AF863459

Lecture Notes in Mechanical Engineering

Lecture Notes in Mechanical Engineering (LNME) publishes the latest developments in Mechanical Engineering—quickly, informally and with high quality. Original research reported in proceedings and post-proceedings represents the core of LNME. Volumes published in LNME embrace all aspects, subfields and new challenges of mechanical engineering. Topics in the series include:

- Engineering Design
- Machinery and Machine Elements
- Mechanical Structures and Stress Analysis
- Automotive Engineering
- Engine Technology
- Aerospace Technology and Astronautics
- Nanotechnology and Microengineering
- Control, Robotics, Mechatronics
- MEMS
- Theoretical and Applied Mechanics
- Dynamical Systems, Control
- Fluid Mechanics
- Engineering Thermodynamics, Heat and Mass Transfer
- Manufacturing
- Precision Engineering, Instrumentation, Measurement
- Materials Engineering
- Tribology and Surface Technology

To submit a proposal or request further information, please contact the Springer Editor of your location:

China: Ms. Ella Zhang at ella.zhang@springer.com
India: Priya Vyas at priya.vyas@springer.com
Rest of Asia, Australia, New Zealand: Swati Meherishi at swati.meherishi@springer.com
All other countries: Dr. Leontina Di Cecco at Leontina.dicecco@springer.com

To submit a proposal for a monograph, please check our Springer Tracts in Mechanical Engineering at http://www.springer.com/series/11693 or contact Leontina.dicecco@springer.com

Indexed by SCOPUS. All books published in the series are submitted for consideration in Web of Science.

More information about this series at http://www.springer.com/series/11236

Caterina Rizzi · Francesca Campana ·
Michele Bici · Francesco Gherardini ·
Tommaso Ingrassia · Paolo Cicconi
Editors

Design Tools and Methods in Industrial Engineering II

Proceedings of the Second International Conference on Design Tools and Methods in Industrial Engineering, ADM 2021, September 9–10, 2021, Rome, Italy

Springer

Editors
Caterina Rizzi
Dipartimento di Ingegneria Gestionale, dell'Informazione e della Produzione
Univesita Degli Studi di Bergamo
Dalmine, Bergamo, Italy

Francesca Campana
Dipartimento di Ingegneria Meccanica e Aerospaziale
Sapienza Università di Roma
Rome, Italy

Michele Bici
Dipartimento di Ingegneria Meccanica e Aerospaziale
Sapienza Università di Roma
Rome, Italy

Francesco Gherardini
Dipartimento di Ingegneria "Enzo Ferrari"
Università di Modena e Reggio Emilia
Modena, Italy

Tommaso Ingrassia
Dipartimento di Ingegneria
Università degli Studi di Palermo
Palermo, Italy

Paolo Cicconi
Dipartimento di Ingegneria
Università degli Studi Roma Tre
Rome, Italy

ISSN 2195-4356 ISSN 2195-4364 (electronic)
Lecture Notes in Mechanical Engineering
ISBN 978-3-030-91233-8 ISBN 978-3-030-91234-5 (eBook)
https://doi.org/10.1007/978-3-030-91234-5

This Springer imprint is published by the registered company Springer Nature Switzerland AG
The registered company address is: Gewerbestrasse 11, 6330 Cham, Switzerland

Preface

The ADM International Conference, promoted by the Italian Association of Design Methods and Tools for Industrial Engineering (ADM), constitutes the biennial international conference organized in the years in which the JCM Conference (joint conference of ADM, INGEGRAF, and S.mart) is not held.

The ADM2021 International Conference, held with a blended formula in Rome, Italy, September 9–10, 2021, and hosted by Sapienza University, represents the opportunity to restart after the pandemic.

On behalf of ADM, I thank the Scientific Societies INGEGRAF and S.mart that have given a very important contribution.

I wish to thank all the members of the ADM board and all the members of the Scientific Council, in particular the Coordinator, Prof. Caterina Rizzi, for their availability, sensitivity, and efforts that made possible to organize the ADM2021 International Conference.

A special thanks to all the members of the scientific and the organizing committees, in particular to Professor Francesca Campana who has solved all the organizational problems.

Thanks a lot to all the authors and to all the international and Italian reviewers. Their activity has allowed the papers to reach such a high scientific level to be published in this prestigious book.

Finally, I thank all the ADM members. Their support has been fundamental for the organization of the ADM2021 International Conference.

Vincenzo Nigrelli
ADM President

Letter to the Authors

Dear Attendees and Authors,

It is a great honor and pleasure to welcome you to the International Conference ADM 2021 promoted by ADM, the Italian Association of Design Methods and Tools for Industrial Engineering, held, with a blended formula, in Rome, Italy, September 9–10, 2021. The conference is hosted by Sapienza University, a worldwide well-known university.

The conference aims at sharing knowledge, experience, and progress in the areas of design and manufacturing with links between industry and academia. It provides a forum for researchers, educators, and professional engineers to present and discuss their latest researches, results, major bottlenecks, and future trends.

Through the exchange of ideas, even virtually due to the pandemic, ADM 2021 is intended to facilitate the creation of multidisciplinary collaborations, new developments and discoveries for new product design and manufacture, health care, transportation, cultural heritage, and sustainability.

The conference offers the opportunity to the delegates to interact and understand how the application of emerging technologies impacts on critical engineering activities such as product design, manufacturing, management, and integration of information along the product life cycle.

This book collects 96 scientific papers across a wide range of session topics that cover a broad spectrum of themes including theoretical issues, methods, tools, processes, and case studies. The topics span from geometric modeling to virtual/augmented reality, human-related and user-centered design, integrated methods for system design, simulation, analysis and optimization, design for additive manufacturing, eco-design, engineering education, and many other subjects applied in different contexts, such as automotive, aerospace, cultural heritage, and health care.

We have also the privilege to host outstanding keynote speakers who will focus on related state-of-the-art technologies in some of main topics of the conference.

To conclude, I would like to thank all the authors for their valuable contribution to the book and each and every one of you for making ADM 2021 successful with your expertise, commitment and active engagement. Special thanks go to Conference Program Chair, Prof. Francesca Campana, and to the members of the organizing committee for their tremendous efforts for making this conference possible, especially during the pandemic.

You are warmly welcome to enjoy the conference and, who will be in presence, the wonderful Rome city.

Caterina Rizzi
ADM2021 Conference Chair

Introduction

This volume collects the proceedings of the ADM2021 International Conference, entitled "Design tools and methods in industrial engineering II," held in Rome (Italy) from September 9–10, 2021, Faculty of Civil and Industrial Engineering of Sapienza.

It is a great honor for me hosting the conference as Chair and coordinating the second edition of this International Conference, which took place for the first time in the University Campus of Modena in September 2019. ADM conference was conceived as the Italian venue to strengthen the network of international research and to meet our colleagues in Italy. The spirit of this venue is similar in conception to that inspiring the Joint Conference on Mechanics, Design Engineering and Advanced Manufacturing (JCM), organized by ADM (Italian Association of Design and Methods for Industrial Engineering) together with the Spanish members of INGEGRAF and the French members of S.mart.

The presence of 96 articles, coming from different European countries and the work of over 100 reviewers, demonstrates the activity of our research community, despite the problems related to the pandemic.

I wish to thank all the colleagues who have collaborated with me as Organizers: Michele Bici and Daniela Pilone of Sapienza, Paolo Cicconi of the University of Roma 3, Tommaso Ingrassia of the University of Palermo, and Francesco Gherardini of the University of Modena and Reggio Emilia to whom I deserve special thanks for having "passed the baton" to me in the conference organizational relay.

I wish to thank also all the colleagues who have supported us as Members of the Scientific Committee and as Reviewers and all Authors who gave their valuable contribution to the conference with their research. Likewise, I thank President of ADM Prof. Vincenzo Antonio Nigrelli and Prof. Caterina Rizzi, Coordinator of the Scientific Council of the ADM.

A further thank is due to Altair and the ASME Italian Section for their patronage, to the Sapienza institutions, the Faculty of Civil and Industrial Engineering and the Department of Mechanics and Aerospace, for hosting us. Last, but not least, the publisher Springer Nature, who honored us by publishing the proceedings of this second edition in the series "Lecture Notes in Mechanical Engineering."

With the hope that ADM2021 and the presence in Rome can also set the restart for new opportunities for social sharing and research.

Francesca Campana
ADM2021 Conference Program Chair

Organization

ADM2021 is organized by ADM—Italian Association of Design Methods and Tools for Industrial Engineering, in cooperation with Dipartimento di Ingegneria Meccanica e Aerospaziale of Sapienza Università di Roma, Italy.

Organizing Committee

Michele Bici	Sapienza Università di Roma, Dipartimento di Ingegneria Meccanica e Aerospaziale
Francesca Campana	Sapienza Università di Roma, Dipartimento di Ingegneria Meccanica e Aerospaziale
Paolo Cicconi	Università degli Studi Roma Tre, Dipartimento di Ingegneria
Francesco Gherardini	Università di Modena e Reggio Emilia, Dipartimento di Ingegneria "Enzo Ferrari"
Tommaso Ingrassia	Università degli Studi di Palermo, Dipartimento di Ingegneria
Daniela Pilone	Sapienza Università di Roma, Dipartimento di Ingegneria Meccanica e Aerospaziale

Scientific Committee

Conference Chair

Caterina Rizzi (ADM Scientific Committee Coordinator)	Università di Bergamo, Italy

Conference Program Chair

Francesca Campana	Sapienza Università di Roma, Italy

Members

Fabio Bisegna	Sapienza Università di Roma, Italy
Francisco Cavas Martínez (INGEGRAF Scientific Council Director)	Universidad Politécnica de Cartagena, Spain
Aitor Cazón (INGEGRAF)	Tecnun Universidad de Navarra, Spain
Vincent Cheutet (S.mart)	INSA Lyon, France
Frédéric Demoly (S.mart)	Université de Technologie de Belfort-Montbéliard, France
Paolo Di Stefano (ADM Scientific Committee)	Università de L'Aquila, Italy
Benoît Eynard (S.mart)	Université de Technologie de Compiègne, France
Salvatore Gerbino (ADM Scientific Committee)	Università degli studi della Campania Luigi Vanvitelli, Italy
Valentín Gómez-Jáuregui (INGEGRAF)	Universidad de Cantabria, Spain
Tim Grunwald	Fraunhofer Institute for Production Technology, Aachen, Germany
Sebastian Hähnel	Fraunhofer Institute for Production Technology, Aachen, Germany
Rafael Hidalgo (INGEGRAF)	Universidad de Córdoba, Spain
Lazhar Homri (S.mart)	Ecole Nationale Supérieure d'Arts et Métiers, France
Rubén Lostado Lorza (INGEGRAF)	Universidad de la Rioja, Spain
Cristina Manchado (INGEGRAF)	Universidad de Cantabria, Spain
Ferruccio Mandorli (ADM Scientific Committee)	Università Politecnica delle Marche, Italy
Cristina Martín Doñate (INGEGRAF)	Universiad de Jaén, Spain
Rikardo Minguez (INGEGRAF)	Universidad del País Vasco, Spain
Ramón Miralbes (INGEGRAF President)	Universidad de Zaragoza, Spain
Maurizio Muzzupappa (ADM Scientific Committee)	Università della Calabria, Italy

Reviewers

Fiorillo Iolanda
Fiorineschi Lorenzo
Fucile Pierpaolo
Furferi Rocco
Germani Michele
Grandi Fabio
Graziosi Serena
Grazioso Stanislao
Greco Alessandro
Guardiani Emanuele
Íñiguez Macedo Saúl
Landi Daniele
Leali Francesco
López-Forniés Ignacio
Mandolini Marco
Marannano Giuseppe
Marconi Marco
Martorelli Massimo
Marzullo Domenico
Morabito Anna
Morosi Federico
Motyl Barbara
Mussi Elisa
Neri Paolo
Paderno Diego
Panariello Dario
Paoli Alessandro
Papetti Alessandra
Pascoletti Giulia
Patalano Stanislao
Pellicciari Marcello
Peruzzini Margherita
Petruccioli Andrea
Pini Fabio
Raffaeli Roberto
Razionale Armando Viviano
Rega Andrea
Regazzoni Daniele
Renzi Cristina
Ricotta Vito
Rossi Marta
Rosso Stefano
Rossoni Marco
Rotini Federico
Santhosh Sandhya
Savio Gianpaolo
Scarfone Luigi
Sequenzia Gaetano
Servi Michaela
Severino Umberto
Sierra-Pérez Jorge
Sponchiado Riccardo
Spreafico Christian
Tamassia Eugenio
Tamburrino Francesco
Tumino Davide
Uberti Stefano
Vaiani Lorenzo
Villecco Francesco
Vitali Andrea
Volpe Yary
Zago Marco
Zanetti Elisabetta M.

Contents

Geometric Modelling and Analysis

On the Geometrical Complexity Index as a Driver for Selecting the Production Technology

Alessandro Greco(✉), Pasquale Manco, and Salvatore Gerbino

Department of Engineering, University of Campania Luigi Vanvitelli, via Roma 29, 81031 Aversa, CE, Italy
{alessandro.greco,pasquale.manco, salvatore.gerbino}@unicampania.it

Abstract. Emerging production technologies, in particular Additive Manufacturing (AM), nowadays are extremely suitable for creating highly complex products, tending towards the concept of 'complexity for free', which is often associated with AM. However, there are no adequate guidelines to provide decision support for the correct selection of the most economically appropriate technology. Indeed, from literature it has been highlighted the need to develop a technology selection methodology based no longer on production volume but on product complexity. This paper investigates this need by presenting an approach to determine the geometrical (or shape) complexity index of a part, which, combined with the assembly complexity, represents the driver for helping to decide the best production technology (traditional or additive). The geometrical complexity index has been determined based on complexity judgments, provided by CAD modelling experts, for a sample of CAD models. In this way, it has been possible to define a preliminary complexity index model, strictly linked to the CAD model information. The results showed that the geometrical complexity metrics from the literature, if individually considered, are not comprehensive. However, a combination of them makes it possible to obtain an index that best reflects the subjective judgement of the experts. In addition, by combining the geometrical and assembly complexity with a cost analysis it is possible to obtain convenience zones for better selecting the production technology.

Keywords: Geometrical complexity · CAD · Survey · Technology selection

1 Introduction

In the era of the fourth industrial revolution, the technological innovation is closely linked to the increasing complexity of products available on the market. Identifying, analyzing and understanding complexity factors represent the first step to manage the complexity at a strategic level to improve company competitiveness. Providing a universal definition of complexity, to date, is still a difficult process because, unlike other physical quantities,

C. Rizzi et al. (Eds.): ADM 2021, LNME, pp. 3–12, 2022.
https://doi.org/10.1007/978-3-030-91234-5_1

complexity is not properly measurable. Therefore, further research is needed to make practical the concept of complexity.

The state of the art about complexity in science is considered multidimensional and it is explored from three points of view: (i) Design and Product Development Complexity, (ii) Manufacturing and Manufacturing System Complexity, (iii) Business and Marketing Complexity [1]. From an engineering point of view, the literature review on complexity can focus on three fundamental aspects: design, manufacturing and assembly.

Since the 1980s, several methodologies for evaluating technological complexity have been introduced. In the first decade of 1980s, the technique of expert systems was widespread, i.e. mathematical programs which, on the basis of input provided by operators, provided an index of complexity. Instead, since the 1990s, mathematics has been simplified, leaving the assessment of certain parameters to the subjective judgment of experts by means of questionnaires. In addition, there are also several more rigorous approaches based on objective system data, as described in the following section.

1.1 Geometrical or Shape Complexity

Complexity in design is generally related to the geometry of the part. In applications such as computer graphics and Finite Element Analysis, polygonal meshes are defined in terms of the geometry and connectivity of the nodes. The shape complexity measures how entrapped the polygon is. In addition to being closely associated with geometry, shape complexity is also associated with organizational and operational aspects of CAD software. Indeed, often it is also called CAD-complexity [2]. About CAD modelling, for each individual the complexity varies; for someone a task may be feasible, while for someone else it may be frustrating. Complexity is associated with the actual shape to be achieved in the project and derives from a strategic use of the functionalities provided by the CAD software. Contemporary CAD software, based on parametric associations, facilitates the creation of fully parameterized products. Thus, the complexity of the design activity is related to the complexity of the product, hence to the geometric complexity of the design.

The complexity of a component has implications in the design and especially in the production phase. In fact, in the mechanical field there is a need to produce increasingly complex and multi-functional parts, so the need for the development of a complexity model as in additive as in the traditional manufacturing is essential.

In literature there are several definitions about geometrical complexity and several metrics have been defined for an objective quantification of it. In particular, about objective metrics, Joshi and Ravi [3] proposed both *Sphere Ratio* (SR) and *Part Volume Ratio* (PVR) for quantifying the complexity. SR represents the ratio between the surface area of a sphere with the same volume of the part and the surface area of the part itself; PVR is the ratio between the volume of the part and the volume of its minimum *Bounding Box*. Lian et al. [4] defined the complexity as the ratio between the volume of the part and the volume of its convex envelope. They talk about *Convex Envelope Complexity* (CEC) and this is the most used metric. Always linked to geometric characteristics, Chougule and Ravi [5] quantified the complexity by proposing the *Cube Ratio* (CR), defined as the ratio between the surface area of a cube with the same volume of the part and the surface area of the part itself. Table 1 describes the mathematical definition of these metrics.

Table 1. Geometrical complexity metrics.

Metrics	Equation	Parameter
Sphere Ratio - SR	$SR = 1-(A_{sphere}/A_{part})$	• A_{sphere}: surface area of sphere* • A_{part}: surface area of part
Part Volume Ratio -PVR	$PVR = 1-(V_{part}/V_{BBox})$	• V_{part}: part volume • V_{BBox}: part's bounding box volume
Convex Envelope Complexity - CEC	$CEC = 1-(V_{part}/V_{CE})$	• V_{part}: part volume • V_{CE}: part's convex envelope volume
Cube Ratio – CR	$CR = 1-(A_{cube}/A_{part})$	• A_{cube}: surface area of cube* • A_{part}: surface area of part

(*) cube and sphere have the same part's volume.

Other metrics, proposed by literature, appear to be linked also to other information characterizing a CAD model, not always related to geometrical properties. Qamar et al. [6] estimate omplexity as function of the ratio of the perimeter of a cross-sectional area to be extruded and the round bar perimeter having same cross-sectional area. Bodein et al. [7] suggest to examine the complexity based on the *number of surface* composing a part. Lastly, Valentan et al. [8] associate the complexity to the number of triangle required for representing an object in a *.stl* file.

From the literature review, it is possible to point out that: i) none of these metrics, taken individually, are exhaustive in describing geometrical complexity; ii) only a few studies tried to compare them with each other, but based on small samples of models [9, 10]; iii) probably a weighted combination of these metrics could provide more comprehensive results than those provided by individual metrics.

1.2 Aim and Outline of the Paper

In continuity with a previous work [10], in which the geometrical complexity metrics described in Table 1 have been compared and correlated with each other, this paper is aimed at filling some gaps of literature by providing a more general complexity metric, based on a linear combination of the metrics described above. Each metric has been correlated against the CAD expert user judgments, pointed out from a survey.

In the next section, the methodology is described. Then, a case study is presented, followed by results and discussion. Finally, conclusions and future work are detailed.

2 Method

The heuristic approach, which uses metrics based on knowledge and personal experience, is easy to apply but the actual complexity metrics remain subjective. W. Elmaraghy and Urbanic [11] present a methodology to assess product and process complexity by considering a set of complexity indices. Fera and Macchiaroli [12] propose a revisitation of the methodology in [11], and introduce a new formula (Eq. 1), which considers the contribution of geometrical complexity, evaluated as: i) Convex Envelop Complexity (CEC), as defined in Table 1; ii) Operational Complexity (OC), which depends on the number and the type of operations necessary to produce the part; iii) the variety of information (IV) and the information entropy (H), which are related to the number, type and diversity of features that have to be manufactured. M is a normalizing factor.

$$PC = \frac{1}{M}(IV + CEC + OC) * H \tag{1}$$

As reported in the Sect. 1.1, the literature proposes different approaches for evaluating the geometrical complexity. So, the purpose is to introduce a complexity index able to account both objective metrics and expert judgement.

The new geometrical complexity contribution, substituted in Eq. 1, provides an overall complexity index that better addresses the selection of the appropriate production technology.

Figure 1 shows the adopted methodological framework. It points out the focus of the present article which is related to the geometrical complexity.

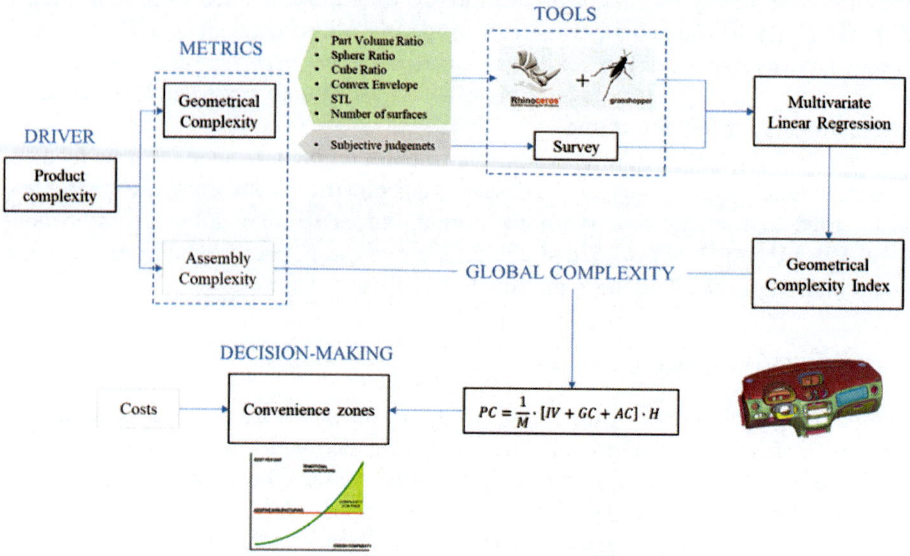

Fig. 1. Methodological framework.

As mentioned above, the driver for technology selection is the Product Complexity (PC), which is defined, in the case of manufacturing field, by both geometric complexity and assembly complexity. The latter, is obtained by using the information contained in the Standard Operating Procedures (SOPs). The measuring procedure as explained in Samy & ElMaraghy's work [13] has been applied. Considerations about handling and insertion attributes of each component have been made to measure the weighted average values of the part assembly complexity factors (C_{part}), according to the following equation:

$$C_{part} = \frac{C_h \sum_1^J C_{h,j} + C_i \sum_1^K C_{i,k}}{\sum_1^J C_{h,j} + \sum_1^K C_{i,k}} \in [0; 1] \tag{2}$$

where $C_{h,j}$ (j = 1,..,J) and $C_{i,k}$ (k = 1,..,K) are respectively the values of relative handling and insertion complexity factors, reported in [13]. The value J and K are the number of the considered attributes. C_h and C_i take on values between 0 and 1. They are respectively the average handling and insertion complexity factor of the part.

About the geometrical complexity, as described in Sect. 1.1, several authors proposed a series of metrics to define it in an objective manner (Table 1). Instead, subjective metrics are based on complexity judgments attributed by experts in CAD modelling, who have a broader vision of the modelling world and who know how to recognize the difficulties that may be encountered during the part creation process. The complexity judgement is attributed following a simple psychometric measurement technique developed by Rensis Likert [14]. Typically, the range of ratings goes from 1 to 5, where 1 stands for "very simple" and 5 stands for "very complex".

In order to evaluate the geometrical complexity metrics, the tools used for evaluating objective and subjective metrics are the plug-in Grasshopper for Rhinoceros® and a survey respectively. In Grasshopper, by means of blocks, it is possible to build real workflows in a Canvas that, starting from the CAD model, provides all the parameters needed to evaluate the objective metrics.

The survey for the acquisition of subjective metrics involves CAD modelling experts (both from industry and academy). Each interlocutor is invited to express an opinion, based on Likert scale, on CAD models, considering some evaluation parameters such as modelling strategies, symmetries, patterns, features, linearity and curvature of surfaces, etc.

Once the objective and subjective metrics, related to a sample of CAD models, have been obtained, an analytical relationship between the "perceived" complexity of a component and its objective characteristics can be obtained by means of multivariate linear regression, by considering the subjective judgements as depended variable and the objective metrics as independent variables.

The overall complexity index (PC), considering the obtained Geometric Complexity (GC) index and the Assembly Complexity (AC), based on Eq. 1, is done by the following equation:

$$PC = \frac{1}{M}(IV + GC + AC) * H \tag{3}$$

The decision-making process involves the complexity of the product and the cost analysis related to the production technology (additive or traditional). So, it allows the identification of areas of convenience for selecting the appropriate technology.

3 Case Study

This section is aimed at demonstrating the effectiveness of the framework proposed in Fig. 1. The case study investigates the technology selection for the production of 26 components, most of which selected from the industrial field. All the components are made of polymeric materials and they have no structural function. Two *near-net-shape* production technologies have been considered: Selective Laser Sintering (SLS) and Injection Molding (IM), but the approach can be easily applied to different technologies.

Figure 2 shows the CAD models of part of the investigated components. The figure appears blurred since it contains multiple copyrighted components. Anyway, it is possible noting the considered models are characterized from very different levels of complexity.

Fig. 2. Investigated components (blurred effect is added for copyright reasons of some components).

According to the framework, a routine has been implemented in Grasshopper plug-in for Rhinoceros® for calculating the objective geometrical complexity metrics of CAD models. The subjective judgements have been collected by means of a survey which involved 50 CAD modelling experts (20 industrial designers and 30 academics). For each item, the overall expert judgment of complexity is obtained from the mean value of the proposed judgments.

Table 2 shows the objective and subjective metrics values. The experts' judgements values have been normalized.

3.1 Geometrical Complexity Index

Multivariate linear regression is performed by using the *stepwise* method [15].

For each metric, maximum and minimum values (highlighted in red in Table 2) have been excluded from the database to avoid extrapolation problems. In addition, 70% of the data has been used for the training phase, while 30% for the testing phase by means of an *offset test*.

The only metrics that have a certain linear relationship with the dependent variable (experts' judgements) are considered for the definition of the linear regression model. For each pair of variables, the Pearson correlation coefficient has been evaluated (Table 3), choosing a lower bound of 0.7.

Table 2. Geometrical complexity metrics.

Item	PVR	CR	SR	CEC	STL	Surfaces	Experts judgements
1	0,8458	0,7762	0,8196	0,7333	417083	1032	0,8200
2	0,9825	0,8911	0,9123	0,9641	3483112	325	0,8800
3	0,9821	0,8396	0,8707	0,9664	16383890	443	0,9400
4	0,9862	0,9093	0,9269	0,9681	372918	12979	0,9400
5	0,9483	0,9161	0,9323	0,9209	10808618	1839	0,8400
6	0,9929	0,8638	0,8902	0,9835	8682983	575	0,9800
7	0,9798	0,9237	0,9385	0,9648	132262	1626	0,9200
8	0,7954	0,6368	0,7073	0,6385	13444	150	0,7600
9	0,9590	0,9003	0,9197	0,9364	36616	330	0,9200
10	0,9816	0,9341	0,9469	0,9639	205847	6422	0,9400
11	0,9365	0,8063	0,8439	0,8895	1691003	203	0,9000
12	0,8925	0,7849	0,8266	0,8246	26056065	1104	0,8400
13	0,9789	0,9041	0,9227	0,9662	9108612	435	0,9400
14	0,9665	0,9214	0,9367	0,9508	28192106	2383	0,9000
15	0,9166	0,7130	0,7687	0,7919	21954769	894	0,8000
16	0,9530	0,9391	0,9510	0,9059	246682	177	0,9000
17	0,5960	0,5797	0,6612	0,6612	7614	101	0,3540
18	0,0397	0,6411	0,7107	0,7107	1952	32	0,2200
19	0,4800	0,0001	0,0890	0,0890	104853	631	0,4460
20	0,1441	0,4792	0,5803	0,5803	1116	15	0,2000
21	0,1072	0,3511	0,4770	0,4770	4444	56	0,2920
22	0,9671	0,8430	0,8735	0,8735	8748	300	0,8300
23	0,5585	0,4250	0,5365	0,5476	5644	101	0,4520
24	0,7561	0,8014	0,8399	0,8399	4940	100	0,5740
25	0,7710	0,6572	0,7237	0,7236	8212	148	0,4420
26	0,8600	0,3389	0,4672	0,4671	11452	69	0,5340

Table 3. Pearson correlation coefficient.

	STL	PVR	CR	SR	CEC	Experts judgements
Surfaces	-0,0010	0,2735	0,2889	0,2889	0,2970	0,3309
STL	-	0,3311	0,2815	0,2815	0,3196	0,3786
PVR	-	-	0,6580	0,6581	0,6720	0,9144
CR	-	-	-	1,0000	0,9795	0,7300
SR	-	-	-	-	0,9795	0,7301
CEC	-	-	-	-	-	0,7538

As shown in Table 3, PVR, CR, SR and CEC has a Pearson coefficient (highlighted in green) higher than the lower bound. So, STL and number of Surfaces metrics are excluded from the analysis. A *partial-F* hypothesis test has been performed to verify that the variables selected from the correlation could be included in the linear regression model. The stepwise method highlighted that PVR and CR are the most significant metrics to better describe the geometrical complexity on the basis of the starting dataset. Hence, it was possible to deduce that the estimation of the Geometrical Complexity (GC) index of a component is a bivariate regression model based on two objective complexity metrics, Part Volume Ratio (PVR) and Cube Ratio (CR), according to the following equation:

$$GC = 0.623 \cdot PVR + 0.277 \cdot CR + 0.025 \quad (4)$$

From the analysis of the residuals of this model, it has been verified that both the assumptions of linearity and homoscedasticity are satisfied.

To finally accept the model, the linear regression equation has been subjected to the *offset test*, considering the 30% of the database previously excluded. It has been found to be valid since the average deviation of the estimated complexity from the subjective judgments was less than 5%.

3.2 Convenience Zones

Having obtained the overall Product Complexity (PC) values, according to Eq. 2, a production cost analysis has been performed, as a function of complexity and considering the production volume as fixed. Two production volume scenarios have been considered: 500 and 1000 pieces. The considered manufacturing technologies are Injection Molding (IM) and Selective Laser Sintering (SLS) as traditional and additive manufacturing techniques respectively. Only 20 of the 26 CAD models have been selected for simulating the production scenarios.

Figure 3 shows the convenience zones for a volume production of 500 and 1000 pieces respectively. The chosen components are ordered on x-axis according to increasing complexity. By the cost analysis, the Additive Manufacturing (AM) production cost turned out to be constant for the considered *build* as the complexity of the produced part increase.

Looking more in detail, regarding the production scenario of 500 pieces (Fig. 3.A), the convenience zone for AM starts from component number 7, which has complexity index equal to 0.5571. At this point, the production cost of a single component is identical for both production technologies (€ 6.69). For components with complexity index higher than 0.5571, it is never convenient to produce in IM, while it is advantageous to move to AM production.

Considering the scenario of 1000 pieces (Fig. 3.B), it is interesting to observe how the area of convenience for AM begins with component number 19, which has complexity index equal to 0.7979. The results show that IM is better suited to a *mass production* scenario, in which the components, although complex, are convenient to be produced with traditional technology.

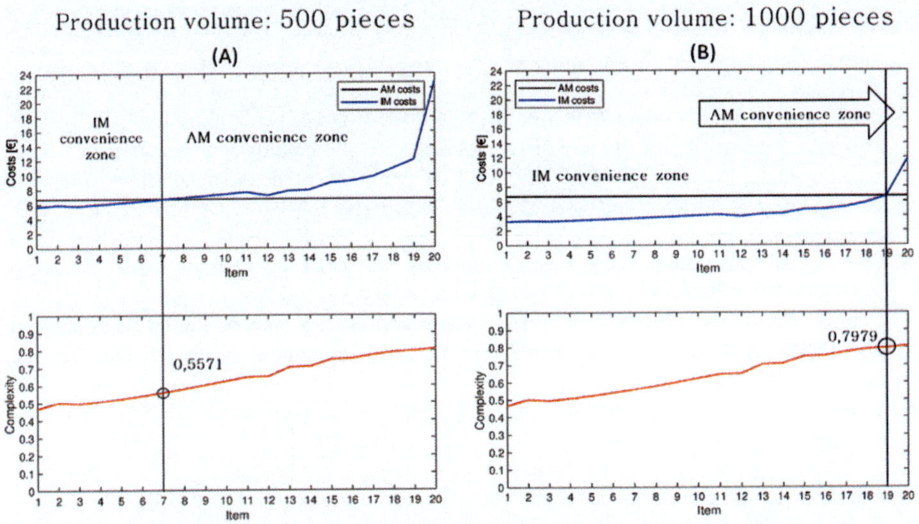

Fig. 3. Convenience zones for production volumes of 500 (A) and 1000 (B) pieces.

4 Conclusions

The analysis of the literature about the metrics for evaluating the geometric complexity of a CAD model has shown that, evaluated individually, these metrics are not exhaustive. The aim of this research was to investigate and combine these geometrical complexity estimation techniques, together with subjective judgements of experts, in order to present a new approach for determining the geometrical complexity index of a part.

Geometrical complexity, combined with assembly complexity and costs analysis, could represent the driver for selecting the best production technology, offering a method for evaluating the convenience of Traditional Manufacturing techniques with Additive Manufacturing ones.

The objective and subjective metrics, obtained from a survey that involved 50 CAD modelling experts, related to 26 CAD models have been analysed and correlated. Then, by means of multi-variate linear regression, a new Geometrical Complexity index has been defined and validated. Part Volume Ratio and Cube Ratio resulted the objective metrics that, linearly combined, better approximate the subjective judgment about geometrical complexity.

However, this research is only a first step in defining a geometric complexity index that can be as effective and generic as possible.

In order to achieve this goal and to enrich the literature about the topic, further research will be conducted, increasing the sample of CAD models for the survey, obtaining more information about the judgements parameters and investigating on the use of other judgement scales and other statistical methods for the definition of the mathematical model.

References

1. ElMaraghy, W., ElMaraghy, H., Tomiyama, T., Monostori, L.: Complexity in engineering design and manufacturing. CIRP Ann. **61**(2), 793–814 (2012)
2. Chase, S., Murthy, P.: Evaluating the complexity of CAD models as a measure for student assessment. In: Proceedings of the 22nd Annual Conference of the Association for Computer-Aided Design in Architecture/1-880250-09-8, Washington DC 19–22 October 2000, pp. 173–182 (2000)
3. Joshi, D., Ravi, B.: Quantifying the shape complexity of cast parts. Comput.-Aided Des. App. **7**(5), 685–700 (2010)
4. Lian, Z., Godil, A.A., Rosin, P.L., Sun, X.: A new convexity measurement for 3D meshes. In: 2012 IEEE Conference on Computer Vision and Pattern Recognition, Providence, RI, 2012, pp. 119–126 (2012)
5. Chogule, R.G., Ravi, B.: Variant process planning of casting using AHP-based nearest neighbour algorithm for case retrieval. Int. J. Prod. Res. **43**(6), 1255–1273 (2005)
6. Qamar, S.Z., Chekotu, J.C., Al-Maharbi, M., Alam, K.: Shape complexity in metal estrusion: definition, classification and applications. Arab. J. Sci. Eng. **44**, 7371–7384 (2019)
7. Bodein, Y., Rose, B., Caillaud, E.: Explicit reference modeling methodology in parametric CAD system. Comput. Ind. **65**(1), 136–147 (2014)
8. Valentan, B., Brajlih, T., Drstvenšek, I., Balič, J.: Development of a part-complexity evaluation model for application in additive fabrication technologies. J. Mech. Eng. **57**(10), 709–718 (2011)
9. Johnson, M.D., Valverde, L.M., Thomison, W.D.: An investigation and evaluation of computer-aided design model complexity metrics. Comput.-Aided Des. App. **15**(1), 61–75 (2018)
10. Greco, A., et al.: Investigation on geometrical complexity techniques for assessing AM feasibility. Macromol. Symposia **396**, 2000309 (2021)
11. ElMaraghy, W.H., Urbanic, R.J.: Modelling of manufacturing systems complexity. CIRP Ann. **52**(1), 363–366 (2003)
12. Fera, M., Macchiaroli, R., Fruggiero, F., Lambiase, A.: A new perspective for production process analysis using additive manufacturing-complexity vs production volume. Int. J. Adv. Manuf. Technol. **95**, 673–685 (2018)
13. Samy, S.N., ElMaraghy, H.: A model for measuring product assembly complexity. Int. J. Comput. Integr. Manuf. **23**(11), 1015–1027 (2010)
14. Likert, R.: A technique for the measure of attitudes. Arch. Psychol. **140**, 5–55 (1932)
15. Yamashita, T.: A stepwise AIC method for variable selection in linear regression. Commun. Statist. Theory. Methods. **36**(13), 2395–2403 (2007)

A Voxel-Based 2.5D Panel Method for Fluid-Dynamics Simulations

Antonio Bacciaglia(✉), Alessandro Ceruti, and Alfredo Liverani

University of Bologna, Department of Industrial Engineering - DIN, Bologna, Italy
{antonio.bacciaglia2,alessandro.ceruti,
alfredo.liverani}@unibo.it

Abstract. The Panel method is an approach for the estimation of the lift of 3D models which is faster than CFD. This can be useful especially in the conceptual design stage where several configurations should be evaluated in a reduced time with a limited computational cost. However, the meshing of the 3D body surface with rectangular panels can be a time-consuming activity because the designer should define from scratch a cloud of points that matches the external surfaces of the tested object to obtain consistent panelling. Therefore, a voxelization-based methodology has been developed to obtain the panels' position, speeding up and automating the model preparation process. The obtained discretization has been integrated into a panel method available in the literature. Four case studies, of increasing complexity, have been analyzed to investigate the capability of the innovative voxel-based panel methodology. A parametric study has been carried out to study the effect of the voxel grid dimension on the accuracy of the results. Benchmarking values of lift coefficient obtained from literature or xFoil software have been used to evaluate the precision that can be achieved with this approach. The results show a good agreement between the voxel-based panel method and the literature when the overall pressure distributions and aerodynamic coefficient values are considered. Higher errors are noticed with drag.

Keywords: Voxelization · Panel method · Fluid dynamics · Potential flow

1 Introduction

The determination of the forces and moments acting on an object (e.g. aircraft, car, truck, trains) due to the surrounding airflow is one of the most significant aerodynamic problems in industrial engineering. On the one hand, testing the component in real-life implies the manufacturing of a new prototype each time a change is introduced: this can be considered the most precise way to evaluate the aerodynamic forces, but it is an expensive methodology both in terms of time and money: such an approach could be unsustainable in the conceptual/preliminary design stage. Another alternative is to test a small-scale object in wind tunnel facilities, but it is still a very expensive and slow procedure to be adopted in the first design stages when different configurations must be tested to find the optimal one. This is the reason why numerical methods are often used to

C. Rizzi et al. (Eds.): ADM 2021, LNME, pp. 13–26, 2022.
https://doi.org/10.1007/978-3-030-91234-5_2

compute the velocity and pressure field around an object in the conceptual/preliminary design stages.

The panel method is one of the simplest numerical approaches in aerodynamics to solve the potential flow around an object if the following assumptions about the flow are verified: inviscid, incompressible, irrotational, and steady [1]. Since the panel method provides inviscid solutions, the resulting flow field is not characterized by viscous effects that can affect the drag in a significant way. However, thanks to the simplicity of the method it is possible to solve most of the common aerodynamic problems. In the last couple of decades the panel method was pretty ignored: the attention was increasingly shifted on Computational Fluid Dynamic (CFD) software that nowadays can be considered the most widely used strategy to numerically solve aerodynamic problems thanks to the always increasing computational power available in Personal Computers. The CFD approach is based on the Navier-Stokes equations discretization by finite volumes elements and is capable to analyze different types of flow. Indeed, the CFD methodology uses a discretized volume domain, as opposed to the panel method where the discretization occurs only on the surface of the body. This reflects on the high computational time spent to find the solution, making the CFD a slower approach, even if more accurate. For this reason, a panel method is still a valuable approach that can be used in a conceptual/preliminary design stage.

However, the 'panelization' process namely the population of the external surface of a 3D object with panels- plays a critical role. The position of the four edge points needed to obtain a rectangle on the object surface is unknown to the user in the case of complex or freeform surfaces. For this reason, the user is asked to design from scratch a cloud point that maps the external surface in these cases. To overcome this problem, voxelization which can be defined as the discretization of an object with cubic cells [2], could be used to set the position of the panels on the object surface before the aerodynamic problem is solved. This can be obtained by matching each panel with the square external surface of every single voxel. Voxelization is a numerical methodology used for graphic manipulation and visualization of complex 3D objects and it has many applications in industrial engineering [3].

The use of a voxel-based representation of an object in the panel method could solve the problem of positioning the panels on the external surface of the part because the edges of the panels could coincide with the voxels' edges. On the other hand, the main drawback is that the adoption of cubic cells creates a stair-effect along the component's surface that could locally affect the pressure distribution.

The scope of this research is to investigate the feasibility of the use of voxels as a discretization unit to easily and automatically panelize 2.5D objects for numerical potential-flow solutions. In the scientific literature, there are few contributions where voxels are used as discretization unit in fluid dynamic problems (i.e. [4] and [5]), but to the best of the authors' knowledge, there is no scientific contribution describing the use of voxels and potential-flow methodology to solve aerodynamic problems. The methodology introduced in this paper uses the APAME open-source panel method program [6] written by Daniel Filković coupled with voxelization to get a feasible panel distribution.

The article is structured as follows. In Sect. 2 the 3D panel methodology and the voxelization process are briefly described. In the following, Sect. 3 shows the integration between the voxel-based model and the available panel code. Section 4 shows the results from the simulation of different case studies used to validate the developed methodology. In the end, Sect. 5 sums up the article with conclusions and future developments of the proposed approach.

2 Theoretical Background

2.1 APAME 3D Panel Method

APAME is a 3D Panel Method open-source software, coded in MATLAB, that is used to compute aerodynamic forces and moments acting on aerospace structures [6]. The scope of this numerical approach is to replace CFD programs for subsonic attached flows where calculation time is important in the context of the conceptual design stage and overall drag estimation can be commissioned to the following stages. Panel methods are used to calculate lift and induced drag forces/coefficients on a body as long as viscous effects are negligible due to the hypothesis at the base of the method. With panel methods, the induced drag can be computed, but the skin friction contribution can't be evaluated because the latter depends on viscous modelling. For airfoils, skin friction is considerable when the angle of attack is low; for high angles of attack or blunt bodies, the shape drag is more significant. For this reason, panel methods shouldn't be relied upon for drag prediction, unless corrections based on the equivalent flat plate skin drag and shape drag are added.

Several assumptions on the type of flow surrounding the object should be respected to apply the panel method (steady, incompressible, and inviscid flow), so that the potential flow method is limited to the resultant pressure forces over the surface of the body [7]. The resulting calculation time is shorter by far if compared with the classic CFD method because the discretization occurs only at the surface of the body and not in the entire control volume. The panel method is essentially a numerical approximation that relies on using discrete elements on the surface of an object, called panels. A potential flow element that could be a vortex, a doublet, a source, or a sink is distributed on each discretization unit to satisfy certain boundary conditions such as the no-slip one. The mutual interaction of these panels is accounted for; far from the object, the velocity flow should be equal to the free stream velocity approaching the object.

If we assume potential flow, the equation that represents continuity (conservation of mass) becomes the Laplace equation where the solution represents the scalar speed potential φ of the velocity field v (Eq. 1).

$$\begin{cases} \frac{\partial^2 \varphi}{\partial x_i^2} = 0 \\ \frac{\partial \varphi}{\partial x_i} = v_i \end{cases} \tag{1}$$

Another important equation at the basis of the panel method comes from the conservation of momentum equation for irrotational, inviscid flow and steady flow, which is called e Euler-Bernoulli equation. It links the velocity field with the pressure distribution (p), being ρ the fluid density (Eq. 2).

$$\frac{v^2}{2} + \frac{p}{\rho} + gz = cost \tag{2}$$

Thanks to the implementation of these equations, the panel method returns the velocity field using the Laplace equation and the pressure field with Bernoulli's formula. Finally, by integrating the pressure over the surface of the object, the total aerodynamic forces and moments acting on the body can be evaluated. A crucial property of the Laplace equation is that a sum of any set of solutions of that equation forms a solution as well. This means that the potential flow solutions can be organized to obtain the sum of singularities whose values are found to satisfy appropriate boundary conditions. This is crucial in the potential flow methodology because the APAME software uses a combination of quadrilateral panel sources and dipoles distributed over the surface of the aerodynamic body. For a more detailed description of the panel method and APAME software, please refer to [6] and [8].

2.2 Voxelization

Voxelization is a geometrical discretization of a 3D model mainly used for graphic visualization of complex models to speed up geometrical and algebraic manipulation. Starting from the B-rep (Boundary representation) of common CAD software, the model is numerically translated into a 3D logical matrix, where each cell represents a "voxel", which is a 3D cubic element that can be defined as a tri-dimensional unitary element equivalent to the pixel in computer graphics. Each voxel is initialized with a binary value based on the fact that that element belongs to the component or it doesn't.

Voxelization allows to speed up all the geometrical operations (boolean, slicing, rotation…) transforming them into matrix operations and can store intricate details about the interior of functional parts, unlike B-rep or constructive solid geometry models which contain only information of the external surface of the part [9]. As a consequence, Voxel models have been exploited to support the realistic modelling of physical parts finding application in medicine, engineering, geography, and computer science. Despite the above-cited advantages, voxelization is a geometry discretization technique leading to approximations in the shape representation: after voxelization, the external shape suffers from stair effect which can be magnified or reduced depending on the dimensions of voxels (Fig. 1).

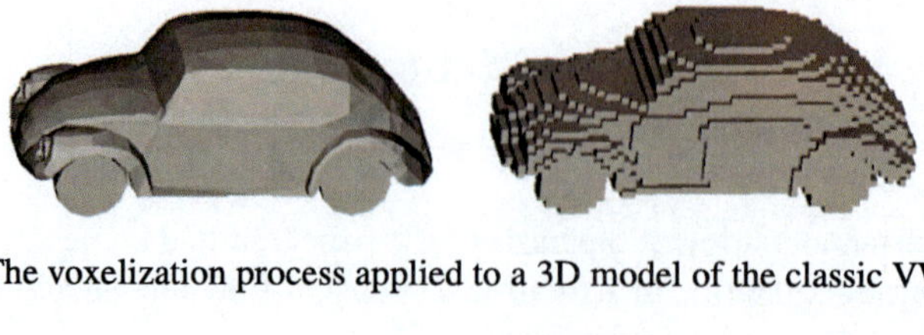

Fig. 1. The voxelization process applied to a 3D model of the classic VW Beetle.

The ray intersection [10] is one of the most widespread voxelization methods. In this voxel methodology, a 3D model described by a Standard Triangulation Language file format (STL) is ray-traced in each of the x,y,z directions, with the overall result being a combination of the result from each direction (Fig. 2). In the STL file, the 3D model is discretized by a set of 3D triangles that approximate the surface. For each triangle of the STL mesh, the ray intersection method can be divided into five steps:

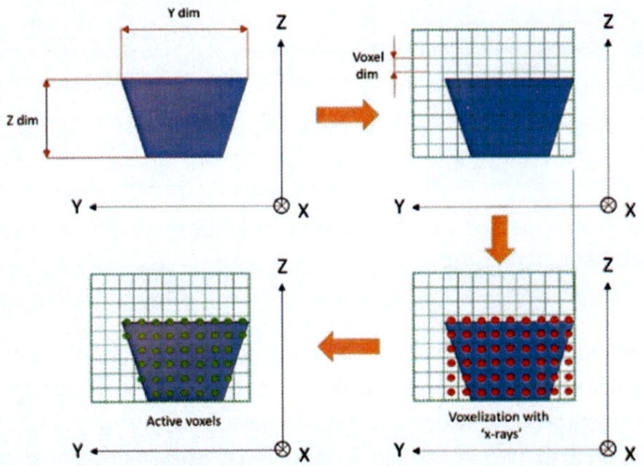

Fig. 2. Ray intersection method for object voxelization; sketched by authors adapting [11].

- Evaluate each edge of the triangle
- Find the position of the opposing vertex to that edge
- Find the position of the ray relative to that edge
- Check if the ray is on the same side of the edge as the opposing vertex
- If this is true for all three edges, then the ray passes through the facet.

In the following steps of our methodology, voxelization is coupled with a panel method to create an innovative and user-friendly tool for preliminary estimation of the aerodynamic forces acting on a body without the need for sketching from scratch a cloud of points matching the surface of the object. The vertices of cubic cells that compose the external layer of voxels will be selected and chosen as vertices of the panels.

3 Voxel-Based Panelization Methodology

In this section, the methodology to couple voxelization and 3D panel approach to solving aerodynamic problems is described. All the operations - better described in the following - are contained in scripts programmed in MATLAB programming environment to easily communicate with APAME codes.

Due to the early stage of this research, the methodology has been implemented up to now only on 2.5D models which can be seen as 2D silhouettes extruded in the third dimension (later called spanwise dimension), but in the future, its applicability will be extended to every kind of 3D models.

The overall methodology is summarized in the flowchart shown in Fig. 3:

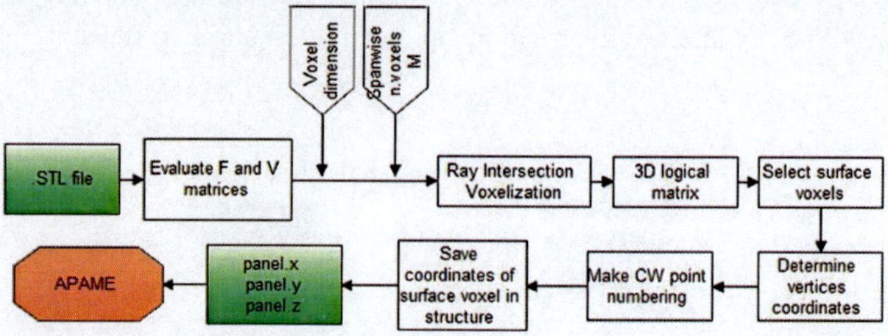

Fig. 3. Voxel-based panel methodology flowchart: coupling of 3D model voxelization and APAME, a 3D panel method platform.

As the first step of the methodology, the user is required to select an STL file of the object he/she would like to analyze (the STL file can be either of binary or ASCII type). From the stereolithography 3D model, the information of vertices and triangle faces are extrapolated and saved in two matrices V and F of dimensions respectively nx3 and mx3, where n is the number of vertices and m is the number of triangular facets.

In the following, the designer is asked to input the voxel dimension useful to discretize the 3D model. This decision affects the number of lateral panels N used in the APAME approach: with reduced voxel dimension the discretization is more precise reflecting on a lower distance between vertices and an increase of the number of lateral panels N. Moreover, the user is tasked to choose the number of voxels he/she would like to obtain in the spanwise dimension. This number represents the number of longitudinal panels M of the APAME approach [6]. The 3D model is oriented in the following way: the 2D profile belongs to the XZ 2D plane, while the spanwise direction follows the Y-axis.

After the active intervention of the designer, the ray intersection voxelization method discretizes the STL file and returns a 3D logical matrix composed of 0 and 1 (black and white representation). If the voxel dimension is known it is possible to return the spatial coordinates of the voxel grid.

Right after the voxelization, the developed methodology is capable to select the surfaces of voxels that are positioned at the surface boundary of the voxel model: this thanks to a function available to the MATLAB community which returns the spatial coordinates of the vertices composing the surface of the voxel model [12].

As stated before, in this stage only 3D models which can be seen as a 2D profile extruded in the spanwise dimension are considered. For this reason, the developed methodology proceeds by isolating the vertices coordinates of the voxel model that belongs to the 2D profile which is extruded in the third dimension (Fig. 4) in the following.

After isolating these points, our code rearranges the points of the 2D profile in a clockwise order to adapt them for the APAME software. In the end, the spatial coordinates of the surface voxel model are rearranged in the specific order described before and saved in a structure called "panel", defined as follows:

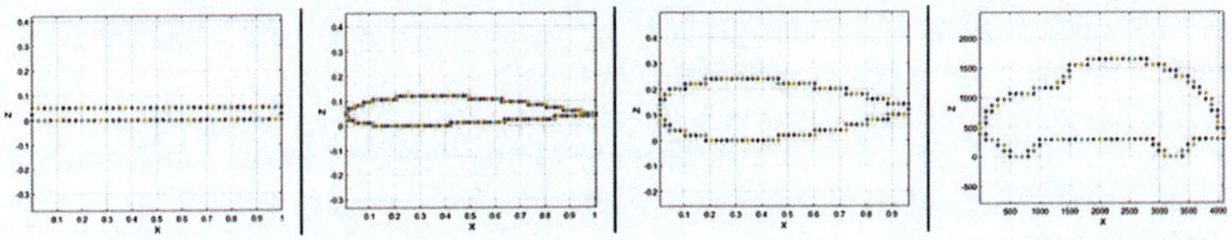

Fig. 4. The 2D profile of voxel models used as case studies to investigate the voxel-based panel methodology performances: from left to right a flat plate, NACA 2412 airfoil, NACA 0024 airfoil, simplified Volkswagen Beetle.

- panel.x = $(N + 1) \times (M + 1)$ matrix containing the $N + 1$ x coordinates of vertices of the 2D profile extracted from the surface voxel model in each column, which are repeated $M + 1$ times to describe the extrusion in the spanwise dimension;
- panel.y = $(N + 1) \times (M + 1)$ matrix containing a spatial grid with $M + 1$ y coordinates in the spanwise dimension in each row, which are repeated $N + 1$ times to describe the extrusion in the spanwise dimension;
- panel.z = $(N + 1) \times (M + 1)$ matrix containing the $N + 1$ z coordinates of vertices of the 2D profile extracted from the surface voxel model in each column, which are repeated $M + 1$ times to describe the extrusion in the spanwise dimension.

Finally, the structure is saved and imported in the APAME software, where each matrix describing the spatial coordinates is extracted and used to populate the surface voxel model with panels.

In the following section, several case studies are presented to evaluate the capability of the voxel-based panel method in the accurate estimation of the lift load on different 3D voxel models; drag can't be accurately estimated using the potential flow method. Several simulations have been performed to understand how the change of the number of longitudinal panels M and the voxel dimension could affect the accuracy of the method. This research also investigates the impact of the stair effect of the external surface of the body (typical of voxelization) on the global aerodynamic performances of the model. However, thanks to the voxelization process, the discretization with panels before the aerodynamic analysis is speeded up since there is no need to design from scratch a cloud of point matching the external surface of the body to analyze. This computational acceleration can be useful in the preliminary design stages where a large number of configurations must be analyzed before freezing the shape of an object.

4 Case Studies

In this section, four case studies will be discussed to investigate the accuracy performances of the voxel-based panel methodology described in this paper. The case studies have been defined as increasing the geometry complexity: A flat plate, which is the simplest geometry discussed in every aerodynamic book is tested at first. In the following, two finite span wings using the NACA 0024 and NACA 2412 airfoils are considered. Finally, a simplified 3D model of the Volkswagen Beetle is evaluated. The simulations are run using a boundary condition given by a free stream velocity with a small angle of attack to avoid flow separation and turbulent effects.

It's important to highlight that the APAME software returns C_X and C_Z aerodynamic coefficients normalized with the XY surface area (the reference is so fixed with the body). Thanks to a simple rotation of the system of reference by the angle of attack (AOA) it is possible to estimate the lift coefficient C_L for the overall 3D model (this coefficient is the integral of the forces in a direction perpendicular to the direction of the flow direction).

In the following the C_L value returned by APAME is compared to the lift coefficient for the 3D model available in the scientific literature or evaluated by xFoil software [13], correcting the results of 2D airfoil characteristics keeping into consideration the aspect ratio.

4.1 Flat Plate

The first geometry selected to test the voxel-based panel method is a flat plate, that is chosen for its geometrical simplicity and easy voxelization process. Moreover, the aerodynamic theory states that using a Joukowski mapping and applying the Kutta-Joukowski Lift Theorem, the lift coefficient for a flat plate at small AOA α can be assumed to be equal to $C_L = 2\pi\alpha$ [14].

Several simulations are performed changing panelization parameters, such as the number of longitudinal panels M and the voxel dimension vox_d to find the optimal trade-off in terms of accuracy.

The following geometrical dimensions and boundary conditions have been applied in the simulations run:

- Free stream velocity V = 27.778 m/s
- Air density $\rho = 1.225$ kg/m^3
- Flat plate airfoil chord c = 0.5 m
- Flat plate airfoil thickness t = 0.05 m
- Flat plate span length b = 2 m
- Angle of attack $\alpha = 0°$ to $6°$ with a $2°$ step
- Reference area for aerodynamic coefficients $S_w = b*c = 4$ m^2

The lift coefficients coming from several simulations performed using different discretization criteria at different AOAs are collected in Table 1. A first remark involves the choice of the numbers of longitudinal panels M, whose influence on C_L can be extrapolated from Table 1 and the plot included in Fig. 5. It can be drawn that a low value of M reflects on a more accurate lift coefficient estimation which is very close to the theoretical result.

The best performance is achieved when $M = 2$ (1,46% of error) with a significant reduction of the computational power required. This could be explained by the fact that a lower dimension of the overall matrix that describes the interaction of panels is obtained. This matrix must be inverted to find the flow solution so that numerical errors can increase with the size. Increasing M could bring singularities in the problem that may affect the final result. Another interesting study involves the choice of the best voxel resolution and how this parameter affects the aerodynamic lift coefficient (Fig. 6). In this

Table 1. Lift coefficient of flat plate at different AOA; comparison of Theory (Kutta-Joukowski theorem) and voxel-based panel method with different number of panels. In green the best discretization. In the last row, an estimation of the lift slope for the 3D model.

α [°]	Theory	M=5, vox_d= 0,025 m	M=10, vox_d= 0,025 m	M=10, vox_d= 0,0125 m	M=20, vox_d= 0,025 m	M=2, vox_d= 0,025 m	M=40, vox_d= 0,025 m	M=2, vox_d = 0,01 m
0	0,000	0,000	0,000	0,000	0,000	0,000	0,000	0,000
2	0,135	0,121	0,114	0,116	0,111	0,131	0,109	0,138
4	0,270	0,243	0,229	0,232	0,221	0,274	0,218	0,276
6	0,405	0,364	0,343	0,348	0,332	0,411	0,321	0,414
$C_{L\alpha_{3D}}$	3,867	3,474	3,273	3,323	3,170	3,923	3,125	3,961

case, an intermediate voxel dimension value should be chosen as a trade-off between good geometry fidelity and the fast run of the simulation. However, at least two layers of voxels are needed in the thickness dimension to avoid problems in the ray intersection voxelization, thus limiting the upper boundary of the voxel dimension interval.

4.2 NACA 0024 and 2412 Finite Wings

Since the evaluation of aerodynamic loads is crucial in the aeronautic and automotive field, the following case studies involve the voxelized 3D model of two finite wings based on airfoils taken from the NACA series which are commonly used in engineering applications.

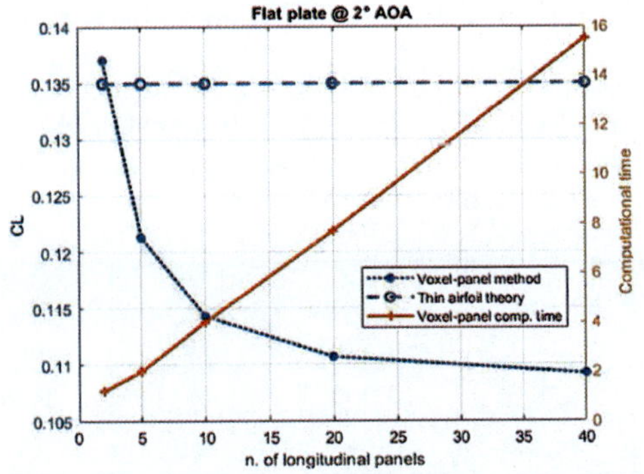

Fig. 5. Lift coefficient value at AOA = 2° for the flat plate for different discretization along the spanwise direction.

In more detail, the NACA 0024 and NACA 2412 which are widely used for low-speed aircraft wings are considered. The xFoil code is used to compute realistic lift coefficients to be used as a reference. xFoil computes the aerodynamic of the 2D airfoil, while the case study involves a finite span wing: a correction for 3D finite wings has

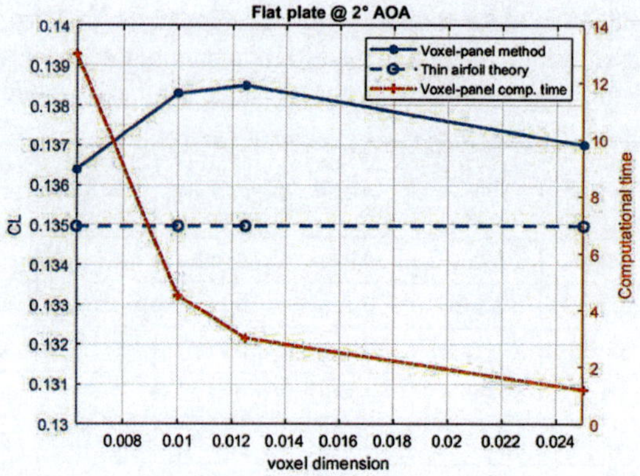

Fig. 6. Lift coefficient value at AOA = 2° for the flat plate for different voxel dimension discretization.

been applied depending on the aspect ratio. The following geometrical dimensions and boundary conditions have been applied in the simulations:

- Free stream velocity V = 27.778 m/s
- Air density $\rho = 1.225$ kg/m^3
- Airfoils chord c = 1 m
- Finite wingspan length b = 10 m
- Angle of attack $\alpha = 0°$ to 6° with a 2° step
- Reference area for aerodynamic coefficients $S_w = b*c = 10$ m^2

The lift coefficients for the NACA 0024 coming from simulations performed using different discretization criteria at a different AOA are collected in Table 2:

Table 2. Lift coefficient of NACA 0024 finite wing at different AOA; comparison of xFoil and voxel-base panel method with different number of panels. In green the best discretization. In the last row, an estimation of the lift slope for the 3D model.

α [°]	xFoil	M=2, vox_d= 0,02	M=2, vox_d= 0,01	M=20, vox_d= 0,02
0	0,000	0,000	0,000	0,000
2	0,204	0,197	0,201	0,178
4	0,408	0,394	0,401	0,357
6	0,611	0,591	0,599	0,536
$C_{L\alpha_{3D}}$	5,840	5,646	5,741	5,113

The lift coefficients for the NACA 2412 coming from simulations performed using different discretization criteria at a different AOA are collected in Table 3.

Table 3. Lift coefficient of NACA 2412 finite wing at different AOA; comparison of xFoil and voxel-based panel method with a different number of panels. In green the best discretization. In the last row, an estimation of the lift slope for the 3D model.

α [°]	xFoil	M=2, vox_d=0,01	M=10, vox_d=0,01
0	0,213	0,200	0,667
2	0,318	0,303	1,129
4	0,521	0,494	1,558
6	0,715	0,667	1,954
$C_{L\alpha_{3D}}$	7,801	7,370	24,439

From these results it follows that a lower amount of longitudinal panels reflects on a more accurate solution, very close to xFoil (1,70% error for NACA 0024 and 5,88% for NACA 2412) reducing the computational costs. Regarding the voxel dimension, in the simulation of the NACA 0024 finite wing, it can be noticed that a more accurate discretization ($vox_d = 0{,}01$) is needed to better capture the shape of the airfoil which affects the aerodynamic performances; a coarser voxel resolution reflects on a slightly higher error (3,31%). It should be remarked that a strange behaviour for the NACA 2412 model appears in the case of a fine voxel resolution, probably due to singularities in the overall system matrix inversion possible after the increase of the mathematical dimension problem.

4.3 Volkswagen Beetle

Finally, a simplified STL model of the classic Volkswagen Beetle is considered. As it can be seen in Fig. 3, the car silhouette in the XZ plane is extrapolated and extruded in the spanwise dimension from the original voxelized 3D model (Fig. 1). In the automotive field, drag estimation is more important than lift, so that this case study has been set up to investigate the evaluation of the overall drag, summing up: the induced drag coming from the APAME software, the friction drag, and the form drag estimated from literature with some geometrical assumptions. The following geometrical dimensions and boundary conditions have been applied in the simulation:

- Free stream velocity V = 20 m/s
- Air density $\rho = 1.225$ kg/m^3
- Car length c = 4 m
- Car height h = 1.6m
- Car width (span length) b = 1.54 m
- Angle of attack $\alpha = 0°$

- Reference area for induced drag coefficient $S_{planar} = b*c = 6.16\ m^2$
- Reference area (frontal area) for form drag coefficient $S_{front} = b*h = 2.46\ m^2$
- Reference area for friction drag coefficient $S_{fric} = c*(2b + 2h) + 2S_{front} = 30\ m^2$
- Reynolds number $Re = (V*c)/\nu = 5.71e6$

The APAME software returns the overall pressure distribution (Fig. 7) and an induced drag coefficient $C_{D_{ind}}$ evaluated in 0.0307. The form drag coefficient $C_{D_{form}}$ can be estimated to be around 0.3, approximating the car shape with an ellipse with the narrow side to the wind [15]. The friction drag coefficient can be estimated with the Prandtl one-seventh power law for turbulent flows [16], returning $C_{D_{fric}} = 0.00292$.

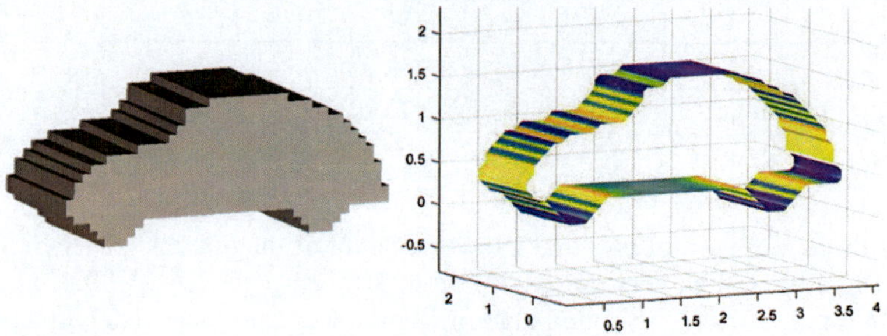

Fig. 7. On the left, the voxelized 3D model of the VW Beetle car, on the right the pressure distributions around the car returned by APAME with hided sides to better show pressure distribution in the front chassis.

After the computation of all the drag coefficient terms [N], it is possible to estimate the overall C_D of the 3D body by summing up all the drag contributions multiplied by the reference surfaces [N × m²], and then dividing by the frontal area [m²] which is the reference value used in the scientific literature dealing with the automotive industry (Table 4).

The drag coefficient which has been obtained in this way is consistent with real-life cars aerodynamic data available in the literature (VW Beetle overall $C_D = 0,48$ [17]), even the value obtained with our methodology should be considered a rough estimation after significant assumptions. However, this test demonstrates that the overall drag coefficient can be estimated from voxel-based panel methodology as well. About

Table 4. Volkswagen Beetle drag coefficient contributions and total drag coefficient evaluation

	C_D	$C_D \cdot S$
Friction	0,0029	0,0439
Induced	0,0307	0,1891
Form	0,3000	0,7392
Total	0,3946	0,9723

lift, as seen in the previous case studies satisfactory accuracy can be achieved: the benchmarking with literature or other proven software such as xFoil confirms it.

5 Conclusion

The panel method is still nowadays a valuable numerical method to estimate in the conceptual and preliminary design stages the aerodynamic forces and moments for subsonic flows: this achieved thanks to lower computational costs compared to CFD-based approaches. However, the main drawback of the potential flow approach is the meshing of the external surface of the 3D model with panels in which a potential flow element (sources and dipoles) must be associated. Nowadays the user must design from scratch a cloud of points that matches the external surface of the object he/she would like to test, which could be a highly demanding task, especially when complex and free form surfaces are considered.

The authors propose an innovative coupling of panel method with voxelization to overcome this issue: in this way it is possible the fast and completely automated panelling of the external surface of the object. Each panel collapse with the external square surface of each voxel belonging to the external surface of the voxelized model. In other words, the vertices of the cubic cells that compose the external layer of voxels are chosen as vertices of the panels.

Four case studies are included to show the accuracy of the methodology and the results are compared with benchmarking values coming from literature or reliable software. A good agreement (less than 6% of error in the coefficient estimation) is found when the lift is considered. This amount of inaccuracy should be considered satisfactory in a context of a preliminary design stage, where a high number of configurations must be tested in a short time to detect the design solution to bring to preliminarily and detailed design stages. Wind tunnel tests need time and economical efforts that can't be sustained in the conceptual design, while CFD tools typically require hours to run, thus reducing the number of configurations that can be evaluated. Parametric studies involving the number of longitudinal panels and voxel grid dimensions are included in the research. As a conclusion, it appears from the case studies analyzed that the lift can be predicted with good approximation ($<10\%$ error), while poorer results can be obtained for the drag (error $< 20\%$) where corrections for shape drag and skin drag have been applied.

In the near future, the proposed methodology will be improved to simulate every type of 3D object: up to now, only 2.5D models which can be seen as a 2D silhouette extruded in the third dimension are considered. Further studies will enhance the methodology capability simulating more complex and interesting bodies, comparing the results with other numerical methods such as the CFD when data from literature aren't available.

References

1. Abbott, I.H., VonDoenhoff, A.E.: Theory of Wing Sections: Including A Summary of Airfoil Data. Dover Publ, New York, NY (2010)
2. Jense, G.J.: Voxel-based methods for CAD. Comput. Aided Des. **21**(8), 528–533 (1989). https://doi.org/10.1016/0010-4485(89)90061-4

3. Bacciaglia, A., Ceruti, A., Liverani, A.: A systematic review of voxelization method in additive manufacturing. Mech. Indus. **20**(6), 630 (2019). https://doi.org/10.1051/meca/2019058
4. Kobayashi, K., Shimizu, H., Kaiho, M.: Numerical simulation of unsteady turbulent flow by voxel method. JSME. Int. J. Ser. B. **47**(3), 477–482 (2004). https://doi.org/10.1299/jsmeb.47.477
5. Onishi, K., Bale, R., Tsubokura, M.: Assessment of Rotating Wheel Vehicle Aerodynamics Simulation Using Cartesian Grid Method and Open-Grill Full Vehicle Models. SAE Technical Paper 2019–01–0660 (2019). https://doi.org/10.4271/2019-01-0660
6. Filkovic D.: APAME – Aircraft Panel Method Tutorial (2008). http://www.3dpanelmethod.com. Accessed 13 Feb 2021
7. Hoeijmakers, H.W.M.: Panel methods in aerodynamics; some highlights. In: Ballmann, J., Eppler, R., Hackbusch, W. (eds.) Panel Methods in Fluid Mechanics with Emphasis on Aerodynamics, pp. 1–34. Vieweg+Teubner Verlag, Wiesbaden (1988). https://doi.org/10.1007/978-3-663-13997-3_1
8. Katz, J., Plotkin, A.: Low-Speed Aerodynamics, 2nd edn. Cambridge University Press, Cambridge (2001)
9. Aremu, A.O., et al.: A voxel-based method of constructing and skinning conformal and functionally graded lattice structures suitable for additive manufacturing. Addit. Manuf. **13**, 1–13 (2017). https://doi.org/10.1016/j.addma.2016.10.006
10. Patil, S., Ravi, B.: Voxel-based representation, display and thickness analysis of intricate shapes. In: Ninth International Conference on Computer Aided Design and Computer Graphics (CAD-CG 2005), pp. 415–422, Hong Kong, China (2005). https://doi.org/10.1109/CAD-CG.2005.86
11. Bacciaglia, A. Ceruti, A. and Liverani, A.: Structural analysis of voxel-based lattices using 1D approach. 3D Printing and Additive Manufacturing. Published Online: 16 Apr 2021, https://doi.org/10.1089/3dp.2020.0178
12. Adam, A.: Converting a 3D logical array into an STL surface mesh. MATLAB Central File Exchange (2021). https://www.mathworks.com/matlabcentral/fileexchange/27733-converting-a-3d-logical-array-into-an-stl-surface-mesh. Accessed 1 Feb 2021
13. Drela, M.: XFOIL: an analysis and design system for low reynolds number airfoils. In: Mueller, T.J. (ed.) Low Reynolds Number Aerodynamics, pp. 1–12. Springer Berlin Heidelberg, Berlin, Heidelberg (1989). https://doi.org/10.1007/978-3-642-84010-4_1
14. Anderson, J.D.: Fundamentals of Aerodynamics in SI Units, 5th edn. McGraw-Hill, Singapore (2011)
15. Standards Australia (Organization) and Standards New Zealand: Structural design actions. Part 2. Sydney, N.S.W.; Wellington N.Z. Jointly published by SAI Global under licence from Standards Australia ; Standards New Zealand (2011)
16. Prandtl, L.: Bericht über Untersuchungen zur ausgebildeten Turbulenz ZAMM. J. Appl. Math. Mech./Zeitschrift für Angewandte Mathematik und Mechanik **5**(2), 136–139 (1925). https://doi.org/10.1002/zamm.19250050212
17. The Mayfield, Coefficient of Drag for Selected Vehicles. http://mayfco.com/dragcd~1.htm. Accessed:16 Oct 2021

Enhanced Statistical Shape Model: A Statistical-Based Tool to design Custom Orthopaedic Devices

Antonio Marzola, Francesco Buonamici(✉), Lorenzo Guariento, and Lapo Governi

Department of Industrial Engineering, University of Florence, Via di S. Marta 3, 50139 Florence, Italy
{antonio.marzola,francesco.buonamici,lorenzo.guariento, lapo.governi}@unifi.it

Abstract. Dealing with the design of personalized medical devices, mass production is not an option that can be hypothesized. Indeed, a cumbersome production process must be considered in such cases, mainly to account for a delicate design phase that needs to take into consideration, as input, an anatomy that vary each time. This article discusses the development of a statistical tool able to support the design of patient-specific devices. By expanding the classical formulation of the Statistical Shape Model (SSM) with the introduction of multiple levels of information within the same model, the authors have experimented with the concept of an "enhanced SSM". While the traditional SSM only provides information on the variations that a class of shapes can manifest, the eSSM may include more levels of information. The article discusses two possible mathematical formulations of such statistical tool. Its application to the design of custom-made pelvic implants is discussed. Such application scenario is described starting from the generation of the eSSM for the pelvis. The features of interest considered in this paper are the centers of the acetabular regions of the pelvis, the segmentation of the anatomy in a series of semantical regions that must be considered when developing a load-bearing implant. Finally, the conclusions of this research are drawn and discussed together with possible future development of eSSMs.

Keywords: Statistical shape analysis · SSM · Biomedical engineering · Pelvis · Human modelling · Custom implant design

1 Introduction

Statistical Shape Analysis (SSA) [1–6] is an established method widely employed to provide a quantitative description of the geometrical characteristics of a specific family of shapes.

When applied to the study of anatomical shapes, SSA allows to synthetize and analyze the information provided by a training set composed by an adequate number of healthy samples of the region of interest (ROI). In its classical formulation, SSA works

C. Rizzi et al. (Eds.): ADM 2021, LNME, pp. 27–38, 2022.
https://doi.org/10.1007/978-3-030-91234-5_3

on a set of training samples provided as discrete models (usually, Point Distribution Models - PDM [5]): the training dataset is then defined as the list of the spatial coordinates of each point of the samples. By assuming all the training samples to be in correspondence (i.e. analogous points are reported in the same positions), SSA exploits powerful mathematical tools, such as multivariate statistical analysis, to interpret and synthesize the information delivered by the training dataset about the possible shape variations from a mean shape. SSA results in a generative parametric model, usually referred to as Statistical Shape Model (SSM), able to describe the family of shapes under consideration and to "summarize" them under a parametric model.

Exploiting its ability to encode an a-priori knowledge of the healthy shape of the ROI, SSM have proved to be particularly suitable as a source of templates for 3D medical image segmentation [5, 7], for the reconstruction of deformed or defective models [1, 8–10] and for dysmorphisms and asymmetries recognition and evaluation [11–13].

With the aim to increase the scope and capabilities of the SSM, this work explores the idea of an implementation of the classical SSA by expanding the information enclosed in the resulting statistical model. In particular, the focus is on the application of SSMs to design patient-specific devices, taking the pelvis as case study, increasing its usefulness as a guide in the design process of personalized orthopedical devices.

Indeed, the design of patient-specific devices would be made easier by an access to a collection of statistical information of geometrical descriptors and landmarks of the ROI that goes beyond the mere surface data. Such database would increase the efficiency of the design process by providing easy-to-access information on various geometric characteristics, improving the typical approach which only exploits surface information.

In the application scenario considered by this publication - the pelvis - the use of an enhanced SSM (eSSM) would be beneficial whenever the anatomical structure of the acetabulum is severely damaged, and an implant is required. In such cases, the first step is typically the retrieval of the missing geometry and the computation of the natural articular Center Of Rotation (COR) to restore the limb functionality. While the first operation can be effectively carried out automatically by a typical SSA (which calculates the missing geometry according to the highest probability), manual operations are required to retrieve the COR [14]. When the contralateral acetabulum is healthy the most common approach is to mirror its COR, though such method fails when an implant or a bi-lateral defect is presented; in addition, the mirroring approach is an important approximation, since the symmetry in human anatomy is only an abstraction and a perfectly symmetric anatomy does not actually exist. The use of an eSSM makes this step automatic, by building an SSM that encompasses the positions of the CORs as learned from the dataset. By following this example, the eSSM could include all the significant landmarks useful in the subsequent designing phases (attachment regions of muscles and tendons, references for the definition of specific planes, etc.).

In addition to landmarks, the eSSM could also include the semantic segmentation of the model starting from the semantic segmentation of the mean shape resulting from the SSA: by assigning each point of the mean model to a specific region, this information will be inherited by the fitted mean shape on the defective target anatomy.

This work aims to explore the capabilities of an enhanced SSM and its implementation, with a focus on the design of personalized orthopedic implants, which could greatly benefit of an automatic procedure able to simultaneously cover several aspects of the design phase, which depend on the specific application. As the presented paper focuses on pelvic personalized implants, the information enclosed within the eSSM presented in the following, are namely the articular centers of rotation, the acetabular orientation and the semantic segmentation.

The paper is organized as follows: Section 2 presents the core concepts of the mathematical framework of the SSA and the authors' proposal to enhance the classic formulation. Section 3 discusses the application of the eSSM to the design of orthopaedic devices. Section 4 presents the conclusions of the authors and possible future developments and applications of this research.

2 Classical Statistical Shape Analysis and proposed Improvements

The mathematical framework of the SSA is strongly affected by the representation of the training samples; in this work, the classical approach based on the Principal Component Analysis (PCA) applied on the training samples provided as PDM is considered.

PDM is the most convenient representation of a complex geometry [15], consisting of a discrete description based on landmarks distributed across the surface [5, 16]. In the following, M will be defined as the number of the training samples, and N as the number of points of each training sample. All the training samples have the same number of points because, as said, it is mandatory that they are in correspondence: in the related literature, many approaches to create a suitable training set are described [17].

With the just described assumptions, each training shape Γ^{i} can be represented as a discrete set of landmarks $\mathbf{x}^{i}$:

$$\Gamma^{i} = \{x_k^i \big| x_k^i \in R^3,\ k = 1,\ \ldots,\ N\} \tag{1}$$

Referring to Eq. 1, in typical SSA approach $\mathbf{x}_k^i$ contains the Cartesian coordinates, in three dimensions, of the k-th point. Γ^{i} is then a $3 \times N$ matrix.

Each shape Γ^{i} is rearranged by stacking the coordinates of each point k in a large ($3N$ elements) column vector $\vec{x}^{\,i}$. Considering the global reference system in x-, y- and z-axis:

$$\vec{x}^{i} = [x_1^i\, x_2^i \ldots x_N^i y_1^i y_2^i \ldots y_N^i z_1^i z_2^i \ldots z_N^i] \tag{2}$$

As a result, the column vector reported in Eq. 2 represents the $3N$ variables of the single observation Γ^{i}.

The whole training set has to be then organized in a $3N \times M$ matrix (**TS**) whose columns are the observations (i.e. the M shapes) and rows are the variables:

$$\mathbf{TS} = [\vec{x}^1 \vec{x}^2 \ldots \vec{x}^M] \tag{3}$$

Once the dataset **TS** is properly defined, and assuming that the family of shapes under consideration is a linear space of R^{3N}, new shapes can be generated by linear combinations of the training samples:

$$x = \bar{x} + \sum_{m=1}^{c} \alpha_m \sqrt{\lambda_m} \varphi_m \tag{4}$$

Where:

- $\bar{\mathbf{x}}$ is the mean shape defined as:

$$\bar{\mathbf{x}} = \frac{1}{M}\sum_{i=1}^{M}\vec{\mathbf{x}}_i \tag{5}$$

- λ_m and φ_m are, respectively, the eigenvalues and eigenvectors resulting from the eigendecomposition of the covariance matrix **S**:

$$\mathbf{S} = \frac{1}{M-1}\sum_{\mathrm{i}=1}^{M}(\vec{\mathbf{x}}_i - \bar{\mathbf{x}}_i)(\vec{\mathbf{x}}_i - \bar{\mathbf{x}}_i)^T \tag{6}$$

A PCA-based approach is typically used to provide the eigendecomposition on **S** delivering the principal *Modes of Variation* (MoVs) φ_{m} (eigenvectors of the matrix **S**) and their respective variances λ_{m} (eigenvalues of the matrix **S**) estimated on each points $\mathbf{x}_k$ of the initial dataset **TS**. PCA returns the plausible deformations, delivered as directions (φ_{m}) and related amplitudes (λ_{m}) of displacement, of every single point of the mean shape. λ_m and φ_m are ordered by their variances so that $\lambda_1 \geq \cdots \geq \lambda_{M-1}$.

- $\boldsymbol{\alpha} \in R^M$ is a coefficient vector that permits to constrain the allowed variation to plausible shape; usually, $\boldsymbol{\alpha}$ is delimited in the range [–3;+3], which corresponds to a deformation that is 3 standard deviations away from the mean ($\boldsymbol{\alpha}$ =0 means that the model is equal to the mean shape).
- c represents the number of significant eigenvalues. In the presented application, c is defined so that the accumulated variance (the numerator of Eq. 7) reaches a certain ratio r of the total variance (the denominator of Eq. 7). Common values of r are between 0.9 and 0.98 [5]. In this work, an r equal to 0.98 has been considered.

$$\frac{\sum_{i=1}^{c}\lambda_i}{\sum_{j=1}^{M-1}\lambda_j} = r \tag{7}$$

As Eq. 4 shows, SSA results in a mathematically convenient representation of the information contained in the training dataset: such representation is delivered as a parametric model able to generate new consistent set of data simply varying the parameters of $\boldsymbol{\alpha}$.

In the typical approach, the set of data is represented, as said, by only the spatial coordinates of the training samples. As a result, Eq. 6 is able to return pathologically unaffected geometries belonging to the same family of shapes forming the training set. Such model can be used, for example, to infer the full shape of the defective anatomy to be restored [9].

In this paper, the described approach is improved expanding the information enclosed in the SSM. Regarding the information contained in the training set, the vector $\vec{\mathbf{x}}^i$ can be expanded by including not only the spatial coordinates of the points describing Γ^{i}, but also those of significant other points such as a joint center, the direction of load applications and so on.

By defining n more points in addition to the N points used to describe the shape, Eq. 1 becomes Eq. 8.

$$\Gamma^{i} = \left\{ \mathbf{x}_k^i \middle| \mathbf{x}_k^i \in \mathrm{R}^3, k_i = 1, \ldots, N, N+1, \ldots, N+n \right\} \tag{8}$$

As discussed, the added points must be in correspondence across all the Γ^{i} too. Starting from the enlarged training samples, the SSA follows the subsequent steps as described above.

An additional approach that could be followed to enhance the SSM could be based on the introduction of additional information in the already built mean shape. While the points composing $\overline{\mathbf{x}}$ remain the same, additional information could be associated to each point using labels, weights, and other parameters. By following this strategy, the mean shape and the SSM remain the same, as they are built using the same dataset.

This information could be, as an example, the indication to which region of the semantic segment each point belongs, or the position of significant landmarks. Such additional information can be effortlessly added after the definition of the SSM directly on the mean shape instead of on each training samples. In this case, and considering the most generic case presented in Eq. 8, the mean shape can be defined as:

$$\overline{\mathbf{x}} = \left\{ \overline{\mathbf{x}}_k | \overline{\mathbf{x}}_k \in \mathrm{R}^d, k_i = 1, \ldots, N+n \right\} \tag{9}$$

where d is defined as an integer greater than or equal to 3.

As a consequence, the so obtained eSSM allows, in the first case, to synthetize more geometric information than the mere shape and, in the second case, to encompass more information about each point of the mean shape in addition of its spatial position. This information will be inherited to the restored shape when the eSSM is used to infer the full shape of the defective model.

It is worth to note that adding rows to the TS can affect the results of the subsequent PCA, but, operatively speaking, the difference introduced in the covariance matrix by adding a few points causes negligible variations to the principal components. Authors added up to 30 new points (the considered training samples had approximately 5000 points) without experiencing significative differences between the two covariance matrices.

3 Custom Pelvic Implants for Severe Bone Defects

To date, custom metal pelvic implants are widely employed to deal with large bone defects because they can be tailored to the very patient's anatomy and can be adapted to fill the missing bone and provide a stable fixation whenever standard implants are not suitable [18]. The design process requires an accurate pre-operative planning to assess the bone deficiency and define the main constraints which will drive the whole process. An accurate virtual reconstruction of the defected anatomy is crucial to determine the best surgical approach and to design an implant which perfectly fits the patient's anatomy with good biological and mechanical performances; furthermore, a well-designed prosthesis should restore the limb functionality with a range of motion as similar as possible to the physiological one.

Due to recent advancements in additive manufacturing technologies, biocompatibility is easily achievable by means of biocompatible powders as Titanium and Chromium-Cobalt alloys. Additive manufacturing technologies allow the design of implants with complex internal lattice structures [21, 22] which could be beneficial to achieve lightweight and to improve bone ingrowth, though the assessment of the actual mechanical behavior requires complex patient-specific FE models.

The performances in terms of mechanical resistance and limb functionality are strictly related to the implant's geometry. Contrarily to other surgeries, e.g. maxillo-facial surgery, pelvic implants does not always reproduce the original anatomical shape to limit the overall volume and weight, though the acetabular kinematics is restored whenever possible (Fig. 1). In addition, it is mandatory to avoid stress concentrations both in the bone and in the implant to avoid fractures and mechanical failure [19, 20].

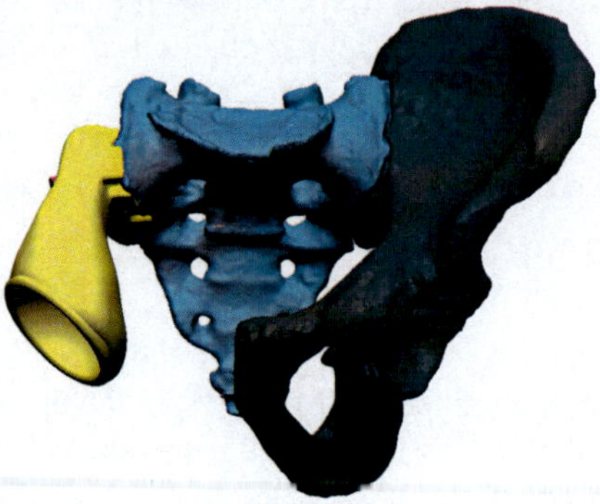

Fig. 1. Patient Specific Implant – Acetabular reconstruction that neglects the restoration of the original anatomy of the iliac crest and pubic ring.

To support the design process of a custom acetabular implant, knowledge of the reconstructed anatomy and the acetabular parameters are of great interest because the first can be exploited to create the implant's shape, while the latter drives the generation of the acetabular cup.

Specifically, the retrieval of the natural acetabular COR and orientation (to set a correct acetabular inclination and anteversion) is essential to restore the limb functionality, guarantee an equal leg length and ensure a good kinematics.

The acetabular orientation can be defined as the normal to the plane of best fit through a series of points on the acetabular rim. Typically, when the contralateral acetabulum is present, the spatial coordinates of the articular COR can be mirrored with respect to the sagittal plane; evidently, in case of bilateral defects, such method cannot be applied. An automatic procedure able to deliver the above cited information would dramatically accelerate the design of custom orthopedic implants and deliver fast and reliable results with little effort. This work aims to implement an eSSM which encloses, besides information regarding the shape of the pelvis, the acetabular parameters and provides details about the location of important landmarks which can be exploited to create patient specific FE models to optimize the implant's design and deliver high performance devices.

3.1 The e-SSM Implementation

The implemented eSSM was trained with 50 healthy pelvises downloaded from an open-source repository of anonymized medical images [23]. For this work, there was no distinction between sex or age, though the reliable automatic procedure will allow to refine the model and differentiate between sex, ethnicity and so on as soon as the number of cases composing the dataset would guarantee an adequate variability to perform such differentiations. The CT scans have been manually segmented by an experienced user with Materialise® Mimics 23.0 to extract the external shape and the acetabular CORs; each COR was considered coincident with the center of a sphere fitted to each femur head.

The SSA is performed by an in-house automatic algorithm developed in MATLAB®. Such algorithm is based on the one presented in [7], properly improved to deliver the eSSM.

The input data of the algorithm are the triangular mesh of the complete pelvis and the coordinates of the CORs. In the first stages, the algorithm automatically finds the correspondences between the training samples and re-arranges the resulting dataset to obtain $\overrightarrow{\mathbf{x}}^i$ (Eq. 8) and **TS** (Eq. 3). In this application, the column vector $\overrightarrow{\mathbf{x}}^i$ contains both the spatial coordinates of the shape descriptors as well as the positions of the CORs for the training sample Γ^i. As a consequence, by defining $[x^i_{RC} y^i_{RC} z^i_{RC}]$ and $[x^i_{LC} y^i_{LC} z^i_{LC}]$, respectively, the right and left COR spatial coordinates, $\overrightarrow{\mathbf{x}}^i \in \mathrm{R}^{3N+6}$ can be defined as:

$$\overrightarrow{\mathbf{x}}^i = \left[x^i_1 x^i_2 \ldots x^i_N x^i_{RC} x^i_{LC} y^i_1 y^i_2 \ldots y^i_N y^i_{RC} y^i_{LC} z^i_1 z^i_2 \ldots z^i_N z^i_{RC} z^i_{LC}\right] \tag{10}$$

After the creation of the matrix **TS**, the algorithm performs the described PCA-based SSA including the CORs. The result of the automatic procedure delivers the mean shape and the mode of variation of the pelvis and CORs as learned from the variability of the initial models. Figure 2 shows the first two modes of variations from the maximum negative ($\alpha = -3$) to the maximum positive ($\alpha = +3$) amplitude. $\alpha = 0$ is equal to the mean shape.

On the obtained mean shape, an expert user has selected significant landmarks and the points pertaining to a semantic segmentation of the pelvis (e.g. Fig. 3a). Such landmarks represent points of application of loads and constraints which could be exploited to create a FE model of the pelvis including ligaments and boundary conditions. Additionally, the user has selected the points shown in Fig. 3b on the acetabular rim, useful to define the acetabular orientation.

Operationally speaking, this additional information will be added to the $\overline{\mathbf{x}}$ vector of the mean shape as in Eq. 9, expanding the dimensionality of each $\overline{\mathbf{x}}_k$. The kind of information delivered is defined by the position in the $\overline{\mathbf{x}}_k$ vector. Each segmented region is saved as a new PDM, while the coordinates of each selected landmark are exported as text files. This way, the developed algorithm can automatically store the additional data on the proper point of the mean shape by comparing the position within the vector of each point with the positions of the points selected by the user.

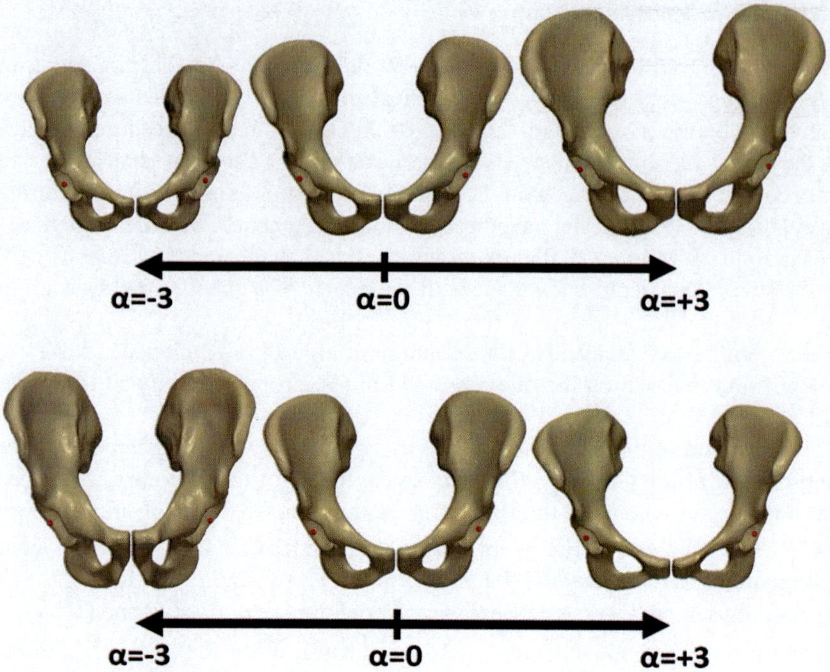

Fig. 2. The first (top row) and the second (bottom row) modes of variation of the pelvis by varying α. The CORs are depicted as red dots.

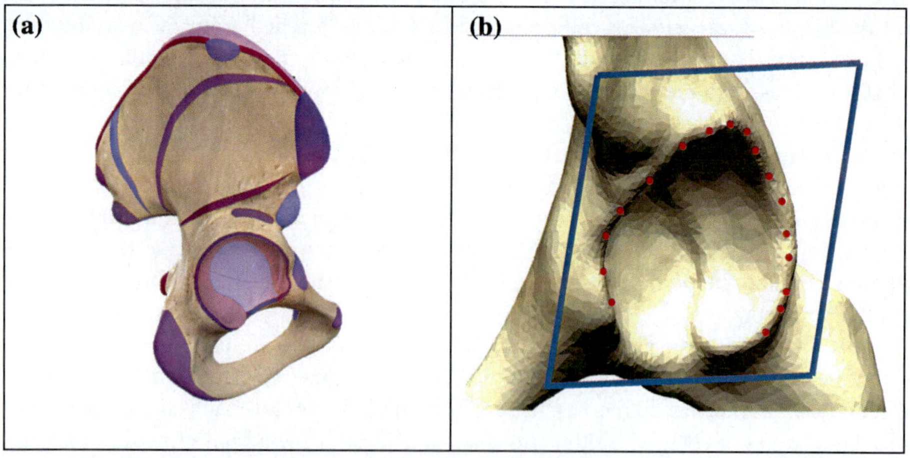

Fig. 3. a) Semantic segmentation of the pelvis: the areas highlighted in the figure define meaningful regions. b) selection of the points on the acetabular rim performed by the user.

In this application, the positions within the $\overline{\mathbf{x}}_k$ vector are defined as follows:

- the first three positions are reserved for the *x*- *y*- and *z*- spatial coordinates of the global reference system;
- in the fourth position, a scalar index defines the attribution of the point to a specific labelled region. As an example, a value of 1 defines all the points belonging to the iliac crest, a value of 21 defines the anterior gluteal line, etc.
- the fifth position is a binary variable that equals to 1 if the point has been selected as an acetabular rim point and to 0 otherwise.

As a consequence, in the case addressed $\overline{\mathbf{x}}_k \in \mathrm{R}^5$.

When this eSSM is used to infer the full shape of the defective model to be restored, the deformation process affects only the first three positions of the $\overline{\mathbf{x}}_k$ (i.e. the spatial coordinates of the point), while the other information remains unchanged. Therefore, all the additional information stored in each point of the mean shape will be inherited by the restored model. The developed eSSM has been used in a real-case scenario characterized by a large acetabular deficiency to design a patient-specific implant that filled the missing bone regions and restored the COR effectively. The result of the design operation is visible in Fig. 4.

In this specific case, the eSSM was employed to semi-automatically reconstruct the pelvic anatomy and to retrieve the COR location and orientation: typically, both the restoration of the anatomy and the retrieval of the COR's information would require cumbersome manual operations. The 3D model of the pelvis was segmented and converted in a STL format by means of Materialise Mimics® software; the obtained model was then used as a target to fit the eSSM. As a result, eSSM provided a coherent healthy shape for the target pelvis to use both as a reference for the prosthesis design as well as for the identification of the COR. The prosthetic acetabulum was generated with simple

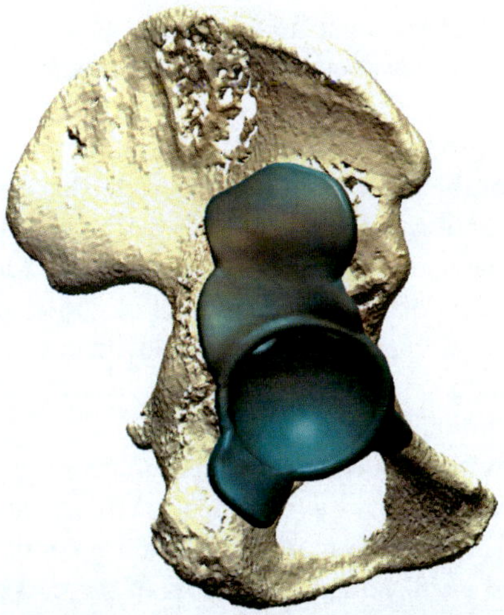

Fig. 4. Patient Specific Implant designed with the support of the eSSM.

CAD operations using the diameter as input and considering the COR as its center. The flanges were obtained by thickening and cropping the triangular mesh obtained with the eSSM.

4 Conclusions

This article has experimented with the idea of developing and applying a specifically built and enhanced SSM to support the development of personalized medical devices: by adding more information than the mere geometry, the data included in the eSSM can be, theoretically, tailored for the specific application that must be carried out. As an example, its application for the design of orthopaedical implants for the pelvic district has been discussed and the results obtained through its application in a real case scenario have been presented. In the reported eSSM only COR location has been included, but it could be implemented by adding information regarding acetabular inclination and anteversion, so that many manual operations can be saved.

Indeed, eSSMs could represent an interesting strategy to support the increasing growth in the need for more effective design methods and techniques in the biomedical field that are able to consider the specific needs of each patient.

As it has been discussed in the article, the introduction of different levels of information within the same statistical model can provide the designer with a more complete tool. The modelling of orthopaedical devices can benefit from a complete mean model of the surfaces composing the anatomy of interest as well as from a series of additional data. On this aspect, this work has proposed to introduce novel information in the eSSM of the pelvis, starting with the introduction of the acetabular COR among the points described by the SSM. As discussed, knowledge on this feature is essential in the design of pelvic implants and its automatic retrieval could, consequently, provide a ground-breaking advantage to a designer empowered with such a tool. Additionally, topological and semantical labels have been included to provide information about significant landmarks and areas that must be taken into consideration when designing a load-bearing implant. These key aspects could be expanded in future work, as different application scenarios, features and landmarks could be considered for the development of eSSM, tailored to answer to the requirements of designers. Staying in the orthopaedical application field, one of the most interesting features that could be studied relates to the bone density across solid bone. Bone density directly influences design choices such as fixation strategies and the shape of the fixation elements. Moreover, it is influenced by several aspects such as age of the patient or pathologies, and it could be evaluated analyzing CT data. All these elements could be introduced in a eSSM, essentially building a 4D model that is able to associate an estimated density value to each point of the solid retrieved with a traditional SSM approach.

Finally, the processes, within the entire design framework of patient-specific devices, that could benefit the most from the development of this type of tools are simulation and analysis. Dealing with load-bearing orthopaedical implants, the identification of the most performing shape and a subsequent validation of the design are key steps in order to obtain a light but safe device. The use of FE models, in this area, is hindered by the high efforts required to build a model that is faithful to the patient's anatomy as well as characterized by a sufficient level of detail and a proper definition of loads, constraints and boundary conditions between all the elements. Up to this day, this type of result can be achieved only by means of time-consuming manual operations; accordingly, most implants are designed with high tolerances (producing a part that weighs more). The introduction of eSSMs could allow to adapt a pre-built high quality FE model to the specific anatomy of the patient. The FE model could be built and fully detailed using the mean shape of the eSSM; the mean FE model with all its information about point of application of forces, ligaments, constraints can be fitted to the patient-specific anatomy. As a result, a high-fidelity analysis could be run in a short time window without any compromise on its quality. In addition, this approach could pave the way towards a ready-to-use models that will make it possible to perform stress analyses without the need of specialized users.

References

1. Fuessinger, M., et al.: Planning of skull reconstruction based on a statistical shape model combined with geometric morphometrics. Int. J. Comput. Assist. Radiol. Surg. **13**(4), 519–529 (2017). https://doi.org/10.1007/s11548-017-1674-6
2. Cootes, T.F., Taylor, C.J., Cooper, D.H., Graham, J.: Active shape models-their training and application. Comput. Vision Image Underst. **61**(1), 38–59 (1995)
3. Cootes, T.F., Edwards, G.J., Taylor, C.J.: Active appearance models. IEEE Trans. Pattern Anal. Mach. Intell. **23**(6), 681–685 (2001)
4. Blanz, V., Vetter, T.: A morphable model for the synthesis of 3D faces. In: SIGGRAPH 1999, pp. 187-194. ACM press (1999)
5. Heimann, T., Meinzer, H.P.: Statistical shape models for 3D medical image segmentation: a review. Med. Image Anal. **13**(4), 543–563 (2009)
6. Heimann, T., et al.: Comparison and evaluation of methods for liver segmentation from CT datasets. IEEE Trans. Med. Imaging **28**(8), 1251–1265 (2009)
7. Audenaert, E.A., et al.: Cascaded statistical shape model based segmentation of the full lower limb in CT. Comput. Methods Biomech. Biomed. Eng. **22**(6), 644–657 (2019)
8. Fuessinger, M.A., et al.: Virtual reconstruction of bilateral midfacial defects by using statistical shape modeling. J. Cranio-Maxillofacial Surg. **47**(7), 1054–1059 (2019)
9. Semper-Hogg, W., et al.: Virtual reconstruction of midface defects using statistical shape models. J. Cranio-Maxillofacial Surg. **45**(4), 461–466 (2017)
10. Marzola, A., Governi, L., Genitori, L., Mussa, F., Volpe, Y., Furferi, R.: A semi-automatic hybrid approach for defective skulls reconstruction. Comput. Aid. Des. App. **17**(1), 190–204 (2019)
11. Mendoza, C.S., Safdar, N., Okada, K., Myers, E., Rogers, G.F., George, M.: Personalized assessment of craniosynostosis via statistical shape modeling. Med. Image Anal. **18**(4), 635–646 (2014)
12. Keustermans, W., et al.: High quality statistical shape modelling of the human nasal cavity and applications. R. Soc. Open Sci. **5**(12), 181558 (2018). https://doi.org/10.1098/rsos.181558
13. Lynch, J.T., et al.: Statistical shape modelling reveals large and distinct subchondral bony differences in osteoarthritic knees. J. Biomech. **93**, 177–184 (2019)
14. Gelaude, F., Clijmans, T., Broos, P.L., Lauwers, B., Vander Sloten, J.: Computer-aided planning of reconstructive surgery of the innominate bone: automated correction proposals. Comput. Aid. Surg. **12**(5), 286–294 (2007)

15. Cootes, T.F., Taylor, C.J., Cooper, D.H., Graham, J.: Training models of shape from sets of examples. In: Hogg, D., Boyle, R. (eds.) BMVC92, pp. 9–18. Springer, London (1992). https://doi.org/10.1007/978-1-4471-3201-1_2
16. Bookstein, F.L.: Landmark methods for forms without landmarks: morphometrics of group differences in outline shape. Med. Image Anal. **1**(3), 225–243 (1997)
17. Marzola, A., Servi, M., Volpe, Y.: A reliable procedure for the construction of a statistical shape model of the cranial vault. In: Rizzi, C., Andrisano, A.O., Leali, F., Gherardini, F., Pini, F., Vergnano, A. (eds.) ADM 2019. LNME, pp. 788–800. Springer, Cham (2020). https://doi.org/10.1007/978-3-030-31154-4_67
18. Baauw, M., Van Hellemondt, G.G., Van Hooff, M.L., Spruit, M.: The accuracy of positioning of a custom-made implant within a large acetabular defect at revision arthroplasty of the hip. Bone. Joint. J. **97**(6), 780–785 (2015)
19. Yan, C., Hao, L., Hussein, A., Young, P.: Ti-6Al-4V triply periodic minimal surface structures for bone implants fabricated via selective laser melting. J. Mech. Behav. Biomed. Mater. **51**, 61–73 (2015)
20. Weinans, H., Sumner, D.R., Igloria, R., Natarajan, R.N.: Sensitivity of periprosthetic stress-shielding to load and the bone density-modulus relationship in subject-specific finite element models. J., Biomech. **33**(7), 809–817 (2000)
21. Guariento, L., Buonamici, F., Marzola, A., Volpe, Y., Governi, L.: Graded gyroid structures for load bearing orthopedic implants. In: Editor, F., Editor, S. (eds.) Proceedings of the 2020 IEEE 10th International Conference on "Nanomaterials: Applications and Properties". IEEE (2020)
22. Zanetti, E.M., Aldieri, A., Terzini, M., Calì, M., Franceschini, G., Bignardi, C.: Additively manufactured custom load-bearing implantable devices: grounds for caution. Austral. Med. J. **10**(8), 694–700 (2017)
23. https://www.cancerimagingarchive.net

Design of a Low Cost 3D Printable Single-Component Compliant Mechanism for FWMAV's Wing Actuation

Giovanni Carollo(✉), Tommaso Ingrassia, and Antonio Pantano

Università Degli Studi Di Palermo, Palermo, Italy
giovanni.carollo01@unipa.it

Abstract. Considering a modern approach to design, one of the viable options for developing innovative projects is the possibility of integrating the effectiveness of the solutions offered by nature and living beings with the latest design methods. With this in mind, the following research exploits the idea of reproducing the natural flexibility inherent in biological structures by combining the advantages of compliant mechanisms with the adaptability of additive manufacturing processes. In the specific, the authors intend to highlight the potential and critical aspects of a possible approach for the application of compliant mechanisms in the development of single-component structures suitable for the actuation of bio-inspired Flapping Wing Micro Air Vehicles (FWMAVs), which can be produced via low cost 3D printing. Some designs conceived by interpreting the movement of insects' wings have been developed with the aim of reproducing the functionality and morphology of their thorax through single-component flexible mechanisms. The results of this research demonstrate the high potentiality of realizing bio-inspired single-component compliant mechanisms through 3D printing.

Keywords: Compliant mechanisms · Flapping wing micro air vehicles · Bio-inspired · Additive manufacturing

1 Introduction

Generally, when it is necessary to design new mechanisms for the transmission of motion and forces, multicomponent structures are designed. These kinds of structures usually are composed of a set of rigid elements connected to each other by joints of various kinds. However, by observing the structures of living organisms present in nature, whether they are plants or animals, it can be seen that these have been able to develop in such a way as to exploit the flexibility of the elements that compose them, rather than their rigidity, so that can perform functions necessary for their survival. Just think of the ability of organs such as the lungs and the heart that are able to sustain pressure gradients by repeatedly expanding and contracting, the function performed by cartilages, tendons and muscles that allow the movement of the limbs, or the ability of the octopuses' tentacles to adapt to the objects to be manipulated or the movement to be made. Based on these premises, in the last years many researchers have tried to find innovative solutions by

C. Rizzi et al. (Eds.): ADM 2021, LNME, pp. 39–49, 2022.
https://doi.org/10.1007/978-3-030-91234-5_4

designing bio-inspired structures. In this context, Compliant Mechanisms (CMs) constitute a tool of considerable importance to try reproducing biological structures, being able to integrate rigid and flexible zones within a single-component structure. Although the main Compliant Mechanisms' application areas are still substantially related to precision manipulators [1, 2], MEMS [3], nano-pliers [4] and micro-grippers [5–9], it can also be found in several recent applications in the field of robotics the growing tendency to simultaneously exploit the possibility of creating flexible structures and to take advantage of the increasingly in-depth knowledge on living organisms [10–12]. For instance, the studies on the biomechanical principles related to the structure of the fins of fishes have made it possible to use the so-called Fin Ray Effect to design soft fingers composed of compliant structures [13], which, without further incorporated drives, are able to maneuver fragile objects simply by deforming themselves, offering unmatched performance. In a similar way, the analysis of the wing structure of ladybugs inspired the creation of compliant origami structures suitable for self-locking and capable of elastic energy storing in large quantities through flexible surfaces [14]. The in-depth observation of living organisms has also allowed the development of innovative propulsion systems for bimodal unmanned vehicles [15] that are inspired by the flying squid's propulsion system in order to expel water at high pressure through a CO2 cartridge. The interest towards the design of this kind of UAV suitable for both aerial and underwater motion [16] it was indeed increased by the possibility of benefiting from the study of morphology of birds capable of plunge-diving. By observing the plunge movements of gannet's Morus, the wing root pivot joint's radial load of a submersible aircraft was studied [17], so that it can be exploited in the development of innovative HUAUV (Hybrid Unmanned Aerial Underwater Vehicles). In that regard, the field of robotics that deals with Micro Air Vehicles (MAVs) undoubtedly constitutes a particularly suitable field both for a bio-inspired approach to design and for the use of Compliant Mechanisms. In fact, bio-inspired CMs used for MAVs could store part of the energy in the form of elastic deformation energy, reduce friction, weight, wear or the need for lubrication, and simplify the assembly process by reducing the amount of components needed, also allowing the development of single-component structures. These advantages offered by CMs are remarkable especially considering the possibility of being able to increase the flight range of the MAV and decrease the construction complexity. Among the various types of MAVs, although rotary-wing and fixed-wing drones are the most developed and widespread on the market, insect-inspired flapping-wing drones have recently gained increasing interest from researchers [18–21]. Compared to rotary-wing drones, these bio-inspired Flapping Wing Micro Air Vehicles (FWMAVs) generate the necessary lift for flight by alternately moving the wings around a vertical axis. FWMAVs are able to fly effectively even with small wings in case of low Reynolds numbers and that makes them suitable for being made with reduced dimensions. Moreover, compared to MAVs, FWMAVs have a greater thrust/total input power ratio with the same thrust/area of wing sweep [22]. The bio-mimetic capabilities of these vehicles together with the possibility of reaching very small dimensions, allow their use also for security purposes, defense, surveillance, search and rescue in confined environments, collapsed structures or in the presence of risks of chemical nature. Based on the above considerations and inspired by the muscular structure of insects, the authors have proposed and analyzed several alternative solutions of single-component compliant mechanisms to be used in bio-inspired FWMAVs and produced through low-cost 3D printing.

2 Materials and Methods

Unlike MAVs, FWMAVs waste a significant amount of the engine power to overcome the inertial forces that act on the wings when their motion is reversed. For this reason, in order to design efficient FWMAVs with reduced energy consumption, in this research the flight mechanisms of flapping wing insects were preliminarily analyzed. Big insects, such as dragonflies, use two direct synchronous muscles (Fig. 1) to fly. These muscles, alternately contracting, actuate the wings to which they are directly connected to carry out the upstroke and the downstroke.

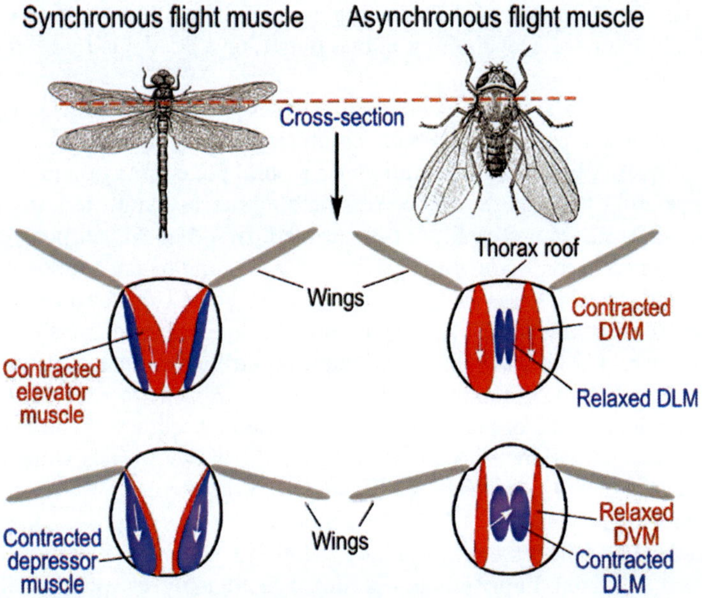

Fig. 1. Insect chest cross section schematic: direct synchronous muscles (left), indirect asynchronous muscles (right).

Small insects, such as flies or bees, instead, use two types of asynchronous indirect muscles whose contraction causes a cyclic deformation of the upper part of the thorax, which generates the alternating movement of the wings [23]. Some studies show that the ability to store energy through internal elastic structures, rather than the use of particularly efficient muscles, allows for lower energy consumption [24].

Based on these observations, it is assumed that the use of structural elements, able to deform and returning energy cyclically, could help to increase the efficiency and flight range of a FWMAV. For this reason, the authors decided to exploit low cost 3D printable single-component Compliant Mechanisms for the design of new bio-inspired FWMAVs. Analyzing different prototypes of FWMAVs in scientific literature, it emerges that some authors have proposed the use of flexible elements in the actuation mechanism, although not produced with additive manufacturing processes.

A significant contribution in this direction is provided by Bejgerowski et al. [25] that proposed to use the injection molding process to obtain a single-component mechanism where flexible regions allow alternating rotation of the flapping wings. Wood et al. [26] have proposed another interesting approach to integrate rigid zones and flexible joints within a component. In this study authors proposed to use the so-called Smart Composite Microstructures (SCM) process, consisting of laser micromachining and lamination operations, through which polymeric material, inserted between rigid sheets of composite material, makes the sheets flexible in the areas where empty spaces are created. Although both previous studies suggest interesting uses of flexible structures for MAVs, their feasibility turns out to be complex due to the manufacturing processes proposed. For these reasons, in this study, Fused Deposition Modeling (FDM) has been chosen for the manufacturing of the new single-component CMs to use in FWMAVS. FDM, in fact, is a low costs technology, allows obtaining a good quality level and offers a wide availability of materials.

This Additive Manufacturing technology was developed by Stratasys Inc. in USA in 1990s, and allows to progressively build complex 3D parts layer by layer, by melting material in a liquefier head and depositing it on a build plate through a moving nozzle. Prior to the printing process, the choice of suitable process parameters is essential to ensure good performance and quality of the printed parts. For this purpose, the choice of optimal parameters can be oriented by referring to some recent studies which have investigated how the thickness of the layers and the build orientation can affect the roughness and the hardness of the surface but also the shape and the dimensional accuracy of the printed parts [27–29]. Another remarkable contribution for the identification of the optimal FDM printing parameters was provided by Boschetto et al. [30] who have developed a model capable of predicting the dimensional accuracy of the FDM printed parts, as a function of parameters such as the thickness of the layers and the deposition angle, considering the dimensions of the filament to be printed.

In order to identify the most suitable materials for our purpose, some common materials used in FDM processes (ABS, PLA, Nylon 645, Ninjaflex, FlexiFil, Onyx) have been preliminarily tested. In particular, non-linear FEM analyzes were initially carried out on various layout of flexible actuation mechanisms to estimate their mechanical behaviors. Based on the existing FWMAVs prototypes, the analyzed layouts have a maximum of 7 cm $\times$ 7 cm planar size, while the minimum thickness of the structure was chosen equal to 1 mm. Among the tested materials, ABS, PLA, Nylon and Onyx, were discarded since they proved to be excessively stiff to be exploited within such a small design space. To obtain the required movement of flapping wings, in fact, many regions of the structure exceed the elastic limit of the material. On the other hand, Ninjaflex, even if it allows to obtain large elastic deformations of the flexible regions, has been discarded because it is not able to guarantee sufficient overall stiffness of the mechanism which could lead to buckling instability.

Therefore, it was decided to use FlexiFil, a partially bio-based TPC (Thermoplastic Co-Polyester) that globally has shown the most suitable mechanical properties for this kind of application.

The main mechanical properties of FlexiFil [31] are reported here below in Table 1.

Table 1. FlexiFil ™: mechanical properties

	Specific gravity	Young's modulus	Tensile strength	Elongation at break	Hardness
FlexiFil ™	*1.14 [g/cc]*	*95 [MPa]*	*24 [MPa]*	*530%*	*45 D*

The overall weight reached by the two layouts of Compliant Mechanisms described in the next chapter is between 1.5 and 3 g, taking into account both the maximum planar size of the layouts (7 cm x 7 cm) and the use of FlexiFil material, which has a specific gravity of 1.14 g/cc. The choice of the MAV's propulsion system fell on electric DC motors, since they are reliable, versatile, cheap and easily available in the market. After considering that the own weight of the CMs reaches the maximum of 3 g, several examples of existing MAVs using DC motors as propulsion system were observed [32–36], in order to compare their main characteristics and better define a category of Micro Air Vehicles able to take advantage of the two CMs here analyzed. Although in this early design stage it is not possible to precisely define all the characteristics of a MAV that could benefit from the analyzed CMs, a total weight of around 15 g can still be approximated for the target MAV category, and a maximum wingspan of 18 cm, with a flapping frequency of at least 25 Hz.

3 Results

Taking inspiration from the thoracic structure of flapping wing insects with indirect muscles (Fig. 1.), two different layouts of single-component CMs, in the following called "1" and "2", have been designed and deeply investigated by FEM analyses to estimate their performances. Both layouts of the single-component CM have been conceptually divided in two parts: the first one would reproduce the role of the flexible area of the thorax connected to the muscles, which have the task of deforming it alternately upwards and downwards, the second part, instead, represents the rigid region of the thorax that supports the wings rotation.

3.1 Single-Component CM for FWMAV: Layout 1

The layout 1 is shown in Fig. 2. The mechanism has a uniform thickness of 4 mm along the Z-axis. A brushless DC motor (whose rotation axis is marked by an orange X in Fig. 2) is connected to the CM by means of a crank (not shown in Fig. 2).

The crank drives the area A of the CM along a circular trajectory (yellow dashed line in Fig. 2) in the XY plane. The area A consists of a circular hole so as to be able to contain a low friction bearing that connects the CM to the crank.

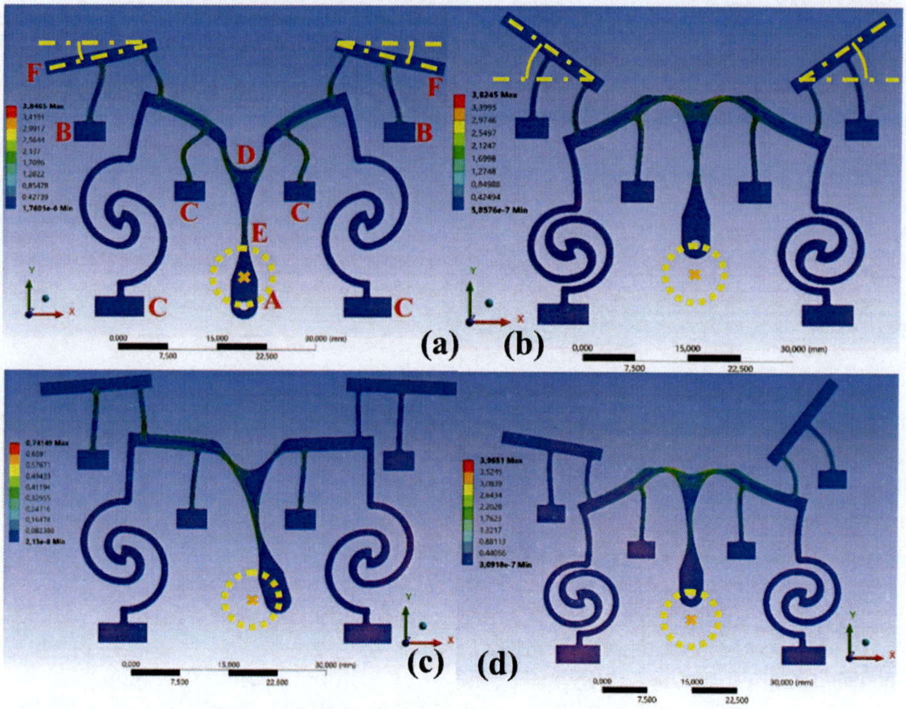

Fig. 2. Layout of bio-inspired flexible mechanism 1: (a) end of downstroke, (b) wings in central position with an angle of 0°, (c) end of upstroke, (d) asymmetrical wings rotation

When the area A moves along the circular yellow dotted path, the area D, which is directly connected to the area A, moves alternately downwards and upwards. These two areas of the single-component flexible structure allow to transform the continuous rotary motion of the electric motor in a translational movement. The movement of the local area D allows, also thanks to the bending of the adjacent areas, the alternating symmetrical rotational movement of the two areas F. The two flapping wings (not shown in Fig. 2) are linked to the single-component CM through the areas F.

The yellow dash-dot lines in Figs. 2a and 2b highlight the angles delimited by the rotation of the wings moving from the central horizontal position. The supports B and C are fixed to the frame and are connected to the mechanism through flexible areas, one of which has spiral shape to allow energy storing by elastic deformations.

The flexible area E has a reduced thickness compared to the adjacent ones in order to decouple the displacement of the area A along the x direction from the other parts of the mechanism. In this way, also when the area A reaches asymmetric positions along x direction (see Fig. 2c), the wing supports mutually maintain symmetrical positions. Figure 2d shows a possible approach to make the two wings rotate different from each other through the opposite displacement of the supports B by a few millimeters upwards and downwards.

Such an approach could potentially be very advantageous making the geometry of the structure parametric and optimizing the various dimensional parameters based on the different design requirements for any FWMAV.

In order to predict the effects of the stiffness changes due to large deformations of the structure, all the FEM analyses were carried out considering non-linear behavior and calculating the large deflection effects. The maximum values of the equivalent von Mises stress calculated in the various configurations analyzed are approximately 6 times lower than the FlexiFil tensile strength (24 MPa).

3.2 Single-Component CM for FWMAV: Layout 2

An alternative interpretation of the insects' indirect muscle structure has led to the development of the layout 2 shown in Fig. 3. In this case, also a prototype, in FlexiFil

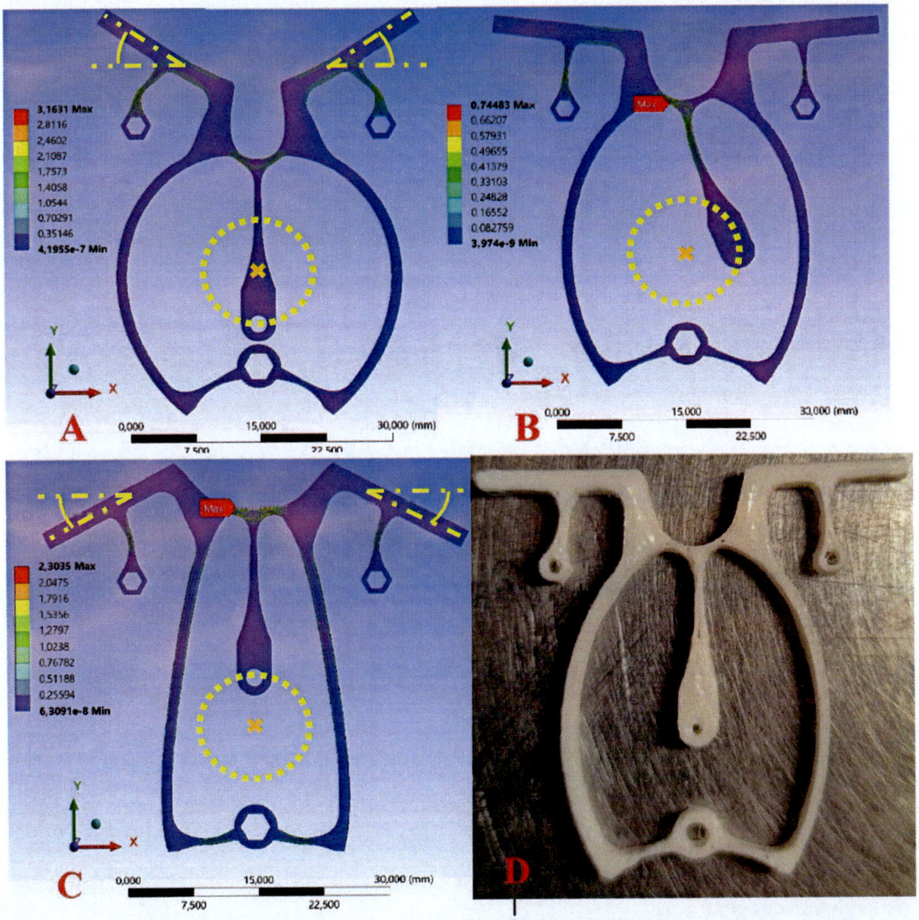

Fig. 3. Layout of bio-inspired flexible mechanism 2: (a) end of upstroke, (b) wings in central position with an angle of 0°, (c) end of downstroke, (d) Flexifil model printed with FDM.

material, has been produced by FDM (Fig. 3). Compared to layout 1 described in the previous Sect. 3.1, a more compact and simple structure has been developed to reproduce as much as possible the shape and the functionality of the thorax of insects with indirect asynchronous muscles. The function of the different areas of the single-component flexible mechanism is similar to that indicated in the Fig. 2 for layout 2, but in this case the component deformation is more distributed over the entire structure. In this way, there is no risk of having highly stressed areas. For the layout 2, in fact, despite the minimum thickness of the structure is 0.6 mm, lower than that of layout 1 (0.8 mm), the maximum value of the von Mises stress is about 3.16 MPa, approximately 8 times lower than the tensile strength of the material (24 MPa). This result should ensure a greater fatigue resistance of the mechanism.

A simplified perspective view of layout 2 is shown in Fig. 4, in order to provide a better representation of the CM layout and the positioning of the wings and engine relative to it. The orange dash-dot line points out the rotation axis of the motor (represented with the cylindrical shape A). The CM is fixed to the frame through areas B, while the two areas C show the connection point of the two wings with the CM.

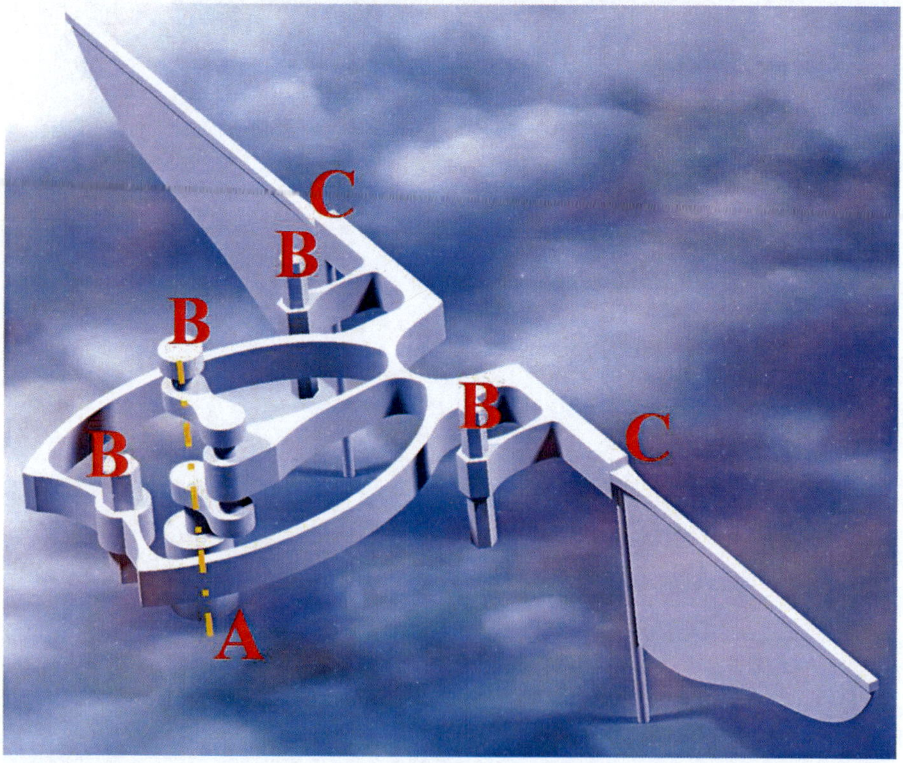

Fig. 4. Layout of bio-inspired flexible mechanism 2: simplified perspective view with positioning of the wings and engine

4 Conclusions

Two different layouts of bio-inspired mechanisms to use in FWMAVs have been proposed. In particular, two single-component compliant mechanisms that reproduce the muscle structure of insects have been designed. A physical prototype of one of the two layouts was also produced to demonstrate its manufacturability through 3D printing. The performances of the developed mechanisms were investigated by FEM analyses in order to evaluate the material's stress state and the ability of the mechanism to reproduce the correct kinematics of flapping wing micro air vehicles.

The obtained results are promising. It was demonstrated the designed layouts may well reproduce the wings movements thanks to the deformations of the structure and they can be manufactured by low-cost 3D printing processes. Regarding the mechanical behavior, it was found that the maximum values of the von Mises stress always are much smaller than the tensile strength of the material.

Considering the designed structures will be subjected to remarkable cyclic loads due to high flapping frequencies, future works should be addressed to fatigue testing.

To do this an experimental mechanical campaign will be performed since currently there are no data regarding the fatigue behavior of the material.

Further attention will also be dedicated to measuring the possible dimensional differences of the 3D printed CMs with the corresponding CAD models, by using 3D scanning techniques. Indeed, dimensional differences could induce unforeseen deformations or buckling situations, especially in local areas which, having a minimum thickness in the order of millimeters, are more affected by high frequency cyclic loads. It will therefore be extremely important to investigate through further accurate FEM analyzes the effects that such possible dimensional deviations would have on the CM's performance.

Additional future works will be addressed to implement topological optimization processes that could lead to optimized layout of the single-component flexible mechanism that could allow to further improve its performance.

References

1. Ghafarian, M., et al.: An XYZ micromanipulator for precise positioning applications. J. Micro-Bio Robot. **16**(1), 53–63 (2020). https://doi.org/10.1007/s12213-020-00124-5
2. Ghafarian, M., Shirinzadeh, B., Al-Jodah, A., Das, T.K., Pinskier, J.: FEA-based optimization of a complete structure of a monolithic z/tip/tilt micromanipulator. J. Micro-Bio Robot. **16**(1), 93–110 (2020). https://doi.org/10.1007/s12213-020-00133-4
3. Schmitt, P., Hoffmann, M.: Engineering a compliant mechanical amplifier for MEMS sensor applications. J. Microelectromech. Syst. **29**(2), 214–227 (2020)
4. Buzzin, A., Cupo, S., Giovine, E., de Cesare, G., Belfiore, N.P.: Compliant nano-pliers as a biomedical tool at the nanoscale: Design, simulation and fabrication. Micromachines **11**(12), 1–13 (2020)
5. Lofroth, M., Avci, E.: Development of a novel modular compliant gripper for manipulation of micro objects. Micromachines **10**(5), 1–22 (2019)
6. Chen, X., Deng, Z., Hu, S., Gao, J., Gao, X.: Design of a compliant mechanism based four-stage amplification piezoelectric-driven asymmetric microgripper. Micromachines **11**(1), 1–25 (2020)

7. Hao, G., Zhu, J.: Design of a monolithic double-slider based compliant gripper with large displacement and anti-buckling ability. Micromachines **10**(10), 665 (2019)
8. Mayyas, M., Mamidala, I.: Prosthetic finger based on fully compliant mechanism for multi-scale grasping. Microsyst. Technol. **27**(5), 2131–2145 (2020). https://doi.org/10.1007/s00542-020-05045-8
9. Liu, C.H., Chiu, C.H., Hsu, M.C., Chen, Y., Chiang, Y.P.: Topology and size-shape optimization of an adaptive compliant gripper with high mechanical advantage for grasping irregular objects. Robotica **37**(8), 1383–1400 (2019)
10. Bilancia, P., Baggetta, M., Berselli, G., Bruzzone, L., Fanghella, P.: Design of a bio-inspired contact-aided compliant wrist. Robot. Comput. Integr. Manuf. **67**, 102028 (2021)
11. Hu, Y., Zhang, L., Li, W., Yang, G.Z.: Design and fabrication of a 3-d printed metallic flexible joint for snake-like surgical robot. IEEE Robot. Autom. Lett. **4**(2), 1557–1563 (2019)
12. Wang, W., Wu, S.: A caterpillar-climbing robot with spine claws and compliant structural modules. Robotica **34**(7), 1553–1565 (2016)
13. Shan, X., Birglen, L.: Modeling and analysis of soft robotic fingers using the fin ray effect. Int. J. Robot. Res. **39**(14), 1686–1705 (2020)
14. Baek, S.M., Yim, S., Chae, S.H., Lee, D.Y., Cho, K.J.: Ladybird beetle inspired compliant origami. Sci. Robot. **5**(41), eaaz62062 (2020)
15. Guo, D., Bacciaglia, A., Ceruti, A., Marzocca, P., Bil, C.: Design and development of a transition propulsion system for bimodal unmanned vehicles. Int. Conf. Unmanned Aircraft Syst. ICUAS **2018**, 549–558 (2018)
16. Drews, P.L.J., Neto, A.A., Campos, M.F.M.: Hybrid unmanned aerial underwater vehicle: modeling and simulation. In: IEEE International Conference on Intelligent Robots and Systems, IROS, 4637–4642 (2014)
17. Liang, J., et al.: Wing load investigation of the plunge-diving locomotion of a gannet Morus inspired submersible aircraft. Sci. China Technol. Sci. **57**(2), 390–402 (2014)
18. Gong, D.H., Lee, D.W., Shin, S.J., Kim, S.Y.: String-based flapping mechanism and modularized trailing edge control system for insect-type FWMAV. Int. J. Micro Air Veh. **11**, 1–10 (2019)
19. Rubio, J.E., Chakravarty, U.K.: Experimental structural dynamic measurements of an artificial insect-sized wing biomimicking a crane fly forewing. Acta Mech. **230**(12), 4273–4286 (2019). https://doi.org/10.1007/s00707-019-02503-x
20. Hudson, O.A., Fanni, M., Ahmed, S.M., Sameh, A.: Bio-inspired jumping maneuver for launching flapping wing micro air vehicles. In: 18th IEEE International Conference on Autonomous Robot Systems and Competitions, ICARSC 2018, pp. 104–109 (2018)
21. Bin Abas, M.F., Bin Mohd Rafie, A.S., Bin Yusoff, H., Bin Ahmad, K.A.: Flapping wing micro-aerial-vehicle: Kinematics, membranes, and flapping mechanisms of ornithopter and insect flight. Chin. J. Aeronaut. **29**(5), 1159–1177 (2016)
22. Phan, H.V., Park, H.C.: Mimicking nature's flyers: a review of insect-inspired flying robots. Current Opin. Insect Sci. **42**, 70–75 (2020)
23. Cao, T., Jin, J.P.: Evolution of Flight Muscle Contractility and Energetic Efficiency. Front. Physiol. **11**(October), 1–14 (2020)
24. Dickinson, M.H., Lighton, J.R.B.: Muscle efficiency and elastic storage in the flight motor of Drosophila. Science **268**(5207), 87–90 (1995)
25. Bejgerowski, W., Ananthanarayanan, A., Mueller, D., Gupta, S.K.: Integrated product and process design for a flapping wing drive mechanism. J. Mech. Des. Trans. ASME **131**(6), 0610061–0610069 (2009)
26. Wood, R.J., Avadhanula, S., Sahai, R., Steltz, E., Fearing, R.S.: Microrobot design using fiber reinforced composites. J. Mech. Des. Trans. ASME **130**(5), 1–11 (2008)

27. Mohamed, O.A., Masood, S.H., Bhowmik, J.L.: Optimization of fused deposition modeling process parameters: a review of current research and future prospects. Adv. Manuf. **3**(1), 42–53 (2015). https://doi.org/10.1007/s40436-014-0097-7
28. Mwema, F.M., Akinlabi, E.T.: Basics of fused deposition modelling (FDM). In: Fused Deposition Modeling: Strategies for Quality Enhancement. (2020)
29. Ingrassia, T., Nigrelli, V., Ricotta, V., Tartamella, C.: Process parameters influence in additive manufacturing. In: Eynard, B., Nigrelli, V., Oliveri, S., Peris-Fajarnes, G., Rizzuti, S. (eds) Advances on Mechanics, Design Engineering and Manufacturing. Lecture Notes in Mechanical Engineering. LNME, 261–270. Springer Cham (2017). https://doi.org/10.1007/978-3-319-45781-9_27
30. Boschetto, A., Bottini, L.: Accuracy prediction in fused deposition modeling. Int. J. Adv. Manuf. Technol. **73**(5–8), 913–928 (2014). https://doi.org/10.1007/s00170-014-5886-4
31. https://www.formfutura.com/downloads.
32. Baek, S.S., Ma, K.Y., Fearing, R.S.: Efficient Resonant Drive of Flapping-Wing Robots, pp. 2854–2860 (2009)
33. Madangopal, R., Khan, Z.A., Agrawal, S.K.: Small Flapping Wing Air Vehicles Using Four-Bar Mechanisms and Quasi-steady Aerodynamics (2014)
34. Mueller, D.: Integrated product and process design for a flapping wing drive mechanism. J. Mech. Des. **131**(6), 1–9 (2009)
35. Sahai, R., Galloway, K.C., Wood, R.J.: Elastic element integration for improved flapping-wing micro air vehicle performance. IEEE Trans. Robot. **29**(1), 32–41 (2013). https://doi.org/10.1109/TRO.2012.2218936
36. Hines, L., Member, S., Campolo, D., Sitti, M., Member, S.: Liftoff of a motor-driven, flapping-wing microaerial vehicle capable of. Resonance **30**(1), 220–232 (2014)

A Bespoke Neck Orthosis for Additive Manufacturing with Improved Design Method and Sustainable Material

Rita Ambu[1](✉), Salvatore Massimo Oliveri[2], and Michele Calì[2]

[1] Department of Mechanical, Chemical and Materials Engineering, University of Cagliari, via Marengo 2, 09123 Cagliari, Italy
ambu@unica.it

[2] Electric, Electronics and Computer Engineering Department, University of Catania, viale A. Doria 6, 95125 Catania, Italy
moliveri@dii.unict.it, mcali@diim.unict.it

Abstract. The use of additive manufacturing (AM) has widespread over the years in different areas, including the biomedical field. In particular, the design of customized orthoses, external medical devices used in the treatment of specific pathologies, was proposed in different studies mainly concerning upper limbs, while few investigations are reported relatively to the cervical area. In this paper a new design of a bespoke neck orthosis is reported. The manufacturing of a light device with a good transpiration allows to increase the patient's comfort and, compatibly with the structural requirements, is a main goal to pursue. With this aim, various aspects were considered in the design and manufacturing of the orthosis. At the design stage, the geometry was conceived with a ventilation pattern based on Voronoi cells, which generally allows a better performance in terms of breathability with respect to a pattern made with uniform geometrical features, keeping at the same time structural requirements, as assessed by numerical finite elements simulations. At the manufacturing stage, a new composite material was used, namely Hemp Bio-Plastic® (HBP) filament, composed by polylactic acid (PLA) and hemp shives which provided lightweight, improved superficial finish and antibacterial properties. In order to assess the thermal comfort, an experimental analysis was finally conducted on a prototype of the orthosis, worn by a volunteer subject, with a thermal imaging camera. The beneficial effect of the ventilation pattern considered in terms of temperature and, accordingly, for the patient's comfort, was highlighted also in relation to a neck orthosis previously designed.

Keywords: Orthosis modeling · Additive manufacturing · Bio-composite

1 Introduction

External medical devices or orthoses are commonly used to support and modify the structural and functional behavior of the musculoskeletal system. They are generally categorized depending on the affected portion of the body such as upper-limb orthoses,

C. Rizzi et al. (Eds.): ADM 2021, LNME, pp. 50–58, 2022.
https://doi.org/10.1007/978-3-030-91234-5_5

spinal orthoses, lower limbs orthoses, and named according to the joint involved. Prefabricated orthoses are readily available on the market and, generally, are less expensive than customized devices. However, these are more beneficial since they have a better fit to the patient's body, contributing both clinically and for the patient's comfort to final user satisfaction [1]. Developments in computer aided design (CAD) and additive manufacturing (AM) potentially give the possibility to design customized external devices. More generally, the specific properties of AM, together with the possibility to create complex geometries, has made this manufacturing method a viable option in the biomedical field for different applications. These include tissue engineering [2–4], patient specific surgical instruments, and preoperative planning [5, 6]. As for external medical devices, the applications have mainly concerned upper-limb, in particular wrist orthoses [7, 8] and lower limbs, specifically, ankle-foot orthoses [9] offering the possibility of successfully creating functional devices and the capability of adapting the design procedures for the use in a hospital setting. However, few research [10, 11] has considered the application of AM for the design of customized cervical orthoses. In particular, a recent approach was relative to a patient-based case study [11] where tentative prototypes of a neck orthosis were progressively improved based on comfort of the patient until the final design. Another study [12] reports an algorithm for generating ventilated casts based on temperature distribution in order to improve thermal comfort, and an application to the neck area is also reported. The limited number of studies, mainly focused on peculiar aspects of the design of these devices, is probably due to the complex functionality of a cervical orthosis which has to support, immobilize or correct the spine. Furthermore, clinical requirements have to be combined with other factors such as the sensation of comfort and aesthetics. Prefabricated semi-rigid cervical collars are available on the market to protect the cervical spine after injuries and, postoperatively, to immobilize it following surgical reconstructions. A primary measure of their effectiveness is the capability to immobilize the spine. As known, some problems are usually associated with the use of these devices including muscle atrophy, skin rush or breakdown, pain, lack of hygiene, heaviness and discomfort. For this reason, the design of a customized cervical orthosis for AM is supposed to combine clinical performance, by adopting a proper geometry, and improved comfort by the reduction of weight and increased breathability.

In this paper the design of a cervical orthosis for 3D printing is reported and some main aspects are studied principally aimed at improving patient's comfort. This study describes a methodology with improved features compared to a previous study of the Authors [13] and, with respect to existing literature, proposes an overall procedure for the design of semi-rigid neck orthoses, potentially adaptable to clinical setting. The design process of a orthosis, generally, comprises the acquisition of the body's part of interest, the CAD modeling of the device based on the acquired geometry and the manufacturing of the final geometry. Each of these steps has to be conducted so as to obtain a global efficient methodology suitable for the medical environment. As for the acquisition of the neck's morphology, different methodologies can be used. Computed tomography (CT) is a technique generally employed to facilitate diagnostics of neck injuries, but the volumetric reconstructions for 3D representation can also be used for the digital representation of the outer surface of the part for the manufacturing of an orthotic device. The feasibility and high accuracy of this method for the reconstruction of the neck area was previously assessed by the Authors [13]. However, during scanning the patient is exposed to radiation which depends on the duration of the scanning besides the fact that the extraction and reconstruction of the external surface can be time consuming. However, this should be a convenient solution if CT scan acquisitions are already available for reasons independently from the manufacturing of the customized orthosis. A more practical method to capture the human external shape is 3D scanning by laser or structured light 3D scanner. This method was used in this study for obtaining the external shape of the human neck, with an equipment that was efficient in terms of accuracy and time of acquisition and reconstruction. This is also beneficial for the patient since it is non invasive and does not require particular effort on his/her part. The modelling of the neck orthosis was conducted with particular attention to the application of a pattern of voids or ventilation pattern, which has an important role both for reducing the overall weight and improving breathability for a better comfort. A ventilation pattern based on Voronoi cells was considered, developed in Rhino-Grasshopper environment whose configuration can be easily changed according to the geometry of the surface and the requirements of the specific orthosis. The feasibility of the final model was structurally assessed by means of Finite Elements (FE) simulations. The manufacturing of a prototype was done by means of a fused deposition modelling (FDM) process. At this stage, among the different aspects, the choice of the material has an important role. An innovative composite 3D printing filament composed by polylactic acid (PLA) and hemp shives, patented as Hemp Bio-Plastic® (HBP) [14] was used. It has a sustainable nature, lightweight, superficial finish, and antibacterial properties [15] which are all characteristics that make it potentially advantageous also for comfort purposes. In order to assess the comfort of the manufactured orthosis, an experimental thermal analysis was finally conducted. Thermal comfort is, indeed, an important issue for 3D printed orthoses [12] and in particular in this case where the device has to fit the neck. This preliminary investigation has highlighted the thermal beneficial effect of the proposed ventilation pattern giving useful indications for further improvements.

2 Methodology

2.1 Acquisition Procedure

Data acquisition was the first phase of the digital reconstruction of human neck surface. For this purpose, the commercial version of Microsoft Kinect sensor device was employed (Fig. 1a) to acquire the neck surface of a volunteer subject. The Kinect device operated with an integrated system of camera sensor and laser.

In addition to a VGA sensor, which had the task of recording the images with a predetermined resolution, two three-dimensional infrared ray depth sensors performed the 3D mapping of the surrounding environment, evaluating the position of objects in space, while a LED sensor assessed the light intensity of the surrounding environment. Data acquired through Microsoft Kinect were processed by the Skanect software (Rare® - Redmond society), generating the neck 3D superficial model. Hardware-software systems operated in real time to obtain a three-dimensional map of the object acquired with

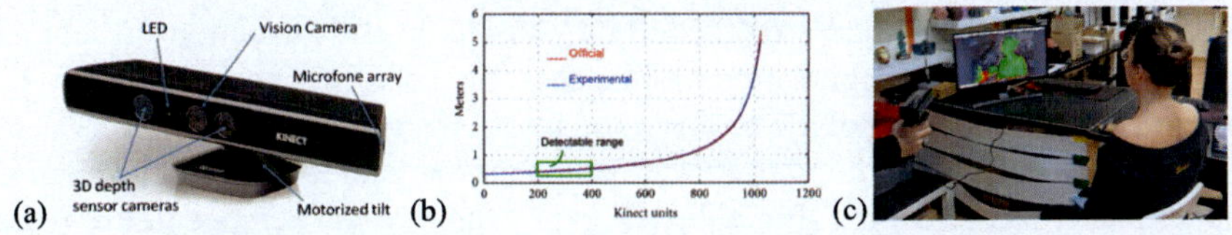

Fig. 1. Acquisition of the neck surface: (**a**) Microsoft Kinect sensor; (**b**) accuracy in depth measurement; (c) digitized surface, colored map of triangular mesh discontinuity is visible.

a preset resolution. The hardware-software system analyzed the data received from the sensors to identify, among the various scanned objects, those of interest. The system used *repetition algorithms*, evaluating the repetition of some features and their position in space. In this way, it was able to fully acquire the foreground object. The acquisition was faster and more precise when the objects that composed the background were clearly distinct from the acquiring object. The experimental measurements provided a sensor sensitivity curve very close to the official one (Rare® - Redmond society) with a maximum error between experimental and official less than 1% (Fig. 1b). The Kinect units corresponded to distances between acquired anatomical landmarks increasing when the distance of the sensor from the object to be acquired increased. The accuracy of the acquired mesh was guaranteed when the relationship $\frac{d_{max}}{S} \leq 1$ is verified [16], where $d_{\max}$ is the maximum mesh dimension and S is the sensibility provided by the curve in Fig. 1b. Therefore, by acquiring the surface of the neck from a distance of less than 400 mm, the surface mesh will have an accuracy greater than 0.5 mm. The integrated use of RGB camera together with an infrared (IR) structured light sensor contribute in fact to the optimal conditions: static conditions, perfect and uniform lighting.

The followed procedure allowed to obtain an adequate precision with a single acquisition (Fig. 1c). The Skanect software, in fact, by managing the information coming from the Kinect device provided in real time a chromatic map of watertight point cloud surface that highlighted any hole, gap and/or tear. Setting the desired resolution accuracy, acquired data was processed in real time by Greedy algorithm (open source PCL1.6.0 libraries) elaborating a colored triangular mesh surface which indicated geometric discontinuities and/or holes, gaps and tears (red zone in Fig. 1c). In these defective zones further scans were performed to integrate the point cloud data [17].

2.2 CAD Modeling and Numerical Assessment of the Orthosis

The design of the CAD model was aimed at obtaining an external device with functional characteristics similar to that of cervical semi-rigid orthoses available on the market. This kind of orthosis is conceived as an external device that fits the neck, anywhere from the jaw to the chest. It is used to restrict movement and support the neck to allow healing when treating a cervical injury, fracture, or surgical procedure. Produced soft cervical orthoses are also available, which have a lighter geometrical design and are used to help in the treatment of minor neck strains, such as whiplash, support the neck and control of pain after an injury.

The first step in the process of modelling consists in the extraction of geometric reference features from the digital model of the neck. First of all it is necessary to establish a top plane for the modelling of the orthosis and its orientation. For this purpose a central axis is created going from the chin to the occipital bone and identify the direction of the upper plane, whose choice should be guided by the medical personnel. From there, the upper plane of the collar is generated automatically and a new reference system is established (Fig. 2a). Subsequent cut planes are defined and in correspondence of each of them profiles with a semi-automatic procedure are extracted. The surface of the orthosis can be obtained by means of the loft operation guided by the profiles properly selected, each made of points which, by means of spline curves, define a contour. An offset is introduced between the surface of the neck and that of the orthosis in order to allow the placement of a soft padding for improved comfort. The procedure determines a reference surface and the final geometry of the orthosis is obtained depending on the specific needs of the patient.

The comfort of the orthosis can be greatly improved by the introduction of a pattern of voids which accomplishes different tasks. The introduction of this pattern is useful for improving breathability of the medical device, thus reducing the risk of irritation. Furthermore, it contributes to weight reduction giving rise to a lightened structure resulting in further benefit for the patient. However, the pattern has to be accurately designed so as to keep an adequate stiffness in relation with the material used and the loads acting on the device. Different approaches can be applied for the generation of a ventilation pattern, but, independently from the method used, an automatic or semi-automatic technique is preferable. A straightforward approach considers a simple CAD feature, such as a hole, ellipse, honeycomb which is evenly or randomly repeated on the model of the orthosis. Another method consists in defining a surface pattern which is mapped on the geometry of the CAD model. The final configuration of the pattern, in this case, depends on the geometry of the surface. The application of this method also allows to introduce geometrical constraints, such setting areas devoid of the pattern, or applying functional requirements. This approach was applied in this study by considering a hollowed Voronoi tessellation. The procedure was implemented in Rhino-Grasshopper environment providing in the block diagram the possibility to exclude the pattern in defined areas, in order to insert hinges and closures or for other purposes and to thicken the surface to obtain the final solid model. The amount of seeds points can be varied within a defined range in order to obtain different distributions of the cells on the surface. The use of Visual Programming Language in CAD environment is useful also with a view to future development of a easy-to use procedure for the application in the clinics as reported in [8] for wrist orthoses. The final model of the neck orthosis for 3D printing consists of two halves which are reported in Fig. 2b.

The model reported in figure has a thickness of 4 mm and allows a weight reduction of about 35% with respect to a model with analogous geometry devoid of the pattern. The assessment of the patterned orthosis can be done by means of Finite Elements (FE) numerical simulations with proper loads and constraints.

At this aim, a one-piece solid model of the neck orthosis was imported in a commercial FE code and meshed by means of eight nodes linear elements. The presence of the Voronoi pattern with sharp edges on the irregular geometry of the model, generally,

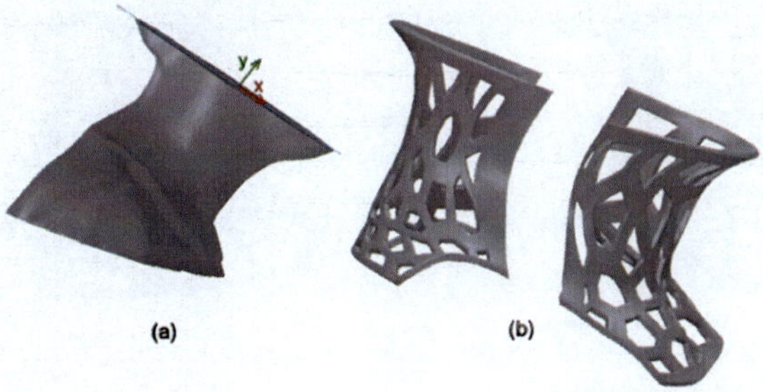

Fig. 2. (a) Model of the human neck; (b) model of the neck orthosis.

produces meshes with a high number of elements. A sensitivity analysis was performed for an optimal compromise between accuracy and computational time which conducted to a mesh of about 2×10^6 elements. The composite material Hemp Bio-Plastic® (HBP), previously described, was chosen for the simulations. The mechanical properties were assumed as follows: elastic modulus 4420MPa, flexural modulus 3833MPa and Poisson's ratio 0.32.

A cervical orthosis is generally subjected to a combination of loads, since from a mechanical and structural point of view, the cervical spine is a very complex mechanism [18]. However, in brief, the movements of the human neck are essentially six, namely, flexion, extension, axial rotation, and lateral bending, and all movements are combinations of these. In this study, the main ones were considered, that is extension, flexion and lateral bending. The loads exerted by the neck of a female subject on the model are approximately derived by the values of three-dimensional moments reported in literature measured experimentally during maximum voluntary contractions of neck muscles in humans, as described in a previous study of the Authors [13].

Table 1 reports the maximum total displacement and the maximum Von Mises stress relative to the loading configurations considered.

Table 1. Results of FE analysis for different loading configurations.

Loading	Max total displacement [mm]	Max Von Mises stress [MPa]
Extension	1.26	41.52
Flexion	1.02	27.88
Lateral bending	1.67	38.85

The highest values of Von Mises stress were obtained for the extension loading. Figure 3 reports an iso-colour representation of these stresses, expressed in MPa, and an analogous representation relative to the total displacement expressed in mm.

The obtained results show that the maximum total displacement is quite limited for each loading configuration considered while the maximum stresses fall under the flexural maximum strength of HPB (69.7MPa).

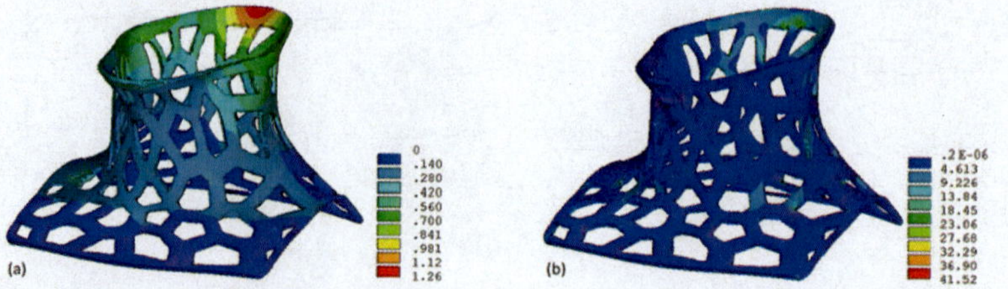

Fig. 3. Iso-colour representation of FE results for extension loading: (a) total displacement; (b) Von Mises stress.

3 Manufacturing and Thermal Comfort Assessment

The prototype model of the bespoke neck orthosis was manufactured with a D300 Technology® 3D Printer for Fused Deposition Modeling (FDM). The main characteristics of this printer are a resolution of 0.1 mm for the Z-axis, and an accuracy of 0.05 mm for the X, Y axes, where X and Y axes identify the plane containing the print bed and the Z-axis represents the vertical direction, perpendicular to the print bed. The printer was equipped with a 0.6 mm nozzle. The filaments used were made of Hemp Bio-Plastic® (HBP), which also offers higher mechanical and thermal dispersion properties as compared to simple PLA.

In order to evaluate the comfort offered by the manufactured prototype, temperature measurements were made by using the thermal Imaging Camera HT-02 Dongguan Xintai Instrument Co., Ltd.. The thermal sensor of the instrument was calibrated on an emissivity value of 0.90. The value of this parameter for PLA plastic is, in fact, between 0.88 and 0.94. The two halves of each prototype were joined by a tape system with double-sided adhesive velcro strips (Fig. 4). The system allows the orthosis to be locked on the neck and continuously adjusted. The measurements were conducted with the orthosis just worn and after one hour. The external ambient temperature during the measurements was 18 °C. In addition, measurements were also repeated by inserting a padding between the neck and the orthosis. For qualitative comparison purposes, similar measurements were also done on an orthosis previously studied by the Authors [13] with a uniform ventilation pattern made of honeycomb cells and with an orthosis devoid of the ventilation pattern. Figure 4 shows the results obtained.

The analysis showed that the temperature decreased by an average of 1.5° passing from the model without pattern to that lightened with the honeycomb pattern and a further decrease of about 1° was observed passing to the model with the Voronoi pattern. Furthermore, in the ventilated models, especially in the model with the Voronoi pattern, the highest temperature values were limited to the shoulder areas and therefore can be tolerated without any discomfort. The model with the Voronoi pattern allowed a complete ventilation in the neck and only a negligible increase of temperature. In all the prototypes, the introduction of a padding, perceived as an additional element of comfort, increased the temperature by only 0.5° on average.

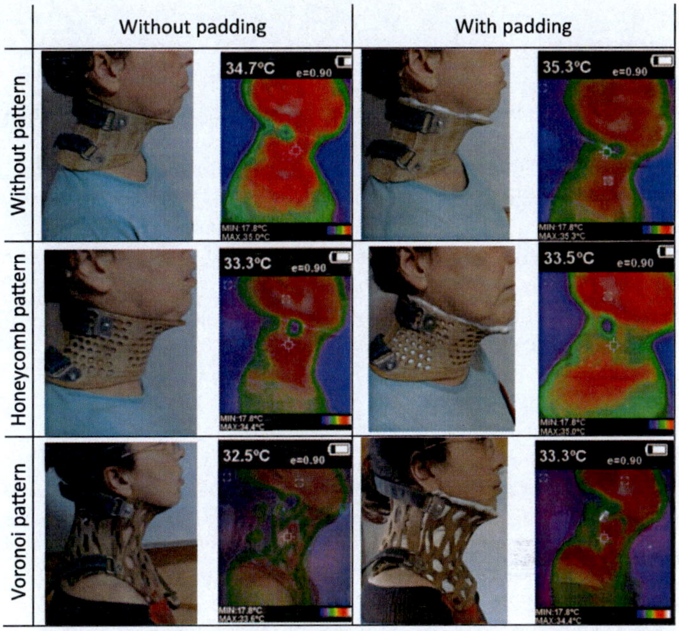

Fig. 4. Thermal images of neck orthoses.

4 Conclusions

In this paper the design and manufacturing of a customized neck orthosis for 3D printing with improved comfort, in terms of lightness and breathability, was studied. The obtained model had an adequate stiffness, as verified by means of FE simulations, and thermal comfort characteristics, as shown by the preliminary thermal analysis on a prototype model 3D printed with HBP filament. The methodology applied encompassed a fast and accurate acquisition of the body's part with a minimal discomfort for the person, while, with regards to the modeling, a breathable ventilation pattern based on Voronoi cells was introduced with a procedure which could be potentially adapted to the use by medical personnel. This main goal of full automation for non-expert CAD users will be implemented with further research.

References

1. Wang, J., Tan, J., Boone, D., Zhang, M.: A review of the application of additive manufacturing in prosthetic and orthotic clinics from a biomechanical perspective. Engineering **6**, 1258–1266 (2020)
2. Gu, B.K., Choi, D.J., Park, S.J., Kim, M.S., Kang, C.M., Kim, C.-H.: 3-dimensional bioprinting for tissue engineering applications. Biomater. Res. **2**(12), 1–8 (2016)
3. Ambu, R., Morabito, A.E.: Modeling assessment and design of porous cells based on Schwartz primitive surface for bone scaffolds. Sci. World. J. **2019**, 1–17 (2019)

4. Percoco, G., Uva, A.E., Fiorentino, M., Gattullo, M., Manghisi, V.M., Boccaccio, A.: Mechanobiological approach to design and optimize bone tissue scaffolds 3D printed with fused deposition modeling: a feasibility study. Materials **13**(3), 648 (2020)
5. Ganguli, A., et al.: 3D printing for preoperative planning and surgical training: a review. Biomed. Microdevice **20**(3), 1–24 (2018). https://doi.org/10.1007/s10544-018-0301-9
6. Speranza, D., et al.: Additive manufacturing techniques for the reconstruction of 3D fetal faces. Appl. Bion. Biomech. **2017**, 1–10 (2017). (Article ID 9701762)
7. Buonamici, F., et al.: A practical methodology for computer-aided design of custom 3D printable casts for wrist fractures. Vis. Comput. **36**(2), 375–390 (2019). https://doi.org/10.1007/s00371-018-01624-z
8. Li, J., Tanaka H.: Feasibility study applying a parametric model as the design generator for 3D–printed orthosis for fracture immobilization. 3D Print. Med. **4**, 1 (2018)
9. Wojciechowski, E., et al.: Feasibility of designing, manufacturing and delivering 3D printed ankle-foot orthoses: a systematic review. J. Foot Ankle Res. **12**(1), 11 (2019). https://doi.org/10.1186/s13047-019-0321-6
10. Prates, A.: Self-adjusting orthoses design. Scripta-Ingenia **3**, 3–8 (2014)
11. Hale, L., Linley, E., Kalaskar, D.M.: A digital workflow for design and fabrication of bespoke orthoses using 3D scanning and 3D printing, a patient-based case study. Sci. Rep. **10**, 7028 (2020)
12. Zhang, X., et al.: Thermal-comfort design of personalized casts. In: UIST 2012 Proceedings, pp.243–254, Quebec City, Canada (2017)
13. Ambu, R., Motta, A., Calì, M.: Design of a customized neck orthosis for FDM manufacturing with a new sustainable bio-composite. In: Rizzi, C., et al (eds.) Design Tools and Methods in Industrial Engineering. ADM 2019. Lecture Notes in Mechanical Engineering, 707–718. Springer International Publishing, Cham (2020). https://doi.org/10.1007/978-3-030-31154-4_60
14. Kanesis. https://www.kanesis.it. Accessed 26 Feb 2021
15. Khan, A.K., Warner, P., Wang, H.: Antibacterial properties of hemp and other natural fibre plants: a review. BioResources **9**(2), 1–18 (2014)
16. Tuvshinjargal, D., Dorj, B., Lee, D.J.: Hybrid motion planning method for autonomous robots using kinect based sensor fusion and virtual plane approach in dynamic environments. J. Sensors **2015**, 471052 (2015)
17. Marin, G., Dominio, F., Zanuttigh, P.: Hand gesture recognition with jointly calibrated leap motion and depth sensor. Multimed. Tools Appl. **75**(22), 14991–15015 (2016)
18. Nightingale, R.W., Myers, B.S., Yoganandan, N.: Neck injury biomechanics. In: Yoganandan, N., Nahum, A.M., Melvin, J.W. (eds.) Accidental Injury, pp. 259–308. Springer, New York (2015). https://doi.org/10.1007/978-1-4939-1732-7_11

Battens Modelling and Optimization in Air-Sail Interaction Analysis

Michele Calì(✉), Ignazio Sapienza, and Salvatore Massimo Oliveri

Department of Electric, Electronics and Computer Engineering, University of Catania, Catania, Italy
mcali@dii.unict.it

Abstract. Numerical codes can play an important function in the nautical design when advanced simulation tools are increasingly adopted for prediction purposes, and in particular for the calculation of the interaction between boat parts and surrounding fluids. Fluid-structure interaction analysis (FSI) can help engineers to set the optimum design of a sail and also the influence of panels and battens sail arrangement. The paper describes a methodology that enhances the understanding of panel and battens arrangement, as well as their stiffness on sail performance. Specifically, through the battens position, shape and stiffness parametrization, the sail non-developable shape was optimized, in order to compensate the variation in the angle of attack that occurs following variations in the speed of the wind at different sail heights. For an efficient and robust implementation of integrated Reynolds Averaged Navier Stokes (RANS) equations with Shear Stress Transport (SST) turbulence model and Computational Structural Mechanics (CSM) analysis, the constitutive sailcloth and battens characterization was conducted experimentally. The case study of a mainsail in Dacron© TNF 240 of a Vaurien dinghy boat, offered tangible results to support the methodology by validating it with experimental data.

Keywords: Triangular membrane modeling · CFD-CSM analysis · SST model · Sail battens arrangement · Flying shape · Twisting

1 Introduction

The correct numerical simulation of the aerodynamics of a sail allows to create a model with which various analyses can be carried out to improve its performance. The analyses conducted using the current Computational Fluid Dynamics methods provide good confidence in the results when based on models possibly validated with an experimental theoretical comparison or with other established theoretical methods [1–3]. On the other hand, the aerodynamics of the sail is rather complex. Usually, it consists in flexible surfaces whose attitude varies rapidly and whose stiffness is strongly influenced by the type of sail fabric, its texture, orientation and thickness. A good sailmaker cuts, sews and overlaps the sailcloth (the strips) in order to optimize its behaviour. Furthermore, the presence of stiffening battens, whose main function is to maintain the shape of the sail, has a significant influence on sail performance.

C. Rizzi et al. (Eds.): ADM 2021, LNME, pp. 59–68, 2022.
https://doi.org/10.1007/978-3-030-91234-5_6

The performance of sailing yachts or boats designed for racing mainly in the context of America's Cup and relevant international sailing races has been studied in several researches for continuous improvements. Higher performances computing environment and very large domains are often selected to analyze the interaction of boat parts with surrounding fluids. It is assumed that methods coupling Computational Fluid Dynamics (CFD) and Finite Element Analysis (FEA) permit this fluid-structure interaction to be investigated through the analysis of sails flying shape.

Traditional CFD tools, however, may prove approximate and need to be validated in relation to predictions concerning the dynamic behaviour of non-developable shape with camber and massively detached flow around thin and flexible membranes. In literature, several works dealing with sail aeroelastic analysis, shapes parametrization as well as materials and manufacturing production technologies can be consulted for specific elaborations [4–6]. Various tools and methods have been proposed for the evaluation of sail flying shape performance during navigation or in real and virtual wind tunnel [7, 8], such as: Coordinate Measuring Machine [9]; the optical profilometry technique; cameras in wind tunnel tests [10]; laser measuring tools like "LIDAR"; a "Time-Of-Flight" radar to detect flying shapes [11]. Some other studies, instead, have focused on sail performance by recording pressure and stress measurements [12]; combined sail and rig shape recognition system using imaging techniques – VSPARS [13]; determination of the pressure distributions on the sails [14, 15] and so on. Nevertheless, none of the above mentioned methods allow to make a quantitative shared evaluation of panel and batten arrangement on sail performance. Moreover, usually sail CFD analyses are relative to isotropic homogeneous membranes and the simulation does not consider any directional characteristics of sailcloth fibre.

The present paper shows the main results of the authors' research work on aerodynamic analysis of the mainsail of a Vaurien dinghy boat, a small 4 m dinghy. The geometry survey was carried out on the real sail, modelling the various sail clothes, their seams and stiffening battens. The mechanical characteristics of the fabric, Dacron © TNF 240, and of the HDPE battens, were obtained experimentally. The work made it possible to construct the polar diagram, lift as a function of drag, of the isolated mainsail, to determine the maximum lift angle and to verify at this angle, through detailed analysis, the mechanical influence of the stiffening battens. A verification of the reliability of numerical data carried out by comparing it with literature data is also presented. The article is organized in 4 sections: in the first section, the modelling methodology and experimental sailcloth characterization is illustrated with a special concern of the panel and battens parametrization strategy used in the FSI problem analysis; in the second section the principal results of the studies are shown. Final considerations and conclusions are discussed in section three.

2 Materials and Methods

2.1 Sail Parametric Modelling

Employing commercial software CATIA V5 R19 (Dassault Systèmes®) the 3D model of the mainsail was created through the loft command, interpolating the R-Spline obtained in correspondence with 6 contours of the sections of the sail equally spaced along the vertical direction (z). The 6 sections were arranged in the space twisted together along the leech profile (red line Fig. 1b) according to the design values.

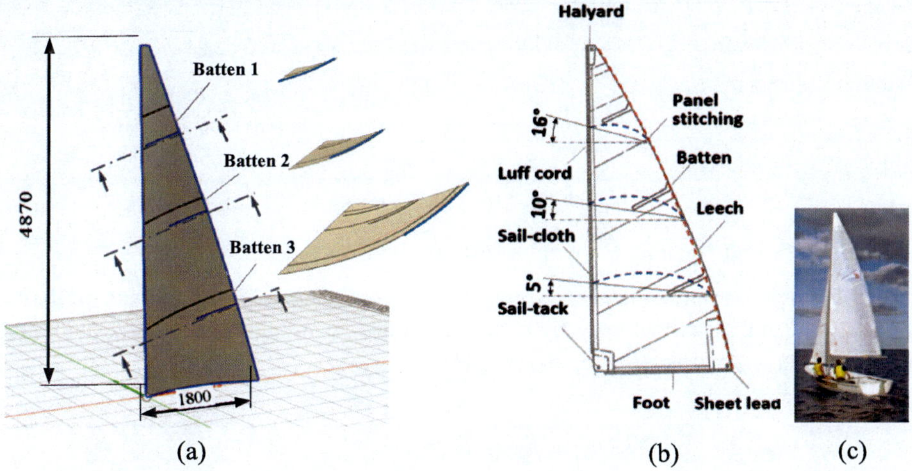

Fig. 1. Vaurien mainsail: (a) design shape; (b) panels and battens arrangement; (c) flying shape.

The sail thus constructed has the percentage curvature and the real incidence α, which varies with respect to the incident wind. The values in the most significant sections are shown in Table 1. Concavity in % was evaluated through the relationship:

$$C = \frac{f}{c}100 \quad (1)$$

where f is the arrow and c the sail chord. The last column indicates the value of the real incidence α of the wind in the section considered if the incidence α at the boom is equal to 22°; the real incidence in the various sections is given by the relation $\alpha = 22° - \beta$. Being β the twisting angle of the sail in the indicated section.

Table 1. Mainsail geometric characteristics.

Section	c [mm]	f [mm]	C [%]	Height [%]	Twisting β [°]	Incidence α [°]
Boma	1800	200	11	0%	0	22
1/4	1494.2	133.5	8.9	25%	5	17
1/2	1132.1	71.6	6.3	50%	10	12
3/4	692.6	23.5	3.4	75%	16	6

The twisting angle values of the battens cross-section dimensions, of the chord c and of the battens position were parameterized together with the stiffness of the battens in order to find their optimal values. In this way it was possible to compensate the variation in the angle of attack that occurred following variations in the speed of the wind at different sail heights (Table 2 and Fig. 2).

Table 2. Sail battens parameters.

Parameters	Design value	Min value	Max value
Position	1141; 2533; 3684 mm	−20%	+20%
Length	620 mm	400	800
Width	15 mm	12	20
Height	3.5 mm	3	5
Stiffness	6.93 GPA	1/5	50×

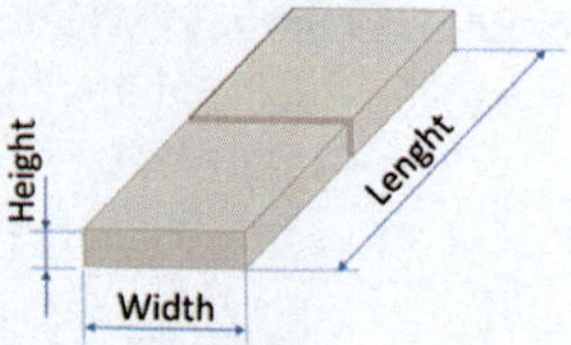

Fig. 2. Batten in polyurethane resin.

2.2 Sailcloth and Battens Characterization

The assessment of sailcloths' anisotropic mechanical behaviour and battens elasticity were carried out experimentally by evaluating the four engineering elastic constants (E_{we}, E_{wa}, $G_{we,wa}$ and $\nu_{we,wa}$) [3, 10] (Fig. 3).

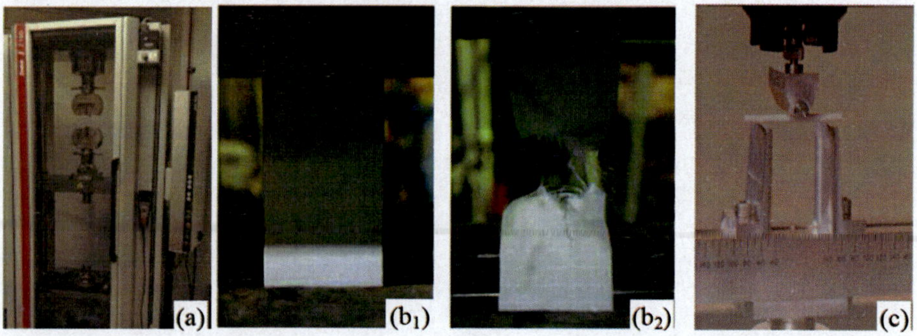

Fig. 3. (a) Zwick & Roell Z100 tensile testing machine; (b) Tensile test on Dacron © TNF 240; (c) Three-point flexion on battens.

Using Zwick & Roell Z100 tensile testing machine Young's modulus in weft (E_{we}) and warp (E_{wa}) directions, Poisson's ratio and shear modulus for the Dacron© TNF 240 ($\nu_{we,wa}$, $G_{we,wa}$) at ambient condition (20 °C − 1 atm) were evaluated (Table 3) [3].

Table 3. Battens and sailcloth structural characteristics.

Battens	E [GPa]		ν	G [GPa]	$\rho[\frac{kg}{m^3}]$	K_f [GPa]
	6.96		0.3	2.68	1400	5.8
Dacron TNF 240	E_{we} [GPa]	E_{wa} [GPa]	0.3	0.26	1300	
	1.05	0.66				

In particular, the Poisson's ratios $\nu_{we,wa}$ was evaluated with the simplified formula $\nu_{we,wa} = \frac{E_{med}}{2G} - 1$, where E_{med} is the average value of E_{we} and E_{wa}; the shear modulus

$G_{we,wa}$ was, instead, calculated using Eq. (2) as Agarwal and Broutman proposed in 1990 [16].

$$G_{we,wa} = \frac{1}{4}\left[\frac{1}{E_{bi}} - \frac{1}{4E_{wa}} - \frac{1}{4E_{we}} + \frac{2\nu_{we,wa}}{E_{we}}\right]^{-1} \quad (2)$$

The Young's modulus, the tangential modulus and flexional stiffness of battens in polyurethane resin were calculated with three-point flexion text. The mass-volume ratio was measured by an electro-optical single-pan precision balance (Gibertini mod. 704/N) with sensibility of 0.1 mg for mass measurement and a graduated cylinder with distilled water for volume measurement.

2.3 Setup of Aeroelastic-Structural Analysis

A Finite Element model using triangular membrane elements for the sail and tetrahedral elements for the battens was developed to calculate the structural deformation of the sail with battens for given flow force acting on it. The model was implemented in a FSI analysis coupling RANS commercial code (ANSYS Fluent® vers 18.0) with the structural code (ANSYS Mechanical® vers 18.0). The iteration of the FSI started from the design shape of the sail (Fig. 1). Flow forces were calculated using the RANS simulation with viscous-SST k-omega turbulence model. Sail deflection was calculated from the flow forces. Flow simulation was repeated with deflected geometry, whereas the structural calculation was repeated in turn with modified flow forces until force convergence was reached (i.e. force residuals become less than 0.001). The maximum number of iterations was set to 1000.

The structural deformation calculation of sail was based on the method of Arcaro, which assumes homogeneous and orthotropic linear elastic material [3, 10]. A 70 m tall non-structured multiblock hexahedral mesh was generated. The domain was extended about 100 m in front and on the sides of the isolated sail. A flat inlet boundary was located 70 m behind the sail (Fig. 4a). The sail was modelled with 1200 triangular membrane elements, varying in size from 35 mm to 100 mm; each batten was modelled with 480 tetrahedral elements, having dimensions of 5 mm; the air was modelled with 4 million of tetrahedral elements, varying in size from 10 mm to 1000 mm. Figure 4b and Fig. 4c show mesh detail in proximity of the sail.

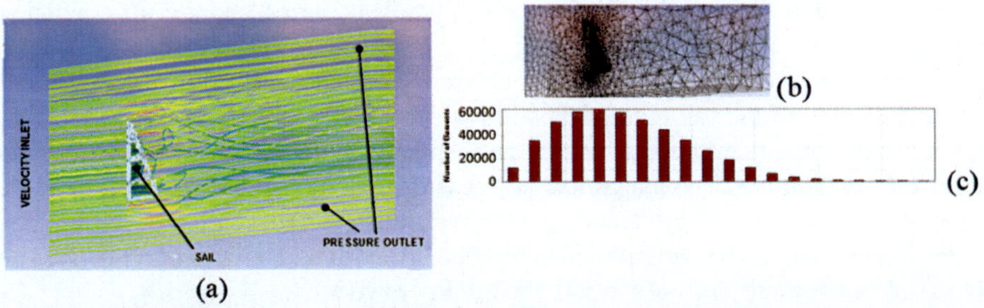

Fig. 4. (a) Wind tunnel; (b) grid air mesh; (c) number of air elements in 20 transversal blocks.

3 Results and Discussion

The numerical simulations carried out in the study allowed to evaluate the influence of the stiffening battens on the deformations of the sail and, therefore, on the lift and resistance values for the different angles of incidence (α) and the different wind speeds. The influence of stiffening battens was evaluated by varying their position, their dimensions (length, width and height) and their stiffness, using the parameters shown in Table 2. In particular, as these parameters varied, the stall angle was assessed, the angle of incidence for which the lift is maximum and the deformations of the sail are at various heights.

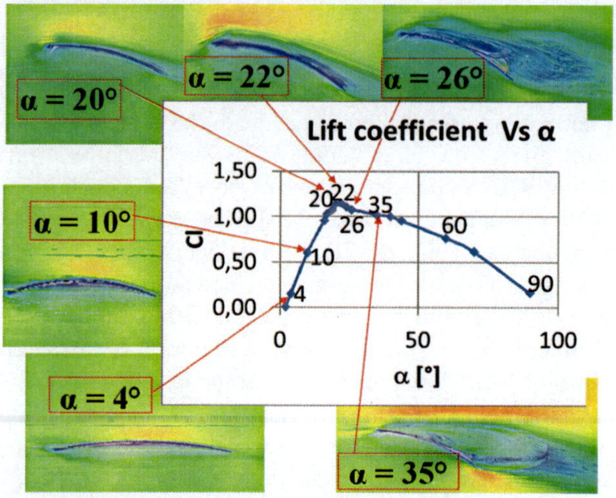

Fig. 5. Lift coefficient Vs incidence α at 2/5 of the sail height (wind = 20 knots).

Thanks to the performed analyses, the phenomenon of the detachment of the threads and the consequent loss of lift were clearly evaluated. Figure 5 shows the behaviour of the fluid threads around the sail at an altitude of almost 2 m, corresponding to 2/5 of the sail height, when using the battens with the design values. Up to an angle of incidence of 22° an increase in the coefficient of lift was registered. Around 22° the detachment of the fluid threads from the surface of the sail (stall) occurred and the lift coefficient decreased quickly (Fig. 6).

It can be observed that the displacements trend of the trailing edge of the mainsail and the flattening of the sail terminal section which greatly attenuated the onset of the flanking with consequent degradation of performance were greatly influenced by the presence of the battens. The length and stiffness of the battens, in fact, influenced the deformations of the sail fabric causing greater resistance to deformation. With the same surface of constraints, the battens were a practical trick to add structure to the one side (the leech) of the three sides of the sail that did not have it.

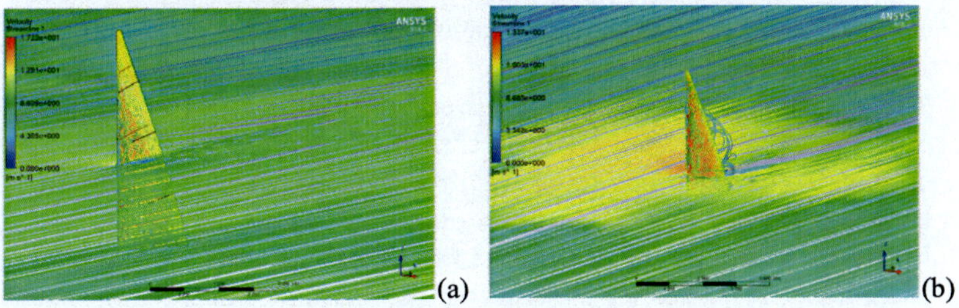

Fig. 6. Streamlines and pressure on sail surface (wind 20 knots): (a) α = 20°; (b) α = 35°.

Simulations were then carried out by progressively increasing the stiffness value and the longitudinal extension of the original battens (design battens). The graphs shown in Fig. 7 highlight the deformations obtained on the three battens as the Y parameter, product of stiffness and longitudinal extension, vary.

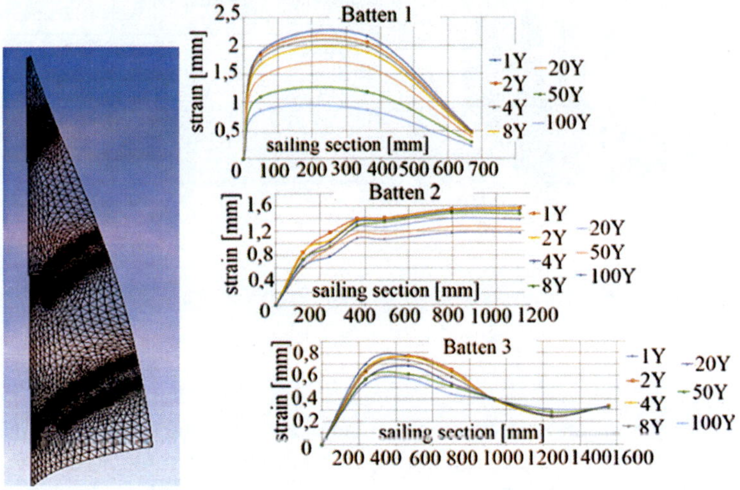

Fig. 7. (a) Sail strain for different battens stiffness (wind = 20 knots).

On batten 1(the superior batten), whose optimal length is close to the entire longitudinal section of the mainsail, a 75% reduction was noted in the initial deformations as the stiffness value and longitudinal extension increased. The phenomenon is attenuated on the center line of intermediate batten (batten 2) where the reduction is about 30% of the initial one. Finally, batten 3 (the inferior batten) is the one that undergoes the lowest percentage reduction (<20%). In the latter, however, a change in local curvature can be observed. All curves show a maximum of deformation near the center line and then decrease near the leech.

The increase in length and stiffness of the battens allows to maintain the concavity of the sail and to limit its localized non-homogeneous deformation. This also has a beneficial effect on warping, the total value of which can thus be reduced. However, it resulted that the increase in the stiffness values of the battens far beyond the design values causes a discontinuity in behaviour in the different sections of the sail and an unacceptable increase in the tensions in the battens themselves. From Fig. 7 it can be deduced that on batten 1, an increase of the stiffness beyond a certain value is not desirable, since keeping rigid the extremity of the sail is not convenient in very strong winds. In battens 2 and 3, on the other hand, the increase in stiffness is beneficial, limiting the rowing and detachment of the fluid threads in the trailing edge of the sail. Figure 8 shows the maximum value of sail lift coefficient of the sail for the best four configurations of twisting angle of the sail and stiffness values of the battens in different wind conditions (intensity and angle of attack). Summarizing the results obtained, it can be deduced that the best solution consists in adopting a differentiated stiffness of the battens: a lower value in batten 3, an average value in batten 2 and an higher in batten 1.

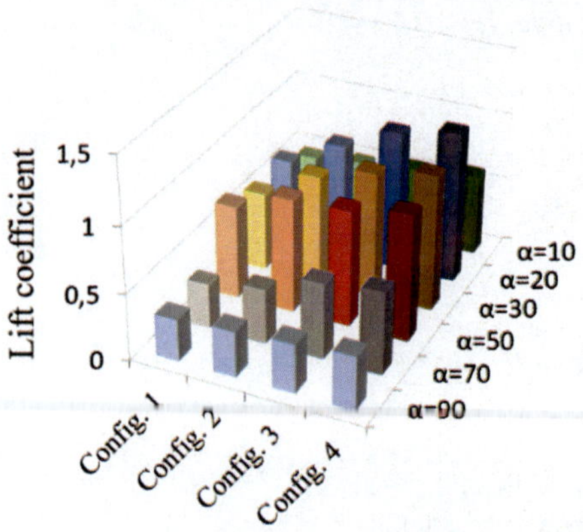

Configuration	Twisting [°]	Battens stiffness		
		Batten1	Batten2	Batten3
Config.1	16	2Y	4Y	20Y
Config.2	18	2Y	4Y	20Y
Config.3	20	2Y	4Y	8Y
Config.4	20	2Y	4Y	20Y

Fig. 8. Optimized lift coefficient.

In the lower part of the sail, where the wind was less, in fact, it was sufficient that the batten 3 controls the filing of the trailing edge of the leech; to achieve this, it was sufficient to place at the exit end of the leech a batten with a length of about one third of the short cross section of the sail. In the intermediate and upper parts of the sail, where the wind is greater, the reduction of turbulent phenomena and the wing's rowing effect can be obtained, however, only by resorting to longer and stiffer battens which extend respectively to about half full length of the cross section of the sail. By adopting battens with quadratically increasing stiffness towards the top of the sail, it was possible to give the sail a correct profile, controlling and maintaining its curvature throughout its height.

4 Conclusions

The paper presents a methodology that enhances panel and batten arrangement on sail's flying shapes. Through a triangular membrane element formulation coupled with a solid mesh of the battens, the aeroelastic effects on the sail were numerically investigated by coupling fluid-dynamic and structural analysis. Experimental structural characterization of sailcloth anisotropic mechanical behaviour (Dacron© TNF 240) and flexional stiffness of battens were carried out experimentally and implemented in the FSI analysis. The use of battens with different stiffness, quadratically increasing towards the top of the sail, proved to be the best solution. In particular, by stiffening the battens in proportion to the speed profile and extending their length to the full width of the sail for the battens at higher altitudes, it was possible to give the sail a correct airfoil, maintaining its curvature even in its upper part where the wind is more intense. Two main advantages in the proposed methodology can be identified: less computational time to evaluate the optimal panel and batten arrangement due to the integration of CAD parametric model CFD and CSM simulation; and correct simulation of the deformation of the sailcloth fabric due to its experimental characterization.

Acknowledgments. Starting Grant 2020–22 Linea 3-Progetto NASCAR-University of Catania Pr. 308811.

References

1. Braun, J.B., Imas, L.: High fidelity CFD simulations in racing yacht aerodynamic analysis. In: Proceedings 3rd High Performance Yacht Design Conference, Auckland, pp. 168–175 (2008)
2. Bak, S., Yoo, J., Song, C.Y.: Fluid-structure interaction analysis of deformation of sail of 30-foot yacht. Int. J. Naval Architect. Ocean Eng. **5**(2), 263–276 (2013)
3. Calì, M., Oliveri, S.M., Cella, U., Martorelli, M., Gloria, A., Speranza, D.: Mechanical characterization and modeling of downwind sailcloth in fluid-structure interaction analysis. Ocean Eng. **165**, 488–504 (2018)
4. Manikandan, A., Rajkumar, R.: Evaluation of mechanical properties of synthetic fiber reinforced polymer composites by mixture design analysis. Polym. Polym. Compos. **24**(7), 455–462 (2016)
5. Gazzo, S., Cuomo, M., Boutin, C., Contrafatto, L.: Directional properties of fibre network materials evaluated by means of discrete homogenization. Eur. J. Mech. A Solids **82**, 104009 (2020)
6. Boutin, C., Contrafatto, L., Cuomo, M., Gazzo, S., Greco, L.: Discrete homogenization procedure for estimating the mechanical properties of nets and pantographic structures. Lecture Notes in Mechanical Engineering, pp. 716–732 (2020)
7. Pearson, W.E.: Textiles to composites: 3D moulding and automated fibre placement for flexible membranes. Marine Appl. Adv. Fibre-Reinforced Composit. **305**, 305–334 (2015)
8. Richter, H.J., Horrigan, K.C., Braun, J.B.: Computational fluid dynamics for downwind sails. In: Proceedings 16th Chesapeake Sailing Yacht Symposium, Annapolis, MD, pp. 19–28 (2003)
9. Ranzenbach, R., Kleene, J.: Utility of flying shapes in the development of offwind sail design database. In: The High Performance Yacht Design Conference, Auckland (2002)
10. Renzsch, H., Graf, K.: An experimental validation case for fluid-structure interaction simulations of downwind sails. In: Proceedings of the 21st Chesapeake Sailing Yacht Symposium, Annapolis, p. 8 (2013)
11. Fossati, F., et al.: A novel full scale laboratory for yacht engineering research. Ocean Eng. **104**, 219–237 (2015)
12. Hochkirch, K.: Design and Construction of a Full-Scale Measurement System for the Analysis of Sailing Performance. Technical University of Berlin, PhD Thesis (2000)

13. Le Pelley, D.J., Modral, O.: VSPARS: a combined sail and rig shape recognition system using imaging techniques. In: Proceedings of 3rd High Performance Yacht Design Conference, Auckland, pp. 57–66 (2008)
14. Graves, W., Barbera, T., Braun, J.B., Imas, L.: Measurement and simulation of pressure distribution on full size sails. In: 3rd High Performance Yacht Design Conference, Auckland, vol. 2, p. 4 (2008)
15. Augier, B., Bot, P., Hauville, F., Durand, M.: Experimental validation of unsteady models for fluid structure interaction: application to yacht sails and rigs. J. Wind Eng. Ind. Aerodyn. **101**, 53–66 (2012)
16. Agarwal, B.D., Broutman, L.J.: Analysis and Performance of Fiber Composites, vol. 2. Wiley, New York (1990)

Industrial Design and Ergonomics

A Detailed Analysis of the Most Promising Concepts of Soft Wearable Robots for Upper–Limb

Dario Panariello[1,2(✉)], Stanislao Grazioso[1], Teodorico Caporaso[1], Giuseppe Di Gironimo[1], and Antonio Lanzotti[1]

[1] Fraunhofer Joint Lab IDEAS, Dipartimento di Ingegneria Industriale, Università degli Studi di Napoli Federico II, P.le Tecchio 80, 80125 Napoli, Italy
dario.panariello@unibg.it

[2] Dipartimento di Ingegneria Gestionale, dell'Informazione e della Produzione, Università degli Studi di Bergamo, Viale Marconi 5, 24044 Dalmine, (BG), Italy

Abstract. In this study we propose a brief analysis of recent soft wearable robots for upper–limb which could have a major impact on future developments and applications. The systems are analysed with respect to: design concepts, actuation systems, sensing systems, control strategies and applications. Finally, a discussion and open issues are presented.

Keywords: Soft robotics · Wearable robotics · Exoskeletons

1 Introduction

A Wearable robot (or exoskeleton) can be defined as "a wearable, external mechanical structure that enhances the power of a human" [1]. They can be divided in: (i) rigid exoskeletons, that transmit forces from robot to human using rigid links [1]; (i) soft exoskeletons, that transmit forces mainly through cable–driven systems or fluid-driven soft actuators (or pneumatic actuators) [2]. Soft wearable robots have emerged in the last years, outperforming the performances of rigid exoskeletons in terms of: wearability, safety and overall human-robot symbiosis. An exhaustive list of soft wearable robots is presented in many recent works [2–4]. Here, the authors have illustrated the soft wearable robots for upper body (i.e. neck, upper limb and trunk) and lower body (i.e. hip, knee and ankle), describing in detail the different types of actuators and power transmissions.

In this work, we propose a detailed analysis of the most promising soft wearable robot for upper–limb assistance, excluding hand wearable robots. In particular, compared to studies already presented in the literature [2,3], we analyze the main features of a soft exoskeleton, by taking into consideration design concept (i.e. degrees of freedom, supported movements, range of motions and mass); actuation (i.e. actuation methods, actuators and power transmissions); sensing system; control strategy and application. Finally, advantages, disadvantages and current limitations of the soft wearable robots are discussed.

C. Rizzi et al. (Eds.): ADM 2021, LNME, pp. 71–81, 2022.
https://doi.org/10.1007/978-3-030-91234-5_7

Table 1. List of the most promising soft wearable robots for upper–limb, presented in the literature from 2016 to 2020.

	Design concept			Actuation		Sensing system	Control strategy	Appl.
	SM	RoM [deg]	Mass [g]	Actuation method	Actuator & transmission			
Shoulder joint								
[5]	α, β	100	-	passive systems	tendon-driven	-	-	assistive
[6]	β	90	35	pneumatic	textile/ fabric	IMU pressure sensor	position control	rehab., assistive
[7]	β	90	350	pneumatic	textile/ fabric	-	pressure control	rehab.
[8]	α, β	90	480	pneumatic	textile/ fabric	-	pressure control	assistive
[9]	α	120	-	pneumatic	tendon-driven	-	-	-
Elbow joint								
[10]	δ	180	1200	electric	tendon-driven	encoder, load cell	admittance control	assistive
[11]	δ	-	-	electric	tendon-driven	EMG, IMU, encoder	EMG–based torque control	rehab., assistive
[12]	δ	75	-	pneumatic	textile/ fabric	EMG, pressure sensor	EMG–based control	rehab.
[13]	δ	-	-	pneumatic	textile/ fabric	-	pressure control	lifting activity
[14]	δ	-	1000	pneumatic	tendon-driven	encoders	position control	assistive
Wrist joint								
[15]	ϵ	160	-	pneumatic	PAM	force sensor, encoder	position control	rehab.
[16]	ϵ	95	200	pneumatic	textile/ fabric	pressure sensor	pressure control	rehab.
Multi–joint								
[17]	α, β, γ, δ, ϵ, ζ, η	-	-	electric	tendon-driven	mocap, load cell	position control	rehab.
[18]	α, β, γ δ, ϵ	-	1300	electric	tendon-driven	IMU	mimetic control	rehab.
[19]	α, δ	-	7500	electric	tendon-driven	encoders	voice control	industry
[20]	α, δ, ϵ	-	-	pneumatic	textile/ fabric	-	-	-
[21]	α,δ, ϵ, ζ	-	2100	pneumatic	PAM	pressure sensor	static control	rehab.

2 Classification of the Main Soft Wearable Robots for Upper–Limb Assistance

The most promising soft wearable robots presented in the literature are reported in Table 1. The table classifies wearable robots based on supported joint angles, i.e. shoulder, elbow, wrist and multi–joint. Moreover, it reports, for each wearable robot: supported movements (SM), range of motion (RoM), mass, actuator and power transmission, sensing system, control strategy and application. The pictures of the analysed soft wearable robots are reported in the following figures: Fig. 1 for shoulder joint exoskeletons, Fig. 2 for elbow joint exoskeletons, Fig. 3 for wrist joint exoskeletons and Fig. 4 for multi–joint exoskeletons.

Upper–Limb Kinematics. The upper limb is composed by three main articulations [22]: shoulder, elbow and wrist joint. The shoulder articulation is the most complex joint and it connects the humerus with the shoulder. The shoulder joint is modeled as a 3 degrees of freedom (DOF) joint, i.e. a spherical joint, simulating the movements of the glenohumeral. The three movements are defined as: shoulder flexion–extension α, shoulder abduction–adduction β and shoulder rotation γ. The elbow is a synovial joint between the humerus, in the upper arm, and the radio and ulna, in the forearm. For a first approximation, the elbow joint is modelled as a 1 DOF joint, i.e. a simple hinge joint. The movement is defined as elbow flexion–extension δ. The wrist joint connects the radius with the ulna and the radius with the hand. It is schematised as a 3 DOF joint and the movements are defined as: pronation and supination ϵ, wrist flexion-extension ζ and wrist radial-ulnar deviation η. Table 1 reports the supported movements (SM) by the wearable robots: α, β and γ for wearable robots supporting the shoulder; δ for wearable robots supporting the elbow; ϵ, ζ and η for wearable robots supporting the wrist; for the wearable robots supporting multiple joints, we have a combination of all the angles.

Design Concept. Apart from the actuation sources and systems, soft exoskeletons are also characterized by: (i) anchor points, that guarantee the execution of correct movement to be supported; (ii) suits, that guarantee the correct position of the anchor points, actuators and power transmissions. In particular, the correct execution of the required movements is guaranteed by the correct positioning of the anchor points of the pneumatic actuators or power transmission, i.e. tendons, on the subjects' body. The positions of the anchor points are not standard and change in each configuration presented in Table 1. However, some common features are: (i) the shoulder movements are carried out placing the anchor points on trunk/upper back/shoulder and upper arm, see Fig. 1; (ii) the elbow movement is assisted placing the anchor points on the upper arm and forearm, see Fig. 2; (iii) the wrist movements are assisted placing the anchor points on forearm and hand, see Fig. 3. In most of the cases reported in Table 1, two anchor points allow to perform the movement in one direction, i.e. 1 DOF; these exoskeletons are defined full–actuated since for each DOF they use an actuator. The wearable robots proposed in [5,8,19], instead, are defined under–actuated since an actuator can assist more than 1 DOF. In particular, a specific configuration of the anchor points allow to perform shoulder flexion–extension α and shoulder abduction–adduction β using a free–to–move anchor point from anterior part to the medial part and soft cruciate linkage, respectively. A mechanical solution, instead, is proposed in [19] to carry out shoulder and elbow flexion–extension (α and δ). In this exoskeleton the anchor points, located on the upper arm and forearm, allow to flex the elbow joint (δ) until 90°C, when δ reaches 90°C the elbow angle is locked and the same cable, using a different position of anchor point (i.e. shoulder and upper arm), assists the shoulder joint (α). The suits, as mentioned above, are used to fix the anchor points on the human and for adjustment the wearable robot to individual users using strap, Velcro, hook–and–look or by reinforcing the anchor areas with flexible plate [8]. Most of the proposed suits use shoulder, elbow, wrist braces and in same cases jackets [8,14,18,20]. Finally, as reported in Table 1, the soft exoskeletons are generally lightweight; the mass of the systems is lower than 500 g, except for wearable robots where the power system is worn [10,18,19,21].

Actuation. The actuators and power transmission systems reported in Table 1 can be classified in: (i) cable–driven systems, which use Bowden cables (or tendons) to generate the required movements; (ii) fluid–driven soft actuators (or pneumatic actuators), which use specific cavities and geometries to generate force when pressurised. In particular, the latter can be divided in: pneumatic artificial muscles (PAM), fabric–based inflatables and textiles (textile/fabric). The PAM are historically the first pneumatic actuators that have been used for soft wearable robots [2]; then, they have been improved by making actuators capable of generating higher forces, such as flat pneumatic actuators (FPA) [15] and pneumatic gel muscles (PGM) [21]. In the last decade, soft actuators made from fabric or a combination of material properties, i.e. textiles, have started having great impact on the design and fabrication of soft wearable robots. The common methods of fabrications include the combination of the following materials: polyurethane (PU) [6,7]; thermoplastic polyurethane (TPU) [12,13,16] and inextensible textile [8]; Twaron with a latex tube (M.T.D13.L1) [20]. The actuation method is electric for cable–driven systems, i.e. based on electric motors, and pneumatic for fluid–driven soft actuators. Hybrid solutions are also proposed: passive (i.e. springs) and pneumatic systems are used to generate the contraction of the tendons in [5] and in [9,14].

Sensing System. The sensors used and reported in Table 1 can be divided in: (i) sensors to monitor human activities, i.e. encoders, motion capture (mocap), inertial measurements units (IMU) and electromyography (EMG); (ii) sensors to monitor the state of the actuators and power transmissions, i.e. pressure sensors and load cells. In particular, encoders are usually used for measuring elbow movement (δ) [10,11,14,24] or for measuring elbow and shoulder movements (α and δ) [19]; motion capture systems are used to estimate shoulder, elbow and wrist movements (α, β, γ, δ, ϵ, ζ and η) [17]; IMUs, placed on the upper arm and forearm, are used to estimate attitude (i.e. roll, pitch and yaw) [18] and measuring

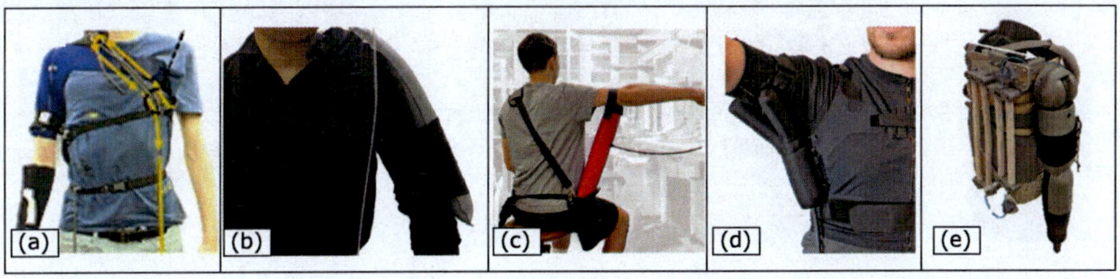

Fig. 1. Main soft wearable robots for assisting the shoulder joint. (a) passive tendon–driven wearable robot from Seoul National University [5]; (b) fabric–based wearable robot from National University of Singapore [6]; (c) fabric–based wearable robot from Stanford University [7]; (d) textile wearable robots from Harvard University at the Whyss Institute for Biologically Inspired Engineering [8]; concept of hybrid soft wearable robot from University of Illinois [9].

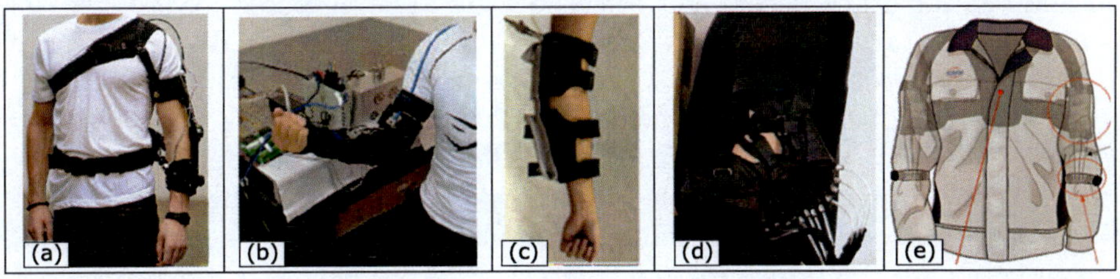

Fig. 2. Main soft wearable robots for assisting the elbow joint. (a) tendon–driven wearable robot from School of Mechanical and Aerospace Engineering at Robotics Research Center in Singapore [10]; (b) tendon–driven wearable robot from University of Aeronautics and Astronautics in China [11]; (c) fabric–based wearable robot from National University of Singapore [12]; (d) fabric–based wearable robot from Arizona State University [13]; (e) hybrid soft wearable robot from Korea Institute of Machinery and Materials [14].

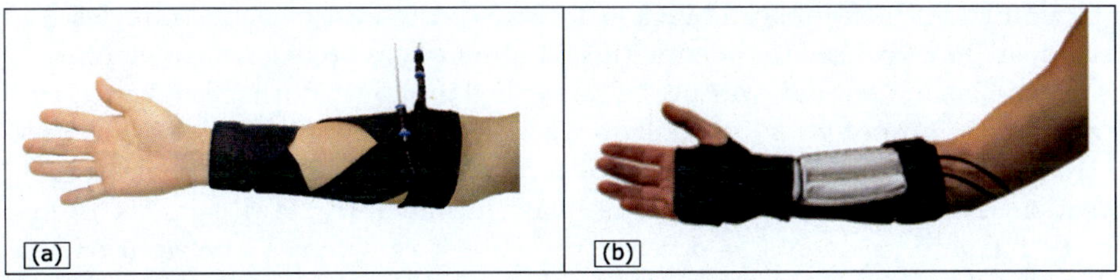

Fig. 3. Main fabric–based wearable robots for assisting the wrist joint, from Soft Robotics Research Center at Seoul National University [15] (a) and from University of Southern California [16] (b). Main fabric–based wearable robots for assisting the wrist joint, from Soft Robotics Research Center at Seoul National University [15] (a) and from University of Southern California [16] (b).

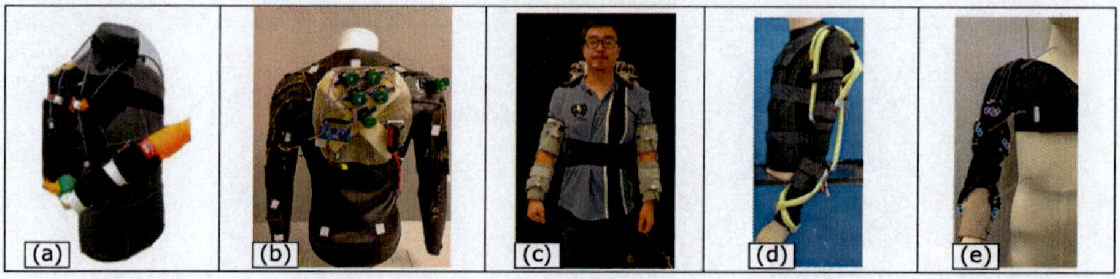

Fig. 4. Main soft wearable robots for multi–joint. (a) tendon–driven wearable robots from Shenyang Institute of Automation [17]; (b) tendon–driven systems from University of California [18]; (c) tendon–driven wearable robot from Nanyang Technological University in Singapore [19,23]; (d) textile wearable robot from Politecnico di Torino [20]; (e) soft wearable robot with pneumatic gel muscles from Hiroshima University [21].

elbow movement (δ) [11] through the extraction of quaternions; EMG, usually placed on biceps brachii (BB) and triceps brachii (TB), are used to estimate the maximum voluntary contractions (MVC) in [12] and, placed on BB, to estimate the elbow joint torque using a sequence of filters including the Kalman filter [11]; encoder and EMG together, placed on BB, TB and brachioradialis, are used to estimate the elbow joint torque using the Hill–type muscle model [24]. Finally, load cells can be integrated in cable–driven systems to measure the tension of the cable [10,17]; pressure sensors, instead, are usually adopted by pneumatic–based systems to monitor air pressure [6,12,16,21].

Control Strategy. The main control strategies reported in Table 1 can be divided in: (i) position control; (ii) pressure control; (iii) intention–detection control; (iv) torque estimation control. The position control (i) and pressure control (ii) are easy–to–use control strategies, but they do not guarantee a perfect symbiosis between human and robot. In particular, in position control strategies (i) the pneumatic actuator pressure or cable length is varied in order to follow a predetermined trajectory, dependent on the task to be performed. Position control is usually performed in closed–loop and encoders, mocap and IMU are implemented in the feedback branch [6,14,15,17]. Pressure control strategies (ii), instead, are usually used for pneumatic actuators where predefined trajectories of the arm define a reference pressure to be applied in input. With this regards, two strategies are proposed: an open–loop control [7,8,13] and closed–loop control where pressure sensors are usually implemented in the feedback branch [16,21]. Intention–detection (iii) and torque estimation control (iv) strategies try to follow the human movements, in order to get closer to a symbiosis between human and robot. Intention–detection controls (iii) are proposed in [19] as open–loop using voice command control and in [12,18] as closed–loop control. In particular, intention–based motions using EMG signals is proposed in [12] and mimetic control using IMU is illustrated in [18] where the actuated arm, i.e. the right arm, follows the movement of the non-actuated arm, i.e. the left arm. The torque estimation control (iv) is proposed in [11] as EMG–based torque estimation control strategy. In this work the authors propose to estimate the elbow joint torque using EMG signals and a Kalman filter; then, an incremental angle is estimated and compared to feedback branch composed by elbow joint (δ) measured using encoder. Finally, a promising control strategy is proposed in [10,24] as gravity compensation torque. In particular, the authors estimate the elbow joint torque required using single–joint model and the assistive torque generated by the soft exoskeleton is evaluated measuring the tension of the cable, using load cell; then the difference between gravity compensation and assistive torque, defined as interaction force, is converted to a reference velocity for the motor in order to have an interaction force equal to zero. In this way the assistive torque is equal to gravity compensation torque of the elbow joint (δ). An improvement of the control strategy is proposed in [24] where the elbow torque is evaluated thought Hill–type muscle model using the EMG and encoder signals.

Application The fields of applications can be divided in: (i) rehabilitation for medical use; (ii) assistive for daily living (ADL) tasks; (iii) industry for industrial tasks. The main applications of examined soft wearable robots, as Table 1 underlines, are rehabilitation and assistive.

3 Discussion and Open Issues

The analysis of the soft wearable robots for upper limb illustrated in Sect. 2 underlines that tendon–driven, fabric–based inflatables and textile are the actuation systems mostly adopted in the last decade. As indicated in [2], the first are controllable and can generate high forces; however, during use, the cables could generate high friction forces and the anchor points could slip. The second, instead, are simple and fast to fabricate, lightweight and can generate high forces; however, during use, delamination of the layers could be generated.

The illustrated advantages of the soft wearable robots are still linked to experiments performed in a laboratory setting. In particular, unlike rigid exoskeletons that have been widely implemented and validated in application contexts, as in industry [1], most of the examined soft wearable robots have been only tested in laboratory settings, as for instance, validation through reduction of muscle activations and fatigue of the subject during the execution of specific tasks. The authors in [8], instead, propose to validate the soft wearable robots across five stroke survivors. However, the involvement of five subjects is still not statistically relevant as underlined in [1], where the authors suggest that eight is the minimum number of subjects to consider the study as statistically significant [1].

Finally, to improve design and fabrication and to observe a complete involvement of soft wearable robots in all fields of application the following limitations should be overcome:

Tools for Design and Simulation. Currently the behaviour of soft actuators, i.e. forces generated and displacements, are studied using (i) finite element method (FEM) [25,26]; (ii) experimental simulations [20]. Then, the joint torques generated by the actuators are evaluated through static or dynamic considerations using kinematic model of the arm. In other cases, i.e. rigid exoskeletons, the systems are modeled and simulated using musculoskeletal models as reported in [27], using Anybody Modeling System[1], and in [28], using OpenSim simulator[2]. Software applications for design and simulation of soft wearable robots integrated with musculoskeletal models have not yet been explored in depth; a recent example of application is proposed in [29]. They could lead to significant improvement in soft wearable robotics, since they would allow to compare different design concepts, materials and geometries of wearable robots before the physical prototyping.

Transportability of Soft Wearable Robots. Tendon–driven systems are defined transportable systems since they wear, usually on the back, the power systems composed of electric motors, gears and batteries. The pneumatic exoskeletons, instead, are not transportable since there are still technological limitations that prevent the wearing of the power system, i.e. air compressor. In this directions, different works have been presented solutions that aim to miniaturize the pumps for soft pneumatic actuators [30]. To date, we can conclude that the system transportability is fundamental for daily living tasks in which the wearable robot assists the subject throughout the day. In other applications, where the subject's position is within predefined areas such as medical and industrial tasks, the portability of the system plays a less relevant role. For these reasons, today tendon-driven systems are preferred for assistive applications and pneumatic systems are more suitable for rehabilitation and industrial contexts.

Wearability of Soft Wearable Robots. The wearability of the systems depend on the size of the exoskeleton. In this regard, one of the main advantages of pneumatic actuators is the possibility to wear them under clothing in order to create a more comfortable and less invasive wearable systems [14]. This feature could bring enormous benefits in industrial sector, where it will no longer be necessary to adapt workstations to workers who wear wearable robots; it will be enough to wear a simple actuated suit to reduce the risks associated to musculoskeletal disorders related to biomechanical overload [31,32]. To enhance the wearability, design approaches where the suits and anchor points are based on the real 3D models and anthropometric characteristics of the human [33] can be explored, as already done in different fields as orthopaedics [34,35]. Another important trend that can be observed in the direction of enhanced wearability is the implementation of flexible sensors (or soft sensors) in order to eliminate bulky components of exoskeletons, i.e. IMU and encoders. Moreover, two different approaches could be used: (i) flexible sensors on the human that evaluate the movements, i.e. joint angles, based on voltage of the soft sensor resistance [36]; (ii) flexible sensors in the actuators that measure the position and behaviour of the actuators [2].

Control of Soft Wearable Robots. The control strategies of tendon–driven wearable robots, i.e. EMG–based torque estimation and admittance control strategies

[1] https://www.anybodytech.com/.

[2] https://simtk.org/projects/opensim/.

[10, 11], have been investigated more thoroughly than control strategies for pneumatic soft wearable robots. Control strategies for soft pneumatic actuators are required to successfully implement soft wearable robots in real environments. In particular, control strategies must be able to follow and adapt to joint movement in a natural and accurate way. To do that, two main approaches could be followed: (i) kinematics–based approach [37]; (ii) synergy–based approach [38–40].

4 Conclusions

In this study, we have presented an analysis of the most promising soft wearable robots for upper–limb that aim to support human movements during the execution of medical, daily or industrial tasks. To date, soft wearable robots have a low level of technological readiness (TRL), as most of the presented systems have not been implemented in a real application context. This allows to further explore soft wearable robots to propose new solutions which maximise the forces transmitted and comfort while reducing the mass and their invasiveness.

References

1. De Looze, M.P., Bosch, T., Krause, F., Stadler, K.S., O'Sullivan, L.W.: Exoskeletons for industrial application and their potential effects on physical work load. Ergonomics **59**(5), 671–681 (2016)
2. Thalman, C., Artemiadis, P.: A review of soft wearable robots that provide active assistance: trends, common actuation methods, fabrication, and applications. Wearable Technol. **1**, e3 (2020)
3. Sanchez, V., Walsh, C.J., Wood, R.J.: Textile technology for soft robotic and autonomous garments. Adv. Functional Mater. **31**(6), 2008278 (2021)
4. Xiloyannis, M., et al.: Soft robotic suits: state of the art, core technologies, and open challenges. IEEE Trans. Robot. (2021)
5. Park, D., Cho, K.J.: Development and evaluation of a soft wearable weight support device for reducing muscle fatigue on shoulder. PloS One **12**(3), e0173730 (2017)
6. Natividad, R., Yeow, C.H.: Development of a soft robotic shoulder assistive device for shoulder abduction. In: 2016 6th IEEE International Conference on Biomedical Robotics and Biomechatronics (BioRob), pp. 989–993. IEEE (2016)
7. Simpson, C., Huerta, B., Sketch, S., Lansberg, M., Hawkes, E., Okamura, A.: Upper extremity exomuscle for shoulder abduction support. bioRxiv (2020)
8. O'Neill, C., et al.: Inflatable soft wearable robot for reducing therapist fatigue during upper extremity rehabilitation in severe stroke. IEEE Robot. Autom. Lett. **5**(3), 3899–3906 (2020)
9. Thompson, N., Sinha, A., Krishnan, G.: Characterizing architectures of soft pneumatic actuators for a cable-driven shoulder exoskeleton. In: 2019 International Conference on Robotics and Automation (ICRA), pp. 570–576. IEEE (2019)
10. Xiloyannis, M., Chiaradia, D., Frisoli, A., Masia, L.: Physiological and kinematic effects of a soft exosuit on arm movements. J. Neuroeng. Rehabil. **16**(1), 29 (2019)
11. Lu, L., Wu, Q., Chen, X., Shao, Z., Chen, B., Wu, H.: Development of a SEMG-based torque estimation control strategy for a soft elbow exoskeleton. Robot. Auton. Syst. **111**, 88–98 (2019)

12. Koh, T.H., Cheng, N., Yap, H.K., Yeow, C.H.: Design of a soft robotic elbow sleeve with passive and intent-controlled actuation. Front. Neurosci. **11**, 597 (2017)
13. Thalman, C.M., Lam, Q.P., Nguyen, P.H., Sridar, S., Polygerinos, P.: A novel soft elbow exosuit to supplement bicep lifting capacity. In: 2018 IEEE/RSJ International Conference on Intelligent Robots and Systems (IROS), pp. 6965–6971. IEEE (2018)
14. Park, S.J., Park, C.H.: Suit-type wearable robot powered by shape-memory-alloy-based fabric muscle. Sci. Rep. **9**(1), 1–8 (2019)
15. Park, S.H., Yi, J., Kim, D., Lee, Y., Koo, H.S., Park, Y.L.: A lightweight, soft wearable sleeve for rehabilitation of forearm pronation and supination. In: 2019 2nd IEEE International Conference on Soft Robotics (RoboSoft), pp. 636–641. IEEE (2019)
16. Realmuto, J., Sanger, T.: A robotic forearm orthosis using soft fabric-based helical actuators. In: 2019 2nd IEEE International Conference on Soft Robotics (RoboSoft), pp. 591–596. IEEE (2019)
17. Li, N., et al.: Bio-inspired upper limb soft exoskeleton to reduce stroke-induced complications. Bioinspiration Biomimetics **13**(6), 066001 (2018)
18. Lessard, S., Pansodtee, P., Robbins, A., Trombadore, J.M., Kurniawan, S., Teodorescu, M.: A soft exosuit for flexible upper-extremity rehabilitation. IEEE Trans. Neural Syst. Rehabil. Eng. **26**(8), 1604–1617 (2018)
19. Kim, Y.G., Little, K., Noronha, B., Xiloyannis, M., Masia, L., Accoto, D.: A voice activated bi-articular exosuit for upper limb assistance during lifting tasks. Robot. Compu. Integrated Manuf. **66**, 101995 (2020)
20. Belforte, G., Eula, G., Ivanov, A., Raparelli, T., Sirolli, S.: Presentation of textile pneumatic muscle prototypes applied in an upper limb active suit experimental model. J. Text. Inst. **109**(6), 757–766 (2018)
21. Das, S., Kurita, Y.: Forcearm: a wearable pneumatic gel muscle (PGM)-based assistive suit for the upper limb. IEEE Trans. Med. Robot. Bionics **2**(2), 269–281 (2020)
22. Pennestrì, E., Stefanelli, R., Valentini, P., Vita, L.: Virtual musculo-skeletal model for the biomechanical analysis of the upper limb. J. Biomech. **40**(6), 1350–1361 (2007)
23. Kim, Y.G., Xiloyannis, M., Accoto, D., Masia, L.: Development of a soft exosuit for industrial applications. In: 2018 7th IEEE International Conference on Biomedical Robotics and Biomechatronics (Biorob), pp. 324–329. IEEE (2018)
24. Lotti, N., et al.: Adaptive model-based myoelectric control for a soft wearable arm exosuit: a new generation of wearable robot control. IEEE Robot. Autom. Mag. **27**(1), 43–53 (2020)
25. Nguyen, P.H., Zhang, W.: Design and computational modeling of fabric soft pneumatic actuators for wearable assistive devices. Sci. Rep. **10**(1), 1–13 (2020)
26. Grazioso, S., Di Gironimo, G., Siciliano, B.: A geometrically exact model for soft continuum robots: the finite element deformation space formulation. Soft Robot. **6**(6), 790–811 (2019)
27. Fournier, B.N., Lemaire, E.D., Smith, A.J., Doumit, M.: Modeling and simulation of a lower extremity powered exoskeleton. IEEE Trans. Neural Syst. Rehabil. Eng. **26**(8), 1596–1603 (2018)
28. Ferrati, F., Bortoletto, R., Pagello, E.: Virtual modelling of a real exoskeleton constrained to a human musculoskeletal model. In: Lepora, N.F., Mura, A., Krapp, H.G., Verschure, P.F.M.J., Prescott, T.J. (eds.) Living Machines 2013. LNCS (LNAI), vol. 8064, pp. 96–107. Springer, Heidelberg (2013). https://doi.org/10.1007/978-3-642-39802-5_9

29. Harbauer, C.M., Fleischer, M., Bandmann, C.E.M., Bengler, K.: Optimizing force transfer in a soft exoskeleton using biomechanical modeling. In: Black, N.L., Neumann, W.P., Noy, I. (eds.) IEA 2021. LNNS, vol. 223, pp. 274–281. Springer, Cham (2022). https://doi.org/10.1007/978-3-030-74614-8_33
30. Cacucciolo, V., Shintake, J., Kuwajima, Y., Maeda, S., Floreano, D., Shea, H.: Stretchable pumps for soft machines. Nature **572**(7770), 516–519 (2019)
31. Panariello, D., Grazioso, S., Caporaso, T., Palomba, A., Di Gironimo, G., Lanzotti, A.: Evaluation of human joint angles in industrial tasks using opensim. In: 2019 II Workshop on Metrology for Industry 4.0 and IoT (MetroInd4. 0&IoT), pp. 78–83. IEEE (2019)
32. Panariello, D., Grazioso, S., Caporaso, T., Di Gironimo, G., Lanzotti, A.: Preliminary requirements of a soft upper-limb exoskeleton for industrial overhead tasks based on biomechanical analysis. In: Black, N.L., Neumann, W.P., Noy, I. (eds.) IEA 2021. LNNS, vol. 223, pp. 317–324. Springer, Cham (2022). https://doi.org/10.1007/978-3-030-74614-8_38
33. Grazioso, S., Selvaggio, M., Di Gironimo, G.: Design and development of a novel body scanning system for healthcare applications. Int. J. Interact. Des. Manuf. (IJIDeM) **12**(2), 611–620 (2018)
34. Grazioso, S., Selvaggio, M., Caporaso, T., Di Gironimo, G.: A digital photogrammetric method to enhance the fabrication of custom-made spinal orthoses. JPO J. Prosthetics Orthotics **31**(2), 133–139 (2019)
35. Grazioso, S., Caporaso, T., Selvaggio, M., Panariello, D., Ruggiero, R., Di Gironimo, G.: Using photogrammetric 3D body reconstruction for the design of patient-tailored assistive devices. In: 2019 II Workshop on Metrology for Industry 4.0 and IoT (MetroInd4. 0 & IoT), pp. 240–242. IEEE (2019)
36. Kim, D., Kwon, J., Han, S., Park, Y.L., Jo, S.: Deep full-body motion network for a soft wearable motion sensing suit. IEEE/ASME Trans. Mechatron. **24**(1), 56–66 (2018)
37. Zhou, Y.M., Hohimer, C., Proietti, T., O'Neill, C.T., Walsh, C.J.: Kinematics-based control of an inflatable soft wearable robot for assisting the shoulder of industrial workers. IEEE Robot. Autom. Lett. **6**(2), 2155–2162 (2021)
38. Grazioso, S., et al.: Assessment of upper limb muscle synergies for industrial overhead tasks: a preliminary study. In: 2019 II Workshop on Metrology for Industry 4.0 and IoT (MetroInd4. 0 & IoT), pp. 89–92. IEEE (2019)
39. Caporaso, T., Grazioso, S., Panariello, D., Di Gironimo, G., Lanzotti, A.: Understanding the human motor control for user-centered design of custom wearable systems: case studies in sports, industry, rehabilitation. In: Rizzi, C., Andrisano, A.O., Leali, F., Gherardini, F., Pini, F., Vergnano, A. (eds.) ADM 2019. LNME, pp. 753–764. Springer, Cham (2020). https://doi.org/10.1007/978-3-030-31154-4_64
40. Averta, G., et al.: On the time-invariance properties of upper limb synergies. IEEE Trans. Neural Syst. Rehabil. Eng. **27**(7), 1397–1406 (2019)

A New Sedan Concept Car in Stylistic Design Engineering (SDE) Enhanced with Augmented Reality

Gian Maria Santi(✉), Leonardo Frizziero, Giampiero Donnici, Daniela Francia, Marco Neri, and Alfredo Liverani

University of Bologna, 40136 Bologna, BO, Italy
{gianmaria.santi2,leonardo.frizziero,giampiero.donnici, d.francia,alfredo.liverani}@unibo.it, marco.neri12@studio.unibo.it

Abstract. Many industrial technologies are developed to optimize products and bring innovation. In particular, the automotive sector is renewing itself according to the rules of green energy and consumption. This huge change requires a reinterpretation of the models on the market updating them to the present and the future needs of automotive industry. In this paper the best compromise between innovation and tradition is found for the Ford brand that has not yet presented electric cars in the sedan segment. Following the SDE method enriched with Quality Function Deployment (QFD), Benchmarking (BM) and Top Flop Analysis (TPA), it is possible to carry out an innovative project. All these technologies must, however, be ordered according to a specific product allowing the best result for the design process. It is therefore necessary identifying the most common stylistic trends in order to draw the external styling of the vehicle using virtual prototyping techniques. To achieve an innovative result, Augmented Reality (AR) is considered to complete the method substituting the static and expensive procedure of making maquettes.

Keywords: Stylistic design engineering (SDE) · Car design · Design engineering · QFD · Benchmarking · Additive manufacturing · Augmented reality

1 Introduction

The initial design of a new car is mandatory for the success of the entire project. In this phase are considered: the shape, the innovations, the costs, the times of production and the impact on the market. Because of the latter, buyers' judgment is crucial and it is not possible to start a new project without a market analysis that highlights the needs of future buyers. Errors in this phase will lead to inefficiencies that will affect the budget. Many approaches to the problem are possible since the design stage isn't simple [1]. The following paper presents a case study based on the different phases of the Industrial Design Structure (IDeS) with the integration of AR. Using IDeS applied to

C. Rizzi et al. (Eds.): ADM 2021, LNME, pp. 82–90, 2022.
https://doi.org/10.1007/978-3-030-91234-5_8

innovative industrial products [2], it is possible to decompose the process in four steps: style, design, optimization and production. The usage of breakthrough technologies such as Augmented Reality (AR) [3] show that non expensive solutions are possible for virtual prototyping and they can be implemented in the IDeS scheme. Figure 1 shows the process divided in three macro-phases: Design Setting, Product Development and Production.

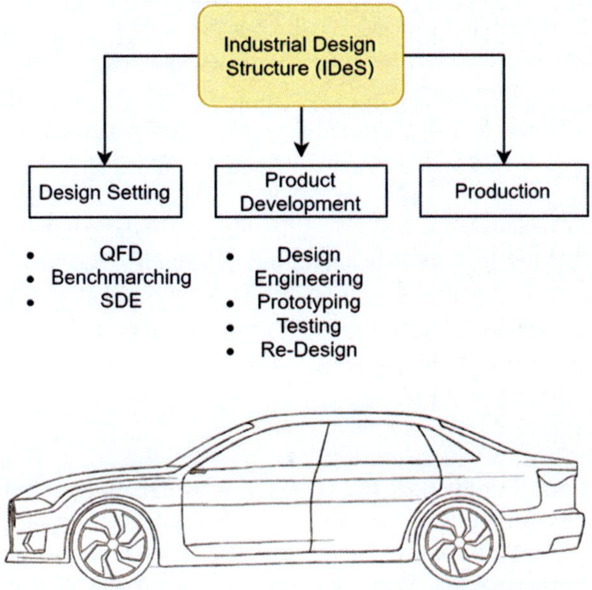

Fig. 1. Paper architecture following IDeS phases and implementing AR.

A Gantt chart is built to manage all phases and the Quality Function Deployment (QFD) [4, 5] is used to define the customer's requests and the target market segment. Secondly, a Benchmarking analysis is performed to identify the number of requirements that will allow the product to be innovative. This phase of the method is enhanced by a Top-Flop analysis that underlines the main characteristics of the best products on the market, which will be considered the limit to achieve innovation in the new model designed. Ford style was initially studied to better understand the characteristics of the brand with its main stylistic properties from his history. The SDE method applied to the automotive sector [6, 7] brought to the sketching of the digital vehicle that improves the classic freehand drawing.

2 Design Setting

Following the Gantt Chart (Fig. 2) and the increasing selling trend of electric cars in the last years, a new project is started using a cost analysis. This firs tool is intended to foresee the budget of all Research and Development costs which are essential for the project setting and product development phases.

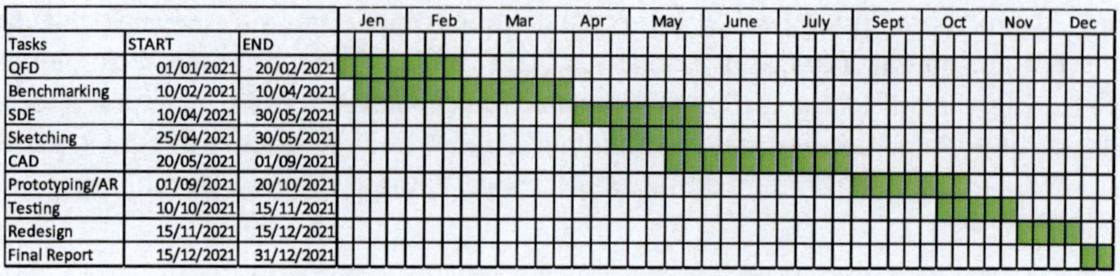

Fig. 2. Gantt Chart of the design phases.

The goal of cost analysis is to minimize risks and maximize earnings for the project. It allows to understand the advantages and disadvantages of various hypothesis including: transactions, activities, business requirements and investments. In the case study, the budget is set on an annual time basis, according to the subdivision of the phases defined in the Gantt chart. Considering as main costs the conceptual design, the sketching phase, the 2D and 3D CAD design, prototype and testing, the budget rises to 456400 EUR for the entire project.

The marketing assumptions described in the following paragraphs are built following the interview of 20 students to mimic the real customer answers.

2.1 QFD

The Quality Function Deployment (QFD) [8] is a methodology that focuses on the aspects that pursue customer satisfaction in order to establish the quality that must be ensured. The QFD forces to connect the key characteristics of the products to the needs of the customer. Because of that, it is applicable to different sectors [9, 10].

The six questions (Who, What, Where, When, Why and How) are used to the case study in the first design phase of the innovative sedan. The answers of the chosen population allowed to find the requirements to be analysed with the interrelation matrices.

It is then possible to rank the most dependent and the most independent needs. The analysis indicates that the items "Low price" (31 points) and "Comfort" (31 points) are the most influenced ones, and they are therefore the most dependent. Rather high values are also recorded for the items "Performance", "Personalization" and "Electric". On the other hand, the item "Electric" reaches a value of 25 points on the columns meaning that this feature has the higher influence on the other specs (Table 1a).

Once the Dependency Matrix is defined, the data relating the various customer product requirements are known, while the information regarding the relative importance of the various items is not yet known. The analysis shows that "Performances", "Electric, "Safety" and "Design" are the most important characteristics for customers as expected.

Table 1. (a) Dependences matrix; (b) What-How matrix.

	Low Price	Performance	Comfort	Electric	Safety	On board space	Design	Smart	Customization	Total
Low Price		9	3	9	3		3	1	3	31
Performance	9		1	3		3	1			17
Comfort	3	3		9	3		1	3	9	31
Electric	3	9						1		13
Safety		3	1			1	1	3		9
On board space			3				1			4
Design	1		3			3		1		8
Smart				1		1			3	5
Customization	3		1	3	3		1	3		14
Total	19	24	12	25	9	8	8	12	15	

(a)

	Length	Width	Height	Weight	Supply	Power	Max Velocity	Acceleration (0-100)	Max Torque	Seat Number	Door Number	Price	Consumption	CO2 Emission	Autonomy	Battery Capacity	Charging Time	
Low Price				1		9	3	9	9	1		9	1	9	3	3	3	60
Performance				3	3	9	9	9	9	3		3	9	3	9	3	3	75
Comfort	9	9	9	1						3	3	3			9		9	55
Electric				3	9	3	1	3	3			9	9	9	9	9	9	76
Safety				1			3	3	3			1		1	1		1	14
Design	9	9	9	3						3	3	1						37
	18	18	18	12	12	21	16	24	24	10		26	19	22	31	15	25	

(b)

2.2 Benchmarking

Strengths and weaknesses of the cars already on the market are analysed through a benchmark analysis.

The cars compared correspond to electric models belonging to the Executive cars segment already in production or prototypes.

The models considered for this analysis are: Tesla Model S and Model 3, Porche Taycan, Audi e-Tron GT, BMW i4, Mercedes Vision EQS, Volkswagen ID Vizzon, Lucid Air and Lightyear One.

Table 2 shows the result of the Benchmarking Analysis for the various competitors underlying the values that would make the product innovative.

Table 2. Innovation value per each competitor's car models.

LENGTH (mm)	WIDTH (mm)	HEIGHT (mm)	WEIGHT (kg)	SIT'S NUMBER
< 4640	< 1850	< 1380	< 1847	5
DOOR'S NUMBER	SUPPLY	HOMOLOGATED POWER (kW)	PEAK POWER – BOOST (kW)	MAX VELOCITY (km/h)
4	E	nd	> 700	> 320
ACCELERATION (0-100) (s)	MAXIMUM TORQUE (Nm)	DRAG COEFFICIENT Cd	PRICE (EUR)	CONSUMPTION (Wh/km)
< 2,6	> 967	< 0,2	< 60000	< 83
EMISSIONS (C02) (g/km)	TEORETICAL AUTONOMY (km)	ACTUAL AUTONOMY (km)	BATTERY CAPACITY (kWh)	CHARGE TIME 2,3 kW (h)
0	> 830	> 630	> 130	< 30
CHARGE TIME 7,4 kW (h)	CHARGE TIME (80%) IN DC (min)			
< 10	< 20			

Not being able to create an innovative model that achieves all 22 objectives of the matrix, chosen is done using the Top-Flop analysis. To understand which parameter is more likely to be improved, reference is made to another tool, the What-How Matrix (Table 1b) which allows to compare the customer's requirements (What) with the designer's answers (How). The most important performances are Price, Autonomy and Charging Time that can be subdivided in six components: Price, Theoretical autonomy WLTP, Real autonomy, Charging time at 2.3kW, Charging time at 7.4 kW and Charging time for 80% battery power in DC. Improving these six parameters means define a new and innovative product since the maximum delta of innovation found in the competitors is five.

The What-How Matrix is therefore applied to evaluate the best type of engine for the future, according to the customer's requests. The needs of consumers are analysed in three main areas: price, performance, and emissions for fee benefits. As expected, electric engine is both the most innovative for future developments and the most requested by the customers.

2.3 SDE

SDE is a design method that consists of the study and analysis of current stylistic trends, their relationship with customer requests and their visualization within the sketch.

For the case study presented, the trends can be classified according to four main stylistic trends of the last few years: Retrò, Natural, Stone, Advanced. The four stylistic trends are generally separate and distinguishable in different product designs, but they can also be mixed together in a completely new style that combines the past, the present and the future of car design. The proposed model aims to be innovative both for characteristics and style.

The cars considered for the stylistic analysis of the Ford brand are the Ford Falcon and the Ford Mondeo with their evolution over the years. For each proposal, a sketch of the car was made in the front three-quarter, rear three-quarter and side views leading to the final sketch (Fig. 3) representing a mix of the four styles.

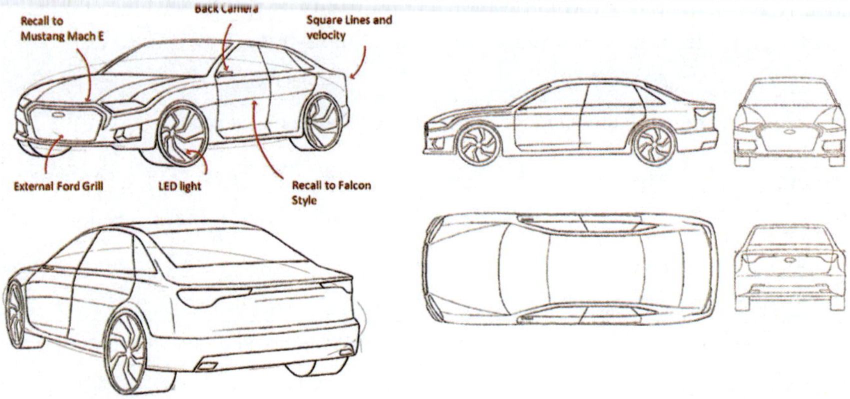

Fig. 3. Final sketches of the proposed car style combining the four possibilities.

3 Product Development

After selecting the final sketch using SDE, the first step of Product Development consists in transforming the digital sketch into 2D and finally 3D models using a CAD environment. Surface modelling was performed in order to have a fine representation both in Augmented reality and in the rendering environment. A first look of the final prototype is presented in Fig. 4 where a superimposition of the CAD model with a realistic environment is performed using a rendering engine.

Fig. 4. 3D representation of the car model and rendering

3.1 Prototyping – Augmented Reality

Because of the 3D model, it is possible to have a realistic view of the car in a digital environment as described above. However, viewing the 3D model on a monitor has some limitations, such as the correct definition of surface curvatures and the limited perception of the proportions. Using Augmented Reality, it is possible to create a physical and a virtual model of the car in its real shape. The model (maquette) have always been a key point in the design phases [11] and is here used to test the final result of the SDE reproducing the "real object (1:1 scale). This representation is extremely useful to begin the technical the industrialization phase minimizing the possible errors.

Figure 5 shows a summary of 3D model and Augmented Reality. This prototyping phase is necessaire to improve the entire design before the tasting and re-design steps which will produce high costs in the total project budget.

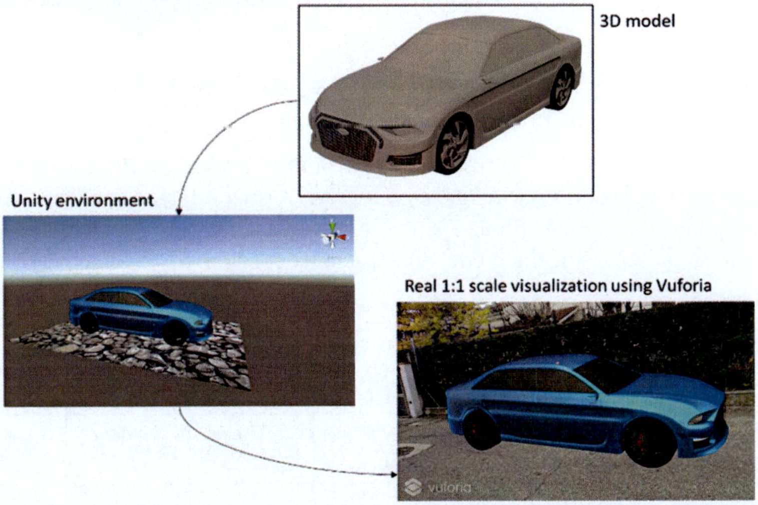

Fig. 5. AR representation in 1:1 scale of the car 3D model.

The model is created in Blender 2.8, the Augmented Reality scene is defined in Unity v2019 and the 1:1 scale view is performed using Vuforia on an Android tablet. This procedure proved to be easy, fast and applicable in many fields. It is important to notice that the 1:1 scale needs a high accuracy of the marker tracking not always reachable.

3.2 CFD Analysis & Re-design

Once the virtual model of Ford Eagle has been designed, it is subjected to virtual tests to confirm the goodness of the product. The vehicle is tested from an aerodynamic point of view and compared with its competitors. The aerodynamic drag coefficient C_d is set as reference.

A Lattice Boltzmann methodology [12] was applied for a fast overview of streamlines and drag coefficient calculation (Fig. 6).

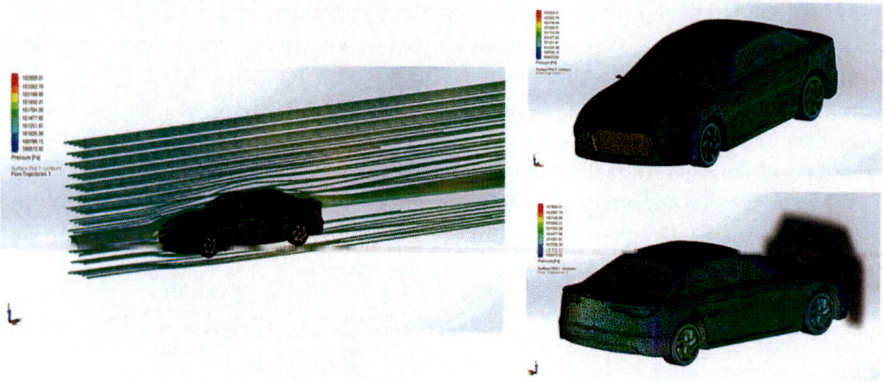

Fig. 6. Simulation of the 3D model to calculate streamlines and drag coefficient.

After a first analysis the Cd resulted to be slightly higher (0.26) than the average value between the chosen cars. A redesign was performed closing the front grille and accentuating the front spoiler. Table 3 shows that the last coefficient calculated (0.23) is inside the average values of the models' coefficients considered for this paper.

Table 3. Comparison of Cd values after optimization phase.

	TESLA		PORSCHE	VOLKSWAGEN	LUCID	LIGHTYEAR	FORD

	Model S	Model 3	Taycan	ID Vizzion	Air	One	Eagle
Cd	0.23	0.24	0.22	0.24	0.21	0.20	0.23

4 Conclusion

This paper was intended to describe a stylistic development of an electric, futuristic, and highly innovative E segment sedan. Through the application of design methods, it was possible to create an innovative car model for the Ford brand that met the demands of consumers identified in a population of 20 students. The construction of the model through sketches allowed to integrate four different car styles within the same prototype. The subsequent 3D realization through surface modeling has allowed the construction of a virtual object useful for many development steps and verification environments. First, a render was made that met the needs of the style. The same model was later used for 1:1 scale view in an Augmented Reality environment. This step is critical to understand the correct aspect ratio and dimensions that can't be found by simply displaying the car on a screen or using sketches. Finally, the 3D model was used for aerodynamic performance verification in order to guarantee an average value within the range defined by competitors. The proposed study shows that the development of a prototype car focuses on new analysis and visualization techniques reducing costs produced by physical maquettes or design errors. Choosing the right design tools in the proposed order leads to linear development of innovative products without costly recursive actions.

References

1. Freddi, A.: Imparare a progettare: principi e metodi del progetto concettuale per lo sviluppo della creatività industriale, 2004th ed.
2. Frizziero, L., Donnici, G., Liverani, A., Alessandri, G., Menozzi, G.C., Varotti, E.: Developing innovative crutch Using IDeS (industrial design structure) methodology. Appl. Sci. **9**(23), 5032 (2019). https://doi.org/10.3390/app9235032
3. Boboc, R.G., Gîrbacia, F., Butilă, E.V.: The application of augmented reality in the automotive industry: a systematic literature review. Appl. Sci. **10**(12), 4259 (2020). https://doi.org/10.3390/app10124259
4. Cox, C.A. ReVelle, J.B., Moran, J.W.: The QFD Handbook.
5. Cristiano, J.J., Liker, J.K., White, C.C.: Key factors in the successful application of quality function deployment (QFD). IEEE Trans. Eng. Manag. **48**(1), 81–95 (2001). https://doi.org/10.1109/17.913168
6. Donnici, G., et al.: A new car concept developed with stylistic design engineering (SDE). Inventions **5**(3), 1–22 (2020). https://doi.org/10.3390/inventions5030030
7. Donnici, G., Frizziero, L., Liverani, A., Aiello, S., Marinelli, L., Salzano, A.: Stylistic design engineering (Sde) framed inside ides (industrial design structure) to conceive a new future citycar, no. August, pp. 2088–2114 (2021)
8. Ginting, R., Ishak, A., Fauzi Malik, A., Satrio, M.R.: Product development with quality function deployment (QFD) : a literature review. IOP Conf. Ser. Mater. Sci. Eng. **1003**(1), 012022 (2020). https://doi.org/10.1088/1757-899x/1003/1/012022
9. Elhegazy, H., Ebid, A., Mahdi, I., Haggag, S., Abdul-Rashied, I.: Implementing QFD in decision making for selecting the optimal structural system for buildings. Constr. Innov. **21**(2), 345–360 (2020). https://doi.org/10.1108/CI-12-2019-0149
10. Adiandari, A.M., Winata, H., Fitriandari, M., Hariguna, T.: Improving the quality of Internet banking services: an implementation of the quality function deployment (QFD) concept. Manag. Sci. Lett. **10**(5), 1121–1128 (2020). https://doi.org/10.5267/j.msl.2019.10.029

11. Brevi, F., Gaetani, F.: The process of surveying maquettes in car design. *diségno*, (7), 141–150 (2020), https://doi.org/10.26375/disegno.7.2020.15
12. Cheylan, I., Fritz, G., Ricot, D., Sagaut, P.: Shape optimization using the adjoint lattice Boltzmann method for aerodynamic applications. AIAA J. **57**(7), 2758–2773 (2019). https://doi.org/10.2514/1.J057955

User Experience Design in Software and Hardware Components Studied in Human-Computer Interaction

Aurora Berni(✉) and Yuri Borgianni

Faculty of Science and Technology, Free University of Bozen-Bolzano, Piazza Università, 3, 39100 Bolzano, Italy
aurora.berni@natec.unibz.it, yuri.borgianni@unibz.it

Abstract. User Experience (UX) is a concept based on the human-product interaction. An increase of UX studies in the Human-Computer interaction (HCI) field was observed in the last decade. Empirical studies based their experimental activity on HCI products, which are characterized by two components: Software and intangible (digital interfaces and web apps) and Hardware and physical (devices). Trough an explorative study, the authors propose a research direction to compares UX studies targeting software and hardware components of HCI products. A preliminary sample of papers was considered. The authors collected contributions where UX in HCI design is investigated through case studies involving devices with software and hardware components. Objectives, methods, and tools of each case study were compared. It emerged that complex systems require both quantitative and qualitative analysis approaches, as the wide variety of tools for data acquisition and processing show. Since Hardware components are more closely related to products such as consumer goods and engineering products, it is possible that methods and tools used to study hardware components could also be applicable to other physical and tangible products, i.e., the main reference for product, engineering, and mechanical design.

Keywords: User experience · HCI design · Software · Hardware · Tangible · Intangible

1 Introduction

The term "User Experience" (UX) was introduced by Norman [1] to extend the study of human-product interaction beyond the consideration of usability and effectiveness. Scholars attempted to find a shared definition to this concept [2–5]. Simultaneously, theoretical models on UX have also been developed over time [6, 7]. Ogunyemi et al. [8] reported a strong growth of UX studies in the last decade. Especially with the advent of new technologies, the domain of Human Computer Interaction has been established. Hassenzahl and Tracktinsky [9] noted a shift in focus within the 'analysis of the study of UX in Human Computer Interaction (HCI). At the beginning contributions focused only on programmatic and task related aspects. Gradually literature on the topic became

C. Rizzi et al. (Eds.): ADM 2021, LNME, pp. 91–99, 2022.
https://doi.org/10.1007/978-3-030-91234-5_9

more conceptual. These articles aimed at a common vision of what characterizes a good UX. These concepts have been applied in empirical studies that involved products with a software and intangible component and, a hardware and tangible component [10, 11]. The former refers to digital products such as Web Apps and digital interfaces; the latter refer to the physical part of the devices (e.g., smartphones, keyboards, and laptops).

Tests and experiments that analyze HCI products are numerous, while empirical studies that examine tangible objects such as consumer goods or engineering products are not likewise widespread or do not rely on adequate research questions [12]. The hardware component can be considered close to the world of consumer goods and engineering products. Thus, it is plausible that methods and tools used to study hardware components could also be suitable for consumer goods and any other product that involves a physical component.

Based on these considerations this work aims at investigating the UX with HCI products characterized by software and/or hardware components. Particularly, objectives, methods, instruments, and typologies of experiences in experimental research have been classified and compared.

The paper is structured as follows. Section 2 provides the selection method applied for the research of contributions, as well as the selection criteria to collect a pertinent sample of papers to be analyzed. Section 3 proposes the classification criteria in detail followed by the resulting table. Section 4 presents the discussions, while Conclusions are drawn in Sect. 5.

2 Methods of Research

To verify the presence of different UX approaches towards software and hardware components in HCI, a sample of convenience of studies should be selected.

The selection of a preliminary reference sample of studies took place by first selecting a pertinent journal of the topic. The "International Journal of Human-Computer Interaction" was considered an authoritative and pertinent source. It was chosen since it addresses, among the others, the cognitive, and ergonomic aspects of interactive computing with a particular attention to human elements related to system and contexts of interaction. The search was performed in Google Scholar, where keywords such as "user experience", "HCI", and "case study" have been used to search pertinent papers limited to articles published in the aforementioned journal. The authors considered papers presenting case studies described in contributions published between 2015 and 2021 only. From an initial sample of more than 100 papers those contributions proposing methodologies and theoretical approaches have been neglected. For brevity, only the papers including "User experience" and "case study" in the title or in the abstract have been considered. 15 articles have been collected to form this preliminary sample. This limited number of sources is clearly not exhaustive. It was nevertheless considered sufficient for extrapolating the existence of differences in UX towards hardware and software. The admitted length of the present paper represented a further constraint for the authors.

3 Classification

The articles belonging to the final sample were classified along the dimensions shown in Table 1, which provides an overview of the result of the categorization work. The most relevant classes were identified in an intuitive way upon after reading the selected contribution. The choice of the classes was also supported by other review works carried out within UX and the use of technological systems within design [13, 14]. Accordingly, contributions have been organized based on their year, source, product typology (software of hardware), objectives, methods, instruments involved in the study and typology of UX. The details of each category follow.

- The "Product" column considers the specific HCI products investigated in each study.
- HCI products have been categorized into two main groups: Software ("S") and "Hardware" ("H"). In the table, an "S" was assigned when the case study involved digital products (e.g., Web apps, websites interfaces, pop-up warnings among the others); "H" when physical products (smartphone hardware, tablet hardware, haptics systems components among the others) were analyzed. "S/H" means that both software and hardware components were considered.
- "Objectives and topic facets" have been expressed readapting the categorization made by Ogunyemi et al. [8] (table 6, p. 13–14,). Since the original categorization was general, the purpose of the adaptation was to focus on the UX domain, including the evaluation of UX and related approaches.
- "Methods" include to the theoretical approach and methodology used to acquire and collect data. Authors focused on data analysis (quantitative and/or qualitative), environment and experiment setting (natural or controlled or non-specified), and participants involved (final users and/or experts).
- "Instruments" relate to product representations presented to participants (prototype or end-use product) and the technology involved in data acquisition as support tools, such as biofeedback (biometric measurements tools such as skin conductance (SC), Electroencephalography (EEG), raw electrocardiogram (ECG)), Virtual Reality (VR) and haptics.
- "Experience" lists the 3 different kinds of experiences identifiable in the design and UX literature; they are Ergonomic, cognitive, and emotional. Ergonomic experiences are related to usability and effectiveness [15]. Cognitive experiences are related to the aesthetic perception of a product [16]. Emotional experiences involve emotions, affective phenomena, feelings, and pleasure [17].

Table 1. Classification UX of HCI products featured by software ("S") and hardware ("H") components or both ("S/H")

Source	Product	S, H, S/H	Objectives and topic facets	Methods	Instruments	Experience
[18]	Android smartphones models	S/H	UX assessment for design and improvement of products; topic facet: UX evaluation and measurement, design innovations	Data analysis: quantitative and qualitative; participants: final users; environment: controlled	Product representation: end-use products; Support tools: Biofeedback	Ergonomic cognitive
[19]	Information Kiosk	S/H	Effective and satisfying user experience design (UXD), product implementation; topic facet: UX evaluation and measurement, Design for Users, User research methods	Data analysis: qualitative; participants: final users; environment: non specified	Product representation: prototype (virtual prot. of the kiosk); support tools: other devices (laptop)	Ergonomic cognitive
[20]	Conversational Agent and Remote Control Unit	S/H	Measurement and comparison of UX of a conversational agent (CA) and remote-control unit (RVU) for TV control; topic facet: UX evaluation and measurement, Design methods and contexts	Data analysis: qualitative and quantitative; participants: final users; environment: controlled	Product representation: end-use products; support tools: Biofeedback	Emotional cognitive
[21]	Web Documentary	S	Discussion of the impact of serialization and interactivity on audience reception and user engagement; topic facet: Collaboration and team communication, Design methods and contexts	Data analysis: quantitative; participants: final users; environment: natural	Product representation: End-use product; support tools: other devices (users' own mobile and non-mobile devices)	Emotional ergonomic
[22]	Learning Management System platform	S	Usability evaluation of the Blackboard system platform; topic facet: Post-deployment activities and software maintenance	Data analysis: qualitative and quantitative; participants: final users; environment: natural	Product representation: End use product; Support tools: Other devices (users 'own mobile or non-mobile devices)	Ergonomic

(*continued*)

Table 1. (*continued*)

Source	Product	S, H, S/H	Objectives and topic facets	Methods	Instruments	Experience
[23]	Petro-chemical manufacturing plant software system	S	Identification of low-level usability-related software issues; topic facet: Post-deployment activities and software maintenance, Collaboration and team communication	Data analysis: qualitative; participants: final users and experts; environment: controlled	Product representation: prototype (static screen of interface); Support tools: other devices (computer screen)	Ergonomic cognitive
[24]	Traffic Supervision System	S	Usability evaluation of user interfaces of a traffic supervision system; topic facet: Design for Users, Design methods and contexts	Data analysis: quantitative and qualitative; participants: final users and experts; environment: controlled	Product representation: end use product; support tools: other devices (computer screen)	Ergonomic
[25]	Smartphone	S/H	Smartphone User experience analysis and categorization, focus on relation between product smartness and user satisfaction; topic facet: UX evaluation and measurement, User research methods	Data analysis: qualitative; participants: no direct users' involvement; environment: natural	Product representation: prototype (Semantic description); support tools: other devices (computer screen)	Emotional cognitive
[26]	Service provider interface	S	User evaluation of interfaces automatically generated by Egoki system; topic facet: Design-driven software architecture	Data analysis: quantitative and qualitative; participants: final users; environment: controlled	Product representation: prototype (interface generation); support tools: other devices (tablet)	Ergonomic Cognitive
[27]	Input methods for VR	H	Analysis of the performance/effectiveness of different input methods for VR; topic facet: Design innovations, Design methods and contexts	Data analysis: quantitative; participants: final users; environment: controlled	Product representation: end use product; support tools: VR and haptics, other devices smartphone) (computer screen)	Ergonomic

(*continued*)

Table 1. (*continued*)

Source	Product	S, H, S/H	Objectives and topic facets	Methods	Instruments	Experience
[28]	Pop-up warnings	S	Examination of positive/negative emotional response when pop-up warnings appear; topic facet: Design tools and techniques, User research methods	Data analyses: quantitative; participants: final users; environment: natural	Product representation: end use product; support tools: other devices (portable/non portable devices)	Emotional
[29]	Cross-platform services	S	Analysis of user experience of cross-platform services; topic facet: UX evaluation and measurement, Collaboration and team communication	Data analysis: quantitative and qualitative; participants: final users and experts; environment: natural	Product representation: prototype; support tools: other devices (portable/non portable devices)	Cognitive
[30]	Smartphone guide technology	S	Usability evaluation of Smartphone-based guide tour and its effect on visitors 'experience in a cultural space; topic facet: Design for Users, Design tools and techniques, Collaboration and team communication	Data analysis: qualitative; participants: final users and experts; environment: natural	Product representation: end use product; support tools: other devices smartphones and a paper guide)	Ergonomic
[31]	Portable hand grip haptic system	H	Implementation of realistic and immersive interaction with virtual environment through hand grips; topic facet: Design methods and contexts, Design tools and techniques, Post-deployment activities and software maintenance	Data analysis: quantitative; participants: final users and experts; environment: controlled	Product representation: prototype; support tools: biofeedback, VR and haptic devices	Ergonomic Cognitive
[32]	Mobile game	S/H	User experience of augmented reality mobile games (increasing the sense of immersion); topic facet: Design tools and techniques, Design-driven software architecture	Data analysis: quantitative and qualitative; participants: final users and experts; environment: controlled	Product representation: prototype; support tools: VR and haptic devices, other devices smartphone)	Emotional cognitive

4 Analysis and Discussions

The main insights from the analysis follow.

In terms of objective of the study, the outcomes confirmed the results highlighted by Ogunyemi [8], in that UX and usability are the main objectives of HCI studies. Usability is predominantly investigated when the software and hardware components are considered separately. UX is studied when these components are evaluated as a whole system.

In terms of analysis approaches, scholars tend to use both quantitative and qualitative data analysis approaches to study HCI products. However, when both software and hardware are considered, qualitative analysis seems to be more leveraged [18–20, 25, 32]. This highlights that a complex system needs to be considered in a more comprehensive way. Therefore, objective data only is not sufficient to describe and analyze the different facets of UX.

In terms of experimental environment and technologies used as support tools, there is a predominance of artificial settings and controlled environments in HCI experiments (presented in 8 papers). It is worth noting that natural environments are preferred to study the software components of a device only [21, 22, 28–30]. This also reflects the diffuse involvement to involve other devices such as smartphones, tablets, or laptops as support tools in software evaluation (omitted in the classification for the sake of brevity, because of them being standard). When hardware or both components are considered, scholars use a wider variety of support tools, which includes biofeedback and VR technology and haptics in addition to the ones mentioned above [18, 20, 27, 32].

In terms of participants, when UX and the investigation of usability are the focus of HCI studies, final users are involved to perform evaluations. Experts' evaluation was employed in addition to final users only in three cases dealing with the software component only.

In terms of typology of product representation, a balance was found as regards the number of contributions using prototypes and end-use products in their experiments. End-use products are slightly more involved (presented in 8 papers) when the software component is considered.

In terms of typology of experience, the ergonomic experience is the most studied one (presented in 10 papers). Cognitive experience follows immediately, while just a few contributions focused on emotional experience, which is generally studied in combination with other kind of experiences. Ergonomic experience is most investigated when software and hardware components are considered separately. In software it is frequently studied alone [22, 25, 27, 30]. Cognitive experience appears more frequently when software and hardware are considered simultaneously [15, 18–20, 32].

To summarize, complex systems require broader evaluation and analysis approaches. HCI involving both hardware and software components is difficult to be analyzed in a quantitative way only. To have a complete analysis, it is necessary to combine both qualitative (less objective but more wide-ranging) and quantitative (objective but more limited) approaches. This is reflected in the use of a greater variety of technologies and supporting tools for data acquisition and processing. Unlike the ergonomic and cognitive experiences, emotional ones seem to be largely overlooked in HCI research; The importance of emotions in HCI is stressed also by Cristescu [33], who notice a deficiency in emotional approaches and a prominence of traditional cognitive ones. This

this aspect might be seen as a limitation considering transferring HCI approaches to consumer goods and engineering products.

5 Conclusions

This work is intended to get a first understanding of UX and usability in the HCI design field as a trigger to develop UX practices for tangibles. The results achieved are not conducive to conclusive considerations, due to the exploratory nature of this study. A method and criteria for categorizing those kinds of studies represent therefore the necessary first step towards this goal. Based on the proposed approach, the authors examined a sample of convenience of articles proposing case studies where UX is analyzed in combination to the studied software and/or hardware components. To confirm or disconfirm the first insights into HCI, a further analysis of a larger sample of papers is needed. Despite the above limitations, this study was able to capture when software and hardware components are investigated separately or as a whole system, as evident in Sect. 4.

References

1. Norman, D., Miller, J., Henderson, A.: What You See, Some of What's in the Future, And How We Go About Doing It: HI at Apple Computer, p. 1. (1995)
2. McCarthy, J., Wright, P.: Technology as Experience, vol. 11. MIT Press, Cambridge, Massachusetts, United States (2004)
3. Law, E.L.-C., Vermeeren, A.P.O.S., Hassenzahl, M., Blythe, M.: Towards a UX Manifesto. In: Proceedings of HCI 2007, the 21st British HCI Group Annual Conference University of Lancaster, UK (Sep. 2007)
4. Karapanos, E., Zimmerman, J., Forlizzi, J., Martens, J.-B.: User experience over time: an initial framework. In: Proceedings of the SIGCHI Conference on Human Factors in Computing Systems, New York, NY, USA, pp. 729–738 (Apr. 2009)
5. Law, E.-C., van Schaik, P.: Modelling user experience – an agenda for research and practice. Interact. Comput. **22**(5), 313–322 (2010)
6. Hassenzahl, M.: Experience design: technology for all the right reasons. Synth. Lect. Hum. Center. Inform. **3**(1), 1–95 (2010)
7. Pucillo, F., Cascini, G.: A framework for user experience, needs and affordances. Des. Stud. **35**(2), 160–179 (2014)
8. Ogunyemi, A.A., Lamas, D., Lárusdóttir, M.K., Loizides, F.: A systematic mapping study of HCI practice research. Int. J. Hum. Comput. Interact. **35**(16), 1461–1486 (2019)
9. Hassenzahl, M., Tractinsky, N.: User experience – a research agenda. Behav. Inform. Technol. **25**(2), 91–97 (2006)
10. Wong, M.L., Khong, C.W., Thwaites, H.: Applied UX and UCD design process. Interface Des. Procedia Soc. Behav. Sci. **51**, 703–708 (2012)
11. Remy, C., Huang, E.M.: Addressing the Obsolescence of End-User Devices: Approaches from the Field of Sustainable HCI. In: Hilty, L.M., Aebischer, B. (eds.) ICT Innovations for Sustainability. AISC, vol. 310, pp. 257–267. Springer, Cham (2015). https://doi.org/10.1007/978-3-319-09228-7_15
12. Robinson, J., Lanius, C., Weber, R.: The past, present, and future of UX empirical research. Commun. Des. Q. Rev. **5**(3), 10–23 (2018)

13. Berni, A., Borgianni, Y.: Applications of virtual reality in engineering and product design: why, what, how, when and where. Electronics **9**(7), 1064 (2020)
14. Berni, A., Borgianni, Y.: From the definition of user experience to a framework to classify its applications in design. In: 23rd International Conference on Engineering Design ICED (Aug. 2021)
15. Alben, L.: Quality of experience: defining the criteria for effective interaction design. Interactions **3**(3), 11–15 (1996)
16. Hassenzahl, M.: The interplay of beauty, goodness, and usability in interactive products. Hum. Comput. Interact. **19**(4), 319–349 (2004)
17. Norman, D.A.: Emotional design. Ubiquity, p. 1 (2004)
18. Ding, Y., Cao, Y., Qu, Q., Duffy, V.G.: An exploratory study using electroencephalography (EEG) to measure the smartphone user experience in the short term. Int. J. Hum. Comput. Interact. **36**(11), 1008–1021 (2020)
19. Ekşioğlu, M.: User experience design of a prototype Kiosk: a case for the Istanbul public transportation system. Int. J. Hum. Comput. Interact. **32**(10), 802–813 (2016)
20. Lee, S., Ryu, H., Park, B., Yun, M.H.: Using physiological recordings for studying user experience: case of conversational agent-equipped TV. Int. J. Hum. Comput. Interact. **36**(9), 815–827 (2020)
21. Ducasse, J., Kljun, M., Čopič Pucihar, K.: Interactive web documentaries: a case study of audience reception and user engagement on iOtok. Int. J. Hum. Comput. Interact. **36**(16), 1558–1584 (2020)
22. Alhadreti, O.: Assessing academics' perceptions of blackboard usability using SUS and CSUQ: a case study during the COVID-19 pandemic. Int. J. Hum. Comput. Interact., 1–13 (2020)
23. Lindgaard, G.: Challenges to assessing usability in the wild: a case study. Int. J. Hum. Comput. Interact. **31**(9), 618–631 (2015)
24. Assila, A., de Oliveira, K.M., Ezzedine, H.: Integration of subjective and objective usability evaluation based on IEC/IEC 15939: a case study for traffic supervision systems. Int. J. Hum. Comput. Interact. **32**(12), 931–955 (2016)
25. Rhiu, I., Yun, M.H.: Exploring user experience of smartphones in social media: a mixed-method analysis. Int. J. Hum. Comput. Interact. **34**(10), 960–969 (2018)
26. Aizpurua, A., Miñón, R., Gamecho, B., Cearreta, I., Arrue, M., Garay-Vitoria, N.: Accessible ubiquitous services for supporting daily activities: a case study with young adults with intellectual disabilities. Int. J. Hum. Comput. Interact. **35**(17), 1608–1629 (2019)
27. Choe, M., Choi, Y., Kim, H.K.: Comparison of gaze cursor input methods for virtual reality devices. Int. J. Hum. Comput. Interact. **35**(7), 620–629 (2019)
28. Buck, R., Khan, M., Fagan, M., Coman, E.: The user affective experience scale: a measure of emotions anticipated in response to pop-up computer warnings. Int. J. Hum. Comput. Interact. **34**(1), 25–34 (2018)
29. Shin, D.-H.: Cross-platform users' experiences toward designing interusable systems. Int. J. Hum. Comput. Interact. **32**(7), 503–514 (2016)
30. Petrie, H., Othman, M.K., Power, C.: Smartphone guide technology in cultural spaces: measuring visitor experience with an iPhone multimedia guide in Shakespeare's Church. Int. J. Hum. Comput. Interact. **33**(12), 973–983 (2017)
31. Kim, M., Kim, J., Jeong, K., Kim, C.: Grasping VR: presence of pseudo-haptic interface based portable hand grip system in immersive virtual reality. Int. J. Hum. Comput. Interact. **36**(7), 685–698 (2020)
32. Sekhavat, Y.A., Zarei, H.: Sense of immersion in computer games using single and stereoscopic augmented reality. Int. J. Hum. Comput. Interact. **34**(2), 187–194 (2018)
33. Cristescu, I.: Motions in human-computer interaction: the role of nonverbal behavior in interactive systems. Inform. Econ. **1**(42), 7 (2008)

Virtual Reality and Interactive Design

Spatial Augmented Reality as a Visualization Support for Engineering Analysis

Federico Morosi(✉), Giandomenico Caruso, and Gaetano Cascini

Department of Mechanical Engineering, Politecnico di Milano, Via La Masa 1, Milan, Italy
federico.morosi@polimi.it

Abstract. Projector-based Spatial Augmented Reality (P-SAR) is a technology that allows to alter the external appearance of a physical object by means of an almost infinite variety of computer-generated contents. Thanks to the adoption of colored lights, a projected image acquires coherent spatial properties with respect to the data it represents; this has been demonstrated to facilitate the users' interpretation of complex information. The current paper presents the development of a P-SAR system aiming at supporting the real-time visualization and the inspection of engineering simulation results. Particular attention is paid on detailing the algorithms necessary for the generation of the color maps to be displayed on the prototype's surface. These are interpolated starting from a discrete array of output data coming from a generic simulation to resemble the configuration of sensors commonly adopted in real experimental setups. An illustrative case study applied to CFD analysis is finally discussed to show the applicability of such immersive environments in engineering fields that require to perform testing activities with equipment like wind tunnels.

Keywords: Spatial augmented reality · Engineering analysis · Simulation

1 Introduction

Experimental and numerical simulations are commonly applied in engineering processes to test if a product or a component under development behave as expected under predefined use conditions. The widespread adoption of simulations is made possible thanks to the availability of sophisticated testing frameworks and powerful software tools that, by means of complex and accurate calculations, handle the huge amount of required data. Despite the continuous improvements in terms of simulation capability of multidomain phenomena, the visualization of results is still linked to classical representations. Simulation software applications, in fact, are generally based on WIMP-style (Window, Icon, Menu and Pointing device) interfaces to present and interact with the output datasets. For this reason, several researchers are investigating the integration of innovative and immersive technologies for enhancing the quality of engineering analysis. Augmented Reality (AR) turns out to be one of the most promising technologies for its capability of merging computer-generated visual contents with the physical environment [1]. Consequently, AR can give a coherent spatial context to any kind of numerical data.

C. Rizzi et al. (Eds.): ADM 2021, LNME, pp. 103–115, 2022.
https://doi.org/10.1007/978-3-030-91234-5_10

This aspect aims at improving the comprehension of the simulation results with respect to purely digital visualizations, thanks to the natural human perception of the physical world. Not only the visualization of the data but also the interaction with the simulation can take advantage of the adoption of AR. According to the technology and the setup, in fact, single or multiple users can intuitively modify the input/output parameters of the simulation by means of interactions with target objects, manipulations of tools or free-hands gestures.

This paper presents the implementation of a visualization tool for simulation results and its integration with a Projector-based Spatial Augmented Reality (P-SAR) system (Fig. 1). As illustrated in the figure below and described in [2], the system was initially developed to support collaborative activities in the field of product and packaging design and for this reason it also integrates an interaction system through which modifying the projected contents. Despite its intrinsic complexity, P-SAR technology has been selected to generate an immersive environment where users are not forced to wear or hold visualization devices. Projectors, in fact, can be installed inside the same room where the experiment takes place and used to overlay colored images directly onto the external surface of the physical object under investigation. Thanks to this setup, the collaboration between multiple engineers working on the project is made more agile by enabling the observation of the same mixed prototype (i.e. the output of the P-SAR technology made by a physical object overlayed with projected images) from their own perspective. Moreover, the adoption of appropriate tracking systems for the detection of the physical object placement offers the potential of natural and direct interactions with the mixed prototype which, in turn, may also behave as an input interface. A change of the physical object's position inside the tracking volume, in fact, can be interpreted by the system not only as means for inspecting hidden sides of the mixed prototype but also as a change of the simulation's input parameters. Thanks to that, the (re)execution of the analysis according to the new settings is simplified and made more understandable also for possible non-expert users.

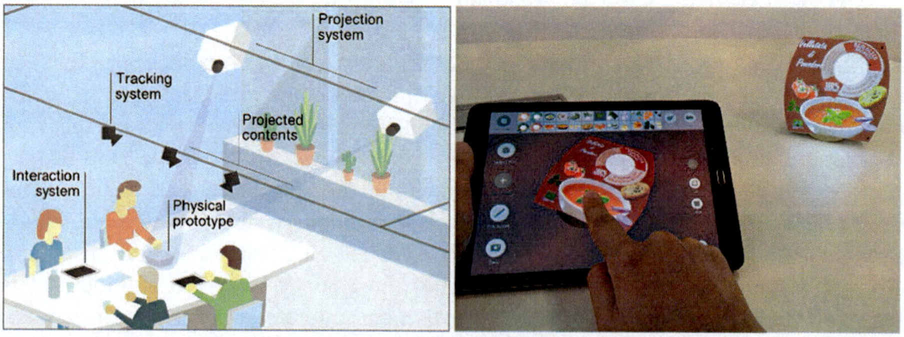

Fig. 1. Schematic representation of a P-SAR system for product design (left); a user interacting with the projected contents of the mixed prototype by means of a tablet interface (right) [2].

The aim of the current research is to study and validate if P-SAR technology can be used as a support tool for the visualization and inspection of the results of engineering analysis. The review of previous studies in Sect. 2, in fact, reveals very few applications of P-SAR in this field while AR examples are more common. The P-SAR system, as well as its architecture, is presented in term of implementation and case study in Sect. 3 and 4 respectively. Section 5 analyses the results and the performance achieved by the developed application. Finally, Sect. 6 discusses limitations and potential impact in the engineering field.

2 Literature Review

There is a growing interest of scholars in the development and validation of AR applications for engineering analysis [1]. This positive trend is mainly due to the democratization of cheap and compact pieces of hardware (e.g. tablet, smartphone, camera, laptop) that are necessary for running such type of applications, as well as to the established advantages provided by AR technologies in human activities: immersive data visualization, intuitive interactions modalities, increased comprehension of complex information and agile collaboration, just to mention a few.

The analysis of the literature shows several works in the field that can be classified in three main categories according to the type of analysis for which they were developed: numerical simulations [3], experimental tests [4] or real monitoring activities [5]. The main difference between the three is related to how the datasets, necessary for the creation of the augmented contents, are generated and structured. In the first case, for instance, CFD mathematical models were used to analytically calculate the flow lines of turbine runner. In the last two cases, instead, strain gauge sensors were directly applied on physical models/buildings to measure their deformation after the application of external forces. Another way to distinguish the huge number of applications available in the literature considers their capability to support human interactions. There were AR systems developed to support the three-dimensional visualization of data only [6] and others where the adoption of non-traditional means of interaction facilitated the setup of the simulation and/or the inspection of the results. This was the case of [7] where engineers manipulated tangible tools to change the properties of the external loads applied onto aluminum frames or [8] where speech and gesture interfaces were used to modify the graphical representations of CFD analysis simulating the heat transfer inside a building room.

The number of publications drastically decreases if the usage of a projector for the generation of the augmented contents is considered. The selection of P-SAR systems as support of engineering analysis, in fact, introduces several technological limitations that reduce the use case scenarios where there are actual benefits. The calibration of the hardware components is difficult and time consuming, the required equipment could be expensive and the installation space should be well structured. Moreover, the adoption of colored light beams for rendering the simulation data requires a physical surface towards which directing the projected images. Because of this, it is evident why the applications available in literature were mainly focused on the engineering fields were P-SAR is used to display superficial features of the target and/or the surrounding objects (e.g. construction or urban planning). An early example of application is the tabletop projection described in [9, 10] where architectural models were used to simulate their shadows, reflections and effects on the wind during the design of urban landscapes. By detecting the placement of the target objects, the system was able to perform the calculations and to automatically display the results on the table surface. A similar setup was more recently implemented in [11] for supporting the interior design of buildings according to different daylight illuminations. Walls and windows were again represented by tangible tools that simultaneously acted as output projection surfaces and input elements by which modifying the simulation setup.

P-SAR technology offers to the users a more intuitive visualization of the digital contents and a more natural interaction with respect to other AR approaches. These are the reasons why projectors have been selected in applications like [12] where a landscape model made of clay or sand were used as input/output systems for GIS (Geographic Information System) simulations. By modifying the geometry of the model, the system was able to detect any variations in its elevation and to project back the simulation results in near-real-time. Despite the declared scope was mostly educational, it is worth mentioning the work by Lundin and Schönborn [13] because of their adoption of sensor data for the generation of the digital contents. In this example projectors were used to superimpose on the external surface of physical objects their temperature-related information, detected in real time by means of thermal cameras.

The review of the literature reveals that, despite the maturity of the technology, it is still missing a real understanding of the potentials of P-SAR when applied in engineering analysis. This includes the implementation of features for the real-time generation of the digital contents according to the data coming from the simulation and the definition of the properties of the projected images in order to effectively represent the results of the specific field of application.

3 Methodology and Implementation

The implementation here described makes use of SAR technology to project the results of engineering analysis directly onto the external surface of a physical prototype that represents the object of the simulation. Since the selected visualization medium allows modifying only the external appearance of an object, this study is focused on those fields where the data are usually presented with color maps. Stress, thermal, and fluid analyses are examples of engineering simulations that fall into this classification. Taking as a reference the schema in Fig. 2, three phases characterize the current approach: (i) Engineering simulation, (ii) Post-processing and (iii) SAR visualization.

3.1 Engineering Simulation

The first stage gathers the experimental or numerical simulations thanks to which the data to be visualized are acquired. It includes, according to the settings of the analysis, elements that are mandatory to achieve the desired functionalities of the 2 subsequent steps: simulation results, physical prototypes and/or virtual 3D models. While the first one needs to be structured inside a.csv file, the last two elements can be used as they are by the developed application if there is a high correspondence between their geometries.

When this requirement is not satisfied or when one of the 2 types of prototype is not available, additive manufacturing or reverse engineering approaches are recommended to comply with the required precision.

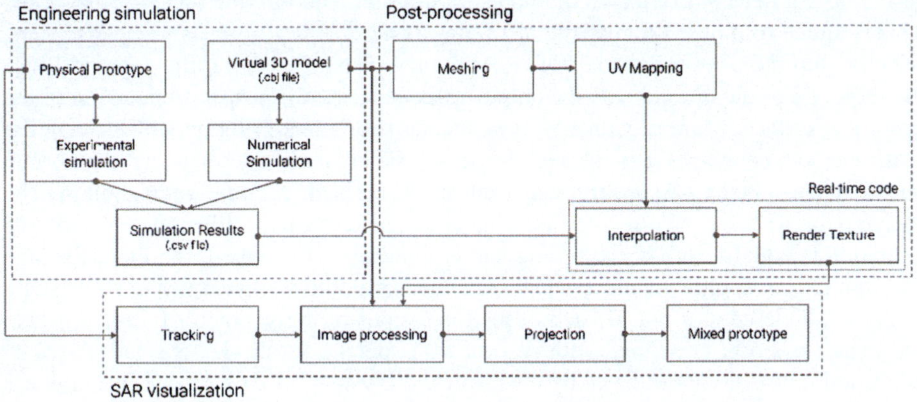

Fig. 2. Flowchart of the entire process developed for the generation of a mixed prototype displaying the results of numerical or experimental engineering simulations. All the steps are clustered within 3 phases: engineering simulation, post-processing and SAR visualization.

3.2 Post-processing

The second stage is the focus of the presented research. It receives the virtual model and the simulation results as input files, and it produces the colored texture to be projected onto the physical prototype by means of real-time interpolations between punctual data distributed across the entire 3D surface. To perform this calculation, the 3D geometry is firstly re-meshed with triangular elements that, by exploiting the optimized computer graphic algorithms, increase the overall efficiency and robustness of the application. The Delaunay triangulation [14] is selected to provide a mesh where the minimum angle of its triangles is maximized. Thanks to this, the errors caused by the numerical precision, while localizing the position of the data onto the surface, are reduced. The new polygonal mesh is then parameterized, an essential geometry processing tool for applications dealing with three-dimensional renderings since it creates a correlation between any 3D point of the mesh ("X, Y & Z") and the 2D coordinates of an image ("U & V"). This UV mapping task is manually or automatically performed by identifying the seam edges on the 3D model where the mesh will be cut to be unfolded and where the texture will be discontinuous. If made properly, it allows to reduce distortions and stretches on the corresponding texture and to obtain a correct wrap of the render texture's colors on each triangular face of the mesh. The final parametrized mesh is exported in Wavefront (.obj) format to encode the information related to the geometry and the UV map.

The main part of the post-processing code is related to the localization of each point data, given as tri-dimensional mesh coordinates, with respect to the bi-dimensional texture reference system and their interpolation. To work even with empirical results, the algorithm is developed to perform the interpolations with a limited amount of data which do not necessarily match the nodal values computed by the analytical solution. This is aimed to mimic the realistic distribution and quantity of measurement sensors, which equip the physical models of experimental tests. Therefore, the paper will use the term sensor data to indicate the output value measured in a specific location of the prototype surface. More specifically, the code starts by identifying on which triangular mesh element each sensor is located. To do so, the cross product between the normal vector of the planes passing through each mesh element and the vector joining the sensor point with a known point of the plane itself is performed. When the result of this operation is equal to 0, i.e. the orthogonality condition is satisfied, the Pineda's edge function [15] is applied to check whether the sensor point is located inside the triangle of the mesh element or not. If each of the three sides of the geometry fulfills this test, the sensor position is defined with respect to a specific mesh element and it can be converted in texture coordinates by means of the UV map. Shape functions and linear interpolations between the sensors data are finally implemented to derive the value of the mesh nodes. In a few words, the nodal value of each vertex of each triangular mesh element is calculated as function of its distance with respect to the 3 closest sensors (i.e. a barycentric coordinate weight).

Even if the results obtained with the previously described procedure are sufficient to generate a color map, and additional computational step is performed with the aim of improving the quality of the final visual output. This consists in a refinement of the mesh structure by adding new vertices, and consequently by creating new triangular elements, in those surfaces of the 3D model where is necessary to obtain a greater resolution of the texture, i.e. where the gradient of the measured data is too high. The mesh subdivision algorithm is iterated until the values interpolated in two contiguous elements have a difference below a threshold or when a predefined minimum size of the mesh is reached. These two conditions are necessary to avoid an exponential increase of the number of faces and to affect the overall computational cost of the application. Finally, the nodal values of the new vertices, now calculated by means of a quadratic shape function, are added at the end of the results array for their RGB conversion. The color of a single triangular element of the UV map is thus established by calculating the mean of the nodal data that delimits the corresponding mesh face, and by interpolating this value onto a blue-to-red color scale. The latter is weighted between the maximum and the minimum numbers that are stored in the results array to have the full rainbow scale available for the generation of the final image that schematize in two dimension the prototype's color map.

The part of the code that starts with the interpolation of the sensors data and ends with the creation of the color map is executed as a real-time process. This means that, whenever a modification of the data stored inside the.csv file is detected, the program automatically updates the visual output according to the new input values. The application can thus easily process transition stages of the simulation when, for instance, the environmental conditions are changed and not only a single-step result.

3.3 SAR Visualization

As soon as a new UV-based render texture is created, or an old one is updated, the application moves to the final step described in Fig. 2 where physical and digital contents are blended to obtain the mixed prototype. This SAR module uses state-of-the-art technologies for projecting colored images directly onto the external surface of the physical prototype which, in turn, has the capability to be moved inside a working volume. The reconstruction of the physical object's shape by means of its virtual 3D model is fundamental to correctly deform the projected images according to the actual position of the target. The latter is triangulated from the images streamed by an array of optical sensors where small IR reflective markers previously attached on the prototype surface are detected. The adoption of this type of tracking solution requires an initial calibration procedure of the main hardware components, i.e. projector(s) and cameras, to derive their configuration inside the installation space.

4 Case Study

The application of the previously described algorithm and workflow, both required for the visualization in SAR environment of the outcomes of experimental testing in engineering field, has been tested by means of an illustrative example based on a numerical CFD analysis. The latter is used for the computation of the fluid dynamics results in a similar form to those that would be obtained from an empirical test inside a wind tunnel. The complexity and the operational costs of this type of experimentation, in fact, leads the current study to identify alternative solutions for the adoption of real-like data in term of magnitude and configuration.

A simplified 3D model of a Tesla CyberTruck is selected as reference for this case study and it is imported in SimScale [16], a CAE software based on OpenFOAM and cloud computing, to simulate the effect of front wind when the car is moving at 100 km/h. Appropriate settings related to the model symmetry, mesh refinement and boundary inflation are chosen to reduce the calculation time and load. Upon completion to convergence of the simulation, the results of the pressure distribution onto the model surfaces (Fig. 3) are exported as node values and post-processed with ParaView [17]. This open-source software enables the visualization and manipulation of the original data to resemble the discrete and limited configuration of the superficial pressure sensors commonly used in wind tunnel tests. Due to the assumption of a symmetric distribution of the pressure, a total of 380 sensors are arranged on one half of the car model; they will be used to perform an identical interpolation also on the second part. A.csv file is finally exported containing the global placement coordinates of each sensor and the pressure values "measured" on the corresponding superficial position. This, together with the 3D model of the car in.obj format, constitute the two input files required by the developed algorithm to generate the colormap to be projected with the SAR platform.

The generation of a render texture that accurately reproduces the pressure distribution on the car surfaces, requires an additional preparatory phase for the 3D model. This task is performed with Gmsh [18] first, to automatically elaborate a mesh based on the Delaunay triangulation, and Blender [19] after, to manually adjust the UV mapping. The result of these operations is illustrated on the right side of Fig. 4 where 16 separated patches have been created on a single UV map starting from the triangular mesh displayed on the left 3D model.

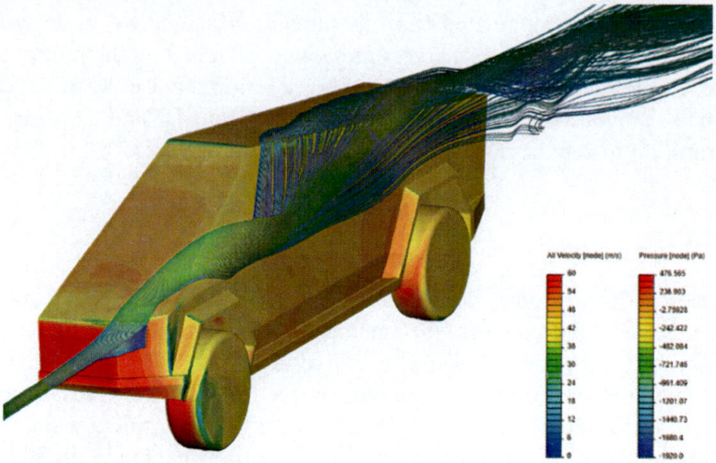

Fig. 3. Results of the CFD analysis on a Tesla Cybertruck 3D model (screenshot from a simulation available on SimScale).

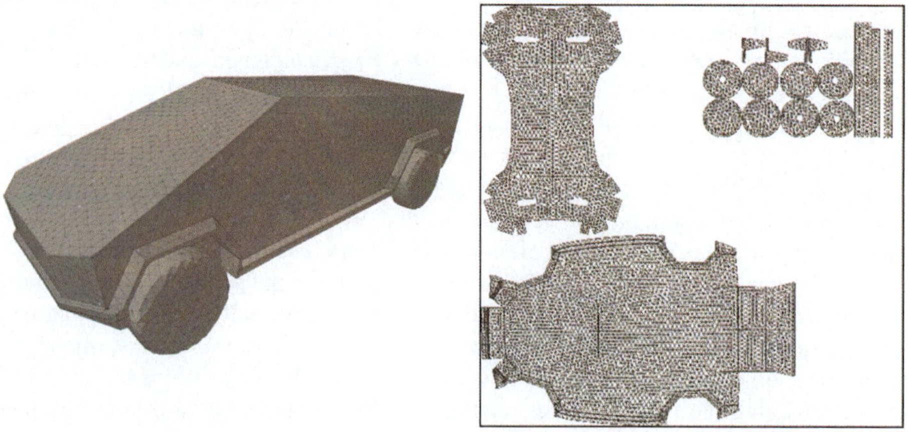

Fig. 4. Elaborated mesh (left) and related UV map (right) of the Tesla Cybertruck 3D model.

The 3D model and the sensors data are finally uploaded on the SAR software. Developed in Unity3D [20], the application implements the interpolation and the augmented reality algorithms necessary to generate a textured mixed prototype. The color map and the object mesh are computed at the startup of the application maintaining the relationship between their reference systems. The outcome of this process is illustrated in Fig. 5 where the right texture is wrapped onto the object mesh thanks to an appropriate render camera. The SAR module, instead, manages the visualization of the projected contents and the simultaneous tracking of the physical object [21]. The first task is performed by means of an overhead projector device while the second one requires an array of 6 IR cameras fixed on the same structure.

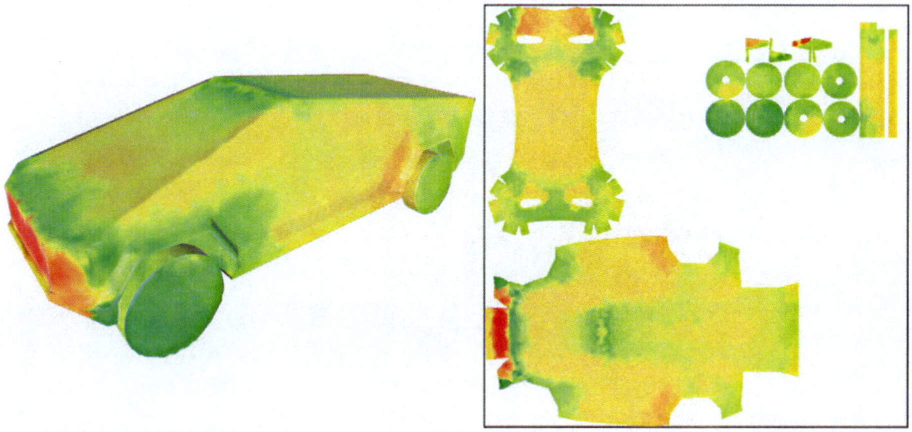

Fig. 5. Colormap results provided by the implemented algorithm in form of textured 3D model (left) and 2D render texture (right).

Crucial for the correct functioning of the SAR platform and the realistic rendering of the mixed prototype is the preparation of the physical target of the projection. In the current use case scenario, 3D printing technology has been selected to manufacture the scaled model of the car (28 × 11.5 × 8 cm) and to obtain an optimal correspondence with the shape and the dimension of its virtual duplicate. For the sake of simplicity, the 3D model is separated and printed in 5 parts (i.e. the body and the wheels) whose surfaces are treated with sand paper to remove the stepping effect and to improve the adhesion of the spray paint. The creation of a white, matte and uniform surface, in fact, is required by the projector device to display the color map with truthful colors. The final step, after re-assembling the prototype, consists of randomly attaching the IR reflective markers necessary for the tracking. 7 small spheres are arranged on the top surfaces to simplify their detection by means of the optical sensors fixed on the ceiling of the room and to reduce the risk of occlusion due to user's direct interactions. The three stages of this procedure are illustrated in Fig. 6 together with the mixed prototype generated by the SAR platform (bottom-right).

5 Results

The developed algorithm and its application with the case study of Sect. 4 have been evaluated in term of both quality of the projected output and computational performance. Except for those areas where the pressure's gradient is high and the sensors number is not sufficient to detect peculiar trends, the texture displayed by the mixed prototype resembles the color configuration of the reference model, the output displayed by the CAE software. This result is well visible on the right side of Fig. 7 where the texture generated starting from the nodal values of the CFD and the one calculated with the sensor data are compared in term of color bands: gray regions indicates where the two images have the same color intensity, magenta and green regions are indicators of differences in the RGB values.

Fig. 6. Procedure for the preparation of the physical 3D model, target of the SAR projection: 3D printing of the body (upper-left), assembly of the printed parts (upper-right), refinement of the surface finishing, color and composition of the marker's constellation (bottom-left), final mixed prototype generated by the SAR with the computed color map (bottom-right).

The measurement of the performances, instead, aims at understanding the capability of the system to display in real time the data streamed by the sensors in a hypothetical experimental analysis. To do so a standard laptop computer with an Intel Core i7-8565U CPU @1.80 GHz processor, 16.0 GB RAM and 64-bit operating system has been used to generate the textures from the data provided by different number of sensors. By measuring the time required by the application to execute the entire code, it is obtained the linear trend illustrated on the left side of Fig. 7. With a reasonable number of sensors for experimental tests (i.e. below 50) a new texture takes about 3 s to be generated. This means that, with the adopted hardware, the delay of the application cannot be considered a strong limitation in its usability in real scenarios. Moreover, the selection of computers with higher performances can further reduce the delay until a condition where it is completely negligible with respect to the one introduced by the same measurement sensors.

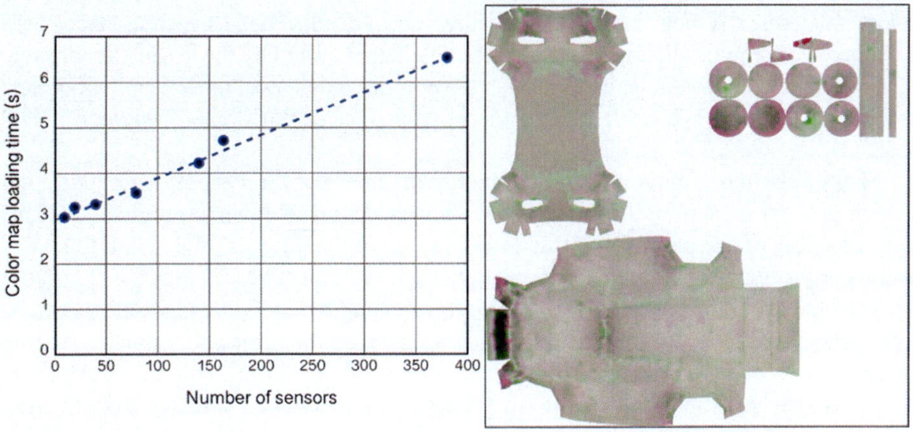

Fig. 7. Performances of the implemented algorithm for the generation of the colormap according to the number of sensors (left) and color bands difference between the generated texture and the one provided by the CFD software (right).

6 Conclusions and Future Works

The current paper presents the development of P-SAR application for the visualization of numerical simulation results in the engineering field. The application has been illustrated with a CFD analysis of a car model. The results and the performance achieved by the algorithm in performing the data interpolation and the color map generation reveal the applicability of such systems during the monitoring and the review phases of both experimental and numerical analysis.

The P-SAR system presented in this research aims at facilitating professional engineers in perceiving the spatial properties of simulation results that would be otherwise visualized by means of flat screens only. This means that, by projecting directly on the external surface of tri-dimensional objects with real or scaled dimensions, the observers' sense of proportion for the digital data is increased with respect to other VR or AR applications based on alternative display devices. This advantage becomes even more meaningful if non-expert participants are involved in the development process due to their poorly trained ability in interpreting digital results. Moreover, the process of idea sharing in face-to-face collaborative scenarios remains natural even during prolonged uses of the technology thanks to the absence of external devices placed on or between the users.

Color maps, and the related interpolation between sensors values, have been selected to visualize the output of the simulation due to the availability of a high number of sensor data extrapolated from the CFD analysis. In this way, the visual result displayed by the SAR platform can be easily compared, in terms of color distribution and hue, with the one calculated by the CAE software. Moreover, this type of visualization offers the advantage to be widely diffused among the experts in the field so they can rely on personal knowledge and experience to correctly interpret the results. Alternative means of visualization can be selected according to the specific use scenario and type of analysis. With the field of application of the current case study, for instance, punctual colors indicating the measured pressure value, can be displayed in correspondence of the sensors' position without applying the interpolations required by the texture generation. This alternative approach leads to reduce the computational cost of the application and to facilitate the users while identifying the value and the disposition of the experimental data.

Future research directions concern the improvement of the system interactivity not only during the inspection of the data but also for the setup of the numerical analysis. The capability of the platform to track the position of the physical object and to consequently adjust the projection leads to the definition of innovative and intuitive interfaces without increasing the complexity of the platform. Tangible tools and instruments can be easily integrated to let users performing adjustments on the projected output, changing the boundary conditions of the simulation and possibly exploring volumetric data (i.e. the current biggest limitation of P-SAR technology). Moreover, a further investigation aiming at validating the applicability of such technology in real engineering fields is necessary. These may include testing activities in wind tunnels, as already mentioned, or in experimental setups involving large and complex models, structures or building where multiple projectors with wide field of view can be easily integrated in the P-SAR system.

References

1. Li, W., Nee, A.Y.C., Ong, S.K.: A state-of-the-art review of augmented reality in engineering analysis and simulation. Multimodal Technol. Interaction **1**(17), 22 (2017). https://doi.org/10.3390/mti1030017
2. Cascini, G., et al.: Exploring the use of AR technology for co-creative product and packaging design. Comput. Ind **123**, 17 (2020). https://doi.org/10.1016/j.compind.2020.103308
3. Niebling, F., Griesser, R., Woessner, U.: Using Augmented Reality and Interactive Simulations to Realize Hybrid Prototypes. In: Bebis, G., et al. (eds.) ISVC 2008. LNCS, vol. 5358, pp. 1008–1017. Springer, Heidelberg (2008). https://doi.org/10.1007/978-3-540-89639-5_96
4. Naets, F., Cosco, F., Desmet, W.: Improved human–computer interaction for mechanical systems design through augmented strain/stress visualisation. Int. J. Intell. Eng. Inform. **5**(1), 50–66 (2017). https://doi.org/10.1504/IJIEI.2017.082565
5. Clothier, M., Bailey, M.: Augmented reality visualization tool for Kings Stormwater Bridge. In: Proceedings of the Fourth IASTED International Conference on Visualization, Imaging, and Image Processing, pp. 754–759. http://web.engr.oregonstate.edu/~mjb/WebMjb/Papers/kings.pdf (2004)
6. Moreland, J., et al.: Integration of augmented reality with computational fluid dynamics for power plant training. In: Proceedings of the International Conference on Modeling, Simulation and Visualization Methods. https://search.proquest.com/docview/1491735652 (2013)
7. Ong, S.K., Huang, J.M.: Structure design and analysis with integrated AR-FEA. CIRP Ann. **66**(1), 149–152 (2017). https://doi.org/10.1016/j.cirp.2017.04.035
8. Lakaemper, R., Malkawi, A.M.: Integrating robot mapping and augmented building simulation. J. Comput. Civil Eng. **23**(6), 384–390 (2009). https://doi.org/10.1061/(ASCE)0887-3801(2009)23:6(384)
9. Underkoffler, J., Ishii, H.: Urp: a luminous-tangible workbench for urban planning and design. In Proceedings of the SIGCHI conference on Human Factors in Computing Systems (CHI 1999), pp. 386–393. Association for Computing Machinery, New York, NY, USA (1999). https://doi.org/10.1145/302979.303114

10. Underkoffler, J., Ullmer, B., Ishii, H.: Emancipated pixels: real-world graphics in the luminous room. In Proceedings of the 26th annual conference on Computer graphics and interactive techniques (SIGGRAPH 1999), pp. 385–392. ACM Press/Addison-Wesley Publishing Co., USA (1999). https://doi.org/10.1145/311535.311593
11. Sheng, Y., Yapo, T.C., Young, C., Cutler, B.: A spatially augmented reality sketching interface for architectural daylighting design. IEEE Trans. Visual. Comput. Graph. **17**(1), 38–50 (2011). https://doi.org/10.1109/TVCG.2009.209
12. Ishii, H., Ratti, C., Piper, B., Wang, Y., Biderman, A., Ben-Joseph, E.: Bringing clay and sand into digital design – continuous tangible user interfaces. BT Technol. J. **22**, 287–299 (2004). https://doi.org/10.1023/B:BTTJ.0000047607.16164.16
13. Lundin Palmerius, K., Schönborn, K.: Visualization of Heat Transfer Using Projector-Based Spatial Augmented Reality. In: De Paolis, L.T., Mongelli, A. (eds.) AVR 2016. LNCS, vol. 9768, pp. 407–417. Springer, Cham (2016). https://doi.org/10.1007/978-3-319-40621-3_29
14. Cheng, S.W., Dey, T. K., Shewchuk, J.R.: Delaunay Mesh Generation, Chapter 2. CRC Press, Boca Raton, FL. http://www.cse.ust.hk/~scheng/meshbook.html (2012)
15. Pineda, J.: A parallel algorithm for polygon rasterization. In: Proceedings of the 15th annual conference on Computer graphics and interactive techniques (SIGGRAPH 1988), pp. 17–20 (1988). https://doi.org/10.1145/54852.378457
16. SimScale Homepage: www.simscale.com. Last accessed 15 Feb. 2021
17. ParaView Homepage: www.paraview.org. Last accessed 15 Feb. 2021
18. Gmsh Homepage: www.gmsh.info. Last accessed 5 Feb. 2021
19. Blender Homepage: www.blender.org. Last accessed 15 Feb. 2021
20. Unity3D Homepage: www.unity.com. Last accessed 15 Feb. 2021
21. Morosi, F., Carli, I., Caruso, G., Cascini, G., Dhokia, V., Ben Guefrache, F.: Analysis of co-design scenarios and activities for the development of a spatial-augmented reality design platform. In: Proceedings of International Design Conference (DESIGN 2018), pp. 381–392 (2018). https://doi.org/10.21278/idc.2018.0504

Investigating the Application of Augmented Reality to Support Wire Harness Activities

Alessandra Papetti(✉), Marianna Ciccarelli, Agnese Brunzini, and Michele Germani

Università Politecnica delle Marche, Via Brecce Bianche 12, 60131 Ancona, Italy
{a.papetti,m.ciccarelli,a.brunzini,m.germani}@staff.univpm.it

Abstract. In the industrial context, the wire harness represents a process with a high degree of manual work and significant customizability of final products. To reduce the workers' mental demand, the augmented reality (AR) can be an effective tool. The existing industrial applications mainly focus on manual assembly assistance, but few solutions have been developed for the wire harness. This paper tries to enhance such a process by supporting operators with an AR application that shows the work instructions and allows reducing their cognitive workload. The main goals are to propose a user-friendly and versatile tool and carry out a structured and complete evaluation of the user experience. Two experimental sessions were conducted in the laboratory by simulating the wire harness assembly and quality inspection. The use of the AR application is efficient and effective especially when instructions far from the workstation are considered as a reference.

Keywords: Augmented reality · Human-centered manufacturing · Wire harness assembly · Industry 4.0 · Hololens 2

1 Introduction

The new production paradigm is based on collaboration and data exchange between different elements of the smart factory of the future (e.g., people, cyber-physical systems, internet of things) eliminating the boundaries between the physical and digital world. The role of the operator will be revolutionized with the final aim of improving his working conditions. If the reduction of humans' physical effort has been extensively addressed, the cognitive workforce capabilities will have to be increasingly enhanced. Rauch et al. [1] highlight how sensorial and cognitive aid systems will play a key role in future manufacturing processes, also exploiting the augmented reality (AR) potential. The research community has shown how the AR, despite its technical limitations, is an effective support tool for operators [2]. An AR system, which allows the superimposition of virtual objects on the real world [3], has to comply with the following three characteristics: combining real and virtual objects, being interactive, and registering real and virtual objects in real-time. In the industrial context, AR represents an effective help for companies that deal with complex processes and have to increase manufacturing

C. Rizzi et al. (Eds.): ADM 2021, LNME, pp. 116–124, 2022.
https://doi.org/10.1007/978-3-030-91234-5_11

flexibility. Nowadays, industrial applications mainly focus on manual assembly or disassembly processes, training activities, and quality inspections. Several research works prove that the chance to display information directly in the field involves reducing task execution time and improving product quality [4]. Indeed, the operator usually performs the assigned tasks reading paper-based instructions, sometimes placed distant from the working area. In this way, he loses time, strains his memory, and commits avoidable errors.

Danielsson et al. identified assembly instructions and human factors as the two main perspectives of AR smart glasses for operators [5]. Workers asked for more individualized and dynamic instructions and some studies showed that multimedia information could reduce mental workload, human errors, and training time. The operator enhances his skills and capabilities by interacting with digital and intangible contents within the working environment. Longo et al. proposed a smart personal digital assistant that integrates digital content and intelligent tutoring systems [6]. In the mixed-reality system proposed by Rodriguez et al., the work instructions are projected directly on the workstation and the operator can follow the procedure step-by-step [7].

Existing approaches mainly focus on manual assembly assistance in terms of sequence instructions, virtual pick-to-light systems, components orientation, etc. However, another interesting application could refer to the wire harness, which is a common problem of several industrial sectors such as automotive, aviation, capital goods, electrical components, etc. This manufacturing process is characterized by a high degree of manual work and increasing customizability of final products, which implies a large diversity of wire harness variants in small batch sizes. One attempt has been made by Rice et al. that realized a prototype system that helps the worker during the electrical wire harness assembly of commercial aircraft by the combined use of mobile, wearable, and location tracking technologies [8]. Considering the relevance of the topic, the development of new solutions is recommended. For this aim, this paper proposes a user-friendly AR application able to enhance the manual process of wire harness by supporting operators with step-by-step work instructions. The application adaptability allows using it in different wire harness activities (assembly, training, and quality inspection).

Ongoing research also needs to establish the effectiveness of AR solutions through multiple measures of performance and user experience (UX), to encourage their diffusion in the industrial world. However, such structured evaluation approaches are usually limited to training applications, by comparing virtual approaches with traditional ones (i.e., flank the trainees with experienced workers) [9]. In this context, the proposed solution tries to simplify the user interface and reduce the needed mental effort to accomplish the tasks. The UX is addressed from different perspectives: design of a user-friendly application for workers, simplification of the development process, and implementation of a structured and complete evaluation protocol.

2 The Augmented Reality System

The proposed AR system is an easy-to-use tool intended for helping the operator in different kinds of wire harness activities such as assembly, quality control, or training. It has been designed to be versatile, adaptable, and simple to customize, both for the operator and the developer.

2.1 Architecture Design of the AR System

Figure 1 shows the design of the AR system's architecture, thought to achieve the suggested adaptiveness to be suitable for different industrial operations.

The selection of the AR content and device mainly depends on the kind of activity that the operator must accomplish (info classification), and on the database that includes all the data related to the CAD. The database collects the product data (i.e., 2D/3D CAD model, product ID, customers, etc.), the wiring's features and routing, the workstation's characteristics, the processes, and the tasks to perform.

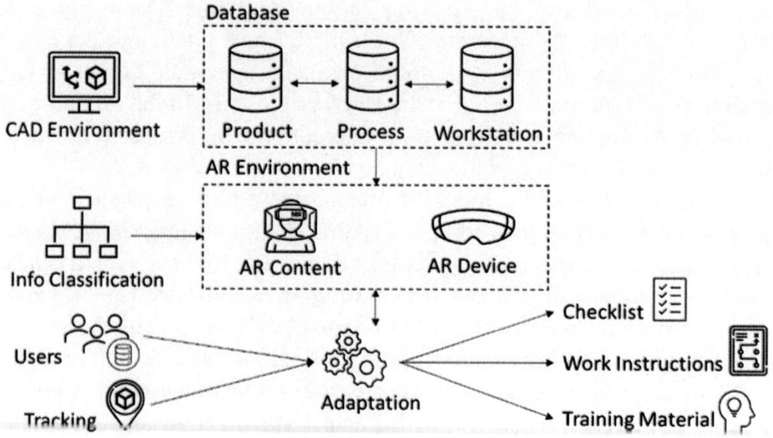

Fig. 1. Architecture design.

The design and development of the AR content refer to four macroblocks: the application's scope, the product's features, the tasks to be performed, and the workstation. Based on the application scope, the AR content may consist of smart checklists and indications for quality control, step-by-step 3D (or 2D) instructions for assembly tasks, or interactive training material to be visualized in front of the product. The number of steps or checks shown at the same time can be customized according to the operator's level of expertise (beginner, expert, etc.).

2.2 Design and Development of the AR Application

The AR application proposed in this study was developed implementing the following requirements: (i) to be user-friendly, (ii) to be used in any environment without markers, and (iii) to help the operator in the harness' assembly and quality control to speed up the process and enhance the performance. It was designed to allow:

- Easy retrieval of process instructions by product data (e.g., name, code), customer (e.g., quality requirements), workstation characteristics (fixed or mobile), etc.
- Possibility of easily handle the 2D/3D object (i.e., the instructions).
- Selection of a "hand-movable menu" or making it follows the user's gaze.

- Minimization of commands to reduce the mental effort and improve the UX.

The application was developed by Unity and Visual Studio. The product CAD model was inserted in Unity where the materials and textures were assigned. The AR application was composed of the object that can alternately show different parts of the product with additional textual instructions, and a specific menu to interact with the 3D object.

From the database of the Mixed Reality ToolKit (MRTK) package, a 2 × 2 standard menu was chosen and customized. Two buttons allow resizing, positioning, and rotating the object, and then keeping the object stationary in the space. The other two are created in Visual Studio for going forward/backward with the tasks. To configure a step-by-step procedure, the position of each object's feature has been assigned starting from "position 0", which corresponds to the first task to be performed. A specific object to visualize is assigned to each step.

The developed AR application can be deployed in different HMD. In this case, Hololens 2 by Microsoft has been selected because of its advanced features in terms of technical specs (optics, field of view, and resolution) and available hands-free interactions (hand tracking, eye tracking, and voice commands). In the application, both eye-tracking and hand-tracking have been implemented for the interaction with the GUI, allowing the marker-less tracking of the physical environment. Through eye-tracking, the AR GUI is automatically placed in front of the user's eyes. Through hand-tracking, the operator can align and adjust the augmented contents with the workstation.

Wearing the AR device, the operator can retrieve the instructions by inserting a product keyword and then move and zoom them in the most convenient position to perform the task. By pressing the "Forward" button, the worker can start the wiring operation, and the object will show its different parts, in the defined sequence. When the operator has completed the task can move on. If there is the need to check back the task performed, with the "Backward" button it is possible to return to the previous steps. With the 3D digital content, the instruction can be provided through textual indications or symbols (such as arrows, circles, etc.) combined with the colouring and animation of the elements to be assembled or checked. With 2D digital content, the instructions mainly rely on images with textual indications or coloured elements and symbols. Figure 2 shows an example of the 3D model respectively in the CAD and AR environments.

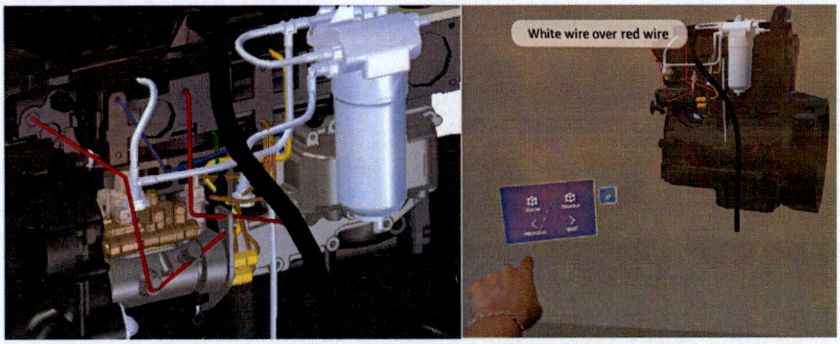

Fig. 2. From CAD environment to AR environment.

3 Experimental Campaign

To preliminarily evaluate the AR application, a pilot test was conducted at the university laboratory. Although 3D instructions would have made the intersections and the fixation points clearer, in this test, 2D pictures were used. It allowed involving users without experience on wire harness, easily simulate tasks, preliminary test the developed design logic, and mainly focus on the UX.

A total of 10 participants without experience with AR technologies were involved in the study. The test was divided into two sessions, 5 users performed the wire harness assembly, and 5 users the quality control (Table 1). To avoid the possibility for the user to memorize the exercise, the tasks that present the same difficulty level were different.

Table 1. Tasks schedule.

Task	Activity	Difficulty	Instruction
T1	Assembly	Simple	Paper-based
T2	Assembly	Complex	Paper-based
T3	Assembly	Simple	AR
T4	Assembly	Complex	AR
T5	Quality control	Simple	Monitor
T6	Quality control	Complex	Monitor
T7	Quality control	Simple	AR
T8	Quality control	Complex	AR

Before starting the test, the user completed a brief training session with HoloLens 2. Figure 3 shows the user during the activities and the workbench configuration both for the assembly and the quality control (respectively with the paper-based and monitor-based instruction used without AR), and the AR application designed for quality control.

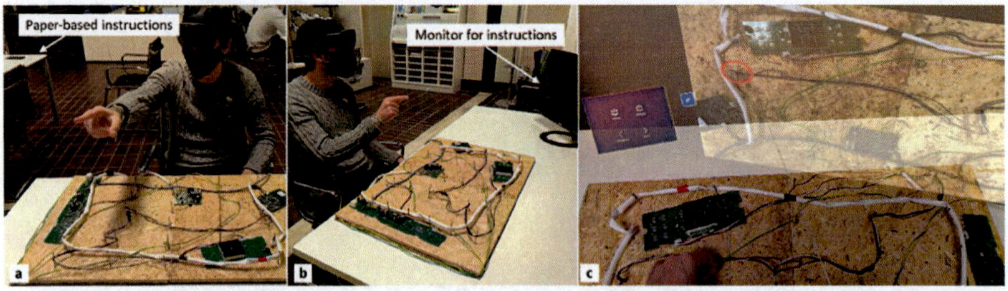

Fig. 3. Wire assembly activity (a), quality control activity (b), and AR application (c).

To simulate the wire harness assembly, it was used a wooden table with some spikes as reference points and electrical wires. The users had to replicate the correct routing, focusing on the cables' intersection and fixations in the reference points. In the simple tasks, the users performed the routing of only one wire, while the complex tasks presented a higher number of cables, fixations, and intersections. In the first two tasks, the participants could read paper-based instructions, placed far from their workstation. With the support of the AR device, they had all the instructions superimposed over their workbench, showing step-by-step consecutive sections of the virtual cables.

In the quality control activity, a final wire assembly was simulated on the wooden table. The users had to check if the sequence was correct and fix the errors. The areas to check were highlighted with virtual red circles shown in sequence. In the simple tasks, the users had to control only the segment route relative to the defined reference points, while the complex tasks included a depth inspection of intersections and fixations. The traditional support modality consists of showing the correct wires assembly on a monitor placed near the workbench.

3.1 Evaluation Protocol and KPIs

As shown in the workflow of the test session (Fig. 4), multiple evaluation methods were adopted. The evaluation protocol aimed to include qualitative, quantitative, subjective, and objective key performance indicators (KPIs).

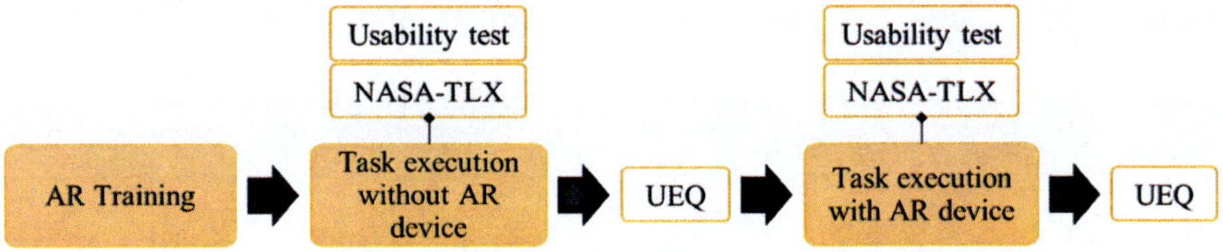

Fig. 4. Test workflow.

During the task execution, the moderator recorded the objective parameters usually considered in usability tests: success rate (i.e., percentage of steps completed for each task), task completion time, errors (i.e., wrong assemblies, undetected defects), and significant events (e.g., expert consultations, problems). The time dedicated to non-value-added activities (NVAA) such as walking, instructions management, etc. was also evaluated. Qualitative parameters refer to users' comments and nonverbal communication, which were monitored by observing users' movements and facial expressions.

To evaluate the cognitive load/stress perceived by users, at the end of each task they filled in the NASA-TLX [10]. It considers the following aspects: physical demand, mental demand, temporal demand, effort, performance, and frustration.

The comparison of supporting materials (traditional work instructions, AS-IS, and AR application) is also based on the user experience questionnaire (UEQ) [11], which is a semantic differential with 10 items (not understandable/understandable, inefficient/efficient, conventional/innovative, etc.).

4 Results and Discussion

All the participants have successfully completed the tasks without the support of the moderator (Table 2). Only one consultation was registered about the AR gestures. The task completion time and the number of errors vary significantly by activity (assembly or quality control). In the first case, time and errors were more than halved with the AR system. However, if the traditional instructions had been closer to the work area it is likely that fewer benefits were obtained. Indeed, from the users' comments, it emerged that proximity and sequencing of the AR instructions were crucial for concentration and not making mistakes. In the second case, the two systems did not significantly affect the made mistakes. On the contrary, the AR application caused an NVAA time increase due to the greater interaction with the instructions by users. Users stated that the closer proximity of the AR instructions to the work area than the screen, although minimal, was very helpful. The opportunity to adjust the positioning and size of the instructions would have played a fundamental role were it not for the users' difficulty with gestures. The halving of the NVAA time between T3/T7 and T4/T8 confirms that becoming familiar with the gestures is crucial. Despite the greater complexity, everyone declared that would have benefited more from the interaction with a 3D model.

Table 2. Usability test results.

KPI	T1	T2	T3 (AR)	T4 (AR)	T5	T6	T7 (AR)	T8 (AR)
Success rate	100%	100%	100%	100%	100%	100%	100%	100%
Errors	2.4 ± 1.7	2.0 ± 1.6	0.2 ± 0.4	0 ± 0	1.0 ± 0.7	1.0 ± 0.7	1.0 ± 0.7	1.2 ± 0.4
Time [s]	258 ± 83	439 ± 138	115 ± 29	187 ± 58	105 ± 74	139 ± 56	268 ± 85	387 ± 175
NVAA time [s]	56 ± 16	72 ± 9	54 ± 16	52 ± 21	11 ± 2	10 ± 1	79 ± 28	98 ± 67

The AR use for assembly generated a reduction of NASA score of 42%, which means a workload reduction, while in the quality control it resulted in a 23% increase. In the former, the high mental demand, which is due to having to remember the instructions, prevailed. In the latter, a greater effort was perceived in obtaining the best instructions positioning due to the unfamiliarity with the gestures.

The UEQ results (Fig. 5) highlight both the great potential of the AR system and the clear difference between the two test sessions, which is also influenced by the fruition modality of the traditional work instructions. Most of the items were evaluated about the work instruction provided and the way of interacting with it. In the quality control, the screen proximity (AS-IS) and the lack of practicality in positioning the AR contents give rise to a minimal difference in the evaluation of the two systems. The obstruction was often associated with the discomfort experienced by users.

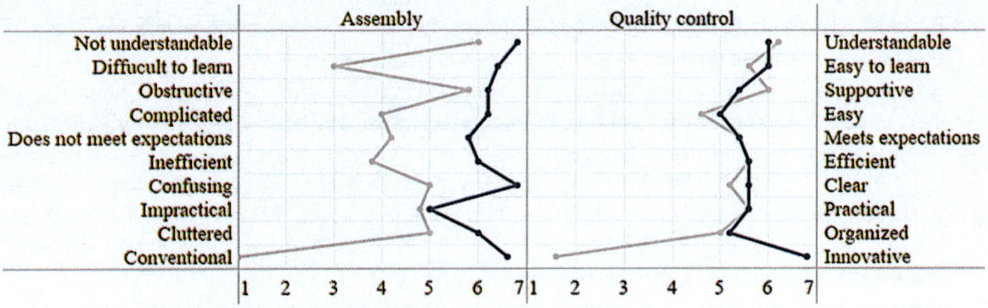

Fig. 5. UEQ results (AS-IS: light gray; AR: dark gray).

5 Conclusions

The paper proposes a user-friendly AR application supporting workers in wire harness activities. Two experimental sessions were performed to evaluate the UX and the AR potentialities for wire assembly and quality control. Benefits in efficiency and effectiveness are significant only if paper-based instructions far from the workstation are considered as a reference. However, good potential emerges when familiarity with gestures increases. More attempts should be made to improve workers' ergonomics and interaction skills. Almost everyone has found the viewers a bit intrusive.

Future works will address the real-time product tracking to automatically manage the wire harness assembly instructions. The AR system will be experimented in a real industrial case study by using more lightweight and comfortable AR devices.

References

1. Rauch, E., Linder, C., Dallasega, P.: Anthropocentric perspective of production before and within Industry 4.0. Comput. Ind. Eng. **139**, 105644 (2020)
2. Danielsson, O., Holm, M., Syberfeldt, A.: Augmented reality smart glasses in industrial assembly: current status and future challenges. J. Ind. Inform. Integr. **20**, 100175 (2020).
3. Trommnau, J., Kühnle, J., Siegert, J., Inderka, R., Bauernhansl, T.: Overview of the state of the art in the production process of automotive wire harnesses, current research and future trends. Procedia CIRP **81**, 387–392 (2019)
4. de Souza Cardoso, L.F., Mariano, F.C.M.Q., Zorzal, E.R.: A survey of industrial augmented reality. Comput. Ind. Eng. **139**, 106159 (2020)
5. Danielsson, O., Holm, M., Syberfeldt, A.: Augmented reality smart glasses for operators in production: survey of relevant categories for supporting operators. Procedia CIRP **93**, 1298–1303 (2020)
6. Longo, F., Nicoletti, E., Padovano, A.: Smart operators in industry 4.0: a human-centered approach to enhance operators' capabilities and competencies within the new smart factory context. Comput. Ind. Eng. **113**, 144–159 (2017)
7. Rodriguez, L., Quint, F., Gorecky, D., Romero, D., Siller, H.R.: Developing a mixed reality assistance system based on projection mapping technology for manual operations at assembly workstations. Procedia Comput. Sci. **75**, 327–333 (2015)

8. Rice, M., Tay, H.H., Ng, J., Selvaraj, S.K., Lim, C., Wu, E.: Extraction, rendering and augmented interaction in the wire assembly of commercial aircraft. In: Proceedings – 2016 IEEE International Conference on Internet of Things (iThings) and IEEE Green Computing and Communications (GreenCom) and IEEE Cyber, Physical and Social Computing (CPSCom) and IEEE Smart Data (SmartData), vol. 349–354, p. 7917111 (2017)
9. Langley, A., et al.: Establishing the usability of a virtual training system for assembly operations within the automotive industry. Hum. Factors Ergon. Manuf. **26**, 667–679 (2016)
10. Hart, S., Staveland, L.: Development of NASA-TLX (task load index): results of empirical and theoretical research. Hum. Mental Workload **1**, 139–183 (1988)
11. Schrepp, M., Hinderks, A., Thomaschewski, J.: Design and evaluation of a short version of the user experience questionnaire (UEQ-S). Int. J. Interactive Multimedia Artif. Intell. **4**, 103–108 (2017)

User-Centered Design of an Augmented Reality Tool for Smart Operator in Production Environment

Emanuele Marino[1(✉)], Loris Barbieri[1], Biagio Colacino[2], and Fabio Bruno[1]

[1] Department of Mechanical, Energy and Management Engineering (DIMEG), University of Calabria, Pietro Bucci 46C, 87036 Rende, CS, Italy
{emanuele.marino,loris.barbieri,fabio.bruno}@unical.it

[2] Baker Hughes Nuovo Pignone Srl, Contrada Le Marinate, 89900 Vibo Valentia, VV, Italy
biagio.colacino@bakerhughes.com

Abstract. In the context of Industry 4.0, Operator 4.0 paradigm represents a key factor when dealing with the integration of new digital technologies into smart factories that are suited for workers with different skills, capabilities, and preferences. In this regard, to encourage the introduction of these new digital solutions and achieve high user acceptance, it is fundamental to consider human factors and put the worker at the center of the development process through the adoption of structured design strategies such as user-centered design (UCD) approaches.

In this perspective, the paper proposes a novel Augmented Reality (AR) tool for supporting operators at the workplace, in real-time, while performing inspection activities on built products. The proposed tool has been developed according to a user-centered design approach by involving end-users in the various design and testing stages. Preliminary tests have been conducted with representative users on a real case study to assess the usability of the proposed solution. The outcomes are very encouraging and lead us towards further investigations for effective and valid implementation of this AR tool in an industrial scenario.

Keywords: Industrial augmented reality · User-centered design · Usability · Operator 4.0 · Industry 4.0

1 Introduction

In the highly technological and innovative scenario of Industry 4.0 humans and resources (enabling technologies, machines, products) are increasingly interconnected for achieving higher efficiency and productivity [1]. There are, in fact, significant implications on the nature of the work in industry that is going to transform the manufacturing workforce in "smarter operators", towards the vision of an Operator 4.0 typology [2, 3]. In this direction, a worker becomes a smart and skilled operator whose human physical, sensory, and cognitive capabilities are enhanced by new systems, machines, and advanced digital technologies. In this way, operators will be able to work alongside digitalized and automated production systems, and use and develop their creative, innovative, and

C. Rizzi et al. (Eds.): ADM 2021, LNME, pp. 125–132, 2022.
https://doi.org/10.1007/978-3-030-91234-5_12

improvisational skills to solve unforeseen challenges [3]. Nevertheless, to successfully embrace the Operator 4.0 paradigm, operators are requested to acquire new skills and knowledge for interacting with new digital technologies and machines, and so achieve a higher level of efficiency [4].

Among the key enabling technologies that can support the development of the Operator 4.0, within the Industry 4.0 framework, Augmented Reality (AR) is one of the most promising and enabling technologies for supporting factory workers and engineers at the workplace. AR, in fact, enhances smart operator's perception of the factory environment by overlaying digital information on his/her field of view through the adoption of handheld or wearable devices, or projectors [5]. In particular, this technology augments the human capabilities of the smart operator, which becomes an Augmented Operator [5], by supporting him/her in real-time during manual operations for reducing human errors and dependency on printed work instructions, computer screen, and operator memory [6]. Thus, industrial workers see their transformation as an opportunity to improve their performances, safety, working environment, and autonomy in organizing their work by continuing to have a central role in the industry [7]. To address such opportunity, new trends in Industry 4.0 are focusing on the importance of applying approaches, principles, and engineering philosophies alongside the concept of Operator 4.0 to ensure a smoother and sustainable transition and accommodate pertaining socio-technical challenges [3]. To this end, many industrial and manufacturing enterprises are increasingly recognizing the need for human-centric production systems where advanced digital and manufacturing technologies are seen as key factors to the further enhancement and augmentation of the human's sensorial and cognitive capabilities.

On the basis of these considerations, the paper proposes a novel AR tool for supporting smart operators to perform real-time inspection activities at their working environment. In particular, this AR tool has been developed through a user-centered design (UCD) approach in order to meet end-users' needs and expectations and facilitate user's acceptability. A preliminary user study has been carried out on a real case study with end-users in order to investigate usability and users' acceptance of the proposed AR tool.

2 Related Works

Over the past few years, the industrial transformation is taking shape in traditional companies through the development of new industry 4.0-enabling technologies. In this context, it is fundamental to take into consideration workers' needs and expectations to promote and achieve closer cooperation between people and machines and a higher level of smart interactions with manufacturing systems. In this regard, UCD approaches help industrial enterprises to effectively integrate the user perspective throughout the design process to develop usable systems. According to the ISO 9241-210 [8] standard, the UCD approach consists of four basic activities that are iterated until achieving user satisfaction. In particular, the design process starts by recognizing the potential users and their context of use. In the second step, users' requirements are defined. The third step produces design solutions to meet users' requirements. In the last step, an evaluation of the product against requirements is conducted to check how well a design is performing on the basis of users' feedbacks. In this context, various researches involved end-users for the development and testing of industrial AR-based systems in different application fields [9–12], such as assembly, maintenance, training, and inspection. Similarly, this paper proposes a novel AR tool that has been developed according to a user-centered design approach, by involving end-users in all the various stages of the development process, in order to support operators at the workplace, in real-time, while performing inspection activities on built products. In particular, the proposed tool overcomes the most common limits of the AR-based systems adopted for inspection activities in industrial contexts [13, 14], mainly due to the adoption of dedicated and expensive hardware, a complex calibration process and limited space of usage, by running on off-the-shelf Android™ devices to allow greater compatibility with commonly used devices, e.g., smartphones and tablets, and higher freedom of movement to the user.

3 UCD Approach

The development of the proposed AR tool was driven by a UCD approach in order to involve users in all the various design stages, from conceptualization to prototype evaluation. As depicted in the following figure (Fig. 1) the adopted UCD approach consisted of four iterative stages. In particular, in the first stage, structured interviews, with workers, engineers, manufacturing and production line managers, were carried out to gather users' needs and desires. Furthermore, scheduled site visits were of fundamental importance in order to have a better understanding of the instruments and procedures adopted to carry out inspection activities. The analysis of the inspection processes allows for identifying two main limitations: inspection activities adopt technical drawings and photographic material that may not be exhaustive and lead to misconceptions; there is a lack of structured and reliable methods and tools to support workers to keep track and formalize information about the detected design changes and errors.

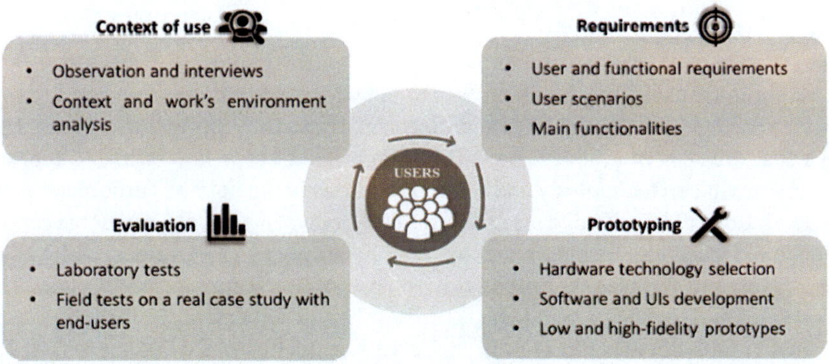

Fig. 1. UCD approach adopted for the development of the proposed AR tool.

All the data collected in the first stage were then adopted to define users' and functional requirements aimed to develop a tool able to support the inspection activities carried out at the workplace and simplify the detection of design discrepancies. To this end, AR technologies were preferred to other solutions because, for this specific context of use, they bring tangible benefits in terms of efficiency when compared to other instruments traditionally adopted for the detection of design discrepancies [15]. Then main functionalities and user scenarios were defined too to visualize aspects of the proposed AR tool which users might appreciate most in their contexts of use and better orient the design stages. In the following prototyping stage, hardware and software technology selections were carried out to identify the best technical solutions that allow addressing requirements. About hardware solutions, handheld devices were preferred to projectors and static screens for their capability to allows users to move in the real environment without any restrictions. While the choice of a tablet device is due to an explicit need expressed by the end-users who prefer displays larger than smartphones' screen. About software development, ARCore™ [16] was preferred to the other solutions because it is open source, and it does not need specific and dedicated hardware components but it can run on any Android™ devices (version 7.0 and above). In particular, this solution ensures, in the first place, very high compatibility with commonly used devices for which workers are already trained and, secondly, to take advantage of the hybrid-tracking techniques that combine vision- and sensor-based methods to improve the reliability of the AR visualization.

On the basis of the hardware and software selection, the main functionalities were implemented in order to allow the user to: perform a real-time augmentation of 3D models, as designed by the technical office, on the physical assembled product; simplify the detection of discrepancies by customizing 3d models' color and transparency settings; report detected discrepancies by adding 3D annotations directly onto virtual models; share the collected data with the technical office. Low- and high-fidelity prototypes were developed and tested with both expert and representative users. All stages of the UCD approach were iterated until satisfying requirements and accommodate as much as possible users' needs and expectations before achieving the current solution.

4 User Study

A first preliminary user study has been carried out with representative users on a real case study at the Baker Hughes Nuovo Pignone Srl plant located in Vibo Valentia (Italy). Baker Hughes is a company of global scale with a high level of know-how, which offers some of the most advanced technology and service solutions in the field of Turbomachinery & Process Solutions (TPS). Nuovo Pignone Srl (TPS) represents in Italy the global center of excellence for research, development, and manufacturing of gas turbines, compressors, pumps, valves, and services. The case study consists of a large auxiliary baseplate system that works as a supporting system for a gas turbine by managing temperature and pressure values of the lubrification oil. Due to the large size of this baseplate system, the test took place on one of its sides (Fig. 2) and the consequent area of experimentation was approximately 6 m long and 2 m wide. In particular, five fiducial markers have been printed on papers and placed, at a distance of about 1.5 m from each other, on the physical components in order to allow users to move freely along the entire length of the case study.

Fig. 2. Case study: assembly inspection of a baseplate system © 2021 Baker Hughes Company – All rights reserved.

The experimentation consisted of four main steps. In the first step, the participants were introduced to the experiment and filled in a demographic information questionnaire. After completing the demographic survey, each user received a brief demonstration by a tutor (about 1/2 min long) about the main functionalities of the AR tool. In the third step, the participants were invited to perform the assembly inspection of the case study by executing a set of six predefined tasks in the following order. First, the user framed a marker placed onto the real prototype to perform the marker detection and the augmentation of 3D models on the real scene. Secondly, the user started the inspection task for detecting potential discrepancies between the designed and built assembly (Fig. 3a). Thirdly, the participant added a 3D annotation in correspondence of the detected discrepancy by taking a picture of the augmented scenario too. Subsequently, the user edited the 3D annotation in terms of content and size. Next, the participant customized the color and transparency level of the augmented 3D model related to the detected discrepancy (Fig. 3b). Finally, the user sent the picture and 3D annotation to the technical office.

Fig. 3. Participants while performing tasks © 2021 Baker Hughes Company – All rights reserved.

After completing all tasks, the participants were invited to perform a one-to-one personal interview in order to gather their opinions, comments, and suggestions about the AR tool. In particular, the SEQ (Single Ease Question) usability metric [17, 18] was adopted to measure task-performance satisfaction. This consists of a standard question: "Overall, how difficult or easy did you find this task?", asked for each task, to which the participant had to rate the ease of use on a 7-step Likert scale from 1 (very difficult) to 7 (very easy). The SEQ has been proved to be a reliable, sensitive, and valid metric, and has been preferred to other ones for this preliminary study because it is an easy-to-understand questionnaire [18] and requires very little time. Sixteen participants recruited among engineers and factory workers were employed for the user study. In particular, they were divided into two groups:

- G1: 8 engineers (4 males and 4 females) with average age 30.5 (min 24, max 46, SD = 6.6);
- G2: 8 factory workers (all males) with average age 40.8 (min 23, max 53, SD = 10.5).

All but one, of the G2 group, of the participants stated that they use very often, several times per day, tablets and smartphones. Furthermore, 62.5% of factory workers and 25% of engineers were inexperienced in the use of AR technologies. On the contrary, 37.5% of factory workers and 75% of engineers had previous experience with the use of AR applications. All of them were novices in the adoption of such technologies in their working environment.

5 Results and Discussions

The data collected in the user study were analyzed through the statistical packages IBM® SPSS® and Microsoft® Excel®. The statistical significance level was set at $p < 0.05$. The following graph (Fig. 4) shows the results of the SEQ questionnaire, error bars represent a 95% confidence interval.

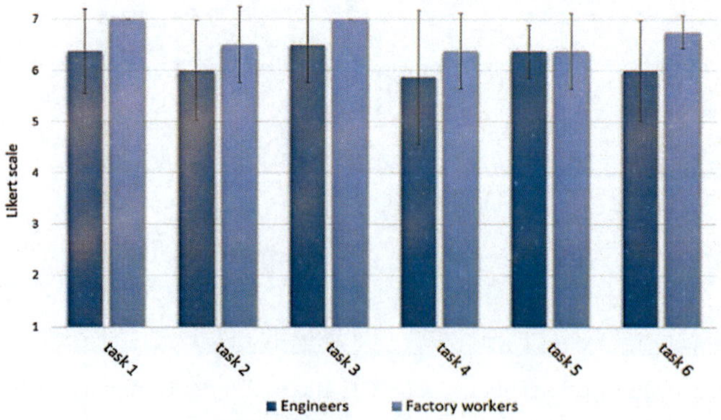

Fig. 4. SEQ results by groups.

The results show that factory workers found tasks easier than engineers, with a mean value respectively of 6.19 (SD 0.26) and 6.67 (SD 0.29), except for task 5 in which both groups perceived the same usability level. Nevertheless, these differences were not confirmed by an independent t-test analysis which results, depicted in Table 1, indicate that there is not a statistically significant difference between the two groups.

Table 1. Independent samples t-test analysis results.

	Task 1	Task 2	Task 3	Task 4	Task 5	Task 6
Sig.	0.180	0.438	0.227	0.524	1	0.176
t value	−1.488	−0.798	−1.323	−0.654	0	−1.426

Then, even if all the participants were novices in the use of industry 4.0 enabling AR technologies, both groups performed relatively well and reached a high level of usability. Furthermore, very positive feedbacks emerged from the one-to-one personal interview about functionalities implemented and intuitiveness of the AR tool. In particular, they appreciated the capability to easily detect the presence of design discrepancies and assembly errors by augmenting virtual 3D models, as defined in the project plan, on the actual design. In summary, the end-users involved in the study perceived a high usability level of the proposed tool and all of them clearly expressed their satisfaction and interest in the possibility to adopt this tool in their working environment to perform inspection activities.

6 Conclusions

In this paper, an AR tool has been proposed aiming to support smart operators in performing inspection activities at the workplace and in real-time. The AR tool has been developed through a UCD approach by involving end-users in all the various stages of the development process, and especially during the experimentation and testing activities. A preliminary user study was conducted at an industrial company that had started a digital transformation journey. In particular, the study was carried out on a real case study with representative users in order to investigate users' acceptance of the proposed AR tool. The outcomes revealed a high level of usability and user acceptance, and all the participants expressed positive opinions and interest for an effective and valid implementation of this AR tool in their working activities.

Future works will focus on extensive field experimentations with end-users in order to assess the usability and effectiveness of the proposed AR tool by means of objective and subjective standard metrics.

Acknowledgments. This research has been partially funded by the project SMILE, funded by the Italian MISE under the PON "Imprese e Competitività" program. Project n. F/190084/03/X44 – CUP: B21B19000530008 COR: 1679960.

References

1. Xu, L.D., Xu, E.L., Li, L.: Industry 4.0: state of the art and future trends. Int. J. Prod. Res. **56**(8), 2941–2962 (2018)
2. Papetti, A., Gregori, F., Pandolfi, M., Peruzzini, M., Germani, M.: A method to improve workers' well-being toward human-centered connected factories. J. Comput. Des. Eng. **7**(5), 630–643 (2020)
3. Romero, D., et al.: Towards an operator 4.0 typology: a human-centric perspective on the fourth industrial revolution technologies. In: Proceedings of the Int. CIE46, Tianjin, China, pp. 29–31 (2016)
4. Kaasinen, E., et al.: Empowering and engaging industrial workers with Operator 4.0 solutions. Comput. Ind. Eng. **139**, 105678 (2020)
5. Gazzaneo, L., Padovano, A., Umbrello, S.: Designing smart operator 4.0 for human values: a value sensitive design approach. Procedia Manuf. **42**, 219–226 (2020)
6. Gattullo, M., Evangelista, A., Uva, A.E., Fiorentino, M., Gabbard, J.: What, how, and why are visual assets used in industrial augmented reality? A systematic review and classification in maintenance, assembly, and training (from 1997 to 2019). IEEE Trans. Visual. Comput. Graph. (2020)
7. Peruzzini, M., Grandi, F., Pellicciari, M.: Exploring the potential of Operator 4.0 interface and monitoring. Comput. Ind. Eng. **139**, 105600 (2020)
8. ISO 9241-11:2018: Ergonomics of human-system interaction – Part 11: Usability: Definitions and concepts (2018)
9. Platonov, J., Heibel, H., Meier, P., Grollmann, B.: A mobile markerless AR system for maintenance and repair. In: 2006 IEEE/ACM International Symposium on Mixed and Augmented Reality, pp. 105–108. IEEE (2006)
10. Webel, S., Bockholt, U., Engelke, T., Gavish, N., Olbrich, M., Preusche, C.: An augmented reality training platform for assembly and maintenance skills. Robot. Auton. Syst. **61**(4), 398–403 (2013)
11. Ferraguti, F., et al.: Augmented reality based approach for on-line quality assessment of polished surfaces. Robot. Comput. Integr. Manuf. **59**, 158–167 (2019)
12. Cascini, G., et al.: Exploring the use of AR technology for co-creative product and packaging design. Comput. Ind. **123**, 103308 (2020)
13. Doshi, A., Smith, R.T., Thomas, B.H., Bouras, C.: Use of projector based augmented reality to improve manual spot-welding precision and accuracy for automotive manufacturing. Int. J. Adv. Manuf. Technol. **89**(5–8), 1279–1293 (2016). https://doi.org/10.1007/s00170-016-9164-5
14. Wasenmüller, O., Meyer, M., Stricker, D.: Augmented reality 3D discrepancy check in industrial applications. In: International Symposium on Mixed and Augmented Reality, pp. 125–134. IEEE (2016)
15. Barbieri, L., Marino, E.: An augmented reality tool to detect design discrepancies: a comparison test with traditional methods. In: International Conference on Augmented Reality. Virtual Reality and Computer Graphics, pp. 99–110. Springer, Cham (2019)
16. ARCore Homepage: https://arvr.google.com/arcore
17. Sauro, J.: 10 things to know about the Single Ease Question (SEQ). Measuring U (2012)
18. Sauro, J., Dumas, J.S.: Comparison of three one-question, post-task usability questionnaires. In: Proceedings of the SIGCHI Conference on Human Factors in Computing Systems, pp. 1599–1608 (2009)

HoloLens 2 for Maxillofacial Surgery: A Preliminary Study

Agnese Brunzini[1] (✉), Marco Mandolini[1], Manila Caragiuli[1], Michele Germani[1], Alida Mazzoli[2], and Mario Pagnoni[3]

[1] Department of Industrial Engineering and Mathematical Sciences, Università Politecnica delle Marche, Via Brecce Bianche 12, 60131 Ancona, Italy
{a.brunzini,m.mandolini,m.caragiuli,m.germani}@univpm.it

[2] Department of Materials, Environmental Sciences and Urban Planning, Università Politecnica delle Marche, Via Brecce Bianche, 12, 60131 Ancona, Italy
a.mazzoli@univpm.it

[3] yourFACE, Clinica Ars Biomedica, Via Luigi Bodio 58, 00191 Roma, Italy
mario.pagnoni@yourface.it

Abstract. Since custom-made 3D printed surgical guides for maxillofacial surgery are usually expensive, Augmented Reality (AR) can be efficiently employed to overcome the high costs.

The proposed work aims to develop and test an AR application for different maxillofacial surgeries. The application consists in overlaying the cutting lines on the patient's mandible to guide the clinician during the procedure. It has been realized in Unity and preliminary tested with HoloLens 2 and a 3D printed mandible. Seven participants performed two consecutive trials. The mandible with the obtained surgical lines has been scanned after each test to digitally reconstruct the traced lines and compare them with the surgical lines previously designed.

The results allowed the preliminary analysis of the developed AR system's accuracy and precision. Mean distances from the designed surgical guides showed good accuracy for the genioplasty (deviation error around 1.03 mm) and orthognathic surgery (deviation error around 1.27 mm), suggesting the applicability of HoloLens 2 for these kinds of surgery. On the contrary, the application was not suitable for the mandibular angle osteotomy (deviation error over 2.50 mm).

Keywords: Augmented reality · HoloLens 2 · Surgical guides · Maxillofacial surgery

1 Introduction and Scientific Background

Today, Maxillo-Facial surgeons rely on 3D printed custom cut guides to accurately reproduce, during the surgical procedure, the preoperative simulation. The main drawbacks of physical surgical guides are the fabrication time, the selection of certified protocols, adequately experienced technicians, the overall cost of the tailored design, and the use of sterilizable and biocompatible materials for the fabrication.

In this context, Augmented Reality (AR) may represent an essential help for the surgeon because it would allow the visualization of additional information (e.g., patient data,

C. Rizzi et al. (Eds.): ADM 2021, LNME, pp. 133–140, 2022.
https://doi.org/10.1007/978-3-030-91234-5_13

internal organs, or the surgical guide itself), maintaining a perfect visualization of the real environment. However, since maxillofacial surgeries are lengthy and sophisticated procedures, AR applications in the operating room are still not widespread [1].

Nevertheless, in the last years, the use of augmented reality in the healthcare sector has increased. It allows, for example, the optimization of the integration between the surgical plan and the surgery itself [2]. Moreover, using head-mounted displays (HMD) instead of cumbersome displays allows overcoming the necessity to move the display during the surgery [3] and keeping the surgical team focused on the patient.

However, some drawbacks related to the technology must be highlighted. The accuracy of the systems, the registration, and the occlusion issues are the most relevant limitations. Moreover, the HMD could be uncomfortable to be worn for long periods and the AR system may also produce nausea and headache for the user [3]. Nevertheless, several scientific literature studies applied AR in the surgery context, particularly in the maxillofacial field, trying to achieve better working experiences, more excellent patient safety, and reduced operational costs.

AR application for mandibular angle split osteotomy (MASO) has been performed by L. Lin et al. [4], assisted by a robotic arm system, with limited errors. Both L. Lin et al. and Y.K. Lin et al. [4, 5] have used AR to superimpose the 3D digital models of colored lines indicating the nerves, or the sinuses, on the real environment to avoid collateral damages. For the registration, they both employed an image target, integrated with the mandible through a splint, as a marker to accurately align the digital and the real world. The head-mounted displays nVisor ST60 [4] and Sony HMZ-T1 [5] have been chosen to deploy the application. However, they both also used the physical surgical guide, respectively, for MASO and dental implant surgery [4, 5].

Another marker-based guidance AR application has been developed by Jiang et al. [6] to drill holes into ten beagle dogs' mandibles. In this case, they did not use the physical guide, and they tested the precision of the system using nVisor ST60 [6]. Even HMD HoloLens 1 has been applied in the craniomaxillofacial field. T. Frantz et al. used HoloLens 1 with the SDK Vuforia to study the hologram's stability and the registration accuracy over a skull [7]. Similar work has also been proposed by T. Itamiya et al., who used HoloLens 1 and Vuforia SDK to develop an easy application for the vision of 3D medical models and surgical instruments [8]. In these last two cases, AR has been used to superimpose anatomical structures.

In this context, the present paper proposes an AR application to guide the surgeon during different maxillofacial operations, replacing 3D printed and expensive surgical guides. The selected HMD is the new Microsoft's HoloLens 2, released in 2019. To the best of the authors' knowledge, this is the first scientific study investigating the applicability and usability of HoloLens 2 for different kinds of maxillofacial surgery.

2 An Augmented Reality Application for Maxillofacial Surgery

The AR application has been designed and developed starting from the CT scan of a skull. Then the surgical guides and the markers' support (for the registration of the AR content) have been created in a CAD environment, and then the full model has been 3D printed for the preliminary validation. Subsequently, the AR application has been developed. Figure 1 shows the difference between the 3D models of the standard tailored surgical guides (a) and the developed AR application with the cutting lines (b).

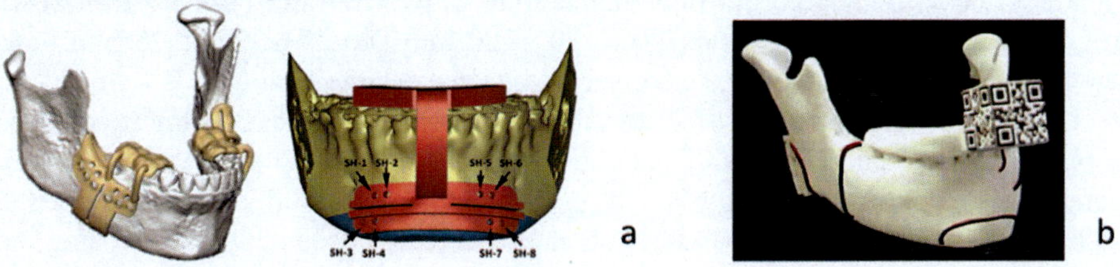

Fig. 1. 3D models of the standard tailored surgical guides (a) and the proposed cutting lines implemented in the AR application (b).

2.1 The CAD Model Design

The marker-based AR application for maxillofacial surgery has been designed and developed, fulfilling the following clinical and mechanical requirements:

- To easily guide the surgeon during different operations to save time and costs:
  - The surgical guides must have a fast and straightforward design (with AR, hooks, drilling holes, and other features can be avoided).
  - The surgical guides should be provided for different kinds of maxillofacial surgeries, such as genioplasty, orthognathic surgery (mandibular advancement) and prominent mandibular angle osteotomy.
- To allow a precise registration of the AR content over the patient's skull, from different angulations, without hindering the surgeon:
  - A three-dimensional marker (i.e., a multi-target) allows the AR content to be correctly visualized based on the surgeon's viewpoint (letting him/her move around the patient).
  - The 3D target must have the smallest dimensions (it must not be cumbersome during the surgery, and, in the meantime, it must be large enough to be recognized by HoloLens 2).
  - The 3D target must be placed in a comfortable and convenient position.
  - The 3D target must be lightweight.

A 3D model of a skull, reconstructed from a CT scan, has been imported with a customized dental splint. For this work, only the lower jawbone has been selected. Based on the surgeon's indications, the cutting lines have been designed to allow genioplasty, orthognathic surgery, and angle mandibular reduction. For this purpose, five curves have been sketched as surgical guides: one on the chin to simulate the genioplasty surgery, and two on each side of the mandible to simulate the orthognathic surgery and the mandibular angle osteotomy. The cutting lines, colored in red, are 1 mm thick.

According to the abovementioned requirements, the 3D multi-target has been realized as a hollow cube, integrated onto the dental splint (Fig. 1b). Four cube sizes of 10 × 10 × 10 mm, 15 × 15 × 15 mm, 20 × 20 × 20 mm and 25 × 25 × 25 mm have been designed, 3D printed and tested to verify the smallest size HoloLens 2 were able to detect. Finally, a hollow cube with 1 mm thickness and 20 × 20 × 20 mm dimensions has been realized. The cube has been rigidly connected to the dental splint through a parallelepiped splint measuring 9 × 8 × 20 mm. The full model, consisting of the mandible, the red cutting lines, and the 3D multi-target, has been 3D printed using the Z Printer 450 (*by 3D-System*) capable of printing also in multiple colors.

2.2 The AR Application Development

The AR application has been developed using Unity 2020.1.17f1 (*by Unity Technologies*) and Visual Studio 2019 (*by Microsoft*).

The development process consists of five main steps:

1. Packages Import (Unity): different QR codes (20 × 20 mm) have been selected as markers to be placed on the cubic multi-target. In Vuforia SDK (*by PTC Inc.*), the cuboid multi-target model has been chosen to realize the target with the QR codes. The created database with the images can be downloaded as unitypackage file and imported into Unity.
2. Settings (Unity): through the Mixed Reality Toolkit (MRTK), some settings have been defined to improve the AR experience. The possibility of visualizing the mesh of the user's hands and the real environment has been removed since it would be annoying and harmful during the surgery.
3. Scene (Unity): The scene is built by adding the multi-target marker and the.obj file of the surgical guides. The surgical guides and the marker have been previously aligned in Rhinoceros to obtain the highest precision. An opaque red material has been assigned to the cutting lines. The surgical guides have been inserted as "child" of the multi-target "game object" to correctly register the AR cutting curves with the physical mandible. Thanks to the multi-target, the cutting lines' projection is adjusted based on the detected marker on the cube.
4. Build (Unity): the developed project must be built and compiled to be run in Visual Studio.
5. Deployment (Visual Studio): the deployment allows distributing the application onto HoloLens 2 through Visual Studio. The smart-glasses need to be connected to the laptop via USB-C. Once the pairing is completed, the deployment begins, and then the AR application can be experienced through Hololens 2.

3 Preliminary Laboratory Validation of the AR Application

Preliminary laboratory validation of the AR application has been performed to assess the system's accuracy and precision and understand if it is usable during different maxillofacial surgeries.

3.1 Trial Participants and Protocol

Seven participants (four students of Biomedical Engineering, one Engineering Professor, one Ph.D. student in Industrial Engineering, and one research fellow in Industrial Engineering) without experience in surgical procedures and familiarity with extended reality technologies were involved in the trial. Figure 2 shows a participant performing the trial.

Fig. 2. A Participant performing the trial

Each participant followed the defined protocol:

- To wear HoloLens 2.
- To perform eye calibration on HoloLens 2 to improve how the user sees the holograms.
- To have a 2-min warm-up session to take confidence with the use of Hololens 2.
- To start the AR application from HoloLens 2.
- To look at the multi-target on the 3D printed mandible.
- To wait until the holograms of the cutting lines are displayed.
- To draw the visualized cutting lines directly on the 3D printed mandible with a pencil.
- To highlight the final drawn lines with a red felt pen.

The 3D printed mandible was covered with white adhesive tape to allow all participants to sketch surgical curves over the mandible directly. It was fixed over a box to simulate the height of the operating table.

The users were asked to perform the task by moving around the mandible and tracing the lines when they felt more comfortable (i.e., when they could better visualize the lines from any point of view). The users were also free to trace a line in a specific position and then move, stand, and trace the other lines.

Each participant repeated the trial twice to verify eventual improvements. The execution time was recorded for each test.

3.2 Evaluation Procedure

For each participant and for each trial, the AR system's accuracy has been assessed by comparing the drawn cutting lines with those previously designed and projected by HoloLens 2. In particular, the assessment's steps can be summarized as follows:

1. 3D scan of the mandible with the drawn cutting lines. The 3D printed mandible has been scanned using Go!SCAN 3D (by Creaform), a white-light scanner with high accuracy (0.05 mm) and resolution (0.10 mm) able to acquire also textures and colors (and thus the drawn cutting lines). A.vrml file was saved and exported.
2. CAD reconstruction of the drawn and scanned cutting lines. The reverse engineering process to reconstruct the three-dimensional surgical guides (thickness of 1 mm) was performed in Rhinoceros.
3. 3D comparison between the cutting lines initially designed and the ones sketched using HoloLens 2 (reconstructed in Rhinoceros 3D). Using CloudCompare, the two 3D models have been aligned and compared, setting a maximum distance of 10 mm.

4 Results and Discussion

The results analysis concerns, on one side, evaluating each participant's overall performance in terms of mean deviations calculated over the five cutting lines, and duration of the trials (thus observing eventual improvements between trials 1 and 2, in terms of precision and timing), and, on the other side, assessing the deviation for each of the three maxillofacial surgeries (genioplasty, orthognathic surgery, and prominent mandibular angle reduction).

Results in Table 1 show that the overall mean deviation between the designed and the drawn cutting lines is below 2 mm (except for one trial) for the five surgical guides. The duration of the tests is between one and five minutes. Thus, it can be asserted that such technology is fast and easy to learn even for users without experience. Furthermore, over 70% of participants took less time to accomplish the second trial than the first one, with a greater accuracy in 57% of cases. This result confirms that a few practices are required to become confident with HoloLens 2 and augmented reality technology. Feedbacks from participants agree with this consideration.

Then, the different cutting lines have been separately considered to assess the feasibility of the AR application in the genioplasty (frontal cutting line on the chin), orthognathic surgery (two middle-lateral cutting lines for mandibular advancement), and mandibular angle osteotomy (two retro-lateral cutting lines for prominent mandibular angle reduction).The Pareto chart has been calculated for the three different cutting lines to analyze the deviations at 75% of the total number of occurrences (last three columns in Table 1). Satisfactory results have been obtained for genioplasty and orthognathic surgery. Instead, the deviations related to the mandibular angle reduction's cutting lines are not acceptable for applying the AR system during the surgery. Indeed, in oral and maxillofacial surgery, the accuracy is considered acceptable if it is between 1 and 2 mm [9]; 1 mm errors at the skeletal level are not visible at the cutaneous level.

Table 1. Trials results: mean deviations, timing, and 75% deviations from the Pareto chart for genioplasty, orthognathic surgery, and prominent mandibular angle reduction.

User	#Trial	Mean deviation [mm]	Time [min]	Genioplasty [mm]	Orthognathic surgery [mm]	Mandibular angle osteotomy [mm]
User 01	Trial 1	1.83 ± 1.58	4:00	1.69	1.66	3.88
	Trial 2	1.33 ± 1.05	4:23	1.99	1.00	2.77
User 02	Trial 1	1.07 ± 0.84	3:00	1.00	1.28	2.09
	Trial 2	1.19 ± 0.87	1:43	2.51	1.26	1.50
User 03	Trial 1	3.26 ± 1.95	3:26	0.62	1.35	4.93
	Trial 2	1.44 ± 1.19	2:06	0.53	1.67	3.28
User 04	Trial 1	1.00 ± 0.77	1:25	0.51	1.78	1.82
	Trial 2	0.95 ± 0.71	1:06	0.59	1.74	1.65
User 05	Trial 1	1.22 ± 1.11	1:00	1.25	0.43	2.96
	Trial 2	1.49 ± 1.21	2:20	1.56	0.97	3.38
User 06	Trial 1	1.54 ± 1.28	2:43	0.55	2.44	3.29
	Trial 2	0.70 ± 0.83	1:42	0.38	0.68	2.30
User 07	Trial 1	0.52 ± 0.58	2:20	0.58	0.39	1.29
	Trial 2	1.00 ± 0.78	1:31	0.69	1.12	2.18
Mean			**2:20**	**1.03**	**1.27**	**2.67**

Moreover, it is impossible to compare the results with scientific literature because previous research concerning the adoption of Hololens 2 in maxillofacial surgery is missing. However, the obtained mean deviations compared with the ones of other tests available in literature, are enhanced respect to those obtained with HoloLens 1 [9–11] and comparable with those obtained using other HMDs [1, 6].

5 Conclusions and Perspectives

The paper proposes an innovative use of augmented reality with HoloLens 2 in maxillofacial surgery. An AR application with the surgical guides suitable for genioplasty, orthognathic surgery, and mandibular angle osteotomy have been designed and developed to be used with HoloLens 2 and tested with the 3D printed model of a mandible.

Seven inexperienced participants were enrolled in the preliminary validation. They were asked to draw the AR surgical guides over the mandible. Good accuracy has been reached for the genioplasty (deviation around 1.03 mm) and orthognathic surgery (deviation around 1.27 mm), suggesting the applicability of HoloLens 2 for these kinds of surgery. On the contrary, the application was not suitable for the mandibular angle osteotomy (deviation over 2.50 mm). Subjects performed the tests without any experience in surgery. Therefore, the calculated deviations include also the error related to the lack of knowledge about surgical procedures. The opportunity to use AR in maxillofacial surgery would lower the surgery time, and the production costs of the physical surgical guides, leading to considerable long-term savings.

Future work will include the design of additional visual aids. The right cutting inclination, or the right mandible reposition could be highlighted, giving interactive real-time visual feedback about the correctness of the action performed. The validation will be performed with a greater sample of users, and, also, in the real clinical setting, to analyze other aspects such as the accuracy of the system in the operating room (with real conditions in terms of blood presence, etc.) and the invasiveness of the multi-target.

References

1. Badiali, G., et al.: Augmented reality as an aid in maxillofacial surgery: validation of a wearable system allowing maxillary repositioning. J. Cranio-Maxillofacial Surg. **42**(8), 1970–1976 (2014)
2. Scolozzi, P., Zulma, C.: Applications de la chirurgie maxillo-faciale assistée par ordinateur. Rev. Med. Suisse **15**, 1226–1230 (2019)
3. Vávra, P., et al.: Recent development of augmented reality in surgery: a review. J Healthcare Eng. **2017**, 1–9 (2017)
4. Lin, L., et al.: Mandibular angle split osteotomy based on a novel augmented reality navigation using specialized robot-assisted arms: a feasibility study. J. Cranio-Maxillofacial Sur. **44**(2), 215–223 (2016)
5. Lin, Y.K., Yau, H.T., Wang, I.C., Zheng, C., Chung, K.H.: A novel dental implant guided surgery based on integration of surgical template and augmented reality. Clin. Implant Dent. Relat. Res. **17**, 3 (2015)
6. Jiang, T., et al.: Precision of a novel craniofacial surgical navigation system based on augmented reality using an occlusal splint as a registration strategy. Sci. Rep. **9**, 501 (2019)
7. Frantz, T., Jansen, B., Duerinck, J., Vandemeulebroucke, J.: Augmenting Microsoft's HoloLens with Vuforia tracking for neuronavigation. IET J. **5**(5), 221–225 (2018)
8. Itamiya, T., Iwai, T., Kaneko, T.: The holographic human for surgical navigation using Microsoft's HoloLens. EPiC Ser. Eng. **1**, 26–30 (2017)
9. Meulstee, J.W., et al.: Toward holographic guided surgery. Surg. Innov. **26**(1), 86–94 (2019)
10. Rae, E., Lasso, A., Holden, M.S., Morin, E., Levy, R., Fichtinge, G.: Neurosurgical burr hole placement using the Microsoft HoloLens. In: Conference: Image-Guided Procedures, Robotic Interventions, and Modeling (2018)
11. Mitsuno, D., Ueda, K., Hirota, Y., Ogino, M.: Effective application of mixed reality device HoloLens: simple manual alignment of surgical field and holograms. Plast. Reconstr. Surg. **143**(2), 647–651 (2019)

CompassbAR: A Technique for Visualizing Out-of-View Objects in a Mixed Reality Environment

Alessandro Evangelista(✉), Vito Modesto Manghisi, Francesco Laera, Michele Gattullo, Antonio Emmanuele Uva, and Michele Fiorentino

Department of Mechanics, Mathematics, and Management, Polytechnic University of Bari, Bari, Italy
alessandro.evangelista@poliba.it

Abstract. The introduction of the new generation of Head Mounted Displays (HMD) makes users' experiences in Mixed Reality (MR) environments more engaging. However, these devices still have a limited field of view, which negatively affects the spatial localization process of virtual objects in the 3D environment. The literature presents several visualization techniques to address this issue, but they currently have several drawbacks, such as visual clutter, occlusion of the real scene, high user workload, and there is still no visualization technique that solves such issues definitively. Therefore, inspired by the gaming industry, we present CompassbAR a visualization technique for out-of-view objects. CompassbAR encodes the position of all out-of-view objects surrounding the user, in a 2D bar positioned at the top of the field of view. In addition, we propose a validation procedure and metrics that aim to evaluate the ability of the CompassbAR visualization technique to guide users towards the out-of-view objects.

Keywords: Mixed Reality · Visualization techniques · Head-mounted display

1 Introduction

In the last two decades, Mixed Reality (MR) technologies achieved significant growth in navigation, interaction, and collaboration application for different types of environments, both real and virtual. The popularity of these technologies has been increased by the latest generation of Head Mounted Displays (HMDs) that offers more engaging user experiences thanks to technological advances such as increased computational power, improved displays, and better wearability. However, the development of these devices represents a constant technological challenge for manufacturers. Indeed, despite the recent advances, they still suffer from technical limitations that slow down the spread of these technologies in real application scenarios. One of the most critical weaknesses of HMDs is the limited Field of View (FOV), especially in Optical See-Through devices. For instance, the Oculus Rift (https://www.oculus.com/rift) one of the most used Virtual Reality (VR) HMD, has a horizontal FOV limited to about 90°, while the Microsoft HoloLens (https://www.microsoft.com/it-it/hololens), an Augmented Reality (AR) HMD, has an even narrower

C. Rizzi et al. (Eds.): ADM 2021, LNME, pp. 141–148, 2022.
https://doi.org/10.1007/978-3-030-91234-5_14

Horizontal FOV of about 40° [1]. Such values are significantly lower than the natural human FOV, which extends about 180° horizontally and 150° vertically.

The restricted FOV of MR systems limits their immersive capability, negatively affecting the situational awareness of the user. Indeed, augmented and virtual environments can theoretically be very complex to explore, as they can include a variety of virtual objects that are often hidden and located out of user's view. This issue is not trivial when the user needs to be conscious of the number and location of out-of-view objects in the environment, e.g., during a maintenance operation in a hazardous industrial scenario where virtual objects are used to signal the presence of hazards. Providing such information can improve the user's comprehension of the environment and allow for quick navigation with targeted search to points of interest, and thus improve user safety.

In the literature, various visualization techniques have been developed to improve situational awareness regarding out-of-view objects in both virtual and augmented HMD-based environments. Nevertheless, existing visualization techniques have numerous disadvantages, such as visual clutter, occlusion of the real scene, and high user workload, and there is still no visualization technique that solves such problems definitively.

Our contribution in this field of research consists in investigating how to represent out-of-view-objects in HMD User Interfaces (UI) and, inspired by commercial video games such as Call of Duty Modern Warfare [2], The Elder Scrolls V: Skyrim [3] and others, we designed and implemented a customized Compass Bar solution that can be used both in AR and VR environments. The wide use of this visualization technique in the videogames industry has led us to think about its effective application to real-life scenarios of industry, military, automotive, and so on, through the MR technology. The CompassbAR technique consists of a graduated bar (in degrees) at the top center of the UI. Each virtual object within the environment is displayed on the graduated bar as a distinct, color-coded proxy icon. In order to provide the user with information about the distance of the virtual object from his/her position, we implemented a specific algorithm that increases the size of the inner point in the center of the annulus as the user moves closer to the virtual object.

After designing and developing our customized solution, we defined an experimental procedure in order to evaluate the effectiveness of such solution by comparing it with the other ones that represent the current state of the art techniques for visualizing out-of-view objects. As soon as the rules on social distancing imposed for fighting the COVID 19 pandemic will allow it, we will carry out this user study.

In the remainder of this work, first we present the current state of the art in this field, then we describe our design approach and the proposed solution, and finally we describe the design of the experiment we planned for the user study.

2 Related Work

In the literature, there are several visualization techniques to display out-of-view objects in MR environments; they can be grouped into three main classes: Overview+Detail, Focus+Context, and Contextual Views [4, 5].

The Overview+Detail technique uses two separate spaces in the UI layout to convey to the user the presence and the position of the out-of-view objects. The first space is dedicated to the Overview window, which typically conveys information about the user's surroundings through a minimap, with proxy icons inside it, indicating points of interest. The second space is used as a detail window, which is typically the user's first-person view of the environment. The main drawback of this technique is the cognitive load required by the user to mentally assimilate both windows (Overview and Detail). The most representative example of this technique is the 2D/3D Radar. This representation technique of the surrounding environment through a minimap is widely used in the videogame industry.

The Focus+Context technique uses a window, typically the user' first-person view of the surrounding environment, with a fisheye distortion of the whole off-screen space [6] or just the points of interest from the off-screen space. In this visualization technique, the context information is positioned along the edges of the window; this allows context information to better integrate with the human frame of reference. The drawback of Focus+Contest is that the user loses information about the proportions of the 3D space, due to the compression of information about objects in the entire 3D space, on the edges of the window. Thus, the user cannot perceive the relative distance among the off-screen objects. AroundPlot [7] and EyeSee360 [8] are two visualization techniques that adopt the Focus+Context approach. AroundPlot uses a window delimited by a rectangle to define the focus area. The whole 3D space surrounding the user is mapped outside of the rectangle using 2D orthogonal fisheye coordinates. If there are many objects in the 3D space, this technique suffers from an object density problem at the four corners of the rectangle. EyeSee360 is a visualization technique based on a grid system between two ellipses. The inner ellipse contains the focus window, which is the user's field of view, and in the area between the outer ellipse and the inner ellipse is mapped the out-of-view 3D space. The horizontal line of the grid represents the altitude of the virtual object, while the vertical curved lines represent the horizontal direction of the virtual object. Unfortunately, EyeSee360 has the disadvantages of having a high user cognitive load and cause clutter to devices with limited fields of view.

The Contextual Views technique uses a detail window and exploits visual indicator (mainly arrows) pointing to an object placed in the off-screen space. In the literature, there are three main solutions belonging to this technique: City Lights, Halo, and sidebARs. City Lights [9] was one of the first techniques to use Contextual Views by highlighting contextual information along the edges of the screen to indicate out-of-view objects. Subsequently, Halo [10] was presented whit an improvement on the previous technique, drawing an intersecting circle on the screen, which encodes the distance from the out-of-view object, with the radius of a circle having the center around the object. Other Contextual Views technique implementation uses other visual indicators such as spheres [11] (in the case of 3D) or arrows [12] (2D and 3D) for pointing towards the out-of-view objects. All the latter techniques have the disadvantage of clutter, which is the accumulation of many visual indicators (halos or arrows) in the corners. The literature pointed out that arrows with fixed dimensions perform worse than Halo, while arrows with scale dimensions (which depend on the distance of the out-of-view object) perform slightly better [13]. Finally, the sidebARs technique [14] uses vertical bars at the right and left edge of the window, inside of which there are proxy icons that show the position and distance (through a numeric indicator below the proxy icon) of out-of-view objects.

The high number of out-of-view objects and the proximity of each other has a negative impact on the performance of this technique in terms of disorder and occlusion of the scene.

3 Our Approach

We investigate how to allow users to be rapidly aware of the number and position of out-of-view objects in a 3D environment using an HMD. We propose the CompassbAR technique.

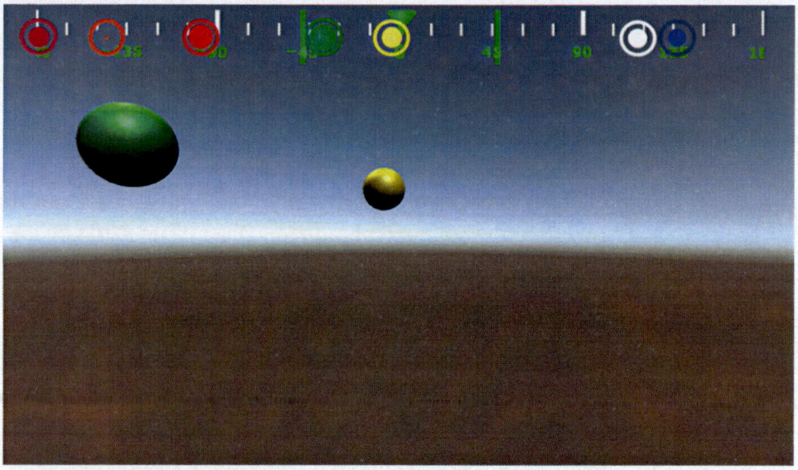

Fig. 1. The CompassbAR visualization technique.

This visualization technique encodes the position of all out-of-view objects surrounding the user, in a 2D bar as shown in Fig. 1.

The choice to use a 2D visual metaphor is based on previous works by Gruenefeld et al. [15] that showed how 2D techniques to visualize out-of-view objects can be used successfully in HMDs. Given the limited space required in the UI, we hypothesize that this visualization technique allows mitigating the typical occlusion issues of the visualization techniques to display out-of-view objects.

In order to evaluate such a capability together with the effectiveness of the proposed solution and its user acceptance, we designed a validation procedure based on a custom user study.

3.1 The CompassbAR Solution

The CompassbAR is mainly constituted by a graduated bar (values expressed in degrees) at the top center of the UI. Bar values range from −180 (extreme left) to +180 (extreme right), the zero value is located in front of the user. The user' FOV width is depicted by two vertical lines on the bar (which can be calibrated according to the FOV of the

HMD used), as shown in Fig. 2. Every single object within the virtual scene is depicted onto the compass bar as a specific color-coded proxy icon (an annulus with a colored dot at its center). We hypothesize that the use of a graduated bar allows the user to immediately identify the position of the object, knowing in advance if the object is on his right or left and how many degrees with respect to the center of his gaze. For example, if CompassbAR is displaying the proxy icon at about +90, users will be aware that the virtual object is to their right (to their left if the value is displaying at about −90), if it is displaying the proxy icon at about +180 or −180, users will be aware that the virtual object is behind them, and so on. Consequently, by turning the head left or right, the proxy icon position is updated in real time resulting in an immediate feedback for the user.

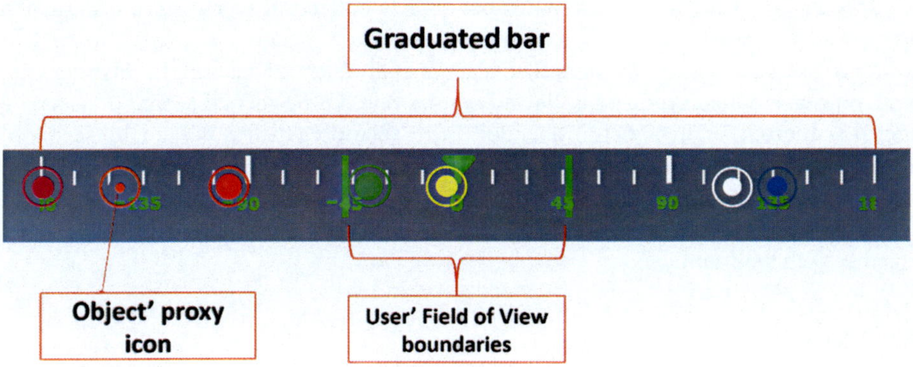

Fig. 2. The CompassbAR components.

The algorithm evaluates the position of the proxy icon on the CompassbAR by assessing, in a plane parallel to the floor, the angle between the vector connecting the object to the player and the player forward direction.

To provide the user with the correct information about the object's distance from his/her location, we designed a visual feedback consisting of an internal dot in the center of the annulus that increases in size as the user approaches the virtual object (when the distance between the annulus and the user is minimal, the dot completely fills the annulus). In order to manage objects that are close together we allowed the proxy icons to overlap. By doing so, if the objects are aligned with respect the user, only the proxy icon of the nearest one is visible on the CompassbAR.

We have developed CompassbAR on the multi-platform graphics engine Unity 3D (https://unity.com) in order to support various VR/AR HMD devices such as Oculus Rift and Microsoft HoloLens.

3.2 The Validation Procedure in a Real Industrial Scenario

Our validation procedure aims at evaluating the capability of the CompassbAR visualization technique for visually guiding the view of HMD users towards out-of-view objects. Similarly to [16], we designed a user study to compare the CompassbAR technique with the state of the art solutions in the literature: sidebARs, 3D Radar, EyeSee360, 3D Arrows, AroundPlot, and Mirror Ball. We plan to conduct the user study using two HMD devices: the Oculus Rift and Microsoft HoloLens. All the scenarios we will use for the user study are comparable to the goal-oriented visual search tasks that we can find in real application scenarios.

A typical industrial application scenario, where CompassbAR can be applied, is that of logistics. In fact, HMDs are of particular interest because they offer a hands-free solution, and operations such as the storage and picking of objects can benefit from this display devices as well as our visualization technique. The aim is to enable picking and storage that is fast, error-free and allows human operators to simplify their work by lowering the overall cognitive load.

Specifically, there will be three distinct objects-collection scenarios in a simulated warehouse operation, for example display the storage location of items. In the first scenario, users are asked to accomplish a sequential object collection. In this scenario, eight objects are displayed and collected one-by-one. In the second scenario, users are asked to collect the eight objects in a random order according to their will. In this scenario, the objects hints are displayed simultaneously, and the user explores the scenario trying to collect target objects as quickly as possible. In the third scenario, users are asked to collect the virtual objects in a specific order meanwhile they are simultaneously displayed together with their hints.

3.3 Metrics

As our user study aims at evaluating the CompassbAR solution effectiveness, we will evaluate three different aspects: the task completion time, the user acceptance, and the cognitive load associated with each visualization technique under test.

As regards the completion time, we will measure the average time for each object collection and the overall time.

As regards user acceptance, after each user test, we will administer the System Usability Questionnaire (SUS) [17]. Furthermore, we will carry out users' interviews to identify specific pros and cons of each visualization technique such as occlusions, clutters, and physical fatigue.

As regards the cognitive load, we will apply two assessing methodologies. The first one will be based on the National Aeronautics and Space Administration – Task Load Index (NASA-TLX) questionnaire [18] that will be administered after each user test. The second one is based on the monitoring of the human-body parameters, under the assumption that when the cognitive load changes, physical changes are observed in the user [19]. In particular, our evaluation will be based on the Heart Rate Variability (HRV) analysis that will be carried out in the time domain by recording the ECG-signal during the user experience and by extracting the NN-tachogram. We will assess the Heart Rate (HR) and the square root of the mean squared differences of successive NN intervals (RMSSD) as such parameters are effective indicators of a stress condition induced by increased cognitive load [20].

4 Conclusion

In this work we presented the main features of the CompassbAR, a novel visualization technique for out-of-view objects that we designed inspired by some solutions belonging to the field of gaming. Our ultimate goal is to bring these solutions into the industrial field, thus improving the effectiveness and usability of new technologies to supporting the operator's renewed role in Industry 4.0 scenario [21]. Given the characteristics of the CompassbAR visualization technique, we expect that it can mitigate some of the common drawbacks of the other techniques and keep performance comparable in terms of search time and selection accuracy. Specifically, we hypothesize that the CompassbAR can lessen both the occlusion problem and the mental workload. In order to test such a hypothesis, we designed a specific validation procedure based on user tests. We illustrated the approach that we will use for the validation procedure with the respective metrics (objective and subjective ones). However, we do not believe that our solution is issues free. Given the high condensation of information in the UI, it could suffer from an increased clutter problem. We believe that the user study we planned, will be able to evidence the pros and cons of our solution. Furthermore, given the comparative design, it will return useful information regarding the best visualization technique to apply for specific application scenarios.

References

1. Xiao, R., Benko, H.: Augmenting the field-of-view of head-mounted displays with sparse peripheral displays. In: Proceedings of the Conference on Human Factors in Computing Systems, pp. 1221–1232 (2016). https://doi.org/10.1145/2858036.2858212
2. Call of Duty 4: Modern Warfare. https://www.callofduty.com/it/modernwarfare. Accessed 09 Sep 2020
3. Skyrim: The Elder Scrolls|Skyrim. https://elderscrolls.bethesda.net/it/skyrim. Accessed 09 Sep 2020
4. Cockburn, A., Karlson, A., Bederson, B.B.: A review of overview+detail, zooming, and focus+context interfaces. ACM Comput. Surv. **41**(1), 1–31 (2008). https://doi.org/10.1145/1456650.1456652
5. Gustafson, S., Baudisch, P., Gutwin, C., Irani, P.: Wedge: clutter-free visualization of off-screen locations. In: Proceedings of the Conference on Human Factors in Computing Systems, pp. 787–796. ACM Press, New York, NY, USA (2008). https://doi.org/10.1145/1357054.1357179
6. Sarkar, M., Brown, M.H.: Graphical fisheye views. Commun. ACM **37**, 73–83 (1994). https://doi.org/10.1145/198366.198384
7. Jo, H., Hwang, S., Park, H., Ryu, J.H.: Aroundplot: focus+context interface for off-screen objects in 3D environments. Comput. Graph. **35**, 841–853 (2011). https://doi.org/10.1016/j.cag.2011.04.005
8. Gruenefeld, U., Ennenga, D., Ali, A.E., Heuten, W., Boll, S.: EyeSee360: designing a visualization technique for out-of-view objects in head-mounted augmented reality. In: Proceedings of the 2017 Symp. Spat. User Interact., SUI 2017, pp. 109–118 (2017). https://doi.org/10.1145/3131277.3132175
9. Zellweger, P.T., MacKinlay, J.D., Good, L., Stefik, M., Baudisch, P.: City Lights: Contextual views in minimal space. In: Proceedings of the Conference on Human Factors in Computing Systems, pp. 838–839 (2003). https://doi.org/10.1145/765891.766022

10. Baudisch, P., Rosenholtz, R.: Halo: a technique for visualizing off-screen locations. Conf. Hum. Factors Comput. Syst. – Proceedings, pp. 481–488 (2003)
11. Trapp, M., Schneider, L., Lehmann, C., Holz, N., Döllner, J.: Strategies for visualising 3d points-of-interest on mobile devices. J. Locat. Based Serv. **5**, 79–99 (2011). https://doi.org/10.1080/17489725.2011.579579
12. Rai, B.: Wedge. In: Rai, B. (ed.) Practical Dentistry, pp. 26–26. Jaypee Brothers Medical Publishers (P) Ltd. (2007). https://doi.org/10.5005/jp/books/10647_8
13. Henze, N., Poppinga, B., Boll, S.: Experiments in the wild: public evaluation of off-screen visualizations in the Android Market. In: NordiCHI 2010: Extending Boundaries – Proceedings of the 6th Nordic Conference on Human-Computer Interaction, pp. 675–678. ACM Press, New York, NY, USA (2010). https://doi.org/10.1145/1868914.1869002
14. Dozeman, T.B.: Sidebars. The Pentateuch. xiii–xviii (2017). https://doi.org/10.2307/j.ctt1ggjhj2.4
15. Gruenefeld, U., El Ali, A., Heuten, W., Boll, S.: Visualizing out-of-view objects in head-mounted augmented reality. In: Proceedings of the 19th International Conference on Human-Computer Interaction with Mobile Devices and Services, MobileHCI 2017. Association for Computing Machinery, Inc. (2017). https://doi.org/10.1145/3098279.3122124
16. Bork, F., Schnelzer, C., Eck, U., Navab, N.: Towards efficient visual guidance in limited field-of-view head-mounted displays. IEEE Trans. Visualization. Comput. Graphics **24**, 2983–2992 (2018). https://doi.org/10.1109/TVCG.2018.2868584
17. Lewis, J.R.: The system usability scale: past, present, and future. Int. J. Hum. Comput. Interact. **34**, 577–590 (2018). https://doi.org/10.1080/10447318.2018.1455307
18. Şeker, A.: Using outputs of NASA-TLX for building a mental workload expert system. Gazi Univ. J. Sci. **27**, 1132–1142 (2014)
19. Charles, R.L., Nixon, J.: Measuring mental workload using physiological measures: a systematic review. Appl. Ergon. **74**, 221–232 (2019). https://doi.org/10.1016/j.apergo.2018.08.028
20. Mansikka, H., Virtanen, K., Harris, D.: Comparison of NASA-TLX scale, modified Cooper–Harper scale and mean inter-beat interval as measures of pilot mental workload during simulated flight tasks. Ergonomics **62**, 246–254 (2019). https://doi.org/10.1080/00140139.2018.1471159
21. Manghisi, V.M., Uva, A.E., Fiorentino, M., Gattullo, M., Boccaccio, A., Evangelista, A.: Automatic ergonomic postural risk monitoring on the factory shopfloor – the ergosentinel tool. Procedia Manuf. **42**, 97–103 (2020). https://doi.org/10.1016/j.promfg.2020.02.091

Cinematic Virtual Reality as a Rehabilitative Tool in Subjects Affected by Schizophrenia

Vito Modesto Manghisi[1](✉), Alessandro Evangelista[1], Antonio Boccaccio[1], Michele Gattullo[1], Michele Fiorentino[1], Domenico Semisa[2], Valeria Latorre[2], and Antonio Emmanuele Uva[1]

[1] Department of Mechanics, Mathematics, and Management, Polytechnic University of Bari, Bari, Italy
vitomodesto.manghisi@poliba.it

[2] Department of Mental Health, Azienda Sanitaria Locale Bari, Bari, Italy

Abstract. Although Virtual Reality Social Skills Training has proven its effectiveness in treating psychiatric disorders, this VR application field is still under-researched for two main reasons. The first one is the unavailability of low-cost VR technologies with sufficient computational capacity needed to render realistic Virtual Environments. The second one consists of the need for specialized VR application developers, usually far from the mental health research field. The recent diffusion of low-cost stereoscopic viewers and the introduction of easy and fast VR content authoring systems, such as Cinematic Virtual Reality (CVR), allow overcoming these limitations. CVR makes it possible to capture real scenes through 360 cameras, augment them with additional virtual objects, and finally immerse the user in these synthetic but highly immersive environments. We present the design and the features of the Entellect360 prototype -an innovative tool supporting the rehabilitation process of subjects affected by schizophrenia. It exploits CVR technology to create Virtual Environments aimed at the rehabilitation of psychiatric patients. The Entellect360 features allow for rehabilitation sessions and patient-performance data-collection even under conditions of social distancing. We also explain the experimental protocol and the validation procedure the prototype will undergo to assess its effectiveness.

Keywords: 360-degree virtual reality · Human-computer interaction · Mental health · Social skills training · Cognitive rehabilitation

1 Introduction

The 'core' characteristics of schizophrenia include cognitive deficits that affect daily life and combine each other resulting in high costs for the individual with schizophrenia, their family members, and society. Rehabilitative interventions aim at recovering patients' cognitive functions primarily related to daily activities. Currently, however, the results of such interventions are still limited, with a consequent increase in disability in subjects suffering from chronic pathology.

C. Rizzi et al. (Eds.): ADM 2021, LNME, pp. 149–156, 2022.
https://doi.org/10.1007/978-3-030-91234-5_15

The literature suggests that Virtual Reality (VR) could improve therapeutic approaches in different medical areas. Over the past decade, VR played a leading role in some works aimed both at studying and treating schizophrenia [1]. In this research field, VR tools allow proposing human behaviors and emotions in virtual contexts where subjects are immersed through a stereoscopic viewer and maintain the perception of being really involved. Simultaneously, the therapist can observe the patients' behavior and manipulate virtual elements presented to subjects in real-time according to the needs. Despite the encouraging results from the research field, some issues hindered the widespread use of VR technologies in medicines. The most relevant were the time-consuming Virtual Environments (VEs) authoring process, the high computational requirements VEs rendering, the unaffordable price, and the necessity of specialized VR developers. Nowadays, it is possible to leverage VR like never before, thanks to technological advances. High performance Stereoscopic viewers are available at affordable prices. VR scene authoring tools evolved, making the creation of VR scenes much easier and faster. The introduction of the Cinematic VR technology (from now on CVR) overcame the poor graphic quality of computer-reconstructed environments. CVR refers to a type of immersive VR experience where individual users can look around synthetic worlds at 360°, often with stereoscopic views. This technology allows capturing real scenes, increasing them with additional elements, thus creating a very realistic feeling of presence. The advantages of CVR become decisive when there is a need to create multiple VR environments, such as in the functional recovery of the subjects suffering from schizophrenia.

As far as the authors know, given the recent implementation of the CVR technology, to date, there is still no data in the literature on the use of such technology for Virtual Reality Social Skills Training (VR-SST) in subjects with schizophrenia.

This paper presents a project developed in collaboration with the Local Department of Mental Health of Bari, applying a transdisciplinary approach to promote the implementation and validate an innovative and original CVR-based VR-SST tool.

The pandemic situation at the time of writing further reinforces the importance of this project. Since the extraordinary social distancing measures applied worldwide to contain the spread of severe acute respiratory syndrome coronavirus 2 (Sars-CoV-2) affected both research and rehabilitation centers activities.

The proposed system overcomes these issues. By using VR technology, it is possible to both carry out the rehabilitation process and adhere to social distancing measures. By implementing a client-server architecture for data collection, the system allows carrying out the user test campaign outside the research laboratories without requiring researchers to access the rehabilitation centers where the VR-SST is administered.

2 Exploiting VR for Mental-Diseases Treatment

Thanks to the recent technological advances, the decreasing costs, and the increasing power of digital media, Mixed Reality technologies are experiencing a remarkable growth of applications, including industry, rehabilitation, and culture [2–5].

Among Mixed Reality technologies, VR has many advantages that make it particularly suited for mental-diseases treatment. One of these advantages is the capability to assess cognition, emotions, and behavior in an ecologically valid environment. The ecological validity of VR derives from the accurate presentation and control of dynamic perceptual stimuli. Indeed, VEs may provide valid assessments through presenting situations that combine the control of laboratory measures with the verisimilitude of everyday experiences [6]. Another significant added value of VR is the sense of presence (the psychological sensation of "being there") that individuals can experience in immersive VR environments [7]. Additionally, VR allows manipulating the environmental triggers that elicit distress in people with mental health problems, thus allowing customized treatments to train towards better management of individual difficulties. The VR capacity of exploiting the impact of perception and presence on emotional reactions has been reviewed by Diemer et al. [8]. All this can directly induce the phenomenon of generalization [9], a crucial element for psychosocial interventions for psychoses.

Given these properties, VR has been used in clinical settings to treat a range of cognitive, emotional, and motor problems in various psychological and psychiatric disorders. Table 1 reports the most relevant ones.

Table 1. Most relevant application of VR technology for mental disease

Mental diseases	Study topic	References
Agoraphobia	Virtual Reality exposure therapy effectiveness evaluation	[10]
Social anxiety and fear of public speaking	Virtual Reality exposure therapy effectiveness evaluation	[11]
Spider phobia	Virtual Reality exposure therapy effectiveness evaluation	[12]
Post-traumatic disorders	VR- job interview training	[13]
Schizophrenia	VR-based Social Skills Training	[14]

In particular, a few clinical trials compared VR-based Social Skills Training (VR-SST) to the classical rehabilitation approach and reported positive results [14]. A recent poll of 70 psychotherapy experts ranked VR and other computerized interventions at the top of interventions predicted to increase in use in the next ten years [15].

3 The Entellect360 Prototype Design

The Entellect360 prototype is an innovative tool to support the rehabilitation process of subjects affected by schizophrenia. The design process started with brainstorming among specialists of psychiatric patients' rehabilitation, VR technologies, and user performance evaluation and led to the identification of the following requirements:

1. Allow a simple process for creating VEs while minimizing authoring effort;
2. Provide a simple and easy to learn interaction system between patient and VE;
3. Collect significant performance data on the subjects in the rehabilitation process;
4. Analyze performance data remotely while respecting the rules of social distancing;
5. Validate the prototype effectiveness as a support to the rehabilitation approach;
6. Provide a specific knowledge in a research field scarcely investigated up to now – the VR and Human-Machine Interactions for psychiatric patients.

3.1 The Architecture

The architecture of the prototype is described in Fig. 1. The core of the application is the VE -a CVR scene- which allows the patients to immerse themself in the familiar environment of everyday experiences. Each VE provides an interaction system that allows users to select specific hotspots of interest for rehabilitation purposes previously defined by psychiatrists.

During each rehabilitation session, the system records task performance metrics. Data from each rehabilitation session is recorded locally and in a remote server.

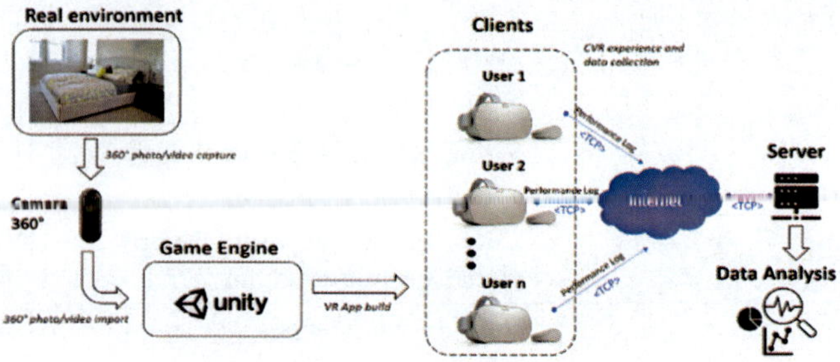

Fig. 1. The Entellect360 architecture.

The VE is entirely built using 360-photos or 360-video of a real environment (https://www.insta360.com/it/product/insta360-one), which has within it tasks that the patient must perform in daily life, such as personal hygiene care, tidying the house, and shopping. The stereoscopic head-worn display used for the rehabilitation program is the Oculus GO (https://www.oculus.com/go/), a low-cost standalone mobile headset based on Android OS.

The Entellect360 prototype was developed using Unity (https://unity.com/). The whole VE interaction system and client-server architecture were implemented in C#.

3.2 The User Interface

User Interactions

The classic approach to rehabilitation consists of bringing patients into the real environment in which they will interact by following the general "command" given by the rehabilitator. The Entellect360 prototype envisions a similar patient experience. By wearing the headset, patients are immersed in a VE corresponding to the rehabilitation domains; thus, they will interact with the VE, as it would in the real environment by following the same "command" from the rehabilitator.

In the early phase of the project, two different types of VE were implemented, both related to the rehabilitation of dysfunctionalities concerning the personal care and the management of daily living spaces: the bathrooms and the bedrooms.

The rehabilitation process aims to support the recovery of patients from dysfunctionalities by stimulating them to identify the correct elements they need to interact with. The implemented modality of rehabilitation involves two types of tasks: (i) research and identification of targets of interest (e.g., untidy bed, toothpaste for brushing teeth, shower gel, and so on); (ii) selection of targets.

The User Interface (UI) provides two different feedbacks according to the type of VE:

1. Bedrooms (management of daily living spaces) - the correct selection of a target triggers a visual feedback in which the VE is changed, e.g., by selecting the untidy bed it turns into the tidy bed (see Fig. 2a and 2b);
2. Bathrooms (personal care) - when a target is correctly selected (e.g., toothbrush or toothpaste), the UI displays a textual label that works according to the mechanism of positive reinforcement (see Fig. 2c and 2d).

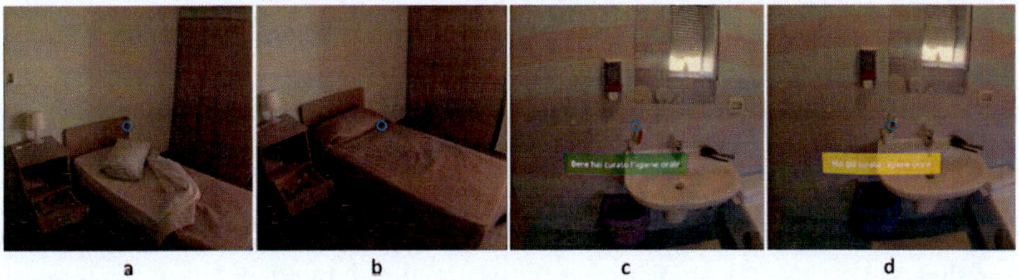

Fig. 2. An example of the visual feedback provided. For management of daily living spaces VE: 2a) pre-selection visualization and 2b) post-selection visualization. For personal care VE: 2c) correct target selection and 2d) multiple selection.

The target selection task consists of two subtasks: focusing on the target and selecting it. The Oculus Go system allows the implementation of two egocentric metaphors for focusing, both based on raycasting. Raycasting can be declined through two metaphors: controlling a gaze pointer through head (or eye) movements or moving a pointer (similar to a laser pointer) through a handheld device (i.e., a controller). We choose to implement a gaze-based UI as the users involved in this study use psychotropic drugs that can affect limb coordination and movement skills. The main element of gaze-based UI is the gaze pointer, where the user controls a pointer with head orientation and interacts with UI objects. Regarding the selection subtask, the metaphors that could be implemented include using either the "dwell" approach or a physical button. We discarded the "dwell" because the implementation of this metaphor is related to the definition of a timer to activate the selection, and, a typical timer value is not available for psychiatric patients. Thus, we implemented for the selection subtask metaphor a physical trigger on the handheld controller.

Rehabilitation Experience
Each rehabilitation session includes two preliminary scenes for recording the patient ID and the experimented VE (Fig. 3). Subsequently, the patient wears the VR headset can begin to interact with the VE. Interaction occurs through hotspots located on points of interest that are transparent to the patient, who must locate them autonomously. During the VE exploration, a crossfire visualizes the pointer, and patients can activate the selection on objects they consider target of interest.

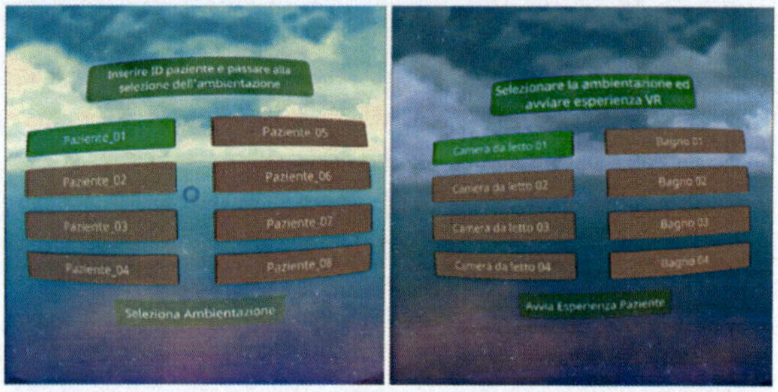

Fig. 3. The start menu composed of two scenes for data collection (Patient ID and VE name)

4 Experimental Protocol

The project is aimed at patients, of both genders, with Schizophrenic Spectrum Disorder according to DSM V. The study's protocol was approved by the Ethics Committee of the Host Institution (Prot. # 173521/2020 12/04/2020).

The inclusion criteria are age between 18 and 65, lack of moderate and severe intellectual disability, and absence of diagnosis of drug use. The test phase involves about 40 voluntary patients. All subjects received information concerning the study and the treatments. Patients are randomly split into two equal groups, the first group undergoing rehabilitation through the CVR-SST-based approach and the second group by the usual rehabilitation practice. The psychologists and doctors of the Department of Mental Health will follow the patient from the clinical point of view.

5 The Validation Procedure

The validation procedure aims to study: (i) the acceptability of the system to end-users; (ii) the effectiveness as a rehabilitation support tool.

Acceptability will be evaluated in terms of usability and user experience. After the rehabilitation session, each user will have to complete two different questionnaires: the System Usability Scale [16] and the User Experience Questionnaire aimed at evaluating also the hedonistic side of the user experience [17].

Effectiveness will be evaluated by analysis of user performance and of patients' dysfunctionality recovery. The user performance evaluation involves real-time collection of the following data: (i) experience duration; (ii) number of correctly selected targets; (iii) number of incorrectly selected targets; (iv) targets selection time; (v) number of times the target was focused before its selection. Evaluation of dysfunctionality recovery will be conducted in strict collaboration with the Local Department of Mental Health of the Metropolitan City of Bari. Psychologists will administer the VADO test [18] to the patients to assess the efficacy of the rehabilitation approach. All patients will be evaluated at the starting time and every month for three months.

6 Conclusion

We designed and implemented an innovative prototype of CVR-SST for psychiatric patients. Using CVR allowed a simplified procedure for VEs creation.

VEs allow experiencing rehabilitation scenarios that do not need to have a consistent spatial placement. Furthermore, with a low effort, it is possible to differentiate the characteristics of the scenarios where rehabilitation occurs. This has two advantages: confronting the patients with similar but not identical stimuli avoids the phenomenon of learning, VR overcomes the mobility limits related to social distancing measures. Together with the client-server architecture implementation, these features allowed us to start the experimental phase with the user test campaign despite the standing social distancing measures involving the research-laboratories lockdown and the inaccessibility of the rehabilitation centers.

Acknowledgements. We thank the psychiatric rehabilitation structures belonging to the EPASSS Foundation for the contribution to the experimental study.

References

1. Freeman, D.: Studying and treating schizophrenia using virtual reality: a new paradigm. Schizophr. Bull. **34**, 605–610 (2008)
2. Evangelista, A., Ardito, L., Boccaccio, A., Fiorentino, M., Messeni Petruzzelli, A., Uva, A.E.: Unveiling the technological trends of augmented reality: a patent analysis. Comput. Ind. 118 (2020). https://doi.org/10.1016/j.compind.2020.103221
3. Gattullo, M., Evangelista, A., Uva, A.E., Fiorentino, M., Boccaccio, A., Manghisi, V.M.: Exploiting augmented reality to enhance piping and instrumentation diagrams for information retrieval tasks in industry 4.0 maintenance. In: Bourdot, P., Interrante, V., Nedel, L., Magnenat-Thalmann, N., Zachmann, G. (eds.) EuroVR 2019. LNCS, vol. 11883, pp. 170–180. Springer, Cham (2019). https://doi.org/10.1007/978-3-030-31908-3_11

4. Bevilacqua, V., et al.: Design and development of a forearm rehabilitation system based on an augmented reality serious game. In: Rossi, F., Mavelli, F., Stano, P., Caivano, D. (eds.) WIVACE 2015. CCIS, vol. 587, pp. 127–136. Springer, Cham (2016). https://doi.org/10.1007/978-3-319-32695-5_12
5. Manghisi, V.M., Uva, A.E., Fiorentino, M., Gattullo, M., Boccaccio, A., Monno, G.: Enhancing user engagement through the user centric design of a mid-air gesture-based interface for the navigation of virtual-tours in cultural heritage expositions. J. Cult. Herit. (2018). https://doi.org/10.1016/j.culher.2018.02.014
6. Parsons, T.D., Courtney, C.G., Arizmendi, B.J., Dawson, M.E.: Virtual reality stroop task for neurocognitive assessment. In: MMVR, pp. 433–439 (2011)
7. Slater, M.: Presence and emotions. Cyberpsychol. Behav. **7**, 121 (2004)
8. Diemer, J., Alpers, G.W., Peperkorn, H.M., Shiban, Y., Mühlberger, A.: The impact of perception and presence on emotional reactions: a review of research in virtual reality. Front. Psychol. **6**, 26 (2015)
9. Shepard, R.N.: Stimulus and response generalization: a stochastic model relating generalization to distance in psychological space. Psychometrika **22**, 325–345 (1957). https://doi.org/10.1007/BF02288967
10. Bethencourt, J.M., Roca-Sánchez, M.J., Acosta, L., Villaverde, M.L., Gracia, R.: The combined use of virtual reality exposure in the treatment of agoraphobia. Actas Esp Psiquiatr. **43**, 133–141 (2015)
11. Anderson, P.L., et al.: Virtual reality exposure therapy for social anxiety disorder: a randomized controlled trial. J. Consult. Clin. Psychol. **81**, 751 (2013)
12. Shiban, Y., Brütting, J., Pauli, P., Mühlberger, A.: Fear reactivation prior to exposure therapy: does it facilitate the effects of VR exposure in a randomized clinical sample? J. Behav. Ther. Exp. Psychiatry **46**, 133–140 (2015)
13. Smith, M.J., et al.: Virtual reality job interview training and 6 month employment outcomes for individuals with schizophrenia seeking employment. Schizophr. Res. **166**, 86–91 (2015)
14. Tsang, M.M.Y., Man, D.W.K.: A virtual reality-based vocational training system (VRVTS) for people with schizophrenia in vocational rehabilitation. Schizophr. Res. **144**, 51–62 (2013)
15. Norcross, J.C., Pfund, R.A., Prochaska, J.O.: Psychotherapy in 2022: a Delphi poll on its future. Prof. Psychol. Res. Pract. **44**, 363 (2013)
16. Lewis, J.R.: The system usability scale: past, present, and future. Int. J. Hum. Comput. Interact. (2018). https://doi.org/10.1080/10447318.2018.1455307
17. Schrepp, M., Hinderks, A., Thomaschewski, J.: Construction of a benchmark for the user experience questionnaire (UEQ). Int. J. Interact. Multimed. Artif. Intell. (2017). https://doi.org/10.9781/ijimai.2017.445
18. Morosini, P., Magliano, L., Brambilla, L.: VADO Valutazione di Abilità e Definizione di Obiettivi. Trento Ed. Erickson, 104–109 (1998)

Mixed Reality in STEM Didactics: Case Study of Assembly Drawings of Complex Machines

Enricoandrea Laviola(✉), Michele Gattullo, Antonio Boccaccio, Alessandro Evangelista, Michele Fiorentino, Vito Modesto Manghisi, and Antonio Emmanuele Uva

Department of Mechanics, Mathematics, and Management, Polytechnic Institute of Bari, via Orabona, 4, 70125 Bari, Italy
enricoandrea.laviola@poliba.it

Abstract. Mixed Reality (MR) could help students in the understanding of complex concepts as well as increase their motivation in the learning process. In this work, our aim is to propose a MR application for the support of engineering students in the understanding of assembly drawings of complex machines. We presented the application of our design methodology for this case study. Then, based on the results of a user study with a sample of students, we tried to improve the usability and the user experience of the MR application, proposing an updated version. The usability of the revised application was in the range "good-excellent" (mean SUS score 77.0). We also presented the lessons learned in this case study, that can be a starting point for a renewal of consolidated didactic processes aiming at future application of MR in other STEM courses.

Keywords: Mixed Reality · Distance education and online learning · Augmented and virtual reality · Improving classroom teaching · Mobile learning

1 Introduction and Motivation

In the last thirty years, there has been numerous attempts of introducing Mixed Reality (MR) in various fields as industry, education, retail, cultural heritage, medical, and so on. However, we still see a rare use in everyday life, except for entertainment applications as social media and television. As to the education field, it could be very important to integrate traditional didactic with emerging technologies such as MR which could help students in the understanding of complex concepts as well as increase their motivation in the learning process [1, 2]. MR can provide added content especially for STEM (Science, Technology, Engineering and Mathematics) subjects which require the observation of real phenomena. This learning activity is traditionally accomplished through drawings, images, videos, practical activities, and applications in a real context. Especially in engineering, students are often involved in the understanding of topics close to the industrial context and far from the university reality. Examples are the representations of pumps and turbines in the fluid dynamic field or drawings of technological processes. Furthermore, university laboratories, if present, often are not easily accessible, such

C. Rizzi et al. (Eds.): ADM 2021, LNME, pp. 157–164, 2022.
https://doi.org/10.1007/978-3-030-91234-5_16

as due to ongoing renovation, lack of laboratory staff, and safety reasons. Thus, MR can help reduce this existing gap to bring students closer to the working environment through the possibility to augment traditional didactic material (e.g., drawings) to show additional useful information that would not be possible to view in classrooms. In this work, our aim is to propose a MR application for the support of engineering students in the understanding of the machine components in assembly technical drawings.

Despite many pilot studies and prototype were proposed in the education field [3–5], there are still few real applications of MR in schools and universities. One of the reasons is the resistance of teachers to the introduction of new practices in their consolidated processes, as pointed out in [6, 7]. However, the epidemiological emergency due to SARS-CoV-2 is offering new opportunities to effectively use new technologies to maintain a high level of teaching through Distance Education (DE). At this purpose, MR gives the possibility to have laboratory activities even with DE. In fact, as revealed by Clark [8], one of the disadvantages of DE is the lack of hands-on practical activities during laboratory lectures.

Mixed Reality is intended as the possibility to use both Augmented Reality (AR) and Virtual Reality (VR) in the same app [9]. AR allows the association between a virtual 3D representation of a product model and its 2D real representation on a printed drawing, usually in the form of the augmented book, as experienced by Cheng [10] and Lytridis et al. [5]. Drawings can also be enriched with every additional information (text, videos, pictures, animations) useful for self-learning, as made by Ali et al. [11] and Huerta et al. [4]. VR has been used for the observation of an experiment in a virtual laboratory in which students can also interact to replicate the tasks done in the experiment, as shown by Dalgarno et al. [12] and Petrakou [13].

MR has been proposed in the past for STEM didactics [4, 5, 10–13], but, as far as the authors know, never with an evaluation in a real context of DE, as the one we are experimenting with SARS-CoV-2 pandemic. Furthermore, in previous studies [14], teachers were less involved in the development of the application, with the consequence that even valid prototypes were hard to use for teachers in a real course. Therefore, we used a methodology that can be valid for all STEM disciplines. In this work, we detailed how this methodology was applied in our case study and how we improved the initial design considering the first evaluation of the usability of the MR application. Furthermore, we presented the lessons learned by this experience that can be useful for the use of MR in other STEM courses.

2 Case Study

The case study of our work refers to the discipline of "Methods for Technical Representation" of the bachelor's degree in Mechanics, Management, and Electrical Engineering of the "Polytechnic Institute of Bari." In particular, the application will help students in the understanding of assembly drawings of complex machines. Traditionally, this skill is acquired in the first years of the bachelor's degree, through laboratory activities where the teacher shows the corresponding components (e.g., bearings, screws, shafts) in a real machine and their representation in the assembly drawing.

2.1 First Implementation

We developed a mobile application using Unity3D, aiming to solve the issues related to this discipline. They derived from a focus group with teachers and interviews to students, who had already attended the course. Respect to showing virtual components on CAD interfaces, our MR application allows to associate them to their drawings on 2D printed assembly drawing. This is possible using natural feature tracking, as made in [15]. The application can be displayed on students' tablet or smartphones.

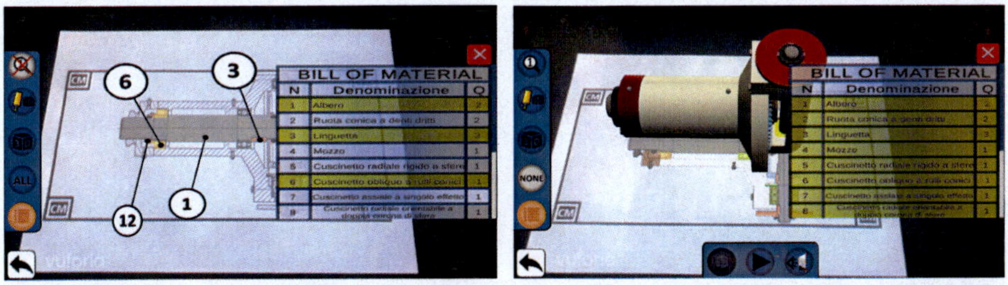

Fig. 1. From left to right: a) AR interface with the interactive BOM (on the right) and the "highlight" and "balloon" visualization; b) AR interface with the "3D model" visualization enabled with the possibility to play an animation or compare 2D with 3D.

The application presents an initial menu where students can choose the learning activity and the relative drawing to augment. Then, an AR interface appears. At the beginning no virtual elements are displayed on the drawing, but only GUI elements. The interactive Bill of Material (BOM) allows users to select the component to highlight the correspondent portion of the 2D technical drawing (see Fig. 1.a). The "Highlight" button allows to change the visualization mode of each component, for example, between 2D semi-transparent planes and 2D realistic identification. The "Balloon" button can display two possible versions of balloons directly on the drawing: the standard one with a component positioning number or that with a preview of the virtual 3D component. The "3D model" button shows 3D model of a single component selected in the BOM. The "List" button allows users to access a VR section of the app with additional components information. Only when all the components are highlighted through the "ALL" button, there is the possibility to display the 3D model of the entire machine and thus the "Animation" and "Compare" buttons appear (see Fig. 1.b). The "Animation" button allows students to play and pause animations, while the "Compare" button displays a half-Sect. 3D model of the entire machine with labels to compare with the 2D section on the drawing. The proposed AR interface makes use of button icons designed to promote "virtual affordance" as well as a great attention was paid to the filtering of information, formatting, and relative arrangement of content.

One of the most important students' needs is the integration of traditional didactic material in the application. For this purpose, we considered that AR is not needed, rather it could cause distractions and increase the mental effort in the using of the application. Then, we created a separated VR section. We did not use an immersive VR environment because it would be hard to use on student's mobile device, it would be rich of distracting graphics, and the authoring would be demanding for teachers. Therefore, we prefer using desktop VR [16], where users interact with virtual components using the touch screen of their mobile device. Students can pan, rotate and zoom the 3D model of a component as done in whatever 3D CAD software. This desktop VR section is accessible from the main menu and from the AR GUI and it allows an in-depth study on all types of components not limited to those visible in the chosen technical drawings (see Fig. 2.a). After having chosen the type of component, the user has access to all the variants of that component. They can be further elaborated through labels, animations, pictures, information filtering, and textual content (see Fig. 2.b).

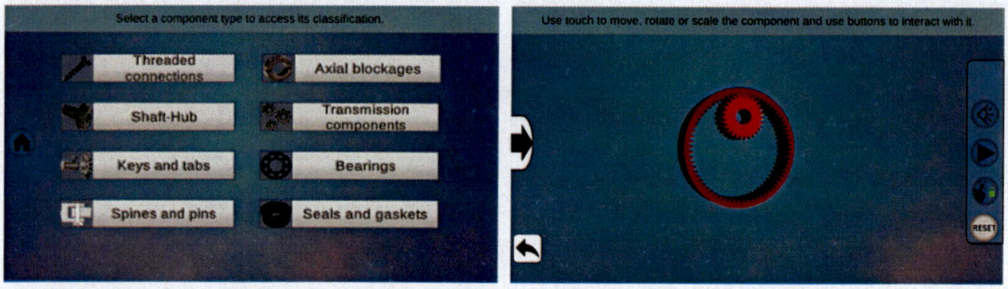

Fig. 2. From left to right: a) selection menu of the desktop VR section; b) GUI elements of the desktop VR section: a navigable 3D model of a component (in the middle), a hidden window with additional pictures (left), buttons for alternative visualizations of the component (right).

2.2 Results of the Usability Test

The initial design of the application, proposed by the course teachers, was evaluated through a user study involving 48 master's degree students (13 females, mean age 24.0 (SD 1.4)). They all were accustomed with the subject because they already passed the exam. In this way, they could provide useful feedbacks on what is really needed for the learning. They all had previous experience with AR applications in other disciplines. After a first training session in which we showed the main features, we asked them to explore all the functionalities of our application. This analysis was carried out through a questionnaire based on the System Usability Scale (SUS) whose results showed that the mean SUS score was 84.7 (SD 11.0, max 100.0, min 55.0) over a maximum of 100. Using the rating scale provided by Bangor et al. [17], we can define the usability of our application as "excellent".

The results from this user study allowed us to obtain students' feedbacks that could be useful to improve both the usability and the user experience of the MR application. We separated comments and issues from additional functionalities and tips that are more relevant for our study. Issues regarded the compatibility with students' own device experienced by four users and tracking lost beyond a certain distance for five users. We further distinguished recommendations into five topics: contents, interaction, GUI, portability, and tracking. Then we continued with an analysis of the extensions that we planned to do in short and mid-term, those that need further evaluation and those out of the learning scopes.

We classified as short term activities the possible implementations that we considered useful in order to ensure better learning and usability of the application. An additional suggested function is a video tutorial integrated directly into the application to provide users with information about the basic features of the system itself. Moreover, simulations of product kinematics and audio descriptions could help students to better understand how the mechanism works. Finally, the possibility to take screenshot could allow users to save pictures taken during their study session in the device to access them even without starting the application.

Then we indicated as mid-term activities those which have not been considered crucial because they do not affect learning performance. An example of a potential implementation consists of the auto-detection of multiple markers at the same time. However, this technical choice will be subjected to a further assessment because we experienced that in Unity3D multiple image targets are able to change only the augmented content and not the canvas related to them especially for the BOM. In addition, some students expressed the need to use a zoom lens in the AR scene to better observe the details of the mechanism. Another users' recommendation was about the device compatibility with iOS and not only with Android. We have planned to extend the accessibility of our prototype to multiple devices as soon as it will be used effectively during the course.

All other recommendations were considered too far from the MR application intent to support engineering students in the understanding of assembly drawings of complex machines. It is the case of the implementation of a more attractive interface with a realistic rendering of CAD models. This solution has been considered out of our scope because rich graphics could distract users from learning. Other activities not taken into account were about the possibility to assemble a product autonomously and to place it in a real context. Finally, the function regarding the use of an external link to access additional material was rejected by us because the necessary teaching content had already been integrated in the MR application as expressed by the teachers and students interviewed in the preliminary phase of the case study.

2.3 Second Implementation

An update of the MR application was also carried out to better meet the users' needs after analyzing their advice regarding the first version of the application. In fact, we added all the functions previously classified in short term activities: simulation of product kinematics, possibility to take screenshots which can be saved in a gallery scene, video tutorials, audio descriptions of products. In Fig. 3, the updated version of the AR scene interface is shown with buttons of new functions and a more ergonomic placement of an interactive bar, moved from the middle zone of the device screen to the left. This new disposition allows users to access all the functionalities of the application without having to perform hands movements not entirely user-friendly.

We made an evaluation of this second implementation through another user study, involving 36 bachelor's degree students (5 females, mean age 20.2 (SD 1.4)) that were currently attending the discipline of "Methods for Technical Representation." They had no previous experience with AR applications. The mean SUS score was 77.0 (SD 12.9, max 100.0, min 47.5) that is in the range "good-excellent" according to the Bangor et al. rating scale [17]. The lower result of the SUS score can be assigned to the low experience with both the subject and AR applications. However, it is a valuable result because users also said that they would use it during the preparation of the exam.

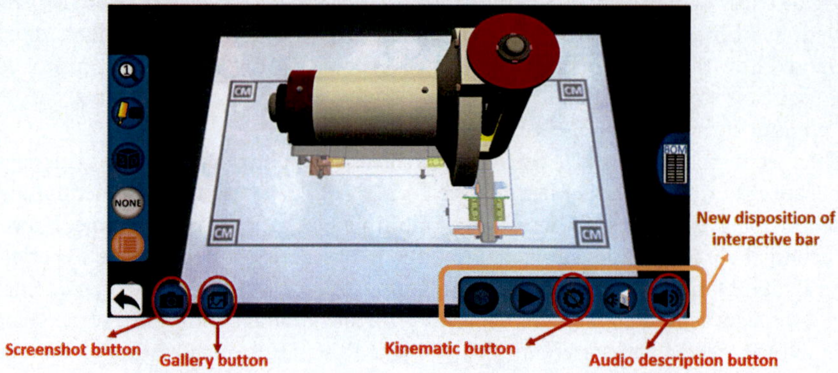

Fig. 3. GUI elements of the updated AR interface: a screenshot button and a gallery one, a new disposition of the interactive bar with kinematic button and audio description button.

3 Lesson Learned

Looking at the implementation described in the previous paragraphs, we can state that our case study is a real example of a MR application development process which could help other researchers to apply MR to other STEM disciplines. In fact, it is possible to extract some insights useful for future implementations, in particular for disciplines that make use of the association between real products and their drawings.

In the AR scene, for example, a great potential is offered by the association between a list of components (BOM in our case study) and different types of virtual highlighting on the real drawing for an immediate understanding of symbolic 2D representations and their locations. The identification of objects in the drawing is possible also through a labelling. We were inspired from the standard balloon with numbers used in technical drawings, but we implemented also a non-standard scalable balloon that shows a 3D model preview of the object. In this way, it is possible to transmit directly the information about the shape of the real object, facilitating students' learning. The use of desktop VR to create a section for the integration of the traditional didactic material, is another feature that can be applied to other courses. In future works, an immersive VR environment could also be considered exploiting VR headsets in the universities. However, factors as the use of distracting graphics and a more demanding authoring should be kept in mind.

We also designed the MR application, following a scalable approach for the interaction. In fact, there is not a preset learning path, but users interact with all the features with the combinations and the order they prefer. This technical choice allows the application to be scalable to students with different learning skills and also to different learning tasks. At this last purpose, we present how the same framework of our application was effectively applied to the course of "furniture design" of the master's degree in Industrial Design of the "Polytechnic Institute of Bari." The learning information provided in this context are related to the design and appearance of a sofa texture. The results of the implementation (see Fig. 4) allow us to say that the proposed functionalities as well as the interaction can be easily applied to other courses and are not limited to the engineering classes.

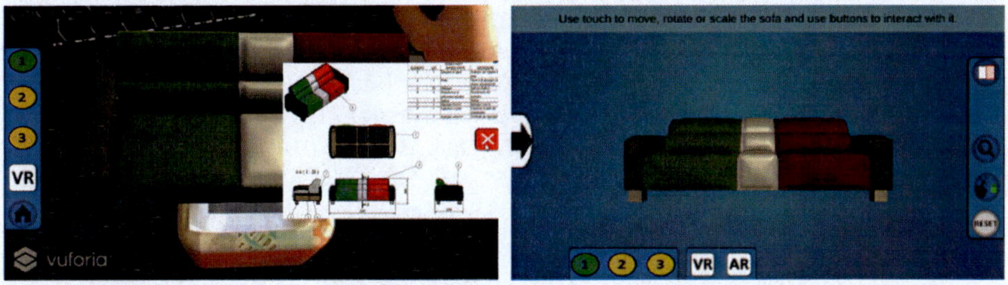

Fig. 4. Example about the using of the same framework of the developed application for another STEM discipline: screenshot of AR interface (left) and desktop VR interface (right).

4 Conclusion and Future Works

In this work, we proposed a case study for the use of MR in STEM disciplines. We presented the application of our design methodology for the understanding of assembly drawings of complex machines. We considered the results from a usability study with students, to improve both the usability and the user experience of the MR application, and to provide useful insights for future developments in other STEM courses. In fact, our case study is a real example of a development process which can be also applied for the continuous improvement of other STEM disciplines with learning activities involving the association between real objects and their drawings. In this way, it is possible to contribute to a renewal of the didactic system making it more compatible with future generations. The use of emerging technologies can contribute to a greater students' involvement through more interactive lectures than the traditional frontal ones.

In future works, we planned to perform a second user study to evaluate the learning effects of the updated MR application also involving students who have not yet attended the course. Our intent is to test as much as possible our MR application to be able to insert it as an official tool to make available to students for learning activities.

References

1. Petrov, P.D., Atanasova, T.V.: The Effect of augmented reality on students' learning performance in stem education. Inf. **11** (2020). https://doi.org/10.3390/INFO11040209
2. del Cerro Velázquez, F., Morales Méndez, G.: Augmented reality and mobile devices: a binominal methodological resource for inclusive education (SDG 4). An example in secondary education. Sustainability **10**, 3446 (2018). https://doi.org/10.3390/su10103446

3. Akçayır, M., Akçayır, G., Pektaş, H.M., Ocak, M.A.: Augmented reality in science laboratories: the effects of augmented reality on university students' laboratory skills and attitudes toward science laboratories. Comput. Hum. Behav. **57**, 334–342 (2016). https://doi.org/10.1016/j.chb.2015.12.054
4. Huerta, O., Unver, E., Aslan, R., Kus, A., Chotrov, D.: Application of VR and AR tools for technical drawing education. In: Proceedings of CAD 2019, pp. 363–366. CAD Solutions LLC (2019). https://doi.org/10.14733/cadconfP.2019.363-366
5. Lytridis, C., Tsinakos, A., Kazanidis, I.: ARTutor—an augmented reality platform for interactive distance learning. Educ. Sci. **8** (2018). https://doi.org/10.3390/educsci8010006
6. Sırakaya, M., Alsancak Sırakaya, D.: Augmented reality in STEM education: a systematic review. Interact. Learn. Environ. **0**, 1–14 (2020). https://doi.org/10.1080/10494820.2020.1722713
7. Viegas, C., et al.: Impact of a remote lab on teaching practices and student learning. Comput. Educ. **126**, 201–216 (2018). https://doi.org/10.1016/j.compedu.2018.07.012
8. Clark, J.T.: Distance education. In: Clinical Engineering Handbook, pp. 410–415. Elsevier (2020). https://doi.org/10.1016/B978-0-12-813467-2.00063-8
9. Speicher, M., Hall, B.D., Nebeling, M.: What is mixed reality? In: Proceedings of the 2019 CHI Conference on Human Factors in Computing Systems, pp. 1–15. ACM, New York (2019). https://doi.org/10.1145/3290605.3300767
10. Cheng, K.H.: Surveying students' conceptions of learning science by augmented reality and their scientific epistemic beliefs. Eurasia J. Math. Sci. Technol. Educ. **14**, 1147–1159 (2018). https://doi.org/10.29333/ejmste/81811
11. Ali, D.F., Omar, M., Mokhtar, M., Suhairom, N., Abdullah, A.H., Halim, N.D.A.: A review on augmented reality application in engineering drawing classrooms. Man India **97**, 195–204 (2017)
12. Dalgarno, B., Bishop, A.G., Adlong, W., Bedgood, D.R.: Effectiveness of a virtual laboratory as a preparatory resource for distance education chemistry students. Comput. Educ. **53**, 853–865 (2009). https://doi.org/10.1016/j.compedu.2009.05.005
13. Petrakou, A.: Interacting through avatars: Virtual worlds as a context for online education. Comput. Educ. **54**, 1020–1027 (2010). https://doi.org/10.1016/j.compedu.2009.10.007
14. Alvarez-Marin, A., Velazquez-Iturbide, J.A., Castillo-Vergara, M.: Intention to use an interactive AR app for engineering education. In: 2020 IEEE International Symposium on Mixed and Augmented Reality Adjunct (ISMAR-Adjunct), pp. 70–73. IEEE (2020). https://doi.org/10.1109/ISMAR-Adjunct51615.2020.00033
15. Gattullo, M., Evangelista, A., Uva, A.E., Fiorentino, M., Boccaccio, A., Manghisi, V.M.: Exploiting augmented reality to enhance piping and instrumentation diagrams for information retrieval tasks in industry 4.0 maintenance. In: Bourdot, P., Interrante, V., Nedel, L., Magnenat-Thalmann, N., Zachmann, G. (eds.) EuroVR 2019. LNCS, vol. 11883, pp. 170–180. Springer, Cham (2019). https://doi.org/10.1007/978-3-030-31908-3_11
16. Lee, E.A.-L., Wong, K.W.: Learning with desktop virtual reality: low spatial ability learners are more positively affected. Comput. Educ. **79**, 49–58 (2014). https://doi.org/10.1016/j.compedu.2014.07.010
17. Bangor, A., Kortum, P.T., Miller, J.T.: An empirical evaluation of the system usability scale. Int. J. Hum. Comput. Interact. **24**, 574–594 (2008). https://doi.org/10.1080/10447310802205776

Defining the Potential of Extended Reality Tools for Implementing Co-creation of User Oriented Products and Systems

Sandhya Santhosh[1(✉)], Francesca De Crescenzio[1], and Bonaventura Vitolo[2]

[1] University of Bologna, Forli, Italy
sandhya.santhosh2@unibo.it
[2] Geven S.p.A, Naples, Italy

Abstract. In the last two decades a huge number of interactive and collaborative applications of Virtual Environments for designing products has been proposed. Such applications have been recommended as tools to implement Human Centered Design Approach in experiments where potential users are involved in participatory design sessions before going for production. In this kind of experiments, we observe that users are mainly involved in the validation of solutions previously elaborated by designers while in the last decade the most innovative approach in the creation of solutions seems to be moving from a user centered design to a co-design or co-creation process. Thus, it is essential to have a platform where the elements of co-creation can be fulfilled in building a successful project. The purpose of this paper is to outline the concept of co-creation and the significance of co-creation platforms alongside of proposing innovative tools for building the co-creative environments. The paper aims to layout a classification of the Extended Reality (XR) tools currently available and of their functionality as valuable means to actually embed co-design and co-creation concepts in Virtual Environments evaluating the advantages that this can bring to Industry through field studies. A review of innovative solutions like Virtual, Augmented and Mixed reality technologies is examined and drawn towards the requirements of the concept through a literature research. Additionally, a co-creative environment for designing aircraft cabin interiors is conceived and discussed a with company representative.

Keywords: Co-creation · Extended reality · Collaboration · Interaction platforms

1 Introduction

In the recent times innovative organizations are shifting their ideation strategies towards the concept of "Co-Creation" in order to improve their creative businesses. Such process is based on the active involvement of final users or customers in the design process of products or services and can be implemented in whatsoever domain. In fact, according to Sanders and Stappers et al. there is no specific definition for co-creation and it can be adapted to the field it can be related and developed to [1]. With regards to new

C. Rizzi et al. (Eds.): ADM 2021, LNME, pp. 165–174, 2022.
https://doi.org/10.1007/978-3-030-91234-5_17

product, co-creation is believed to be the collaborative development of new value, the value being concepts, solutions, products, and services [2]. For product development, it is well known that early collaboration of various users is essential to work towards a shared goal reducing the time taken in completing the task. To this aim, co-creation meetings can be challenging due to diverse participants, be it designers or non-designers [3].

In this context, it is essential to identify the technological tools that allow users to go beyond the validation of already designed products and to play an active role in generating ideas and directly building solutions. Such tools must be capable of providing the key elements in developing co-creation platforms which are: Collaboration, Interaction and Experience [4]. How and where to collaborate, how to interact and what will be the experience are the basic questions every user would have when the concept of co-creation is introduced in product development phase.

To this note, Extended Reality (XR) technology, a term referring to all real-and-virtual combined environments & a rapid growing field, is paving its way in to numerous grounds. Furthermore, XR serves as an umbrella term for Virtual, Augmented and Mixed Reality technologies (VR, AR and MR) offering high potential in generating solutions to common constraints as collaborating and interacting platforms [5]. There are very few studies actually correlating the co-creation process with the conception of VR, AR and MR platforms as tools to successfully implement the process. Nevertheless, recent advancements in Head Mounted Display's (HMD's), Goggles, CAVE and mobile technology lead us to believe that companies will be facilitated in the actual evolution towards the co-creative approach.

In this paper, we review the virtual platforms available for building co-creating environments and their suitability towards the concept alongside of the hardware available and previous works performed using XR tools.

2 The Concept of Co-creation and Importance of Co-creation Platforms

Co-creation is still an emerging concept and is usually linked to business activities with customer playing a crucial role from beginning to end in a product or service design process. Prahalad and Ramaswamy et al. illustrates co-creation as interaction between companies and customers in order to co-create value experience with product and to deliver a unique co-created value [6]. These experiences signify the quality of interactions and lead to positive relationship between the customer and the company. Two famous companies which adapted customer co-creation and co-creation in new product development are LEGO [7] and BMW Group's Co-Creation Lab [8] respectively.

Mandolfo et al. has elaborated co-creation variables affecting consumers and companies [9]. In a study by Hitachi on "Co-creating the Future" [10], 61% of respondents have mentioned that adopting co-creative approach in their businesses to innovation has reduced the cost of developing products and services. Overall, co-creation is believed to bring together users and designers to work towards a shared goal. In fact, Prahalad et al. stated that introducing co-creation in to design practice will change the way we design and will also affect the tools and methods [6]. The study even relates co-designing and participatory design to co-creation [11].

The power of co-creation lies in to how well the users are collaborating and communicating in the environment [12]. Developing a co-creative environment or a platform has been unprecedented and has not been given much of importance till now. Designers, stakeholders, customers, engineers, suppliers, and other actors should interact on a common platform in order to adapt to the concept of co-creation and avoid misinterpretation and communication challenges. So, the experience they perceive also becomes a key factor in maintaining relationship and work collectively. Since co-creation can be regarded as a new way of practicing innovation within organizations in multiple fields of study, interactions and collaborative innovation experience factors play a vital role in the success of the concept. This recalls the need for social structures to support the quality of interactions in succeeding co-creation. A deeper definition to co-creation thus focuses on connections/interactions between people and collaboration rather than just involvement and creativity [13, 14].

An empirical study performed by Fuller et al. has highlighted the importance of co-creation experience and proves that it impacts user's quality of work [15]. Companies involved in co-creation bear the risk of less participants due to poorly managed collaboration platforms and consequently discouraging innovation process through valuable contributions. There are numerous existing technologies used as tools for co-creating environments. Most of them are driven to be web and network based to connect the users [14] such as: Crowdsourcing (a software or online platform with unrestricted access) and Netnography (webforms, vlogs, blogs etc.) [16]. A pattern which can be noted through these existing platforms is that collaboration and interaction between the users although happen in a same environment, user experience to the project is still limited.

With the growth of internet and innovative solutions towards communication technologies, there has been a lime light towards virtual platforms for developing co-creation environments. Kohler et al. explored virtual worlds for co-creation systems and highlights the importance of user experience to improve connections, creativity and thus co-creation [17]. It is not only bringing two or more users to perform an activity, but requires various types of interactions among the users, technological systems and environments.

Little research exists on how to design co-creation platforms for a user-centric experience and, to address this concern, we propose immersive environments for collaborating, interacting and user experience. Technologies such as Virtual, Augmented and Mixed Realities offer high potential to virtually collaborate, visualize and co-create in an imaginary unknown world. The following section will extend the types and definition to these technologies and explore the suitability towards the concept.

3 Potential Co-creation XR Platforms

Deep interactions with conversation and additional information introduces user to the product features that are more convincing. Thus, it is important to develop interaction-based platforms for co-creation environments. Extended Reality is such technology which offer high promising challenges to collaborate, interact, visualize, co-create/co-design. The technology have evolved to use avatars to offer appearances, gestures, voice and ability to interact with others using virtual artifacts [18]. XR can go beyond reality, providing users with elements that are not possible in the real world. With these objectives in consideration, the following section will provide an overview of the technologies and what the world is up to in developing the elements of co-creation with XR.

3.1 Virtual Reality

Virtual Reality (VR) is a complete 3dimensional digital experience. Unlike traditional interfaces, VR places the user inside an experience without any realization and contact with real world [19]. Numerous VR technologies based on display systems have emerged enabling a user to experience the virtual world [20] such as: Passive Stereoscopic Theatre (PST), Cave Automatic Virtual Environment (CAVE) [21], and finally, Head-Mounted Displays (HMD's) like HTC Vive and Oculus Rift.

Interacting with a virtual environment is the key constituent of many virtual world applications and co-creation. In order to improve the effectiveness of development projects, VR enabled-interactive objects are designed by Kostis and Ritala et al. to improve digital co-creation practices [22].

Interactions in the form of non-verbal communications (voice and gestural commands) have been successfully incorporated to VR environments. A set of hand-based tools such as wands, hand-controllers or other input devices such as goggles & gloves are used to manipulate virtual objects for the users to interact has been designed in developing collaboration in virtual reality [23].

A number of projects have been developed in VR which illustrate the elements of co-creation. For instance: *Coco Verse* is an immersive VR experience allowing interaction and collaboration for learning and creativity [24], an initial iterative design process in VR using Oculus Medium was developed by El-Jarn et al. for artists in order to collaborate with other creative leads on sketching slaying the 2D sketching step with a leap to 3D modelling [5], a unique social VR platform using HMD's like Oculus Rift has been explored by Saffo et al. for conducting user studies, *Facebook Spaces* is another VR app which allows interaction between friends using a VR device such as Oculus Rift and HTC Vive [26].

3.2 Augmented Reality

Augmented Reality (AR) unlike Virtual Reality isn't fully immersive, artificial experience. Instead, it adds digital information such as graphics, sound and video on top of an existing physical world. Augmented Reality can be implemented through various kinds of visualization devices such as Heads-up displays, Holographic displays, smart glasses and hand-held devices [27]. The excellent thing about AR is it does not always require dedicated hardware but also phones, tablets, computers can be utilized. With advent growth in technology, it can be predicted that phones, AR glasses and headsets will coexist in the future. Some of the popular AR HMDs are Microsoft HoloLens, Magic Leap One & Google Glass. Collaboration in Augmented Reality can be designed to be either co-located or remote.

There are several examples in the last two decades of collaborative AR environments. *The Studierstube* is an AR project using HMD's and allows multiple users to examine similar data in the same room [28]. Another such example is the use of Microsoft HoloLens by NASA's Jet propulsion Laboratory allowing the scientists to work on their Mars rover missions [29]. *Spatial* is one another augmented reality conference application with shared experiences for multiple remote users using HMD's [30]. Mourtzis et et al. developed AR collaboration for product design [31]. A complete remote collaboration has been designed by Zillner et al. in order to support the local helper by a remote expert [32]. Thus, interaction is considered smooth and easy in AR as it can be pre-designed across multiple devices.

3.3 Mixed Reality

Mixed Reality (MR) is an extension of Augmented Reality allowing real and virtual elements to interact in an environment. MR is capable to interact with and manipulate both physical and virtual items using next-generation sensing and imaging techniques. It provides the ability to have one foot (or hand) in the real world and the other in an imaginary place offering an experience that changes everything. It is also called as enhanced AR, as it is similar to AR but provides more physical interaction [33]. It requires an MR headset device such as Microsoft HoloLens or Magic Leap One.

Collaborative experiences in MR utilizes gestural and speech recognitions allowing multi user interaction between virtual and physical environment. There are many systems designed for this purpose. *SharedSphere* is one such system developed by Lee et al. allowing remote collaboration using See-through HMD, Epson BT-200 equipped with a camera and sharing captured video with remote user using a VR HMD, HTC Vive [34]. Bait et al. uses a similar strategy but captures live 3D panorama using a sensor cluster installed in the room. With the combination of gaze and gesture cues, the system has been proven to provide a stronger feeling of co-presence for both users [35]. *CoVar* is a platform supporting VR and MR technologies developed by Piumsomboon et al. where users can collaborate with same local real-world environment or remotely [36]. A recent illustration of *Microsoft Mesh* has broken all barriers demonstrating MR platform to be delivering collaborative experiences [37]. The application has been tried by the laboratory and affirms the presence and shared experiences from anywhere.

A wide variety of technologies are available which can be integrated with the existing techniques to make collaboration a future possibility. Some of the main aspects for building a user-centric successful co-creation and the suitability of the mentioned tools have been studied in the following section.

4 XR and Its Suitability to Co-creation

The potential for XR platforms has highly altered the way to work, collaborate and interact. As specified above, the technologies have adopted to collaborate, visualize, co-design or co-create providing experiences that had never been achieved before. However, the ability of these tool characteristics for building co-creative platforms anyways need to be verified whenever a co-creative session involving multi-users is built. Studies highlighting user experience to co-create in virtual world have concluded that with right approach, virtual worlds improve innovation, engagement, and co-creation [38]. These have been mapped in a traditional L-matrix diagram for easier understanding. The level of the XR equipment to cover the co-creation requirements is recorded as High (H), Medium (M) and Low (L) and the resulting matrix is as shown below.

		CAVE	Immersive HMD	See-through HMD	Hand/Held Display
EXTENDED REALITY	MR				
	AR				
	VR				
CO CREATION PLATFORM REQUIREMENTS	Collaboration [1] [2]	H	M	H	L
	Interaction [2] [4]	M	H	H	L
	Co-Presence [20] [30]	M	L	H	L
	Co-Located [23]	H	L	H	L
	Remote Sharing [5]	L	M	M	L
	Situational Awareness [26]	L	L	H	M
	Portability	L	M	H	H
	Intrusiveness	H	L	H	H
	Collaborative Digital content generation [18] [21]	M	H	H	L

Fig. 1. How does the XR platform cover the co-creation requirement? H: High; M: Medium; L: Low.

The simple matrix diagram visualized as in Fig. 1 is linked to the literature review and the study of innovation platforms. Besides the key elements necessary for building a successful co-creative platform, a number of other factors have been taken into consideration based on the studies and the aspects the tools offer. Co-creation has to be organized, managed and facilitated and it usually involves users from different passions and interests. Considering the on-going pandemic situation, apart from face-to-face collaboration, remote collaboration is regarded as high priority and thus the tools are also marked with consideration of remotely working capabilities.

5 Field Study – A Co-creative XR Environment for Designing Aircraft Cabin Interiors

From the previous section it can be observed that see-through HMD's can enhance the co-presence features of a XR platform. In order to provide an example of a co-creative shared space we outline an application in the aircraft cabin interiors design evaluation field. The idea is emerged from the work done by authors in the framework of a Horizon 2020 project, the CASTLE (CAbin Systems design Toward passenger welLbEing). In CASTLE, a Human Centered Design approach has been followed for the validation of design solutions through Human in the Loop Experiments performed in the CAVE and Head Mounted Displays (HoloLens1). These experiments aim at validating design solutions through the involvement of voluntary participants representing potential users [39, 40]. Several aspects of comfort, such as visual comfort, interaction comfort or living space comfort have been measured and used to iteratively validate design solutions.

During such tests, each participant has been asked to experience the cabin interiors scenario and eventually to interact with it as a single user, representing, depending on the scenario to be validated, a potential passenger or a potential crew member. However, in a real aircraft several users are involved with different roles in the same scenario and they usually interact at the same time with the cabin environment.

For instance, in view of a Flight Attendant Seat (FAS) validation of a regional aircraft in a CAVE, two users enacting the roles of passenger and crew member cannot share the same environment mutually facing each other as in a real aircraft due to the fact that in the CAVE multiple users can share similar views and the virtual camera can have only a single point of view which is usually the point of view of one of the users, if tracked. Therefore, CAVE users obliged to stay on the same side of the screen limiting the front and rear view of the aircraft concurrently. Instead, using Head Mounted Displays multiple users can share the same virtual environment with independent point of views. The presence of another user can be developed to be physically visible in the same environment as a virtual avatar. With the aim of solving this issue we have conceived the co-creative XR environment interface based on the co-location of two different users, with two different roles, facing each other and living the same Virtual Environment wearing two synchronized See-Through HMDs (Microsoft HoloLens2). This scheme allows to have a shared and co-locating experience with an independent point of view in a virtual aircraft environment (Fig. 2). The design can also be extended to bring remotely located multiple users in to same virtual environment to interact and collaborate to work towards a shared goal.

Currently, the application is under technical refinement. Simultaneously, the authors are working at the tuning of the simulation scenario and of the validation procedure.

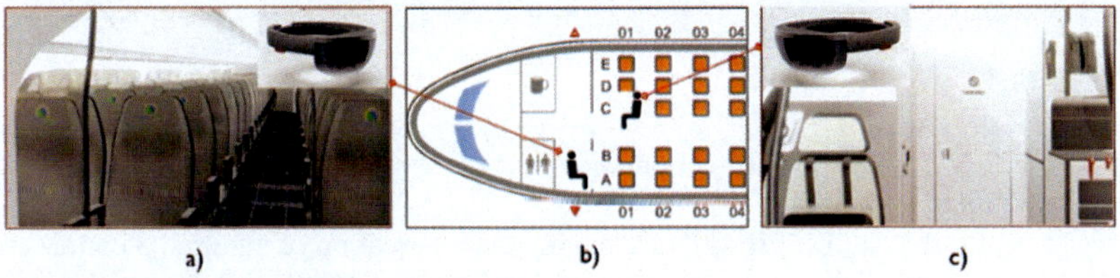

Fig. 2. Layout for multiple HoloLens users in cabin interior virtual environment. a) view form the crew seat; b) layout; c) view from the passenger's seat.

Furthermore, a reputed organizations manufacturer has been consulted to collect a practical point of view in the conception of multi-user co-located virtual environment in co-creative design validation and product development using XR tools. The manufacturer's industrial knowhow has provided an insight to extend similar environment to other figures of a company such as a designer (producing cabin interior and cabin configuration) and certification engineer (who provides the main requirements) along with positive expectations of the results.

6 Conclusions and Future Work

This conceptual paper is an attempt to provide an insight to the idea of co-creation and explore the innovative tools and technologies available to build a successful co-creation platform. Whilst, co-creation is being evoked in every industry other than business world, it needs to be further investigated to be understood and applied correctly. Many works have been performed highlighting the importance of the user's perspective in a co-creative environment, but no prominence has been provided towards developing an innovative platform for practicing the approach. Thus, this paper provided potential innovative solutions like Virtual Reality, Augmented Reality and Mixed Reality as means to collaborate, to interact and to make true the experience for a user in a co-creative environment. We have listed out existing works with a literature review and have drawn a level-based matrix on to what level the XR platform/device cover the co-creation requirements.

A wide range of industries are showing interest in these new technologies and using the XR tools for co-creation will have a huge impact on them. The future works can investigate the development of co-creation concept with the tools mentioned for any kind of testcases to practically prove the potential of XR tools in building co-creation.

Finally, a field study scenario has been presented in the aeronautics domain. Such scenario is being studied in the framework of CASTLE European project further to demonstrate the use of XR tools to co-create in industry.

Acknowledgements. The work presented in this paper has received funding from the Clean Sky 2 Joint Undertaking under the European Union's Horizon 2020 research and innovation program under grant agreement No. 807083.

References

1. Sanders, E.B.N., Stappers, P.J.: Co-creation and the new landscapes of design. Co-design **4**(1), 5–18 (2008)
2. Payne, A.F., Storbacka, K., Frow, P.: Managing the co-creation of value. J. Acad. Mark. Sci. **36**(1), 83–96 (2008)
3. O'Hare, J.A., Dekoninck, E., Giunta, L., Boujut, J.F., Becattini, N.: Exploring the performance of augmented reality technologies in co-creative sessions: Initial results from controlled experiments. In: DESIGN 2018 15th International Design Conference, vol. 15, pp. 405–416 (2018)
4. Alimamy, S., Deans, K.R., Gnoth, J.: The role of augmented reality in the interactivity of co-creation: a critical review. Int. J. Technol. Hum. Interact. (IJTHI) **14**(3), 88–104 (2018)
5. El-Jarn, H., Southern, G.: Can co-creation in extended reality technologies facilitate the design process? J. Work Appl. Manag. (2020)
6. Prahalad, C.K., Ramaswamy, V.: Co-creation experiences: the next practice in value creation. J. Interact. Mark. **18**(3), 5–14 (2004)
7. LEGO to launch production of fan-created Saturn V moon rocket model | collectSPACE, collectSPACE.com. http://www.collectspace.com/news/news-060916a-saturn-v-lego-ideas.html. Accessed 02 Feb 2021
8. LAUNCH of the BMW Group Co-Creation Lab | The Making-of Innovation. https://michaelbartl.com/2010/09/11/launch-of-the-bmw-group-co-creation-lab-2/. Accessed 14 Feb 2021

9. Mandolfo, M., Chen, S., Noci, G.: Co-creation in new product development: which drivers of consumer participation? Int. J. Eng. Bus. Manag. **12**, 1847979020913764 (2020)
10. Lane, S.C.: Co-creating the Future, p. 24. https://social-innovation.hitachi/en-eu/about/whitepaper/. Accessed 10 Nov 2021
11. Ruhl, E., Richter, C., Lembke, J., Allert, H.: Beyond methods: co-creation from a practice-oriented perspective (2014)
12. Humphreys, A., Grayson, K.: The intersecting roles of consumer and producer: a critical perspective on co-production, co-creation and prosumption. Sociol. Compass **2**(3), 963–980 (2008)
13. Ind, N., Coates, N.: The meanings of co-creation. Eur. Bus. Rev. (2013)
14. Piller, F.T., Walcher, D.: Toolkits for idea competitions: a novel method to integrate users in new product development. R&d Manag. **36**(3), 307–318 (2006)
15. Füller, J., Hutter, K., Faullant, R.: Why co-creation experience matters? Creative experience and its impact on the quantity and quality of creative contributions. R&D Manag. **41**(3), 259–273 (2011)
16. Bartl, M., Jawecki, G., Wiegandt, P.: Co-creation in new product development: conceptual framework and application in the automotive industry. In: Conference Proceedings R&D Management Conference–Information, Imagination and Intelligence, Manchester, vol. 9, pp. 1–9, June 2010
17. Kohler, T., Fueller, J., Matzler, K., Stieger, D., Füller, J.: Co-creation in virtual worlds: the design of the user experience. MIS Q., 773–788 (2011)
18. Ramaswamy, V., Ozcan, K.: What is co-creation? An interactional creation framework and its implications for value creation. J. Bus. Res. **84**, 196–205 (2018)
19. Berg, L.P., Vance, J.M.: Industry use of virtual reality in product design and manufacturing: a survey. Virtual Real. **21**(1), 1–17 (2016). https://doi.org/10.1007/s10055-016-0293-9
20. Piumsomboon, T., et al.: Mini-me: an adaptive avatar for mixed reality remote collaboration. In: Proceedings of the 2018 CHI Conference on Human Factors in Computing Systems, pp. 1–13, April 2018
21. Cruz-Neira, C., Sandin, D.J., DeFanti, T.A.: Surround-screen projection-based virtual reality: the design and implementation of the CAVE. In: Proceedings of the 20th Annual Conference on Computer Graphics and Interactive Techniques, pp. 135–142, September 1993
22. Kostis, A., Ritala, P.: Digital artifacts in industrial co-creation: how to use VR technology to bridge the provider-customer boundary. Calif. Manag. Rev. **62**(4), 125–147 (2020)
23. Berni, A., Borgianni, Y.: Applications of virtual reality in engineering and product design: why, what, how, when and where. Electronics **9**(7), 1064 (2020)
24. Greenwald, S.W., Corning, W., Maes, P.: Multi-user framework for collaboration and co-creation in virtual reality. In: 12th International Conference on Computer Supported Collaborative Learning (CSCL), June 2017
25. Saffo, D., Di Bartolomeo, S., Yildirim, C., Dunne, C.: Remote and collaborative virtual reality experiments via social VR platforms (2021)
26. Facebook Spaces. https://about.fb.com/news/2017/04/facebook-spaces/. Accessed 10 Nov 2021
27. Bagassi, S., De Crescenzio, F., Piastra, S.: Augmented reality technology selection based on integrated QFD-AHP model. Int. J. Interact. Des. Manuf. (IJIDeM) **14**(1), 285–294 (2019). https://doi.org/10.1007/s12008-019-00583-6
28. Schmalstieg, D., et al.: The studierstube augmented reality project. Presence Teleoperators Virtual Environ. **11**(1), 33–54 (2002)
29. Mars Virtual Reality Software Wins NASA Award. https://www.jpl.nasa.gov/news/mars-virtual-reality-software-wins-nasa-award. Accessed 11 Feb 2021
30. Ali, A., Glackin, C., Cannings, N., Wall, J., Sharif, S., Moniri, M.: A framework for augmented reality based shared experiences. In: Immersive Learning Research Network-iLRN (2019)

31. Mourtzis, D., Siatras, V., Angelopoulos, J., Panopoulos, N.: An augmented reality collaborative product design cloud-based platform in the context of learning factory. Procedia Manuf. **45**, 546–551 (2020)
32. Zillner, J., Mendez, E., Wagner, D.: Augmented reality remote collaboration with dense reconstruction. In: 2018 IEEE International Symposium on Mixed and Augmented Reality Adjunct (ISMAR-Adjunct), pp. 38–39. IEEE, October 2018
33. Sobota, B., Korečko, Š., Hudák, M., Sivý, M.: Mixed reality: a known unknown. In: Mixed Reality and Three-Dimensional Computer Graphics. IntechOpen (2020)
34. Lee, G.A., Teo, T., Kim, S., Billinghurst, M.: Mixed reality collaboration through sharing a live panorama. In: SIGGRAPH Asia 2017 Mobile Graphics & Interactive Applications, pp. 1–4 (2017)
35. Bai, H., Sasikumar, P., Yang, J., Billinghurst, M.: A user study on mixed reality remote collaboration with eye gaze and hand gesture sharing. In: Proceedings of the 2020 CHI Conference on Human Factors in Computing Systems, pp. 1–13, April 2020
36. Piumsomboon, T., Lee, Y., Lee, G., Billinghurst, M.: CoVAR: a collaborative virtual and augmented reality system for remote collaboration. In: SIGGRAPH Asia 2017 Emerging Technologies, pp. 1–2 (2017)
37. Introducing Microsoft Mesh | Here can be anywhere. https://www.microsoft.com/en-us/mesh. Accessed 20 Mar 2021
38. Kohler, T., Fueller, J., Stieger, D., Matzler, K.: Avatar-based innovation: consequences of the virtual co-creation experience. Comput. Hum. Behav. **27**(1), 160–168 (2011)
39. De Crescenzio, F., Bagassi, S., Asfaux, S., Lawson, N.: Human centred design and evaluation of cabin interiors for business jet aircraft in virtual reality. Int. J. Interact. Des. Manuf. (IJIDeM) **13**(2), 761–772 (2019). https://doi.org/10.1007/s12008-019-00565-8
40. Bagassi, S., Lucchi, F., De Crescenzio, F., Piastra, S.: Design for comfort: aircraft interiors design assessment through a human centered response model approach. In: 31st Congress of the International Council of the Aeronautical Sciences, ICAS (2018)

A New Approach for CAD Modelling of Customised Orthoses by Generative Design

Laura Bragonzoni[1], Tommaso Ingrassia[2], Giuseppe Marannano[2], Vincenzo Nigrelli[2], and Vito Ricotta[2](✉)

[1] Alma Mater Studiorum, Università di Bologna, Bologna, Italy
[2] Università degli Studi di Palermo, Palermo, Italy
vito.ricotta@unipa.it

Abstract. The standard method of design and manufacturing customised orthoses is still very time-consuming due to their often very complex shape. Different authors have tried to solve this problem but, unfortunately, the proposed approaches cannot be easily used in clinical practice because they require substantial interaction among medical staff and engineers or technicians.

The aim of this work is to present the framework of a new design approach that could allow clinicians to easily model a customised orthosis, without a skilled technician develops the entire procedure.

In particular, an automatic process based on Generative Design has been implemented. The obtained results have demonstrated that the implemented algorithm is simple to use and could allow also not-skilled users to design customised orthoses.

Keywords: Reverse engineering · Generative design · CAD · Customised orthosis

1 Introduction

Personalised medicine is a constantly evolving approach in patient care. Unlike the traditional approach [1], the personalised medicine takes more into account some needs of the patients and, thanks to the use of modern reverse engineering and additive manufacturing techniques [2–5], it is spreading more and more in recent years.

Personalised approaches are most used in rehabilitation medicine, especially physical rehabilitation, in which a specific treatment is given to each patient [1, 6].

One important branch of rehabilitation medicine is related to the production of orthoses, which are external medical devices, used in orthopaedic and traumatology, having the main function of restraining articulation movements after a trauma, a surgery, a distortion or in arthrosis patients [7].

For patients with disabilities that induce functional limitations, orthoses are employed to apply forces on the body for biomechanical needs by helping patients during walking and limb movement.

C. Rizzi et al. (Eds.): ADM 2021, LNME, pp. 175–182, 2022.
https://doi.org/10.1007/978-3-030-91234-5_18

In the literature there are different works about elbow [8, 9], foot [10] and hip orthoses [11]. The present work focuses on the study of wrist orthoses. Wrist orthoses provide different treatment outcomes to patients through immobilisation of wrist while allowing mobility of all fingers [12].

Although the rehabilitation medicine field has moved away from prefabricated orthoses, which in some cases reduced the success rate of therapy and/or patient acceptance of the device, towards more personalised devices, the standard method of design and manufacturing of customised orthoses is still a time consuming, arduous and wasteful process and often not completely personalized [1].

Different authors have tried to improve the design process of personalised devices by speeding up the modelling process.

Palousek et al. [13] proposed to simplify the design and manufacturing process by scanning only one side of the arm and generating a splint, thus gaining some advantages such as ventilation, cleanability and lightness, but resulting in a limited protection for the patient.

Huhn Kim et al. [14] proposed a hybrid model with a wrist orthosis divided in two components, consisting of an inner structure, perfectly fitting the patient, and a standardised outer cover.

Górski et al. [15] proposed another approach to simplify the process by trying to automate the procedure. All these approaches, although interesting, still require reverse engineering and CAD experts to generate the models of the orthosis [16, 17].

Other authors proposed several solutions regarding the 3D acquisition or CAD modelling phases of orthoses or prostheses [18–20]. However, the proposed approaches are not ready to be used directly in clinical practice since the design of these devices requires significant interaction among medical staff and technical experts.

In this work a new approach to rapidly design personalised orthoses is proposed. For this purpose, the CAD modelling process was designed to be handled by non-expert users by means of an automatic process based on Generative Design (GD).

2 3D Acquisition of the Wrist by Photogrammetry

For the 3D acquisition there are many technologies with different feature in terms of accuracy, costs and ease of use [21–23].

To scan the patient's wrist without the use of complex and expensive 3D scanners, it was decide to use the photogrammetry and, in particular, a camera of a smartphone (Fig. 1-left).

Photogrammetry is a very versatile technique that allows the generation of 3-dimensional CAD model of any object from a series of photos. Compared to other reverse engineering techniques, photogrammetry is a more user friendly technique for non-expert users.

The pictures must be taken from different point of views and, usually, some hundreds of photos are generally enough to guarantee a good result.

For the case study analysed in this work, about 130 photos of the patient's wrist were acquired from different positions. This phase did not require much time and it was not necessary to use an external device to keep the wrist and the fingers into a fixed position.

However, according to the pathology of each patient, a support system can be provided for positioning and fixing the patient's wrist.

The reconstruction of the 3D model of the patients' wrist was made by the Agisoft software.

The obtained 3D mesh is shown in Fig. 1-right.

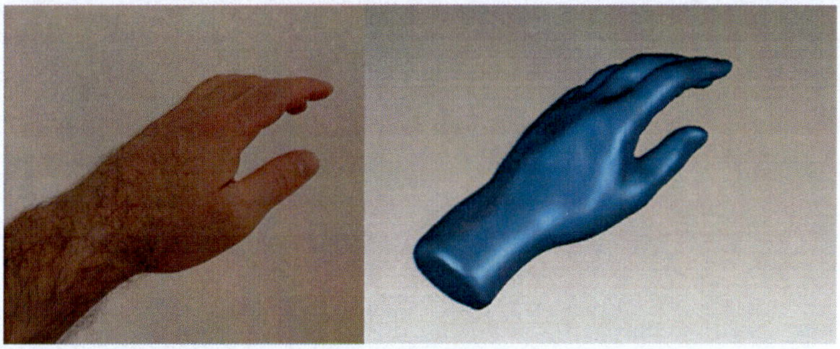

Fig. 1. Wrist and hand (on the left) and reconstructed mesh (on the right).

3 CAD Modelling of the Wrist Orthosis

3.1 CAD Modelling by Generative Algorithm

A new modelling approach based on generative algorithms was proposed to model the wrist orthosis by means of an automatic process.

The new algorithm was developed using Grasshopper, a graphical algorithm editor that can be used as a plug-in to Rhinoceros.

The block diagram of the generative algorithm developed in Grasshopper for the realization of the wrist orthosis is shown in Fig. 2.

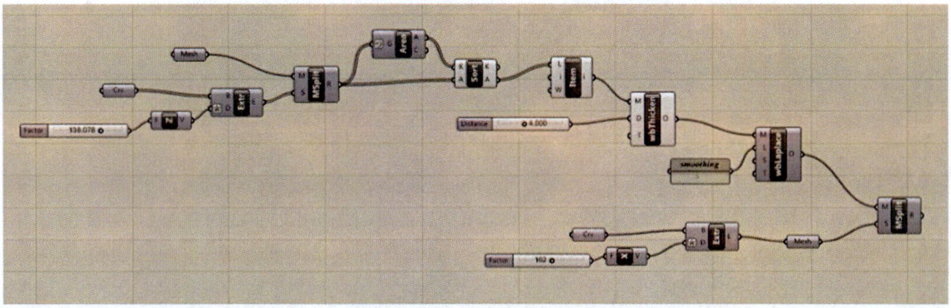

Fig. 2. Implemented function blocks diagram in Grasshopper.

To design the customised orthosis, after importing the mesh of the patient's wrist/hand into the software, the user must define the initial and final contours of the orthosis by simply sketching two lines on the 3D model (Fig. 3).

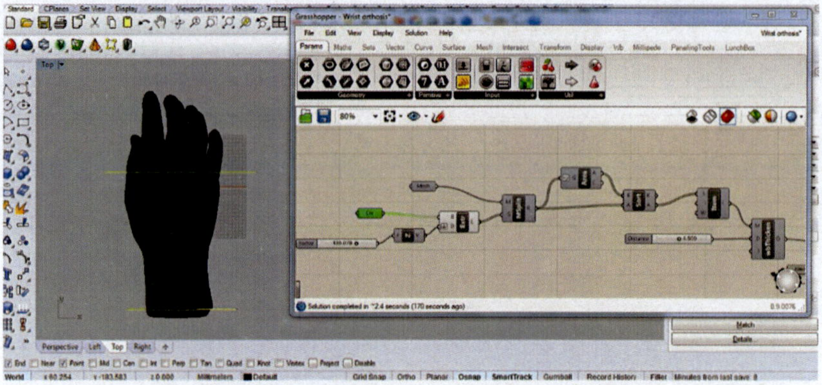

Fig. 3. Indication of the ends of the orthosis.

Then the algorithm automatically creates the mesh perfectly adapted to the patient's wrist.

In particular, the algorithm generates a mesh starting from the initial and final contours on the patient's wrist/hand one. Then, the two meshes are split, extracting the mesh of the orthosis (Fig. 4).

A thickness of 4 mm of the structure is preliminary set through an offset function. This value of thickness is usually adopted in commercial orthoses and can be modified by the user.

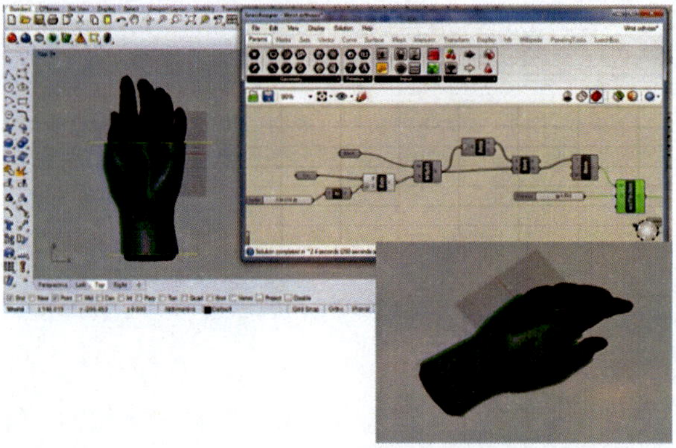

Fig. 4. Mesh of wrist orthosis.

The designed orthosis has a longitudinal gap to allow the devices be easily worn and adjusted according to medical recommendations. The orthosis is worn through the longitudinal opening thanks to some flexibility. This aspect could be improved after a better trial with clinicians.

The position of the gap can be indicated by the user by moving a "cutting" plane (Fig. 5), after the algorithm creates the gap automatically (Fig. 6).

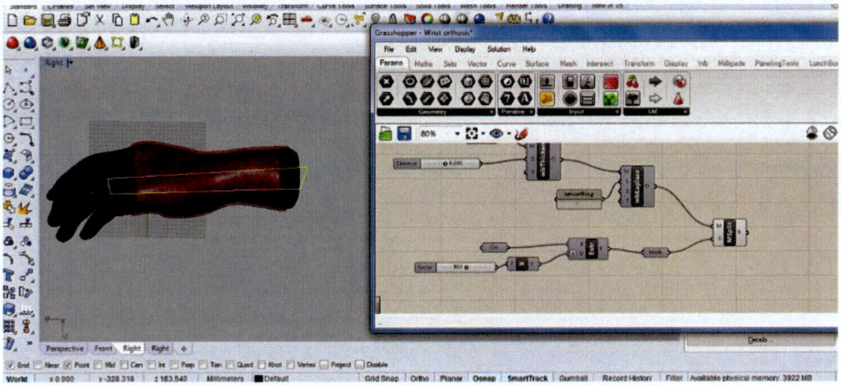

Fig. 5. Part of wrist orthosis to open.

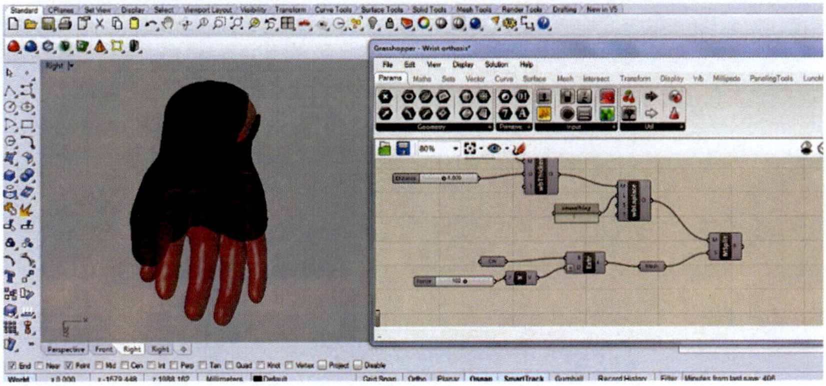

Fig. 6. Opening of wrist orthosis.

After defining the position of the gap, the algorithm creates the final mesh and smooths the orthosis structure so to eliminate the sharp edges.

To complete the design of the orthosis, also another important aspect has been considered in the algorithm: the transpiration of the patient's limb.

The orthoses model can be perforated to increase ventilation of the treated area, so reducing the risk of tissue inflammation and the overall device weight, but must maintain the mechanical properties of the device.

Figure 7 shows the final CAD model of the orthosis that could be directly used for manufacturing purposes, for example for its realization by AM.

To increase the aesthetic appeal of the device, the orthosis can be also customised with different patterns depending on patients liking (Fig. 8).

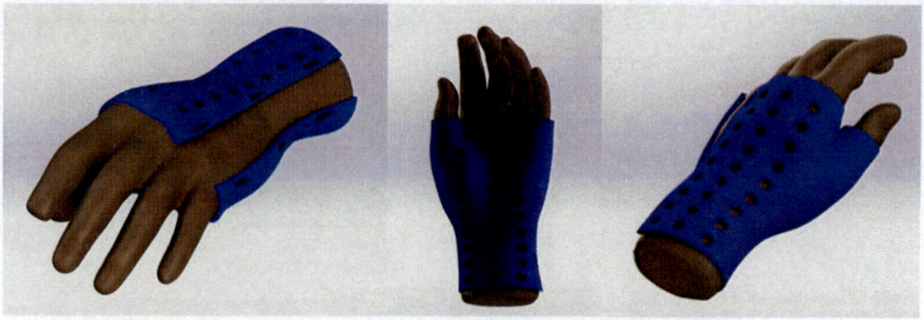

Fig. 7. Final CAD model.

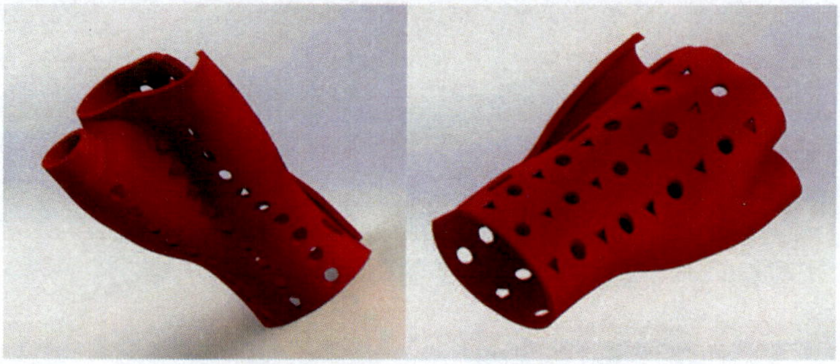

Fig. 8. Final CAD model with different pattern.

4 Conclusions

In this work a new approach that can allow medical staff to carry out the 3D acquisition and easily model the orthosis without the necessity of an expert following the entire procedure has been developed.

The preliminary phase of the proposed approach is related to the 3D CAD reconstruction of the wrist of the patient that can be easily completed by a smartphone, by means of a simple reverse engineering process based on photogrammetry.

An ad hoc modelling algorithm based on generative algorithms was proposed to allow user to model the wrist orthosis by means of an highly automated process.

The developed algorithm was tested by clinicians and students. The future step will be to perform a trial with clinicians to better understand the time needed for the orthoses design and the easy-of-use of the tool, creating an interface more user-friendly for all users.

However the obtained results are promising contributing to make the design process of customised orthoses accessible even to non-expert users so reducing the time needed to design and manufacturing these medical devices.

References

1. Leite, M., Soares, B., Lopes, V., Santos, S., Silva, M.T.: Design for personalized medicine in orthotics and prosthetics. Procedia CIRP **84**, 457–461 (2019)
2. Ingrassia, T., Nigrelli, V., Ricotta, V., Tartamella, C.: Process parameters influence in additive manufacturing. In: Eynard, B., Nigrelli, V., Oliveri, S., Peris-Fajarnes, G., Rizzuti, S. (eds.) Advances on Mechanics, Design Engineering and Manufacturing. LNME, pp. 261–270. Springer, Cham (2017). https://doi.org/10.1007/978-3-319-45781-9_27
3. Li, C., Pisignano, D., Zhao, Y., Xue, J.: Advances in medical applications of additive manufacturing. Engineering **6**, 1222–1231 (2020)
4. Lal, H., Patralekh, M.K.: 3D printing and its applications in orthopaedic trauma: a technological marvel. J. Clin. Orthop. Trauma **9**, 260–268 (2018)
5. Sheoran, A.J., Kumar, H., Arora, P.K., Moona, G.: Bio-medical applications of additive manufacturing: a review. Procedia Manuf. **51**, 663–670 (2020)
6. Halem, A., Javaid, M.: 3D printed medical parts with different materials using additive manufacturing. Clin. Epidemiol. Glob. Health **8**, 215–223 (2020)
7. Buonamici, F., et al.: A CAD-based procedure for designing 3D printable arm-wrist-hand cast. Comput. Aided Des. Appl. **16**(1), 25–34 (2019)
8. Ricotta, V., Campbell, R.I., Ingrassia, T., Nigrelli, V.: A new design approach for customised medical devices realized by additive manufacturing. Int. J. Interact. Des. Manuf. (IJIDeM) **14**(4), 1171–1178 (2020). https://doi.org/10.1007/s12008-020-00705-5
9. Campbell, R.I., Ingrassia, T., Nigrelli, V., Ricotta, V.: New customized elbow orthosis made by additive manufacturing. In: Cavas-Martínez, F., Eynard, B., Fernández Cañavate, F., Fernández-Pacheco, D., Morer, P., Nigrelli, V. (eds.) Advances on Mechanics, Design Engineering and Manufacturing II. LNME, pp. 473–483. Springer, Cham (2019). https://doi.org/10.1007/978-3-030-12346-8_46
10. Fantini, M., De Crescenzio, F., Brognara, L., Baldini, N.: Design and rapid manufacturing of a customized foot orthosis: a first methodological study. In: Eynard, B., Nigrelli, V., Oliveri, S., Peris-Fajarnes, G., Rizzuti, S. (eds.) Advances on Mechanics, Design Engineering and Manufacturing. LNME, pp. 457–467. Springer, Cham (2017). https://doi.org/10.1007/978-3-319-45781-9_46
11. Munhoz, R., Da Costa Moraes, C.A., Tanaka, H., Kunkel, M.E.: A digital approach for design and fabrication by rapid prototyping of orthosis for developmental dysplasia of the hip. Res. Biomed. Eng. **32**(1), 63–73 (2016)
12. Paterson, A.M., Bibb, R., Campbell, R.I., Bingham, G.: Comparing additive manufacturing technologies for customised wrist splints. Rapid Prototyp. J. **21**(3), 230–243 (2015)
13. Palousek, D., Rosicky, J., Koutny, D., Stoklásek, P., Navrat, T.: Pilot study of the wrist orthosis design process. Rapid Prototyp. J. **20**, 27–32 (2014)
14. Kim, H., Jeong, S.: Case study: hybrid model for the customized wrist orthosis using 3D printing. J. Mech. Sci. Technol. **29**, 5151–5156 (2015)
15. Górski, F., Wichniarek, R., Kuczko, W., Żukowska, M., Lulkiewicz, M., Zawadzki, P.: Experimental studies on 3D printing of automatically designed customized wrist-hand orthoses. Materials **13**, 4091 (2020)
16. Marannano, G., Pasta, A., Giallanza, A.: A model for predicting the mixed-mode fatigue crack growth in a bonded joint. Fatigue Fract. Eng. Mater. Struct. **37**(4), 380–390 (2014)
17. Fragapane, S., Giallanza, A., Cannizzaro, L., Pasta, A., Marannano, G.: Experimental and numerical analysis of aluminum-aluminum bolted joints subject to an indentation process. Int. J. Fatigue **80**, 332–340 (2015)
18. Buonamici, F., et al.: A practical methodology for computer-aided design of custom 3D printable casts for wrist fractures. Vis. Comput. **36**(2), 375–390 (2019). https://doi.org/10.1007/s00371-018-01624-z

19. Fernandez-Vicente, M., Escario Chust, A., Conejero, A.: Low cost digital fabrication approach for thumb orthoses. Rapid Prototyp. J. **23**(6), 1020–1031 (2017)
20. Ingrassia, T., Nalbone, L., Nigrelli, V., Pisciotta, D., Ricotta, V.: Influence of the metaphysis positioning in a new reverse shoulder prosthesis. In: Eynard, B., Nigrelli, V., Oliveri, S., Peris-Fajarnes, G., Rizzuti, S. (eds.) Advances on Mechanics, Design Engineering and Manufacturing. LNME, pp. 469–478. Springer, Cham (2017). https://doi.org/10.1007/978-3-319-45781-9_47
21. Marannano, G., Pasta, A., Parrinello, F., Giallanza, A.: Effect of the indentation process on fatigue life of drilled specimens. J. Mech. Sci. Technol. **29**(7), 2847–2856 (2015). https://doi.org/10.1007/s12206-015-0613-0
22. Marannano, G., Parrinello, F., Giallanza, A.: Effects of the indentation process on fatigue life of drilled specimens: optimization of the distance between adjacent holes. J. Mech. Sci. Technol. **30**(3), 1119–1127 (2016). https://doi.org/10.1007/s12206-016-0216-4
23. Dal Maso, A., Cosmi, F.: 3D-printed ankle-foot orthosis: a design method. Mater. Today Proc. **12**, 252–261 (2019)

Semi-automatic Analysis of Spoken Interaction Dynamics in Collaborative Design Sessions

Niccolò Becattini(✉) and Luca Alberio

Politecnico di Milano, Milan, Italy
niccolo.becattini@polimi.it

Abstract. The paper proposes an original approach for design protocol analysis that reduces the time and the effort required to extract relevant insights about the interaction dynamics in co-design sessions. Audio recordings are processed with a newly tailored algorithm that recognizes speakers during collaborative design and extracts indexes that describe the behavior of the designers, such as their degree of involvement in the session, their role (e.g. leadership), the length of their verbal interventions as well as the frequency with which they take the floor. The robustness of the approach is estimated by means of the application of the algorithm to three co-design sessions that are different from each other in terms of number and profile of the participants, language spoken and design task. The algorithm proved to be effective: the results of the analysis of spoken interactions from audio recordings are comparable to those obtained with quantitative techniques (precision up to 80% and faster processing time up to 100x).

Keywords: Co-design · Verbal interactions · Speech · Team dynamics · Human behavior in design

1 Introduction

To study the stages, the routines, the processes and the dynamics of design activities, one of the most acknowledged method is undoubtedly protocol analysis. This typically requires the study of the transcription of the design session (i.e. the protocol), under the assumption that verbal data is the most suitable best proxy to understand cognition and unveil the reasoning process that took place during the design session [1]. Recent literature on design protocol analysis relied on a wider set of sources of acquisition (e.g. [2]). Different facets of design sessions have been studied according to the physiological state of the subjects involved, e.g. using galvanic skin resistance [3], gaze [4], heartbeat [5] as well as brain activity [6]. Visual data acquisitions allow for the analysis of body language characteristics: e.g. facial expression [7] and gestures [8]. On the other hand, audio/verbal data acquisitions also allows the investigation of the interaction dynamics among co-designers [9], but their analysis is still based on the contents of verbal interactions, e.g. [10], which requires all the protocol to be transcribed before any analysis can be carried out. Unfortunately, this kind of investigation is extremely demanding of time and human resources: each hour of recording needs up to 1000 h of analysis [11] (a

C. Rizzi et al. (Eds.): ADM 2021, LNME, pp. 183–195, 2022.
https://doi.org/10.1007/978-3-030-91234-5_19

typical range is 1:10–100 [12]). In fact, before the analysis, the protocol requires transcription (up to 10 times the duration of the recordings) and segmentation into 'chunks'. First, these are labeled with a coding scheme by multiple coders (recording:coding time up to 1:1000). Second, the Inter-Rater Reliability estimation (e.g. [13]) checks the coherence and confirm the codified contents of the protocol. This data allows for qualitative and/or quantitative assessments (e.g. time spent on a topic, sequences of activities, etc.) [14].

With reference to the abovementioned areas of interest and the limitations of this kind of approach (time and HR demands), this research explores an approach for the semi-automatic analysis of audio recordings of co-design sessions. This aims at unveiling the dynamics of the verbal interactions among co-designers, with minimum investment of human resources and time to generate results.

The next section of the paper briefly presents the background by means of relevant contributions from the literature in order to formalize the research question(s). The third section summarizes the research approach, including the overall architecture of the algorithm, the case studies considered to check its suitability as well as the main estimators that enable a comparison with the outcomes of standard design protocol analysis. Then, the following section presents and critically discusses the results according to the objectives of the research, showing the implications of the proposed solution. Eventually, the concluding section summarizes the findings and depicts opportunities for the future developments of the approach.

2 Relevant Background

2.1 Approaches Towards Automatic Design Protocol Analysis

As mentioned in the introduction, the demanding requirements for time resources and coders' expertise are the main obstacles preventing the quick analysis of design protocols and, hence, to understand better Human Behavior in Design. The relatively recent developments of the technology show that automatic transcriptions are becoming easily accessible almost for free, as most of the platforms for online meetings enable the recording and its automatic transcription (depending on the language) [15]. Youtube, as well, enables the automatic transcription of speech once a video is uploaded to the platform. Nevertheless, this is the least time- and human resources- consuming task towards the completion of the analysis: segmentation and coding, as well as the quantification of observations according to the coding are still consuming a lot of resources. Tools for coding the sessions while it takes place, thus without a transcribed and segmented protocol, are now available to analyze gestures in collaborative design sessions [8]. On the other hand, other approaches aim at softening the needs for the coding stage, e.g. by making this task easier also for non-expert coders are becoming popular and progressively diffused (e.g. [15]). Whenever an ICT tool supports the interaction among co-designers, it is also possible to use its log files to obtain a general overview on the session dynamics and its activities, depending on the data processing pipeline and the metrics for the assessment [16]. This reduces the time required for the whole of more than 95%, despite this induces a partial loss of the information that a standard design protocol analysis is capable of producing.

2.2 Approaches to Extract Information About Team Dynamics from Verbal Data Captured in Spoken Interactions

Bhattacharya et al. [17] studied dynamics of collaboration in teams with a comprehensive viewpoint that ranges from the contents of verbal interaction to body language, through the recording of multimodal signals. However, they manage to reach high precisions for the investigation of group behavior (e.g. major contributors and group leaders in design sessions) just through fusing audio and video data and image and natural language processing algorithms. Through spectral analysis of vocal features Dubey et al. [18] studied dominance, participation, emphasis, curiosity and engagement within peer-led team sessions. They showed that speech signal processing is effective to understand group dynamics; despite they checked the precision of their algorithms just against the qualitative interpretation of the above characteristics by session participants. Jayagopi et al. [19] extracted behavioral cues characterizing the speaking and looking patterns in face-to-face interactions by fusing sound and visual expressions, leading to interesting correlations with interpersonal perception (e.g. dominance hierarchy and leadership) and performance, measured through questionnaires. Sanchez-Cortes et al. [20] presented a similar approach with leadership analysis via the processing of audio recordings and body motion, showing good correlations with speaking turn features. Cowie et al. [21] showed that changes in the fundamental frequency of the voice, its intensity as well as word rates are associated with different behavioral and emotional responses, highlighting a correlation with boredom, anxiety, anger, happiness, sadness, etc. In all the above cases, however, questionnaires filled in by session participants were the main source to validate the results, without any specific crosschecking with objective or objectified quantitative data.

Independently from the approach, the background literature [e.g. 18, 22] also underlines that noise and sounds reverberating from different sources towards the different data recording locations are among the main issues preventing the achievement of reliable results, as they could impair the identification of the right source, potentially leading to wrong speaker identification or speech recognition.

2.3 Research Objectives

According to the challenges and the limitations described in Sect. 2.1 and 2.2, this research aims at exploring the performance of an originally developed algorithm-based semi-automatic computational pipeline for the identification of speakers from a multi-source acquisition context. More specifically, the outcomes of the automatic analysis carried out with the algorithm (whose technical development goes beyond the purposes of this paper) focus on the following objectives:

1. Estimating the accuracy of speaker identification in co-design session, compared to the outcomes of standard human-based approaches for design protocol analysis;
2. Assessing, against objective data, the algorithm precision to describe team dynamics in co-design (e.g. leadership and turn taking in spoken interactions);
3. Comparing the time and human resources required for a human-powered design protocol analysis with the ones needed to generate results with the algorithm.

3 Research Method

3.1 Semi-automatic Algorithm for the Analysis of Co-design Team Dynamics

This section briefly provide a description of the computational pipeline of the algorithm for the (semi-)automatic identification of speakers in co-design sessions and the computation of relevant estimators that describe team dynamics in such contexts. The algorithm has three main logical blocks (Fig. 1); each of them entitled to deliver a specific function on the audio files collecting spoken interactions among co-designers.

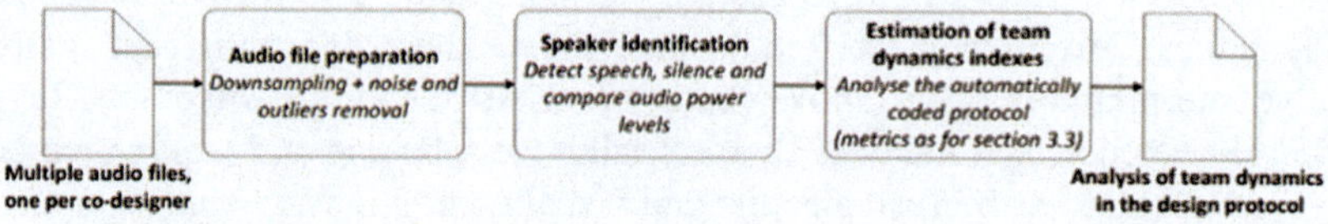

Fig. 1. The functional sequence of actions that the algorithm performs.

The audio files preparation includes signal-processing techniques to clean the audio tracks (one per file) from environmental noise and to filter disturbances as coughing, sneezing or other kind of outliers that are not representative of spoken interactions. This is the only part requiring the intervention of the user (e.g. for downsampling).

The identification of speakers from different audio sources leverages different Matlab® routines. This detects the speaker in a time window of one second, which is compatible with the identification of speakers from design protocols with traditional human-based approaches. A series of concurrent and sequential checks on different signal characteristics also enable distinguishing spoken and silent segments.

The estimation of indexes about team dynamics extracts relevant parameters from the sequence of speakers. The time-based nature of the signal also enables the monitoring of the progression of the co-design session (process-wise) as well as to estimate more comprehensive characteristics of the session (outcome-wise).

3.2 Case Studies Description

Three co-design sessions served as Case Studies (CS) for the assessment of the 3 research objectives presented in Sect. 2.3. Their characteristics (Table 1) span a variety of different conditions to reduce the potential bias of uncontrolled variables on the results and prove the robustness of the algorithm to generate meaningful results.

These CS were also selected because their quantitative analyses were already carried out through traditional design protocol analysis (for the coding method, please check [8, 15, 23]) which facilitate the comparison with the algorithm-based analysis and the estimation of research objective attainment. The three CS took place in real operational environment and involved professional designers as well as other stakeholders. Each session has a different design task to address. CS1 and 2 were oriented to product interface design, respectively of a barbecue and of a personal rescue beacon. CS3, on the contrary focused on the development of a new packaging for baby rice cereal.

Table 1. Characteristics of the co-design sessions examined with the algorithm.

Case study	Language	Co-designers [#]	Duration [min]	Audio file [GByte]	Design domain [Product/packaging]
1	English	4	95	2,1	Product
2	Catalan	4	83	1,8	Product
3	Italian	6	102	3,4	Packaging

CS2 and 3 took place inside a standard room each (without additional acoustic insulation beyond the one provided by walls). While CS2 has a fixed layout for participants positions (all of them seated around a table), CS3 had most of the participants seated, while the session leader was free to move around the table. CS 1, on the contrary, took place into an open space arranged with different barbecues showcased to foster the discussion among co-designers. This setting also triggered the participants to move from their seats and walk around the room to comment different product features that characterized the showcased barbecues.

Each co-designer was equipped with a lapel microphone in CS1, 2 and 3, but just in CS1 and 2 there was an additional environmental microphone for audio capturing.

3.3 Metrics for the Algorithm Effectiveness Estimation

As mentioned, the design protocols of the three CS were already analyzed with standard techniques. This means that the sequence of speakers that took the floor during the co-design sessions was a-priori known. Such list of speakers is, therefore, a vector of categorical variables (name of the speaker). Each data point in the vector corresponds to the main speaker in that time window; i.e. the co-designer who took the floor for most of the time in that 1-s-long segment. This duration for the segments is compatible with human capabilities of discerning speakers, as shorter frames do not allow the coders to identify the speaker with sufficient precision. These 3 lists of speakers (one per each CS) serve as the reference to estimate the performance of the algorithm to properly determine the speaker per second (independent variable) and the attainment of the objectives as for the metrics of Table 2 and 3 (dependent variables).

Last, but not least, is the estimator to check the attainment of the third research objective. As said, the standard approach for design protocol analysis is extremely time consuming because all the activities it requires (except for the transcription of the protocol), need to be cross validated by multiple subjects in order to ensure the reliability of the investigation and quantify its outcomes. This multiplies the manpower (e.g. in terms of man-hours) required to complete the analysis and delays the availability of the results. As the algorithm does not require manpower (or just a few to launch it and to set a few parameters at the beginning of the computational pipeline), the estimator for objective #3 regards just the ratio between the duration of the protocol and the time required to generate the analysis.

Table 2. Metrics for the assessment of research objective #1: accuracy of speaker identification

What to measure	Description of the metrics	Estimator
Process-wise accuracy (second-by-second in 60 s moving window)	Instant error of speaker identification	Difference between the algorithm and the reference [logical: 0→correct; 1→ wrong, averaged over a 1-min time window]
Process-wise accuracy (whole session)	Average error of speaker identification	Sum of errors divided by the whole duration of the session
Outcome-wise accuracy (whole session)	Accuracy by speaker	Ratio: correct assignments of the algorithm and sum of assignments in the reference

Table 3. Metrics for the assessment of research objective #2: team dynamics estimators for checking the coherence between the reference and the algorithm

What to measure	Description of the metrics	Estimator
Number of interventions per co designer	Overall count of the times a speaker takes the floor	Sum, by speaker (i), of the number of times it is different from the person speaking during the previous second
Duration of the average intervention per co-designer	Sum of durations of interventions by the overall count of times a speaker takes the floor	Ratio between the overall time a co-designer is speaking during the session and its number of interventions

4 Results and Discussion

4.1 Accuracy of Speaker Identification in Co-design Sessions (Obj #1)

As per what presented in Table 2, the estimation of the capabilities of the algorithm accuracy, meant as its capability of replicating the outcome of a standard design protocol analysis carried out by humans, should focus on three estimators. Figure 2 shows the process-wise accuracy of the algorithm for CS1 in terms of the instant and average error. The blue line in the graph shows how the error in the estimation of the speaker changes within a time window of 1 min. The horizontal red line, on the contrary, shows the average value of the error computed over the whole session duration. For reasons of space, Table 4 reports the average error by CS as well as the maximum error recorded after the application of the algorithm.

Table 4. Summary of process-wise accuracy of the algorithm

Estimator	CS1	CS2	CS3
Maximum instant error	48%	56%	72%
Average Error	16,2%	20,9%	39,7%

Despite these results show that the range of error variation could be significantly high, depending on the specific timeframe considered along the whole process, the trend depicted with the blue line show that error peaks, despite common, are typically of smaller magnitude than the maximum instant error recorded in Table 4.

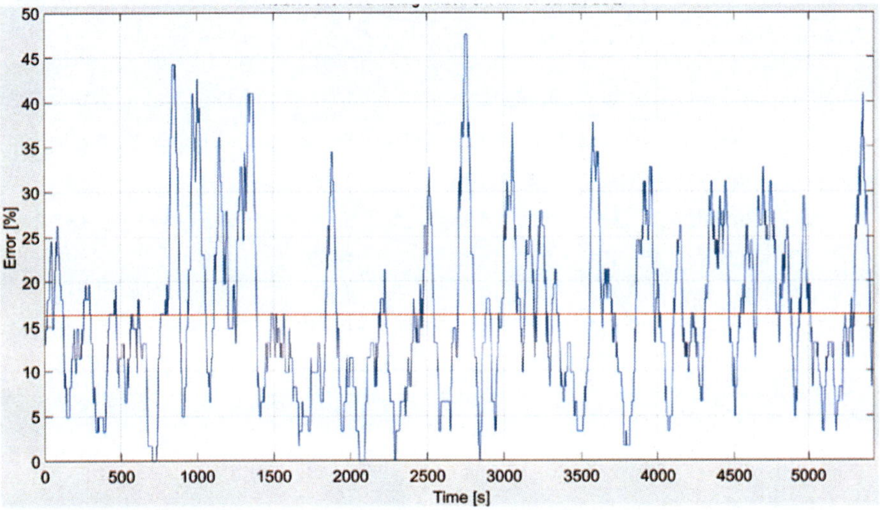

Fig. 2. Process-wise accuracy of the algorithm for CS1: average error and instant error (averaged over a moving window of 1 min)

More specifically, data about the error distributions in the three CS also show that the accuracy is higher for co-design sessions carried out with a smaller team size. In fact, CS1 and 2, which were carried out in teams of 4 co-designers, show lower values for the error and smaller dispersion of the error within the protocol compared to the CS3 co-design session (6 people). Those higher values can also find a potential justification on the lack of a static position for the session leader, which what free to move around the table while interacting with other seated co-designers.

To complete the analysis of the accuracy of the algorithm to identify the right speaker, Table 5, 6 and 7 respectively show the overall accuracy of the algorithm by speaker. The diagonal elements of these contingency-like tables collect successful speaker identifications by second of spoken interaction. The other cells, by column, show the wrong associations the algorithm does with reference to the human-based approach, while the final row “Accuracy” shows the algorithm success ratio.

Table 5. Summary of the outcome-wise accuracy of the algorithm, by speaker, for CS1.

		Reference (CS1)				
	Speaker	Subject 1	Subject 2	Subject 3	Subject 4	Silence
Algorithm	Subj #1	**367**	18	15	2	41
	Subj #2	21	**1631**	102	2	51
	Subj #3	75	160	**2245**	21	172
	Subj #4	4	24	20	**51**	50
	Silence	8	47	57	3	**314**
	Accuracy	77,3%	86,8%	92%	64,6%	50%

Table 6. Summary of the outcome-wise accuracy of the algorithm, by speaker, for CS2.

		Reference (CS2)				
	Speaker	Subject 1	Subject 2	Subject 3	Subject 4	Silence
Algorithm	Subj #1	**841**	140	146	41	54
	Subj #2	31	**738**	50	18	18
	Subj #3	81	122	**1439**	39	39
	Subj #4	26	33	43	**619**	34
	Silence	15	11	30	6	**69**
	Accuracy	84,6%	70,7%	84,3%	85,6%	32,2%

Table 7. Summary of the outcome-wise accuracy of the algorithm, by speaker, for CS3.

		Reference (CS3)						
	Speaker	Subj #1	Subj #2	Subj #3	Subj #4	Subj #5	Subj #6	Silence
Algorithm	Subj #1	**917**	82	66	31	9	17	128
	Subj #2	91	**572**	58	9	0	14	104
	Subj #3	92	103	**571**	21	1	14	265
	Subj #4	147	114	91	**417**	3	34	104
	Subj #5	16	16	19	6	**39**	2	27
	Subj #6	43	30	27	16	7	**65**	117
	Silence	132	185	110	38	5	31	**1095**
	Accuracy	63,8%	51,9%	60,6%	77,5%	60,9%	36,7%	59,5%

With just one exception (CS3, Subj #6), the algorithm proves to be effective in speaker recognition with a success rate that is generically above 50% and on average above 67%. The exception also has an experimental justification for this drop of accuracy: when the co-design session got close to two thirds of its duration, the lapel microphone of Subject #6 in CS3 went offline for lack of power. Moreover, the results also show that the algorithm allows for further improvements for the identification of silent segments, as these are typically the least effective assignments done automatically. These results also allow for the preliminary estimation of the role the different co-designers had during the sessions, as the columns summarize the duration of their spoken interaction and the figures in bold highlight what the algorithm properly computes. Subjects #2 and #3 were the more active ones in CS1. Data from CS2 suggests that Subject #3 acted as the session leader, as he has a significantly stronger prevalence over the other co-designers, which interacted with similar durations of spoken interaction. Eventually, the analysis of CS3 shows that Subject #1 was the session leader and that Subjects #2, #3 and #4 were approximately contributing very similarly to the whole session (while their role can obviously change depending on the considered time frame of the whole design protocol).

All in all, the accuracy of the algorithm for speaker identification appears to be compatible with what emerges after the coding stage of a standard design protocol analysis process, carried out by different humans. Indeed, an approximate result 0,67 for Inter-Rater Reliability [range 0 to 1] is more than sufficient to deem the coherence of the results among coders (moderate to substantial agreement) [13].

4.2 Estimation of Relevant Elements for the Analysis Out of the Automatically Generated Design Protocol (Obj #2)

The estimation of relevant indexes that are capable of describing team dynamics is currently limited to the number of interventions per co-designer and their average duration, as these could be a good proxy to describe leadership and more in general, emergent behaviors by one or more co-designers. However, the identification of speakers presented in Sect. 4.1 might enable the addition of new indexes to extend the considerations and the investigation of interaction dynamics in co-design teams.

Table 8 shows, for the three CS, the values determined with standard approaches for design protocol analysis (reference) and what measured by the algorithm.

Table 8. Comparison of the estimators describing the team dynamics in CS1, CS2, CS3

		Number of interventions [#]		Average duration of interventions [s]	
CS	Speaker	Reference	Algorithm	Reference	Algorithm
1	Subject #1	77	52	6,17	8,52
	Subject #2	229	159	8,21	11,36
	Subject #3	288	225	8,47	11,88
	Subject #4	23	43	3,43	3,47

(continued)

Table 8. (*continued*)

CS	Speaker	Number of interventions [#]		Average duration of interventions [s]	
		Reference	Algorithm	Reference	Algorithm
2	Subject #1	214	179	4,64	6,94
	Subject #2	219	131	4,77	6,68
	Subject #3	261	198	6,54	9,06
	Subject #4	105	90	6,89	8,57
3	Subject #1	474	298	3,03	4,21
	Subject #2	389	233	2,83	3,66
	Subject #3	380	305	2,48	3,51
	Subject #4	152	191	3,54	4,79
	Subject #5	29	48	2,21	2,66
	Subject #6	78	98	2,27	3,28

A quick analysis by CS confirms the early conclusions about the team dynamics commented at the end of Sect. 4.1. CS1 was mostly a dialogue between Subject #2 and #3, with a much more limited role by the other co-designers. CS2 shows a much more balanced role among the co-designers with a prevalence of Subject #3 and the least participation by Subject #4, despite it speaks for longer time, compared to the remaining two other members. Data about CS3 show a harder situation to interpret, probably because of the different nature of the co-design session (packaging design, instead of product design) and the higher number of participants. However, the most evident result is that the average duration of interventions for CS3 is definitely shorter than for the other two CS, which confirms the different nature of this observation.

In terms of statistical validity of the analysis, the outcomes generated with the standard technique have been compared against the semi-automatic approach enabled by the algorithm. The correlation analysis for the of number of interventions yields high values: (Pearson's correlation: CS1 = 0,984; CS2 = 0,892; CS3 = 0,940). Similarly, correlation between the two approaches for the duration of interventions shows that the human- and the algorithm-based analysis are describing the same phenomena (Pearson's correlation: CS1 = 0,998; CS2 = 0,948; CS3 = 0,960).

4.3 Algorithm-Based vs Traditional Protocol Analysis: Comparison of the Time and Human Resources Needed (Obj #3)

While the attainment of Objectives #1 and #2 provide a comprehensive viewpoint on the performance of the algorithm-based analysis of the protocol, the attainment of Objective 3 estimates the costs of running algorithm-based analysis. In other words, the potential lack of accuracy discussed with reference to the first two objectives have to be weighted with the time and human effort savings this semi-automatic approach in order to estimate the convenience of the proposed trade-off.

The algorithm is semi-automatic as it enables the analysts, during the first stage of the computational pipeline (Fig. 1), to fine-tune the algorithm execution by adjusting some of its parameters according to audio features describing moments in which each speaker is silent or speaks, if the analysis relies on already existing protocols. This requires 5 min per co-designer/audio track of the session. On the contrary, the algorithm can proceed automatically if new co-design sessions will enable the recording of silent and spoken episodes for each speaker (e.g. 30 s of silence and 30 s of reading before the session starts), so that the algorithm can self-calibrate microphones for speaker identification, thus reducing the human efforts.

Beyond this difference, Table 9 shows the run-time for the execution of the algorithm for the next two stages presented in Fig. 1 with a standard commercial laptop (Intel® Core™ i5-4690@ 3.50 GHz and 8 GB of RAM (DDR3 @ 2400 MHz).

Table 9. Estimation of the time required for the analysis with reference to the protocol duration

	CS1	CS2	CS3
Algorithm run-time [s]	85,5	97	130
Session duration/run-time ratio [#]	1: 0,015	1: 0,019	1: 0,021

As the time required for the analysis of a design protocol with standard techniques required several times the duration of the protocol itself (typically in the range of 10–100 times), these results clearly highlight that the proposed approach significantly capsizes the situation observed in decades of experience. The application of the algorithm to the three case studies presented in Sect. 3.2 shows that it takes approximately 2% of the protocol duration to extract relevant parameters about team dynamics in collaborative design sessions. This element, according to the authors' opinion, definitely compensate the lack of precision for the analysis mentioned in Sect. 4.1 and 4.2. The convenience of the approach, in fact, needs to consider that this enables the analysts to overcome several limitations (e.g. checking the agreement among coders for high IRR) because it leverages objective and quantitative data. Moreover, the short run-time also suggests the opportunity to run the algorithm in real-time, opening up several opportunities of application that, so far, are underexplored (e.g. the live analysis of online meetings which are becoming more and more common under the pressure of the COVID-19 pandemic, even for co-design sessions across geographical borders).

The approach, then, has to be considered promising also in case of the analysis of previously carried out design protocols whose recordings are available, as the protocol duration vs algorithm runtime ratio shows that results for the analysis come available in less than 1/3 of the co-design session time.

5 Conclusions

This paper presents the application of an originally developed algorithm for an almost human-free analysis of the interaction dynamics among designers and other stakeholders in collaborative design sessions. It leverages the audio recordings captured during the protocols with personal microphones. The purpose of reducing the time needed to complete the analysis is fully satisfied, as the algorithm capsized the ratio between the protocol duration and the time required for the analysis. In the three observed cases studies, the ratio dropped from 1000–10000% typical of human-based techniques to 2–33% of the protocol duration needed by the semi-automatic algorithm. However, the application of this semi-automatic approach also resulted into a lack of precision for the data generated for the analysis. Yet, the overall distribution of the same enables the analysts to infer the same conclusion of a standard investigation as checked via statistical hypotheses testing. Besides, these time savings release the analysts from time consuming activities with small added value and set her/him free to invest part of her/his efforts to focus on specific episodes of the protocol that would require higher attention. Coherently with these findings, the author aims at further checking the goodness of the results by applying the approach to other co-design sessions, e.g. running analysis live, while the design protocol is still taking place, or on single audio tracks collecting the voices of multiple participants at a time. Moreover, they will explore the opportunities generated by the definition of new indexes to provide a more comprehensive viewpoint on team dynamics in collaborative contexts.

References

1. Hay, L., Duffy, A.H., McTeague, C., Pidgeon, L.M., Vuletic, T., Grealy, M.: Towards a shared ontology: a generic classification of cognitive processes in conceptual design. Des. Sci. **3** (2017)
2. Nguyen, P., Nguyen, T., Zeng, Y.: Empirical approaches to quantifying effort, fatigue and concentration in the conceptual design process. Res. Eng. Des. **29**(3), 393–409 (2018)
3. Hu, W.L., Booth, J., Reid, T.: Reducing sketch inhibition during concept generation: psychophysiological evidence of the effect of interventions. In: ASME 2015 IDETC-CIE Conference (2015)
4. Ruckpaul, A., Nelius, T., Matthiesen, S.: Differences in analysis and inter-pretation of technical systems by expert and novice engineering designers. In: International Conference on Engineering Design 2015. Design Society, Glasgow (2015)
5. Steinert, M., Jablokow, K.: Triangulating front end engineering design activities with physiology data and psychological preferences. In: DS 75-7: Proceedings of the 19th International Conference on Engineering Design (ICED13). Design Society, Glasgow (2013)
6. Liu, Y., Ritchie, J.M., Lim, T., Kosmadoudi, Z., Sivanathan, A., Sung, R.C.: A fuzzy psycho-physiological approach to enable the understanding of an engineer's affect status during CAD activities. Comput. Aided Des. **54**, 19–38 (2014)
7. Balters, S., Steinert, M.: Capturing emotion reactivity through physiology measurement as a foundation for affective engineering in engineering design science and engineering practices. J. Intell. Manuf. **28**(7), 1585–1607 (2015). https://doi.org/10.1007/s10845-015-1145-2
8. Masclet, C., Poulin, M., Boujut, J.F., Becattini, N.: Real-time coding method and tool for artefact-centric interaction analysis in co-design situations assisted by augmented reality. Int.

J. Interact. Des. Manuf. (IJIDeM) **14**(4), 1141–1157 (2020). https://doi.org/10.1007/s12008-020-00683-8
9. Wulvik, A., Menning, A., & Steinert, M.: A computational approach to expose conversation dynamics in engineering design activities. In: DS 87-2 Proceedings of the 21st International Conference on Engineering Design (ICED 17). Design Society, Glasgow (2017)
10. Georgiev, G.V., Georgiev, D.D.: Enhancing user creativity: Semantic measures for idea generation. Knowl. Based Syst. **151**, 1–15 (2018)
11. Jiang, H. & Yen, C.-C.: Protocol analysis in design research: a review. In: Rigor and Relevance in Design: IASDR 2009, Seoul, Korea, 18–22 October, pp. 147–156. International Association of Societies of Design Research (2009)
12. Blessing, L.T., Chakrabarti, A.: DRM, a Design Research Methodology. Springer Science, London (2009). https://doi.org/10.1007/978-1-84882-587-1
13. Fleiss, J.L.: Measuring nominal scale agreement among many raters. Psychol. Bull. **76**(5), 378 (1971)
14. . Someren, M.V., Barnard, Y.F., Sandberg, J.A.: The Think Aloud Method: A Practical Approach to Modelling Cognitive Processes. Academic Press, Cambridge (1994)
15. Becattini, N., Cascini, G., O'Hare, J., Boujut, J,F.: Improving the efficiency of design protocol analysis: an approach to speed up the coding stage. In: Rizzi, C., Andrisano, A.O., Leali, F., Gherardini, F., Pini, F., Vergnano, A. (eds.) ADM 2019. LNME, pp. 612–624. Springer, Cham (2019). https://doi.org/10.1007/978-3-030-31154-4_52
16. Becattini, N., Cascini, G., O'Hare, J.A., Morosi, F., Boujut, J.F.: Extracting and analysing design process data from log files of ICT supported co-creative sessions. In: International Conference on Engineering Design, vol. 1, no. 1, pp. 129–138. CUP (2019)
17. Bhattacharya, I., Foley, M., Zhang, N., Zhang, T., Ku, C., Mine, C., et al.: A multimodal-sensor-enabled room for unobtrusive group meeting analysis. In: Proceedings of the 20th ACM International Conference on Multimodal Interaction, pp. 347–355 (2018)
18. Dubey, H., Sangwan, A., Hansen, J.: Using speech technology for quantifying behavioral characteristics in peer-led team learning sessions. Comput. Speech Lang., 343–366 (2017)
19. Jayagopi, D., Sanchez-Cortes, D., Otsuka, K., Yamato, J., Gatica-Perez, D.: Linking speaking and looking behavior patterns with group composition, perception, and performance. In: Proceedings of the 14th ACM International Conference on Multimodal Interaction (ICMI 2012), 433–440 (2012)
20. Sanchez-Cortes, D., Aran, O., Gatica-Perez, D.: An audio visual corpus for emergent leader analysis. In: Workshop on Multimodal Corpora for Machine Learning: Taking Stock and Road Mapping the Future, ICMI-MLMI (2011)
21. Cowie, R., Douglas-Cowie, E., Tsapatsoulis, N., Votsis, G., Kollias, S., Fellenz, W., et al.: Emotion recognition in human-computer interaction. IEEE Signal Process. Mag. **18**(1), 32–80 (2001)
22. Moritz, N., Adiloglu, K., Anemuller, J., Goetze, S., Kollmeier, B.: Multi-channel speech ehancement and amplitude modulation analysis for noise robust automatic speech recognition. Comput. Speech Lang, 558–573 (2017)
23. Becattini, N., Masclet, C., Ben Guefrache, F., Prudhomme, G., Cascini, G., Dekoninck, E.: Characterisation of a co-creative design session through the analysis of multi-modal interactions. In: International Conference on Engineering Design 2017, pp. 479–488 (2017)

Reverse Engineering, Digital Acquisition and Inspection

A CAD-Based Tool for Tissue-Mimicking Replica of Human Costal Cartilage

Michaela Servi(✉), Elisa Mussi, Rocco Furferi, and Yary Volpe

Department of Industrial Engineering, University of Florence, Firenze, Italia
michaela.servi@unifi.it

Abstract. Ear reconstruction surgery is a very demanding intervention for plastic surgeons. The results rely heavily on the "artistic skills" of the surgeon who has the task of creating a three-dimensional sculpture using harvested costal cartilage capable of replicating the shape of a physiological ear. By exploiting a combination of Reverse Engineering, CAD modelling, and Additive Manufacturing tools, it is possible to create efficient and highly personalized simulators to train and support the surgeon before and during surgery. In fact, the flexibility of these techniques allows the development of custom-made medical devices, designed directly on the patient's anatomy. In order to make the simulation as realistic as possible, this work focuses specifically on the creation of accurate costal cartilage replicas, both in terms of geometry and mechanical properties of the material, addressing two aspects: the systematization and automation of the CAD mould modelling procedure and the study of the most efficient mould casting materials.

Keywords: Medical simulation · Costal cartilage mould · Ear reconstruction

1 Introduction

In recent years, medical simulation in healthcare has been increasingly used for both training and preoperative planning to ensure greater patient protection. The use of both physical and virtual simulation as a training tool [1] is essential to reduce the risk of errors, increase accuracy, and improve manual skills thanks to the possibility of repeating clinical procedures without injuring the patient. As a preoperative planning tool, simulation is useful for the detailed analysis of the clinical case and for the identification of a specific treatment strategy for the patient: unconventional surgical and access procedures can be analyzed by the surgeon before surgery.

The reduced costs, the speed, and the extreme freedom in terms of treatable geometries offered by modern rapid prototyping techniques, such as Reverse Engineering (RE) and Additive Manufacturing (AM), have revolutionized the field of medical simulation by facilitating the creation of physical simulators that accurately follow patient-specific anatomies and allow for high-fidelity simulation.

Autologous ear reconstruction (AER) surgery makes no exception in taking considerable advantage from the use of physical simulation. This surgery involves the restoration of malformed or absent ear anatomy by modelling costal cartilage harvested from the

C. Rizzi et al. (Eds.): ADM 2021, LNME, pp. 199–208, 2022.
https://doi.org/10.1007/978-3-030-91234-5_20

patient. The ear reconstruction surgery is a very challenging procedure for plastic surgeons. In fact, the results strongly depend on the "artistic skills" of the surgeon in creating a three-dimensional sculpture, which resembles the shape of a physiological ear.

The necessity of achieving an accurate reconstruction and the complexity of the task have pushed the research on this subject [2]. Long and effective training is the prevailing response presented in the literature for making this surgery more accessible to surgeons and, at the same time, to improve its outcome [3]. In this perspective, thanks to the close collaboration with the plastic surgeons of the Meyer Children's Hospital of Florence, preoperative simulation tools and intraoperative tools were developed to improve AER results [4]. The proposed study focuses on the fabrication of custom costal cartilage replicas capable of mimicking the typical characteristics of human cartilage such as haptic return and flexibility. Physical manipulation of such models has the great advantage of providing the physician with the ability to observe the spatial relationship of sites of interest allowing familiarization with the anatomy and the creation of mental maps for use in the operating room.

The manufacturing of accurate simulators typically involves the use of additive manufacturing techniques; the geometry of anatomical parts, such as bone tissue, can be directly printed due to the availability of suitable materials [1]. In the case of soft tissue, however, anatomical models are manufactured through mould making and casting techniques. In fact, the limited range of materials available for 3D printing presents challenges to the direct reproduction of human tissues using this technology. Some soft materials are available on the market for direct 3D printing of human tissues [5], but their high cost and limited material choice still renders mould-building and casting techniques the gold standard. Accordingly, the manufacturing of costal cartilage replicas also require the production of moulds and subsequent casting of suitable materials. Therefore, the cartilage fabrication requires a CAD design phase of the mould starting from the model of the patient specific anatomy.

On the basis of the above considerations, the aim of this work is twofold: i) to define of a systematic and interactive CAD procedure capable of semi-automatically model the replica of the costal cartilage mould and ii) to study and qualitatively identify materials suitable to replicate the physical characteristics of human cartilage.

With reference to the first aim, the automatic procedure will allow to streamline the process and make the mould design phase independent from expert CAD user and from the inter-individual variability of the anatomy of interest.

2 Materials and Methods

As mentioned in the introductory Section, an ideal simulator for training and preoperative planning of the AER procedure consists of an accurate replica of the patient's cartilage that reproduces the specific volumes and shapes. In fact, one of the goals of the simulator is to train the surgeon in identifying where to sculpt the components of the ear structure from the costal cartilage, thus defining the best cutting strategy for an optimization of the available tissue.

The first step in creating the patient-specific cartilage replica is the acquisition of anatomical data through a high-resolution CT scan of the subject's thoracic region.

The diagnostic images in DICOM format need to be post-processed with dedicated software packages in order to isolate and reconstruct the geometry of interest, through the application of segmentation algorithms. Figure 1 shows an example of a segmentation of the costal cartilage performed with Mimics Materialise® software [6] using thresholding algorithms, Boolean operations and manual segmentation. The resulting 3D model is exported in a tessellated file (Standard Tassellated Language format).

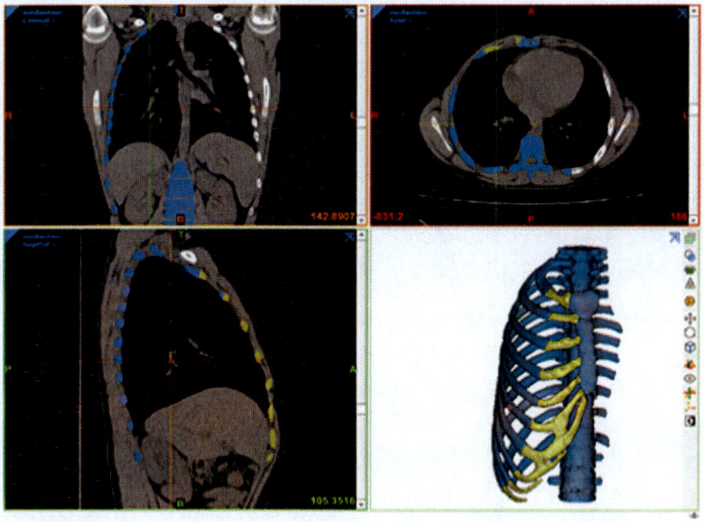

Fig. 1. Example of costal cartilage segmentation from CT scan.

The three-dimensional geometry of the cartilages is used as input for the systematic CAD modelling procedure, consisting of a predefined sequence of CAD operations, as described in the next session.

In addition to geometric accuracy, the costal cartilage replica must comply with the mechanical properties of human cartilage such as flexibility, elasticity, and hardness. Moreover, the material is required to be sutured with 5–0 nylon wires to allow assembly of the individual ear elements. For these purposes, the authors researched and analyzed the required materials to be used for the manufacture of costal cartilage replicas, according to the related literature.

2.1 Mould Systematic Modelling Procedure

The steps of the systematic CAD procedure for the mould modelling are detailed in Table 1. They essentially consist in 1) the creation of a parallelepiped from which the geometry of the cartilages is subtracted and 2) the split of this parallelepiped into two halves so that the physical model can be extracted once the casted material is solidified. The identification of such a dividing surface is the most delicate phase of the procedure, since the geometry of the cartilages is very irregular; in this perspective, it is necessary to find a surface able to fit the shape of the cartilages and, at the same time, simplifies

Table 1. Main steps of the systematic CAD procedure for the design a patient-specific costal cartilage mould.

	The first step of the CAD procedure is the orientation of the model. Accordingly, the origin of the global reference system is moved to coincide with the center of gravity of the model and the principal axes of inertia are aligned with the global reference system (Z-X plane).
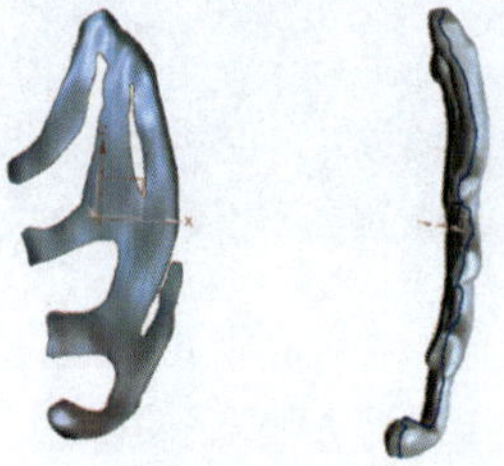 (a) (b)	The best-fit surface of the starting model is evaluated using the NX function "Fit Surface" directly on the STL file. Subsequently, through an intersection Boolean operation between this surface and the STL model, the surface shown in Fig. a is retrieved. Since the extracted surface is extremely complex to be extended and directly used as a parting surface for splitting the mould, it is necessary to simplify it. For this reason, the curve that defines the boundary of such a surface is extracted (Fig. b) using a dedicated operation of Siemens NX (called "extracted geometry") and further processed as described below.
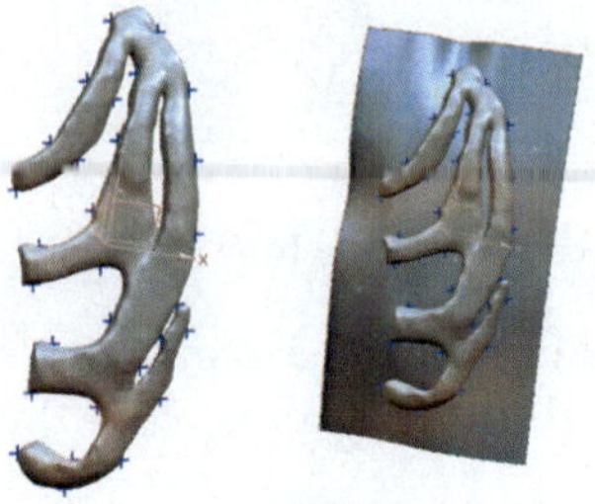 (a) (b)	The curve, extracted in the previous step, is therefore sampled with a set of equal-spaced 30 points (Fig. a). The number of points was defined as a trade-off between the computational cost and the accuracy of the result. The points are used in the next step to create the profiled parting surface as shown in Fig. b. To ensure the performance of the subsequent split process, each edge of the so-calculated surface is extended by 15 mm (this value is a cautionary choice to ensure that in the next step the solid is correctly divided by such surface).
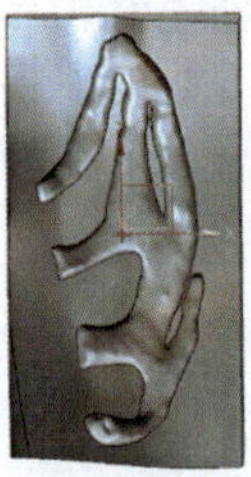	The solid bounding box of the cartilage model is calculated using the NX function "Bounding Body" and applying an offset of 5 mm along each direction to create a valid mould. The costal cartilage model is subtracted from the solid and split into two parts by using the previously created parting surface
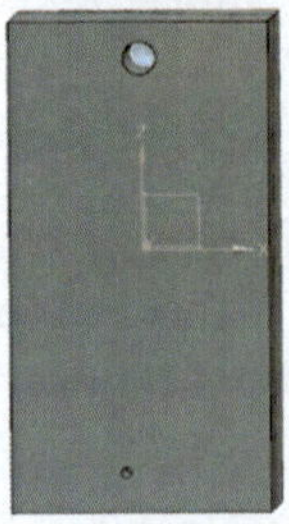	One through-all hole, with a radius of 7 mm is made on the top of the upper shell of the mould in correspondence of the rib cartilages. This hole will be used for the injection of the material. An additional hole, with a radius of 2 mm will be used as air vent. To calculate the position of the holes, an automatic control has been implemented: the program determines the position of the point with maximum Y value of the cartilage model and positions the first injection hole in the surrounding area. The air vent hole is positioned at the point with minimum Z value of the model.

it in order to optimize the manufacturing process. The procedure is carried out through built-in functions available in Siemens NX [7].

To allow the medical staff to generate the patient-specific mould without the need of involving CAD experts, the described procedure was automated using the NX programming API and embedded in a software interface programmed in C++. The application, implemented with the Qt framework [8], allows to manage the user interaction with the routines implemented for CAD modelling. The idea behind the implementation was to provide the surgeon with an intuitive process to manage the creation of the moulds, simplifying as much as possible the modelling phase. The interface, shown in Fig. 2, consists of a form for entering patient information, a tab for accessing the mould modelling procedure, and a 3D navigator.

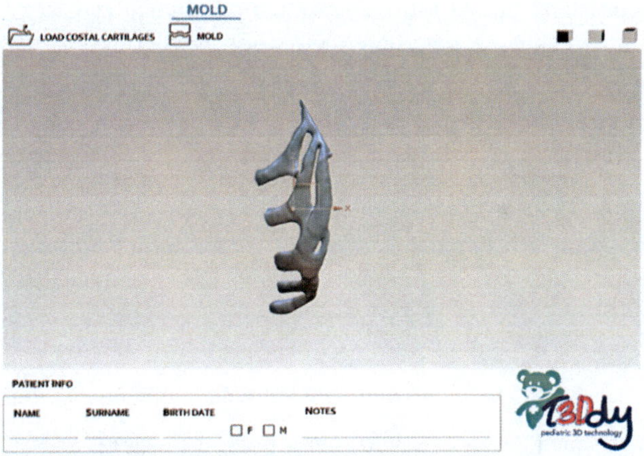

Fig. 2. Graphical User Interface to load the costal cartilage anatomy and call the mould modelling routine.

Through the button "Load costal cartilages", the user is able to load the patient's anatomy. The modelling phases of the mould are then managed with a macro procedure invoked through the "Mould" button.

Once the CAD model of the mould is available, the process of producing the physical model involves two additional steps: 1) manufacturing the mould using AM technologies and 2) casting inside the mould a material that can mimic the costal cartilage tissue.

With regard to the mould manufacturing, FDM (Fused Deposition Modelling) technology can be used employing low-cost 3D printers (such as the Makerbot Replicator + [9] 3D printer) and low-cost materials (such as PLA [10], polylactic acid). With regard to the synthetic material to be cast, an analysis of the most appropriate materials carried out in collaboration with surgeons from the Meyer Children's Hospital is presented in the next paragraph.

2.2 Tissue Mimicking Materials for Costal Cartilage Replica

Synthetic materials, which mimic the characteristics of costal cartilages, are extensively studied in recent literature. In 2009, Yamada et al. used dental impression silicone to create a realistic and accurate replica of cartilage [11]. The authors reported that the replica allowed for a very realistic ear reconstruction using real surgical instruments. In 2011, a rib cartilage replica made of acrylic polyurethane was proposed [12]. The material had cutting and sculpting behaviour similar to that of real cartilage, and thus it was suitable for realistic simulation. In 2016, Berens et al. [13] compared two different materials: 1) a composite of silicone and starch in different ratios, and 2) a dental impression material (vinyl polysiloxane). These studies suggest that dental impression silicone is neither flexible enough nor easily suturable but are adequate for carving operations. On the other hand, the starch-silicone mixture gives satisfactory results in terms of flexibility and carving ability but is less suitable when dealing with suturing ability.

Accordingly, with the aim of identifying the best material for moulding the cartilage, four different materials are here investigated: 1) alginate impression material, 2) mixture of silicone and cornstarch in 1:2 ratio, 3) mixture of silicone and wood powder in 2:1 ratio and 4) mixture of silicone, cornstarch and wood powder in 1:1:1 ratio (as shown in Fig. 3).

Fig. 3. Samples of material: 1) alginate impression material, 2) mixture of silicone and corn starch in proportion 1:2, 3) mixture of silicone and wood powder in proportion 2:1, 4) mixture of silicone, corn starch and wood powder in proportion 1:1:1

Inspired by the study conducted in [13], to evaluate the four materials, three plastic surgeons at Meyer Children's Hospital filled out a questionnaire based on a Likert scale [14]. Surgeons had the opportunity to judge the materials on five specific items (cutting, carving, suturing, bending and general impact) by assigning scores from 1 to 5, where 1 indicates a low similarity and 5 indicates a strong similarity of the material with respect to the actual cartilage. To give the reader a better understanding of the anatomical terminology used below and the choice of items on which the questionnaire was conducted, Fig. 4 shows the auricular elements involved in the procedure and an example of the final result.

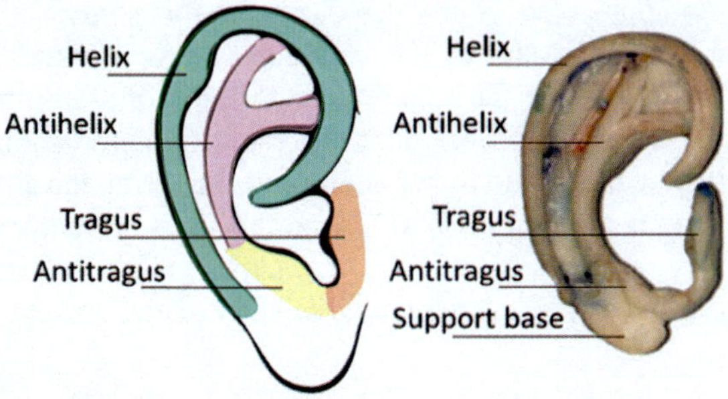

Fig. 4. Anatomical elements of the ear involved in AER surgery and an example of the reconstructed ear framework.

The items were determined in collaboration with the surgeons based on the specific operations that are performed during surgery: 1) cutting the elements, 2) bending the helix, 3) carving the base to recreate the typical grooves in the ear structure, and 4) suturing the 3D structure (see Fig. 5).

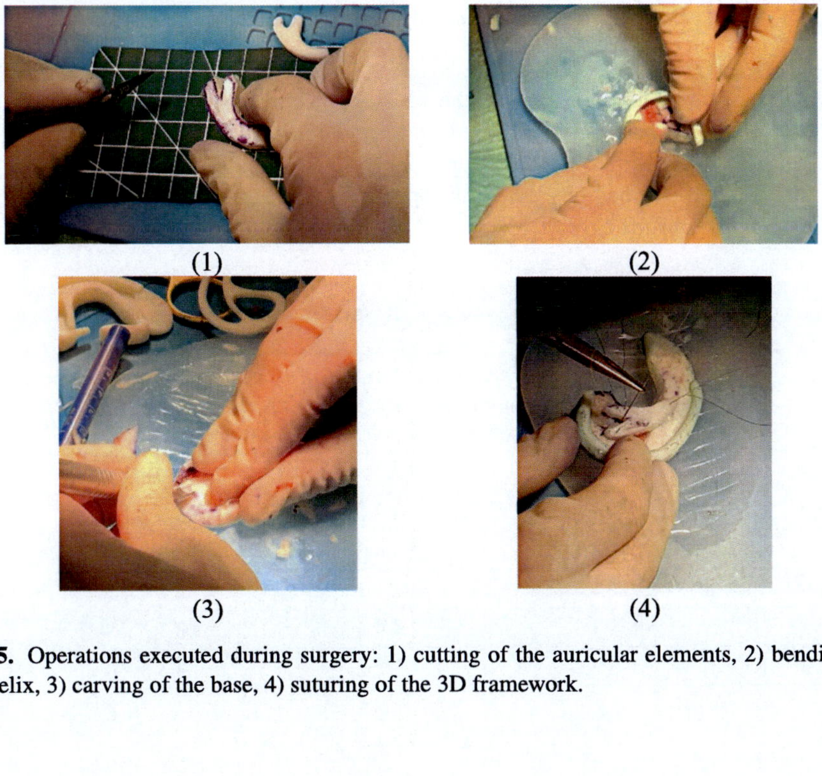

Fig. 5. Operations executed during surgery: 1) cutting of the auricular elements, 2) bending of the helix, 3) carving of the base, 4) suturing of the 3D framework.

3 Results

In collaboration with the Meyer Children's Hospital, 10 tomographic sequences of the thoracic region of patients affected by microtia were collected. Each patient, or legal guardian, signed an informed consent for the use of the data for scientific purposes and the diagnostic examinations were anonymized prior to this study.

In order to test the robustness of the software to model the moulds of different anatomies the automatic routine (launched through the proposed interface) was tested for each costal cartilage. The authors assessed the results on a fundamental aspect, namely the possibility of opening the mould in two shells and extracting the physical replica after casting. In all the examined cases, the moulds models were correctly generated without any complication and, once casted, the physical cartilage model was easily extracted. Figure 6 shows two examples of the generated CAD models.

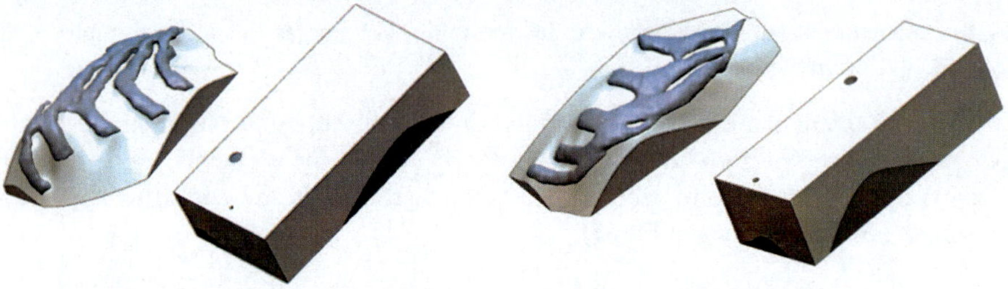

Fig. 6. Example of two automatically generated patient-specific moulds.

As already mentioned, regarding the study of the tissue mimicking material, the evaluation questionnaire was submitted to three pediatric surgeons. For each casted cartilage, each surgeon performed independently the tasks depicted in Fig. 5, without prior knowledge of the composition of the specimen under examination. Table 2 reports the questionnaire results.

Table 2. Results of the questionnaire.

	Alginate Impression material	Mixture of silicone and corn starch	Mixture of silicone and wood powder	Mixture of silicone, corn starch and wood powder
Surgeon	A B C	A B C	A B C	A B C
Cutting	4 3 4	4 4 4	3 2 2	4 3 4
Bending	2 2 1	3 3 2	4 4 3	4 3 3
Carving	4 4 5	2 3 2	2 1 1	2 2 2
Suturing	1 1 1	1 1 1	4 4 4	3 3 3
Overall	3 2 3	3 3 2	3 2 2	3 3 3

The questionnaire revealed that alginate impression material was considered the more similar to human cartilage in terms of carving and cutting; however, this material is non-suturable and unbending. The mixture of cornstarch and silicone was found to be better in terms of carving ability and flexibility, but again the material is not suture-compliant. The mixture of silicone and wood powder is suture resistant but at the same time loses out in terms of carving ability as the material is too elastic. Finally, the fourth material, while suture resistant and flexible, is less carvable than the alginate. On the basis of literature studies, supported by the outcomes of this study, it is not possible to identify an ideal material able to accurately mimic the costal cartilages. However, the best results are obtained with the mixture of silicone, corn starch and wood powder. The material is in fact a good compromise between alginate, which offers excellent carvability but is neither suturable nor flexible, and the mixture of wood powder and silicone, which is easily sutured and sufficiently flexible but cannot be carved.

4 Conclusions

Patients affected by microtia can be successfully treated with autologous reconstruction technique, by restoring the auricular anatomy using the patient's autologous cartilage. The need to accurately plan the intervention to optimize time and results has highlighted the demand for faithful replicas of the anatomy of the patient's costal cartilage. RE and AM technologies have proven to allow a more effective approach to clinical treatment, with important steps forward also in the field of preoperative simulation. It is important to note, however, that CAD modelling of personalized medical devices cannot dispense with the abilities of CAD operators. From this point of view, the automation of CAD procedures, enabling the creation of personalized devices to be independent from the intervention of CAD operators, is a crucial step.

In this work, a systematic procedure for the modelling of moulds is proposed with the aim to reproduce the costal cartilages of each patient treated with AER. The development of a dedicated application allows medical staff (not CAD modelling experts) to be independent in the creation of simulators, enabling surgical planning to be more straightforward.

Another objective of this study was to investigate the most efficient materials to obtain a more realistic simulation.

The automatic modelling procedure has been successfully tested in 10 cases and a material has been identified through a qualitative assessment carried out by means of a specially designed questionnaire. Although not capable of mimicking all the mechanical characteristics of the costal cartilage, at least one of the tested materials proved to be effective in providing a sufficiently realistic simulation. Future developments will be addressed to the automation of the cartilage segmentation from the tomographic sequence of the patient's chest so that the entire process can be performed automatically.

References

1. Mussi, E., et al.: Current practice in preoperative virtual and physical simulation in neurosurgery. Bioengineering **7** (2020)

2. Firmin, F., Marchac, A.: A novel algorithm for autologous ear reconstruction. Semin. Plast. Surg. **25**, 257–264 (2011)
3. Mussi, E., Furferi, R., Volpe, Y., Facchini, F., McGreevy, K.S., Uccheddu, F.: Ear reconstruction simulation: From handcrafting to 3D printing. Bioengineering **6** (2019)
4. Facchini, F., Morabito, A., Buonamici, F., Mussi, E., Servi, M., Volpe, Y.: Autologous ear reconstruction: towards a semiautomatic CAD-based procedure for 3D printable surgical guides. CAD Solutions, LLC, pp 1–5 (2020)
5. J750 digital anatomy 3D printer for lifelike medical models | Stratasys. https://www.stratasys.com/3d-printers/j750-digital-anatomy. Accessed 2 Oct 2021
6. Materialise Mimics | 3D medical image processing software. https://www.materialise.com/en/medical/mimics-innovation-suite/mimics. Accessed 28 Sept 2020
7. Siemens PLM Software Inc NX: Siemens PLM Software.
8. Qt for developers by developers | Cross-platform development. https://www.qt.io/developers. Accessed 17 Apr 2020
9. Replicator+ Desktop 3D Printer | MakerBot. https://www.makerbot.com/3d-printers/replicator/. Accessed 10 Feb 2021
10. MakerBot PLA: Rigid and Glossy Parts for Quick and Easy Prototyping | MakerBot. https://www.makerbot.com/3d-printers/materials/method-pla/. Accessed 10 Feb 2021
11. Yamada, A., Imai, K., Fujimoto, T., Morimoto, K., Niitsuma, K., Matsumoto, H.: New training method of creating ear framework by using precise copy of costal cartilage. J. Craniofac. Surg. **20**, 899–902 (2009)
12. Miyamoto, J., Miyamoto, S., Nagasao, T., Kasai, S., Kishi, K.: Preoperative modelling of costal cartilage for the auricular reconstruction of microtia. Plast. Reconstr. Surg. **128** (2011)
13. Berens, A.M., Newman, S., Bhrany, A.D., Murakami, C., Sie, K.C.Y., Zopf, D.A.: Computer-aided design and 3D printing to produce a costal cartilage model for simulation of auricular reconstruction. In: Proceedings of the Otolaryngology - Head and Neck Surgery (United States), SAGE Publications Inc., vol. 155, pp. 356–359 (2016)
14. Robinson, J.: Likert Scale. In Encyclopedia of Quality of Life and Well-Being Research, pp. 3620–3621. Springer Netherlands (2014). https://doi.org/10.1007/978-94-007-0753-5

A Rapid Prototyping Strategy for Manufacturing of Personalized Bolus

Elisa Mussi[1(✉)], Gabriele Simontacchi[2], Cinzia Talamonti[3], Michaela Servi[1], Rocco Furferi[1], Yary Volpe[1], Lorenzo Protesti[4], Monica Carfagni[1], and Stefania Pallotta[3]

[1] Department of Industrial Engineering, University of Florence, Firenze, Italia
elisa.mussi@unifi.it

[2] Radiation Oncology Unit - Oncology Department, Azienda Ospedaliero Universitaria Careggi, Florence, Italy

[3] Department of Experimental and Clinical Biomedical Sciences, "Mario Serio" University of Florence, Florence, Italy

[4] PIN S.C.R.L. - Polo Universitario Città Di Prato, Prato, Italy

Abstract. In radiotherapy, the treatment of skin tumors requires the application of tissue-equivalent devices in order to avoid the skin-sparing effect, and thus maximize and equalize the dose on the target. In the common clinical practice these devices, called bolus, are standard pads that do not perfectly adapt to the geometry of the patient generating empty spaces at the skin-bolus interface. This phenomenon hinders an efficient planning of the dose to be administered due to the non-adherence and the unpredictability of the position of the air gaps. In this work a workflow for the production of patient-specific bolus in Ecoflex® 00–10 silicone rubber (soft, biocompatible and sterilizable material) is proposed. The process exploits modern Reverse Engineering and Additive Manufacturing technologies to guarantee an accurate compliance with the involved geometries. Significant improvement, with respect to flat boluses, in the adhesion of the medical device to the anatomical region were found through in-vitro testing using replicas of the involved anatomy. Repeatability tests of the positioning performed on the customized bolus led to satisfying results demonstrating the possibility of accurate and predictable treatment planning.

Keywords: Custom bolus · Radiotherapy · CAD modeling · Molding

1 Introduction

In recent years, advances in Reverse engineering (RE) and Additive manufacturing (AM) techniques have led to the development of innovative solutions for personalized medicine. 3D technologies are rapidly becoming a new treatment concept driven by the ability to model patient-specific, more effective devices. In fact, these technologies have revolutionized personalized medicine, as they can be effectively applied to acquire information about the patient's anatomy, to design CAD models based on specific anatomical

C. Rizzi et al. (Eds.): ADM 2021, LNME, pp. 209–219, 2022.
https://doi.org/10.1007/978-3-030-91234-5_21

references, and to produce parts with sufficient precision for various medical applications. The versatility of these techniques makes them suitable for use on a wide range of applications, such as for the fabrication of anatomical models [1], surgical implants [2], surgical cutting templates [3], external supports such as prostheses and orthoses [4], or even the fabrication of biofabrics [5].

In radiotherapy these techniques can be used for the realization of customized bolus for the treatment of superficial tumors [6]. In fact, it is well known that the superficial treatment of tumors with megavoltage-photon beams is affected by the skin-sparing effect, i.e. the absorption of the maximum dose occurs at a certain depth below the surface of the skin, which makes the treatment unsuitable for superficial targets [7]. To overcome this problem, soft tissue equivalent material pads called bolus are placed on the patient's skin to maximize the surface dose and improve uniformity [8]. In the common clinical practice bolus are presented as standard pads that, positioned on irregular surfaces, create unwanted air gap [9]. The scientific community has therefore turned to the use of rapid prototyping techniques to produce boluses that can adapt to the patient's specific anatomy, demonstrating the advantages [10]. The production of custom bolus can exploit two different processes, namely i) the direct printing with plastic materials or ii) molding and casting techniques. In [6] the authors investigated additive manufacturing with plastic materials such as ABS and PLA by modifying fill rates to achieve ideal density and median Hounsfield Unit for treatment. Although the created boluses have been shown to improve therapy over commercial boluses, patient discomfort and considerable air gaps are noted from the use of these techniques [11]. In addition, the rigidity of these boluses has revealed their inability to adequately fit deformable anatomies. Soft boluses allow to overcome these limits: in [12] Park et al. propose a study of bolus made of rubberlike material (Tango plus [13]) realized with the Objet500 Connex 3 3D printer [14]. Two comprehensive studies have been carried out in [11, 15] on customized bolus exploiting molding and casting techniques. In [11] two materials were tested, Clear Flex® 30 and Dragon Skin® 10, by performing a biological evaluation on the biocompatibility of the materials, a dosimetric evaluation, and an evaluation of the physical properties. After identifying the best material (i.e. Dragon Skin® 10), they tested the bolus on four patients by analyzing the reduction of the air gap between the bolus and the skin. The study was repeated in [15] comparing Dragon Skin® 10 with Ecoflex® 00–30. Ecoflex® 00–30 was found to be better solution in terms of adaptability to irregular surfaces, fabrication time and for ease of casting, given the lower viscosity of the material before the curing time.

The great advantage in the realization of customized bolus consists in the possibility, provided by these devices, of an extremely accurate treatment planning. In the common clinical practice, the use of flat bolus implies two main issues: the lack of reproducibility of the bolus positioning between treatment cycles and the unpredictability of the location of the air gaps. The use of custom bolus helps to reduce these problems by allowing reproducible positioning based on the patient's anatomical features and greater predictability of air gaps. This allows for a reliable treatment plan which is a critical aspect in radiotherapy.

The aim of this work was to realize customized bolus for the treatment of skin cancer. For this purpose different silicone materials were investigated taking into account those already researched in the literature and a methodology for the realization of molds was defined, in view of the two phases of 3D printing and the subsequent casting of the silicone materials. The work evaluated the efficacy and possible influence of the adherence of the device to the skin for different geometries of the anatomical region of interest.

2 Materials and Methods

The ideal properties of a bolus for the treatment of superficial lesions are mainly the tissue equivalence, the ability to adapt to different anatomies and the biocompatibility. However, characteristics such as durability, sterilizability and resistance are also fundamental aspects. The integration of Reverse Engineering and Additive Manufacturing techniques allows the realization of patient specific molds with which, through the casting of suitable materials, it is possible to create customized boluses able to respect all requirements. Accordingly, the common workflow for the production of custom bolus, summarized in Fig. 1, involves two main phases: 1) CAD modeling and mold production and 2) bolus fabrication by casting the most suitable material. The following paragraphs discuss in detail these two phases.

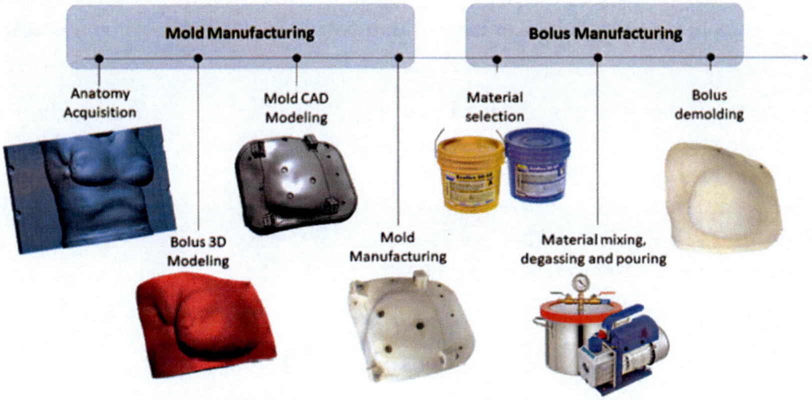

Fig. 1. Workflow for customized bolus realization.

2.1 Mold Manufacturing

The first step of this phase consists in the acquisition of the patient's anatomical data, which can be performed with diagnostic imaging techniques (CT scan, MRI), with optical scanners or with photogrammetric techniques. In this phase the accuracy of the acquisition is fundamental to ensure the geometric accuracy of the final device, in fact the three-dimensional reconstruction of the anatomy of interest is the starting point for the creation of the mold and the subsequent manufacture of the bolus.

Using different digital postprocessing techniques, depending on the technology employed for the acquisition, the raw data are processed in order to generate a three-dimensional model of the anatomical region of interest, without noise and errors; in the case of a tomographic sequence, for example, it is necessary to segment the region of interest with thresholding techniques. The anatomical model is imported into a CAD software for the 3D mold modeling. The modeling requires as first step the definition of the exact contours of the region interested by the treatment, therefore the contours of the bolus. To manage this aspect the intervention of the physician is required, in order to evaluate the extension of the pathological region and the safety margins to be adopted. Figure 2a shows an example of indication of the bolus boundaries, drawn by the physician.

(a) (b)

Fig. 2. Example of the definition of personalized bolus contours in the mammary region.

The model of the device is obtained by cutting the reference mesh according to the contours indicated by the physician and thickening the obtained entity by the desired thickness. In this way, the digital model of the object to be manufactured is obtained (Fig. 2b). Through a sequence of well-known CAD operations of fitting, extrusion, fillets and booleans, the geometry of the mold is obtained, visible in Fig. 3. The entire CAD procedure is carried out with Geomagic Design X software.

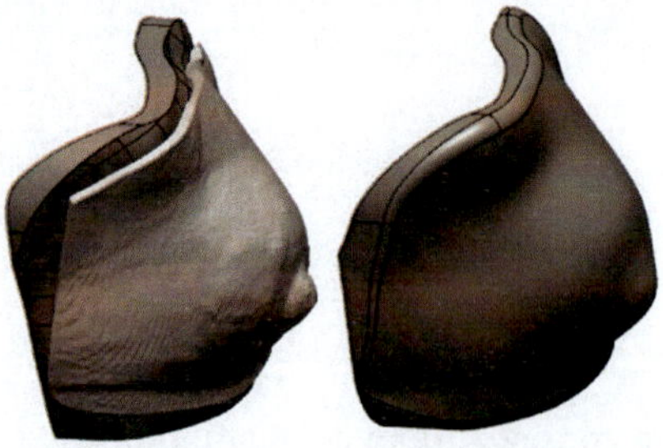

Fig. 3. Initial CAD model of the mold for a customized bolus.

In this work the thickness of the bolus is set equal to 5 mm in accordance with other studies in the literature [12].

To be used operationally for casting silicone materials, the mold must be further refined. In a second modeling phase, the operator provides the realization of injection holes, to allow the casting of the material inside the mold, vent holes, to avoid the formation of air bubbles during the casting, and pass-through holes with screw-nut junction to hold together the two halves during the curing time (Fig. 4). Considering the significant size of the anatomical region to be treated, in this work a mold manufacturing strategy was devised to divide the upper shell into three parts. As shown in Fig. 4, the upper shell of the mold is composed of three pieces that are held together during casting by screw-nut junctions to facilitates the opening phase during the demolding. The digital model of the mold can be manufactured with additive manufacturing techniques, these technologies offer great advantages in terms of both time and cost for the rapid manufacturing of highly customized prototypes that do not involve serial production.

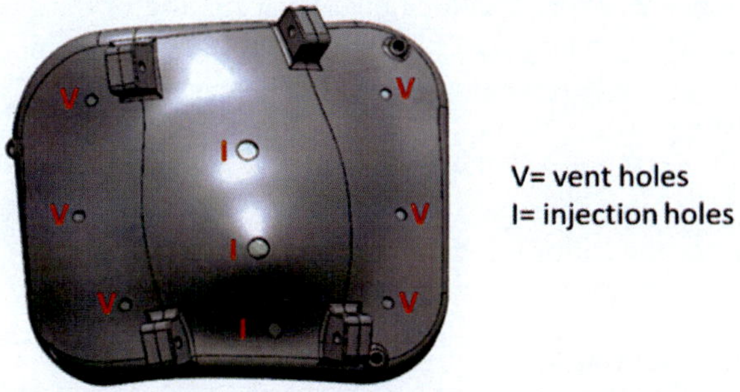

Fig. 4. Final printable model of the mold.

2.2 Bolus Manufacturing

The identification of the most suitable material involved an initial phase of literature review concerning the use of soft biocompatible materials for the fabrication of patient-specific bolus. The most important studies have been conducted by Park et al. [11, 15]; the authors investigated a set of silicones offered by Smooth-on® company, assessing the possibility to use these materials in radiotherapy treatment, certifying the necessary biocompatibility: skin irritation, skin sensitization and cytotoxicity according to ISO 10993–5. Among the materials offered by the company, Park et al. tested the following materials (listed in descending order of hardness): Clear Flex® 30 (shore hardness 30-A), Dragon Skin® 10 (shore hardness 10-A), Ecoflex® 00–50 (shore hardness 50–00) and Ecoflex® 00–30 (shore hardness 30–00). The analysis of the results shows an improvement, as the hardness decreases, in the adhesion of bolus to the surface to be treated. In agreement with the Department of Experimental and Clinical Biomedical Sciences of the Azienda Ospedaliera Universitaria di Careggi, Florence (Italy), observing the conclusions reported in the above mentioned papers, it was decided to investigate a material of the Ecoflex® line, which, from previous studies, resulted to be the best in terms of tissue equivalence and adhesion. Accordingly, it was decided to use a material with a lower hardness than those already investigated, i.e. Ecoflex® 00–10 (shore hardness 10–00) in order to create customized bolus able to better adapt to the specific anatomy of the patient. The rationale behind this choice is represented by the need to create a bolus that, thanks to its high deformation capacity, can offer the possibility to cover any small changes in the anatomy due, for example, to the progressive healing between treatment sessions, or small changes in the pose. This material presents a density equal to 1.04 g/cc, a tensile strength equal to 120 psi, an elongation at break equal to 8 psi.

In order to confirm [16] the suitability in terms of tissue equivalence, a first test was conducted on a circular sample (thickness 5 mm, radius 30 mm) by scanning with a CT, obtaining values od Hounsfield Unit equal to 159, comparable to the one of human skin. Having defined this aspect, Ecoflex® 00–10 represents a super-soft material that can be used for the realization of personalized bolus, able to respect all the necessary properties.

After identifying the material (in this study the Ecoflex® 00–10), the production workflow of the bolus provides for three phases: 1. mixing of the two components of the silicone rubber (base and catalyst) 1:1 ratio by weight or volume, 2. degassing the mixture in a vacuum chamber to eliminate air bubbles incorporated during the previous phase, and 3. casting inside the mold. Finally, after the curing time specific for the silicone rubber (4 h for Ecoflex® 00–10) the bolus is extracted opening the mold.

The following chapter describes and examines the parameters for evaluating the custom boluses produced following the workflow defined in this study.

3 Results and Discussion

The bolus has the main function of creating a layer of replacement tissue in the anatomical region of the pathology in order to shift the dose on the patient's skin to avoid the skin sparing effect, that is the deposit of the higher dose in deep tissues. In order to optimize the efficacy of treatment, it is therefore of primary importance that the bolus precisely matches the profile of the anatomical area on which it is positioned, reducing the creation of air gaps; such air gaps have the potential to disrupt the distribution of the surface dose. In addition, it is of paramount importance to ensure that the dose distribution established during treatment planning coincides with the one administered in actual radiotherapy sessions. Accordingly, it is critical that the location of the bolus is repeatable between sessions.

The effectiveness of a bolus can therefore be measured taking into account two main aspects: i) the ability to adhere to the skin and ii) the repeatability of the positioning phase on the anatomical region being treated.

These two aspects were evaluated by considering boluses of two different anatomical regions, the mammary area and the nasal district, using diagnostic images.

The patients signed the informed consent form to allow data usage for research purposes, after being anonymized by the physician.

To evaluate adhesion and repeatability of placement, replicas of the anatomical district of interest were made using additive manufacturing techniques. In detail, the customized bolus was placed on the replica and the CT sequence was acquired; the tomographic images thus obtained were analysed to isolate and reconstruct the 3D model of the bolus and of the anatomical replica on which it was placed. To quantify the air gap, the two models were compared with a deviation map in the contact surfaces evaluating the mean deviation and the points of maximum deviation. To assess repeatability, these comparisons were performed on three successive placements of the bolus on the anatomical replica; each test involved complete removal and subsequent repositioning of the bolus on the replica by an experienced operator. The mean, maximum and standard deviation values for the bolus deviation of the two examined anatomical regions for the three tests conducted are shown in Table 1.

Table 1. Mean, maximum and standard deviation deviation values for the repeatability assessment.

	Nasal district		Mammary region	
I	μ: 1.19 mm σ: 1.01 mm	Max distance 4.99 mm	μ: 1.32 mm σ: 2.17 mm	Max distance 4.99 mm
II	μ: 0.69 mm σ: 0.56 mm	Max distance 4.98 mm	μ: 0.36 mm σ: 0.67 mm	Max distance 4.99 mm
III	μ: 0.93 mm σ: 0.78 mm	Max distance 4.98 mm	μ: 0.77 mm σ: 1.3 mm	Max distance 4.98 mm

The values reported in Table 1 provide an indication of the global mean deviation between the two surfaces. In order to qualitatively assess the local repeatability of the air gaps, and therefore the predictability of their localization in the planning phase, Fig. 5 reports a comparison of the deviation maps in which it is observed that the local trend of the deviation is comparable.

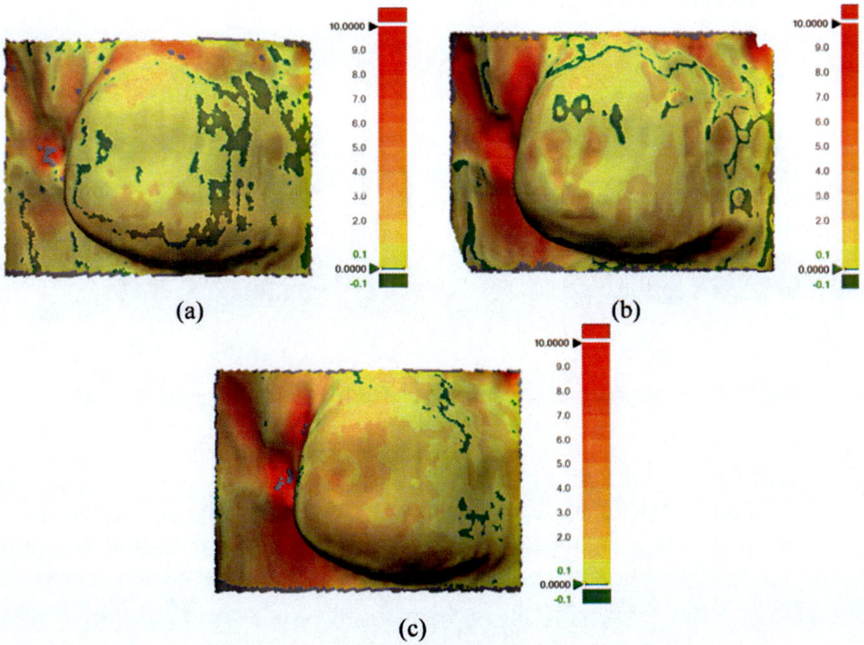

Fig. 5. Qualitative evaluation of repeatability.

The overall evaluation of the effectiveness of the custom bolus was performed by comparing the deviation of the bolus from the anatomic surface for both custom and flat devices (commonly used in clinical practice). Table 2 compares the mean values of the three tests presented in the previous table and the deviation values obtained with flat bolus.

Table 2. Comparison of mean deviation in case of flat and custom bolus.

	Nasal district	
Flat bolus	μ: 4.38 mm σ: 4.12 mm	Max distance 23.73 mm
Custom bolus	μ: 0.94 mm σ: 0.78 mm	Max distance 4.98 mm
Mammary region		
Flat bolus	μ: 5.15 mm σ: 4.84 mm	Max distance 48.82 mm
Custom bolus	μ: 0.81 mm σ: 1.38 mm	Max distance 4.99 mm

As an example, Fig. 6 shows the deviation map between the anatomical region of the nose and the bolus, both in case of custom bolus (Fig. 6a) and standard bolus (Fig. 6b). The same for the mammary area is shown in Fig. 7.

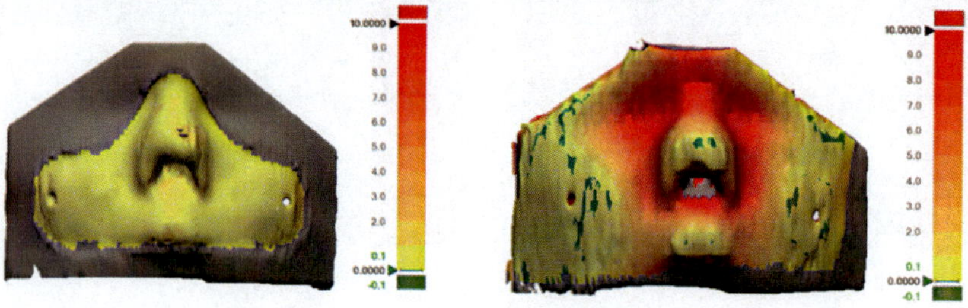

(a) (b)

Fig. 6. Deviation map for custom (a) and flat (b) bolus in the nasal district.

For a qualitative evaluation by the reader, Fig. 8 shows the tomographic sections and related 3D reconstruction of the flat bolus of the mammary region (Fig. 8a) and the custom bolus (Fig. 8b). From the image, it can be observed the better fit of the custom bolus to the irregular surface compared to the flat bolus that is not able to conformally cling to skin folds and crests.

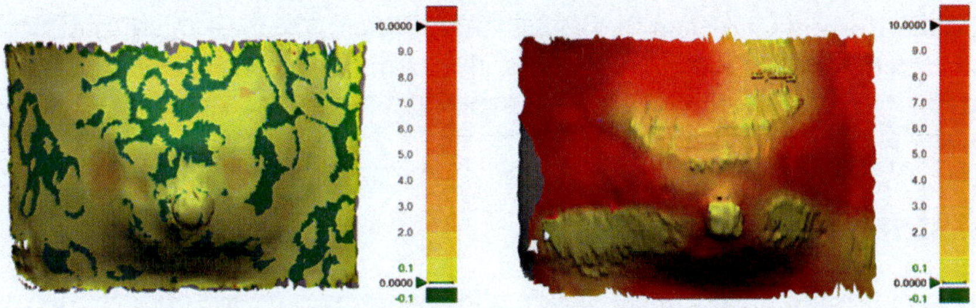

(a) (b)

Fig. 7. Deviation map for custom (a) and flat (b) bolus in the mammary region.

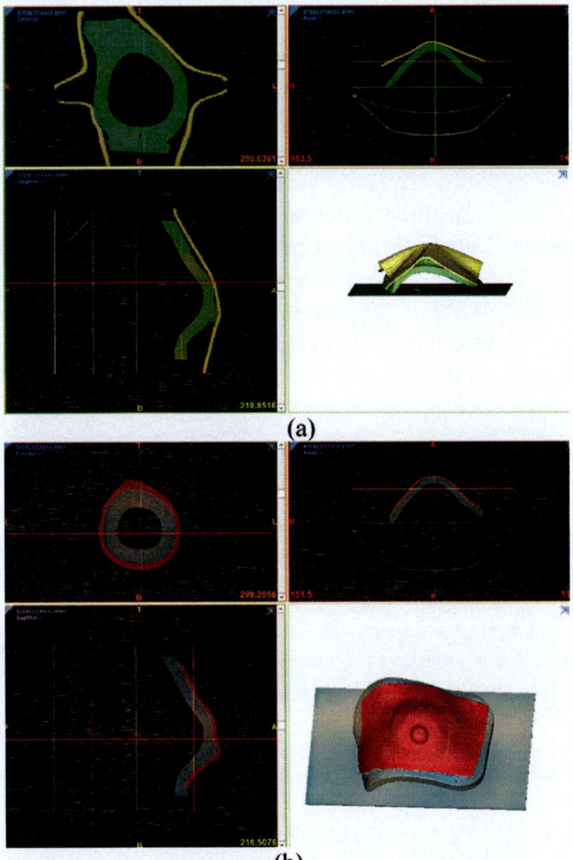

(a)

(b)

Fig. 8. Example of CT scan and 3D reconstruction for flat (a) and custom (b) bolus.

The obtained deviation values emphasize how personalized bolus allow for significantly higher adhesion performance than standard ones. While there is a significant decrease in the mean deviation values (from 4.38 mm to 0.94 mm for the nasal region and from 5.15 mm to 0.81 mm for the mammary region), the most significant decrease occurs in terms of the maximum distance that corresponds to the maximum air gap (from 23.73 mm to 4.98 mm for the nasal region and from 48.82 mm to 4.99 mm for the mammary region).

4 Conclusions

In radiation therapy, the treatment of skin tumours involves the use of rubber-like pads (flat bolus) to increase the dose to the skin and improve dose uniformity. The use of such medical devices in recent years has been reconsidered due to poor adhesion to the tissues interested by the treatment. Thanks to the advent and consolidation of Reverse Engineering and Additive Manufacturing technologies in the medical field, patient-specific alternatives to standard bolus have emerged. The customization of these medical devices allows a better adhesion and therefore an increase in treatment efficacy. In this study, a workflow for the fabrication of custom malleable bolus with a soft, biocompatible material, not yet tested in the literature, was presented. Specifically, the proposed workflow is composed of two main phases: a first phase of realization of the custom mold and a second phase consisting in the realization of the bolus through casting methods. Two types of bolus, manufactured following the proposed procedure, were evaluated in terms of adhesion and repeatability of in-vitro positioning to pave the way for real in-vivo applicability. The tests showed a clear qualitative and quantitative improvement of adhesion and a repeatability of the positioning that ensures accurate and reproducible dose planning. Future developments include in-vivo positioning studies, which can then take into account tissue deformability and geometry changes between different treatment sessions.

References

1. McMenamin, P.G., Quayle, M.R., McHenry, C.R., Adams, J.W: The production of anatomical teaching resources using three-dimensional (3D) printing technology. Anat. Sci. Educ. **7**, 479–486 (2014)
2. Willemsen, K., Nizak, R., Noordmans, H.J., Castelein, R.M., Weinans, H., Kruyt, M.C: Challenges in the design and regulatory approval of 3D-printed surgical implants: a two-case series. Lancet Digit. Heal. **1**, e163–e171 (2019)
3. Leng, S., et al.: Customized cutting template to assist Sternotomy in Pectus Arcuatum. Ann. Thorac. Surg. **107**, 1253–1258 (2019)
4. Dodziuk, H.: Applications of 3D printing in healthcare. Kardiochirurgia i Torakochirurgia Pol. **13**, 283–293 (2016)
5. Hollister, S.J., et al.: Integrating image-based design and 3D biomaterial printing to create patient specific devices within a design control framework for clinical translation. ACS Biomater. Sci. Eng. **2**, 1827–1836 (2016)
6. Burleson, S., Baker, J., Hsia, A.T., Xu, Z.: Use of 3D printers to create a patient-specific 3D bolus for external beam therapy. J. Appl. Clin. Med. Phys. **16**, 166–178 (2015)
7. Loevinger, R.: A formalism for calculation of absorbed dose to a medium from photon and electron beams. Med. Phys. **8**, 1–12 (1981)
8. Vyas, V., et al.: On bolus for megavoltage photon and electron radiation therapy. Med. Dosim. **38**, 268–273 (2013)

9. Butson, M.J., Cheung, T., Yu, P., Metcalfe, P.: Effects on skin dose from unwanted air gaps under bolus in photon beam radiotherapy. Radiat. Meas. **32**, 201–204 (2000)
10. Canters, R.A., et al.: Clinical implementation of 3D printing in the construction of patient specific bolus for electron beam radiotherapy for non-melanoma skin cancer. Radiother. Oncol. **121**, 148–153 (2016)
11. Park, J.M., et al.: Bio-compatible patient-specific elastic bolus for clinical implementation. Phys. Med. Biol. **64** (2019)
12. Park, J.W., Oh, S.A., Yea, J.W., Kang, M.K.: Fabrication of malleable three-dimensional-printed customized bolus using three-dimensional scanner. PLoS One **12** (2017)
13. Tango: A Soft Flexible 3D Printing Material | Stratasys. https://www.stratasys.com/materials/search/tango. Accessed 16 Feb 2021
14. J8 Series 3D Printers Designed for Designers: J850, J835, & J826 | Stratasys. https://www.stratasys.com/3d-printers/j8-series. Accessed 16 Feb 2021
15. Son, J., Jung, S., Park, J.M., Wu, H., Kim, J.: Evaluation of platinum-catalyzed silicones for fabrication of biocompatible patient-specific elastic bolus (2020)
16. Steinmann, A., et al.: Developing and characterizing MR/CT-visible materials used in QA phantoms for MRgRT systems. Med. Phys. **45**, 773–782 (2018)

CNN Approach for Monocular Depth Estimation: Ear Case Study

Roberto Magherini(✉), Michaela Servi, Elisa Mussi, Rocco Furferi, Francesco Buonamici, and Yary Volpe

Department of Industrial Engineering of Florence, Via di S. Marta 3, 50139 Florence, Italy
roberto.magherini@unifi.it

Abstract. The realization of medical devices to assist the surgeon in autologous auricular reconstruction (personalized surgical guides) requires the use of depth maps images obtained from 3D scans of the patient's profile. In order to make the process of ear geometry acquisition and depth map realization faster, more comfortable for the patient and easily accessible by hospital staff, this work proposes a system able to create depth maps from a single RGB image. The proposed approach involves the integration of tools based on convolutional neural networks to build a modular system capable of isolating the ear from the profile, creating the corresponding depth map, and refining it to correct inaccuracies. The system was trained and tested using a database of human profile images and corresponding depth maps made available by the University of Notre Dame. To evaluate the result in this preliminary study, standard metrics such as mean square error and structural similarity were used, yielding results suitable for the targeted application.

Keywords: CNN · Depth map estimation · 3D vision · Ear reconstruction

1 Introduction

In recent years, personalized medicine is rapidly transforming the healthcare sector by overcoming the paradigm of standardized medicine and offering treatments built on the specific characteristics of the individual. This approach has proven to make treatment more effective and outcome more predictable in many medical fields [1]. In particular, the surgical field has significantly benefited from the creation of medical devices tailored to each patient using modern 3D modeling and manufacturing techniques [2]. In recent years, autologous ear reconstruction (AER) surgical procedure is also taking advantage of these technologies in order to personalize and improve the treatment. The AER procedure is performed in case of total or partial absence of the external ear and involves: 1) the harvest of cartilaginous material from the costal region of the patient 2) the manual modeling of the tissue in order to create a framework that reproduces the features of the ear and 3) the insertion of this framework in correspondence of the pathological auricular region in a subcutaneous pocket [3]. To ensure the harmony of the patient's face after surgery, the framework is modelled using the contralateral mirrored ear as guide. The procedure is very complex for the surgeon who has to manually cut and sew the cartilages

C. Rizzi et al. (Eds.): ADM 2021, LNME, pp. 220–228, 2022.
https://doi.org/10.1007/978-3-030-91234-5_22

to reproduce the auricular anatomy and the final outcome is therefore strongly related to the "artistic skills" of the physician performing the surgery. For this reason, in the last years several techniques are being investigated to realize surgical aids able to support and help the physician in this procedure [4]. Recently 3D acquisition and CAD-based modeling approaches have allowed great progress in the realization of guides for AER outperforming the results obtained using two-dimensional aids [5, 6].

To assist the construction of the framework, previous works from the authors have focused on the definition of a workflow for the realization of patient-specific surgical guides using Reverse Engineering and Additive Manufacturing techniques and dedicated algorithms [5]. The workflow involves the acquisition of the three-dimensional geometry of the patient's healthy ear and the modeling and the manufacturing of 3D printed surgical guides from the depth map of the ear obtained from 3D scanning. Currently, the acquisition is made by means of a high-precision optical scanner, the Artec Eva [7] and the depth map is retrieved through appositely devised algorithms. The process therefore involves, in addition to the availability of the optical scanner mentioned above, two main steps: i) the acquisition of the ear by an experienced user who can perform the acquisition of the 3D model, and ii) a post-processing of the model that includes the orientation and the realization of the depth map [8].

With the idea of making the process of acquisition and realization of the depth map more accessible to non-expert users such as hospital staff, this work focuses on the development of a system based on deep learning techniques that, through a single RGB image of the ear, can create the depth image which by definition is an image that contains information about the distance of the surfaces of objects in the scene from a point of view.

Searching for geometric relationships between objects in the scene by using neural network-based approaches is a topic extensively investigated in the literature [9, 10]. This is because depth perception is a key component for example for autonomous systems that interact in the real world, such as warehouse robots, self-driving cars, etc. [10]. Many systems study the arrangement of objects in the scene, but they are usually applied to external scenario contexts that require a much lower resolution compared to the one required for the application under consideration. For this reason, the systems available in the literature are not directly applicable for the realization of depth maps of the auricular anatomy. Accordingly, the present work focuses on the study of alternative approaches suitable to create depth maps with higher resolution. In addition, to make the realization of the depth map as simple as possible this work focuses on approaches based on monocular depth estimation [11, 12], in which the depth image is created from a single photo. This aspect is of considerable relevance if we contextualize the acquisition of the ear on pediatric subjects (generally uncooperative) and in a hospital environment (not always equipped with modern acquisition technologies).

The devised approach, detailed in the following chapters, takes as input the side picture of the face of subject and returns the depth map of the ear using three different architectures: 1) a Faster Region-based Convolutional Neural Network (R-CNN) [10] that has the task of isolating the ear region from the face image; 2) a cycle Generative Adversarial Network (GAN) [13] that takes as input the cropped anatomical region and creates the depth map; 3) a semantic-aware approach [14] to refine the depth map created in the previous step. If on the one hand the use of Faster R-CNN is well established for this type of task (object recognition), to determine the architecture for the creation of the depth map it was necessary to study the recent literature in order to identify the best approach. The cycle GAN architecture was identified as the best choice as it supports relatively high resolutions compared to other approaches [12], not negligible aspect for the success of the workflow.

2 Materials and Methods

The present work is intended to use machine-learning systems for the creation of a depth map of a human ear from a single RGB image with a level of accuracy acceptable for the creation of custom surgical guides for auricular reconstruction.

In order to train and test the system, a subset of the ND-Collection F [15] and ND-Collection G [16] made available by the University of Notre Dame was employed, containing depth maps and corresponding 2D images of 302 and 235 human subject profiles, respectively.

Figure 1 shows the general structure defined to create the depth map of the ear region from a profile photo.

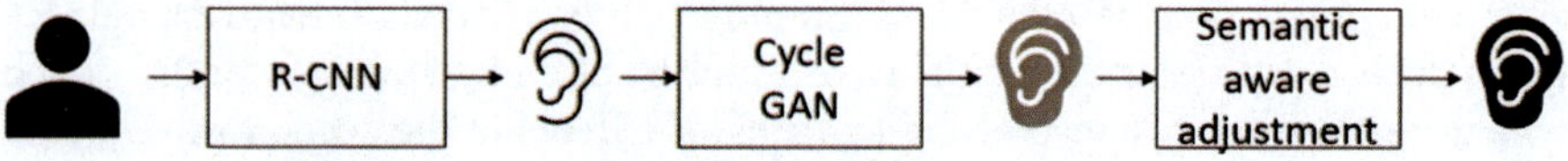

Fig. 1. General structure diagram: profile image of a person given as input to the R-CNN that extracts the ear, that it is used by the Cycle GAN to make the depth map, which is finally adjusted using a semantic-aware module.

All the architectures used are explained in more detail in the following paragraphs.

2.1 Faster R-CNN

The first step of the workflow involves the extraction of the ear region from the RGB images and from the depth map images in order to use all the resources on the creation of the ear depth map without considering the remaining portion of the scene. To do this, the algorithm is based on the use of a R-CNN, which was trained specifically for the detection of ears, implementing procedures known in the literature to train this type of network to specific tasks [17].

R-CNNs are a class of techniques that share the common trait of using deep models to pursue object detection. An R-CNN model first selects several proposed regions from an image and then labels their categories and bounding boxes. Then, a CNN is used to perform features extraction for each region and use them to predict the regions' categories. In our case, a Faster R-CNN Inception v2 [18], with the architecture illustrated in Fig. 2, was used; the specific CNN architecture used is the Inception v2. This paper does not report implementation details of these architectures as they are well known in the literature.

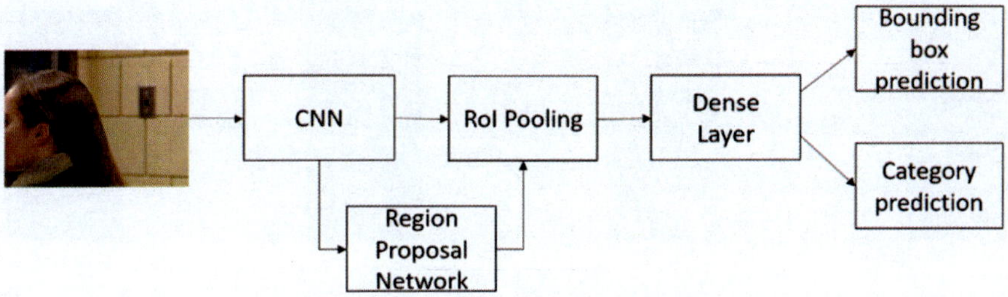

Fig. 2. Faster R-CNN ear detection

2.2 Cycle GAN

The region of the ear cropped from the original image is used to train a cycle GAN that has the task of creating the depth image from the single photo. The GAN is a class of machine learning frameworks, in which two neural networks contest with each other in a game (zero-sum game) [19]. The cycle GAN is a variation that combines the properties of condition constraints and cycle consistencies to use the GAN architecture in Image-to-Image translation task effectively.

Cycle-GANs can be used to alter a given input image to a distribution of the target domain. Specifically, an image is taken from input domain X and transformed into an image of the target domain Y without a one-to-one mapping between the images. A generator is used to map the image to the target domain and the quality of the result image is checked using a discriminator which pushes the generator to perform better.

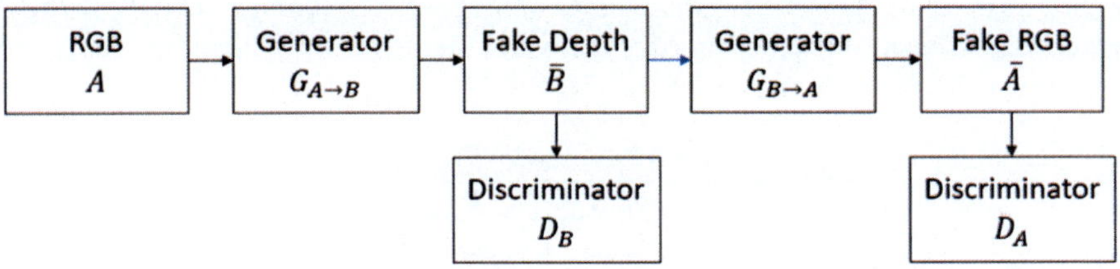

Fig. 3. Simplified scheme of a cycle GAN architecture

2.3 Depth Map Refinement

Finally, to increase the definition and accuracy of the generated depth maps, a semantic-aware module is used, which receives as input both the RGB image and the estimated depth map. The module uses the RGB image to get the semantic context, that is used to create several masks which are used to perform a max pooling operation over the pixels inside the same mask within the estimated depth map. This approach has already been used in the literature to correct errors generated in depth images [14].

Figure 4 shows an example of the entire process applied to a profile photo, showing the results for each step.

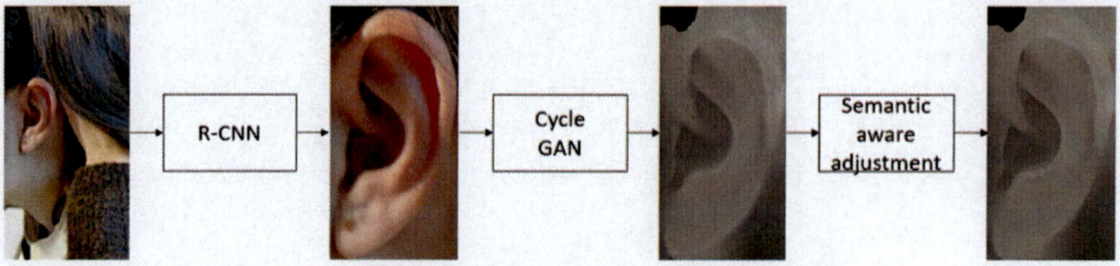

Fig. 4. An example of the output of each module.

3 Results

In this section the results obtained through the proposed architecture are shown, divided in two sections: the first one related to the ear detection task, and the second one related to the depth map generation from RGB images. For both tasks the dataset was randomly split into two parts, 70% for training, and 30% for testing.

3.1 Faster R-CNN

The Faster R-CNN architecture was implemented exploiting the Tensorflow APIs [20] and the model was fine-tuned over the ear dataset. To generate the labels for the training, it was used the free software LabelImg [21].

The obtained results are shown in Fig. 5. The first graph shows how the loss of the network decreases with the increasing number of training epochs. In the second graph is represented the function AP@0.75IOU [22]. The AP represents the Average Precision (AP) of the network, i.e. the average of the precision scores. The AP@0.75IOU corresponds to the AP considering only the bounding boxes for which the Intersection Over Union (IOU) is greater than 75%, where IOU is computed as follow:

$$IOU = \frac{Area(B_{predicted} \cap B_{ground\_truth})}{Area(B_{predicted} \cup B_{ground\_truth})}$$

At the end of the overall training process an AP of 97.63% is reached over the testing set.

Figure 6 shows an example of the output at this stage where, from the profile, the ear region is detected.

3.2 Cycle GAN and Depth Map Refinement

The architecture of the cycle GAN was realized by exploiting a standard PyTorch implementation of the 4 necessary components, the two Generators and the two Discriminants. The Adam optimizer was exploited as suggested by Goodfellow et al. [19]. Regarding GAN results, it is quite complex to evaluate the accuracy of the generated images since there is no standard measure described in literature [23].

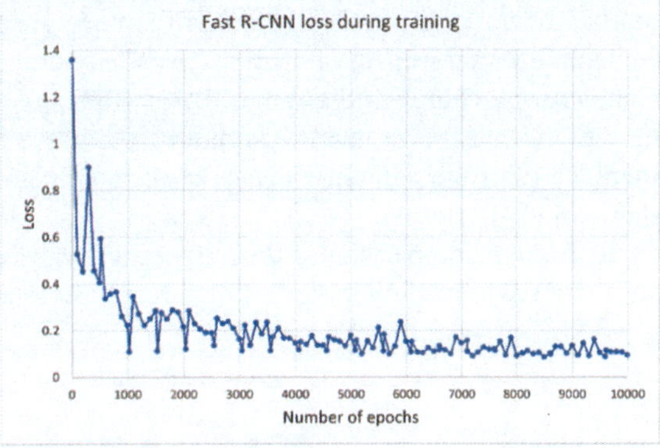

(a)

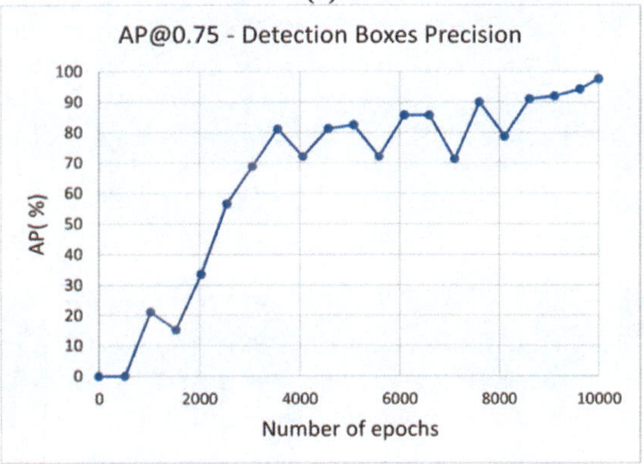

(b)

Fig. 5. Graph of the loss (a) and of the average precision (b) achieved by the Faster R-CNN

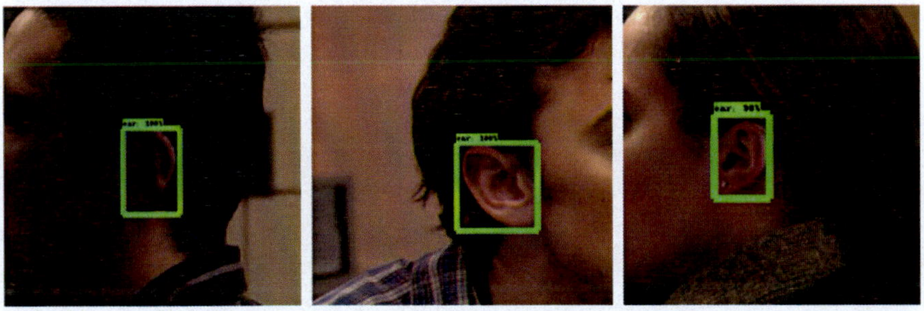

Fig. 6. Output of the Faster R-CNN module.

Considering the difficulty in evaluating the results of GAN architectures, in this work it was decided to use two standard metrics, the mean square error (MSE) and the structural similarity (SSIM), often used to assess the quality of images after they have undergone changes. The MSE indicates the cumulative error between the generated image and the original image therefore lower values indicate more similar images [24].

SSIM in recent years has become an accepted standard among image quality metrics. This technique evaluates the visual impact of changes in image luminance, contrast and structure [25], and can assume values in the range [0,1]. MSE and SSIM values were evaluated comparing the original image and the output image at the final step, i.e. after the depth map refinement with the semantic-aware adjustment module implemented using the code available in [14].

Figure 7 shows a qualitative comparison of the output and original depth maps. The MSE and SSIM values were averaged over the test data resulting in an average MSE of ~ 0.07 and an average SSIM of ~0.80.

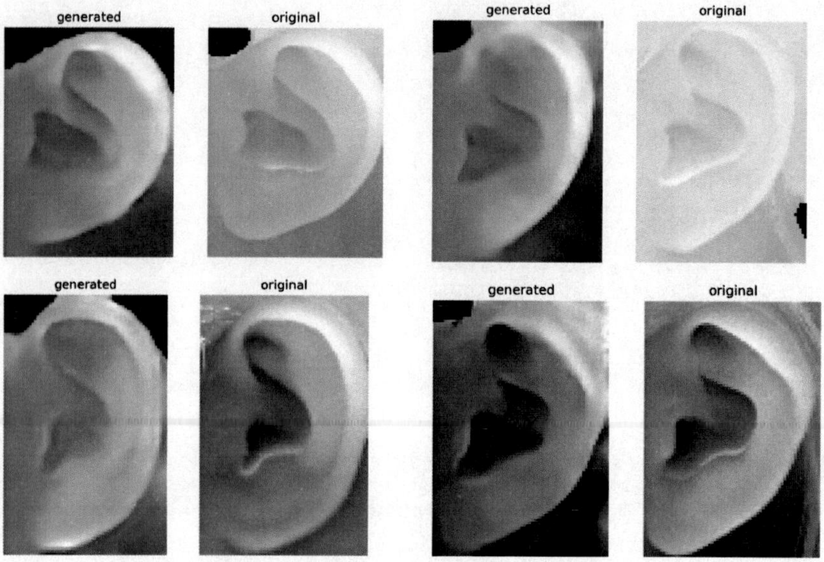

Fig. 7. A subset of the depth map generated on the test set.

4 Conclusions

Autologous ear reconstruction has been the subject of several studies in the literature involving medical and engineering fields. The great interest is due on one hand to the importance from the psychological point of view of a satisfactory outcome for the patient, on the other hand to the intrinsic difficulty in the manual construction of the ear reported by surgeons. The authors are working on the development of a user friendly and fast system for the fabrication of patient-specific devices that can assist the physician during surgery. In this context, this work aimed to define a system to create depth map images of the patient's ear from a single profile photo. The analysis of the literature revealed the lack of approaches that can provide a sufficiently detailed depth map. In this work the most suitable architectures for the purpose were first identified, they were then implemented and trained on a specific dataset and the interfaces between the different architectures were defined in order to obtain an efficient workflow. The results obtained for the single steps were shown and the quality of the final result was quantitatively evaluated in terms of MSE and SSIM, obtaining satisfying results considered suitable for the final application. The implemented system constitutes a disruptive innovation that considerably streamlines the workflow currently foreseen for the realization of surgical guides. Future developments envisage the evaluation of the depth maps produced by the network by performing an acquisition campaign on an adequate number of subjects, reconstructing the corresponding point clouds from the depth images by controlling the intrinsic acquisition parameters.

References

1. Dodziuk, H.: Applications of 3D printing in healthcare. Kardiochirurgia i Torakochirurgia Pol. **3**, 283–293 (2016)
2. AlAli, A.B., Griffin, M.F., Butler, P.E.: Three-dimensional printing surgical applications. Eplasty **15**, e37 (2015)
3. Nagata, S.: A new method of total reconstruction of the auricle for microtia. Plast. Reconstr. Surg. **92**, 187–201 (1993)
4. Storck, K., et al.: Total reconstruction of the auricle: our experiences on indications and recent techniques. Biomed Res. Int. (2014)
5. Facchini, F., Morabito, A., Buonamici, F., Mussi, E., Servi, M., Volpe, Y.: Autologous ear reconstruction: towards a semiautomatic CAD-based procedure for 3D printable surgical guides. In: CAD Solutions, LLC., pp. 1–5 (2020)
6. Zhu, P., Chen, S.: Clinical outcomes following ear reconstruction with adjuvant 3D template model. Acta Otolaryngol. **136**, 1236–1241 (2016)
7. Professional 3D Scanners I Artec 3D I Best 3D scanning solutions. https://www.artec3d.com. Accessed 16 Jun 2021
8. Mussi, E., Servi, M., Facchini, F., Furferi, R., Governi, L., Volpe, Y.: A novel ear elements segmentation algorithm on depth map images. Comput. Biol. Med. **129**, 104–157 (2021)
9. Alonso, M.: Y-GAN: a generative adversarial network for depthmap estimation from multi-camera stereo images. In: arXiv (2019)
10. Lore, K.G., Reddy, K., Giering, M., Bernal, E.A.: Generative adversarial networks for depth map estimation from RGB video. In: IEEE CVPRW, pp. 1258–12588 (2018)
11. Eigen, D., Puhrsch, C., Fergus, R.: Depth map prediction from a single image using a multi-scale deep network. In: Advances Neural Information Processing System, vol. 3, pp. 2366–2374 (2014)
12. Kwak, D.H., Lee, S.H.: A novel method for estimating monocular depth using cycle gan and segmentation. Sensors (Switzerland) **20** (2020)
13. Kwon, Y.-H., Park, M.-G.: Predicting future frames using retrospective cycle GAN. In: IEEE/CVF, pp. 1811–1820 (2019)
14. Niklaus, S., Mai, L., Yang, J., Liu, F.: 3D ken burns effect from a single image, vol. 38, p. 15. In: arXiv (2019)
15. Yan., P., Bowyer, K.: Empirical evaluation of advanced ear biometrics. In: IEEE, p. 41 (2006)
16. Yan, P., Bowyer, K.W.: An automatic 3D ear recognition system. In: 3DPVT, pp. 326–333 (2006)
17. Bizjak, M., Peer, P., Emeršic, Ž.: Mask R-CNN for ear detection. In: MIPRO, pp. 1624–1628 (2019)
18. Szegedy, C., Vanhoucke, V., Ioffe, S., Shlens, J., Wojna, Z.: Rethinking the inception architecture for computer vision. In: IEEE/CVPR, pp. 2818–2826 (2016)

19. Goodfellow, I.J., et al.: Generative adversarial nets. In: NIPS, vol. 2, pp. 2672–2680 (2014)
20. Chen, X., Gupta, A.: An implementation of faster RCNN with study for region sampling. In: arXiv (2017)
21. Tzutalin. LabelImg. Git code (2015). https://github.com/tzutalin/labelImg. Accessed 19 Feb 2021
22. Zhang, E., Zhang, Y.: Average precision. In: Encyclopedia of Database Systems, pp. 192–193. Springer, Boston (2009). https://doi.org/10.1007/978-0-387-39940-9
23. Shmelkov, K., Schmid, C., Alahari, K.: How good is my GAN? In: arXiv (2018)
24. Zhang, L., Wu, X.: Color demosaicking via directional linear minimum mean square-error estimation. IEEE Trans. Image Process. **14**, 2167–2178 (2005)
25. Dosselmann, R., Yang, X.D.: A comprehensive assessment of the structural similarity index. Signal Image Video Process **5**, 81–91 (2011)

A Fast Scanning System for Automatic 3D Object Reconstruction

Umberto Severino[1(✉)], Fabrizio Fuoco[2], Felix Manfredi[2], Loris Barbieri[1], and Maurizio Muzzupappa[1]

[1] Department of Mechanical, Energy and Management Engineering (DIMEG), University of Calabria, P. Bucci 46C, 87036 Rende, CS, Italy
{umberto.severino,loris.barbieri,muzzupappa}@unical.it
[2] 3DResearch S.r.l., P. Bucci 45C, 87036 Rende, CS, Italy
{fabrizio.fuoco,felix.manfredi}@3dresearch.it

Abstract. During field 3D reconstruction activities and indoor reverse engineering processes, there is often a need for a portable and easy-to-deploy system that can autonomously conduct a full 3D scanning operation, aligning and registering point clouds even in a low-light environment and considering different target sizes. Currently, this process can be obtained through various technological solutions, such as photogrammetric systems and laser scanners, that come with some limitations like portability, affordability, and degree of automation.

In this context, the paper presents a feasibility study of a low-cost solution that, by integrating a commercial depth camera into an easy to unfold frame design, autonomously operates to perform a fast 3D scanning and reconstruction of small and medium-sized objects. In particular, the proposed system features a rotating two-dimensional light detection and ranging system and integrates an Inertial Measurement Unit sensor to automatically performs a pre-alignment and registration of the acquired scan data. According to the experimental results, this system may become a cost-effective solution for fast 3D scanning when high levels of accuracy are not required.

Keywords: 3D acquisition · Depth camera · LiDAR · Product development

1 Introduction

3D scanning is a useful technique that can be applied in a variety of contexts where dimensional information about an object or scene needs to be acquired. In particular, it is largely used in the industrial/robotics field, environmental sciences, and in medical and archaeological fields. 3D scanning, in fact, along with other techniques, allows robots to be equipped with three-dimensional perception [1]. In industry, 3D scanning is used for reverse engineering and inspection processes [2]. In the context of cultural heritage, it is useful for documenting and creating a digital copy of unique assets [3–5]. In environmental sciences, it is used for surveys and collecting qualitative and quantitative data [6]. In almost all cases, when very large surfaces and areas have to be acquired,

C. Rizzi et al. (Eds.): ADM 2021, LNME, pp. 229–236, 2022.
https://doi.org/10.1007/978-3-030-91234-5_23

the related equipment is expensive or non-portable and does not address the need for a lightweight solution for field research activities. Indeed, at the moment, there is not a low-cost solution for automatic 3D scanning acquisition and reconstruction activities, in particular for outdoor campaign.

Because of these limitations, the paper presents a feasibility study of a low-cost depth camera for in-place fast 3D reconstruction of small- medium-size objects. In particular, the proposed system features a rotating two-dimensional Light Detection and Ranging (LiDAR) system and integrates an Inertial Measurement Unit (IMU) sensor to automatically performs a pre-alignment and registration of the acquired scan data. Furthermore, the proposed system features an easy to unfold frame design in order to be portable and easy to assemble. A preliminary laboratory test was carried out to assess the accuracy of the proposed scanning system.

2 Related Works

The process of collecting data from the shape of an object is called 3D scanning. It can be achieved by employing various 3D scanning technologies, based on different physical principles, and can span to many applications such as close-range, aerial, terrestrial [7–11], but generally, the classification is based on contact and non-contact methods. The most common are non-contact methods, which can be active or passive, referring to the possibility to emit some kind of radiation or light and then detect the reflection of radiation from the surface of the target object. The most used passive non-contact 3D scanning method is photogrammetry. Through photogrammetry is possible to acquire larger scene shooting photos from a different position and matching through the features of the captured scene. However, trying to obtain the highest resolution possible expensive solution are needed, and is necessary to strictly follow precise rules and expedients. Moreover, to achieve a good level of detail is necessary to cover multiple poses around the target [12, 13]. Automated photogrammetric systems are widely used to reconstruct small assets through a high-resolution camera coupled with a rotating platform [14]. Nevertheless, this acquisition requires a high number of poses for the reconstruction process and constant and diffuse lighting conditions to avoid shadows on the object surface that can affect feature matching. Another solution consists of a set of low-cost camera modules deployed around the target object that allow to achieve a lower accuracy with a mean distance of 13 $\pm$ 9 mm [15]. Due to optics aberration, photogrammetric systems need a precise calibration process that could be tricky and time-consuming but necessary to achieve sub-millimeter accuracy [16].

Active non-contact technologies allow overcoming some of the abovementioned limitations [17]. In fact, if on the one hand photogrammetric data require a reference to scale the acquired scene to its actual size, on the other hand, active systems are capable to collect full-scale dense 3D point data and accomplish a 3D reconstruction with a smaller number of acquisitions. Furthermore, in time-of-flight systems [18], such as LiDAR scanners, a signal in the infrared wavelengths is emitted and then received from the sensor, allowing a direct measure of positional information in a single shot. The latest state-of-the-art laser scanners can capture millions of highly accurate data points in minutes [19] and the overall time to make a digital 3D model representation of the scene is faster than a full photogrammetry procedure [14]. Moreover, continued improvements in the performance and accuracies of these systems in recent years have enabled the use of LiDAR data as a source of control suitable for photogrammetric applications.

About low-cost scanning technologies, the most common commercial solution is represented by the Microsoft Kinect™ [20] sensor, which produces an RGB-D image [21] through which it is possible to achieve an average RMSE of about 3 mm [22]. However, its performance degrades rapidly when increasing the distance from the object to acquire [20]. A more accurate low-cost (1000 USD) laser scan solution achieves an accuracy of 0.59 mm and a precision of 0.1–0.2 mm [23] for indoor applications using a not fast-to-deploy setup, constituted by several markers. There are several commercial solutions based on structured light or photogrammetric principle capable to provide an automatic or semi-automatic 3D scanning but usually they need a high-power supply (e.g., high brightness projectors, high resolution camera) or they are expensive (1500–30000 USD).

What emerges from these considerations is that at the moment there are no low-cost portable solutions for automatic 3D scanning acquisition and reconstruction activities.

This paper then proposes a portable low-cost scanning system for automatic 3D reconstructions of small- and medium-size objects.

3 HW/SW Architecture

As depicted in the following figure (Fig. 1), the proposed scanning system consists of two main hardware components: a motion control module (MCM) equipped with a LiDAR camera, and a folding support structure.

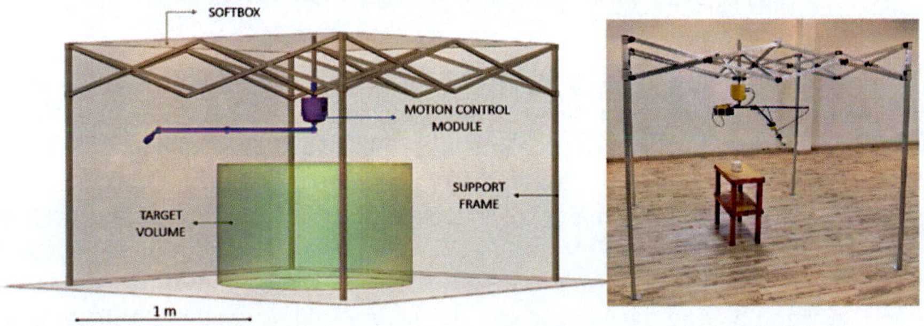

(a) (b)

Fig. 1. CAD model (a) and physical prototype (b) of the LiDAR scanning system.

MCM is the core part of the proposed system, which contains all the elements deputed to the 3D acquisition process. It is composed of two subsystems: the LiDAR camera, and its actuation sub-system which consists of the mechanism, sensors and electronics to control it. About the folding support frame, it is composed by four symmetrical telescopic legs extendable from 1 m to 2 m and a gusseted roof. The legs of the support frame can be adjusted individually to better adapt to the unevenness of the ground. The workspace at its maximum extension defines a cylindric target volume of 1.5 m diameter and 1.75 m height, that is adequate for the scanning of small- and medium-sized objects, such as mechanical components and assemblies, archaeological artifacts, or it can be adopted as body scanner too. Furthermore, the support frame is equipped with a softbox, comprised of a translucent cloth of ballistic nylon fabric, that can be adopted in those cases in which the environmental conditions can affect the scanning process negatively. And specifically, to prevent the interference from the atmospheric environment [24] such as the infrared components of the sunlight to the emitter signal due to the sensitivity of the receiver sensor at the infrared wavelength, and to avoid blurring effects due to oscillations by the wind in the open field [25, 26].

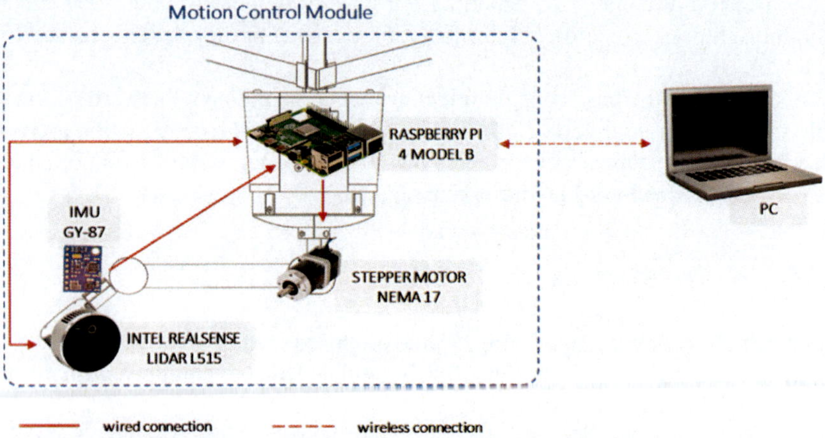

Fig. 2. MCM main components and communication channels.

As depicted in figure (Fig. 2), the MCM consists of: a control unit with a single-board microcomputer; a stepper motor; a multi-sensor board; a LiDAR camera; and an articulated arm. The articulated friction arm is assembled to the top-center of the support frame roof, while its end-effector is equipped with an Intel® RealSense™ L515 LiDAR camera. Compared to stereo-camera-based technologies, a LiDAR camera is preferred because of its capacity to scan object in poor light conditions and it does not require additional light source. Furthermore, the arm has 2 DOF motion: tilt of the camera and 360° rotation around the vertical axis through a stepper motor. On the arm take place also a GY-87 multi-sensor board, which incorporates IMU, magnetometer, and barometer sensors. The control unit of the MCM is a Raspberry Pi® 4 Model B, and the LiDAR and the IMU sensors are wired to it respectively via USB and I2C protocols. The control unit is placed in a cone-shaped case whose dimensions are approximately 140 mm diameter and 160 mm height. This case has been designed to be waterproof to counteract weathering when the system is used outdoor. The MCM is powered by a LiPo battery pack which allows an autonomy of approximately 3 h.

The software architecture consists of two main parts implemented in Python 3: the motion control software, and the 3D reconstruction software. The motion control software actuates the arm equipped with the LiDAR and automatically perform a 360°

acquisition. The 3D reconstruction software consists of a web app and a registration module. In particular, the web app allows to command and receive data from the motion control software. While, the registration module, adopts ICP (Iterative Closest Point) algorithms [19] to perform alignment, registration and 3D object reconstruction.

4 Experimental Tests

A preliminary laboratory test was carried out to assess the accuracy of the proposed scanning system. A common approach to determine the precision and resolution of a 3D acquisition system is to test its performance with geometrically simple calibration objects of known dimensions (e.g. cubes or spheres) [14, 27]. Accordingly, this preliminary test was carried out by using a cube as a ground-truth for evaluating the 3D model produced through the proposed scanning system.

The support frame was mounted considering a square workspace with a 2 m side and with a baseline of 0.7 m from the center of the scene to the camera optics. Referring to the manufacturer datasheet [28], the Intel® RealSense™ LiDAR L515 features a depth accuracy of 5 ± 2.5 mm and 14 ± 15.5 mm respectively at a distance of 1 m and 9 m from the target. The following figure (Fig. 3) depicts the scene acquired with the developed scanning system. It is characterized by an area of 0.50 m $\times$ 0.46 m in which are placed a cube and three fiducial square markers, with a side of 0.13 m long, on its corners.

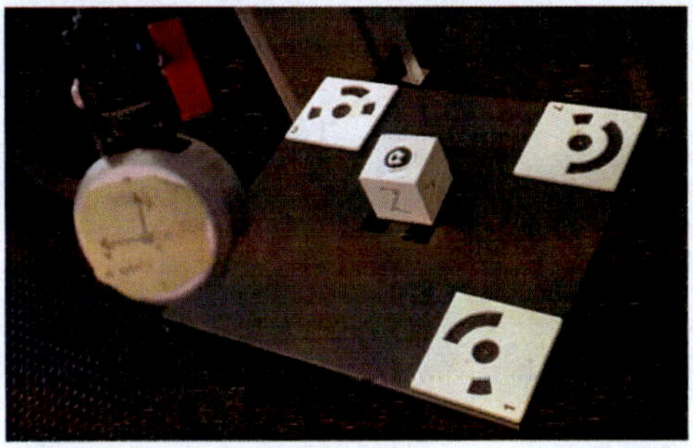

Fig. 3. Experimental scene acquired by the LiDAR system.

A cube with a side 8 cm long is placed in the center of the rectangular scene. The cube was made with Fused Deposition Modeling (FDM) technology and markers in the form of letters are embossed on its lateral faces while additional fiducial planar markers were placed on the remaining faces. Furthermore, the cube was coated using an opaque spray primer to smooth its surface and reduce its reflectivity [29]. A total of twelve acquisitions, every 30 degrees, were automatically carried out in about a minute. The information content of each acquisition is composed by: the timestamp, the pose of the LiDAR camera, the point cloud, the RGB image of the scene.

The reference frame of each point cloud was then automatically aligned to the world reference frame applying a pose-related transformation matrix automatically computed by the software. Using the CloudCompare [30] customized script, the point clouds were automatically aligned to the world reference frame and cropped to remove all the non-relevant objects outside the workspace. In particular, the first transformation is the translation from the frame origin of the target {T} to the frame origin of the camera {C} moving by the length of the baseline along the z-axis of the camera. Then, for each point cloud, it is necessary to orient properly the point cloud frame to the world frame. The point clouds are then automatically registered in few seconds through the ICP algorithm.

In figure (Fig. 4a) is depicted the scene automatically reconstructed through the LiDAR system, while the nearby figure (Fig. 4b) depicts the results of a cloud-to-mesh distance operation performed between the reconstructed object and the CAD model adopted as reference. In particular, the results are represented in the RGB scalar field and show an average distance, between the actual geometry and the nominal geometry, of 2.60 mm with a standard deviation of 1.89 mm.

These results are consistent with the accuracy of 5 mm below 1 m range, reported in the manufacturer datasheet, with the low-cost 3D reconstruction systems discussed in Sect. 2. Furthermore, the proposed system offers the advantage to be suitable for both indoor and outdoor activities and perform an automatic acquisition and reconstruction process.

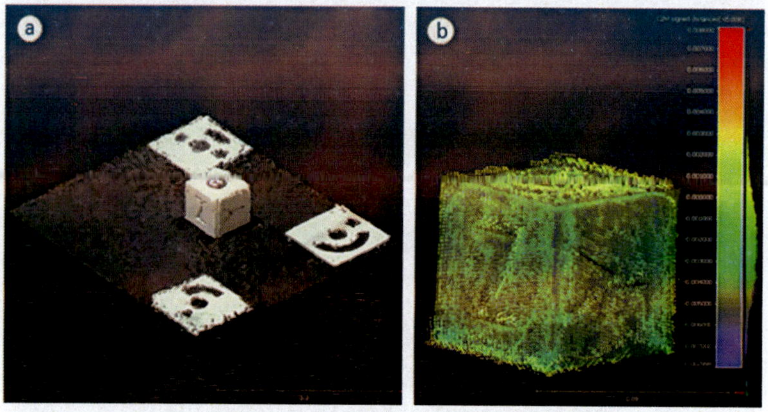

Fig. 4. 3D reconstructed scene (a) and test's results (b).

5 Conclusions

This paper has presented a feasibility study for the development of a low-cost portable scanner system that can automatically perform a fast 3D reconstruction of small-medium-size objects. It is characterized by a rotating two-dimensional LiDAR system that integrates an IMU sensor to automatically performs a fast pre-alignment and registration of the acquired scan data. Furthermore, an unfold frame design allows to make it portable and easy to deploy in field and indoor activities.

A preliminary laboratory test was carried out to assess the accuracy of the proposed scanning system. And the results showed the proposed system can be a cost-effective solution for fast 3D scanning when a sub-millimeter accuracy is not required.

In the next future, further tests will be undertaken both indoor and outdoor environments, to assess the accuracy and reliability of the proposed system with objects of different size and complexity and in different lighting conditions.

References

1. Surmann, H., Nuchter, A., Hertzber, J.: An autonomous mobile robot with a 3D laser range finder for 3D exploration and digitalization of indoor environments. Robot. Auton. Syst. **45**(3–4), 181–198 (2003)
2. Son, S., Park, H., Lee, K.H.: Automated laser scanning system for reverse engineering and inspection. Int. J. Mach. Tools Manuf **42**(8), 889–897 (2002)
3. Wachowiak, M.J., Karas, B.V.: 3D scanning and replication for museum and cultural heritage applications. J. Am. Inst. Conserv. **48**(2), 141–158 (2009)
4. Remondino, F.: Heritage recording and 3D modeling with photogrammetry and 3D scanning. Remote Sens. **3**(6), 1104–1138 (2011)
5. Bruno, F., Bruno, S., De Sensi, G., Luchi, M.L., Mancuso, S., Muzzupappa, M.: From 3D reconstruction to virtual reality: a complete methodology for digital archaeological exhibition. J. Cult. Herit. **11**(1), 42–49 (2010)
6. Naesset, E., Okland, T.: Estimating tree height and tree crown properties using airborne scanning laser in a boreal nature reserve. Remote Sens. Environ. **79**, 105–115 (2002)
7. Daneshmand, M., et al.: 3D scanning: a comprehensive survey. arXiv preprint, n. 1801.08863 (2018)
8. Haleem, A., Javaid, M.: 3D scanning applications in medical field: a literature-based review. Clin. Epidemiol. Global Health **7**(2), 199–210 (2019)
9. Vitali, A., Regazzoni, D., Rizzi C., Molinero, G.: 3D scanning procedure for the evaluation of lymphedema of upper limbs using low-cost technology: a preliminary study. In: International Conference on Design, Simulation, Manufacturing: The Innovation Exchange, pp. 177–188. Springer, Cham (2019)
10. Cucinotta, F., et al.: On the morphology of the abrasive wear on ploughshares by means of 3D scanning. Biosys. Eng. **179**, 117–125 (2019)
11. Ingrassia, T., Nalbone, L., Nigrelli, V., Ricotta, V., Pisciotta, D.: Biomechanical analysis of the humeral tray positioning in reverse shoulder arthroplasty design. Int. J. Interact. Des. Manuf. (IJIDeM) **12**(2), 651–661 (2017)
12. Curless, B., Seitz, S.: 3D Photography. Course Notes. SIGGRAPH (2000)
13. Wilfried, L.: Digital Photogrammetry: Theory and Applications. Springer (2013). https://doi.org/10.1007/978-3-662-06725-3
14. Collins, T., Woolley, S., Ch'ng, E., Gehlken, E.: Automated low-cost photogrammetric acquisition of 3D models from small form-factor artefacts. Electronics **8**(12), 1441 (2019)
15. Santise, M., Thoeni, K., Roncella, R., Sloan, S.W., Giacomini, A.: Preliminary tests of a new low-cost photogrammetric system. Int. Arch. Photogramm. Remote Sens. Spatial Inform. Sci. **42**, 229 (2017)
16. Galantucci, L.M., Lavecchia, F., Percoco, G., Raspatelli, S.: New method to calibrate and validate a high-resolution 3D scanner, based on photogrammetry. Precis. Eng. **38**(2), 279–291 (2014)
17. Beraldin, J. A., Blais, F., Cournoyer, L., Godin G., Rioux M.: Active 3D Sensing. National Research Council of Canada (2003)
18. Cui, Y., Schuon S., Chan, D., Thrun, S., Thèobalt, C.: 3D shape scanning with a time-of-flight camera. In: IEEE Conference Computer Vision and Pattern Recognition, pp. 1173–1180 (2010)

19. El-Omari, S., Moselhi, O.: Integrating 3D laser scanning and photogrammetry for progress measurement of construction work. Autom. Constr. **18**(1), 1–9 (2008)
20. Khoshelham, K.: Accuracy analysis of Kinect depth data. In: ISPRS Workshop Laser Scanning, vol. 38, no 1 (2011)
21. Henry, P., Krainin, M., Herbst, E., Ren, X., Fox, D.: RGB-D mapping: using kinect-style depth cameras for dense 3D modeling of indoor environments. Int. J. Robot. Res. **31**(5), 647–663 (2012)
22. Jaiswal, X., Sun, Y.: 3D object modeling with a Kinect camera. In: Asia IEEE Signal and Information Processing Association Annual Summit and Conference, pp. 1–5 (2014)
23. Akyol, O., Duran, Z.: Low-cost laser scanning system design. J. Russ. Laser Res. **35**(3), 1–8 (2014)
24. Gupta, M., Yin, Q., Nayar, S.K.: Structured light in sunlight. In: Proceedings of the IEEE International Conference on Computer Vision (2013)
25. Blais, F., Picard, M., Godin, G.: Accurate 3D acquisition of freely moving objects. In: 2nd International Symposium on 3D Data Processing, Visualisation, and Transmission, pp. 422–429. IEEE Computer Society (2004)
26. Salil, G., Bharat, L.: A motion correction technique for laser scanning of moving objects. IEEE Geosci. Remote Sens. Lett. **1**(11), 225–228 (2014)
27. Boheker, W., Vicent, M.B., Marbs, A.: Investigating laser scanner accuracy. Int. Arch. Photogramm. Remote Sens. Spatial Inform. Sci. **34**(part 5), 696–701 (2003)
28. Intel RealSense LiDAR Camera L515. https://www.intelrealsense.com/lidar-camera-l515/. Accessed 12 Feb 2021
29. Pereira, J.R.M., Penz, I., Da Silva, F.P.: Effects of different coating materials on three-dimensional optical scanning accuracy. Adv. Mech. Eng. **11**(4), 1–6 (2019)
30. CloudCompare: https://www.danielgm.net/cc/. Accessed 10 Feb 2021

Online 3D Reconstruction in Underwater Environment Using a Low-Cost Depth Camera

Luigi Scarfone(✉), Rosario Aiello, Umberto Severino, Loris Barbieri, and Fabio Bruno

Department of Mechanical, Energy and Management Engineering (DIMEG), University of Calabria, via Pietro Bucci 46C, 87036 Rende, CS, Italy
{luigi.scarfone,rosario.aiello,umberto.severino,loris.barbieri, fabio.bruno}@unical.it

Abstract. Although computer vision techniques are widely and effectively used for terrestrial applications, the adoption of these techniques in underwater scenarios is often a challenging task. This is due to the water medium that degrades the quality of the acquired images requiring the use of specific methodologies for their correction and improvement. In this context, this paper presents a low-cost system for underwater 3D reconstruction and 3D measurements based on an off-the-shelf depth camera. The system consists of a Stereolabs ZED stereo camera and its waterproof housing designed to be fabricated through additive manufacturing technologies. A dedicated software has been developed to manage all the steps of the acquisition process, starting from the calibration phase and continuing for rectification, depth map computation, 3D scene reconstruction, and 3D measurement. A preliminary laboratory test has been carried out in a controlled environment to quantify system performance in accuracy of measurement.

Keywords: Stereo Camera · Underwater 3D Reconstruction · Underwater Inspection · Stereolabs ZED · Calibration Techniques

1 Introduction

Applications of computer vision are rapidly spreading across a broad spectrum of underwater operations requiring highly accurate environmental information. Vision systems, in fact, are increasingly being used as the primary tool for inspection of underwater sites, in disciplines ranging from archaeology and biology to offshore engineering and pipeline inspection [1]. This has been facilitated by the increasing industry adoption of remotely operated vehicles (ROV) and autonomous underwater vehicles (AUV), which led to a variety of new challenges and applications for machine vision.

Instead of just a two-dimensional projection, as a standard camera does, underwater 3D vision systems allow operators to obtain quantitative 3D data and capture 3D shapes of underwater structural elements that can be used in many applications, such as underwater robotic manipulation [2], and computational fluid dynamics of reconstructed 3D geometries obtained from the underwater survey. All the above-mentioned scenarios

C. Rizzi et al. (Eds.): ADM 2021, LNME, pp. 237–244, 2022.
https://doi.org/10.1007/978-3-030-91234-5_24

and application fields require high-quality 3D data of the underwater operating environment. These kinds of data can be obtained only by using 3D computer vision systems because traditional acoustic detection systems are not able to provide the level of detail and accuracy required for such applications.

Among the various computer vision solutions, binocular stereo vision technology has proven to be a valid and reliable solution that has proved its efficacy in many areas such as 3D scene reconstruction, robot vision navigation, target recognition and tracking, industrial detection, etc. [3]. Furthermore, it could be a cost-effective solution too, since, in the last years, this technology has benefited by the presence on the market of various low-cost off-the-shelf cameras [4]. Although the use of such low-cost solutions has been consolidated for terrestrial applications, in underwater scenarios their adoption is still a challenging task due to the intrinsic limitation of the medium. Its severe condition, in fact, degrades the quality of the acquired images requiring the use of specific methodologies for their correction and improvement. In this regard, it is important to point out that the images captured by any camera underwater do not follow the principles of perspective projection [5] because of the refraction effect at the transparent interface of the housing of the camera.

Some attempts have been made to use low-cost off-the-shelf cameras underwater. In particular, a Stereolabs® ZED Stereo camera has been adopted for the development of a real-time underwater onboard vision sensing system for robotic gripping [6] and a stereo vision system for underwater object detection and 3D reconstruction [7]. But both works provide only qualitative results.

In this context, the paper proposes a low-cost system for online 3D reconstruction in underwater environments. In particular, the proposed system consists of an off-the-shelf camera, i.e., a Stereolabs® ZED Stereo camera, an ad-hoc dedicated software, and a waterproof case made with additive manufacturing technologies. A first prototype of the proposed system has been characterized in a controlled environment in order to measure its performances, in terms of accuracy, and verify then if it could be suitable for 3D measurements and inspection activities carried out in the underwater environment.

2 System Development and Prototyping

The hardware consists of a self-contained camera device and its waterproof housing (Fig. 1a). In particular, the housing has been designed to minimize the overall dimensions and to be used underwater up to 30 m depth. Two main removable parts allow access to the camera when outside the water, while a cable penetrator is designed for connecting the Camera via USB 3.0 to an external PC.

The camera is a ZED stereo camera from Stereolabs® [8]. This is a passive stereo depth camera that reproduces the way human vision works by using two 4MP RGB cameras which can provide high-definition images up to 2K at 15 fps with a 110° FOV.

In order to adopt the standard photogrammetric model by avoiding the introduction of systematic errors [9], mainly due to the refraction effect at the air-water interface, the camera has been placed in the case on a support (Fig. 1a), specifically designed to minimize the distance between its entrance pupil and the flat port (thickness 4 mm).

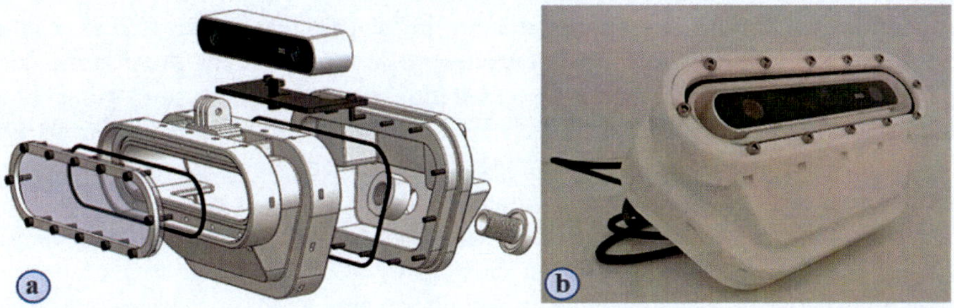

Fig. 1. Depth camera and exploded view of its waterproof housing (a) and its physical prototype (b).

About the software architecture (Fig. 2), it consists of five main modules that starting from the acquired raw images allow to perform underwater 3D measurement and reconstruction activities.

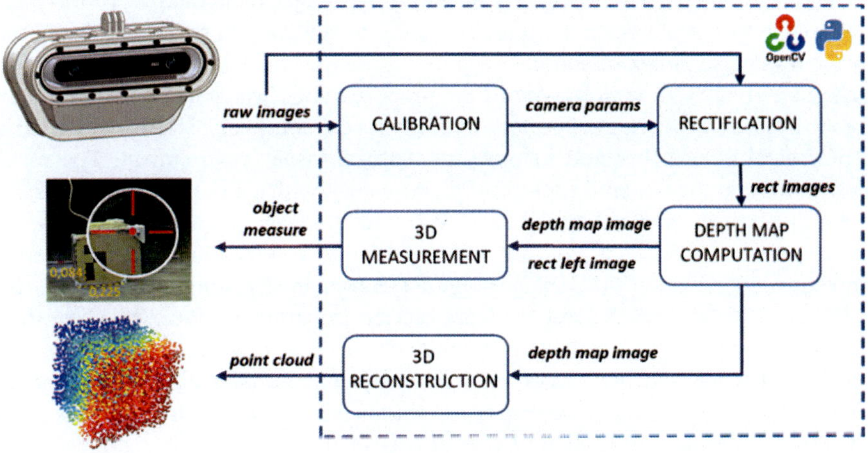

Fig. 2. SW architecture.

The software modules have been programmed in Python, by integrating open source OpenCV libraries [10], to perform an online 3D object reconstruction and 3D measurements of the acquired underwater scene. In particular, the software grabs raw image pairs from the stereo device, at a defined frame rate, that are then adopted to execute image rectification and depth map estimation procedures before to generate the resultant 3D object point cloud. About the calibration module, it requires chessboard image pairs as input to compute the interior and exterior orientation parameters (i.e. intrinsic and extrinsic parameters) [11] of the camera according to the pinhole model [12]. The calibration procedure is automatically carried out by implementing filtration operations and a sub-pixel finder function [13] provided by OpenCV that allow to detect points with an accuracy higher than the pixel size. The detected points are then used to perform the camera calibration process based on the Bouguet algorithm [14], finding the interior and exterior orientation parameters of the camera, that are adopted as input by the rectification module. This module corrects raw images, sending them to the depth map computation module for depth map generation. In particular, the semi-global block matching (SGBM) algorithm for stereo correspondence [15] is used to convert both left and right rectified images into a single disparity map. To avoid significant noise, a WLS (weighted least square) method and a guided post-filtering algorithm, provided by OpenCV, have been implemented. In particular, the WLS filter allows to remove holes in the disparity map due to half-occlusions by filling the image to get a uniform segmentation between layers, then the guided filter is applied for image edge-preserving smoothing. The disparity map is then reprojected to generate the depth map, obtaining the value of the depth Z for each pixel in the image. Finally, the resultant two-dimensional depth map and the corresponding RGB image are sent to the 3D reconstruction module, for generating the desired 3D point cloud, and to the 3D measurement module that allows to compute 3D-distance between two points picked on the 2D images. In particular, the three-dimensional coordinates of the two points are smoothed using the nearest points in a defined circular area, then the z score of each point is computed to reject outliers and the remaining consolidated points coordinates are averaged. The filtering of the points picked by the user reduces the inaccuracy caused by a locally mismatching depth map or by a non-precise points selection.

The Fig. 1b shows the prototype of the proposed underwater stereo vision system. Before to proceed with physical prototyping, a finite element analysis (FEA) was carried out to assess its reliability and integrity in the underwater environment. The results demonstrated that the system keeps a full functionality, with a minimum factor of safety of 1.3 at a depth of 30m below the sea level, with a water pressure of 0.4 MPa. The housing was then produced in PA2200 material through SLS (Selective Laser Sintering) technologies by means of an EOS Formiga P110 printer. The print job required 15 h, then the parts were cleaned using a media blaster. In particular, PA2200, a non-filled powder on basis of PA12, allows to print parts characterized by strength, rigidity, good chemical resistance and low water resistance capabilities. Nevertheless, its porosity can be reduced by optimizing the process parameters or performing postprocessing operations. Furthermore, the degradation effects of the seawater environment can be alleviated by sealing the porous surface of the plastic parts by means polyurethane coatings.

3 Experimental Test

An experimental test was carried out in the laboratories of the DIMEG (Department of Mechanical, Energy and Management Engineering) of the University of Calabria to assess the camera's performance. The test was performed both underwater and in air in order to assess the performances degradation due to the water medium. The underwater test was undertaken within a controlled environment, i.e., a circular pool of 3 m diameter and 1 m deep (Fig. 3a). The stereo vision system is mounted on a sliding bracket, allowing it to stay in a fixed position during the image acquisition process. The camera is connected via USB 3.0 to an external PC for process debugging and post-processing operations. All the images during tests were acquired at a resolution of 2K pixels and 15 frames per second.

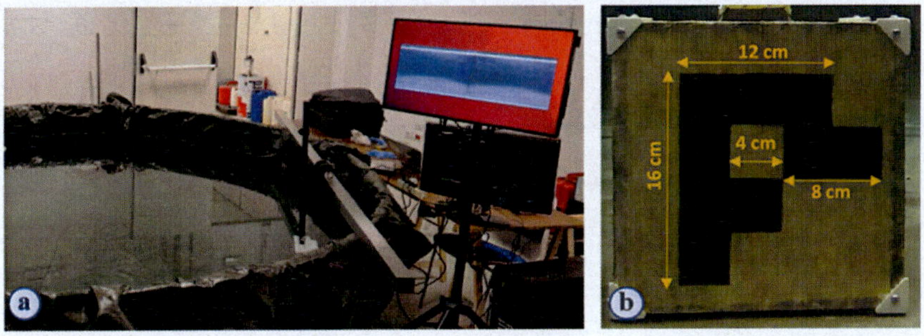

Fig. 3. Experimental setup (a) and the cubic object adopted for the 3D reconstruction test (b).

In particular, the test was carried out by using a cubic object (Fig. 3b), with black patterns consisting of square figures with a 4 cm side long, on each face.

The procedure adopted for the experimental test, both underwater and in air conditions, is depicted in the following figure (Fig. 4). Several acquisitions, with different object's positions and orientations, were made and then processed through the developed software to perform the 3D reconstruction. For each one of the four acquisition distances, i.e. 1 m, 1.5 m, 2 m, and 2.5 m, the cube was oriented at angles of 0°, 25°, 50°, and 75°.

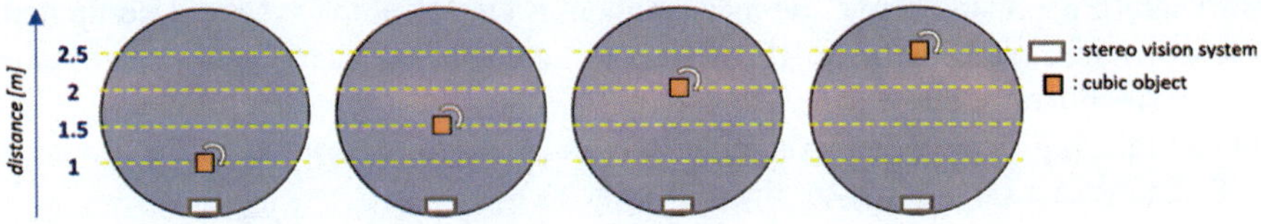

Fig. 4. The experimental test procedure.

The calibration process was carried out in the submerged environment by means of a flat chessboard pattern. Fifty chessboard image pairs were detected by the calibration module and then automatically processed: the location of internal chessboard corners was performed, and stereoscopic parameters information was saved for the next rectification stage. The resultant stereo camera parameters were used to perform 3D object reconstruction acquiring, for each cubic target configuration, an image pair from the stereo device and then sending it to the software for offline processing. In particular, the rectification module took as inputs the calibration parameters and the acquired raw images to perform the rectification process. The corrected images were then transmitted to the depth map computation module, which performed stereo correspondence to calculate the disparity map required for the depth map. Then, the resultant depth map was used by the 3D measurement module to perform object 3D measurements.

The following table (Table 1) shows the error analysis results achieved for the 3D measurements performed underwater and above water. In particular, the error analysis was carried out by calculating accuracy, computed as the percentage of the mean error between object feature measurements and their true dimensions, and Mean Absolute Error (MAE).

Table 1. 3D measurement results for each acquisition distance in underwater and air conditions.

Distance [mm]	Accuracy [%] (underwater)	MAE [mm] (underwater)	Accuracy [%] (air)	MAE [mm] (air)
1000	6.91	11.035	Disregarded	Disregarded
1500	5.06	5.217	2.37	3.385
2000	5.97	7.142	4.068	4.614
2500	5.73	6.121	3.43	4.385

The results show that a mean error smaller than a centimetre can be obtained underwater for distances greater than one meter, arriving at an accuracy of 5% for objects at 1.5 m. The error trend does not follow a monotonous function over distances. This is due to the aleatory nature of the 3D measurement module algorithm used to reduce the inaccuracy caused by a non-precise point selection. The measurement algorithm provides acceptable errors for an operating distance range from 1.5 m to 2.5 m; this cannot be verified for an operating distance of 1 m, which makes the proposed algorithm inefficient. Therefore, outside water, the measurement accuracy of the system is computed at the same distances starting from 1.5 m; as shown in the previous table, the measurement accuracy significantly improves.

The following figure (Fig. 5a) depicts the resultant object point cloud computed by the 3D reconstruction module for the underwater condition.

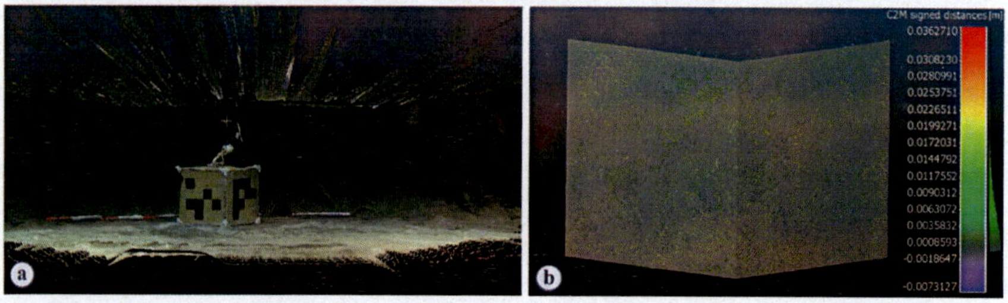

Fig. 5. Underwater 3D object reconstruction (a) and resulting scalar field's colour scale representing distance error in meters (b).

This point cloud was then compared with a reference model, i.e., a CAD model with the real dimensions of the cubic object, to verify the 3D reconstruction performance of the proposed system. The reconstruction error was measured by means of the open-source CloudCompare software [16]. The alignment process was carried out by picking equivalent point pairs in both entities. The resulting scalar field's colour scale (Fig. 5b) of the distance error between the generated point cloud and the reference mesh object, and related distribution (Fig. 6a), is finally computed. A mean error of 5.5 mm is achieved underwater, validating the generated point cloud accuracy.

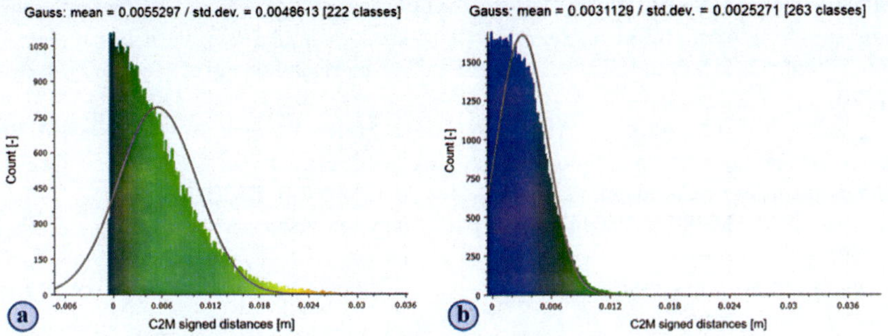

Fig. 6. Gauss distribution representing distance error in meters in underwater condition (a) and in air (b).

Figure 6b shows the error distribution computed outside water conditions; in particular, the 3D reconstruction quality significantly improves with minimal planar distortions and a mean error of 3.12 mm. On the basis of these results, it is possible to point out that the proposed system is suitable for object 3D measurement and 3D reconstruction activities to be carried out underwater for industrial and scientific applications. For example, the proposed system could be mounted on ROVs and AUVs for pipeline inspection activities, with an operating distance up to 3 m due to water turbidity. Or it can be adopted for the documentation and 3D reconstruction of submerged environments for applications in the environmental and archaeological fields.

4 Conclusions

The paper has proposed a low-cost solution for underwater 3D reconstruction and 3D measurements based on an off-the-shelf depth camera. In particular, a dedicated software for online processing was developed using open-source libraries.

A laboratory test was conducted in a controlled environment to characterize the proposed system and quantify its performance in accuracy of measurement, varying the relative distance between the camera and framed objects. A quantitative analysis was conducted: accuracy and Mean Absolute Error (MAE) were employed in the evaluation of the 3D reconstructed object and the error analysis results are provided, obtaining the lowest error for 1.5 m operating distance.

Future developments will focus on improvements to strengthen the calibration procedure and additional experimentations will be carried out at sea in real operating conditions.

References

1. Bruno, F., Bianco, G., Muzzupappa, M., Barone, S., Razionale, A.V.: Experimentation of structured light and stereo vision for underwater 3D reconstruction. ISPRS J. Photogramm. Remote. Sens. **66**(4), 508–518 (2011)

2. Bruno, F., Lagudi, A., Barbieri, L., Rizzo, D., Muzzupappa, M., De Napoli, L.: Augmented reality visualization of scene depth for aiding ROV pilots in underwater manipulation. In: Ocean Engineering, vol. 168, pp. 140–154 (2018)
3. Vitali, A., Regazzoni, D., Rizzi, C., Lupi, G.: Low cost markerless motion capture systems: a comparison between RGB cameras and RGB-D sensors. In: ASME International Mechanical Engineering Congress and Exposition, vol. 84539. American Society of Mechanical Engineers (2020)
4. Digumarti, S.T., Chaurasia, G., Taneja, A., Siegwart, R., Thomas, A., Beardsley, P.: Underwater 3D capture using a low-cost commercial depth camera. In: IEEE Winter Conference on Applications of Computer Vision (WACV) 2016, pp. 1–9. IEEE (2016)
5. Andrew, A.M.: Multiple View Geometry in Computer Vision, Kybernetes, pp. 1333–1341 (2001)
6. Wang, Y., et al.: Real-time underwater onboard vision sensing system for robotic gripping. IEEE Trans. Instrum. Meas. **70**, 1–11 (2020)
7. Wang, C., et al.: Research and experiment of an underwater stereo vision system. In: OCEANS Marseille 2019, pp. 1–5. IEEE (2019)
8. Stereolabs Homepage: https://www.stereolabs.com/zed/. Accessed 13 Mar 2021
9. Menna, F., Nocerino, E., Remondino, F.: Photogrammetric modelling of submerged structures: influence of underwater environment and lens ports on three-dimensional (3D) measurements. Latest developments in reality-based 3D surveying and modelling. MDPI Basel, Switzerland, pp. 279–303 (2018)
10. OpenCV Homepage: https://opencv.org/. Accessed 13 Mar 2021
11. Granshaw, S.: Photogrammetric Terminology. 3rd edn, pp. 210–252. The Photogrammetric Record (2016)
12. Forsyth, D. A., Ponce, J.: Computer Vision: A Modern Approach. 2nd edn. Pearson (2012)
13. Forstner, M.A., Gulch, E.: A fast operator for detection and precise location of distinct points, corners and centers of circular features. In: Proceedings of the ISPRS Intercommission Conference on Fast Processing of Phonogrammic Data, Interlaken, pp. 281–305 (1987)
14. Zhang, Z.: A flexible new technique for camera calibration. IEEE Trans. Pattern Anal. Mach. Intell. **22**(11), 1330–1334 (2000)
15. Hirschmuller, H.: Stereo processing by Semiglobal matching and mutual information. IEEE Trans. Pattern Anal. Mach. Intell. **30**(2), 328–341 (2008)
16. CloudCompare. http://www.cloudcompare.org/. Accessed 13 March 2021

A Depth-Camera Based System for the Real-Time Scanning of Upper Limb Anatomy

Paolo Neri(✉), Sandro Barone, Alessandro Paoli, Armando Viviano Razionale, and Francesco Tamburrino

University of Pisa, Largo L. Lazzarino 1, 56122 Pisa, Italy
paolo.neri@dici.unipi.it

Abstract. The 3D reconstruction of upper limb anatomy plays a significant role in many biomedical fields as ergonomics, motion rehabilitation, prosthesis design. Conventional manual measurements have been progressively replaced by 3D optical scanning in collecting and storing 3D anatomical data, thus increasing reliability and data accuracy, shortening, at the same time, the overall acquisition process. However, the real-time scanning of human body parts still represents a complex task since it is challenging to keep the arm in a stable position and avoid artifacts in the collected data. Also, optical undercut geometries often impair the 3D reconstruction's completeness. In this paper, a compact and low-cost 3D scanning system has been developed by integrating three D415 Intel RealSense cameras. The three depth cameras have been assembled in a circular rig to define a lightweight handheld scanner capable of carrying out 3D data acquisition in different scenarios. The optical system has been validated through anthropometric measurements on different subjects.

Keywords: 3D scanning · Upper limb anatomy · Depth-camera · D415 Intel realSense

1 Introduction and Background

In the medical field, the 3D digital information of the surface patient's anatomy plays a significant role in designing highly customized medical devices [1]. More effective and comfortable therapies can be enabled by bespoke orthoses, customized prostheses, or active controllers. The rehabilitation of the upper limb functionalities, for example, may involve working with patients who have musculoskeletal disorders due to strokes, dystonia, or sports injuries. Creating an advanced patient's biomechanical profile is thus crucial for developing effective bespoke assistive controllers, and, for this purpose, the 3D scanning of the patient's upper limb anatomy represents its starting point.

Body sizes are conventionally taken by tape measurements characterized by invasive and time-consuming processes [2]. Also, accuracy is a significant concern since a considerable level of subjectivity characterizes manual measurements. Instead, the patient's morphology is usually acquired through plaster molds. Hand-crafting processes are then

C. Rizzi et al. (Eds.): ADM 2021, LNME, pp. 245–255, 2022.
https://doi.org/10.1007/978-3-030-91234-5_25

adopted to customize the medical device and enhance comfort once the patient wears the prototype. The effectiveness of the final appliance, therefore, depends on both skills and expertise of the specialist.

In the last two decades, manual measurements have been replaced by 3D optical scanning in the acquisition of human anatomy, thus increasing reliability and data accuracy, shortening, at the same time, the overall process. The digital anatomical model, created as an output of the acquisition process, represents an effective support in the design of bespoke devices, also allowing the extraction of measurement of interest at any time. The standard procedure has been progressively replaced using computer-aided design (CAD) and computer-aided engineering (CAE) tools. However, a significant step forward was possible thanks to the development of Additive Manufacturing (AM) technologies that allowed the direct fabrication of fully functional parts starting from a 3D model, without involving the traditional manufacturing process [3–5].

Capturing human upper limbs' 3D shape and motion has been the scientific focus for many researchers due to its applications in many engineering fields as computer graphics, human-computer interaction, rehabilitation, and robotics. The recent technological advancement in off-the-shelf RGB-D cameras has further increased this interest, especially for virtual and augmented reality applications. Many technologies exist, and countless are the scientific community's approaches [6, 7].

In this paper, a compact and low-cost 3D scanning system, composed of multiple RGB-D sensors, has been developed to perform real-time acquisitions of upper limb anatomy. Three D415 Intel RealSense cameras have been assembled in a circular rig to define a lightweight handheld scanner. The system effectiveness has been assessed by acquiring upper limb anatomies of different subjects and comparing anthropometric measurements with conventional tape data collection.

2 Upper Limb 3D Scanning: Main Issues

The 3D scanning of human upper limbs (i.e., hands, fingers, wrists, and forearms) is not a trivial task since it is not easy for the patient to keep the arm in a stable and static position for the duration of the whole scanning process. Therefore, artifacts due to involuntary movements are usually present in the 3D scanned data and must be handled [8]. For this reason, popular phase shift-based structured light methods are not suitable since they require at least three images, and the use of real-time single-shot techniques would be preferable [9]. However, critical issues remain since fingers have many degrees of freedom, move quickly, and look similar in color and shape, thus making difficult the tracking process. Moreover, the hand and fingers shape lead to self-occlusions, which make ambiguous their pose inference.

Technical literature has investigated the use of low-cost sensors, as the Kinect sensor (Microsoft, Redmond, WA, USA) to track the full-body motion by depth cameras and skeleton stream and the Leap Motion Controller (LMC, Leap Motion, San Francisco, CA, USA) for hand tracking [10].

2.1 Scanning Layout

Obtaining a 360° view of the scanned arm raises two issues: long acquisition time and registration issues between different clouds. A handheld multi-sensor layout was developed, exploiting three D415 Intel RealSense sensors placed at 90° on a circular rig to solve both these issues (Fig. 1). A full metrological characterization of the D415 Intel RealSense was performed in [11], demonstrating its effectiveness as a low-cost device for biomedical 3D-scanning applications. In the present work, the three optical sensors are mounted on the rig to have both a radial and an axial orientation. This allows acquiring target surfaces placed in front of the sensor, even if they do not go through the ring (e.g., in the case of a patient with impaired mobility). The two diametrically opposite sensors allow obtaining a wide acquisition angle around the arm axis. In contrast, the third sensor is oriented to acquire some arm's typical optical undercut, such as the thumb back surface. A similar approach was followed in [7] where eight D415 Intel RealSense were assembled in two circular rigs composed of 4 sensors each. This configuration, however, defines a static architecture, which represents a limiting factor for the acquisition of patients suffering from partially inhibited musculoskeletal system function (i.e., dystonia disorders).

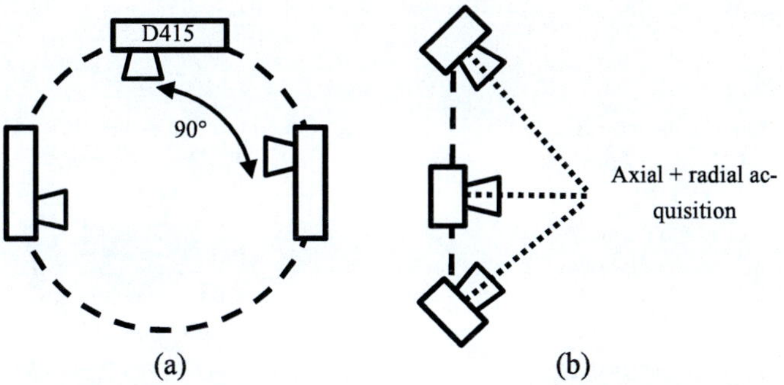

Fig. 1. The layout of the proposed scanning device: a) front view and b) lateral view.

2.2 Device Calibration

The relative placement of the sensors needs to be calibrated to automatically register the three point clouds acquired in each frame during the scanning stage. To achieve this result, a calibration specimen was designed, and 3D printed (Fig. 2). The specimen has three color-coded markers, which allow to automatically detect and isolate the specimen with respect to the background in the scanned data. The specimen is placed in front of the device, approximately in the rig's center, and acquired by the three sensors.

Fig. 2. Calibration specimen.

The markers are then used to obtain a rough alignment of the three point clouds with respect to a reference geometry obtained through the CAD model of the specimen, Fig. 3(a). Finally, the registration of the three-point clouds with respect to the reference geometry is refined through an Iterative Closest Point (ICP) algorithm, Fig. 3(b), to obtain the three transformations defining the relative placement of the sensors. It is worth noting that the described calibration procedure is fully automated; thus, the user must only acquire the specimen taking care that all the markers are visible by all the sensors.

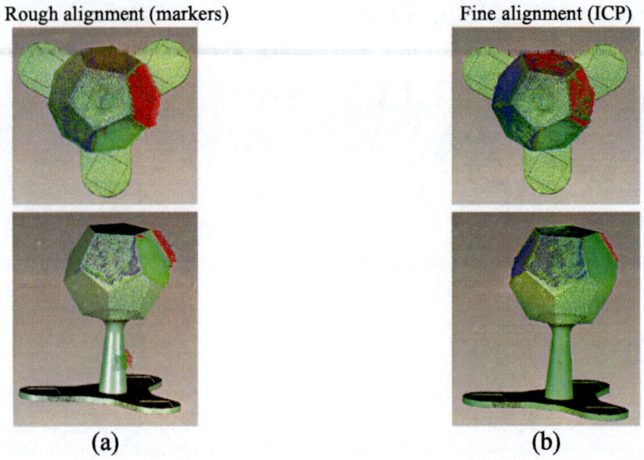

Fig. 3. Registration procedure during calibration: a) rough estimation based on markers and b) fine registration based on ICP algorithm.

2.3 Scanning Workflow

The scanning process can be carried out by the user (e.g., therapist), handling the structure and rotating it around the extended patient's arm's central axis. Preliminary acquisition trials demonstrated that, a small number of viewpoints ($5 \div 8$) is enough to achieve the

360° description of the arm surface. Thus, all the acquisitions shown in this paper were carried out with six different viewpoints, requiring an acquisition time of about 55 s. In patients with impaired mobility, some issues could arise since it could not be possible for the patient to extend his arm fully. In this case (e.g., the arm is folded close to the body), it is possible to exploit the sensors' axial orientation to frontally scan the patient, as shown in Fig. 4, even if it is no more possible to obtain a 360° scan of the arm.

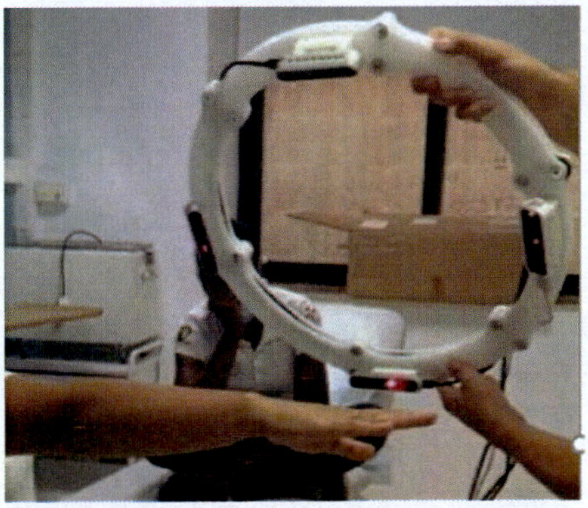

Fig. 4. Scanning process around the patient's arm.

2.4 Point Clouds Elaboration

Once the acquisition stage is completed, three series of different point clouds are obtained, coming from each of the sensors. The elaboration stage requires to: 1) register the three separated point clouds deriving from the sensors to a common reference frame, 2) remove noise from the clouds, and 3) register the subsequent acquisition with respect to each other. To this extent, a semi-automatic procedure was developed. Firstly, the point clouds acquired by different sensors can be registered to a common reference frame by exploiting the roto-translation determined during the calibration stage. This is a straightforward step, which only requires applying three known rigid transformations to the acquired data. Subsequently, this registration can be further refined using an automatic ICP algorithm to the three clouds of the single viewpoint, thus obtaining their final relative placement. After this step, the three sensors' points clouds are appropriately aligned, but the subsequent scans (e.g., the six viewpoints) are still noisy and not properly registered, as shown in Fig. 5(a).

The point clouds are then elaborated to remove noise, i.e., detect and delete outliers. This is performed through an automatic algorithm that detects clusters in the point clouds, separated by a given distance. The points belonging to each detected cluster are counted, and all the clusters with numerosity lower than a given threshold are deleted. Both the separation distance and the cluster numerosity can be set to achieve more or less aggressive noise reduction. Finally, the point clouds need to be registered to obtain the full reconstruction of the patient's arm. This is achieved by the conventional 3–2–1 pairwise registration approach [12]: firstly, three corresponding points are selected on

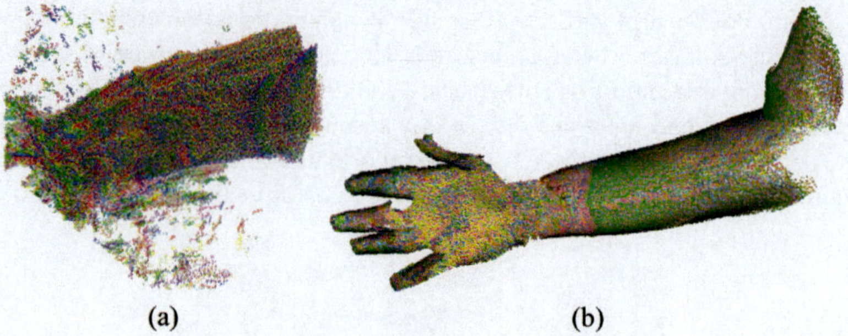

Fig. 5. Scanning raw outcomes: a) noisy and overlapped point clouds and b) de-noised and register arm scanning.

consecutive scans to obtain a rough registration. Secondly, an ICP algorithm is used to refine the registration. This process is repeated for all the six viewpoints, and a final global registration algorithm is then applied to refine the results further, obtaining the 360° point cloud as shown in Fig. 5(b).

After the elaboration, the data consist of 3D coordinates of the point cloud (de-noised, registered), RGB texture for each point, and surface normal. This information is stored in arrays: three columns of data for the coordinates, three columns for the texture, and three columns for the normal, being the number of rows equal to the number of measured points. The data can then be saved in several formats, such as.txt or ASCII files.

3 Assessment of Anatomical Scanning

The developed scanning system was tested, performing several acquisitions to assess hardware and software functionality, the system's usability, the effectiveness of the measurement, and the training time needed to become confident with the equipment. To this extent, three different researchers performed a complete scan of each other arms. This first assessment also allowed to verify the scanning outcome with respect to anthropometric measurements taken through rulers. These tests demonstrated that, after about five trials, each researcher was able to perform a successful complete scanning of the human arm. Figure 6 shows the scanning results for one of the researchers from the six different viewpoints. As stated, each viewpoint is the combination of the acquisition of three sensors, thus providing the surface of a wide-angle. The figure highlights that the full 360° acquisition was achieved, with no gaps or relevant holes in arm and forearm geometry. Some noise and surface holes can be noted between the fingers, which are the most challenging region to be acquired. Nevertheless, arm and forearm surfaces are smooth. After the training trials, all the researchers could complete the acquisition procedure in less than 55 s.

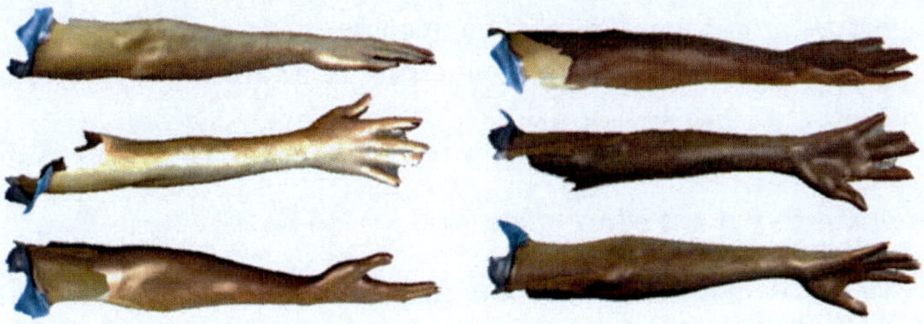

Fig. 6. Example of 3D scanning outcomes: 360° view of the human arm

The good qualitative results of Fig. 6 were quantitatively evaluated by comparing the measurement of some key points of the arm on the 3D scanning and on the actual arm. The comparison results are shown in Table 1.

Table 1. Anthropometric measurements for scanning validation

Measurement	Scanning (mm)	Tape (mm)	Tape σ	Diff. %
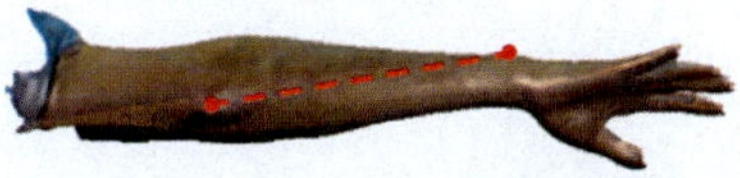	286	275	8	3.8
	94	97	4	3.2
	56	58	2	3.6
	179	181	7	1.1
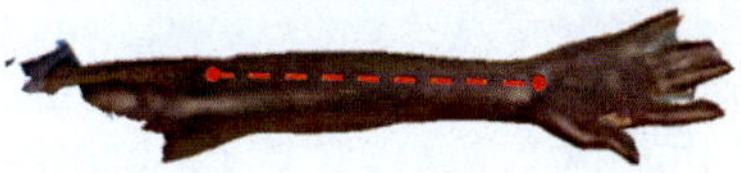	309	294	12	4.8

The first column shows an image of the performed measurements, the second column reports the measurement value as obtained from the 3D scanning (rounded to the nearest unit), the third column reports the mean value of the corresponding measurement taken through a tape with five repetitions, the fourth columns shows the standard deviation of the tape measurement, while the fifth column reports the percentual difference between 3D scan and tape measurements. The assessment produced promising results concerning the usability and reliability of the system. It is worth noting that a high standard deviation was found concerning the tape measurements due to significant difficulties for the operator to select the proper measurement point on the arm repeatedly. This also justifies the absolute measurement discrepancies, which are to be ascribed to uncertainty in the tape measurement.

3.1 Issue Related to Patient Movement

The previous section demonstrated that 360° acquisition of the patient's arm is possible for a stretched and still arm. On the other hand, some issues may arise if the patient cannot stretch the arm or be perfectly still. This phenomenon was simulated by repeating the scanning with different grades of arm movement and different arm positions. The first trial results are shown in Fig. 7, highlighting an acquisition issue since the last two frames of the scanning showed the wrist movement. Indeed, the subject slightly changed the hand and the forearm's angle, thus impairing the registration process. Nevertheless, two separate surfaces could be defined, which can be processed separately. Indeed, even if the last two views could not be aligned with the previous, all the frames were successfully acquired, meaning no sides of the arm are missing in the acquisition. Thus, the arm geometry's 360° information is still preserved if the two separate clouds are considered.

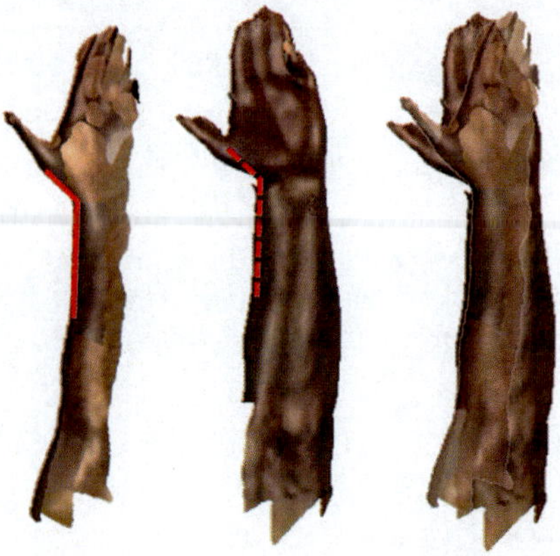

Fig. 7. Example of partial 3D scanning: the wrist angle changed during the acquisition.

Some more trials were then performed, simulating the typical movement of dystonia patients. To this extent, several tests were repeated, mimicking tremors of the hand and/or of the arm, to assess the movements' impact on the scanning quality. In particular, two examples will be reported. The first case was simulated, assuming that the patient could keep the arm almost stretched, even if he could not be still. The second case was simulated considering that the patient could not stretch the arm at all and could not be still either. In the first case, the six acquisitions around the arm could be performed as for a healthy subject, while in the second case, the arm was folded too close to the subject's body. Thus a 360° view was not possible, but several different orientations were acquired to describe as much geometry as possible. The same post-processing procedure (cloud registration and denoising) applied to healthy patients was used in both cases. Figure 8(a) shows the registered point clouds for the first case, while Fig. 8(b) shows the second case results. As can be noted, in both cases, the proper registration of different viewpoints to obtain a unique point cloud is not possible because of the uncontrolled movement, which leads to inconsistent clouds between acquisitions. Nevertheless, the single views were separately elaborated to assess if they are still usable for measurement purposes. Some examples are reported in Fig. 9(a) and (b), which show two different viewpoints of the first and second simulated cases, respectively. The images demonstrate that, even if the full 360° point cloud cannot be achieved, the clouds corresponding to single viewpoints are not impaired by the patient's movement since the device's frame rate is sufficiently high to avoid blur corrupted or noisy acquisitions. As can be seen, the single frame appears consistent and can be used to assess some specific anthropometric measurements or describe some portions of the geometry.

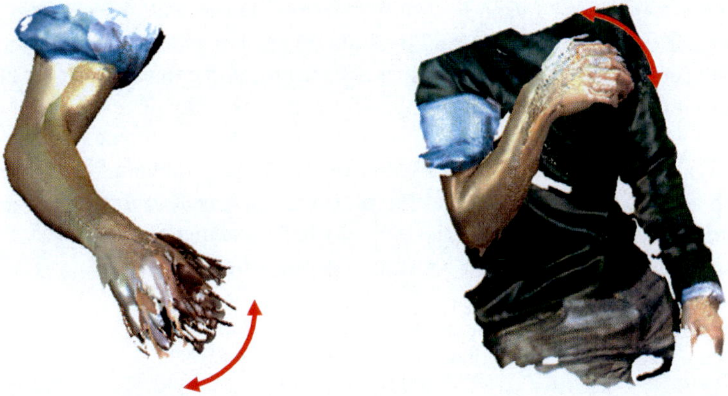

(a) (b)

Fig. 8. Scanning results for simulated dystonia patients: (a) stretched arm and (b) folded arm.

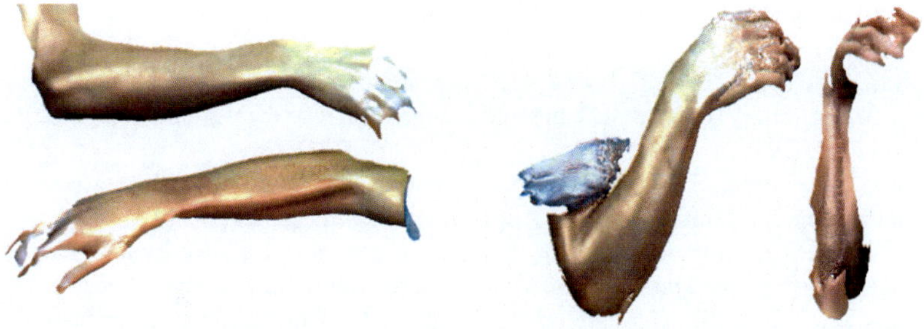

(a) (b)

Fig. 9. Scanning results for simulated dystonia patients: (a) stretched arm and (b) folded arm

The issues related to patient's movement may be mitigated by using fixed supports to constrain the patient's arm. Nevertheless, this could impair the usability of the system for most fragile users (such as dystonia patients), thus it was discarded at the present research stage.

4 Conclusion

The 3D scanning of upper limb geometry was studied and discussed in this paper. The main issues concerning the acquisition process were discussed, and a scanning device was proposed to accomplish the task. To this extent, a compact, low-cost handheld scanner was assembled based on D415 Intel RealSense cameras. Three sensors were mounted on a rig at 90°, to allow for a wide-angle scanning in a single frame. The device's automatic calibration procedure was proposed, basing on a simple calibration specimen and color-coded markers. Consequently, the scanning workflow was presented, consisting of acquisition, clouds denoising, and registration. The device was then tested, showing promising results in terms of usability (only five training acquisitions were need, acquisition time (360° acquisitions obtained in about 55 s), and accuracy (good comparison with tape measurements). A more extensive validation campaign with a larger number of patients and a more robust tape measurement procedure is under development for further characterization of the system. Finally, some critical issues of the acquisition process were investigated, which are mainly related to involuntary movements. Nevertheless, the research activity demonstrated that the procedure is robust to this noise source: the full 360° acquisition of moving subjects could not be achieved. However, the single viewpoints resulted in being usable for further processing.

Acknowledgement. This project has received funding from the European Union's Horizon 2020 research and innovation programme under grant agreement No 856998.

References

1. Barone, S., Neri, P., Paoli, A., Razionale, A.V.: Design and manufacturing of patient-specific orthodonticappliances by computer-aided engineering techniques. Proc. Inst. Mech. Eng. **232**, 54–66 (2018)
2. Xia, S.B., Guo, S.M., Li, J.Y., Istook, C.: Comparison of different body measurement techniques: 3D stationary scanner, 3D handheld scanner, and tape measurement. J. Text. Inst. **110**, 1103–1113 (2019)
3. Baronio, G., Harran, S., Signoroni, A.: A critical analysis of a hand orthosis reverse engineering and 3D printing process. Appl. Bionics Biomech. **2016** (2016)
4. Barrios-Muriel, J., Romero-Sanchez, F., Alonso-Sanchez, F.J., Rodriguez Salgado, D.: Advances in orthotic and prosthetic manufacturing: a technology review. Materials **13** (2020)
5. Paterson, A.M., Bibb, R., Campbell, R.I., Bingham, G.: Comparing additive manufacturing technologies for customised wrist splints. Rapid Prototyping J. **21**, 230–243 (2015)
6. Zollhofer, M., et al.: State of the art on 3D reconstruction with RGB-D cameras. Comput. Graph. Forum **37**, 625–652 (2018)
7. Buonamici, F., Carfagni, M., Puggelli, L., Servi, M., Volpe, Y.: A Fast and Reliable Optical 3D Scanning System for Human Arm. In: Roucoules, L., Paredes, M., Eynard, B., Morer Camo, P., Rizzi, C. (eds.) JCM 2020. LNME, pp. 268–273. Springer, Cham (2021). https://doi.org/10.1007/978-3-030-70566-4_43

8. Paoli, A., Neri, P., Razionale, A.V., Tamburrino, F., Barone, S.: Sensor architectures and technologies for upper limb 3D surface reconstruction: a review. Sensors-Basel **20** (2020)
9. Wang, Z.Z., Zhang, C.S.: Three-dimensional hand reconstruction by single-shot structured light line pattern. IEEE Access **6**, 59881–59890 (2018)
10. Guzsvinecz, T., Szucs, V., Sik-Lanyi, C.: Suitability of the kinect sensor and leap motion controller—a literature review. Sensors-Basel **19** (2019)
11. Carfagni, M., et al.: Metrological and critical characterization of the Intel D415 stereo depth camera. Sensors-Basel **19** (2019)
12. Zhu, H., et al.: A review of point set registration: from pairwise registration to groupwise registration. Sensors-Basel **19** (2019)

Geometrical Product Specification and Tolerancing

The ASME Y14.43 Standard and Its Root in the American Pragmatism

Sergio Rizzuti and Luigi De Napoli(✉)

Università della Calabria, DIMEG - Ponte Pietro Bucci 46/C, 87030 Rende, CS, Italy
{sergio.rizzuti,luigi.denapoli}@unical.it

Abstract. ASME Y14.5 and Y14.43 standards share the task to define a coherent context in which geometric tolerances may find a rational approach. Even ASME Y14.5 defines the rules, meanings, and descriptions of geometric tolerances, ASME Y14.43 applies them aiming to design and build functional gages able to check if the geometric characteristics, assigned on a drawing, have been reached in a physical component. Many of the assumptions given in the ASME Y14.5 may be understood only when the design of a functional gage is approached. The need to build such kind of device is strictly connected with the need of industry that requires best-practices able to check geometric characteristics of products in a short time. The basic elements of ASME Y14.43 are summarized in the paper and a functional gage, taken from the edition of 2011, is elaborated and discussed. The root of the American standards has been associated with the philosophical school of pragmatism in which more attention is paid to practical examples and how every theoretical conceivement must be verified by scientific experimentation.

Keywords: Geometric dimensioning and tolerancing · Functional gages · Pragmatism

1 Introduction

ISO and ASME never diverged in the conception of Geometric Tolerances as in this last period. The interest of both organizations is to define and clearly pose the conditions to solve the problem of the best representation of geometric errors inside industrial products. Even GPS (ISO) and GD&T (ASME) strategies seem to treat the same subject and share the same symbols, they really differ from the cultural point of view. ISO researchers invented the GPS master plan to build a structure able to incorporate all the standards that have been emanated along time (or that will be in the future) regarding the problem of industrial products conformity. ASME professionals conceived GD&T methodology in order to allow mechanical engineers to check the validity of their projects, controlling the geometric errors inside their products. When the geometric tolerances topic is presented to students it is not easy to stimulate the audience on the difference between ISO and ASME avoiding losing the main task to introduce them in the hard and tricky culture of geometric control of components. The comparison between ISO and ASME has been discussed in the literature, recently by Stefano Tornincasa [1], even probably the best

C. Rizzi et al. (Eds.): ADM 2021, LNME, pp. 259–268, 2022.
https://doi.org/10.1007/978-3-030-91234-5_26

way to treat the question is to discuss both "theories" independently of each other, and at the end compare them. ISO, in the recent 1101 standard of 2017 [2], exceedingly stressed the metrologists' point of view inside the Committee that guides the study of such norms. ASME technicians, in the latest norm Y14.5 of 2018 [3] stressed the applicability of the norm to its real employment in the design and the final check of a part. But the importance of the ASME Y14.5 standard is not clearly evident if it is not merged to the ASME Y14.43 [4] that dictates the rules to design a functional gage able to check the geometric errors specified on a component. The name of ASME Y14.43, the latest edition has been in 2011, is in fact "Dimensioning and Tolerancing Principles for Gages and Fixtures. Engineering Drawing and Related Documentation Practises". It has been proposed for the first time in 2003, as argued by Meadows in [5] and maybe it will be reviewed in the near future to adhere more precisely to the latest edition of the ASME Y14.5–2018. These two ASME standards (Y14.5 and Y14.43), studied in conjunction, give a coherent ambient, in which many of the established rules become absolutely clear.

The paper discusses the employment of the ASME Y14.43 standard, its main characteristics with the task to build a functional gage. The final section will try to identify such ASME approach in the philosophical theory of the American pragmatism, that has in John Dewey its main representative in a pedagogical context.

2 The ASME Y14.43-2011 Standard

This Standard gives the design practices for dimensioning and tolerancing of gages and fixtures for the inspection of geometric tolerances. This standard not only govern the principles for the appropriate procedures for the creation of gages for geometric tolerances but also continues the practices to measure maximum material condition with GO gages and least material conditions with NOGO gages.

The foreword of the ASME Y14.43-2011 standard contains a set of sentences that outlines its main characteristics:

- *This Standard contains information showing methods for creating gages and fixtures for features that use principles found in ASME Y14.5, Dimensioning and Tolerancing.*
- *This Standard shows the principles and choices available to design, dimension, and tolerance gages and fixtures in compliance with the principles in ASME Y14.5-2009 and previous editions. The gages and fixtures displayed in this Standard represent the physical embodiment of the theory shown in ASME Y14.5 for the simulation of virtual condition boundaries and proper datum feature simulation.*
- *The understanding of gages and fixtures is the key to understanding dimensioning and tolerancing of products in accordance with ASME Y14.5. This Standard is intended to serve the needs of those professionals who are designing gages and fixtures for workpieces dimensioned and toleranced per ASME Y14.5.*

Since ASME Y14.5 is not a gaging standard, ASME Y14.43 shows the practical embodiment of the theory displayed in ASME Y14.5 by illustrating how workpieces can be fixtured and gaged for tolerance verification.

Gages and fixture are defined as:

- *gage element*: a physical feature of the gage used in the verification of workpiece compliance to the associated tolerance requirement. These physical features represent datum feature simulators, virtual condition, or datum material boundaries.
- *GO gage*: a fixed limit gage that checks a feature of size for acceptance within MMC perfect form boundary.
- *NOGO gage*: a fixed limit gage that checks a feature of size for violation of the LMC actual local size.
- *functional gage*: a fixed limit gage used to verify virtual condition boundaries (MMC concept) generated by the collective effect of the feature's MMC and the applicable geometric tolerance at the MMC size.
- *fixture*: a device used to hold parts securely in the correct position in a tool or gage during manufacturing, assembly, or inspection.

Most part of ASME 14.43 is addressed to gages design, considering the design of fixtures as something that have contact with part datum feature and that typically holds parts as they would be held when assembled. Fixtures represent simulated datum features of the part.

3 Gage Characteristics

The basic premise of all gages is to reject bad parts (those that violate the tolerances) and accept all of the good parts (those parts that are in compliance with specified tolerances). But since all gages need to be toleranced, it is understood that they will fail to achieve these lofty goals. They will either reject a small percentage of technical in-tolerance parts or they will accept a small percentage of technically out-of-tolerance parts.

The characteristics of a gage are based on how the designer chooses to apply the different principles available, such as gaging policy, percent of workpiece tolerance used, material condition modifier, and wear factor allowance. With all these choices available, it is possible to provide the basis for several gages to be designed with different characteristics for a single workpiece drawing.

3.1 Gaging Policy

The gaging policy should be the first decision made, as this will define the functional acceptance characteristic of the gage.

The choices are absolute, practical absolute, tolerant, and optimistic policies.

The *absolute* tolerancing is a policy of tolerancing gages that ensures complete random assemblability of parts. Gages produced under this policy will accept most part features the are within tolerance, reject all part features not within tolerance, and reject a small percentage of borderline part features that are technically within tolerance.

The *practical absolute* tolerancing is a policy of tolerancing gages that predicts most part features within tolerance will be accepted by the gage, some borderline part feature within tolerance will not be accepted by the gage, and that there is a very low probability that some borderline part feature not within tolerance will be accepted by the gage.

The *tolerant* tolerancing is the policy of tolerancing gages that ensure most part features within tolerance that are gaged are accepted, and most part features not within tolerance that are gaged are rejected.

The *optimistic* tolerancing is the policy of tolerancing gages that ensures all part features within tolerance that are gaged are accepted by the gage.

3.2 Percent of Workpiece Tolerance Used by Gage

The gage tolerance is based on a percentage of the workpiece tolerance. This percentage value is determined by the gage designer and may vary from one gage to another as function changes. Each feature of the gage that represents a feature on the workpiece is recommended to receive a tolerance between 5% or 10% of the tolerance assigned to that particular workpiece feature. This Standard recommends that the gage designer consider 5% of the part tolerance used as gage tolerance with an additional 5% considered for wear allowance.

The selection of appropriate material condition modifiers is important in determining where the gage size elements fall within the gage tolerance range. Each modifier contributes a different characteristic to the gage. Consideration should be given to understanding where the gage pin size occurs within the tolerance band.

When gage element wear is considered a factor in the gage design, a percentage of the gage tolerance can be applied to increase the gage element's lower size limit to account for wear. Wear allowance applied to a new gage will reject a larger number of good parts, and as it wears toward the lower size limit, will tend to accept more good parts.

3.3 Design Constraints

As with any measurement tool, consideration shall be given to the advantages and disadvantages of gages as they pertain to the design, manufacture, use, and maintenance of the gage.

Useful Life. Gages wear as they are used. Eventually, the gage will wear beyond acceptable limits and begin to accept parts that are not within tolerance. Therefore, gages shall be closely monitored for wear to determine when it is appropriate to replace or refurbish the gage. Where possible, the original gage design should facilitate both monitoring and maintenance.

Availability of Commercial Components. When it is possible to purchase off-the-shelf components for gages, they should be considered for use. This practice has the potential to reduce the original and refurbishment cost of the gages.

Size and Weight. Whenever possible, gages should be made at a physical size and weight that allows the gage to be easily handled for optimal use. A gage that is unnecessarily heavy may be difficult to maneuver and use.

Physical Properties. The material used for gages shall be selected with consideration to stability, durability, and rigidity.

(a) Material. Gaging elements shall normally be manufactured from high-quality tool steel suitably selected to provide a high degree of wear resistance after heat treatment.
(b) Hardness. The hardness of the gaging surface shall be at least 700 HV (Vickers Hardness scale)/ RHC60 (Rockwell C Hardness).
(c) Stabilization. The gage manufacturer shall ensure that the gages are adequately stabilized by a method appropriate to the material, shape, and size.
(d) Surface Texture. The surface texture shall be consistent with the accuracy of the gage desired. The maximum roughness values are expressed in roughness average values, Ra, for the preferred classes, ranging from 0.2 to 0.05 μm.

4 Case of Study

The design of a functional gage is based on the strategic choice of some parameters, such as the tolerancing policy, the percentage of tolerance on the piece to be transferred on the geometric counterpart of the gage, the modifiers of the material condition, and lastly the percentage of tolerance that takes into account wear allowance.

The case of study proposed here is an example of the ASME Y14.43 (see [4] fig B-14 pg. 96) and the task consists of designing a functional gage able to check the assemblability of the component represented in Fig. 1, under the *practical absolute* tolerancing policy.

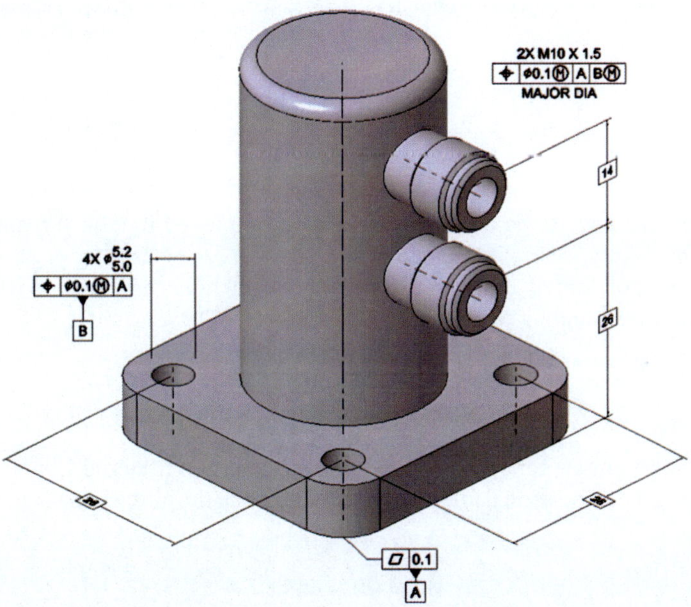

Fig. 1. The component for which a functional gage must be designed.

The component may be fixed to an external support by a set of 4 screws that will be inserted in the corresponding pattern of holes present in the plate. In the upper part, two pipe unions are present, controlled by position tolerance with respect to the base, taken as a datum.

The geometric tolerances taken into account begin with a datum reference (A) on the lower surface of the plate, where planarity of 0.1 mm is assigned. A second datum reference (B) is assigned to the pattern of holes. All four holes are manufactured as a pattern with a dimensional tolerance of 0.2 (Ø5.0–5.2) and controlled by a position error of Ø0.1, at the MMC. Then there is the position of the two pipes union, where the axes of their threads are controlled by a position error of Ø0.1, under the MMC modifier, with respect to datum reference frame [A, B]. The datum B is located at the center of the pattern of holes, which establishes the true geometric counterpart used to derive the second and third datum reference plane.

In Table 1 the error values to be employed in the gage are summarized, to compute the direct percentage for the pins tolerance on the gage.

Table 1. Direct percentage gage tolerance method

Practical absolute policy	Direct percentage gage tolerance **MMC**
No wear allowance: (10%) = total percent of workpiece size tolerance used by gage	Direct Percentage Gage Tolerance: 0.02 (10% of workpiece tolerance): Pin Ø4.90 – 4.92 $Ø(MMC) = 4.92 \rightarrow Ø0.01\,position$ $Ø(LMC) = 4.90 \rightarrow Ø0.03\,position$
With 5% wear allowance (added to pin inner boundary): (5%) = total percent of workpiece size tolerance used by gage before wear allowance	Direct Percentage Gage Tolerance: 0.01 (5% of workpiece tolerance): Pin Ø4.90 – 4.91 $Ø(MMC) = 4.91 \rightarrow Ø0.01\,position$ $Ø(LMC) = 4.90 \rightarrow Ø0.02\,position$

Assuming the gage to be designed will not be employed in a hostile environment, wear is not necessary, 10% of tolerance workpiece size can be used by gage, as reported in the upper part of Table 1. The pin that will check the holes on the plate will be made with the size corresponding to the Virtual Condition (VC) evaluated as

$$Ø(\mathrm{VC})_{pin} = Ø(\mathrm{MMC}) - \delta = 5.0 - 0.1 = 4.9 \quad (1)$$

where δ is the geometric tolerance applied to the Maximum Material Condition.

Employing the dimensional tolerance, the diameter of the pins will have the following range

$$Ø_{pin}4.90 - 4.92.$$

The second check is for the pipes union, for which the gage must provide two holes in which the threaded surfaces can be inserted. The Virtual Condition for such holes is

$$\text{Ø}(VC)_{hole} = \text{Ø}(MMC) + \delta = 10.0 + 0.1 = 10.1 \tag{2}$$

where δ is the geometric tolerance applied to the Maximum Material Condition (in this case the maximum diameter) and further when datum B is in its Maximum Material Boundary.

The holes' tolerance will be at 10% of the geometric tolerance assigned to pipes union (0.01) and their size tolerance will be in the range

$$\text{Ø}_{hole} 10.09 - 10.10$$

The position of both pipe union holes' axes must be at the perfect form with respect to the bottom surface. The gage surface where the component will be put must be manufactured with a 10% of the planarity tolerance of that assigned to the component (0.01).

Figure 2 represents the gage able to check such a component. The gage has been conceived in two parts: the first that allows the component to be positioned and checked with the pattern of pins; the second that checks the position of the threaded surface of pipes union. Both parts are provided with a dovetail grooving that allows part #2 to slide with respect to part #1.

On such kind of gages where stack-up is present, a further check must be made to verify that the errors due to accumulation of errors do not override the 50% globally assigned tolerancing, that would prejudice gage functionality (see [4] Sect. 4.5.1).

Figure 3a shows the way in which both gage components are assembled, and Fig. 3b shows as the functional gage may check the component.

A lot of work must be done in order to guarantee that functional gage operates properly. The example on ASME Y14.43 does not explain how to design the dovetail surfaces that allow mating the components. In Fig. 2 additional tolerances of orientation, profile and position have been added to the surfaces on part #1 that must guide the datum reference on part#2.

The dovetail slot on part#1 has been made lowering the level of datum A by 3 mm. The parallelism of such surfaces is checked by the profile of a surface with a range of 0.01. The dovetail coincides on both parts: the theoretical slot width is fixed at 22 mm and the angularity at 45°.

The datum reference frame on part#2 is established with the first plane on the lateral surface (C), the second one on the lower surface (A) with a check of perpendicularity of 0.01, and the third datum as the mid-plane of slot width (B), with a further check of perpendicularity. This latter is positioned on the theoretical dimension of the slot, before remove the chamfer.

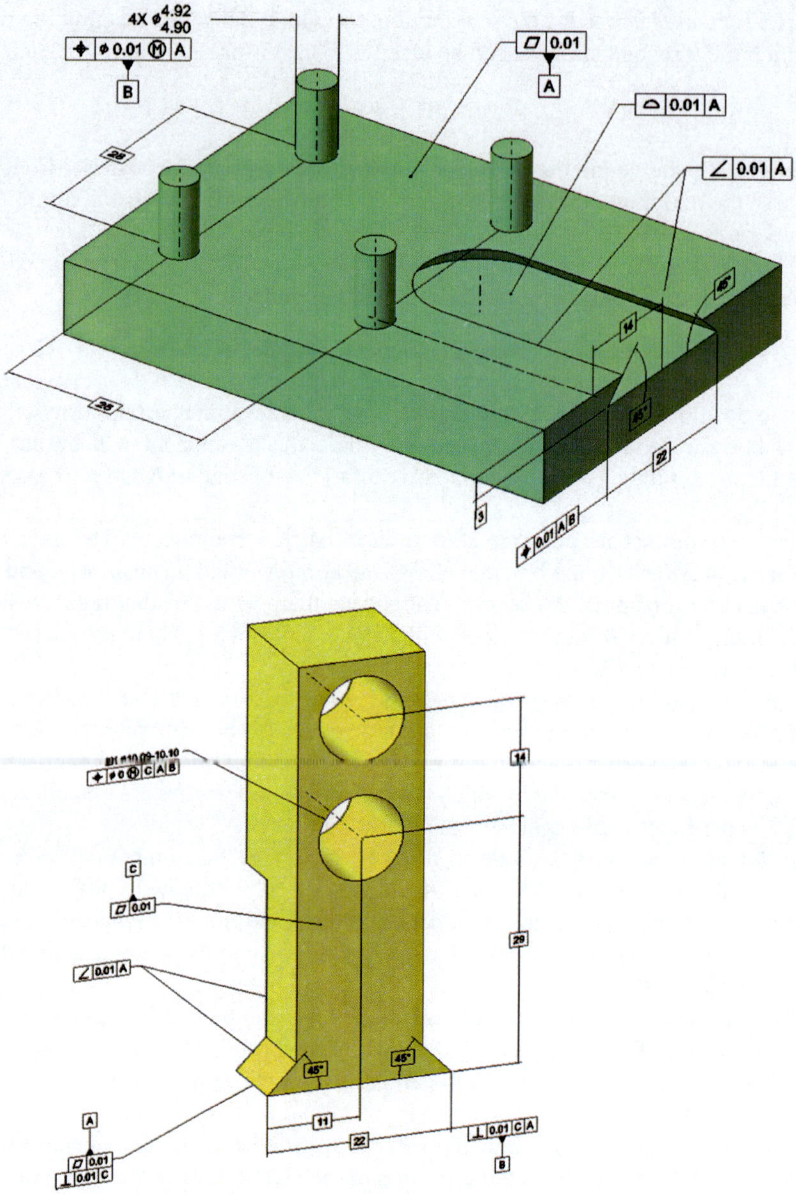

a)

b)

Fig. 2. Functional gage able to check geometric tolerances decomposed in two parts: a) part #1 for checking the pattern of holes; b) part #2 for checking the position of pipe unions.

The position of both pipe union threads is checked by a couple of holes positioned at a theoretical distance of 29 mm from the bottom and 14 mm between their axes.

Lubrication must be provided in order to guarantee that sliding of part#2 on part#1 occurs properly.

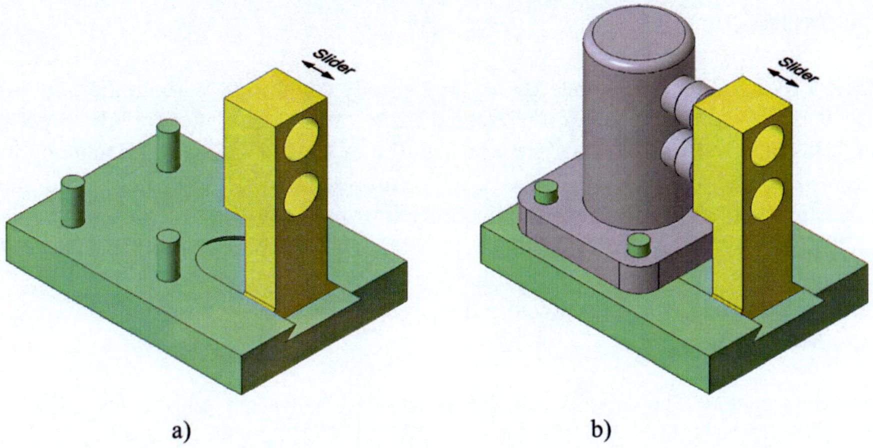

Fig. 3. a) Functional gage assembled; b) the functional gage that checks the satisfaction of the geometric condition imposed on the component.

5 ASME Roots in American Pragmatism

The different approach between ISO and ASME can be identified in the different mentality by which it is not only read and interpreted a technical drawing, even every aspect of real life. The American approach pays more attention to practical examples and how every theoretical conceivement must be verified by scientific experimentation.

This way of thinking is defined as pragmatism [6] and originated in the USA in the late '800 due to Charles S. Peirce and William James. John Dewey was the principal activist of the second generation and turned pragmatism towards politics, education, and other dimensions of social improvement.

Dewey based his philosophy on school considered as a social institution, mean to form future leaders' generation. His pedagogy, called also structuralism, is based on observation and on the direct experimentation of the theory he conceived so that these never remained confined at the idealistic level. Pragmatism is the core of his thinking where the practice is absolutely fundamental and ideas are not other than tools to produce actions. According to Dewey, the place where the real-world (ambient) and the subject (human) that faces it meet is the experience, where the human is not a passive viewer but interacts with what surrounds him/her. Human thought grows with experience; education must open the way to new experiences. To solve a problem is essentially the search for tools that give consciousness to the human.

Just to compare such a philosophical approach with the topic of the paper it can be said that the functional gage represents, for a strategy of industrial production, the tool that meets the need for the functionality check with something that immediately gives an answer about the conformity required to a physical component that must adhere to the requirements assigned to it.

6 Conclusion

ASME Y14.5 and Y14.43 standards are very strictly connected. Without their conjoint study, it is not really possible to understand many of the assumptions made in ASME Y14.5. Just considering the news in the last edition of ASME Y14.5 (2018) that removes concentricity and symmetry tolerances, as meanings, checks, and symbols, the motivation may be found in their not applicability in functional gage designing. A radical choice. But surely because the check of symmetry and concentricity can be made validly by a gage employing the geometric tolerances of position. Considering that a kind of geometric tolerance is not employable for gage design, it has no reason to exist, it is only a theoretic question.

Acknowledgements. The authors would like to thank Mr. Annibale Spizzirri for his work on the ASME Y14.43 standard and the Department DIMEG at UNICAL for the financial support.

References

1. Tornincasa, S.: Technical Drawing for Product Design: Mastering ISO GPS and ASME GD&T. Springer, Cham (CH) (2021)
2. ISO 1101. Geometrical Product Specifications GPS – Geometrical Tolerancing – Tolerances of form, orientation, location and run-out, Switzerland (2017)
3. ASME Y14.5-2018, Dimensioning and Tolerancing – Engineering Product Definition and Related Documentation Practices, ASME Two Park Avenue, New York (2019)
4. ASME Y14.43-2011, Dimensioning and Tolerancing Principles for Gages and Fixtures– Engineering Drawings and Related Documentation Practices, ASME Two Park Avenue, New York (2011)
5. Meadows, J.D.: Dimensioning and tolerancing of gages per ASME Y14.43-2003. In: Meadows, J.D. (Ed.) Geometric Dimensioning and Tolerancing Handbook: Applications, Analysis & Measurement Chap. 22. ASME Press (2009)
6. Stanford Encyclopedia of Philosophy Homepage. http://plato.stanford.edu/entries/pragmatism. Accessed 19 Feb 2021

Proposal for a Method for Advanced Search of Information and Standards in the Field of Geometrical Product Specification

Ileana Bodini[1], Gabriele Baronio[1], Diego Paderno[1,2,3], Valerio Villa[1,2,3], Paolo Martinelli[1,2,3], Roberto Frizza[2], Alessandro Balsamo[3], and Stefano Uberti[1(✉)]

[1] DIMI, Università Degli Studi di Brescia, Brescia, Italy
stefano.uberti@unibs.it
[2] MG Marposs, Brescia, Italy
[3] INRIM, Torino, Italy

Abstract. The Geometrical Product Specification (GPS) is one of the most powerful tools available to link the "perfect" geometrical world of the models to the imperfect world of manufactured parts and assemblies. GPS aims to be an unambiguous common language between designers, process engineers, and Coordinate Measuring Machines operators (CMM) in the ISO (International Organization for Standardization) environment. GPS standards are 150 and further 26 are under development. The need to have a tool to search through the standards, to optimize the work of the designer and to minimize the design, production and control costs is great. A database and the structure for a search engine, called GPS Navigator, has been studied and developed, and the requirements for the following coding phase have also been defined, in order to realize a powerful, efficient, fast, robust and rigorous tool to navigate through the GPS standards. Final aim of this tool is to help and guide the designer to quickly consult the correct standard or the most appropriate set of standards.

Keywords: Geometrical product specifications (GPS) · Database · Advanced search criteria · Design methods

1 Introduction

The Geometrical Product Specification (GPS) and the Geometrical Dimensioning and Tolerancing (GD&T) languages are the most powerful tools available to link the "perfect" geometrical world of the models to the imperfect world of manufactured parts and assemblies. GPS and GD&T should be unambiguous common languages between designers, process engineers, and Coordinate Measuring Machines operators (CMM) in the ISO (International Organization for Standardization) and ASME (American Society of Mechanical Engineers) environments, respectively [1].

This works only deals with GPS standards, because they are the ones adopted in Italy and in almost all Europe and because they are also the most numerous. Consequently,

C. Rizzi et al. (Eds.): ADM 2021, LNME, pp. 269–276, 2022.
https://doi.org/10.1007/978-3-030-91234-5_27

it is easier “to get lost” among them. The whole GPS standards set is under continuous development through the cooperation of 27 participating members and 26 observer, and currently includes 150 standards, while 25 others are currently being developed [2].

GPS standards have been considered in several works as an integral part of the production chain of a product, from its design to its final control, passing through production and processing. Their knowledge and correct application has often been considered as fundamental medium to reduce the overall cost of a certain product [3–6]. Even if GPS standards have been increasingly taught at engineering schools or in basic tolerancing classes, many circulating drawings are still prone to ambiguity of interpretation and do not communicate true functional requirements of the parts. Potential savings related to the correct implementation of GPS language are generated by eliminating these types of losses. All companies involved in designing or manufacturing mechanical parts can benefit from a correct usage of GPS, but advantages are greatest for those companies outsourcing the manufacturing of parts because these companies, more than any others, need their drawings to clearly communicate functional requirements, univocally, not ambiguously [4–7].

There have been many attempts in synthesizing these standards [4, 8–10], providing guidelines or simplifications [1, 3, 4, 7–10] or otherwise creating a new language [11]; in all cases, authors of these studies highlighted the need to provide guidance in choosing and applying standards, in order to reach the best communication level during the first phase of the product lifecycle..

In this moving chain of information, designer is often the weakest link since he must formalize and express the desired prescriptions with clear terms and a proper language that can sometimes be cryptic. Whatever the complexity of a drawing, not all the 150 standards are needed to understand it, therefore having a tool that allows the designer to find the group of appropriate standards to reach the functional requirements of the component can be extremely useful.

Present work originates from all the aforementioned needs, first of which is the correct choice of the most appropriate set of rules to convert the functional requirement into an appropriate graphic sign on the drawing. This paper presents the development of an advanced search engine, named “GPS Navigator”, to help designers in untangling themselves among ISO GPS standards. The database and the structure of this system have been developed and made available, and the specifications for the coding of the GPS Navigator search engine were studied and provided as well.

The operational steps were the following: Database development, Technical specifications definition, and Implementation.

2 Methods

Whatever the complexity of a design, not all the 150 standards are needed to understand it. The aim of this work is to find a method to choose between the more suitable standards, starting from a simple keyword, e.g. the functional aim of the component, or the required form. In this regard, GPS Navigator is an advanced search engine that facilitates designers in their search for GPS information or standards.

2.1 Database Development

Currently, the database is essentially an Excel table whose rows contain ISO GPS standards identification numbers while the columns represent the attributes that will be exploited during the search process which define the standard from different points of view: (i) ISO codification; (ii) UNI codification; (iii) Title; (iv) Class (Fundamental, General, Complementary in accord to ISO 14638); (v) Relationship to matrix model (ISO 14638).

Each standard of the ISO GPS has been analysed so far, and consequently different attributes representing probable search criteria have been inserted in an Excel spreadsheet for each standard: (i) ISO coding; (ii) UNI coding; (iii) title; (iv) type; (v) position in the GPS matrix; (vi) 3 levels of topics from general to specific; (vii) keywords; (viii) index; (ix) purpose; (x) notes.

GPS MATRIX	A Symbols and indications	B Feature requirements	C Feature properties	D Conformance and non-conformance	E Measurement	F Measurement equipment	G Calibrations
1 Size							
2 Distance							
3 Form							
4 Orientation							
5 Location							
6 Run-out							
7 Profile surface texture							
8 Areal surface texture							
9 Surface imperfections							

Fig. 1. GPS matrix structure.

Figure 1 shows the GPS matrix structure as defined in the ISO 14638:2015: the GPS Matrix is composed of 9 rows and 7 columns. The different geometric properties are arranged in the rows while the position of the product in the production chain is defined through the columns. This matrix is divided into three sections: specification (first 3 columns, continuous line in Fig. 1); comparison and decision rules (fourth column, dashed line Fig. 1); verification (remaining 3 columns, dotted line in Fig. 1).

ISO 14638 specifies the classification of all the GPS International Standards into three categories, according with a matrix model:

- fundamental standards, which fill all boxes of the matrix model (Fig. 2a)
- general standards, which take up one or more boxes of the matrix model (Fig. 2b)
- complementary standards, which sit in one or more boxes of a row (Fig. 2c) occasionally added at the bottom of matrix model.

GPS MATRIX	A Symbols and indications	B Feature requirements	C Feature properties	D Conformance and non-conformance	E Measurement	F Measurement equipment	G Calibrations
1 Size	•	•	•	•	•	•	•
2 Distance	•	•	•	•	•	•	•
3 Form	•	•	•	•	•	•	•
4 Orientation	•	•	•	•	•	•	•
5 Location	•	•	•	•	•	•	•
6 Run-out	•	•	•	•	•	•	•
7 Profile surface texture	•	•	•	•	•	•	•
8 Areal surface texture	•	•	•	•	•	•	•
9 Surface imperfections	•	•	•	•	•	•	•

(a)

GPS MATRIX	A Symbols and indications	B Feature requirements	C Feature properties	D Conformance and non-conformance	E Measurement	F Measurement equipment	G Calibrations
1 Size	•	•					
2 Distance							
3 Form							
4 Orientation							
5 Location							
6 Run-out							
7 Profile surface texture							
8 Areal surface texture							
9 Surface imperfections							

(b)

GPS MATRIX	A Symbols and indications	B Feature requirements	C Feature properties	D Conformance and non-conformance	E Measurement	F Measurement equipment	G Calibrations
1 Size							
2 Distance							
3 Form							
4 Orientation							
5 Location							
6 Run-out							
7 Profile surface texture							
8 Areal surface texture							
9 Surface imperfections							
Standards on mouldings	•	•					

(c)

Fig. 2. (a) Relationship between ISO 1:2016 and matrix model. (b) Relationship between 286–1,2:2010 and matrix model. (c) Relationship between 286–1,2:2010 and matrix model

The position in the GPS matrix is one of the attributes listed in the database, and thus it is one of the possible search criteria.

In the following, an example of how UNI EN ISO 286:2010 standard have been classified within the database:

- UNI Codification: UNI EN ISO 286:2010.
- Class: General.
- Relationship to the GPS matrix model as in Fig. 2(b).
- 1st level topic: ISO code system for tolerances on linear sizes.
- 2nd level topic: Basics of tolerances, deviations, and fits.

- 3rd level topic: Establishes the ISO code system for tolerances to be used for linear sizes of features of the following types: cylinder, two parallel opposite surfaces.
- Keywords: dimensional tolerancing, IT tolerance grades, upper tolerance limit, lower tolerance limit, fundamental deviations, tolerance interval, hole, shaft, diameter, size, fit.
- Scope: This document establishes ISO code system for tolerances which define linear sizes of features of the following types: cylinder, two parallel opposite surfaces. It defines the basic concepts and the related terminology for this code system, and provides a standardized selection of tolerance classes for general purposes

2.2 Definition of the Search Criteria

User's interest in a specific subject, according to GPS "philosophy", could have two prevailing connotations that lead to aspects related to: (i) the specification and (ii) the verification. In the classification phase, GPS system splits the standards into this couple of categories adding a third one, less populated, concerning the issue of establishing rules to define conformity/non-conformity. This classification will be exploited as one of the possible means of search the database.

The output of the search activity could be essentially a standard or a list of standards presented to the user with a configurable list of the attributes stored in the database.

The following search modalities have been hypothesized: (i) free search; (ii) simple guided search; (iii) advanced guided search.

Free Search. This search is the simplest one and can be accomplished in two ways: (i) free keywords search, aimed to having an overview on all the standards containing the required word in the title, in the scope, or in the other "attributes" of the database; (ii) ISO codification search, directly targeting the standard of an already known ISO code.

Simple Guided Search. Different types of searches are possible in the following modality: (i) guided keywords search, in which the alphabetic list of keywords are showed as suggestion to help the user; (ii) search through symbols, where the list of symbols found in the various standards is shown to the user; (iii) Macro-topics, macro-topics have been assigned to attributes through each standard with the aim of summarize covered arguments. There are 3 level of topics, from 1st to 3rd, which gradually detail more deeply the theme of a standard; (iv) GPS standards classification search, in which user can browse into three lists of standards, subdivided according to the hierarchical structure of the GPS standards (fundamental, general, complementary); (v) search through the relationship to GPS matrix.

Advanced Guided Search. The intent is to guide the user through different routes or collections of standards with a pre-established order, which should enable a specific subject to be dealt with. The suggested order should consent a didactic approach to the user, who gradually is introduces to the standards of the collection, looking at each topic with the necessary preliminary knowledge. Different types of search are possible: (i) the "preparatory route" aims to enable the user to gain access to the introductory knowledge to understand how GPS system is structured with the specific content of each standards;

(ii) the "drawing comprehensive route" aims to enable the user to be able to decode the requirements expressed in blueprints; (iii) the "form errors" route aims to guide the user in deepening form errors comprehension, considering both specification and verification aspects; (iv) the "location errors route" aims to show the user the available standards to comprehend location errors considering both specification and verification aspects; (v) the "surface micro-geometric errors" route aims to show the user the available standards to understand surface texture errors considering both specification and verification aspects, through two possible strategies (the profile method and the areal method).

3 Results and Discussion

In the following, examples of possible research methods are detailed.

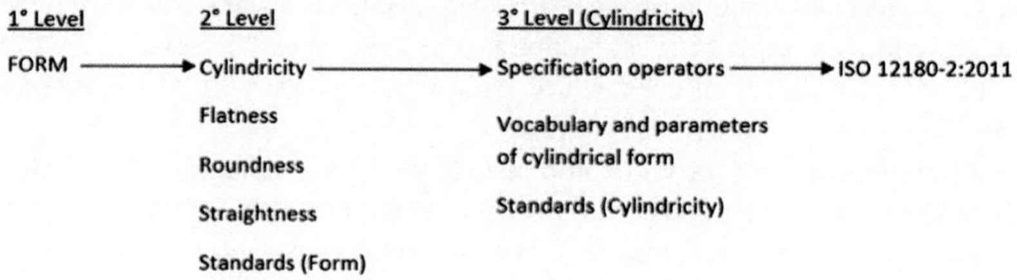

Fig. 3. Example of a complete topic search.

Figure 3 shows a complete topic search, included in the simple guided search mode. This research flows through three different levels. The 1st level topic sums up the main search argument in less of a line, and a list of alphabetically ordered 1st level topics are presented to the user. In the presented case the user chooses "FORM", as 1st level topic, then a list of 2nd level topics related to the chosen ones is shown. The user chooses "Cylindricity", then a list of 3rd level topics appears. The user chooses "Specification operators" and finally the related standard is retrieved.

	A	B	C	D	E	F		G
1	286-1,2 1119 2538-1,2 2692 3040 5459 14405-1,3 17863	286-1,2 1119 2538-1,2 2692 3040 5459 14405-1,3 17863	2538-2 2692 5459 14253-4 14405-1,3 14406 16610-series 17863	14253-1,2,3,4 16015	1938-1 14253-2,3,4,6 16015	463 1938-1,2 3611 9493 10360 13102 13225 13385-1,2	14253-2,3,4,5,6 14406 14978 16015 16610* 17865 23165	1938-1,2 3650 13385-1 14253-2,3,4,6 14978 15530-1,3,4 16015 16610-1

Fig. 4. Example of a research through the relationship to GPS matrix.

Figure 4 shows an example of search through GPS matrix, included in the "simple guided search" mode. The columns and rows are indexed with letters from A to G and numbers from 1 to 9, respectively, and user can request standards indicating bidimensional coordinates. In the example, the two-dimensional coordinate A1, provides the list of standards dealing with "Symbols and Indications" on the subject "Size".

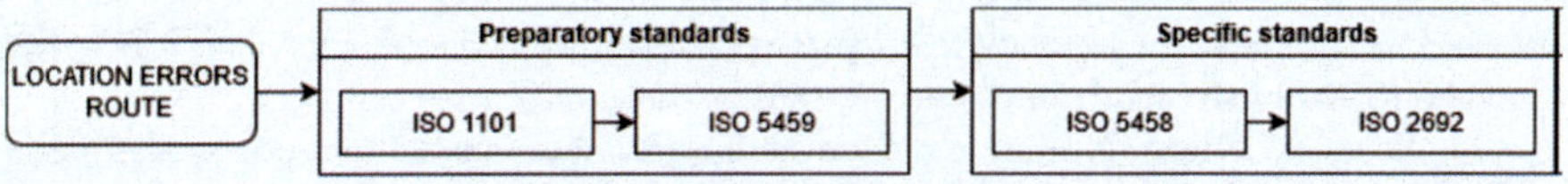

Fig. 5. Path for location errors.

Figure 5 shows an example of "advanced guided search", in particular the path for location errors is presented. User chooses the path "LOCATION ERRORS ROUTE", as a consequence the propaedeutic standard ISO 1101 (Geometric tolerances) and ISO 5459 (Datum) are shown, both including general concepts and definitions; then, specific standards are presented, to deepen the knowledge about the topic. So, ISO 5458 is about location tolerances and ISO 2692 is about maximum material, least material, reciprocity requirements concepts.

4 Conclusions

The created database contains attributes to characterize each GPS Standard. Thanks to this it is possible to create an indexing strategy to cover the various searching methods of the GPS Navigator. The illustrated coding strategies (flowchart of search strategies) can create an environment where users can perform different searches according to the needs, from the simplest to the most structured one.

Current work results can be exploited for: (i) updating information about new standards or updated standards; (ii) data enrichment, especially the section of key-words and symbols; (iii) vocabulary enlargement, for example with the translation into other technical languages about product specifications.

The final result will be soon online as a search engine supporting users in orientating themselves in GPS system to find and apply the standards of interest.

References

1. Tornincasa, S.: Technical Drawing for Product Design. Springer, Cham (2021)
2. ISO/TC 213: Dimensional and geometrical product specifications and verification Home Page. https://www.iso.org/committee/54924.html. Last accessed 17 March 2021
3. Dantan, J.Y., Ballu, A., Mathieu, L.: Geometrical product specifications – model for product life cycle. Comput. Aided Design **40**, 493–501 (2008)
4. Bennich, P., Nielsen, H.: An Overview of GPS – A Cost Saving Tool, 1st ed. Institute for Geometrical Product Specifications, Vaerloese (2005)
5. Mandolini, M., Campi, F., Favi, C., Germani, M., Raffaeli, R.: A framework for analytical cost estimation of mechanical components based on manufacturing knowledge representation. Int. J. Adv. Manuf. Technol. **107**(3–4), 1131–1151 (2020). https://doi.org/10.1007/s00170-020-05068-5
6. Violante, M.G., Vezzetti, E., Alemanni, M.: An integrated approach to support the requirement management (RM) tool customization for a collaborative scenario. Int. J. Interact. Design Manuf. **11**(2), 191–204 (2015). https://doi.org/10.1007/s12008-015-0266-3
7. Schuldt, J., Hofmann, R., Gröger, S.: Introduction of a maturity model for assessment of the integration of the GPS system in companies. Procedia CIRP **92**, 129–133 (2020)

8. Nielsen, H.S.: The ISO Geometrical Product Specification Handbook. Find your way in GPS, 1st edn. ISO/Danish Standards, Denmark (2012)
9. Qi, Q., Scott, P.J., Jiang, X., Lu, W.: A hierarchical category model for geometrical product specification (GPS). Procedia CIRP **43**, 214–219 (2016)
10. Humlenny, Z.: State of art in standardization in GPS area. CIRP J. Manuf. Sci. Technol. **2**, 1–7 (2009)
11. Loureiro, G.B., Ferreira, J.C.E., Messerschmidt, P.H.Z.: Design structure network (DSN): a method to make explicit the product design specification process for mass customization. Res. Eng. Design **31**(2), 197–220 (2020). https://doi.org/10.1007/s00163-020-00331-y

Model-Based Approach for Optimal Allocation of GD&T

Andrea Petruccioli(✉), Fabio Pini, and Francesco Leali

"Enzo Ferrari" Department of Engineering, University of Modena e Reggio Emilia, Via Vivarelli 10/1, 41125 Modena, Italy
{andrea.petruccioli,fabio.pini,francesco.leali}@unimore.it

Abstract. Model-based Definition (MBD) is a known design approach that aims to an effective integration of Product Manufacturing Information (PMI) within geometrical data. By means of MBD, product requirements and specifications based on Geometric Dimensioning and Tolerancing (GD&T) can be directly associated to 3D models, improving interoperability between design and simulation virtual environments. However, especially in industrial settings, many challenges still limit MBD diffusion, such as limited knowledge and application of GD&T rules, inconsistent representation of PMI, lack of methodological and organizational approach based on PMI. As a consequence, the Dimensional Management practice based on GD&T cannot be systematically applied, and the full potential of Computer-Aided specific tools remains unexpressed. In this paper, the effective implementation of MBD for PMI during both product and process design is proved through its direct application on tolerance-cost optimization. Thanks to 3D semantic annotations, a model-based framework is suggested to validate functional requirements of a mechanical assembly and to assess production efforts, enhancing the integration between tolerance analysis and manufacturing cost tools. The interrelation of GD&T schemes enables the automated transfer of the data linked to annotations toward Computer-Aided Tolerancing (CAT) and Product Cost Management (PCM) virtual environments. Consequently, PMI guides the simulations during the multi-disciplinary optimization, proving its effectiveness in communicating engineering information and enabling the transition to digital manufacturing though MBD.

Keywords: Product manufacturing information · Model-based definition · Geometric dimensioning and tolerancing (GD&T) · Tolerance-cost optimization

1 Introduction

The demand for an effective development of design processes is directly related to a methodical integration and management of the information associated with the products to develop. Specifically, the Model-based Definition (MBD) is the engineering approach that addresses this purpose [1]. It suggests to include the 3D model representation by means of Product and Manufacturing Information (PMI), that is engineering information related to the product [2]. PMI consists of 3D annotations associated with CAD features,

C. Rizzi et al. (Eds.): ADM 2021, LNME, pp. 277–284, 2022.
https://doi.org/10.1007/978-3-030-91234-5_28

including material specification, component lists, process related information, inspection requirements and Geometric Dimensioning and Tolerancing (GD&T) [3]. Geometric and dimensional variations have a direct impact on performance and production costs. Consequently, GD&T annotation of Datum Reference Frames (DRFs), tolerance types and values, should be carefully managed along the main phases of product and process design, from Tolerance analysis to Inspection [4]. Regarding GD&T presentation and semantics, it should be noted that ASME and ISO Standards are implemented in PMI [5]. With regards to interoperability capabilities, the Standard for the Exchange of Product Model Data (STEP) defines the rules for 3D neutral format ISO STEP AP 242 [6], that represents a useful means to spread the GD&T rules and approach over the industrial field. In addition, PMI addressed in 3D CAD models provides a considerable benefit for information exchange between Computer-Aided tools like Computer-Aided Tolerancing (CAT) [7, 8] and Computer-Aided Manufacturing (CAM) [9], as well as Coordinate Measuring Machine (CMM) software [5, 10]. Moreover, it should be noticed as the attention of industrial companies is growing around multi-objective optimizations [11]. In particular, tolerance-cost optimization seems to be very promising to combine the tolerance effects on both functionality and production cost [1, 12]. Concurrently, it is also increasing the development of cost manufacturing aided tools, like Product Cost Management (PCM) software, also known as Computer-Aided Cost Estimating (CACE) or Product-Cost Estimation (PCE) [13]. Nevertheless, despite the growing popularity, industrial application of tolerance-cost optimization is still limited due to the lack of integration and, consequently, of direct data exchange with CAT tools [14]. Therefore, the potential of integrated frameworks for tolerance-cost optimization can be highly increased by the integration of PMI as an information vendor.

Furthermore, as reported by [15], a systematic application of PMI is still missing. In fact, manufacturers frequently use different standards and file formats [16], resulting in decreased interoperability. In addition, GD&T is not completely assessed and applied [17]. Most of the time, tolerance specification and allocation are directly done according to individual experience, experimental data or information from previous drawings [1]. A lot of research has been done focusing on the reduction of typical issues of interoperability and data transfer between different environments [18], as well as on the definition of frameworks to enhance MBD across product lifecycle [6, 7, 10, 19]. Consequently, industrial settings are still far from getting full advantages of MBD approach [20], mainly due to methodological, organizational and "cultural" obstacles and challenges, rather than technical ones [21]. The result is that, despite the integration of PMI into 3D CAD models improves efficiency providing human and machine-readable file to be machined and simulated, reducing development time and costs compared to traditional approach (i.e., based on 2D drawings), a widespread application of PMI in industry is not completely met. Consequently, the Dimensional Management approach based on GD&T cannot be systematically applied, and the full potential of Computer-Aided specific tools is still unexpressed.

For this reason, the present paper aims to highlight the central role of MBD and PMI in the context of tolerance-cost optimizations based on Computer-Aided tools integration. A mechanical assembly is chosen as case study to be representative of the typical requirements of industrial applications. PMI tolerances, chosen as variables for the multi-disciplinary optimization and updated at each iteration, guide the simulations through the automatic transfer of GD&T information between tolerance analysis and manufacturing cost tools. By this way, the communication of engineering information is effectively enhanced by PMI, becoming the central node of the optimization framework. Therefore, PMI can improve the knowledge of GD&T rules and effects on both product and process design simulation, as well as during the inspection phase.

Rest of the paper is structured as follows: in Sect. 2, the PMI-based tolerance-cost methodology is presented. Section 3 reports the case study application, with PMI integration and the simulation results. Section 4 presents the final discussion and some conclusive remarks.

2 Methodology

Tolerance-cost optimization frameworks and methodologies have been constantly developed and improved to integrate manufacturing and design, enhancing concurrent engineering [2]. A promising approach is based on the integration of specific Computer-Aided tools within an optimization environment to overcome the methodological difficulties. This framework, based on state-of-the-art tolerance-cost optimization and structured within a Multi-Disciplinary Optimization platform (MDO), can be effectively improved by MBD through the integration of PMI.

Following the main phases of the tolerance-cost optimization framework (see Fig. 1), respectively, Modelling and Annotation, Tolerance Analysis, and Manufacturing Cost Estimation, GD&T expressed with PMI becomes the central node from which the iteration between simulation tools is managed.

Modelling and Annotation. Starting from the CAD models, it is the first phase that provides all the input for the optimization, including the GD&T definition. The functional analysis of the components is formalized by the identification of the functional features to be controlled and the definition of DRFs and tolerance scheme. Therefore, the tolerance types are identified for each feature and the first tolerance values are assigned. The GD&T information is properly integrated on 3D CAD models in 3D machine-readable semantic annotations (i.e., PMI representation) [8]. Once defined, GD&T PMI can be automatically extracted from the 3D models and set as input variables of the optimization.

Tolerance Analysis. CAD geometry features and assembly structure, including PMI, are inputs of this phase. Thanks to PMI, GD&T is automatically read and included in the tolerance simulation model. The set-up of tolerance distribution types and the measuring control points is followed by tolerance stack-up simulation. The results of simulation are compared with respect to functional requirements and process capability values. If output values are not acceptable, a synthesis of tolerances is needed, modifying tolerance values and, if necessary, types and scheme.

Manufacturing Cost Estimation. It is the final phase. It requires information related to product assembly and individual components to assign production processes and assembly operations, as well as PMI to retrieve tolerances and assign the corresponding manufacturing process. Set-up of materials, machine parameters and production rates complete the modelling and enable the cost estimation. The evaluated costs are checked with targets and, if necessary, corrective actions are applied on tolerances values, types and scheme. Otherwise, the product is validated and product design is complete.

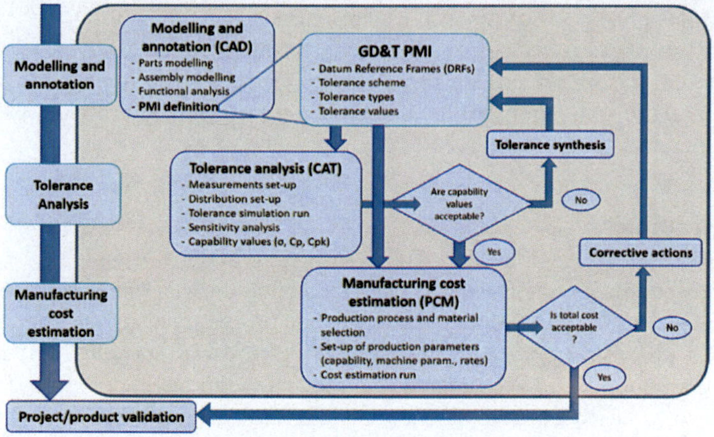

Fig. 1. Tolerance-cost optimization framework with PMI.

As described above, PMI tolerances are selected as variables for multi-disciplinary optimization. The tolerance values drive the simulations during the optimization steps. At each step, the optimization algorithm updates the variables, which are automatically transferred to the simulation models, and runs tolerance and cost simulations. After each iteration, input variables are changed and updated considering the global constraints and the defined objectives. At the end of the optimization, the best configurations are selected, as well as the corresponding 3D model with the optimized tolerance values.

3 Case Study

The presented framework is evaluated on a general mechanical assembly, where the tolerance values must optimize the trade-off between functional requirements and manufacturing costs. The assembly is composed of three components: a plate, a support and a shaft, respectively referred as #1, #2 and #3 as in Fig. 2. The main function of the assembly is the accurate positioning of the shaft and of the plate on the support: functional requirements to be checked is the axis alignment with the corresponding radial gap between the shaft and the plate hole. By this way, the case study is representative of many industrial products, such as gearbox assemblies and engine block assemblies, where the positioning of the main components on a case must allow the required performance. According to the assembly sequence, the shaft is inserted into the support, then the plate is positioned on the support by means of two pins.

Modelling and Annotation. With respect to the assembly sequence, the DRFs for the components are identified, functional features are selected and PMI semantic annotations are inserted (see Fig. 2b, Fig. 2c and Fig. 2d).

- Support: The contact plane with the shaft is the first Datum A; the hole axis is the second Datum B, controlled with a perpendicularity tolerance; the plane opposite to A is controlled by a dimensional tolerance and a parallelism tolerance; the plane, identified as Datum C, is used in conjunction with B to control the position of the holes (*H1* and *H2*) for the assembly with the plate; all the holes are controlled by dimensional tolerances.
- Shaft: The contact plane with the support is Datum A; the cylindrical surface, identified as Datum B, is controlled by perpendicularity tolerance, dimensional tolerances on the main diameter and on the axial length.
- Plate: The contact surface with the support is Datum A; the first pin (*P1*) is identified as Datum B, perpendicular with respect to A; the second pin (*P2*), the Datum C, is controlled by a position tolerance with respect to A and B; the central hole is controlled by a position tolerance with respect to A, B and C; all the holes are controlled by dimensional tolerances.

Once the tolerance scheme has been defined, first values of tolerances are chosen as starting configuration to be optimized and inserted 3D semantic annotations. For the pin/hole joints, a clearance fit is considered to avoid physical interferences.

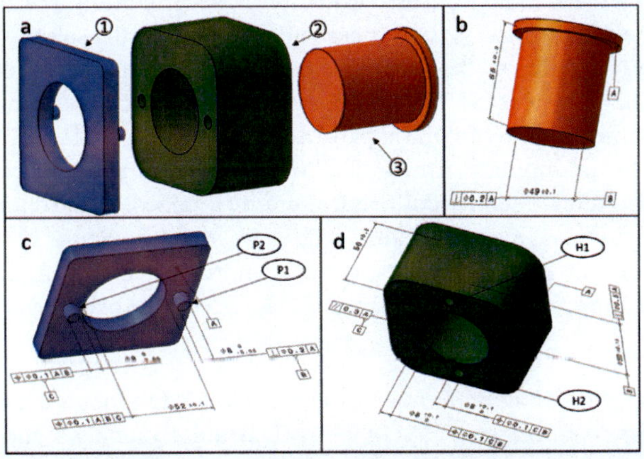

Fig. 2. Mechanical product: Assembly (a), shaft (b), plate (c) and support (d).

Tolerance Analysis. 3D models and assembly are imported into the CAT environment, in this case the CAT software Cetol 6σ© by Sigmetrix, LLC. Once inserted the components in the correct order following the assembly sequence, PMI GD&T annotations of each part are automatically read. The tolerance model is then completed defining the joints between components, the tolerance distribution types and setting the measurement to be controlled by the simulation: the radial gap, *Rg*, between shaft and central hole of the plate, whose range of acceptance is ±0.1 mm. In addition, the gaps between pins and holes, respectively *H1g* and *H2g*, are checked as an assembly requirement, to avoid interference. To verify if the process meets these requirements, the process capability index Cpk is chosen to be optimized [22]. Observing the VDA 6.3 automotive standard as a quality target, the target value is set at Cpk = 1.33, considered as the minimum requirement for long term process capability. Then, the simulation with the initial configuration is run. As reported in Table 1, the stack-up simulation shows a low value of σ for *Rg*, corresponding to a Cpk of 0.62. The *H1g* shows an optimal capability, while *H2g* is strongly influenced by the geometrical tolerances. Tolerance main contributors of variation are identified and reported in Table 1.

Table 1. Tolerance simulation output (initial configuration)

Measurement	Value and range	Capability	Tolerance main contributors
Rg	1.5 ± 0.1 mm	$\pm 1.87\sigma$, Cpk = 0.62	⊥ (#3), ∥ (#2), Ø DatumB (#3)
H1g	$0 + 0.2$ mm	$\pm 4.17\sigma$, Cpk = 1.33	Ø *H1*, Ø *P2*
H2g	$0 + 0.2$ mm	$\pm 1.71\sigma$, Cpk = 0.45	⌖ *H1* and *H2*, ⌖ DatumC (#1)

Manufacturing Cost Estimation. Cost evaluation is set importing the 3D models and assembly into the PCM software, in this case aPriori©. After the setting of components main machining operation, of the batch size (set as 5500 annual units for 5 years), and of production database (western Europe), the material for each component is selected: ANSI 6012 Aluminium Alloy for the support and plate, and AISI 4140H Alloy Steel for the shaft. After this preliminary setting, the PMI manufacturing tolerances are imported and their corresponding geometrical features are recognized, identifying the available operations to achieve the tolerance values. The estimation of manufacturing cost for the initial configuration is run and the costs for each component are given, that respectively are $ 5.34 for the support, $ 8.24 for the shaft and $ 2.25 for the plate, for a total assembly cost of $ 15.83.

Once the starting configuration has been set, the CAD models are inserted in the MDO optimization environment. CATIA V5 by Dassault Systemes® and ModeFrontier by Esteco© are the two tools used for the previous steps. PMI tolerances are extracted from the CAD models and selected as input variables: their limit of variation is defined, as well as their distribution type, if necessary. The optimization algorithm (i.e., in this case the multi-strategy self-adapting pilOPT), starting from a preliminary design, generates new populations to reach Pareto front while satisfying the constraints (i.e., feasible designs). By this way, PMI is updated at each iteration exploiting the available designs and refining the Pareto optimal set, with the new values constantly transferred to tolerance and cost models. The simulation results provide a set of 16 optimal configurations, where the target value of Cpk is reached both for functional and assembly requirements. As reported in Table 2, the least cost configuration is identified as the n°534 (with an assembly cost of $ 17.16 and lowest value of Cpk equal to 1.37), and the best Cpk (considering the lowest value) is n°446, where the higher cost is compensated by higher values of Cpk (respectively, $ 19.62 and Cpk = 1.61).

Table 2. Simulation results: best configurations.

ID	Cost (#1)	Cost (#2)	Cost (#3)	Cost (Tot)	Cpk (*Rg*)	Cpk (*H1g*)	Cpk (*H2g*)
446	$ 4.51	$ 6.80	$ 8.31	$ 19.62	1.61	2.11	1.67
534	$ 2.25	$ 6.61	$ 8.30	$ 17.16	1.37	2.67	1.37

The optimal configuration is evaluated thanks to the trade-off between production cost and the cost of rejected parts. Consequently, 3D models with optimized PMI tolerance values can be directly selected and then used for the inspection phase.

4 Conclusive Remarks

As proved by the high number of research works concerning MBD, PMI is a promising technique to enhance concurrent engineering and to enable the transition to digital manufacturing. Although considerable effort to overcome technical limitations has been made, PMI diffusion in industry has to contend with the lack of a methodological approach on design integration and, in particular, with the limited application of GD&T rules during product/process design. As a result, the full potential of Computer-Aided tools is still unexpressed. This paper has shown how PMI plays a central role on tolerance-cost optimization, transferring GD&T information across simulation environments. The case study application and the optimization results prove the approach to be very promising. At each iteration PMI tolerances are updated providing a set of CAD models within tolerance configurations to be directly used. The automation of PMI upload on 3D models increases flexibility, with time reduction for redesign, i.e., with reduction of product design associated costs. In addition, data exchange between product/process design and Quality Control Plan can be effectively enhanced. In fact, the optimal tolerance configuration from the optimization can be directly transferred to the CMM software: measurement routines are automatically generated from the 3D model with best PMI, with tolerance scheme and values to be controlled. By this way, PMI becomes the enabler to integrate product design, manufacturing and inspection. Future work will further improve the model to include and automate information related to material selection and manufacturing processes.

References

1. Hallmann, M., Schleich, B., Wartzack, S.: From tolerance allocation to tolerance-cost optimization: a comprehensive literature review. Int. J. Adv. Manuf. Technol. **107**(11–12), 4859–4912 (2020). https://doi.org/10.1007/s00170-020-05254-5
2. Quintana, V., Rivest, L., Pellerin, R., et al.: Will model-based definition replace engineering drawings throughout the product lifecycle? A global perspective from aerospace industry. Comput. Ind. **61**(5), 497–508 (2010)
3. Hedberg, T., et al.: Testing the digital thread in support of model-based manufacturing and inspection. J. Comput. Inform. Sci. Eng. **16**(2) (2016)

4. Schleich, B., Wartzack, S.: A discrete geometry approach for tolerance analysis of mechanism. Mech. Mach. Theory **77**, 148–163 (2014)
5. Ricci, F., et al.: PMI: a PLM approach for the management of geometrical and dimensional controls in modern industries. Comput. Aided Design Appl. **11**(sup1), S36–S43 (2014)
6. Feeney, A.B., Frechette, S.P., Srinivasan, V.: A portrait of an ISO STEP tolerancing standard as an enabler of smart manufacturing systems. J. Comput. Inform. Sci. Eng. **15**(2) (2015)
7. Ramnath, S., et al.: Interoperability of CAD geometry and product manufacturing information for computer integrated manufacturing. Int. J. Comput. Integr. Manuf. **33**(2), 116–132 (2020)
8. Sarigecili, M.I., Roy, U., Rachuri, S.: Interpreting the semantics of GD&T specifications of a product for tolerance analysis. Comput. Aided Des. **47**, 72–84 (2014)
9. Bijnens, J., Kellens, K., Cheshire, D.: Accuracy of geometry data exchange using STEP AP242. Procedia CIRP **78**, 219–224 (2018)
10. Morse, E., et al.: Interoperability: linking design and tolerancing with metrology. Procedia CIRP **43**, 13–16 (2016)
11. Favi, C., Germani, M., Mandolini, M.: A multi-objective design approach to include material, manufacturing and assembly costs in the early design phase. Procedia CIRP **52**, 251–256 (2016)
12. Peng, H.P., Jiang, X.Q., Liu, X.J.: Concurrent optimal allocation of design and process tolerances for mechanical assemblies with interrelated dimension chains. Int. J. Prod. Res. **46**(24), 6963–6979 (2008)
13. Ehlhardt, H.: Computer aided cost estimating. In: Proceedings of the 16th E&PDE14, Design Education and Human Technology Relations, Twente, NL, pp. 320–326 (2014)
14. Li, H., Zhu, H., Zhou, X., Li, P., Yu, Z.: A new computer-aided tolerance analysis and optimization framework for assembling processes using DP-SDT theory. Int. J. Adv. Manuf. Technol. **86**(5–8), 1299–1310 (2016). https://doi.org/10.1007/s00170-015-8266-9
15. Peng, Z., et al.: A new method for interoperability and conformance checking of product manufacturing information. Comput. Electr. Eng. **85**, 106650 (2020)
16. Schleich, B., Wartzack, S.: Challenges of geometrical variations modelling in virtual product realization. Procedia CIRP **60**, 116–121 (2017)
17. Chen, M.S.: Optimising tolerance allocation for mechanical components correlated by selective assembly. Int. J. Adv. Manuf. Technol. **12**(5), 349–355 (1996)
18. Hallmann, M., Goetz, S., Schleich, B.: Mapping of GD&T information and PMI between 3D product models in the STEP and STL format. Comput. Aided Des. **115**, 293–306 (2019)
19. Fang, F.Z., et al.: Closed loop PMI driven dimensional quality lifecycle management approach for smart manufacturing system. Procedia CIRP **56**, 614–619 (2016)
20. Goher, K., Shehab, E., Al-Ashaab, A.: Challenges of model-based definition for high-value manufacturing. In: Proceedings of the 17th International Conference on Manufacturing Research, Belfast (2019)
21. Pippenger, B.S.: Three-dimensional model for manufacturing and inspection. In ASME Turbo Expo 2013: Turbine Technical Conference and Exposition, American Society of Mechanical Engineers Digital Collection, San Antonio, Texas (2013)
22. Pyzdek, T., Keller, P.: Six sigma handbook. McGraw-Hill Education, New York (2014)

Design for Manufacturing and Assembly

A Method to Assess Design for Assembly Efficiency of Aircraft Cabin Concepts

Giovanni Formentini[1](✉), Claudio Favi[1], Francois Bouissiere[2], Claude Cuiller[2], Pierre-Eric Dereux[2], and Cedric Jurbert[2]

[1] University of Parma, Parco Area delle Scienze 181/A, 43124 Parma, Italy
giovanni.formentini@unipr.it

[2] Airbus SAS, 1 Rond-Point Maurice Bellonte, 31700 Blagnac, France

Abstract. The design for assembly and installation of aircraft systems is a challenging topic to tackle in the conceptual design phase. This paper presents the definition of a Conceptual Design for Assembly (CDfA) methodology for cabin architecture concept of a commercial aircraft. The cabin equipping includes the assembly of many interior components (here called modules) such as toilets, galleys, seats, etc. The method has been developed and experimented on a civil aircraft cabin installation. Results provide interesting insight in the identification of the most complex items to install, enabling to understand, in terms of design, were criticalities lie and where improvements can be implemented. Results highlight how the new cabin architecture design performs better (approximately 23%) than the previous one in terms of assembly, which has been confirmed by the workload measurement performed on the assembly line.

Keywords: Conceptual design · Aircraft · Design for assembly · Design for manufacturing · DFA · DFMA · Cabin architecture

1 Introduction

The aircraft manufacturers are continuously trying to improve the industrial performances at all levels, keeping the industrial investment low [1]. To achieve this goal, a close collaboration with all departments along the engineering process is required [2]. Product engineering and manufacturing engineering need to collaborate since the very beginning of the Product Development Process (PDP) [2]. During the design process, different methodologies are available to improve product's features. These methodologies are collected under the umbrella of *Design for X*, where the X is replaced with the method optimization goal (i.e., Design for Environmental, Design for Manufacturing, etc.) [3]. The Design for Assembly (DFA) is a family of methods whose aim is to improve the product concept to reach better assembly performances. Different methods are available in literature, for instance the Boothroyd-Dewhurst (B&D) [4], the Hitachi method (AREM) [5] or the Lucas method [6]. The majority of the DFA methods require detail information such as the number of screws used to fix a component or the orientation. Thus, even though the assembly complexity is reduced, a product re-design is required.

C. Rizzi et al. (Eds.): ADM 2021, LNME, pp. 287–297, 2022.
https://doi.org/10.1007/978-3-030-91234-5_29

Few attempts are available in literature where DFA methods are applied to highly complex products, like aircraft. For instance, [7] applied DFA techniques to improve the wiring system installation, [8] developed a method to evaluate assembly processes of the aircraft fuselage in the final assembly line (FAL), using digital manufacturing techniques, [9] derived a set of design guidelines (i.e., good practices) applicable to aircraft to improve assembly and manufacturing performances. However, only a few attempts were done to apply DFA principles at the conceptual level when information of lower granularity is present. Among them, [10] laid the foundation proposing a method to link functional modules to assembly aspects with limited capabilities to the number of functions. In the aerospace field, [11] proposed a tool to retrieve information at the conceptual design from previous airplane designs, while other researchers moved the research angle from product to the FAL [8, 12]. The way to improve product assembly performance at the conceptual design phase is strictly related to the concept of product architectures [13]. Product architectures are usually used to create modular products, leading to many advantages such as improved product maintenance and reduced development costs [14]. The main challenge with working at early phase design (conceptual phase) is the information quality which is characterized by a low granularity information [15]. The authors of this paper tried to assess the assembly complexity at the product architecture level for aircraft [16–18]. The method has been preliminary developed and experimented on the nose-fuselage of a civil aircraft which has a high number of modules and interfaces. The main objective of this paper is to present the Conceptual Design for Assembly (CDfA) methodology and its experimentation for the assessment of commercial aircraft cabin installation. The cabin installation is related to the installation within aircraft airframe of many and various components (here called modules) such as toilets, galleys, seats, hat racks, linings, oxygen system etc. The improvement of cabin assembly performances leads to an overall assembly workload reduction and potentially a reduction of the aircraft manufacturing lead-time. The original CDfA method was adapted to the cabin case study with the aim to catch critical issues observed in the cabin installation process. The overall framework of the CDfA methodology was kept (hierarchical structure based on level, domains and attributes) but the definition of each assessment parameter was reviewed and adapted to address the issue related to the Cabin installation. The method was tested on a civil aircraft cabin installation, demonstrating the added value of the most advanced cabin architecture concepts. The paper is structured as follow: after this introduction (1), the Cabin System is described, and the application of the method is explained in detail (2). Then, results are presented and discussed (3). Finally, conclusions are presented (4).

2 Material and Methods

Cabin system analyzed in this study is presented in Fig. 1.

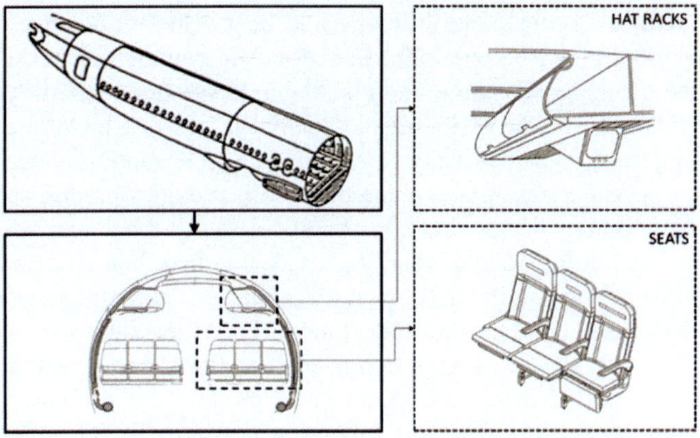

Fig. 1. Aircraft cabin and main modules

An aircraft is composed of several systems that interact with each other in order to fulfil different functions [19]. From the manufacturing point of view, six main systems characterize an aircraft: i) a nose-fuselage, ii) a typical fuselage, iii) a rear end fuselage, iv) a cabin, v) wings, and vi) empennages. The main function of the cabin is to accommodate passengers and hand luggage during the flight, to provide comfort for passengers (e.g., fresh air, toilets, water, etc.), while ensuring user safety (e.g., seats-belts, oxygen, etc.). The analysis focused to the cabin modules installation, indeed systems interfaces (i.e., electrical cables and harnesses, air ducts, etc.) are out the scope of this work because installed before the cabin equipping. The CDfA methodology is composed of seven (7) steps (Fig. 2).

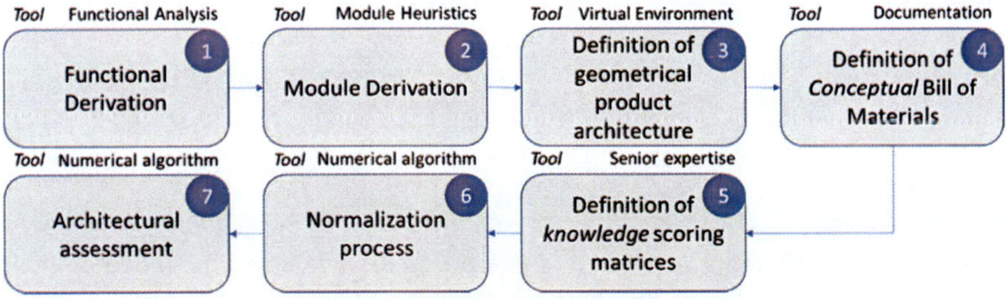

Fig. 2. CDfA methodology steps

The core of the method is the definition of a hierarchical structure composed of levels, domains, and attributes. Attributes are defined as key driving parameters to address the product assembly performance; domains are groups of attributes related to the same assembly topic (i.e., ergonomic aspects, assembly tools, etc.), while levels are used to cluster domains to accommodate different level of information granularity. Then, by using specific mathematical algorithm, the information is processed and a single score for each analyzed module feature is obtained. Scores express the assembly complexity of each module feature. Using this information, it is possible to identify and, subsequently, improve module features considered critical inside the product architecture. The first step of the CDfA methodology is to obtain the Functional derivation followed by the second step which is the Module derivation. The Functional Derivation related to the proposed case study (cabin) was obtained with the method proposed by [2], while modules were derived following the module heuristics proposed by [6]. The modules heuristics method was chosen as a modularization technique because it provides a wide solution space. Indeed, it enables to consider several constraints related to the aerospace sector (e.g., safety, maintenance, etc.). Modules identified with the three heuristic were compared with the ones already existing in the cabin and a one-to-one correspondence was found. A list of modules identified after the application of the second step is shown in Table 1. For the sake of brevity, the full description of the functional analysis and the modules derivation are not reported within this work. These modules are difficult to merge and, therefore, to delete due to their specific functions, position allocation constraints and aircraft building process. For example, the module Toilet is limited to its basic function because its location would not enable to encompass another cabin module function. In addition, it cannot be deleted because its need is required by the number of passengers and associated cabin layout.

Table 1. Aircraft cabin modules derived by heuristics [6]

Identified modules	Acronym
Toilet	T
Seats	S
Galley	G
Cabin crew rest compartment	CCRC
Passengers service units	PSU
Hat racks	HR
Doors entrance	DE
Side walls	SW

From an assembly point of view, it is interesting to understand the module ranking in relation to their complexity and therefore to their installation constraints driving the assembly performance, and then to assess different cabin architecture in relation to assembly performance. This information enables to design new module concepts and associated physical integration concepts to increase the assembly performance. The third step of the approach is the definition of the simplified Digital Mock Up (sDMU). The sDMU is built by using the functional and the module derivation, indeed it aims at transforming that information into geometrical ones by translating functions and interfaces into simple geometries (e.g., boxes, cylinder, etc.). In particular, modules are represented with their bounding box (parallelepiped) and interfaces are represented by cylinder (i.e., cables, pipes). An extract of the cabin sDMU is shown in Fig. 3.

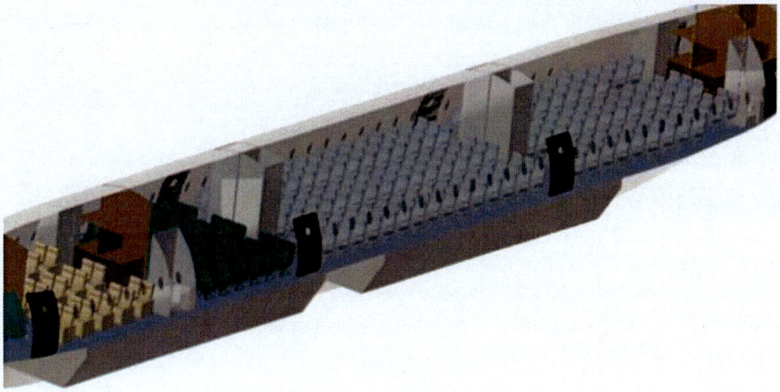

Fig. 3. Cabin sDMU

The fourth step is the creation of the hierarchical structure to model the cabin modules installation process, called *conceptual Bill of Material* (cBoM). The cBoM is used to translate information from the sDMU into numerical data to be processed with mathematical algorithms. The modelling process consists of defining attributes, domains, and levels. The definition process was performed using a concurrent approach. For this task, five meetings were organized, involving cabin manufacturing engineering department and product architecture department with the aim to identify parameters characterizing the assembly process and their effects on the assembly lead-time and workload. Parameters considered critical from an assembly point of view, such as the module shape, attachment points location, etc. were defined and justified with the help of manufacturing engineers. These attributes were then clustered into domains considering their thematic (e.g., characterization of mechanical interfaces, of systems interfaces, of handling difficulty, etc.) and, domains collected into levels according to the information granularity (i.e., at what time of the conceptual design phase the attribute information is available). At the end of the modelling process, the hierarchical structure was obtained: a total of 17 attributes were identified, clustered into 4 domains (Table 2).

Table 2. cBoM structure (attributes and domains for cabin equipping)

Domain	Attributes
System interface domain	Type of system interfaces
	Number of system interfaces
	Position
	Number of plates
	Number of interfaces on plate
	Number of interfaces stand-alone

(*continued*)

Table 2. *(continued)*

Domain	Attributes
Furnishing domain	Number of furnishing
	Screwed
Mechanical domain	Number of mechanical interfaces
	Position
	Standardization
	Design principle
	Rigging
	Gaps
Handling domain	Tool/assistant
	Clearance
	Weight

Only one level was considered for the cabin case-study. Moreover, a template was included to consider metadata such as: (i) module family, (ii) name, and (iii) identifying label (ID). After the definition of the cBoM structure, the fifth step consist of collecting manufacturing knowledge using scoring matrices. Scoring matrices were built concurrently with assembly attributes identification. A scoring matrix is a table that enables to link module characteristics into a common scale of scores representative of the assembly performance (i.e., score ranging from 1 to 5, 1 being the most efficient and 5 the less efficient). A scoring matrix was obtained for each attribute identified in the cBoM through brainstorming sessions involving manufacturing and architectural department. Scoring matrices of attributes not directly measurable were obtained with the help of senior engineers. Scoring matrices obtained for attribute "Tool/Assistant" and attribute "Weight" of the Handling domain are shown in Fig. 4. Once the scoring matrices are defined, the sixth step called normalization process, was performed. All data collected inside the cBoM were normalized using the defined scoring matrices. This enabled to obtain a score ranging from 1 to 5 for each attribute previously defined and for each module of the cabin.

Attribute 1		Score
No Jig and No Assistant	=	1
No Jig with Assistant	=	2
Simple Jig with Assistant	=	4
Complex Jig with Assistant	=	5

Attribute 3			Score
	x <=	15	1
15	< x <=	40	2
40	< x <=	80	3
80	< x <=	120	4
	x >	120	5

Fig. 4. Scoring matrices

Finally, within the seventh step, scores were collected using the root-mean-square and the mean operators. The former was used to collect score of attributes belonging to the same domain, while the latter to collect scores of each domain to obtain one single score for each module, expressing the overall assembly performance. The model was developed to obtain two scores: i) a global score for the overall cabin architecture assessment, and ii) specific score for cabin modules and module families. The global score is a single score expressing the overall cabin assembly performance. It is useful to compare, for a given aircraft category, different cabin architectures. The specific score is composed by two scores which expresses: i) Module family score, and ii) Single module score. Modules family score represents the assembly process performance of a given module's family, while the Single module describes the assembly process performance of a specific module. For instance, *Toilet "T1"* is a single module that belongs to the *Toilet* module family. The Module Family score is useful to understand which family is providing the lowest assembly performance, while the Single module score identifies critical modules and where their critical assembly issues lie.

3 Results and Discussions

The CDfA analysis previously described was experimented on the *Cabin Concept 1* of a civil aircraft. The specific scores (i.e., module family score and single module score) are shown in Fig. 5. Details about the normalization process through the use of scoring

Module Family	Family Score	Module Name	Module Score
Doors Entrance	3,19	DE 1	3,19
Cabin Crew Rest Compartment	3,41	CCRC 1	3,82
		CCRC 2	3,82
Side Walls	2,13	SW 1	2,13
Galley	**4,46**	G1	4,40
		G2	4,52
		G3	4,40
		G4	3,82
		G5	**4,57**
		G6	4,53
Toilet	4,08	T1	4,07
		T2	3,99
		T3	3,99
		T4	3,99
		T5	3,99
		T6	4,47
Hat Racks	2,33	HR1	2,35
		HR2	2,31
Passengers Service Units	3,08	PSU1	3,08
		PSU2	3,08
Seats	3,65	S1	3,71
		S2	3,93
		S3	3,34
		S4	3,62

Fig. 5. Cabin modules assessment (family and modules scores)

matrices and score assessment for different modules are not provided within this work for confidentiality reasons. However, module family scores show that the most critical module family is the Galley one with a score of 4.46, followed by the module Toilet with a score of 4.08 then Seats with a score of 3.65. A more detailed analysis of results is given by the single module score (last column of the Fig. 5). Among all, the most critical module is the Galley G5 with a score of 4.57, while the second most critical module is the Galley G2 with a score of 4.53. In order to understand where design criticalities are, it is possible to proceed backward in the analysis of the cBoM.

By looking at the overall score for each domain, the highest score for the module Galley G5 is the *Furnishing Domain* with a score of 5, followed by the *Mechanical Domain* with a score of 4.76. Moving inside the *Furnishing Domain* it is possible to understand which attributes are critical. For Galley G5 both attributes (number of furnishing and screwed) are critical, indeed they present a score of 5. Looking at the whole *Furnishing Domain* scores, 70% of scores are above 4 (Fig. 6). This result suggests that a modification of the furnishing technology would result in a more efficient assembly process by improving the score of all modules. For instance, from the scoring matrix associated to the attribute "Screwed", it is possible to understand that changing the attachment principle by removing screws and make use of hook and loop closure would bring a score of 3. The same process can be repeated for other domains and other modules, to identify possible improvements on the current cabin architecture. The *Cabin*

Module Family	Name	Number of furnishing	Screwed	SCORE
Toilet	T1	4	5	4,53
Toilet	T2	4	5	4,53
Toilet	T3	4	5	4,53
Toilet	T4	4	5	4,53
Toilet	T5	4	5	4,53
Toilet	T6	5	5	5,00
Seats	S1	3	5	4,12
Seats	S2	3	5	4,12
Seats	S3	3	5	4,12
Seats	S4	3	5	4,12
Galley	G1	5	5	5,00
Galley	G2	5	5	5,00
Galley	G3	5	5	5,00
Galley	G4	5	5	5,00
Galley	G5	5	5	5,00
Galley	G6	5	5	5,00
Side Walls	SW1	3	3	3,00
Passengers Service Units (PSU)	PSU1	3	3	3,00
Passengers Service Units (PSU)	PSU2	3	3	3,00
Hat Racks	HR1	3	3	3,00
Hat Racks	HR2	3	3	3,00
Cabin Crew Rest Compartment (CCRC)	CCRC 1	3	5	4,12
Cabin Crew Rest Compartment (CCRC)	CCRC 2	3	3	3,00
Doors Entrance	DE 1	3	3	3,00

Fig. 6. Furnishing domain scores

Concept 1 architecture was modified according to the information retrieved from the CDfA analysis, for instance the screwing process was changed for the modules Galley and it led to a reduction of the attribute *Number of furnishing* from a score of 5 to a score of 3. Moreover, improvements were performed on the *Passenger Service Units (PSU)* modules. Here, the *Standardization* and *Rigging* attributes (Mechanical Domain) were modified respectively, by changing non-standard parts to standard ones and by changing the attachment design principle to minimize the riggings.

With the aim to compare the different cabin architecture concepts, the global score was computed (Table 3).

Table 3. Global score comparison

Cabin concept 1	Cabin concept 2	Difference
3.01	2.31	23%

Data reported in Table 3 shows the global score of the *Cabin Concept 1* (before modification) and *Cabin Concept 2* (after modification). By measuring the performance of the two cabin concepts in terms of workload, a reduction of the 27% between *Cabin Concept 1* and *Cabin Concept 2* were found. The CDfA result is consistent with the reality of the production, which validates its use at conceptual phase to identify assembly criticalities and assess benefits brought by new design options. It is interesting to notice that the method can be used also to compare aircraft cabin architectures of different aircraft size (i.e., different range of application), by including in the global score the: i) CDfA scores, ii) number of modules present in each module family and iii) overall number of modules present in the cabin architecture. In fact, the number of modules is linked to the aircraft size. For instance, the number of toilets, seats, etc. is related to the number of passengers the cabin can accommodate. Even though the method proposed is powerful, it is important to notice that it may require adaptations according to the type of System analysed.

4 Conclusion

The paper aims at applying the CDfA methodology on a new System of Interest (i.e., the Cabin System of a civil aircraft). The application of the methodology is presented, and the obtained results analyzed. The application of the CDfA methodology provides three scores representing the assembly complexity of: i) each cabin module family, ii) each cabin individual module, and iii) the overall cabin architecture. The first two scores were used to provide root causes of the cabin modules assembly performance. It was shown where criticalities lie (i.e., in the *Furnishing Domain*) and how a modification might lead to a global assembly improvement. The score assessing the overall architecture was used to compare different cabin concepts. Results shown that the cabin concept created starting from the CDfA results, provided a significant optimization of the assembly performances confirmed at manufacturing level. The current CDfA approach presents some difficulties;

first the modelling process (i.e., creation of the cBoM) is time-consuming and requires to be adapted according to the field of application (e.g., airframe assembly), second the definition of attributes domains and scoring matrices requires the involvement of several expertise, third it requires specific calibration process with huge amount of data to collect. The demonstrated method efficiency on the cabin System perimeter is promoting its extension to the full scope of the aircraft components. Future development will focus on the application of the method on different case-studies to improve the overall method modelling efficiency.

References

1. Maropoulos, P.G., Muelaner, J.E., Summers, M.D., Martin, O.C.: A new paradigm in large-scale assembly—research priorities in measurement assisted assembly. Int. J. Adv. Manuf. Technol. **70**(1–4), 621–633 (2013). https://doi.org/10.1007/s00170-013-5283-4
2. Pahl, G., Beitz, W., Feldhusen, J., Grote, K.H.: Engineering design: a systematic approach. Engineering Design: A Systematic Approach. Springer London (2007). https://doi.org/10.1007/978-1-84628-319-2
3. Favi, C., Germani, M., Mandolini, M.: A multi-objective design approach to include material, manufacturing and assembly costs in the early design phase. Procedia CIRP **52**, 251–256 (2016). https://doi.org/10.1016/j.procir.2016.07.043
4. Boothroyd, G., Dewhurst, P., Knight, W.A.: Product design for manufacture and assembly. CRC Press (2011)
5. Suzuki, T., Ohashi, T., Asano, M.: Assembly reliability evaluation method (AREM). CIRP Ann. Manuf. Technol. **52**(1), 9–12 (2003). https://doi.org/10.1016/S0007-8506(07)60518-6
6. Stone, R.B., Mcadams, D.A., Kayyalethekkel, V.J: A product architecture based conceptual DFA technique. (2003). https://doi.org/10.1016/j.destud.2003.09.001
7. Lockett, H., Fletcher, S., Luquet, N.: Applying design for assembly principles in computer aided design to make small changes that improve the efficiency of manual aircraft systems installations. SAE Int. J. Aerosp. **7**(2), 284–291 (2014). https://doi.org/10.4271/2014-01-2266
8. Butterfield, J., et al.: Optimization of aircraft fuselage assembly process using digital manufacturing. J. Comput. Inf. Sci. Eng. **7**(3), 269–275 (2007). https://doi.org/10.1115/1.2753879
9. Barbosa, G.F., Carvalho, J.: Guideline tool based on design for manufacturing and assembly (DFMA) methodology for application on design and manufacturing of aircrafts. J. Braz. Soc. Mech. Sci. Eng. **36**(3), 605–614 (2013). https://doi.org/10.1007/s40430-013-0103-1
10. Stone, R.B., McAdams, D.A., Kayyalethekkel, V.J.: A product architecture-based conceptual DFA technique. Des. Stud. **25**(3), 301–325 (2004). https://doi.org/10.1016/j.destud.2003.09.001
11. Domeshek, E.A., Herndon, M.F., Bennett, A.W., Kolodner, J.L.: Case-based design aid for conceptual design of aircraft subsystems. In: Proceedings of the Conference on Artificial Intelligence Applications, pp. 63–69 (1994). https://doi.org/10.1109/caia.1994.323691
12. Gómez, A., Ríos, J., Mas, F., Vizán, A.: Method and software application to assist in the conceptual design of aircraft final assembly lines. J. Manuf. Syst. **40**, 37–53 (2016). https://doi.org/10.1016/j.jmsy.2016.04.002
13. Ulrich, K.: The role of product architecture in the manufacturing firm. Res. Policy **24**(3), 419–440 (1995). https://doi.org/10.1016/0048-7333(94)00775-3
14. AlGeddawy, T., ElMaraghy, H.: Reactive design methodology for product family platforms, modularity and parts integration. CIRP J. Manuf. Sci. Technol. **6**(1), 34–43 (2013). https://doi.org/10.1016/j.cirpj.2012.08.001

15. Gomes, M., Ogliari, A., Fernandes, R.B., Marques, K.: Systematic approach to develop physical models as creative stimuli in conceptual design phase. Smart Innov. Syst. Technol. **198 SIST**, 228–238 (2021). https://doi.org/10.1007/978-3-030-55374-6_23
16. Bouissiere, F., Cuiller, C., Dereux, P.-E., Malchair, C., Favi, C., Formentini, G.: Conceptual design for assembly in aerospace industry: a method to assess manufacturing and assembly aspects of product architectures. In: Proceedings of the International Conference on Engineering Design, ICED (2019). https://doi.org/10.1017/dsi.2019.303
17. Formentini, G., et al.: Extrapolation of design guidelines during the conceptual design phase: a method to support product architecture design. In: Proceedings of the Design Society: DESIGN Conference, vol. **1**, pp. 857–66. Cambridge University Press (2020). https://doi.org/10.1017/dsd.2020.26
18. Favi, C., Formentini, G., Bouissiere, F., Cuiller, C., Dereux, P.-E., Malchair, C.: Design for assembly in the conceptual development of aircraft systems. Lect. Notes Mech. Eng., 268–278 (2020). https://doi.org/10.1007/978-3-030-31154-4_23
19. Chakraborty, I., Mavris, D.N.: Heuristic definition, evaluation, and impact decomposition of aircraft subsystem architectures. In: 16th AIAA Aviation Technology, Integration, and Operations Conference (2016). https://doi.org/10.2514/6.2016-3144

Design and Integration of Automation Systems with Manual Operation: Small and Medium Enterprises Issues

Rocco Salvato[1], Giovanni Marra[2], Paola Scardamaglia[1], Giuseppe Di Gironimo[2], Domenico Marzullo[3], and Rocco Mozzillo[4](✉)

[1] PROMEC SRL, Research and Development Department, Naples, Italy
[2] Department of Industrial Engineering, University of Naples Federico II, Naples, Italy
[3] Department of Engineering and Architecture, University of Trieste, Via Alfonso Valerio, 6/1, 34127 Trieste, Italy
[4] School of Engineering, University of Basilicata, Potenza, Italy
rocco.mozzillo@unibas.it

Abstract. Today it is more and more mandatory for all commercial companies to comply with the principles and methodologies of Industry 4.0 and to achieve the related capabilities protecting their competitiveness and taking a leading-edge position on market as regards technologies. Specifically, the whole production and sale system must achieve the fundamental characteristics of Industry 4.0 approach, but specially the manufacturing companies must also change and update their management procedures, internal organization, resource training, assets and all production process to keep safe their current business capacities. This evolution process is even more critical for Small and Medium Enterprises (SME), that traditionally tend to be conservative and to protect their way of operation, usually characterized by a low level of automation. The work presented focuses on the design and integration of a semi-automatic welding cell of train bolster in a SME which is currently realizing a project aimed to the acquisition of Industry 4.0 capabilities, with special focus on manufacturing processes. Among them, one of the most important is the production of welded-steel critical structures, that the Company supplies to prime manufacturer of railway rolling stock systems. The experience gained during the activity, the criticalities due to the integration processes and the adopted design methodologies are here described. The work has been carried out consistently with the Systems Engineering principles, starting from the requirements elicitation and analysis to the systematic approach for the design and integration activities.

Keywords: Industry 4.0 · Internet of things (IoT) · Design methodologies · Systems engineering · CAD · Welding process

1 Introduction

To keep up production standards and the whole sale system, companies must achieve the four following fundamental characteristics of Industry 4.0 [1]:

C. Rizzi et al. (Eds.): ADM 2021, LNME, pp. 298–307, 2022.
https://doi.org/10.1007/978-3-030-91234-5_30

- Vertical integration between all hierarchical levels of the value chain
- Horizontal integration between all the players in the supply chain
- End-to-end design that links all the phases of product conception, production and distribution
- Technological integration.

The manufacturing companies must also change and update their management procedures, internal organization, resource training, assets and all other production process components to safe their current business capacities [2]. Therefore for SME, traditionally tending to be protective towards their way of operation, these changes might be critical. The digital transformation process is involving all business realities, be they large, medium or small. This transformation deeply reshapes organizations which, called upon to take up the challenge linked to new technologies, must innovate the processes, the management of information flows and often their business models. The fourth industrial revolution [3] and its technologies have undoubtedly contributed to this evolution; these last ones are an essential element for new factories or in general for smart businesses.

The transition to the Industry 4.0 paradigm is determining a new product supply logic. Manufacturing is no longer a mere process to obtain the product but becomes a service available by different actors. Factories are no longer static and immutable, they become modular and can be rearranged depending on the order [4]. The long and laborious product research and development activity is now enriched with Big Data analysis reducing the over time, increasing product quality and customer satisfaction. In the vast nomenclature of Industry 4.0, the concept of "cyber-physical systems", in short CPS [5], is the one that probably arouses the greatest astonishment and is among the least known. Industry 4.0 aims essentially to maximize the adaptability of Europe as a production site in general and of individual production plants. Ultimately, the aim is to maintain international competitiveness, jeopardized by countries with low labor costs and countries with emerging markets. In this scenario, the present work is focused on the experience gained with the design and integration of semi-automatic welding cell of train bolster assembly, activity developed in close collaboration between the School of Engineering of Basilicata University, the Department of Industrial Engineering of University of Naples Federico II, Department of Engineering and Architecture of Trieste University and PROMEC Srl.

In the following paragraphs the conceptual design of the semi-automatic welding cell will be explained. The reasons that led to this mixed configuration, the functional diagram of the cell and the flow of information that the different stations exchange with the central server will be presented in detail. All that has been developed according to the principles of the Industry 4.0 and through the Systems Engineering approach [6].

2 Motivation and Objective

PROMEC is a SME that is currently realizing a project aimed to the acquisition of Industry 4.0 capabilities, with special focus on manufacturing processes [7]. Among them, one of the most important is the creation of welded-steel critical structures, that the company supplies to prime manufacturer of railway rolling stock systems (i.e. trains, vehicles and other major components).

The welding process is a critical task due to:

- the high safety requirements that parts shall comply [8];
- the consideration that – as well known – welding is a Special Process;
- very binding inspections and traceability requirements [9];
- the large amount of workmanship that the process currently employs (almost all the process operations are traditionally not automatized but realized by qualified workers);
- the high grade of geometrical complexity of the product, combined with low production rates.

For these reasons it seems very difficult to make innovation in this specific area. Nevertheless, the company has bought a robotized welding cell, trying to automate as much phases as possible, but the largest part of them (about 50%) will remain necessarily manual. Due to this it has become mandatory integrate the robotic cell with manual processes. Some tasks (part positioning, tack welding, Non-Destructive Tests) could be performed by one or more operators while other ones (as welding of heavy joints) could be realized by the robot, always respecting the sequences provided by the work schedule/work order. In any case, it has been mandatory for the design to guarantee the automatic acquisition (through the Company Information System) of all the parameters and information (planned and actual) that must be reported in the Manufacturing and Control Plan (MCP). The MPC resumes and certifies characteristics and values of the whole manufacturing process of each part, ensuring the total compliance to the Quality and Safety Requirements of the part itself.

3 Material and Methods

This work has been carried out following the Systems Engineering approach which allows to understand how systems fit into the larger context of day-to-day life, how they behave, and how they are managed. This process has an iterative nature that supports continuous learning and improvement; indeed, it involves looking at a problem in its entirety, considering all the facets and all the variables relating social and technical aspects.

As a matter of fact, Systems Engineering and Industry 4.0 imply the implementation of technological advances (in some cases already used in the traditional manufacturing system) in a way that gives birth to a fully integrated, automated, and optimized production flow, also defined as Smart Manufacturing [10]. Adopting a SE approach allows for the requirements elicitations (e.g. quality and safety requirements) from the early stage of the design and it provides the techniques for the Verification and Validation process [11], representing a key phase for the approval of the welding operations [12]. The aim of the work concerns the integration of the existing technologies into real business contexts and not only mere technological installations in the manufacturing process. The integration of these technologies and approaches in SME is the critical point of the development process since typically, the environment of SME has a low level of automation with processes not very structured. Sometimes the knowledge about the critical manufacturing process (be technical, logistics, commercial etc.) oversees few employers or operators and it is not managed by enterprise information system. In these industrial contexts it has become difficult to have a good traceability of materials and processes state along the entire manufacturing process [13]. The focus is on specific macro phases of production regarding welding. Therefore, in addition to the traceability of the material, the designed cell guarantees, as an information output, other essential data such as: operator traceability, welding parameters, processing start and end time, current and voltage values. The expected results will be to facilitate the production flow and above all to control it in the manual phases, which nowadays does not occur. With the design of a semiautomatic welding cell combined with a robotic welding cell, an attempt has been made to overcome all the problems relating to the transmission and compliance with parameters and data.

Very often, in a working day, structured over several shifts and with different operators, the directives imposed by a welding procedure specification (WPS) are not properly respected because they are subject to variations, based on the evaluations of the single operator; these problems can be overcome using an appropriate data control and management architecture. The robotic cell has been integrated with a "manual welding cell"; both cells cooperate for the manufacturing of the welded component. In this configuration the manual tasks carried out by the operators are integrated with the tasks of the welding robot. Figure 1 illustrates schematically the layout of the two cells, both in the simple configuration and in a more evolved form (manual cell and robot cell).

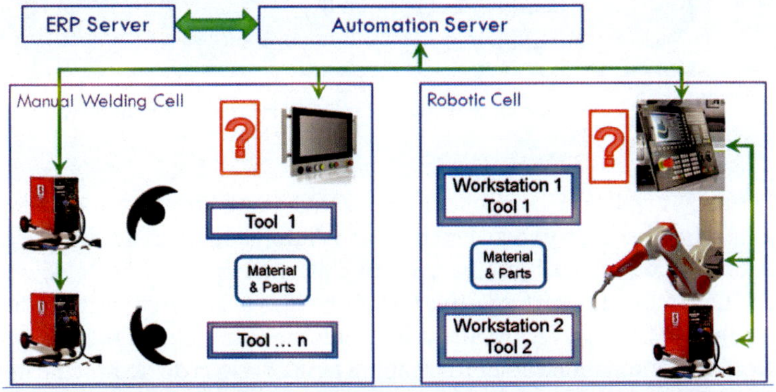

Fig. 1. Functional scheme of the integrated welding cell.

In this communication flow the voltage, the current and the wire speed are transmitted. There is a forward feed that lists the parameters to use, and a feedback that records the ones used. The kinematics between the unit and the robot is defined, the control unit is thus able to control everything and provides all the necessary information. If the robot were the only one to oversee the welding phase, we would already be in a situation of industry 4.0, but at present not all information is collected. Both cells are managed by a control unit exchanging information with the Automation Server. The control units manage the cells working areas and the related inputs which can be materials, tools and operators. These units manage the access in working areas in two different ways: it can be physical, by an access barrier, or consists of a series of photocells detecting or acquiring welding data, in direct connection through a port with the welding machines. According to SE approach this architecture assures the traceability and control, also in real time, of the process parameters both for manual and robotic tasks. For our project concerning the train Typical Moving Bolster, the cell must be equipped to perform two of the five welding phases plus all the control phases. The Fig. 2 schematically illustrates the data flow expected to be managed, the flow chart has been superimposed on the layout, previously presented, to highlight separately the characteristic data of each cell.

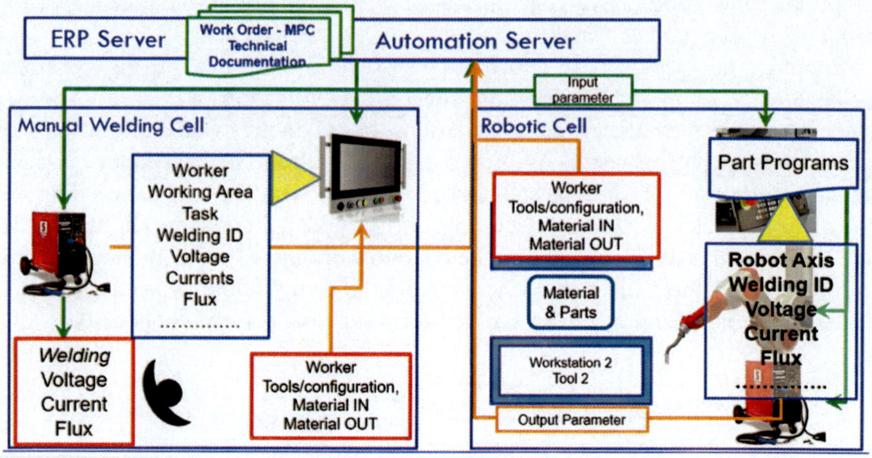

Fig. 2. Information flow between central server and workstations

Work Order, Technical Documentation and MCP data are to be considered shared (fully or as much as necessary) between ERP and Automation Server; the green connectors indicate the flows to and from the Control Units and/or the work cells, while those in orange return to the automation server from devices also different from the Control Units themselves. The large yellow arrow indicates the extraction of the aggregated data received from the automation server for sending to the various devices in the field, while the red boxes list the parameters that will need to be detected. To understand in detail what information the server needs compiling and updating the MCP, a typical one made in PROMEC was analysed.

4 Case Study

Currently PROMEC is dealing with the modernization of both its organization and manufacturing assets to protect its competitiveness and to achieve new market opportunities. Two years ago, it launched a program aimed to introduce significant modifications in the management and realization of all its processes maximizing the implementation of the principles of Industry 4.0 in its production. One of the most important manufacturing programs is a train *Typical Moving Bolster* (TMB), shown in Fig. 3.

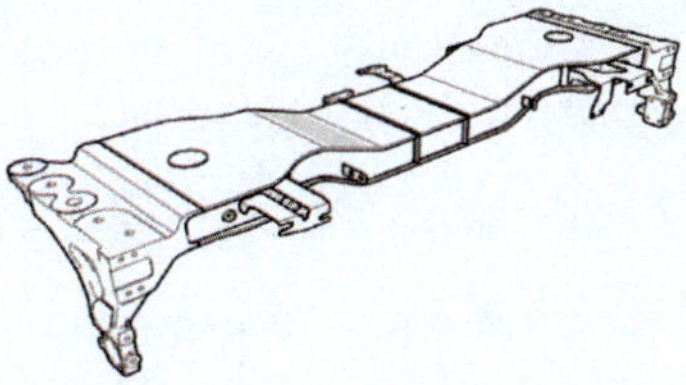

Fig. 3. Typical moving bolster

The TMB is a typical welded assembly, used as intermediate element linking the bogie and the train body shell. It is very important to allow the required degree of freedom between the two parts, transmitting in a suitable way the forces. TMB runs stably on both straight and curved track; it absorbs vibration generated by track irregularities and minimizes the impact of centrifugal forces when train runs on curves at high speed. The production flow of a typical TBM order has been analysed for understanding how resources are located and what is the technological level of the individual production steps Fig. 4.

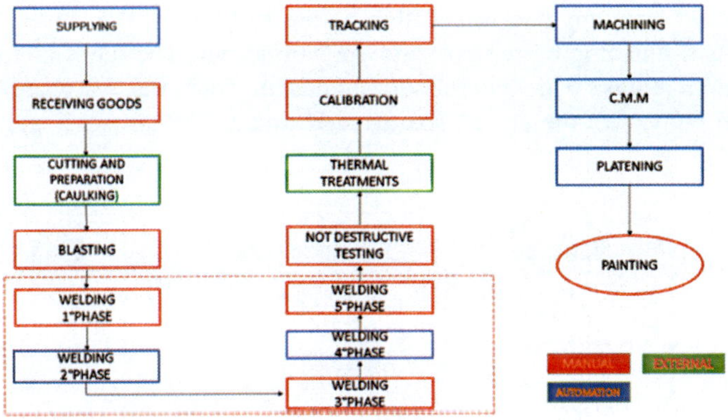

Fig. 4. Complete TBM production flow diagram

Figure 4 schematically represents the manufacturing process performed in PROMEC for a TMB fabrication. It must be highlighted that such kind of process is characterized by a significant alternation of automated and manual phases. In order to achieve the optimal integration between manual welding phases with robotic welding operation and maintaining at the same time unaltered the production flow, the following steps have been carried out: collection of a series of information and parameters on the manual operations not currently considered; integration of all information for the whole cycle until the MCP is automatically completed [14]. According to SE principles, the concept design of work cells has started from the collection of the information (parameters and needs) related to the welding operation and from the reporting of such information into functional requirements. This stage builds the guidelines for the next design stage as well as for the verification and validation activity. Using the layouts provided by the robot manufacturer, the overall dimensions have been considered to adapt the cell with the space available in the company and integrate it with the semiautomatic cell.

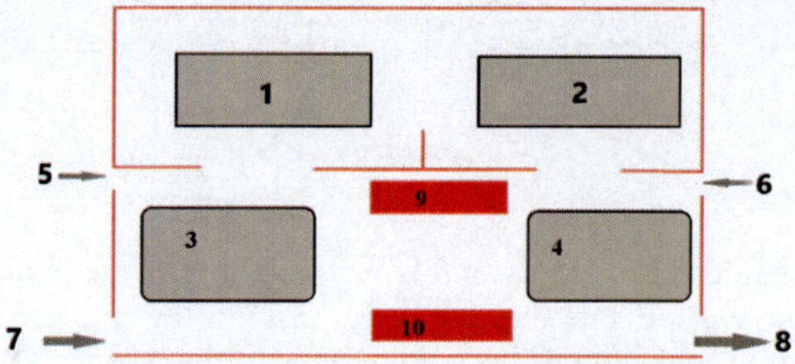

Fig. 5. Final layout of the two cells

Considering the subdivision of the process phases, the available area has been analysed and implemented to optimize the construction of two cells. Figure 5 shows the production flow and the logistic organization of the available space. The areas indicated with the numbers 3 and 4 are part of the manual welding cell in which the first welding steps, the bolster checking and verification before authorizing the subsequent phases, the welding and finishing of certain parts are carried out. The areas indicated by the numbers 1 and 2 represent the two robotic stations, in which the two phases of welding of the central body with the closed box are performed. The arrows 5 and 6 represent

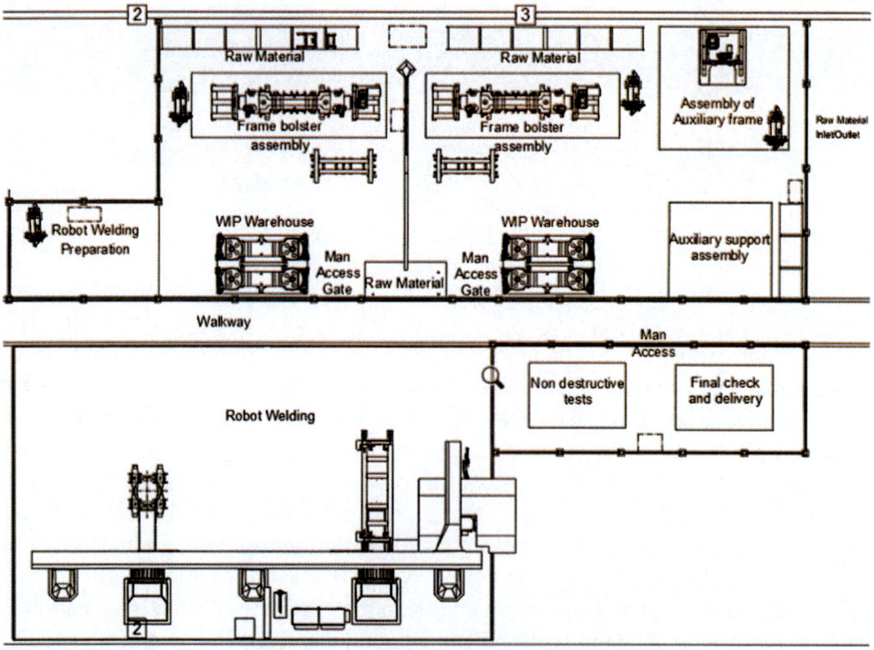

Fig. 6. The 2D layout of the two cells

the two independent entrances for operators equipped with badges (without which they would not have access to the cell), while arrows 7 and 8 indicate, respectively, the input of the semi-finished pieces and the exit of the welded components. The semi-finished pieces entering the cell, in the form of kit, are identified with a code that is transmitted to the cell control unit and then to the servers. The areas labelled 9 and 10 are reserved for the quality control. According to the flow described in Fig. 5 and the available area in the PROMEC context a detailed layout of the robotic and semiautomatic welding cell has been developed Fig. 6.

The final design of the two cells is visible in Fig. 7. The entire working area has been modelled in virtual environment (Fig. 7) [15].

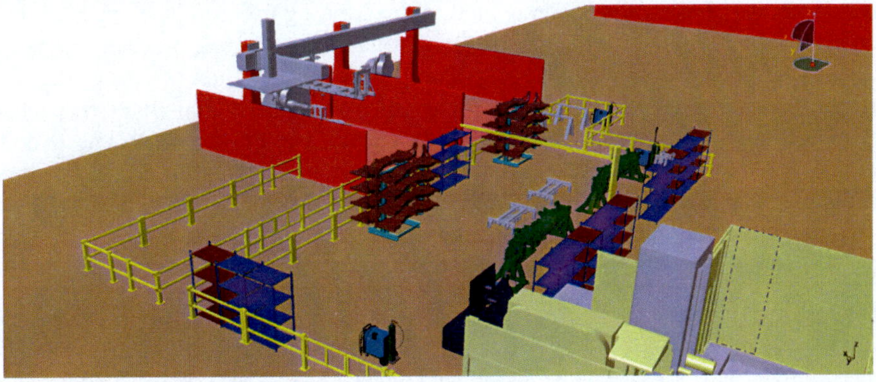

Fig. 7. Working area in virtual environment

For the detection of data on machines not equipped with processing units, the use of appropriate sensors (relative low cost) connected directly to the cell control unit is envisaged. Even if the robotic cell has a central unit for the data ex-change, an additional unit which communicates directly with the automation server and therefore with the ERP server would be helpful.

This architecture enables a dense network of data exchange in real time, which allows to gather excellent results with a view to saving time and reducing errors. This architecture helps in saving materials, in more reliable and economic management of their traceability, in the automatic compilation of MCP and all manufacturing documentation. The interface with the company management (ERP) allows the pre-processed production plans to be distributed in the factory; the orders progress in real time are returned via control terminals to the ERP management system, which also updates materials and finished products warehouse. The efficiency of the new production process is significantly improved compared with the previous one and it represents a big step forward from the industry 4.0 point of view.

5 Conclusion

The design and development of an innovative architecture for a semi-automatic welding cell in a SME based on main principles of Industry 4.0 has been presented in this paper.

The new architecture, based on IoT and human-robot cooperative environments, leads to a time saving in the product completion and a greater interaction between human and robotic work with excellent results in terms of final product quality. Furthermore, the new architecture allows to know and store previously neglected information, such as the welding machine's working parameters, the beginning and end of the welder's working hours and useful information for the automatic compilation of the manufacturing and control plan. All that contributes to an increased quality of the production processes and products since all the parameter related to them can be managed and controlled on time. This method, allowing the storage and the management of the industry knowledge, will bring also numerous benefits on process planning for sub-sequent orders. The semiautomatic welding cell is now under construction, the cell can guarantee the management and traceability of the information related to the welding process. The approach here described does not allow for the global automation of the welding process, this aspect could be a limitation of the potential benefit of the Industry 4.0 approach but on the other hand the work developed demonstrate its feasibility in not structured industrial context. The authors are aware a test activity to finalize and validate the approach is needed, the experience gained in a real context highlighted the issues related to the implementation of design approach and design solution in industrial context, due to the constraints related to the product and process complexity.

References

1. Dalenogare, L.S., Benitez, G.B., Ayala, N.F., Frank, A.G.: The expected contribution of Industry 4.0 technologies for industrial performance. Int. J. Prod. Econ. **204**, 383–394 (2018)
2. Finance, A.T.C.C.: Industry 4.0 challenges and solutions for the digital transformation and use of exponential technologies, pp. 1–12. Finance, Audit Tax Consulting Corporate, Zurich, Swiss (2015)
3. Schwab, K:. The fourth industrial revolution. Currency (2017)
4. Deloitte: Industry 4.0. Challenges and solutions for the digital transformation and use of exponential technologies (2014)
5. Tarallo, A., Mozzillo, R., Di Gironimo, G., De Amicis, R.: A cyber-physical system for production monitoring of manual manufacturing processes. Int. J. Interact. Des. Manuf. **12**(4), 1235–1241 (2018). https://doi.org/10.1007/s12008-018-0493-5
6. Haskins, C., Forsberg, K., Krueger, M., Walden, D., Hamelin, D.: Systems engineering handbook. In: INCOSE, Vol. **9**, pp. 13–16 (June 2006)
7. Davies, R.: Industry 4.0 digitalisation for productivity and growth. In: European Parliament Briefing (Sept. 2015)
8. Hu, J., Xie, L., Sun, W., Li, X., Lv, X., Meng, W.: Structural safety and fatigue reliability assessments of a metro-train bolster. Adv. Mech. Eng. **11**(1), 1687814018821755 (2019)
9. ISO 3834-5:2015: Quality requirements for fusion welding of metallic materials.
10. Zheng, P., et al.: Smart manufacturing systems for Industry 4.0: conceptual framework, scenarios, and future perspectives. Front. Mech. Eng. **13**(2), 137–150 (2018)
11. Marzullo, D., Di Gironimo, G., Lanzotti, A., Mazzone, G., Mozzillo, R.: Design progress of the DEMO divertor locking system according to IPADeP methodology. Procedia CIRP **34**, 56–63 (2015)
12. ISO 9001:2015: Quality management systems requirements

13. Iaccarino, P., Inserra, S., Cerreta, P., Mozzillo, R.: Determinant assembly approach for flat-shaped airframe components. Int. J. Adv. Manuf. Technol. **108**(7–8), 2433–2443 (2020). https://doi.org/10.1007/s00170-020-05459-8
14. Jin, W., Liu, Z., Shi, Z., Jin, C., Lee, J.: CPS-enabled worry-free industrial applications. In: 2017 Prognostics and System Health Management Conference (PHM-Harbin), pp. 1–7. IEEE (July 2017)
15. Miccichè, G., Lorenzelli, L., Frascati, F., Di Gironimo, G., Mozzillo, R.: Remote handling refurbishment process for the European IFMIF target assembly: concept design, simulation and validation in virtual environment. IEEE Trans. Plasma Sci. **45**(7), 1824–1830 (2017)

Integrated Product and Process Design

Modular Design Assessment: Development and Application of a Product Characteristic-Based Tool

Laura Asión-Suñer(✉) and Ignacio López-Forniés

Universidad de Zaragoza, Saragossa, Spain
lauraasion@gmail.com, ignlopez@unizar.es

Abstract. Agile manufacturing and product renovation are achieved in industry by methods such as modular design that has been investigated to improve product development and manufacturing processes. However, it has been detected that there is some confusion in its actual application in product design due to the lack of consensus on its definition. As a consequence, it is common to confuse it for other related terms, making its correct evaluation and application difficult. To facilitate its understanding, this work offers a characterization of modular design through a series of essential and complementary characteristics that define it. A tool is proposed to evaluate modular products through said characterization. This tool proposes a series of questions that help to verify compliance with each characteristic. The objective is that the user could respond to each one with *Yes*, *Partially* or *No*. Compliance with essential characteristics allows the product to be considered modular. As a result, this work offers an accessible and usable tool capable of evaluating modular products and improving their future design.

Keywords: Modular design · Assessment tool · Product characterization

1 Introduction

Modular design is frequently used to adapt the manufacturing process to change due to it is an excellent base for product renovation [1]. Modularity, versatility and adaptability are highly appreciated characteristics today, so various authors reflect the importance of its development in the literature. They claim that modularity responds to large industrial needs [2] and that its development is a goal in the industrial production [3]. In addition, the growing demand for personalized products is pushing companies to incorporate modular design to create customized products easily and economically [4]. The terms *module* and *modularity* have been used generically and have been open to interpretation for years. The only existing consensus is that a modular product consists of modules understood as building blocks, so the current definition of modular design is based on the definition of modules [5]. However, there is no consensus on the definition of modular design in the literature [2, 6], which gives rise to confusion since the term *modular design* lacks its own definition and is often confused with *modularity*. In this regard, modularity is only one characteristic that a design can fulfill, while modular design is a

C. Rizzi et al. (Eds.): ADM 2021, LNME, pp. 311–319, 2022.
https://doi.org/10.1007/978-3-030-91234-5_31

design method used intentionally in the design and development process of a product to obtain modular products. As a consequence, this terminological confusion has affected the identification, evaluation and application of modular design.

It is detected the need to facilitate the evaluation of modular design, which will also allow to improve its application by offering a better understanding of the term. This work proposes an evaluation tool by filtering a series of characteristics that have been considered inherent to modular design. These characteristics have been identified through a bibliographic review carried out during the research phase of a doctoral thesis. Nevertheless, the work presented here does not focus on this review and characterization, but on its applicability to evaluate modular design in design concepts. For this, the work continues with the proposal of a tool based on the characterization. To make the tool accessible and easy to use, a series of questions are written for each characteristic. The user can respond to each one with *Yes*, *Partially* or *No* indicating whether the product fulfills or not with each characteristic (*Yes*/*No*) or only some modules fulfill with it (*Partially*). Finally, the work concludes with the analysis of a real example to understand and justify the defined characteristics, and the assessment of six current products considered modular design to evaluate the tool and verify its usability. As a result, the work offers an assessment tool to evaluate whether a product can be considered modular design with the objective of apply it in design concepts in future work.

2 Recalls from the State of the Art

The characterization is based on a study that analyzes numerous publications, research articles and theses of the main authors who have defined *modular design*, *module*, *modularity* and *product architecture*. This review includes the search and analysis of 31 modular design methods and 29 real cases [7]. To provide an applicable characterization, the characteristics were counted (Table 1) and subsequently defined. In this section, it is presented a summary of the review though the definition of module, product architecture, modular design and its relationship to obtain modular products.

A *module* is a component of the product that performs a specific function to achieve a general objective [8]. Among its main characteristics, we find its independence and ability to combine with other modules [2]. In this way, a module is a fragment of the product with an identifiable function [9]. Regarding to *product architecture*, it can be seen as a construction principle where the functional elements are arranged in physical components [10]. Modular design has the potential to incorporate life cycle considerations into the product architecture configuration, achieving benefits in development, assembly, reuse and other goals [11]. Most definitions of *modular design* focus on modularity to characterize it, and many of them derive from the basic notion of a block-based design [3]. For this reason, the term *modularity* is often confused with *modular design*. Modularity can be defined as the degree to which a system is made up of independent but interactive elements in which each one houses a set of functionalities. This flexibility is also reflected in the customer as modularity gives rise to flexible designs capable of responding to changes in functional requirements [12].

Analyzing these definitions, a link is established between *module*, *product architecture* and *design process*. Modules are identified as individual and independent elements that work together as part of a whole. Therefore, the system must provide a product architecture that selects and arranges those modules to achieve the desired result. In this aspect, modular design is a model integrated into the design process to obtain a product with modular characteristics. The connection between the three terms can be understood in the representation of the module as a component, product architecture as the organization of the elements and modular design as a process. Each module has its own characteristics and their combination leads to modifications on the product architecture within a process based on achieving an effective modular design.

3 Approach

This section proposes a series of characteristics that allow to evaluate whether a product can be considered modular design or not. These characteristics are the premise for the development of an analysis and evaluation tool that will be presented later. Table 1 shows the definition and number of appearances of each characteristic in the analyzed definitions. The first three characteristics have been classified as *essential*, not only because they have the highest number of appearances, but also because they have been considered necessary to achieve a modular design according to the authors' definitions and the real cases analyzed. In this way, essential characteristics are those that suppose a *sine qua non* condition, so they should be present at the same time to get a modular design. The rest of characteristics have been classified as *complementary* and they can improve the degree of modularity of a product, but their application has not been considered necessary to obtain a modular design. Moreover, their presence depends on the fulfillment of the essential characteristics. Table 1 indicates these dependencies with respect to independence (I), connectivity (C) and variability (V). Strong dependencies are indicated with capital letter and weak dependencies in small letter.

Next, we propose a checklist tool with which to evaluate modular design through the defined characterization. Some modular design methods are based in checklists [13], such as the Holonic Product Design method [14]. Table 2 proposes some questions for the checklist that help to verify compliance with each characteristic. The objective is that the user can answer each one with Yes (✓), Partially (-) or No (X).

The proposed tool assesses through a table the fulfillment of the essential and complementary characteristics. To do this, each characteristic should be assessed through the defined questions and be answered with Yes (✓), Partially (-) or No (X). This assessment can be applied to support product design in the evaluation and optimization of concepts. It also can be applied to finished products in order to update and optimize them, so that the tool serves to improve the application of modular design in any phase.

Table 1. Summary of the appearance of characteristics in the revised definitions.

Characteristic	Description	Related characteristics	Dependence	Sum
Independence (I)	A modular system consists of independent modules that function as a whole. Modular design is based on the relationship between achieving functional independence and reducing interactions between modules, allowing to change components without affecting the rest.	Separable	-	18
Connectivity (C)	To connect construction elements in groups from which to assemble variants. Thanks to compatibility between connections, modular product architectures and platforms can use common elements, resulting in interchangeability and standardization.	Collaboration, union, interface	-	18
Variability (V)	Modular design is a basic design element in promoting product diversity. This makes it possible to generate a variety of products through the combination of platforms and modules, resulting in product families or variations of the same product.	Combination, configuration	-	18
Economy	Cost and time reduction in the design, development and manufacture of complex products. Modular design contributes to the achievement of an economy of scale through the reuse of proven elements.	Quick, easy assembly, added value	*objective*	12
Multifunctionality	To cover and expand the number of functionalities, resulting in a function-oriented design that can be integrated into different systems for the same functional purpose, but without modifications.	Complex	I C V	11
Adaptability	To replace individual modules to improve functionality with minimal impact to the rest of the system, allowing it to adjust and adapt to change, both temporary and permanent.	Replaceable, changeable, versatile	I C V	11
Interchangeability	Division and decomposition of the product architecture in modules to facilitate their interchangeability in a reduced cycle time, facilitating in turn the combination of components.	-	I C	10
Standardization	Component, interface and process standardization allows for different sizes or versions of a module to be interchanged, resulting in potential benefits of variety.	Compatible	i C	10
Flexibility	Ease of incorporating future modifications in the system and allowing the expansion of elements through redesign, change or evolution of the modules, favoring innovation in the product.	-	I C v	10
Upgrades	Continuous product renewal through system update when new technology is developed to improve a specific function, affecting only one product module.	Evolutionary, redesigned	I C	9

(*continued*)

Table 1. (*continued*)

Personalization	To choose, mix and match modules according to specific customer preferences to create a final product that suits their wishes and needs.	Customizable	c V	9
Reusability	The interchangeability of the module, both within a single product and between several different ones, makes it easy for a product to be reused, even when it stops working.	Recyclable	i c V	9

Table 2. Questions for the checklist.

Characteristic	Question
C1 Independent	Does this product have modules that can be physically separated and used in an analog product without changing its function?
C2 Connectable	Does this product allow to change its modules through a compatible interface that connects them without changing their functionality?
C3 Variable	Does this product have modules whose combination allows the creation of a family of products and/or variants of the same product?
C4 Economical	Does this product have a reduction in the cost and/or time of design, development, production or sale thanks to its modularity?
C5 Multifunctional	Can this product achieve a greater number of functions or extended functionality by adding or changing modules?
C6 Adaptable	Does this product have modules that allow it to be functionally versatile in different situations (temporary or permanent)?
C7 Interchangeable	Does this product allow the combination of different components of the same product and/or other types of products?
C8 Standardized	Does this product allow the integration of compatible modules or components from other products, families or manufacturers?
C9 Flexible	Does this product have modules that favor the incorporation of future modifications and innovations in the product design?
C10 Upgradeable	Can this product evolve through the renovation and/or redesign of a specific module without changing the complete system?
C11 Customizable	Does this product allow the end user to make custom configurations prior to purchase and/or during its life cycle?
C12 Reusable	Does this product have modules that can be reused in another version of the product and/or in another product of the same family?

4 Test Cases and Results

In order to justify the choice of the characteristics and make them more understandable, an example of a good application of modular design has been analyzed against another that could be considered modular design but is not. The chosen examples are two types of pens from the BIC company [15]: the common one-ink pen (#1) and the four-ink pen (#2). Three common basic modules have been identified between them: barrel, cartridge and cap (Fig. 1). It should be taken into account that some modules group several components inside them. Finally, Table 3 shows which characteristics each example fulfills. The symbol "✓" means that the product fulfills the characteristic completely, "-" means that it only partially fulfills it and "X" indicates that it does not fulfill it.

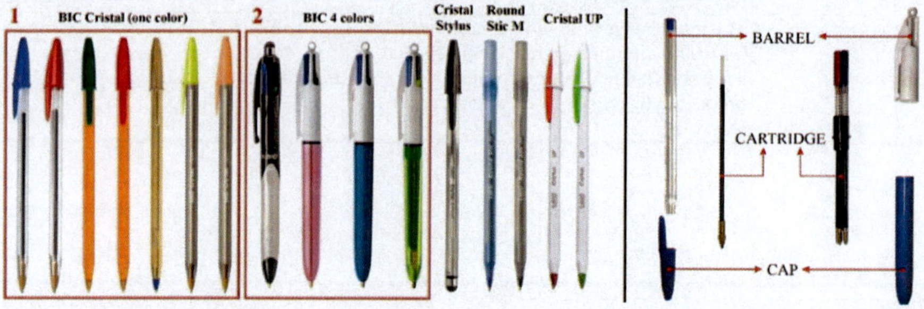

Fig. 1. Product range of BIC company and main modules of the two examples to analyze.

Table 3. Summary of compliance with the characteristics in each case.

Product	C1	C2	C3	C4	C5	C6	C7	C8	C9	C10	C11	C12
One-ink pen (#1)	✓	✓	✓	✓	-	-	✓	✓	-	✓	✓	✓
Four-ink pen (#2)	-	X	-	X	✓	-	-	-	-	X	-	-

These results clarify why essential characteristics must be met to consider a product as modular design. In this case, pen #1 *connects* all its modules by pressure, which allows each one to separate from the rest and continue to function. This gives them *independence*, allowing the creation of a family of products through modular *variants* (see Fig. 1). However, pen #2 has a highly integrated system. The cartridge is embedded in the barrel and it can't work independently. However, the connection to the cap is by thread, which gives the product its unique modular facet. The results also justify why the presence of the essential characteristics leads to the inclusion of complementary characteristics, but the same does not happen the other way around. For example, the presence of adaptability and multifunctionality in the pen #2 does not make it modular. Moreover, the lack of independency and good connectivity in its modules affects directly to other complementary characteristics such as interchangeability, customization, upgrade and reusability.

In Table 4 the modular design of six real products is evaluated. They have been chosen for being defined as a modular design or its good application. The evaluation was carried out by two evaluators. The first with 23 years of experience and 17 of teaching in design engineering. And the second with 5 years of experience and investigation in the same scope. The evaluation consisted of the assessment of the 6 cases individually and then there was an agreement session to reach a consensus on the results. The six products to evaluate are: (1) concept of a lamp with different support and lampshade modules [16]; (2) lamp created with identical hexagonal modules [17]; (3) Dyson vacuum cleaner with various heads and accessories [18]; (4) concept of kitchen with electrical appliances that can be fitted by their standardized measurements [19]; (5) mixer with multiple heads and accessories [20]; and (6) identical portable batteries connectable between them [21].

Table 4. Experimentation with the proposed tool to evaluate modular design.

Product	C1	C2	C3	C4	C5	C6	C7	C8	C9	C10	C11	C12
Doolight by Doosan Baek [16]	✓	✓	✓	✓	-	✓	✓	✓	✓	✓	✓	✓
Helios Touch [17]	✓	✓	X	X	X	-	-	✓	X	-	✓	✓
Dyson Cyclone V10 [18]	✓	✓	✓	✓	✓	✓	✓	✓	✓	✓	X	✓
Addition by Heewoong Chai [19]	✓	-	X	X	✓	-	✓	✓	✓	X	✓	✓
PowerGear 1500XL [20]	✓	✓	✓	✓	✓	✓	✓	✓	✓	✓	X	✓
Google PixelBloc [21]	✓	✓	X	✓	X	-	X	✓	X	X	X	✓

As can be seen, only half of the products have been considered modular design. This may be because the rest were not based on a modular design process, but on the design of identical modules. In this way, despite meeting the independence and connectivity characteristics, they could not provide variability to the product. As a consequence, compliance with the complementary characteristics has been lower.

5 Discussion

Most of the characteristics defined are compatible between them, but some are antagonistic depending on the context in which they are applied. For example, a module can be reusable if the product is disassembled and attached to a new one. However, by allowing products to be upgradeable, modular design can make some modules obsolete and therefore not reusable. We must take into account this antagonism from the beginning of the process to ensure that the characteristics are finally compatible.

From an external point of view, it is difficult to know whether a product has been designed under a modular design process or not. We can only know it where manufacturers name their process or method as modular. However, the proposed tool can assess whether a product meets the characteristics that give rise to a modular design at the time of evaluation. In this aspect, there is a confrontation between the goals of modular design focused on the company and the user. The first seeks to optimize the manufacturing process, and the second allows the end user to use modular products to their own benefit. The proposed tool offers an assessment from the point of view of the end user.

Having modules does not imply that the result is a modular product, being necessary to differentiate between process and result. If we conceive modular design as a result, we obtain products made up of identical modules [17, 19, 21]. Instead, modular design understood as a process integrates module and architecture to create a modular products [16, 18, 20]. For this reason, the proposed tool only considers modular design as a process that integrates module and architecture, and not as a result of the design of modules. This aspect gives rise to discussion in practice, where it is common to attribute the term modular design to products based on design of modules.

6 Conclusions

The work presents a modular design evaluation tool based on a series of predefined characteristics. The results show a valid first approximation to understand, apply and correctly evaluate the modular design. No incongruities are detected in the results and the evaluations allow to detect points of improvement in the evaluated products. Nevertheless, the characteristics identified have been worked on to present a new evaluation tool as a test that must continue to evolve with the aim of improving the definition of the characteristics, the questions for the checklist and their application in real cases. The work currently offers limitations such as usability by non-specialized users or applicability in other types of products that must be resolved in the future work.

Multiple definitions of modular design have hampered its understanding. A more integrated approach facilitates its evaluation and application. Modular design is a process to achieve technical advantages through configurable modules within a product architecture. For this reason, a product made up of modules is not always designed under a modular design process. Products with internal objectives for the company, such as economic savings, generally are not based on modular design. On the contrary, those that are conceived from the beginning to include modules that allow the end user to obtain advantages, are conceived under the idea of modular design.

Recent publications focus on the development of methods intended for specific purposes where modular design is only a tool and not the goal. In the real application, the correct use of the essential characteristics conditions the presence of the complementary ones. The use of these characteristics in product design increases the level of modularity by providing qualities that benefits the life cycle. Conceptualization as an initial design phase foresees the possibility of a product being modular thanks to the integration of the desired characteristics. Used during the conceptual phases, these characteristics are translated into Product Design Specifications (PDS's).

References

1. Baldwin, C.Y., Clark, K.B.: Design rules: the power of modularity. Acad. Manag. Rev. **26**, 471 (2000)
2. Marshall, R.: Design modularisation: a systems engineering based methodology for enhanced product realization. Doctoral dissertation, Loughborough University (1998)
3. Lehtonen, T.: Designing Modular Product Architecture in the New Product Development. Tampere University of Technology (2007).
4. Shamsuzzoha, A.: Modular product development for mass customization. Vaasan yliopisto (2010)
5. Gershenson, J.K., Prasad, G.J., Zhang, Y.: Product modularity: definitions and benefits. J. Eng. Des. **14**(3), 295–313 (2003)
6. Smith, R.J.: The impact of modular design on product use and maintenance. Doctoral dissertation, Georgia Institute of Technology (2009)
7. Asión-Suñer, L., López-Forniés, I.: Modular Design: Product Design Opportunities and a Case Analysis. In: Cavas-Martínez, F., Eynard, B., Fernández Cañavate, F.J., Fernández-Pacheco, D.G., Morer, P., Nigrelli, V. (eds.) Advances on Mechanics, Design Engineering and Manufacturing II. LNME, pp. 596–609. Springer, Cham (2019). https://doi.org/10.1007/978-3-030-12346-8_58

8. Sinha, K., Suh, E.S.: Pareto-optimization of complex system architecture for structural complexity and modularity. Res. Eng. Des. **29**(1), 123–141 (2017). https://doi.org/10.1007/s00163-017-0260-9
9. Hölttä-Otto, K.: Modular product platform design. Helsinki University of Technology (2005)
10. Løkkegaard, M., Mortensen, N.H., McAloone, T.C.: Towards a framework for modular service design synthesis. Res. Eng. Des. **27**(3), 237–249 (2016). https://doi.org/10.1007/s00163-016-0215-6
11. Gu, P., Hashemian, M., Sosale, S., Rivin, E.: An integrated modular design methodology for life-cycle engineering. CIRP Ann. **46**(1), 71–74 (1997)
12. Tate, D., Lindholm, D., Harutunian, V.: Dependencies in axiomatic design. J. Integr. Des. Proc. Technol. **3**, 159–166 (1998)
13. Gershenson, J.K., Prasad, G.J., Zhang, Y.: Product modularity: measures and design methods. J. Eng. Des. **15**(1), 33–51 (2004)
14. Marshall, R., Leaney, P.G.: Holonic Product Design: a process for modular product realization. J. Eng. Des. **13**(4), 293–303 (2002)
15. BIC Homepage. https://www.bicworld.com/es. Accessed 18 Jan 2021
16. Doolight by Doosan Baek. https://www.yankodesign.com/2018/09/25/meet-doolight-the-modular-smart-chandelier-of-2018/. Accessed 18 Jan 2021
17. Helios Touch Homepage. https://heliostouch.com/. Accessed 18 Jan 2021
18. Dyson Cyclone V10. https://www.dyson.es/aspiradoras/sin-cable/dyson-cyclone-v10/absolute. Accessed 18 Jan 2021
19. Addition by Heewoong Chai. https://www.yankodesign.com/2018/04/04/best-of-modularity. Accessed 18 Jan 2021
20. Cecotec PowerGear 1500XL. https://www.storececotec.com/es/batidoras-de-mano/792-powergear-1500-xl-mash-pro.html. Accessed 18 Jan 2021
21. Google PixelBloc. https://www.yankodesign.com/2020/09/18/google-pixelbloc-modular-power-bank-lets-you-stack-up-external-chargers-for-a-bigger-battery/. Accessed 18 Jan 2021

Concept Assessment Using Objective-Based Metrics on Functional Models

Ignacio López-Forniés(✉)

Zaragoza University, 50018 Zaragoza, Spain
ignlopez@unizar.es

Abstract. The measurement of design concept variables for future products is essential to assess the functionality, durability and feasibility, as well as the novelty and usefulness that they can offer to product innovation. Creativity can be evaluated in the conceptual phase through the creative dimensions, but there is no consensus on which method is more appropriate. The need to measure creativity by objective means is detected, leaving aside the subjectivity of some metrics that depend on the experience of the evaluators.

The experience presented here deals with the design and construction of a model to perform functional tests measured with multidimensional metrics. The results of a test carried out with 37 functional models are presented, observing that the measurement with hybrid metrics allows obtaining continuous variables that improve the classification and discrimination of positions. The models allow to observe continuous, measurable and quantifiable variables based on functional objectives.

Keywords: Conceptual models · Conceptual design · Creativity metrics · Objective-based metrics

1 Introduction

Measuring creativity is a necessary activity, since innovation and the development of new products depends on its results [1]. It is complex due to the physical and psychological factors that come into play such as design and manufacture, satisfactory operation, market value, impact on society, among others. Engineering and design identify creativity as the generation of effective novelty [2], said effectiveness is linked in the literature with utility [3] expanding its meaning by being associated with the value or impact it produces in society [4] which is difficult to measure in a conceptual phase.

The way to show a functional design can be very rudimentary and basic with artifacts or very precise and detailed with fully functional prototypes; election depends on the phase of the project, the resources available and the detail or amount of information that the test can offer. The type and level of detail in the construction of models corresponds to the design phase, in the divergent phases of idea generation, drawings and diagrams are enough to represent the operation. In the convergent phases, in which concepts are defined, proof of concept POC, functional models or prototypes (low or high detailed)

C. Rizzi et al. (Eds.): ADM 2021, LNME, pp. 320–327, 2022.
https://doi.org/10.1007/978-3-030-91234-5_32

are used and called concept models. Concept models could be focused prototypes that implement one or a few attributes and answer specific questions. Or else, comprehensive prototypes that implement many or all attributes and offer opportunities for rigorous testing [5].

A creative design experiment is presented. Concept models are validated with metrics that assess novelty, usefulness and the achievement of the proposed objectives. This experience only request to meet the objectives and demonstrate the functionality and feasibility. The choice for the type of model is free and there is no minimum level of detail. Viability is not requested as the economic and formal characteristics are not evaluated. The data obtained with the conceptual models can be used in various metrics, since a single metric lacks the discriminatory power necessary to analyze and compare the results [6].

The objective of the work is to create a metric that determines the validity of a concept model, measuring its creative dimensions by the fulfillment of certain objectives. The creative dimensions, novelty and usefulness, are defined in the metric by means of the satisfaction of measurable objectives such as conceptual differentiation and operational effectiveness; and they are combined with performance dimensions.

2 Literature Review. Theoretical Background

Comparing creativity assessment methods and metrics is found that novelty is a key component of creativity [7], that could be a prerequisite [4] to assess creativity [1, 8]. In some cases, it is defined with originality, infrequent, non-obviousness, rarity or unusualness [3]. Utility and Feasibility are characteristic dimensions defined openly with terms such as resolution, elaboration, synthesis, purpose, implementation, relevance, workability [3]. A generalized metric is that based on novelty (N), usefulness (U) and feasibility (F) [3, 9]. Other dimensions are based on: satisfaction of objectives through indicators or fulfillment of some conditions [10], how close to the correct solution the method reaches or how much resources are consumed in reaching its solution.

The classic metrics measure in ranges with discrete values using Likert scales, being imprecise and subjective, the inter-rater reliability is relatively low, the most reliable results refer to individual characteristics [11]. Regarding the type of measurement, various approaches are observed: answers (yes/no) for satisfying a condition or compared to a datum, Likert scales ranging from 3 up to 10 point-scale, evaluation with semantic differential, among others [4]. So hybrid metrics are proposed in this work, balancing NUF dimensions with others based on measurable objectives or satisfaction of conditions [12]. A concept as a result of a creative process must meet some properties [13]: Anticipatory expressed by novelty, differentiation or value contribution. Well-founded evidencing the performance. Focused on an objective to be achieved. Understandable to explain the functionality and operation of the object.

Conceptual models serve to validate concepts, being focused prototypes that implement one or some attributes to answer specific questions, or else, comprehensive prototypes that implement many or all attributes offering opportunities for rigorous testing. Concept models in a divergent stage can test functionality, they represent a safe opportunity to fail and learn, set challenges and provide immediate feedback, some changes can be done, so those models are called editable models [14]. Testing a concept model is a way out of the uncertainty, it is a demonstration that the idea works and the outcome that give us confidence to overcome the final design and to achieve the design requirements [15], so test various concept models is recommended to avoid idea fixation [16].

3 Experience to Test the Functionality of a Concept Model

The experience consists of the design and construction of a conceptual model to collect marbles that simulate a particle system, in a space delimited by walls and with a marked collection area, the designed device must be built by hand. University students have participated, practicing with fundamentals of physics and materials, such as rigidity and flexibility, deformation, friction, etc. In addition, they must develop a certain skill in the management of the model and its effective and efficient use it is conditioned in relation to the given time. Similar examples of model and prototype test lead to a learning of creativity in which designers think with their hands [17] and the problem is solved in a physical way, since everything holds up on paper and a completely different thing to put it into practice.

The experience has the rules grouped into objectives, difficulties and limitations, just as if it were a design problem to find a valid concept. The objectives of the experience are: Validate the concept or idea through a physical model that has to collect 50 marbles in the shortest time possible, counting the number of marbles in the first thirty seconds, and using the fewest parts and materials to build the model. The difficulties of the experience are: Collect the marbles of different diameters (Ø16–25 mm, 35 and 15 pieces respectively) by one person using only one hand, in a delimited area of 400 × 400 mm marked on the ring. The limitations are: (1) container ring of 660 × 660 mm, with 150 mm vertical walls. (2) Only recycled materials, parts or components are allowed. (3) Rapid prototyping technologies or FabLabs machines cannot be used. (4) In case of model breakage, a second attempt is allowed, permanent breakage is considered elimination. (5) Five minutes, maximum time, to collect all the marbles, otherwise it is considered eliminated. (6) The time to design and develop the model is 4 weeks.

3.1 Creativity Metrics Applied to the Experience

A comparison between two metrics that evaluate creative results through conceptual models is proposed. The first M_1, is multidimensional assessing the satisfaction of the basic elements of creativity (novelty and usefulness) and the specific performance objectives. And the second M_2, is also a multidimensional metric where novelty and functionality is a prerequisite and the score of the metric is based on continuous and quantifiable variables linked to the performance objectives to differentiate the concepts. Both metrics evaluate the creative product, in an advanced conceptual phase in which the operation can be evidenced by means of a conceptual model. The purpose of the metrics is not to evaluate personal creativity or functional novelty with respect to what already exists, since functional differentiation is sought between the models that participate in the experience and the optimal performance results for particular conditions.

Existing studies systematically identify operational variables and key ideation components of design methods and relate them to performance metrics [18]. The data obtained with the conceptual models is used in both metrics, since a single metric lacks the discriminatory power necessary to analyze and compare the results [6].

The first metric M_1 includes novelty and utility as factors for the evaluation, is calculated with the formula (1). Where N_1 novelty evaluates the conceptual differentiation of the models presented among them, with a maximum value of 1 in a range [0,1], it is calculated with formula (2). M is the total number of models and C_n is the number of conceptually equal models. It favors the unique characteristic and penalizes that of similarity. U_1 Usefulness evaluates the effectiveness of its operation in the mode of collecting marbles compared to each other, with a maximum value of 1 in a range [0,1] and is calculated with the formula (3). Where B is the total number of marbles and G_n is the group of marbles collected in each attempt, grouped as follows: all at once (G50 scores 50) or in groups of 11 to 49 (G11 scores 25), 6 to 10 (G6 scores 10), 2 to 5 (G2 scores 5) and 1 each attempt (G1 scores 1).

In the second metric M_2, novelty and utility are set as sine qua non conditions, it is calculated with formula (4). Where N_2U_2 is an *AND* type Boolean factor that can only be 0 or 1 since it is the product of both conditions. Where N_2 is 0 if there is a number of conceptually equal models greater than 50% and 1 if less. U_2 is 1 if collects all the marbles in the given time and 0 if not achieved.

Performance (Q, T, P) is a factor that evaluates the objectives of the experience for both metrics, expressed in a range [0, 1] with respect to the percentile of the sample ($Q_{\%}$, $T_{\%}$, $P_{\%}$). Q is the quantity of marbles collected in the first thirty seconds; T is the time required to collect all the marbles, scoring the maximum time (300 s) if they are not collected after five minutes; and P the number of pieces, the minimum number of pieces being the most optimal and taking into account the possibility of manufacturing said pieces; bearing in mind that in the model there may be several pieces that can be integrated into one in the manufacture process.

$$M_1 = (N_1 \times U_1) \times \text{Performance}\ (Q,\ T,\ P) \tag{1}$$

$$N_1 = ((M - C_n) + 1) \div M \tag{2}$$

$$U_1 = 1 - ((B - G_n) \div B) \tag{3}$$

$$M_2 = N_2U_2 \times \text{Performance}\ (Q,\ T,\ P) \tag{4}$$

3.2 Evaluation

The evaluation was carried out by two evaluators. The first with 23 years of professional experience in product design and 17 years of teaching in design engineering. And the second with 18 and 3 years respectively. The evaluation consisted of the classification and grouping of the functional principle of each model and the evaluation of the number of pieces used. It was carried out individually and then there was an agreement session to reach a consensus on the results. The rest of the values are obtained from the objective data of the test that were recorded on video for later verifications.

3.3 Conceptual Differentiation

There are a total of 37 models who have done the experience. Table 1 lists the different concepts, the number of similar concepts (Sum) and their operating principle. There is a great diversity of concepts with few repetitions, and although in some cases the principle of operation coincides conceptually, they have differences. The largest grouping of principles is for drag, 13 models, which groups "Wedge drag", "Container with ramp", "Excavator with container" and "Container with shovel".

Table 1. Conceptual models (CM), similarity and working principle

#	Concept	Sum	Principle
CM1	Centrifugal	1	Vertical rotation
CM2	Wedge drag with container	1	Drag
CM3	Sliding sheet	1	Slide
CM4	Fishing net with pusher	2	Enveloping
CM5	Container with ramp	3	Drag
CM6	Sphincter (point/linear)	3	Material deformation
CM7	Harvester / noria	4	Horizontal rotation
CM8	Excavator with container	4	Drag
CM9	Gripper with funnel with container	4	Angle closure
CM10	Scissors with deposit	4	Angle closure
CM11	Container with shovel	5	Drag
CM12	Tennis racket (checkered/linear mesh)	5	Material deformation

4 Results and Discussion

The sample of 37 models is representative to make a critical analysis and obtain valid data (Fig. 1). Furthermore, measurements at thirty seconds (Q), at the end of collecting all marbles (T) or counting of pieces (P) are precise measurements and with a wide range to list the positions (POS_1 and POS_2) of each model according to the metrics (M_1 and M_2 respectively). The metrics are considered consistent based on the congruence of results, there are seven models that are among the top ten positions in both metrics (70%, in green 7 of 10 models: 10, 15, 16, 20, 22, 26, 30 and 35), in the same way, 6 are among the worst positions (87% in red color 6 of 7 models: 6, 12, 13, 31, 32 and 36). Both metrics evaluate correctly using the test results, whether or not they have the novelty and usefulness dimensions. Models that have collected 50 marbles in less than 30" are in the top 6 positions with both metrics (POS_1 and POS_2). The N_1U_1 factor gives greater precision to the metric and allows better positioning of each concept. The N_2U_2 factor expected as restrictive and acting as a filter is not, it is observed that it only affects directly to the models that have not finished the test. However, the N_2U_2 factor can be configured and made more restrictive by the similarity of concepts that affects novelty or by making the cut in a shorter time. Comparing both metrics (Fig. 1), it is observed that both serve to select the best concepts and eliminate the ineffective ones, M_1 can be more useful in measurements where greater discrimination is necessary, obviously by including more factors, greater differentiation is obtained between concepts, but it implies establishing a measurable criterion for novelty and usefulness away from Likert scales that have a certain load of subjectivity. Novelty is the most critical factor in evaluating creativity due to its subjectivity. According to recent studies, even the most objective metrics [18] maintain the subjectivity of the evaluators [19]. However, the metrics presented here propose to move away from subjectivity by linking the novelty with functional indicators (discriminated by similarity or repetition) individually evaluated and consensual a posteriori. The M_2 can be more useful in measurements in which the achievement of the design conditions surpasses novelty and utility, knowing that they act as filter factors, if the conditions are not met, the concept is considered invalid.

#	Q	T	P	CONCEPT	N_1	U_1	C_n	G_n	N_2	U_2	$Q_\%$	$T_\%$	$P_\%$	M_1	M_2	POS_1	POS_2	CM
1	34	90	2	C 01	1,00	0,10	1	G02	1	1	0,56	0,36	0,97	0,020	0,195	16	17	Ma
2	29	115	3	C11	0,89	0,10	5	G02	1	1	0,39	0,25	0,64	0,006	0,062	23	23	Ma
3	32	66	4	C 05	0,95	0,10	3	G02	1	1	0,50	0,58	0,11	0,003	0,033	27	26	Ma
4	20	96	2	C 09	0,92	0,10	4	G02	1	1	0,17	0,33	0,97	0,005	0,054	24	25	Ma
5	44	68	3	C 04	0,97	1,00	2	G50	1	1	0,78	0,56	0,64	0,269	0,276	4	12	Ma
6	16	300	3	C 06	0,95	0,10	3	G02	1	0	0,08	0,06	0,64	0,000	0,000	35	34	Ma
7	30	69	2	C 06	0,95	0,02	3	G01	1	1	0,44	0,53	0,97	0,004	0,228	26	16	Mf
8	40	47	3	C10	0,92	0,10	4	G02	1	1	0,67	0,72	0,64	0,028	0,308	14	11	Ma
9	23	225	3	C 07	0,92	0,10	4	G02	1	1	0,28	0,08	0,64	0,001	0,015	29	29	Ma
10	43	36	2	C 09	0,92	0,10	4	G02	1	1	0,75	0,81	0,97	0,054	0,588	10	4	Ma
11	21	300	3	C 04	0,97	1,00	2	G50	1	0	0,19	0,06	0,64	0,007	0,000	22	34	Ma
12	10	114	3	C 08	0,92	0,10	4	G02	1	1	0,03	0,28	0,64	0,000	0,005	32	32	Ma
13	24	120	5	C12	0,89	0,20	5	G06	1	1	0,33	0,22	0,03	0,000	0,002	33	33	Ma
14	22	58	1	C 05	0,95	0,50	3	G11	1	1	0,22	0,61	1,00	0,064	0,136	9	20	Ma
15	50	13	3	C 03	1,00	1,00	1	G50	1	1	0,89	1,00	0,64	0,567	0,567	1	5	My
16	49	32	3	C11	0,89	0,20	5	G06	1	1	0,86	0,86	0,64	0,085	0,474	8	9	Ma
17	31	74	3	C11	0,89	0,10	5	G02	1	1	0,47	0,47	0,64	0,013	0,143	18	19	Ma
18	29	73	2	C11	0,89	0,10	5	G02	1	1	0,39	0,50	0,97	0,017	0,189	17	18	Ma
19	36	44	4	C11	0,89	0,20	5	G06	1	1	0,64	0,78	0,11	0,010	0,056	19	24	Mf
20	50	20	3	C12	0,89	0,50	5	G11	1	1	0,89	0,95	0,64	0,239	0,536	5	7	Mf
21	34	77	2	C10	0,92	0,20	4	G06	1	1	0,56	0,45	0,97	0,044	0,240	12	15	My
22	50	21	2	C12	0,89	0,50	5	G11	1	1	0,89	0,92	0,97	0,353	0,792	3	2	Ma
23	35	50	3	C 09	0,92	0,02	4	G01	1	1	0,61	0,67	0,64	0,005	0,260	25	14	Mf
24	40	51	3	C 08	0,92	0,10	4	G02	1	1	0,67	0,64	0,64	0,025	0,272	15	13	Ma
25	27	109	2	C 07	0,92	0,10	4	G02	1	1	0,36	0,31	0,97	0,010	0,107	20	21	Mf
26	50	22	2	C10	0,92	0,20	4	G06	1	1	0,89	0,89	0,97	0,141	0,768	6	3	Mf
27	23	86	3	C 02	1,00	0,10	1	G02	1	1	0,28	0,39	0,64	0,007	0,069	21	22	Ma
28	48	35	3	C 05	0,95	0,10	3	G02	1	1	0,83	0,83	0,64	0,042	0,444	13	10	Ma
29	44	45	2	C 09	0,92	0,10	4	G02	1	1	0,78	0,75	0,97	0,052	0,567	11	6	Ma
30	50	18	2	C12	0,89	0,50	5	G11	1	1	0,89	0,97	0,97	0,375	0,841	2	1	Mf
31	32	77	6	C 08	0,92	0,20	4	G06	1	1	0,50	0,45	0,00	0,000	0,000	36	34	My
32	11	160	3	C 07	0,92	0,10	4	G02	1	1	0,06	0,14	0,64	0,000	0,005	31	31	Ma
33	19	132	3	C 07	0,92	0,10	4	G02	1	1	0,14	0,20	0,64	0,002	0,017	28	27	Mf
34	16	140	3	C12	0,89	0,10	5	G02	1	1	0,08	0,17	0,64	0,001	0,009	30	30	Ma
35	41	47	2	C10	0,92	0,20	4	G06	1	1	0,72	0,72	0,97	0,093	0,508	7	8	Ma
36	2	300	4	C 08	0,92	0,20	4	G06	1	0	0,00	0,06	0,11	0,000	0,000	36	34	Ma
37	22	164	3	C 06	0,95	0,02	3	G01	1	1	0,22	0,11	0,64	0,000	0,016	34	28	Mf

Fig. 1. Results after the test with the conceptual models

Looking at the finish of the models, there is no difference between making an artifact, a working model, or a prototype. The three cases with a better built and detailed model have been positioned in the best, worst and average zone. There are 24 artifacts, not very robust, without details and with poor finishes whose results in M_2 are worse, the effectiveness is reduced by the construction. The functional models obtain better results in the M_2, they are oriented to the fulfillment of objectives and the effectiveness without penalizing in N and U.

Regarding the number of pieces and use of materials, despite being an objective, it has not been a critical factor. The case with one-piece design achieves a good result. 31 models (84%) have used 2 or 3 pieces, scoring good and bad results. And models with 4, 5 or 6 parts did not achieve better results by using more parts or materials.

The values of gathering marbles in groups (G_n) show slight differences. The models that tried all-in-one or one-by-one obtained similar positions, grouping in smaller groups (G02 and G06) penalized, and grouping in larger groups (G11 and G50) benefited with respect to the time variable.

Three conceptual models present differences among metrics (models 7, 11 and 14) of about ten positions, due to effectiveness in grouping marbles (U_1). Showing that metrics measuring continuous variables are more reliable than those using Likert scales.

The limitations of this work focus on: (1) even working with an objective-based metric, it is difficult to eliminate subjectivity due to the intervention of the evaluator, (2) not all designers feel comfortable moving from idea to model, this affects the results in the test by the inequality between the models (due to the materials, the way of construction, the details and the finishes) due to the decisions of the designer and his skill in the construction of models, and finally (3) the recommendation to build various models or editable models has not been followed by all the participants, so the idea or conformity has appeared.

5 Conclusions

The different ways of measuring concepts are adjusted to specific needs, metrics that can be custom-designed, but some aspects can be left unexamined. Hybrid metrics can be an option to improve the evaluation of concepts and design proposals. Hybrid are those metrics that meets a series of essential conditions and then satisfies some objectives in a measurable way with computable variables that can discriminate positions in a ranking.

Assessing a design concept, the score can be more coherent depending on the metric (M_1 or M_2) used and the objectivity of the evaluators, the dimensions of the metric are those that mark the objectivity of the results. Dimensions such as novelty or level of finish are subject to the criteria of the evaluator and lose objectivity. Dimensions such as time or number of pieces collected are objective and only it is necessary to establish the ranges with their limits and determine how the calculation is done by the observable, measurable or quantifiable variables.

Details of the construction related to the effectiveness may affect its performance, but there are models that, despite not having a good presence, achieve good results, showing an optimal design made quickly and with few resources.

References

1. Jagtap, S.: Design creativity: refined method for novelty assessment, Int. J. Des. Creat. Innov., **7**(1–2), 99–115 (2019)
2. Cropley, D.H., Kaufman, J.C.: Rating the creativity of products, in Handbook of research on creativity, Edward Elgar Publishing (2013)
3. Dean, D.L., Hender, J.M., Rodgers, T.L., Santanen, E.L.: Identifying quality, novel, and creative ideas: constructs and scales for idea evaluation. J. Assoc. Inf. Syst. **7**(10), 646–698 (2006)
4. Oman, S.K., Tumer, I.Y., Wood, K., Seepersad, C.: A comparison of creativity and innovation metrics and sample validation through in-class design projects. Res. Eng. Des. **24**(1), 65–92 (2013)
5. Ulrich, K.T., Eppinger, S.D.: Product design and development. Irwin/McGraw-Hill, Boston (2000)
6. Verhaegen, P.-A., Vandevenne, D., Peeters, J., Duflou, J.R.: Refinements to the variety metric for idea evaluation. Des. Stud. **34**(2), 243–263 (2013)
7. Sarkar, P., Chakrabarti, A.: Assessing design creativity. Des. Stud. **32**(4), 348–383 (2011)
8. Chulvi, V., Mulet, E., Chakrabarti, A., López-Mesa, B., González-Cruz, C.: Comparison of the degree of creativity in the design outcomes using different design methods. J. Eng. Des. **23**(4), 241–269 (2012)
9. Kudrowitz, B.M., Wallace, D.: Assessing the quality of ideas from prolific, early-stage product ideation. J. Eng. Des. **24**(2), 120–139 (2013)
10. Bansiya, J., Davis, C.G.: A hierarchical model for object-oriented design quality assessment, IEEE Trans. Softw. Eng. Softw. Eng. IEEE Trans. on, IIEEE Trans. Softw. Eng., **28**(1), 4–17 (2002)
11. Srivathsavai, R., Genco, N., Hölttä-Otto, K., Seepersad, C.C.: Study of existing metrics used in measurement of ideation effectiveness. In: 22nd International Conference Design Theory Methodology Specific Conference Mechanical Vibration Noise, Vol. 5, pp. 355–366 (2010)
12. López-Forniés, I., Sierra-Pérez, J., Boschmonart-Rives, J., Gabarrell, X.: Metric for measuring the effectiveness of an eco-ideation process. J. Clean. Prod. **162**, 865–874 (2017)
13. Keinonen, T., Takala, R.: Product Concept Design: A Review of the Conceptual Design of Products in Industry. Springer, London (2006)
14. Lennings, A.F., Broek, J.J., Horváth, I., Sleijffers, W., De Smit, A.: Editable physical models for conceptual design. In: Proceedings of the TMCE 2000 Symposium, p. 10 (2000)
15. Cohen, A., et al.: Prototype to product: a practical guide for getting to market. O'Reilly Media, Inc (2015)
16. Carfagni, M., Fiorineschi, L., Furferi, R., Governi, L., Rotini, F.: The role of additive technologies in the prototyping issues of design, Rapid Prototyp. J., **24**(7), 1101–1116 (2018)
17. Brown, T.: Tales of creativity and play. USA: TED TALKS I SERIOUS PLAY 2008 (2008)
18. Shah, J.J., Smith, S.M., Vargas-Hernandez, N.: Metrics for measuring ideation effectiveness, Des. Stud., **24**(2), 111–134 (2003)
19. Fiorineschi, L., Frillici, F.S., Rotini, F.: Impact of missing attributes on the novelty metric of Shah et al., Res. Eng. Des., **31**(2), 221–234 (2020)

An Integrated Approach to Ensure Requirements Traceability During the Product Development Process

Domenico Viapiana[1](✉), Giovanni Riggio[2], Loris Barbieri[3], and Fabio Bruno[1,3]

[1] 3D Research s.r.l., via P. Bucci 46C, 87036 Rende (CS), Italy
domenico.viapiana@3dresearch.it

[2] Dassault Systemes Italia s.r.l., via dell'Innovazione 3, 20126 Milano, Italy
giovanni.riggio@3ds.com

[3] Department of Mechanical, Energy and Management Engineering (DIMEG), University of Calabria, P. Bucci 46C, 87036 Rende (CS), Italy
{loris.barbieri,fabio.bruno}@unical.it

Abstract. Today's Product Lifecycle Management (PLM) platforms are a fundamental strategy for the development of complex products because of their capability to integrate processes, business systems, and information getting a complete and clear vision of all the various stages of the product development process and supporting the design and management of an extensive set of requirements.

Nevertheless, common PLM platforms lack efficient traceability of design requirements within the product development process. Rather than a stand-alone tool for requirements management, it is desirable to take a design method that places requirements in the spotlight of the design process.

For this reason, the paper presents a method, based on the Requirement - Functional - Logical - Physical (RFLP) design approach, that ensures requirements traceability during each stage of the product development process and provides decision support in a multidisciplinary collaborative digital environment. In particular, Dassault Systèmes' 3DEXPERIENCE platform has been adopted for the case study of the suspension and tyre systems of a vehicle to assess the efficacy of the proposed method and validate the integration within the collaborative business experience platform.

Keywords: Design requirements · Requirements traceability · Product lifecycle management · 3DEXPERIENCE Platform · Vehicle suspension system · Collaborative business experience platform

1 Introduction

Social, technological, and ecological changes are challenges met by many companies producing even more complex products in today's scenario. Developing complex products requires that product and process knowledge is available to several actors in the

C. Rizzi et al. (Eds.): ADM 2021, LNME, pp. 328–335, 2022.
https://doi.org/10.1007/978-3-030-91234-5_33

value network such as designers, partners, suppliers, and customers. The tendency is to use collaborative platform strategy to integrate processes, business systems, and information to manage the product development and support its lifecycle. Thanks to the business collaboration of all company roles involved in the product lifecycle, a collaborative business platform can manage all product data along the entire lifecycle, from the initial phase of product requirements specification, design, production, delivery, and evolutionary modification phase [1].

In this context, defining clear specifications is important for succeeding in developing and supporting the right and innovative product [2, 3], but is equally important to keep track of the system requirements during the development process and the upgrade of the system.

The requirements traceability refers to the ability to describe and follow the life of a requirement, in both a forward and backward direction [4]. Traceability of requirements is essential because throughout the system lifecycle numerous design changes are made to components and functions.

Nevertheless, requirements management tools, integrated within common Product Lifecycle Management (PLM) systems, lack efficient traceability of design requirements within the product development process. In fact, current traceability techniques are deficient and may cause problems [4–7]. Rather than a stand-alone tool for requirements management, it is desirable to take a design method that places requirements in the spotlight of the design process. This paper presents an innovative method, based on the RFLP design approach, to keep track of the requirements. Through the proposed method, following the requirements specification, performed in the Requirement phase of the RFLP approach, a parameter is defined for each specified requirement. A simulation scenario is created after the system modelling phase, it represents all the procedures, measures, and tests to verify that all the requirement has been met. To determine whether each requirement is met or not, the measurements made during the simulation are compared with the corresponding requirement parameter. While numerous system changes are needed during the product development process, the simulation scenario will always remain the same. For this reason, the proposed method ensures the traceability of requirements during each phase of the product development process and provides decision support in a collaborative business platform.

Dassault Systèmes incorporates the basic RFLP approach within their business experience platform 3DEXPERIENCE® [8]. For this reason, the 3DEXPERIENCE platform has been adopted for the improvement of the suspension and tyre systems of a vehicle in order to validate and assess the efficacy of the design requirements traceability feature following the method based on the RFLP approach described in this paper.

2 The Integrated Approach

2.1 Methodology

The method presented in this paper extends the RFLP approach capabilities. The RFLP approach is a specific V-model derived method [9]. The conventional V-model includes only two main structures, i.e., requirements (R) and parts (P). The lack of a common framework between the requirements and part structures represents a gap in the product concept definition [10]. By inserting two structures, i.e., the functional (F) and logical (L) structures, between requirements (R) and physical part (P) structures, it transforms the traditional V-model to the requirement, functional, logical, and physical model, namely the RFLP approach.

The innovation brought to the RFLP approach through the methodology described in this paper consists in the creation of a simulation scenario that represents all the procedures, measures, and tests to verify that the performance requirement has been met. Through the proposed method, following the requirements specification, performed in the Requirement stage of the RFLP approach, a parameter is defined for each specified requirement. This requirement parameter represents the fulfillment of the same. During the functional, logical, and physical stages of the RFLP approach all the product subsystems are designed following the engineering activities related to the method, i.e., the functional design, the logical design, and the physical design. Within the simulation scenario, the system to be tested and all the parameters associated with the requirements are embedded.

During the simulation, to determine whether each requirement is met or not, the measurements made are compared with the corresponding requirement parameter. If all the requirements are met, then it can be concluded that the system was successfully designed. If at least one requirement is not met, then the system configuration must be changed and the non-compliant subsystem must be redesigned. To identify the cause that led to the not requirement fulfilment it could be useful to apply a statistical technique such as the design of experiments (DOE). The proposed method is fully compatible with DOE techniques. A conceptual scheme of the method proposed in this paper is shown in the following figure (Fig. 1).

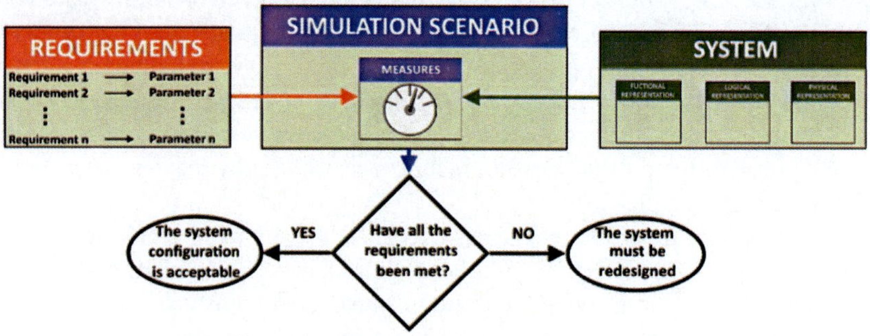

Fig. 1. RFLP-based method conceptual scheme

It is easy to understand that following the modification of the system configuration and therefore the redesign of one of the subsystems, the simulation scenario does not change and all the procedures, measures, and tests to verify that the performance requirements have been met are not altered. For this reason, the method proposed in this paper ensures requirements traceability during each stage of the product development process and provides decision support in a collaborative business platform.

2.2 3DEXPERIENCE® Applications to Support Methodology

Within the Dassault Systèmes' 3DEXPERIENCE platform it is possible to incorporate this RFLP-based approach in order to provide a collaborative engineering methodology that can capture, manage, and track all the stages of the product development process. In particular, the 3DEXPERIENCE platform provides the following applications to support the RFLP-based approach:

- ENOVIA® Traceable Requirement Management (TRM) application allows to define and structure customers' needs and requirements [11].
- CATIA® Functional & Logical Design application adopts the requirements created in the TRM application to define functional and logical architectures through a model-based approach [12].
- CATIA® Dymola® Behavior Modelling application is a physical modeling and simulation environment, for model-based design of dynamic systems, that can be then added to functional and logical architectures to perform a virtual product test execution [13].
- CATIA® 3D Modelling applications consist of dedicated solutions like CATIA Part Design, CATIA Sheet Metal Design, CATIA Generative Shape Design, that can be used for the physical representation.

3 Case Study

The above-mentioned RFLP-based approach has been adopted in a case study where performance improvement is needed for a vehicle. The vehicle that was analyzed is a formula student car, weighing 216 kg, and with a suspension subsystem of double wishbone type [14]. Through physical experimental observations carried out on the vehicle under analysis, during a cornering manoeuvre in which a lateral acceleration of 1g-force has been measured on the vehicle body, the rear inner wheel rise has been detected. In-depth inspections on this aspect it has been concluded that the cause of this phenomenon is an excessive roll motion.

To improve the cornering performance of the vehicle the proposed method has been applied to modify the suspension subsystem. As mentioned in Sect. 2.1, to keep track of requirements during the system redesign activity, the creation of a simulation scenario is needed. For this reason, a simulation scenario was built in which the vehicle is subjected to a variable lateral acceleration between the values of 1g-force and -1g-force. This simulation scenario reproduces the physical experiment done on the vehicle and represents the digital test to verify that the performance requirements have been met. Moreover, the redesign engineering activity has been automated by applying the DOE technique to the car suspension subsystem, particularly the spatial coordinates of the eight suspension attachment points were made to vary on two levels. Finally, during the simulation, many measurements are made on the vehicle subsystems to determine whether the requirements are met.

3.1 Requirement Specification

The first step of the RFLP-based approach is to translate needs into a set of all requirements to ensure traceability during the product development process. The lateral vehicle performance was most carefully examined as the problem arises during a cornering manoeuvre. Thus, as a car goes around a corner, the outside wheels have more load than the inside ones. Lateral weight transfer produces undesirable effects, it is independent of suspension configuration, but it produces different normal loads on the inner and outer wheels that generate a rolling motion [15]. To examine the roll behaviour was considered the roll centre position. It is the point at which forces in the suspension links are reacted by the body and if it is situated on the centre of gravity there will be no roll at all. Initially, this sounds desirable, but in fact, it is not. If the roll centre is too high off the ground level the jacking effect will occur and the body will lift. So, the roll centre position is generally placed between the centre of gravity and the ground, the exact point is a trade-off between the jacking forces and roll motion [15].

Following the previous considerations, it was possible to specify three requirements through the TRM application on the 3DEXPERIENCE Platform. With the theories above mentioned, the suspension geometry must be changed to ensure that the roll centres of the front and rear suspensions are closer to the centre of gravity than the current configuration. The new peak of the roll angle should be lower than the current configuration (*Roll*). A requirement deriving from the above is the Camber Thrust. An effect of the roll motion is to alter the angles of the wheels. Camber is the angle of the wheel relative to vertical line to the road, as viewed from the front or the rear of the car. If the wheel leans in toward the chassis, it is called negative camber and if it leans away from the car, it is called positive camber. The cornering force that a tyre can develop is highly dependent on camber angle, particularly in a roll movement of the car a little negative camber angle on the outer tyre is desirable to increase the car road-holding, so the new configuration of the suspension must guarantee a negative camber angle on the external wheel to benefit from the effect of the camber (*CamberThrust*). Another requirement that ensures vehicle road-holding under 1g-force lateral acceleration is to check that all four wheels produce enough lateral force. To check if this force is acceptable, the value of the lateral slip coefficient is taken under control. It must be ever less than the maximum lateral slip coefficient and higher than the minimum lateral slip coefficient (*TyresGrip*).

To establish whether a requirement is achieved or not, a parameter is associated with it. It is a boolean parameter that establishes whether that requirement is fulfilled. The value of the boolean parameter will be changed running the simulation scenario.

3.2 System Model

Following the RFLP-based approach, once all the requirements were set up, the logical and functional architecture representing the global view of the system, in which the requirement specification is embedded, was then created.

The vehicle complex system consists of heterogeneous subsystems. On the first hand, the subsystems behave completely different and are represented by distinct mathematical models, but, on the other hand, they cooperate to define vehicle performance.

As depicted in the following figure (Fig. 2), through the CATIA Functional & Logical Design application, the elementary functions and the characteristic subsystems of the product were identified.

The CATIA Functional & Logical Design application allows the system developer to monitor the relations existing between the vehicle subsystems, but the Functional and Logical architecture alone does not permit representing the system physical behavior. Therefore, it is needed a physical representation of the components making up the system. This was done by embedding Dymola behavior models inside the vehicle schematic representation.

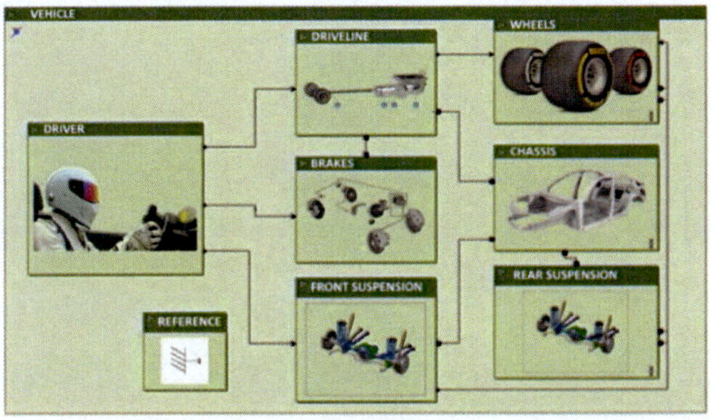

Fig. 2. Vehicle schematic representation

3.3 Simulation Scenario

A simulation scenario was built in which the vehicle is subjected to a variable lateral acceleration between the values of 1g-force and -1g-force.

To keep track of the requirements, it was necessary to measure some variables. The maximum value of the roll angle, the variation of the camber angle during compression and elongation of the suspension, and the maximum and minimum slip coefficient calculated with the Pacejka model [16] were recorded.

The new configurations that the designer is searching for must satisfy all requirements. After a simulation, if the requirement is not satisfied, the boolean parameter value associated with it becomes false and the designer can be concluded that the new configuration is not an acceptable solution. The suspension geometry parameters were changed through the DOE technique. An experiment matrix was created in which the spatial coordinates of the eight suspension attachment points were made to vary on two levels in order to create 256 candidate configurations.

3.4 Results

By the illustrated method based on the RFLP approach, and using the Dassault Systèmes 3DEXPERIENCE platform, 6 configurations were determined to accomplish the requirements. The following figures (Fig. 3a-d) show the results obtained with one of the suspension subsystem configurations deemed acceptable and compared with the start reference configuration.

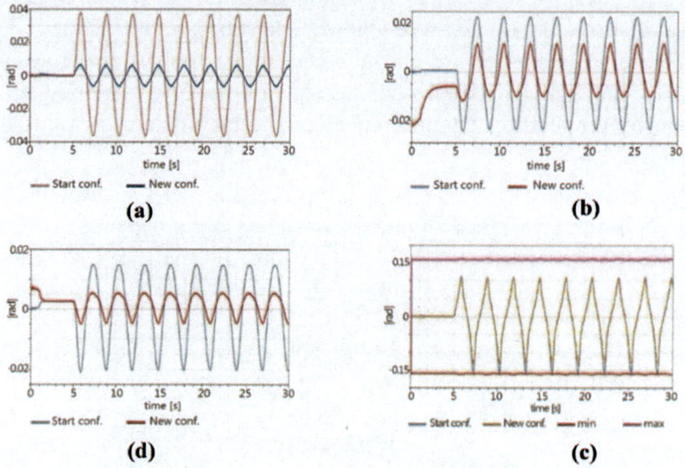

Fig. 3. Body roll angle (a); front (b) and rear (c) wheel camber angles; and lateral slip angle (d)

Figures 3b and 3c show the camber angle for front and rear wheels taking negative value when the corresponding suspension mechanism is in the bump state, this measure was compared with the *CamberThrust* requirement. The new suspension configuration ensures that the tyre never reaches the lateral slip angle limits (Fig. 3d), this measure refers to the *TyresGrip* requirement. The body roll motion is drastically reduced (Fig. 3a), as explained in the *Roll* requirement. By ensuring requirements traceability through the RFLP-based method, improvements were made to the vehicle system.

4 Conclusions

This paper has shown how requirements can be tracked during each stage of the product development process to provide decision support for engineering business process. In particular, Dassault Systèmes' 3DEXPERIENCE platform was adopted for the case study of the suspension and tyre systems of a vehicle to assess the efficacy of design requirements traceability and the integration of the presented method within the multidisciplinary collaborative digital environment. The RFLP-based method was validated by performing design changes to the vehicle subsystems. Following these modifications, the requirement specification has not lost consistency and through the requirements traceability feature, it has been possible to improve the vehicle system. For the above reasons it was possible to confirm that the proposed method ensures the design requirements traceability, and it is possible to conclude that the 3DEXPERIENCE platform provides effective and suitable applications to support this methodology.

References

1. Stark, J.: Product Lifecycle Management. In: Product Lifecycle Management. DE, pp. 1–35. Springer, Cham (2016). https://doi.org/10.1007/978-3-319-24436-5_1
2. Violante, M.G., Vezzetti, E.: A methodology for supporting requirement management tools (RMt) design in the PLM scenario: an user-based strategy. Comput. Ind. **65**, 1065–1075 (2014)
3. Violante, M.G., Vezzetti, E., Alemanni, M.: An integrated approach to support the Requirement Management (RM) tool customization for a collaborative scenario. Int. J. Interact. Des. Manuf. (IJIDeM) **11**(2), 191–204 (2015). https://doi.org/10.1007/s12008-015-0266-3
4. Gotel, O.C.Z., Finkelstein, A.C.W.: An analysis of the requirements traceability problem. In: Proceedings of 1st International Conference on Requirements Engineering, pp. 94–101. IEEE. Colorado (1994)
5. Kirkman, D.P.: Requirement decomposition and traceability. Requirements Eng. **3**, 107–114 (1998)
6. Ramesh, B., Powers, T., Stubbs, C., Edwards, M.: Implementing requirements traceability: a case study. In: Proceedings of 1995 IEEE International Symposium on Requirements Engineering, pp. 89–95. IEEE. Newyork, UK (1995)
7. Bashir, M.F., Qadir, M.A.: Traceability techniques: a critical study. In: 2006 IEEE International Multitopic Conference, pp. 265–268. IEEE. Islamabad, Pakistan (2006)
8. 3DEXPERIENCE platform, https://www.3ds.com/3dexperience, last accessed 2021/03/21.
9. Kleiner, S., Kramer, C.: Model based design with systems engineering based on RFLP Using V6. In: Abramovici, M., Stark, R. (eds.) Smart Product Engineering. Lecture Notes in Production Engineering. Springer, Berlin, Heidelberg (2013). https://doi.org/10.1007/978-3-642-30817-8_10
10. Baughey, K.: Functional and logical structures: a systems engineering approach. In: SAE 2011 World Congress, SAE Technical Paper 2011–01–0517 (2011)
11. ENOVIA Traceable Requirements Management. https://www.3ds.com/products-services/enovia/products/3dexperience/. Accessed 21 Mar 2021
12. CATIA Functional & Logical Design. https://www.3ds.com/products-services/catia/. Accessed 21 Mar 2021
13. CATIA Dymola Behavior Modelling. https://www.3ds.com/products-services/catia/products/dymola/. Accessed 21 Mar 2021
14. UNICAL Reparto Corse. https://www.unicalrepartocorse.it/?page_id=14716. Accessed 16 Jan 2018
15. Balkwill, J.: Performance Vehicle Dynamics Engineering and Applications, 1st edn. Butterworth-Heinemann, Oxford (2018)
16. Pacejka, H.B., Bakker, E.: The magic formula tyre model. Veh. Syst. Dyn. **21**(sup001), 1–18 (1992)

Design of a Bioinspired Multi-fingered Soft Pneumatic Gripper with Embedded Suckers

Stanislao Grazioso(✉), Teodorico Caporaso, Giuseppe Di Gironimo, and Antonio Lanzotti

Fraunhofer Joint Lab IDEAS, Dipartimento di Ingegneria Industriale Università degli Studi di Napoli Federico II, P.le Tecchio 80, 80125 Napoli, Italy
stanislao.grazioso@unina.it

Abstract. Applications as robotic harvesting or pick and place in the agrifood domain require robotic grippers able to gently manipulate delicate products, while guaranteeing high gripping power and adhesion forces on smooth surfaces. Existing soft grippers are mainly based on pneumatic bending actuators which can guarantee a gentle manipulation, but they suffer from low gripping power and possibility of slip of the manipulated object. This paper describes a novel design concept of soft robotic pneumatic gripper with embedded suckers. The concept consists of four soft fingers, each one comprising an elastomeric structure with two separate air paths, one for pressurizing the finger for generating bending motion, one for vacuum–based adhesion to the object's surface via suction pads distributed along the surface of the finger. In this work we highlight the concept design of the mechanical system and the pneumatic control unit.

Keywords: Bioinspired design · Concept design · Soft grippers · Soft robotics

1 Introduction

Traditional robotic grippers consist of a discrete set of rigid links connected through rigid joints. Flexibility and dexterity in grasping and manipulation using rigid grippers can be achieved only through complex mechanical design and control strategies.

A different design approach involves the use of compliant elements embedded within their mechanical structure [1]. This results in simpler design concepts, where the control complexity is reduced by the "active" role of compliance in achieving effective grasping and manipulation. Robotic grippers made of compliant elements are called soft robotic grippers [2].

One of the most adopted approach for the design and development of soft grippers involves the use of fluid–driven actuators [3], namely elastomeric chambers which shape in a bending motion when inflated thanks to the adoption of embedded reinforcing layers. Many design architectures of robotic grippers make

C. Rizzi et al. (Eds.): ADM 2021, LNME, pp. 336–341, 2022.
https://doi.org/10.1007/978-3-030-91234-5_34

use of fluid–driven actuators, organized as multi–fingered systems or single–fingered, tentacles–like systems. The first typology allows soft grasping and manipulation of the object through multiple contact points. The second typology allows grasping and manipulation by enrolling the object through a bending or even helix motion; this is referred to as whole–arm manipulation. Gripping by pneumatic actuation of inflatable soft structures has many advantages as simplicity of realization, robustness and low cost. Despite fluid–driven actuators have been used for successful grasping of a large variety of objects, they suffer from low gripping power and low adhesion forces on smooth surfaces. These limitations can represent a problem for enabling applications, as in the agricultural sector, where a soft gripper can be used simultaneously for harvesting or pick and place applications. While most of design solutions, especially the multi–fingered concepts, have been successfully used for pick and place of fruits, no concepts have been provided for realizing, with the same systems, pick and place and harvesting of fruits [4]. This could be due to the fact that during the harvesting process, the gripper should guarantees, besides safe interaction with the fruit, also higher grasping power and adhesion to the surface. Furthermore, robotic grippers in the agrifood domain should also avoid the problem of potential slip of the product during the handling process.

One possible solution to this problem can by endowing the soft actuators with a biologically inspired mechanism of adhesion. Bioinspiration has always been present in soft robotics, aiming at transferring the working principles of biological soft–bodied living organisms to artifical machines [5]. Examples of living organisms which are source of inspiration for controllable adhesion mechanisms are the gecko and the octopus. Geckos are able to climb on vertical surfaces thanks to special microfibers, located on their foot, able to attract the object surface by the van der Waals force, resulting in the generation of shear forces [2]. The adhesion in robotic grippers exploiting this concept is guaranteed by the pressure of gecko–inspired microfibers in the direction normal to the object's surface [6]. Octopus is probably the most well–known biological source of inspiration for soft robotics [7]. Such incredible animals have impressive manipulation capabilities; as a matter of fact, their soft bodies produce complex motions which allow to easily conform to different objects, while the suckers–like structures distributed along their arms allow to attach to multiple surfaces [8]. This organism has been the source of inspiration of two examples of octopus–inspired grippers. The first example is the TentacleGripper by FESTO (https://www.festo.com/), a pneumatically actuated tentacle gripper with distributed suction cups that can produce 2D motion (inward bending), analysed in detail in recent research papers [9] and endowed with distributed sensors [10]. The second example is a tentacle gripper with distributed suction cups that can produce 3D motion, useful for grasping and manipulation of complex–shaped objects in confined environments as industrial pipes [11]. According to the literature, the combination of fluid–driven actuators and suckers for the development of soft grippers is reported only for tentacle–like structures.

In this paper we propose a novel architecture for bioinspired soft grippers, based on multi–fingered fluid–driven actuators, each one with bending motion capabilities endowed with distributed suction cups along their length. The integration of bending and suction in multi–fingered soft grippers is expected to provide improved gripping power and possibility to stick to multiple surfaces. This combination can allow to safely grasp and manipulate delicate objects which require high adhesion and gripping power. To the best of our knowledge, the synergistic combination of bending motion and vacuum–based adhesion via suction cups is proposed here for the first time in a multi–fingered soft gripper architecture. Indeed, so far the combined usage of pneumatic shaping and vacuum–based adhesion via suction cups is only reported for tentacle–like structures with a single finger [9,11]. As an application example, the implementation of this concept in the agrifood domain, can allow for combined harvesting and pick and place, without potential slip, of spherical–shaped fruits with smooth surface, as apples. As a matter of fact, while commercial solutions already exist for pick and place of fruits and food products (as the pneumatic grippers from Soft Robotics Inc. (https://www.softroboticsinc.com/), no commercial solutions exist for harvesting robot. In the rest of this paper, we present the concept of the gripper (Sect. 2) as well as the design description of the mechanical system (Sect. 2) and of the pneumatic control unit (Sect. 3). Conclusions are reported in Sect. 5.

2 Concept

The concept of the multi–fingered soft pneumatic gripper with embedded suckers in illustrated in Fig. 1. Here, we can appreciate the combined function of bending and suction in a simple example, i.e. safe grasping of an apple. Bending of the soft actuators, as generated by inflating compressed air within the pneumatic chambers, allows the gripper to conform to the apple's shape in a safe manner. Suction cups made of compliant materials are used indeed for enhanced adhesion to the apple's (usually smooth) surface, when vacuum is applied. The resulting adhesion forces, normal to the surface of the apple, might enable higher gripping power (useful, for example, in harvesting applications) and enhanced handling of the fruit (useful, for example, in pick and place with reduced possibility to slip).

3 The Mechanical System

The CAD model of the soft gripper is illustrated in Fig. 2. The system mainly comprises four soft fingers with embedded suckers and one gripper base. Each finger is composed by an elastomeric structure with two separate air paths, one for inflating the chambers on the back side to allow the bending motion, one for vacuumizing the channels on the front side for vacuum–based adhesion via the suction cups. In the isometric view of the finger (bottom right of Fig. 2), we can see that two air channels are dedicated to vacuum (one for each line of suction cups), while one channel is dedicated to compressed air for bending motion.

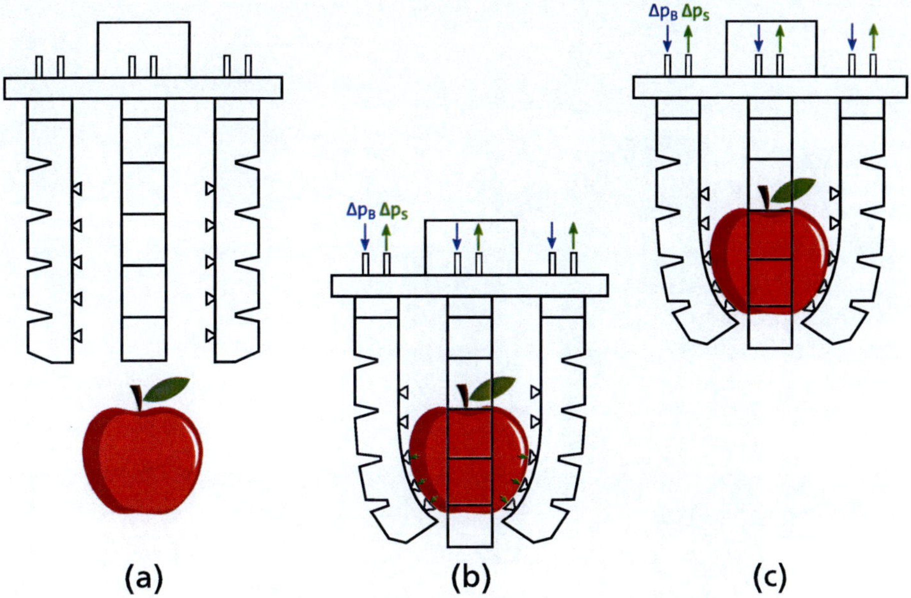

Fig. 1. Concept of multi–fingered soft pneumatic gripper with embedded suckers. Positive pressure generates bending of the actuators, allowing for shape adaptation to the object (Δp_B). Negative pressure generates suction, and thus better adhesion to the surface (Δp_S). (a) = soft gripper approaching the apple; (b) = soft gripper grasping the apple via bending motion and vacuum–based adhesion; (c) = soft gripper firmly handling the apple during its transportation.

To allow for grasping of objects with different sizes, we design a gripper base with three series of positioning holes; in this way the gripper has three opening distances. Furthermore, the gripper base has an internal opening for air tubes passage and it is equipped with a flange for coupling with robotic arms.

4 The Pneumatic Control Unit

The architecture of the pneumatic control unit is shown in Fig. 3. It comprises the control paths for vacuum and compressed air. For each finger, the architecture foresees: pressure regulator, solenoid valve, pressure sensor. It allows for independent control of the bending angles of the various fingers. For the vacuum path, we just consider one vacuum regulator and one solenoid valve for the simultaneous activation and deactivation of the suction cups. Indeed, in our concept, the suction cups are all activated simultaneously after the object to be grasped is enrolled by the bending motions of the fingers. The vacuum can be generated using a vacuum ejector downstream of the compressor or using a vacuum pump.

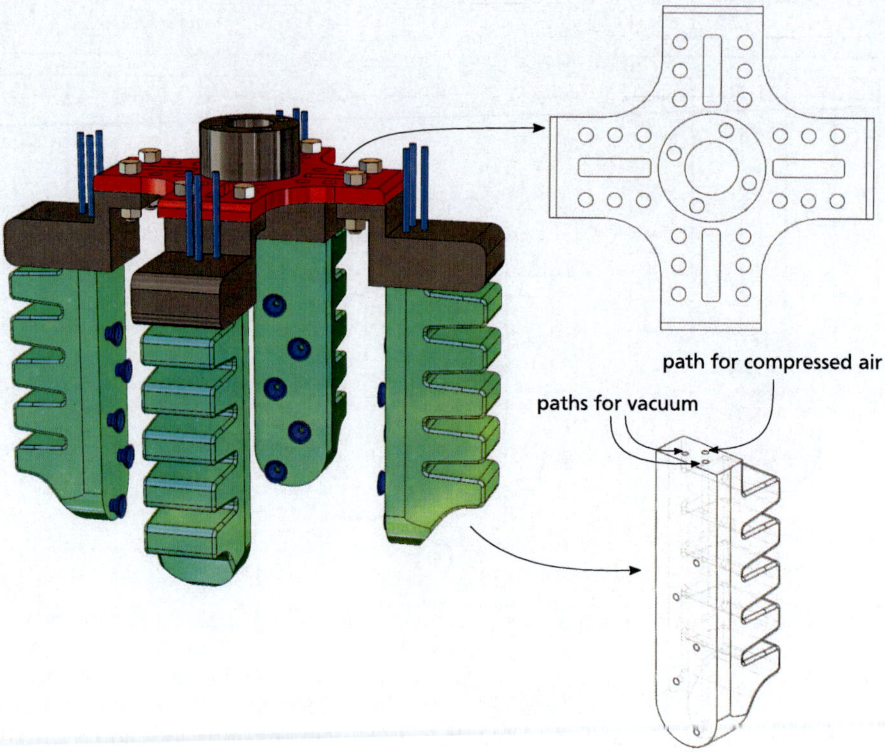

Fig. 2. CAD model of the multi–fingered soft pneumatic gripper with embedded suckers.

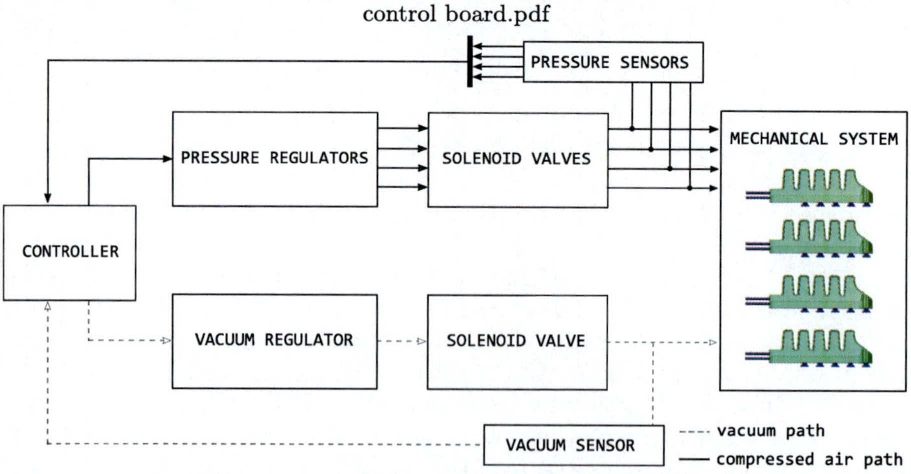

Fig. 3. Architecture of the pneumatic control unit.

5 Conclusions

A novel design concept of soft robotic gripper comprising a multi–fingered architecture with suction capabilities was presented in this paper. The design concept might improve the gripping power and reduce the objects' slip during trasport of existing solutions. These features are suitable for agrifood applications; more specifically, in the harvesting and pick and place of spherical–shaped fruits with smooth surface, as apples. However, the concept can also be valid for applications in different domains, which require soft grasping without slip and high gripping power. This is just a preliminary work dealing with the design of the gripper. Future works will be dedicated to fabrication and testing of the overall system.

Acknowledgments. This work was supported by the BIOIC project (Bioinspired soft robotic systems for cognitive production) https://www.bioic.unina.it/.

References

1. Della Santina, C., Catalano, M.G., Bicchi, A.: Soft robots. Ang, M., Khatib, O., Siciliano, B. (Eds.) Encyclopedia of Robotics. Springer (2020). https://doi.org/10.1007/978-3-642-41610-1
2. Shintake, J., Cacucciolo, V., Floreano, D., Shea, H.: Soft robotic grippers. Adv. Mater. **30**(29), 1707035 (2018)
3. Polygerinos, P., et al.: Soft robotics: review of fluid-driven intrinsically soft devices; manufacturing, sensing, control, and applications in human-robot interaction. Adv. Eng. Mater. **19**(12), 1700016 (2017)
4. Zhang, B., Xie, Y., Zhou, J., Wang, K., Zhang, Z.: State-of-the-art robotic grippers, grasping and control strategies, as well as their applications in agricultural robots: a review. Comput. Electron. Agric. **177**, 105694 (2020)
5. Kim, S., Laschi, C., Trimmer, B.: Soft robotics: a bioinspired evolution in robotics. Trends Biotechnol. **31**(5), 287–294 (2013)
6. Hawkes, E.W., Christensen, D.L., Han, A.K., Jiang, H., Cutkosky, M.R.: Grasping without squeezing: Shear adhesion gripper with fibrillar thin film. In: 2015 IEEE International Conference on Robotics and Automation (ICRA), pp. 2305–2312. IEEE (2015)
7. Margheri, L., Laschi, C., Mazzolai, B.: Soft robotic arm inspired by the octopus: I. from biological functions to artificial requirements. Bioinspiration Biomimetics **7**(2), 025004 (2012)
8. Tramacere, F., Beccai, L., Kuba, M., Gozzi, A., Bifone, A., Mazzolai, B.: The morphology and adhesion mechanism of octopus vulgaris suckers. PLoS One **8**(6), e65074 (2013)
9. Xie, Z., et al.: Octopus arm-inspired tapered soft actuators with suckers for improved grasping. Soft Robot. **7**(5), 639–648 (2020)
10. Xie, Z., Yuan, F., Liu, Z., Sun, Z., Knubben, E.M., Wen, L.: A proprioceptive soft tentacle gripper based on crosswise stretchable sensors. IEEE/ASME Trans. Mechatron. **25**(4), 1841–1850 (2020)
11. Mazzolai, B., et al.: Octopus-inspired soft arm with suction cups for enhanced grasping tasks in confined environments. Adv. Intell. Syst. **1**(6), 1900041 (2019)

Investigation of Warpage and Tolerances in Injection Moulding Components Based on Simulation and Experimental Validation

Giovanni Garlaschelli[1], Francesco Gherardini[1](✉), Simone Malagola[2], and Francesco Leali[1]

[1] Department of Engineering "Enzo Ferrari", University of Modena and Reggio E., Modena, Italy
francesco.gherardini@unimore.it
[2] Cattini srl, San Martino in Rio, RE, Italy

Abstract. This paper presents a systematic approach to model and simulate the influence that the variation of process parameters has on the final quality of an injection moulded component. In the first phase, we define a multi-steps procedure to develop a reliable digital model of the injection moulding process by the fine-tuning of the part cavity and the mould elements (e.g. detail simplification, discretization and mesh density, elements modelling, etc.) using real data as validation. In the second phase, we investigate the correlation between selected process parameters and the final tolerances of the moulded component, based on a Design of Experiments. As a case study, we selected the body of a mass airflow sensor for an automotive high performance engines, made of polybutylene terephthalate reinforced with glass-fibre, which presents roundness issues on functional features. The effects of the injection velocity and packing pressure on the deformation of the component are investigated, identifying the best combination of their values that leads to compliance with the roundness tolerances on its functional features. The injection moulding Computer Aided Engineering (CAE) software Moldex3D (CoreTech) is used to run the simulations, and the results are finally validated by comparing the experimental data obtained from the injection moulding machine that produces the component.

Keywords: Simulation-based design approach · Design of Experiments · Injection moulding · Warpage · Tolerances

1 Introduction

The injection moulding (IM) manufacturing process of a polymeric component leads to a final shape that differs from the ideal one due to a large number of parameters that rule the manufacturing process itself. Many works deal with the study and optimization of IM parameters [1–3] to comply with the dimensional and geometric tolerances on functional elements of injection moulded components [4–7]. On the one hand, IM simulations [8, 9] are used in industries mainly for preliminary validations and first attempts in filling

C. Rizzi et al. (Eds.): ADM 2021, LNME, pp. 342–353, 2022.
https://doi.org/10.1007/978-3-030-91234-5_35

the cavity but, on the other hand, the most recent trends show that researchers are moving towards smart manufacturing [10, 11] and digital twin approaches [12, 13]. Among the models required in the digital twin approach, the main one is about the simulation of the injection process, considering all the fluid-dynamic and thermal effects. Computer Aided Engineering (CAE) applications for IM are widespread and work correctly if the starting model of the mould system is reliable.

In this work, we want to carry out an in-depth study of the correlation between the simulated mould model and the real mould used within the IM machine, thanks to a close alignment between the simulated data and the real ones. In the modelling phase, we investigate the effects of the number of mesh elements, of the modelling of the elements of the cooling system as well as its mesh elements, up to convergence on target values [14–16]. Once the model has been calibrated and validated with experimental tests, this can be used to correlate the process parameters with the deformations (warpage) of the part, and therefore to further define new values that reduces the deformation. In order to tackle this task, a Design of Experiments (DOE) [17–19] is applied. The study does not consider the influence of the characteristics of the material (which is kept unchanged with respect to current manufacturing process and considered as an input to the analysis) as well as the design choices that led to the creation of the mould. The only fields of action available are the process parameters that could be easily modified in production.

The paper is structured as follows: Sect. 2 presents the method, Sect. 3 presents the case study, Sect. 4 highlights the results and Sect. 5 closes the paper.

2 Method

The design approach to model, simulate and study the IM process consists of two main phases. In the first (i.e. modelling phase), the system elements are modelled, simplified and discretized as follows:

1) Defeaturing of the part: Before the mesh creation, the features that may have a negative influence on the quality of the mesh, or are not relevant for the goal, are eliminated from the model.
2) Modelling and discretization of the mould elements: Before creating the whole mesh of the mould system, the mould elements needs to be simplified and preliminary evaluation of the mesh density of these elements are required, in particular, of the entire mould and the cooling system.
3) Definition of mesh size and density: It is a preliminary phase in which some considerations on the discretization of the domain were made. However, the mesh convergence eliminates the influence of discretization from the deformation results, allowing an objective assessment of the correlation with the process parameters subjected to variation. These assessments have been made considering the available computing power with respect to the mesh size of the part.

All these steps are validated by comparing the simulations with IM experiments.

In the second phase (systematic simulation), the developed model is finally used as input to CAE simulations, to correlate the IM parameters to be investigated according to a factorial plan (DOE). The parameters as well as their values involved in the DOE are previously selected thanks to experience or a screening phase. In this way a full factorial design involving a low number of factors, each with a high number of levels (i.e. values) can be performed. This is important to simulate and analyse the correlation between inputs (i.e. factors) and outputs (i.e. responses).

3 Case Study

The component selected as a case study is the body of a mass airflow sensor (mass flow meter) for an automotive high-performance engine. The complexity of the geometry (Fig. 1) is mainly due to the presence of the housing of the sensor for detecting the intake airflow. The component is made of polybutylene terephthalate (PBT) reinforced with 30% glass fibre. PBT has excellent mechanical (i.e. stiffness and toughness), thermal, tribological, and aesthetic characteristics. It also absorbs a very low moisture content, so the moulded part present a high dimensional stability. The runner has 3 equally spaced gates (Fig. 1, left), and a cooling system with a central baffle (Fig. 1, right). The geometry of the part and the mould are considered design inputs and they cannot be changed.

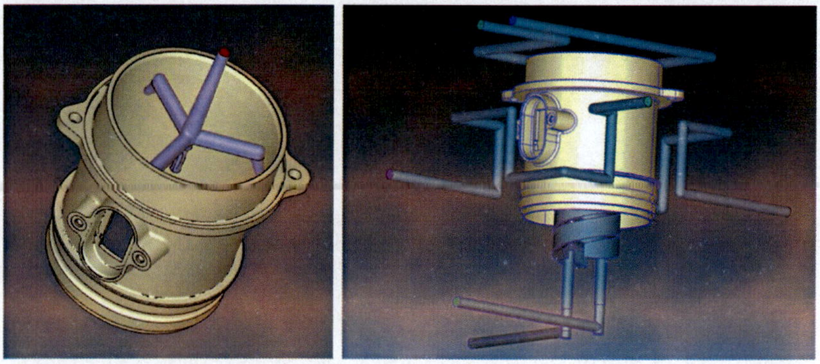

Fig. 1. Runner geometry and location (left); 3D model of the cooling system in Moldex3D (right).

In the current manufacturing process, the design specifications are generally achieved, satisfying most of the dimensional and geometric requirements established in the design phase but some latent defects occur due to local deformations that causes roundness (circularity) issues in particular on the side opposite to the gates location. On the other hand, we selected this case study thanks to the high level of knowledge of the component and the manufacturing process, and the large amount of data available. This allows for accurate evaluations on the variation of the process parameters and an experimental validation of the simulation results.

4 Results

4.1 Modelling Phase

In the first phase, we modelled the part according to the part defeaturing, simplifying the mould elements, setting the mesh size and density, etc. To achieve the mesh convergence, we had to run a reference simulation many times. To this aim, we set the values of the IM parameters as in the real process. In the following text, we refer to them as "initial parameters" as they represent the starting point of the DOE investigation. The values of parameters are listed in Table 1. Among them, we focused on two parameters that will be further investigated in the second phase: the injection velocity, set constantly to 80 mm/s in the first part of the injection cycle (up to 62% of the injection stroke), and then reduced to 10 mm/s in the second part. The packing pressure is set constantly to the value of 85 MPa for 15 s.

Table 1. Initial process parameters.

Parameter	Value	Parameter	Value
Melt temperature (°C)	265	Max. Pack. pressure (MPa)	85
Mould temperature (°C)	85	Cooling time (s)	30
Max. Inj. pressure (MPa)	175	Mould-open time (s)	15
Injection volume (cm^3)	153.562	Eject temperature (°C)	170
Packing time (s)	15	Cycle time (s)	60.59

Defeaturing of the Part (1). Before creating the mesh, the features of the component are analysed to remove unnecessary details. For this purpose, the text and symbols were eliminated, since these are in relief on the component surface and present reduced thickness compared to the one of the main body. Then, we analysed the dress-up features of the component, such as chamfers and fillets, which cannot be simplified for their importance in the manufacturing process. We maintain all fillets above 0.2 mm, which positively affect the injection of the molten polymer and avoid stress localization.

Modelling and Discretization of the Mould Elements (2). The mould elements such as runners, cooling channels, inserts, etc., may require to be simplified in their geometry and then discretized by an adequate number of mesh elements. About the mould base, as there was no appreciable difference in reproducing all its external dimensions and shape, it was modelled with a simple prism, which returned fully comparable results. Similarly, we simplified the two mould inserts, which are placed in a not-critical area (i.e. the sensor housing), creating holes corresponding to their external diameter.

Besides, the mesh size of the cooling system is investigated. The analysis of the cooling efficiency (i.e. the parameter that establishes the percentage of thermal energy exchanged by each component of the cooling system) of the various parts of the cooling system shows that the central baffle has the most important role in cooling the molten material inside of the cavity (Fig. 2, left). The discretization of the boundary layer of the cooling channels may have a significant influence on the cooling performance. A recommended number of elements (at least 3 layers of prismatic elements) is suggested by Moldex3D software. Therefore, four different configurations of mesh elements in the cooling system were investigated to check whether there was a real advantage in deviating from the number of elements recommended by the software, as well as in increasing the number of mesh elements in the baffle (4 mesh sizes). A simulation was carried out with each of these mesh models, using the initial values of the process parameters (Table 1). The results of the four simulations show how increasing the number of elements of the boundary layer (from 2 to 4 elements, with meshes 2 and 3 kept at the recommended value equal to 3) does not lead to a consistent difference in the responses, i.e. the number of Reynolds along the cooling system and the cooling heat flux. The maximum value of the number of Reynolds assumes the following values: 8.825 e^{+04}; 8.806 e^{+04}; 9.000 e^{+04} in the changes of direction of the channels. The minimum value of the number of Reynolds assumes the following values: 3.485 e^{-09}; 2.174 e^{-07}; 2.090 e^{-11}; 1.586 e^{-11} in some critical areas of the baffles, i.e. those in which stagnation phenomena of the cooling water occur (in the upper part and in correspondence with the inlet and outlet areas of the baffle). In these areas, the Reynolds number also varies by orders of magnitude as the mesh varies. However, the differences, in terms of exchanged heat flow, between the various simulations are very limited and are concentrated in the upper area of the baffle. In these areas, the low velocity of the water (used as refrigerant fluid) causes the refrigerant to reach a high temperature, but have a little influence for the study of cooling effect. In all the other regions, the results of the four simulations are absolutely comparable in terms of heat exchanged. As there was no advantages in refining the mesh size of the baffle, as well as there was no advantage in using a number of boundary elements higher than the recommended value (3), we set the lowest number of elements for the cooling system mesh.

Furthermore, we checked the impact on the results of using a simplified model of the baffle instead of its real geometry. Moldex3D offers the possibility of modelling complex parts of the cooling system in a simplified shape, through the management of a few parameters. This simplified model (Fig. 2, right) returns a reliable result regarding the deformation of the part, as demonstrated by the experimental validation, but presents a different percentage of thermal energy subtracted from the cavity. The difference in the total percentage of energy removed from the baffle with respect to the totality of the cooling systems is 31% (real geometry) vs 25% (simplified geometry). For this reason, the real geometry of the baffle was used for the second phase of simulation.

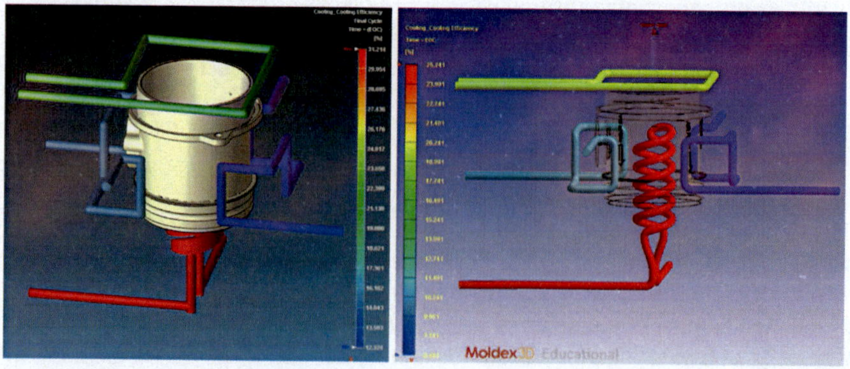

Fig. 2. Cooling efficiency of the cooling system: the real baffle (left); the simplified one (right).

Mesh Size and Density (3). We finally generated a surface mesh for the cavity (part), the runners and the cooling system. In particular, the small-sized fillets were discretized with at least two surface elements, to ensure that they kept characteristics closer to those of a fillet than to those of a chamfer. For the preliminary phase of the study, aimed at identifying the most suitable mesh size and density to achieve the goal, four different meshes were created, always doubling (indicatively) the number of total volume elements. Mesh 1: 829,140 elements; Mesh 2: 1,590,440 elements; Mesh 3: 4,049,484 elements; Mesh 4: 9,552,816 elements. The comparison of the mesh density required to identify a parameter to assess mesh convergence. We opted for four values of the roundness tolerance of as many circumferences on both the part sides (Fig. 3). Using different mesh sizes, a variation in the temperature distributions occurs along the circumferences, which is in the order of a few degrees, which can significantly affect the part deformation. However, the charts in Fig. 4 show a growing trend of the roundness value in correspondence with each circumference. This trend settles down, generating insignificant differences between the results obtained with meshes 3 and 4. As the tolerance values are in line with those found by the quality control carried out on the moulded part, mesh 3 was chosen as discretization for the implementation of the next phase.

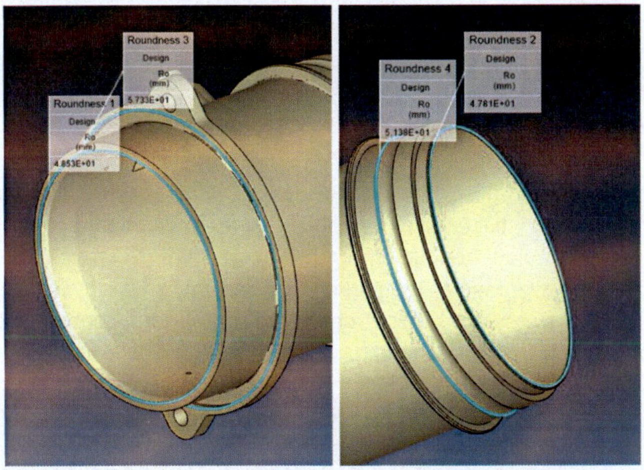

Fig. 3. Roundness tolerances for the comparison.

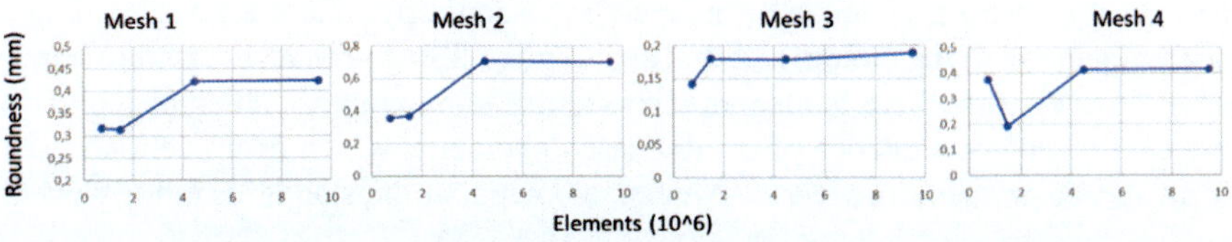

Fig. 4. Roundness trend in the 4 edges corresponding to different mesh sizes.

4.2 Simulation Phase

In the second phase, we aim to simulate the behaviour of the injection moulded part as functions of the process parameters. In particular, a set of 16 (2^4) simulations, considering the injection velocity and the packing pressure parameters as inputs (Table 2), were carried out following a full factorial design. The outputs consist in the correlation between the two process parameters and the part deformation and shrinkage. In particular, we investigated a range of the packing pressure values that includes the value (85 MPa) normally used for the part production, by increasing and reducing its value (Table 2). This choice was made to evaluate the effectiveness with which the volumetric shrinkage of the material can be influenced, by varying the contribution of the compensating material during the packing phase, to ensure a lower distortion of the final geometry compared to that of the cavity. In fact, the shrinkage is one of the main causes of the part deformation in the side opposite the flange.

Conversely, as regards the injection velocity, it could be increased from the initial value of 80 mm/s, since the sprue pressure graph shows that the injection pressure limit is never reached during the injection phase.

Table 2. Factors and levels of the full factorial design (DOE).

Factors	Level 1	Level 2	Level 3	Level 4
Injection velocity (mm/s)	100	120	140	160
Packing pressure (MPa)	42.5	63.75	106.25	127.5

A four-level analysis was then carried out for the two parameters, using the uniform volumetric shrinkage (calculated according to the thermal effects at each point of the discretization) and the warpage as the parameters (responses) for evaluating the quality of the results. The software then carried out sixteen simulations, representative of all the possible combinations of the selected parameters.

The curves in Fig. 5 link each parameter with the volumetric shrinkage. There is no shrinkage variation due to the injection velocity variation, while the packing pressure has a much greater effect on it, showing an important reduction of volumetric shrinkage at the maximum level of the packing pressure (127.5 MPa, i.e. the 72.8% of the maximum packing pressure of the injection machine in the real process).

As regard the injection velocity, the slight negative effect on the shrinkage conditions as the velocity increases is due to the increase in viscous effects at the gates, leading to a higher temperature of the molten material at the end of filling with respect to the standard process. Furthermore, there is also an increase in the maximum values of the viscous stresses, a logical consequence of the increase in velocity, which negatively affects the uniform shrinkage, therefore it is convenient to keep it at the standard (initial) level.

As regard the part deformation (warpage), the results show a substantial improvement of the geometric conditions of the part at the side opposite the flange, with a significant reduction of the displacements during the cooling phase (Fig. 6, left). This is due to the better compensation of the shrinkage of the region below the sensor housing. Figure 6 (left) shows the results with the maximum value of the packing pressure (127.5 MPa), while Fig. 6 (right) shows the initial process. The improvement in terms of roundness (but also of dimensional tolerances) on the most critical circular edges (among those subjected to a specific design tolerance) is relevant. Conversely, there is an increasing deformation (displacements) in the flange region, due to the overpacking phenomenon, shown by the yellow areas in Fig. 6, left, which worsen the flatness condition that is obtained with the initial process parameters. However, the resulting values in the flange region are acceptable, so the value of the packing pressure generates the best uniform shrinkage, since there is an improvement in the deformation in areas subjected to tolerance, while the deterioration is localized in less important areas. This result was validated thanks to experiments on the real mould and the IM machine.

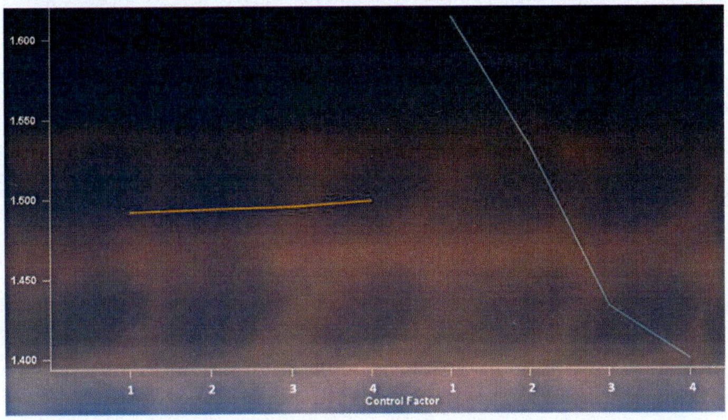

Fig. 5. The correlation between the volumetric shrinkage and the two control factors: (left) injection velocity and (right) packing pressure.

Additional simulations were carried out to evaluate if a further increasing of the packing pressure may have positively affected the general deformation. The injection velocity was fixed at its initial value, and we ran three additional simulations in which the packing pressure increased (143.5 MPa, 159 MPa and 175 MPa, which corresponds to the maximum pressure limit allowed by the IM machine). However, due to bad results in the flange zone, and the risk of damages to the mould and the IM machine, these last three simulations were not validated with the real process.

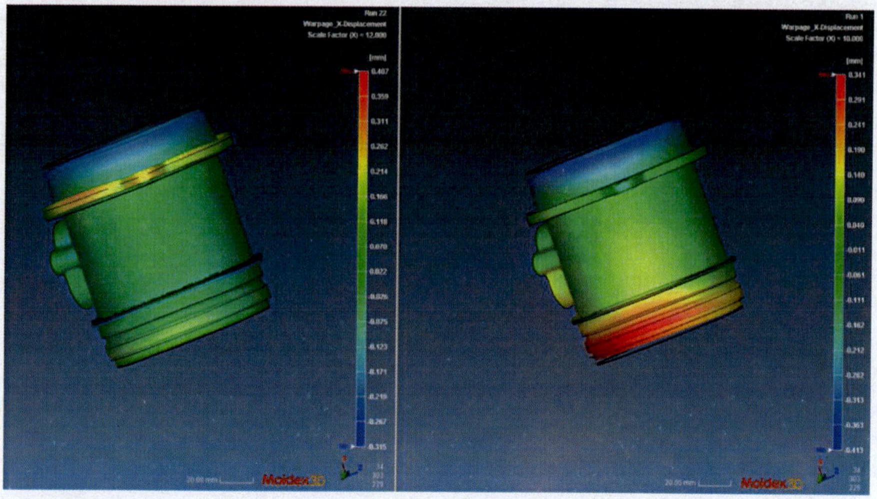

Fig. 6. Displacements along the axis of the component, using (left) the highest value of packing pressure (127.5 MPa); (right) the initial value of packing pressure.

5 Discussion and Conclusions

This work presents a study on injection moulded components to investigate the influence that the variation of the process parameters has on the part deformation, and therefore to predict and reduce the warpage, with particular respect to the functional features.

The study is based on two main design phases: Firstly, we develop a model of the mould system (cavity, runner, cooling system, mould, etc.) that can be reliable in the following simulation phase, and so capable of simulating the real behaviour of the system, thanks to the experimental validation with real data. This fine-tuning of the model considered the defeaturing of the part (removing unnecessary details), the modelling strategies of the mould elements, the mesh size and density. The analysis of the cooling efficiency reveals that the central baffle has the most important role in cooling the molten material inside of the cavity. Four different configurations of mesh elements in the cooling systems were investigated, but no advantages was found when using an increased number of elements with respect to what suggested by the CAE software. Finally, thanks to a convergence evaluation, the best number of mesh elements of the mould elements was determined and the resulting model was evaluated thanks to experimental data.

Secondly, to systematically assess the influence of the process parameters on the part deformation, we carried out a DOE factorial plan considering two process parameters as factors. The variation of the injection velocity (in particular its increase) did not return significant results, both from the point of view of shrinkage and deformation. To obtain positive results by acting on this parameter, changes to the mould would be necessary, designed to mitigate the increase in viscous effects. On the other hand, the packing pressure proved to be very effective in reducing the effects of volumetric shrinkage linked to thermal effects. In particular, by increasing the pressure up to about 72.8% of the maximum pressure limit allowed by the injection machine (127.5 MPa), a worsening of the overall deformation of the product was found, but a significant improvement in compliance with the tolerances of all the functional elements. In particular, in all areas subject to design tolerances, the range of the measured tolerance field is reduced compared to the initial process. The worsening of the deformation is concentrated in less relevant areas from the dimensional point of view, such as the flange. Therefore, although the software points out that there was an overall worsening of the part deformation, it also showed an improvement in tolerances in all the functional features.

Then, the simulations were carried out by returning to the injection velocity value used during the initial process, thus focusing only on the variation in the packing pressure. The model results reliable to predict the volumetric shrinkage and the warpage as revealed by the comparison with experimental data. The part obtained with the modified value of packing pressure already complies with the tolerances when it is extracted from the mould.

The described workflow is fully applied to a case study but can be easily generalized to other injection moulding components to predict and analyse their warpage and tolerances following the general two-phase procedure described in the Method section.

Some limitations of the current study are listed below. Firstly, the geometry of the mould system and the part were considered as inputs; therefore, we achieved an improvement of the existing production process acting only on the process parameters. Variations of the design parameters, such as the section of the gates, which could have influence on the results, were not considered.

Secondly, the cooling system geometry has not been changed, even if differentiating the flow rates of the cooling system on the two sides of the mould could ensure higher temperatures of the material in the area diametrically opposite to the sensor housing at the beginning of the cooling phase. This could lead to a lower differential shrinkage between the two sides of the component. In fact, the increase in the packing pressure has proved to be effective for reducing deformation, but it cannot be brought above certain values without worsening the elements that are not subject to problems related to shrinkage.

Thirdly, a further increase in the packing pressure, from the largest value used in the DOE calculations up to the maximum pressure limit, showed no significant benefits in terms of improvement of the results. The improvement of the volumetric shrinkage continues to be present but the effects of the overpacking become so significant in the flange area as to increase the width of the tolerance range of the circular features, worsening the overall quality. Furthermore, the resistance of the mould could set an insurmountable limit to the investigation of the highest values of packing pressure. For this reason, no experimental validation of these highest values was carried out.

Finally, the use of a smaller increment step of the packing pressure could have given better overall results (average between the condition that showed the best uniform shrinkage and the one that led to the best uniform deformation). It would therefore be possible to deepen the study by looking for an optimum condition for both these two parameters.

Acknowledgment. The authors acknowledge Cattini srl, San Martino in Rio (RE), Italy, and Moldex3D Italia srl, Lecco, Italy, for their valuable support.

References

1. Fernandes, C., Pontes, A.J., Viana, J.C., Gaspar-Cunha, A.: Modeling and optimization of the injection-molding process: a review. Adv. Polym. Technol. **37**(2) (2016). https://doi.org/10.1002/adv.21683
2. Kashyap, S., Datta, D.: Process parameter optimization of plastic injection molding: a review. Int. J. Plast. Technol. **19**(1), 1–18 (2015). https://doi.org/10.1007/s12588-015-9115-2
3. Singh, G., Verma, A.: A brief review on injection moulding manufacturing process. In: Materials Today: Proceedings, vol. **4**(2), pp. 1423–1433 (2017). https://doi.org/10.1016/j.matpr.2017.01.164
4. Fan, Z., Speight, R.: Deformation and stress prediction of injection molded components after being mounted into designed position. In: Annual Technical Conference – ANTEC 2018, Conference Proceedings, pp. 151541, Orlando (2018)
5. Kurt, M., Kamber, O.S., Kaynak, Y., Atakok, G., Girit, O.: Experimental investigation of plastic injection molding: assessment of the effects of cavity pressure and mold temperature on the quality of the final products. Mater. Des. **30**(8), 3217–3224 (2009). https://doi.org/10.1016/j.matdes.2009.01.004
6. Bici, M., Campana, F., Petriaggi, S., Tito, L.: Study of a point cloud segmentation with part type recognition for tolerance inspection of plastic components via reverse engineering. Comput. Aided Des. Appl. **11**(6), 640–648 (2014). https://doi.org/10.1080/16864360.2014.914382
7. Bici, M., Campana, F., Trifirò, A.: Automatic post-processing for tolerance inspection of digitized parts made by injection moulding. Comput. Aided Des. Appl. **13**(6), 835–844 (2016). https://doi.org/10.1080/16864360.2016.1168231
8. Miranda, D.A.D., Nogueira, A.L.: Simulation of an injection process using a CAE tool: assessment of operational conditions and mold design on the pro-cess efficiency. Mater. Res. **22**(2) (2019). https://doi.org/10.1590/1980-5373-mr-2018-0564M
9. Berti, G., Monti, M.: A virtual prototyping environment for a robust design of an injection moulding process. Comput. Chem. Eng. **54**, 159–169 (2013). https://doi.org/10.1016/j.compchemeng.2013.04.005
10. Lee, H., Ryu, K., Cho, Y.: A framework of a smart injection molding system based on real-time data. Proc. Manuf. **11**, 1004–1011 (2017). https://doi.org/10.1016/j.promfg.2017.07.206
11. Bici, M.B., Campana, G.B., F. Dughiero A.: Computer aided inspection procedures to support smart manufacturing of injection moulded components. Proc. Manuf. **11**, 1184–1192 (2017). https://doi.org/10.1016/j.promfg.2017.07.243
12. Liau, Y., Lee, H., Ryu, K.: Digital twin concept for smart injection molding. In: IOP Conference Series Material Science Engineering, vol. 324, pp. 012077 (2018)
13. Bibow, P., et al.: Model-Driven Development of a Digital Twin for Injection Molding. In: Dustdar, S., Yu, E., Salinesi, C., Rieu, D., Pant, V. (eds.) CAiSE 2020. LNCS, vol. 12127, pp. 85–100. Springer, Cham (2020). https://doi.org/10.1007/978-3-030-49435-3_6
14. Candal, M.V., Morales, R.A., Gorrin, K.: Influence of the mesh density on injection molding simulation results. J. Plast. Technol. **3**(3), 1–11 (2007)
15. Marin, F., Fagali de Souza, A., Pabst R.G., Ahrens, C.H.: Influences of the mesh in the CAE simulation for plastic injection molding. Polímeros, **29**(3) (2019). https://doi.org/10.1590/0104-1428.05019
16. Huszar, M., et al.: The influence of flow and thermal properties on injection pressure and cooling time prediction. Appl. Math. Model. **40**(15–16), 7001–7011 (2016). https://doi.org/10.1016/j.apm.2016.03.002
17. Heinisch, J., Lockner, Y., Hopmann, C.: Comparison of design of experiment methods for modeling injection molding experiments using artificial neural networks. J. Manuf. Process. **61**, 357–368 (2021). https://doi.org/10.1016/j.jmapro.2020.11.011

18. Dowlatshahi, S.: An application of design of experiments for optimization of plastic injection molding processes. J. Manuf. Technol. Manag. **15**(6), 445–454 (2004). https://doi.org/10.1108/17410380410547852
19. França, D.: Polymer injection molding parameters optimization into computational analysis using design of experiments methodologies. SAE Technical Papers 2018–36–0093 (2018). https://doi.org/10.4271/2018-36-0093

Ecodesign

Investigating a Circular Economy Application: Reuse of End of Life Tire Fibers in a Plastic Compound

Daniele Landi[1](✉) and Marco Marconi[2]

[1] Department of Engineering Management, Information and Production, University of Bergamo, Bergamo, Italy
daniele.landi@unibg.it

[2] Department of Economics, Engineering, Society and Business Organization, Università Degli Studi Della Tuscia, Viterbo, Italy

Abstract. According to the European Waste Codes 19.12.08, fibers derived from end-of-life tires (ELT) are classifies as a special waste to be sent to landfill or incineration with energy recovery. However, these activities would pose additional risks including soil pollution, and groundwater contamination. A change in the current ELT waste management practices is needed to reduce the environmental impacts. The aim of this paper is to present and investigate the technical and environmental feasibility of a circular economy path for ELT fibers. Several PP-based compounds have been manufactured and tested to verify the possibility of reusing ELT fibers in such an application. Then a Life Cycle Assessment (LCA) study has been carried out to compare the proposed reuse scenario with the two standard scenarios for ELT fibers. Reuse scenario leads to environmental savings for several impact categories, even if for some indicators the incineration is preferrable due to the additional resources and energy required needed to treat the dirty ELT fibers before reuse.

Keywords: End of life tires · Reinforced compound · Circular economy · Life Cycle Assessment

1 Introduction

Modern urbanization during the last years has contributed to increase the vehicle demand. Consequently, the annual tire demand was increased, with about 1500 million tires manufactured globally and about 1000 million tires that reach their end of life (EoL) each year [1, 2]. In this context, in Europe, about 320.000 tons of dirty fibrous material must be disposed as special waste. This produces negative impacts on the environment, in terms of green-house gases (GHG) emissions, economic losses and public costs. Nowadays, several studies and solutions have been implemented to address the ELT issue: reuse (5–23%), recycling (3–15%), landfilling (20–30%), and recovery (25–60%) [3]. Unfortunately, about 50% of used tires still end up in landfills or are discarded as untreated waste, and insufficiently valued [4].

C. Rizzi et al. (Eds.): ADM 2021, LNME, pp. 357–364, 2022.
https://doi.org/10.1007/978-3-030-91234-5_36

Generally, three main materials are produced from ELT: rubber, steel wires and textile fibers [5]. Rubber granule is the main material derived from the treatment process of ELT. Such material is then used as a basis in manufacturing new materials, such as different kinds of tires, plastic parts, concrete, asphalt, etc. [6]. Separated steel wires are generally used by steel producers for the manufacturing of new virgin steel [7]. The ELT fibers, instead, are currently classified as a unique waste according to the European Waste Code 19.12.08, thus they have to be burnt and/or buried [8].

In scientific literature, different uses for these fibers have been considered. In particular, it is possible to find applications as reinforcements in cementitious composites [9], or as indicated in Mucsi et al. [10] they are used to reinforce the soil. In recent years, researchers have tried to reuse these wastes and prevent them from being burnt or buried. In general, various types of fibers, both natural and synthetic, can be used as reinforcements. Examples of natural fibers used as reinforcements are coir [11], bagasse [12], sisal [13], jute [14] and palm [15]. Although natural fibers are normally cheaper and more sustainable than synthetic fibers, they are less durable, and synthetic fibers, specifically those composed of polymeric substances, are highly resilient as exposed to the natural environment. Synthetic fibers are widely used as reinforcement of polyethylene, polypropylene polyester, carbon, glass, and steel [16, 17]. The recycling and reuse of polypropylene (PP) have been investigated in several literature studies. It is possible to increase the properties of waste PP by adding fibers [18].

The ELT fibers mainly consist of a blend of polyester, rayon and nylon fibers, these materials have yielded positive results in different industrial applications. This study proposes a valorization strategy for ELT fibers by adopting a circular economy approach. More specifically, the paper aims to demonstrate that is possible to reuse the ELT fibers to produce new PP-based compounds with superior technical features and an improved environmental sustainability. A comparative Life Cycle Assessment (LCA) evaluation has been performed to finally verify the positive influence of the proposed circular economy.

The paper is structured as follows. Section 2 illustrates the reinforced PP-based compound properties. Section 3 presents the case study and the first two step of the LCA study and Sect. 4 shows and discusses the LCA results. Finally, Sect. 5 reports conclusions and proposal for future developments.

2 Materials

To reduce the environmental impact of ELT fibers this paper studies and compares different second life applications. Three EoL scenarios have been considered: landfill (#1), incineration with energy recovery (#2), and reuse in plastic compounds (#3). The first two options are standard scenarios, currently used in the waste management chain. Instead, the last option, is the proposed innovative applications for the ELT fibers. ELT fibers generally include Nylon 6,6, Nylon 6, Dacron and Rayon, with fibers of approximately 3– 8 mm in length and 30–50 μ in diameter [3]. Such fibers were used to prepare a PP-based compound with ELT fibers content in the range 0 – 45%. A DSM Micro compounder 15 cc, twin screw with co-rotating screws has been used to create the material samples.

Different types of blends have been obtained, by suitably mixing master and post-consumer PP. The percentages have been identified in steps of 5% increase from 0% to 20%, then it is passed to 30% and 45% by weight of fibers. For understanding the benefit of the ELT fiber in the plastic compound, tensile and compression tests have been carried out. Tensile tests were performed using the Lloyd LR30k dynamometer equipped with a 30 kN load cell. The test was conducted at a deformation rate equal to 5 mm/min. Young module and yield stress decrease as the percentage of fiber increases. However, the total deformation is reduced, with an increase of the material stiffness. As regards the mechanical impact characterization, the Izod methodology was used. The UNI EN ISO 180 standard provide a boundary condition of the test. For the impact a 1,2 J sledgehammer has been used. The ELT fibers do not produce a significative difference in the impact values. The energy values are independent from the percentage of fiber. For understanding the behavior of different fiber percentage in the plastic compound, Fig. 1 shows a sample test morphology after impact tests. There is a reduction of fractured surface as the percentage of fiber increases. The increase of fiber percentage reduces the fracture surface and two distinct parts are not obtained at the end of test. This last result suggests that the actual energy required to generate the same fracture surface increases as the amount of fiber increases.

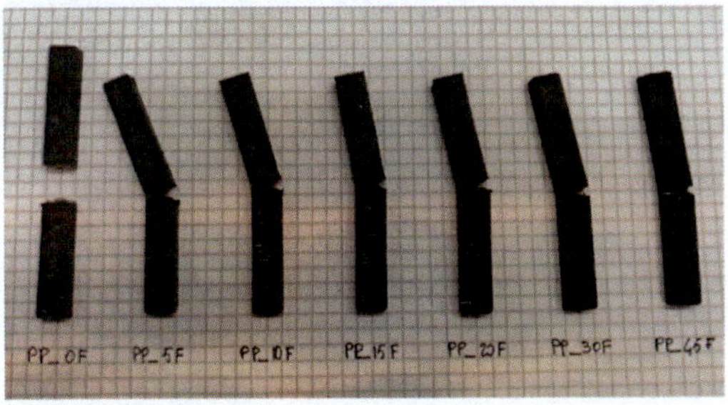

Fig. 1. Sample test morphology after impact tests

The presence of the ELT fibers in plastic compounds, reduces tensile strength but improves impact resistance. The increase of impact resistance is directly proportional to the percentage of the fiber. The new plastic compound can be used for pallets, storage bins, etc., and in general products that have to resist at impacts. In the following LCA study, a plastic compound with 45% of ELT fiber (PP- F45%) was considered.

3 Methods

The LCA methodology (ISO 14040–14044) has been used in this study by following the, four phases: (i) Goal and scope definition, (ii) Life cycle inventory (LCI), (iii) Life cycle impact assessment (LCIA), and (iv) Results interpretation.

3.1 Goal and Functional Unit

The goal of this study is to calculate the environmental impacts related to the EoL management of the textile fibers obtained from ELTs according to Life Cycle Assessment methodology (ISO 14040–14044). The activity has been carried out in collaboration with two Italian company: STECA S.p.a., an authorized recycler of ELTs, and TECNOFILM S.p.a., a producer of plastic compounds. The functional unit is defined as "disposal of the annual production of ELT fibers in STECA". The reference flow is equal to the amount of fibers that STECA manages to treat in a year, that is 1200 tones, with a pilot equipment capacity of about 300 kg/h of ELT fibers processed.

3.2 System Boundary

The study is a "gate to gate" type, this means it refers only to a specific step of the life cycle of fiber. All processes starting from the clean fibers in STECA company until landfill, incineration or reuse in plastic compound. Therefore, the various transports within these phases are also included. The system boundaries are summarized in the following Fig. 2.

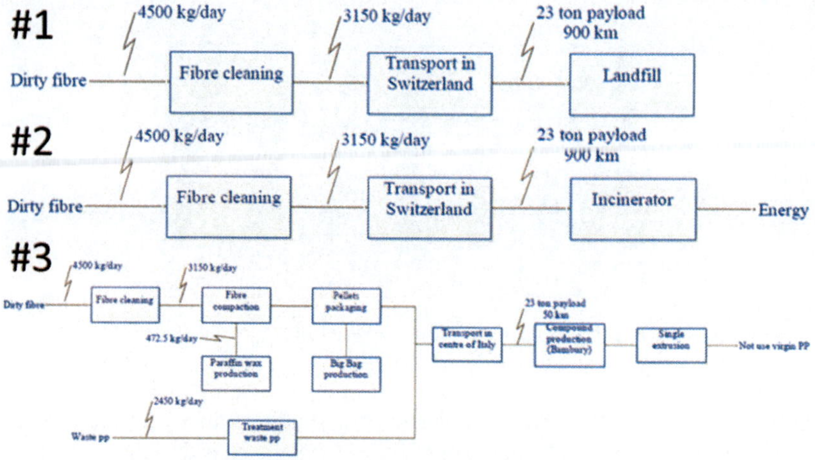

Fig. 2. Schematization of the system boundaries

While in the landfill and incineration scenarios the ELT fibers do not need further processing, in the reuse scenario it is necessary to compact the ELT fiber for the successive reuse. Though a hopper, the fiber is compacted into pellets to allow better transport and storage and a dosage in the compound plastic production. The paraffinic wax allows good lubrication and low temperature during the pelletizing phase [3].

3.3 LCI Definition

The set of inventory data used in this study for each scenario are shown in Table 1. Primary and secondary data are included on the life cycle inventory.

Table 1. Inventory data

Scenario	Input	Quantity	Unit
#1	Textile material	1700	[ton/year]
	Clean fiber	1200	[ton/year]
	Electricity of cleaning machine	29	[MWh/year]
	Euro 5 trucks load	23	[ton/travel]
	Distance	900	[km]
#2	Textile material	1700	[ton/year]
	Clean fiber	1200	[ton/year]
	Electricity of cleaning machine	29	[MWh/year
	Euro 5 trucks load	23	[ton/travel]
	Distance	900	[km]
#3	Textile material	1700	[ton/year]
	Clean fibers	1200	[ton/year]
	Electricity of cleaning machine	29	[MWh/year]
	Electricity of compaction machine	229	[MWh/year]
	Paraffinic wax for pellets production	0.15	[kg wax/kg fiber]
	PP fibers for big bag production	340	[kg]
	Euro 5 trucks load	23	[ton/travel]
	Distance	50	[km]
	Waste polypropylene	603.75	[ton/year]
	Electricity for treatment of waste pp	169	[MWh/year]
	Electricity for compound production	528.281	[MWh/year]

3.4 Cut-Off and Assumptions

A cut-off to all energy-consuming processes with an incidence lower than 1% of the total electricity consumption has been applied. An approximate distance of 900 km is assumed between the factory STECA and Switzerland for the incinerator. The big bags are PP bags with a weight of 20 kg, that are used to carry both fibers in the natural state and reduced into pellets. A reduction of 2,5 times of volume compared to the fibers in the natural state is obtained after the compaction and pelletization.

4 Results and Discussion

The third LCA step involves the measurement of the environmental impacts related to each process and flow considered. The LCIA method used for the calculation of the environmental impacts is the ReCiPe midpoint - Hierarchist (H) version – Europe. The obtained results are summarized in Table 2.

Table 2. Split of contributions of the life cycle impact assessment for each scenario and process

Scenario	Process	Climate change [kg CO2-Eq.]	Fossil depl [kg oil eq]	Freshwater eutr [kg P eq]	Ozone depl [kg CFC-11 eq]	Particulate [kg PM10 eq]	Terrestrial acid. [kg SO2 eq]
#1	Diesel	5,89E + 03	1,99E + 04	2,85E−01	4,93E−08	1,12E + 01	3,97E + 01
	Landfill	9,18E + 05	1,07E + 04	1,29E + 02	−1,80E−06	3,99E + 02	1,62E + 02
	Transport	5,46E + 04	0,00E + 00	0,00E + 00	0,00E + 00	1,59E + 01	3,62E + 01
	Electricity	1,30E + 04	3,89E + 03	6,24E-02	4,91E−07	6,05E + 00	2,03E + 01
	TOT	**9,91E + 05**	**3,45E + 04**	**1,29E + 02**	**−1,26E-06**	**4,32E + 02**	**2,58E + 02**
#2	Electricity generated	−2,07E + 05	−4,65E + 04	−9,47E−01	−5,17E−04	−1,03E + 02	−3,21E + 02
	Heat gen	−4,55E + 04	−1,05E + 04	−1,31E + 01	−2,47E−06	−6,90E + 02	−1,89E + 03
	Diesel	5,89E + 03	1,99E + 04	2,85E−01	4,93E−08	1,12E + 01	3,97E + 01
	Incineration	2,74E + 06	2,95E + 04	1,17E−01	5,05E−07	1,41E + 03	3,70E + 03
	Transport	5,46E + 04	0,00E + 00	0,00E + 00	0,00E + 00	1,59E + 01	3,62E + 01
	Electricity	1,30E + 04	3,89E + 03	6,24E−02	4,91E−07	6,05E + 00	2,03E + 01
	TOT	**2,56E + 06**	**−3,84E + 03**	**−1,35E + 01**	**−5,18E-04**	**6,55E + 02**	**1,58E + 03**
#3	PP production (av.)	−3,66E + 06	−3,34E + 06	−5,66E + 00	−5,80E−06	−1,76E + 03	−5,57E + 03
	Diesel	5,73E + 02	1,94E + 03	2,78E−02	4,80E−09	1,09E + 00	3,86E + 00
	PP fiber	5,45E + 03	4,33E + 03	8,95E−03	5,49E−08	3,57E + 00	1,06E + 01
	Wax paraffin	6,38E + 04	7,38E + 04	1,74E−01	2,48E−07	5,31E + 01	1,89E + 02
	Transport	5,31E + 03	0,00E + 00	0,00E + 00	0,00E + 00	1,55E + 00	3,53E + 00
	Electricity	5,50E + 05	1,65E + 05	2,64E + 00	2,08E−05	2,56E + 02	8,60E + 02
	TOT	**−3,03E + 06**	**−3,09E + 06**	**−2,81E + 00**	**1,53E-05**	**−1,45E + 03**	**−4,50E + 03**

Table 2 demonstrate that reusing ELT fibers in plastic compound is a feasible solution that significantly reduces the environmental load related to textile fibers. Going into more details, the reuse is the best scenario in terms of Climate Change, while incineration with energy recovery scenario leads to very high impacts. The incineration process (burning of plastic materials) produces toxic emissions in the air that strongly contributes to the rise of the world temperature. Instead, the landfilling of plastic materials, does not cause relevant emissions in the air, thus the influence on GWP impact category is quite limited.

Also considering the Fossil Depletion indicator, the situation is the same. The use of fibrous material as reinforcement in plastic compounds allows avoiding production of virgin PP, with considerable advantages in environmental terms. In the same way, in the incineration scenario, the ELT fibers allows saving fossil fuels (coal, oil, gas, etc.) commonly used for the generation of energy and/or heat. In the reuse scenario, most of the impacts are related to the paraffinic wax (produced by using fossil resources), needed to make the clean fibers reusable in the plastic compound.

Regarding impacts related to the Freshwater Eutrophication indicator, energy recovery is the most environmentally friendly, while the plastic compound scenario is penalized by the production of the wax paraffine needed to produce ELT fiber pellets.

As for Eutrophication indicator, also Ozone Depletion Potential is convenient in the energy recovery scenario. Ozone Depletion has been defined as a relative measure of the ozone depletion capacity of an ODS and uses CFC-11 (trichlorofluoromethane) as a reference. So, this indicator is very high in landfills while less significative in the paraffine production. This is due to the emissions of trichlorofluoromethane.

Fine Particulate Matter (PM10) represents a complex mixture of organic and inorganic substances. PM10 causes health problems as it reaches the upper part of the airways and lungs when inhaled. Landfill and electricity consumption are the main causes of high levels of PM10. There is a drastic reduction, about 300% of PM 10 in cases of reusing fibers in PP compounds.

Terrestrial Acidification is characterized by changes in soil chemical properties following the deposition of nutrients (namely, nitrogen and sulfur) in acidifying forms. The terrestrial acidification mainly depends on the production/consumption of energy. As for terrestrial acidification, the reuse of fibrous material produces a reduction of this impact of about 300%.

5 Conclusions

This paper investigates a reuse scenarios of ELT fibers in plastic compounds (PP-based). Firstly, the technical feasibility was evaluated to assure that the reuse path leads to a material with acceptable properties, as well as to choose the best alternative (PP-based compound with 45% fiber content in weight). Successively, such reuse scenario has been compared from the environmental point of view with two other common dismantling scenarios (landfill and incineration with energy recovery) by using the standard LCA methodology. Results demonstrate that it is not possible to univocally identify the best scenario, although reuse scenarios appear to be very promising: reuse scenario is the best alternative in case of four considered impact categories (Climate Change, Fossil Depletion, Particulate Matter Formation, Terrestrial Acidification), while incineration with energy recovery is preferrable in case of Freshwater Eutrophication and Ozone Depletion. Landfill that is currently the most common EoL scenario for ELT fibers, due to economic reasons, can be never considered the best option from the environmental perspective. Future work should be focused on optimizing the LCA study (e.g. sensitivity analysis), as well as to find improvement strategies for the most critical phases (e.g. use of paraffine wax).

References

1. Williams, P.T.: Pyrolysis of waste tyres: a review. Waste Manage. **33**, 1714–1728 (2013)
2. Jun, Z., Xiang-ming, W., Jian-min, C., Kai, Z.: Optimization of processing variables in wood–rubber composite panel manufacturing technology, Bioresour. Technol. **99** (2008)
3. Landi, D., Gigli, S., Germani, M., Marconi, M.: Investigating the feasibility of a reuse scenario for textile fibres recovered from end-of-life tyres. Waste Manag. **75**, 187–204 (2018)
4. Thomas, B.S., Gupta, R.C.: A comprehensive review on the applications of waste tire rubber in cement concrete. Renew. Sustain. Energy Rev. **54**, 1323–1333 (2016)
5. Landi, D., Marconi, M., Meo, I., Germani, M.: Reuse scenarios of tires textile fibers: an environmental evaluation. Proc. Manuf. **21**, 329–336 (2018)
6. Fornai, D., Sangiorgi, C., Mazzotta, F., Bermejo, J.M., Saiz, L.: A new era for rubber asphalt concretes for the green public procurement in road construction. In: 1st European Road Infrastructure Congress (Leeds, United Kingdom) (2016)
7. European Tyre Rubber Manufactures Association (ERTMA), 2015. The European Tyre Industry e Our Vision for 2030 (2015)
8. Yilmaz, Y.: Compaction and strength characteristics of fly ash and fiber amended clayey soil. Eng. Geol. **188**, 168–177 (2015)
9. Caggiano, A., Folino, P., Lima, C., Martinelli, E., Pepe, M.: On the mechanical response of hybrid fiber reinforced concrete with recycled and industrial steel fibers. Construct. Build. Mater. **147**, 286–295 (2017)
10. Mucsi, G., Szenczi, A., Nagy, S.: Fiber reinforced geopolymer from synergetic utilization of fly ash and waste tire. J. Clean. Prod. **178**, 429–440 (2018)
11. Sivakumar Babu, G.L., Vasudevan, A.K.: Strength and stiffness response of coir fiberreinforced tropical soil. J. Mater. Civ. Eng. **20**(9), 571–577 (2008)
12. Dang, L.C., Fatahi, B., Khabbaz, H.: Behaviour of expansive soils stabilized with hydrated lime and bagasse fibres. Proc. Eng. **2016**(143), 658–665 (2016)
13. Kafodya, I., Okonta, F.: Effects of natural fiber inclusions and pre-compression on the strength properties of lime-fly ash stabilised soil. Constr. Build. Mater. **170**, 737–746 (2018)
14. Wang, Y.X., Guo, P.P., Shengbiao, S., Haiping, Y., Binxiang, Y.: Study on strength influence mechanism of fiber-reinforced expansive soil using jute. Geotech. Geol. Eng. **34**(4), 1079–1088 (2016)
15. Azadegan, O., Kaffash, A.E., Yaghoubi, M.J., Pourebrahim, G.R.: Laboratory study on the swelling, cracking and mechanical characteristics of the palm fiber reinforced clay. Electron. J. Geotech. Eng. **17**, 47–54 (2012)
16. Akbulut, S., Arasan, S., Kalkan, E.: Modification of clayey soils using scrap tire rubber and synthetic fibers. Appl. Clay Sci. **38**(1–2), 23–32 (2007)
17. Consoli, N.C., Montardo, J.P., Donato, M., Prietto, P.D.: Effect of material properties on the behaviour of sand-cement-fibre composites. Proc. Instit. Civ. Engineers Ground Improve **8**(2), 77–90 (2004)
18. Marconi, M., Landi, D., Meo, I., Germani, M.: Reuse of tires textile fibers in plastic compounds: is this scenario environmentally sustainable? Proc. CIRP **69**, 944–949 (2018)

Ephemeral Products: Opportunities for Circularity Based on Ideation for Reuse. An Experience

Ignacio López-Forniés(✉) and Jorge Sierra-Pérez

Zaragoza University, 50018 Zaragoza, Spain
ignlopez@unizar.es

Abstract. Ephemeral products have a high potential for improvement within the framework of the Circular Economy since, due to their short cycle of use, their state at the end of their life is usually optimal and available for other uses. Ephemeral products have wide possibilities of being introduced into another cycle of use, either as a product, a disassembled part or as a material. This work presents a metric that evaluates the opportunity indicator applicable to ephemeral products as a product of two dimensions: the economic one based on Cost-Benefit Analysis and an environmental one based on Life Cycle Assessment. To present the proposed metric, crown caps are used as a case study. A physical characterization of these products is carried out, and later, through a creative process, some concepts are obtained for reuse. The new concepts are compared in the environmental and economic dimensions to assess the level of opportunity.

Keywords: Ephemeral products · Circular economy · Metrics · Eco-ideation

1 Introduction

The concept of ephemeral product (EP) has undergone a drastic change since its first definition in 1969 by Buckminster Fuller [1] of the term ephimeralization, doing more with less in a continuous process of efficiency thanks to technological progress. Current definitions are related to short life cycles, single-use products, planned obsolescence, trends and fashions, etc. The validity in the market of EP depends on the product satisfying the needs and desires of the consumer at a reason-able cost; in part it depends on trends and sociocultural aspects. Historically, some industrial sectors have promoted the culture of discard, the overproduction of articles of ephemeral or disposable life over repairable durable goods. That focuses on short life cycles and rapidly changing trends to stimulate consumer desire for novelty and change, encouraging frequent acquisition of new styles and premature removal of old ones [2].

The longer a product lasts, the better for the environment; the old ability to repair products, extend their life or give them new functions once the main one is finished, is lost with EP. The most conventional term EP is related to disposable product to define single-use products applied in perishable products in the food sector, disposable containers and

C. Rizzi et al. (Eds.): ADM 2021, LNME, pp. 365–372, 2022.
https://doi.org/10.1007/978-3-030-91234-5_37

packaging in industry, medical-sanitary and hygiene products or kitchen and dining room products [3]. Ephemeral products can benefit from Circular Economy (CE) strategies since their final state after their useful life is good and they can be reused, recovered or recycled. This can open up possibilities for it to be reintroduced into another cycle of use, either as a product, disassembled part or as a material.

The objective is the evaluation of design concepts from EP to apply circularity strategies. The opportunity indicator is assessed by a metric that measures depending on the environmental and economic dimensions. The environmental dimension is the difference between the environmental impact of the old and the new application, while the economic dimension is calculated based on the residual value for the material, parts or components, the added value of the new product and the production costs.

2 Literature Review

2.1 Ephemeral Products

The definition of EP is barely defined in the scientific literature, it refers the life of the product due to its longevity, durability and resilience [4]; for their survival of fashions and trends that make obsolescence arrive in its various forms [5] and for attachment, nostalgia or emotional attachment [6]. Ephemeral is a trait related to seasonal, trendy and fashionable. An EP is short-life and often carries some connotation of excess or waste [7]. Janssen [7] proposes a checklist to determine whether the products are ephemeral or not, based on the ephemerality characteristic. It establishes five possible levels that relate them to functional use lifespan and fashion, ranging from the highest level in which the product suddenly disappears to what we could consider a classic, a product that we do not want to get rid of by continuing to operate, with technological validity and emotional attachment.

To define ephemeral products, the literature highlights three characteristics, albeit in isolation. These are: product lifespan in full functional use [8] validity in the market [7], and the emotional attachment [9]; highlighting the lack or low level of each characteristic, or the opposite of said characteristic. Functional use is defined by two factors the time that the product can perform its function and by the number of use cycles. The market validity is conditioned by trend and fashion factors, loss of attractiveness or replacement by a fresher one and by the introduction of novelties as factor of change in technologies, functionalities, benefits, etc. Emotional attachment is conditioned by the emotional bond factor, lack of generation of emotional attachment and easily changing tastes and by the factor of the representativeness or image that it transmits that normally it is not linked to recognized brands and does not usually represent a status or social position.

2.2 Circularity

Circularity is structured in levels ranging between macro (national, regional), me-so (city, industrial park and supply chain), micro (company) and nano (product with different methods and techniques [10]. Linder et al. [11] define product-level circularity as "the fraction of a product that comes from a used product" and reveal the urgent need for metrics to critically evaluate product-level circularity to facilitate a transition to a more circular economy.

EC strategies that aim to extend the useful life of materials and products focus on: Narrowing (Use less), Slowing (Use longer) and Closing (Use again) [12, 13]. In the case of EP, the introduction of the Slowing strategy would render the definition of the EP meaningless. Regarding the implementation of the strategy Narrowing, expresses, the quantity of products, components, materials and energy during the different life cycle stages of product, so the EP should suffer a decrease in material and energy intensity. Regarding to Closing strategy, it refers to what was previously said about the reintroduction of the product, part or material in another cycle of use.

Moreover, we have selected the environmental and economic dimensions that de-fine the opportunity indicator, helping designers in the selection of concepts and improving decision-making through a result, indicative rather than definitive, of the strategic and innovative changes applied to product design [14].

3 Creating Opportunities for Ephemeral Products

To create a metric that helps select concepts that integrate circularity strategies generating an Opportunity indicator as a continuation of previous work [15]. For that, it is used the economic dimension, Cost-Benefit Analysis (CBA) indicator [16] and the environmental dimension, using an environmental indicator based on Life Cycle Assessment methodology (LCA). That compares the current situation of an ephemeral product with respect to the new conceptual proposals. The CBA aims at identifying and valuing the impacts of a project in terms of their effects on social well-being, comparing the positive effects (benefits) with the negative effects (costs). Current reality shows that the marriage between ecology and economics is universally accepted [16]. The CBA is used in the economic evaluation and it is desirable that the environmental factor prevails over the economic one, so it is necessary to propose tools that facilitate decision-making towards circularity.

The Environmental Dimension (ED) will be obtained with the development of LCA of the concepts developed, comparing it with similar products on the market with different materials. To do this, once the concepts have been defined, a market study is carried out to identify products with functions and characteristics similar to those proposed in the conceptualization process. The first step to carry out an LCA is to design the inventory of materials that make up the product to be analyzed, both for the concept and for the products to be compared. Once all the elements that make up each product, its materials and manufacturing processes have been listed, they will be characterized environmentally. For this, OpenLCA is used, an open access software [17].

The Opportunity indicator proposed, which using the Environmental (ED) and the Economic Value (EV) dimensions does not determine whether one dimension should prevail over the other, but rather provides actions based on the opportunity defined by Eq. (1).

$$\text{Opportunity} = \text{EV} * \text{ED} = [\text{RV}-\text{RP}] * \text{ED} \tag{1}$$

Where Residual Value (RV) for the circularity metric is the measure of value change after usage, is a term fixed by market value, as is the case with the market for second-hand cars, recycled metals (steel, cooper, aluminum), recovered textiles, buildings, etc. As a circularity strategy waste may be used as a resource. The RV is calculated with the difference between the product of the cost of the raw material by its weight and the product of the residual value of the recycled material by the weight of the new concept.

Recovery and Processing (RP) is the cost of all the actions necessary for the transformation and production of the new product, In the case of an existing product, the cost can be analyzed in detail. However, in a concept we apply a factor in which the costs are a third of the market price, taking into account the cost per kilogram.

The ED indicator is the balance between the environmental impact of the reference product and that of the new application, it is calculated with the difference of the impact measured in Kgs equiv. CO_2, with the aforementioned software.

Figure 1, in Sect. 4, shows the four quadrants generated by these two dimensions and the different opportunities for ephemeral product concepts. The first quad-rant, white background, encompasses the concepts with the greatest development opportunity, high level, since both the environmental and economic dimensions are positive, they represent products that will allow a reduction in environmental impact as well as economic savings. They will serve to disseminate the strategies adopted in CE, green marketing and as examples to apply in academia and industry.

The second and fourth quadrants, a light grey background, with a low level of opportunity, encompass concepts with one of the two positive dimensions. Both quad-rants represent concepts that must be supported so that the environmental interest prevails over the economic one. The second quadrant with the positive environmental dimension needs institutional support, financial aids and promotion of a good environmental response, to prevent the economic dimension leads to abandon these initiatives. The fourth quadrant, with the positive economic dimension, needs strategies to improve the environment with policies to support change and redesign or conceptualization actions with environmental objectives. The concepts of the third quadrant, dark grey, with no opportunity represent concepts that are not economically viable and generate a greater environmental impact than existing products.

Once the values of the two dimensions have been calculated for each concept, it can be represented graphically so that its level of opportunity is easily observed visually and what actions should be taken in each case.

4 Case Study: Crown Cap

We will use as an example an EP with a very short life and a single use, it is the metallic crown-type cap of soft drink bottles. This type of stopper guarantees a hermetic and hygienic closure at a low cost, as well as an advertising and brand support. Its life cycle is very short, it begins at the time of bottling and ends with the opening of the bottle and its subsequent disposal. It is a waste that is not recycled and that in the best of cases ends up in the packaging container. The choice of this residue serves to evaluate the proposed metric with the possibility of reusing the material, in order to extend its life, providing them with a second functional opportunity to obtain another product. The first step is to characterize the product: Weight (M/SD) = 2,097 g/0,029 g (measured in 100 units); Weight metal/plastic ratio (PE) = 1.9036 g/0,1709 g; Diameter (max/min) = 32,1 ± 0,2 mm - 26,75 + 0,15 mm; Height (h) = 6,0 ± 0,15 mm. Solid granulated PE-based compounds suitable for liquid foods with an alcohol volume of up to 15%. Granulated solid PVC-based compounds suitable for non-fatty liquid foods with an alcohol volume of up to 10%.

The concepts developed for the reuse of the caps were three concepts with a common characteristic, their modularity with the use of different amounts of caps. The selected concepts are: #1 room divider, #2 bottle rack, and #3 tile. All the products would be formed with pressed and cut caps, and placed on steel rods, with the exception of concept #3 that exclusively uses transformed caps to be hooked together.

Based on these concepts, three products with different materials were selected. The market costs of both the recovered steel and the raw materials of the original products have been obtained based on approximate values of different prices on web pages. Table 1 collects the information used to calculate the environmental value (ED) of the selected concepts and products. This table includes the Functional Unit used in each concept, the data on materials and the weight of the products with their original material and being replaced by the plates. All these data are those used in the LCA inventory and the equivalence between them, and the environmental implications for the manufacture of all products. In the case of products made of sheet metal, as it is a reused product, the impacts of its manufacture were not included. In this case, the sheet metal transformation process (pressing and cutting) and treatment as waste were included. In the case of the rest of the product, the production of the materials, their transformation and their treatment as waste were included in the LCA.

Table 1. Inventory data and results of LCA for concepts and substitute products.

	Functional Unit (FU)	Product material - name	Raw material	Crown cap as substitute
			Quantity (kg) / Kg CO_2 –eq	
#1 Room divider	1 m^2	Plastic (PVC) - O1p	5,00/21,50	16,20/95,20
		Metal - O1m	21,50/774,2	5,50/32,34
		Wood - O1w	2,50/145,0	5,20/30,57
#2 Bottle rack	Contain 6 bottles	Plastic (PC) - O2p	0,82/7,24	0,50/2,90
		Metal - O2m	0,99/35,80	0,40/2,30
		Wood - O2w	0,46/27,50	0,30/1,70
#3 Floor tile	1 m^2	Plastic (LDPE) - O3p	3,10/9,00	4,10/24,10
		Porcelain - O3g	12,30/10,10	3,10/18,20
		Ceramic - O3c	58,30/48,1	5,60/32,90

Table 2 shows the results of the economic and environmental value and the final result of the opportunity to be able to compare the different concepts and original products that they are going to replace.

Table 2. Metric results for case study with Crown Caps.

Concept		RV			COST				RP		ED		Opportunity	
		Vo (€/Kg)	Vc (€/kg)	DIF (Vo-Vc)	Raw (€/Kg.)	Vo/ (3*kg)	Original	Crown cap	EV [-1,1]		ED [-1,1]		OP [-1,1]	
#1 Room divider	O1p	0,89	0,27	0,08	119,00	2,45	21,5	95,2	-2,37	-0,00	-73,7	-0,10	-174,86	-0,007
	O1m	2,00	0,27	41,52	137,00	8,30	774,2	32,3	33,21	0,04	741,9	1,00	24638,63	1,000
	O1w	0,03	0,27	-1,34	135,00	8,65	145,00	30,6	-10,00	-0,01	114,4	0,15	-1143,77	-0,046
#2 Bottle rack	O2p	2,30	0,27	1,75	7,90	5,27	7,20	2,90	-3,52	-0,00	4,30	0,01	-15,26	-0,001
	O2m	2,00	0,27	1,87	24,00	20,00	35,8	2,30	-18,13	-0,02	33,50	0,05	-607,29	-0,025
	O2w	0,03	0,27	-0,07	31,00	34,44	27,5	1,7	-34,51	-0,04	25,80	0,03	-890,46	-0,036
#3 Floor tile	O3p	1,35	0,27	3,08	54,90	4,46	9,00	24,10	-1,39	-0,00	-15,10	-0,02	-20,92	-0,001
	O3g	5,00	0,27	60,66	14,00	1,51	10,10	18,20	59,16	0,07	-8,10	-0,01	-479,18	-0,019
	O3c	15,00	0,27	872,99	22,29	1,33	48,10	32,90	871,66	1,00	15,20	0,02	13249,25	0,538

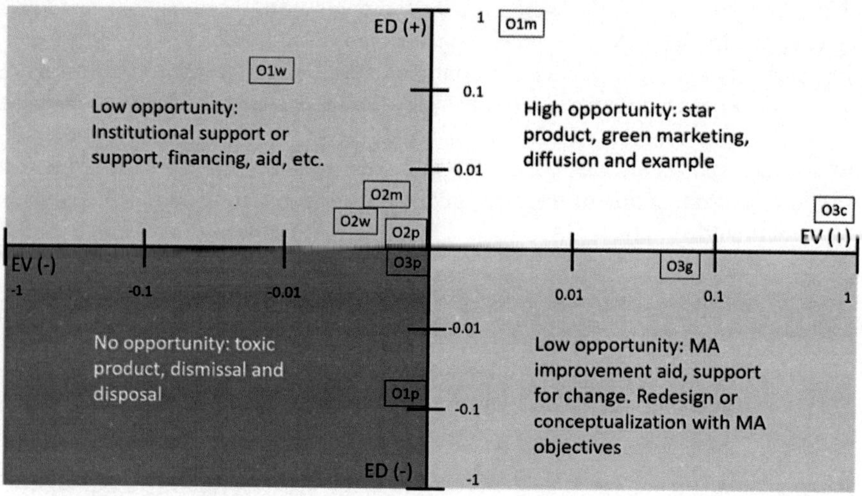

Fig. 1. Concept opportunities for substitution based in dimension EV and ED

Figure 1 shows the results of the economic and environmental value from a graphical approach that is easier and faster to interpret. The concepts and products to be replaced are distributed along the two axes, their position within the quadrant also indicates their potential to move to the adjacent quadrant through actions and decision making that support these ideas to improve their opportunity. In this regard, only two concepts could have a chance to replace the original products, these two concepts coincide with the two extreme values for EV (O3c) and for ED (O1m). The O3c concept corresponds to the replacement of the product with the highest market value (tile) but more data must be obtained to show that the costs of reprocessing crown taps are not high due to the energy required to produce a tile. The O1m concept represents the metal room divider whose environmental impact is the best. There are two concepts that are excluded because of the negative values for EV and ED. The concept #2 improves the environmental dimension of all products, albeit only slightly, so it is necessary to look at the economic cost assessment to see if the proposed concept has a chance and can be developed as a more environmentally friendly object from a waste and has an alternative of use. Most of them are very close to the vertical axis, so efforts in economic improvement represent opportunities.

The metric shows that there are large differences between the concepts and that there is dependence on the material used. The metric discriminates the concepts well, and the EV and ED dimensions are quantifiable, and the data for evaluation can be obtained from reliable sources. The test with the metric seems correct but is not conclusive and should be contrasted with other metrics and indicators for ED and CE, and with other experiences with other ephemeral products. We have relied on CBA and LCA, but they can be changed for other already known indicators [18, 19], the only requirement is that the dimension can take positive and negative values to be able to identify in which quadrant it is located.

5 Discussion and Conclusions

The proposed metric indicates that there may be difficulty in finding ephemeral products that can be replaced by new concepts, since the differences in the economic and environmental dimensions are not highlighted.

It is necessary to continue studying which ephemeral product can be recovered to give a chance of a second life, recovery of parts or materials. However, metrics can help redefine ephemeral product concepts that have a better timeline indicator. In addition, it would be necessary to compare the different ephemeral products and determine their differences. The differences between them lie in their materiality, due to the number of cycles for which they have been designed, and which have finally been used.

The metric and the graphical representation explain well the opportunity of the EP, which is low, but that there are options mainly improving the EV indicator. In this sense, using other assessment tools of the Environmental Dimension, more qualitative compared to a quantitative tool such as the LCA, can improve decision-making. The use of other indicators of generalized circularity in the scientific com-munity can provide us with other opportunity values not included in this metric.

References

1. Chu, H.-Y., Trujillo, R.G.: New Views on R. Buckminster Fuller. Stanford University Press, Palo Alto (2009)
2. Joyner Armstrong, C.M., Kang, J., Lang, C.: Clothing style confidence: the development and validation of a multidimensional scale to explore product longevity. J. Consum. Behav. **17**(6), 553–568 (2018)
3. Muranko, Z., Tassell, C., van der Laan, A.Z., Aurisicchio, M.: Characterisation and environmental value proposition of reuse models for fast-moving consumer goods: reusable packaging and products. Sustainability **13**(5), 2609 (2021)

4. Dewberry, E.L., Sheldrick, L., Sinclair, M., Moreno, M.A., Makatsoris, C.: Developing scenarios for product longevity and sufficiency. In: PLATE: Product Lifetimes and the Environment (2017)
5. Cooper, T.: Inadequate life? Evidence of consumer attitudes to product obsolescence. J. Consum. Policy **27**(4), 421–449 (2004)
6. Ceschin, F., Gaziulusoy, I.: Evolution of design for sustainability: from product design to design for system innovations and transitions. Des. Stud. **47**, 118–163 (2016)
7. Janssen, C., Vanhamme, J., Lindgreen, A., Lefebvre, C.: The catch-22 of responsible luxury: effects of luxury product characteristics on consumers' perception of fit with corporate social responsibility. J. Bus. Ethics **119**(1), 45–57 (2013). https://doi.org/10.1007/s10551-013-1621-6
8. Marchand, A., Walker, S.: Product development and responsible consumption: designing alternatives for sustainable lifestyles. J. Clean. Prod. **16**(11), 1163–1169 (2008)
9. Shimp, T.A., Madden, T.J.: Consumer-object relations: a conceptual framework based analogously on Sternberg's triangular theory of love. ACR North Am. Adv. (1988)
10. de Oliveira, C.T., Dantas, T.E.T., Soares, S.R.: Nano and micro level circular economy indicators: assisting decision-makers in circularity assessments. Sustain. Prod. Consum. **26**, 455–468 (2021)
11. Linder, M., Sarasini, S., Loon, P.: A metric for quantifying product-level circularity. J. Ind. Ecol. **21**(3), 545–558 (2017)
12. Konietzko, J., Bocken, N., Hultink, E.J.: A tool to analyze, ideate and develop circular innovation ecosystems. Sustainability **12**(1), 417 (2020)
13. Bocken, N.M.P., de Pauw, I., Bakker, C., van der Grinten, B.: Product design and business model strategies for a circular economy. J. Ind. Prod. Eng. **33**(5), 308–320 (2016)
14. Cayzer, S., Griffiths, P., Beghetto, V.: Design of indicators for measuring product performance in the circular economy. Int. J. Sustain. Eng. **10**(4–5), 289–298 (2017)
15. López-Forniés, I., Sierra-Pérez, J., Boschmonart-Rives, J., Gabarrell, X.: Metric for measuring the effectiveness of an eco-ideation process. J. Clean. Prod. **162**, 865–874 (2017)
16. Hundloe, T., McDonald, G.T., Ware, J., Wilks, L.: Cost-benefit analysis and environmental impact assessment. Environ. Impact Assess. Rev. **10**(1), 55–68 (1990)
17. OpenLCA, The open source Life Cycle and Sustainability Assessment software (2020)
18. Mesa, J., Esparragoza, I., Maury, H.: Developing a set of sustainability indicators for product families based on the circular economy model. J. Clean. Prod. **196**, 1429–1442 (2018)
19. Saidani, M., Yannou, B., Leroy, Y., Cluzel, F., Kendall, A.: A taxonomy of circular economy indicators. J. Clean. Prod. **207**, 542–559 (2019)

A Design for De-manufacturing Methodology to Improve the Product End of Life Environmental Sustainability

Marta Rossi[1](✉), Federica Cappelletti[1], Marco Marconi[2], and Michele Germani[1]

[1] Università Politecnica delle Marche, Ancona, Italy
marta.rossi@staff.univpm.it
[2] Università degli Studi della Tuscia, Viterbo, Italy

Abstract. De-manufacturing and re-manufacturing are well-known solutions for recovering value from products that have reached their End of Life (EoL) and thus reducing resource exploitation. Although such scenarios are implemented after the use phase, they must be considered since the very early stage of design. The paper proposes a methodology that can be applied at the design stage to detect space for product design improvements, also representing a baseline for organizations approaching de-manufacturing for the first time. The methodology consists of four main steps, in which firstly target components are identified according to their environmental impact; then the disassembly sequence is qualitatively evaluated, and successively quantitatively too. This leads to the identification and evaluation of different EoL scenarios. The application of the methodology to a professional espresso coffee machine highlighted a reduction of impacts up to 52% if re-using and re-manufacturing strategies are implemented.

Keywords: Environmental sustainability · De-manufacturing · Ecodesign

1 Introduction

Environmental sustainability increases companies' competitiveness in global market contexts, enhances the reduction of consumption of energy, resources, key materials, and decreases the product lifecycle footprint [1]. Circular Economy (CE) concept and strategies aim at considering each life cycle step functional to the subsequent one with the objective to reduce the product burden on the environment.

This work proposes a methodology to support the implementation of design for de-manufacturing strategies, detecting the most impacting components and investigating how they will be disassembled. Indices are proposed to evaluate their re-usability, recyclability, and re-manufacturability, so different scenarios and design alternatives can be evaluated. The novelty of the proposed methodology against previous literature studies [2, 3], stands in the possibility to select target components to be re-used, re-manufactured or re-cycled according to quantitative results of environmental analysis and quantitatively evaluate the potential benefits of the proposed new scenarios.

C. Rizzi et al. (Eds.): ADM 2021, LNME, pp. 373–380, 2022.
https://doi.org/10.1007/978-3-030-91234-5_38

2 State of the Art

Circular systems must be self-sustainable, [4]; they focus on high value and high-quality material cycles [5] and related design practices are focused on maintaining materials at their highest value at any point of the products' lifecycles [6]. In this context, design has the power to influence all downstream phases, affecting their probability to happen, their efficiency and success rate [7]. The analysis and continuous improvement of the disassembly and End of Life (EoL) phases, together with a correct design, are key actions to obtain the maximum result in terms of fatigue, cost, and time, using only the essential resources [8]. Stavropoulos et al. [4] proposed a work aiming at bridging the gap between the multi-material component design and the a priori knowledge of the component disassembly features and by extension. Their work only prospects for re-usability and re-cycling. Sabaghi et al. [9] proposed disassembly-related parameters with controllable characteristics at the design phase, then modelled a polynomial regression model for estimating the disassemblability index. Zhang et al. [10] focused on the re-manufacturing process and verified the prediction model of each evaluation index. Favi et al. [11] used LeanDFD to assess the economic feasibility of the different EoL scenarios, combining it with LCA parameters to have a global vision of the product lifecycle. Disassembly processes usually include the largest number of employees and the highest complexity in EoL treatments [12]. Numerous research works focused on disassembly sequence and time calculation using analytical hierarchy process (AHP) [13], AND/OR graph [14], oriented graphs and constraint matrices [15].

3 Methodology

The methodology supports the identification of components more suitable for re-use or re-manufacturing strategies at their EoL and the quantification of benefits, in terms of environmental impact, related to the implementation of these strategies. At the same time, it allows identifying where product EoL criticalities are and how their resolution can contribute to the reduction of environmental impact at product EoL.

3.1 Product Environmental Analysis

The first step aims to identify the most impactful components, defined as "target components", which represent the object of the next step. It is carried out starting from the product Bill of Materials (includes materials and manufacturing processes), CAD model, (includes information about product structure), and information about dismantling and EoL. The environmental analysis can be realized through simplified tools (e.g., Simplified LCA) or full LCA tools, according to the level of details available for the product. During the goal and scope definition, the selection of the most representative impact categories is realized according to their representativeness for the specific product under analysis. The system boundaries include the material, the manufacturing and the EoL phases, the only ones affected by de-manufacturing strategies. Transport phase can be included according to the quality of the available data.

3.2 Qualitative Disassembly Analysis

The second step has the objective to analyse the disassembly level for the identified target components, through a top-down approach. First, sub-assemblies and gathering components are grouped according to their disassembly level, and then all types of connections (e.g. fixed/movable, permanent/temporary, etc.) are examined with a dual scope: (i) to understand how to disassemble the groups, and (ii) to pave the way to the quantitative analysis. Each sub-assembly is analysed to identify the disassembly sequence for the removal of each component, which can be summarized in an oriented graph. Furthermore, the authors suggest also using the constraint matrix [16]. Such matrix is populated according to the Boolean rule shown in Eq. 1:

$$\mathrm{x_{ij}} = \begin{cases} 1\ \textit{if the}\ i_{th}\ \textit{component is constrained by}\ j_{th}\ \textit{component} \\ 0\ \textit{if}\ i_{th}\ \textit{component is NOT constrained by}\ j_{th}\ \textit{component} \end{cases},\quad i \neq j \tag{1}$$

Figure 1 shows an example of graph-oriented and level matrix of a product made of 6 components/sub-assemblies: before removing component 2 you must separate components 1, 3, and 4. Then, the disassembly level definition is possible. This consists of re-organizing the rows of the level matrix in a way that the sub-totals of all values in a row are arranged in ascending order. Parts belonging to the i_{th} level can be disassembled after all parts till level $(i-1)_{th}$ are removed.

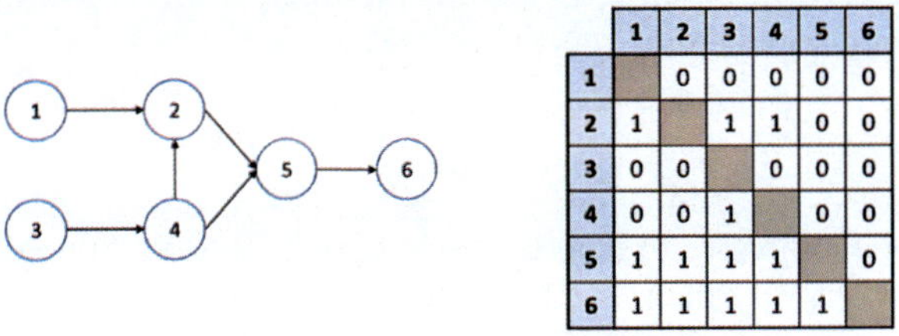

Fig. 1. Example of an oriented graph and a level matrix.

3.3 Quantitative Analysis of EoL Strategies

This step aims at quantifying several indices useful to understand the most suitable EoL strategies for the target components previously identified. Besides the outcomes of the previous phase, this also requires the disassembly level and disassembly time.

Disassembly Time. The authors followed the procedure suggested by Mandolini et al. [16] in quantifying the manual disassembly time.

EoL Indices. Indices represent a useful tool to compare and select the best EoL strategies. Two indices [17] are calculated for all target components. The first index (Eq. 2) is the reusability rate and represents the percentage (in weight) of the product that is potentially reusable:

$$\mathrm{I_{Reuse}} = \frac{\sum_i \sum_k m_{reuse_{i,k}}}{m_{tot}} \cdot 100 = \frac{\sum_i \sum_k m_{i,k} D_{i,k} M_{D_{i,k}}}{m_{tot}} \cdot 100 \tag{2}$$

where:

m_{tot}: the total mass of the target component [kg];

$m_{reuse\,i,k}$: the potentially reusable mass of k_{th} material of i_{th} component [kg];

$m_{i,k}$: the reusable mass of k_{th} material of i_{th} component [kg];

$D_{i,k}$: disassemblability index of k_{th} material of i_{th} component [%];

M_D: degradation index of k_{th} material [%].

According to Ardente and Wolf [17], the recyclability index is defined as shown in Eq. 3 and represents the percentage (in weight) of the product that is potentially recyclable:

$$\mathrm{I_{Recycle}} = \frac{\sum_i \sum_k m_{recycle_{i,k}}}{m_{tot}} \cdot 100 = \frac{\sum_i \sum_k m_{i,k} D_{i,k} M_{R_{i,k}} \cdot C'_{1_{i,k}}}{m_{tot}} \cdot 100 \tag{3}$$

m_{tot}: the total mass of the target component [kg];

$m_{recycle_{i,k}}$: the potentially recyclable mass of material of i_{th} component [kg];

m_{ik}: the recyclable mass of material of i_{th} component [kg];

D_{ik}: disassemblability index of k_{th} material of i_{th} component [%];

M_R: recyclability index of k_{th} material [%];

$C'_{1i,k}$: complementary contamination index of recyclable material [%].

Two parameters which affect the applicable EoL strategies for components are taken into account in the following. The degradation index (M_D), which considers the level of reusability of materials with concerning to their chemical properties and use, and the contamination index (C_1), which considers the presence of contaminants which reduces the recyclability of components.

3.4 EoL Strategies and Environmental Benefits

The last step has the objective to identify the best suitable EoL strategy for each target component and to quantify the related environmental benefits; parameters and related rates considered to select the best EoL options are shown in Table 1.

Table 1. Indices and rates for EoL strategies

Indices and Rates	EoL strategy
$I_{Reuse} \geq 50\%$ and $M_D \geq 75\%$	Re-use
$I_{Reuse} > 50\%$ and $20\% \leq M_D \leq 75\%$	Re-manufacturing
$I_{Reuse} \leq 50\%$ and $M_D \geq 75\%$	Re-manufacturing
$I_{Recycle} > 30\%$ and $C'_1 \geq 50\%$	Recycling
$I_{Reuse} < 50\%$, $M_D < 20\%$, $C_1' < 50\%$, or $I_{Recycle} < 30\%$	Landfill

At this stage, it is possible to assign to each target component, a specific EoL strategy, to calculate (Eqs. 4–6) the product mass suitable for each EoL strategy (i.e. re-using, re-manufacturing and re-cycling mass) and to assess the potential environmental benefits.

$$\mathrm{M_{Reuse}} = \frac{\sum_i m_{reuse_i}}{m_{tot_i}}, \tag{4}$$

$$\mathrm{M_{reman}} = \frac{\sum_i m_{reman_i}}{m_{tot_i}}, \tag{5}$$

$$\mathrm{M_{recycle}} = \frac{\sum_i m_{recycle_i}}{m_{tot_i}}. \tag{6}$$

m_{reuse_i} is the mass of the i_{th} component intended to be re-used [kg];
m_{reman_i} is the mass of the i_{th} component intended to be re-manufactured [kg];
$m_{recycle_i}$ is the mass of the i_{th} component intended to be re-cycled [kg];
$m_{tot\,i}$ is the total mass of the i_{th} component [kg].

In particular, to quantify the environmental benefits deriving from the implementation of the EoL strategy, the percentage variation is calculated as the difference of environmental impacts of a certain EoL strategy in the future (V_f) and the current EoL scenario (V_i) over the impact of the same category of the current EoL scenario. In this way, several improvement strategies can be evaluated and compared, to identify the best solution from the environmental point of view.

4 Case Study

The proposed methodology has been applied in the case of an Italian company, leader in the production of professional espresso coffee machines, which are complex products, containing more than 500 elements. The company provided the bill of materials (BOM) of the product. For confidentiality reasons, in this case study the components are identified with BOM codes, while their names are neglected.

4.1 Product Environmental Analysis

The LCA analysis has been performed through SimaPro software tool (database Ecoinvent 3), and impacts were calculated with the ReCiPe Midpoint (H) method (Fig. 2).

Concerning assumptions, all elements whose weight is lighter than 0,005 kg have been cut off. The use phase has been neglected since it is assumed equal for current and future scenarios. Concerning the EoL phase, it is assumed that 50% of aluminum, 25% of brass, and 70% of copper, steel and iron are re-cycled. Remaining materials and components are treated as Waste from Electric and Electronic Equipment (WEEE).

Figure 2 shows the split of impacts for the material and manufacturing phases in six environmental impact categories. For each most impacting group further deployment has been carried out, and 16 target components have been selected and analysed in the successive methodology steps.

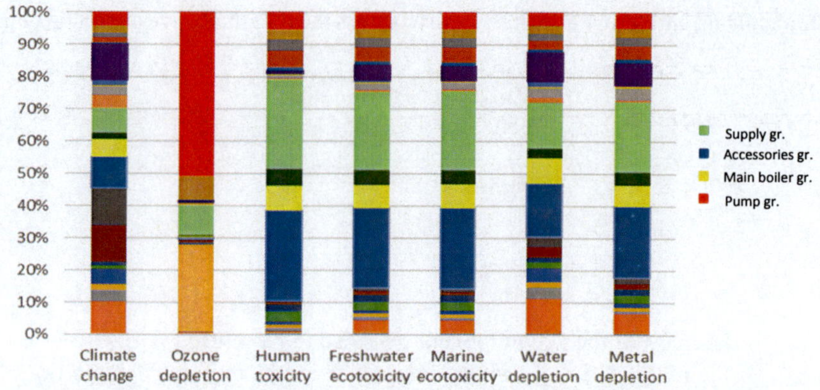

Fig. 2. Results of material and manufacturing phase of the espresso coffee machine.

4.2 Qualitative and Quantitative Disassembly Analysis

After removing the sides parts and the upper cups-heating elements it is possible reach the internal components of the machine. From the analysis of the CAD model and connectors, the oriented graph has been created. It has been converted to the constraint matrix (Table 2). Disassembly times were quantified by using the LeanDfD tool [11].

Table 2. Constraint matrix of disassembly of the main boiler group.

	5B	8B	9B	12B	14B	16B	17B	18B	19B	20B	21B	22B	23B	24B	25B	26B	27B	TOT.	LEVEL
9B	0	0	-	0	0	0	0	0	0	0	0	0	0	0	0	0	0	0	0
12B	0	0	0	-	0	0	0	0	0	0	0	0	0	0	0	0	0		
17B	0	0	0	0	0	0		0	0	0	0	0	0	0	0	0	0		
22B	0	0	0	0	0	0	0	0	0	0	0		0	0	0	0	0		
23B	0	0	0	0	0	0	0	0	0	0	0	0		0	0	0	0		
25B	0	0	0	0	0	0	0	0	0	0	0	0	0	0		0	0		
8B	0	-	1	0	0	0	0	0	0	0	0	0	0	0	0	0	0	1	1
21B	0	0	0	1	0	0	0	0	0	0	-	0	0	0	0	0	0		
24B	0	0	0	0	0	0	0	0	0	0	0	0	1	-	0	0	0		
26B	0	0	0	0	0	0	1	0	0	0	0	0	0	0	0	-	0		
27B	0	0	0	0	0	0	1	0	0	0	0	0	0	0	0	1	-	2	2
16B	0	0	0	0	0	-	1	0	0	0	0	1	1	1	1	1	1	7	3
14B	0	0	1	0	-	1	1	0	0	0	0	1	1	1	1	1	1	9	4
18B	0	1	1	0	1	1	1	-	0	0	0	1	1	1	1	1	1	11	5
5B	-	1	1	0	1	1	1	1	0	0	0	1	1	1	1	1	1	12	6
19B	1	1	1	0	1	1	1	1	-	0	0	1	1	1	1	1	1	13	7
20B	1	1	1	1	1	1	1	1	1	-	1	1	1	1	1	1	1	16	8

4.3 EoL Strategies

According to the criteria suggested in Table 1, only components 16A, 17A and, 16B are reusable; 24B, 27B, 20B and 4D are suitable for re-manufacturing; target components 15A, 14B, 4C and 7C can undergo recycling processes. When modelling future scenarios in SimaPro, it is supposed that re-manufacturing activities consist of working with general metal processes up to 50% of the weight of the components. Both for re-used and re-manufactured parts, there is no material phase because any resources aren't required to be extracted and worked. Three scenarios are modelled: i) the first is the one described above; ii) components supposed to be re-used are re-manufactured (17A, 16A, 16B); iii) all target components are re-cycled. The results (Fig. 3) prove that re-using and re-manufacturing are much more sustainable than the current EoL strategy. The impacts can be improved, more than 45% for certain categories. In general, scenarios 1 and 2, which include re-using and re-manufacturing of certain products, reduce the most the overall impact. The third scenario is less impacting and has between 7 and 15% lower impacts, because the recycling process requires a high quantity of energy.

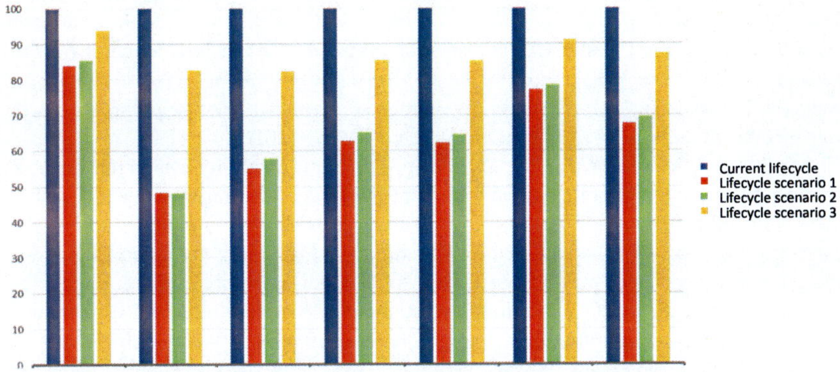

Fig. 3. Comparison between actual lifecycle and future scenarios

5 Conclusions

The proposed methodology supports the design decision-making process toward the optimization of product EoL. The four steps aim to evaluate the performance of the product EoL, to identify product criticalities and favour re-use, re-manufacturing and re-cycle. The experimentation in the context of an espresso coffee machine case study demonstrated the ability of the proposed methodology in supporting the definition of strategies (e.g., re-use, re-manufacturing) to maximize components/groups value. Despite the potential of the proposed approach, further work is needed to optimize it. The methodology, which guide in the optimization of single parameter (environmental impact of EoL phase) should be integrated with other design drivers, to have a multi-objective metric able to guide the product improvement.

References

1. Sabaghi, M., Cai, Y., Mascle, C., Baptiste, P.: Sustainability assessment of dismantling strategies for end-of-life aircraft recycling. Resour. Conserv. Recycl. **102**, 163–169 (2015)
2. Ong, S.K., Fang, H.C., Nee, A.Y.C.: A design feature-based approach for product remanufacturability assessment and analysis. Procedia CIRP **53**, 15–20 (2016)
3. Francia, D., Ponti, S., Frizziero, L., Liverani, A.: Virtual mechanical product disassembly sequences based on disassembly order graphs and time measurement units. Appl. Sci. **9**(17), 1-21 (2019)
4. Stavropoulos, P., Spetsieris, A., Papacharalampopoulos, A.: A circular economy based decision support system for the assembly/disassembly of multi-material components. Procedia CIRP **85**, 48–53 (2020)
5. Sumter, D., Bakker, C., Balkenende, R.: The role of product design in creating circular business models: a case study on the lease and refurbishment of baby strollers. Sustainability **10**(7), 2415 (2018)
6. Franco, M.A.: A system dynamics approach to product design and business model strategies for the circular economy. J. Clean. Prod. **241**, 118327 (2019)
7. Bracke, S., Yamada, S., Kinoshita, Y., Inoue, M., Yamada, T.: Decision making within the conceptual design phase of eco-friendly products. Procedia Manufact. **8**, 463–470 (2017)
8. Frizziero, L., Liverani, A., Caligiana, G., Donnici, G., Chinaglia, L.: Design for disassembly (DfD) and augmented reality (AR): case study applied to a gearbox. Machines **7**(2), 7–29 (2019)
9. Sabaghi, M., Mascle, C., Baptiste, P.: Evaluation of products at design phase for an efficient disassembly at end-of-life. J. Clean. Prod. **116**, 177–186 (2016)
10. Zhang, M., Ao, X., Cai, W., Jiang, Z., Zhang, H.: A sustainability evaluation method integrating the energy, economic and environment in remanufacturing systems. J. Clean. Prod. **239**, 118100 (2019)
11. Favi, C., Germani, M., Mandolini, M., Marconi, M.: Promoting and managing end-of-life closed-loop scenarios of products using a design for disassembly evaluation tool. In: Proceedings of the ASME 2012 International Design Engineering Technical Conferences and Computers and Information in Engineering Conference, pp. 1339–1348 (2019)
12. Poschmann, H., Brüggemann, H., Goldmann, D.: Disassembly 4.0: a review on using robotics in disassembly tasks as a way of automation. Chemie Ingenieur Technik **92**(4), 341–359 (2020)
13. Cong, L., Zhao, F., Sutherland, J.W.: A design method to improve end-of-use product value recovery for circular economy. J. Mech. Des. **141**(4), 044502 (2019)
14. Harivardhini, S., Murali, Krishna, K., Chakrabarti, A.: An integrated framework for supporting decision making during early design stages on end-of-life disassembly. J. Cleaner Prod. **168**, 558–574 (2017)
15. Belhadj, I., Khemili, I., Trigui, M., Aifaoui, N.: Time computing technique for wear parts dismantling. Int. J. Adv. Manufact. Technol. **103**(9–12), 3513–3527 (2019). https://doi.org/10.1007/s00170-019-03692-4
16. Mandolini, M., Favi, C., Germani, M., Marconi, M.: Time-based disassembly method: how to assess the best disassembly sequence and time of target components in complex products. Int. J. Adv. Manufact. Technol. **95**(1–4), 409–430 (2017). https://doi.org/10.1007/s00170-017-1201-5
17. Ardente, F., Wolf, M.-A., Mathieux, F., Pennington, D.: Integration of resource efficiency and waste management criteria in the implementing measures under the Ecodesign Directive. In: European Commission. Joint Research Centre. Institute for Environment and Sustainability (2011)

An Approach for the Responsible Design of the Packaging Employed in Household Appliances

Daniele Landi[1(✉)] and Paolo Cicconi[2]

[1] Department of Engineering Management, Information and Production, University of Bergamo, 24044 Dalmine, Italy
daniele.landi@unibg.it

[2] Dipartimento di Ingegneria, Università degli Studi Roma Tre, 00146 Roma, Italy

Abstract. The packaging is responsible for the production of a great amount of waste in the world. Every product comes with different levels of packaging to protect the product during shipping, store the content in the warehouse, and show the product to customers in retail shops. Therefore, the designer of packaging is more and more involved in a responsible analysis while defining the package configurations for a product. This paper proposes an approach to support the packaging configurations considering life cycle data, analytical structural analysis, and parametric cost modeling. Rules, formulas, and specific standards are formalized into a Knowledge Base. As a case study, the methodological approach has been applied to design the packaging of a household appliance. The results show the possibility to reduce the cost and environmental impacts of packaging by a responsible approach.

Keywords: Eco-design · Packaging · EPS · Corrugated cardboard

1 Introduction

The main function of the packaging is to protect the product during the different transportation phases. Packaging should have characteristics that add convenience in distribution, handling, stacking, using, reusing, recycling, and ease of disposal [1]. Globalization and worldwide market have been increasing the need for good transportation between the different actors in the production chain, producers, customers, and end-users [2]. In the pre-and post-consumer logistics activities, the packaging is considering one of the principal sources of consumption and environmental impact [3]. In the traditional packaging, the main materials employed are corrugated paper board [4], card, wood, and plastics such as polyurethane (EPU), polyethylene (EPE), polypropylene (EPP), and expanded polystyrene (EPS) [5]. Paper and corrugated boards are among the most widely used materials in the packaging [6, 7]. Corrugated board boxes are used to protect products against damage that may arise from handling, transportation, storage, hazards, and environmental conditions [8]. For the interior parts packaging, considering the need to

C. Rizzi et al. (Eds.): ADM 2021, LNME, pp. 381–389, 2022.
https://doi.org/10.1007/978-3-030-91234-5_39

have lightweight, easy manufacturability, and cushioning properties that guarantee good energy absorption, the EPS is the most used material [9]. The packaging configurations with corrugated board boxes and interiors in EPS are widely applied in hi-tech products to reduce damage related during the transportation phase as vibration and impacts [10]. In the scientific literature, there are various studies that investigate non-fossil-origin materials to find alternative solutions to EPS. In Europe, the design of packaging must consider the directives provided by the European Commission (EC). These EC directives have the target of achieving a market where all plastic placed is either reusable or recyclable in a cost-effective manner by 2030 [11]. Even if a lot of plastics used in the packaging of consumer goods are recyclable, an approach based on responsible design is suitable to achieve a higher reduction of environmental impacts [12]. In the literature, responsible design activities for packaging are most focused on the selection and testing of environmental-friendly materials. In this context, a specific normative has been also provided in Europe for testing the packaging recoverable through composting and biodegradation [7]. As discussed by Tencati et al. [3], it is important to adopt solutions that aim at improving the packaging design by minimizing the relative environmental impacts with a life cycle perspective. Several scholars adopt the Life Cycle Assessment (LCA) method to support the Design for the Environment activities. However, these LCA-based approaches do not propose design support in the sizing and configuration of new packaging solutions. To overcome the highlighted limits, the paper proposes an approach to configure the scheme of packaging evaluating cost, environmental impacts, and structural behavior during the conceptual phase. The outcome method is a tool to support the engineer in the responsible design of the packaging. The remainder of the paper describes the research approach in Sect. 2. A test case with results is reported in Sect. 3 to validate the proposed approach. Finally, Sect. 4 shows discussions and conclusions.

2 Materials and Method

The definition of the early packaging configuration implies the definition of the primary product protection, the cushioning selection, the pallet for the logistics scheme, etc. After this level of analysis, the second step regards the analytical analysis of three important characteristics such as structural behavior, cost, and environmental impacts.

The approach aims at optimizing the main analytical parameters related to the objectives that maximize the structural performance and the reduction of cost and environmental impacts Fig. 1. The next design phase regards the validation of the packaging study. After that, there is the phase of the details design called Modeling. Here the packaging is modeled using a 3D CAD tool before realizing the prototype for the final test phase.

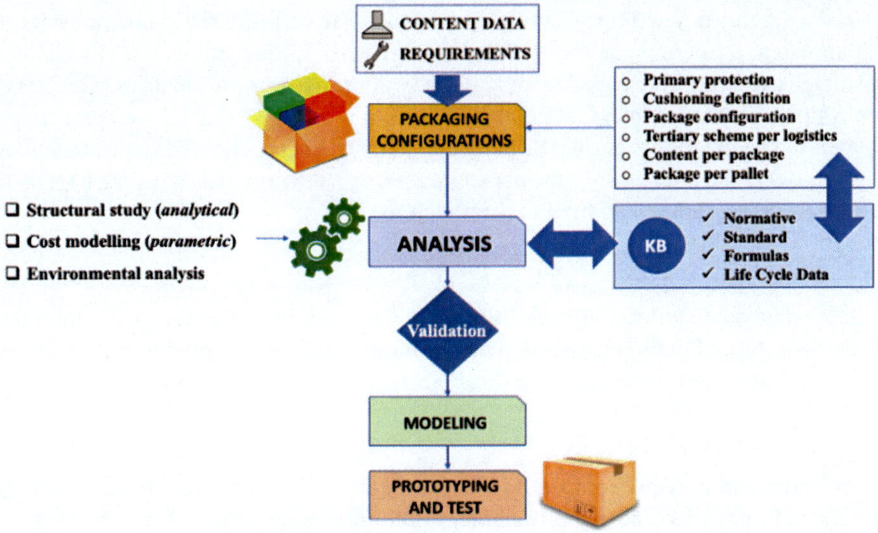

Fig. 1. The proposed design approach.

2.1 Structural Study

The structural analysis regards the study of the external cardboard box and the interior parts with cushioning properties. The cardboard box [13] has been verified considering the analytical calculation of the Box Compression Test (BCT), which has been defined by the McKee test [14]. The BCT test includes the calculation of Edge Crush Test (ECT), which is a test method used to measure the cross-direction crushing of a sample of corrugated board. The simplified form of the McKee formula (1) is most used by practitioners.

$$BCT = 5.877 \times ECT \times P \tag{1}$$

BCT is specific test for the corrugated board to obtain data and information on the ability of a board configuration to resist crushing. The ECT resistance (2) is expressed in kilonewtons per meter, and it is calculated by the following equation where F is the maximum force (measured in newtons).

$$R = 0.001 \times F \tag{2}$$

The interior parts of the packaging increase the level of product protection during the transportation and handling phases. EPS parts are very common cushioning materials in packaging. The design of a cushioning part is related to the fragility rating, represented by the Fragility factor. The Fragility factor (G) is evaluated by analyzing the deformation energy related to a drop. Equation 3 describes how the Fragility factor is defined. While the terms a and g represent the deformation and gravity accelerations (m/s^2), the terms h and d represent the drop height and the deformation value (mm).

$$G = a/g = h/d \tag{3}$$

In the literature, medium electronic appliances are considered moderately fragile with an average Fragility factor of about 80–120 G's. In this study, 115 G's has been considered as a Fragility factor for cooker hoods. As a check, the ultimate tensile stress of the cushioning material (σ_{max}, MPa) has been compared in (4) with the maximum stress achieved after a drop considering the Fragility factor of the product. The stress calculated in (4) considers the total mass of the packaging (*m*, kg), the gravity acceleration (*g*, m/s^2), the contact area (*A*, m^2), and the Fragility factor (*G*).

$$\sigma_{max} \geq (m \times g) \times \mathrm{G}/a \tag{4}$$

After this verification, the Fragility factor has been used for sizing the minimum thickness of the cushioning material. The minimum thickness (*s*, m) has been evaluated by the maximum deformation after a drop. This value is calculated considering Eq. (5).

$$s = (h/G) \times \mathrm{C} \tag{5}$$

This equation derives from Eq. (3), but it includes the buffer factor (C) which has been added to considers the compressibility of the cushioning material. The buffer factor is a value between 2.5 and 3.5 for EPS packaging. In this study, the buffer factor is 3.5 and the drop height is 0,8 m.

3 Test Case

To investigate the packaging redesigning process and shows the advantages of proposed methodology, hood family (vertical hoods, specifically, a series of hoods with the same external shape) were selected. The starting scheme of hood packaging is described in Fig. 2. The external part of packaging is a corrugated board. The internal parts consist of four lateral heads (2), one package base (1) in EPS, one corrugated board U-bar for the chimney protection (5).

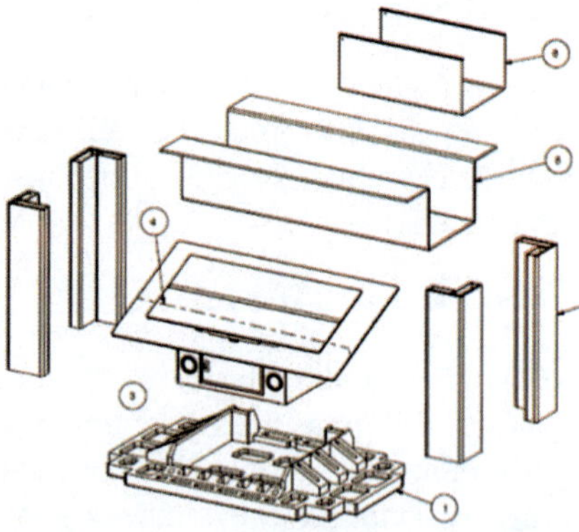

Fig. 2. Starting packaging used for a standard vertical cooker hood.

Test case study includes three different models. The first model, called *Case A*, is the reference and current packaging. Case B and Case C have been analyzed to demonstrate the advantages of the proposed methodology. These two cases regard the redesigning of the EPS base to measure the advantages (if any). Since each cooker hood belongs to a product family, each packaging structure has been analyzed for two deferment sizes: 60 cm and 90 cm, the length of the product. Figure 3 describes the geometry of the EPS base used in the packaging model called *Case A*. This model is the reference packaging already described in Fig. 2. The main components of this packaging consist of an EPS base (Fig. 3), four cardboard lateral heads (Fig. 2), and an external cardboard box.

Fig. 3. The EPS base analyzed in Case A for the sizes 60 cm (on the left) and 90 cm (on the right).

Figure 4 shows the EPS base redesigned for the model *Case B*. Even if the maximum dimensions on the base plan XY are the same, the vertical geometry related to the *Case B* model is different. The objective of this redesign is to eliminate the employment of the four lateral heads made in cardboard. This EPS base also shows a simplified geometry that reduces undercuts improving the positioning and fixing of the product. The advantages of this EPS model are the reduction of cardboard material and the use of the same sizing for the external box.

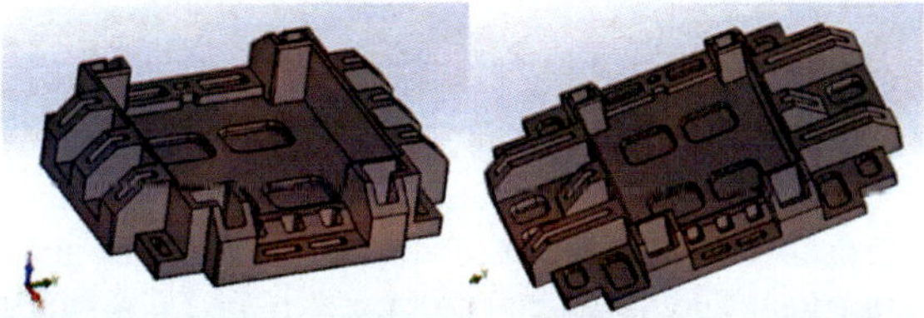

Fig. 4. The EPS base analyzed in Case B for the sizes 60 cm (on the left) and 90 cm (on the right).

The 60-cm base achieves a weight of 336 g (before it was 272 g). The 90-cm base achieves a weight of 516 g (before it was 396 g). This third case (*Case C*) shows a complete redesign of the packaging base, substituted with a molded pulp solution to optimize environmental impacts and cost. One of the advantages of this solution is the elimination of EPS parts, about 40% of the total cost. On the other hand, molded pulp is a recycled material and comes from the waste of paper, carton boards, etc. This recycled material also shows properties such as high recyclability and several times, acceptable mechanical properties, and the possibility to be molded in different shapes (Fig. 5).

Fig. 5. The redesigned packaging base in molded pulp (Case C) for the 60-cm size.

3.1 Results

For the structural behavior, according to UNI EN ISO 4180:2010 and ISTA (International Safe Transit Association), three different tests have been performed on physical prototypes. These tests are stacking/compression (evaluated by BCT), fall test, and vibration test (UNI EN ISO 2233:2001). Table 1 shows the results of these packaging tests. The value of BCT was obtained according to standard FEFCO N° 50. This value indicates the maximum vertical load expressed in kilograms for cardboard. The Fall test has been carried out according to ISO 2248:1985 standard. Each model passed the fall and vibration test. Moreover, the achieved BCT value was similar for each model. The BCT values are close because the box dimension is the same and only a little difference in the cardboard composition has been applied in each model.

Table 1. Result of the structural tests applied to each packaging model.

Model	BCT	Fall	Vibration
A	370 kg	Ok	ok
B	368 kg	ok	ok
C	367 kg	ok	ok

A life cycle approach has been used to evaluate the environmental impacts and estimate the packaging cost. The functional unit used in the LCA study is "to protect the product". In particular, the analyzed protection of cooker hoods is complied with EN and ISO standard. Considering the aim of the study, the system boundaries include the production of raw material, all transport of material and components, and the process for the realization of the packaging component [15]. The end of life has not been considered.

The primary data has been collected through measurements and surveys with experts, the secondary data has been retrieved from the literature and LCA databases such as Ecoinvent. Table 2 reports the flows related to the different packaging models. Each model has a similar external cardboard box but a different interior base. The method ReCiPe 2016 mid-point - Hierarchist (H) version has been chosen and used to assess the environmental impacts. The assessment (Fig. 6) shows the greater impact contribution of the starting packaging model over the redesign models (B and C) in all categories. Moreover, Case C shows a reduction of environmental impacts. The average reduction is about 10% in each category.

Table 2. Component list for each packaging model with the description of the flows.

Category	Flow A	Quantity A	Cost € A	Flow	Quantity B	Cost € B	Flow C	Quantity C	Cost € C
External box	Corrugated board	2950 g	0.92	Corrugated board	2950 g	0.92	Corrugated board	2950 g	0.92
	Printing ink	0.48 g		Printing ink	0.48 g		Printing ink	0.48 g	
	Water	2.32E-03 l		Water	2.32E-03 l		Water	2.32E-03 l	
	Printer slotter	0.02 kWh		Printer slotter	0.02 kWh		Printer slotter	0.02 kWh	
	Cut and fold machine	1.95E-02 kWh		Cut and fold machine	1.95E-02 kWh		Cut and fold machine	1.65E-02 kWh	
	AGV handling	2.24E-04 kWh		AGV handling	2.24E-04 kWh		AGV handling	1.38E-04 kWh	
	Press and wrapping	3.00E-04 kWh		Press and wrapping	3.00E-04 kWh		Press and wrapping	1.98E-04 kWh	
U-bar	Kraft paper	395 g	0.64	Kraft paper	395 g	0.64	Kraft paper	395 g	0.64
	Cut and pre-creasing	1.71E-02 kWh		Cut and pre-creasing	1.71E-02 kWh		Cut and pre-creasing	1.71E-02 kWh	
4 lateral heads	Corrugated board	644 g	0.25	Corrugated board	/	/	Corrugated board	/	/
	Cut and pre-creasing	3.25E-02 kWh		Cut and pre-creasing	/		Cut and pre-creasing	/	
Package base	Expandable Polystyrene (EPS)	396 g	0.89	Expandable Polystyrene (EPS)	516 g	0.96	Molded pulp	786 g	0.67
Paper clips	Steel	15 g	0.02	Steel	15 g	0.02	Steel	15 g	0.02
Total cost			2,72			2,54			2,25

The redesign of the packaging base described in Case B and C reduces the use of EPS materials and avoids the employment of the four lateral heads. Moreover, the proposed solutions also achieve the same mechanical performance as the reference model (Case A). These solutions show a cost reduction of 6,7% for Case B and 17% for Case C (Table 2).

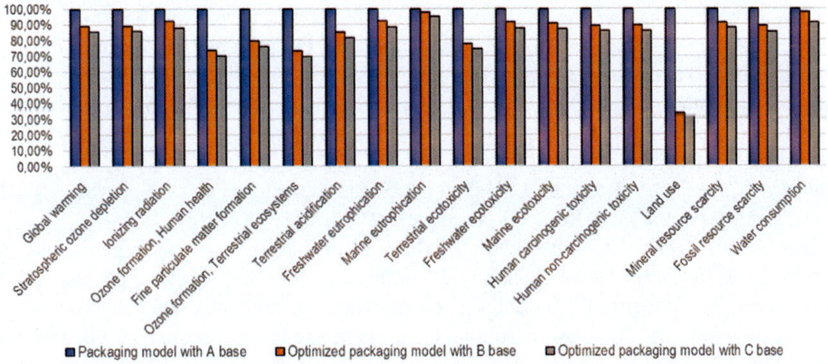

Fig. 6. The comparison of the three model of packaging in terms of ReCiPe characterization.

4 Conclusion

This paper proposed a study, development, and application of a methodological approach to support the sustainable design of packaging for household appliances. The aim of this approach is to fill the distance between the eco-design guidelines and the packaging requirements. The final objective is to reduce the environmental impacts by introducing a new design method, new shape, new material.

This paper compares three packaging solutions for cooker hoods in terms of environmental impacts, cost, and structural behavior. While structural testing has been performed on physical prototypes, LCA and LCC studies have been carried out using life cycle tools. Before the physical prototyping, the BCT values have been estimated using an analytical approach. Focusing on the proposed test case, two solutions of packaging have been proposed. These solutions achieve the same level of the structural behavior of the reference model (Case A); however, Case B and C show some advantages in terms of environmental impacts and cost reduction. The results described in this paper may be used by appliances manufacturers to improve performances and design solutions of their different packaging.

References

1. Fadiji, T., Coetzee, C.J., Berry, T.M., Opara, U.L.: Investigating the role of geometrical configurations of ventilated fresh produce packaging to improve the mechanical strength – Experimental and numerical approaches. J. Food Packag. Shelf Life **20**, 100312 (2019)
2. Reuters: Alibaba's Singles' Day Sales Smash Record With $25 Billion Haul. http://fortune.com/2017/11/11/china-singlesday-alibaba-2/. Accessed 25 Jan 2021
3. Tencati, A., Pogutz, S., Moda, B., Brambilla, M., Cacia, C.: Prevention policies addressing packaging and packaging waste: some emerging trends. Waste Manage. **56**, 35–45 (2016)
4. Chen, J., Zhang, Y., Sun, J.: An overview of the reducing principle of design of corrugated box used in goods packaging. Procedia Environ. Sci. **10**, 992–998 (2011)
5. Landi, D., Postacchini, L., Cicconi, P., Ciarapica, F.E., Germani, M.: Study and design of sustainable packaging for household hoods. In: Proceedings of the ASME Design Engineering Technical Conference (2018)
6. Zaheer, M., Awais, M., Rautkari, L., Sorvari, J.: Finite element analysis of paperboard package under compressional load. Procedia Manufact. **17**, 1162–1170 (2018)
7. Eriksson, D., Korin, C., Thuvander, F.: Damage to carton board packages subjected to concentrated loads. In: 19th IAPRI World Conference on Packaging, pp. 172–182 (2014)
8. Navaranjan, N., Johnson, B.: Modelling and experimental study of creep behaviour of corrugated paperboard. In: 60th Appita Annual Conference and Exhibition, pp. 43–50, Melbourne, Australia (2006)
9. Ling, C., Cardiff, P., Gilchrist, M.D.: Mechanical behaviour of EPS foam under combined compression-shear loading. Mater. Today Commun. **16**, 339–352 (2018)
10. Kun, G., Xi, W.: Design and analysis of cushioning packaging for home appliances. Procedia Eng. **174**, 904–909 (2017)
11. EC 2018: A European strategy for plastic in a circular economy (2018). http://ec.europa.eu/environment/circular-economy/pdf/plastics-strategy.pdf
12. Landi, D., Cicconi, P., Germani, M.: Analyzing the environmental sustainability of packaging for household appliances: a test case. Procedia CIRP **90**, 355–360 (2020)

13. ISO 3037:2013: Corrugated fibreboard —Determination of edgewise crush resistance (unwaxed edge method)
14. McKee, R.C., Gander, J.W., Wachuta, J.R.: Compression strength formula for corrugated boxes. Paperboard Packag. **48**(8), 149–159 (1963)
15. Wikström, F., Williams, H., Verghese, K.: The influence of packaging attributes on consumer behaviour in food-packaging life cycle assessment studies - a neglected topic. J. Clean. Prod. **73**, 100–108 (2014)

Implementation of Eco-Design Actions in the Development of a Sustainable Heat Exchanger in a Life Cycle Perspective

Núria Boix Rodríguez[1](✉), Marta Rossi[2], Federica Cappelletti[2], and Claudio Favi[1]

[1] Department of Engineering and Architecture, Università degli Studi di Parma, Parco Area delle Scienze 181/A, 43124 Parma, Italy
nuria.boixrodriguez@unipr.it

[2] Department of Industrial Engineering and Mathematical Sciences, Università Politecnica delle Marche, Via Brecce Bianche, 60131 Ancona, Italy

Abstract. According to the European energy consumption reports, the highest energy consumption in residential sector is due to space heating, followed by water heating. Generally, the product used to warm water in residential building is the boiler where heat exchanger is the core of the system. The paper aims to develop a novel concept of heat exchanger by following eco-design actions retrieved by the analysis of life cycle performance. Several eco-design actions were put into practice to reduce the environmental issues in each phase of the life cycle. Concerning the materials and manufacturing phase, a novel design based on different material (i.e., stainless steel) was developed to replace a mix of materials (i.e., copper and aluminum alloy). Concerning the use phase, the overall product efficiency was increased allowing important savings in terms of gas/energy consumptions. Finally, concerning the end-of-life phase, brazing processes was replaced by other joining processes to increase component's disassembly and varnishing process was avoided due to the better corrosion resistance performance of the stainless steel. The new heat exchanger shows better environmental performance in each Life Cycle Assessment indicator, saving more than 40% in CO_2 emissions (GWP) in the whole product life cycle.

Keywords: Heat exchanger · LCA · Ecodesign · Life cycle · Energy consumption

1 Introduction and Research Context

Heat exchangers play a key role in devices such as boilers. According to the European energy consumption reports, the highest energy consumption in residential sector is due to space heating, followed by water heating [1]. Space heating refers to the energy necessary to provide heat in an interior of a dwelling, different types of space heating equipment can be identified (e.g., central steam/hot water space-heating, heat pumps, reversible heat pumps, electric storage heater, etc.) [2]. On the other hand, water heating

C. Rizzi et al. (Eds.): ADM 2021, LNME, pp. 390–398, 2022.
https://doi.org/10.1007/978-3-030-91234-5_40

refers to the energy necessary to heat water for different purposes (e.g., cleaning, hot running water, etc.), being able to carry out this process together with the space heating (e.g., using combi boiler) or separately (e.g., water heater/boiler) [2]. Due to the great relevance of these devices in residential energy consumption, the EU has adopted various directives to improve efficiency and reduce energy demand in the building sector, which is responsible for 36% of the EU CO_2 emissions [3]. The global market for heat exchangers is expected to grow significantly [4], which makes clear the need of researches of these devices to make them more sustainable. There are not many studies on heat exchangers, due to the fact that researches in literature calculate the impact of the boiler as a whole, or the impact of the dwelling. Shah et al. [5] have carried out an analysis of different heating systems in different climate regions; with identical building dimensions but specific construction characteristics for each territory. This research shows the importance of climate, construction (i.e., insulation) and energy mix in environmental impacts, since there is no heating system that performs best for all scenarios simultaneously. Studies like those by Vignali [6] and Tremonti et al. [7] have exposed that latest generation of boilers (i.e., condensing boilers) have better environmental performances than traditional ones. Šulga [8] demonstrates the benefits of applying eco-design actions to two boilers, however this study is focus again on boilers and does not offer the possibility of analysing and applying improvements to the heat exchanger in particular. Considering the few Life Cycle Assessment (LCA) studies carried out on this topic, the main purpose of this paper is to conduct an LCA analysis of traditional heat exchangers used in domestic boilers that allows the development of eco-design measures, increasing the overall sustainability of heat exchangers. The novelty of this article is to overtake the main limitation of general studies, providing a research regarding the environmental performance of a traditional heat exchanger (individually, not with a boiler), which permits the improvement of its environmental sustainability and facilitate eco-design decision-making. Section 2 describes the methods adopted for the development of this research. Section 3 discusses the results and how to use them for eco-design initiatives. Finally, Sect. 4 presents conclusions and future works.

2 Materials and Methods

This research work is based on three phases, as can be seen in Fig. 1.

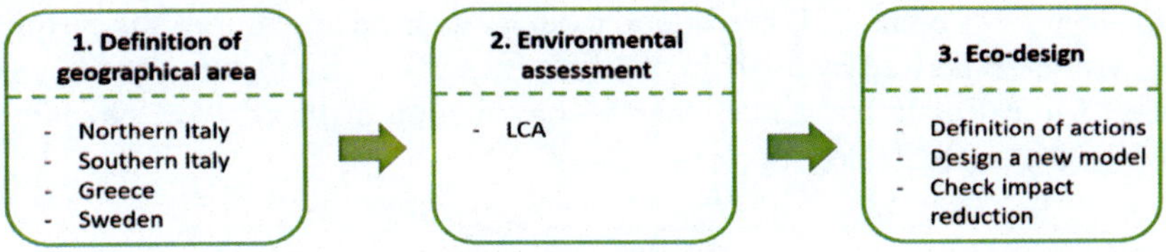

Fig. 1. Project methodology

In the first phase, the study region must be selected. For this project has been decided to focus the analysis on Italy, divided into north (N-IT) and south (S-IT), Greece (GR) and Sweden (SE). For each area, the average number of hours that the boiler is on has been obtained from the literature [9–11]. Table 1 shows these results. In the second phase, environmental analyses have been conducted following the LCA methodology to determine the impact of a traditional heat exchanger in the four different scenarios described above. The analysis was performed in line with the ISO 14040:2006 [12], which identifies four phases: (i) Goal and Scope definition, (ii) Life Cycle Inventory (LCI), (iii) Life Cycle Impact Assessment (LCIA), and (iv) Interpretation. According to this methodology, the functional unit was defined as: "*To increase the water temperature to fulfil the domestic hot water demand of a flat where three persons live during a useful life of 7 years*". This functional unit has as boundary condition the use of a 24 kW power heat exchanger. The system boundaries included the following phases: (i) material extraction, (ii) manufacturing, and (iii) use. The maintenance, transport and End-of-Life phases fell outside the system boundaries.

Table 1. Average number of hours with the boiler on per region.

Season	Unit	N-IT	S-IT	GR	SE
Winter	[h/day]	9	6	5	10
Summer	[h/day]	0	0	0	0
Spring/Autumn	[h/day]	7	4	5	9
Winter	[month/year]	3	3	3	3
Summer	[month/year]	0	0	0	0
Spring/Autumn	[month/year]	6	6	2	6

For the LCI of the manufacturing phase a virtual approach was adopted. Bill of Materials (BoM) and CAD have been used to obtain data (e.g., dimensions, weights, etc.). Ecoinvent database was used for secondary data. For the LCI of the use phase, it has been considered that the studied product, a traditional finned-tube heat exchanger (Trad. HE), works at maximum power (24 kW) 20% of time, while the other 80% works at 10 kW. The useful life of 7 years has been established based on warranty average [13] in addition to expert's opinion. These values, together with the estimation of use explained above, have been derived by the help of a heat exchanger manufacturer and have been considered to calculate the annual energy consumption of this device, obtaining the following results (see Table 2).

Table 2. Gas and heat needed during the use-phase.

Type		Unit	N IT	S IT	GR	SE
Space heating	Gas needed for 7 years	m3	2,0E + 04	1,2E + 04	7,1E + 03	2,4E + 04
	Heat needed for 7 years	MJ	7,0E + 05	4,3E + 05	2,5E + 05	2,5E + 05
Water heating	Gas needed for 7 years	m^3	1,5E + 03	1,3E + 03	1,3E + 03	1,8E + 03
	Heat needed for 7 years	MJ	5,3E + 04	4,5E + 04	4,8E + 04	6,3E + 04

For the Life Cycle Impact Assessment (LCIA) specialized software (SimaPro 9.0.0.49) was used and the ReCiPe Midpoint (H) was adopted, selecting the most representative midpoint indicators for this study. Table 3 shows the 6 indicators analysed. Finally, for the eco-design phase, the Life Cycle Thinking (LCT) approach has been followed together with the methodology presented in the standard ISO 14006:2011 [14] (updated in 2020). Which is based on the following phases: (i) Define product functions, (ii) environmental analysis, (iii) environmental improvement strategies, (iv) develop environmental objectives, (v) environmental product specification, and (vi) develop technical solutions. Thus, in Sect. 3 an alternative heat exchanger model (New HE) is proposed to compare the environmental impacts of both devices and check if the eco-design actions taken lead to benefits on the environmental performance of the product.

Table 3. ReCiPe Midpoints (H) indicators.

Name	Acronym	Unit
Global warming potential	GWP	[kg CO2 eq]
Ozone depletion potential	ODP	[kg CFC11 eq]
Photochemical oxidant formation potential	OFP	[kg NOx eq]
Particulate matter formation potential	PMFP	[kg PM2.5 eq]
Human toxicity potential: carcinogenetic	HTPc	[kg 1,4-DCB]
Fossil fuel potential	FFP	[kg oil eq]

3 Results Discussion

This section shows the results of LCA analysis of a traditional heat exchanger (Fig. 2).

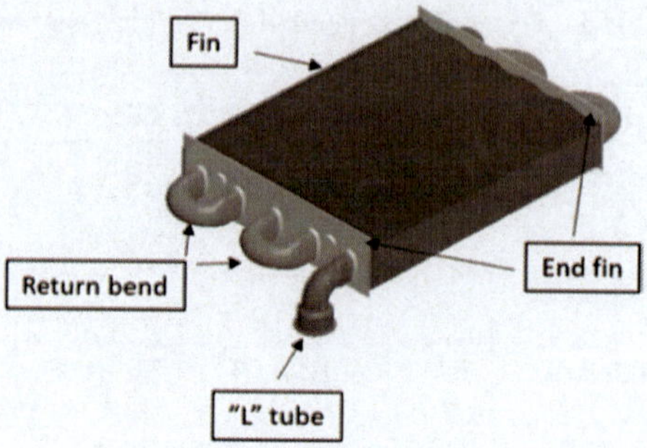

Fig. 2. Traditional finned-tube heat exchanger components

Figure 3 A) shows the values for each indicator studied during the assembly and manufacturing phase. Figure 3 B) displays results of the assembly phase, being fins the component that causes the greatest impact in all categories (more than 60%), since it is the one that requires more material for its production. Manufacturing processes are less relevant compared to the impact of raw material, which is mainly copper. In this way, the total impacts of heat exchanger components for this phase are proportional to the amount of material required for its production. The different parts of the heat exchanger have been grouped considering their function: (i) water transport, composed of tube, return bend and "L" tube assembly; (ii) improvement of heat exchange, made up of fin assembly; and (iii) assembly and protection, composed of end-fin assembly. Thus, facilitating the analysis and interpretation of the results. As expected, assembly fin has the greatest values of environmental impacts in all indicators. So, it is evident that efforts of eco-design actions should be focused on this part of the heat exchanger.

A)

Midpoint	Trad. HE
GWP	6,9E+00
ODP	1,1E-05
OFP	4,5E-02
PMFP	6,3E-02
HTPc	2,5E+00
FFP	1,7E+00

B)

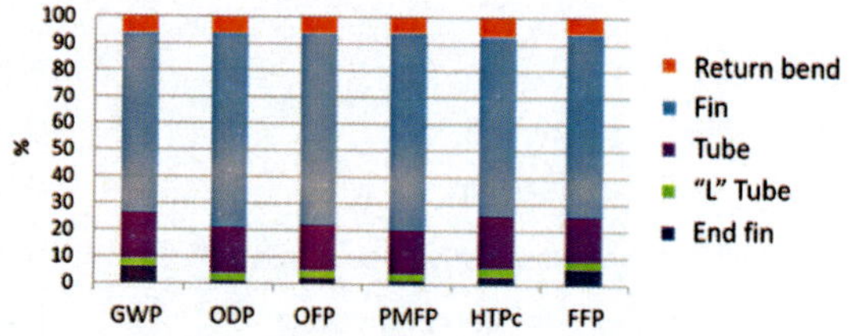

Fig. 3. A) ReCiPe Midpoints (H) indicators (assembly and manufacturing); **B)** Component's impact in each Midpoint indicator for the assembly and manufacturing

Results related to the entire life cycle (materials, manufacturing and use phase) for the different scenarios studied are shown below (see Table 4).

Comparing the case of Italy, it can be seen how in all indicators the impact of N-IT scenario is always greater than S-IT, with an average difference of 35%. This is due to colder temperatures and higher yearly use of the boiler in the first scenario. GR values are lower in all categories, and SE scenario has the highest levels in all indicators. Once LCA analysis has been carried out, eco-design actions can be established in order to reduce

Table 4. ReCiPe Midpoints (H) indicators for each scenario.

Midpoint	N-IT	S-IT	GR	SE
GWP	8,4E + 03	5,3E + 03	3,4E + 03	1,0E + 04
ODP	1,1E-02	6,7E-03	4,3E-03	1,3E-02
OFP	1,9E + 01	1,2E + 01	7,4E + 00	2,2E + 01
PMFP	7,1E + 00	4,5E + 00	2,9E + 00	8,6E + 00
HTPc	1,4E + 02	8,7E + 01	5,7E + 01	1,7E + 02
FFP	2,1E + 04	1,3E + 04	8,5E + 03	2,6E + 04

its environmental impact. An extract of these actions is shown in Table 5. Following the guidelines exposed above (obtained by applying the LCA methodology to HE) together with meetings with industry specialists, a new heat exchanger can be developed. The New HE is a stainless-steel serpentine heat exchanger with oval section (see Fig. 4). In this case, serpentine allows water transportation and thermal exchange simultaneously. This design was performed in collaboration with the heat exchanger manufacturer and it seeks to optimize efficiency during the use phase. As in the previous case (Trad. HE) it works at maximum power (24 kW) 20% of time, but now the other 80% works at 3 kW instead of 10 kW.

Table 5. Eco-design guidelines.

Lifecycle phases	Criticalities	Eco-design actions
Material and manufacturing	High impacts due to the use of copper as raw material for the heat exchanger	To use a different material with high thermal conductivity and high corrosion resistance (i.e., stainless steel)
	Several components with different materials	To design components with more than one functionality and same material
Use	The efficiency of this product is not the maximum value that it can be achieved	Optimize efficiency, achieving better space and water heating performance
End-of-Life	Difficulties in separating components and materials	To avoid welded components, designing parts that can be disassembled To avoid varnish components using stainless steel, which already has good resistance to corrosion

Figure 5 shows how this New HE has higher impacts in all categories. The average difference between both models is approximately 70%, due to the fact that it uses a greater number of components and entails bigger consumption of raw materials. These greater impacts during the materials and manufacturing phase will be compensated throughout its entire life cycle.

Fig. 4. New HE.

The behavior during the use phase in the different scenarios is analogous to Trad. HE (see Table 6). Normalized results of the entire life cycle are shown in Fig. 6. Figure 6 reports the percentage of GWP indicator obtainable with the New HE, being the trend analogous in the other indicators studied. For the specific case of Italy, even the worst scenario (N-IT) has lower environmental impacts with the New HE than the best scenario (S-IT) with the Trad. HE. With this New HE, impacts of SE (the most unfavorable scenario) are lower than Trad. HE in N-IT region, with a difference of 18% in indicators such as GWP or FFP. The importance of the geographical area (i.e., average outside temperature) is shown by the GR results, that already in its Trad. HE version presents lower impacts than the optimized New HE for SE and N-IT scenarios, having also very similar values to those of New HE in S-IT (with a difference of 13% in indicators such as GWP or ODP).

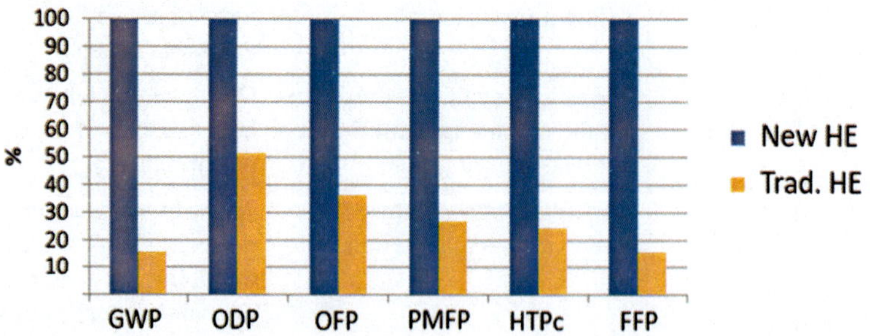

Fig. 5. Normalized Midpoints for the materials and manufacturing (New HE vs Trad. HE).

Table 6. ReCiPe Midpoints (H) indicators use-phase for each scenario (New HE).

Midpoint	N-IT	S-IT	GR	SE
GWP	4,6E + 03	2,9E + 03	2,0E + 03	5,6E + 03
ODP	5,8E-03	3,7E-03	2,5E-03	7,1E-03
OFP	1,0E + 01	6,5E + 00	4,4E + 00	1,2E + 01
PMFP	4,0E + 00	2,7E + 00	1,9E + 00	4,8E + 00
HTPc	8,3E + 01	5,7E + 01	4,2E + 01	9,9E + 01
FFP	1,2E + 04	7,4E + 03	4,9E + 03	1,4E + 04

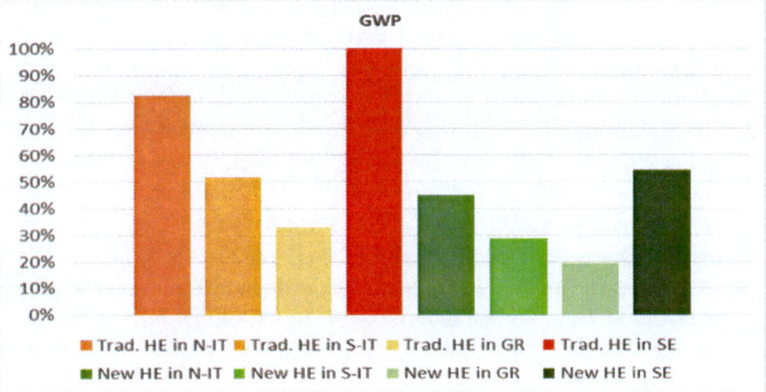

Fig. 6. Normalized GWP Midpoint for the entire life cycle (Trad. HE vs New HE).

4 Conclusions

The present paper deals with the adoption of eco-design actions for the development of environmentally sustainable heat exchangers. The approach used in this work followed the well-known LCA methodology and outcomes from the life cycle analysis were used to derive eco-design actions. Results highlight how the increment of environmental impacts for the materials and manufacturing phase was recovered by the improvement in terms of product efficiency which exhibits a reduction of life cycle GWP and ODP of approx. 45%, and a variable reduction of HTPc up to 37% based on the analyzed scenarios. The overall performance in a life cycle perspective seems very promising to reduce the greenhouse gas emissions in residential sector which is a critical topic in the EU agenda (European Green Deal). Moreover, the new product shows important features in terms of disassembly, increasing the product recycling at the end-of-life. Some questions are still uncovered, such as the possibility to develop the same product concept for different power and the development of specific features to increase the product performance (i.e., efficiency) based on different geographic scenarios.

References

1. Energy consumption in households – Statistics Explained – Eurostat. https://ec.europa.eu/eurostat/statistics-explained/index.php?title=Energy_consumption_in_households. Accessed 19 Mar 2021
2. EUROSTAT: Manual for statistics on energy consumption in households. Publications Office of the European Union, Luxembourg (2013)
3. Directive (EU) 2018/844 of the European Parliament and of the Council of 30 May 2018 amending Directive 2010/31/EU on the energy performance of buildings and Directive 2012/27/EU on energy efficiency (Text with EEA relevance)
4. Heat Exchangers Market Size and Share with Industry Forecast By 2026, https://www.alliedmarketresearch.com/heat-exchanger-market. Accessed 19 Mar 2021
5. Shah, V.P., Debella, D.C., Ries, R.J.: Life cycle assessment of residential heating and cooling systems in four regions in the United States. Energy Build. **40**(4), 503–513 (2008)
6. Vignali, G.: Environmental assessment of domestic boilers: a comparison of condensing and traditional technology using life cycle assessment methodology. J. Clean. Prod. **142**, 2493–2508 (2017)
7. Tremonti, M., Breedveld, L., Padovan, G., Bosio, S., Corsini, M.: LCA di caldaie a condensazione nell'ambito delle Direttiva EuP1. Ambiente, Risorse e Salute 117 (2008)
8. Šulga, M.: Life Cycle Assessment (LCA) based environmental impact minimization of solid fuel boilers in Lithuanian industry. Environ. Res. Eng. Manag. **58**(4), 74–84 (2011)
9. ISTAT.: I Consumi Energetici Delle Famiglie. ISTAT, 120 (2014)
10. Davaki, M.: Analysis of energy use in typical Greek residential buildings and proposed retrofit strategies. Ph.D. thesis. Georgia Institute of technology (2011).
11. National Research Council.: District Heating And Cooling in the United States: Prospects and Issues. National Academies Press (1985)
12. ISO 14040:2006 Environmental management - Life cycle assessment - Principles and framework. International Organization for Standardization: Geneva, Switzerland (2006)
13. Boiler Warranties & Guarantees – A Guide to Good Value. https://www.theheatinghub.co.uk/worcester-bosch-guarantee-ideal-boilers-warranty. Accessed 19 Mar 2021
14. ISO 14006:2011 Environmental Management Systems - Guidelines for Incorporating Ecodesign. International Organization for Standardization: Geneva, Switzerland (2011)

Design for Additive Manufacturing

The Role of Rapid Prototyping Devices in the Design and Manufacturing Practices of FabLab Visitors: A Survey

Aurora Berni(✉), Florian Dallago, Lorenzo Maccioni, Franco Concli, and Yuri Borgianni

Faculty of Science and Technology, Free University of Bozen-Bolzano, Piazza Università, 3, 39100 Bolzano, Italy
{aurora.berni,florian.dallago}@natec.unibz.it, {lorenzo.maccioni,franco.concli,yuri.borgianni}@unibz.it

Abstract. The diffusion of Fab Labs and the continuous development of Additive Manufacturing technologies are undoubtedly two relevant phenomena nowadays. The former fuels the latter and vice versa, but their mutual relationship has not been systematically explored so far. This paper presents an exploratory study based on a survey, in which various aspects of the use of 3D printers in Fab Labs are investigated. The results show how different scales of Fab Labs are present in terms of investments made in 3D printing, as well as work purposes are not negligible for many attendees. In addition, the outcomes of the survey show that 3D printers are considered easy to use, as well as manufactured parts are deemed satisfactory. Despite this, there are still some barriers to boosting the market of domestic 3D printers for Fab Lab visitors.

Keywords: Fab Labs · 3D printers · Skills · Design for Additive Manufacturing · CAD

1 Introduction and Background

The rise of laboratories with cutting-edge technologies accessible to local communities is certainly one of the most evident signs of the process of democratizing manufacturing. By the way, a strong correlation is observed between fabrication laboratories, their tools, and their community. According to [1], those can be classified in Hackerspaces, Maker Spaces and Fab Labs. In Hackerspaces, members are more focused on electronics projects, while Maker Space is the general term to indicate any shared workshop. Fab Labs are places equipped with different kind of tools such as 3D printers, laser cutters, and drills, among others, for peer-to-peer production. Here people can work, learn, and share information or skills. Fab Labs have a common identity but each one is free to determine its own projects or activities and its own target user group. Santos et al. [2] limited their research on the concept of Fab Lab and compared different Fab Labs all over the world to understand if the economic context affects their development and features.

C. Rizzi et al. (Eds.): ADM 2021, LNME, pp. 401–409, 2022.
https://doi.org/10.1007/978-3-030-91234-5_41

Other scholars categorized these environments taking into account Maker spaces as a macro-category to be compared to public Print centers. According to [3], a substantial difference exists between public Print Centers and traditional Maker Spaces. Visitors usually find more experienced people in the latter, which make Maker Spaces more popular than simple public Print Centers.

As it can be noticed, it is almost impossible to separate completely Fab Labs and Maker Spaces from their target users, but the reasons behind the use of 3D printers, supposedly devices targeting Rapid Prototyping to the largest extent, in these laboratories might differ substantially. In addition, while the use of Additive Manufacturing (AM) in industrial settings is established and much information is available about its pros and cons [4], insufficient scientific research has been carried out to study the appropriateness of 3D printers for other scopes. By the way, with regard to industrial settings, the perceived advantages and disadvantages ascribable to AM as reported in [4] are substantially aligned with those claimed in textbooks and commonly described in educational experiences. More precisely, while advantages are seen in allowed geometric complexity (consistently with principles of Design for Additive Manufacturing, DfAM, [5]) and time reduction to produce small batches, accuracy and mechanical properties are still seen as weaknesses of AM processes.

Between the 1990s and the early 2000s, just few curious people and enthusiasts approached and explored the world of 3D printing, but the numbers rose quickly especially starting from 2005 [6]. In order to acquire preliminary knowledge on the topic, the authors focused on Fab Labs and Maker Spaces, where, as aforementioned, 3D printers are expectedly a major asset for prototyping and fabricating items. However, as these laboratories are acquiring visibility and improving their equipment, their 3D printing services might be directed to manifold stakeholders. Indeed, although AM technologies were originally developed to target the industrial production, many companies might avoid investing on 3D printers because of the availability of Fab Labs and Maker Spaces. Professional 3D printer machines and their maintenance costs are indeed not always affordable for some companies. Other barriers that hinder the complete adoption of this technology are: a) the low awareness of 3D printing and its capabilities; b) the difficulties to access the 3D printing supply chain; c) the limited choice of materials; d) the required investment in professional 3D CAD systems [7].

It follows that the use of 3D printers in Fab Labs and Maker Spaces deserves specific attention, as AM technological developments might address uncovered needs related to these specific environments. The overview of previous studies provides potentially useful information on the topic, but it clearly emerges that the subject has not been studied systematically yet. According to [3], motivation and advanced skills make the difference between the so-called "Makers" and the "Casual users" approach to the 3D printing process. The former has already experienced such a technology. They are usually people who need 3D printing machines for work reasons and include designers, engineers, architects, students, or enthusiasts who are genuinely interested in learning the process. Casual makers, instead, find their motivation in the final output and they see the 3D printing process as a means-to-an-end. They do not have a knowledge background about the hardware and software related to 3D printing. Their learning time is generally slower compared to experienced makers. That is why casual makers usually need the external help of a Print Center Operator. As aforementioned, these places are not for peer-to-peer production and the operators guide and help the user during the whole process. However, even though Makers' learning processes are faster, and their motivation is higher, they share some difficulties with the Casual makers. Such difficulties are found especially in the 3D modelling phase and in software and hardware interfaces. This happens because 3D printing is rooted in the industrial field and professional skills are often needed. 3D printers' interfaces turned not to be user friendly as rapidly as their widespread private use.

With reference to non-industrial use, it is worth remarking that high value is given to self-designed and self-printed objects, thanks to an engendered affective bond [8]. By the way, although objects created with low-performing AM technologies are overall deemed inappropriate to replace commercial products, the perception gap is significantly lower than the actual quality gap [9]. This further supports the potential of Fab Labs as alternative production niches for end-use 3D-printed products beyond attendees' mere hobby and curiosity motivation.

As a result, motivations, skills, learning processes, the presence of industrial players using 3D printers in Fab Labs and Maker Spaces, typically available outfit, and barriers to domestic use are the main issues to be studied to the full understanding of the topic. The following Section presents an investigation aimed at clarifying these aspects within Fab Labs and Maker Spaces (the term "Fab Labs" will be used hereinafter to indicate both kinds of laboratories). The final Section discusses the results and draws conclusions.

2 Investigation into the Use of 3D Printers in Fab Labs

The authors launched an eight-question online survey to acquire information related to the use of 3D printers in Fab Labs. The questionnaire was prepared in three languages (English, Italian and German) and the invitation to fill it in was sent out to email addresses found in websites of online communities coordinating Fab Labs. The popular tool Surveymonkey was used for the survey, as it allows proponents to formulate questions requiring different forms of answers (open answers, Likert scales, multiple choice, etc.). In particular, 5-point Likert scales (see Fig. 1) were used for answers meant to

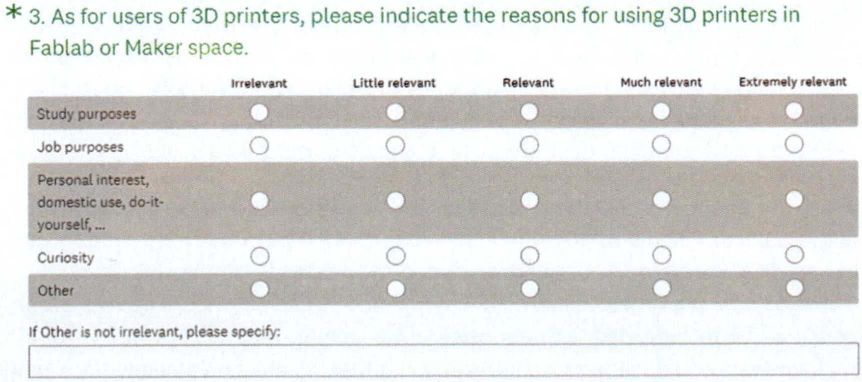

Fig. 1. Illustrative screenshot of the survey representing the 5-point Likert scale for evaluating the relevance of options.

evaluate relevance of options, as a common instrument employed in engineering and design when it comes to rating people's stated importance and agreement, e.g. [9–11].

30 Fab Labs replied to all the questions of the survey, which can be accessed through https://it.surveymonkey.com/r/5P58LM3. The questions are presented in the subsections that follow along with results and comments. An illustrative screenshot of the survey is provided in Fig. 1.

Availability of 3D Printers. The first question aimed to test the actual relevance of 3D printers in Fab Labs. Respondents were asked to indicate the "Total number of 3D printers available in the lab". Numbers of 3D printers varied between 2 and 26 (median 6; mean 6.2; SD 4.6). Despite the large variability, the results support the hypothesis that 3D printers are part of the fundamental equipment a Fab Lab should offer.

Investments in 3D Printing Technologies. The second question was intended to investigate the expected quality level of 3D printers available in Fab Labs based on their costs. Respondents were asked to indicate the "Approximate price of the most expensive 3D printer". Prices were then converted in Euros. Here, the differentiation was even larger, as costs varied between 1200€ and 220000€ (median 3500€; mean 17000€; SD 42000€). While investments are oriented to get mid-quality 3D printers as the best machine in many cases, it clearly emerges that a few Fab Labs have succeeded introducing professional AM technologies.

Motivations Behind the Use of 3D Printers. Through the third question (details in Fig. 1), the respondents were asked to "indicate the" users' "reasons for using 3D printers in Fab Lab". In particular, the perceived relevance had to be indicated of a standard list of possible reasons, including study purposes, working needs, personal interests or do-it-yourself, curiosity or other. Figure 2 presents the results as for the third question, showing how not only study purposes and personal interests, but also job reasons are overall relevant for using 3D printers in Fab Labs. This confirms that the evolution of the Fab Labs has attracted enterprises that can benefit from their equipment and expertise. The fifth question clarifies this aspect.

Skills in Using 3D Printers. The fourth question aimed to verify whether the Fab Lab attendees using 3D printers are able "to quickly acquire sufficient skills to create objects with complex geometries". The respondents should use a five-level Likert scale to indicate the intensity and the rapidity of this learning process. The outcomes show that the abilities required for a proficient use of 3D printers are diffusedly acquired with a medium to moderately high speed. Indeed, more than two thirds of the Fab Labs replied answered this question with the third or fourth level of the Likert scale.

Firms Using Fab Labs' 3D Printers. The fifth question had to be answered just by those respondents (28 out of 30) who believed that work purposes have some relevance in the use of 3D printers. This question aimed to investigate the reasons behind using the Fab Lab's "3D printers instead of buying one". The relevance of five proposed rea-sons had be to indicated, namely price, time saving, rarity of use (that would not justify investments in 3D printers), the need to explore the technology better (advantages and disadvantages are still unclear), and other. The results (see Fig. 3 for details) show

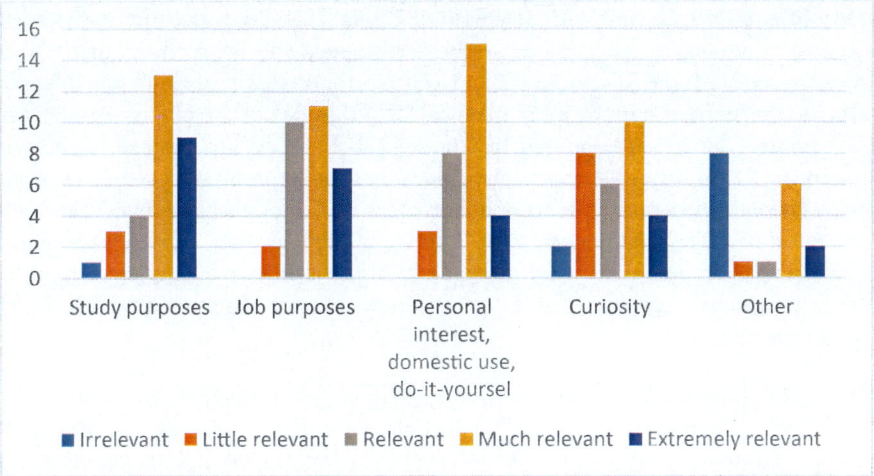

Fig. 2. Occurrences of answers on the perceived relevance of the motivations behind the use of 3D printers in Fab Labs.

that, in contrast to other hypothesized reasons, costs play a little role in firms using Fab Labs' 3D printers. These outcomes stress the importance of AM technologies for rapid prototyping purposes, while the benefits provided by 3D printers are not clear yet to all industrial players. This aspect partially conflicts with what was stressed in the Introduction, i.e. industrialists are well aware of actual AM's strengths and weaknesses. It can be inferred that, while companies who have introduced AM in their manufacturing or prototype processes have a clear picture of these technologies, the same does not apply for firms that have not implemented AM yet. Time resulted another important reason behind firms' use of Fab Labs' 3D printers, which can be interpreted as the presence of some difficulties in mastering AM technologies.

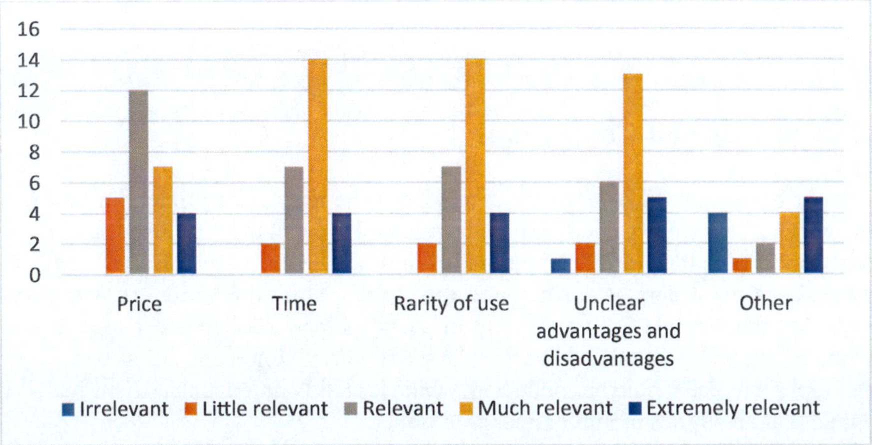

Fig. 3. Occurrences of answers on the perceived relevance of the reasons behind the firms' use of 3D printers in Fab Labs.

Quality of and Satisfaction with 3D-printed Items. The sixth question targeted the perception of the users' satisfaction "with the aesthetics and the surface quality of the objects created with the 3D printers". In addition, the seventh question was meant to investigate the satisfaction "with the mechanical properties of the objects created with the 3D printers". As it is supposed that many Fab Labs do not dispose of the most professional 3D printing systems, a question about the appropriateness of 3D-printed items was deemed necessary. The respondents answered through a five-level Likert scale to both questions. The quality perception in terms of both aesthetics and surface quality, and mechanical properties resulted moderately high (fourth level) for more than half of the respondents. Therefore, Fab Labs' equipment is seemingly appropriate for the scopes of their users.

Private 3D Printers. The eighth question moved from the consideration that Fab Lab visitors have acquired skills in using 3D printers and might be tempted to buy one for private use. More in details, it was asked the perceived reason behind the fact that "Fab-lab users do not purchase a 3D printer for domestic use". The respondents were asked to check the reasons that supposedly apply, which include price, limited user-friendliness of the required software, the lack of complementary tools (e.g. CAD), problems with the acquisition of materials, absence of an appropriate environment, and other. Costs resulted the main motivation behind users' relinquishment of buying a 3D printer for domestic use (24 out 30 answers). The lack of complementary tools (11 answers) and an appropriate environment (8 answers) play a relevant role too. Among the answers "Other", respondents have highlighted potential problems in terms of skills and maintenance. The results show that, despite the existence of 3D printers sold at accessible prices, costs are not considered in line with families' budget. It is also worth noting that Fab Labs' equipment is seldom of low quality (see the first and second questions) and users have consequently got used to good 3D-printed items. It can be derived that users might like to print objects of akin quality at home, but this would imply unsustainable costs. Moreover, the frequency of other answers indicate that the purchase of a 3D printer is not sufficient to make what is possible in Fab Labs, but additional re-sources should be channeled, e.g. tools like CAD systems, maintenance, preparation of an appropriate environment.

3 Discussions and Conclusions

The paper has presented exploratory research on the use of 3D printers in Fab Labs. Some insights have been gained despite the relatively low number of respondents to the survey described in the preceding section. In addition, the results are affected by the fact that Fab Labs' managers have expressed their perception on a variety of user-related aspects, but those might not correspond to users' perceptions. In order to over-come this limitation, while keeping a worldwide perspective, time-consuming and lengthy ethnographic studies should be conducted, which is clearly unpractical and un-necessary in a field that undergoes preliminary exploration.

The main findings can be summarized in the following bullet list.

- Fab Labs are featured by a markedly uneven availability of equipment, at least in terms of 3D printing devices.
- Although the ultimate scope of Fab Labs is to let people tinker and create by involving communities of private users, the accumulated expertise in 3D printing is considerably attracting industrial players. To this respect, Fab Labs can be seen as substitutes for prototyping and experimenting AM capabilities. This is pushed by the overall good quality of 3D-printed items that can be fabricated in Fab Labs.
- Firm representatives attending Fab Labs and using their 3D printers are probably not fully aware of AM technologies' main pros and cons. This reinforces the utility of Fab Labs in terms of spreading fresh knowledge of technological advancements and possibilities, markedly those offered by 3D printing.
- While Fab Labs offer acceptable quality for infrequent industrial users of 3D printing technologies, such results could not be achieved by the majority of attendees if they switched to domestic use. Here, costs and issues related to the whole process that involves 3D printing play a relevant role in discouraging the purchase of 3D printers for domestic use.

Therefore, 3D printers well feature Fab Labs' tendency to capture users' segments with intermediate exigencies: while professional results are not expected, low-end 3D printers could be considered unsuitable, or their use deemed too time-consuming for getting satisfying results. To this respect, future work is intended to analyze the results of the survey further and verify whether more equipped Fab Labs can generate higher user satisfaction and attracting a larger number of industrial players.

To keep attracting relevant users' segments, skills of Fab Labs' operators represent a further competitive factor. On the one hand, because of the relevance of skills, it is not surprising to notice the growing interest in AM education [12, 13] in academia.

On the other hand, it emerges that required skills are quite easily acquired by Fab Labs' users, but this is not sufficient to fuel the domestic market of 3D printers. Here, it is difficult to state whether acquired skills concern the mastering of 3D printing devices only, or can be generalized to the whole process that starts with modeling objects tailored to 3D printing, i.e. DfAM. Undoubtedly, supplementary tools required for 3D printing, such as CAD, represent an obstacle, supposedly because of costs and limited ease of use especially for wielding AM potential. To this respect, it is also worth noting that DfAM-oriented CAD systems or add-ins, e.g. those for facilitating the realization of lattice structures, are particularly costly and clearly not accessible by a large non-specialized audience. As Fab Labs are spreading and mid-range 3D printers represent a part of their fundamental equipment, tailored CAD systems with fewer functions, thus expectedly cheaper, might be seen as a key to empower more and more people in the whole design and/or DfAM process.

Overall, while the study has clarified some questions with respect to the topic, some points are still to be addressed and other research issues have been opened up, which are worth stressing. Beyond satisfaction and appreciation of 3D-printed parts, the actual effect of the learning process about 3D-printing taking place in Fab Labs is still obscure. This applies particularly to firms benefitting from Fab Labs' 3D printing services and it would be interesting to know whether the decision to introduce AM technologies in the enterprises ever takes place. In some cases, companies might have already considered the convenience and economic advantage of keeping attending Fab Labs for prototyping purposes and/or production of small batches. In other cases, the already mentioned learning process could push companies to acquire their own 3D printers once advantages and limitations get clearer and clearer.

From the educational viewpoint, Fab Labs and their digital manufacturing equipment have proved to be a trigger for interest in technical disciplines, see [14] in particular. However, it is unknown whether the awakening of this interest is enduring and affects future learners' decisions, e.g. the choice of Bachelor or Master Studies. Likewise, the disciplines (e.g. design, engineering) that mostly benefit from hands-on 3D printing in Fab Labs could be investigated.

As for other categories of users, who do not exploit 3D printers in Fab Labs for study or work reasons, the effects of this experience should also be studied. In particular, it is worth investigating whether the interest in tinkering was innate or pushed by other circumstances, e.g. the attendance of Fab Labs was first guided by socialization drivers and an interest in technical aspects followed. Still, the way people exploit newly acquired skills in 3D printing, e.g. for new jobs or career advancements, can be also beneficially researched.

Acknowledgements. The study is conducted in the frame of the project "Additive Manufacturing FDM: Dimensional Accuracy and Product Acceptability" (AMDAPA) funded by the Free University of Bozen-Bolzano with the call CRC2016.

References

1. Kohtala, C.: Making "Making" critical: how sustainability is constituted in Fab lab ideology. Design Journal **20**(3), 375–394 (2017). https://doi.org/10.1080/14606925.2016.1261504
2. Santos, G., Murmura, F., Bravi, L.: Fabrication laboratories: the development of new business models with new digital technologies. J. Manuf. Technol. Manag. **29**(8), 1332–1357 (2018)
3. Hudson, N., Alcock, C., Chilana, P.K.: Understanding newcomers to 3D printing: motivations, workflows, and barriers of casual makers. In: Conference on Human Factors in Computing Systems CHI, pp. 384–396 (2016). https://doi.org/10.1145/2858036.2858266
4. Pradel, P., Zhu, Z., Bibb, R., Moultrie, J.: Investigation of design for additive manufacturing in professional design practice. J. Eng. Des. **29**(4–5), 165–200 (2018). https://doi.org/10.1080/09544828.2018.1454589
5. Thompson, M.K., et al.: Design for additive manufacturing: trends, opportunities, considerations, and constraints. CIRP Ann. **65**(2), 737–760 (2018). https://doi.org/10.1016/j.cirp.2016.05.004
6. Moilanen, J., Vadén, T.: 3D printing community and emerging practices of peer production. First Monday **18**(8), 1–14 (2013). https://doi.org/10.5210/fm.v18i8.4271
7. Dorrington, P., Lacan, F., Bigot, S.: How are micro enterprises adopting emergent technologies? In: Setchi, R., Howlett, R.J., Liu, Y., Theobald, P. (eds.) Sustainable Design and Manufacturing 2016. SIST, vol. 52, pp. 215–226. Springer, Cham (2016). https://doi.org/10.1007/978-3-319-32098-4_19
8. Kudus, S.I.A., Campbell, R.I., Bibb, R.: Customer perceived value for self-designed personalised products made using additive manufacturing. Int. J. Ind. Eng. Manag. **7**(4), 183–193 (2016)

9. Borgianni, Y., Maccioni, L., Basso, D.: Exploratory study on the perception of additively manufactured end-use products with specific questionnaires and eye-tracking. Int. J. Interact. Des. Manufact. (IJIDeM) **13**(2), 743–759 (2019). https://doi.org/10.1007/s12008-019-00563-w
10. Cascini, G., Graziosi, S., Montagna, F., Rotini, F.: On the factors affecting design education within a multi-disciplinary class. J. Integr. Des. Process. Sci. **21**(2), 21–44 (2017). https://doi.org/10.3233/jid-2017-0001
11. Maccioni, L., Borgianni, Y., Basso, D.: Value perception of green products: an exploratory study combining conscious answers and unconscious behavioral aspects. Sustainability **11**(5), 1226 (2019). https://doi.org/10.3390/su11051226
12. Borgianni, Y., Maccioni, L., Russo Spena, P., Shunmugavel, M.K.: University education in additive manufacturing and the need to boost design aspects. In: Proceedings of the Design Society: International Conference on Engineering Design ICED, vol. 1(1), pp. 629–638. Cambridge University Press (2019). https://doi.org/10.1017/dsi.2019.67
13. Stern, A., Rosenthal, Y., Dresler, N., Ashkenazi, D.: Additive manufacturing: an education strategy for engineering students. Addit. Manuf. **27**, 503–514 (2019). https://doi.org/10.1016/j.addma.2019.04.001
14. Togou, M.A., Lorenzo, C., Lorenzo, E., Cornetta, G., Muntean, G. M.: Raising students' interest in STEM education via remote digital fabrication: an Irish primary school case study. In: EDULEARN18 Proceedings, pp. 2835–2840. IATED (2018)

Geometrical Issues in Design for Binder Jetting – The Effect of Anisotropic Dimensional Change on Sintering

Marco Zago[1(✉)], Nora Francesca Maria Lecis[2], Maurizio Vedani[2], and Ilaria Cristofolini[1]

[1] Department of Industrial Engineering, University of Trento, Via Sommarive 9, 38123 Trento, Italy
marco.zago-1@unitn.it

[2] Department of Mechanical Engineering, Politecnico di Milano, Via Privata Giuseppe La Masa 1, 20156 Milan, Italy

Abstract. Recent research has been focused on the binder jetting (BJ) additive manufacturing technique due to the high potential possibilities in industrial applications. The actual limitation of BJ process can be attributed to the difficult control of the product quality. In fact, a high dimensional variation occurs on sintering, which can detrimentally affect dimensional and geometrical precision, when not properly considered in the design step.

This paper aims at investigating the influence of sintering on the dimensional change of through holes, with different diameter size and different axis orientation with respect to the building direction. Samples were measured in the green and sintered state by means of a coordinate measuring machine in order to calculate the diameter shrinkage. The empirical data were successfully compared with the prevision of an analytical model demonstrating that diameter shrinkage is influenced by: the anisotropic dimensional change, the axis orientation and the position of the two diametral opposite points used to identify the diameter. A deep analysis of the results showed a non-negligible effect of the gravity-induced load and of the inhomogeneous shrinkage on sample geometry. This study highlighted that the analytical model may serve as a basis in the design step for improving the dimensional quality of BJ product.

Keywords: Binder jetting · Design for additive manufacturing · Dimensional and geometrical precision

1 Introduction

Binder jetting additive manufacturing technique has currently demonstrated promising results due to the high production rate and low manufacturing cost [1]. Even though the performance of BJ product might be lower in comparison to laser powder bed fusion product, BJ can theoretically process any type of metal and ceramic materials, and it does not require controlled atmosphere, neither high energy heat source in the printing

C. Rizzi et al. (Eds.): ADM 2021, LNME, pp. 410–421, 2022.
https://doi.org/10.1007/978-3-030-91234-5_42

machine [2]. The BJ working flow is split in two fundamental operations: the printing operation and the sintering process. The printing step consists in spreading a layer of powder on a working plate, then injecting a binder agent in correspondence of the section of the CAD file of the product. After layer-by-layer shaping, the green part is obtained. However, metal powders are weakly bonded, therefore the green part is cured, to partially enhance the mechanical resistance, aiming at safely handling the product in the further operation. The residual powders of the printing operation are then carefully removed from the green part. Finally, part is debinded and sintered in order to obtain the final product.

The successful development of BJ and other AM technologies needs for specifically oriented design methods and guidelines, known as Design for Additive Manufacturing (DfAM) [3, 4]. Since the great advantage of the AM is the free form design approach, large part of research has been focused on the shape, investigating both topology optimization [5] and geometrical product specification [6, 7].

The assessment and control of the geometrical quality of AM parts are fundamental for the standardization of the manufacturing process, and particularly in BJ, because the sintering operation can detrimentally affect the dimensional and geometrical precision of sintered-based product [8]. For this reason, the effect of sintering process has to be considered in the design step. Few studies experimentally measured the anisotropic dimensional change on sintering [9–11]. Kafara et al. studied the influence of binder quantity on the dimensional accuracy and mechanical strength [12]. Islam et al. focused on the dimensional accuracy of as-built BJ cylinders [13]. Ollison and Berisso investigated the cylindricity of green parts with different axis orientation with respect to the fabrication direction [14]. However, the influence of the anisotropic dimensional change on sintering on the quality of sintered BJ products is not specifically investigated in depth in the scientific literature.

In previous works the anisotropic dimensional change on sintering and the effect on the dimensional precision have been investigated in depth [15–17]. This work aims at investigating the influence of sintering on dimensional variation of through holes with different diameter size and different orientation. An analytical model was derived, aimed at predicting the dimensional change of diameters. The results were successfully compared with the experimental measurements.

2 Method and Model

In the Sect. 2.1, the dimensions and geometrical characteristic of samples are reported, in addition to the material and experimental procedure adopted. Section 2.2 concerns the statement of the assumptions and the analytical formulation of the model for the prediction of diameter shrinkage.

2.1 Experimental Procedure

Five samples were fabricated in order to study the dimensional precision of cylindrical holes. Each sample has four through holes equally spaced in a plane of 20 × 70 mm, which is inclined at angles of 0°, 30°, 45°, 60° and 90° with respect to the fabrication direction as shown in Fig. 1. The dimensions of such plane and the diameter of the holes were kept constant in all of the samples. The other sample dimensions were optimized in order to yield the depth of the hole in a range close to 10 mm, to directly compare the measurements results.

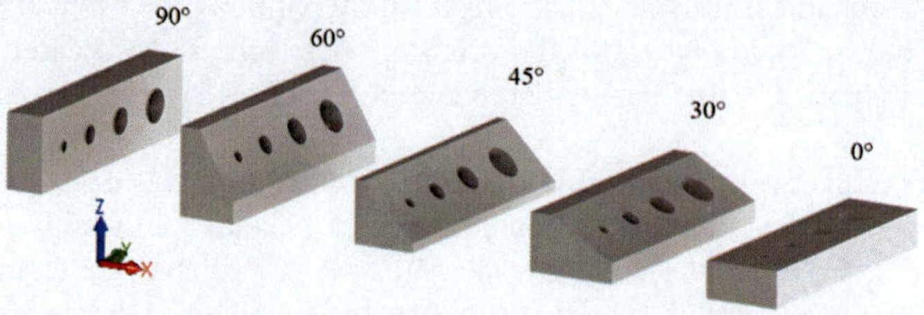

Fig. 1. Nominal geometry and name of the 5 samples.

The nominal dimensions of the five samples are reported in Table 1, referring to dimensions shown in Fig. 2.

Table 1. Nominal dimensions of the linear dimensions and the cylindrical holes of the five geometries (all dimensions are in millimeters).

	0°	30°	45°	60°	90°
LX1	20	23.19	16.62	15.41	10
LX2	–	5.87	2.47	5.41	–
LY1	70	70	70	70	70
LZ1	10	15.41	16.62	23.19	20
LZ2	–	5.41	2.47	5.87	–
W	–	20	20	20	–
$\phi 3$	3	3	3	3	3
$\phi 5$	5	5	5	5	5
$\phi 7$	7	7	7	7	7
$\phi 9$	9	9	9	9	9
$H_{\phi 3}$	10	11.15	12	11.15	10
$H_{\phi 5}$	10	10.58	11	10.58	10
$H_{\phi 7}$	10	10	10	10	10
$H_{\phi 9}$	10	9.42	9	9.42	10

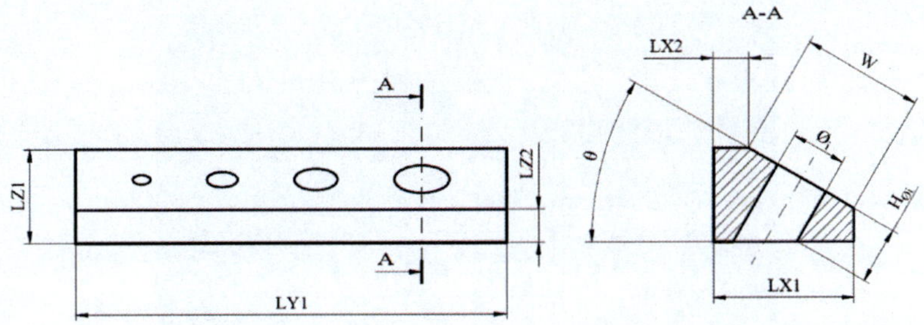

Fig. 2. Section view and conventional name of the main features in inclined samples.

The samples were fabricated by an ExOne Innovent+ printing machine, set-up parameters are reported in Table 2. Gas atomized AISI 316L powder was used, being one of the most promising material currently studied in scientific literature and applied in industrial production [18].

Table 2. BJ process parameters.

Layer thickness	Saturation level	Recoat	Roller	Ultrasonic intensity
50 μm	55%	90 mm/s	500 rpm	100%

All samples were printed, cured and sintered in a single batch in order to minimize the variability of the manufacturing process. The larger dimension of all the samples was set parallel to the printhead movement direction, so that the variability of the fabrication process would affect all the samples in the same way. In this work it is assumed that the sample location does not affect the dimensional and geometrical precision, despite some variability was observed by Vitolo et al. [19].

After the printing operation, samples were cured at 180 °C for 180 min and successively samples were extracted by the printing chamber and extra powders were removed.

The measurement of the green samples was performed with a Hexagon DEA Global Image 07-07-07 coordinate measuring machine. The CMM mounted a Renishaw SP600M touch-probe which has a maximum permissible error of 1.5 + L/333 μm in touch acquisition mode according to ISO 10360-2 [20]. The diameter of the tip used is 1 mm and the effective working length of the stylus is 20.5 mm, to guarantee the inspection of the holes through all the depth, also considering the shrinkage on sintering.

In the set-up of the measurement procedure, the sample was carefully blocked to the CMM working plane as shown in Fig. 3. The clamping system allowed accessing all the planes and holes, also limiting the shadow effect of the supports.

A direct-computer-control measurement routine was developed for each sample. A 3-2-1 alignment procedure was initially performed to define the three datum planes. The primary datum plane (A) was identified on the bottom plane, which corresponds to the first plane fabricated by the layer-by-layer construction. The secondary datum plane (B)

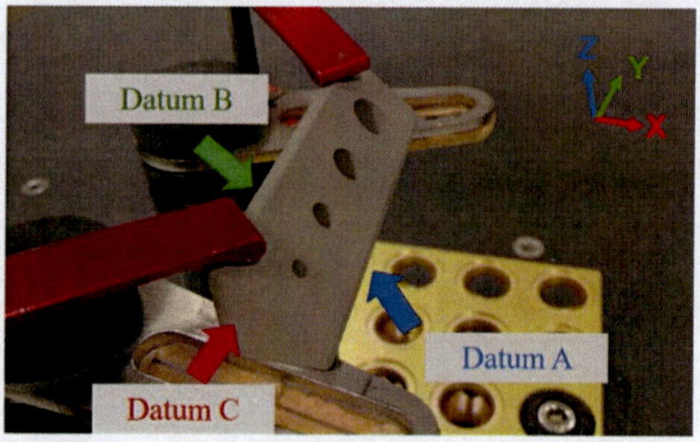

Fig. 3. Example of the clamping system adopted for the measurement of green and sintered samples and the planes used as datum reference.

corresponds to the plane parallel to YZ plane, and the third datum plane (C) to the plane parallel to XZ plane, as shown in the Fig. 3.

A pattern of 50 to 150 points was acquired on each plane to maximize the measurement data points for the reconstruction of the plane by a gaussian best fit least squares method. The linear dimensions of samples were derived as distances between planes. Moreover, the effective depth of the hole ($H_{\phi i}$, where i subscript refers to the different nominal diameters, as in Table 1) can be computed by the trigonometric relationships expressed by Eq. (1) for samples 30° and 45°, and Eq. (2) for sample 60°. The effective depth corresponds to the depth which includes a complete cylinder, as displayed in Fig. 2.

$$H_{\phi i} = \frac{LZ1}{\cos\theta} - \left(\frac{\phi_i}{2} + \frac{W}{2}\right)\tan\theta \tag{1}$$

$$H_{\phi i} = \frac{LX1}{\sin\theta} - \left(\frac{\phi_i}{2} + \frac{W}{2}\right)\cot\theta \tag{2}$$

On the basis of the effective depth of the holes, the circular section of the cylinder was measured at five equidistant levels, as shown in Fig. 4a. On each level, 12 points were acquired on the circular section, 30° spaced, according to the reference system shown in Fig. 4b.

Consequently, on each circular section six values of the diameter can be calculated by Eq. (3), as the distance of the diametral opposite points, where x_j, y_j, z_j are the coordinates of each point, subscript *j* referring to values from 1 to 6, and subscript *i* to the nominal diameter.

$$\phi_{i-j} = \sqrt{\left[\left(x_j - x_{j+6}\right)^2 + \left(y_j - y_{j+6}\right)^2 + \left(z_j - z_{j+6}\right)^2\right]} \tag{3}$$

After measurement, green parts were debinded at 470 °C for 4 h in Argon atmosphere, then heated at 5 °C/min up to the isothermal sintering temperature of 1360 °C for 3 h in a vacuum furnace at the pressure of 10^{-1} mbar (HT-S1 LPC by HTS, Mozzanica BG, Italy).

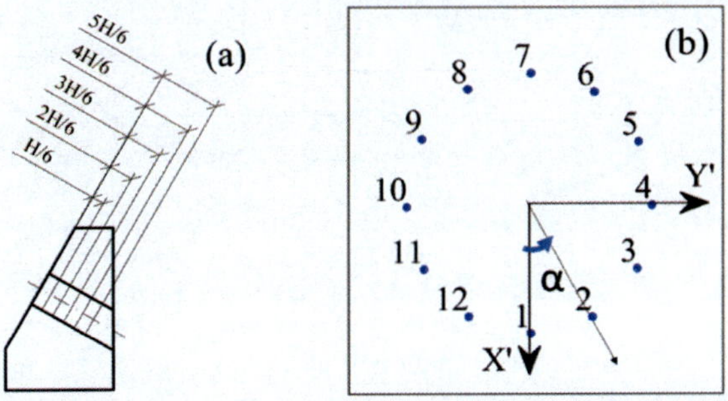

Fig. 4. Schematic representation and labeling of the position of the levels corresponding to the circle measurement (a), and the 12 nominal points acquired on each level (b).

Sintered samples were measured according to the same routine developed for green parts. The measurement routine was the same used for the green parts, the points acquired were theoretically located at the same fraction of the effective depth, even if not in the same identical position due to the large shrinkage occurred on sintering.

The dimensional changes of the linear dimensions and diameters were calculated according to Eqs. (4) and (5), respectively.

$$\varepsilon_L = \frac{L_s - L_g}{L_g} \tag{4}$$

$$\varepsilon'_{\phi i} = \frac{\phi_s - \phi_g}{\phi_g} \tag{5}$$

2.2 Theoretical Model

The analytical model presented hereafter aims at predicting the dimensional change on sintering and the geometrical variation of cylindrical holes. The main points of the model are here briefly recalled, description in depth can be found in [21]. As documented in scientific literature [9–11], the dimensional change on sintering is supposed to be anisotropic. Moreover, dimensional change is assumed totally dependent on the direction in the printing reference system, and independent on the sample geometry. On the basis of these assumptions, the reference system is defined according to Fig. 5. XYZ reference system is the printing reference system. X′Y′Z′ reference system is related to the hole, Z′ corresponds to the cylinder axis, Y′ is overlapped with Y direction and θ angle defines counterclockwise rotation around Y′ axis of X′Z′ plane. Within X′Y′Z′, the reference system in cylindrical coordinates (R, α, Z″) is identified (positive value of α corresponds to counterclockwise rotation moving from X′ to Y′ axis).

Two diametral opposite points (P1, P2) belong to plane X′Y′ and to the cylinder, therefore their radial coordinates are equal to the nominal radius of the hole (R). Assuming the angular coordinate of P1 equal to α_0, the coordinate of the diametral point P2 would be equal to α_0 plus 180°, by definition. The two points lie on the cylindrical surface at a generic coordinate z''_0.

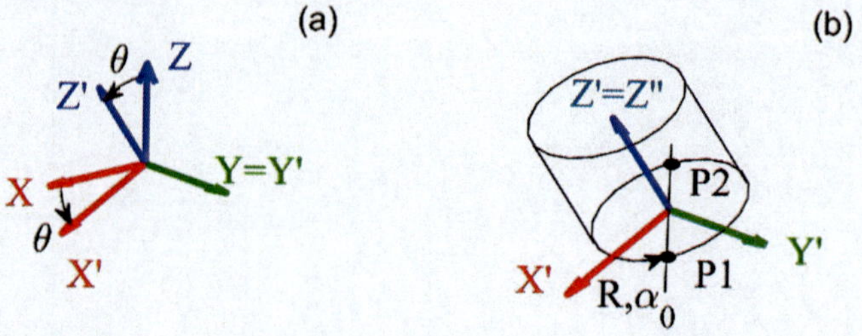

Fig. 5. Conventional reference systems of the cylindrical hole: (a) XYZ printing reference system, X′Y′Z′ rotated reference system and (b) R, α, Z″ cylindrical reference system.

The coordinates of the points can be translated in the X′Y′Z′ reference system by Eqs. (6) and (7).

$$\begin{Bmatrix} x'_{P1} \\ y'_{P1} \\ z'_{P1} \end{Bmatrix} = \begin{Bmatrix} R\cos\alpha_0 \\ R\sin\alpha_0 \\ z''_0 \end{Bmatrix} \tag{6}$$

$$\begin{Bmatrix} x'_{P2} \\ y'_{P2} \\ z'_{P2} \end{Bmatrix} = \begin{Bmatrix} -R\cos\alpha_0 \\ -R\sin\alpha_0 \\ z''_0 \end{Bmatrix} \tag{7}$$

Moreover, the P1 and P2 coordinates can be expressed in the fabrication reference system (XYZ) by Eq. (8).

$$\begin{Bmatrix} x_{g-i} \\ y_{g-i} \\ z_{g-i} \end{Bmatrix} = \begin{bmatrix} \cos\theta_g & 0 & -\sin\theta_g \\ 0 & 1 & 0 \\ \sin\theta_g & 0 & \cos\theta_g \end{bmatrix} \begin{Bmatrix} x'_{g-i} \\ y'_{g-i} \\ z'_{g-i} \end{Bmatrix} \tag{8}$$

Where the subscript i refers to the point number and θ_g represents the angle between the cylinder axis and the building direction in the green state.

Considering the shrinkage occurring on sintering, the new coordinates of the points in the sintered state can be calculated by Eq. (9).

$$\begin{Bmatrix} x_{s-i} \\ y_{s-i} \\ z_{s-i} \end{Bmatrix} = \begin{bmatrix} 1+\varepsilon_x & 0 & 0 \\ 0 & 1+\varepsilon_y & 0 \\ 0 & 0 & 1+\varepsilon_z \end{bmatrix} \begin{Bmatrix} x_{g-i} \\ y_{g-i} \\ z_{g-i} \end{Bmatrix} \tag{9}$$

Where ε_x, ε_y, ε_z represent the dimensional changes along the XYZ fabrication reference system, which are defined according to Eq. (4). The subscripts g and s refer to the green and sintered state, respectively.

In agreement with the measurement strategy, the diameter of the hole can be estimated by the point-by-point distance. In this model, the green diameter is theoretically equivalent to the nominal value defined in the CAD file, while the sintered diameter can be evaluated by the distance of the two opposite points as by Eq. (10).

$$\phi'_s = \sqrt{\left(x'_{s-P2} - x'_{s-P1}\right)^2 + \left(y'_{s-P2} - y'_{s-P1}\right)^2 + \left(z'_{s-P2} - z'_{s-P1}\right)^2} \tag{10}$$

The sintered coordinates of the points can be expressed as function of the XYZ coordinates by the rotation matrix in Eq. (11).

$$\begin{Bmatrix} x'_{s-i} \\ y'_{s-i} \\ z'_{s-i} \end{Bmatrix} = \begin{bmatrix} \cos\theta_s & 0 & \sin\theta_s \\ 0 & 1 & 0 \\ -\sin\theta_s & 0 & \cos\theta_s \end{bmatrix} \begin{Bmatrix} x_{s-i} \\ y_{s-i} \\ z_{s-i} \end{Bmatrix} \tag{11}$$

Where θ_s is calculated by Eq. (12).

$$\theta_s = \tan^{-1}\left(\frac{x_s}{z_s}\right) = \tan^{-1}\left[\tan\theta_g \frac{(\varepsilon_x + 1)}{(\varepsilon_z + 1)}\right] \tag{12}$$

By Eqs. (10) and (11) the sintered diameter is derived, as defined by the coordinates in XYZ reference system, to apply the dimensional change defined in Eq. (9). The coordinates are then converted in the X′Y′Z′ reference system by Eq. (8), and in the cylindrical reference system by Eqs. (6) and (7).

Finally, the dimensional change of the hole can be predicted by Eq. (5), which turns to Eq. (13) using the terms derived above.

$$\varepsilon'_{\phi i} = \sqrt{\cos^2\alpha_O\left[(\varepsilon_x + 1)^2\cos^2\theta_g + (\varepsilon_z + 1)^2\sin^2\theta_g\right] + \sin^2\alpha_O\left(\varepsilon_y + 1\right)^2} - 1 \tag{13}$$

In agreement with the starting hypothesis, the analytical model is independent on the cylindrical size (diameter and the depth), it depends on:

- the dimensional change along the fabrication reference system (ε_x, ε_y, ε_z);
- the orientation of the cylinder axis with respect to the building direction (θ_g);
- the angular position of the selected points on the circumference (α_O).

Summarizing, this model described the evolution of the circular section of the cylinder to an elliptical shape, as affected by the anisotropic dimensional change on sintering and by the cylinder orientation.

3 Results and Discussion

All the dimensional changes on sintering of the linear dimensions, as calculated by Eq. (4) in all the samples, are reported in Table 3, as mean value and standard deviation.

Table 3. Average and standard deviation of the dimensional change on sintering of the linear dimensions of the five samples.

Printing direction	Average ε_L [%]	Std. Dev. ε_L [%]
X	−15.5	0.8
Y	−14.9	0.2
Z	−17.8	0.8

An almost isotropic behavior is shown in plane XY, while shrinkage is considerably higher along Z direction. The analysis of the results of linear dimensional changes is exhaustively discussed in [21].

The mean values of the linear shrinkage were used as ε_x, ε_y, ε_z input parameters in the analytical model.

Figure 6 shows the dimensional change of the diameters of the four cylindrical holes of sample 0°. Plots from A to D in Fig. 6 refer to the holes having the nominal diameter of 3, 5, 7, 9 mm, showing the data of the experimental measurements, along with the corresponding values predicted by the analytical model. Empirical data are divided in five groups, related to the different levels (H/6, 2H/6, 3H/6, 4H/6, 5H/6 fraction of the effective depth). Each group contains the diameter shrinkage corresponding to α angular coordinate (0°, 30°, 60°, 90°, 120°, 150°).

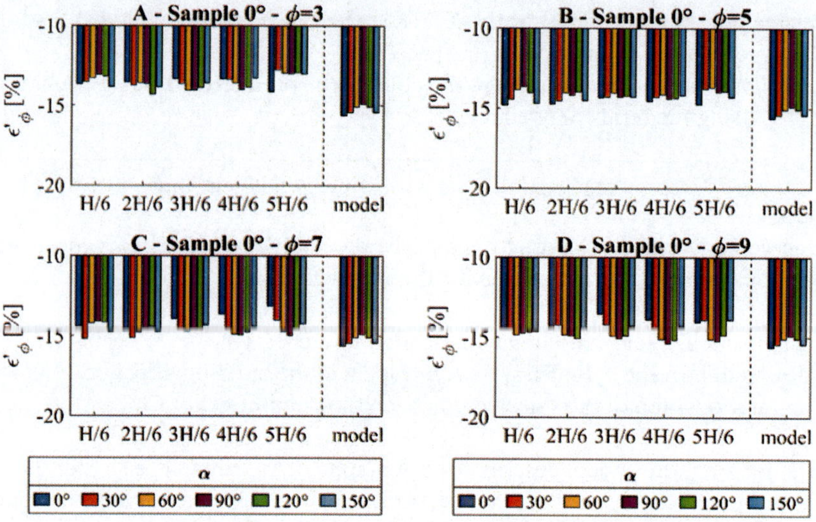

Fig. 6. Experimental and predicted dimensional change on sintering of the hole diameters evaluated at different depth and angular positions. Sample 0°: (A) hole 3 mm, (B) hole 5 mm, (C) hole 7 mm, (D) hole 9 mm.

In sample 0°, the analytical model predicts slightly lower diameter shrinkage corresponding to α angle equal to 90°, due to the slight difference between ε_y and ε_x used as input parameters. The empirical results generally show the same shrinkage at different α angles, practically unaffected by depth. Therefore, the experimental shrinkage is in line with the model prevision, concerning the influence of α angle. Nevertheless, in the biggest holes (7 and 9 mm), deeper scan levels, empirical data show an opposite trend compared to the model prevision.

Moreover, mean values of experimental data look to be affected by hole size, and larger shrinkage is observed in larger holes. The influence of hole size might be related to the amount of material surrounding the different holes: as by Fig. 1, the larger the hole, the smaller the amount of surrounding material. Considering the thermal gradient during sintering, the lower amount of material surrounding the larger holes should heat up faster than the higher amount of material surrounding the smaller holes, determining different time at the isothermal sintering temperature, and consequently different dimensional change. Future work will model and empirically study the possible temperature gradient occurring during the sintering cycle in different samples, in order to confirm this hypothesis.

Moreover, the influence of green density should be considered, and in particular the inhomogeneous green density due to irregular powder spreading in Y direction. As reported in scientific literature, lower powder bed density determines higher dimensional variation on sintering [22]. The influence of green density will be investigated in future work, referring to different positions of samples in the working plane.

Figure 7 shows the experimental measurements, along with the corresponding values predicted by the analytical model, for sample 90° (same grouping as in Fig. 6). The evolution of data plots from sample 0° to sample 90° is represented by the data corresponding to samples 30°, 45° and 60°, showing intermediate values according to the predicted trend. They are not shown here due to space limitations.

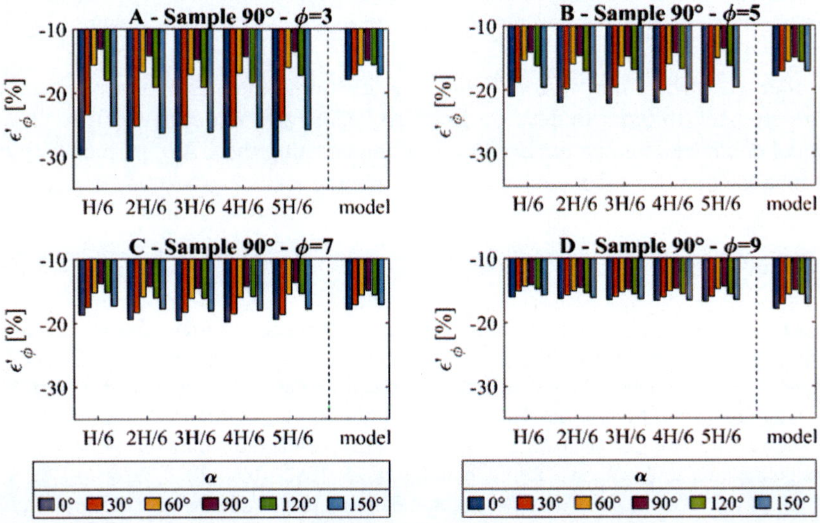

Fig. 7. Experimental and theoretical dimensional change on sintering of the hole diameters evaluated at different depth levels and angular positions. Sample 90°: (A) hole 3 mm, (B) hole 5 mm, (C) hole 7 mm, (D) hole 9 mm.

In sample 90° the trend of experimental data, as affected by α angle, fully confirms the trend predicted by the model. Shrinkage decreases on increasing α from 0° to 90°, then increases on increasing α up to 150°. Significant influence of hole size is also observed, which might be ascribed to the influence of gravity force during sintering. Considering the orientation of the sample in the working space, the mass of weakly bonded particles above the holes might have determined the distortion of the cylindrical shape of the holes. The effect is more pronounced in the smaller holes, which are subject to higher gravity force due to the higher mass of powder above the hole surface. The influence of gravity is not considered in the analytical model, and consequently the dimensional change is underestimated. Future work will consider the load induced by gravity in order to enhance the current model formulation.

4 Conclusion

This works investigated the dimensional precision of cylindrical holes produced by binder jetting process. Five samples were fabricated by AISI 316L powder feedstock and sample geometry was measured before and after the sintering operation by means of a coordinate measuring machine. The experimental results were compared with the analytical solution of a model previously developed showing that:

- the analytical trend is generally confirmed. The shrinkage of diameter depends on: the anisotropic dimensional change, the angle between the cylinder axis and the building direction, and the position of the selected diametral points on the circumference.
- The difference between the measured and predicted values can be ascribed to the gravity effect and to a temperature gradient within the sample during sintering.

Further work will improve the model, also including the effect of gravity and temperature gradient, in order to develop an effective design approach for BJ. In this way, the model could be effectively adopted for compensating the CAD geometry, in order to improve the quality and standardization of BJ product.

References

1. Ziaee, M., Crane, N.B.: Binder jetting: a review of process, materials, and methods. Addit. Manuf. **28**, 781–801 (2019)
2. Li, M., Du, W., Elwany, A., Pei, Z., Ma, C.: Metal binder jetting additive manufacturing: a literature review. J. Manuf. Sci. Eng. **142**, 1–17 (2020)
3. Wiberg, A., Persson, J., Ölvander, J.: Design for additive manufacturing – a review of available design methods and software. Rapid Prototyp. J. **25**, 1080–1094 (2019)
4. Kumke, M., Watschke, H., Vietor, T.: A new methodological framework for design for additive manufacturing. Virtual Phys. Prototyp. **11**, 3–19 (2016)
5. Rosso, S., et al.: Optimization approaches in design for additive manufacturing. In: Proceedings of International Conference on Engineering Design (ICED), August 2019, pp. 809–818 (2019)
6. Ameta, G., Lipman, R., Moylan, S., Witherell, P.: Investigating the role of geometric dimensioning and tolerancing in additive manufacturing. ASME. J. Mech. Des. **137**(11), 111401 (2015). https://doi.org/10.1115/1.4031296
7. Moroni, G., Petrò, S., Polini, W.: Geometrical product specification and verification in additive manufacturing. CIRP Ann. - Manuf. Technol. **66**, 157–160 (2017)
8. Zago, M., Larsson, M., Cristofolini, I.: An improved design method for net-shape manufacturing in powder metallurgy. In: Rizzi, C., Andrisano, A.O., Leali, F., Gherardini, F., Pini, F., Vergnano, A. (eds.) ADM 2019. LNME, pp. 257–267. Springer, Cham (2020). https://doi.org/10.1007/978-3-030-31154-4_22

9. Mostafaei, A., Rodriguez De Vecchis, P., Nettleship, I., Chmielus, M.: Effect of powder size distribution on densification and microstructural evolution of binder-jet 3D-printed alloy 625. Mater. Des. **162**, 375–383 (2019)
10. Kumar, A., Bai, Y., Eklund, A., Williams, C.B.: Effects of hot isostatic pressing on copper parts fabricated via binder jetting. Procedia Manuf. **10**, 935–944 (2017)
11. Miyanaji, H., Rahman, K.M., Da, M., Williams, C.B.: Effect of fine powder particles on quality of binder jetting parts. Addit. Manuf. **36**, 101587 (2020)
12. Kafara, M., Kemnitzer, J., Westermann, H.H., Steinhilper, R.: Influence of Binder Quantity on Dimensional Accuracy and Resilience in 3D-Printing. Procedia Manuf. **21**, 638–646 (2018)
13. Islam, M.N., Sacks, S.: An experimental investigation into the dimensional error of powder-binder three-dimensional printing. Int. J. Adv. Manuf. Technol. **82**(5–8), 1371–1380 (2015). https://doi.org/10.1007/s00170-015-7482-7
14. Ollison, T., Berisso, K.: Three-dimensional printing build variables that impact cylindricity. J. Ind. Technol. **26**, 1–10 (2010)
15. Baselli, S., Torresani, E., Zago, M., Amirabdollahian, S., Cristofolini, I., Molinari, A.: Sintering shrinkage of uniaxial cold compacted iron: influence of the microstructure on the anisothermal and isothermal shrinkage of uniaxial cold-compacted iron. Powder Metall. **61**, 276–284 (2018)
16. Zago, M., Cristofolini, I., Molinari, A.: New interpretation for the origin of the anisotropic sintering shrinkage of AISI 316L rings based on the anisotropic stress field occurred on uniaxial cold compaction. Powder Metall. **62**, 115–123 (2019)
17. Cristofolini, I., et al.: Design for powder metallurgy: predicting anisotropic dimensional change on sintering of real parts. Int. J. Precis. Eng. Manuf. **20**(4), 619–630 (2019). https://doi.org/10.1007/s12541-019-00030-2
18. Mirzababaei, S., Pasebani, S.: A review on binder jet additive manufacturing of 316L stainless steel. J. Manuf. Mater. Process. **3**, 8–12 (2019)
19. Vitolo, F., Martorelli, M., Gerbino, S., Patalano, S., Lanzotti, A.: Controlling form errors in 3D printed models associated to size and position on the working plane. Int. J. Interact. Des. Manuf. (IJIDeM) **12**(3), 969–977 (2017). https://doi.org/10.1007/s12008-017-0441-9
20. ISO, I.S.O.: 10360-2: 2009—Geometrical Product Specifications (GPS)—Acceptance and Reverification Tests for Coordinate Measuring Machines (CMM)—Part 2: CMMs Used for Measuring Linear Dimensions. International Organization for Standardization, Geneva, Switzerland (2009)
21. Zago, M., Lecis, N.F.M., Vedani, M., Cristofolini, I.: Dimensional and geometrical precision of parts produced by Binder Jetting process as affected by the anisotropic shrinkage on sintering. Addit. Manuf. **43**, 102007 (2021)
22. Do, T., et al.: Improving structural integrity with boron - based additives for 3D printed 420 stainless steel. Procedia Manuf. **1**, 263–272 (2015)

A Hollowing Topology Optimization Method for Additive and Traditional Manufacturing Technologies

Emmanuele Barberi, Filippo Cucinotta, Marcello Raffaele, and Fabio Salmeri(✉)

Department of Engineering, University of Messina, Contrada Di Dio (S. Agata), 98166 Messina, Italy
{emmanuele.barberi,filippo.cucinotta,marcello.raffaele, fabio.salmeri}@unime.it

Abstract. Recently, thanks to the evolution of rapid prototyping, the interest in designing topologically optimized components have grown, to maintain good mechanical characteristics while reducing their weight. This work proposes a new method of topological optimization that associates the information obtained through a finite element analysis and a grid of prefixed points in space. The software used for this study is a Rhinoceros plugin called Grasshopper, which is composed of a parametric environment schematized by graphic algorithms. The Finite Element analysis is carried out through the Ansys Workbench software. The obtained stresses are the basis for the algorithm to parameterizes the hollowing with a variation of the diameter of the holes. The ability of the algorithm to directly modify the CAD, avoiding post-processing and generating directly the CAD topologies, represents the true potential of the method. Furthermore, the method lends itself to being used for both additive and subtractive manufacturing. In the presented case study, once the beam model was designed, it was printed using a Fused Deposition Modeling 3D printing technique and then a 3-point bending test was carried out on it. Finally, a comparison was made between the original non-optimized beam and the optimized one by the algorithm, observing an increase in the strength/weight ratio of about 20% but, conversely an increase in printing time of about 50%.

Keywords: Hollowing · Topology Optimization · Additive Manufacturing · Lightweight design

1 Introduction

Additive Manufacturing is a production methodology with which it is possible to create three-dimensional objects through the deposition of material. The deposition is generally made layer by layer, proceeding oppositely to traditional subtractive manufacturing methods [1], such as those based on chip removal. Developed in the 1980s as a method for rapid prototyping [2], Additive Manufacturing has evolved, so much so that it can be used in the production of finished pieces. The mechanical properties that these final

C. Rizzi et al. (Eds.): ADM 2021, LNME, pp. 422–430, 2022.
https://doi.org/10.1007/978-3-030-91234-5_43

products have essentially depend on the method by which the layers of material are deposited, as well as on the material itself. By focusing on the Fused Deposition Modeling (FDM) Additive Manufacturing technique used in the experimental part of this paper, it is possible to see how both the surface finish and the mechanical properties of the piece depend on the thickness of the layer and of the printing direction [3, 4]. The realization of a three-dimensional object with the FDM technique takes place through the use of a thermoplastic polymer filament which, after being heated and brought to the molten state, it is deposited using an extruder. Therefore, for the correct realization of the piece, it is necessary to know the stress direction that will act on it during its use and, consequently, to define an appropriate printing orientation that gives the object the mechanical properties it needs: this is allowed by the possibility to spatially position 3D CAD model and to be able to print, at the same time, the necessary support structures [5] (which will be later removed) for the slicing of the geometries in the desired direction. Among the main advantages of this production method, there is the possibility of creating objects with a complex shape, even if limited in size by the dimensions of the machine with which they are produced. This specific advantage is linked to the concept of Topological Optimization which is a family of mathematical methods which aim at designing objects maximizing or minimizing an objective function, such as weight or costs [6]. This leads to components safer, lighter, or cheaper than those they would have if they were made with traditional technologies; these objects, optimized in shape, weight, and cost, continue to address the target mechanical characteristics required by their particular application. There are several platforms with which Topological Optimization can be carried out (e.g. Rhinoceros-Grasshopper [7], ANSYS [8], etc.) and the areas in which it is applied are many: from the civil sector [9] to that of transport (e.g. automotive [10, 11], naval [12] and aerospace [13, 14]), to the medical field [15], also passing through the sports competition sector [16]. Topological Optimization can be based on Finite Element Analysis (FEA), Computational Fluid Dynamic (CFD), or also Fluid-Structure Interaction (FSI) [17].

In this work, an FDM technology, which consists of a thermoplastic polymer filament deposition, has been used. It has been decided to make use of PolyLactic Acid (PLA) both for its good mechanical characteristics and its environmental compatibility since it comes from renewable sources such as starch [18]. PLA furthermore is biodegradable material [19], therefore it can replace polymers petrochemicals. Seol et al. [20] get an overview of the mechanical properties of different dog-bone specimens made of PLA with the FDM technique showing how the mechanical characteristics of the specimen vary according to the geometry and density of the filling.

The potentiality of the method consists not only in its simplicity and in the practicality of the readiness of the CAD geometries but also in the high versatility of the hollowing technique in terms of manufacturing technologies. Indeed, hollowing is very adapt to be manufactured both with additive and subtractive technologies, such as drilling or milling machines.

2 Materials and Methods

2.1 The Topology Optimization Method

The proposed method consists of emptying the volume of the model, varying the size of the holes as an inverse function of the stresses evaluated through the FEA carried out through the Ansys Workbench software.

As case study, a simple rectangular-shaped beam subjected to a 3-point bending test has been chosen. The non-optimized model has been populated with an equispaced grid of points. These points represent the seeds from which the hollowing of the component will start.

Once the results from the FEA are obtained, they are inserted as input data within the algorithm, which, evaluates the maximum stress that acts around each of them.

The algorithm created for this work is monoparametric and acts by varying the diameter of the holes that hollow the component with an inverse function of the intensity of the stresses acting. The chosen mathematical function is shown in (1)

$$D \propto \left(1 - k \cdot \frac{\sigma}{\sigma_p}\right) \tag{1}$$

In which D, which is the local hole diameter, is proportional to a linear trend in which k is a fixed parameter, σ the local von Mises stress and σ_p the permissible stress.

The algorithm was developed in the Rhinoceros-Grasshopper environment. Grasshopper is a visual programming language that allows parameterizing the drawing tools of Rhinoceros with block programming. The algorithm is shown in Fig. 1 in which each box represents a specific function (an input or an operation) and each connection line a flow of information. In particular, the algorithm read the geometry and the associated FEA and perform the optimization by assigning the new diameters for each seed, according to Eq. (1).

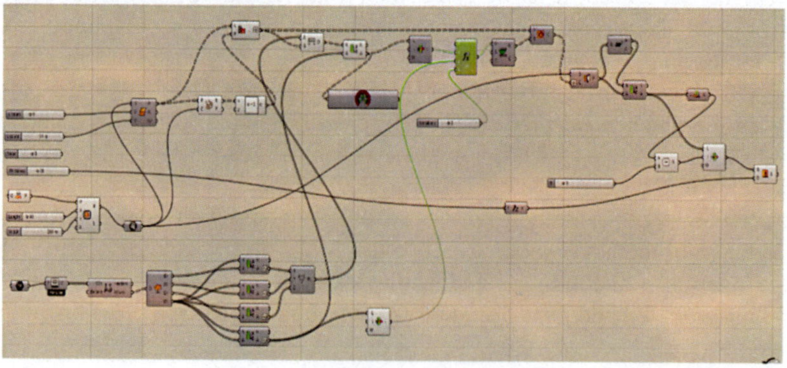

Fig. 1. The Grasshopper algorithm.

After the optimization process, the software creates a solid CAD geometry ready to be exported in a standard format (such as StL) and prototyped without the need for post-processing work.

2.2 The FDM Manufacturing

The material used for printing the specimens is PLA. This material is currently booming, since it has interesting features such as good mechanical characteristics, low cost and it is more environmentally friendly than the other most used plastic materials. The mechanical properties of the material are shown in Table 1.

Table 1. Mechanical properties of PLA.

Property	Unit	Value
Density	kg/m^3	1165
Young's Modulus	MPa	2300
Poisson's ratio	–	0.4
Tensile yield strength	MPa	50
Tensile ultimate strength	MPa	60

The beam specimen (280 × 20 × 40 mm) has been printed in PLA using a 100% infill, for a weight of about 0.260 kg. The direction of slicing has been oriented along an axis directed along with the thickness. The same dimensions and slicing direction have been used also for the optimized specimen.

At this stage, it is particularly important to print with an adequate temperature and speed to obtain adequate adhesion between the various printing layers. In particular, a temperature of 200 °C has been set for this print, giving good results. The printing time varies according to the complexity of the model to be printed and to the printing parameters. In our case study, a nozzle of 0.4 mm, a printing speed of 50 mm/s, and an infill of 100% have been used, for a total time of 24 h and 40 min to print the non-optimized specimen.

2.3 The Bending Tests

All the tests were performed with a servo-hydraulic axial load machine INSTRON 8854 with a maximum load capacity of 250 kN. The 3-point bending tests were performed with a distance between the support rollers equal to 220 mm. The rollers had a diameter of 20 mm. The upper roller lowering speed has been set at 5mm per minute. The tests were carried out until the breaking (Fig. 2). Figure 3 shows the non-optimized specimen after the bending test.

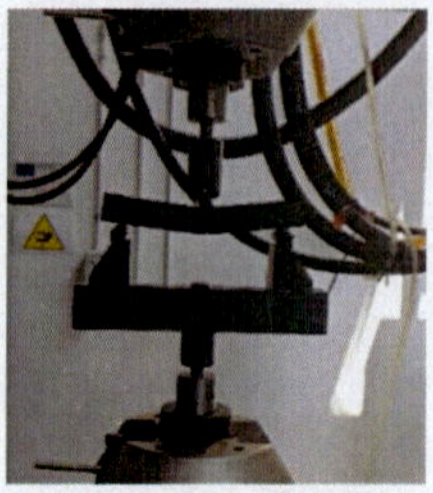

Fig. 2. The experimental setup.

Fig. 3. The broken sample.

2.4 The FE Model

FEA simulations were performed using Ansys workbench 2020 with the linear elastic model. The geometric regularity of the specimen allowed to create a regular structured mesh. the three rollers were modeled to have the same size as those used in the experimental test and were positioned in the same way. The material was set by inserting the mechanical characteristics of the PLA as validated from the experimental tests, while the rollers were modeled in steel. The lower rollers have been constrained with fixed supports. The upper roller, instead, has been set with a lowering equal to that obtained during the bending test at the end of the elastic part of the experimental curve.

3 Results and Discussion

A Topology Optimization study was carried out on a rectangular-shaped beam. The FEA of the non-optimized model provided the stress values distributed within the specimen to be inserted into the algorithm. The model obtained using the optimization algorithm is shown in Fig. 4, and the effect due to the outer fibers, neutral axis and contact pressures on the supports is well visible.

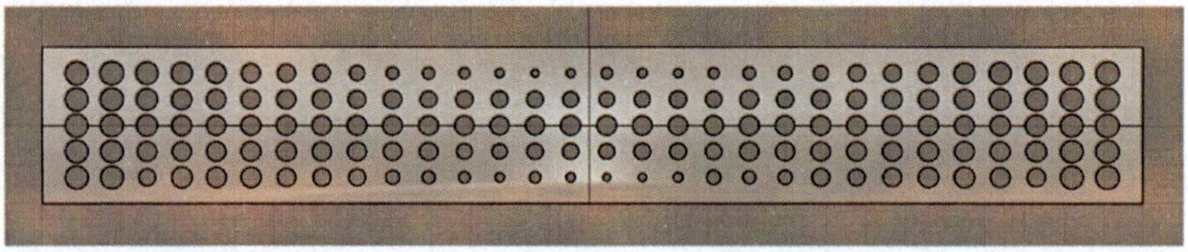

Fig. 4. The hollowed specimen, after the algorithm optimization.

The optimized specimen has been printed. In this case, despite the weight is lowered (0.205 kg), the printing time has increased up to 38 h, i.e. it has increased by about 54% compared to the non-optimized specimen (Fig. 5), maintaining the same printing parameters.

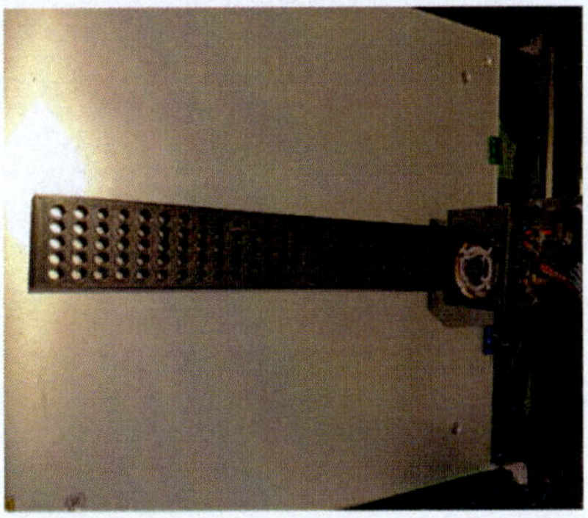

Fig. 5. The printing of the optimized specimens.

A comparison of the different behavior of the two samples is reported in Fig. 6 and Fig. 7. The limit of the elastic phase has been calculated when the derivative of the curve changes value, reducing itself. This value doesn't change significantly in the two samples. The specimen broke at a load equal to about 6000 N, for the non-optimized sample and of almost 5000 N for the optimized one. However, the deflection at the ultimate strength is more or less equal for both the samples. The optimized sample has, obviously, minor stiffness and tensile strength but has a higher stress/weight ratio. Furthermore, the two ways of breaking are quite different since the non-optimized sample has a wide plastic area, while the optimized sample collapses suddenly, after the ultimate tensile strength.

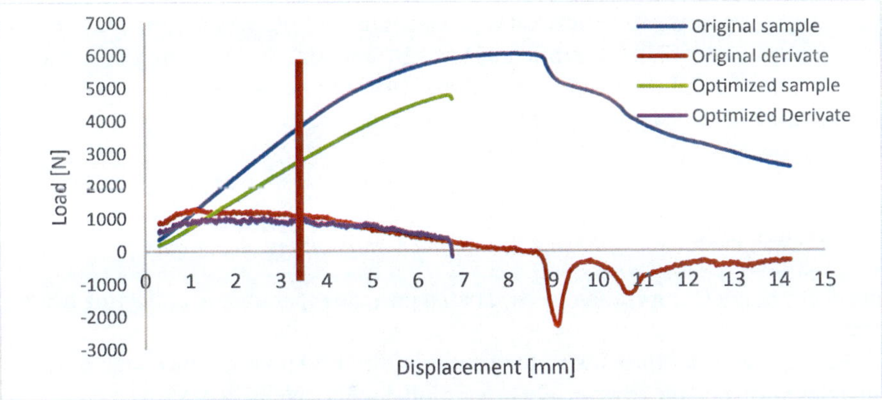

Fig. 6. The two samples Load-Displacement curves.

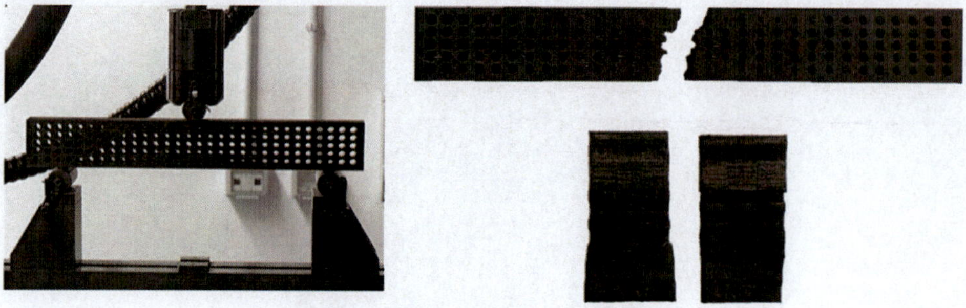

Fig. 7. The broken specimen.

By superimposing the result of the FE analysis on the optimized specimen, it can be observed how the variation in the diameter of the holes is related to the trend of stress (Fig. 8).

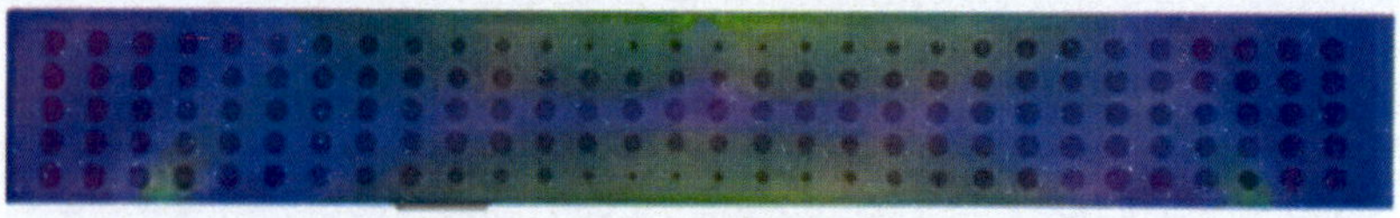

Fig. 8. A comparison of the stress field superimposed on the hollowing pattern.

The weight decreases by about 21%, the stress/weight ratio increases by about 20% in the optimized specimen. An interesting aspect is that the utilization coefficient, defined as the average stress on the sample, increases by about 37% under the same displacement. This means that each part of the sample works more on average. Conversely, the manufacturing time increases by about 54%, due to the constant changes in direction of the print head.

4 Conclusions

A novel Topology Optimization method based on a monoparametric hollowing has been presented and tested.

The algorithm interfaces with the results of the FEA and hollows the component with holes with a diameter inversely proportional to the intensity of the stress.

A rectangular-shaped beam has been optimized, as case study, with a good increase of stress/weight but also of time/weight ratios.

The same approach can be used with components of any shape and nature and it is useful for both additive or traditional manufacturing technologies, by using CNC drilling or milling machines since the hollowing is suitable also for traditional subtractive technologies.

References

1. Subramaniam, S.R., et al.: 3D printing: overview of PLA progress. In: AIP Conference Proceedings (2019). https://doi.org/10.1063/1.5085958
2. Gardan, J.: Additive manufacturing technologies: state of the art and trends. Int. J. Prod. Res. **54**, 3118–3132 (2016). https://doi.org/10.1080/00207543.2015.1115909
3. Yao, T., Deng, Z., Zhang, K., Li, S.: A method to predict the ultimate tensile strength of 3D printing polylactic acid (PLA) materials with different printing orientations. Compos. Part B Eng. **163**, 393–402 (2019). https://doi.org/10.1016/j.compositesb.2019.01.025
4. Ahn, D., Kim, H., Lee, S.: Fabrication direction optimization to minimize post-machining in layered manufacturing. Int. J. Mach. Tools Manuf. **47**, 593–606 (2007). https://doi.org/10.1016/j.ijmachtools.2006.05.004
5. Gardan, N., Schneider, A.: Topological optimization of internal patterns and support in additive manufacturing. J. Manuf. Syst. **37**, 417–425 (2015). https://doi.org/10.1016/j.jmsy.2014.07.003
6. Muzzupappa, M., Barbieri, L., Bruno, F., Cugini, U.: Methodology and tools to support knowledge management in topology optimization. J. Comput. Inf. Sci. Eng. **10**, 417–425 (2010). https://doi.org/10.1115/1.3518386
7. Cucinotta, F., Raffaele, M., Salmeri, F.: A stress-based topology optimization method by a Voronoi tessellation Additive Manufacturing oriented. Int. J. Adv. Manuf. Technol. **103**(5–8), 1965–1975 (2019). https://doi.org/10.1007/s00170-019-03676-4
8. Cucinotta, F., Guglielmino, E., Longo, G., Risitano, G., Santonocito, D., Sfravara, F.: Topology optimization additive manufacturing-oriented for a biomedical application. In: Cavas-Martínez, F., Eynard, B., Fernández Cañavate, F.J., Fernández-Pacheco, D.G., Morer, P., Nigrelli, V. (eds.) Advances on Mechanics, Design Engineering and Manufacturing II. LNME, pp. 184–193. Springer, Cham (2019). https://doi.org/10.1007/978-3-030-12346-8_18
9. Wonoto, N., Blouin, V.: Integrating grasshopper and matlab for shape optimization and structural form-finding of buildings. Comput. Aided. Des. Appl. **16**, 1–12 (2018). https://doi.org/10.14733/cadaps.2019.1-12
10. Sudin, M.N., Tahir, M.M., Ramli, F.R., Shamsuddin, S.A.: Topology optimization in automotive brake pedal redesign. Int. J. Eng. Technol. **6**, 398–402 (2014)
11. Abdi, M., Ashcroft, I., Wildman, R.D.: Design optimisation for an additively manufactured automotive component. Int. J. Powertrains **7**, 142–161 (2018). https://doi.org/10.1504/IJPT.2018.090371
12. Jia, D., Li, F., Zhang, C., Li, L.: Design and simulation analysis of trimaran bulkhead based on topological optimization. Ocean Eng. **191**, 106304 (2019). https://doi.org/10.1016/j.oceaneng.2019.106304
13. Zhu, J.-H., Zhang, W.-H., Xia, L.: Topology optimization in aircraft and aerospace structures design. Arch. Comput. Meth. Eng. **23**(4), 595–622 (2015). https://doi.org/10.1007/s11831-015-9151-2
14. Tomlin, M., Meyer, J.: Topology optimization of an Additive Layer Manufactured (ALM) Aerospace Part. In: 7th Altair CAE Technology Conference, vol. 2011 (2011)
15. Sutradhar, A., Park, J., Carrau, D., Nguyen, T.H., Miller, M.J., Paulino, G.H.: Designing patient-specific 3D printed craniofacial implants using a novel topology optimization method. Med. Biol. Eng. Comput. **54**(7), 1123–1135 (2015). https://doi.org/10.1007/s11517-015-1418-0
16. Cucinotta, F., Raffaele, M., Salmeri, F.: A topology optimization of a motorsport safety device. In: Rizzi, C., Andrisano, A.O., Leali, F., Gherardini, F., Pini, F., Vergnano, A. (eds.) ADM 2019. LNME, pp. 400–409. Springer, Cham (2020). https://doi.org/10.1007/978-3-030-31154-4_34

17. Cirello, A., Cucinotta, F., Ingrassia, T., Nigrelli, V., Sfravara, F.: Fluid–structure interaction of downwind sails: a new computational method. J. Mar. Sci. Technol. **24**(1), 86–97 (2018). https://doi.org/10.1007/s00773-018-0533-7
18. Valerga, A.P., Batista, M., Salguero, J., Girot, F.: Influence of PLA filament conditions on characteristics of FDM parts. Materials (Basel) **11**, 1322 (2018). https://doi.org/10.3390/ma11081322
19. Jullanun, P., Yoksan, R.: Morphological characteristics and properties of TPS/PLA/cassava pulp biocomposites. Polym. Test. **88**, 106522 (2020). https://doi.org/10.1016/j.polymertesting.2020.106522
20. Seol, K.-S., Zhao, P., Shin, B.-C., Zhang, S.-U.: Infill print parameters for mechanical properties of 3D printed PLA parts. Korean Soc. Manuf. Process Eng. **17**, 9–16 (2018). https://doi.org/10.14775/ksmpe.2018.17.4.009

A Preliminary Study on the Effect of Strut Waviness on the Mechanical Properties of BCC Lattice Unit Cells

Davide Tumino(✉), Andrea Alaimo, Calogero Orlando, and Stefano Valvano

Facoltà di Ingegneria e Architettura, Università degli Studi di Enna Kore, Cittadella Universitaria, 94100 Enna, Italy
davide.tumino@unikore.it

Abstract. In this paper, the effect of strut waviness on the mechanical properties of a BCC unit cell has been studied by means of numerical FEM simulations. Different waviness amplitudes have been introduced on variable relative density unit cell and resulting rigidity has been calculated. Two different loading configurations have been setup in order to simulate the uniaxial and the shear behavior. Periodic repetition of the unit cell in the three cartesian directions has been simulated with the proper definition of kinematic constraints on the boundaries. From the two loading configurations, the uniaxial modulus, the shear modulus and the Poisson ratio of the unit cell have been extracted and these entities have been compared with the corresponding of a straight strut BCC unit cell. Results have evidenced for the waved struts a considerable increase in the uniaxial modulus, a slight reduction of the shear modulus and a highly variable Poisson ratio, depending on the waviness amplitude.

Keywords: 3D lattice · Waviness · BCC · FEM

1 Introduction

Developments in Additive Manufacturing (AM) processes have allowed the industrial application of a wide range of meta-materials, with simple to complex topologies. The use of these materials can be found in large civil engineering applications as well as in micro biomechanical devices [1]. In this field, an important role play lattice structures, constituted by 2D or 3D repeated open cell structures with an ordered topology. The success of these new meta-materials is mainly the possibility to tune the response of the unit cell in terms of mechanical properties as a function of connectivity and geometry of the cells [2]. Moreover, lattices are lightweight structures if compared with the parent material and this quality is very attractive for application, for example in the aerospace framework, where the final weight of the product is of paramount importance. Modern AM technologies have no difficulties in printing complex shapes, with vertical or horizontal elements, especially those based on Powder Bed Fusion methods where powder bed can act as support to overhang geometries during manufacturing.

C. Rizzi et al. (Eds.): ADM 2021, LNME, pp. 431–441, 2022.
https://doi.org/10.1007/978-3-030-91234-5_44

Most diffused lattice units are strut based architectures, in particular the Body-Centered Cubic (BCC) and the Face-Centered Cubic (FCC) with the improvements obtained by including z-struts (BCCZ and FCCZ respectively). These lattice geometries have been extensively studied in literature to obtain the equivalent mechanical properties of the unit cell [3]. Analytical [4], numerical and experimental [5] approaches have given several tools to investigate on the rigidity and strength of various lattice structures and to predict their properties starting from the cell geometry. In [2, 5] and [6] it is demonstrated that the strut connection topology and the relative density are the main factors that influence the mechanical response of the lattice structure. These factors are correlated via the Gibson-Ashby model where a different deformation behavior is evidenced between a stretch dominated and a bending dominated unit cell. For a fixed topology, mechanical properties ideally depend only on the relative density, i.e. on the strut diameter [7].

The possibility to introduce a regular waviness on the cell struts has been recently investigated in [8] for 2D lattices as stretch dominated triangular and bending dominated hexagonal structures. In this paper, the influence of strut waviness and undercuts has been studied for 0.1 relative density cells by means of experimental and numerical analysis, obtaining that waviness enhances ductility especially in hexagonal structures.

In the present paper, the idea to introduce a waved shape in the struts of a 3D lattice structure has been implemented in a BCC unit cell. The aim of this study is to evaluate the mechanical response of the lattice to the waviness amplitude in the struts. To achieve this result, numerical FE simulations have been performed on a single cell subjected to uniaxial vertical load and to in plane shear load, changing the relative density of the cell and the waviness amplitude. Results of simulations are given in terms of ratio between the unit cell modulus (axial and shear) and the bulk modulus and in terms of Poisson ratio. A final comparison is proposed between straight strut lattice cells and waved ones to demonstrate the improving effects that waviness leads to.

2 Lattice Unit Cells

The behavior of a lattice cell mainly depends on its relative density, calculated as the density of the lattice structure divided by the density of the cubic unit. Different lattice geometries lead to an internal deformation of the structures that can be ideally stretch dominated or ideally bending dominated or a combination of both [6]. In literature [3, 6], such behaviors are associated with the Maxwell's stability criterion where the Maxwell number M can be defined as:

$$M = s - 3n + 6 \quad (1)$$

where s is the number of struts and n the number of nodes in the unit cell.

For those lattice geometries where $M < 0$, there are too few struts to equilibrate external moment loads at nodes and struts react with internal bending. On the contrary, where $M \geq 0$, the number of struts is sufficient to react with only axial tension or compression at nodes. This simple criterion works fine for different simple strut-based geometries: for example body-centered cubic has $M = 8 - 3 \times 9 + 6 = -13$ and face-centered cubic has $M = 16 - 3 \times 12 + 6 = -14$ meaning that for these structures a bending dominated deformation can be predicted. For an octet-strut unit cell $M = 36 - 3 \times 14 + 6 = 0$ meaning that the structure is stretch-dominated [3]. In Fig. 1, previously defined cell structures are shown.

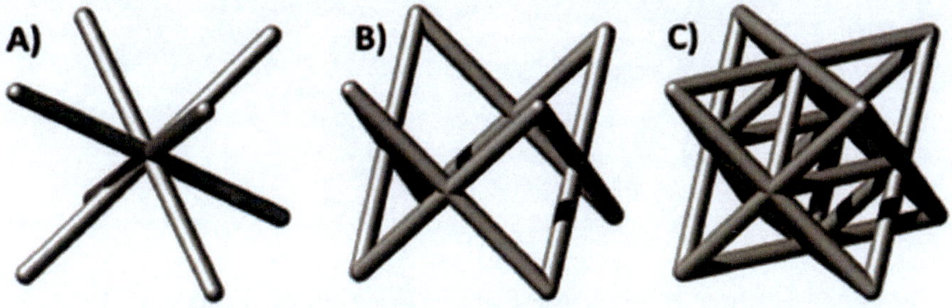

Fig. 1. Geometry of typical lattice cells: BCC (A), FCC (B) Octet (C).

According to the Gibson-Ashby [2] power criterion, modulus of the lattice unit cell can be predicted with simple equations:

$$\frac{E^*}{E_s} = C\left(\frac{\rho^*}{\rho_s}\right)^2 \tag{2}$$

for bending dominated structures, and

$$\frac{E^*}{E_s} = C\left(\frac{\rho^*}{\rho_s}\right) \tag{3}$$

for stretch-dominated structures, where E^* is the static modulus of the cellular structure, E_s is the static modulus of the structure bulk material, ρ^* is the density of the cellular structure, ρ_s the density of the bulk material and C is the Gibson-Ashby constant that depends on the unit cell geometry and topology and can be determined by experimental tests.

A typical Body-Centered Cubic (BCC) cell is depicted in Fig. 2 left. Edge length are equal in the three cartesian directions and the reference system is assumed to be located in correspondence of a corner node with axes x and y horizontal and the axis z vertical. It can be noticed that at corner nodes strut ends are cut with horizontal and vertical planes to allow the periodic repetition of the cell in all the directions. For these analyses it has been used $c = 10$ mm, but all following results are given in terms of dimensionless entities (i.e. divided by the correspondent of the bulk material).

To introduce the strut waviness, a rotated plane $\chi - z$ is defined, aligned with the diagonals of the cube, see Fig. 2 right. Due to the equal-size edges, the length of the horizontal diagonal is simply given by $\sqrt{2}c = 14.142$ mm. The straight struts are replaced with sine waved struts with axes on the main diagonals of the cube. This sine function has a period equal to $l = \sqrt{3}c = 17.32$ mm and the amplitude a.

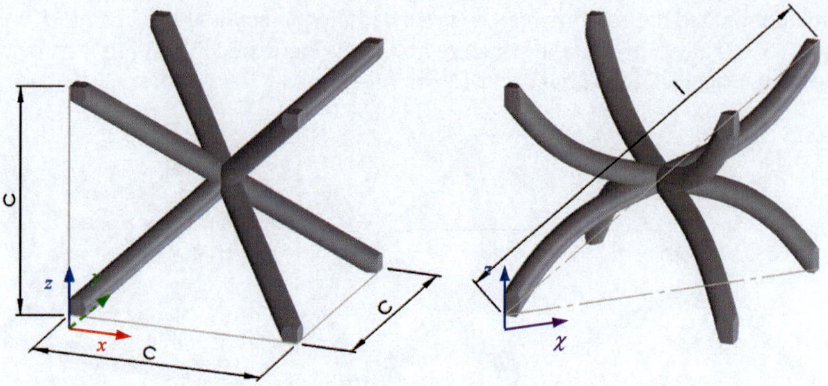

Fig. 2. Geometry of the BCC models: straight strut (left) and waved strut (right).

The influence of the sine function amplitude is shown in Fig. 3 left, from a minimum of $a = 0$ (corresponding to a straight strut) to a maximum of $a = 2.5$ mm. It is important to notice that, as the amplitude grows, the slope of the strut at mid length decreases and it becomes zero for $a = 2$ mm. This behavior can be better observed in Fig. 3 right where the slope of the waved struts is plotted for different a values. At $a = 2$, corresponding to ratio $a/l = 0.115$, the slope is zero at mid length and it would become negative for greater values of a. However, it is not preferable to adopt too large sine amplitudes because negative slopes cause multiple intersections between struts in the same rotated plane and, consequently, difficulties in the manufacturing process. For this reason, in this paper it is decided to limit the analysis to amplitudes that do not lead to negative slope, that is up to $a = 2$ mm.

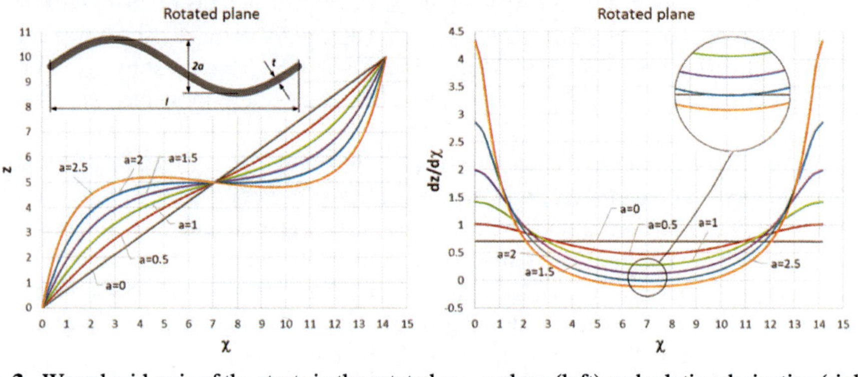

Fig. 3. Waved mid-axis of the struts in the rotated $\chi - z$ plane (left) and relative derivative (right), dimensions in [mm].

The mathematics of the sine wave is imported in the CAD software SOLIDWORKS where different relative densities of the unit cell have been obtained by simply increasing the diameter t of the strut.

The waviness has influence on the relative density as well, as reported in Table 1.

Table 1. Relative cell density [dimensionless] as function of *a* and *t* in [mm] (cell size length *c* = 10 mm).

a ↓ t →	0.5	0.75	1	1.25	1.5	1.75	2	2.25	2.75	3.5
0	0.0130	0.0285	0.0495	0.0755	0.1059	0.1404	0.1785	0.2197	0.3096	0.4565
0.5	0.0131	0.0287	0.0496	0.0755	0.1059	0.1401	0.1779	0.2187	0.3074	0.4514
1	0.0132	0.0289	0.0498	0.0754	0.1052	0.1386	0.1752	0.2146	0.2995	0.4353
1.5	0.0134	0.0288	0.0492	0.0738	0.1022	0.1338	0.1683	0.2049	0.2833	0.4071
2	0.0125	0.0269	0.0458	0.0684	0.0944	0.1232	0.1543	0.1872	0.2574	0.3666

3 Numerical Simulations

Parametric CAD models defined in SOLIDWORKS have been imported via STEP files into the FEM structural module of ANSYS. In this section results of numerical simulations of uniaxial compression and of in plane shear on the unit lattice cell are reported.

3.1 FEM Model

The mesh is obtained with tetrahedral quadratic elements with a typical edge length of 0.2 mm, in order to have a sufficiently fine mesh also for the thinner struts (cell edge length $c = 10$ mm). The total number of elements varies considerably because of the different volumes obtained by changing the strut diameter t: the maximum number of elements has been found as 253296 for $t = 3.5$ mm, while the minimum is 5727 for $t = 0.5$ mm. In Fig. 4, two meshed models are shown as example. FEM models can be used both to simulate uniaxial compression and in plane shear but boundary constraints and symmetry conditions must be modified for the two cases.

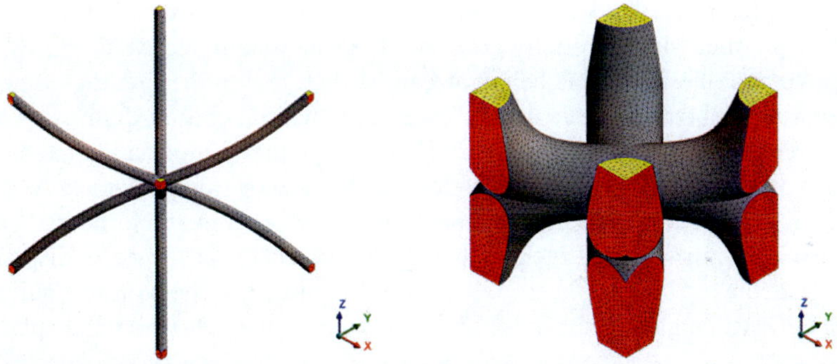

Fig. 4. Meshed models for two different a and t combinations: $a = 0.5$, $t = 0.5$ (left) and $a = 2$, $t = 3.5$ (right).

3.2 Uniaxial Compression

Uniaxial compression is a well-established loading configuration to analyze the elastic properties of the unit cell of a lattice structure [7, 9]. To correctly reproduce the deformation that a unit cell is subjected to, the multiple linear repetition of it must be correctly simulated in the FEM code. When the unit cell is compressed along the z axis on top surfaces (yellow in Fig. 4), deformations in the directions x and y happen but they are not completely free because the lateral faces of the cell are boundaries to the adjacent cells. To mimic the phenomenon, lateral faces of the cell must remain planar and vertical. This result is obtained by imposing a remote point on each of the external cell nodes: these points, because of the double symmetry in x and y, can move only in z and along the cell diagonals. Displacements of lateral faces (red in Fig. 4) in their normal direction are coupled with each associated remote point.

FEM simulations have been performed for different combinations of strut diameter t and of waviness a. Numerical results are given in terms of relative uniaxial modulus vs. relative cell density, see Fig. 5. This figure includes four plots for different waviness a. In each plot, the orange line and the cyan line represent the behavior of an ideal stretch-dominated cell and of a bending-dominated cell respectively. By following Eq. 2 and 3, in a log-log diagram the behavior of these cells appears as a straight line with different slope that converge at $x = y = 1$. In literature [3], evidence is reported that real cell structures follow the above mentioned rules especially for relative density lower than 0.1; for higher densities the curve can move with different slopes. In the same plots, a light cyan band is reported that represents the domain where the behavior of most diffused bending-dominated cells is located [5]. This band is described by Eq. (2) where the Gibson-Ashby constant C varies in the range [0.1–4]. Furthermore, in each diagram the relative uniaxial modulus calculated for the straight BCC is plotted with a solid gray line: the curve starts at the right bound of the band for very low relative density, as in [9], but its slope monotonically grows up to the last computed density $\rho^*/\rho_s = 0.4565$. The entire curve remains below the ideal bending-dominated one.

The effect of waviness is evident especially for $a \geq 1$. The increase in the relative modulus with respect to the straight cell at the same density is much more perceived as the waviness grows. For the waved cells, calculated points follow curves that show an inflection point: their slopes initially grow and then, at certain relative densities, start to decrease. An hypothesis for this behavior can be that, due to the fact that the slope of ideal stretch-dominated cell structures is lower than that of bending-dominated ones, as the relative density of waved cells grows, struts change their deformation from bending (in the case of very slender aspect) to stretch (in the case of stocky elements).

In Fig. 6, contour maps of displacement are reported as example for the cell with waviness $a = 1.0$ mm and strut thickness $t = 1.25$ mm. On the left, vertical displacement u_z is shown; at the top blue surfaces $u_z = -1$ mm is applied, at the bottom a null vertical displacement is calculated on the red surfaces. On the right, the horizontal displacement, calculated as $u_H = \sqrt{u_x^2 + u_y^2}$ is shown; it can evidenced that the central point has no horizontal displacement and that a double symmetric behavior is correctly obtained.

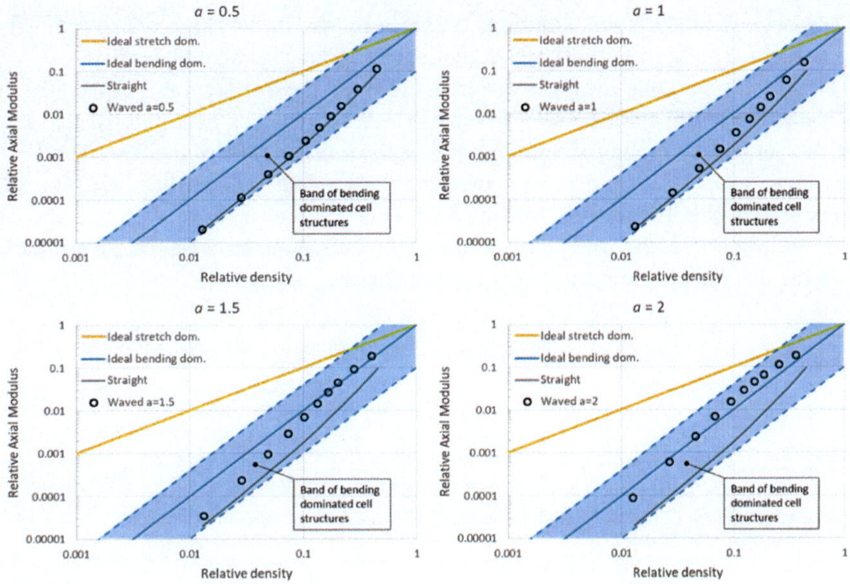

Fig. 5. Log-log diagrams of relative axial modulus vs. relative density for different wave amplitudes.

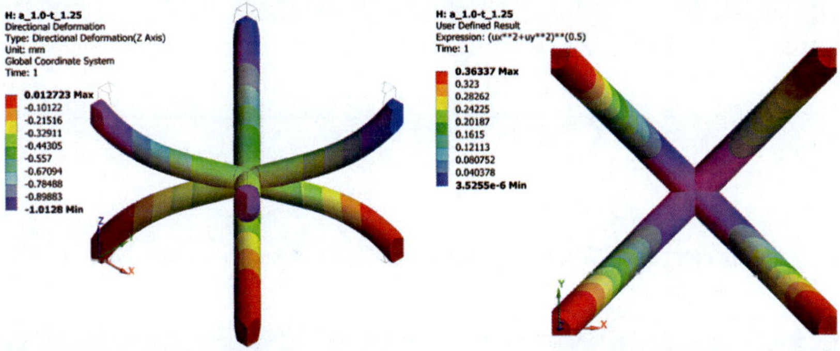

Fig. 6. Contour map of displacements for the cell $a = 1$ mm, $t = 1.25$ mm: u_z displacement (left) and horizontal displacement (right).

Another important parameter that can be extracted from the uniaxial compression is the transversal deformation of the unit cell. The lateral expansion in the x (or equally in the y) direction has been calculated during compression in order to evaluate an equivalent Poisson ratio that is a function of the relative density and of the strut waviness. These results are reported in Fig. 7 for different values of a. In each plot, the cyan horizontal line represent the Poisson ratio of the bulk material (in this case, 0.3). Moreover, it is also shown the curve obtained for the straight strut cell with a solid gray line. This line starts, for low relative density, from 0.5 and decreases very slowly up to $\rho^*/\rho_s = 0.1$, in accordance with analytical solutions obtained in [4, 10]. For high relative density, it decreases faster and arrives at 0.365 at the higher computed relative density.

The waved strut cells exhibit a different effect on the Poisson ratio especially for $a \geq 1$. The effect on the Poisson ratio given by the relative density are emphasized by the waviness: for similar ranges of relative density, the Poisson ratio of the waved cell varies in the range [0.27–0.50] for $a = 1$ and even in the very wide range [0.13–0.76] for $a = 2$. This result gives high versatility to waved strut cells, in fact with an adequate choice of density and waviness, the overall response of the lattice material can be functionalized to fit to the desired elastic properties of the material.

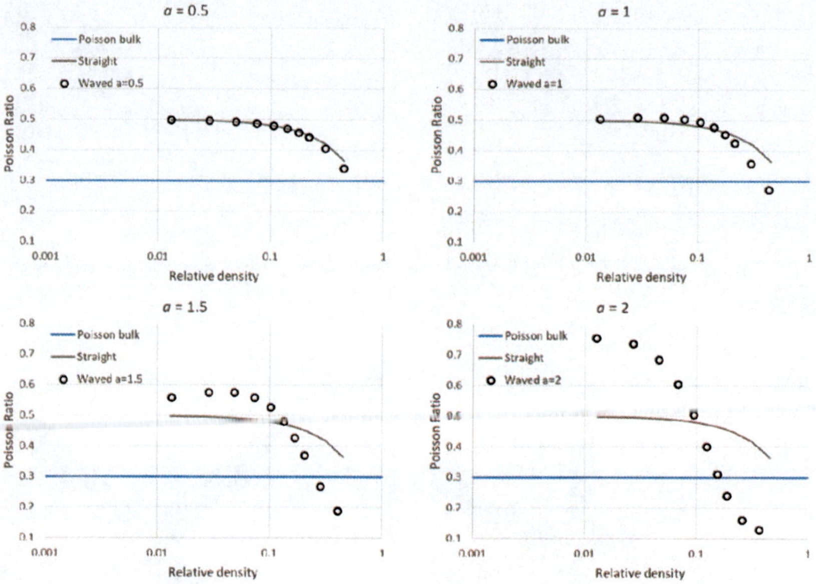

Fig. 7. Semi-log diagrams of Poisson ratio vs. relative density for different wave amplitudes.

3.3 In Plane Shear

To determine the shear characteristics of the BCC cell, both for straight and for waved struts, the same FEM model described in Sect. 3.2 has been used but different constraints have been defined on the boundaries. It is assumed now that the upper face has a uniform displacement in the x direction while the displacement in the z direction is assumed null for the upper and the bottom face and for the lateral faces normal to the x direction. Symmetry is imposed on the lateral faces normal to the y direction. This constraint system ensures a pure in plane shear deformation on the unit cell.

In Fig. 8, log-log diagrams of the relative shear modulus vs. relative density are reported for different values of waviness. For these plots, the relative shear modulus is calculated as the ratio between the shear modulus of the unit cell and the shear modulus of the bulk material. In each diagram, the relative shear modulus for the straight BCC is plotted with a solid gray line: the curve has a positive, slightly increasing slope and tends towards the point $x = y = 1$.

Computed points in the four plots for different a values give evidence that the waviness has an impact to the shear modulus of the unit cell. In fact, for all the values of a, calculated points are always below the solid line of the straight BCC and this behavior increases with the waviness. At each value of a, points follow an increasing trend with a decreasing slope and the solid line of the straight BCC acts as an asymptote for the points of the waved BCC. The explanation can be found in the strain mechanism that acts on the unit cell when shear is imposed. For straight BCC, it has been found that the struts are always stretch-dominated also for low relative density (under the assumption of small displacements) while waved BCC show large deflections due to bending in cases of slender struts. For waved strut, this behavior gradually vanishes as the relative density increases and struts become thicker.

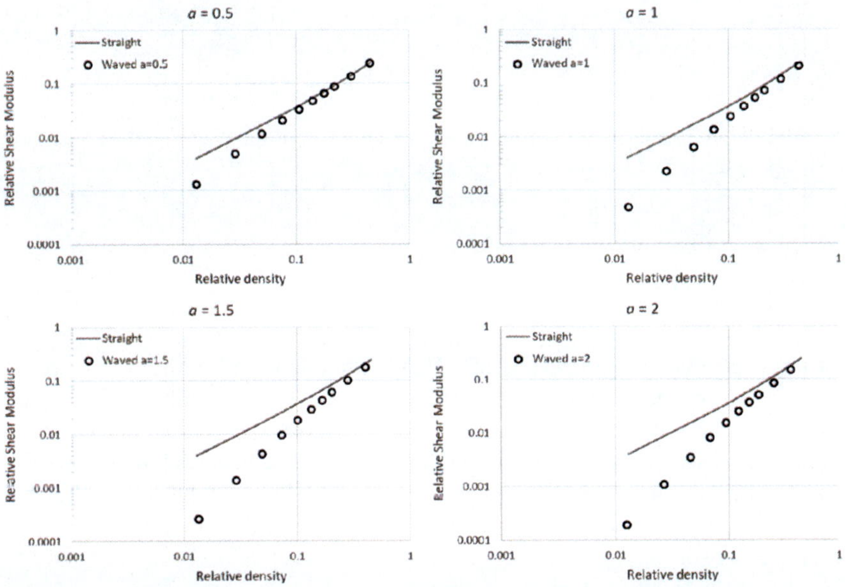

Fig. 8. Log-log diagrams of relative shear modulus vs. relative density for different wave amplitudes.

3.4 Comparison

To better quantify the effect of waviness with respect to straight strut cells, the ratio between the modulus of the waved cell and the corresponding straight cells have been calculated for axial load and for shear load. These results are reported in Fig. 9.

By looking on axial properties in Fig. 9 left, it can be noticed that the ratio is always above 1. For each curve the relative density corresponding to the maximum point decreases as the waviness increases. It must be remarked that at $a = 2$ mm and a relative density of 0.1 the modulus of the waved cell is 8.7 time higher than the modulus of the straight cell.

In terms of shear properties, the comparison between waved and straight cell is quite different. Curves admit no maximum point and a consistent loss of shear modulus happens already at low waviness. The ratio is an increasing function of the relative density, is always below 1 and approaches to 1 as the density grows. For values of *a* greater than 1.5 mm the curves overlap.

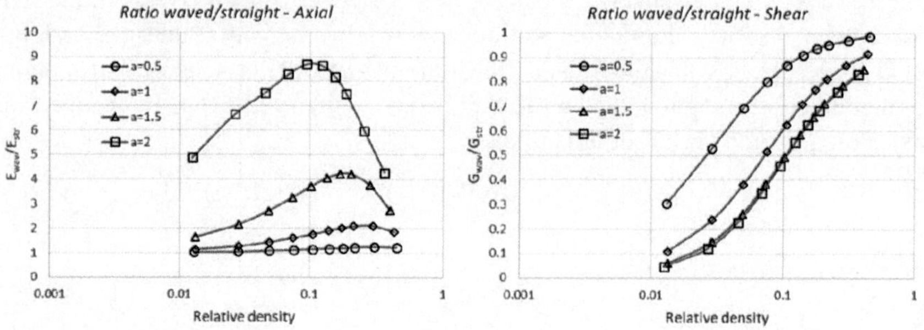

Fig. 9. Semi-log diagrams of the ratio between the modulus of the waved cell and the modulus of the straight cell: in compression (left) and shear (right).

4 Conclusions

In this paper, mechanical properties of a BCC lattice unit cell with waved struts have been studied by means of numerical FEM simulations. A cubic unit cell has been considered where the struts have a sinusoidal shape and different wave amplitudes have been considered. Axial and shear relative properties have been obtained for different relative densities of the cell. Moreover, transverse strain has been calculated during uniaxial compression in order to calculate the cell Poisson ratio. Main results obtained are summarized as follows:

- from the uniaxial compression analysis it can be noticed that the strut waviness always increases the axial modulus with respect to the straight cell; this increment is much more evident as the waviness grows and, at the limit value of
- $a = 2$ mm, it has been calculated an increment of the modulus of 8.7 time higher than the straight cell modulus;
- the uniaxial compression analysis has also shown that waviness has a big impact on the transversal cell deformation; the Poisson ratio of the unit cell varies with the relative density in a range that becomes wider as the waviness grows;
- in terms of shear behavior, the strut waviness always reduces the shear modulus with respect to the straight cell; this happens even with very low waviness and the reduction grows with the waviness.

These preliminary results have given evidence that such waved unit cell can be fully customized by the designer to achieve the required rigidity of the unit cell and of the entire multi cellular lattice. In the authors opinion, results obtained for BCC cells could be similarly obtained for other bending dominated lattice cells, but specific investigations need to be performed. On the other hands, waviness could be cause of reduced mechanical performance for stretch dominated lattice cells, because curvature of struts could promote local buckling phenomena.

Future developments will concern the study of the orthotropic properties of the unit cell and the effect of the period of the sine wave in the strut length for BCC cells. Furthermore, these analyses will be spread also to other lattice cell geometries.

Acknowledgments. This research activity has been financially supported by the Italian Ministry of University and Research (MUR) within the project SIADD (Innovative Solutions for quality and sustainability of Additive Manufacturing processes), ARS01_00806, funding program PNR 2015-2020.

References

1. Vyatskikh, A., Delalande, S., Kudo, A., Zhang, X., Portela, C.M., Greer, J.R.: Additive manufacturing of 3D nano-architected metals. Nat. Commun. **9**(1), 593 (2018)
2. Gibson, L.J., Ashby, M.F.: Cellular Solids: Structure and Properties. Cambridge University Press, Cambridge (1999)
3. Maconachie, T., et al.: SLM lattice structures: properties, performance, applications and challenges. Mater. Des. **183**, 108137 (2019)
4. Yang, Y., Shan, M., Zhao, L., Qi, D., Zhang, J.: Multiple strut-deformation patterns based analytical elastic modulus of sandwich BCC lattices. Mater. Des. **181**, 107916 (2019)
5. Ashby, M., Evans, T., Fleck, N.A., Hutchinson, J.W., Wadley, H.N.G., Gibson, L.J.: Metal Foams: A Design Guide, 1st edn. Butterworth-Heinemann, Oxford (2000)
6. Ashby, M.F.: The properties of foams and lattices. Philos. Trans. R. Soc. A **364**, 15–30 (2006)
7. Umer, R., Barsoum, Z., Jishi, H.Z., Ushijima, K., Cantwell, W.J.: Analysis of the compression behaviour of different composite lattice designs. J. Compos. Mater. **52**(6), 715–729 (2018)
8. Seiler, P.E., Li, K., Deshpande, V.S., Fleck, N.A.: The influence of strut waviness on the tensile response of lattice materials. J. Appl. Mech. **88**(3), 031011 (2021)
9. McKown, S., et al.: The quasi-static and blast loading response of lattice structures. Int. J. Impact Eng. **35**(8), 795–810 (2008)
10. Ptochos, E., Labeas, G.: Elastic modulus and Poisson's ratio determination of micro-lattice cellular structures by analytical, numerical and homogenisation methods. J. Sandwich Struct. Mater. **14**(5), 597–626 (2012)

Effects of Coating Post-processing on the Compressive Properties of Strut-and-Node-Based FDM Lattice Structures

Sandro Barone, Paolo Neri, Alessandro Paoli, Armando Viviano Razionale, and Francesco Tamburrino(✉)

Department of Civil and Industrial Engineering, University of Pisa, Largo L. Lazzarino 1, 56122 Pisa, Italy
francesco.tamburrino@ing.unipi.it

Abstract. Lattice structures have many outstanding properties, and their use in diversified industrial and biomedical fields is widely studied. The advent of additive manufacturing (AM) technologies has further pushed the design of these cellular structures allowing for the fabrication of complex trusses and tailored local geometries. However, geometrical defects introduced by the AM process into printed lattice structures significantly affect their mechanical properties. In this work, the effect of chemical post-processing on the compressive properties of FDM-PLA strut-and-node-based lattice structures is evaluated. A UV resin has been used as a coating film on samples fabricated using Simple Cubic (SC) and Face-Centered Cubic (FCC) unit cells. Results demonstrated a 65% increase in compressive strength for SC unit cells and a 12% increase for FCC unit cells with respect to as-printed samples. Resin coating demonstrated to represent an effective approach to minimize defects of strut-and-node-based lattice structures, thus enhancing mechanical properties.

Keywords: Lattice structures · Additive manufacturing · Coating post-process · Compressive properties

1 Introduction and Background

Lattice structures made by additive manufacturing (AM) technologies are a growing trend for vibration isolation [1], heat dissipation, energy absorption, and, more generally, lightweight design [2]. They are characterized by high stiffness and strength, high energy absorption, high damping. A lattice structure is defined by the unit cell, which can be typically classified as strut-and-node-based or surface-based. Strut-and-node-based unit cells consist of a network of thin prismatic struts, having constant cross-sections, connected at nodes, similar to truss structures. Polymer lattice structures are commonly produced by using VAT photopolymerization, Material Extrusion (ME) and even Selective Laser Sintering (SLS). However, among AM technologies, ME is probably the most popular and low cost. This technology has a high potential in printing

C. Rizzi et al. (Eds.): ADM 2021, LNME, pp. 442–450, 2022.
https://doi.org/10.1007/978-3-030-91234-5_45

lattice structures, since a wide range of thermoplastic polymers is available and also many composite materials can now be printed. Furthermore, ME machines equipped with a dual extruder system are increasingly frequent. This characteristic can allow the use of soluble supports or multi-material printing, providing the possibility to further customize geometries and the lattice structures' mechanical behavior. However, AM parts are generally characterized by defects that can be more pronounced than those typical of traditional manufacturing methods (e.g., anisotropy, porosities). Additionally, also layer stratification, which is peculiar of these manufacturing techniques, can lead to defects as delamination or poor adhesion between adjacent layers. Strut elements of lattice structures particularly suffer from poor surface characteristics, which negatively affect mechanical properties, lowering strength and toughness and resulting in a brittle behavior.

For this reason, post-processing treatments aimed to repair the external surface of lattice structures represent a practical approach to enhance 3D printed parts. According to the ISO/ASTM 52900:2015, post-processing treatments are defined as the steps taken after the completion of an additive manufacturing (AM) build cycle to achieve the desired properties in the final product.

In the last few years, a wide variety of post-processing treatments have been developed to enhance the effectiveness of AM polymers [3–5]. Among them, mechanical and chemical treatments are certainly the most used [4]. Mechanical treatments include all the techniques based on the use of abrasive media and/or mechanical tools to remove material from the part. They represent the most traditional post-processing treatments since derived from conventional metal finishing techniques, even if their use on polymeric parts is impaired by greater limitations if compared to metals (e.g., lower temperature resistance, lower surface hardness, lower strength). Also, mechanical treatments may damage weak features (e.g., strut-and-node-based structures). On the other hand, chemical treatments replace the mechanical contact with the part with a chemical contact carried out by substances in different forms (vapor, spray, bath). Chemical treatments are more effective than mechanical treatments in processing complex shapes, blind holes, and hardly accessible areas. Moreover, they can process small thin structures. Among chemical treatments, coating is one of the most used since characterized by fast processing time and minimal equipment requirements. The coating post-processing technique consists of applying a thin film of material on the external surface of the treated part by the *dip-coating* technique [6, 7], by simple *immersion* in a tank or by *manual brushing* or *spraying* [8–10]. In the last few years, coating treatments have been extensively used to provide smoother surfaces [6], increase mechanical properties and/or reduce water absorption of polymeric 3D printed parts [8–11]. However, no evidence exists about the use of coating treatments to enhance polymeric strut-and-node-based lattices structures.

In this work, coating-based treatments applied to Simple Cubic (SC) and Face-Centered Cubic (FCC) FDM-PLA strut-and-node lattice structures were experimented. The effects of the coating on the compressive strength of SC and FCC structures were investigated and the results compared to the respective as-printed structures. The coating effects were evaluated by comparing force-displacement curves, strength and displacement at break and, finally, by retrieving some observations on the failure behavior of the tested structures.

2 Materials and Methods

2.1 Materials and Specimen's Fabrication

In this study, PLA strut-and-node lattice structure specimens, having a size of 45 mm × 45 mm × 85 mm and 5 unit cells along Y and Z, and 10 along X, were manufactured with a Creality CR10-S5 FDM machine. A nozzle with a diameter of 0.4 mm was used. Layer height, printing speed and extrusion temperature, instead, were set respectively at 0.2 mm, 50 mm/s and 200 °C. The unit cell size is 10 mm, all the struts have a diameter of 1.4 mm and at the nodes a fillet radius of 0.7 mm was applied. The 3D model of the specimen was created in Solidworks®, exported as StL file, and then imported into Cura 4.6.1, an open-source slicer software provided by Ultimaker company, to create the G-code necessary for specimen's manufacturing. Two different specimen types were investigated, one based on a SC unit cell and the other based on a FCC unit cell. Both were inspired by the crystal structure of metals and are largely used when strut-and-node lattice structures are produced by Additive Manufacturing (AM) technologies [12]. An example of the unit cells adopted is shown in Fig. 1. A total number of 20 specimens were fabricated, 10 per unit cell type.

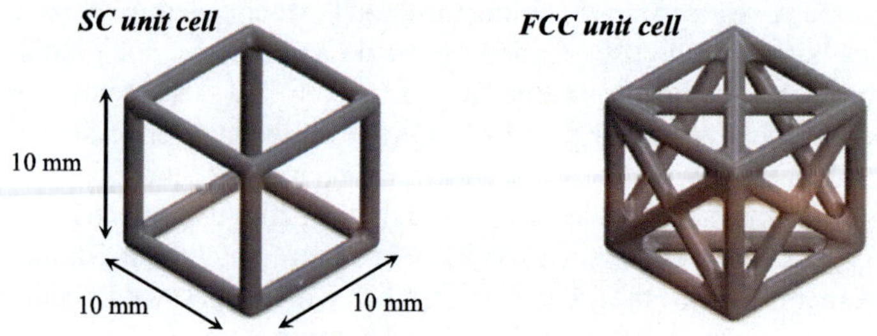

Fig. 1. Studied unit cells.

2.2 Specimen's Post-processing

Ten of the 20 printed specimens were treated and post-processed by applying a coating. The choice of this treatment was aimed to improve the surface characteristics of the lattice structures, which particularly suffer for the inhomogeneity, the presence of defects and for the so-called layer-by-layer appearance.

The coating was carried out using "UV DLP Poliglass" resin, supplied by Photocentric Group® and consisting of a mixture of acrylates, methacrylates, and photoinitiators as declared in the product safety data sheet. Each specimen was immersed in a tank containing the resin and kept still for 5 min (see Fig. 2). Then, the specimens were drained to remove the extra resin and achieve a fairly uniform coating distribution over the external surfaces. The resin coated on the specimen was then cured using the Wash & Cure Machine 2.0 supplied by Anycubic Company. Each sample was UV-cured for 10 min per side at a wavelength of 405 nm, for a total time of 60 min of UV curing. The characteristic solid-void geometric organization of the lattice structures and the 360° rotating curing platform allow the UV rays to reach all the lattice structures' elements and complete the curing of the coated resin. The described post-processing was carefully replicated for each specimen to guarantee the results' comparability. Furthermore, all the coated specimens were weighed before and after the coating process, to assess weight variation and have the same conditions for each specimen.

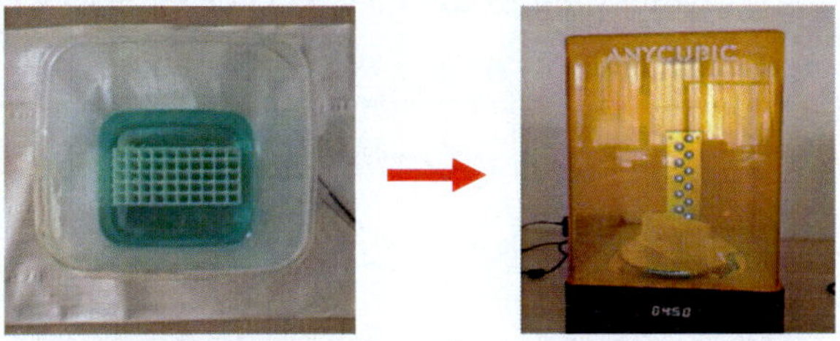

resin coating *resin UV curing*

Fig. 2. Coating and curing processes.

2.3 Mechanical Testing

Compressive tests were carried out on all the printed specimens. For each unit cell typology, five specimens were coated, and five were left "as-printed". Compressive tests were carried out at room temperature on a MTS testing machine with a load cell of 25 kN, a cross-head speed of 0.05 mm/s, and a data acquisition rate of 25 Hz (Fig. 3). Force-displacement curves were plotted, and the mechanical behavior of SC and FCC lattice structures was analyzed for uncoated and coated specimens. Furthermore, observations on their failure mechanisms were retrieved.

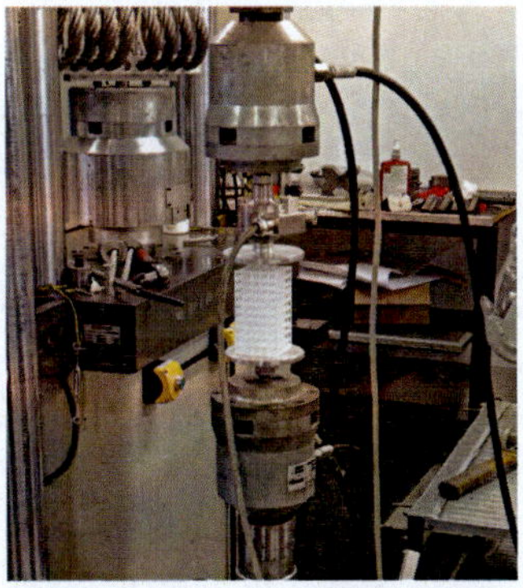

Fig. 3. Mechanical testing of a SC unit cell specimen.

3 Results and Discussion

The coated specimens based on SC unit cell showed an average weight gain of 7 g if compared to the same specimens weighted before the post-processing, corresponding to a 35% weight increase. Instead, the coated specimens based on FCC unit cell showed an average weight increment of 12 g, corresponding to a 25% weight variation.

Figure 4 reports the SC and FCC structures' behavior by plotting one representative force-displacement curve for as-printed specimens and one for coated ones.

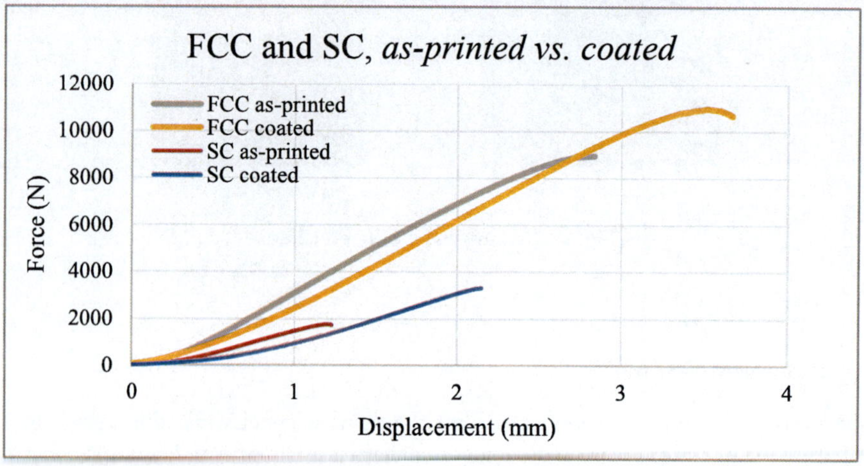

Fig. 4. Force-displacement curves for as-printed and coated SC and FCC specimens.

The representative curves show in the case of SC structure a notable improvement of the compressive strength and a similar enhancement for the displacement reached at the break. The increase of compressive strength and displacement is also confirmed for FCC structure. However, it is less significant than for SC structure. The slope of the curves is similar for "as-printed" and coated specimens, for both SC and FCC structure, confirming that only slight changes occur for the specimens' stiffness.

The mean values and the standard deviation of force and displacement at break, for the five tested specimens of each type, are reported in Table 1.

The results show that the coating provides an increase for SC and FCC structures, both in force and displacement reached at break. This increase is significant in the case of SC structure, where the mean value of the force rises from 1835 N to 3026 N (i.e., 65% increment with respect to "as-printed" specimens), and the mean value of the displacement rises from 1.44 mm to 2.36 mm (i.e., 64% increment with respect to "as-printed" specimens). The increase is still occurring for FCC specimens, but it is less impressive than for SC ones. The mean value of the force rises from 8814 N to 9914 N (i.e., 12% increment) and the mean value of the displacement rises from 3.23 mm to 3.66 mm (i.e., 13% increment).

Table 1. Mean values and standard deviation (STDV) of force and displacement at break for SC and FCC specimens, before and after coating.

Specimen	F (N)	STDV	Coated vs. as-printed	Displacement (mm)	STDV	Coated vs. as-printed
SC as-printed	1835	85.39	+65%	1.44	0.18	+64%
SC coated	3026	178.02		2.36	0.24	
FCC as-printed	8814	404.14	+12%	3.23	0.12	+13%
FCC coated	9914	464.57		3.66	0.37	

The enhancement of the mechanical behavior observed for coated specimens can be attributed to the reduction of the surface defects, lower roughness, and a possible slight increase of the struts' diameter.

The improvement related to the surface characteristic is shown in Fig. 5.

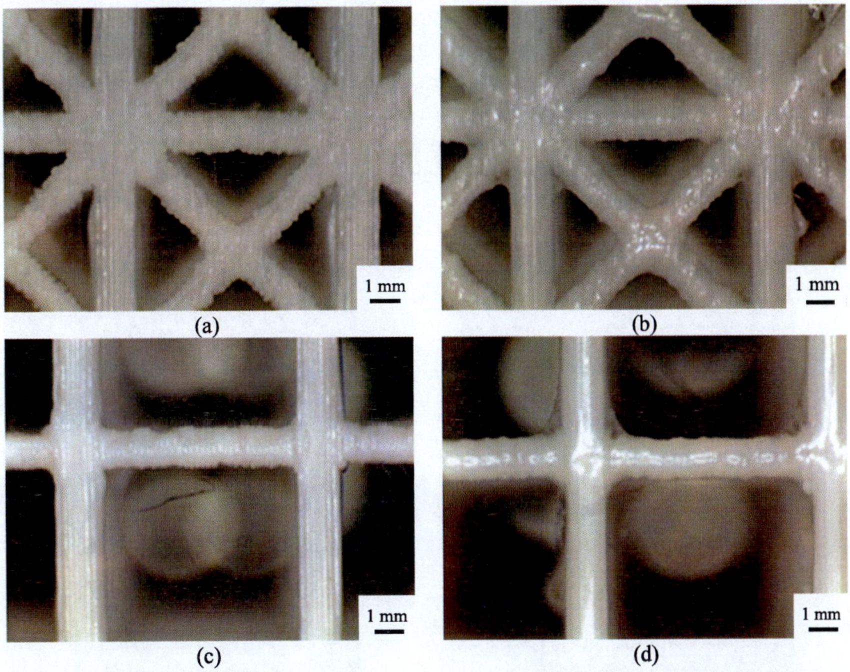

Fig. 5. Digital microscopy of "as-printed" (a and c) and coated (b and d) SC and FCC specimens.

The comparison of the digital microscopy of "as-printed" and coated specimens highlights that the layer-by-layer appearance is mitigated by the coating layer on the surface of the 3D printed specimen. Furthermore, the cohesive forces existing between the coating resin molecules (when it is in a liquid state, before curing) associated with its surface tension shrink the liquid at the nodes created by the struts and generate geometric fillets. Although the geometric fillets were already present when the StL file was generated (see Fig. 1), the fillets cannot always be easily 3D printed and resolution and nozzle diameter can be a limit when a small fillet radius must be printed. However, as reported in [13], nodal fillets can enhance the compressive strength of strut-and-node-based lattice structures and increase the stress required for buckling. In particular, this last aspect probably had a positive and significant impact on SC structures. The SC unit cell geometry, according to Maxwell's criterion [14], can be defined as bending-dominated ($M < 0$). The deformation of this kind of structure, differently from stretch-dominated unit cells ($M = 0$), is predominantly bending because the thin features that make it up (the struts) face buckle easily. The presence of nodal fillets [13, 15] may have had a most substantial and positive impact on SC structures, whose geometry is more prone to nodal rotations, strain energy localization at the nodes, and buckling than FCC structures [16].

This consideration can be further supported by observing Fig. 6, which shows a change in the failure behavior between "as-printed" and coated SC specimens. Uncoated specimens behave as expected for bending-dominated structures. The predominant deformation mode is bending, and part of the vertical struts face buckle easily, as evidenced by red dotted lines in Fig. 6-a. On the contrary, coated specimens do not show vertical strut's buckling, while the main break mechanism is the failure of the horizontal struts (orange dotted lines reported in Fig. 6-b), and vertical planes get partially or entirely detached. This different failure behavior could mean that after coating, because of the nodal fillets' presence and a slight change in struts' thickness, the bending deformation mode and buckling are not more predominant for SC structure, and the failure occurs because the horizontal struts are stretched, and they break up.

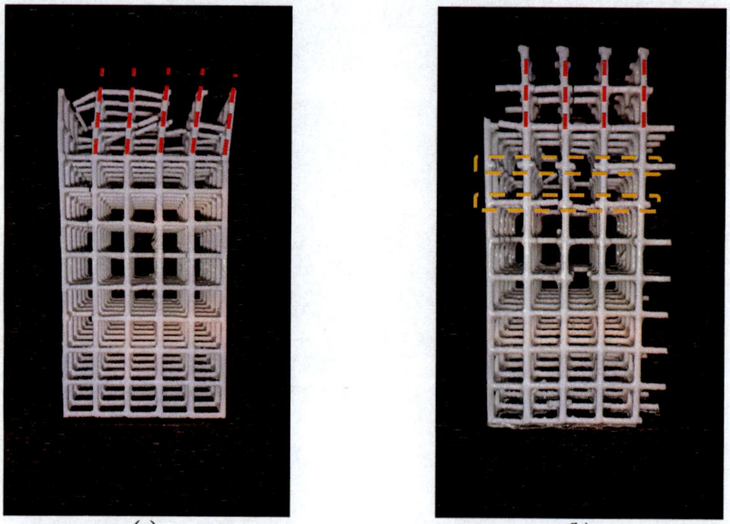

(a) (b)

Fig. 6. Failure behavior of as-printed (a) and coated (b) SC specimens.

4 Conclusions

Additively manufactured lattice structures are increasingly studied and adopted as solutions for lightweight design, vibration isolation, energy absorption, and much more. The interest in lattice structures is motivated by their vast potential to reduce material waste and part weight and to customize the mechanical behavior of the designed component through aspects such as design, selection and sizing of the unit cell geometry, the approach used to populate the design space and the use of density gradients. The presence of manufacturing defects related to the use of AM for fabricating the characteristics thin feature which constitute strut-ad-node structures, can have a significant and negative impact on the considerable potential of additively manufactured lattice structures. A chemical post-processing method based on a coating treatment has been developed and experimented in this work. The results show a general enhancement of the mechanical strength of lattice structures tested after coating. The improvement is impressive in the case of SC unit cells, where for a 35% weight increase, a 65% increase in compressive strength with respect to as-printed specimens is registered. The enhancement of the compressive strength (12% increment) is still significant, albeit minor, in case of FCC unit cells and the weight increase is of 25% after coating. The reported results demonstrated that resin coating could be an effective solution for post-processing treatments of strut-and-node lattice structures when manufacturing defects should be minimized, and reliable and advanced mechanical properties are needed. Some differences about the level of effectiveness of the coating process applied to bending-dominated and stretch-dominated lattice structures were observed and should be further investigated. In particular, the introduction of performance indexes, which consider aspects such as maximum force and/or displacement reached at break in relation to the weight increase observed after the coating process, can represent a step forward. Their introduction could be a valid tool to properly design coating post-processing of strut-and-node-based FDM lattice structures. Furthermore, future developments could be addressed to experiment coating-based post-processing for a wider range of unit cell geometries and to investigate new and different coating products, aiming to enhance mechanical properties and get novel and more advanced ones.

References

1. Syam, W.P., Wu, J.W., Zhao, B., Maskery, I., Elmadih, W., Leach, R.: Design and analysis of strut-based lattice structures for vibration isolation. Precis. Eng. **52**, 494–506 (2018)
2. Pan, C., Han, Y.F., Lu, J.P.: Design and optimization of lattice structures: a review. Appl. Sci.-Basel **10**, 6374 (2020)
3. Kumbhar, N.N., Mulay, A.V.: Post processing methods used to improve surface finish of products which are manufactured by additive manufacturing technologies: a review. J. Inst. Eng. (India): Ser. C **99**(4), 481–487 (2016). https://doi.org/10.1007/s40032-016-0340-z
4. Chohan, J.S., Singh, R.: Pre and post processing techniques to improve surface characteristics of FDM parts: a state of art review and future applications. Rapid Prototyping J. **23**, 495–513 (2017)
5. Wickramasinghe, S., Do, T., Tran, P.: FDM-based 3D printing of polymer and associated composite: a review on mechanical properties, defects and treatments. Polymers-Basel **12**, 1529 (2020)

6. Zhu, J., Chen, J.L., Lade, R.K., Suszynski, W.J., Francis, L.F.: Water-based coatings for 3D printed parts. J. Coat. Technol. Res. **12**(5), 889–897 (2015). https://doi.org/10.1007/s11998-015-9710-3
7. Lee, K.M., Park, H., Kim, J., Chun, D.M.: Fabrication of a superhydrophobic surface using a fused deposition modeling (FDM) 3D printer with poly lactic acid (PLA) filament and dip coating with silica nanoparticles. Appl. Surf. Sci. **467**, 979–991 (2019)
8. Leite, M., Varanda, A., Ribeiro, A.R., Silva, A., Vaz, M.F.: Mechanical properties and water absorption of surface modified ABS 3D printed by fused deposition modelling. Rapid Prototyping J. **24**, 195–203 (2018)
9. Miguel, M., Leite, M., Ribeiro, R., Deus, A.M., Reis, L., Vaz, M.F.: Failure of polymer coated nylon parts produced by additive manufacturing. Eng. Fail. Anal. **101**, 485–492 (2019)
10. Barone, S., Neri, P., Orsi, S., Paoli, A., Razionale, A.V., Tamburrino, F.: Two coatings that enhance mechanical properties of fused filament-fabricated carbon-fiber reinforced composites. Addit. Manuf. **32**, 101105 (2020)
11. Vicente, C.M.S., Fernandes, J., Reis, L., de Deus, A.M., Vaz, M.F., Leite, M.: Effect of protective coatings on the water absorption and mechanical properties of 3D printed PLA. Frattura Ed Integrita Strutturale **13**, 748–756 (2019)
12. Tamburrino, F., Graziosi, S., Bordegoni, M.: The design process of additively manufactured mesoscale lattice structures: a review. J. Comput. Inf. Sci. Eng. **18**, 6374 (2018)
13. Latture, R.M., Rodriguez, R.X., Holmes, L.R., Zok, F.W.: Effects of nodal fillets and external boundaries on compressive response of an octet truss. Acta Mater. **149**, 78–87 (2018)
14. Ashby, M.F.: The properties of foams and lattices. Philos. Trans. R. Soc. A **364**, 15–30 (2006)
15. Zok, F., Begley, M., Latture, R.: Design and performance of periodic trusses. In: Proceedings of the IUTAM Symposium Architectured Materials Mechanics, pp. 1–2. Purdue University Libraries Scholarly Publishing Services (2018)
16. Portela, C.M., Greer, J.R., Kochmann, D.M.: Impact of node geometry on the effective stiffness of non-slender three-dimensional truss lattice architectures. Extreme Mech. Lett. **22**, 138–148 (2018)

Examination of Additively Manufactured Auxetic Components Using a Novel Testing Setup

Lewis William Robert Urquhart[1(✉)], Francesco Tamburrino[2], Andrew Wodehouse[1], Paolo Neri[2], and Armando Viviano Razionale[2]

[1] University of Strathclyde, Glasgow, UK
lewis.urquhart@strath.co.uk
[2] University of Pisa, Pisa, Italy

Abstract. This paper reports upon the results of an initial test cycle using a bespoke testing rig designed expressly to examine additively manufactured auxetic components. Firstly, the key problems facing practical researchers in the field of auxetics is explored with the treatment of the boundary condition identified as a key issue. The testing setup that is then introduced utilises a novel method of part mounting and facilitates optical analysis and real-time force measurements. The study analyses three different auxetic structures (re-entrant, chiral, and semi-rigid), a set of samples of which were additively manufactured in TPU material. A range of parameters were varied across the three designs including interior geometry and wall thicknesses in order to demonstrate the effectiveness of the setup for the examination of different structures. Several key results were distilled from the tests that were then further analysed through numerical modelling and discussed with respect to future testing. Our investigation shows a close alignment between the physical testing results and the simulations, indicating that the testing configuration is rigorous and may be used to explore the mechanical behaviour of more complex auxetic componentry.

Keywords: Auxetics · Mechanical analysis · Simulation · Additive manufacturing

1 Introduction and Background

Auxetics are interesting materials that display deformation properties with a negative Poisson's ratio and have been studied extensively over the last few decades within material and engineering science [1]. Common designs include re-entrant, chiral, and semi-rigid [2] (Fig. 1). Testing auxetics and other deformable structures with a view to advancing understanding of the behaviour of additively manufactured parts, exploring applications, or expanding the scope of possible testing strategies presents a number of challenges for researchers. Notably, control and monitoring of boundary conditions and buckling behaviour have proved to be difficult to implement rigorously due to the requirements of direct interference with the structure. For this reason, achieving setups

C. Rizzi et al. (Eds.): ADM 2021, LNME, pp. 451–461, 2022.
https://doi.org/10.1007/978-3-030-91234-5_46

that can replicate simulated results with reliability and accuracy is difficult. This work seeks to address some of these issues by exploring the short comings of various practical testing approaches proposed throughout the literature and presenting results from a new testing setup developed by the authors. The new testing system explores a novel method for controlling the boundary conditions to a much higher degree. Boundary conditions - the way in which the components are held or fixed at their perimeters - are a fundamental concern in this kind of mechanical testing as they affect how the properties of the overall structure respond to change such as large deformations. Furthermore, it is to highlight that the method presented for testing auxetics structures is not explicitly addressing the derivation of mechanical properties such as strength at yield or break or Poisson's ratio. The aim is to explore a novel testing approach based on a new setup for boundary and constraints conditions under a bi-axial loading condition. This paper will be split into three key stages in which we can explore the effectiveness of the new testing approach. Firstly, we will consider the literature on physical testing approaches undertaken by other researchers examining auxetics. Secondly, the experimental work undertaken by the authors as part of a wider research project is set out. Thirdly, experimental results will be compared to numerical results, to highlight the differences between the specimen behaviour in actual experimental conditions with respect to ideal simulated conditions. The discussion of the results will then be provided, describing the effectiveness of the approach and next steps for future researchers.

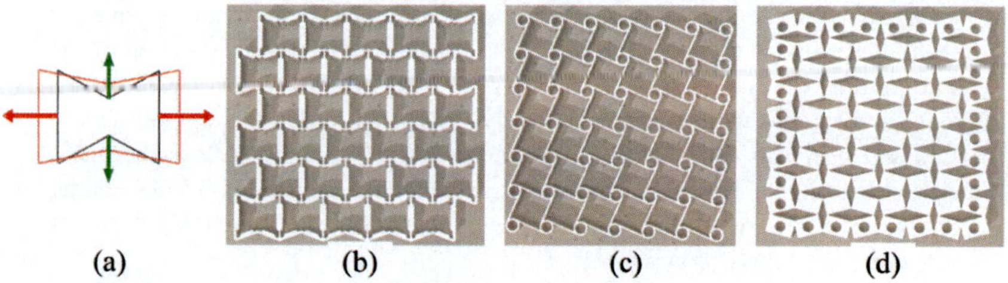

Fig. 1. Force-deformation diagram for an auxetic structure (a). Typical re-entrant (b), chiral (c) and semi-rigid (d) auxetic geometry.

2 Auxetics: Key Issues in Testing Set-Ups

2.1 Conventional Testing Setups

An "auxetic" is a structure that produces a negative Poisson's-ratio (NPR) in terms of its mechanical deformation when subjected to loads [1]. Within the literature focusing on the examination of 3D printed auxetic components, a number of setup variations have been identified that we could classify as more conventional, requiring less advanced equipment. Work from [3] exemplifies how simple jigs can be developed in order to examine the behaviour of parts under loading. A simple configuration exploits a rudimentary cuboidal transparent plastic structure to encase the auxetic, allowing the structure to compress without buckling across the z-axis. The transparent plastic allows for examination of the part as it is compressed although there is no formalised way to assess the changes in dimensional structure. Additionally, the boundary problem is addressed only by applying a uniform pressure to the top surface; the part is not held within a vice. While this offers some practical flexibilities, it leaves open the possibilities of slippages and inaccuracies in any given analysis. Other researchers have explored the use of slightly more advanced setups with readily available componentry. There are notable examples in the literature of different methods of holding the parts within custom made jigs. Research from [4] for example used a fixed jaw and a movable jaw in conjunction with a high-speed camera for their optical analysis. Markers are placed directly on the part to track the relative position of the points as the part is subjected to tension. The change in the relative position of the points is then interpreted by a computer, providing an assessment of how the part behaves. A similar approach was taken by [5] who examined the behaviour of thinner auxetic samples. The work used simple time lapsed photos to record the changes of the structure, resulting in a simple and effective method. [6] utilised a more thorough optical analysis method. The set up uses a jaw style jig that holds the component which is then pulled to tension using a series of rail-mounted stepper motors. The optical analysis method involves tracking the relative position of many points in the structure. Interestingly, a set of nodes have been added to the core structure that are then integrated into the jig, thus allowing the central structure to behave somewhat independently, moving the boundary problem away from the section being studied directly. In another interesting setup, the authors [7] secured the sample with use of a set of supporting truss structures. The pins are placed uniformly within the auxetic structure itself. This means that the boundary condition can be more carefully controlled, and the setup potentially facilitates more complex local deformation tests although the boundary condition remains uncontrolled on two sides of the component. The researchers used a similar optical analysis method to determine the results.

2.2 Advanced Testing Setups

Beyond the optical analysis tools that are characteristic of the "conventional" testing methods, other researchers have utilised what we can categorise as more advanced methods of analysis. In work by [8] for example, an auxetic sample is mechanically tested using a low velocity hydraulic press. Used in conjunction with an accelerometer, the analysis method determines material behaviour by examining the speed rate of material compression. The boundary condition is set by uniform surface contact of the striker and the bottom plate with the sample. Other work [2] explores a less conventional approach by compressing the auxetic sample off-axis. This creates a complex boundary condition where one side of the component (in this case a 3d printed metallic part) is subjected to compression while the other side is held in a static position. In a highly developed study [9], the researchers tested a range of auxetic structures by physical experimentation and numerical analysis with FE modelling. The researchers made a number of observations including that some structures exhibited properties of high stress resistance. A similar study explored the reliability of FE modelling using SolidWorks finding mixed results: strain behaviour could be accurately predicted but forces and stresses were less accurate [10]. Other research [11] has studied the mechanical dynamics of additively produced lattices through FE analysis. In a comprehensive review they concluded that there is a significant mismatch between the understanding of the design process for additively manufactured lattice structures and the modelling of these parts, leading to inefficiencies in the design process. Physical testing validated by numerical analysis has also been undertaken by [12] who used optical analytics and FE modelling to study uniaxial loading and by [13] in a comprehensive study. Less conventional auxetic forms have been studied by [14] in a similar multi-format approach concluding that "double-U" structures can reduce manufacturing damage and stress concentration under elastic loading.

3 Development of a Novel Testing Configuration

3.1 Testing Setup

The main boundary condition problem has been addressed through the creation of a bespoke jig. Figure 2 shows how the samples were mounted during testing. A series of adjustable pins were set into the sample on rails – this allowed both a small degree of movement but crucially held the sample firmly enough in order that the tension and compression forces could be applied. A basic lubricant was added to the rails to reduce the effects of friction. Taking inspiration from other research efforts (notably [7]) mounting points were set into the samples at uniformly distributed perimeter positions on all four sides. Recognizing that there are clear benefits to both the optical analysis and a more technical mechanical analysis, our setup was configured to both record time lapsed images revealing displacements and to measure the mechanical forces. As shown in Fig. 2a, a high-definition camera was placed above the sample, recording the physical changes in the structure. Sensitive force gauges were placed on two sides of the jig, recording real-time forces in the X and Y directions. Figure 2c shows the final testing rig configuration.

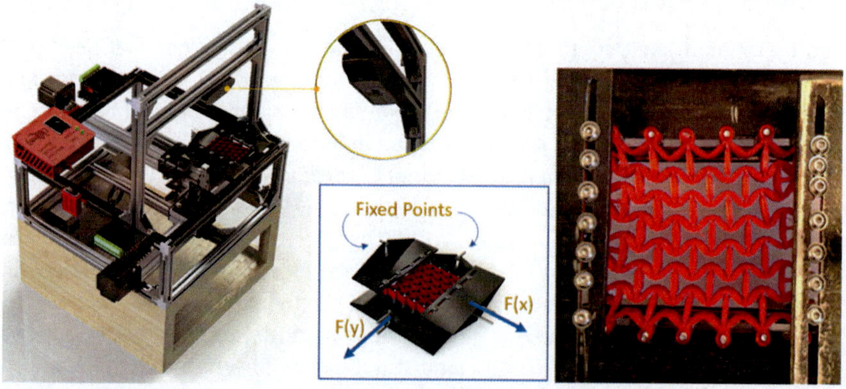

Fig. 2. Design of fully integrated testing rig with camera detail (a). Mounting design with support trusses and rails detail (b). Physical re-entrant sample mounted in testing jig (c).

3.2 Summary of Key Results

As this was the first run of test utilising this newly developed testing system, we decided to focus on conventional auxetic designs allowing us to thoroughly test the system with a set of well-understood structures. Figure 1 shows the sample designs that we focused on; re-entrant, chiral and semi-rigid. Various design parameters were edited in CAD including the interior angle (IA), and the wall thickness (WT) and then printed in TPU using a FlashForge 3D printer. Each sample was 10 mm in depth. The results were analysed by means of load-displacement curve, which are shown in Figs. 3, 4, 5, 6 and 7. As a point of information, the graphs track a complete cycle from tension to compression.

Re-entrant Sample Tests

Much of the derived data produced consistent and usable results despite a difficult boundary condition setup. Starting with the lowest interior angle variable – 15°, several results are of note when comparing these structures against their own variants in wall thickness and the other interior angle variants. For instance, Fig. 3 shows only small changes in recorded tension and compression forces at the two wall thickness extremes. This suggests that the largest overriding factor allowing the sample to sustain these loads may be the smaller interior angle, providing more mechanical freedom within the given deformation. There are some noted differences between the measured compressive forces, but this is only approximately 5–10 Newtons (N) across both axes.

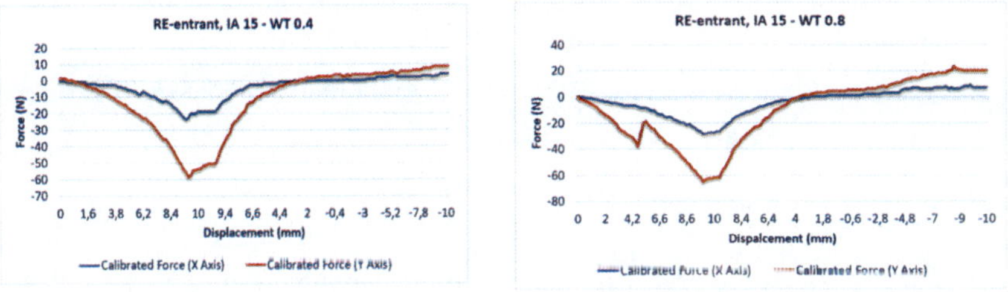

Fig. 3. Results of force-displacement study for two re-entrant structures characterized by different IA and WT.

If we now consider these in comparison to the other sample tests, some interesting conclusions can potentially be drawn. Figure 4 shows the distinction between the recorded forces of the 0.8 mm wall thickness extreme for the re-entrant sample with a 20° interior angle as compared to 25° at 0.6 mm. Looking at the data for an interior angle of 25°, the lower recorded forces suggest even greater flexibility for that structure. Again, the higher wall thicknesses of this design led to marginally higher force readings as flexibility is reduced when compared with other tests (not shown for brevity).

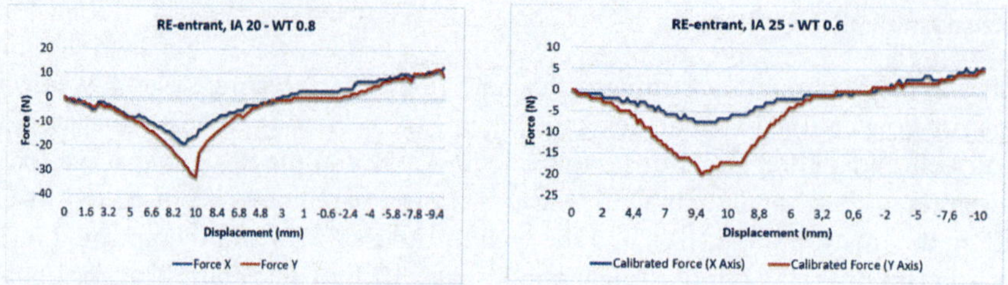

Fig. 4. Results of force-displacement study for two re-entrant structures characterized by different IA and WT.

Chiral Sample Tests

The chiral results differed from the re-entrant results in one distinct way since they showed greater uniformity in recorded mechanical behaviour across the X and Y axes. Again, the interior angles and the wall thicknesses of the build were varied. Considering the results at an interior angle of 60°, we can see a high level of uniformity in the recorded forces across both axes.

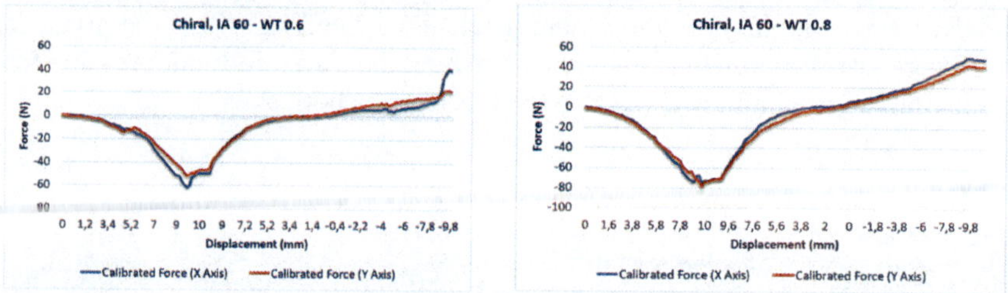

Fig. 5. Results of force-displacement study for two chiral structures characterized by different IA and WT.

A similar pattern is also seen with an increase in recorded forces as the wall thickness is increased. Interestingly, the results for the chiral samples show a distinctly larger increase in measured forces across both axes than the re-entrant samples. Figure 5 shows an approximate increase by 20 N in tension and compression when different wall thicknesses are compared. When looking at a sample variant with a larger interior angle in Fig. 6, the recorded forces are reduced markedly when compared to the 60-degree sample (approximately 60 N) suggesting that these samples have a higher degree of mechanical flexibility, echoing the re-entrant results (Figs. 3 and 4). Interestingly however, doubling the wall thickness from 0.4 to 0.8 mm has a very large influence on the recorded forces, contrasting much of the re-entrant test data. Looking at Fig. 6 and comparing the graphs, a difference of about 60 N is recorded suggesting that this configuration is greatly affected by changes in wall profile, particularly in the tension phase.

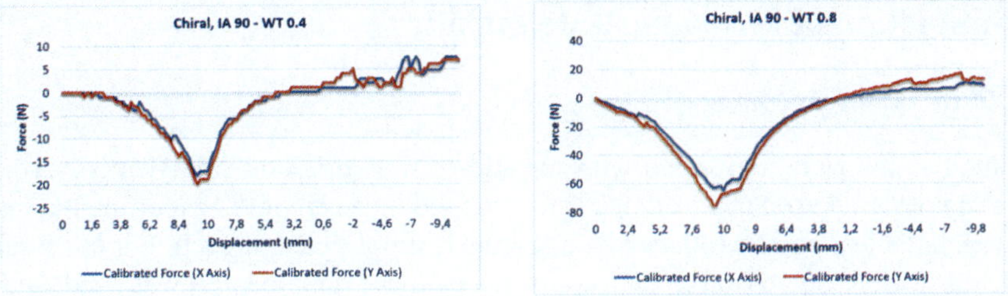

Fig. 6. Results of force-displacement study for two chiral structures characterized by different IA and WT.

To add more complexity to the discussion, the data for an interior angle of 120° record results very similar to that of the 90° samples with an increase in wall thickness hugely influencing the recorded forces although these results are not graphed for brevity. Only small differences are recorded when the parts are in compression, where a marginally larger compressive force is recorded for the larger 0.8 mm wall thickness. The similarities between the 90° and the 120° results (not shown) suggest that the interior angle above a certain threshold does not make a huge impact of the amount of force sustained by the sample as it is deformed.

Semi-rigid Sample Tests

The semi-rigid samples showed the same lack of uniformity as the re-entrant samples in terms of x/y behaviour but showed much greater recorded forces for both tension and compression when compared with most of the chiral and re-entrant sample variations. Considering the results shown in Fig. 7, it is clear how much force the samples are sustaining as they are deformed. This is mostly due to the increased rigidity of the samples due to their implicit geometric makeup, i.e., the samples are less skeletal and mostly solid compared to the chiral and re-entrant variants. While the chiral and re-entrant samples recoded forces at approximate maximums of 80 N, the semi-rigid samples sustained more than 100 N on the x-axis across the change in wall thickness, predictably showing greater flexibility with the smaller wall thickness. The higher compression forces show that the sample is becoming more structurally uniform as the material is pressed together – significantly more uniform than the chiral and re-entrant samples.

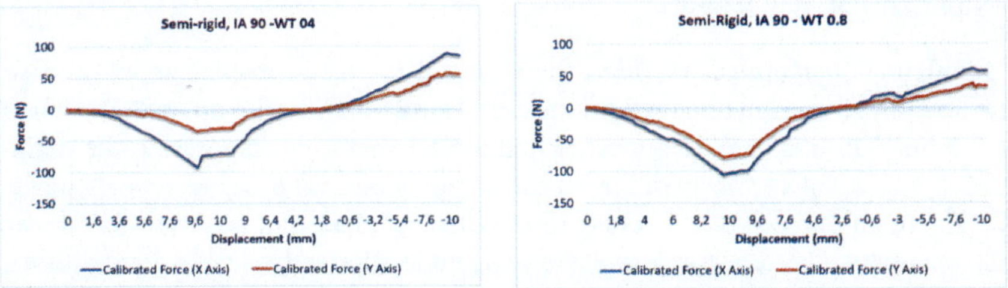

Fig. 7. Results of force-displacement study for two semi-rigid structures characterized by different IA and WT.

4 Simulation of Mechanical Behaviour

4.1 Finite Element Model

The simulations were performed with the add-in *simulation* of SolidWorks 2020. The mounting design, based on a series of pins trapped on a rail of the experimental testing rig, was modelled, and reproduced by adopting a simplified approach. For constraining and loading the simulated auxetic structures, a cylinder (i.e., the pin) was used, and two cubes were added at its top and bottom. The choice to utilize the cube was motivated by the possibility to have five free faces (one is attached to the cylinder) to constrain the auxetic structure on some axes and loading it on other axes. The method adopted is shown in Fig. 8a overleaf and a chiral structure is adopted as an example. The material applied to the auxetic structure was the same utilized for the experimental tests; TPU 95A. It was edited basing on the technical data sheet provided by the Ultimaker company (see https://ultimaker.com/materials/tpu-95a). The material applied to the fixtures, instead, was the AISI 316L. This was already included in the material library of SolidWorks and the choice was motivated by the requirement to use a stiffer material than TPU 95A for the fixtures. The simulation was run by using, for both materials, a linear elastic isotropic model with all the geometrical entities of the study treated as solid bodies. Finally, the mesh factor was set to "fine". This choice increased the calculation time, but it facilitated a higher accuracy in predicting the mechanical behaviour of the tested structure. All the auxetic structures characterized for the experimental tests (re-entrant, chiral, and semi-rigid structures) were tested. For each type, one or two variants with a different interior angle or a different wall thickness were simulated. The magnitude of loads, for every single structure, was derived by the corresponding experimental test. Such an approach was adopted for comparing physical and virtual testing based on the respective displacement results. For each structure, the magnitude of the force applied was derived by using the load corresponding to the higher displacement (10 mm) measured during the experimental test. If the displacement of the virtual test was close to the one of the experimental tests, the simulations were validated, and the results considered robust. Finally, the results of simulations and experimental tests were also assessed by a qualitative visual comparison through observations of the deformation modes occurring.

4.2 Summary of Key Results

The simulations performed provided the charting of stress, displacement, and strain. Figure 8b shows the result obtained for the chiral structure (interior angle: 60°; wall thickness: 0.6 mm). In this case, the undeformed model was superimposed to the deformed shape for a more direct observation of the strain mode. The figure reports the linear displacement on the y-axis. It is close to the displacement that occurred for physically tested specimens (10 mm for x and y axes), under the same loading conditions and magnitude of the force (55 N on the x and y axis). The result was also confirmed for the chiral structure that, on the same loading and constraints conditions, the prediction of the displacement obtained through virtual test is reliable. Results obtained for the

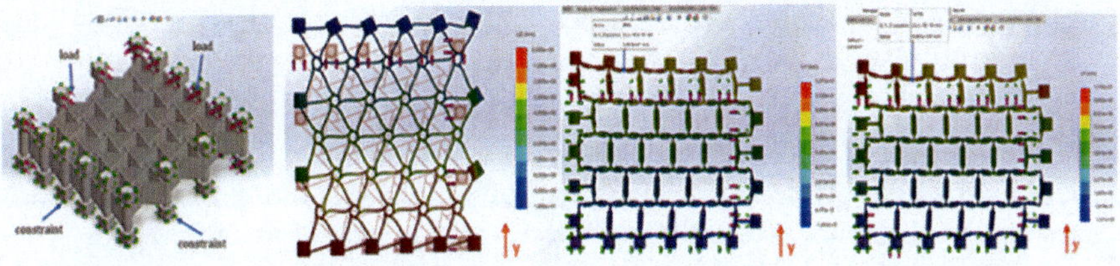

(a) (b) (c) (d)

Fig. 8. Simulation setup with loads and constraints (a). Chiral structure (IA 60°; WT 0.6 mm), the undeformed is superimposed to the deformed result (b). Simulated result of re-entrant structure (IA 20°; WT 0.8 mm) (c). Simulated result of re-entrant structure (IA 25°; WT 0.6 mm) (d).

semi-rigid sample (interior angle: 60°; wall thickness: 0.8 mm) recorded a similar level of continuity between the physical behaviour and the simulated behaviour.

Figure 8 also shows the displacement chart of the re-entrant structure with an interior angle of 20° and a wall thickness of 0.8 mm. The magnitude of the load set on the x-axis was 23 N and the magnitude on the y-axis was 34 N (analogous to the physical test). For this structure, the maximum displacement observed on the y-axis is larger than 10 mm. For the re-entrant structure, an additional variant was simulated, with a different interior angle (25°) and wall thickness (0.6 mm). The magnitude of the loads applied is in this case was 7 N on the x-axis and of 19 N on the y-axis. The mechanical behaviour confirms the observations made for the variant, in both cases the displacement is close to a value of 10 mm, the peak is at the corner on the left and the strain mode is aligned with it. Figure 9 shows a qualitative superimposed comparison between physical and virtual testing of chiral (a) and re-entrant structures (b) at 10 mm of displacement. The overlap between physical and virtual tests, in terms of displacement, was very good for both structures. However, it is to highlight that some slight deviations are present for the re-entrant structure, where on the horizontal axis the displacement is moderately larger for simulated structure compared to the experimental test.

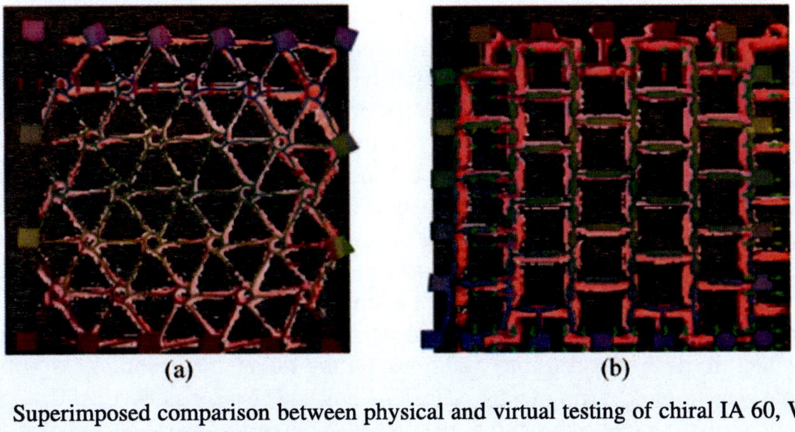

(a) (b)

Fig. 9. Superimposed comparison between physical and virtual testing of chiral IA 60, WT 0.6 (a) and re-entrant IA 20, WT 0.8 (b) structures.

5 Discussion

From our initial analysis, we can derive several key results that we can discuss in this section. With respect to a more general conclusion, we can clearly show that all sample variations showed differences in recorded forces, demonstrating that interior angle geometry and wall thicknesses influence these mechanical behaviours. More specifically, it was shown that an increase in wall thickness always led to an increase in the recorded deformation forces indicating higher stiffnesses. A larger interior angle correlated strongly with a decrease in measured forces for chiral and re-entrant samples, suggesting that a smaller interior angle creates greater flexibility in the structure. With respect to the differences shown between the X and Y forces, large deviations were recorded for the re-entrant and semi-rigid samples but by contrast there is a close correlation in the chiral samples suggesting that the design facilitates behaviour that is more mechanically uniform. Differences in X/Y behaviour in re-entrant and semi-rigid samples is most likely due to the directional differences in geometry creating different mechanical effects. The chiral samples sustained greater forces than the re-entrant samples and semi-rigid samples sustained the largest forces of all the samples. Semi-rigid samples consistently sustained larger forces under compression than the re-entrant or chiral variants, most likely due to the material becoming denser and more uniform under compression. With respect to the simulation efforts, we showed a robust match between the physical and virtual behaviour of the auxetic structures under deformation and the results were aligned for all the tested geometries. The method proposed can be used not only for the validation of the auxetic structures, but also as a supporting tool for designing more optimized and advanced structures. Future work can be more focused on the study of stress distribution occurring when the auxetic structures are strained. This would facilitate the retrieval of useful information which could potentially drive critical design tasks such as material selection.

6 Conclusion

This work presents the results from a novel setup to mechanically test auxetic components with a view to unifying a number of testing approaches into one umbrella system. Even if the proposed setup was limited to "off-the-shelf" componentry, we sought to produce results similar to previous research efforts but provide novel solutions for the boundary condition question and unify particular methods such as optical analysis and force-displacement analysis. This paper presented the results from an initial study focusing on a range of small scale auxetic components. The data derived from the tests allowed for several conclusions to be drawn relating to the behaviour of auxetics of different build characteristics. These results were subsequently validated through finite element simulations. The close match between force-displacement analytics and the optical comparisons demonstrates the effectiveness and efficiency of the testing setup, and its ability to reproduce the desired constraint conditions. Future studies could address aspects such as fatigue behaviour occurring with cyclical loading and also consider stress distribution occurring when the auxetic structures are strained. With respect to the design of auxetics, a testing setup such as ours presents the opportunity for the quick analysis of additively manufactured componentry that is both useful and reliable, stimulating the creation of more elaborate auxetic structures.

References

1. Lakes, R.: Foam structures with a negative Poissons ratio. Science **235**, 1038–1040 (1987)
2. Kolken, H.M.A., Zadpoor, A.A.: Auxetic mechanical metamaterials. RSC Adv. **7**, 5111–5129 (2017)
3. Abdel-Rahman, A., Tafrihi, E.: Heat-actuated auxetic facades. In: Noble, D., Kensek, K. Facade Tectonics, 2018, World Congress, Los Angeles. Tectonic Press, LA, USA (2018)
4. Zulifqar, A., Hua, T., Hu, H.: Development of uni-stretch woven fabrics with zero and negative Poisson's ratio. Text Res. J. **88**, 2076–2092 (2018)
5. Dubrovski, P.D., Novak, N., Borovinsek, M., Vesenjak, M., Ren, Z.R.: In-plane behavior of auxetic non-woven fabric based on rotating square unit geometry under tensile load. Polymers-Basel **11**(6), 1040 (2019)
6. Brighenti, R., Spagnoli, A., Lanfranchi, M., Soncini, F.: Nonlinear deformation behaviour of auxetic cellular materials with re-entrant lattice structure. Fatigue Fract. Eng. M **39**, 599–610 (2016)
7. Auricchio, F., Bacigalupo, A., Gambarotta, L., Lepidi, M., Morganti, S., Vadala, F.: A novel layered topology of auxetic materials based on the tetrachiral honeycomb microstructure. Mater. Design **179**, 107883 (2019)
8. Jiang, L.L., Hu, H.: Finite element modeling of multilayer orthogonal auxetic composites under low-velocity impact. Materials **10**(8), 908 (2017)
9. Alomarah, A., Masood, S.H., Sbarski, I., Faisal, B., Gao, Z.Y., Ruan, D.: Compressive properties of 3D printed auxetic structures: experimental and numerical studies. Virtual Phys. Prototy. **15**, 1–21 (2020)
10. Ardebili, M.K., Ikikardaslar, K., Chauca, E., Delale, F.: Behavior of soft 3D-printed auxetic structures under various loading conditions. In: ASME International Mechanical Engineering Congress and Exposition, 2019, Salt Lake City, Utah. American Society of Mechanical Engineers, NY, USA (2019)
11. Tamburrino, F., Graziosi, S., Bordegoni, M.: The design process of additively manufactured mesoscale lattice structures: a review. J. Comput. Inf. Sci. Eng. **18**, 040801 (2018)
12. Li, X., Wang, Q.S., Yang, Z.Y., Lu, Z.X.: Novel auxetic structures with enhanced mechanical properties. Extreme Mech. Lett. **27**, 59–65 (2019)
13. Yang, C., Vora, H.D., Chang, Y.: Behavior of auxetic structures under compression and impact forces. Smart Mater. Struct. **27**, 025012 (2018)
14. Yang, H., Wang, B., Ma, L.: Mechanical properties of 3D double-U auxetic structures. Int. J. Solids Struct. **180**, 13–29 (2019)

Heterogeneous Materials Additive Manufacturing: An Overview

Riccardo Sponchiado[1(✉)], Luca Grigolato[1], Stefano Filippi[3], Gianmaria Concheri[1], Roberto Meneghello[2], and Gianpaolo Savio[1]

[1] ICEA Department, University of Padova, Via Venezia, 1, 35131 Padova, Italy
riccardo.sponchiado@unipd.it
[2] DTG Department, University of Padova, Stradella S. Nicola, 3, 36100 Vicenza, Italy
[3] DPIA Department, University of Udine, Viale delle Scienze, 206, 33100 Udine, Italy

Abstract. Advancements in additive manufacturing technology have made it possible to create machines that allow the use of a wider range of materials, even simultaneously in the production of a single piece. The production of heterogeneous objects allows to include multifunctionality within the domain by varying the composition in a gradual or net fashion. This paper analyzes the AM technologies that allow multi-material, emphasizing the constrains and the possible applications with the goal of identifying guidelines for design methods development. From the analysis we observe important innovations that permitted to easily process polymeric materials, especially with material extrusion and material jetting. However, the use of ceramic powders and metallic materials for the creation of heterogeneous objects requires the development of methods which remain very limited by the process conditions.

Keywords: Multi material AM · Functionally graded materials · Heterogeneous objects

1 Introduction

Additive manufacturing (AM) was first introduced during the 1980s to serve the highly specialized needs of model making and rapid prototyping [1]. AM is a solid freeform technology capable of delivering parts of very intricate and complex geometries with a minimum need for post-processing, by placing material layer by layer at set positions within the boundaries of a design domain [2, 3].

With the evolution of technology over the last three decades, important improvements in spatial resolution, materials, and physical properties of the parts produced were introduced. These advancements and a general increase in awareness of the potentials of AM, led to production of end-use items in sectors such as aerospace, automobile, biomedical and defense. Often, due to this phenomenon, we have moved from the use of the terms rapid prototyping or 3D printing to that of additive manufacturing, specifically to

C. Rizzi et al. (Eds.): ADM 2021, LNME, pp. 462–473, 2022.
https://doi.org/10.1007/978-3-030-91234-5_47

emphasize the transition to production [4–6]. Until recently AM has been generally relegated to the prototyping of single-material structures for form fit evaluation. However, a possible disruptive step in the evolution of AM will be to increase the functionality of the manufactured components [5, 7].

Functional objects that need variable properties or multi functionality can be produced by varying composition inside a single domain to achieve tailored characteristics for highly specific applications [8]. An object with heterogeneous composition could have confined spaces in the domain associated with different functions or a gradient between two or more materials or microstructures that can be designed and engineered for a specific set of functions, for instance being responsive towards specific stimuli (such as a 4D-printed part [9]), or for mechanical shock resistance, thermal insulation, catalytic efficiency and relaxation of thermal stresses [10, 11].

Materials designed with gradient composition are so called Functionally Graded Materials (FGM) and their manufacturing Functionally Graded Additive Manufacturing (FGAM) to differentiate them from classic FGMs introduced firstly more than 30 years ago by Niino et al. [12]. Usually, FGMs are obtained via bulk or surface treatments of pre-designed part [13]. FGAM techniques modify process parameters and gradationally vary either the densification of a homogenous composition or the combination of two or more materials through gradual transition [14]. This last type of FGAM will be the focus of the article.

Design of functional distributions of materials encompasses dispersed site-specific properties with transitions in geometry, chemical compositions, constituents or microstructures [15].

This paper aims at defining the foundation for design and produce innovative components using functionally graded materials by varying the chemical composition within the piece. Especially, the available techniques for producing heterogenous objects and FGMs by AM, their manufacturing limitations/constrain, and industrial/research applications are summarized.

2 Multi Material AM

AM manufacturing methods are classified based on the ISO/ASTM 52900 standard, that organizes the techniques by the type of material used (polymer, metal, ceramic or composite), the principles applied for fusion or bonding (melting, curing, sintering etc.), the form of the feedstock that is used for the addition of material (liquid, powder, suspension, filament etc.), and the way that the material is brought together. The seven categories in which AM is subdivided are: material extrusion, vat photopolymerization, binder jetting, material jetting, sheet lamination, directed energy deposition, and powder bed fusion [16].

Not all AM methods are capable or optimized for producing multi material parts [3], so, for the purpose of this paper, binder jetting and sheet lamination were not considered because of the lack of significative advancements in recent years. Figure 1 shows the classification of additive manufacturing technologies faced in the article, with some significative pros and cons for each one. To better understand the multi material capabilities, the techniques will be evaluated according to the way they can be used to produce heterogenous objects:

- Discrete spaces in the domain with dissimilar material composition.
- Gradient composition among zones in the domain.

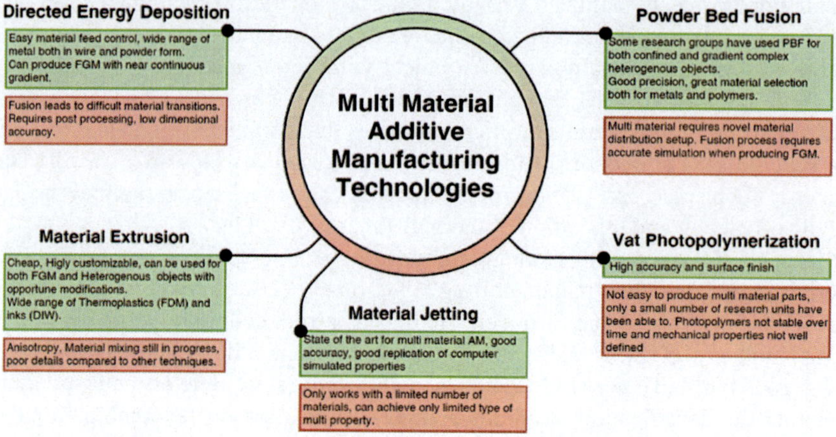

Fig. 1. Multi material AM technologies faced in the paper with pros and cons for each one.

2.1 Direct Energy Deposition

Direct Energy deposition (DED) is a process in which focused thermal energy is used to fuse materials by melting as they are being deposited [16]. DED systems use a concentrate energy source (laser beam, Electron beam, plasma or electric arc) and a stream of raw material (powder or wire) that intersect at the same focal point, where the material is being deposited [17]. Laser metal deposition (LMD) consist on a laser beam (most commonly a CO2 or Nd:YAG laser) as the main energy source to melt metal powders that can be prior mixed to obtain graded composition [18]. LMD is the most present procedure in the literature to produce heterogenous materials. Wire arc additive manufacturing (WAAM) instead, can use multiple metal wires whose feeding speed can be individually controlled to obtain graded composition [19]. WAAM processes provide a lower resolution as compared to laser-beam powder based processes but have a higher deposition rate and the ability to build larger structure.

As stated before, it is possible to use multi material feeding both by controlling the metal wires, as in the work of Heralic et al. [20], and the powder volumetric fraction, Greer et al. [21]. Hence it may seem possible to obtain both confined spaces and gradients with different composition.

However, since DED is a fusion-based process, the development of unwanted intermetallic phases in the gradient zone can be detrimental for mechanical properties during cooling. To investigate the feasibility of multiple material DED, Hofmann et al. [22] presented a roadmap for developing gradient alloys that uses multi-component phase diagrams as maps for composition selection to avoid unwanted phases. The aim was to successfully design the gradient path such that the unwanted brittle phases could be avoided in the microstructure, which would eventually give better mechanical properties for the printed parts. The results confirmed that a gradient in material composition is feasible with DED.

Thus far, the production of gradient materials with DED systems in literature has been limited to methods for obtaining a near-continuous gradual composition (as low as 1 wt.-% per layer, even if additional benefits are not necessarily observed for finer grading [23]) in the building direction or radially if the system is equipped with rotating parts [24]. Talking about industrial applications, an example came from the work of Morvan et al. [25] who produced a FGM flywheel by positioning Cobalt and Tungsten carbide inside the volume to optimize density.

2.2 Powder Bed Fusion

Similar to DED, powder bed fusion (PBF) uses lasers to selectively fuse regions of a powder bed [16]. This technology can produce metal or polymer parts and is involved in various methods among which selective laser sintering (SLS) and selective laser melting (SLM).

SLS is a widely used AM method that can produce a variety of objects based on polymer powders. One way of producing graded compositions with SLS is to mix varying fractions of ceramic nanoparticles in the powder bed. Chung et al. [26] investigated the influence of filler volume fractions on nylon-based composites, adding silica particles to nylon-11 powder and observing a one-dimensional gradient in the final piece.

Stated that it is possible to obtain gradient composition along Z axis, true three-dimensional gradient may also be possible in PBF if more advanced powder distribution and positioning systems are implemented [16]. For instance, the design of new selective powder dispensing methods introduced the possibility to have true gradients and confined multi-functional areas. Zhang et al. [27] developed a new approach of combining point-by-point selective powder delivery with powder bed fusion for multiple material SLM. The procedure consists of a dual ultrasonic vibration dispenser that can achieve stable flowrates of different material powders over the bed. Gu et al. [28] developed an integrated modelling framework for multi-track, multi-layer and multi material SLM in which special attention was given in modelling the molten pool evolution and track morphology of multiple materials deposited on the same and across layers.

Thanks to these, it is possible to obtain multifunctional objects using PBF, with gradients and with confined areas dedicated to different compositions and functions.

2.3 Material Extrusion

Extrusion based additive manufacturing consists in a layer-by-layer deposition of molten and semi-molten polymers, ceramic pastes, polymer solutions, and dispersions selectively dispensed through a nozzle or orifice [16]. Depending on the temperature required for the extrusion, it can be classified into two main sub groups: fused filament fabrication (FFF) (also known as fused deposition modeling (FDM)) for extrusion of melted thermoplastic polymers and direct ink writing (DIW) for extrusion without melting [29].

Material extrusion technologies can be easily extended to multi material 3D printing system by adding several material flows to the same machine [29], either by adding more nozzles or by adding more extruders that converge in a single nozzle or orifice [30, 31]. The main difficulties concern CAD modeling, data exchange and numerical control (CNC) pattern computation.

Fused deposition modeling consists of the extrusion of thermoplastic material, which is supplied as a thin filament (the most common diameters are 1.75 and 3.0 mm). The material is pushed towards the hotend, either directly or through PTFE tubes. The hotend heats up to the appropriate temperature for the different types of polymer: above the melting temperature for semicrystalline polymers or above the glass temperature for amorphous polymers.

Multi material FDM is commercially accessible and available in various prosumer and industrial grade printers [32]. In the most common multi material systems, the printing head carriage could have two or more identical extruders, each processing a single material, spaced by an offset, such as Ultimaker S3 [33] or multiple independent extruders each one moved by an independent motion system, such as Bolt Pro by Leapfrog or cylindrical coordinate systems [34]. Another way to add multi material capabilities to an FDM printer is represented by the multi material units, from Prusa3d [35] and Mosaic [36]. Those are stand-alone systems that can support several filament feeds and combine them to work with most desktop machines. The functional principle consists in splicing the filaments into segments calculated by the software and combining these together to provide a continuous supply to the FDM system. During the print, a purge tower is used to unload the filament, in the interval where the change between one material and another takes place. The above systems can be used to create prints with discrete material variations; it is therefore impossible, in this way, to create material gradients. FGMs in 3D printing can be achieved with the use of a multiple material, single nozzle hotend. An example comes from the RepRap project with the diamond hotend [31]. The hotend system features three independently controlled filament lines that are combined into a single nozzle output. This type of printing head is often termed as mixing nozzle. However, due to high viscosity resulting in laminar flow, homogeneous mixing of the polymers is difficult to obtain by existing commercial mixing nozzles [37]. To better understand the physics of a multiple material single nozzle system, several studies have been proposed. Han et al. [38] designed an extruder by Ansys Fluent simulations to study the influence of the temperature field on the flow of materials inside the nozzle. Their results highlighted that temperature control is a key factor in mixing material, achieving different optimal temperature values depending on the flow rate. Khondoker et al. [39] used a static intermixer placed inside the hotend to induce mechanical interlocking of pairs of otherwise immiscible materials: PLA, ABS, and HIPS. They subsequently characterized the parts produced with and without intermixers and observed a noticeable improvement in intralayer adhesion. Kennedy et al. [40] produced hardware to mix in situ two filaments of similar or dissimilar composition. The mixing takes place by rotating tips inside the hotend, whose speed is controlled by software. The sample specimens, produced blending TPU and PLA, nylon and PLA, and conductive carbon nanotube (CNT)-PLA, were characterized and the quality of the mix were expressed as a function of the rotational speed and the geometry of the hardware blending.

Direct ink writing is a technique in which paste-like materials showing suitable viscoelastic and rheological characteristics are extruded. Most of the preparations used for DIW show a rheological behavior of shear thinning, characterized by a decrease in viscosity as the shear rate increases [29]. In the DIW, the print head consists of a nozzle and a cartridge, which can be moved via a computer-controlled system. The deposition of the material is usually carried out pneumatically [41]. In multi material DIW the speed of the extrusion of each single material will correspond to the volumetric fraction of the material in the final product.

One of the first systems suitable for MMAM with DIW was proposed by Oxman et al. [42]. The printing platform was capable of mixing and extruding two materials, which were in a liquid state at room temperature, and solidified after mixing and deposition, such as certain epoxies, ultra-violet (UV)-cure polymers, drying adhesives, and thixotropic fluids. Three different mixing strategies were explored: diffusive, static and active. The diffusive strategy simply consisted of an additional length of tubing between the nozzle and a "Y" connector joining the material feed tubes. The active mixing strategy used a small reservoir immediately before the nozzle into which material could be fed. A series of blades driven by a small motor inside the tank facilitates mixing. The static mixing strategy used static mixers for in-line pipes inserted between the nozzle and material feed tubes. The gradients were still one-dimensional, even if methods for the development of multidimensional gradients are in development.

Later, Bakarich et al. [30], with a technology similar to that of Oxman's group, have manufactured an artificial tendon that simulates the heterogeneous composition of the human tendon with smooth gradients between a tough alginate/poly (acrylamide) ionic covalent entanglement hydrogel and an acrylated urethane based UV-curable adhesive material. The composite materials were mechanically characterized and their performance could be modeled with standard composite theory.

One of the most significative efforts to produce real 3d complex multifunctional objects with extrusion technique comes from Skylar-Scott et al. [43]. Their research exploited the potential of DIW to pattern a broader range of materials. Their group reported the design and fabrication of voxelated soft matter using multi material multi-nozzle 3D (MM3D) printing which can be seen in Fig. 2 in which the composition, function and structure of the materials are programmed at the voxel scale. Co-printing multiple epoxy and silicone elastomer inks with stiffness in several orders of magnitude, some functional results can be achieved.

To summarize, extrusion technology allows for both heterogeneous objects with confined areas dedicated to different properties and functions, and material gradients. Components with simplified or discretized gradients have been produced. Despite studies focus on the estimation of the delay for the material change due to the geometry of the extruder [37] and other researches aimed at overcame the existing limits of MMAM, complex 3D FGM production with extrusion machines is still not mature at the hardware level and most accessible extruders can make coextrusion of filaments more than real mixed extrusion. Moreover, the bottleneck in complex 3D gradients is mainly at the software level: effective design methods and CNC pattern computation tools do not exist. File exchange formats are in development, to integrate the information concerning the material with the geometric ones [6, 41].

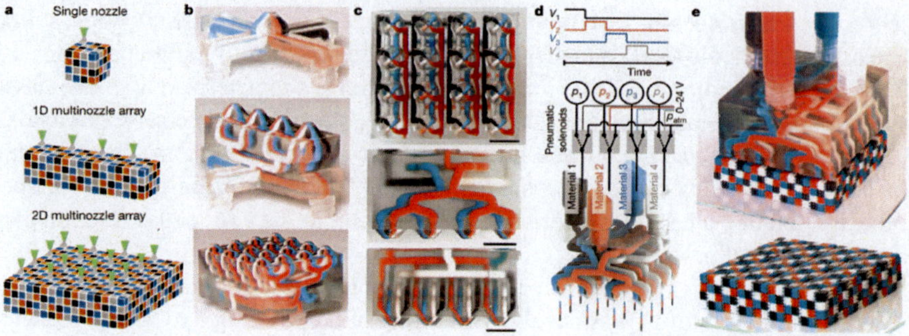

Fig. 2. Schematic of voxelated architectures printed using MM3D printheads [43]

2.4 Material Jetting

Material jetting is a process where droplets of material are selectively deposited on a platform by piezoelectric nozzles, using either a continuous or drop on demand (DOD) approach. Printable materials are mostly polymers, which come in the form of photosensitive inks or waxes [16, 29], but also metal and ceramic nanoparticles wit Xjet NanoParticle Jetting technology [44].

MJ is often referred to as Polyjet, from the name of the most known technology in this sector produced by Stratasys. The technology is considered as state of the art for the production of multi-material objects with confined spaces or graded properties such as color, transparency and stiffness. Various compositions can be obtained by mixing different fractions of resins before the curing step. These types of composites are called digital materials (DM).

Salcedo et al. [45] validated with mechanical tests the performances of an FGM produced with different gradations of DM made with different percentages of rubber-like material (Tango black+) and ABS-like material. The group concluded that their previous simulations were able to predict both local and global mechanical properties with a 5% error on maximum elongation.

Several other authors have used Polyjet printers to produce multifunctional objects, both FGM and confined. Childs et al. [46] proposed a new method, which requires partial assembly but prevents the work of finishing and removing the supports, for the creation of microfluidic components (Fig. 6) capable of retaining an internal pressure of 189 kPa before breaking. Similar works were previously done to create valves capable of precisely controlling the flow [47].

Doubrovski et al. [48] presented a new printing method called Bitmap printing and a digital workflow based on the voxelization of the model and the printing with local variation of stiffness. To prove the accuracy of the method, they produced a prosthetic leg socket where pressure sensors were installed to evaluate the socket's fit.

To conclude, Polyjet printing is probably the unique mature technology for the production of heterogeneous objects and several applications are already exploited at an industrial level. The real disadvantage of material jetting remains the scarce availability of usable materials.

2.5 VAT Photopolymerization

Vat photopolymerization is a process in which liquid photopolymer in a vat is selectively cured by light activated polymerization. Photopolymerization process usually is not designed for multi-material manufacturing. Indeed, using multiple materials within the same vat could be complicated. The quality and ease of use meant that multi-vat systems were developed to produce heterogeneous objects, each containing a different UV curable photopolymer. The change of material between the different vats during printing make the printing process very slow. Moreover, gradients in the growing direction can be obtained.

Choi et al. [49] presented a multi material stereolithography machine (MMSL) based on multiple vats. New rotating vat carousel, platform assembly, automatic leveling system, and less viscous resins were developed. A scheme of MMSL can be seen in Fig. 3. Complex multi material objects such as different types of chess pieces have been produced thanks to MMSL which are functional and visually illustrative.

Han et al. [50] developed a new SLA process called multi-material projection micro-stereolithography using dynamic fluidic control of multiple liquid photopolymers within an integrated fluidic cell. The pressure-tight and leak-free fluidic cell enables an active and fast switch between liquid photopolymers, even including micro-/nano-particle suspensions, which could potentially lead to easy manufacturing of multi-material metallic/ceramic structures or heterogeneous biomaterials.

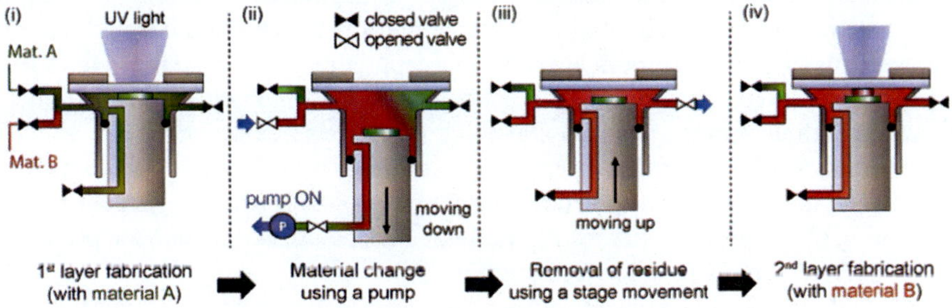

Fig. 3. Material exchange process in multi-material projection micro-stereolithography [50]

2.6 Multiprocess AM

The research in multi material functional objects design will lead to the development of new manufacturing technologies, which will make possible to create components that are not manageable through a single manufacturing process, produced in a unified, tooling-free, multi-process manufacturing environment. In this context, multi-process (or hybrid) AM (MPAM) is defined as an additive process enhanced by complementary technologies [5]. An example of MPAM derives from the need to integrate electronic circuits within printed parts. Previously, the method of interrupting the printing process was used to carry out the embedding of electronic circuits [51, 52]. More recently, a MPAM where two 3D printers create a multi-material structure in combination with a gantry that provides complementary processes such as wire embedding and micromachining was developed. Everything is controlled by a 6-axis robotic arm that moves the build plate between the various processing stations [53, 54].

Chueh et al. [8] recently presented a system that integrates FDM and PBF to produce hybrid metal-polymer parts. The junction of the metal parts with the polymeric parts was carried out by creating physical interconnections with the PBF. The particular type of powder bed fusion that they use in this technique includes a distribution of the material through ultrasonic dispensing, which allows to deposit different metal powders, even creating gradient compositions. With this technique, despite the weakness of interlocking, the group has managed to create multi-material parts that consist of hybrids between stainless steel, copper and polymer.

3 Conclusions and Outlook

This paper provides a broad view of the possibilities of producing heterogeneous objects and FGM by AM. These indications can be useful in choosing the production process and the most suitable materials in multifunctional design.

MJ is the more established technology in terms of precision and quality, even if the limited choice of materials restricts the field of application.

A significant progress in material extrusion highlights a broad range of properties. Polymers that can be processed together range from common thermoplastics to high performance technopolymers. Metals too can be manufactured exploiting sintering in the furnace processes. The study of the compatibility between materials and material blending extrusion AM, must be deepened by incorporating knowledge that comes from material science and the theory of blends, not only from fluid dynamics, even if it remains a fundamental topic moving from coextrusion to real mixing extrusion.

Significant results came also from DIW, especially by extending aspects coming from MJ. The new methods of powder distribution applied to the most recent powder bed fusion systems promise interesting findings that could, be crucial for producing metal or ceramic parts with tailored properties.

Multi-material printing is opening up new frontiers for designers and engineers to create FGM and heterogeneous structures useful in the conceptualization of multifunctional components, which could be crucial for technological evolution in the near future.

Acknowledgement. This work was partially funded by the Regional Operational Program F.S.E. 2014–2020 Veneto Region and the European Regional Development Fund POR 2014–2020, project code 2105-0036-1463-2019, "Design and manufacturing by additive technologies of innovative components using functionally graded materials".

References

1. Ligon, S.C., Liska, R., Stampfl, J., Gurr, M., Mülhaupt, R.: Polymers for 3D printing and customized additive manufacturing. Chem. Rev. **117**, 10212–10290 (2017)
2. Bikas, H., Stavropoulos, P., Chryssolouris, G.: Additive manufacturing methods and modeling approaches: a critical review. Int. J. Adv. Manuf. Technol. **83**, 389–405 (2016)

3. Loh, G.H., Pei, E., Harrison, D., Monzón, M.D.: An overview of functionally graded additive manufacturing. Addit. Manuf. **23**, 34–44 (2018)
4. Li, Y., et al.: A review on functionally graded materials and structures via additive manufacturing: from multi-scale design to versatile functional properties. Adv. Mater. Technol. **5**, 1900981 (2020)
5. MacDonald, E., Wicker, R.: Multiprocess 3D printing for increasing component functionality. Science **80**, 353 (2016)
6. Leung, Y.-S., et al.: Challenges and status on design and computation for emerging additive manufacturing technologies. J. Comput. Inf. Sci. Eng. **19**, 021013 (2019)
7. Conner, B.P., et al.: Making sense of 3-D printing: creating a map of additive manufacturing products and services. Addit. Manuf. **1**, 64–76 (2014)
8. Chueh, Y.H., Wei, C., Zhang, X., Li, L.: Integrated laser-based powder bed fusion and fused filament fabrication for three-dimensional printing of hybrid metal/polymer objects. Addit. Manuf. **31**, 100928 (2020)
9. Pei, E., et al.: A study of 4D printing and functionally graded additive manufacturing. Assem. Autom. **37**, 147–153 (2017)
10. Oxman, N.: Variable property rapid prototyping: inspired by nature, where form is characterized by heterogeneous compositions. Virtual Phys. Prototyp. **6**, 3–31 (2011)
11. Oxman, N., Keating, S., Tsai, E.: Functionally graded rapid prototyping. In: Innovative Developments in Virtual and Physical Prototyping: Proceedings of 5th International Conference on Advanced Research in Virtual and Rapid Prototyping, pp. 483–489 (2012). https://doi.org/10.1201/b11341-78
12. Niino, M.: Functionally gradient materials as thermal barrier for space plane. J. Jpn. Compos. Mater. **13**, 257–264 (1987)
13. Saleh, B., et al.: 30 years of functionally graded materials: an overview of manufacturing methods, Appl. Future Challenges Compos. Part B Eng. **201**, 108376 (2020)
14. Cen, P.D.: PD CEN/TR/ISO/ASTM 52912 : 2020 BSI Standards Publication Additive manufacturing — Design — Functionally graded additive manufacturing (2020)
15. Liu, Z., Meyers, M.A., Zhang, Z., Ritchie, R.O.: Functional gradients and heterogeneities in biological materials: design principles, functions, and bioinspired applications. Prog. Mater. Sci. **88**, 467–498 (2017)
16. ISO/ASTM International: BSI Standards Publication Additive manufacturing — General principles — Terminology, pp. 1–19 (2017)
17. Dass, A., Moridi, A.: State of the art in directed energy deposition: from additive manufacturing to materials design. Coatings **9**, 418 (2019)
18. Shamsaei, N., Yadollahi, A., Bian, L., Thompson, S.M.: An overview of Direct Laser Deposition for additive manufacturing; Part II: mechanical behavior, process parameter optimization and control. Addit. Manuf. **8**, 12–35 (2015)
19. Wang, J., et al.: Characterization of wire arc additively manufactured titanium aluminide functionally graded material. Mater. Sci. Eng. A **734**, 110–119 (2018)
20. Heralić, A., Christiansson, A., Lennartson, B.: Height control of laser metal-wire deposition based on iterative learning control and 3D scanning. Optics Lasers Eng. **50**(9), 1230–1241 (2012)
21. Greer, C., et al.: Introduction to the design rules for metal big area additive manufacturing. Additive Manufact. **27**, 159–166 (2019)
22. Carroll, B.E., et al.: Functionally graded material of 304L stainless steel and inconel 625 fabricated by directed energy deposition. Acta Mater. **108**, 46–54 (2016)
23. Lin, X., Yue, T., Yang, H., Huang, W.D.: Laser rapid forming of SS316L/Rene88DT graded material. Mater. Sci. Eng. **391**, 325–336 (2005)
24. Hofmann, D.C., et al.: Developing gradient metal alloys through radial deposition additive manufacturing. Sci. Rep. **4**, 5357 (2014)

25. Morvan, S., Fadel, G.M., Keicher, D.: Manufacturing of a heterogeneous flywheel on a LENS apparatus
26. Chung, H., Das, S.: Functionally graded Nylon-11/silica nanocomposites produced by selective laser sintering. Mater. Sci. Eng. A **487**, 251–257 (2008)
27. Zhang, X., Wei, C., Chueh, Y.-H., Li, L.: An integrated dual ultrasonic selective powder dispensing platform for three-dimensional printing of multiple material metal/glass objects in selective laser melting. J. Manuf. Sci. Eng. **141**, 011003 (2019)
28. Gu, H., et al.: Multi-physics modelling of molten pool development and track formation in multi-track, multi-layer and multi-material selective laser melting. Int. J. Heat Mass Transf. **151**, 119458 (2020)
29. Rafiee, M., Farahani, R.D., Therriault, D.: Multi-material 3D and 4D printing: a survey. Adv. Sci. **7**, 1–26 (2020)
30. Bakarich, S.E., et al.: 3D printing of tough hydrogel composites with spatially varying materials properties. Addit. Manuf. **14**, 24–30 (2017)
31. Diamond Hotend - RepRap. https://reprap.org/wiki/Diamond_Hotend
32. Wohlers Report 2020: 3D Printing and Additive Manufacturing State of the art
33. Ultimaker 3. https://ultimaker.com/it/3d-printers/ultimaker-3
34. Monteiro, D.L., Vicente, C.M.S., de Oliveira Leite, M.A., Ribeiro, A.M.R.: Development of a cylindrical coordinate-based fused filament fabrication machine with multiple print heads. Int. J. Adv. Manufact. Technol. **110**(11–12), 3129–3143 (2020). https://doi.org/10.1007/s00170-020-05900-y
35. Original Prusa i3 Multi Material 2.0 - Prusa3D - 3D Printers from Josef Průša. https://www.prusa3d.com/original-prusa-i3-multi-material-2-0/
36. Palette 2S – Mosaic Manufacturing. https://www.mosaicmfg.com/products/palette-2s
37. Garland, A., Fadel, G.: Design and manufacturing functionally gradient material objects with an off the shelf three-dimensional printer: challenges and solutions. J. Mech. Des. Trans. ASME **137**, 1–11 (2015)
38. Han, S., Xiao, Y., Qi, T., Li, Z., Zeng, Q.: Design and analysis of fused deposition modeling 3D printer nozzle for color mixing. In: Advances in Materials Science and Engineering (2017)
39. Khondoker, M.A.H., Asad, A., Sameoto, D.: Printing with mechanically interlocked extrudates using a custom bi-extruder for fused deposition modelling. Rapid Prototyp. J. **24**, 921–934 (2018)
40. Kennedy, Z.C., Christ, J.F.: Printing polymer blends through in situ active mixing during fused filament fabrication. Addit. Manuf. **36**, 101233 (2020)
41. Li, J., Wu, C., Chu, P.K., Gelinsky, M.: 3D printing of hydrogels: Rational design strategies and emerging biomedical applications. Mater. Sci. Eng. R Rep. **140**, 100543 (2020)
42. Oxman, N., Tsai, E., Firstenberg, M.: Digital anisotropy: a variable elasticity rapid prototyping platform. Virtual Phys. Prototyp. **7**, 261–274 (2012)
43. Skylar-Scott, M.A., Mueller, J., Visser, C.W., Lewis, J.A.: Voxelated soft matter via multimaterial multinozzle 3D printing. Nature **575**, 330–335 (2019)
44. XJet Nano Particle Jetting. https://www.xjet3d.com/
45. Salcedo, E., Baek, D., Berndt, A., Ryu, J.E.: Simulation and validation of three dimension functionally graded materials by material jetting. Addit. Manuf. **22**, 351–359 (2018)
46. Childs, E.H., Latchman, A.V., Lamont, A.C., Hubbard, J.D., Sochol, R.D.: Additive assembly for PolyJet-based multi-material 3D printed microfluidics. J. Microelectromech. Syst. **29**(5), 1094–1096 (2020). https://doi.org/10.1109/JMEMS.2020.3003858
47. Keating, S.J., et al.: 3D printed multimaterial microfluidic valve. PLoS One **11**, e0160624 (2016)
48. Doubrovski, E.L., et al.: Voxel-based fabrication through material property mapping: a design method for bitmap printing. CAD Comput. Aided Des. **60**, 3–13 (2015)

49. Choi, J.W., Kim, H.C., Wicker, R.: Multi-material stereolithography. J. Mater. Process. Technol. **211**, 318–328 (2011)
50. Han, D., Yang, C., Fang, N.X., Lee, H.: Rapid multi-material 3D printing with projection micro-stereolithography using dynamic fluidic control. Addit. Manuf. **27**, 606–615 (2019)
51. Kataria, A., Rosen, D.W.: Building around inserts: Methods for fabricating complex devices in stereolithography. Rapid Prototyp. J. **7**, 253–261 (2001)
52. Weiss, L.E., et al.: Shape deposition manufacturing of heterogeneous structures. J. Manuf. Syst. **16**, 239–248 (1997)
53. Espalin, D., Muse, D.W., MacDonald, E., Wicker, R.B.: 3D Printing multifunctionality: structures with electronics. Int. J. Adv. Manuf. Technol. **72**, 963–978 (2014)
54. MacDonald, E., et al.: 3D printing for the rapid prototyping of structural electronics. IEEE Access **2**, 234–242 (2014)

Conformal Lattice Structures: Modeling and Optimization

Pierandrea Dal Fabbro[1](✉), Stefano Rosso[2], Alessandro Ceruti[3], Roberto Meneghello[2], Gianmaria Concheri[1], and Gianpaolo Savio[1]

[1] Department of Civil, Environmental and Architectural Engineering, Laboratory of Design Tools and Methods in Industrial Engineering, University of Padova, Via Venezia, 1, 35131 Padova, Italy
pierandrea.dalfabbro@unipd.it

[2] Department of Management and Engineering, Laboratory of Design Tools and Methods in Industrial Engineering, University of Padova, Stradella S. Nicola, 3, 36100 Vicenza, Italy

[3] Department of Industrial Engineering, University of Bologna, Viale Risorgimento, 2, 40136 Bologna, Italy

Abstract. One of the open issues in additive manufacturing is the design of conformal lattice structures, leading to an optimal layout of the struts in the design domain. This paper aims to compare different struts distributions in conformal lattices via low computational power methods in a CAD environment. Four approaches for a wireframe virtual model definition are presented for a simple cubic conformal lattice structure. An iterative variable diameter optimization method and two linear structural analyses based on mono-dimensional elements and different theories are compared. These verification methods widen the capability of checking the results so the user can compute the deformation of 3D periodic structures, or other visual results, without spending a huge amount of time and computational power. Results show that both the analysis methods give reliable results and the struts layout based on trivariate NURBS shows the most flexible solution allowing for a real-time variation of the boundary condition.

Keywords: Conformal lattice structure · Size optimization · Additive manufacturing · Virtual modeling · Design for additive manufacturing

1 Introduction

Nowadays, the developments in additive manufacturing (AM) technologies give more freedom in the design and fabrication of optimized structures, overcoming the limits of other manufacturing technologies [1]. Thanks to the recent evolution in materials and AM processes, the design of the structure morphology allows for full customization of the geometry and functional proprieties of the products [2]. Indeed, the mesoscale characteristics of a lattice component govern its macroscale behavior [3] and structural gradation contributes to a smooth transition between properties [4]. Currently, the main limitations lie in the tools packages available for the designer [5]. The traditional Computer Aided Design (CAD) and Computer Aided Engineering (CAE) software request

C. Rizzi et al. (Eds.): ADM 2021, LNME, pp. 474–485, 2022.
https://doi.org/10.1007/978-3-030-91234-5_48

high computational resources to design and simulate complex structures. When the component dimension is significantly larger than the scale of the lattice structure unit cell, a classic modeling approach for designing each cell is time-consuming [6]. To overcome this limit, several design tools are available in the market, like nTopology, Grasshopper, Ansys Spaceclaim, Altair Inspire, etc. In any case, a huge amount of physical power is required to perform tasks such as Boolean operation, blends, rendering, and visualization [7]. Typically, in AM the models are based on triangular meshes [8, 9] and the discretization of the model requires a lot of vertices and faces, resulting difficult to handle for further analyses and slicing operations [10]. For these reasons, several geometric modeling approaches for non-stochastic structures were developed [9, 11–13]. Another challenge in the design of lattice structures that conform to the boundary of the parts or to the principal stress directions [14–16]. Moreover, the structural analysis of lattice structures is another current limitation, due to the high number of elements needed to discretize the model inside the finite element analysis (FEA) software. In this scenario, modeling and simulating the cellular structures in one dimension, as a wireframe, is suitable and time-saving. Regarding structural linear analyses, for a small deformation in the elastic region, both solid and beam-based models can properly describe the mechanical behavior of the lattice structures [17]. In literature there are several approaches for structural analyses on lattice structures with the aim of process time reduction; the most common method is the asymptotic homogenization (AH) [18]. AH approach works by substituting the lattice with a fully dense material which have equivalent mechanical proprieties; in this way the polyhedral mesh presents a lower number of nodes and the resources required by the analysis are greatly reduced. As a disadvantage, AH may lose the geometrical lattice characteristics by using a bulk component instead of a periodic structure [19]. The mono-dimensional structural analysis, instead, is based on two types of beam elements model: Euler–Bernoulli and Timoshenko [17]. The most appropriate beam theory depends on the load conditions and the slenderness ratio of the lattice structure, defined as the ratio between the length and the thickness of the beams. The Timoshenko theory allows the rotation between the cross-section and the neutral axis of the bending curve, whereas the Euler-Bernoulli theory assumes that the cross-section is perpendicular to the neutral axis, disregarding the shear deformation in bending [17]; neglecting this causes underestimation of the deflection results. Ptochos et al. [20] investigated the elastic modulus and Poisson's ratio of micro cellular structures by analytical and numerical beam theories, and homogenization method; for micro-BCC lattices in compression condition and for low slenderness ratio, the results of the different methods are similar. Ahmadi et al. [21] involved the same methods for studying the mechanical behavior of diamond lattices. They changed the slenderness ratio varying the beam diameter to increase the relative density of the lattice structure. They noticed that for low slenderness ratio the analytical solution based on the Euler-Bernoulli theory exhibits larger differences from the experimental results whereas the analytical solution based on the Timoshenko theory is closer to the experimental results. Another important parameter is the relative density of a lattice structure, defined as the ratio between the cellular material density and the full material density. Gibson and Ashby stated that for an all open-cell foam the relative density depends to the inverse of slenderness ratio squared [22], and governs the structure elastic modulus. FEA can be iteratively performed to optimize the size of the elements according to a target (stiffness, maximum stress, utilization, etc.) and to respect the boundary conditions of the problem [23]. In a previous work [24] a method for optimizing regular lattice structures was proposed. After the construction of the wireframe, the elements size was defined via an iterative size optimization process, based on a utilization target. Once the wireframe has been optimized, the lattice structure needs to be modeled by NURBS or meshes [25].

In this paper, four modeling methods for conformal lattice structures were proposed and compared. Three methods for the wireframe modeling are based on curves division whereas the other is based on NURBS free form deformation. A size optimization through mono-dimensional FEA is performed on the obtained wireframes according to the boundary conditions. These methods require low computational power and physical memory since they are based on wireframe modeling and 1D structural linear analysis approaches: almost 80% of the computational time is expected to be saved if compared to a FEA based on mesh [19]. Being these experimental methods for structural linear analysis in a preliminary design environment, results are then checked and confirmed by another FEM software.

2 Materials and Method

The proposed method is based on the following steps:

- Wireframe modeling of 4 conformal lattice structures;
- Size optimization of the wireframes [24];
- Linear structural analysis based on beam theory of the optimized structures [19].

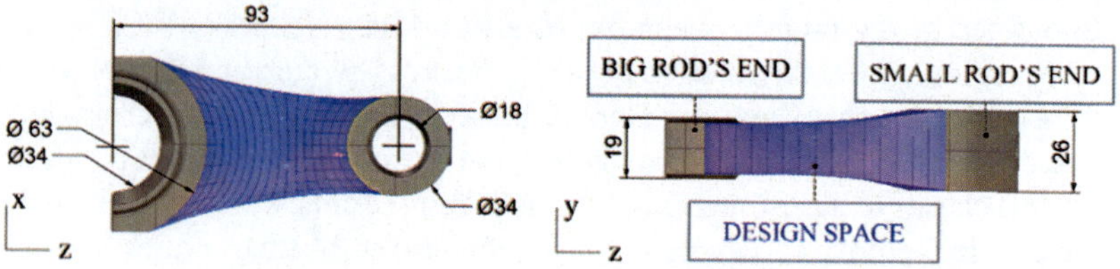

Fig. 1. Piston rod: the case of study main dimensions [mm].

The aforementioned method is applied to a piston rod (Fig. 1) previously investigated by Rosso et al. [6].

2.1 Design Methodologies of Conformal Wireframe Model

There are several ways to model the wireframe of a lattice structure: from the generation of regular cells and the conformal morph into a design space, to the construction of the conformal cells through basic operations with lines, curves, and arcs. A suitable technique for morphing a lattice structure into a curved domain is identified in the Free-Form Deformation (FFD) approach [26]. FFD consists in placing and "imprisoning" the object inside a trivariate tensor volume, which will be deformed through its nodes causing the deformation of the captive object. This technique can be used by many modeling systems as well as in virtual reality and biomedical fields [27]. Several evolutions of the FFD idea are well known in literature [28], from improvements in the flexibility of the control volume, combining several tricubic Bezier volumes [29], to the enhancement of the control of the deformations through control points weighting, via NURBS-based approach [30]. Software like Maya, Rhinoceros, and Blender implemented NURBS FFD tools. They consist of the creation of a trivariate NURBS cage around the object and in the transformation of its control points [31]. As previously mentioned, four design methodologies were developed to obtain a conformal lattice structure based on a simple cubic cell inside a design domain.

The software used is Rhinoceros 7 (Robert McNeel & Associates, release 7.8.2) and the methodologies have been automated in Python's codes inside Grasshopper, a visual algorithm editor inside Rhinoceros. In the proposed case study, the wireframe replaces the central part of the piston rod, with a lattice made by 14 cells along the z-direction, 10 cells along the x-direction, and 4 cells along the y-direction. To simplify the comprehension, the 4 wireframe modeling methods are shown in 2D, even if developed for 3D (Fig. 2).

Tween Curve Division Method (TCD)

Starting from 2 opposite edges of the design domain, it is possible to create a set of intermediate curves according to the number of the cells in x- and z-direction, respectively. Moving along the coordinate system axes, a net of curves is obtained, and by mutual intersections, a list of points is obtained, which is used to construct the wireframe model. Depending on the position of the curve between the two primitives, the curve will assume a behavior influenced by the nearest primitive.

Arc Division Method (AD)

The method subdivides the design space by arcs. The vertical centerline is drawn together with the side curves that define the design space. These curves are subdivided by the desired instances along Z, obtaining a set of points. Arcs are drawn through the 3 points at each subdivision level and then they are subdivided as many times as the instances along the x-direction, obtaining the points for the wireframe definition. This method can only be applied if the edges of the design domain can be approximated by arcs.

Curvature Division Method (CD)

This method is based on linearly variating the curvature of the arcs that make up the wireframe. The curvatures of the upper and lower arc are obtained. Then, a list of n curvatures, where n is the number of instances along the z-direction, is created by linearly interpolating the curvature of the upper and lower arc. The vertical centerline is created starting from the midpoint of the upper arc to the midpoint of the lower arc and is divided into n points. Then, an arc through each point, centered along the vertical centerline, and with the previously computed curvatures is drowned. The same procedure is repeated considering the arcs of the left-right profile, using the number of instances along the x-direction. The intersections of the arcs are finally used for the wireframe

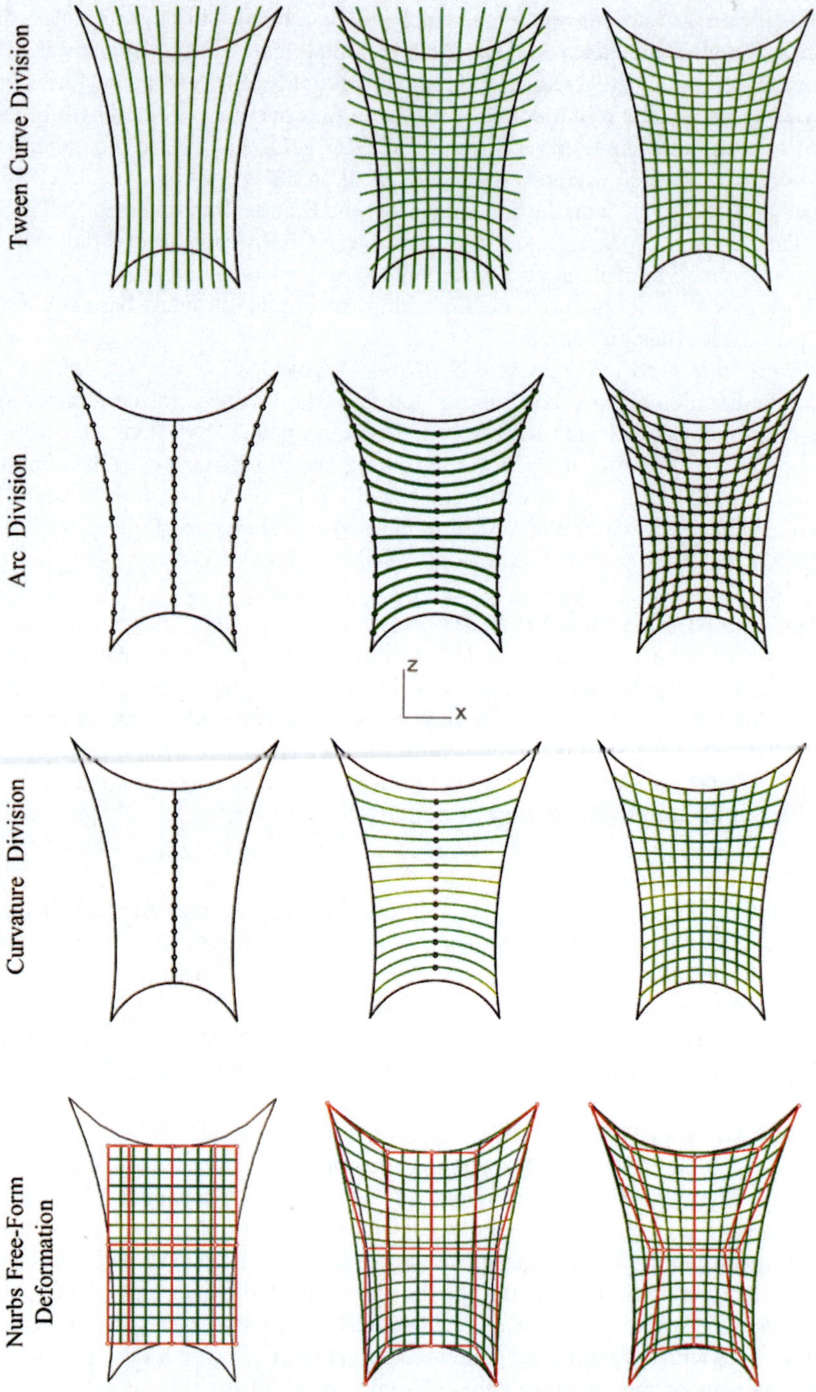

Fig. 2. Design methodologies for conformal wireframe modeling.

modeling. This method, too, can be applied only if the boundary of the design domain can be approximated by arcs.

Nurbs Free-Form Deformation Method (NFFD)
A regular wireframe based on the repetition of a simple cubic cell along the x-, y- and z-axis is generated. By the Cage-Edit command [31], choosing the number of the control points in each direction, a 3D NURBS cage is built around the wireframe. In the case study, the cage is governed by 5 control points in the x-direction, 3 control points in the z-direction, and 2 control points in the y-direction. The regular wireframe, with the control cage around it, has been overlapped to the design space The vertices of the control cage is then adapted to the vertices of the design space. Then, the cage is adpted to the design space, by moving the internal control points.This approach can be automated by defining a specific transformation of the cage and it is more adaptable to different geometric configurations of the design space. Moreover, it allows for warping transformation if densification of the cells is needed in a specific part of the design space.

2.2 Size Optimization and Structural Analysis

The size optimization process is performed by consecutive linear elastic structural analyses where the diameter of the beams is iteratively upgraded to reach the target utilization.

The method was developed in Grasshopper via Karamba3D (release 1.3.3) [32], a parametric structural engineering plug-in, and an ad-hoc algorithm developed by Savio et al. [24]. The diameter of each beam is optimized according to a utilization factor, defined as the ratio between the equivalent stress (including buckling) and the strength stress, as defined by the Eurocode3 [33]. The target utilization ratio was set to (90 ± 1%).

Then, another beam-based structural analysis was performed via the 1D approach developed by Bacciaglia et al. [19], in Patran/Nastran software (MSC Software Corporation) for validating the results. The validation procedure starts from the creation of a text file that contains the geometry wireframe information. The model is imported in Patran and the material, the cross-sections, and the constraints are defined. The main difference between the two structural analysis approaches lies in the type of beam element used: Karamba3D works with the Timoshenko beam element, whereas the Patan/Nastran 1D approach adopts the Euler Bernoulli beam theory [19, 34]. Considering producing the part using Selective Laser Melting (SLM) technology, the material is the aluminum AlSi10Mg (Table 1). The material was assumed as a linear elastic isotropic material, without considering the influence of the printing direction, and the deformations are assumed to be small.

Table 1. Proprieties of AlSi10Mg [35].

Density [kg/m^3]	Young modulus [GPa]	Tensile strength [MPa]	Yield strength [MPa]	Poisson ratio
2700	68	336.5	192	0.3

Two load cases, applied to the big rod's end along the z-direction, are investigated. A traction load case of 7.5 kN is considered to compare the results with the work of Rosso et al. [6] and a compression load case of 10 kN is applied for evaluating the elastic bucking. The model is constrained by locking all the rotations and displacements of all the nodes belonging to the face of the small rod end and by locking the displacements along x- and y-directions of all the nodes of the big rod end zone, where the load is applied (Fig. 3). Other applied constraints are the maximum and minimum beam diameter, which are based on geometrical limitations due to the AM technologies and chosen material [6]. A range between 0.5 mm to 1.5 mm was selected, respecting the minimum printable dimension and for facilitating the powder evacuation process [36].

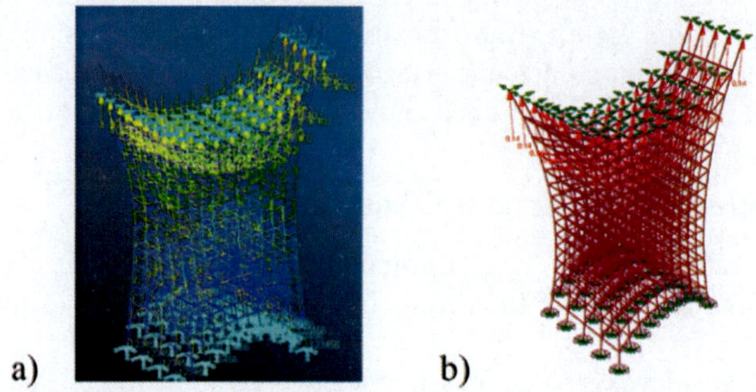

Fig. 3. Boundary condition in traction: a) Nastran/Patran, and b) Karamba3D.

3 Result and Discussion

3.1 Wireframe Design Comparison

The wireframes obtained by the four modeling methods are presented in Fig. 4, where each one is compared to the CD method, used as a reference since it was previously proposed [6]. The methods mostly differ in the xz-plane, whereas in the xy-plane minor differences appear, mainly among the position of transversal nodes. Consequently, results are discussed concentrating in the xz-plane. The TCD method, as well as the NFFD method, differ primarily on the horizontal lines, near the right-left boundaries. The AD method diverges in both directions, but still close to the borders of the design space. In the central area of the structures, instead, the cells remain undeformed.

The beam's length depends on the position in the design space. Looking at the barplot in Fig. 4, the minimal length is similar in all the configurations. Higher differences are noted in the maximum length curves, associated with the lines at the boundary of the design space, close to the big rod end; the CD method presents the longest beam dimension whereas, at the opposite, the AD method presents the lowest due to the regular division of the design space edges.

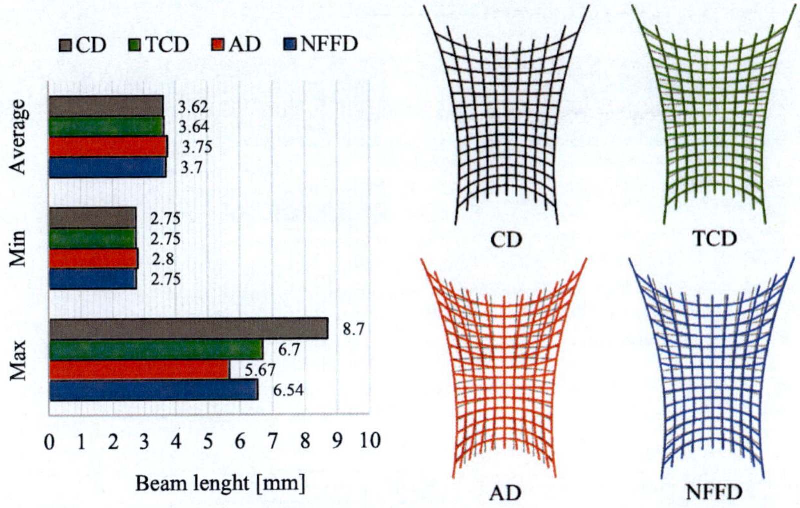

Fig. 4. Wireframe comparison.

3.2 Optimization and Structural Analyses

In Fig. 5 are reported the main results of the optimization procedure and in Fig. 6 the result of the structural analysis carried out by Karmaba3D and Nastran. For all the configurations, the average diameter is similar (about 0.70 mm for the traction and 0.76 mm for the compression load case) due to the similar beam distribution in the central portion. The TCD lattice structure has the highest diameter in both traction and compression load cases. NFFD and CD methods have the lowest maximum diameter in traction with the same value, 1.05 mm; moreover, the CD method reaches the lowest diameter for the compression load case.

The minimum slenderness ratio of each lattice structure and in both the load cases is around 3 whereas the average values are close to 6. This could lead to a loss of accuracy in the results of the structural analyses but at the same time is more conservative.

In all the configurations, the optimization algorithm converged at the first iteration. Each configuration has 2141 beams; in the traction load case, 1316 beams would require an optimized cross-section below the lower size limit (0.5 mm); in the compression load case, this happens for 1240 beams. Those beams are the ones along the x-direction and they result to be over-dimensioned in the final model.

The volume of the design space is 38779 mm^3 and all the lattice configurations showed a volume smaller than more the 90%, as can be seen from the relative density values in Fig. 5. Moreover, the AD configuration presents the highest relative density in both the load cases and the lattice structure is less compliant (Fig. 6). Finally, the CD configuration has the best compromise between relative density and structure stiffness.

The structural verification carried out with Nastran solver confirms the validity of the results and shows that for low values of the slenderness ratio in axial load condition, the difference between the two beam approaches adopted in the structural linear analysis are limited.

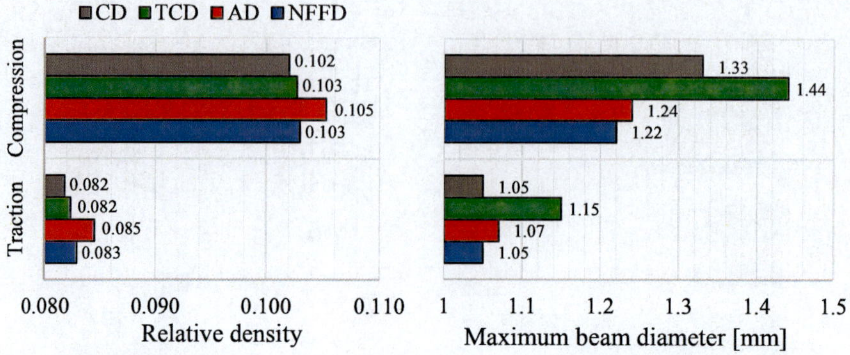

Fig. 5. Size optimization result in base of the modeling methods.

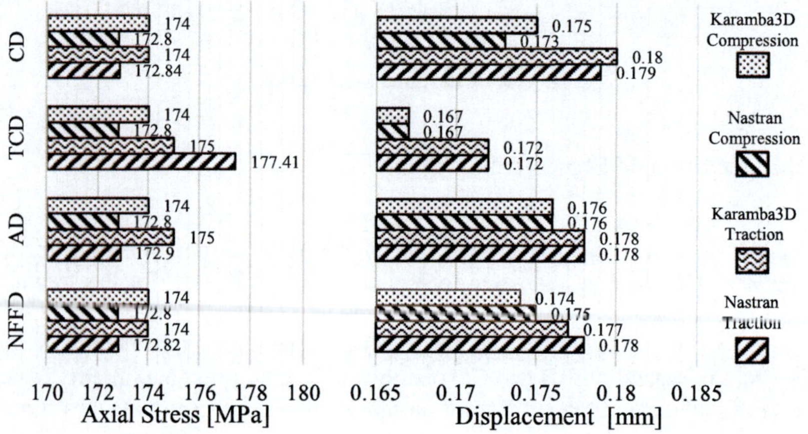

Fig. 6. Result of the linear structural analysis in base of the modeling methods.

4 Conclusion

The most important contributions of this study are the proposed modeling methodologies for obtaining the wireframe of conformal lattice structures based on simple cubic cells. When used to fill a bulk design space, the modeling approaches together with a subsequent size optimization show similar results in terms of mass reduction. Even if the curvature division method gives the best results, the NURBS Free-Form Deformation method is more adaptable to different geometric configurations of the design domain. Moreover, this approach can be automated by defining a specific transformation of the cage and allows for warping transformation when cell densification is needed. Also, the designer can iteratively modify the geometry in the simulation environment obtaining a prediction of the mechanical behavior thanks to the mono-dimensional FEA based on the beam element. Both the Euler-Bernoulli and the Timoshenko beam element theories were used showing a good agreement in the results. An advantage of using these approaches is the calculation and visualization of the structural results as the deformation and the field of tension with the requirement of a relatively low computational and graphic power. Future works will be addressed to study other cell shape and other optimization criteria.

Acknowledgement. This work was partially funded by the Regional Operational Programme F.S.E. 2014–2020 Veneto Region and the European Regional Development Fund POR 2014–2020, project code 2105-0036-1463-2019, "Design and manufacturing by additive technologies of innovative components using functionally graded materials", and a Ph.D. grant of Fondazione Cassa di Risparmio di Padova e Rovigo (CARIPARO).

References

1. Bacciaglia, A., Ceruti, A., Liverani, A.: Additive manufacturing challenges and future developments in the next ten years. In: Rizzi, C., Andrisano, A.O., Leali, F., Gherardini, F., Pini, F., Vergnano, A. (eds.) ADM 2019. LNME, pp. 891–902. Springer, Cham (2020). https://doi.org/10.1007/978-3-030-31154-4_76
2. British Standards Institution CEN/TR/ISO/ASTM 52912:2020: Standards Publication Additive manufacturing: Design Functionally graded additive manufacturing (2020)
3. Vigliotti, A., Pasini, D.: Stiffness and strength of tridimensional periodic lattices. Comput. Meth. Appl. Mech. Eng. **229–232**, 27–43 (2012). https://doi.org/10.1016/j.cma.2012.03.018
4. El-Galy, I.M., Saleh, B.I., Ahmed, M.H.: Functionally graded materials classifications and development trends from industrial point of view. SN Appl. Sci. **1**(11), 1–23 (2019). https://doi.org/10.1007/s42452-019-1413-4
5. Oxman, N., Tsai, E., Firstenberg, M.: Digital anisotropy: A variable elasticity rapid prototyping platform. Virtual Phys. Prototyp. **7**, 261–274 (2012). https://doi.org/10.1080/17452759.2012.731369
6. Rosso, S., Savio, G., Uriati, F., Meneghello, R., Concheri, G.: Optimization approaches in design for additive manufacturing. Proc. Int. Conf. Eng. Des. ICED **1**(1), 809–818 (2019). https://doi.org/10.1017/dsi.2019.85
7. Savio, G., Rosso, S., Meneghello, R., Concheri, G.: Geometric modeling of cellular materials for additive manufacturing in biomedical field: a review. Appl. Bionics Biomech. **2018**, 1–14 (2018). https://doi.org/10.1155/2018/1654782
8. Pei, E., Ressin, M., Campbell, R.I., Eynard, B., Xiao, J.: Investigating the impact of additive manufacturing data exchange standards for re-distributed manufacturing. Prog. Additive Manufact. **4**(3), 331–344 (2019). https://doi.org/10.1007/s40964-019-00085-7
9. Savio, G., Meneghello, R., Rosso, S., Concheri, G.: 3D model representation and data exchange for additive manufacturing. In: Cavas-Martínez, F., Eynard, B., Fernández Cañavate, F.J., Fernández-Pacheco, D.G., Morer, P., Nigrelli, V. (eds.) Advances on Mechanics, Design Engineering and Manufacturing II. LNME, pp. 412–421. Springer, Cham (2019). https://doi.org/10.1007/978-3-030-12346-8_40
10. Ceruti, A., Ferrari, R., Liverani, A.: Design for additive manufacturing using LSWM: a CAD tool for the modelling of lightweight and lattice structures. In: Campana, G., Howlett, R.J., Setchi, R., Cimatti, B. (eds.) SDM 2017. SIST, vol. 68, pp. 756–765. Springer, Cham (2017). https://doi.org/10.1007/978-3-319-57078-5_71
11. Pasko, A., Fryazinov, O., Vilbrandt, T., Fayolle, P.-A., Adzhiev, V.: Procedural function-based modelling of volumetric microstructures. Graph. Models. **73**, 165–181 (2011). https://doi.org/10.1016/j.gmod.2011.03.001

12. Bacciaglia, A., Ceruti, A., Liverani, A.: A systematic review of voxelization method in additive manufacturing. Mech. Ind. **20**, 630 (2019). https://doi.org/10.1051/meca/2019058
13. Loh, G.H., Pei, E., Harrison, D., Monzón, M.D.: An overview of functionally graded additive manufacturing. Addit. Manuf. **23**, 34–44 (2018). https://doi.org/10.1016/j.addma.2018.06.023
14. Nguyen, J., Park, S.I., Rosen, D.W., Folgar, L., Williams, J.: Conformal lattice structure design and fabrication. In: 23rd Annual International Solid Freeform Fabrication Symposium - An Additive Manufacturing Conference, SFF 2012, pp. 138–161 (2012)
15. Wu, J., Wang, W., Gao, X.: Design and optimization of conforming lattice structures. IEEE Trans. Vis. Comput. Graph. **27**, 43–56 (2021). https://doi.org/10.1109/TVCG.2019.2938946
16. Liang, Y., Zhao, F., Yoo, D.-J., Zheng, B.: Design of conformal lattice structures using the volumetric distance field based on parametric solid models. Rapid Prototyp. J. **26**, 1005–1017 (2020). https://doi.org/10.1108/RPJ-04-2019-0114
17. Alomar, Z., Concli, F.: A review of the selective laser melting lattice structures and their numerical models. Adv. Eng. Mater. **22**, 2000611 (2020). https://doi.org/10.1002/adem.202000611
18. Kalamkarov, A.L., Andrianov, I. V., Danishevs'kyy, V. V.: Asymptotic Homogenization of Composite Materials and Structures. Appl. Mech. Rev. **62**, 1–20 (2009). https://doi.org/10.1115/1.3090830
19. Bacciaglia, A., Ceruti, A., Liverani, A.: Structural analysis of voxel-based lattices using 1D Approach. 3D Print. Addit. Manuf. (2021). https://doi.org/10.1089/3dp.2020.0178
20. Ptochos, E., Labeas, G.: Elastic modulus and Poisson's ratio determination of micro-lattice cellular structures by analytical, numerical and homogenisation methods. J. Sandw. Struct. Mater. **14**, 597–626 (2012). https://doi.org/10.1177/1099636212444285
21. Ahmadi, S.M., et al.: Mechanical behavior of regular open-cell porous biomaterials made of diamond lattice unit cells. J. Mech. Behav. Biomed. Mater. **34**, 106–115 (2014). https://doi.org/10.1016/j.jmbbm.2014.02.003
22. Ashby, M.: Materials Selection in Mechanical Design, 4th edn. (2010)
23. Kirsch, U.: Problem statement. In: Kirsch, U. (ed.) Structural Optimization. Springer, Heidelberg (1993) https://doi.org/10.1007/978-1-4842-3051-0_5
24. Savio, G., Gaggi, F., Meneghello, R., Concheri, G.: Design method and taxonomy of optimized regular cellular structures for additive manufacturing technologies. In: Proceedings of the International Conference on Engineering Design, ICED, pp. 235–244 (2015)
25. Savio, G., Meneghello, R., Concheri, G.: Geometric modeling of lattice structures for additive manufacturing. Rapid Prototyp. J. **24**, 351–360 (2018). https://doi.org/10.1108/RPJ-07-2016-0122
26. Sederberg, T.W., Parry, S.R.: Free-form deformation of solid geometric models. In: Proceedings of the 13th Annual Conference on Computer Graphics and Interactive Techniques - SIGGRAPH 1986, pp. 151–160. ACM Press, New York, New York, USA (1986)
27. Andreas, S., et al.: A first step towards cage-based deformation in virtual reality. Smart Tools Appl. Graph. (2020). https://doi.org/10.2312/stag.20201246
28. Fernandez, J.W., Mithraratne, P., Thrupp, S.F., Tawhai, M.H., Hunter, P.J.: Anatomically based geometric modelling of the musculo-skeletal system and other organs. Biomech. Model. Mechanobiol. **2**, 139–155 (2004). https://doi.org/10.1007/s10237-003-0036-1
29. Coquillart, S.: Extended free-form deformation: a sculpturing tool for 3D geometric modeling. In: Proceedings of the 17th Annual Conference on Computer graphics and Interactive Techniques - SIGGRAPH 1990, pp. 187–196. ACM Press, New York, New York, USA (1990)
30. Lamousin, H.J., Waggenspack, N.N.: NURBS-based free-form deformations. IEEE Comput. Graph. Appl. **14**, 59–65 (1994). https://doi.org/10.1109/38.329096
31. Prochazkova, J.: Free form deformation methods - the theory and practice. In: 16th Conference on Applied Mathematics, APLIMAT 2017 – Proceedings, pp. 1276–1282 (2017)

32. Preisinger, C.: Linking structure and parametric geometry. Archit. Des. **83**, 110–113 (2013). https://doi.org/10.1002/ad.1564
33. BSI. Bs en 1993-1-1: 2005: Eurocode 3. Design of steel structures. General rules and rules for buildings (2005)
34. Mäenpää, J.: Algorithm-Aided Structural Engineering of Steel-Framed Warehouse (2018)
35. Renishaw: AlSi10Mg-0403 powder for additive manufacturing. https://resources.renishaw.com/en/details/data-sheet-alsi10mg-0403-400-w-powder-for-additive-manufacturing--73122. Accessed 09 Nov 2021
36. Ferro, P., Fabrizi, A., Berto, F., Savio, G., Meneghello, R., Rosso, S.: Defects as a root cause of fatigue weakening of additively manufactured AlSi10Mg components. Theor. Appl. Fract. Mech. **108**, 102611 (2020). https://doi.org/10.1016/j.tafmec.2020.102611

Optimization of an Engine Piston Through CAD Platforms and Additive Manufacturing Based Systematic Product Redesign

Enrico Dalpadulo(✉), Fabio Pini, and Francesco Leali

"Enzo Ferrari" Department of Engineering, University of Modena e Reggio Emilia, Via Vivarelli 10/1, 41125 Modena, Italy
{enrico.dalpadulo,fabio.pini,francesco.leali}@unimore.it

Abstract. The present work describes an automotive component design optimization process through a systematic approach. The redesign aims to improve product performance by Powder Bed Fusion metal Additive Manufacturing. The approach allows to match Topology Optimization and Design for Additive Manufacturing by exploiting benefits provided by CAD platforms that integrate CAD, CAE and CAM tools. The Systematic Concept-Selection-Based Approach aims to make redesign simple and effective, allowing design solutions exploration while containing product design lead time. Topology Optimization is the key phase to achieve lightweight design by a double-level optimization approach. In particular, the technique is setup to produce different design variants, whose subsequently undergo a Trade-off study to perform the concept selection step. Finally, one final redesign occurs and a design refinement step is performed to achieve product optimization. The case study is a high performance internal combustion engine piston, which has been redesigned to be produced by Selective Laser Melting process with benefit of weight reduction.

Keywords: Design for additive manufacturing · Topology optimization · Re-design · Design method · Automotive

1 Introduction

Modern high performance Internal Combustion Engines (ICE) are subjected to intense thermal and mechanical loadings. Engine crank mechanism components have to be strength and lightweight since inertial loads have to be minimized in order to improve machines performance and efficiency. Thus, since pistons represent the main moving mass, their study and optimization is fundamental. Additive Manufacturing (AM) technologies enable construction of complex shapes and structures that perfectly fit with high specific stiffness and specific strength required to design lightweight components such as high performance engine pistons [1].

Literature provides some interesting piston design approaches based on Topology Optimization, that is a common techniques used in conjunction with AM. Du et al. [2] first investigated structural behaviour, Zhao et al. [3] and, more recently, Gaidur

C. Rizzi et al. (Eds.): ADM 2021, LNME, pp. 486–493, 2022.
https://doi.org/10.1007/978-3-030-91234-5_49

et al. [4] introduced also thermal considerations. Barbieri et al. first combined TO and AM for a steel piston redesign application [5]. Lately, Mahle and Porsche demonstrated even industrial feasibility and endurance capability of such innovative solutions [6]. Nevertheless, none of these actually integrated approaches while facilitating key points such as design exploration, variants selection, linkage of validation and redesign steps. In fact, Design for Additive Manufacturing (DfAM) collects sundry guidelines and techniques but actual implementation requires effective approaches to exploit technology potentials in design practices [7]. Therefore, both design methods and available tools are continuously evolving in order to bridge the gap between theory, applications and integration, as confirmed by recent reviews on AM [8, 9].

The present paper stresses the application of a systematic approach to integrate TO and DfAM supported by CAD platforms for product-process design. The case study is a complex component such as an ICE piston to be re-designed for AM-based enhanced product performance.

2 Method

Many researches are focused on TO advances and AM applications. DfAM implementation methods and description on how systematic approaches can provide benefits in TO techniques and DfAM integration is discussed in [10]. Moreover, in the DfAM context, advanced tools and CAD platforms for product-process design can be exploited to improve approaches by data consistency, features linkage, information retrieval, supported by advanced associativity. The systematic approach (Fig. 1) used in this work aims at supporting DfAM, product optimization, design solutions exploration, design variants selection and design lead time reduction, while exploiting potentials of CAD platforms [11].

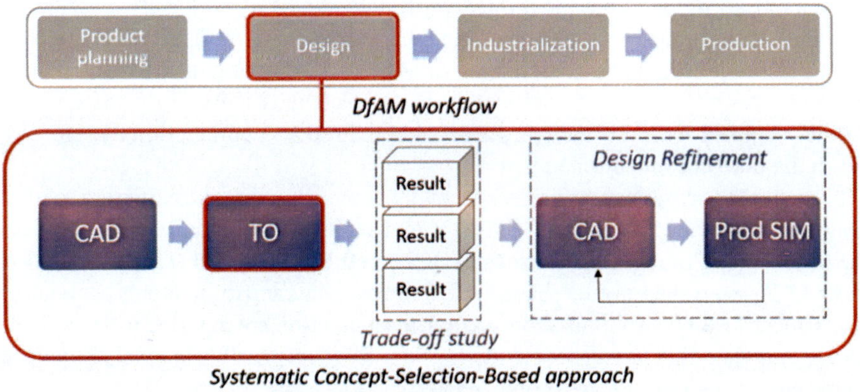

Fig. 1. Systematic concept selection-based approach for product optimization.

2.1 Systematic Concept-Selection-Based Approach

The Systematic Concept-Selection-based approach [11] belongs to a more broadly DfAM workflow based on CAD platforms, consisting of Design and Industrialization core phases. Within the Design phase, TO represent the key step to achieve product optimization on a CAD model after Product Planning phase, thus, a conceptual solution exploration step is performed through TO by generating sundry concept results. Subsequently, they can be evaluated to drive design variants selection with respect to design requirements (KPIs) through a Trade-off study, which represents a significant support for the designer [9]. These driven solution generation steps become simple in CAD based environments integrating TO and Product Simulation (ProdSIM) and supporting automatic geometry construction features. Afterwards, just one manual final redesign is required for design interpretation. This work brings a plain improvement compared to the original design method case [11]. In fact, on such complex case studies, a final Design Refinement step based on ProdSIM and CAD can be performed to further improve model performance. In this way, a double-level optimization through integrated environment is made possible: the first involving design automation to drive the Trade-off study step, the second requiring manual design interpretation to improve the concept solution by Design Refinement step. The product optimization approach thus based on driven solution generation and subsequent design refinement has been applied to the case study by means of the DS 3DExperience CAD platform for product-process design.

3 Case Study

The redesign subject is an ICE piston from a 250 cc single cylinder racing motorcycle. The piston is made of Al 2618 T6 alloy, with a whole weigh of 138 g (see Fig. 2). The expected outcomes of redesign are mass reduction of the piston while keeping the same or improved structural performances, assuming the loads related to heaviest working condition of the engine as the functional constrains.

Product Planning phase involves original part analyses as well as design objectives and constraints definition. Structural analysis is performed to obtain stress and displacement references for redesign. Subsequently, to complete feasibility study, material study and functional and assembly analyses are required. Product optimization therefore aims to reach the outcomes mentioned before.

Reference loadcases expressing respectively combustion pressure, inertial loads and lateral skirts thrust are: Top Dead Centre during Combustion (TDCC); Top Dead Centre at the beginning of the Induction stroke (TDCI); Left Piston Thrust (LPT); Right Piston Thrust (RPT). A tetrahedral mesh based on 1.2 mm elements and refinements has been used for the structural analysis. Loads from analytical calculation are reported in Table 1. Each loadcase is modeled in worst case considering the maximum possible load related to engine working conditions (e.g. 12 MPa comb. pressure, 14000 rpm).

Particular attention has been paid to proper loads and DOF boundary conditions modeling. TDCC is modeled by resulting force on piston crown surface, axial fixed displacement on upper surfaces of piston pin hole, radial fixed displacements on piston ringland. TDCI is modeled by downward force on lower surfaces of piston pin hole, axial fixed displacement on piston crown surface, radial fixed displacements on piston

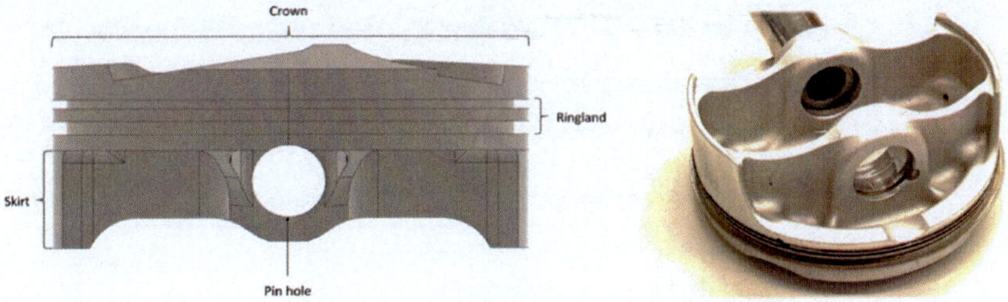

Fig. 2. Original piston geometry.

ringland. LPT and RPT are modeled by thrust force on lateral surfaces of piston pin hole, two nodes axial fixed displacements, fixed horizontal displacement on piston ringland and correspondent skirt. Figure 3 respectively returns behavior of the original piston in TDCC and TDCI.

Table 1. Loadcases summary.

Loadcase	φ	Fx	Fy	Fz
TDCC	15°		–	–56862.5 N
TDCI	0°		–	9745.5 N
LPT	25°	3483.5 N	–	
RPT	335°	–3483.5 N	–	

The material study returns the selection of AlSi10Mg alloy suitable for additive fabrication. Based on literature mean material data of both alloys, temperature-dependent mechanical properties have been compared, considering AlSi10Mg along AM build direction, in favor of safety. In particular, at elevated temperatures (200 °C), a lower gap between the strength of the alloys can be observed [12]. Table 2 reports and compares materials properties.

Considering these data, design objectives and constraints can be defined. The redesign through TO aims to reduce component mass, reducing stress peaks (≤Ys) by smoothening its distribution, while keeping almost the same deformation values.

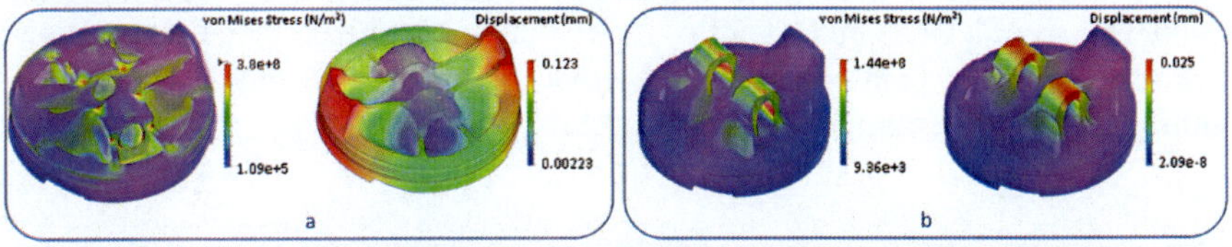

Fig. 3. Original piston structural behavior in TDCC (a) and TDCI (b).

Table 2. Materials properties.

Material	ρ	E	ν	Ys	UTs
AlSi10Mg	2.75 g/cm^3	74.5 GPa	0.33	250 MPa	400 MPa
Al 2618 T6	2.7 g/cm^3	72 GPa	0.33	380 MPa	430 MPa

4 Results

Referring to the workflow presented above, the Design phase is mainly composed by TO, Trade-off study and Design Refinement. CAD model preparation for TO involves the definition of Design Space (DS) and Non-Design Space (NDS), retrieving information from previous analyses. In particular, piston crown keeps the original part minimum thickness, other functional elements keep a 1.5 mm feasible thickness according to the manufacturing process. Finite element model and loadcases are retrieved from the original component analysis setup. TO is setup to maximize part stiffness with a mass reduction value as constraint. In this case, design solution exploration for further Trade-off study is achieved through generation of sundry models from TO working on model settings, finite elements size, computed minimum element dimension, algorithm settings, additional controls such as part symmetry, etc. Results are computed considering at the same time all the four loadcases. A tool for automatic generation of solid models is used to create the design variants from computed shapes selecting ISO-surface values. In this way, each result can be directly validated by FEA ProdSIM, retrieving information from TO setup. Figure 4 depicts design variants involved in Trade-off study to perform the Concept Selection step.

Fig. 4. Topology optimization design variants for concept selection.

Considering structural KPIs such as mass, mean stress, maximum principal stress, maximum displacement, concept selection is performed. Figure 5 depicts the stress distribution map in TDCC and TDCI of the topology optimized model presenting the highest score. At this point, Design Refinement can start and thus a detailed CAD modelling involving design interpretation occurs (Fig. 6).

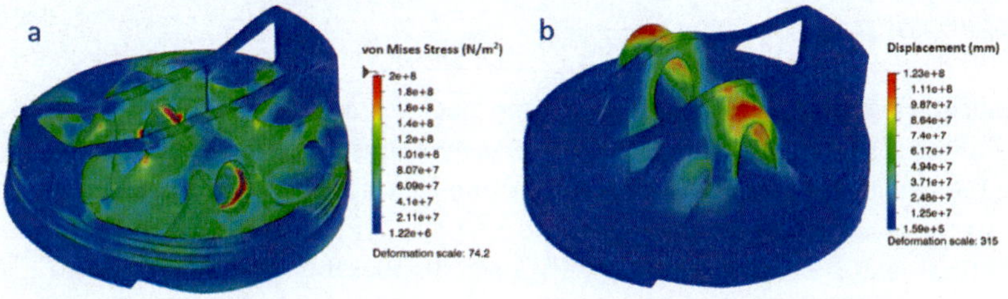

Fig. 5. Stress distribution fields of selected concept in TDCC (a), TDCI, (b).

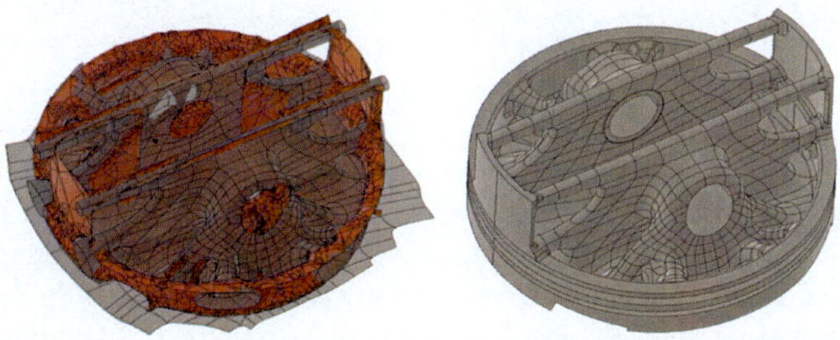

Fig. 6. On the left, design interpretation process through freeform surface modelling; on the right, topology optimized piston final design after Design Refinement.

Subsequently, FEA ProdSIM is performed using the aforementioned setup. The Design Refinement aims to improve optimized component behavior, focusing in particular on part mass, stress concentration and deformation. The optimization loop based on just one manual re-design returns a model whose most significant results (TDCC, TDCI) are depicted in Fig. 7.

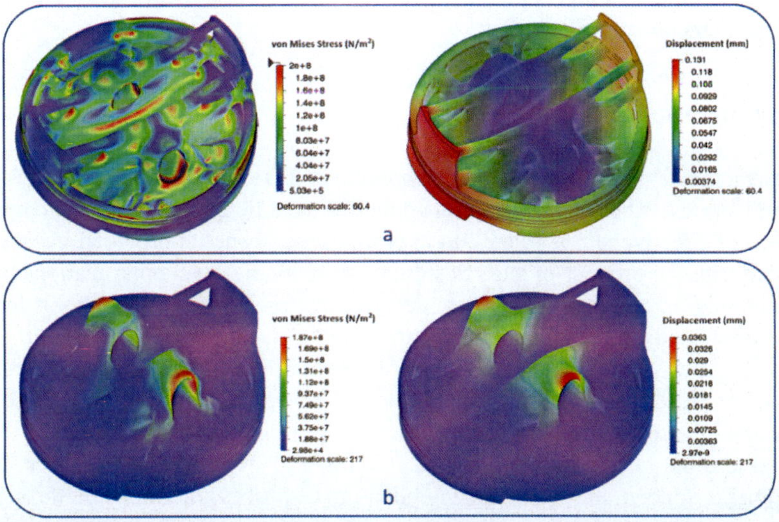

Fig. 7. Topology optimized piston structural behavior in TDCC (a) and TDCI (b).

5 Discussion

The Concept-Selection-based systematic approach for integration of TO in the re-design of the engine piston to be produced by AM provided fine results.

On the side of product requirements, from a functional and structural point of view, firstly stress distribution, component deformations and mass saving can be analyzed. The original piston showed von Mises stress considerably above 300 MPa (Ys/1.25), in particular with maximum pressure during combustion in TDCC. Other loadcases produced less critical stress, with maximum values around 140 MPa in TDCI. The optimized piston, due to its branched and organic shape, actually presents stress hardly above 200 MPa (Ys/1.25) in TDCC. In TDCI, stress values around 170 MPa are in favor of safety due to actual mass reduction, whereas for lateral thrusts (i.e. LPT, RPT) they are reduced again. Concerning deformations, maximum displacement is the same in TDCC, it is slightly higher in TDCI, but however low in magnitude, and it is lower in LPT and RPT. Regarding the main objective, or rather mass saving, after the final optimization loop the Design Refinement step produced a component with the impressive weight of 109 g. Mass saving is therefore more than 20% compared to the original component. Nevertheless, additional remarks can concern fatigue life, since currently SLM produced components provide around 30% less fatigue strength, even though it is highly influenced by process parameters and post-processing treatments [13, 14]. On the other hand, such racing application requires original piston replacement after 40 h evidently providing a short lifecycle.

Anyhow, on the side of method and application, a remark is that the systematic approach structure allows to integrate additional product or process studies (e.g. buckling, fatigue life, thermal, industrialization, printing etc.) into the design phase. In particular, they could exploit the integrated CAD-based environment into the Trade-off study and/or the Design Refinement processes. However, beyond the aforementioned flexibility, the systematic approach can be applied to effective redesigns of AM-based manufacturing high performance metal products.

6 Conclusions

A design method based on a systematic approach for TO and DfAM integration for product optimization has been presented and applied to a challenging automotive component redesign. Its effectiveness is related to exploitation of benefits provided by integrated CAD platforms for product-process design, using a double-level optimization approach and combining automation and manual design. Integration of advanced tools for product validation and design variants selection can be used to drive the design process steps to obtain high performance components design. Moreover, flexible integration of additional product simulations can be performed as well as KPIs for concept selection step can be adapted to meet different design requirements. The ICE piston redesign produced optimal results in terms of mass saving and structural performance. Further developments concern the subsequent application of process optimization through analogue approaches to perform equally the design and the industrialization steps of DfAM and the physical prototyping to validate case studies.

References

1. Plocher, J., Panesar, A.: Review on design and structural optimisation in additive manufacturing: towards next-generation lightweight structures. Mater. Design **183** (2019)
2. Du, F., Tao, Z.: Study on lightweight of the engine piston based on topology optimization. Adv. Mater. Res. **201–203**, 1308–1311 (2011)
3. Zhao, J., Du, F., Yao, W.: Structural analysis and topology optimization of a bent-bar-frame piston based on the variable density approach. In: ASME 2014 Dynamic Systems and Control Conference, vol. 2 (2014)
4. Gaidur, M., et al.: Analytical study regarding the topological optimization of an internal combustion engine piston. In: IOP Conference Series: Materials Science and Engineering, vol. 997. The 9th International Conference on Advanced Concepts in Mechanical Engineering - ACME 2020 (2020)
5. Barbieri, S.G., Giacopini, M., Mangeruga, V., Mantovani, S.: Design of an additive manufactured steel piston for a high performance engine: developing of a numerical methodology based on topology optimization techniques. SAE Int. J. Engines **11**(6), 1139–1150 (2018)
6. Abele, D., Ickinger, F., Schall, V., Klampfl, M.: Additive manufacturing of high-performance powertrain components. MTZ Worldw. **82**, 14–19 (2021)
7. Wiberg, A., Persson, J., Ölvander, J.: Design for additive manufacturing – a review of available design methods and software. Rapid Prototyp. J. **25**(6), 1080–1094 (2019)
8. Alfaify, A., Saleh, M., Abdullah, F.M., Al-Ahmari, A.M.: Design for additive manufacturing: a systematic review. Sustainability (Switzerland) **12**(19), 7936 (2020)
9. Vaneker, T., Bernard, A., Moroni, G., Gibson, I., Zhang, Y.: Design for additive manufacturing: framework and methodology. CIRP Ann. **69**(2), 578–599 (2020)
10. Dalpadulo, E., Pini, F., Leali, F.: Systematic integration of topology optimization techniques in design for additive manufacturing methodologies applied to automotive settings. In: ASME International Mechanical Engineering Congress and Exposition, Proceedings (IMECE), vol. 6 (2020)
11. Dalpadulo, E., Gherardini, F., Pini, F., Leali, F.: Integration of topology optimisation and design variants selection for additive manufacturing-based systematic product redesign. Appl. Sci. **10**(21), 7841 (2020)
12. Uzan, N.E., Shneck, R., Yeheskel, O., Frage, N.: High-temperature mechanical properties of AlSi10Mg specimens fabricated by additive manufacturing using selective laser melting technologies (AM-SLM). Addit. Manuf. **24**, 257–263 (2018)
13. Domfang Ngnekou, J.N., et al.: Fatigue properties of AlSi10Mg produced by additive layer manufacturing. Int. J. Fatigue **119**, 160–172 (2018)
14. Rhein, R.K., Shi, Q., Arjun Tekalur, S., Wayne Jones, J., Carroll, J.W.: Effect of direct metal laser sintering build parameters on defects and ultrasonic fatigue performance of additively manufactured AlSi10Mg. Fatigue Fract. Eng. Mater. Struct. **44**(2), 295–305 (2021)

A Numerical Model for Simulating the Selective Laser Melting of Closed Impellers

Marco Mandolini[1](✉), Federica Santucci[1], Paolo Cicconi[2], Eleonora Santecchia[1], Michele Germani[1], Marco Palladino[3], Claudio Favi[4], and Marco Marconi[5]

[1] Department of Industrial Engineering and Mathematical Sciences, Università Politecnica delle Marche, Ancona, Italy
{m.mandolini,f.santucci,e.santecchia,m.germani}@univpm.it

[2] Department of Engineering, Università degli Studi Roma Tre, Roma, Italy
paolo.cicconi@uniroma3.it

[3] Baker Hughes, Florence, Italy
marco.palladino@bakerhughes.com

[4] Department of Engineering and Architecture, Università degli Studi di Parma, Parma, Italy
Claudio.favi@unipr.it

[5] Department of Economics, Engineering, Society and Business Organization, Università degli Studi Della Tuscia, Viterbo, Italy
marco.marconi@unitus.it

Abstract. The Selective Laser Melting (SLM) is accelerating the adoption of Additive Manufacturing (AM) technologies in the industry. One of the most critical benefits concerns the possibility of manufacturing complex-shaped components, which are not feasible or too expensive using traditional processes. Recent studies are evaluating the SLM manufacturability of closed impellers through laboratory tests. The adoption of numerical simulation models for achieving this goal is still limited due to the complexity of the additive process and the number of phases to be considered. The paper presents a numerical model developed in the ANSYS workbench platform for simulating the SLM process of closed impellers. This work is one of the first studies available in the literature for such a particular kind of components. The paper describes the overall simulation model and the steps required for its definition. Furthermore, boundary conditions and process parameters are provided for a better understanding of the model. A case study illustrates its application to a specific part, intending to evaluate (i) impact between the recoater and the component, (ii) maximum von Mises stress and (iii) maximum displacement during the printing phase and following post-processing. These evaluations will support design and manufacturing engineers during product and process engineering. The promising results of this study are encouraging further research about the application of SLM for closed impellers.

Keywords: Selective laser melting · Impeller · Numerical simulation · Additive manufacturing

C. Rizzi et al. (Eds.): ADM 2021, LNME, pp. 494–501, 2022.
https://doi.org/10.1007/978-3-030-91234-5_50

1 Introduction and Literature Review

Impellers, made of metallic materials such as light alloys, stainless steels, and superalloys, find application in various products, such as centrifugal compressors and pumps. Impellers can be classified according to their shape, which could be very complicated, in closed (shrouded), open or semi-open. Conventionally, impellers are either made from a solid piece of metal, whereby material is removed by 5-axis milling or produced by welding or brazing a pre-processed hub (wax-casted or milled) and a pre-processed shroud [1]. The second solution is required for impellers with a high diameter to channel height ratio. The technological limits of these processes (e.g. minimum thicknesses, the ratio of impeller diameter to channel height) negatively impact the quality of impellers or design freedom. Besides, traditional technologies could negatively affect the structural and modal behaviour of the component being manufactured (e.g. the need to weld two shells determines a significant mechanical strength reduction).

The limitations just introduced in recent years are pushing companies, engineers, and researchers towards the study and application of additive technologies (mainly Selective Laser Melting – SLM) to support impellers' manufacturing process. Research is moving towards additive manufacturing to directly create metal impellers to be used inside machines after post-processing and machining. The most crucial challenge for 3D printing a shrouded impeller is linked to the design of the supports inside the channels. These supporting structures, required for preventing the collapse of overhanging surfaces, must be eliminated after printing, therefore they should be accurately designed for facilitating their removal through specific post processing operations.

This problem can be solved through a continuous modulation of the process parameters (e.g. scan speed, laser power) however an appropriate design of the build part and supports is necessary. The SupportFree™ technology proposed by VELO3D [2] permits near-horizontal features to be printed entirely without supports. Conversely, through a careful design of the impeller shape, it is possible to correctly 3D print this kind of products. Meli et al. [3] presented a general framework for re-designing and manufacturing 3D impellers installed on centrifugal compressors by exploiting Topology Optimisation and Additive Manufacturing potential. From the literature, it is also possible to identify patents of impellers optimised to be made in AM [1]. In the last three years, researchers pushed their effort in optimising the AM process for impellers. The first studies date back to 2014, when Allison et al. [4] presented their experience with manufacturing and testing multiple closed impeller designs constructed from Inconel 718, 17-4 PH Stainless Steel and Titanium Ti-6Al-4V. More recently, Sulzer AG [5, 6], evaluated the adoption of AM for small pumps impellers, with a the lower limit of castability, characterized by complex shapes. Its goals were to define the optimal build orientation and design suitable support structures. Many works have been based on experimental studies, based on trial and error approach. Recently, the research is moving towards the SLM process characterisation to simulate its thermal and structural effects. Many works are focused on developing numerical simulators aimed at studying the melt-pool [7, 8]. Based on these simulators, other scholars [9] have developed finite-element (FE) numerical simulators of the heat transfer analysis of AM processes such as SLM and SLS. Given the SLM process complexity, recent numerical simulation software tools (ANSYS Workbench Additive, Siemens Simcenter) are allowing engineers and researchers to engineer and optimise mechanical components. These simulators can simulate the entire printing process (build, heat treatment, supports removal) to evaluate distortions, shrinkage, stress state, thermal load, and any failures (e.g. impact of the party with the recoater). These tools are recently gaining a lot of interest thanks to their possibility of exploring many design solutions while reducing physical experiments costs. Such software can be used, for example, to quantify the strain and residual stress after printing [10]. Among the studies aimed at numerically simulating the impeller SLM, the first one was by Van Belle [11], who presented the residual stresses and distortion of a turbine impeller. More recently, Jia et al. [12] used a computer-aided design technique to simulate a lightweight compressor impeller manufacturing process by evaluating the residual deformation and residual stress.

From the scientific literature analysis, most of the studies concerning the adoption of SLM for 3D printing impellers are focused on experimental tests. There are only limited applications of numerical simulation tools. Among those present, all are focused on open or semi-open impellers, whose production complexity is much lower than closed (shrouded) ones. To the best of our knowledge, this is the first study that aims to define a numerical model for the simulation of the entire SLM process to study closed impellers.

2 SLM Simulation Model

The numerical model developed in this study for simulating the SLM process is based on ANSYS Workbench Additive. The simulation method used by this platform is called "*Lumped Layer Approach*". It consists of grouping multiple printing layers (around twenty) into a single simple (super) layer. This approach assumes that the thermal histories of neighbouring physical layers are comparable and can be lumped into one representative layer. This solution can be considered as appropriate to quickly obtain the tensile and deformative state of the workpiece during the printing process. The geometry is approximated with a cartesian or tetrahedral mesh and with the "*dead and birth*" technique to simulate the build progress.

The overall methodology consists of:

1. Closed impeller 3D CAD modelling.
2. Positioning and orientating the impeller into the build chamber.
3. Modelling the supporting structures.
4. Setting up the simulation model and material.
5. Evaluation and discussion of the results.

Concerning the ANSYS suite, the first three steps are performed using Additive Mechanical Workbench and Additive Prep, while the last two ones with Workbench Additive. The impeller orientation should be defined to reduces build time, material usage, and supports removal time. This phase is achieved utilising the orientation maps, which help to determine the best printing configuration considering build time, support volume, and distortion tendency. Another parameter to set is the Z-Offset, i.e. the distance between the lowest point of the part and the base plate.

Furthermore, as regards supports, they are structures which can be either surfaces or body lines. They can be defined based on overhang angle (the angle measured from the base plate to the surface of the part). For any predefined support type, a set of parameters must be identified (e.g., strut dimension, cell size). Specific parameters that regulate every structure characteristics can be also defined (e.g., dimensions of the top and bottom intrusion, width and height of contour). Figure 1 presents the modules of the simulation model developed in ANSYS Workbench Additive:

- *Geometry*: 3D CAD model of the impeller.
- *Transient thermal analysis of printing phase*. This module simulates the thermal behaviour of the component during the print phase. The tool applies energy by using temperature and power. This behaviour means that the simulation is not conforming to time. The thermal simulation requires the set layer thickness, hatch spacing, scan speed, recoating time (the time recoater takes for a complete cycle), preheat build temperature, gas and powder temperatures and convection coefficients. By solving transient thermal, it is possible to display the part temperature as the printing process progresses. This module is used to characterize the material, modelled with a bilinear isotropic hardening behaviour. The most relevant properties are density, coefficient of thermal expansion, melting temperature, specific heat, thermal conductivity, Young's modulus and Poisson's ratio.
- *Transient thermal analysis of heat treatment*. This module simulates the heat treatment applied once the printing process ends. The temperature vs time curve is the primary input required for this simulation. Other material and process information are retrieved from the previous module.
- *Static structural analysis*. This module takes results previously calculated by the two thermal analyses and it is used to compute the deformation and tension state during the overall printing steps. Material and process information are retrieved from the *transient thermal* module.

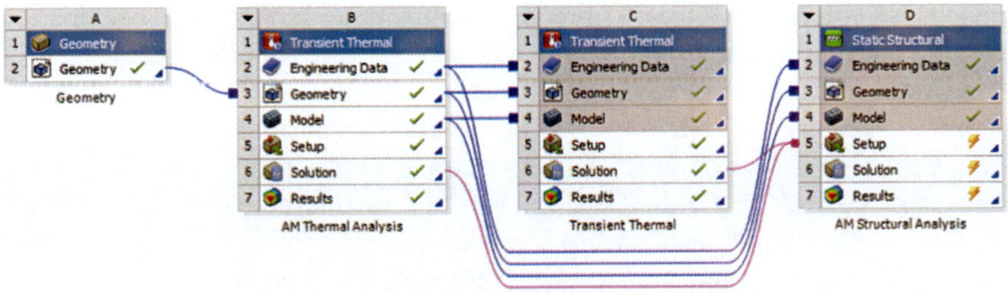

Fig. 1. Simulation model in ANSYS Workbench

The overall process simulated in ANSYS Workbench Additive is the following:

1. *Build*: energy application to the part, layer after layer, until the print completion.
2. *Cooldown*: component cools down to room temperature after printing.
3. *Heat treatment*: heat treatment for stress relieving.
4. *Build plate removal*: directional removal of the base plate and related supports.
5. *Supports removal*: removal of internal supports.

3 Case Study

The component here analysed is an Inconel 718 closed impeller. The hub, inlet and outlet diameters are respectively 45, 137 and 312 mm. The 3D CAD model was downloaded from GrabCAD (www.grabcad.com). To reduce as much as possible the computational effort without losing results quality, the impeller (twelve channels) was split with radial two surfaces to isolate a single channel (Fig. 2). To reduce the edge-effects on the channel analysed, two splitting surfaces have been defined to include a portion of the surrounding channels. The resulting model includes a complete channel that is representative of the full impeller, as represented in Fig. 2.

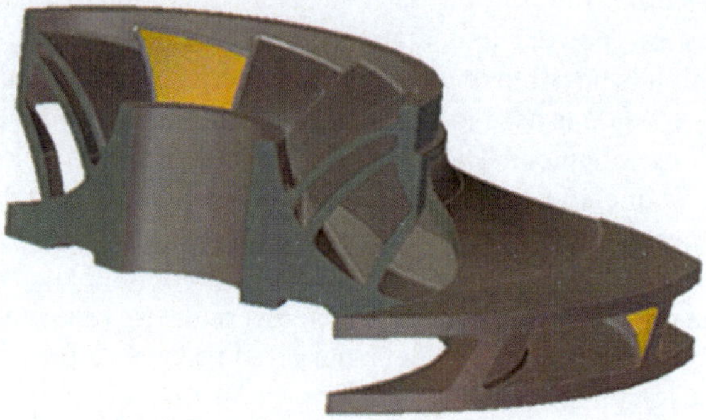

Fig. 2. A portion of the impeller used for the numerical simulation

The component was placed horizontally with the suction channel facing upwards. The machine considered for 3D printing is EOS M400. It has a printing volume of 400 [mm] × 400 [mm] × 400 [mm]. The process parameters are layer thickness: 40 [μm], hatch spacing: 110 [μm], scan speed: 960 [mm/s], recoating time: 7.5 [s], preheat temperature: 80 [°C], gas and powder convection coefficient: 1e−5 [W/m^2 °C]. There parameters have been retrieved from [13] and ANSYS database. Concerning the material, the material properties are density: 8220 [kg/m^3], coefficient of thermal expansion: 1.43e−5 [1/°C] @427 [°C]; 1.83e−5 [1/°C] @1084 [°C], melting temperature: 1260 [°C], Young's modulus: 165 [GPa] @20°[C]; 34 [GPa] @ 1100 [°C], thermal conductivity: 12 [W/m°C] @20 [°C]; 26 [W/m°C] @ 866 [°C], specific heat: 421 [J/kg°C] @20 [°C]; 651 [J/kg°C] @110 [°C]. There parameters are retrieved from ANSYS database.

Two support structures were used: a block structure at the base, to reduce deformations, and a tree structure in the channel, to allow easy removal. The first type has an x and y spacing of 0.5 [mm]. The tree structure has 55° conex apex, 0.5 mm tip height, 0.5 tip scale and 0.15 [mm] tip intrusion. As regards the simulation of the AM process, defined in the Additive Workbench, the following boundary conditions have been used:

- Mesh size (voxel) of supports and impeller: 1.24 [mm] (31 lumped layers).
- Mesh size (cartesian) base plate: 10 [mm].
- Build plate boundary conditions: fixed support on the lower surface during the build phase and twelve longitudinal springs at the four edges during the heat treatment.

The 3D virtually printed impeller has undergone a heat treatment (HT) according to AMS 5662 and 5664 (HT specific for 3D printed components in Inconel 718).

4 Results and Discussion

The numerical simulation of the SLM process allows manufacturing engineers to evaluate the manufacturability of the build parts. The achievement of such a scope is evaluated through the following results:

- Collision between the component and recoater during the printing phase. Hard recoater is preferable for its advantages in terms of part quality and repeatability. However, a silicon soft recoater can reduce the problems due to collisions.
- Equivalent von Mises stress lower than that material yield.
- Maximum displacement of the component at the end of the printing process is within the tolerance range.

Regarding the evaluation of the recoater impact with the component, the directional deformation along the printing direction was analysed. The latter depends on the kind of recoater (soft or hard) used. For hard recoater, directional deformation must be lower than the layer thickness (40 μm) plus a tolerance (20–80 μm). As shown in Fig. 3, this condition is respected at the 30% (−29 μm) of build progress, while a failure will be met at 60% (185 μm) and 90% (127 μm). These results mean that a soft recoater is necessary for printing the closed impeller.

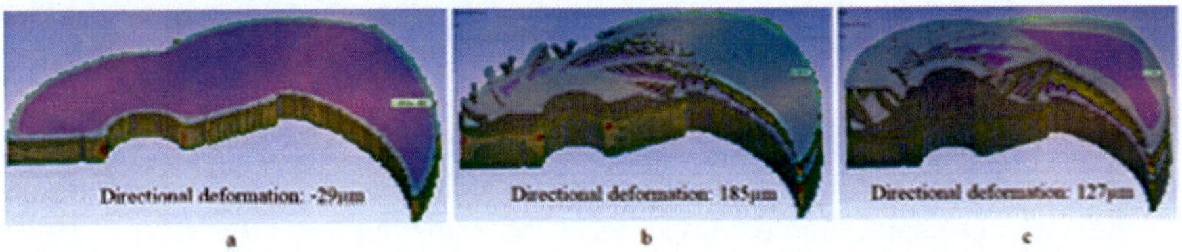

Fig. 3. Directional deformation at 30% (a), 60% (b) and 90% (c) of build progress.

Concerning the evaluation of the deformation and tensional state after the heat treatment, Fig. 4 presents the simulation results. Excluding the stress peaks due to geometry simplification, the maximum value of stress reached is approximately 360 [MPa]. It is located in the area of the inlet walls. As for the displacement, at the end of the cooldown, there is a maximum equivalent deformation equal to 1.01 [mm]. The deformation peaks due to the geometry simplification have not been considered. The maximum displacement observed is still too high for being considered acceptable (further machining operations will be required to respect critical to quality – CtQ – requirements).

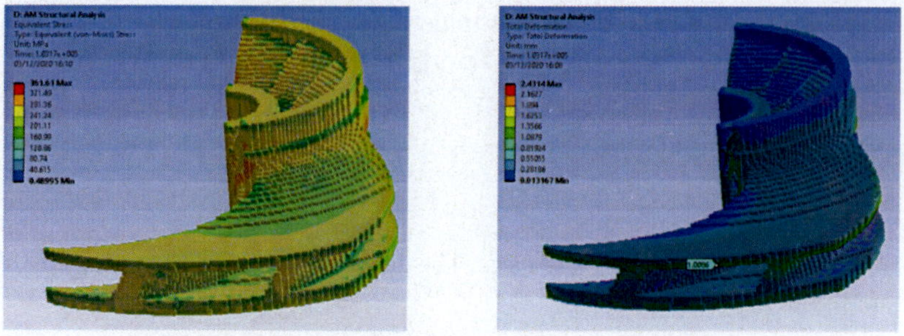

a b

Fig. 4. Equivalent von Mises stress (a) and total deformation (b)

5 Conclusions

The paper presented a numerical model for simulating the SLM process of closed impellers, typically used in centrifugal pumps and compressors. Manufacturing companies are studying the adoption of such a 3D printing process to overcome the limitations (e.g. minimum thickness) of traditional techniques (casting, milling and welding). The simulation model, here described, shows all the steps of a typical AM process, from the build plate warm up to supports removal. Process parameters and boundary conditions here provided will allow other scholars and practitioners to use the proposed model for engineering and research further activities. The results of this study cannot confirm the SLM printability of the closed impeller horizontally oriented into the build chamber. According to the simulation results, there could be impacts with a hard recoater (suggested for high-performance components). Furthermore, deformations will be higher than the typical critical to quality requirements. For these reasons, future investigations will be required. The full-shape impeller should be simulated, in addition to the single-channel geometry here analysed. Different supporting structures should be further investigated to evaluate their effects on the components stress and deformation fields. Besides, the shape should be optimised according to the results to meet the CtQ requirements. The last step will help designing a compensated geometry and include machining allowance where CtQ will not be respected.

References

1. Pulnikov, A.: Shrouded impeller made by additive manufacturing and including voids in the hub and in the shroud (2017)
2. VELO3D: Turbopumps. https://www.velo3d.com/wp-content/uploads/2021/02/velo3d-am-solution-for-turbomachinery.pdf. Accessed 14 Feb 2021
3. Meli, E., Furferi, R., Rind, A., Ridolfi, A., Volpe, Y., Buonamici, F.: A general framework for designing 3D impellers using topology optimization and additive manufacturing. IEEE Access **8**, 60259–60269 (2020). https://doi.org/10.1109/ACCESS.2020.2982841
4. Allison, T., Rimple, A., Pelton, R., Moore, J., Wilkes, J., Wygant, K.: Manufacturing and testing experience with direct metal laser sintering for closed centrifugal compressor impellers. In: 43rd Turbomach. 30th Pump Users Symposium (2014)

5. Meboldt, M., Klahn, C. (eds.): AMPA 2017. Springer, Cham (2018). https://doi.org/10.1007/978-3-319-66866-6
6. Kraenzler, T.: Process method for manufacturing impellers by Selective Laser Melting (SLM). Sulzer Tech. Rev. (2018)
7. Cao, L., Yuan, X.: Study on the numerical simulation of the SLM molten pool dynamic behavior of a nickel-based superalloy on the workpiece scale. Materials (Basel). **12**, 2272 (2019). https://doi.org/10.3390/ma12142272
8. Wu, Y.-C., et al.: Numerical modeling of melt-pool behavior in selective laser melting with random powder distribution and experimental validation. J. Mater. Process. Technol. **254**, 72–78 (2018). https://doi.org/10.1016/j.jmatprotec.2017.11.032
9. Chiumenti, M., et al.: Numerical modelling and experimental validation in Selective Laser Melting. Addit. Manuf. **18**, 171–185 (2017). https://doi.org/10.1016/j.addma.2017.09.002
10. Cicconi, P., Mandolini, M., Favi, C., Campi, F., Germani, M.: Metal additive manufacturing for the rapid prototyping of shaped parts: a case study. Comput. Aided. Des. Appl. **18**, 1061–1079 (2021). https://doi.org/10.14733/cadaps.2021.1061-1079
11. Van Belle, L.: Numerical simulation of complex part manufactured by selective laser melting process. AIP Conf. Proc. **1896** (2017). https://doi.org/10.1063/1.5008039
12. Jia, D., Li, F., Zhang, Y.: 3D-printing process design of lattice compressor impeller based on residual stress and deformation. Sci. Rep. **10**, 1–11 (2020). https://doi.org/10.1038/s41598-019-57131-1
13. Chen, Q., et al.: An inherent strain based multiscale modeling framework for simulating part-scale residual deformation for direct metal laser sintering. Addit. Manuf. **28**, 406–418 (2019). https://doi.org/10.1016/j.addma.2019.05.021

Additive Manufacturing Design of an Argon Condenser Made with Pure Copper Powder for High-Purity Physics Applications: Technological Issues

Daniele Cortis(✉), Alessandro Lalli, and Donato Orlandi

National Institute for Nuclear Physics, Gran Sasso National Laboratory, 67100 L'Aquila, Italy
daniele.cortis@lngs.infn.it

Abstract. Additive Manufacturing (AM) techniques give to designer great freedom as the possibility to rethink the classic geometrical shapes of common objects and it is well suitable for experimental physics research, where components production is based on unique prototype with high specialization and characteristics. In this paper, an AM design process of a heat exchanger for the condensation of argon using liquid nitrogen under cryogenics condition is proposed. The development of the condenser is part of the activities related to the "DarkSide" experiment. The work describes the AM approach considering the re-design of the condenser geometrical shape, the functional and performance constraints together with a focus on the technological issues of the processing of a pure copper powder with Selective Laser Melting (SLM) technology. Unfortunately, it was found that there are still technological issues related to the SLM process of pure copper due to its low-density and its high internal porosity.

Keywords: Additive Manufacturing · Pure copper powder · Heat exchanger · Selective Laser Melting

1 Introduction

1.1 Design for Additive Manufacturing

"Additive Manufacturing (AM) encompasses a range of technologies that allows physical components to be made, from virtual 3D models by building the component layer-upon-layer until the part is complete" [1]. These kinds of new technologies give to designer great freedom as the possibility to rethink the classic geometrical shapes of common objects: the constraints due to the Computer Numerical Control (CNC) manufacturing technologies are not important as before and they are mainly replaced by constraints linked to the specific AM manufacturing process (i.e., part supports generation, geometric and dimensional tolerances, material properties, heat dissipation, etc.). AM technologies have thus allowed the development of new design methods mainly based on virtual simulations (i.e., Topological Optimization, etc.), starting from new geometrical shape

C. Rizzi et al. (Eds.): ADM 2021, LNME, pp. 502–513, 2022.
https://doi.org/10.1007/978-3-030-91234-5_51

to a new material distribution inside components, both spatial and related to multi-scale dimensions (i.e., lattice structures, etc.).

Because of the study of Design for Additive Manufacturing (DFAM) is just at the beginning, several studies are being made in order to identify the framework of application of these new methods [2, 3], considering the entire life cycle of a product, from the initial design strategy to the environmental sustainability compared to CNC applications [4]. Procedures based on such framework to guide the designer in the identification of opportunities, strategies, materials, technologies, shapes and design tools to come to feasible solutions are nowadays under investigation. For examples, some evaluation index to measure the compatibility and suitability of AM technologies have been proposed [5]. The principal challenge for a designer is the identification of components that with a freeform design create added value, justifying the investment in AM machines. Usually, components that can benefits from AM can be defined by following two different design strategies: the "manufacturing driven" and the "functional driven" strategy [6]. In the first the designer maintains the original form the object and the benefits derive from substitution of the manufacturing process, while in the second ones the function of the components, with the AM technology characteristics, guides the entire re-design process.

1.2 High-Purity Physics Applications

Because of AM can be considered as a process without the need of specific machines or tools, the production of an individual object with high customization can be made with relative low costs respect to subtractive processes. Consequently, these technologies are well suited to the field of basic and experimental physics research, where components production is based on unique prototype with high specialization and characteristics that cannot be found on the commercial market and realizable by the common manufactures. An example is the use of high-purity materials, both plastic and metal, for the manufacturing of components used in physics experiments with particular equipment like particle detectors. In this case, a low background is essential for measurements of rare or unknown experimental events. The components can have different characteristics and can be simple objects like frame, support or functional ones such as heat exchanger.

In this paper, an AM design process of a heat exchanger for the condensation of argon using liquid nitrogen under cryogenics condition is proposed. The development of the condenser is part of the activities related to the "DarkSide" experiment [7–9]. Nowadays, cosmological and astrophysical research support in an overwhelmingly way the existence of dark matter as a yet unknown form of matter with high energy density. The research involved Weakly Interacting Massive Particles (WIMPs), whose presence would be indicated by the detection of a recoiling nucleus in a large volume detector. "DarkSide" will attempt to detect WIMP-induced nuclear recoils using two-phase depleted argon Time Projection Chambers (TPCs). Argon TPCs measure both the deposited energy and the ionization charge produced by particle interactions in their liquid active volumes. The biggest challenge is the suppression of the rate of background events to below the very low WIMP interaction rates (a few events per ton-year) to which current dark matter experiments are sensitive. For this reason, "DarkSide" detectors use materials with lowest possible levels of intrinsic radioactivity [10].

1.3 Powder Bed Fusion of Pure Copper

Thanks to the "Hub for Additive Manufacturing, Materials, Engineering and Re-search" (HAMMER) [11] of "National Institute of Nuclear Physics" (Istituto Nazionale di Fisica Nucleare - INFN), it was possible to design the argon condenser considering the powder bed fusion with Selective Laser Melting (SLM) technology of pure copper powder. This is a 99.8% pure copper that responds well to the radioactive purity requirements of physics experiments like "DarkSide". This material is widely used in cryogenics and thanks to the small a content of extraneous elements, the properties of elemental copper, like high ductility, electrical and thermal conductivity and low relative volatility under high vacuum are enhanced. On the other hand, this material has technological issues related to the powder bed fusion process that are still under investigation [12] like the high reflectivity with the use of conventional SLM laser sources and the remarkable thermal conductivity. These characteristics reduce the energy available for the melting, producing final components with highly porous and low relative density which reduce the excellent properties of the above-mentioned elemental copper. The existing literature propose only few positive works about the powder bed fusion of this pure material [13, 14]. Some of them have reach a density well above of 90% but still distant from the typical values of other metallic materials close to 99,8%.

The aim of this work is the is the use of pure copper for the production with AM of a technologically advanced component, like a heat exchanger. The design approach takes into account the re-design of the geometrical shape of a classic heat exchanger for gas condensation, considering the well-known dimensional, functional and performance constrains, together with a focus on the technological issues of the processing of a pure material with SLM, which can be extended to other applications. The first part describes the preliminary design of gas condenser which take into account the technical and functional requirements, while the second and the third ones, address the re-design of the geometrical shape with the 3D modelling and the design validation by means Computer Fluid Dynamics (CFD) simulation of the argon condensation. Eventually, the manufacturing of the first condenser prototype is presented together with the SLM technological parameters used in the printing process and the results of the preliminary experimental tests.

2 Argon Condenser Development

2.1 Functional Requirements

The design workflow of a component always starts with the list of requirements which specify the technical functionalities and establish the main performance to achieve. Usually, the requirements can be dived into internal and external properties: the first ones describe technical aspects (i.e., material properties, dimensional tolerance, etc.) while the second ones refer to functionalities of component and are directly connected to the operation (i.e., general performance, reliability, safety, etc.).

In our case study, we can summarize all the necessary condenser requirements with the data listed in Table 1. Essentially, they concern the outlet temperature of the argon and nitrogen at the specific operative pressure (1.5 bar), the condensate flow rate (L/h), the main dimensions (see Fig. 1) and general needs like: a coupling surface to easily connect the condenser with an external cryocooler, necessary to initially cool down the system to cryogenic temperatures and some holes to install electrical resistances to be used for the fine tuning of the overall temperature.

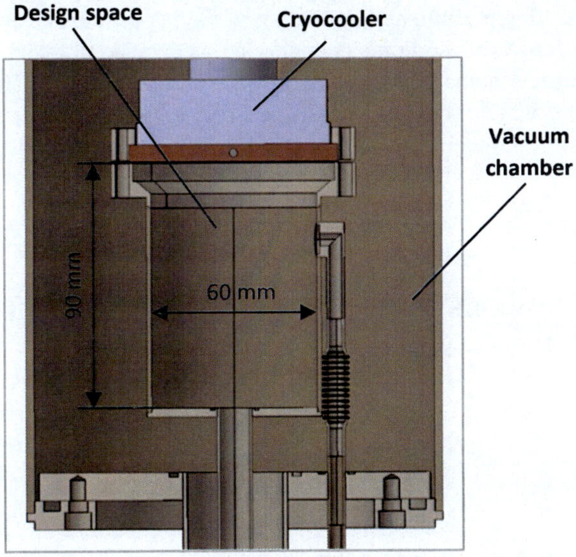

Fig. 1. Design space for the condenser: the image shows the inside of the vacuum chamber with the main dimensions for the installation. Mechanical constraints were not taken into account at this preliminary stage.

Table 1. Functional requirements of the condenser.

N.	Functional requirements	
1	Operative pressure [bar]	1.5
2	Argon outlet temperature* [K]	83.8 ÷ 87.3
3	Nitrogen outlet temperature* [K]	63.2 ÷ 78.2
4	Argon condensate flow rate [L/h]	≥1
5	Dimensions [mm]	Ø60 × 90
6	Coupling surface for an external cryocooler	
7	Holes for the installation of electrical resistances	

(*) Range of temperature for the existence of the argon and nitrogen in the liquid phase at the operating pressure of 1.5 bar.

2.2 Preliminary Design

The aim of a heat exchange design is to realize the heat exchange between two fluids when are known their mass flow rate, their inlet temperatures and the outlet temperature of one of the two. Therefore, the calculation is based on the selection of a heat exchanger type and on the determination of the exchange area necessary to obtain the desired outlet temperature. To a preliminary estimation of the total exchange area required and, consequently, the length of the fluids channel, it was decided to adopt the well-known linear heat exchanger scheme with countercurrent flows (see Fig. 2). The linear scheme allowed to evaluate which geometry it was possible to adopt considering the vacuum chamber dimensions and the estimated length (L) of the heat exchanger.

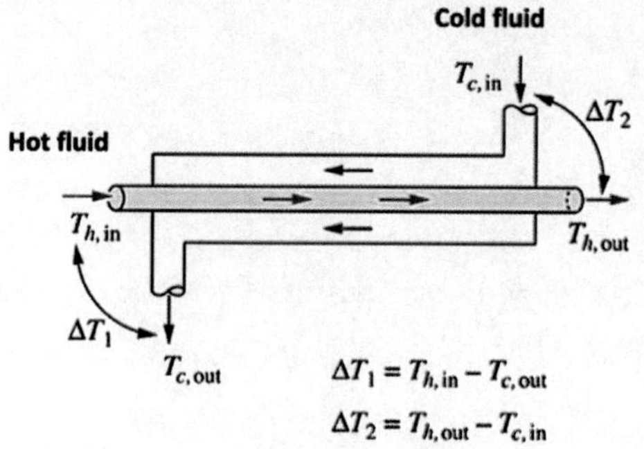

Fig. 2. Linear countercurrent flow heat exchanger scheme [15]

The preliminary design has considered the following assumptions: stationary operating conditions, constant fluids properties (see Table 2), fully developed flows, adiabatic conditions and no variation of kinetic and potential energy.

Table 2. Physical properties of the fluids: nitrogen (liquid phase) and argon (gas phase).

	Nitrogen (T = 77 K)	Argon (T = 300 K)
Density [kg/m^3]	806	1.6
Specific Heat [J/Kg K]	2042	523
Thermal Conductivity [W/m K]	0.1458	0.0177
Dynamic viscosity [Pa s]	1.61 E-04	3.77 E-05
Cinematic viscosity [m^2/s]	1.99 E-07	2.36 E-05

The channels diameter was assumed the same for both fluids and equal to 5 mm with a wall thickness of 1 mm. Due to the high conductivity of copper and the small wall thickness, it was possible to neglect the resistance of the copper and consider the areas of the channels surface equal. The input data and the results are listed in Table 3 and 4.

Table 3. Preliminary design – Input data.

Input data	Nitrogen	Argon
Inlet temperature [K]	77.0	300.0
Outlet temperature [K]	*Unknown*	86.0
Volumetric flow rate [L/min.]	40	20

Table 4. Preliminary design – Results.

Results	
Nitrogen outlet temperature [K]	77.1
Heat exchange length [mm]	486

2.3 Geometrical Shape Re-design

Results about the preliminary design, like the heat exchange length (L), together with the space design available for the definition of the final shape of the condenser, suggest two main well-known solutions: the first one is the "vertical tubes condenser" (see Fig. 3a) and the second one is the "helical coils condenser" (see Fig. 3b). Both would guarantee a correct design considering the functional requirements. The benefit of the

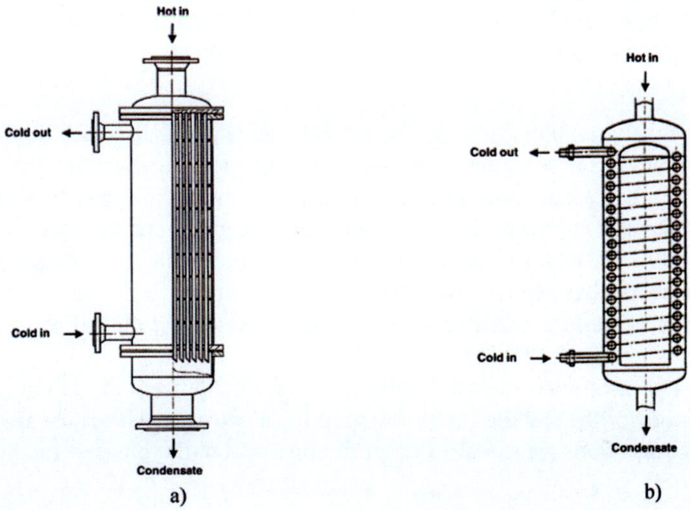

Fig. 3. "Vertical tube condenser" a) and "helical coils condenser" b) schemes [16].

helical condenser respect to the vertical one can be summarized with the maximization of the heat exchange thanks to the centrifugal effect and the turbulence that is established within the fluid vein: this lead to a smaller length pipes respect to the preliminary design indications.

Considering these suggestions, the re-design of the geometrical shape of the condenser starts from the helical solution. Because of there are no more manufacturing constrains due to the standard CNC technologies, the fluid channels project was carried out by reversing the empty space of the usual solution with the full one: in the classic helical condenser, the pipes are produced bending a rough pipe until the desired geometry is reached, then they are installed inside the chamber where the heat exchange between the fluid takes place. In this case, the AM technologies, gives the opportunities to build a new helical pipes geometry directly embedded in the copper volume matrix (see Fig. 4a and 4b). The condenser was modeled with CATIA v5 software.

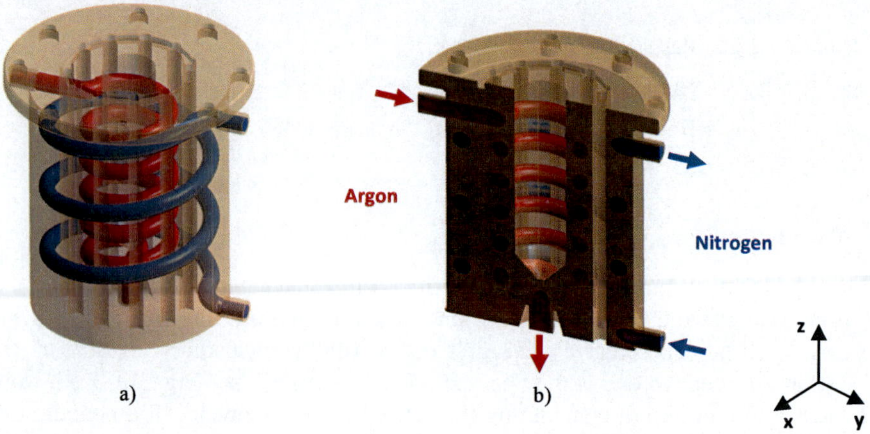

Fig. 4. New geometrical design of the condenser: a) 3D model and b) vertical section. The red arrows indicate the argon flow and the blue the nitrogen one. Nominal dimensions of condenser are height 75 mm, flange diameter Ø 70 mm and body Ø 60 mm.

From the functional point of view, the helical cold channel, in which liquid nitrogen flows, has the purpose of bringing the copper walls to cryogenic temperature; afterwards the helical hot channel has the aim of bringing the argon from gas to liquid phase. Accordingly, the heat exchange takes place between the argon and the highly conductive copper solid wall. The angle of the helical channels respect to the build direction (z-axis) of the SLM machine has been defined with the aim of not creating supports inside them. In Table 5 are reported the main geometrical characteristics of the helical geometry of the condenser channels for both fluids.

Moreover, several holes circularly arranged were made from the top to the bottom of the condenser for the installation of the required electrical resistances, together with the external connections for the welding of the inlet and outlet pipes of the fluids.

Table 5. Geometrical characteristics of the helical channels.

	Nitrogen channel	Argon channel
Helical height [mm]	57.5	51
Helical pitch [mm]	20	10
Helical diameter [mm]	53	35
Helical length [mm]	438	425
Helical number	3	5
Diameter [mm]	5	5

2.4 CFD Simulations

CFD steady simulations were carried out with ANSYS Fluent software. The analysis has considered several different aspects, like the implementation of a multiphase model (VOF - Volume of Fluid method) with surface tension modelling for the simulation of argon condensation: gaseous argon was considered as the first phase, while liquid argon the second one. Energy and viscous model (k-ε) were also applied. Regarding the boundary conditions, the flow rate for the inlet channels (20 ÷ 40 L/min.) and the pressure for the outlet ones (1.5 bar) were set. The external condenser walls have been considered as adiabatic surfaces and the operative conditions included also the gravity.

The trend of temperature for the fluids inside the helical channels is shown in Fig. 5 and 6, while the numerical results are listed in Table 6.

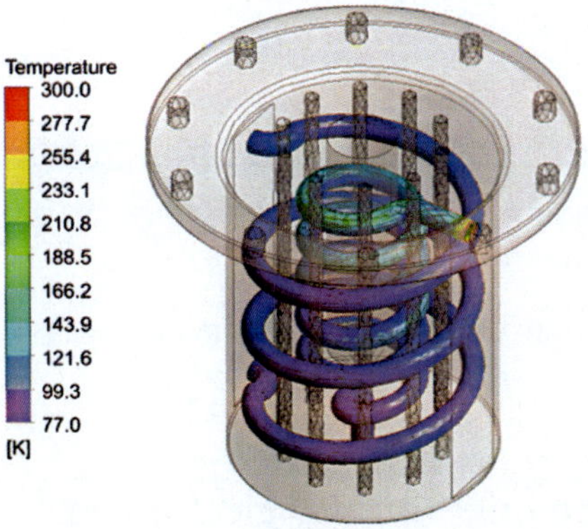

Fig. 5. Map of temperature for both fluids. Flow rate: Argon 20 L/min., Nitrogen 40 L/min.

Data points refer to selected points within the fluid domain along the helical axis. The plot shows how in all situations the argon condensation temperature is reached. This fact proves that the condenser design guarantees a good regulation during the operative conditions by the control of the flow rate and in addition by means the electric resistances: in this kind of application, the main problems is the freezing of argon or nitrogen gasification with consequent abrupt increase of its volume. The estimated condensed argon flow rate is well above the desired 1 L/h.

Table 6. CFD simulation results: the argon flow rate was changed from 15 to 25 L/min., while that of nitrogen was kept constant (40 L/min.) during each run.

Fluid	Inlet flow rate [L/min.]	Inlet temperature [K]	Outlet temperature [K]	Condensate flow rate [L/h]
Argon	15	300.0	82.8	1.5
	20	300.0	84.6	2.3
	25	300.0	86.6	2.2
Nitrogen	40	77.0	77.1	/

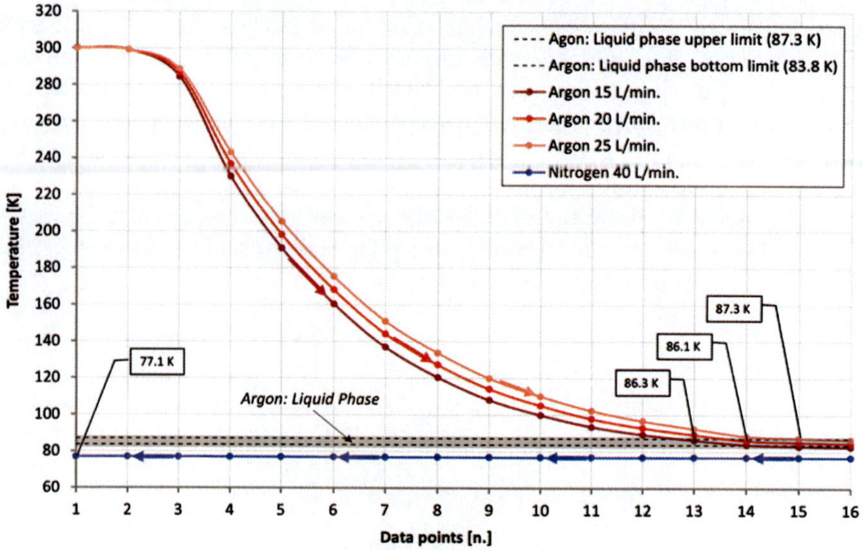

Fig. 6. CFD simulation results: trend of temperature for both fluids inside the helical channels.

3 Prototype Production: Technological Issues

The first prototype of the argon condenser was manufactured with a MYSINT 100 RM SLM machine model produce by SISMA S.p.A (see Fig. 7).

The machine is equipped with a building volume of Ø 100 mm × 100 mm, a fiber laser source of 200 W and a laser spot diameter of 30 μm. It can work with the following

Fig. 7. First prototype of the argon condenser produced by SLM process.

inert gases: Nitrogen, Argon and Helium. The SLM machine parameters used for the 3D printing of the condenser are listed in Table 7. As an initial reference, it has been taken previous works with similar laser power and characteristics [14]. For the printing, a pure copper powder (99.8%), produced by the LWP company, with a grain size of 5 ÷ 32 μm, has been used. The expected copper density should be approximately 90%. The slicing elaboration process has been carried out by Materialize MAGICS software.

Table 7. MYSINT100 RM SLM machine parameters for the 3D printing of the condenser.

Parameter	Value	Energy density
P - Laser Power [W]	175	E = P/SVH 218.8 J/mm^3
S - Layer size [μm]	20	
V - Speed Scanning [mm/s]	1000	
H - Hatch distance [μm]	40	
Inert Gas	Nitrogen	

As said in the introduction, pure copper has technological issues related to the powder bed fusion process connected to its high reflectivity with conventional laser sources and its remarkable thermal conductivity. These aspects reduce the energy available for the melting process producing components with a highly porous state.

For this reason, experimental tests were aimed at verifying how this residual porosity can really affect the functionality of the condenser. Therefore, a water bath test with pressured air (about 6 bar) was arranged. Figure 8 clearly shows the leak of the air from the condenser walls.

This simple experimental evidence demonstrates how a material density of about 90% is still not sufficient to guarantee a perfect seal of that kind of components manufactured

Fig. 8. Experimental water bath tests: compressed air flow inside the argon helical channel.

with the SLM process. This evidence also affects other functional problems, such as the impossibility of bringing the component to a sufficient vacuum level for some physical experimental applications.

To overcome these technological issues related to the powder bed fusion by means SLM technique, some solutions could be applied. For example, it could be possible to further optimize the machine parameters such as those affects applied the energy density (see Table 7) like the speed scanning (the speed reduction should facilitate the achievement of a higher density) and those which influence the heat exchange during the melting process (i.e., scanning strategies, type of inert gas and gas recirculation rate speed). Also, it could be possible to evaluate the use of a different laser source (i.e., green laser light) and the application of specific post heat treatments or the use of metal or copper alloys powders compatible with high-purity physics applications. Eventually, in order to prevent the leaks, it could be possible to apply a surface coating with pure materials (i.e., gold film). In this last case, the relevant surface roughness of the copper components produced with SLM (about 15 μm) [17] would favor the process. Some of these proposed solutions will be taken into consideration in the future developments of the work.

4 Conclusions

In this paper, an Additive Manufacturing (AM) design process of a heat exchanger for the condensation of argon has been presented. The process described the AM approach from the preliminary design to the rethink of the geometrical shape of component. Technical and functional requirements were considered, and the validation of the new design was carried out by dedicated Computer Fluid Dynamics (CFD) simulations. The process shows the great freedom that the AM technologies gives to the designer especially in experimental physics research, where components are unique prototype with high specialization that cannot be found on the commercial market. The achieved results show how it is possible to use high-purity material, like copper, for the production with AM of an advanced component, overcoming the high reflectivity and the low energy density absorption. Usually, literature reports printing example with this material only for small specimens with simple geometrical shape (i.e., cubes, etc.). Moreover, the research demonstrates the feasibility of printing a complex and massive geometry overcoming some the common technological AM issues like the boundary effects when printing components reach the size of the building volume and the minimization of supports.

Unfortunately, it was found that there are still technological issues related to the low-density and the high internal porosity of printing pure copper, which is still well below 99.0%. This aspect produced in the heat exchanger leaks of fluid from the internal material structure. Eventually, some further solutions to overcome this issue has been proposed like further manufacturing parameter optimization, post heat treatments, different metal powders and surface coating. However, the results achieved can be considered a good starting point for future investigations.

References

1. Diegel, O., Nordin, A., Motte, D.: A Practical Guide to Design for Additive Manufacturing. Springer, Singapore (2019). https://doi.org/10.1007/978-981-13-8281-9
2. Alfaify, A., Saleh, M., Abdullah, F.M., Al-Ahmari, A.M.: Design for additive manufacturing: a systematic review. Sustainability **12**, 7936 (2020)
3. Vaneker, T., Bernard, A., Moroni, G., Gibson, I., Zhang, Y.: Design for additive manufacturing: framework and methodology. CIRP Ann. **69**(2), 578–599 (2020)
4. Gebisa, A.W., Lemu, H.G.: Design for manufacturing to design for additive manufacturing: analysis of implications for design optimality and product sustainability. Procedia Manuf. **13**, 724–731 (2017)
5. Lettori, J., Raffaeli, R., Peruzzini, M., Schmidt, J., Pellicciari, M.: Additive manufacturing adoption in product design: an overview from literature and industry. Procedia Manuf. **51**, 655–662 (2020)
6. Klahn, K., Leutenecker, B., Meboldt, M.: Design strategies for the process of additive manufacturing. Procedia CIRP **36**, 230–235 (2015)
7. Picciau, E., et al.: Latest results of dark matter detection with the DarkSide experiment. In: Proceedings of IFAE (2020)
8. Agner, P., et al.: First results from the DarkSide-50 dark matter experiment at Laboratori Nazionali del Gran Sasso. Phys. Lett. B **743**, 456–466 (2015)
9. Meyers, P.D., et al.: DarkSide-50: a WIMP search with a two-phase argon TPC. Phys. Procedia **61**, 124–129 (2015)
10. Wright, A.: The DarkSide program at LNGS. In: Proceedings of 13th ICATPP Conference (2011)
11. Hub for Additive Manufacturing, Materials Engineering and Research (HAMMER) website Homepage. https://hammer.lngs.infn.it
12. Tran, T.Q., et al.: 3D printing of highly pure copper. Metals **9**, 756 (2019)
13. Colopi, M., Caprio, L., Demir, A.G., Previtali, B.: Selective laser melting of pure Cu with a 1 kW single mode fiber laser. Procedia CIRP **74**, 59–63 (2018)
14. Lykov, P.A., Safonov, E.V., Akhmedianov, A.M.: Selective laser melting of copper. Mater. Sci. Forum **843**, 284–288 (2016)
15. Yunus, A.Ç.: Thermodynamics and Heat Transfer. McGraw-Hill, New York (2009)
16. Nitsche, M., Gbadamosi, R.O.: Heat Exchanger Design Guide. Elsevier, Amsterdam (2016)
17. Mazzei, A.: Surface quality analysis of oxygen-free copper components made using SLM technologies (Analisi della qualità superciale di componenti in rame oxygen-free realizzati mediante tecnologie SLM). Bachelor's degree thesis in Industrial Engineering, University of L'Aquila (2019)

A Further Investigation Toward the Design of Topology Optimized Solid-Lattice Hybrid Structures for Biomedical Applications

Chiara de Crescenzo[1(✉)], Maria Richetta[1], Massimo Martorelli[2], Antonio Gloria[3], and Antonio Lanzotti[2]

[1] Department of Industrial Engineering, University of Rome Tor Vergata, Rome, Italy
[2] Department of Industrial Engineering, Fraunhofer JL IDEAS, University of Naples Federico II, Naples, Italy
[3] Institute of Polymers, Composites and Biomaterials – National Research Council of Italy, Naples, Italy

Abstract. Titanium alloys (e.g., Ti6Al4V) have been widely considered for the design of biomedical implants. To avoid stress shielding effects, bone atrophy and implant loosening bone, 3D porous devices with controlled geometry and architecture should represent a promising solution. Several cellular structures were already investigated to obtain a wide range of mechanical properties. Many studies focused on the mechanical performance of diamond and body-centered-cubic. Different kinds of porous and semi-porous femoral stems were also proposed and analyzed. Accordingly, the aim of the current research was to provide further insight into the design of solid-lattice hybrid structures through a two-step process involving the classical and lattice topology optimization. A cementless femoral stem was considered as a case study. The solid isotropic material with penalization (SIMP) was used at varying values of the penalty factor and the effect of the geometrical features of each beam forming the lattice structure was also determined. Differences were found in terms of functional and structural performances according to the selected strategy for the design of the solid-lattice hybrid structures, as a consequence of the material distribution/layout and geometrical features.

Keywords: Topology optimization · Solid-lattice hybrid structures · Biomedical applications

1 Introduction

Titanium alloys (e.g., Ti6Al4V) have been often considered to develop orthopedic and dental implants, benefiting from their excellent biocompatibility, good corrosion resistance, and high strength [1–4].

As an example, total hip replacement generally involves the use of cementless femoral stems, which are proximally inserted in the medullary canal of the femur. The stem must be firmly attached to the tone in order to ensure the physiological stress transfer to the bone during daily activities. In this scenario, Ti6Al4V (elastic modulus of

C. Rizzi et al. (Eds.): ADM 2021, LNME, pp. 514–523, 2022.
https://doi.org/10.1007/978-3-030-91234-5_52

114 GPa) has been frequently selected to fabricate cementless stems [1–4], also causing several complications [1, 4, 5]. In particular, stress shielding effects, bone atrophy and implant loosening are related to devices fabricated using a material with an elastic modulus which is much higher than that of the cortical bone (i.e., 10–30 GPa) [1, 4, 5].

Finite element (FE) models may be successfully used to study the biomechanical behavior of natural tissues and implants [6].

The great need of increasing and optimizing the implant life, as a consequence of population ageing, diffusion of extreme sports, revision surgery and costs of implant replacement, has led to the development of long-lasting implants, also using materials with mechanical properties which are close to those of human tissues [1–4].

Composite and porous implants were also designed with the aim to overcome the above reported drawbacks [7–11]. The use of novel materials and designs through the integration of porous structures, several geometrical profiles and functionally graded materials allowed for the development of strategies to reduce the implant stiffness [12–14].

Contextually, 3D porous devices with controlled geometry and architecture were designed, also trying to promote the possibility of tissue ingrowth, which should stabilize the implant [1]. In recent years, some studies have also reported novel designs of porous and semi-porous stems for reducing the stiffness of the stems [6, 15–17]. It has been frequently reported how the use of low stiffness porous stems provides lower stress shielding and bone resorption in comparison to the dense stems [18].

Several cellular structures were analyzed to obtain a wide range of mechanical properties. Concerning torsional, bending and compression, excellent mechanical performances were found for diamond and body-centered-cubic (BCC) structures [6, 14, 19]. For this reason, the diamond and BCC types were further employed to generate porous and semi-porous femoral stems, which were then properly analyzed.

Several investigations have reported that porous cellular microstructures possess a fatigue strength which is a fraction to that of the effective yield strength (about $0.3\sigma_{ys}$) of the cellular microstructure [13–19].

A recent study on a novel design, analysis and 3D printing of Ti6Al4V alloy bio-inspired porous femoral stem was performed using a wide range of porosity for the inner core [6].

Accordingly, the aim of the current study was to provide further insight into the design of solid-lattice hybrid structures with optimized properties for biomedical applications. A cementless femoral stem was considered as a case study.

2 Materials and Methods

SolidWorks®2017 (Dassault Systemes, Paris, France) computer-aided design (CAD) system was used to create 3D models of a hip stem. A total hip replacement using an alumina head and Ti6Al4V stem was considered (Fig. 1).

A Young's modulus of 114 GPa and a Poisson's ratio of 0.33 were used for Ti6Al4V alloy [1].

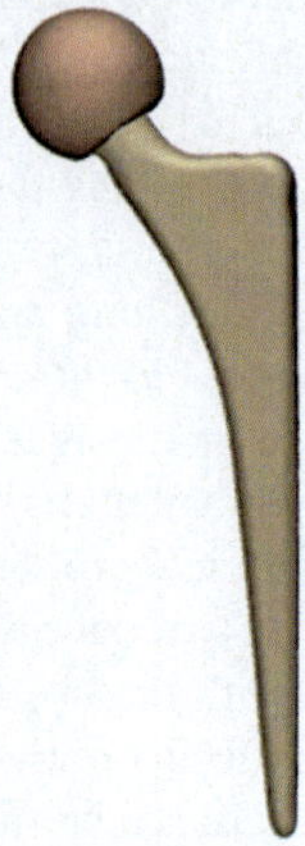

Fig. 1. Total hip replacement using alumina head and Ti6Al4V stem: 3D model.

As reported in a previous work [6], the bodyweight was assumed as 800 N and a wide range of physiological loads was considered.

3D models of the femur with implant were also designed. With regard to the femur bone, cortical bone and cancellous bone were considered A Young's modulus of 17 GPa and a Poisson's ratio of 0.30 were used for cortical bone, whereas values of 0.52 GPa and 0.29 were considered for cancellous bone [20].

The entire model of the femur with the implant was imported into HyperMesh® (HyperWorks®, Altair Engineering Inc., Troy, MI, USA). A 3D mesh was suitably generated, adequate mesh size and mesh refinement techniques were employed. Further technical features for the investigated model were the total number of structural grids [10,340], elements excluding contact [34,142], and degrees of freedom [28,818]. Moreover, "freeze" type elements were used as the contact condition.

Finite element analysis (FEA) was performed on the models. The von Mises stress distributions were evaluated in the different component of the model.

The stem compliance was evaluated for the starting model. As reported in the literature, the compliance (C) is evaluated using the vectors of applied loads (**f**) and nodal displacements (**u**) and the stiffness matrix (**K**), as well as the strain (ε) and stress (σ) vectors.

The starting model was properly optimized from a topological point of view for the given boundary conditions, constraints and loading condition, focusing on the compliance and a material mass fraction of 0.30 and 0.50.

An integrated approach involving classical and lattice topology optimization was used to generate different technical solutions.

As an example, benefiting from our previous studies (data not shown), a specific lattice type (i.e., "LT 3" in the lattice optimization panel - HyperWorks®, Altair Engineering Inc., Troy, MI, USA) was individuated consisting of beams with a fixed radius to length ratio, constant radius (i.e., 0.1 mm) or variable radius and tapered formulation with circular cross sections (i.e., two different radii at the nodes, with values spanning in a range from 0.02 to 0.2 mm).

The topological optimization problems were solved using a density method known as the solid isotropic material with penalization (SIMP), where the density of an element is generally correlated to its stiffness using the following equation:

$$\mathbf{K}(\rho) = \rho^p \mathbf{K}_0 \tag{1}$$

$\mathbf{K}(\rho)$ and $\mathbf{K_0}$ are the penalized stiffness matrix of an element and the real stiffness matrix of an element (i.e., when the element "is filled up" with material; the stiffness of the material of the initial design space), whilst ρ represents the (relative) density and p is the penalization factor.

A typical formulation of topology optimization problems was first considered with a special focus on material distribution for the design space and on the equivalent density of each element as a design variable. In the classical topology optimization, the elements with high density are replaced by full dense matter, whereas the elements with low density are replaced by void. The mid-density elements may be interpreted as matter or void and the aim should be to limit their amount.

Differently from a classical topology optimization approach, in the lattice topology optimization the elements with intermediate density are replaced by lattice structure.

A preliminary (classical) topology optimization was integrated with lattice topology optimization. The penalty factor controls the intermediate densities, also defining the stiffness of the element. With regard to the lattice topology optimization, the penalty factor (p) was also set at 1.0, 1.25 and 1.8, also exploring high, medium and low values of porosity.

3 Results and Discussion

The increasing applications of additive manufacturing technologies in the biomedical field have currently led to product reimagination from a new standpoint.

Additive manufacturing allows to reproduce complex shapes and to design customized prostheses with tailored and improved properties as well as advanced scaffolds for tissue engineering [1].

In general, the performances of scaffolds and prostheses can be an obstacle in terms of satisfaction of the functional requirements. Concerning the manufacturing process, the relationship among the process parameters, microstructure and mechanical properties is also crucial and involves innovative and traditional fabrication techniques [1, 2, 21, 22]. The knowledge of such relationship is clearly fundamental in designing functional prostheses and scaffolds for the regeneration of complex tissues.

In this context, selective laser melting is widely analyzed to develop metal components and lattice structures as lightweight devices with enhanced functional properties for biomedical applications [1]. Ti6Al4V lattice components were manufactured by selective laser melting and then analyzed in terms of compressive properties, surface roughness, microhardness, and dimensional accuracy [5]. A further theoretical-experimental analysis was also performed on selective laser melted Ti6Al4V lattice structures [1].

Over the past few years, many efforts have been devoted to develop strategies for finding the design which is best suited to the requirements (i.e., "optimum" design). Topology optimization may be clearly considered as a tool to implement the design process, providing advanced prostheses and scaffolds, which may most efficiently be fabricated using additive manufacturing techniques. In particular, topology optimization represents a method which is able to optimize the material layout within a defined design space, for a given set of loads, constraints and boundary conditions, with the aim of improving the performance of the device.

The classical topology optimization does not generally take into account buckling, fatigue or further nonlinear kinds of analysis [23]. Moreover, the conventional topology optimization still presents some limitations even with the use of additive manufacturing technologies. One important limitation is represented by the fact that the optimized design often changes the original shape, unless the designer employs some constraints which may be able to "freeze" the external shape of the structure [23].

In addition, the discretization of the design space in elements with relative densities between 1 (solid) and 0 (void) leads to large intermediate relative densities that the conventional algorithms cannot take into advantage. Furthermore, the optimized design may contain overhang structures, also leading to the need for a redesign [23].

On the other hand, nowadays lattice structures in conjunction with additive manufacturing technologies are of great interest because of their design flexibility and mechanical properties [23].

The lattice structure is a kind of cellular structure with interconnected struts and nodes in a 3D space [17]. The lattice structure allows for a more regular configuration in comparison to normal cellular materials. Thus, the mechanical and physical properties of the lattice structure are usually more tailorable and controllable. As an example, the mechanical properties of homogeneous lattice structures may be tailored by changing the topology or the relative density. In the case of heterogeneous lattice structures, it results possible to optimize the mechanical properties by properly tailoring the material distribution in the lattice space [17]. For this reason, the lattice structure has been increasingly considered in the design of scaffolds for tissue engineering and prostheses.

Accordingly, in the current study a further insight into the design and analysis of solid-lattice hybrid structures was reported for biomedical applications. A cementless femoral stem was selected as a case study.

As an example, at a fixed loading condition (i.e., 2400 N, contact forces on the hip joint as 300% of the bodyweight [6]) the results obtained from a preliminary topology optimization (mass fraction of 0.50) for the analyzed femoral stem are reported in Fig. 2, in terms of contour plot of element densities.

In terms of element densities, values closest to 1 are related to important elements from a structural point of view, whereas small density elements may be considered not structurally relevant.

In order to avoid a reduction of the number of low density elements, which does not generally preserve the integrity of the external shape of the stem, the classical topology optimization was integrated with a further lattice topology optimization process.

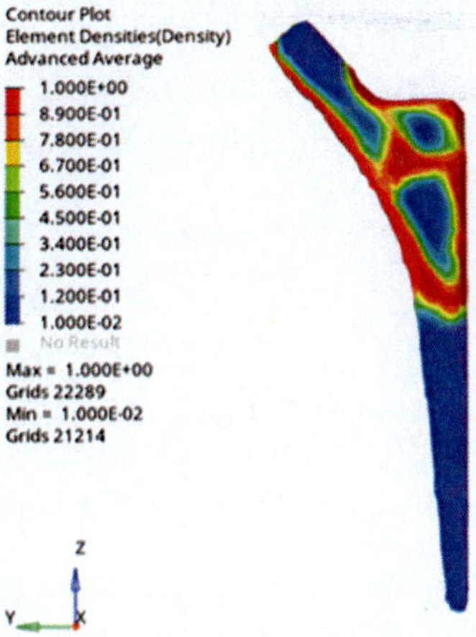

Fig. 2. Preliminary topology optimization: contour plot of element densities.

Starting from the results in terms of element densities (Fig. 2), a lattice topology with specific parameters was defined. The beams represent the elements giving stiffness to the structure, the nodes are the locations where the beams are jointed.

Specifically, a defined lattice type with a tapered beam formulation was selected and the penalty factor (p) was set to 1.0, 1.25 and 1.8 (Fig. 3).

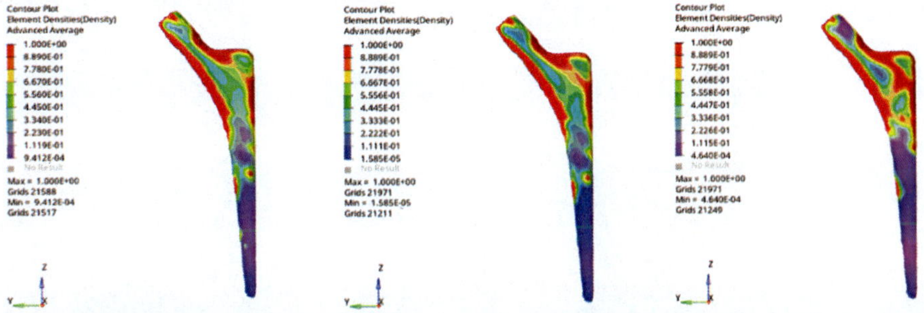

Fig. 3. Lattice topology optimization process: contour plot of element densities for p = 1.0 (left), p = 1.25 (center) and p = 1.8 (right).

The results from the further lattice topology optimization led to different technical solutions (Fig. 4).

Even though the mass fraction was fixed (0.50), the solutions obtained for the three cases (p = 1.0, 1.25, 1.8) provided some differences in terms of material distribution/layout within the design space.

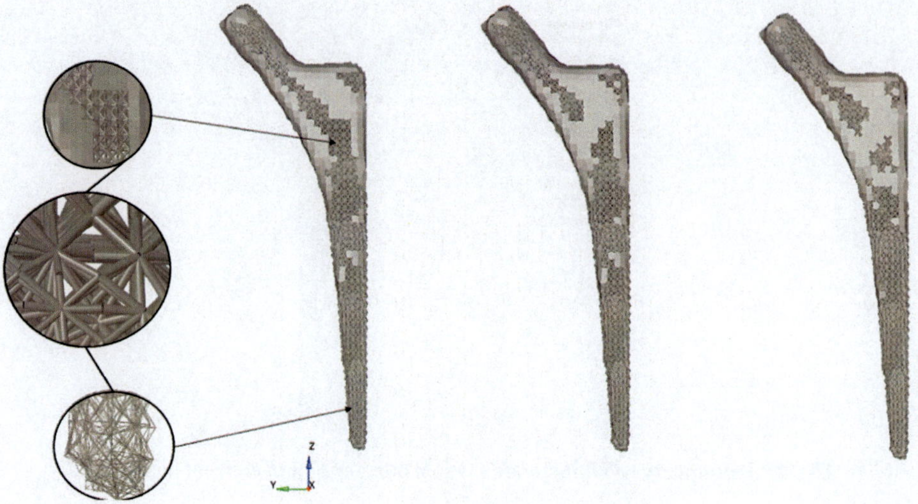

Fig. 4. Lattice topology optimization: example of the generated solid-lattice hybrid structure for p = 1.0 (left), p = 1.25 (center) and p = 1.8 (right). As an example, the insets (left) show some regions of the lattice parts characterized by a tapered beam formulation and beams with constant values of radius.

Differently from the above reported case, when the radius of beam of the lattice was constant and equal to 0.1 mm, at a fixed value of the radius to length ratio, no differences were found in terms of material distribution/layout and compliance value at varying values of the penalty factor (1.0, 1.25, 1.8) (Fig. 5).

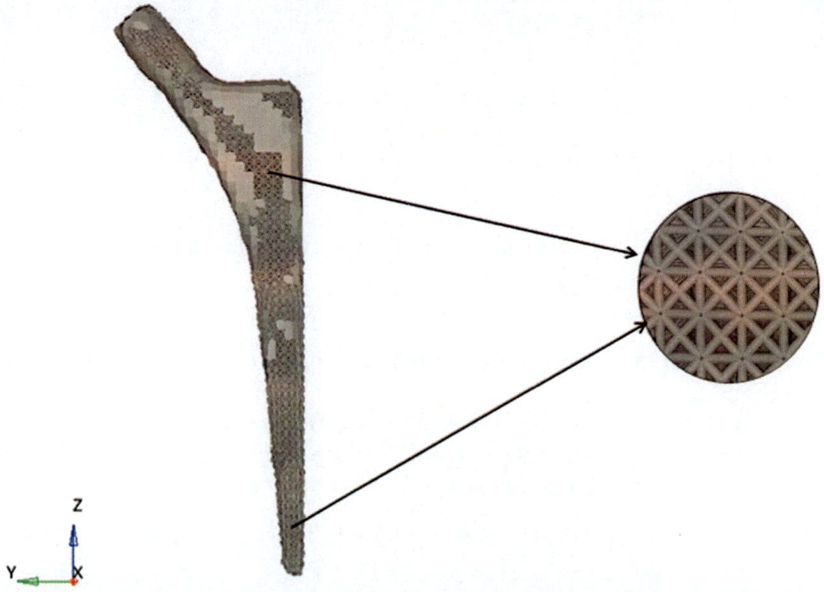

Fig. 5. Lattice topology optimization: an example of the generated solid-lattice hybrid structure in the case of a constant beam radius (0.1 mm) and a fixed radius to length ratio.

However, with regard to the FE re-analysis, considering the von Mises stress distributions, the solid-lattice hybrid structures generated using the tapered beam formulation generally showed maximum values (e.g., 446.9 and 804.1 MPa) in the lattice part, which were higher than $0.3\sigma_{ys}$ as well as than those obtained using a constant radius (i.e., 0.1 mm) and a fixed radius to length ratio for the beam. Specifically, in such latter case maximum stress values of 103.6 MPa and 87.5 MPa were found in the solid and lattice part, respectively.

On the other hand, in the case of topology optimization where the constraint on mass fraction was set to 0.30, in the lattice part of the hybrid structure the tapered beam formulation led to stress values which were much more higher than those achieved for the corresponding case at a mass fraction of 0.50. However, the generated solid-lattice hybrid structures using a constant radius of 0.1 mm and a fixed radius to length ratio for the beam showed a maximum stress value of 136.0 MPa for the solid part, whereas 299.0 MPa was achieved in the lattice part.

4 Conclusions

The current research provided a further investigation toward the re-design of device for biomedical applications. Briefly, a design strategy of cementless femoral stems with optimized properties was reported. Classical and lattice topology optimization were properly explored and integrated, with some focus on the step related to a lattice size optimization, trying to determine the effect of the "formulation" of each beam forming the lattice structure. At a fixed mass fraction (0.50 or 0.30), a constant radius and a fixed radius to length for the beam, variable radius and tapered formulation with circular cross sections were also explored at varying values of the penalty factor.

As demonstrated by the FE re-analysis, regarding the von Mises stress distributions in the solid and lattice part of the hybrid structures, differences were found according to the mass fraction (i.e., 0.50 or 0.30) and beam formulation (tapered or constant radius with a fixed radius to length). In particular, when the mass fraction was set to 0.50, in the lattice part of the hybrid structure the tapered beam formulation led to: i) stress values which were higher than those obtained using a constant radius (i.e., 0.1 mm) and a fixed radius to length ratio for the beam; ii) stress values which were much more lower than those found for the corresponding case at a mass fraction of 0.30.

With regard to a constant radius of 0.1 mm and a fixed radius to length ratio for the beam, at a mass fraction of 0.30 the stress values evaluated in the solid and lattice part were higher than those assessed for a mass fraction of 0.50. However, differently from the corresponding case at a mass fraction of 0.50, when the mass fraction was set to 0.30 higher values of von Mises stress were achieved in the lattice part in comparison to those found in the solid part.

Furthermore, in the case of tapered beam formulation, the solutions found for the different values of the penalty factor evidenced some differences in terms of material distribution/layout within the design space.

Anyway, even if many cases were theoretically explored, only few examples were reported in the current investigation with the aim to focus on a strategy for the development of lightweight devices with optimized properties. The same methodology may be clearly revised and in many cases extended taking into account the manufacturability constraints according to the selected technology.

In this scenario, design for manufacturability (often known as design for manufacturing) plays an important role as it represents the general engineering practice of designing devices in such a way that they are easy to manufacture.

Even though this concept exists in the different engineering disciplines, its implementation clearly differs depending upon the selected manufacturing technology. Several formulations of the design problem may be possible and should be introduced in the different features of the fabrication process. Based on the objectives of the problem, the design formulation may consist of a single problem or a sequential problem according to the selected manufacturing process. In general, due to the problem complexity, further solution approaches may be also considered.

The lattice type with its geometrical features and cell parameters could be "in conflict" with the selected manufacturing technology (e.g., selective laser melting, electron beam melting), as restrictions could exist in terms of both the minimum size of a cell in a lattice structure and the size of the device. The designer should solve the given problems creating feasible solutions rather than exploring novel perspectives on the analyzed problems. For this reason, strategies have to be adopted in order to redefine the problem exploration space for the design innovation. Thus, among the technical solutions which are best suited to the requirements, the "optimum" design should be found taking into consideration the manufacturing constraints due to the adopted technology.

The current research may be also considered as the first step of a future complex study, also aiming at the strict correlation between the stress shielding in the femur and the several solid lattice hybrid structures which may be generated and explored, in order to provide re-design guidelines.

References

1. Maietta, S., Gloria, A., Improta, G., Richetta, M., De Santis, R., Martorelli, M.: A further analysis on ti6al4v lattice structures manufactured by selective laser melting. J. Healthc. Eng. **2019**, 3212594 (2019)
2. Zhu, M., Xu, G., Zhou, M., Yuan, Q., Tian, J., Hu, H.: Effects of tempering on the microstructure and properties of a high-strength bainite rail steel with good toughness. Metals **8**(7), 484 (2018)
3. Campanelli, S., Contuzzi, N., Ludovico, A., Caiazzo, F., Cardaropoli, F., Sergi, V.: Manufacturing and characterization of Ti6Al4V lattice components manufactured by selective laser melting. Materials **7**(6), 4803–4822 (2014)
4. Bandyopadhyay, A., Espana, F., Balla, V.K., Bose, S., Ohgami, Y., Davies, N.M.: Influence of porosity on mechanical properties and in vivo response of Ti6Al4V implants. Acta Biomater. **6**(4), 1640–1648 (2010)
5. Spoerke, E.D., Murray, N.G., Brinson, L.C., Li, H., Dunand, D.C., Stupp, S.I.: A bioactive titanium foam scaffold for bone repair. Acta Biomater. **1**(5), 523–533 (2005)
6. Mehboob, H., et al.: A novel design, analysis and 3D printing of Ti-6Al-4V alloy bio-inspired porous femoral stem. J. Mater. Sci.: Mater. Med. **31**(9), 1–14 (2020). Article no. 78
7. Mehboob, H., Chang, S.-H.: Application of composites to orthopedic prostheses for effective bone healing: a review. Compos. Struct. **118**, 328–341 (2014)

8. Sridhar, I., Adie, P.P., Ghista, D.N.: Optimal design of customised hip prosthesis using fiber reinforced polymer composites. Mater. Des. **31**(27), 67–75 (2010)
9. Rezaei, F., Hassani, K., Solhjoei, N., Karimi, A.: Carbon/PEEK composite materials as an alternative for stainless steel/titanium hip prosthesis: a finite element study. Australas. Phys. Eng. Sci. Med. **38**(4), 569–580 (2015). https://doi.org/10.1007/s13246-015-0380-3
10. Oshkour, A.A., Abu Osman, N.A., Yau, Y.H., Tarlochan, F., Wan Abas, W.A.B.: Design of new generation femoral prostheses using functionally graded materials: a finite element analysis. Proc. Inst. Mech. Eng. H J. Eng. Med. **227**(1), 3–17 (2013). https://doi.org/10.1177/0954411912459421
11. Ramos, A., Completo, A., Relvas, C., Simoes, J.A.: Design process of a novel cemented hip femoral stem concept. Mater. Des. **33**, 313–321 (2012)
12. Limmahakhun, S., Oloyede, A., Sitthiseripratip, K., Xiao, Y., Yan, C.: 3D-printed cellular structures for bone biomimetic implants. Addit. Manuf. **15**, 93–101 (2017)
13. Kadirgama, K., Harun, W.S.W., Tarlochan, F., Samykano, M., Ramasamy, D., Azir, M.Z., et al.: Statistical and optimize of lattice structures with selective laser melting (SLM) of Ti6Al4V material. Int. J. Adv. Manuf. Technol. **97**, 495–510 (2018)
14. Mehboob, H., Tarlochan, F., Mehboob, A., Chang, S.H.: Finite element modelling and characterization of 3D cellular microstructures for the design of a cementless biomimetic porous hip stem. Mater Des. **149**, 101–112 (2018)
15. Gonzalez, F.J.Q., Nuno, N.: Finite element modelling approaches for well-ordered porous metallic materials for orthopaedic applications: cost effectiveness and geometrical considerations. Comput. Methods Biomech. Biomed. Eng. **19**, 845–885 (2016)
16. Kim, C.-H., Lee, S.J., Aditya, K., Kim, H.Y., Yoon, K.S., Yoon, P.W.: Incidence of a stem sitting proud of a proximally coated cementless tapered wedge stem. J. Orthop. Transl. **19**, 118–125 (2019)
17. Dong, G., Tang, Y., Li, D., Zhao, Y.F.: Design and optimization of solid lattice hybrid structures fabricated by additive manufacturing. Addit. Manuf. **33**, 101116 (2020)
18. Arabnejad, S., Johnston, B., Tanzer, M., Pasini, D.: Fully porous 3D printed titanium femoral stem to reduce stress-shielding following total hip arthroplasty. J. Orthop. Res. **35**, 1774–1783 (2017)
19. Wang, L., Kang, J.F., Sun, C.N., Li, D.C., Cao, Y., Jin, Z.M.: Mapping porous microstructures to yield desired mechanical properties for application in 3D printed bone scaffolds and orthopaedic implants. Mater Des. **133**, 62–68 (2017)
20. Chethan, K.N., Shyamasunder Bhat, N., Zuber, M., Satish Shenoy, B.: Finite element analysis of different hip implant designs along with femur under static loading conditions. J. Biomed. Phys. Eng. **9**(5), 507–516 (2019)
21. Fiorese, E., Bonollo, F., Battaglia, E.: A tool for predicting the effect of the plunger motion profile on the static properties of aluminium high pressure die cast components. Metals **8**(10), 798 (2018)
22. Jeon, G., Kim, K., Moon, J.-H., Lee, C., Kim, W.J., Kim, S.: Effect of Al 6061 alloy compositions on mechanical properties of the automotive steering knuckle made by novel casting process. Metals **8**(10), 857 (2018)
23. Tyflopoulos, E., Steinert, M.: A comparative study between traditional topology optimization and lattice optimization for additive manufacturing. Mat. Des. Process Comm. **2**, e128 (2020)

A Preliminary Analysis of the Effects of Process Parameters on the Impact Resistance of 3D Printed PETG and HIPS

Massimo Martorelli[1], Vito Gallicchio[1,2](✉), Antonio Gloria[3], and Antonio Lanzotti[1]

[1] Department of Industrial Engineering, Fraunhofer JL IDEAS, University of Naples Federico II, Naples, Italy
vito.gallicchio@unina.it

[2] Department of Neurosciences, Reproductive and Odontostomatological Sciences, University of Naples Federico II, Naples, Italy

[3] Institute of Polymers, Composites and Biomaterials, National Research Council of Italy, Naples, Italy

Abstract. Additive manufacturing (AM) allows the development of novel and customized products with tailored properties. However, the application of extrusion-based AM techniques (i.e., fused deposition modeling – FDM) in the design of functional parts is often limited because of the poor mechanical performance as a consequence of the nature of the process to build the object in a layer-by-layer fashion. In the current study, the impact resistance of the 3D printed polyethylene terephthalate glycol-modified (PETG) and high impact polystyrene (HIPS) was evaluated as a function of three printing parameters (i.e., printing temperature, layer height and line width). Izod-type test specimens were fabricated and analyzed according to the ASTM D256. The contribution of each factor was properly analyzed. The results also indicated that the printing temperature was the most significant parameter for the impact resistance of 3D printed PETG and HIPS. The obtained findings may be considered as valid only within the limit of parameters and ranges investigated in the current study.

Keywords: Fused deposition modeling · Design of experiments · Impact resistance

1 Introduction

The need to increase flexibility and speed up the design and fabrication process of novel products has led to the development of rapid technologies, including the additive manufacturing (AM) techniques. The design and manufacture steps during a new product development process are critical because any decisions at this point can have a great impact on the final cost and quality of the product [1].

Additive manufacturing (AM) has the potential of revolutionizing the way parts are made. It allows the direct manufacturing via tool-less manufacturing while reducing the cost related to the product development and material wastage [2, 3].

C. Rizzi et al. (Eds.): ADM 2021, LNME, pp. 524–534, 2022.
https://doi.org/10.1007/978-3-030-91234-5_53

Extrusion based AM techniques involve the use of a thermoplastic material to fabricate a 3D component. The Scott Crump developed and patented the extrusion based AM technique which was introduced to the market as fused deposition modeling (FDM). This term and the related acronym were trademarked by Stratasys Inc. As a consequence of the expiration of the essential patents, the RepRap community started the open-source revolution of the material extrusion-based technology, thus leading to fused filament fabrication (FFF). FFF may be considered as a largely employed AM technique due to a wide range of available materials, low-cost equipment, a great variety of equipment and user-friendliness [2].

FFF is a 3D printing process where a continuous filament of a thermoplastic material is extruded through a heated nozzle and then deposited on the build platform for creating the part in a layer-by-layer fashion [2, 4].

Over the past years, FFF has played a fundamental role in enabling rapid prototyping to develop novel product concepts. However, the poor mechanical performance of the FFF-printed parts has strongly limited the production of finished products.

Moreover, FFF is not able to benefit from the total advantage of the raw materials in terms of mechanical properties. This may be ascribed to the fact that the raw material is extruded as rasters deposited in rows and layers. No extra pressure is employed to bond the deposited rasters and only a short time compression occurs during the deposition stage [2, 5].

As reported in the literature, great efforts have been done for improving the mechanical properties of the built part with a special focus on the possibility of tailoring the printing parameters. Specifically, a poor impact resistance was found for the manufactured parts, thus limiting their use. The impact performance of the 3D printed parts was also compared to that of the injection-molded parts. Contextually, some investigations were performed to enhance the impact performance of the 3D print components.

Specific core–shell polymer filaments were studied evidencing the effect of the amount of the shell material on the impact resistance [2, 6], whereas an experimental analysis on the mechanical behavior of 3D printed and injection molded acrylonitrile butadiene styrene (ABS) parts was carried out [7].The effect of the nozzle temperature and layer height and on the impact strength of the polypropylene (PP) was also investigated [8].

Moreover, the effect of the layer height and bed temperature on the impact resistance of 3D printed PLA parts was demonstrated [9]. The effect of raster angle, build orientation and infill pattern was also analyzed considering different materials (e.g., Nylon 12, PLA, ABS, polycarbonate (PC)-ABS, Ultem) [10–19].

Further works already reported the influence of several FDM build parameters (e.g., build orientation, infill density, layer thickness) on the properties of 3D printed parts using mostly ABS and PLA [13–15, 17, 20–25], especially using ABS and PLA.

Many authors generally considered an infill density of 100% and reported [20] the effect of the infill on the resistance to the deformation.

The change in the impact energy of 3D printed specimens was also investigated through Izod impact tests, using several polymers and build orientations [17].

Recently, the effect of raster angle, layer height and raster width on the impact resistance of 3D PLA parts printed using an open source FFF-based 3D printer was properly investigated [2].

On the other hand, the impact resistance of further polymers such as polyethylene terephthalate glycol-modified (PETG) and high impact polystyrene (HIPS) may play an important role in several applications.

For this reason, the aim of the current study was to analyze the effect of three printing parameters (printing temperature- PT, layer height - LH and line width - LW) on the Izod impact resistance of 3D printed PETG and HIPS parts.

2 Materials and Methods

SolidWorks®2017 (Dassault Systemes, Paris, France) computer-aided design (CAD) system was used to create 3D models of Izod-type test specimens, according to the ASTM D256 [26].

The notched specimens were fabricated by Fused Deposition Modeling, using a commercially available 3D printer (Creality3D Ender-3 PRO), polyethylene terephthalate glycol (PETG, FILOALFA®) and high impact polystyrene (HIPS, FILOALFA®).

The test specimens were fabricated varying building conditions. In particular, the specimens were printed at a fixed raster angle (i.e., bead deposition with respect to the x-axis of the printing bed) of 0°, whereas three printing parameters (i.e., printing temperature - PT, layer height – LH, line width – LW) were varied at three levels to assess their influence on the impact resistance. The printing parameters and their changes along with the fixed parameters are reported in Tables 1 and 2.

Table 1. List of control and held factors.

Factor	Type	Value
PT(°C)	Control	Table 2
LH (μm)	Control	Table 2
LW (μm)	Control	Table 2
Bed temperature (°C)	Held-constant	70
Printing speed (mm/s)	Held-constant	60
Wall thickness (μm)	Held-constant	800
Bottom/Top thickness (μm)	Held-constant	700
Infill density (%)	Held-constant	80
Raster angle (°)	Held-constant	0
Infill pattern	Held-constant	Linear

Table 2. Three levels selected for the control factors.

Control factor	Levels		
	1	2	3
PT(°C)	215	230	245
LH (μm)	100	200	300
LW (μm)	500	600	700

As reported in previous study on PLA [2], the impact tests were carried out on specimens (i.e. notched rectangular blocks - 63.5 mm × 12.7 mm × 12.7 mm) which were held as a vertical cantilever beam and broken by a single swing of the pendulum. The specimen was positioned so that the notch was on the side facing the pendulum. During the test, the striking edge made initial contact with the specimen on a line 22 mm above the top surface of the vise (Fig. 1).

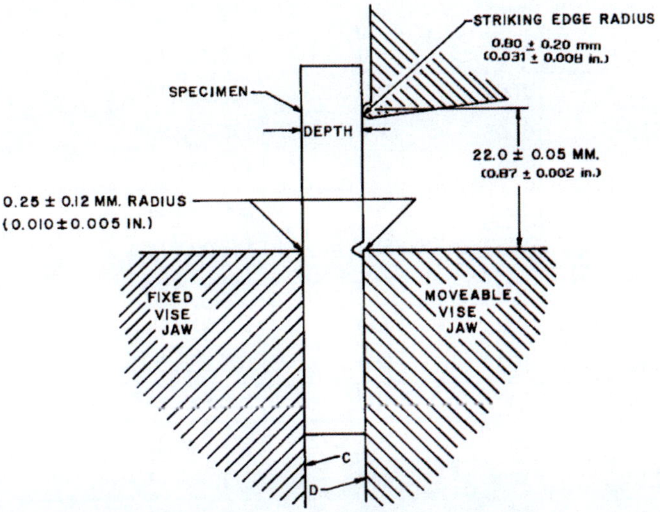

Fig. 1. Schematic representation of fixed and movable vise jaw, specimen, and striking edge to each other for Izod Test Method according to the ASTM D256 [26].

A 5.42 J hammer [2] was employed to break all 3D printed PETG and HIPS specimens. The breaking energy for the specimen and the impact resistance (J/m) were measured.

A full factorial design was adopted in the current research, based on three printing parameters at the 3 levels having 27 experimental runs. In the case of both PETG and HIPS, three replicates were planned for each treatment, resulting in a total of 81 specimens and, hence, experimental runs.

3 Results and Discussions

The influence of the several parameters on the impact resistance of the 3D printed PETG and HIPS specimens was analyzed.

Tables 3 and 4 show the experimental results obtained for the employed printing conditions, in terms of impact resistance (mean value ± standard deviation) and signal-to-noise ratio (SNR).

Table 3. Experimental results obtained for PETG through 3^3 full factorial design: impact resistance (mean value ± standard deviation) and signal-to-noise ratio (SNR).

Run	PT (°C)	LH (μm)	LW (μm)	Impact resistance (J/m)	SNR
1	215	100	500	132.5 ± 11.5	42.4
2	215	100	600	157.6 ± 8.3	43.9
3	215	100	700	130.4 ± 12.9	42.2
4	215	200	500	169.1 ± 6.8	44.5
5	215	200	600	184.3 ± 11.2	45.3
6	215	200	700	210.1 ± 18.2	46.4
7	215	300	500	175.2 ± 9.9	44.8
8	215	300	600	182.6 ± 4.2	45.2
9	215	300	700	200.2 ± 3.4	46.0
10	230	100	500	152.2 ± 12.4	43.6
11	230	100	600	161.6 ± 8.5	44.1
12	230	100	700	152.0 ± 15.8	43.5
13	230	200	500	188.9 ± 10.1	45.5
14	230	200	600	190.6 ± 10.0	45.6
15	230	200	700	229.7 ± 20.1	46.8
16	230	300	500	211.0 ± 13.1	46.4
17	230	300	600	220.0 ± 11.0	46.8
18	230	300	700	241.1 ± 4.5	47.6
19	245	100	500	200.3 ± 13.8	46.0
20	245	100	600	242.6 ± 8.0	47.7
21	245	100	700	199.9 ± 18.0	45.9
22	245	200	500	251.2 ± 13.1	48.0
23	245	200	600	280.6 ± 14.4	48.9
24	245	200	700	323.0 ± 14.0	50.2
25	245	300	500	261.2 ± 12.9	48.3
26	245	300	600	273.6 ± 4.9	48.7
27	245	300	700	352.0 ± 5.0	50.9

Table 4. Experimental results obtained for HIPS through 3^3 full factorial design: impact resistance (mean value ± standard deviation) and signal-to-noise ratio (SNR).

Run	PT (°C)	LH (μm)	LW (μm)	Impact resistance (J/m)	SNR
1	215	100	500	138.3 ± 10.4	42.8
2	215	100	600	154.8 ± 7.2	43.8
3	215	100	700	133.7 ± 13.1	42.4
4	215	200	500	170.5 ± 6.2	44.6
5	215	200	600	187.3 ± 10.1	45.4
6	215	200	700	211.3 ± 17.4	46.4
7	215	300	500	181.1 ± 9.5	45.1
8	215	300	600	187.2 ± 3.6	45.4
9	215	300	700	206.8 ± 2.2	46.3
10	230	100	500	159.8 ± 11.1	44.0
11	230	100	600	171.0 ± 8.9	44.6
12	230	100	700	160.0 ± 14.9	44.0
13	230	200	500	192.3 ± 9.3	45.6
14	230	200	600	194.0 ± 9.7	45.7
15	230	200	700	231.1 ± 19.2	47.2
16	230	300	500	213.4 ± 12.2	46.6
17	230	300	600	227.2 ± 9.9	47.1
18	230	300	700	244.0 ± 4.2	47.7
19	245	100	500	201.2 ± 14.9	46.0
20	245	100	600	237.0 ± 7.4	47.5
21	245	100	700	204.7 ± 17.5	46.2
22	245	200	500	262.3 ± 12.2	48.4
23	245	200	600	281.4 ± 15.2	49.0
24	245	200	700	325.2 ± 13.5	50.2
25	245	300	500	265.8 ± 13.2	48.5
26	245	300	600	275.6 ± 5.7	48.8
27	245	300	700	353.3 ± 4.8	51.0

Figures 2 and 3 reports the mean value of SNR for each of the control factors at each level. In the current research, the larger-the-better criterion was selected to find the optimal printing condition, which should provide the highest impact resistance as the ideal situation.

Considering the criteria of the highest impact resistance and larger-the-better for SNR, Figs. 2 and 3 were used for the determination of the optimal set of printing parameters from the employed experimental design.

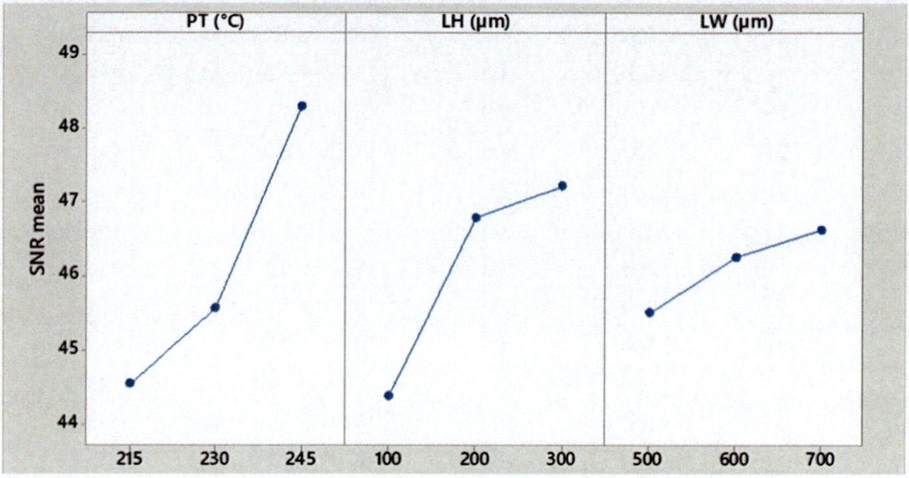

Fig. 2. Main effects plot of SNR for PETG impact resistance.

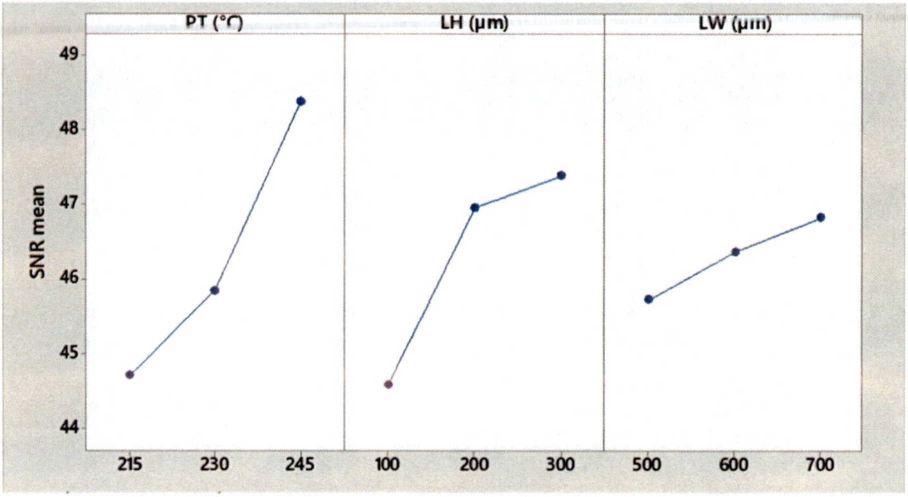

Fig. 3. Main effects plot of SNR for HIPS impact resistance.

Thus, the main effects plots show the optimum values of the control factors to achieve the highest impact resistance. In the case of both PETG and HIPS, the highest impact resistance was achieved for the printing temperature (PT) at level 3 (245 °C), layer height (LH) at level 3 (300 μm) and infill line width (LW) at level 3 (700 μm). It appears clear that the printing temperature (PT) has the highest impact, whereas the line width (LW) has the lowest impact for both PETG and HIPS.

The analysis of variance (ANOVA) was also used to corroborate the obtained findings as well as to assess the percentage contribution of each involved factor. The significance level was set at 5%. Thus, the significance of the control factors was also determined. Therefore, based on the p-values, the printing temperature (PT), layer height (LH) and line width (LW) were significant parameters for both 3D printed PETG and HIPS (Tables 5 and 6).

Table 5. ANOVA table for PETG impact resistance.

Source	DF	Adj SS	Adj MS	p-value	% Contribution
PT (°C)	2	129513	64756.5	0.000	53.6
LH (μm)	2	66030	33015.2	0.000	27.3
LW (μm)	2	13800	6899.8	0.000	5.7
PT × LH	4	4152	1037.9	0.001	1.7
PT × LW	4	3538	884.4	0.002	1.5
LH × LW	4	13114	3278.5	0.000	5.4
Error	62	11517	185.8		4.8
Total	80	241664			

Table 6. ANOVA table for HIPS impact resistance.

Source	DF	Adj SS	Adj MS	p-value	% Contribution
PT (°C)	2	124876	62438.0	0.000	53.2
LH (μm)	2	67494	33746.9	0.000	28.7
LW (μm)	2	13607	6803.5	0.000	5.8
PT × LH	4	4348	1087.0	0.000	1.8
PT × LW	4	2765	691.2	0.005	1.2
LH × LW	4	11554	2888.6	0.000	4.9
Error	62	10316	166.4		4.4
Total	80	234960			

Looking at the ANOVA Tables 5 and 6, it is worth noting the magnitude of the influence of the parameters. Specifically, the contribution of the printing temperature (PT), layer height (LH) and line width (LW) for maximizing the PETG impact resistance was 53.6%, 27.3% and 5.7%, respectively. In the case of HIPS, the contribution of PT, LH and LW was 53.2%, 28.7% and 5.8%, respectively.

Moreover, the (PT × LH), (PT × LW) and (LH × LW) interactions were also significant.

However, the analysis confirmed that the most significant parameter for the impact resistance was the printing temperature (PT).

The obtained results evidenced that for both PETG and HIPS specimens the impact resistance generally increased as the printing temperature increased from 215 to 245 °C.

An increase in the printing temperature clearly facilitates the adhesion of a newly deposited filament on a previously deposited layer, thus resulting in an increased bond strength and mechanical properties. However, it is also worth remembering that in general there is a limit up to which the printing temperature may be increased without negatively affecting the mechanical properties.

An increase of the layer height from 100 to 300 μm usually provided an increase of the impact resistance. A higher impact resistance was found for a layer height of 300 μm, whereas a lower impact strength was achieved for both PETG and HIPS specimens in the case of a layer height of 100 μm. This effect could be probably ascribed to the fact that thicker layers (i.e., higher values of layer thickness) may offer more resistance to the impact blow delivered to the test specimen by means of a pendulum-type hammer. In addition, the presence of thicker layers improves the interface bonding between layers, thus enhancing the impact resistance. By increasing the layer height, a reduction in the number of the deposited layers is obtained, thus reducing the number of bonding lines between the layers which may represent a weak point for the 3D printed systems. Accordingly, it has been frequently demonstrated how the presence of more joining line and voids may cause a premature failure resulting in a lower value of impact resistance [2].

An improvement in the impact strength was also evident by increasing the infill line width, as the higher and lower values were found for LW of 700 and 500 μm, respectively. The increase in line width usually provides more resistance to the impact blow. The obtained findings were also consistent with those already reported by previous works [2, 27], which demonstrated a higher impact resistance as a consequence of a higher bonding strength between the adjacent rasters at a higher infill line width.

4 Conclusions

In the current study, three printing parameters (i.e., PT, LH, LW) were explored to analyze their effect on the Izod impact resistance of 3D printed PETG and HIPS.

The analysis of the experimental results through the SNR allowed to optimize the response and was in agreement with the ANOVA results which further corroborated the experimental investigation. In particular, the contribution of PT, LH and LW for maximizing the PETG impact resistance was 53.6%, 27.3% and 5.7%, respectively. With regard to HIPS, the contribution of PT, LH and LW was 53.2%, 28.7% and 5.8%, respectively. The (PT × LH), (PT × LW) and (LH × LW) interactions were also significant. The results indicated that PT was the most significant parameter for the impact resistance of 3D printed PETG and HIPS.

Anyway, the present study may be also considered as a first step of a future research in which other values of PT, LH and LW as well as further control factors will be considered, thus implementing the design of experiments.

References

1. Medellin-Castillo, H.I., Zaragoza-Siqueiros, J.: Design and manufacturing strategies for fused deposition modelling in additive manufacturing. Chin. J. Mech. Eng. **32**(53) (2019)
2. Rajpurohit, S.R., Dave, H.K.: Impact strength of 3D printed PLA using open source FFF-based 3D printer. Prog. Addit. Manuf. **6**(1), 119–131 (2020). https://doi.org/10.1007/s40964-020-00150-6
3. Bamiduro, O., Owolabi, G., Haile, M.A., Riddick, J.C.: The influence of load direction, microstructure, raster orientation on the quasi-static response of fused deposition modeling ABS. Rapid Prototyp. J. **25**, 462–472 (2019)
4. McLouth, T.D., Severino, J.V., Adams, P.M., Patel, D.N., Zaldivar, R.J.: The impact of print orientation and raster pattern on fracture toughness in additively manufactured ABS. Addit. Manuf. **18**, 103–109 (2017)
5. Costa, A.E., da Silva, A.F., Carneiro, O.S.: A study on extruded filament bonding in fused filament fabrication. Rapid Prototyp. J. **25**(3), 555–565 (2019)
6. Peng, F., Zhao, Z., Xia, X., Cakmak, M., Vogt, B.D.: Enhanced impact resistance of three-dimensional-printed parts with structured filaments. ACS Appl. Mater. Interfaces **10**(18), 16087–16094 (2018)
7. Dawoud, M., Taha, I., Ebeid, S.J.: Mechanical behaviour of ABS: an experimental study using FDM and injection moulding techniques. J. Manuf. Process. **21**, 39–45 (2016)
8. Wang, L., Gardner, D.J.: Effect of fused layer modelling (FLM) processing parameters on impact strength of cellular polypropylene. Polymer **113**, 74–80 (2017)
9. Wang, L., Gramlich, W.M., Gardner, D.J.: Improving the impact strength of Poly (lactic acid) (PLA) in fused layer modelling (FLM). Polymer **114**, 242–248 (2017)
10. Upadhyay, K., Dwivedi, R., Singh, A.K.: Determination and comparison of the anisotropic strengths of fused deposition modeling P400 ABS. In: Wimpenny, D., Pandey, P., Kumar, L. (eds.) Advances in 3D Printing and Additive Manufacturing Technologies, pp. 9–28. Springer, Singapore (2017). https://doi.org/10.1007/978-981-10-0812-2_2
11. Kamoona, S.N., Masood, S.H., Mohamed, O.A.: An investigation on impact resistance of FDM processed Nylon-12 parts using response surface methodology. AIP Conf. Proc. **1859**(1), 020120 (2017)
12. Górski, F.I., Kuczko, W.I., Wichniarek, R.A.: Impact strength of ABS parts manufactured using Fused Deposition Modeling technology. Arch. Mech. Technol. Autom. **31**(1), 3–12 (2014)
13. Es-Said, O.S., Foyos, J., Noorani, R., Mendelson, M., Marloth, R., Pregger, B.A.: Effect of layer orientation on mechanical properties of rapid prototyped samples. Mater. Manuf. Process. **15**(1), 107–122 (2000)
14. Ziemian, C., Sharma, M., Ziemian, S.: Anisotropic mechanical properties of ABS parts fabricated by fused deposition modelling. In: Gokcek, M. (ed.) Mechanical Engineering, pp. 159–180. InTech Open United Kingdom (2012)
15. Tsouknidas, A., Pantazopoulos, M., Katsoulis, I., Fasnakis, D., Maropoulos, S., Michailidis, N.: Impact absorption capacity of 3D-printed components fabricated by fused deposition modelling. Mater. Des. **102**, 41–44 (2016)
16. Fatimatuzahraa, A.W., Farahaina, B., Yusoff, W.A.: The effect of employing different raster orientations on the mechanical properties and microstructure of Fused Deposition Modeling parts. In: IEEE Symposium on Business, Engineering and Industrial Applications (ISBEIA) 2011, pp. 22–27(2011)
17. Roberson, D.A., Perez, A.R., Shemelya, C.M., Rivera, A., MacDonald, E., Wicker, R.B.: Comparison of stress concentrator fabrication for 3D printed polymeric izod impact test specimens. Addit. Manuf. **7**, 1–11 (2015)

18. Gurrala, P.K., Regalla, S.P.: Part strength evolution with bonding between filaments in fused deposition modelling. Virtual Phys. Prototyp. **9**(3), 141–149 (2014)
19. Sood, A.K., Ohdar, R.K., Mahapatra, S.S.: Parametric appraisal of mechanical property of fused deposition modelling processed parts. Mater. Des. **31**(1), 287–295 (2010)
20. Torres, J., Cotelo, J., Karl, J., Gordon, A.P.: Mechanical property optimization of FDM PLA in shear with multiple objectives. JOM **67**(5), 1183–1193 (2015). https://doi.org/10.1007/s11837-015-1367-y
21. Torres, J., et al.: An approach for mechanical property optimization of fused deposition modeling with polylactic acid via design of experiments. Rapid Prototyp. J. **22**(2), 387–404 (2016)
22. Patterson, A.E., Pereira, T.R., Allison, J.T., Messimer, S.: IZOD impact properties of full-densityFDM polymer materials with respect to raster angle and print orientation. J. Mech. Eng. Sci. **203–210**, 1–13 (2019)
23. Caminero, M., Chacon, J., Garcia-Moreno, I., Rodriguez, G.: Impact damage resistance of 3D printed continuous fibre reinforced thermoplastic composites using fused deposition modelling (2018)
24. Lanzotti, A., Grasso, M., Staiano, G., Martorelli, M.: The impact of process parameters on mechanical properties of parts fabricated in PLA with an open-source 3-D printer. Rapid Prototyp. J. **21**, 604–617 (2015)
25. Vega, V.: Effect of layer orientation on the mechanical properties and microstructure of a polymer. J. Mater. Eng. Perform. **20**(6), 978–988 (2011)
26. ASTM D256-10: Standard Test Methods for Determining the Izod Pendulum Impact Resistance of Plastics
27. Coogan, T.J., Kazmer, D.O.: Bond and part strength in fused deposition modeling. Rapid Prototyp. J. **23**, 414–422 (2017)

3D Printing of Prototypes Starting from Medical Imaging: A Liver Case Study

Robinson Guachi[1,2](✉), Michele Bici[1], Fabiano Bini[1], Marcelo Esteban Calispa[3], Cristina Oscullo[2], Lorena Guachi[2,4], Francesca Campana[1], and Franco Marinozzi[1]

[1] Department of Mechanical and Aerospace Engineering, DIMA – Sapienza University of Rome, 00184 Rome, Italy

[2] Department of Mechatronics, Universidad Internacional del Ecuador - UIDE, 170411, Av. Simón Bolívar, Quito, Ecuador

roguachigu@uide.edu.ec

[3] Department of Mechanical Engineering, Escuela Superior Politécnica de Chimborazo - ESPOCH, Panamericana Sur km 1 1/2, 060155 Riobamba, Ecuador

[4] SDAS Research Group, Ibarra, Ecuador

http://www.sdas-group.com

Abstract. Hepatic diseases are serious condition worldwide, and several times doctors analyse the situation and elaborates a preoperative planning based exclusively on the medical images, which are a drawback since they only provide a 2D vision and the location of the damaged tissues in the three-dimensional space cannot be easily determined by surgeons. Nowadays, with the advancement of Computer Aided Design (CAD) technologies and image segmentation, a digital liver model can be obtained to help understand the particular medical case; even with the geometric model, a virtual simulation can be elaborated. This work is divided into two phases; the first phase involves a workflow to create a liver geometrical model from medical images. Whereas the second phase provides a methodology to achieve liver prototype, using the technique of fused deposition modelling (FDM). The two stages determine and evaluate the most influencing parameters to make this design repeatable in different hepatic diseases. The reported case study provides a valuable method for optimizing preoperative plans for liver disease. In addition, the prototype built with additive manufacturing will allow the new doctors to speed up their learning curve, since they can manipulate the real geometry of the patient's liver with their hands.

Keywords: Image segmentation · Convolutional neural network · 3D printing · FDM · Liver disease

1 Introduction

In 2017, the number of adults diagnosed with liver disease was, approximately 4.5 million in USA [1], that means that 1.8% of adults have been diagnosed with liver disease. In many cases, surgery is a procedure widely used for facing a variety of liver

C. Rizzi et al. (Eds.): ADM 2021, LNME, pp. 535–545, 2022.
https://doi.org/10.1007/978-3-030-91234-5_54

pathologies [2]. In selected patients, the advent of new technologies, novel neo-adjuvant methods and liver regenerative properties have permitted effective resection. Although alternative techniques as percutaneous and laparoscopic procedures have been developed, enhanced and implemented over the past few years; liver surgery is still a complex operation involving a high mortality rate and risk of morbidity [3–5]. Surgical training and preoperative planning are required for liver surgery, particularly when minimally invasive surgery (MIS) procedures are involved, since the liver tissues of interest in MIS are not directly visualized by the surgeon during the procedures. Preoperative planning (based on 3D virtual models or rapid prototypes) plays an important role in understanding the liver's complex anatomical structure, identifying the damaged areas, determining the best path to insert the instruments while minimizing risk and thereby preserving the liver's functions [4, 6, 7].

Medical images from Computed Tomography (CT) and Magnetic Resonance Imaging (MRI) have been used in recent years as a valuable method for anatomical liver structure identification [4, 8], medical diagnosis and preoperative planning. But interpreting medical images is often a challenging task and needs specific skills and clinical experience. For this reason, some works in medical field have been done to support the work of the doctors. Therefore, several improvements were made, as proposed in [9–12], to identify, segment or classify the various tissues in healthy, diseased, benign or malignant tumoral. Nevertheless, the images are presented in 2D format, making difficult to distinguish tissues and organs involved. 3D modelling has become a modern toolkit for diagnosis, surgical tool design, preoperative planning, surgical training and simulation of surgical procedures with the use of enhanced medical images through the advancement of computational tools and CAD systems. A 3D geometrical model of anatomical structures obtained from medical images of patients provides details and highlights the location and distance between the organs and the different tissue. Using the additive manufacturing process (3D printing), the 3D geometric model can also be transferred from a virtual environment to a physical environment [13, 14]. Applications with silicone material and printable elastomers may also reproduce physical perception of the biological material. Different colours may increase the recognition of different elements (e.g., vascularization of an organ). Moreover, changing the 3D printing filling structure different levels of stiffness may be reproduced.

With the rapid diffusion and easy access to 3D printing desktop systems, prototypes of the 3D models can be obtained with low cost, with good accuracy and in a relatively short time [15]. This paper reports the fundamental set-up and the key aspects of 3D printing via a desktop system, for free shapes derived from Digital Imaging and Communications in Medicine (DICOM). According to the aim, Sect. 2 outlines a general workflow to obtain a prototype from medical images by 3D printing technology; Sect. 3 shows the experimental results considering the set-up previously. The results are discussed in Sect. 4 and, finally, conclusions are outlined in Sect. 5.

2 Workflow for Printing Patient-Specific 3D Model

Virtual geometrical models have become very helpful tools in the medical field. In Virtual simulations, Finite Element Analysis (FEA) is used to reproduce the interaction among surgical instruments and soft tissues (organs) [16, 17]. In addition, according to unique patient requirements, virtual simulations can be used as a tool to design new instruments. In Virtual surgery, it allows using realistic anatomical models to evaluate and simulate different surgical plans [18]. This procedure is often considered to be a tele-surgery simulator.

With 3D printing processes, [6, 13, 14], the realistic geometric model could help in identifying zone with diseases for surgery planning and also in improving manual skills during surgery. There is not a general workflow to obtain an accurate geometrical model so far, this document provides our experience with the creation of a patient-specific liver geometric model implemented in a case study, and the procedure is depicted in Fig. 1.

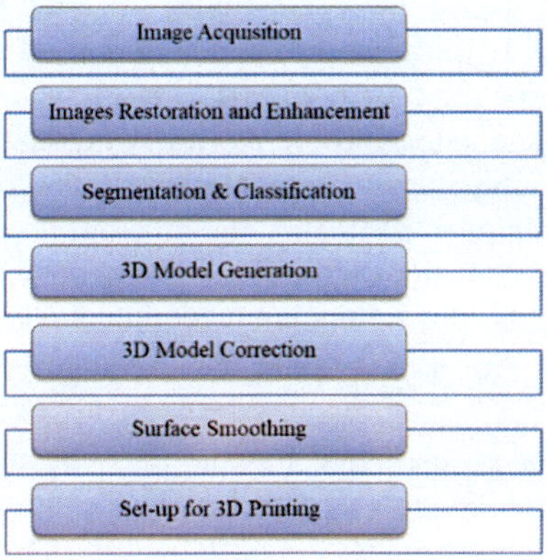

Fig. 1. Workflow for printing patient-specific 3D models

Image acquisition allows to create a visual representation of inner organs, bones and soft tissues. The most common techniques are the Magnetic Resonance Imaging (MRI), Ultrasound, and Computed Tomography (CT). Different techniques are applied depending on the internal area of interest, the interested tissues and the desired accuracy. The images are saved in the standard format called Digital Imaging and Communications in Medicine (DICOM) for the communication and management of medical imaging information and related data.

Pre-processing of the images, done through *Image Restoration and Enhancements*, facilitates the digitalization of the 3D models and it is also widely used to increase the automatic machine perception, avoiding image defects that may increase difficulties of understanding. Techniques used in image acquisition present some drawbacks due to the noise presence and to the equipment's own configuration. For this reason, pre-processing of the medical images should be done to recover image information from images degraded by noise and blur.

Segmentation and Classification aim at distinguishing objects of interest such as organs, cells, tissues, pathologies, among others, from background pixels. An effective segmentation extracts the shapes with their respective borderlines from the objects of interest present in an input sequence. Several semi-automatic and automatic methods exist for tissue segmentation purposes. Some software tools such as 3DSlicer, Mimics, Amira are available to perform semi-automatic segmentation, where the ability to accurately segment an image is restricted to the choice of an accurate threshold. The semi-automatic process is often a time-consuming task, particularly if the boundaries are not clearly defined, and the process should therefore be carried out slice by slice. On the other hand, several approaches based on mathematical and theoretical concepts, such as fuzzy logic, genetic algorithms, artificial neural networks, clustering, and so on, have been introduced to automatically perform a segmentation [21–25].

3D Model Generation concerns with surface modeling from the point cloud derived. It can be done interactively or in a semi-automatic way via STL tessellation. Although the segmentation process has been improved, the created 3D model may still have problems, thus *3D Model Correction* may be necessary. STL tessellation uses several connected triangles that correspond to a representation of a solid model's surface geometry, but in the 3D model generation process, commonly some triangles result missing. The typical task performed to correct the actual problems in the geometrical model, includes the following: a) Correct the triangles that are pointed; b) Remove overlapping triangles; c) Fill the missing holes or triangles; d) Rectify the inverted normal; e) Rectify bad edges (bad contours, near edges); f) Re-mesh process, g) Create a single shell (tessellation). It is also necessary to consider during the holes correction process if the hole is a: hole normal, hole ruled, hole freeform or whether it is a bridge in order to obtain a regular and continuous geometry.

The *Surface Smoothing* process is a flexible task meaning that this step can be performed before or after the "3D model correction". It is anyway necessary to conduct an optimization and smoothing process to correct some sharp areas. The optimization increases the number of triangles, the smoothing process reduces irregularities.

Finally, the *Set-up of the 3D Printing* includes the selection of the process parameters to print the STL model. Among different additive manufacturing techniques, the Fused Deposition Modelling (FDM) is currently the most widely used technology, so that it is considered a domestic desktop system for its easy access. This technique uses thermoplastic polymers as a raw material.

3 Case Study

The case study applies the workflow to print a liver disease taken from CT images. They are taken from [19] and represent medical images a liver belonging to woman born in 1987 (Fig. 2).

Image restoration and enhancement were made to reduce blur. It adopts a bicubic interpolator to obtain a smoother surface. It determines the colour value of the pixel through the 16 closest pixels to the input coordinate. Then, the blur kernel is estimated by modelling statistical irregularities in the power spectrum of blurred natural images [10]. Then, the deconvolution process is performed adjusting the blur kernel employing a fast total variation deconvolution FTVd (Fast Total Variation deconvolution) method [20]. After deconvolution process, a sharpening filter is used to mitigate the smoothest, to enhance the borders and to obtain an image free of impulsive noise. Then, histogram equalization is performed to increase the global contrast of the given image and to

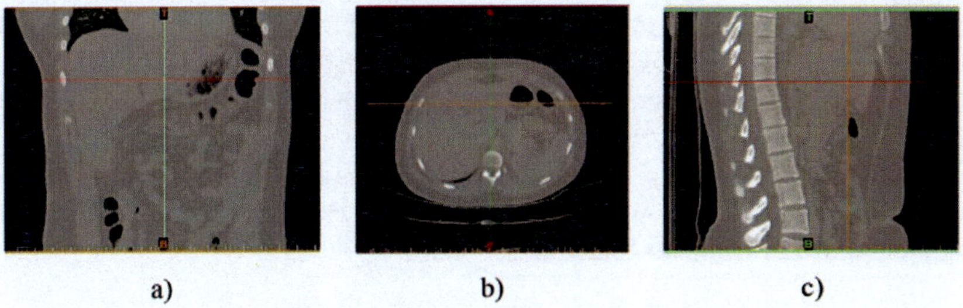

Fig. 2. CT medical images: a) coronal view b) axial view and c) sagittal view

distribute the intensities on the histogram for getting a contrast-enhanced image. Finally, two erosions and one dilation with a structure of 5 × 5 are done in order to separate the liver tissues from the rest of the tissues.

Image Segmentation and Classification has been approached by Convolutional Neural Networks (CNNs), since they have gained the ability to successfully support applications in pattern recognition, computer vision, and artificial intelligence. More in details, it has been adopted to explore the ability of CNNs to automatically distinguish the liver (foreground) from no liver (background) pixels [9] to facilitate the development of accurate 3D geometric models. To predict the probability that a pixel might be part of the liver tissues, the proposed approach first splits each image into patches of 28 × 28 centred on each pixel. Then, each patch is classified through a CNN. Finally, classified patched are arranged to obtain the final segmented image. The general workflow is depicted in Fig. 3.

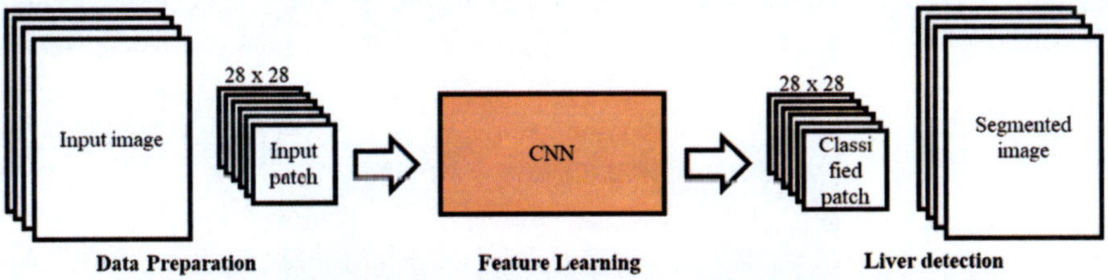

Fig. 3. Scheme of the segmentation approach

Prior to liver tissues segmentation, 80% of the grayscale slices of CT liver images (each CT image has approximately 1600 slices) are pre-processed to generate a dataset of patches that learns the CNN. Dataset preparation uses ground truth images manually labelled by the operator to assign the corresponding label value to input image patches (1 if the central pixel belongs to the liver tissues, 2 if the central pixel belongs to cancer tissues, otherwise 0). After obtaining the set of patches, LMDB dataset file is generated to be used as input to the CNN, considering that LMDB is a compressed format for working with large dataset. The CNN was implemented starting with a convolutional layer, where each patch is convolved with 20 learnable filters of size 5 × 5. After that, a down-sampling is performed, through a pooling layer of size 2 × 2 with a stride of 2 to reduce the spatial size of representation to 25% of its previous size. Then, a convolutional and pooling layer with 50 filters are used. At the end, two fully connected layers with 500 and 3 neurons, respectively act as patch classifier. The outputs from the convolutional and pooling layers represent the high-level features of the input patches, and the fully connected layers use these characteristics to classify the input patch into different classes/labels based on the training dataset. The details of the CNN architecture and its training parameters are depicted in Table 1.

Table 1 CNN architecture and settings for liver segmentation

Structure	Training parameters
Input: *n* patches of size 28 × 28	Weights were initialized randomly
Conv1: 20 filters of size 5 × 5	mini-batch of size 128 training sample patches
Pool1: 20 filters of size 2 × 2	43×10^3 iterations
Conv2: 50 filters of size 5 × 5	
Pool2: 50 filters of size 2 × 2	
Fc3: 500 neurons	
Output: 3 neurons	

Deep learning framework CAFFE was used for training purposes. It allows to process data that comes from in normal image format (.jpg,.png, and so on) and LMDB format. The Fig. 4 shows the liver tissues (white pixels) isolated from another soft tissues and background (black pixels). Preliminary results reached an accuracy value of 91% referring to the ground truths manually generated by an expert for experimental purposes.

Fig. 4. Liver tissues classification

After the application of the CNN, the images belonging to liver (mask) are imported for building the 3D model. Automatic interpolation of segmented images (mask) was made using Mimics. Then the 3D model is generated as an STL file. The 3D model correction has been done through an "Adaptive Re-mesh" with a target triangle edge length of 2 mm, followed by a "Uniform Re-mesh" with a target triangle edge length of 2 mm. Figure 5 shows the 3D final model (Fig. 5a) and before starting the correction (Fig. 5b).

a b

Fig. 5. a) Liver 3D model; b) Liver geometrical model before correction.

Surface smoothing, shown in Fig. 6a, has been applied on the corrected tessellation with a uniform mesh of 2 mm, a smoothing factor of 0.7 (3 iterations). Local smoothing can be applied only to the area of interest (smaller area with tumoral portion on the case being tested) to avoid an increase in the above-mentioned resolution in areas where it does not offer additional benefits to the presented case study.

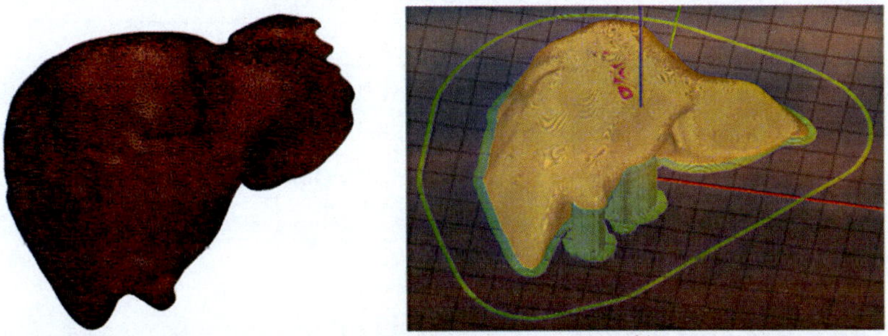

a) b)

Fig. 6. a) Liver geometrical model for printing; b) Liver geometrical model with the configuration of Table 2 into Slic3r.

The liver geometrical model has been built using a FDM device, "TEVO Delta Little Monster" that uses an open source Smoothieboard clone, raw material brown PLA filament (polylactic acid or polylactide) with a diameter of 1.75 mm. The path (G-code) for depositing the melted material has been determined using the open source Slic3r software (Fig. 6b). Table 2 displays the parameters considered for the printing of the geometric model of the liver, which were set to obtain a reliable printed model.

Table 2. Parameter to print using Slic3r software

Parameter	Description
Scale of liver geometrical model	0.5
First layer height	0.2 mm
Layer height	0.1618 mm
Perimeters	4
Solid layer top/bottom	4
Infill density	15%
Infill pattern	3D honeycomb
Top/bottom fill pattern	Concentric
Skirt	Loops:4; Distance from object: 20 mm; Skirt height: 1 layer
Brim	4 mm
Support material/Overhang threshold	45°
Speed for perimeters	70 mm/s
Speed for small perimeters	25%
Speed for external perimeters	40 mm/s
Speed for infill	120 mm/s
Speed for support material	80 mm/s
Speed for bridges	80 mm/s
Speed for non-print moves	650 mm/s
Temperature extruder	First layer: 230 °C; Other layers: 215 °C
Temperature bed	First layer: 80 °C; Other layers: 55 °C
Diameter filament	1.75 mm

4 Results and Discussion

Figure 7 shows half liver with a scale of 0.5 obtained with the parameters described previously. The geometric model has a good surface finish and can illustrate the main characteristics and region of interest of the liver (tumour) as highlighted in Fig. 7. To reduce printing time, the liver was split into two halves.

With the configuration presented in the previous section, the warping effect was reduced, and the support structure was easily removed. Warping effect is a common defect in the FDM technique. It is generated when the printing process goes forward and during the solidification process the previous layer extruded cools. As each region of the printed model cools at different temperature rates, its dimensions change at different rates as well (contraction). Internal stress is therefore produced as a result of the above-mentioned phenomena, and the final effect is to pull the underlying layer upwards (warping).

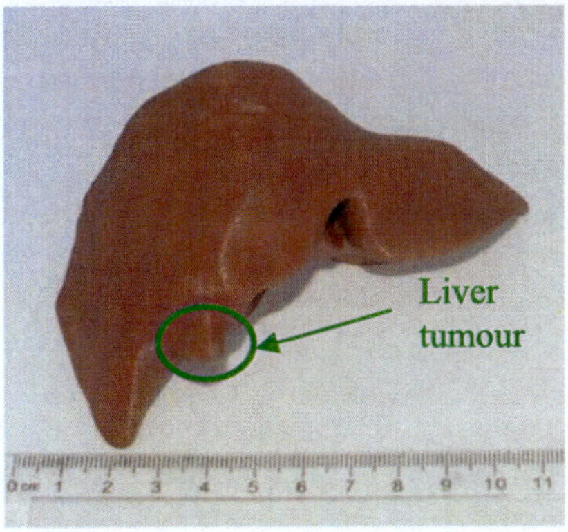

Fig. 7. Liver geometric prototype with the configuration of Table 2 in Slic3r

To prevent the issue of warping, it is crucial to control the print temperature and increase the adhesion between the print bed and the printed geometric model, therefore, in the test case, the parameter 'Brim' was set to 4 mm in order to create additional (external) loops on the first layer of the printed geometric model. This arrangement improves the adhesion of the printed model to the print bed.

For geometric models with irregular shapes (free form), support structures for the overhang portions must be generated, otherwise the overhang areas or bridges cannot be printed. In this work, the "overhang threshold" parameter has therefore been set to 45°. In the medical field, a good accuracy of the printed model is required, but the time taken to produce a printed model also plays an important role. The 15 percent infill (minimum possible) combined with a 3D honeycomb pattern ensured that the structure does not collapse. The external finishing was more significant in the case study established in this paper, so the speed of the infill and the inner perimeter were set at 120 mm/s and 70 mm/s, respectively, while the speed for external perimeters was set at 40 mm/s.

In accordance with these reasonings and parameters, the final result of the two halves was remarkable, confirming the good accuracy the author found in [15] in another test case. The final prototype, although suffered from the precision of the mounting of the two halves, that could be reduced designing specific pins for insertion and closing, confirmed its good usability for 3D assessing of the medical problem. Although FDM with PLA does not mimic biological deformation at all, it can economically solve the problem of the 3D perception of the disease, aiding the mentoring process at school and the medical preliminary assessment.

5 Conclusions

This work introduces and applies set-up rules for a workflow suitable to build a 3D liver model integrating 2D image segmentation based on CNNs and FDM technique. The output of the CNN is therefore a binary mask where the pixels are recognized and divided into two different classes (liver and background pixels).

The FDM technique provides a good trade-off between quality and time, in addition, the FDM is a low-cost alternative to generate prototypes to help the doctor establish preoperative planning. However, prototypes with the same mechanical properties of liver tissues cannot be created by the FDM system, which may be a drawback when the prototype is used for training new surgeons. Nevertheless, it can economically solve the problem of the 3D perception of the disease, aiding the mentoring process at school and the medical preliminary assessment.

To prevent certain impurities from occurring in the final printed model, it is recommended to consider the 'Skirt' parameter that is useful for cleaning the nozzle before beginning model printing. In cases where, given the complex free-form surfaces, the half liver can be printed, the orientation of the slice should be done in such a way that less structural support can be generated to improve the quality of the finished surface. This consideration prevents potential fractures or other problems when the support structures are detached.

Currently, to improve and continue the development of this research, feasibility analyses and studies are focused on the usage of flexible 3D printable elastomers and also on the design of compliant structure that integrated into the prototype can simulate the behaviour of the organ in terms of deformation.

References

1. U.S. National Library of Medicine, "MedlinePlus," Liver Diseases (2019). https://medlineplus.gov/liverdiseases.html#cat_79. Accessed 11 Apr 2021
2. Billingsley, K.G., Jarnagin, W.R., Fong, Y., Blumgart, L.H.: Segment-oriented hepatic resection in the management of malignant neoplasms of the liver. J. Am. Coll. Surg. **187**(5), 471–481 (1998)
3. McColl, R.J., Shaheen, A.A.M., Brar, B., Kaplan, G., Myers, R., Sutherland, F., Dixon, E.: Survival after hepatic resection: impact of surgeon training on long-term outcome. Can. J. Surg. **56**(4), 256–262 (2013)
4. Catalano, O.A., Singh, A.H., Uppot, R.N., Hahn, P.F., Ferrone, C.R., Sahani, D.V.: Vascular and biliary variants in the liver: implications for liver surgery. Radiographics **28**(2), 359–378 (2008)
5. Helling, T.S.: Liver failure following partial hepatectomy. HPB **8**(3), 165–174 (2006)
6. Ikegami, T., Maehara, Y.: Transplantation: 3D printing of the liver in living donor liver transplantation. Nat. Rev. Gastroenterol. Hepatol. **10**(12), 697–698 (2013)
7. Zein, N.N., et al.: Three-dimensional print of a liver for preoperative planning in living donor liver transplantation Liver Transplant. **19**(12), 1304–1310
8. House, M.G., et al.: Survival after hepatic resection for metastatic colorectal cancer: trends in outcomes for 1,600 patients during two decades at a single institution. J. Am. Coll. Surg. **210**(5), 744–752 (2010)
9. Guachi, L., Guachi, R., Bini, F., Marinozzi, F.: Automatic colorectal segmentation with convolutional neural network. Comput.-Aided Des. Appl. **16**(5), 836–845 (2019)
10. Kim, S., Lim, H: Method of background subtraction for medical image segmentation, In: CITSA 2006 - 3rd International Conference on Cybernetics and Information Technology, Systems and Applications jointly with the 4th International Conference on Computing, Communications and Control Technologies, CCCT 2006 - Proceedings, vol. 1, pp. 87–91 (2006)
11. Gayathri Devi, K., Radhakrishnan, R.: Automatic segmentation of colon in 3D CT images and removal of opacified fluid using cascade feed forward neural network. Comput. Math. Meth. Med. **2015**(670739), 1–15 (2015)

12. Kainz, P., Pfeiffer, M., Urschler, M.: Semantic segmentation of colon glands with deep convolutional neural networks and total variation segmentation. ArXiv, pp. 1–15 (2015)
13. Uccheddu, F., Carfagni, M., Governi, L., Furferi, R., Volpe, Y., Nocerino, E.: 3D printing of cardiac structures from medical images: an overview of methods and interactive tools. Int. J. Interact. Des. Manuf. **12**(2), 597–609 (2018)
14. Madurska, M.J., Poyade, M., Eason, D., Rea, P., Watson, A.J.M.: Development of a patient-specific 3D-printed liver model for preoperative planning. Surg. Innov. **24**(2), 145–150 (2017)
15. Bici, M., et al.: Digital design of medical replicas via desktop systems: shape evaluation of colon parts. J. Healthcare Eng. **2018**(3272596), 1–10 (2018)
16. Guachi, R., Bici, M., Guachi, L., Campana, F., Bini, F., Marinozzi, F.: Geometrical modelling effects on FEA of colorectal surgery Comput.-Aided Des. Appl. **16**(4), 778–788 2019
17. Guachi, R., Bini, F., Bici, M., Campana, F., Marinozzi, F., Guachi, L.: Finite element analysis in colorectal surgery: non-linear effects induced by material model and geometry. Comput. Meth. Biomech. Biomed. Eng. Imaging Visual. **8**(2), 219–230 (2020)
18. Kühnapfel, U.G., Kuhn, C., Hübner, M., Krumm, H.-G., Maass, H., Neisius, B.: The Karlsruhe Endoscopic Surgery Trainer as an example for virtual reality in medical education. Minimally Invasive Therapy Allied Technol. **6**(2), 122–125 (1997)
19. Ircad, 3D-IRCADb-01 database. https://www.ircad.fr/research/3d-ircadb-01/. Accessed 11 Apr 2021
20. Yu, H., Huang, T.-Z., Deng, L.-J., Zhao, X.-L.: Super-resolution via a fast deconvolution with kernel estimation. Eurasip J. Image Video Process. **2017**(1), Article no. 3 (2016)
21. Christinal, H.A., Díaz-Pernil, D., Real Jurado, P.: Segmentation in 2D and 3D image using tissue-like P system. In: Bayro-Corrochano, E., Eklundh, J.-O. (eds.) CIARP 2009. LNCS, vol. 5856, pp. 169–176. Springer, Heidelberg (2009). https://doi.org/10.1007/978-3-642-10268-4_20
22. Vorontsov, E., Tang, A., Pal, C., Kadoury, S.: Liver lesion segmentation informed by joint liver segmentation. In: Proceedings - International Symposium on Biomedical Imaging, April 2018, pp. 1332–1335 (2018)
23. Yang, J., Dvornek, N.C., Zhang, F., Chapiro, J., Lin, M., Duncan, J.S.: Unsupervised domain adaptation via disentangled representations: application to cross-modality liver segmentation. In: Shen, D. et al. (eds.) Medical Image Computing and Computer Assisted Intervention – MICCAI 2019, MICCAI 2019. LNCS, vol. 11765, pp. 255–263. Springer, Cham (2019). https://doi.org/10.1007/978-3-030-32245-8_29
24. Bici, M., Campana, F., Petriaggi, S., Tito, L.: Study of a point cloud segmentation with part type recognition for tolerance inspection of plastic components via reverse engineering. Comput.-Aided Des. Appl. **11**(6), 640–648 (2014)
25. Bici, M., Campana, F., Trifirò, A., Testani, C.: Development of automatic tolerance inspection through Reverse Engineering. In: 2014 IEEE International Workshop on Metrology for Aerospace, MetroAeroSpace 2014 - Proceedings, Article no. 6865903, pp. 107–112 (2014)

Integrated Methods for System Design, Simulation, Analysis and Optimization

Fracture Toughness Characterization of Copper-Resin Interface in Power Electronics Application

Michele Calabretta[1(✉)], Alessandro Sitta[1], Salvatore Massimo Oliveri[2], and Gaetano Sequenzia[2]

[1] Automotive and Discrete Group, R&D Department, STMicroelectronics, 95121 Catania, Italy
michele.calabretta@st.com

[2] Dipartimento di Ingegneria Elettrica, Elettronica ed Informatica (DIEEI), Università degli Studi di Catania, 95125 Catania, Italy

Abstract. The use of molding compound as encapsulating material is nowadays increasing in semiconductor industry. Such component guarantees excellent thermal and reliability performances than the current silicone-based gel, enabling higher working temperature for semiconductor device and mitigating the solder joint reliability bottleneck. The adhesion of package interfaces between copper components and molding compound is one of the key aspect for optimized durability. Dedicated experiments and theoretical framework based on fracture mechanic are needed for this purpose. The presented activity proposes the fracture toughness characterization of copper-resin interface in a power semiconductor package. Double Cantilever Beam (DCB) test has been executed on dedicated bimaterial coupon with an initial crack at interface. The aim of this test has been to enhance the fracture propagation mode-I (opening). Strain energy release rate (SERR) and mode-mixity have been estimated from this experiment developing a finite element analysis that is able to predict the crack length during the experimental DCB trials and to predict the energy release rate by virtual crack closure technique (VCCT). Mode-mixity has been estimated collecting displacements near the crack tip by crack surface displacement method (CSD). The proposed methodology for fracture toughness characterization represents a strong pillar to predict fracture behavior due to any load conditions and it is needed to describe interface adhesion by cohesive zone method (CZM).

Keywords: Fracture · Interfacial delamination · Virtual crack closure technique · Finite element analysis · Power electronics

1 Introduction

The current diffusion of Hybrid and Full Electric Vehicles is boosting the development of innovative Power semiconductor products which couple the semiconductor device with the mechanical assemblies and the logic control. This new

C. Rizzi et al. (Eds.): ADM 2021, LNME, pp. 549–556, 2022.
https://doi.org/10.1007/978-3-030-91234-5_55

trend results crucial to manage high power density using power package with low dimensions, reduced cost and enough reliable to satisfy the mission profile targets [6]. The encapsulation in transfer molding packaging solution is one of the pillar of the current development trend [2,3,13,15]. Molding compound mitigates the thermal stress on solder layers induced by mismatch between the coefficient of thermal expansion of the different package components. Solder degradation is one of the main reliabilty bottleneck for power package [5,7,17]. Furthermore, molding compound allows to exploit the excellent temperature performances of the innovative wide-band gap semiconductor, as Silicon Carbide [14,16]. With respect to the silicone gel that are thermally limited by silicon degradation, the innovative high glass temperature resins can withstand at temperature higher than 200 °C. Moreover, molding compound allows to tighten the package out-of-plane tolerance that is an important parameter to control manufacturability and compatibility between modules and external components, such as the printed control board [4]. Nevertheless, adhesion should be guaranteed at the interfaces between molding compound and other package components, not only after manufacturing process. In this context, copper-resin interface is the most important inside the package. Furthermore, there is the need to employ dedicated experiments to characterize the phenomenology, selecting the proper theoretical framework for a quantitative description. Fracture mechanic comprises several approaches to study quantitatively crack initiation at the interfaces, identifying the actual fracture mode-mixity [1]. The mentioned approached could be based on the stress analysis around the crack tip [12] or on by energetic calculation, like the virtual crack closure technique (*VCCT*) in the field of linear fracture mechanic. This last methodology has been considered in the presented work.

In the present paper, it has been characterized the fracture toughness of the interface between copper frame layer and high temperature glass transition molding compound. Dedicated samples have been prepared, as described in Sect. 2.1, to be tested according to *Double Cantilever Beam* test reported in Sect. 2.2. Two dimensional finite element model has been developed as described in Sect. 2.3 in order to enable the strain energy release rate (SERR) calculation by the VCCT method explained in Sect. 2.4. Experiment and theoretical results have been linked and commented in Sect. 3.

2 Materials and Methods

2.1 Test Vehicles and Sample Preparation

Dedicated copper-resin samples have been prepared to perform double cantilever beam (DCB) test. The transfer mold has been fixed on a copper leadframe and it has been filled by epoxy molding compound (EMC). The resin has been cured at 180 °C, then the bi-materials beams have been removed from the mold and chilled to the ambient temperature. Specimen dimensions are 80 $\times$ 10 mm, with EMC thickness of 2 mm and Cu thickness of 0.25 mm (10 mills). An example of realized coupon is shown in Fig. 1-(a).

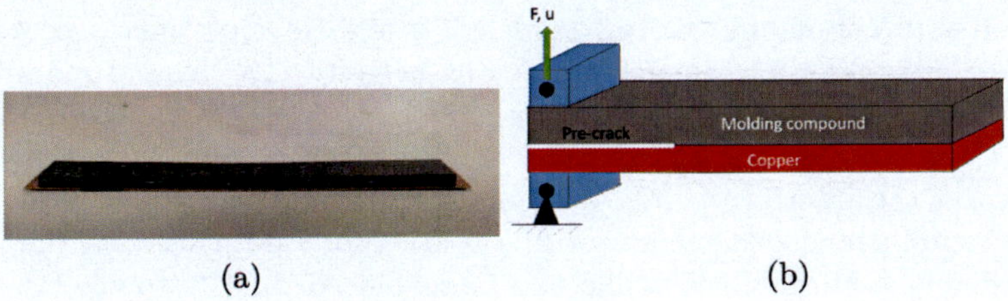

Fig. 1. Optical images of bimaterial sample (a) and schematic of dual beam cantilever test (b).

A initial notch has to be created in the samples before to start the tests in order to enhance crack propagation. The initial damage has been induced in the specimen by blocking the sample at a certain distance and bending the beam in order to start delamination.

2.2 Double Cantilever Beam Experiment

The purpose of double cantileer beam (DCB) test is to enhance the opening crack propagation (mode I) at the interface of a bi-material coupon. In fact, the mode-mixity of such kind of test is close to pure mode I, as later calculated. Test schematic has been shown in Fig. 1-(b). A tensile test equipment has been considered to execute DCB tests. The pre-cracked side of test sample has been connected to the equipment clamps by an epoxy glue. Trials have been done in displacement-controlled mode at 0.5 min/min. In the first stage, the load increases linearly with the displacement. At a certain testing time, the force tends to decrease due to crack propagation. In order to calculate the SERR value correlated with DCB test, exact crack length during the test must be known. Repetitive loading-unloading during DCB has been considered for this purpose. As later shown, force during unloading phase decreases linearly with displacement. As calculated by simulation, the slope in the unloading stages is correlated with the crack length. In order to estimate SERR at different crack length, loading-unloading has been repeated four times.

2.3 Finite Element Analysis for Strain Energy Release Rate Calculation

Modeling activities based on finite element analysis (FEA) have been employed to estimate the SERR value correlated with the employed DCB test. Before SERR calculation, a preliminary finite element model has been developed to determine the crack length in the different unloading phase of DCB trial. As mentioned before, the slope of deformation during unloading phase is correlate with crack length and the model serves to calculate this length. According to the specimen structure and loading types, 2D analysis based on plane strain has been assumed. Considering that relative small deformations occur, molding compound and copper have been modeled as elastic materials. According to material characterization done for previous activity [17], Young modulus has been considered 120 and 15 GPa, respectively for copper and resin. Poisson ratio is assumed 0.364 and 0.2. Initial crack has been reproduced uncoupling the nodes at the interface in the delaminated area with parametric length. In order to reproduce the schematic of Fig. 1-(b), the node placed on the bottom left of model has been fixed while the node placed on the top left has been move upward. Several simulations have been carried out considering a crack length between 5 to 15 mm to correlate the loading slope with crack length by polynomial function. According to this relation, crack length at different loading-unloading steps of experimental DCB trial has been found. The model before described has been replicated more times considering the real crack dimension in the different loading-unloading step. Considering that model is bi-dimensional, the critical force measured during experiments has been divided by the sample width. The strain energy release rate has been finally calculated considering the virtual crack closure technique (VCCT).

2.4 Virtual Crack Closure Technique (VCCT) and Mode Mixity Determination

VCCT is a well-assessed numeric methodology to calculate energy release rate and crack growth in the field linear fracture mechanics in which deformation are purely elastic or, according to Irwin extension [9], plastic behavior is confined in a very small area near the crack tip [11]. VCCT assumes that a crack extension of a quantity Δl does not modify the crack tip. As consequence, the energy rate required to close the crack along the distance Δl is the same to the quantity required to open the crack along the same distance. The total strain energy release rate is given by the sum of mode-I (opening) and mode-II (sliding) fracture toughness $G = G_I + G_{II}$, which could be calculated as reported in Eq. 1 [18]:

$$G_I = -\frac{F_y u_y}{2\Delta l} \qquad G_{II} = -\frac{F_x u_x}{2\Delta l} \tag{1}$$

in which F_x and F_y are the nodal forces. Due to the oscillatory nature of stress and displacements near the crack, mode-I and mode-II components calculated by VCCT depends on considered crack extension [10]. Crack surface displacement (CSD) method has been considered to mode-mixity calculation [12]. According to this, displacements u_x and u_y are considered as function of the distance from crack tip r. For a given distance r_0, Eq. 2 will be satisfy:

$$\frac{u_x^2 + u_y^2}{r_0} = 64\frac{G}{2\pi\left(1 + 4\epsilon^2 E^*\right)} \tag{2}$$

in which $\varepsilon = 0.063$ is a constant related to copper and resin materials [10] and E^* is the harmonic average of resin and copper Young modulus. It has been highlighted that the right term of Eq. 2 does no depend on crack distance r. The resulting u_x and u_y are considered to calculated mode-mixity φ in Eq. 3:

$$\varphi = \arctan\left(\frac{u_x}{u_y}\right) - \arctan(-2\varepsilon) - \varepsilon \ln\left(\frac{l}{r_0}\right) \quad (3)$$

in which l is a characteristic reference length (the sample thickness in this case).

3 Results

The force-displacement curve of DCB test described in Sect. 2.2 has been shown in Fig. 2. The curve initially growths linearly and reaches a maximum force of about 4.9 N. According to theoretical expectations, crack propagation starts at maximum force value. During force decreasing, the four loading-unloading steps are visible. As discussed before, the slope dF/du in each loading/unloading phase is directly linked with the current crack extension.

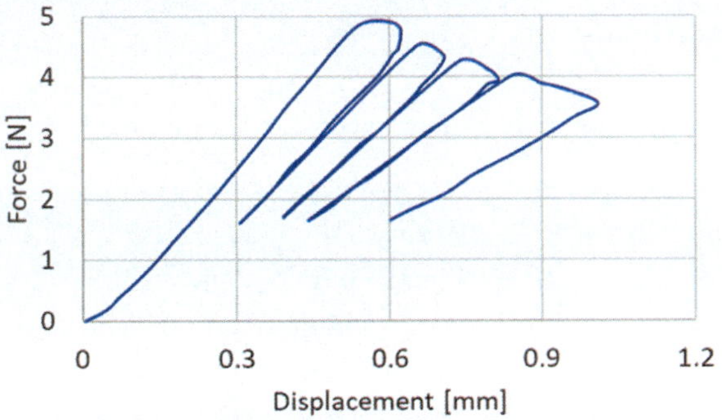

Fig. 2. Diagram of force and displacement in DCB test.

Finite element model presented in Sect. 2.3 has been considered to calculate the slope dF/du with respect to different crack length ranging from 5 to 15 mm. Table 1 has reported the slope dF/du at different crack length Δl: polynomial function $dF/du = -0.0882\Delta l^3 + 3.2291\Delta l^2 - 39.611\Delta l + 166.82$ fits relation with reasonable accuracy ($R^2 = 0.9967$).

Table 1. Calculated slope in loading/unloading stage at different crack length.

Crack length [mm]	5	7	9	11	13	15
Slope [N/mm]	38.8	16.4	8.5	4.9	3.0	2.0

Crack length during the loading/unloading steps reported in Fig. 2 can be estimated replacing the slope value dF/du the above mentioned polynomial fit with the correspondent one in each step. Four models, one for each found crack length, have been built and the VCCT method described in Sect. 2.4 has been applied to calculate the final SERR values. Finally, mode-mixity has been calculated found the critical crack distance r_0 for which interface displacements allow to satisfy the Eq. 3. The main experimental output and the mentioned results in terms of critical load at each loading/unloading step (F), loading/unloading slope (dF/du), crack length (Δl), SERR (G_c) and mode-mixity (φ) have been collected in Table 2. The averaged values for strain energy release rate and mode-mixity are respectively $G_c = 43.7\,\mathrm{J/m^2}$ and $\varphi = 11.6°$. It has been highlighted that resulting mode-mixity angle, close to 0°, has confirmed the capability of DCB test to enhance the fracture mode I. Comparing the found values with other research that followed a similar experimental approach [10], fracture toughness G_C is higher (43.7 than 35.6 J/m^2) while mode-mixity φ is quite similar (11.6° than 11.9°.). The detected differences ($\tilde{2}3\%$) in terms of SERR could be addressed to the differences in terms of sample dimension and epoxy molding compound chemical formulation. According to above observations, the presented results has confirmed that Double Cantilever Beam test is suitable for fracture mode I characterization.

Table 2. Output parameters concerning experiment, finite element and VCCT calculations.

Step	F [N]	dF/du [N/mm]	Δl [mm]	G_c [J/m^2]	φ [°]
1	4.39	9.13	8.84	43.9	11.5
2	4.14	7.82	8.92	43.2	11.2
3	3.87	6.26	9.53	43.1	11.8
4	3.55	4.67	10.66	44.5	11.7

4 Conclusion

The presented work has shown a experimental and numerical technique to characterize the fracture toughness of copper-resin interface according to a specific loadcase, double cantilever beam test, that has stressed mainly the fracture mode I (opening). It has been demonstrated that the obtained results are coherent with references that follow a similar method [10], therefore the proposed activity represents also a validation study on new test samples based on novel molding compound formulation optimized for high temperature applications. The performed activity represents a milestone for the complete fracture characterization of interface: adding the result from experiments which involve fracture mode II (sliding) is possible to predict the fracture toughness for any mode mixity according to the formulation proposed by [8] : $G_C = G_{I,C}\left[1 + \tan^2\left((1-\lambda)\,\varphi\right)\right]$ in which λ shall be found empirically. Considering copper-resin is the most diffused interface inside a power semiconductor package, the capability to predict fracture occurrence could be considered to analyze and forecast package reliability performance by finite element models. The direct application of fracture mechanics results to be tricky considering that starting crack geometry and location have to be assumed as boundary conditions. Such methodology, considering the complexity of package geometry and that multiple cracks can occur, requires a great effort in terms of computation resources. Cohesive-zone theory is a suitable method to predict crack initiation and propagation without the "a priori" knowledge of the initial fracture shape [19]. This method assumes that the traction between surfaces are correlated with the crack tip displacement by a dedicated traction-separation law. The presented double beam cantilever test with strain energy release rate is one of the needed experiment for the calibration of the above mentioned traction-separation law. The next step of the presented activity should be the development of finite element model based on cohesive-zone theory and the employment of experiments to validate the numerical approach.

References

1. Agrawal, A., Karlsson, A.M.: Obtaining mode mixity for a bimaterial interface crack using the virtual crack closure technique. Int. J. Fract. **141**(1–2), 75–98 (2006). https://doi.org/10.1007/s10704-006-0069-4
2. Calabretta, M., Renna, M., Vinciguerra, V., Messina, A.A.: Power packages interconnections for high reliability automotive applications. In: ESSDERC 2019–49th European Solid-State Device Research Conference (ESSDERC), pp. 35–39. IEEE (2019). https://doi.org/10.1109/ESSDERC.2019.8901742
3. Calabretta, M., Sitta, A., Oliveri, S.M., Sequenzia, G.: Design and process optimization of a sintered joint for power electronics automotive applications. In: International Conference on Design, Simulation, Manufacturing: The Innovation Exchange, pp. 470–480. Springer (2019). https://doi.org/10.1007/978-3-030-31154-4_40
4. Calabretta, M., Sitta, A., Oliveri, S.M., Sequenzia, G.: Experimental characterization and modeling of power module warpage during assembly process. Int. J. Interact. Des. Manuf. (IJIDeM) **15**(1), 113–115 (2020). https://doi.org/10.1007/s12008-020-00735-z
5. Calabretta, M., Sitta, A., Oliveri, S.M., Sequenzia, G.: Power semiconductor devices and packages: Solder mechanical characterization and lifetime prediction. IEEE Access **9**, 22859–22867 (2021). https://doi.org/10.1109/ACCESS.2021.3056281
6. Catalano, A.P., Scognamillo, C., d'Alessandro, V., Castellazzi, A.: Numerical simulation and analytical modeling of the thermal behavior of single-and double-sided cooled power modules. IEEE Trans. Compon. Packaging Manuf. Technol. **10**(9), 1446–1453 (2020). https://doi.org/10.1109/TCPMT.2020.3007146
7. Durand, C., Klingler, M., Coutellier, D., Naceur, H.: Power cycling reliability of power module: a survey. IEEE Trans. Device Mater. Reliabil. **16**(1), 80–97 (2016). https://doi.org/10.1109/TDMR.2016.2516044
8. Hutchinson, J.W., Suo, Z.: Mixed mode cracking in layered materials. Adv. Appl. Mech. **29**, 63–191 (1991). https://doi.org/10.1016/S0065-2156(08)70164-9
9. Irwin, G.: Linear fracture mechanics, fracture transition, and fracture control. Eng. Fract. Mech. **1**(2), 241–257 (1968). https://doi.org/10.1016/0013-7944(68)90001-5

10. Krieger, W.E., Raghavan, S., Sitaraman, S.K.: Experiments for obtaining cohesive-zone parameters for copper-mold compound interfacial delamination. IEEE Trans. Compon. Packaging Manuf. Technol. **6**(9), 1389–1398 (2016). https://doi.org/10.1109/TCPMT.2016.2589223
11. Krueger, R.: The virtual crack closure technique for modeling interlaminar failure and delamination in advanced composite materials. In: Numerical Modelling of Failure in Advanced Composite Materials, pp. 3–53. Elsevier (2015). https://doi.org/10.1016/B978-0-08-100332-9.00001-3
12. Matos, P., McMeeking, R., Charalambides, P., Drory, M.: A method for calculating stress intensities in bimaterial fracture. Int. J. Fract. **40**(4), 235–254 (1989). https://doi.org/10.1007/BF00963659
13. Messina, A.A., et al.: The "first and european sic eight inches pilot line": a project, called reaction, that will boost key sic technologies upgrading (developments) in europe, unleashing applications in the automotive power electronics sector. In: 2020 AEIT International Conference of Electrical and Electronic Technologies for Automotive (AEIT AUTOMOTIVE), pp. 1–6. IEEE (2020). https://doi.org/10.23919/AEITAUTOMOTIVE50086.2020.9307385
14. Schmidt, R., Werner, R., Casady, J., Hull, B., Barkley, A.: Power cycle testing of sintered sic-mosfets. In: PCIM Europe 2017; International Exhibition and Conference for Power Electronics, Intelligent Motion, Renewable Energy and Energy Management, pp. 1–8. VDE (2017)
15. Schuderer, J., et al.: High-power sic and si module platform for automotive traction inverter. In: PCIM Europe 2019; International Exhibition and Conference for Power Electronics, Intelligent Motion, Renewable Energy and Energy Management, pp. 1–8. VDE (2019)
16. Sitta, A., Renna, M., Messina, A.A., Sequenzia, G., D'Arrigo, G., Calabretta, M.: Thermal measurement and numerical analysis for automotive power modules. In: 2020 21st International Conference on Thermal, Mechanical and Multi-Physics Simulation and Experiments in Microelectronics and Microsystems (EuroSimE), pp. 1–3. IEEE (2020). https://doi.org/10.1109/EuroSimE48426.2020.9152204
17. Sitta, A., et al.: An integrated approach to optimize solder joint reliability. In: 2020 21st International Conference on Thermal, Mechanical and Multi-Physics Simulation and Experiments in Microelectronics and Microsystems (EuroSimE), pp. 1–5. IEEE (2020). https://doi.org/10.1109/EuroSimE48426.2020.9152688
18. MSC software: Marc 2017.1 Volume A: Theory and User Information. User documentation, MSC Software, Newport Beach, CA, USA (2017). https://www.mscsoftware.com/
19. Tambat, A., Subbarayan, G.: Isogeometric enriched field approximations. Comput. Methods Appl. Mech. Eng. **245**, 1–21 (2012). https://doi.org/10.1016/j.cma.2012.06.0062

A Framework for Virtual Evaluation of Body-Attached Sensor Networks

Giuseppe Sanseverino(✉), Dominik Krumm, and Stephan Odenwald

Department of Sports Equipment and Technology, Chemnitz University of Technology, Reichenhainer Str. 70, 09126 Chemnitz, Germany
giuseppe.sanseverino@mb.tu-chemnitz.de

Abstract. In this work, we propose a framework that can be used for virtual evaluation of Body-Attached Sensor Networks. Normally, the evaluation of Body-Attached Sensor Networks requires numerous subject tests under laboratory conditions. However, since it is difficult to perform the same motion repeatedly without minute deviations, numerous replicate measurements are required to obtain statistically meaningful measurements. To overcome this limitation, we propose the use of virtual environments. These provide both a high degree of flexibility, since a movement can be repeated in the same way each time, and the ability to test many different sensor setups quickly and with little effort. To this end, we modeled the human body parts of interest using the MATLAB tool Simscape Multibody. Digital twins were then implemented in this model to represent real sensors along with their sensor properties at arbitrary locations. This makes it possible to check many different sensor types and their position on the body in a short time without having to perform subject tests. This framework creates a solid basis for the development of effective Body-Attached Sensor Networks.

Keywords: Virtual environment · Multibody simulation · Wearable · Sensor networks · Digital twins

1 Introduction

This work aims to develop a methodology that allows the assessment of Body-Attached Sensor Networks (BASN) using virtual environment. The core idea behind this framework is to model human body parts and digital twins of sensors and run simulations of predefined set of gestures in order to find the best sensors set-up.

To this end, we first review the state of the art in BASN, digital twins, and virtual models of the human body. We then present the steps required to build the framework. In particular, we focus on the steps to model the human hand and digital twins of a uniaxial accelerometer and a gyroscope. Finally, we demonstrate the capabilities of the proposed framework by simulating the following basic gesture: Gesture for the number 1 from American Sign Language (ASL).

C. Rizzi et al. (Eds.): ADM 2021, LNME, pp. 557–568, 2022.
https://doi.org/10.1007/978-3-030-91234-5_56

1.1 Body-Attached Sensor Networks

The impressive improvements in electronic devices that we have seen in recent years have made it possible to develop sensors that are small and light enough to be worn. In addition, the proliferation of wireless networks has offered the opportunity to connect many different sensors together, creating wearable sensor networks.

The idea of such networks was first mentioned in 2001 from Van Dam et al. [1], who used the term Wireless Body Area Network (WBAN). Ullah et al. [2] conducted an extensive review on WBANs. Their work showed that these networks were mainly used in the medical field to monitor the health status of patients and in some cases even to prevent problems at an early stage.

Furthermore, wearable sensor networks are very commonly used in sports and rehabilitation. For example, accelerometers, gyroscopes, electro myographs and inertial measurement units (IMU) [3] can be combined and used to perform field tests [4], measure performance in sports [5], or to enable wireless motion sensing [6].

In recent years BASNs have also been used in gesture recognition and human-robot interaction. Human-robot collaboration is a common problem in industrial environments. For instance, the work of Neto et al. [7] showed how IMUs can be used to recognize human movements of the upper body. This allows the worker to control a robotic arm with simple gestures. Fang et al. [8] also used inertial and magnetic sensors by embedding them in a glove, allowing them to recognize both static and dynamic gestures. Abualola et al. [9] also used an IMU-based glove to recognize sign language.

Although the above solutions provide good results, they all consider specific gestures and do not provide a generic methodology to recognize a wide variety of gestures. If there is a need to recognize other gestures, the type, number and position of sensors on the body may vary, leading to the need to perform new and time-consuming laboratory tests.

The framework proposed in this work aims to overcome the need of extensive laboratory testing by replacing it with testing in a virtual environment. This allows a large number of different tests to be performed in a short time, so that the effectiveness of different types of sensors and their boundary conditions can be evaluated.

1.2 Digital Twins of Sensors

With the advent of industry 4.0, the concept of digital twin has become prevalent. A complete overview of the various forms in which digital twins have evolved is given by Liu et al. [10]. In general, a digital twin is understood to be a virtual replica of a physical component or environment. In this paper the focus is set on simulating the electro-mechanical properties of real sensors in a digital realm.

This work deals with digital twins of physical sensors, such as accelerometers or gyroscopes, which together form a BASN. Only one similar work on this topic was found in literature. Raveendranathan et al. [11] developed a virtual accelerometer to be used in gait analysis to recognize the activities and postures in real-time. However, they proposed only one type of digital sensor that cannot be easily reused for other purposes and also cannot be directly implemented in multibody models. Therefore, there is a need to develop plug and play digital twins of different sensors that can form diverse BASNs to make their evaluation faster and cheaper.

1.3 Human Body Models

In recent years, much academic and industrial research has been conducted in the field of human body modeling leading to interesting results. A good example is OpenSim, an open-source software developed by Stanford University [12] and widely used to perform dynamic simulations of movements. It offers the possibility to estimate the internal load of the musculoskeletal system. Another solution is the AnyBody Modeling System (AnyBody Technology A/S, Niels Jernes Vej 10, Aalborg, Denmark). It is a commercial software that allows dynamic motion simulations and estimation of muscle forces, joint contact forces and moments [13].

Both models described have been used extensively for biomechanics and both give a good approximation of the human musculoskeletal system. However, their software does not allow to develop digital twins of physical sensors or to import them from external environments. Thus, there is a need to find an integrated environment where human body parts and physical sensors can be modeled and used together. The procedure is designed to first get real motion data for predefined gestures using motion capturing systems. The second step is to animate the virtual multibody-model of the human body with virtual sensors attached by the previously captured motion data. Eventually, the digital twin of the Sensor generates virtual electrical signals by an underlying electro-mechanical coupling model. With this digital setup, different simulations for different positions and combinations of sensors will be tested. The output of the digital sensors are the used to identify the original gestures. The best configuration among all the simulated ones will be used to build the physical BASN.

2 Methods

2.1 Simulation Environment

As mentioned earlier, no biomechanics software was found that could combine modeling of human body and digital twins of sensors. Therefore, the MATLAB Simulink environment (The MathWorks, Inc., Natick, MA, USA) and its package Simscape Multibody were selected.

This software enables block modelling of mechatronic systems and finds application in various fields, particularly robotics. For example, some researchers used Simscape Multibody to model a robotic arm and perform a simulation to understand the effectiveness of implementing an innovative joint actuator on this arm [14]. However, this environment has been used only sporadically to model human body parts. Almost all existing work in this direction focuses on human lower body and gait analysis [15, 16].

Simscape Multibody was used in this work to model the human hand and the digital twins of accelerometers and gyroscopes. This choice guaranteed high flexibility due to the possibility of modeling both mechanical and electronic systems in an integrated environment that is powerful and user-friendly. It also offered the possibility of importing data directly from MATLAB and exporting simulation results for quick analysis.

2.2 Multibody Modelling of Human Body Parts

The first step of the methodology was to model the human body parts needed for the simulation. For this purpose, a multibody model of the human hand is designed.

The framework proposed by Sebastian Castro serves as the basis for the modeling [17]. In it, Castro has given an overview of how it is possible to model human-like legged robots using Simscape Multibody.

To get a true reproduction of the human hand, dimensions used for the model were measured on a subject. According to this, the thumb is 20 mm wide, while all other fingers are 15 mm wide, the thickness of the palm and fingers is 10 mm, and the length of each phalanx is between 20 and 52 mm (Table 1). A very simplified schematic of the palm was considered and its cross section was defined in terms of Cartesian coordinates

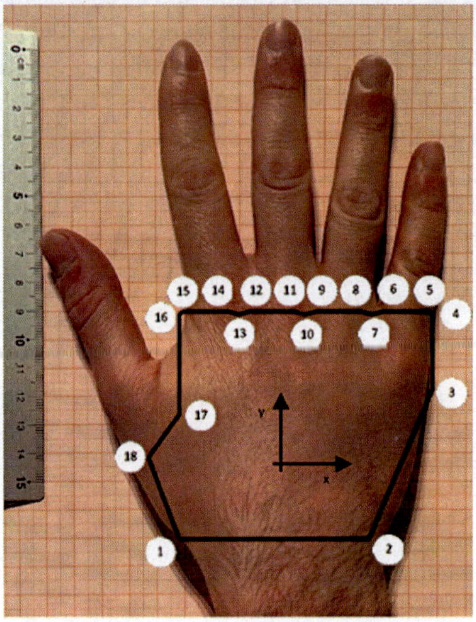

Fig. 1. Photo of measured hand and palm cross section sketch.

Table 1. Fingers phalanx lengths in mm.

Finger	Proximal phalanx	Middle phalanx	Distal phalanx
Index	50	30	20
Middle	52	35	20
Ring	50	30	20
Little	40	25	20
Thumb	35	25	20

in MATLAB. Figure 1 shows the hand of the considered subject and the cross section sketch.

Using the Extruded Solid Block from Simscape library, it was possible to extrude the defined cross section to obtain a solid representing the schematized palm. The fingers were modeled using Solid Bricks and connected to the correct position on the palm.

To allow the finger to move, joints with two degrees of freedom (Revolute Joints), were placed between each phalanx and between each finger and the palm. Each Revolute Joint can be animated using joint angles. The joint angles required for gesture number 1 in ASL can be determined through laboratory experiments (e.g. vision-based or sensor-based motion capture) and imported in MATLAB, or they can be defined arbitrarily. Table 2 displays the joint angles used in this study. These are subject-specific and were determined from a visual analysis of the gesture under consideration. Only three fingers are listed in the table, with the ring finger and little finger animated with the same joint angles as the middle finger. Figure 2 shows an example of how a finger block is modeled.

Table 2. Joint angles for gesture number 1 in ASL. (PP, MP and DP stand respectively for proximal, middle and distal phalanx).

Time [s]	Thumb			Index			Middle		
	PP [deg]	MP [deg]	DP [deg]	PP [deg]	MP [deg]	DP [deg]	PP [deg]	MP [deg]	DP [deg]
0	0	0	0	0	0	0	0	0	0
0.25	11.25	0	0	0	0	0	11.25	11.25	0
0.50	22.50	0	0	0	0	0	22.50	22.50	0
0.75	33.75	5	0	0	0	0	33.75	33.75	0
1	45	10	5	0	0	0	45	45	0
1.25	56.25	15	15	0	0	0	56.25	56.25	0
1.50	67.50	20	25	0	0	0	67.50	67.50	0
1.75	78.75	25	35	0	0	0	78.75	78.75	0
2	90	30	45	0	0	0	90	90	0

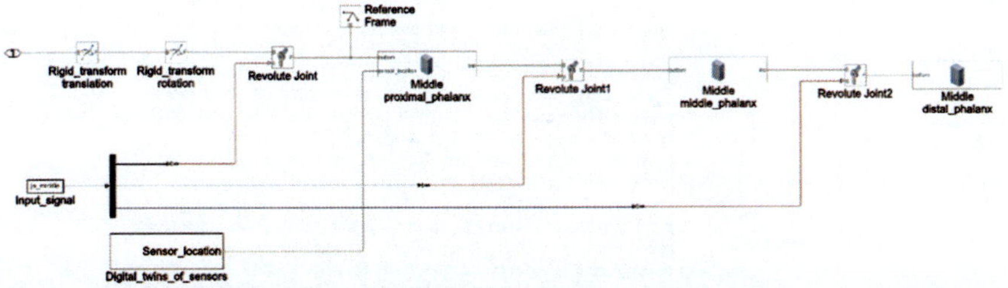

Fig. 2. Simscape block model for middle finger.

2.3 Modelling Digital Twins of Sensors

Simscape Multibody allows the physical quantities present in a joint, such as force and torque, to be read out as parameters in the corresponding settings of the joint. Although many parameters can be captured, these physical quantities do not reflect what is actually happening in the real environment at the location of the body where the sensor is attached. However, Simscape Electrical and Simulink libraries contain several components that enable the modeling of digital twins of physical sensors.

Digital Twin of a Uniaxial Accelerometer. The Simscape Electrical library contains the Accelerometer block, which emulates a uniaxial MEMS accelerometer [18] whose parameters can be set according to the user's specifications. Figure 3 shows the graphical user interface (GUI) mask where the accelerometer parameters can be set.

The physical sensor used as reference is CXL04GP1. This is a single axis MEMS from Willow Technologies (Willow Technologies Ltd., Copthorne, West Sussex, UK). Its main characteristics are also shown in Fig. 3.

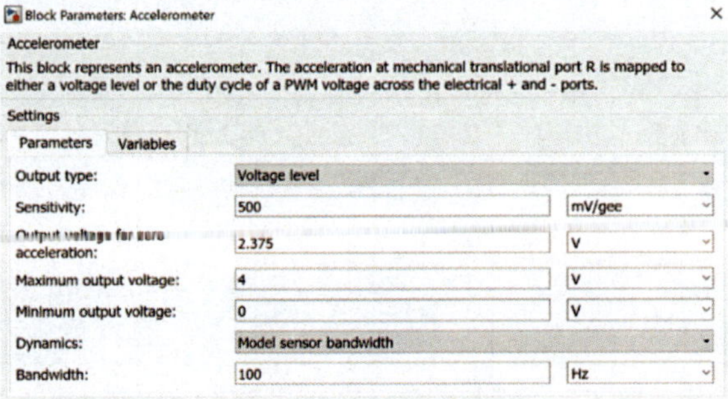

Fig. 3. GUI mask of Accelerometer block from Simscape Electrical library.

A physical accelerometer converts a time-dependent displacement as input and provide voltage as output. While the above-mentioned Simscape block needs a mechanical translational signal that can be generated employing a block called Ideal Translational Velocity Source, it and the GUI mask that describes its features are shown in Fig. 4. The

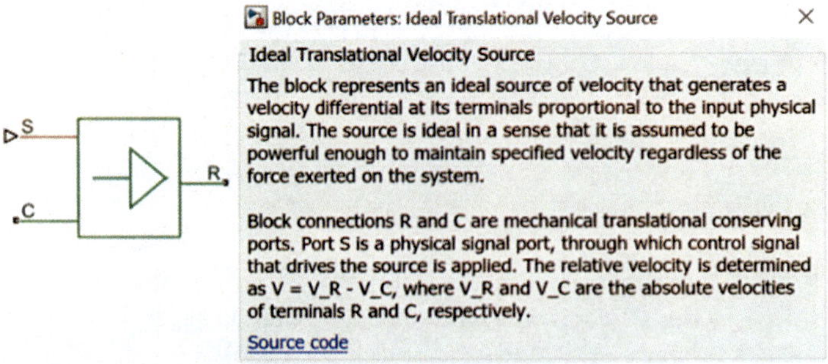

Fig. 4. Ideal translational velocity source block and its GUI mask.

signal that drives the source is the velocity of the point where the sensor is attached, while port C is connected to a Mechanical Translational Reference. The output, on the other hand, is already a voltage signal, but to read it, a volt-meter is needed, consisting of a Voltage Sensor, a Resistor and a ground (Electrical Reference).

Even if accelerometer digital twin is ready, there is still the need to get the velocity signal of the point where the sensor is attached. To overcome this issue a Transform Sensor Block should be used. It needs to be connected to a base and a follower frame and allows to sense displacements, velocities and accelerations, both translational and rotational. Figure 5 shows the block model of the digital twin of the accelerometer with all the above-mentioned components.

The follower port of the Transform Sensor Block is connected to a Brick Solid that gives a shape to the digital twin and allows its visualization during the simulation. The sensor location port (Digital twins of sensors subgroup in Fig. 2) needs to be connected to the Brick Solid that represents the body part where the physical sensor is attached.

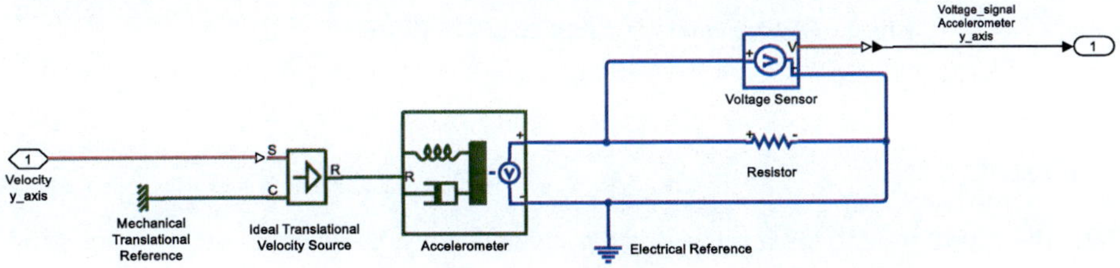

Fig. 5. Block model for digital twin of MEMS accelerometer.

Digital Twin of a Uniaxial Gyroscope. Simscape Electrical library contains also a block model for MEMS gyroscope [18]. Figure 6 shows the block GUI mask that reports the different parameters that can be adjusted.

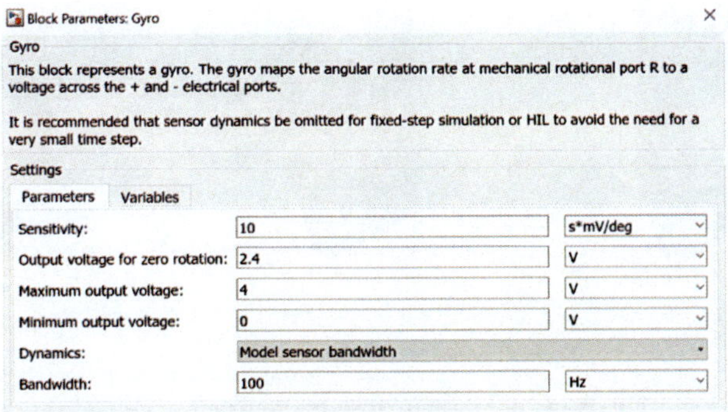

Fig. 6. GUI mask of gyroscope block from Simscape Electrical library.

The physical gyroscope used as reference is a single axis MEMS (CRH02-200) from Silicon Sensing (Silicon Sensing Systems Ltd., Plymouth, Devon, UK), whose main characteristics are displayed in Fig. 6.

As for the accelerometer, this block is not able to work alone and needs additional components. An angular velocity signal is taken from the Transform Sensor Block and given as input to an Ideal Angular Velocity Source Block, which acts like the translational block but is capable of elaborating angular velocities. A Mechanical Rotational Reference is connected to its port C. Thus, digital twin of single axis MEMS gyroscope is obtained. Its block model is presented in Fig. 7.

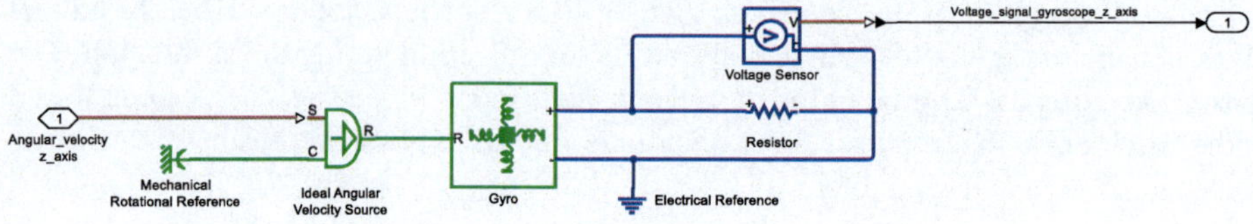

Fig. 7. Block model for digital twin of MEMS gyroscope.

3 Results

Once the block models of human body parts and physical sensors are complete and motion data are fed into the respective joint, simulation can be performed and results can be analyzed in MATLAB. Mechanics Explorer [19] is the Simscape Multibody tool that enables graphical visualization and exploration of the model. Figure 8 shows the designed model both at the beginning and at the end of the simulation. The small block placed on the middle finger, represents the position and shape of the sensing unit.

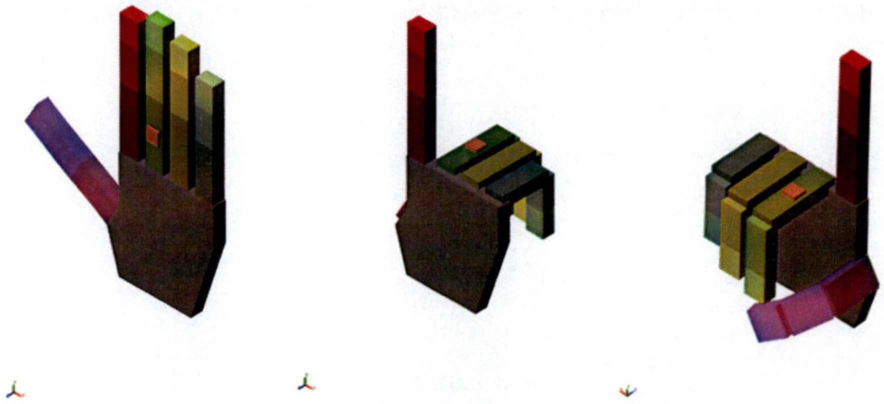

Fig. 8. Simulation of gesture number 1 in ASL.

Figure 9 displays the simulation output with regards to the linear accelerations along the y-axis. The first two signals come directly from the Transform Sensor Block and represent the translation and the acceleration of the point where the digital twin of the accelerometer is attached. While, the voltage signal shown at the very bottom represents the output of the digital twin of the accelerometer.

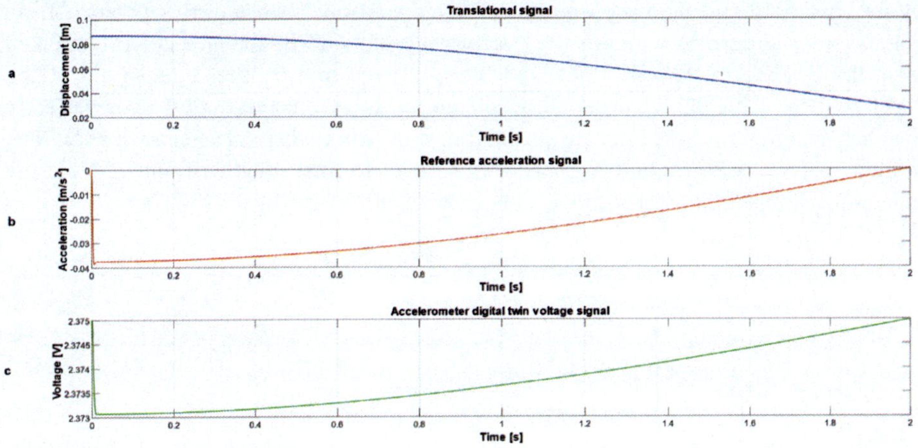

Fig. 9. Plots of translational signals along the y-axis. (a) translation; (b) acceleration taken from the Transform Sensor Block; (c) voltage output of the digital twin of the accelerometer.

The simulation results in terms of angular accelerations about the x-axis of the world frame, are shown in Fig. 10. Here, the first two plots represent the rotational and the angular velocity signals coming from the Transform Sensor Block. The output signal of the digital twin of the gyroscope is shown in the plot at the bottom of Fig. 10.

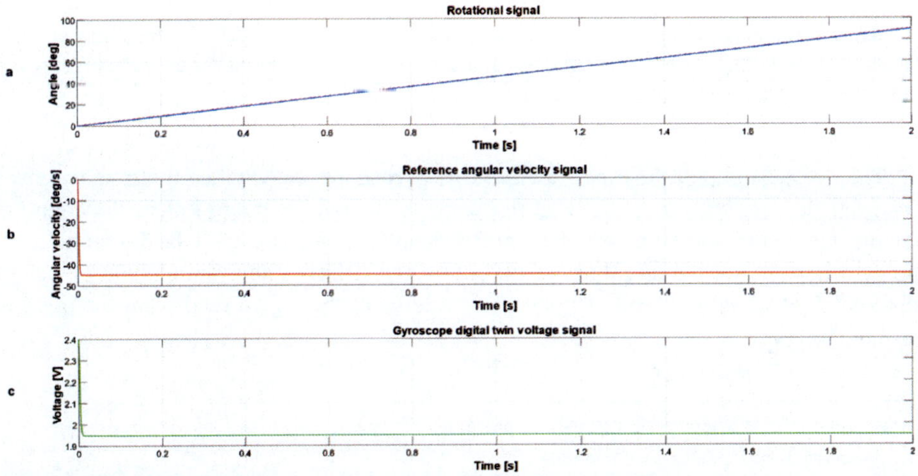

Fig. 10. Plot of rotational signals around the x-axis. (a) rotation; (b) angular velocity taken from the Transform Sensor Block; (c) voltage output of the digital twin of the gyroscope.

4 Discussion

Additional analysis were performed on the output signals from the digital twins of the sensors. The output signal of the digital twin of the accelerometer was first divided by its sensitivity and then multiplied by the gravitational constant to convert it from volts to m/s^2. Figure 11 contains the overlapping plots of the linear acceleration taken from the Transform Sensor Block and the one measured by the digital twin of the accelerometer. This shows how the acceleration read from the digital twin perfectly matches the reference one after the initial transient, where there is a negative peak. The average difference of the two signals with respect to time is 0.0012 m/s^2. While the maximum difference of 0.0142 m/s^2 is registered at the time of 0.0051 s.

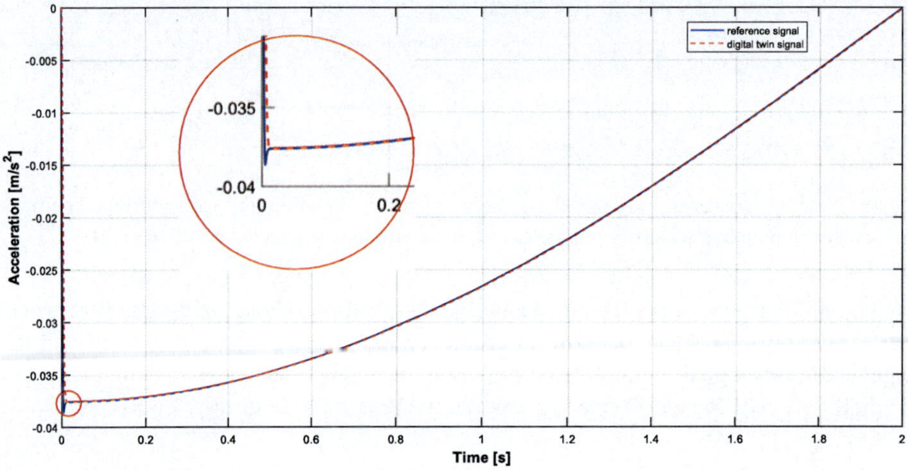

Fig. 11. Overlapping plots of the reference acceleration signal taken from the Transform Sensor Block and acceleration signal read from the digital twin of the accelerometer over the time, with zoom on the negative peak of the reference signal.

The output signal of the digital twin of the gyroscope was divided by its sensitivity to convert it from volts to deg/s. Figure 12 shows the overlapping plots of the angular velocity taken from the Transform Sensor Block and the one read from the digital twin of the gyroscope. After the initial transient, both signals show a constant angular velocity of 45 deg/s. The only notable difference is a delay in the signal read from the digital twin of 0.0074 s.

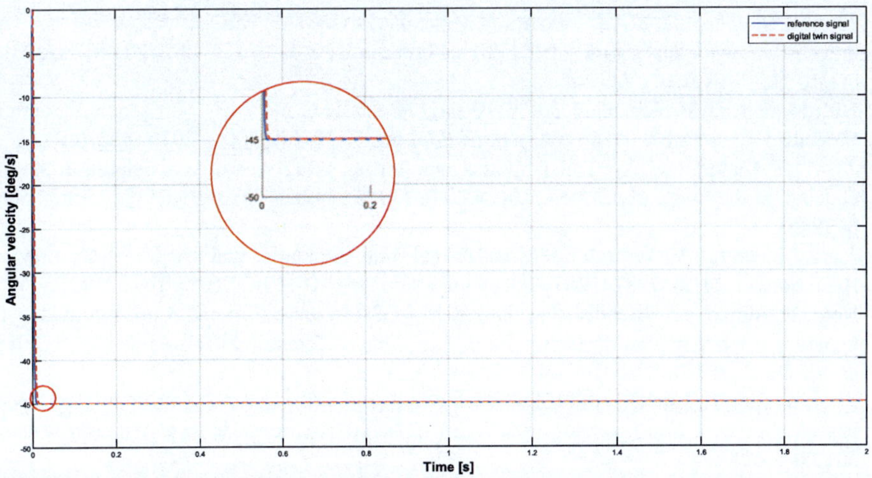

Fig. 12. Overlapping plots of the reference angular velocity signal taken from the Transform Sensor Block and angular velocity signal read from the digital twin of the gyroscope over the time, with zoom on the delay between the signals.

5 Conclusion and Outlook

This work has demonstrated that a virtual environment is a powerful tool for evaluating Body-Attached Sensor Networks. The human hand model presented was able to translate several basic gestures into realistic simulations. For this purpose, only the motion data collected during subject tests had to be implemented into the simulation environment. Furthermore, it was shown that the newly introduced digital twins of sensors, which can be easily attached to the human hand model, generated realistic measurement signals during the simulation. In the future, the library of digital twins of sensors will be enlarged to enable simulation of more complex sensor networks. Eventually, this approach will be used to vary the position and orientation of the sensors and collect measurements for a variety of conditions to systematically evaluate different setups and find the best architecture to develop physical Body-Attached Sensor Networks for gesture recognition.

Acknowledgements. Funded by the Deutsche Forschungsgemeinschaft (DFG, German Research Foundation) – Project-ID 416228727 – SFB 1410.

References

1. Van Dam, K., Pitchers, S., Barnard, M.: Body area networks: towards a wearable future. In: Proceedings of WWRF Kick off Meeting, Munich, Germany (2001)
2. Ullah, S., Higgins, H., Braem, B., et al.: A comprehensive survey of wireless body area networks: on PHY, MAC, and network layers solutions. J. Med. Syst. **36**(3), 1065–1094 (2012). https://doi.org/10.1007/s10916-010-9571-3

3. Ahmad, N., Ghazilla, R.A.R., Khairi, N.M., et al.: Reviews on various inertial measurement unit (IMU) sensor applications. Int. J. Signal Process. Syst. **1**(2), 256–262 (2013). https://doi.org/10.12720/ijsps.1.2.256-262
4. Hill, M., Hoena, B., Kilian, W., et al.: Wearable, modular and intelligent sensor laboratory. Procedia Eng. **147**, 671–676 (2016). https://doi.org/10.1016/j.proeng.2016.06.270
5. Arif, M., Kattan, A.: Physical activities monitoring using wearable acceleration sensors attached to the body. PLoS ONE **10**(7), 1–16 (2015). https://doi.org/10.1371/journal.pone.0130851
6. Rana, M., Mittal, V.: Wearable sensors for real-time kinematics analysis in sports: a review. IEEE Sens. J. **21**(2), 1187–1207 (2021). https://doi.org/10.1109/JSEN.2020.3019016
7. Neto, P., Simão, M., Mendes, N., Safeea, M.: Gesture-based human-robot interaction for human assistance in manufacturing. Int. J. Adv. Manuf. Technol. **101**(1–4), 119–135 (2018). https://doi.org/10.1007/s00170-018-2788-x
8. Fang, B., Sun, F., Liu, H., et al.: 3D human gesture capturing and recognition by the IMMU-based data glove. Neurocomputing **277**, 198–207 (2018). https://doi.org/10.1016/j.neucom.2017.02.101
9. Abualola, H., Ghothani, H.A., Eddin, A.N., et al.: Flexible gesture recognition using wearable inertial sensors. In: 59th International Midwest Symposium on Circuits and Systems (MWSCAS), Abu Dhabi, UAE, pp. 1–4. IEEE (2016)
10. Liu, M., Fang, S., Dong, H., et al.: Review of digital twin about concepts, technologies, and industrial applications. J. Manuf. Syst. **58**, 346–361 (2020). https://doi.org/10.1016/j.jmsy.2020.06.017
11. Raveendranathan, N., Galzarano, S., Loseu, V., et al.: From modeling to implementation of virtual sensors in body sensor networks. IEEE Sens. J. **12**(3), 583–593 (2012). https://doi.org/10.1109/JSEN.2011.2121059
12. Delp, S.L., Anderson, F.C., Arnold, A.S., et al.: OpenSim: open-source software to create and analyze dynamic simulations of movement. IEEE Trans. Biomed. Eng. **54**(11), 1940–1950 (2007). https://doi.org/10.1109/TBME.2007.901024
13. AnyBody Technology Homepage. https://www.anybodytech.com/software. Accessed 29 Jan 2021
14. Lee, K., Lee, J., Woo, B., et al.: Modeling and control of a articulated robot arm with embedded joint actuators. In: 2018 International Conference on Information and Communication Technology Robotics (ICT-ROBOT), Busan, South Korea, pp. 1–4. IEEE (2018)
15. Eldirdiry, O., Zaier, R.: Modeling biomechanical legs with toe-joint using simscape. In: 11th International Symposium on Mechatronics and its Applications (ISMA), Sharjah, UAE, pp. 1–7. IEEE (2018)
16. Milanowski, H., Pilat, A.K.: Comparison of Identified and SimScape model of human leg motion. In: 2020 International Conference Mechatronic Systems and Materials (MSM), Bialystok, Poland, pp. 1–6. IEEE (2020)
17. Walking Robot Modeling and Simulation » Racing Lounge - MATLAB & Simulink. https://blogs.mathworks.com/racing-lounge/2019/12/20/walking-robot-modeling-and-simulation. Accessed 01 Feb 2021
18. Nawrocki, W.: Measurement Systems and Sensors, 2nd edn. Artech House, Norwood (2016)
19. Visualize and explore multibody models - MATLAB. https://www.mathworks.com/help/physmod/sm/ref/mechanicsexplorer-app.html. Accessed 08 Feb 2021

Mobile Robot Navigation in Indoor Environments: Comparison of Lidar-Based 2D SLAM Algorithms

Riadh Dhaoui[1](✉) and Amine Rahmouni[2]

[1] Ruhr University Bochum, Universitätsstr. 150, 44801 Bochum, Germany
riadh.dhaoui@ruhr-uni-bochum.de
[2] LISPEN Laboratory, Arts et Métiers, 8 Bd Louis XIV, 59000 Lille, France
mohamed_amine.rahmouni@ensam.eu

Abstract. In this paper, a comparative study on lidar-based 2D Simultaneous Localization and Mapping (SLAM) algorithms in Robot Operating System (ROS) is addressed. These mapping algorithms are tested and compared in both simulations in ROS and experimentally on TurtleBot3 mobile robot. The experimental results are collected under the same conditions to evaluate the performance of the considered techniques. A generalized performance metric which is based on the Iterative Closest Point method, structure similarity index and Hausdorff distance are evaluated to conduct this comparison study. In addition, the Central Processing Unit (CPU) load and memory usage of each algorithm is considered as well. The present work provides a better insight on the performances of each provided SLAM algorithm to adopt the better solution for a given application.

Keywords: Simultaneous Localization and Mapping (SLAM) · Mobile robot · Indoor navigation · Robot Operating System (ROS)

1 Introduction

Facing the evolution of autonomous mobile robots, SLAM techniques are in the focus of interest within theoretical and practical application. It refers to a process in which the robot builds a map of its environment while simultaneously localizes itself relatively to that map [1]. Several SLAM algorithms have been developed in the past years to solve the navigation issues. Some of them have been published in ROS [2, 3]. However, there is no systematical approach to find the best-fit algorithm for a particular problem. For this aim, these algorithms need to be compared firstly. Moreover, there is no standard method that gives exact information about the accuracy of each algorithm. While some SLAM algorithms are presented and compared in other works, see for instance [1, 4–6], there is no standard method that gives exact information about the accuracy of each algorithm. However, the previously mentioned papers address hardware, evaluation metrics and development environments that differ from present work. Furthermore, there is no explicit agreement on which benchmarks to use, since SLAM algorithms are based

C. Rizzi et al. (Eds.): ADM 2021, LNME, pp. 569–580, 2022.
https://doi.org/10.1007/978-3-030-91234-5_57

on different techniques. Therefore, it is up to each user to choose the appropriate method. However, the performance aspects such as map quality, localization errors and computer performance should be considered to select suitable benchmarks. Thus, ROS-based 2D SLAM algorithms need to be compared and studied first through their performances in a simulation and preferably in a real environment.

The problem of creating maps is just one of the typical problems encountered in mobile robotics navigation. Models of the environment are needed for a variety of applications such as transportation, cleaning, rescue, and various other robotic tasks. Many of the probabilistic designs utilized to solve the trouble of mapping rely upon Bayes filter [7]. For instance, Kalman Filters (KFs), Extended Kalman Filters (EKFs), and Particle Filters (PFs) are among the most popular implementations of Bayes filters [7]. The KF is an algorithm that approximates the state of a system from measured data. The algorithm encompasses a two-step procedure [8]. Firstly, there is the prediction step in which the state of the system in the next step is predicted taking into consideration the previous measurements. Secondly, there is an update step in which the current state of the system is estimated using the measurement during that point in time. The filter iterates the previously presented procedure, steadily enhancing the accuracy. PFs are an additional application of Bayes filters. This filter is one of the consecutive Monte-Carlo techniques [9] and also is a recursive state estimator. To predict the system state, a chosen variety of particles is initially scattered evenly over the entire application room. In the following step, each particle is compared to the measurement of the system, for instance through sensors, and depending upon the congruence, a particular weighting is assigned. By utilizing a so-called resampling algorithm, the particles are filtered according to their probabilities. In the following cycle, the particles are resampled, considering the placements of the fragments with high probabilities in the previous cycle. The state estimation making use of the particles is executed until, ideally, all particles remain in one position and the definition of system state is unambiguous [10]. A modified PF technique is proposed in [11] to deal with a large number of landmarks. Equally vital are graph-based SLAM algorithms, as they cover some weaknesses of PF and EKF strategies [12, 13]. In these SLAM algorithms, the extracted data is employed to create a graph. This type of algorithms was first provided by Lu [12], in which the position constraints were retrieved by the scan matching procedure. In the last few years, the number of SLAM algorithms has risen and the necessity to delineate different SLAM techniques from each other expanded significantly.

On the other hand, visual inspection of the resulting maps does not enable an adequate comparison. So, the requirement to accurately evaluate the results demands a more precise technique, hence a quantitative range. For example, in [15], a metric for juxtaposing SLAM algorithms was developed, wherein the outcome is not assessed using a reference, however rather by thinking about the poses of the robot throughout data acquisition.

The SLAM evaluation methods are mostly dependent on standard datasets. Nevertheless, they are not fully compatible with ROS. However, most 2D SLAM algorithms in ROS are firstly exposed and then compared. The occupancy grids at the final output of a given SLAM algorithm are evaluated using the map similarity metric. The map quality is chosen over the pose estimation errors for issues related to localization. The aim is to provide a complete quantitative comparison study concerning these algorithms for mobile robot navigation. Thus, the ROS user community is enabled to select the suitable algorithm for their requirements.

2 Evaluated 2D SLAM Algorithms

Four SLAM algorithms are considered in this study:

- **Gmapping**: This algorithm uses a horizontal laser scan of the environment, with which it builds an occupancy grid map. It is specifically designed for indoor environments [3]. Gmapping is a Rao blackwellized particle filter-based method that maintains a separate grid map in memory for each particle. Using the motion model and the odometry data of the robot, the most probable subsequent position can be determined for each particle. This position is optimized with a Gaussian curve and a probability calculation is performed in which the sensor data is compared with the corresponding map and a probability value is determined for each particle to produce the final particle list [16].
- **Hector-SLAM:** It is a SLAM method developed by Kohlbrecher et al. (2011). It uses the Gauss-Newton method for mapping and the EKF for positioning. For pose estimation, the EKF requires the estimated position of the scan matcher, i.e. the Gauss-Newton method. This latter is a scan matcher that attempts to minimize the error between two different scans so that they can be overlaid with the least variance. The position is updated based on its old state, the measured and the estimated position of the scan matcher. The advantage of Hector-SLAM is that no odometry is required but it may be added to the model to make the estimation more accurate [17].
- **Karto-SLAM**: It is developed by Karto Robotics of SRI International based on graph technique. One may note that a node from the graph represents the robot position and sensor measurements. For each new node, the map considers constraints on the edges which are then computed via the node's configuration [12]. However, the graph-based approach is generally proved to be more reliable for mapping a large-scale environment.
- **Cartographer:** This algorithm enables real-time SLAM via multiple sensor configurations [18]. The main idea is to divide SLAM problem into two different sub-problems: local SLAM and global SLAM. The main goal here is not to deal with large maps or representations when mapping a new area. Instead, a sub-map is created for the local area and updated with each new scan. Each scan is also tested against the sub-map using a Ceres-based scan matcher to perform position optimization. The global SLAM thread will then have a collection of sub-maps to compute the entire map [19].

3 Comparison Study

All SLAM algorithms above are tested in 3D simulation in ROS-Gazebo environment and experimentally for a mobile robot navigation. Regarding the optimal conditions in

Gazebo environment, such as noise-free odometric, SLAM algorithms consider uncertainty in measurement. The mobile robot was teleoperated in experimental tests. The update rate of the LSD01 used in both simulations and experiments is 5 Hz, which ensures low angular and linear velocities for the robot motion. The output of each SLAM algorithm was generated by 2D occupancy grid map. To conduct the comparison study based on the quality of the maps obtained, an analysis of the error between the generated and the ground truth maps is established. The conducted analysis process is as follows: the resulting map of each algorithm is firstly binarized. This latter contains only boundaries and obstacles information of the chosen scenario. Afterwards, the binarized map is aligned narrowly with the respective ground truth using Iterative Closest Point (ICP) [20]. Another considered metric for assessing the quality of SLAM algorithms is the Structural Similarity Index (SSIM), which attempts to quantify the visible difference between a distorted image (SLAM map) and a reference image (ground truth map) is used. The SSIM algorithm used is provided by Imatest Software [21]. SSIM determines the actual perceptual difference between two similar images. In this case, the localization error is given by the difference between the estimated position and the actual position. Thus, the Hausdorff distance method [22] is used for this purpose.

3.1 Simulation Tests

Gazebo simulation is performed using a ground truth map which is used in both simulation and real-world experiments. The simulation environment is shown in Fig. 1.

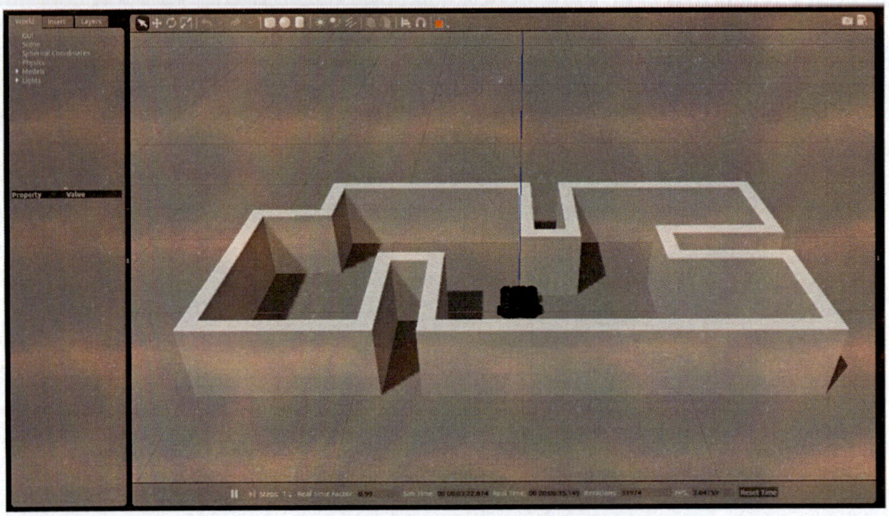

Fig. 1. Simulation environment in Gazebo simulator.

The map enables the analysis of the behavior of the SLAM algorithms in a compact environment with less features. This is important to analyze the dependency on landmarks of each approach. In the simulation tests, the model of the range sensor was defined as the one which is used in real-world experiments: the LDS01 which has a maximum range of about 3.5 m. All the sensing and actuators data issued from the robot (odometry, velocity commands, etc.) was recorded previously and then played back for each SLAM algorithm. One may note that all SLAM packages were tested under the same conditions in the aim to conduct a better comparison study. In experiment tests, the default parameters were used. For example, the default number of particles for the Gmapping SLAM algorithm is 100. The generated SLAM maps from simulation tests are shown in Fig. 2.

Fig. 2. Generated SLAM maps from Gazebo simulator.

Table 1 shows the ICP results after 24 iterations. The rotation matrix and translation vector are computed for each iteration, which both have to be applied to the SLAM map and have to match the ground truth map. Karto-SLAM and Hector-SLAM generated the map with the lowest error. On the other hand, Cartographer has a slightly greater error,

Table 1. ICP-Error between Ground-Truth and SLAM maps.

SLAM method	ICP
Gmapping	max = 2.0099
	mean = 1.8489
	min = 1.8352
Hector-SLAM	max = 1.5881
	mean = 1.4923
	min = 1.4471
Karto-SLAM	**max = 1.3612**
	mean = 1.2981
	min = 1.6628
Cartographer	max = 1.7418
	mean = 1.6671
	min = 1.6628

while the results of Gmapping presents the highest error value. Karto-SLAM is usually more efficient in maintaining a map of an indoor environment, and this justifies the quality of the resulting map. Also, the scan matching process of Hector-SLAM showed its efficiency.

In the next step, the respective maps were compared with ground truth using SSIM index. The SSIM can have a value between 0 and 1. The higher the SSIM value, the more similar the structure of the SLAM map is to the structure of the ground truth map. Table 2 lists the measured SSIM index for each SLAM map. As can be seen in Table 2, the Cartographer algorithm has the lowest SSIM value of 0.9280, followed by Gmapping with 0.9141. As with the ICP test, Karto-SLAM and Hector-SLAM perform best in the SSIM test. With a value of 0.9348, Hector-SLAM has a slightly better SSIM value than Karto-SLAM. It can be concluded that Hector and Karto-SLAM also produce the best maps with the lowest possible noise.

Table 2. SSIM Index between Ground-Truth and SLAM maps.

SLAM method	SSIM value
Gmapping	0.9141
Hector-SLAM	**0.9348**
Karo-SLAM	0.9345
Cartographer	0.9280

In the SLAM comparison, it is also important to examine not only the map quality, but also the localization error. This is measured with the Hausdorff distance method. The results in Table 3 show that Hector-SLAM has the smallest localization error of all algorithms. Gmapping came second in this test, followed by Cartographer and finally Karto-SLAM. In conclusion, one may note that in all three tests Hector-SLAM provides the best results of all four algorithms.

Table 3. Adjustment error between Ground-Truth and SLAM paths using Hausdorff distance

SLAM method	Hausdorff distance
Gmapping	0.0307
Hector-SLAM	**0.0294**
Karo-SLAM	0.0534
Cartographer	0.0323

Fig. 3. Indoor environment.

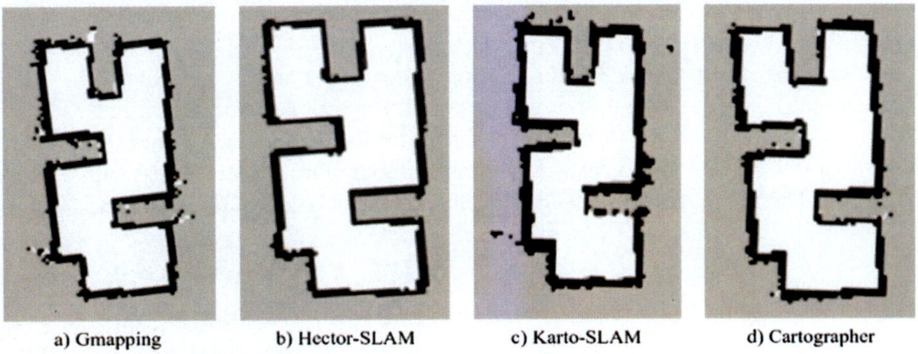

Fig. 4. Generated SLAM maps of the real environment.

3.2 Real-World Tests

In the real-world experimental tests presented in Fig. 3, the mapping was performed using a Turtlebot3 mobile robot which is equipped with an LDS-01 and a Raspberry Pi 3 B +, while running Ubuntu 16.04 and ROS Kinetic. The algorithm parameters used in the simulation tests were again adopted. Figure 4 shows that all four techniques were able to successfully map the scenario.

Since visual observation alone is not sufficient, the next step was an ICP test for SLAM algorithms. The results of the ICP test are shown in Table 4.

Table 4. ICP-Error between Ground-Truth and SLAM maps.

SLAM method	ICP
Gmapping	max = 2.0221
	mean = 1.6484
	min = 1.6231

(continued)

Table 4. (*continued*)

SLAM method	ICP
Hector-SLAM	max = 2.2629
	mean = 1.7120
	min = 1.6566
Karto-SLAM	max = 2.0752
	mean = 2.0706
	min = 2.0294
Cartographer	**max = 1.4900**
	mean = 1.4885
	min = 1.4619

Contrary to our visual evaluation, Cartographer scores best in the ICP test.

After the ICP test, the SSIM test was performed on the respective algorithms. Table 5 shows the results of the SSIM test. In this test procedure, Karto-SLAM scores best with a value of 0.9232, as in our visual inspection. Cartographer is second best with a value of 0.9143. The last places with the lowest values are occupied by Gmapping and Hector-SLAM.

Table 5. SSIM index between Ground Truth and SLAM maps.

SLAM method	SSIM value
Gmapping	0.9109
Hector-SLAM	0.9121
Karo-SLAM	**0.9232**
Cartographer	0.9143

In the last step, the localization error of the respective algorithms is calculated. The results in Table 6 show that Hector-SLAM has the smallest localization error of all algorithms in this test. While the Gmapping algorithm achieves the second-lowest localization error with a minimum deviation of 0.0025 compared to Hector-SLAM.

Beyond the error analysis, the comparison study on the CPU load and memory usage in a computer equipped with an Intel Core i7-3630QM and 8GB RAM running each algorithm is presented in Fig. 5. The Gmapping algorithm presented the highest percentages of CPU usage and memory usage. Moreover, the values obtained are quite distant from the other three SLAM algorithms. This can be explained by the process that Gmapping performed an additional positioning step (ICP algorithm) for 100 particles. The resources occupied by the other three algorithms during the experiments are similar, as shown in Fig. 5.

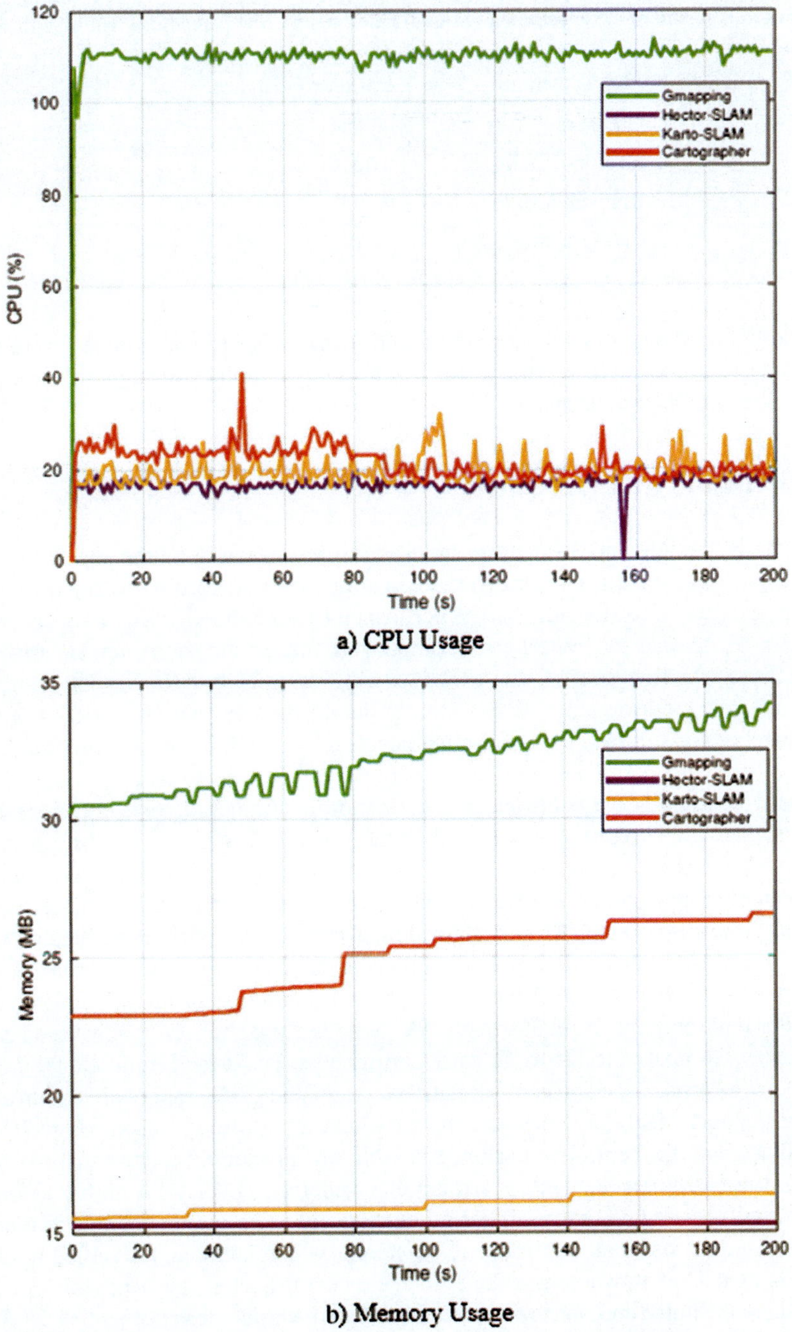

Fig. 5. Real-world CPU load and memory usage evolution of each SLAM method.

Table 6. Adjustment error between Ground-Truth and SLAM paths using Hausdorff distance

SLAM method	Hausdorff distance
Gmapping	0.0084
Hector-SLAM	**0.0056**
Karo-SLAM	0.0093
Cartographer	0.0159

The CPU and memory analyses reveal that all three algorithms analyzed are quite efficient in terms of required resources and can be adopted online during field experiments to map generic 2D scenarios.

4 Discussion and Outcome

In this study, the ICP and SSIM tests are used for the evaluation of the generated SLAM maps and Hausdorff distance for the localization error. According to the results from our experimental tests, Hector-SLAM performed better generally. The map constructed in Hector-SLAM has the lowest error compared to the ground truth map. Hector-SLAM also achieved the lowest error rate in the localization tests, both in Gazebo simulations and in the real environment. This is mainly due to the fact that Hector-SLAM uses a scan matcher and odometry to determine positions. In addition to the lowest error rate, the algorithm also had the lowest computational effort. On the other hand, Gmapping provides the second-best solution for localization in simulation tests and in real tests. The drawback of Gmapping compared to the other algorithms is the higher computational effort. The computing power is because this algorithm uses a combination of scan matching and odometry data to reduce the number of particles in the mapping process. Similar to Hector-SLAM algorithm, Karto-SLAM performed better in terms of map quality in both simulation and real-world environment tests. One may note that the graph-based SLAM algorithm is usually more efficient in maintaining a map of an indoor environment. Both Karto-SLAM and Cartographer are graph-based SLAM approaches. Compared to Karto-SLAM, Cartographer performed significantly worse in the conducted tests. Especially in simulation tests, where Cartographer algorithm always performed poorly. Only in real environments where it performed reasonably well in ICP and SSIM tests. As mentioned earlier, Karto-SLAM produces accurate maps with low CPU and memory usage, while Cartographer generates maps with higher errors and is identified with significantly higher memory usage. The reason for this is that Cartographer requires several sub-maps to be stored, which are then assembled to build a complete map. One may note that the critical element in such a case is the robot environment where the noise and perturbation could affect the input measurement of the SLAM algorithms and therefore the quality of the map.

5 Conclusion

In this work, four representative 2D SLAM algorithms available in ROS are tested through Gazebo simulation and experimentally in real-world environment on a mobile robot. The focus was on the selection of the optimal SLAM algorithm for robot navigation in static indoor environments which is achieved through the presented comparative study.

In future work, some ideas can be retrained. On one hand, there is a wide range of other algorithms for 3D-SLAM, such as the Cartographer algorithm which can be tested. On the other hand, this algorithm offers a variety of configurations regarding scan matching and position optimization. Whereas Gmapping offers the option to set the number of particles and the re-scan threshold individually. Furthermore, several laser scans could be used instead of only one to achieve an even more accurate map. Sensor fusion is another aspect in which IMU and odometry data are combined to be able to calculate an even more precise position or localization of the robot. These are all aspects that enable further important insights to set up mobile robots even more specifically for their use in special environments.

References

1. Yagfarov, R., Ivanou, M., Afanasyev, I.: Map comparison of Lidar-based 2D SLAM algorithms using precise ground truth. In: 2018 15th International Conference on Control, Automation, Robotics and Vision (ICARCV), 18–21 November 2018. 2018 15th International Conference on Control, Automation, Robotics and Vision (ICARCV), Singapore, 11/18/2018–11/21/2018, pp. 1979–1983. IEEE, Piscataway (2018). https://doi.org/10.1109/ICARCV.2018.8581131
2. Joseph, L.: Learning robotics using Python. Design, simulate, program, and prototype an autonomous mobile robot using ROS, OpenCV, PCL, and Python. Packt Publishing, Birmingham (2018)
3. Stachniss, C., Frese, U., Grisetti, G.: OpenSLAM. Give your algorithm to the community (2007). https://openslam-org.github.io/. Accessed 26 July 2021
4. Santos, J.M., Portugal, D., Rocha, R.P.: An evaluation of 2D SLAM techniques available in Robot Operating System. In: 2013 IEEE International Symposium on Safety, Security, and Rescue Robotics (SSRR), pp. 1–6 (2013). https://doi.org/10.1109/SSRR.2013.6719348
5. Fang, L., et al.: Comparative evaluation of time-of-flight depth-imaging sensors for mapping and SLAM applications. In: IEEE International Conference on Robotics and Automation (ICRA) (2016)
6. Filatov, A., Filatov, A., Krinkin, K., Chen, B., Molodan, D.: 2D SLAM quality evaluation methods. In: Balandin, S.I., Tyutina, T. (eds.) Proceedings of the 21st Conference of Open Innovations Association FRUCT. Helsinki, Finland, 6–10 November 2017. 2017 21st Conference of Open Innovations Association (FRUCT), Helsinki, 11/6/2017–11/10/2017, pp. 120–126. IEEE, Piscataway (2017). https://doi.org/10.23919/FRUCT.2017.8250173
7. Thrun, S., Burgard, W., Fox, D.: Probabilistic Robotics. Intelligent Robotics and Autonomous Agents Series. MIT Press, Cambridge; London (2006)
8. MathWorks: What are Kalman-Filters? https://www.mathworks.com/discovery/kalman. Accessed 26 July 2021
9. Metropolis, N., Ulam, S.: The monte carlo method. J. Am. Stat. Assoc. **44**, 335–341 (1949)
10. Sievers, C., Scharfenberg, F., Schneider, U.: Projekt 70c: Labyrinth SLAM mit EV3. https://wiki.hshl.de/wiki/index.php/Projekt 70c: Labyrinth SLAM mit EV3. Accessed 26 July 2021

11. Konolige, K., Grisetti, G., Kümmerle, R., Burgard, W., Limketkai, B., Vincent, R.: Efficient sparse pose adjustment for 2D mapping. In: 2010 IEEE/RSJ International Conference on Intelligent Robots and Systems, pp. 22–29 (2010)
12. Lu, F., Milios, E.: Globally consistent range scan alignment for environment mapping. Auton. Robot. **4**, 333–349 (1997)
13. Kümmerle, R., et al.: On measuring the accuracy of SLAM algorithms. Auton. Robot. **27**, 387–407 (2009)
14. Kummerle, R., Grisetti, G., Strasdat, H., Konolige, K., Burgard, W.: G2o: a general framework for graph optimization. In: Bicchi, A. (ed.) IEEE International Conference on Robotics and Automation (ICRA), Shanghai, China, 5/9/2011–5/13/2011, pp. 3607–3613. IEEE, Piscataway (2011). https://doi.org/10.1109/ICRA.2011.5979949
15. Steux, B., El Hamzaoui, O.: tinySLAM: a SLAM algorithm in less than 200 lines C-language program. In: 2010 11th International Conference on Control Automation Robotics & Vision, pp. 1975–1979 (2010)
16. Werner, S.: Variability Modelling in Localization and Mapping Algorithms. Thesis, Technical University Dresden (2017). https://nbn-resolving.org/urn:nbn:de:bsz:14-qucosa-172457. Accessed 26 July 2021
17. Kohlbrecher, S., Stryk, O. von, Meyer, J., Klingauf, U.: A flexible and scalable SLAM system with full 3D motion estimation. In: IEEE International Symposium on Safety, Security, and Rescue Robotics (SSRR), Kyoto, Japan, 11/1/2011–11/5/2011, pp. 155–160. IEEE, Piscataway (2011). https://doi.org/10.1109/SSRR.2011.6106777
18. Google Cartographer ROS: Algorithm walkthrough for tuning. https://google-cartographer-ros.readthedocs.io/en/latest/algo_walkthrough.html#overview. Accessed 27 July 2021
19. Hess, W., Kohler, D., Rapp, H., Andor, D.: Real-time loop closure in 2D LIDAR SLAM. In: Okamura, A., Menciassi, A. (eds.) 2016 IEEE International Conference on Robotics and Automation (ICRA), Stockholm, Sweden, 5/16/2016–5/21/2016, pp. 1271–1278. IEEE, Piscataway (2016). https://doi.org/10.1109/ICRA.2016.7487258
20. Kak, A.: Version: 2.1.0 (2017). https://engineering.purdue.edu/kak/distICP/ICP-2.1.0.html. Accessed 26 July 2021
21. Imatest LLC: SSIM: Structural Similarity Index. https://www.imatest.com/docs/ssim/. Accessed 26 July 2021
22. Zhang, J.-W., Han, G.-Q., Wo, Y.: Image registration based on generalized and mean Hausdorff distances. In: 2005 International Conference on Machine Learning and Cybernetics, pp. 5117–5121 (2005)

An Integrated Modular Design and Group Decision-Making Method in Complex Design Environments

Cristina Renzi(✉) and Francesco Leali

Department of Engineering Enzo Ferrari, University of Modena and Reggio Emilia, Modena, MO, Italy
cristina.renzi@unimore.it

Abstract. This paper deals with the design of novel products, conceived in a complex design environment, under strict constraints and several disciplines involved in. As an application, the design of micro-bioreactors for tissue-engineering is proposed. To date, in-vitro two-dimensional (2D) culture systems have been widely used to study the mechanisms underlying cell biology. However, 3D culture systems and platforms have showed to better mimic the real tissue environment. The design of a 3D culture system of bioreactor is challenging due to several criteria derived from several disciplines, as biotechnology, engineering, and manufacturing. To this aim, a two-step integrated modular design and group multi-criteria decision-making method is proposed. In step one, a functional analysis is carried out, first, to point out the interfaces between parts and the function of each part in the assembly. Generated design alternatives have been selected by means of a group Fuzzy-TOPSIS technique, in step two. The proposed integrated design method has shown efficient in the early stages of design of novel products in complex design environments.

Keywords: Conceptual design · Modular design · Group decision making · Micro bio reactors · Fuzzy-TOPSIS

1 Introduction

1.1 Background

The design of microbioreactors (MBR) is challenging and complex, due to several criteria, often conflicting which need to be considered concurrently, as well as several actors involved. The use of 3D platforms and bioreactors in tissue engineering is aimed at recreating in vitro the structures of the native tissue, overcoming some drawbacks of the bidimensional environment. As an example, 3D platforms are showed to reduce the gap between the cell culture and the tissue environment [1] Further advantages and drawbacks have been pointed out in [2]. In the literature, bioreactors have been largely investigated, for what concerns fluid dynamics, design of turbulence, shear stress and their correlation with the enhancement of growth of the tissue inside the chamber. The

C. Rizzi et al. (Eds.): ADM 2021, LNME, pp. 581–592, 2022.
https://doi.org/10.1007/978-3-030-91234-5_58

major quest is to recreate the natural cellular niche and the impact on cell behavior. As an example, the VITVO bioreactor [3] is compared to microfluidic based techniques for tumor cells growth, with good results [2].

In the literature, few papers deal with the methodology followed in the early design stages to conceive this kind of biomedical devices. Hence, in this paper, starting from the design requirements, involving experts from different disciplines in the design of bioreactors. the steps required for the generation and selection of conceptual design alternatives have been carried out, in a two-step integrated design and decisional environment. The conceptual design of the device is defined under a modular design approach. The functional analysis of the device is carried out first, to identify the main parts of an MBR. Hence a morphology study is carried out to evaluate possible connections between parts. Among nine design concepts which have been generated, the most suitable alternative has been selected by means of a group multicriteria design making approach. The uncertainty of the verbal judgments of the decision makers has been handled by means of fuzzy numbers.

2 Related State of the Art

MCDM have been subdivided into the multi attribute and the multi objective decision-making methods. In particular, the Multi attribute decision-making methods include the Value measurement methods, the Outranking methods and the Distance-based methods. TOPSIS (Technique for Order Preference by Similarity to Ideal Solution) is a Distance based method [4] which is commonly used to select among conceptual design alternatives in the early design stages (e.g. [5–8]). The idea laying at the basis of the TOPSIS method is that the chosen alternative should have the shortest distance from the ideal solution and the farthest from the negative-ideal solution. Hence one of the most evident advantages of TOPSIS in relation to other MCDM outranking techniques is the quick identification of the best alternative.

For group decision making, Fuzzy sets are efficient for capturing the uncertainty in translating the decision makers' preferences in the selection of design products as largely reported in the literature and remarked also by the authors (e.g. [8]). Several examples in literature use combined techniques to improve computational efficiency. In this paper, the fuzzy-TOPSIS algorithm is the one reported in [9], in which ranking is evaluated with less computations than original fuzzy TOPSIS.

3 Method

An integrated modular design and group decision making approach is carried out in this paper. Integrated design methods have been often used in complex design environments (e.g. [10]). A modular product is a product with some parts groupings performing a single function or a group of functions. This is aimed at enhancing the design variety, reducing cost and leading time, decomposing complex problems into simpler ones, enhancing design flexibility and promoting green life cycle engineering and reusing [11]. Modularity is aimed at increasing the number of common parts, according to the function each part holds within the assembly. Design for Manufacturing and Assembly (DFMA) rules have been adopted in the design of the connections to satisfy safety requirements and to easy up opening/closing operations. Moreover, DFMA is aimed at lowering the total cost of the product by reducing the number of components, assembly time and the complexity of assembly operations. A "two-steps" integrated modular design and group decision making approach has been carried out (Fig. 1). The first step is the modular design approach, starting with the functional analysis to point out the interconnections between the parts of the assembly. This step ends with the generation of the design alternatives. The second step is the group decision making, aiming at considering all the possible criteria in the selection of the most suitable design alternative, among the generated ones. This step ends with the ranking of the design alternatives and the selection of the most suitable one. In the following subsections, these steps have been described.

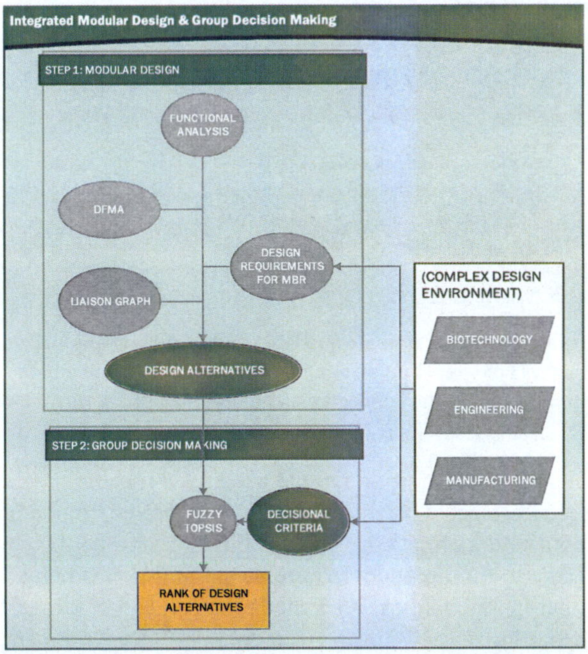

Fig. 1. A two-step integrated modular design and group decision making approach

3.1 Step 1: Modular Design for the Generation of Alternatives

A functional analysis is carried out first, to point out the functions of each part in the assembly and their interaction with the environment. In particular, the components of the MBR have been listed first. Then, their relative interfaces have been defined, and finally, the interface with the environment in which the MBR is included have been defined. Hence, connections related to the interfaces between parts and the environment have been defined. The MBR assembly can be described by means of a modular design approach, as depicted in Fig. 2. In a modular design approach, the main structure remains the same, while the components change according to the methods used for connecting the parts. Five parts have been defined for a generic MBR design. In particular: A1 is the lower cover, A2 is the upper cover, A3 is the oxygenating membrane, A4 is the gasket between the membrane and the external environment, A5 is the gasket between the lower and the upper cover. Cij (for all [i, j = 1:5]) are the connections/contacts between the parts in the assembly. In this context, the connecting method C12 between A1 and A2, as well as the one C23 between A2 and A3, define the possible alternatives to be generated for the MBR.

The lower cover A1 has four exchanges channels with the external environment, dedicated to the inlet/outlet of the culture liquid. Inlets and outlets have been provided with microchannels at different heights to promote fluid shear stresses on the surface of the tissue. In Fig. 2, the liaison graph [12] depicting the connections between the parts of the MBR has been reported.

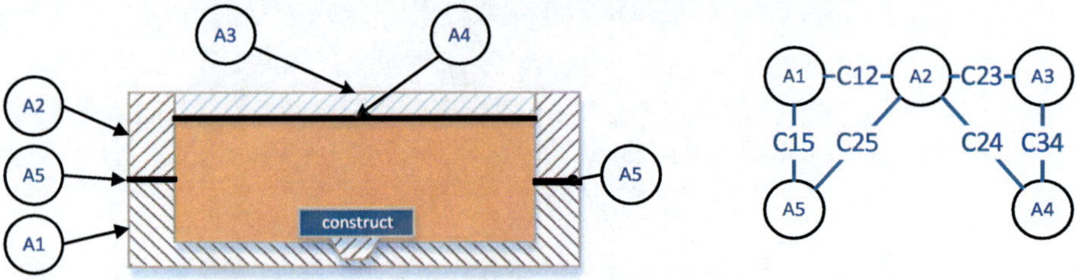

Fig. 2. Modular design of a micro bioreactor (MBR) and corresponding liaison graph. A1 lower cover, A2 upper cover, A3 oxygenating membrane, A4 gasket at the membrane, A5 gasket at the cover. Cij (for all i, j = [1:5]) are the connections/contacts between parts of the assembly.

Connections between parts respond to the need for a specific product functionality. The type of connection selected need to respond to the product functionality requested. The choice of a connection affects the product functionality, so that connections have been selected in order to fulfil product requirements. In this viewpoint, connections are correlated to product functionalities. A1 include the cell construct and host the culture terrain with A2 separating the cell culture from the external environment. A3 guarantees the oxygenation to the cells, A4 and A5 provide protection of the cell culture terrain from the contamination avoiding contact with the external environment.

3.2 Step 2: Fuzzy-TOPSIS for the Selection of Design Alternatives

The verbal judgments of the decision makers have been described using Trapezoidal Fuzzy Numbers. A Fuzzy Number is described by its membership function $\mu_A(x)$ with the following characteristics:

$\mu_A(x) = 0, \forall x \in (-\infty, \alpha] \cup [\delta, \infty)$

$\mu_A(x) = 1, \forall x \in [\beta, \gamma];$

$\mu_A(x)$ increases monotonically in [α, β] and decreases in [γ, δ].

or it may also take these values: $\alpha = -\infty$ or $\alpha = \beta\ or\ \beta = \gamma\ or\ \partial = \infty$. The fuzzy number is described by a 4-tuple [α, β, γ, δ] with straight line segments for $\mu_A(x)$

in [α, β] and [γ, δ]. The membership function of a Trapezoidal Fuzzy Number has the mathematical form: [0,1] as in (1).

$$\mu_A(x) = \begin{cases} \frac{(x-a)}{(b-a)} & for\ a \leq x \leq b \\ 1 & for\ b \leq x \leq c \\ \frac{(d-x)}{(d-c)} & for\ c \leq x \leq d \\ 0 & otherwise \end{cases} \tag{1}$$

The procedure used by the TOPSIS to evaluate the design alternatives is described below and depicted in Fig. 3. In particular, the decision makers express their preference by pairwise comparing the criteria: these linguistic assessments are handled using fuzzy numbers. Criteria are then put in order of priority and a matrix is provided with the values of the mean aggregated preference weights. The suitability of each alternative is evaluated, with respect to the criteria, by means of linguistic variables (e.g. Trapezoidal Fuzzy Numbers). Finally, fuzzy score evaluation and a defuzzification procedure is carried out, to generate a ranking vector of the alternatives. A software tool based on Fuzzy-TOPSIS process was implemented by the authors [13] including all the steps of the method. In the adopted fuzzy-TOPSIS approach, similarity is used in place of closeness criteria, as in [9].

In a Fuzzy environment, the similarity between two normalized fuzzy numbers $\tilde{A}$ and $\tilde{B}$ is given by (2), with $S\left(\tilde{A}, \tilde{B}\right) \in [0, 1]$:

$$S\left(\tilde{A}, \tilde{B}\right) = \left(1 - \frac{\sum_{i=1}^{4} |a_i - b_i|}{4}\right) \cdot \frac{\min\left(P\left(\tilde{A}\right), P\left(\tilde{B}\right)\right)}{\max\left(P\left(\tilde{A}\right), P\left(\tilde{B}\right)\right)} \tag{2}$$

$$P\left(\tilde{A}\right) = \sqrt{(a_1 - a_2)^2 + 1} + \sqrt{(a_3 - a_4)^2 + 1} + (a_3 - a_2) + (a_4 - a_1)$$

$$P\left(\tilde{B}\right) = \sqrt{(b_1 - b_2)^2 + 1} + \sqrt{(b_3 - b_4)^2 + 1} + (b_3 - b_2) + (b_4 - b_1)$$

In the fuzzy approach for multicriteria decision-making, the importance weights of the criteria and the ratings of qualitative criteria are verbal judgments that is linguistic variables to be translated into fuzzy numbers.

A. *Preferences and Suitability.* In the first phase, the decision makers provide their preferences for the criteria, using verbal judgments taken from a previously defined Term Set for "preferences". The weights are allocated for each criterion j, (j = 1, ..., i), by the rth decision maker (r = 1, ..., q). In this way, a matrix of (i x q) elements is produced.

B. *Suitability indices (S_{jm}).* Each weight w_{jr} is a four-element vector, representing the trapezoidal number w_{jr}= (a_{jr}, b_{jr}, c_{jr}, d_{jr}). The linguistic values provided by the decision makers are translated into the corresponding numerical values, in a (ixqx4)

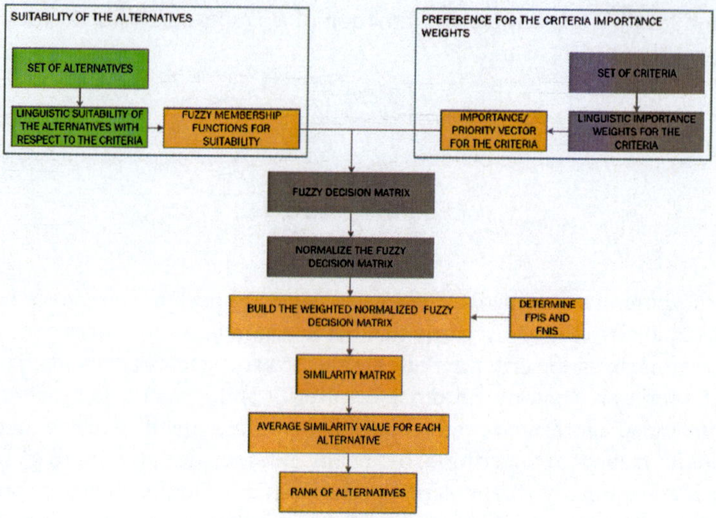

Fig. 3. Fuzzy-TOPSIS procedure

matrix). The q decision makers provide judgments on the suitability of each alternative, with respect to the criteria, among the linguistic variables from the Term Set of "suitability". If $S_{jmr} = (a_{jmr}, b_{jmr}, c_{jmr}, d_{jmr})$ is defined as the suitability assigned to the mth alternative (m = 1, 2…, p), against the jth criterion (j = 1, 2, ..., i), by the rth decision maker (r = 1, 2, ..., q), the suitability is given by i matrices of size (p x q). The suitability indices (S_{jm}) are evaluated for each criterion, as in (3):

$$S_{jm} = q^{-1} \cdot \sum_{r=1}^{q} S_{jmr} \quad (3)$$

C. ***Fuzzy decision matrix*** the matrices of the suitability indices can be collected in a single matrix that has the form of (4), with i criteria and p alternatives, which is defined as the fuzzy-decision matrix.

$$S_{indices\_m,j} = \begin{pmatrix} \tilde{S}_{11} & \dots & \tilde{S}_{1j} & \dots & \tilde{S}_{1i} \\ \dots & \dots & \dots & \dots & \dots \\ \tilde{S}_{m1} & \dots & \tilde{S}_{mj} & \dots & \tilde{S}_{mi} \\ \dots & \dots & \dots & \dots & \dots \\ \tilde{S}_{p1} & \dots & \tilde{S}_{pj} & \dots & \tilde{S}_{pi} \end{pmatrix} \quad (4)$$

D. *Normalization of the fuzzy decision matrix.* First, the set of criteria can be divided into benefit criteria (the larger the rating, the greater the preference) and cost criteria (the smaller the rating, the greater the preference). Therefore, the normalized fuzzy decision matrix can be represented as (5):

$$\hat{R} = [\hat{r}_{ij}]_{mxn} \quad (5)$$

where the members of the matrix are evaluated in a different way, provided they are referred to benefit or cost criteria. If B and C are the sets for benefit criteria and cost criteria respectively, the members of the decision matrix are:

$$r_{ij} = \left(\frac{a_{ij}}{d_j^{\oplus}}, \frac{b_{ij}}{d_j^{\oplus}}, \frac{c_{ij}}{d_j^{\oplus}}, \frac{d_{ij}}{d_j^{\oplus}}\right), j \in B \text{ in which } d_j^{\oplus} = \max_i(d_{ij}), j \in B \text{ and}$$

$$r_{ij} = \left(\frac{a_j^{\ominus}}{a_{ij}}, \frac{a_j^{\ominus}}{b_{ij}}, \frac{a_j^{\ominus}}{c_{ij}}, \frac{a_j^{\ominus}}{d_{ij}}\right), j \in C \text{ in which } a_j^{\ominus} = \min_i(a_{ij}), j \in C.$$

Considering the importance of each criterion, the weighted normalized fuzzy decision matrix is:

$\hat{V} = (\tilde{v}_{ij})_{mxn}$, where $\hat{v}_{ij} = \hat{r}_{ij}(\cdot)\hat{w}_j$.

normalized positive trapezoidal fuzzy numbers can also approximate the elements v_{ij}, for all i, j, according to the weighted normalized fuzzy decision matrix.

E) The positive and negative ideal solutions Hence, the fuzzy positive-ideal solution is: $A^{\oplus} = (\hat{v}_1^{\oplus}, \hat{v}_2^{\oplus}, \ldots, \hat{v}_n^{\oplus})$, in which $\hat{v}_j^{\oplus} = \max_i(v_{ij4})$ and the fuzzy negative ideal solution is: $A^{\ominus} = (\hat{v}_1^{\ominus}, \hat{v}_2^{\ominus}, \ldots, \hat{v}_n^{\ominus})$, where $\hat{v}_j^{\ominus} = \min_i(v_{ij1})$.

F) The similarity matrix. The similarity matrix is evaluated calculating fuzzy similarity. Hence the average similarity is computed by Eq. (6), and the resulting similarity is used as a measure for the rank of the alternatives.

$$S_i^{\oplus} = \frac{1}{i}\sum_{j=1}^{n} S_v(\hat{v}_{ij}, \hat{v}_j^{\oplus}) \tag{6}$$

For each alternative, according to the value of the average similarity, the assessment status for the alternative can be found, as in Table 1. Six conditions have been defined, (from the "not recommended" to the "approved and preferred" status).

Table 1. Assessment status for the alternatives, according to the similarity value.

S_i value	$S_1 \subset [0, 0.2)$	$S_i \in [0.2, 0.4)$	$S_i \in [0.4, 0.6)$	$S_i \in [0.6, 0.8)$	$S_i \in [0.8, 1.0]$
Assessment status	Do not recommend	Recommend with high risk	Recommend with low risk	Approved	Approved and preferred

4 Results and Discussion

4.1 Decisional Criteria and Design Alternatives

In [14] some design requirements for the optimal design of a perfusion bioreactor have been pointed out, as: oxygenation, shear stress, mechanical or chemical stimuli. The hermetic top cover is provided by an oxygenating membrane. The shear stresses are provided by means of two microchannels at different heights, with respect to the position of the construct. The following design criteria have been defined for the selection of the most suitable design alternatives. Three decision makers (an engineer, a biotechnologist, and a manufacturer) have met to list the design criteria for the microbioreactor. Fundamental design requirements as reliability of the assembly connections, stiffness and quality of the oxygenation percentage have been already satisfied in the generation of all the design alternatives. Decision makers listed the criteria for the MBR. A fundamental criterion is the ease of open for the introduction of the construct, the ease of assembly is fundamental for minimizing the operations within sterile environments. The manufacturing ease as well as cost is critical for batch production. Hence, the following five criteria have been defined: C1 Easy opening, C2 Ease of assembly, C3 Cost, C4 Manufacturing ease, C5 Aesthetics. The modular design as well as the functional analysis is fundamental for reducing the number of parts in the assembly, in a DFMA approach. In this approach, nine conceptual design alternatives for the concept design have been generated as a combination of the functional connections C12 between the lower cover A1 and the upper cover A2 and C23 between the upper cover A2 and the membrane A3, as depicted in Fig. 4. Hence, the conceptual design alternatives have all the same number of modules as depicted in Fig. 2, but each module has a different number of parts, according to the specific geometry defined for joining the parts in the module. The alternatives have been generated by means of a functional matrix as the result of all the possible combination of the functional connections C12 and C23 considering the screw joint, dovetail joint, and snap fit joint.

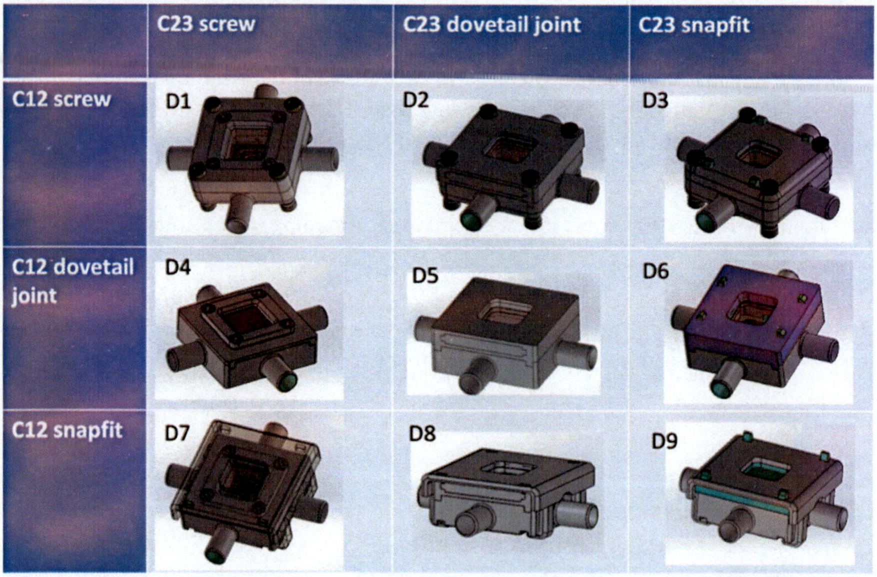

Fig. 4. Generated alternatives for the MBR by means of a modular design approach

4.2 Decisional Steps: Application of the Fuzzy-TOPSIS

In this section, an integrated multi-attribute decision technique and Fuzzy logic is applied to select the most suitable, among nine proposed design alternatives. To this aim, the preferences of three decision makers have been translated into fuzzy numbers, to capture the uncertainty of verbal judgments. Moreover, the TOPSIS technique has been used to rank the alternatives. Fuzzy Linguistic Variables describe the preferences of the decision makers. Here follow the steps related to the outranking approach. *A)Preference and Suitability.* The variables used to indicate the "preference" are taken from the Term Set {Very Low (VL). Low (L). Medium (M). High (H). Very High (VH)}. The variables describing the Fuzzy Variable for "suitability", are taken from the set {Very Poor (VP). Poor (P). Fair (F). Good (G). Very Good (VG)}. Figure 5 shows the membership functions for "Preference" and "Suitability". Linguistic values are provided by the decision makers (Table 2), to evaluate the suitability of each alternative in relation to the criteria. Trapezoidal fuzzy numbers are used for describing the verbal judgments for the importance of each criterion

$VL = [0., 0., .05, .25]; L = [.05, .2, .25, .40]; M = [.25, .4, .45, .60]; H = [.5, .65, .7, .9]; VH = [.75, .9, 1., 1.];$

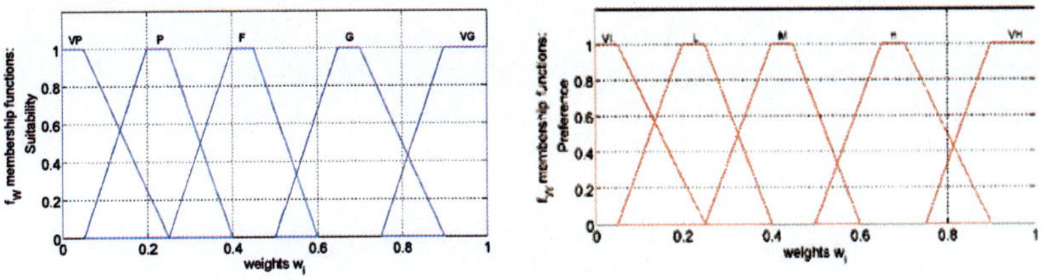

Fig. 5. Membership functions for "Preference" and "Suitability".

Table 2. Linguistic values for the preferences expressed by the decision makers.

		DM1	DM2	DM3
C1	Easy opening	VH	VH	L
C2	Ease of assembly	M	VH	H
C3	Cost	H	L	VH
C4	Manufacturing ease	H	L	M
C5	Aesthetics	M	H	L

Decision makers are asked to analyze the suitability of the alternatives by linguistic terms. $S_{jmr} = (a_{jmr}. b_{jmr}. c_{jmr}. d_{jmr})$ is defined as the suitability assigned to the mth alternative ($m = 1. 2.. p; p = 9$), evaluated against the jth criterion ($j = 1. 2....i; i = 5$), by the rth decision maker ($r = 1. 2... q; q = 3$),

B) Suitability indices are evaluated for all the criteria and collected in a decision matrix. The suitability indices (S_{jm}) are evaluated for each criterion. As an example, for the first criterion and the first alternative, the suitability index is calculated as in Eq. (3).

C) Fuzzy decision matrix. These matrices have been collected into a single matrix defined as the decision matrix for the problem. The positive and negative ideal solutions and similarity matrix. The evaluation of the Fuzzy Positive Ideal Solution and Fuzzy Negative Ideal Solutions (FPIS and FNIS) are reported in Table 3, with related similarity matrix. Hence, the vector of the average similarity for the alternatives is reported in Table 4, following the definition of Eq. (6). The resulted preference, that is the alternative with the maximum value of the average similarity, according to Table 1, is alternative D7, approved and recommended, D5 approved, while D8 and D9 could be approved with low risk.

Table 3. Fuzzy points of positive and negative ideal solution FPIS and FNIS with the Similarity matrix

$$FPIS = (0.75\ 0.90\ 1.00\ 1.00\ 0.67\ 0.82\ 0.90\ 0.97\ 0.75\ 0.90\ 1.00\ 1.00\ 0.75\ 0.90\ 1.00\ 1.00\ 0.75\ 0.90\ 1.00\ 1.00)$$

$$FNIS = (0.02\ 0.07\ 0.12\ 0.30\ 0.05\ 0.20\ 0.25\ 0.40\ 0.17\ 0.27\ 0.32\ 0.47\ 0.05\ 0.20\ 0.25\ 0.40\ 0.12\ 0.27\ 0.32\ 0.47)$$

$$SimilarityMatrix = \begin{pmatrix} 0.21 & 0.32 & 0.57 & 0.61 & 0.31 \\ 0.26 & 0.37 & 0.38 & 0.59 & 0.38 \\ 0.21 & 0.33 & 0.32 & 0.57 & 0.37 \\ 0.82 & 0.42 & 0.60 & 0.56 & 0.35 \\ 1.00 & 1.00 & 0.29 & 0.39 & 0.88 \\ 0.37 & 0.56 & 0.33 & 0.30 & 0.57 \\ 0.91 & 0.92 & 0.89 & 0.83 & 0.99 \\ 0.82 & 0.88 & 0.19 & 0.24 & 0.55 \\ 0.82 & 0.83 & 0.23 & 0.15 & 0.56 \end{pmatrix}$$

Table 4. Average similarity of the alternatives.

Average Similarity	0.41	0.40	0.36	0.55	0.71	0.43	**0.91**	0.54	0.52
Related alternative	D1	D2	D3	D4	D5	D6	D7	D8	D9
rank	7°	8°	9°	3°	2°	6°	1°	4°	5°

4.3 Discussion and Conclusions

In this paper, an integrated modular design and group decision making approach is applied for the generation and selection of the design alternatives of MBR. First, a modular design has been adopted for the evaluation of design alternatives considering requirements from different research disciplines. Generated alternatives and decisional criteria were input data for the group decision making step for selecting the most suitable alternative. This integrated design and selection method has proven effective in the generation and selection of design alternatives in complex design and decisional environment. The experts have been asked to fill in a questionnaire with their verbal judgments for preference and suitability. Since the same experts have been previously involved in the design of an existent MBR described in [3], so their level of knowledge on the topic could be defined satisfying. The design solution D7 which resulted as "approved and recommended" has been described as novel, by the experts, with respect to other existent solutions in the same field. As a further development, in the decisional step, uncertainty in verbal judgments is handled by means of fuzzy numbers. In addition, a weighted sum approach is used to aggregate functions for the overall preference representation. As remarked in the literature (e.g. [8]) a limitation of the proposed selection process within the group decision-making approach is the method for aggregating the alternatives, related to the choice of suitable preference aggregation functions. In particular, the weighted sum approach is a compensatory strategy. Thus, the use of this kind of function to aggregate decision makers' preferences could provide a partial view of the group's overall preference. Hence, as a further development of this integrated approach is the application of the Method of Imprecision (MoI) for aggregating preferences as in [15], which uses the trade-off functions for aggregating decision maker's preferences in an overall one.

References

1. Pennarossa, G., Arcuri, S., De Iorio, T., Gandolfi, F., Brevini, T.A.: Current advances in 3D tissue and organ reconstruction. Int. J. Mol. Sci. **22**(2), 830 (2021)
2. Corallo, D., et al.: Emerging neuroblastoma 3D in vitro models for pre-clinical assessments. Front. Immunol. **11**, 584214 (2020). https://doi.org/10.3389/fimmu.2020.584214
3. Candini, O., et al.: A novel 3D in vitro platform for pre-clinical investigations in drug testing, gene therapy, and immuno-oncology. Sci. Rep. **9**(1), 1–12 (2019)
4. Yoon, K.P., Hwang, C.L.: Multiple Attribute Decision Making: An Introduction. Sage Publications, Thousand Oaks (1995)
5. Renzi, C., Leali, F.: Multicriteria Decision Making Methods in Engineering Design: An Application on Medical Devices. In: Roucoules, L., Paredes, M., Eynard, B., Morer Camo, P., Rizzi, C. (eds.) JCM 2020. LNME, pp. 406–411. Springer, Cham (2021). https://doi.org/10.1007/978-3-030-70566-4_64
6. Renzi, C., Leali, F.: Decision-making methods in engineering design: a designer-oriented approach. Direccion y Organizacion **63**, 21–49 (2017)
7. Renzi, C., Leali, F., Di Angelo, L.: A review on decision-making methods in engineering design for the automotive industry. J. Eng. Des. **28**(2), 118–143 (2017)
8. Renzi, C., Leali, F., Pellicciari, M., Andrisano, A.O., Berselli, G.: Selecting alternatives in the conceptual design phase: an application of fuzzy-AHP and Pugh's controlled convergence. Int. J. Interact. Des. Manuf. (IJIDeM) **9**(1), 1–17 (2013). https://doi.org/10.1007/s12008-013-0187-y
9. Luukka, P.: Fuzzy similarity in multicriteria decision-making problem applied to supplier evaluation and selection in supply chain management. In: Advances in Artificial Intelligence (2011)
10. Panari, D., Renzi, C., Vergnano, A., Bonazzi, E., Leali, F.: Integrated design method for optimal tolerance stack evaluation for top class automotive chassis. In: Eynard, B., Nigrelli, V., Oliveri, S., Peris-Fajarnes, G., Rizzuti, S. (eds.) Advances on Mechanics. Design Engineering and Manufacturing. Springer, Cham (2017). https://doi.org/10.1007/978-3-319-45781-9_101

11. Bryant, C.R., Sivaramakrishnan, K.L., Van Wie, M., Stone, R.B., McAdams, D.A.: A modular design approach to support sustainable design. In: International Design Engineering Technical Conferences and Computers and Information in Engineering Conference, vol. 46962, pp. 909–918, January 2004
12. Xia, P.: A hybrid approach for haptics-based intelligent assembly planning. Adv. Mech. Eng. **8** (2016). https://doi.org/10.1177/1687814016668708
13. Renzi, C., Leali, F.: A multicriteria decision-making application to the conceptual design of mechanical components. J. Multi-Criteria Decis. Anal. **23**(3–4), 87–111 (2016). https://doi.org/10.1002/mcda.1569
14. Selden, C., Fuller, B.: Role of bioreactor technology in tissue engineering for clinical use and therapeutic target design. Bioengineering **5**(2), 32 (2018)
15. Scott, M.J., Antonsson, E.K.: Aggregation functions for engineering design trade-offs. Fuzzy Sets Syst. **99**(3), 253–264 (1998)

Firefly Algorithm for Structural Optimization Using ANSYS

Giuseppe Marannano(✉) and Vito Ricotta

Engineering Department, University of Palermo, Viale delle Scienze, 90128 Palermo, Italy
giuseppe.marannano@unipa.it

Abstract. In the mid-1980s, several metaheuristic methods began to be developed for solving a very large class of computational problems with the aim of obtaining more robust and efficient procedures. Among them, many metaheuristic methods use bio-inspired intelligent algorithms. In recent years, these methods are becoming increasingly important and they can be used in various subject areas for solving complex problems.

Firefly Algorithm is a nature-inspired optimization algorithm proposed by Yang to solve multimodal optimization problems. In particular, the method is inspired by the nature of fireflies to emit a light signal to attract other individuals of this species. In this work, a numerical study for solving a structural problem using the Firefly Algorithm as optimization method is conducted.

In particular, the implementation of the Firefly Algorithm in several input files realized in the ANSYS Parametric Design Language has allowed the definition of the optimal stacking sequence and the laminate thickness of a composite gear housing used to enclose the components of a mechanical reducer.

Keywords: Firefly Algorithm · Structural optimization · Finite element analysis · ANSYS

1 Introduction

In the last twenty years, optimization techniques have reached a high degree of maturity and they can be used in solving any engineering problem, in the automotive, aeronautical and manufacturing fields. Thanks to the use of the latest computerized design software and the integration with innovative optimization techniques, it is possible to improve the conceptual design process of any engineering system [1–4]. Among the different optimization techniques, general-purpose heuristic approaches, called metaheuristics, have gained increasing importance in recent years. Metaheuristic algorithms are approximate solution methods that, by using appropriate properties in the search mechanism, allow to find a good enough solution in reasonable times.

There are several metaheuristic techniques in the literature: Simulated annealing [5], neural networks [6], genetic algorithm [7], particle swarm [8, 9], ant colony optimization [10], artificial bee colony [11], bacterial foraging [12], cuckoo search [13], firefly algorithm [14–17], leaping frog algorithm [18], bat algorithm [19], flower pollination

C. Rizzi et al. (Eds.): ADM 2021, LNME, pp. 593–604, 2022.
https://doi.org/10.1007/978-3-030-91234-5_59

algorithm [20, 21], artificial plant optimization [22, 23], group search optimizer [24], etc.

Among these techniques, the Firefly Algorithm (FA) is one of the metaheuristic optimization methods that draws inspiration from the ability of fireflies to use of bioluminescence and in particular continuous light flashes during twilight to attract mates or prey and, in general, for the communication between members of its species. The degree of attraction of a firefly is proportional to its brightness and, therefore, the less bright one will move towards the brighter one and their attraction will grow as their mutual distance decreases.

In this work, a numerical study of structural optimization of the composite gear housing of a mechanical reducer has been conducted. The implementation of the firefly algorithm in appropriate ANSYS APDL macros has led to the definition of the thickness of the single lamina and of the optimal stacking sequence of the composite gear housing.

2 Firefly Algorithm

The basic idea in the initial formulation of the method proposed by Yang in 2009 [14] is that any firefly will be attracted to the one that produces a higher light intensity (queen firefly), and this attraction is stronger if the mutual distance is smaller. The algorithm is based on a physical formulation of light intensity I (Eq. 1) and on the attractiveness β (Eq. 2) that decrease as the value of distance d increases.

$$I \prec \frac{1}{d^2} = I_0\, e^{-\gamma\, d^2} \tag{1}$$

In the Eq. 1, I_0 is the light intensity at the source or initial light intensity, d is the distance between the two considered fireflies and γ è the light absorption coefficient that controls the light intensity. The attractiveness β is proportional to the light intensity as described in Eq. 2:

$$\beta = \beta_0\, e^{-\gamma\, d^2} \tag{2}$$

where $\beta_0 = 1$ is the attractiveness when the distance d is equal to zero. Therefore, the light intensity I and attractiveness β are in some way synonymous.

The distance d_{ij} between two fireflies (Euclidean distance) can be defined as follows:

$$d_{ij} = \left\| u_i - u_j \right\| = \sqrt{\sum_{k=1}^{n} \left(u_{i,k} - u_{j,k}\right)^2} \tag{3}$$

where n is the total number of fireflies.

The i-th firefly is attracted by another brighter firefly j. The movement of the firefly from one position to another is expressed by the following equation:

$$u_i^{t+1} = u_i^t + \beta_0\, e^{-\gamma\, d_{ij}^2} \cdot \left(u_j^t - u_i^t\right) + \alpha \cdot \varepsilon_i \tag{4}$$

where $\alpha = 0.2$ and ε_i is a random number in the range $[-1, 1]$. The parameter γ has a crucial effect on the convergence speed and it depends on the specific optimization problem. Typically, its value ranges from 0.1 to 10 [15]. In this study, the value of γ was assumed to be 1. The operations carried out during the execution of the Firefly Algorithm are described below:

1) Generate initial population of feasible fireflies x_i (i = 1, 2,..., n). The initial population consists of at least 10 fireflies.
2) Sort the n feasible fireflies according to the increasing Objective Function Value (OFV). Consequently, the first firefly will be assumed as the "queen firefly" of the initial population;
3) Determine the attractiveness β for xi by using Eq. (2) and (3);
4) Update the position of fireflies by using Eq. (4);
5) Compute the OFV for the new positions;
6) Sort the feasible fireflies according to the increasing OFV and redefine the queen firefly;
7) Iterate starting from point 3). The cycle stops when the maximum number of iterations or the convergence condition is reached.

3 Case Study

3.1 Cad Modeling

In this work, a structural optimization for the determination of the thickness of each lamina and of the optimal stacking sequence of a composite gear housing was carried out. Figure 1 (A) shows the CAD model of the assembled reducer while, in Fig. 1 (B), the cad model of the composite gear housing. To realize gear housing, a carbon fiber/epoxy composite laminate made of 16 unidirectional layers of Cycom 5320 carbon fiber prepreg has been considered [26]. Table 1 shows the mechanical properties of the single layer of the used Carbon Fiber Reinforced Polymer (CFRP) laminate.

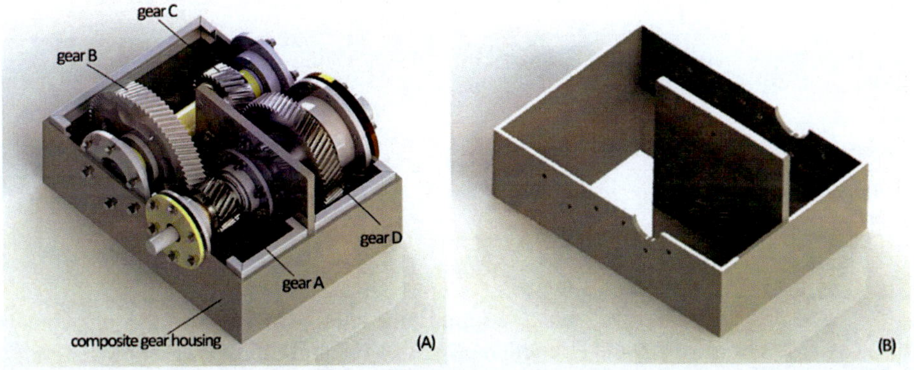

Fig. 1. (A) Cad model of the assembled reducer; (B) cad model of the composite gear housing.

Table 1. Mechanical properties of Cycom 5320 carbon fiber prepreg

E_{11} (GPa)	$E_{22} = E_{33}$ (GPa)	$G_{12} = G_{13}$ (GPa)	G_{23} (GPa)	$\nu_{12} = \nu_{13}$	ν_{23}
141	9.7	5.2	3.4	0.34	0.44
F_{1t} (MPa)	F_{1c} (MPa)	F_{2t} (MPa)	F_{2c} (MPa)	F_6 (MPa)	
2703	1737	81	312	57	
$F_{3c} = -F_{3t}$	F_4	F_5			
10^6	10^6	10^6			

In the Table 1, the numerical subscript (*1*) denotes the direction of the fiber, (2) in-plane transverse to the fibers and (*3*) through the thickness of each lamina. The letter subscript denotes tensile (*t*) and compressive (*c*). With the capital letter *F*, the ultimate stress values are denoted. The stacking sequence is symmetric about the mid-plane $[\alpha_1, \alpha_2, \alpha_3, \alpha_4]_{2s}$, where α symbol denotes a generic orientation of the fibers.

Considering a composite laminate with 68% by volume of carbon fiber and 32% epoxy resin, a density of the carbon fiber equal to 1.79 g/cm^3 and a density of the epoxy resin equal to 1.20 g/cm^3, it is possible to determine an average density of the composite equal to 1.6 g/cm^3.

As shown in Fig. 1(A), four helical gears are used to transmit motion between two parallel shafts. The reducer is driven by an electric motor with a maximum power of 10 kW and a rotation speed of 900 rpm. The motor shaft runs in the anticlockwise direction. Table 2 shows the geometric characteristics of the helical gears.

Table 2. Geometric properties of the helical gears

	Gear A (left-hand helix)	Gear B (right-hand helix)	Gear C (right-hand helix)	Gear D (left-hand helix)
Number of teeth	24	72	34	62
Normal pressure angle [degrees]	20	20	20	20
Helix angle [degrees]	25	25	25	25
Normal module [mm]	2	2	2	2

Tapered roller bearings were used to support the shafts. The reducer shafts are made of 39NiCrMo3 steel with yield strength $\sigma_s = 735$ MPa and ultimate tensile strength $\sigma_r = 930$ MPa. Figure 2 shows the linear dimensions and the axial positioning of main components of the gearbox.

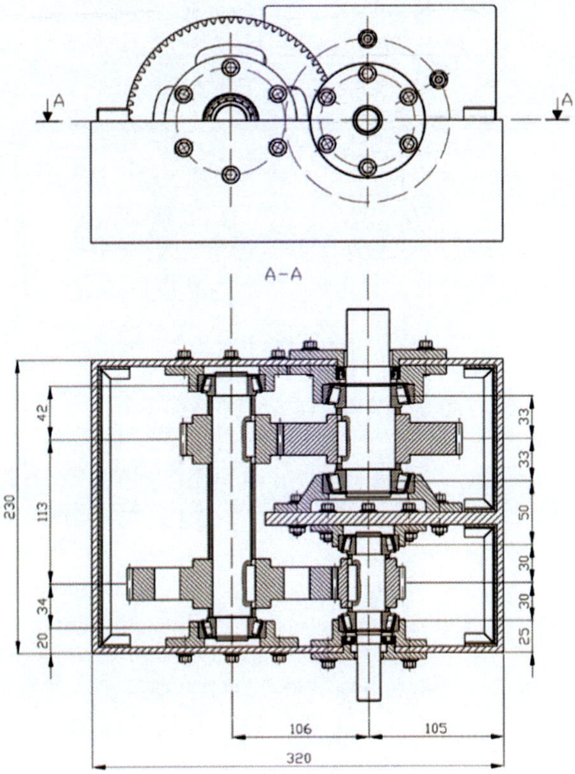

Fig. 2. Dimensions and the axial positioning of main components of the gearbox.

The magnitudes of the reactions at the supports are shown in Table 3.

Table 3. Magnitudes of the reactions at the supports.

Constraint A			**Constraint B**		
R_{AX} [N]	R_{AY} [N]	R_{AZ} [N]	R_{BX} [N]	R_{BY} [N]	R_{BZ} [N]
1868.4	1629.2	2003.4	0	20.1	2003.4
Constraint C			**Constraint D**		
R_{CX} [N]	R_{CY} [N]	R_{CZ} [N]	R_{DX} [N]	R_{DY} [N]	R_{DZ} [N]
2088.2	2076.9	1400.4	0	2939.8	5878.6
Constraint E			**Constraint F**		
R_{EX} [N]	R_{EY} [N]	R_{EZ} [N]	R_{FX} [N]	R_{FY} [N]	R_{FZ} [N]
0	1699.7	4242.5	3956.6	1707.9	4242.5

Figure 3 shows the directions of the reactions at the supports.

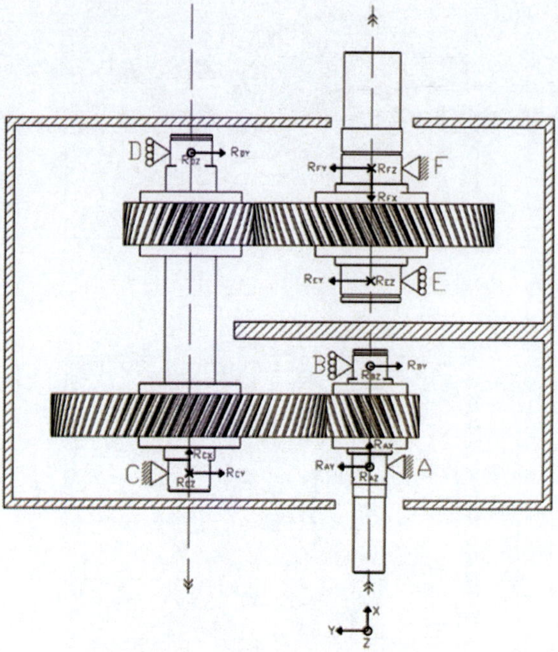

Fig. 3. Reactions at the supports.

3.2 Finite Element Analysis

Numerical analyses were carried out in the ANSYS APDL environment [27–29]. Several macros have allowed to realize a surface modelling of the composite gear housing and of the flanges connected to it. *Shell181* elements, suitable for analyzing thin to moderately-thick shell structures, are used to discretize the parametrically generated surfaces.

A convergence test has been conducted to determine the size of elements in finite element modeling [30]. In particular, the average element size is set to $d = 4$ mm. The Fig. 4 (A) shows the surface cad model and the Fig. 4 (B) shows the discretization realized on the composite gear housing and on the steel flanges.

In the preprocessor, during the definition of the laminate, it is very important to know the orientation of the laminate coordinate system. Material properties and the relative lamina orientation are defined in the laminate coordinate system. For the analysis of laminated composite shells, it is very important to define clearly a reference direction with respect to which one can specify the fiber direction of each lamina. The cad model was constrained to the base and loaded in the same way as described in paragraph Sect. 3.1.

In particular, an input file was created in order to determine, in each node and in each layer of the composite laminate, the failure index I_F defined as (Eq. 5):

$$I_F = \frac{stress}{strength} \tag{5}$$

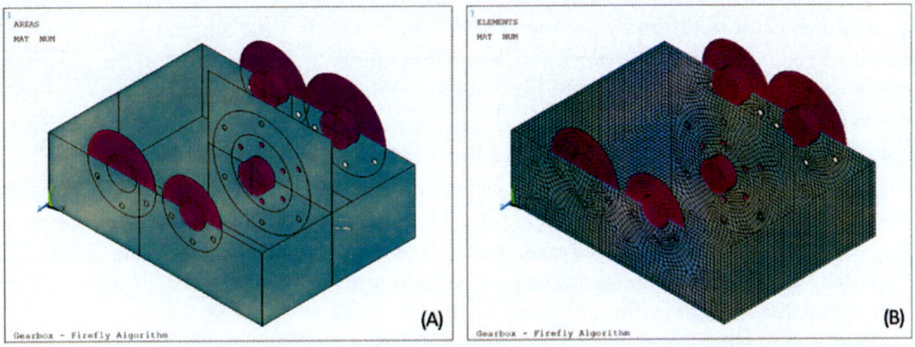

Fig. 4. (A) Surface cad model; (B) Detail of the discretization realized on the model.

Failure is predicted when $I_F \geq 1$. In particular, in the paper, the failure index I_F was calculated by using the Tsai-Wu criterion [31].

3.3 Optimization Study

The following variables were defined for the structural optimization problem:

Five Design Variables (DV), corresponding to the different angles α_i of the fibers in the composite, according to the stacking sequence provided for the laminate $[\alpha_1, \alpha_2, \alpha_3, \alpha_4]_{2s}$. The fifth design variable corresponds to the thickness *Tk* of the single composite layer. The thickness *Tk* is equal to a multiple of 0.05 mm, with $0.35 \leq Tk \leq 1.10$ mm. Two State Variables (SV), corresponding to the value of the Tsai-Wu failure index $I_F \leq 1$ and the allowable displacement *D* in the composite gear housing which must be less than $D_{max} = 0.6$ mm.

Two different optimization analyses were conducted by defining as OFV to be minimized the weight and, in second analysis, the value of the maximum displacement of the composite gear housing. The optimization study involves the determination of a first set of feasible fireflies through the execution of random simulations. In fact, through the definition of randomly generated design variables, the analyses allow to explore the entire search space. Specifically, in each individual analysis, the design variables are defined as follows:

$$\alpha_i = RAND\_i \cdot 180 \quad (i = 1, 2, 3, 4) \tag{6}$$

where *RAND_i* (i = 1,2,3,4) is a random number in the range [0, 1]. Moreover, a fifth random variable *RAND_5* allows the determination of the thickness of the single composite layer. In more detail, the thickness of the single lamina $Tk = 0.35 + (0.75 \cdot RAND\_5)$, a value that will be approximated in order to consider only multiples of 0.05mm. Consequently, there is a one-to-one correspondence between the values of *RAND_i* ($i = 1, 2, 3, 4, 5$), the angles of the fibers α_i and the lamina thickness *Tk*.

In subsequent iteration cycles, an ANSYS macro allows the automatic selection of design variables, the parametric generation of the model, the definition and the positioning of constraints, the application of loads with directions shown in Fig. 3 and magnitudes shown in Table 3. At the end of each analysis, the routine determines the Tsai-Wu failure index I_F value, the displacement D and the weight of the gear housing. These values will be stored in an array of size $n \times m$, where n corresponds to the number of iteration cycles and $m = 9$. In particular, the column 9 contains the Objective Function Value.

At the end of the random analyses, the rows of the array corresponding to unfeasible sets will be automatically deleted. The array is then sorted in ascending order with respect to the objective function value (column 9). The first row of the array corresponds to the first firefly of the initial population, the one that is currently the brightest (queen firefly).

The optimization cycle ends when the convergence condition or the maximum number of iterations is reached. Between one optimization cycle and the others, the software eliminates fireflies that provide OFV values greater than a specified percentage from the OFV value of the queen firefly. For the study, 200 independent runs of random analyses and two optimizations cycles were started using the firefly algorithm. The analyses confirm that layup has a large effect on strength and that the value of the minimum weight of the composite gear housing is equal to 3511.7 g, corresponding to a value $Tk = 0.65$ mm of the thickness of the single layer. The optimal results obtained from the analyses are shown in Table 4.

Table 4. Optimization results

	RAND_1	*RAND_2*	*RAND_3*	*RAND_4*	*RAND_5*	I_F	*D* [mm]	*WEIGHT* [g]	*OFV*
F1	0.46222503	0.4194685	0.44212239	0.44480992	0.3802325	0.3711215	0.554589183	3511.7	3511.7

A second optimization analysis allowed us to determine, for the same weight shown in Table 4, the stacking sequence that determines the minimum value of the displacement D (see Table 5).

Table 5. Optimization results

	α_1	α_2	α_3	α_4	*Tk*	I_F	*D* [mm]	*WEIGHT* [g]
F1	88.93	89.92	87.75	82.86	0.65	0.35805561	0.54321875	3511.7

Figure 5 shows the displacement values in the composite gear housing.

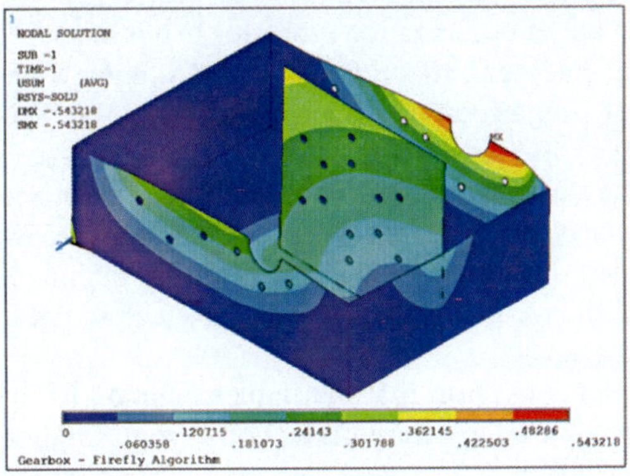

Fig. 5. Displacement values in the composite gear housing.

Table 6 shows the weight of the gear housing, considering to use as material:

The optimized CFRP laminate;
An aluminum alloy Al 6061 T6 ($\sigma_y = 276$ MPa; $\sigma_u = 310$ MPa);
A steel for general structural purposes AISI 1040 ($\sigma_y = 593$ MPa; $\sigma_u = 779$ MPa).

In the analyses, in the case of gear housings made of aluminum alloy or steel, the thickness that determines a displacement $D = D_{max} = 0.6$ mm was considered. The I_F value was obtained using the Von Mises criterion.

Table 6. Comparison results for gear housing realized in different materials.

	Density [g/cm^3]	Total thickness [mm]	I_F	D [mm]	Weight [g]
Optimized CFRP laminate	1.6	10.4	0.35805561	0.54321875	3511.7
Aluminum alloy Al6061-T6	2.7	9.76	0.39409420	0.60551490	5561.3
Steel AISI 1040	7.8	6.64	0.44256365	0.60573587	10930.2

It possible to assert that the use of a composite gear housing allows a weight reduction of about 37% if compared to an aluminum alloy case and of about 49% if compared to a steel case.

4 Conclusions

The firefly algorithm uses the process of attraction based on the brightness in fireflies to optimize an objective function. Several literature studies have shown that the algorithm can solve a large number of optimization problems in dominant subject areas: computer science, engineering, electrical, robotic, mathematics, energy, physics and astronomy. In this work, the firefly algorithm was implemented in ANSYS APDL environment in order to solve structural optimization problems. In particular, the FA algorithm has been tested to optimize the stacking sequence of a composite gear housing used to enclose the components of a mechanical reducer. The analysis allows to evaluate, moreover, the weight reduction when the gear housing is realized in aluminum alloy or steel. With the same stress level, the weight reduction can be even close to 50% in the case of gear housing realized in steel.

The numerical results confirm that the implementation of the firefly algorithm in ANSYS environment allowed to determine a good enough solution in reasonable times. In conclusion, FA proved to be an interesting alternative method for solving complex optimization problems. Moreover, the implementation in an ANSYS environment of the presented code has a perspective of extending to optimization of large scale structures, considering linear or nonlinear effects.

References

1. Ingrassia, T., Nigrelli, V.: Design optimization and analysis of a new rear underrun protective device for truck. In: Proceedings of the 8th International Symposium on Tools and Methods of Competitive Engineering, TMCE 2010, vol. 2, pp. 713–725 (2010)
2. Giallanza, A., Marannano, G., Pasta, A.: Structural optimization of innovative rudder for HSC. In: NAV International Conference on Ship and Shipping Research (2012)
3. Marannano, G., Pasta, A., Parrinello, F., Giallanza, A.: Effect of the indentation process on fatigue life of drilled specimens. J. Mech. Sci. Technol. **29**(7), 2847–2856 (2015). https://doi.org/10.1007/s12206-015-0613-0
4. Marannano, G., Parrinello, F., Giallanza, A.: Effects of the indentation process on fatigue life of drilled specimens: optimization of the distance between adjacent holes. J. Mech. Sci. Technol. **30**(3), 1119–1127 (2016)
5. Kirkpatrick, S., Gelatt, C.D., Jr., Vecchi, M.P.: Optimization by simulated annealing. Science **220**(4598), 671–680 (1983)
6. Grossberg, S.: Nonlinear neural networks: Principles, mechanisms, and architectures. Neural Netw. **1**(1), 17–61 (1988)
7. Holland, J.H.: Adaptation in natural and artificial systems: an introductory analysis with applications to biology, control, and artificial intelligence. University of Michigan Press (1975)
8. Shi, Y., Eberhart, R.C.: Parameter selection in particle swarm optimization. In: Porto, V.W., Saravanan, N., Waagen, D., Eiben, A.E. (eds.) EP 1998. LNCS, vol. 1447, pp. 591–600. Springer, Heidelberg (1998). https://doi.org/10.1007/BFb0040810
9. Shi, Y., Eberhart, R.C.: Empirical study of particle swarm optimization. In: Proceedings of the 1999 IEEE Congress on Evolutionary Computation, Washington, pp. 1945–1950. IEEE (1983)
10. Dorigo, M., Birattari, M., Stützle, T.: Ant colony optimization. IEEE Comput. Intell. Mag. **1**(4), 28–39 (2006)
11. Gao, W.F., Liu, S.Y.: A modified artificial bee colony algorithm. Comput. Oper. Res. **39**(3), 687–697 (2012)

12. Biswas, A., Dasgupta, S., Das, S., Abraham, A.: Synergy of PSO and bacterial foraging optimization - a comparative study on numerical benchmarks. In: Corchado, E., Corchado, J.M., Abraham, A. (eds.) Innovations in hybrid intelligent systems, pp. 255–263. Springer, Heidelberg (2007). https://doi.org/10.1007/978-3-540-74972-1_34
13. Bhandari, A.K., Singh, V.K., Kumar, A., Singh, G.K.: Cuckoo search algorithm and wind driven optimization based study of satellite image segmentation for multilevel thresholding using Kapur's entropy. Expert Syst. Appl. **41**(7), 3538–3560 (2014)
14. Yang, X.-S.: Firefly algorithms for multimodal optimization. In: Watanabe, O., Zeugmann, T. (eds.) SAGA 2009. LNCS, vol. 5792, pp. 169–178. Springer, Heidelberg (2009). https://doi.org/10.1007/978-3-642-04944-6_14
15. Yang, X.S.: Firefly algorithm, stochastic test functions and design optimisation. Int. J. Bio-Inspired Comput. **2**(2), 78–84 (2010)
16. La Scalia, G., Micale, R., Giallanza, A., Marannano, G.: Firefly algorithm based upon slicing structure encoding for unequal facility layout problem. Int. J. Ind. Eng. Comput. **10**, 349–360 (2019)
17. Micale, R., Marannano, G., Giallanza, A., Miglietta, P.P., Agnusdei, G.P., La Scalia, G.: Sustainable vehicle routing based on firefly algorithm and TOPSIS methodology. Sustain. Futures **1**, 100001 (2019)
18. Elbeltagi, E., Hegazy, T., Grierson, D.: A modified shuffled frog-leaping optimization algorithm: applications to project management. Struct. Infrastruct. Eng. **3**(1), 53–60 (2007)
19. Gandomi, A.H., Yang, X.S., Alavi, A.H., Talatahari, S.: Bat algorithm for constrained optimization tasks. Neural Comput. Appl. **22**(6), 1239–1255 (2013)
20. Yang, X.-S.: Flower pollination algorithm for global optimization. In: Durand-Lose, J., Jonoska, N. (eds.) UCNC 2012. LNCS, vol. 7445, pp. 240–249. Springer, Heidelberg (2012). https://doi.org/10.1007/978-3-642-32894-7_27
21. Bekdas, G., Nigdeli, S.M., Yang, X.S.: Sizing optimization of truss structures using flower pollination algorithm. Appl. Soft Comput. **37**, 322–331 (2015)
22. Cui, Z., Yang, H., Shi, Z.: Using artificial plant optimization algorithm to solve coverage problem in WSN. Sens. Lett. **10**(8), 1666–1675 (2012)
23. Cui, Z., Cai, X.: Artificial plant optimization algorithm. In: Swarm Intelligence and Bio-Inspired Computation: Theory and Applications, pp. 351–365 (2013)
24. He, S., Wu, Q.H., Saunders, J.R.: A novel group search optimizer inspired by animal behavioural ecology. In: 2006 IEEE Congress on Evolutionary Computation, CEC 2006, Vancouver (2006)
25. Abdel-Basset, M., Abdel-Fatah, L., Sangaiah, A.K.: Metaheuristic algorithms: a comprehensive review. In: Sangaiah, A.K., Sheng, M., Zhang, Z. (eds.) Intelligent Data-Centric Systems, Computational Intelligence for Multimedia Big Data on the Cloud with Engineering Applications, pp. 185–231 (2018)
26. Technical data sheet of CYCOM® 5320-1 Prepreg (2020). https://www.solvay.com
27. Ingrassia, T., Nigrelli, V., Ricotta, V., Tartamella, C.: Process parameters influence in additive manufacturing. In: Eynard, B., Nigrelli, V., Oliveri, S., Peris-Fajarnes, G., Rizzuti, S. (eds.) Advances on Mechanics, Design Engineering and Manufacturing, pp. 261–270. Springer, Cham (2017). https://doi.org/10.1007/978-3-319-45781-9_27
28. Baron Saiz, C., Ingrassia, T., Nigrelli, V., Ricotta, V.: Thermal stress analysis of different full and ventilated disc brakes. Frattura ed Integrita Strutturale **9**(34), 608–621 (2015)
29. Ingrassia, T., Nalbone, L., Nigrelli, V., Pisciotta, D., Ricotta, V.: Influence of the metaphysis positioning in a new reverse shoulder prosthesis. In: Eynard, B., Nigrelli, V., Oliveri, S., Peris-Fajarnes, G., Rizzuti, S. (eds.) Advances on Mechanics, Design Engineering and Manufacturing, pp. 469–478. Springer, Cham (2017). https://doi.org/10.1007/978-3-319-45781-9_47

30. Marannano, G., Pasta, A., Giallanza, A.: A model for predicting the mixed-mode fatigue crack growth in a bonded joint. Fatigue Fract. Eng. Mater. Struct. **37**(4), 380–390 (2014)
31. Barbero, E.J.: Finite Element Analysis of Composite Materials Using ANSYS®, 2nd edn. CRC Press, Boca Raton (2014)

An Analytical Model for Robot-Based Grinding of Axisymmetric Mold Inserts Using a Rotary Unit

Eugenio Tamassia[1(✉)], Sebastian Hähnel[1], Fabio Pini[2], Tim Grunwald[1], Thomas Bergs[1,3], and Francesco Leali[2]

[1] Department of Fine Machining and Optics, Group of Fine Machining, Fraunhofer Institute for Production Technology IPT, Steinbachstraße 17, 52074 Aachen, Germany
{eugenio.tamassia,sebastian.haehnel,tim.grunwald, thomas.bergs}@ipt.fraunhofer.de

[2] Department of Mechanical Engineering "Enzo Ferrari" (DIEF), Integrated Design and Engineering Application Laboratory (IDEA Lab), University of Modena and Reggio Emilia, Via Pietro Vivarelli 10, 41125 Modena, Italy
{fabio.pini,francesco.leali}@unimore.it

[3] Machine Tool Laboratory WZL, Chair of Manufacturing Technology, RWTH Aachen University, Campus-Boulevard 30, 52074 Aachen, Germany
t.bergs@wzl.rwth-aachen.de

Abstract. The grinding of mold inserts used for injection molding aims to improve the surface roughness according to precise quality standards. The insert surface must also have a surface topography that facilitates the release of the plastic material at the end of the injection process. In particular, fine machining lines must be parallel to the extraction direction from the mold to avoid the sticking of plastic material and subsequent surface damages compromising the functionality of the finished product. However, this step in the production chain is most often conducted manually. This paper presents an analytical model to grind a truncated cone-shaped mold insert for the mass production of plastic cups. The automated solution consists of a flexible robotic system equipped with a rotating external axis to improve the accessibility of the tool to the surface to be machined. The tool path programming requires the development of an analytical model considering the simultaneous motion of the insert and the robot joints. The effectiveness of the developed model is evaluated in terms of final surface quality, grinding lines direction, and total process time. The automated strategy developed can be easily implemented with machine tools and applied to inserts with different axisymmetric geometries.

Keywords: Automated fine machining · CAM programming · Grinding lines · Mold insert · Robot manufacturing

C. Rizzi et al. (Eds.): ADM 2021, LNME, pp. 605–614, 2022.
https://doi.org/10.1007/978-3-030-91234-5_60

1 Introduction

The manufacturing of molds and dies embodies several steps [1], ranging from initial coarse material removal to finishing post-processing operations, such as grinding and polishing, to achieve a surface quality consistent with the work-piece functionality and possibly assign a precise pattern for aesthetic purposes. Automating finishing operations that are most often done manually potentially reduces production times, increases productivity [2], and lowers manufacturing costs in the long term [3]. Besides, automation allows for more precise control of process parameters, resulting in greater repeatability in the final characteristics of machined parts [4]. Grinding operations can be carried out with computer numerical controlled (CNC) machines or industrial robots (IR) [5]. Compared to machine tools, the working space of a robot cell yields better accessibility to the surfaces of medium-large-sized molds and the average investment costs of an IR for manufacturing purposes are significantly lower [6]. Although, the less rigid kinematic structure, the greater sensitivity to external vibrations and disturbances, and the possibility of running into singular configurations make an IR less suitable for fine machining processes than a machine tool [7]. The above issues can be reduced by minimizing the robot movements around the surface to be machined in the case of medium-large-sized mold inserts. The approach presented in this paper consists of an external axis that assigns a continuous rotation to the mold insert. However, the rotary unit is controlled independently from the robot axes and the grinding lines must be parallel to the extraction direction of the finished product from the mold to facilitate its ejection at the end of the injection molding process [8]. Therefore, an analytical model is developed to compute a trajectory that simultaneously handles the insert rotation and the robot-guided tool movements. The impression of grinding lines consistent with the extraction direction from the mold is achieved with a specific tool orientation to the machined surface. Section 2 describes the methods, specifically the analytical model, robot cell setup, and grinding tests. Section 3 shows the measurements of profile roughness and surface topography for each test. Section 4 discusses the results and draws the conclusions of the research conducted.

2 Materials and Methods

2.1 Analytical Model

This paper considers a truncated cone-shaped geometry with the characteristic parameters shown in Fig. 1. The workpiece reference system (WRS) (4), $O_w x_w y_w z_w$, coincides with the Cartesian coordinate system used to calculate the tool path. The parametric equations of the curve $P(t) = (x(t), y(t), z(t))$ are derived as continuous functions of time, t, so that the robot-guided tool (5) follows the lateral surface of the insert during its revolution. Table 1 summarizes the input and output parameters of the analytical model. First, the angular velocity of the mold insert (1), ω_w, is derived from its rotational speed, n_w. The insert

Table 1. Input (a) and output (b) parameters of the analytical modelInput (a) and output (b) parameters of the analytical model

(a) Inputs

Description	Symbol	Unit
Insert bottom radius	$r_{w,1}$	mm
Insert top radius	$r_{w,2}$	mm
Insert height	h_w	mm
Insert rotational speed	n_w	min^{-1}
Tool feed rate	v_f	mm s^{-1}
Number of tool passes	N_t	
Number of machining steps	N_m	

(b) Outputs

Description	Symbol	Unit
Insert apothem length	l_w	mm
Insert opening angle	δ_w	°
Insert angular velocity	ω_w	rad s^{-1}
Tool path coordinates	(x, y, z)	mm
Tool path time	T_t	s
Machining time	T_m	min
Number of tool revolutions	N_r	

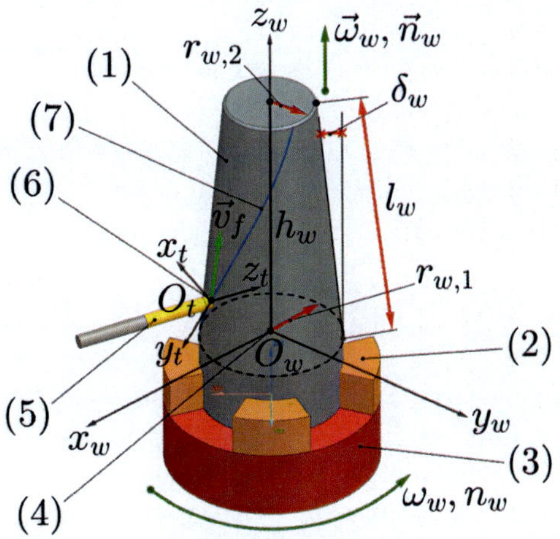

Fig. 1. Dimensions of the truncated cone-shaped geometry and kinematics of the analytical model: (1) Mold insert, (2) Clamping jaw, (3) Rotary unit, (4) WRS, (5) Tool, (6) TCP, (7) Tool path

apothem, l_w, represents the segment that generates the lateral surface through complete rotation around the z-axis:

$$l_w = \sqrt{h_w^2 + (r_{w,1} - r_{w,2})^2}. \tag{1}$$

The opening angle of the truncated cone, δ_w, is the angle between the apothem and the insert axis of rotation:

$$\delta_w = \tan^{-1}\left(\frac{h_w}{r_{w,1} - r_{w,2}}\right). \tag{2}$$

The time required for a tool pass, T_t, can be obtained by imposing a certain feed rate value, v_f, parallel to the insert apothem and assigning to the tool a uniform rectilinear motion:

$$T_t = \frac{l_w}{v_f}\,. \tag{3}$$

The number of tool revolutions, N_r, describes how many times the tool and, therefore, the robot wrist rotate around the insert axis of symmetry:

$$N_r = \frac{\omega_w T_t}{2\pi}. \tag{4}$$

By inserting Eqs. (1), (2), (3) and (4) into the parametric equations of a truncated-conical spiral [9], the tool path trajectory (7) is derived:

$$x(t) = \left[r_{w,1} - \frac{t}{T_t}(r_{w,1} - r_{w,2})\right]\cos(\omega_w t), \tag{5}$$

$$y(t) = \left[r_{w,1} - \frac{t}{T_t}(r_{w,1} - r_{w,2})\right]\sin(\omega_w t), \tag{6}$$

$$z(t) = v_f \frac{h_w}{\sqrt{h_w^2 + (r_{w,1} - r_{w,2})^2}} t. \tag{7}$$

The same equations remain valid for grinding cylinder-shaped ($r_{w,1} = r_{w,2}$) and cone-shaped ($r_{w,2} = 0$) inserts. The number of machining steps, N_m, depends on the desired final surface quality and enables the estimation of the grinding process time:

$$T_m = T_t N_t N_m. \tag{8}$$

This output allows evaluating the effectiveness of the method, because the same surface quality can be achieved with lower machining times than the manual process or other traditional approaches with a static workpiece. Since T_m does not consider the approach and retract tool movements but only those of contact with the insert, Eq. (8) slightly underestimates the real total process times. The model described works under certain boundary conditions. First, the axisymmetric geometry must be generated by the complete rotation of a straight line around a vertical axis. Then, the bottom circumference must be greater than or equal to the top one ($r_{w,1} \geq r_{w,2}$). Finally, the tool must follow the trajectory only during the grinding process ($t \in [0; T_t]$), remain in contact with the insert lateral surface at each time step ($r \in [r_{w,2}; r_{w,1}], \forall t \in [0; T_t]$), start to machine from the bottom circumference at $P(0) = (x(0), y(0), z(0)) = (r_{w,1}, 0, 0)$ and finish on the top one at $P(T_t) = (x(T_t), y(T_t), z(T_t)) = (x(T_t), r_{w,1}, h_w)$.

2.2 Experimental Setup

Figure 2(a) shows the experimental cell setup at Fraunhofer IPT and the relative position between the TCP (6), WRS (8), and rotary unit (9). The grinding tests were performed using a 6-degree-of-freedom (DOF) ABB *IRB 4400-60/1.96* IR (1) and a Walter *RAD-335-A* rotary unit (9) with jaws to clamp the mold insert (5). The tool (4) used was a rubber-bonded mounted point *Daiwa K320* manufactured by Novapax Kunstofftechnik GmbH & Co. KG with corundum abrasive grains. The bonded grits were extra fine and had an average diameter of $d_g = 36\,\mu\text{m}$ according to the FEPA grit designation. The visualization of the tool

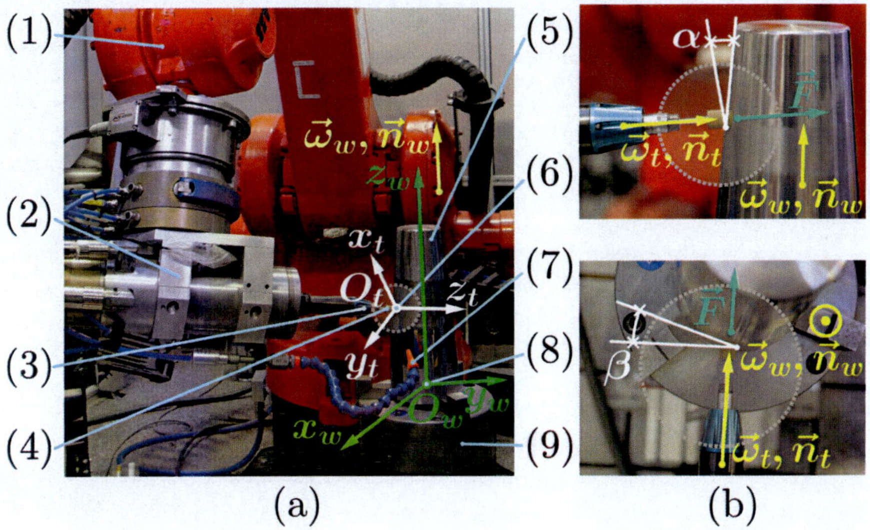

Fig. 2. Robot cell setup (a) and tool orientation with lead or tilt angle (b): (1) IR, (2) Spindle, (3) Tool holder, (4) Tool, (5) Mold insert, (6) TCP, (7) Air cooling, (8) WRS, (9) Rotary unit

contact area in Figs. 2(b) and 2(c) clarifies the relationship between the machining angles α or β, $\boldsymbol{v}_f$, and the transportation direction of the abrasive grains. In particular, a strategy with a tilt angle, β, should yield grinding lines coherent with the ejection direction of the finished product from the insert [10]. The tool direction of rotation was set clockwise to have the transportation direction of the abrasive grains consistent with $\boldsymbol{v}_f$ when using β. The average cutting speed was $v_{c,avg} = 4.08\,\text{mm}\,\text{s}^{-1}$. The mounted point was guided by the robot using an axial compliant tooling system (iCTS) from Hi.automize GmbH. The tool holder (3) compensated for robot inaccuracies in tool positioning by exerting an axial force $F = 12\,\text{N}$ that depended on the feed pressure value of $p = 1\,\text{bar}$ transmitted to the robot spindle (2). The demonstrator material was ISO X40Cr14, which is pre-hardened chromium alloyed stainless steel suitable for manufacturing injection mold inserts. The tool–insert contact zone was cooled with air (7) to lower local temperatures and avoid burn marks. The input parameters were chosen so that $N_r < 1$ in all grinding tests to improve robot stability, avoid joint singularities, and prevent collisions while machining. Table 2 reports on the left (a) the numerical values of the process parameters for this specific study, with the same meaning of the symbols in Table 1, and the common elements adopted for all grinding tests on the right (b). Creating the tool path in a computer-aided design (CAD) environment required to discretize the spiral and create a new interpolating curve. The interpolation points were derived with a time step $\Delta t = 0.5\,\text{s}$ and imported into the *Siemens NX* software through a script programmed in *Visual Basic*. The adoption of a 7th-order spline as an interpolation curve allowed to minimize the root mean square (RMS) error between the continuous spiral and discretized curve. The tool path and process parameter values were transferred to the robot using the *ABB RobotStudio* software.

Table 2. Parameter values of the analytical model (a) and experimental setup for all the grinding tests (b)

(a) Parameters	
Inputs	**Outputs**
$r_{w,1} = 50.00\,\text{mm}$	$l_w = 200.75\,\text{mm}$
$r_{w,2} = 32.65\,\text{mm}$	$\delta_w = 5°$
$h_w = 200\,\text{mm}$	$\omega_w = 0.11\,\text{rad}\,\text{s}^{-1}$
$n_w = 1\,\text{min}^{-1}$	$N_r = 0.11$
$v_f = 10\,\text{mm}\,\text{s}^{-1}$	$T_t = 20.08\,\text{s}$
$n_t = 12\,000\,\text{min}^{-1}$	$T_m = 49.18\,\text{min}$
$N_m = 1$	

(b) Experimental setup	
Feature	**Description**
Abrasive material	Corundum
Bond	Rubber
Tool type	Mounted point
Tool rotation	Clockwise
Tool holder	Axial compliant
Cooling system	Air
External axis	Rotary unit
IR type	6-DOF serial arm

2.3 Test Plan

Table 3 collects the experimental parameters for each grinding test. In Test 1 the tool was tilted with α and followed a sequence of vertical paths while the rotary unit was not activated. The other tests differed from it because they used β as a machining angle. In Test 2 the insert did not rotate, as in Test 1, whereas in Tests 3 and 4 the table was activated with $n_w = 1\,\text{min}^{-1}$. Specifically, the analytical model was only used in the last test. The purpose of Test 1 was to show that the adoption of α did not generate grinding lines in the desired direction and, thus, it was not necessary to further investigate its utility. On the contrary, Test 2 was conducted to confirm that a tool inclination with β allowed the generation of vertical grinding lines. If Test 3 showed grinding lines tilted to the extraction direction, the need to develop the mathematical model would be highlighted. Therefore, Test 4 served to validate the effectiveness of the combined use of the machining angle β and the analytical model concerning the correct transportation of the abrasive grains along the calculated tool path. Each test was performed on different portions of the lateral surface to optically distinguish the machined areas. The test surfaces had sufficient axial, $h_s = 50\,\text{mm}$, and radial, $\delta_s = 80°$, extensions to allow measurements of profile roughness and surface topography. The value of v_f compensated material removal rate (MRR) and total process times. The low stepover value, a, depended on the extension of the tool–insert contact area. The number of tool passes, N_t, consented to grind the entire test area based on the value of a. Only one machining step, N_m, was sufficient to achieve the desired surface quality. Each test was run several times without changing parameters and strategy to verify the stability of the process. Additional tests with different parameters were carried out to check the reliability of the conclusions using a design of experiment (DOE) approach, although only the results from the above tests are shown.

Table 3. Values of common (a) and different (b) parameters of the grinding tests

(a) Common parameters

Parameter	Value
a	0.1 mm
p	1 bar
F	12 N
N_t	39
N_m	1

(b) Different grinding strategies

Feature	Test 1	Test 2	Test 3	Test 4
n_w	0	0	1 min^{-1}	1 min^{-1}
α	10°	0	0	0
β	0	10°	10°	10°
Eqs. (5), (6), (7)	No	No	No	Yes

3 Results

3.1 Surface Quality

Finishing procedures are standardized to obtain a specific surface quality based on the materials of the mold and final product. This study adopts the SPI B-3 quality standard as a reference, which is proper for a wide range of plastics and indicates a normal semi-glossy finish of the mold surface with an average surface roughness in the range 0.15 µm < Ra < 0.32 µm. The profile roughness values (Ra, Rz, Rt) were measured with a Taylor Hobson *Form Talysurf series 2* tactile device both in the axial direction, i.e. moving on the lateral surface along the insert axis of symmetry, and in the radial direction, i.e. orthogonally to the axial direction. The measurement method, the cutoff values, and the sample lengths chosen were consistent with the EN ISO 4288 norm. Before the grinding tests, the average roughness values were Ra = 0.06 µm, Rz = 0.36 µm, Rt = 0.90 µm in the radial direction and Ra = 1.32 µm, Rz = 5.55 µm, Rt = 5.86 µm in the axial one. The latter values were greater than the former because the insert demonstrator was manufactured by external cylindrical turning. The average profile roughness at the end of the tests fell within the range established by the above-mentioned quality standard, in particular Ra = 0.26 µm (Test 1), 0.06 µm (Test 2), 0.27 µm (Test 3), 0.19 µm (Test 4) in the radial direction and Ra = 0.22 µm (Test 1), 0.22 µm (Test 2), 0.30 µm (Test 3), 0.21 µm (Test 4) in the axial one.

3.2 Direction of Grinding Lines

The profile roughness values after the tests did not allow to discard any of the strategies adopted, so the surface topography had to be analyzed to visualize the orientation of the grinding lines. Figure 3 shows the measurements taken with a Bruker *NPFlex3D surface metrology system* white light interferometer (WLI) and a 10x magnification objective before and after grinding. The image also displays the relative orientation between the extraction direction and grinding lines, γ, the size of the analyzed area, and the color scale relative to the surface roughness. The lower roughness values in the axial direction and the absence of regularity in the surface topography after grinding confirm the successful removal of the turning lines. The orientation of the grinding lines is consistent with the assumptions of Sect. 2.3. Test 1 shows a surface pattern parallel to the turning lines ($\gamma = 90°$) since the active abrasive grains are those in the contact area resulting from the α angle. On the contrary, Test 2 displays grinding lines in the axial direction ($\gamma = 0°$), confirming the efficacy of adopting $\beta = 10°$ and a clockwise rotation of the tool. Test 3 shows grinding lines directed obliquely to both the insert axis of symmetry and the turning lines ($\gamma \neq 0°, 90°$), proving the need to develop a mathematical model that compensates for the insert rotation. Test 4 presents a surface topography similar to Test 1 ($\gamma = 0°$) despite the activation of the rotary unit. In conclusion, the adoption of a strategy with β is not sufficient to obtain grinding lines parallel to the extraction direction and the mathematical model presented fulfills this requirement in the case of rotating mold inserts.

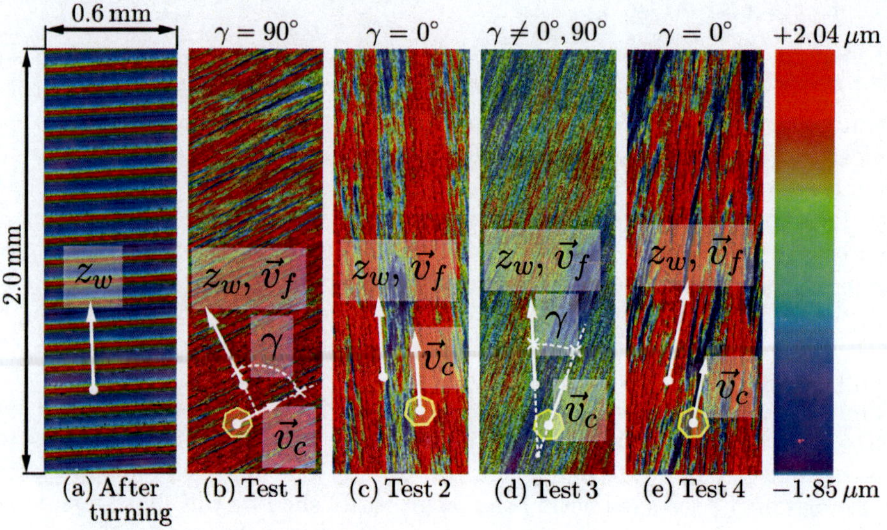

Fig. 3. Surface topography before (a) and after (b–e) grinding with visualization of the insert axial direction, z_w, tool feed rate, $\boldsymbol{v}_f$, transportation direction of the abrasive grains, $\boldsymbol{v}_c$, and relative orientation between extraction direction and grinding lines, γ

3.3 Process Improvements

The insert rotation allowed the robot to avoid singularities and to access all points of the insert lateral surface with the restriction $N_r < 1$. With the demonstrator and the process parameters adopted, the total time for grinding the lateral surface was $T_m \approx 50$ min, which is considerably less than the average time required by manual processes or traditional approaches with a static workpiece.

4 Conclusions

This paper investigated a robot-based approach for grinding injection mold inserts. The mold insert represents a case study for the production of plastic cups. The insert was assigned a continuous rotation with an external rotary unit not synchronized with the robot axes. Different strategies were investigated to generate grinding lines parallel to the extraction direction of the finished product and obtain a surface quality consistent with industrial standards. In particular, the analysis focused on the elaboration and validation of a mathematical model for programming the tool path so that the robot followed the rotation of the insert lateral surface during grinding. The study showed the effectiveness of combining the analytical model with the proper tool inclination angle on the final surface characteristics. The use of a rotary unit instead of a synchronized external axis considerably reduces the investment costs of a robot cell. The approach was validated on a truncated cone-shaped demonstrator with specific dimensions, though it can also be used with cylindrical or conical mold inserts and with different dimensions. In conclusion, the method shown allows to increase productivity and exploit the working space of an IR to machine medium-large components.

Acknowledgments. This research has been funded by the Federal Ministry for Economic Affairs and Energy (Bundesministerium für Wirtschaft und Energie, BMWi) based on a resolution of the German Parliament (Deutsche Bundestag) in the context of the CORNET project "Cobotic Polishing of Moulds" (CoPolMould) grant number HBC.2017.0643.

References

1. Dieste, J.A., Fernandez, A., Roba, D., Gonzalvo, B., Lucas, P.: Automatic grinding and polishing using spherical robot. Procedia Eng. **63**, 938–946 (2013). https://doi.org/10.1016/j.proeng.2013.08.221
2. Brecher, C., Storms, S., Ecker, C., Obdenbusch, M.: An approach to reduce commissioning and ramp-up time for multi-variant production in automated production facilities. Procedia CIRP **51**, 128–133 (2016). https://doi.org/10.1016/j.procir.2016.05.027
3. Barosz, P., Gołda, G., Kampa, A.: Efficiency analysis of manufacturing line with industrial robots and human operators. Appl. Sci. **10**(8), 2076–3417 (2020). https://doi.org/10.3390/app10082862
4. Wilbert, A., Behrens, B., Zymla, C., Dambon, O., Klocke, F.: Robotic finishing process - an extrusion die case study. CIRP J. Manuf. Sci. Technol. **11**, 45–52 (2015). https://doi.org/10.1016/j.cirpj.2015.07.002
5. Hähnel, S., Grunwald, T., Bergs, T., Pini, F., Leali, F.: Vibration-assisted face grinding of mould steel. In: Rizzi, C., Andrisano, A., Leali, F., Gherardini, F., Pini, F., Vergnano, A. (eds.) Design Tools and Methods In Industrial Engineering. ADM 2019, LNME. Springer, Cham (2020). https://doi.org/10.1007/978-3-030-31154-4_25
6. Leali, F., Pini, F., Ansaloni, M.: Integration of CAM off-line programming in robot high-accuracy machining. In: Proceedings of the 2013 IEEE/SICE International Symposium on System Integration, Kobe, Japan (2013). https://doi.org/10.1109/SII.2013.6776741
7. Ji, W., Wang, L.: Industrial robotic machining: a review. Int. J. Adv. Manuf. Technol. **103**, 1239–1255 (2019). https://doi.org/10.1007/s00170-019-03403-z

8. Bryce, D.M.: Plastic Injection Molding: Mold Design and Construction Fundamentals. Society of Manufacturing Engineers (1998). ISBN: 0872634957/9780872634954
9. Lazureanu, C.: Spirals on surfaces of revolution. VisMath **16**(2), 1–10 (2014). ISSN: 1821-1437
10. Koprowski, S., Uhlmann, E., Weingaertner, W.: Influence of tilt and lead angles on 5-axis grinding with spherical mounted points. Prod. Eng. **12**, 449–455 (2018). https://doi.org/10.1007/s11740-018-0812-5

Towards the Upscaling of Biomanufacturing Process Enhanced by Human-Robot Collaboration

Andrea Rega[1,3](✉), Castrese Di Marino[2], Ferdinando Vitolo[3], Stanislao Patalano[3], and Antonio Lanzotti[3]

[1] Department of Neurosciences, Reproductive and Odontostomatological Sciences, University of Naples Federico II, Naples, Italy
andrea.rega@unina.it

[2] Department of Management, Information and Production Engineering, University of Bergamo, Bergamo, Italy

[3] Department of Industrial Engineering, University of Naples Federico II, Naples, Italy

Abstract. Nowadays, the commercial translation of additive manufacturing technologies for scaffolds fabrication is still a challenge. The production methodology of 3D scaffolds for tissue regeneration is a complex and discontinuous process involving several stages, from the isolation of the stem cells to the dynamic cell culture in vitro. Even though in this scenario industries are increasingly implementing automated robotic systems, current technologies are not enough to realize a large industrial scale scaffold fabrication. Accordingly, a relevant improvement could raise from the implementation of a modern collaborative workplace in an existing production line, combining strength endurance and accuracy of cobots, with intelligence, flexibility, and adaptability of the human being. Such a solution overcomes limits related to the low level of process control, low productivity, and risk of contaminations. Therefore, the current work proposes a systematic approach to the design of a collaborative workplace for biomanufacturing of 3D scaffolds. Starting from an overview of basic concepts on scaffolds for tissue engineering and additive manufacturing, as well as from an analysis of automation solutions in cell culture applications, a design methodology section is reported. The paper provides a further insight into the potentials to upscale the scaffolds manufacturing process, taking advantage of the huge possibilities given from the Human-Robot Collaboration and gives evidence of critical features for workplace definition.

Keywords: Workplace design · Human-Robot Collaboration (HRC) · Automatic layout generation · Biomanufacturing

1 Introduction

Additive manufacturing (AM) technologies led to important developments in the fabrication of advanced medical devices. In the field of regenerative medicine, these technologies allow the production of customized device with complex shape, lightweight and

C. Rizzi et al. (Eds.): ADM 2021, LNME, pp. 615–622, 2022.
https://doi.org/10.1007/978-3-030-91234-5_61

tailored properties. These constructs replicate more closely the heterogeneity and complexity of tissues and organs than the current biomedical solutions. Latest approaches in Tissue Engineering (TE) are based on the use of temporary printed scaffolds for supporting cells throughout the tissue formation [1]. Human tissue-engineered products are mostly generated in a laboratory bench-scale process, by culturing and seeding cells from a patient or donor, onto the scaffolds. It is an expensive process of variable quality, if performed manually, which prevents these products entering the market and consequently limits their actual clinical application [2]. The upscaling of the laboratory-based process is a real need and an exciting engineering challenge. The main challenge to address is translating the conventional production techniques of tissue-engineered products into a large-scale process, fast, cost-effective, safe, repeatable, and clinically efficient.

Fully automated robotic manufacturing platforms represent a possible solution. Such a solution could decrease the contamination risks and the variability due to the human error, ensuring a high level of process control, the generation of more data, and a higher throughput [3]. In addition, the standardisation of the production process would facilitate the regulatory specification compliance.

Over last years, researchers applied automation solutions from the production industries into cell culture applications [4]. AUTOSTEM [5] is a modular production line for stem cells, combining robotics, process control and cell production. It minimizes the manual interventions, enhances reproducibility, and assures strict adherence to regulations and standards. StemCellFactory [6] is the result of a research project aiming to develop an automatic production platform for reprogramming, cultivation, and differentiation of induced pluripotent stem cells. BioCell Printing [7] is an integrated automated system for manufacturing and in vitro dynamic culture of tissue engineering constructs. The system produces three-dimensional scaffolds of polymer-based material in an aseptic, continuous, and fully automated way, avoiding contaminations and promoting a broad clinical application.

However, while on the one hand, fully automated systems guarantee robustness and repeatability of the process, on the other they show a lack of flexibility. Furthermore, the complexity of such systems, composed of a large number of interconnected devices, increases the risk for failure of the process. Therefore, the human being proves to be still indispensable in procedures that need a certain degree of flexibility, not achievable using conventional automation concepts. Accordingly, this paper proposes the use of a Human-Robot Collaboration (HRC) solution which combines speed, accuracy, and repeatability of robots with flexibility and adaptability of human being, in order to meet the demand for high quality and customized products.

In the following section, an overview of progress in Tissue Engineering and Human-Robot Collaboration is briefly presented aiming to clarify the context and the vast opportunities. Then, a design methodology for Human-Robot Collaborative workplaces is described. The authors proposed adaptations and improvements to an existing method by introducing a specially developed algorithm. Discussion and conclusions are finally given in the last section.

2 Background

2.1 Advances in Tissue Engineering

Tissue Engineering aims to replace, improve, or repair biological tissue. TE is a multidisciplinary science which combines the efforts of scientist from different fields aiming to restore, maintain or enhance tissue function. Over the last years, material scientists, biologists, geneticists, engineers, and clinicians focused on the development of functional tissue substitutes for damaged tissues and organs.

One of the most used approaches involves the use of stem cells seeded on three-dimensional constructs. These structures, called scaffolds, provide support for the formation of extracellular matrix (ECM) and allow the formation of the new tissue; scaffolds offer the right microenvironment for cell adhesion, proliferation, growth, and differentiation. Therefore, the capability to preserve cells and growth factors, the diffusion of oxygen and cell nutrients, the vascularisation, are the crucial aspects related to the scaffolds features since they affect the outcome of the tissue regeneration process [8]. An ideal scaffold should possess some basic requirements: (i) it should be biocompatible and biodegradable with a controllable kinetics; (ii) it should have interconnected porous structure and tailored pore size and shape; (iii) it should possess mechanical properties similar to those of the tissue to be repaired; (iv) it should have appropriate geometry and size to reproduce the anatomic defect.

Additive manufacturing technologies led to an important development in this field. AM allows realizing three-dimensional constructs with complex morphology. Precisely controlling the interconnectivity, distribution, and shape of the pores, means being able to tune the mechanical and mass transport properties of the scaffolds according to different needs. Furthermore, additive technologies allow to process the polymer-based composites in a quite simple way. These materials are generally used as scaffold constituents since they can be designed with specific properties for a specific application. In the field of bone tissue regeneration, for example, polymeric matrices reinforced with calcium phosphate-based materials are usually adopted, due to their biocompatibility and their capability to undergo a tailored degradation process.

2.2 Human-Robot Collaboration

Several production fields are migrating from batch to flexible production. This shifting is dictated by the high demand for customized products (i.e., high number of variants for the same products, high number of different products). Production lines must be highly responsive for adapting and meeting with the new products. The collaboration between human and robot is a promising solution used to meet the demand for more flexible production processes. This approach aims to combine the strengths of human and robot improving product quality and providing greater job satisfaction.

Collaborative robots (cobots) have flooded the market. Cobots are designed to work with humans in a shared space or proximity without safety cages. Generally, cobots are designed to be inherently safe, lightweight, with rounded edges and limited in power and force. Nonetheless, truly collaborative workplaces are still rare. Several issues must be faced to reach an efficient collaboration between humans and cobots. The main issue concerns, obviously, the safety of the operators in the workplace. Ensuring safety is not related exclusively to the interaction between human being and cobot. It is more complex and concerns the design of the entire workplace.

In general, there are three main categories of standards regarding robotic solutions:

- Type A standards contain the basic safety specifications and requirements to apply to machinery.
- Type B standards include two sub-categories, B1 and B2, which contain generic safety specifications. B1 standards concern specific safety aspects. B2 standards concern safeguard measures, interlocking devices, optical or pressure sensors.
- Type C standards specify safety requirements for specific machinery. ISO 10218-1/2 are dedicated to the safety of industrial robots, but they provide the first indications on the collaborative operations between human and robot. ISO/TS 15066, the most recent standard for collaborative robotics, defines requirements for collaborative robotic systems and reinforces the indications given in ISO 10218-1/2.

It is easy to see that moving among the standards is not very simple. It could be very tiring considering the huge number of rules which usually provide only guidelines and few indications [9]. Accordingly, an approach based on the graph theory was developed and proposed in previous work by the authors [10] to simplify the preliminary HRC design phase. In this scenario, graphs were used to easily manage the relations among the requirements of a collaborative workplace. The result was a multi-level graph that organizes the contents of standard in order to make them usable in algorithms.

3 An Approach for the Collaborative Workplace Design

Designing collaborative workplaces requires a structured approach to gain significant benefits from the HRC. A feasible solution was proposed by F.Pini et al. [11] which adopted a structured method to design a HRC workplace for drip chambers assembling. It consists of four phasis. Firstly, the manufacturing process is thoroughly analysed and described in terms of tasks. Then, the tasks are assigned using a qualitative allocation criterion: the tasks are distributed among the available resources, i.e., human operators, cobots and machine tools, through the evaluation of the safety level, the required quality and the workload of each activity. The task planning phase concerns the definition of the time sequence of the tasks. Finally, the engineering phase of the collaborative workplace is performed according to the input coming from the previous steps, leaving the layout definition to the perception and experience of the designer.

Based on the solution proposed by F. Pini et al. [11], the current paper presents an enriched structured approach driven by a developed algorithm for layout optimisation.

3.1 Towards the Upscaling of Biomanufacturing Process

The design process proposed by the current work consists of two main steps: (i) *task analysis* and (ii) *HRC workplace layout generation.*

The task analysis step involves the *task description*, the *task allocation,* and the *task sequencing definition.*

More in details, *task description* aims to describe the whole manufacturing process in terms of elementary tasks. Two scenarios are possible: (i) renovation of existing process; (ii) design of new process. In both cases, the aim of this step is to define the tasks and the set of resources needed to carry them out. The resources are classified as active and passive resources [12]. Referring to well-known resources in a production line, humans and cobots are active resources, whereas working tables, conveyors and other machines are considered passive resources.

In the *task allocation phase*, each task is assigned to the most suitable resource according to the qualitative selection criterion proposed in [11]. Each task is assessed on the basis of four categories: (i) Impact of the single task on the whole process; (ii) Risk related to the single task; (iii) Fatigue caused by every task in terms of effort and repetitiveness of movements; (iv) Accuracy needed for every task. The qualitative assessment is carried out by means of a scale consisting of three levels for each category: low, medium, and high. Then, based on the ratings, the most suitable resource is assigned to a specific task.

Finally, the *task sequencing definition* sets the task schedule according to time constraints and process requirements.

In the *HRC workplace layout generation,* an in-house algorithm is used to help the designer during the definition of the layout of the workplace. The algorithm disposes the passive resources in the available space in a semi-automatic way, minimizing the occupied floor space. To minimize the occupied space, the resources are placed as close as possible while respecting the regulations' minimum and maximum distances. More in details, a scalar function was defined as the sum of the Euclidean distances among the passive resources; then, a set of nonlinear constraints were identified and translated in Matlab language. The proposed algorithm finds the local minimum of the function subject to the identified constraints, returning the Cartesian coordinates of the centre of gravity of the passive resources in a global reference frame. The tool considers the spatial constraints established by the reference standards and generates a standard compliant solution according to all the inputs obtained by the previous phases. The identification of the regulatory framework and the critical studying of the contents of standards were crucial steps in the algorithm development. The in-force standards were used to identify functional requirements and design parameters for collaborative workplaces. The relations between parameters and requirements were organised by means of a multi-level graph-based approach, in order to simplify their implementation in the code. The algorithm is organised in four main phases as depicted in Fig. 1: (P1) input definition – the designer defines manually the inputs collected during the task analysis, i.e., features of passive and active resources, task typology, task sequence, available space; (P2) Generation of the layout of the collaborative workplace in a 3D and 2D environment. The locations of the resources are defined and displayed by running the algorithm. The 2D plan view is very useful to easier control the visual obstruction and the overall dimensions. The 3D view is enabled by exploiting the Matlab 3D World Editor Toolbox and the V-Realm Builder. A library of parametric CAD objects is available in Virtual Reality Modeling Language (VRML) format; (P3) User editing: the designer can evaluate and modify the rough layout according to his experience. After the necessary modification, the compliance with standards is checked again. (P4) The generated 3D layout is exported as VRML file.

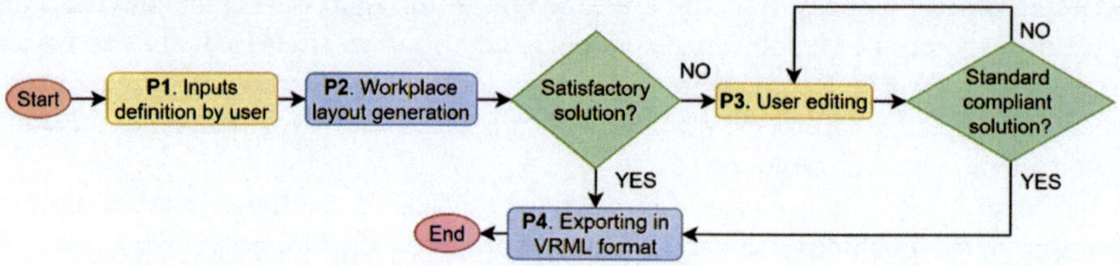

Fig. 1. Flowchart representing the main phases of the proposed algorithm

4 Application to the Preliminary Design of HRC Workplace for Advanced Biomanufacturing

The proposed approach was used to preliminary design the layout of the HRC workplace for scaffolds manufacturing. Task analysis steps are clarified in the following.

Task description: scaffolds are usually produced through multi-head bioprinter which can manufacture acellular and cellular constructs depending on the chosen strategy. Then, the scaffolds are sterilised by physical processes through UV radiation or by chemical processes by means of ethylene oxide or supercritical CO_2. Cells seeding is performed manually or using a robotic dispenser. The culture of matrix-cells systems is performed in dynamic conditions through perfusion flow bioreactor.

Task allocation step consists in the combination of tasks and resources according to the qualitative evaluation criterion previously described. Scaffold manufacturing, sterilisation, cell seeding, and the culture of the tissue are performed by means of specific devices. The cobot is used as a flexible handling unit for moving the scaffolds between the devices. This task is assigned to the cobot due to its repetitiveness and high risk of contamination. The highly skilled operator has the role of supervisor. When necessary, he can adjust the process steps and load or unload the supply materials of the different machines.

Task sequence definition was simple considering that the manufacturing of tissue-engineering constructs is a deeply rooted process. Therefore, the sequence of tasks did not change and remained as follows: (i) scaffold manufacturing; (ii) sterilisation; (iii) cells seeding and (iv) culture of the seeded cells.

HRC workplace layout generation phase leverages the specially developed algorithm to define the layout of the collaborative workplace. Figure 2 shows a first attempt layout that can be modified by the user by changing, for example, the orientation of the Human Machine Interface (HMI). The cobot is integral with a mobile platform which moves on a rail in order to increase the reach and allow access to every device. 3D Bioprinter, steriliser, automatic liquid handling units and bioreactor are the basic components of this system. Most of these are already commercially available, even though some need adaption to be used efficiently in the HRC workplace.

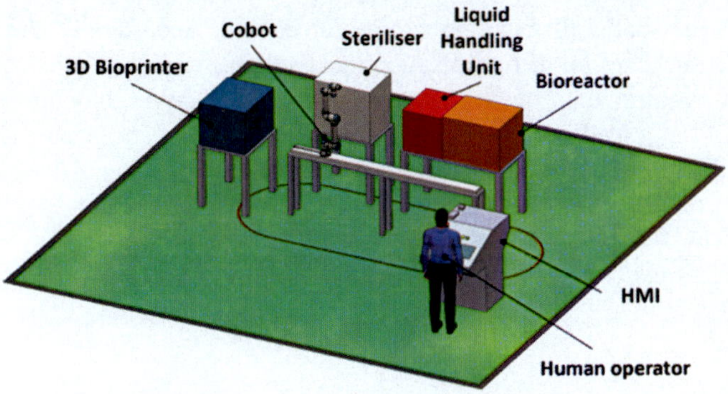

Fig. 2. The HRC workplace layout generated by using the proposed approach

5 Discussion and Conclusions

Starting from an overview of automation solutions in cell culture applications, as well as from the analysis of data reported in the literature, the present paper identifies the actual technological challenge in the necessity to upscale the biomanufacturing process.

The proposed approach would allow the implementation of a modern collaborative robot in an organized workplace with the aim of improving the clinical applications of scaffolds and tissue-engineered products. A relevant improvement would raise from combining the strengths of humans and robots in a collaborative environment that ensures flexibility, high throughput and high-quality products, overcoming the limitations of manual production and the shortcomings of fully automated systems. The design of a HRC workplace clearly requires the use of an organized approach which allows to: (i) breakdown the whole manufacturing process into a sequence of elementary tasks; (ii) allocate the identified tasks to the suitable resources; (iii) define the workplace layout according to one or multiple optimisation objectives. Therefore, the best practices generally adopted for designing the layout of a collaborative workplace have been integrated with an in-house algorithm developed in Matlab environment.

Thus, the final layout of a collaborative workplace for biomanufacturing was obtained through the proposed approach leveraging the developed algorithm.

The advantages obtained from the algorithm are:

- the tool helps the designer to take into account the maze of rules and prescriptions during the designing of the workplace layout.
- the layout planning is no longer left to the perception of the designer. The resources are organized in the available space according to an optimisation criterion aiming to reduce the distances to cover and the material handling costs.
- the layout generated in a semi-automatic way is a rough solution. The designer can edit it according to his experience and knowledge.

The current idea also involves future developments concerning the implementation of a real collaborative workplace for 3D scaffold manufacturing in order to validate the tool and the proposed approach. Furthermore, the employment of multiple criteria for optimising the spatial distribution can be considered. Accordingly, the user will be able to choose the criterion of optimisation most suitable for his purposes. In addition, the regulatory requirements regarding sterilisation, clean room, biocompatibility, and packaging will be considered to enhance the developed tool and ensure that both the workplace and the products comply with reference standards.

Acknowledgment. This study was developed with the economic support of MIUR (Italian Ministry of Universities and Research) under the remit of project ARS01_00861, 'Integrated collaborative systems for smart factory – ICOSAF".

References

1. Fucile, P., et al.: Strategies for the design of additively manufactured nanocomposite scaffolds for hard tissue regeneration. ACTA IMEKO **9**(4), 53–59 (2020)
2. Archer, R., Williams, D.J.: Why tissue engineering needs process engineering. Nat. Biotechnol. **23**(11), 1353 (2005)
3. Ochs, J., Barry, F., Schmitt, R., Murphy, J.M.: Advances in automation for the production of clinical-grade mesenchymal stromal cells: the AUTOSTEM robotic platform. Cell Gene Therapy Insights **3**(8), 739–748 (2017)
4. Kulik, M., Ochs, J., König, N., Schmitt, R.: Automation in the context of stem cell production–where are we heading with Industry 4.0? Cell Gene Therapy Insights **2**(4), 499–506 (2016)
5. Jung, S., Ochs, J., Kulik, M., König, N., Schmitt, R.H.: Highly modular and generic control software for adaptive cell processing on automated production platforms. Procedia CIRP **72**, 1245–1250 (2018)
6. Marx, U., et al.: Automatic production of induced pluripotent stem cells. Procedia CIRP **5**, 2–6 (2013)
7. Bartolo, P., Domingos, M., Gloria, A., Ciurana, J.: BioCell printing: integrated automated assembly system for tissue engineering constructs. CIRP Ann. **60**(1), 271–274 (2011)
8. Bártolo, P.J., Almeida, H.A., Rezende, R.A., Laoui, T., Bidanda, B.: Advanced processes to fabricate scaffolds for tissue engineering. In: Virtual Prototyping & Bio Manufacturing in Medical Applications, pp. 149–170. Springer, Boston (2008). https://doi.org/10.1007/978-0-387-68831-2_8
9. Di Marino, C., Rega, A., Vitolo, F., Patalano, S., Lanzotti, A.: The anthropometric basis for the designing of collaborative workplaces. In: 2019 II Workshop on Metrology for Industry 4.0 and IoT (MetroInd4. 0&IoT), pp. 98–102. IEEE (2019)
10. Di Marino, C., Rega, A., Vitolo, F., Patalano, S., Lanzotti, A.: A new approach to the anthropometric design of human-robot collaborative environments. ACTA IMEKO **9**(4), 80–87 (2020)
11. Pini, F., Leali, F., Ansaloni, M.: A systematic approach to the engineering design of a HRC workcell for bio-medical product assembly. In: 2015 IEEE 20th Conference on Emerging Technologies & Factory Automation (ETFA), pp. 1–8. IEEE (2015)
12. Tsarouchi, P., et al.: A decision making framework for human robot collaborative workplace generation. Procedia CIRP **44**, 228–232 (2016)

A Coarse-Grained Lattice Spring Model to Characterize Nanoindented Stem Cells

Lorenzo Vaiani(✉), Michele Fiorentino, Michele Gattullo, Vito Modesto Manghisi, Antonio Emmanuele Uva, and Antonio Boccaccio

Dipartimento di Meccanica, Matematica e Management, Politecnico di Bari, Bari, Italy
lorenzo.vaiani@poliba.it

Abstract. In this work, a Coarse-Grained Lattice Spring Model to characterize the mechanical behavior of human mesenchymal stem cells subjected to nanoindentation measurements is presented. The model simulated the action of adhesive structures acting on cells, necessary for attaching them to a substrate, and a nanoindentation process, performed by means of an atomic force microscope with a spherical tip. Cells were hypothesized to behave as elastic materials and the model included several subcellular components such as cell cortex and cytoskeleton. The lattice spring model was integrated within an optimization algorithm that iteratively compared the force-indentation curve numerically predicted to the data experimentally obtained, until a best fit condition was reached. The computed mechanical properties of the cell were compared to those obtained via the Hertz contact theory and finite element modelling, showing a good agreement. The proposed lattice spring model appears as a promising tool that can be used, with a very low computational cost, to characterize cell materials and other biological materials.

Keywords: Lattice spring models · Nanoindentation · Stem cells · Cell mechanical properties

1 Introduction

Numerical simulations actually represent a powerful tool, based on engineering disciplines such as computer science, numerical computing and computer graphics, to support scientific research in many areas. They can be considered as an extension of analytical modelling and a versatile and highly scalable link between theories and experimentation. A significant advantage is given by the possibility of exploring physical phenomena in particular conditions that are partially or totally impossible to achieve in the real world or are difficult to observe by using classical experimental setups, often culminating in the synthesis of general parametric predictive models. Computational methods have led to the identification of new disciplines such as Computational Physics or Computational Biology, characterized by the potential of managing complex multi-domain and multi-scale phenomena, recreating interactive environments that appear very close to real conditions [1].

C. Rizzi et al. (Eds.): ADM 2021, LNME, pp. 623–629, 2022.
https://doi.org/10.1007/978-3-030-91234-5_62

Many physical-mathematical modelling techniques can be used, as the observation scale varies in space and time. Ranging from the smallest to the largest dimensions, we can find the atomistic methods, in which the considered masses and force fields coincide with the representation of the individual atoms. In mesoscale approaches we can find the well-known Smoothed-Particle Hydrodynamics (SPH) and Lattice-Boltzmann (LB) methods, up to the more common mesh-based methods such as Finite Elements Method (FEM) and Computational Fluid Dynamics (CFD) [1].

In this simulation landscape, Lattice Spring Models (LSMs) emerge as a class of lagrangian methods for simulating mechanical behaviors of materials through a discretization of physical deformable systems into a series of particles, interconnected by means of spring elements, based on simple physical laws relating the acceleration of masses to the reaction forces given by the spring network [2], which are easy to implement through common programming languages. Even if LSMs are not as accurate as FEM approaches for mechanical case studies, they present a relatively fast and stable formulation for problems involving highly deformable objects even in case of topological changes (e.g. fracture), and thus are currently a consolidated approach for virtual reality and computer graphics applications, where the needs of real-time interaction and feedback are essential requirements [3]. In the biomedicine environment, the emergent computer-based surgery practices involve geometrical deformable models which must present a stable and realistic mechanical response during real-time haptic manipulations [2]. In the biological field, many coarse-grained LSMs have been used to model the structural dynamics of proteins and compute linear normal modes around a given reference structure, with great reductions in computational cost compared to all-atom models [4].

In this paper, the Atomic Force Microscopy (AFM) nanoindentation of a mesenchymal stem cell was considered as a case study. AFM is a common experimental practice in Biology fields, consisting in moving a spherical, conical or pyramidal nanoindenter positioned at the tip of a cantilever of known stiffness, into the surface of the material to investigate. This technique was previously utilized to characterize different cell materials [5–7]. Several studies were also conducted to evaluate the effect of the tip shape and of the residual stress acting on the cell during the nanoindentation process [8–10]. The report of an AFM test is a force-indentation curve which contains all the information associated to the stiffness of the investigated material. To obtain metrics of the internal behavior of the material, it is necessary to replicate the real deformation conditions of the AFM experiment by means of an analytical treatment or a virtual simulation environment, and then an extraction procedure for the elastic constants must be applied to the model. Common choices to do this are to consider the nanoindentation process as a linear hertzian problem in order to compute an averaged global Young's modulus for the entire cell [11, 12] or, in the case of finite elements modelling, an iterative procedure for Young's moduli fit is a solution adopted by many researchers [6], but this approach results computational demanding and often unstable, due to mesh distortions for large indentation magnitudes or in presence of nonlinear materials with low stiffness, typical for biological tissues and living cells [7, 12]. Different alternatives to the previous methods for studying the mechanical behavior of living cells were proposed, e.g. Vassaux and

Milan [13] investigated the effect of the curvature of the substrate where the adhesion of cells takes place, through a particle method.

In this work we propose a Coarse-Grained [14] LSM formulation for the AFM nanoindentation process on mesenchymal stem cells, with the principal focus of overcoming the issues associated to the previously described methods and to establish a fast and stable procedure for the abovementioned class of problems, aiming at extracting the elastic parameters of subcellular components. A LSM of a stem cell was generated to set up a complete computer simulation for extracting the elastic parameters (Young's moduli) of the main components of the cell. This procedure is the necessary starting point for analysing mechanobiological relations in living cells.

2 Materials and Methods

2.1 Modelling, Meshing and Spring Network Generation

The experimental procedure consisted in a series of AFM nanoindentation tests performed towards the soma of a mesenchymal stem cell. All data has been processed and presented as a report curve in force-indentation coordinates, in the range 0–5 nN for the measured forces and 0–200 nm for nanoindentations.

The entire experimental setup was modelled in a virtual simulation environment [15]. A CAD model of a quarter of the stem cell under investigation was built by using the Rhinoceros software (Version 7, McNeel) integrated with the Grasshopper plugin, and the real internal structure was simplified into several subcellular components: the nucleus, the cytoskeleton [16], the cell cortex (a 200 nm thick layer enclosing the whole cell) [7, 10], the stress fibers and the related tensioning regions connecting them to the cytoskeleton [17, 18]. As depicted in Fig. 1a and c, the main geometry was essentially a revolved surface having a diameter of 50 μm and a height of 13 μm [19].

The solid model generated in the above described way was firstly transformed in a tetrahedral mesh (see Fig. 1b) with the open source meshing software Gmsh (version 4.5.2), by choosing a Frontal-Delaunay algorithm for surface meshing and a refined Delaunay algorithm for volume meshing with an element size equal to 0.350 μm, always keeping in mind the need for a uniform and isotropic network of discrete elements to properly describe the mechanical behavior during deformation.

A second phase of the geometrical modelling consisted in the isolation of the edges of the tetrahedrons generated in the meshing phase and the conversion of them, by means of a dedicated Grasshopper script, into a series of springs, as shown in Fig. 1c, thus constituting the expected coarse-grained LSM.

The total number of springs was 887605 (thus leading to a spring density of about 320 springs per cubic μm), the average rest length was 0.345 μm (chosen to obtain a suitable resolution for the deformation field in the indented region) and they were interconnected by spherical hinges at the end points, thus ensuring a high structural compliance during deformation, as expected for soft biological materials.

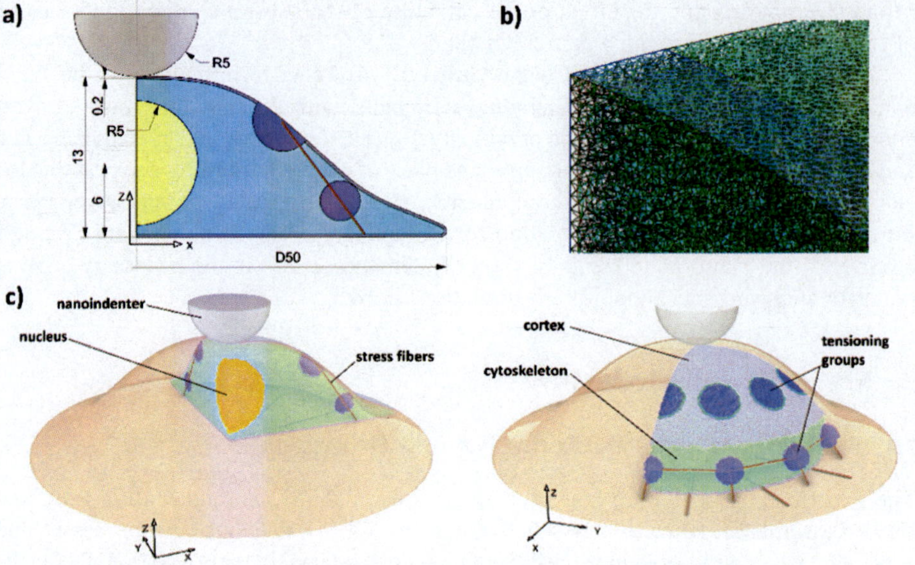

Fig. 1. a) CAD model of the cell (measures in μm), b) detail of the tetrahedral mesh, c) spring network and subcellular components.

2.2 Simulation, Optimization and Extraction of Young's Moduli

In the simulation environment (Abaqus, version 6.14, Dassault Systèmes), the LSM of the stem cell was accurately positioned under a rigid spherical nanoindenter with a 5 μm radius, to simulate a frictionless contact between the two objects.

The springs were treated as truss elements, presenting a cross section A equal to 1 μm^2, rest length L_0 equal to the averaged rest length of the whole lattice model, and axial stiffness $E_{loc}A/L_0$ defined by setting two different local Young's moduli, $E_{loc,cyt}$ and $E_{loc,cor}$ for cytoskeleton and for cortex respectively. The simulation was carried out throughout two consequent phases: the tensioning phase of the stress fibers, to describe the adhesion to a flat substrate and the indentation phase, consisting in the penetration of the spherical indenter into the pre-tensioned deformable cell, as illustrated in Fig. 2. The two unknown Young's moduli were determined by exploiting a numerical routine performing an iterative curve fitting between the known experimental data and the force-indentation curves coming from the simulations, by means of an optimization algorithm coded in the Matlab framework. After this curve matching driven by the experimental set of data, the local optimized values for cytoskeleton and cortex stiffnesses obtained through the LSM model were converted, by means of calibration functions, in global Young's moduli, comparable with the values presented in classical engineering and scientific literature for materials.

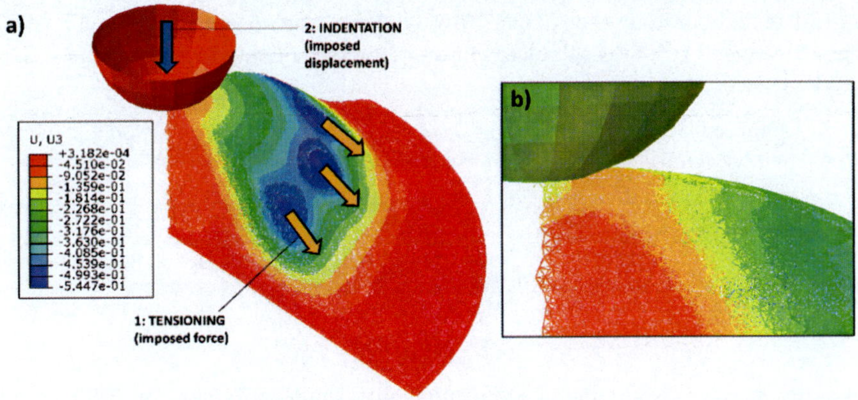

Fig. 2. a) Vertical displacement field on the pre-tensioned and nanoindented LSM of the stem cell, b) tetrahedral spring network detail.

3 Results and Discussion

The output of the LSM optimization for local Young's moduli is depicted in Fig. 3, where a good agreement between the experimental curve, obtained as the report of the AFM nanoindentation experiments and the best fit curve, coming as the output from the optimized computer simulation, can be noticed (the correlation coefficient R^2 is greater than 0.99).

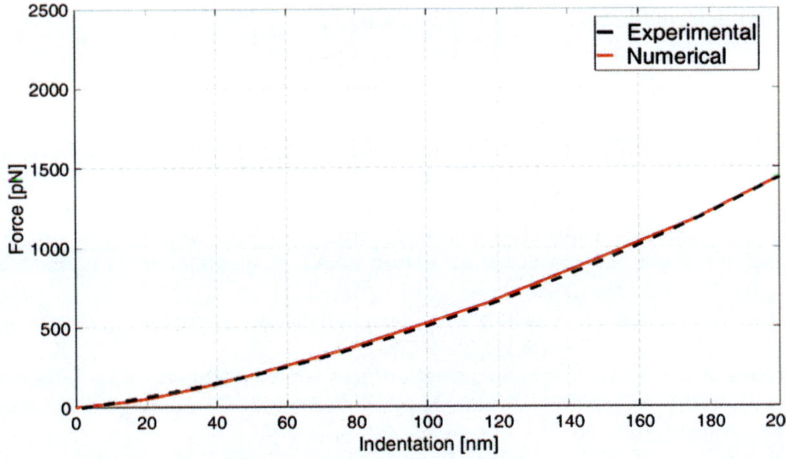

Fig. 3. Experimental curve obtained by AFM nanoindentations and optimized force-indentation curve computed by numerical simulations.

The optimal local Young's moduli for the spring lattice and the results of the extraction of the global Young's moduli E_{cyt} for cytoskeleton and E_{cor} for cortex are presented in Table 1. They are compared to the moduli calculated in a previous study [20] by means of a FEM model based on the Arruda-Boyce hyperelastic constitutive law [11, 12, 21] and to an averaged modulus calculated through the Hertz contact theory [22].

Table 1. Optimized local and global Young's moduli extracted by means of calibration functions compared to those computed through FEM and Hertz contact theory.

Coarse-grained LSM		Finite element method		Hertz theory
E_{cyt} [Pa]	E_{cor} [Pa]	E_{cyt} [Pa]	E_{cor} [Pa]	E_{Hertz} [Pa]
2911 (Local: 160)	9216 (Local: 508)	3519	9754	4110

The proposed coarse-grained LSM resulted in a simple to implement, fast and stable approach to compute elastic parameters of subcellular components in a mesenchymal stem cell, even when a pre-stress field due to a tensioning phase is previously superimposed to the stress field coming from a subsequent nanoindentation phase. The magnitude of the tensioning pre-stress field can increase up to 15% the magnitudes of the elastic moduli of cytoskeleton and cortex [20]. The values for E_{cyt} and E_{cor} computed through the proposed LSM are evidently proportional to those computed through FEM and to the averaged value resulting from the application of the Hertz contact theory in other studies [23–25], and it is possible to state that the refined Delaunay algorithm for tetrahedral discretization and the chosen element sizes were adequate and robust to represent the deformation field induced by AFM nanoindentation. The most interesting feature of the proposed coarse-grained LSM is the intrinsic compliance of the structure, due to the missing bending stiffness of the spring lattice, that can be considered as suitable to describe the deformation of soft biological materials.

References

1. Steinhauser, M.O., Hiermaier, S.: A review of computational methods in materials science: examples from shock-wave and polymer physics. Int. J. Mol. Sci. **10**, 5135–5216 (2009)
2. Golec, K., Palierne, J.F., Zara, F., Nicolle, S., Damiand, G.: Hybrid 3D mass-spring system for simulation of isotropic materials with any Poisson's ratio. Vis. Comput. **36**, 809–825 (2020)
3. Kot, M., Nagahashi, H., Szymczak, P.: Elastic moduli of simple mass spring models. Vis. Comput. **31**, 1339–1350 (2015)
4. Togashi, Y., Flechsig, H.: Coarse-grained protein dynamics studies using elastic network models. Int. J. Molecul. Sci. **19**(12), 3899 (2018)
5. Boccaccio, A., et al.: A hybrid characterization framework to determine the visco-hyperelastic properties of a porcine zona pellucida. Interface Focus **4**(2), 20130066 (2014). https://doi.org/10.1098/rsfs.2013.0066
6. Boccaccio, A., et al.: Nanoscale characterization of the biomechanical hardening of bovine zona pellucida. J. R. Soc. Interface **9**, 2871–2882 (2012)
7. Boccaccio, A., et al.: Nanoindentation characterisation of human colorectal cancer cells considering cell geometry, surface roughness and hyperelastic constitutive behavior. Nanotechnology **28**, 1–18 (2017)
8. Boccaccio, A., Lamberti, L., Papi, M., De Spirito, M., Pappalettere, C.: Effect of AFM probe geometry on visco-hyperelastic characterization of soft materials. Nanotechnology **26**, 325701 (2015)

9. Boccaccio, A., Papi, M., De Spirito, M., Lamberti, L., Pappalettere, C.: Effect of the residual stress on soft sample nanoindentation. Appl. Phys. Lett. **102**, 1–5 (2013)
10. Vargas-Pinto, R., Gong, H., Vahabikashi, A., Johnson, M.: The effect of the endothelial cell cortex on atomic force microscopy measurements. Biophys. J. **105**, 300–309 (2013)
11. Liu, K.-K.: Deformation behaviour of soft particles: a review. J. Phys. D Appl. Phys. **39**(11), R189–R199 (2006). https://doi.org/10.1088/0022-3727/39/11/R01
12. Lin, D.C., Shreiber, D.I., Dimitriadis, E.K., Horkay, F.: Spherical indentation of soft matter beyond the Hertzian regime: numerical and experimental validation of hyperelastic models. Biomech. Model. Mechanobiol. **8**, 345–358 (2009)
13. Vassaux, M., Milan, J.L.: Stem cell mechanical behaviour modelling: substrate's curvature influence during adhesion. BioMech. Model. Mechanobiol. **16**, 1295–1308 (2017)
14. Ingólfsson, H.I., et al.: The power of coarse graining in biomolecular simulations. Wiley Interdiscip. Rev. Comput. Mol. Sci. **4**, 225–248 (2014)
15. Vaiani, L., et al.: Coarse-grained elastic network modelling: A fast and stable numerical tool to characterize mesenchymal stem cells subjected to AFM nanoindentation measurements. Mater. Sci. Eng. C **121**, 111860 (2021)
16. Blanchoin, L., Boujemaa-Paterski, R., Sykes, C., Plastino, J.: Actin dynamics, architecture, and mechanics in cell motility. Physiol. Rev. **94**, 235–263 (2014)
17. Deguchi, S., Ohashi, T., Sato, M.: Tensile properties of single stress fibers isolated from cultured vascular smooth muscle cells. J. Biomech. **39**, 2603–2610 (2006)
18. Deguchi, S., Sato, M.: Biomechanical properties of actin stress fibers of non-motile cells. Biorheology **46**, 93–105 (2009)
19. Su, X., et al.: Nanomorphological and mechanical reconstruction of mesenchymal stem cells during early apoptosis detected by atomic force microscopy. Biology Open (2020). https://doi.org/10.1242/bio.048108
20. Migliorini, E., et al.: Nanoindentation of mesenchymal stem cells using atomic force microscopy: effect of adhesive cell-substrate structures. Nanotechnology **32**(21), 215706 (2021)
21. Arruda, E.M., Boyce, M.C.: A three-dimensional constitutive model for the large stretch behavior of rubber elastic materials. J. Mech. Phys. Solids **41**, 389–412 (1993)
22. Harding, J.W., Sneddon, I.N.: The elastic stresses produced by the indentation of the plane surface of a semi-infinite elastic solid by a rigid punch. Math. Proc. Cambridge Philos. Soc. **41**, 16–26 (1945)
23. Chen, Q., et al.: AFM studies of cellular mechanics during osteogenic differentiation of human amniotic fluid-derived stem cells. Anal. Sci. **26**, 1033–1037 (2010)
24. Kiss, R., et al.: Elasticity of human embryonic stem cells as determined by atomic force microscopy. J. Biomech. Eng. **133**, 1–10 (2011)
25. Liang, X., Shi, X., Ostrovidov, S., Wu, H., Nakajima, K.: Probing stem cell differentiation using atomic force microscopy. Appl. Surf. Sci. **366**, 254–259 (2016)

Analysis of Possible Algorithms for Active Noise Control of Siren Noise into an Ambulance

Massimo Generoso Buttarazzi[1](✉), Chiara Bartalucci[1], Francesco Borchi[1], Monica Carfagni[1], and Libero Paolucci[2]

[1] Department of Industrial Engineering (DIEF) – Florence, University of Florence, Via di Santa Marta 3, 50139 Firenze, Italy
{massimogeneroso.buttarazzi,chiara.bartalucci,francesco.borchi, monica.carfagni}@unifi.it
[2] Department of Information Engineering (DINFO) – Florence, University of Florence, Via di Santa Marta 3, 50139 Firenze, Italy
libero.paolucci@unifi.it

Abstract. This paper discusses a potential application of Active Noise Control (ANC) algorithms for the reduction of ambulance siren noise inside the vehicle. The study starts from the analysis of ANC techniques available in literature and currently used in industrial and consumer applications. The concept of ANC is based on the introduction of a canceling "antinoise" wave through a suitable speaker array known as secondary sources. These generators are managed through a specific signal-processing algorithm tailored according to the characteristics of the noise generated by the primary source that has to be cancelled. The primary signal is recorded by the control microphones and then processed by the ANC algorithm. An "antinoise" is then generated to reduce the primary source noise in a specific target zone. Based on the fundamental principles of wave superposition and wave cancellation, ANC technique is a very effective solution from a theoretical point of view; nevertheless, this technique faces numerous practical implementation problems and despite the wide number of available industrial and domestic applications, it is still poorly exploited. In this paper, the state of the art about most advanced ANC techniques and applications is analyzed. In particular, the implementation of adaptive signal processing algorithms and digital signal processing (DSP) in ANC systems is analyzed.

Keywords: Active Noise Control · Comfort zone · Feed-forward ANC · Virtual ANC · Acoustical · FIR · IIR · X-LMS · U-LMS · Genetic algorithms

1 Introduction

Active Noise Control (ANC) is an effective way to reduce noises that can otherwise be very difficult and expensive to control. It could be effective also when dealing with low frequency disturbances where physical barriers may fail. Lueg proposed the concept of acoustic ANC in 1936, using a microphone and an electronically driven loudspeaker to generate a canceling sound [1]. In general, the properties of the acoustic noise source

C. Rizzi et al. (Eds.): ADM 2021, LNME, pp. 630–640, 2022.
https://doi.org/10.1007/978-3-030-91234-5_63

and the environment change over time; hence the frequency content, amplitude, phase, and sound velocity of the undesired noise are nonstationary. Consequently, it is not possible to operate an ANC with static algorithms. A possible solution is the use of adaptive filters [2–5] that adjust their coefficients to reduce an error signal. Finite impulse response (FIR), infinite impulse response (IIR), and transform-domain filters (TDF) can be used to realize these algorithms. In the following section four ANC strategies based on adaptive filters and a possible application to ambulance noise abatement is presented.

Section 2 describes the basic concept of active noise control techniques and discusses the Least Mean Square (LMS) and FxLMS algorithms. The third section delves into the idea of using active noise control in an ambulance.

2 Active Noise Control Architecture

Concerning the ambulance siren noise, it is a stationary signal characterized by a single tone at a medium-high frequency (2 or 3 kHz). Hence it is possible to estimate its spectral behavior, and the control concerns mainly the amplitude of the disturbance. In the following section a LabVIEW simulation tool is used to test the behavior of different ANC strategies and evaluate their response as a function of a certain noise type. Most of the ANC algorithms are based on the Least-Mean-Square (LMS) filter because of its simplicity of implementation and adaptability. The system is usually composed of a reference microphone which records the primary signal noise, one or more secondary sources that generate the "antisound", one or more error microphones which record the resulting from the superposition of the primary and secondary sound and a processing unit. A general scheme of an ANC system is presented in Fig. 1. Most of these systems are based on digital controllers, as a result, signals from electroacoustic or electromechanical transducers are sampled and processed in real time using Digital Signal Processing (DSP) systems before being synthetized, amplified, and delivered to the loudspeakers. Clearly the entire process must be carried in real time, where the sound is generated and the moment it reaches the target zone which is the physical place where the primary noise must be reduced. Hence, especially in the past years, it has been necessary to achieve an optimal trade-off between processing time and algorithm complexity. In recent years the development of improved high-speed digital signal processing hardware allowed the implementation of complex ANC algorithms [6, 7]. However, the mathematical implementation of the cancellation algorithm has still to be optimized to take into account the causation principle and the convergence time.

The choice of the ANC algorithm type in general depends on the primary noise characteristics in terms of signal bandwidth and amplitude. The goal of adaptive algorithms is to keep filter coefficients up to date in order to achieve optimal noise cancellation. The optimization process of the filter coefficients can usually be stochastic or deterministic. LMS and Normalised-LMS (NLMS) algorithms use the stochastic approach. The deterministic approach, which is used with the RLS (Recursive Least Squares) algorithm, necessitates the calculation of the characteristics of many samples. The Filtered-x LMS (FxLMS) filter is a modified version of the LMS algorithm, and it is computationally simple where secondary path effects are also included, but its convergence speed is low. The fundamental issues in the afore mentioned approaches are inherent to stability issues in IIR-based structures, increases in computational requirements, and numerical instability issues in RLS-based ANC systems. For these reasons, FxLMS is still a viable option for ANC applications [8].

The simulation scenario makes use of the single-channel duct-acoustic ANC system depicted in Fig. 1 with a single reference sensor, single secondary source, and single error sensor.

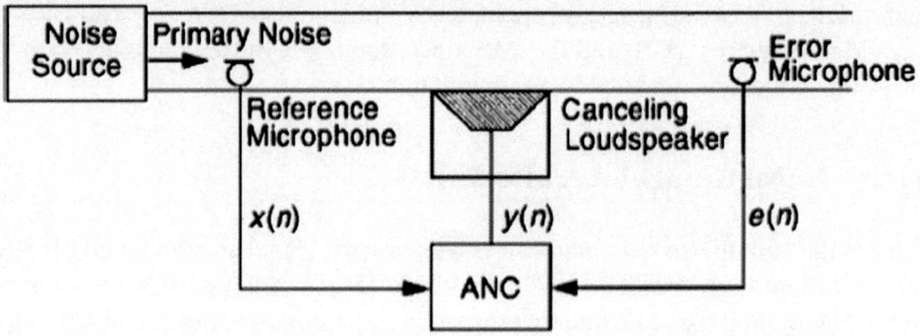

Fig. 1. A duct-based single-channel broad-band feedforward ANC system.

2.1 LMS Algorithm

In Fig. 2 a scheme of an LMS adaptive filter used in system identification applications is shown. The scheme highlights the behavior of the filter where x(n) is the primary source signal, d(n) is the disturbance in the target zone, y(n) is the 'antisound' and e(n) is the sum of d(n) and y(n). The Adaptive Filter in this case is an LMS adaptive filter which has to adapt to an unknown channel to minimize the resulting error e(n). It represents an LMS adaptive filter block diagram where each iteration of LMS involves three steps: filter output y(n), estimation error e(n) and tap-weight adaptation w(n + 1).

The frequency-domain analysis of the residual error signal e(n) can be used to determine the performance of ANC. The e(n) autopower spectrum is given by:

$$S_{ee}(\omega) = [1 - C_{dx}(\omega)]S_{dd}(\omega) \tag{1}$$

where Cdx(ω) is the magnitude-squared coherence function between two stationary random processes with wide-sense d(n) and x(n), while Sdd (ω) is the autopower spectrum of d(n). The term [1 − Cdx(ω)] represents an ANC system's noise reduction at frequency ω, but the maximum noise reduction in decibel is given by −10log10[1 − Cdx(ω)].

In Fig. 3 S(z) is introduced in order to model the secondary path effects, representing the secondary path connecting the canceling loudspeaker and the error microphone.

The z-transform of the error signal is:

$$E(z) = [P(z) - S(z)W(z)]Xr(z) \tag{2}$$

where it is constrained by the coherence of the reference signal to obtain the residual error (ideally equal to zero), the optimal transfer function W(z) must be defined:

$$Wo(z) = \frac{P(z)}{S(z)} \tag{3}$$

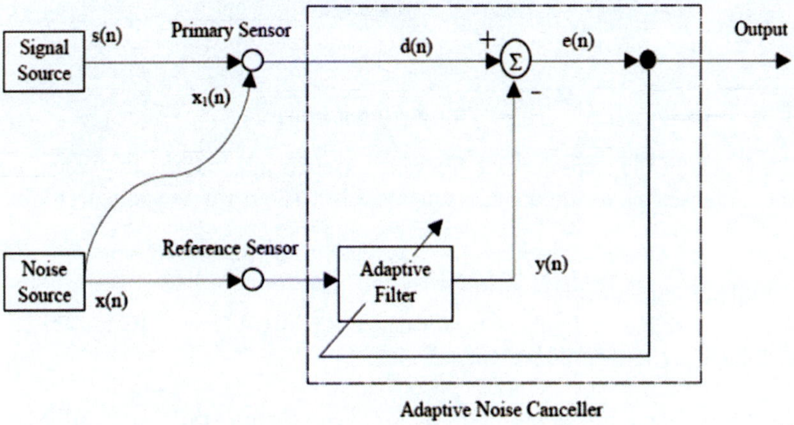

Fig. 2. LMS adaptive filter block diagram used in system identification applications.

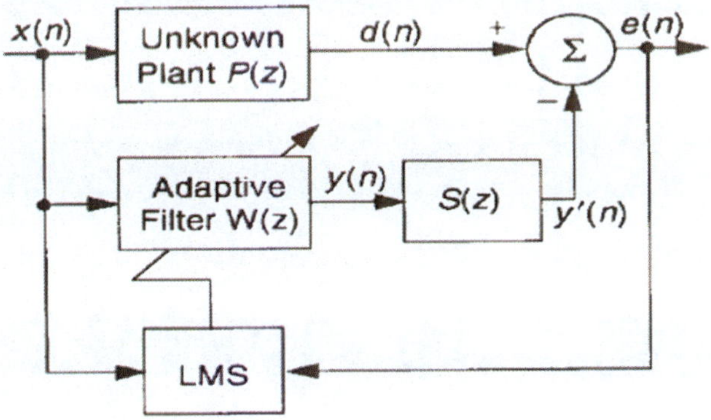

Fig. 3. LMS adaptive filter used with secondary path transfer function.

In other words, the adaptive filter W(z) must model both P(z) and S(z) simultaneously. The expression of the filter output is the following:

$$y_s(j) = \sum_{i=0}^{N-1} x_r(j-i)w_i(j) = x_r^T(j)w(j) \tag{4}$$

Filter weights updating is performed according to Eqs. (5),

$$w(j+1) = w(j) + \mu e(j)x(j) \tag{5}$$

Where:

1. μ = convergence coefficient (step size)
2. x(n) = reference signal
3. p(n) = primary signal
4. w(n) = filter coefficients

the condition for stability is:

$$0 < \mu < \frac{2}{input\ signal\ power} \tag{6}$$

With large values of the step-size, the filter increases its adaptation rate (faster convergence) and increases the residual mean-squared error. The major advantages of this filer are:

1. it is simple in implementation
2. it guarantees stable and robust performance against different signal condition a disadvantage is the slow convergence.

In Fig. 4 it is shown the variation of Mean Square Error (MSE) with different step-size.

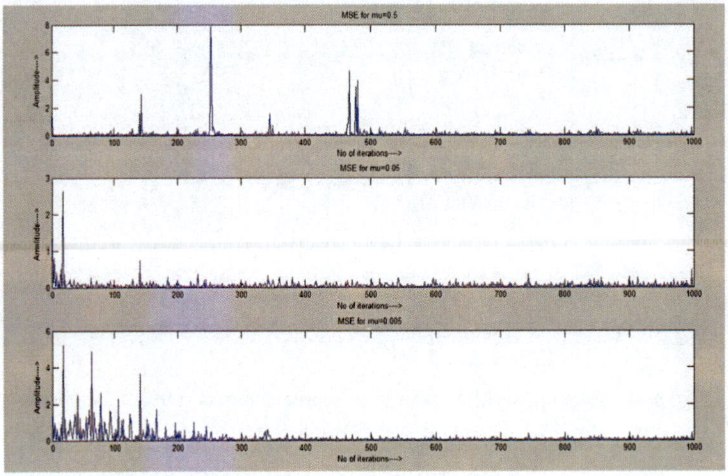

Fig. 4. Variation of MSE with different step-size.

In Fig. 5, instead, it is shown the response of the LMS filter to a broadband noise. It is possible to highlight that given a step size of 0.5 the system is not stable, and the error diverges.

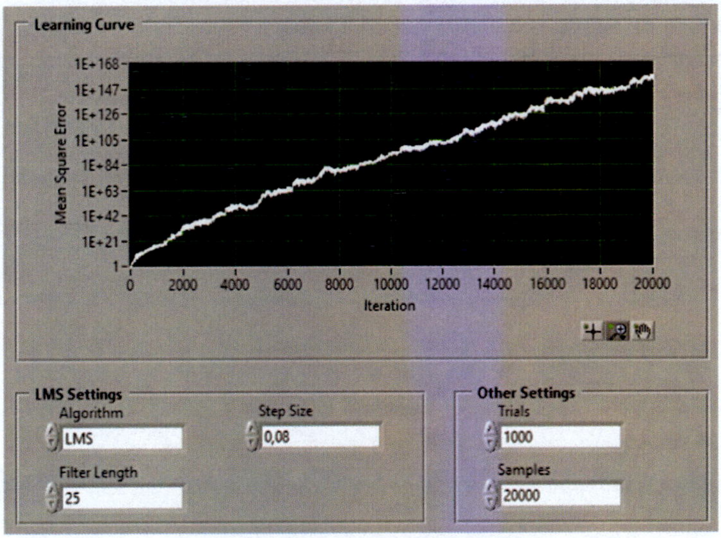

Fig. 5. Instability LMS Algorithm.

2.2 FxLMS Algorithm

The integration of a secondary path model in the reference signal path was proposed by Widrow, Shur, and Shaffer (from the speaker to the error microphone). Figure 6 depicts a block diagram of the FXLMS algorithm, which Kuo and Morgan used to demonstrate how the noise reduction algorithm works.

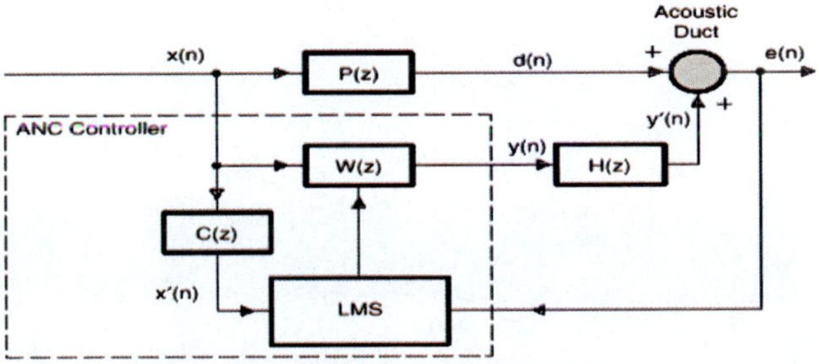

Fig. 6. FxLMS algorithm block diagram.

The FXLMS algorithm is a simple variation on the LMS algorithm [9]. The dashed box (Fig. 6) depicts the functional block diagram of a FxLMS algorithm adjusting an ANC controller, where C(z) represents an estimated model of the secondary path. The reference signal is filtered by C(z) before being used by the standard LMS algorithm, as shown. The reference signal can be determined in two ways: feedforward or feedback. An upstream microphone was used in the feedforward structure to provide information about the unwanted noise propagating down the system. The feedforward ANC algorithm is depicted in Fig. 7. As shown in Fig. 7, a microphone, called 'reference microphone', is located at a position where the signal of the primary source is dominant and the influence of the anti-noise source is negligible.

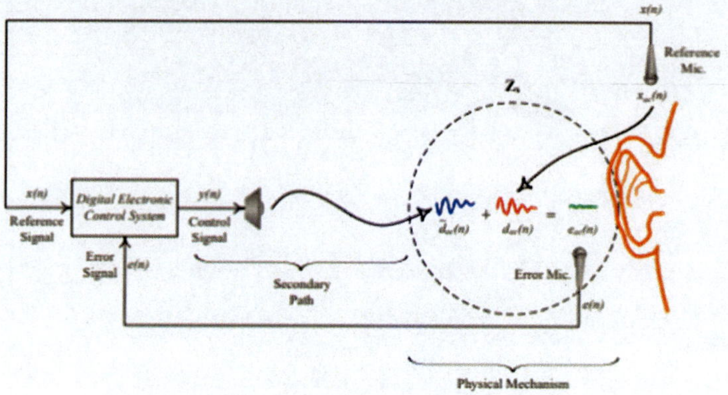

Fig. 7. General diagram for feedforward ANC.

Eriksson proposed ANC controllers with a feedback structure as an alternative to the feedforward structure in 1991. In the feedback structure the error microphone is directly used to evaluate the reference signal (Fig. 8) [10]. Because a reference microphone is not required in this structure, feedback ANC controllers are more compact and cost effective than feedforward ANC controllers.

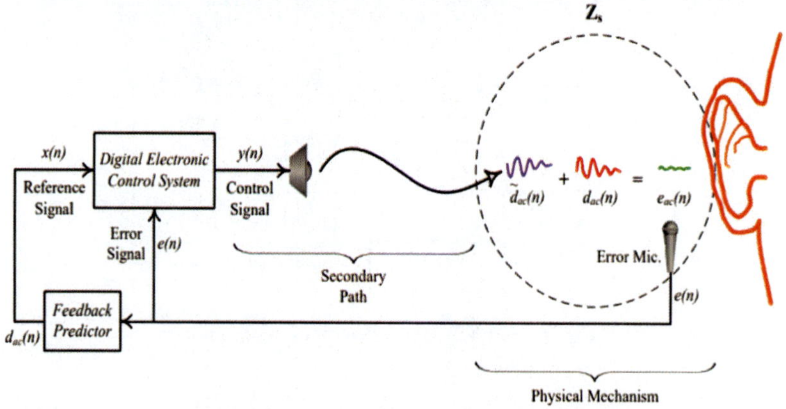

Fig. 8. General diagram for feedback ANC.

Aiming to test the effectiveness of the ANC system varying the step-size, authors have implemented into the Matlab-Simulink environment a single channel Adaptive Feedback ANC using a Filtered-X LMS [11]. Starting from this model of a Filtered-X LMS ANC system, it is possible when the error microphone signal is simulated as the sum of the noise source filtered by the primary acoustic path and the ANC output filtered by the secondary acoustic path.

In Fig. 9 the output of the error microphone is shown referring to step-size 0.5, 0.05 and 0.005. It is possible to highlight that given a step size of 0.5 the system is not stable, and the error diverges, while for step-size of 0.05 and 0.005 the system is stable with a better convergence rate for the step-size of 0.05.

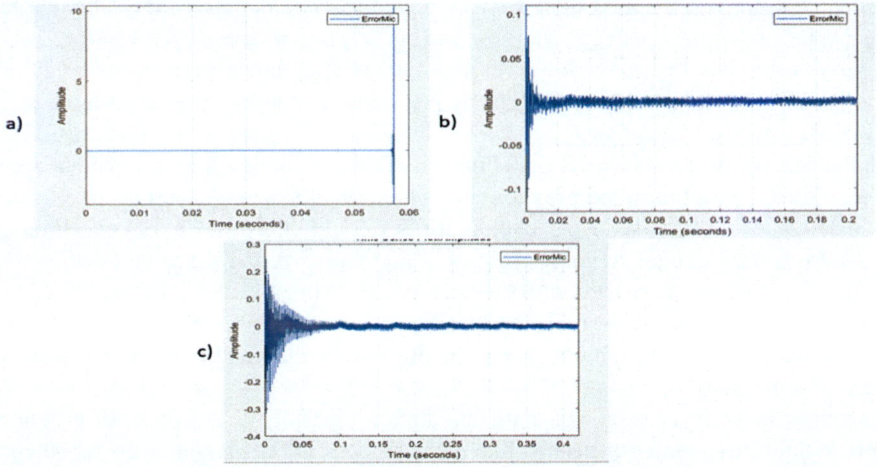

Fig. 9. Output of FXLMS Algorithm Simulation (Single channel - Broadband noise signal) utilizing various step sizes (a) Step size = 0.5, (b) Step size = 0.05, and (c) Step size = 0.005.

In Fig. 10 the results of the simulation, in terms of original noise signal and attenuated one, are shown considering the step-size 0.05.

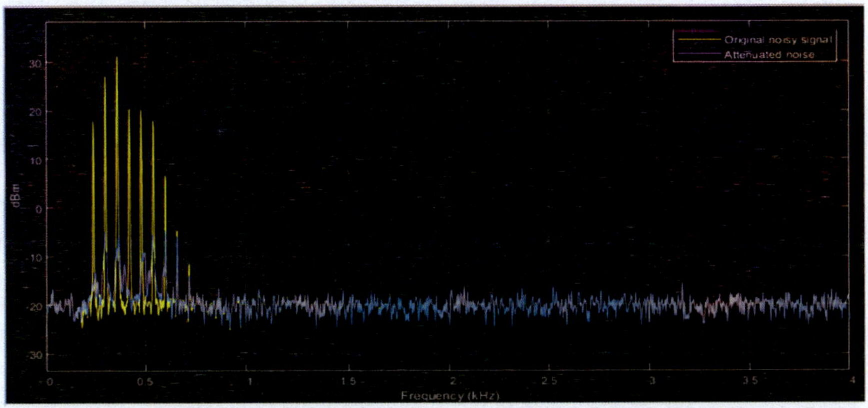

Fig. 10. Output of FxLMS algorithm simulation considering a broadband noise signal: original noise signal (yellow line) and attenuated signal (blue line).

3 Noise Control in Ambulance

Emergency vehicles often use high amplitude screeching sirens. The European normative only allows for the use of a high/low series of tones in emergency vehicles. The ambulance siren signal alternates between two tones at frequencies of 392 Hz (Sol natural) and 660 Hz (Mi natural). The sounds must be played in a continuous sequence with no breaks or noticeable overlap. The duration of the complete cycle is 3 s ± 0.5 s. Between a cycle and the next there may be a pause whose duration should not exceed 0.2 s [12].

Many estimates peg the noise level of a typical ambulance siren in the neighborhood of 120–130 decibels (dB) at roadside positions. The noise of an ambulance siren penetrates the ambulance doors and reaches the ears of the patient and staff. Inside the ambulance, sound pressure levels of about 87 dB(A) are common in the patient zone, while in the driver zone sound pressure levels up to 90 dB(A) are present [13].

Since exposure at these levels, over 80 dB(A), could be a problem of concentration for the driver, also due to the long exposure to siren noise during the same working day, according to the numerous trips with the siren switched on, and also for the patient [14]. In fact, the patient has a short time of exposure to the siren noise, which however could be sufficient to generate problems given the difficult health conditions he may have to face at that moment.

During the rescue operation the patient is already critically ill, and the noise produced can be reduced by passive measures like increasing sound insulation of the ambulance and installing sound absorbing material between doors. However, these interventions require several modifications reducing the free space into the ambulance and involving extra costs. Another possible solution can be that the patient put on ear muffs but based on his condition this solution is considered not feasible [15]. Also for the driver, the use of ear muffs that reduce the overall environmental noise is not considered beneficial, as safe driving conditions must be ensured.

The solution based on the active noise control (ANC) system can be used to solve the problem. It does not need physical modifications of the ambulance and does not affect the safe driving conditions for the driver. A local zone of silence is created around the heads of the staff and of the patient, leading to a significant noise reduction [16].

This solution creates a controller unit that consists of a filter and DSP with use of a fixed step-size, where the system's stability is jeopardized because the convergence rate is slow. The simulation and testing of the two-tones ambulance siren signal is performed for both LMS and FXLMS algorithm, by using Matlab and Simulink software, considering a fixed step-size. The noise signal used in the simulation consists of an ambulance siren signal (Fig. 11).

In Fig. 11 it is shown the examined signal has 51.2 kHz sample/s, 32 bit.

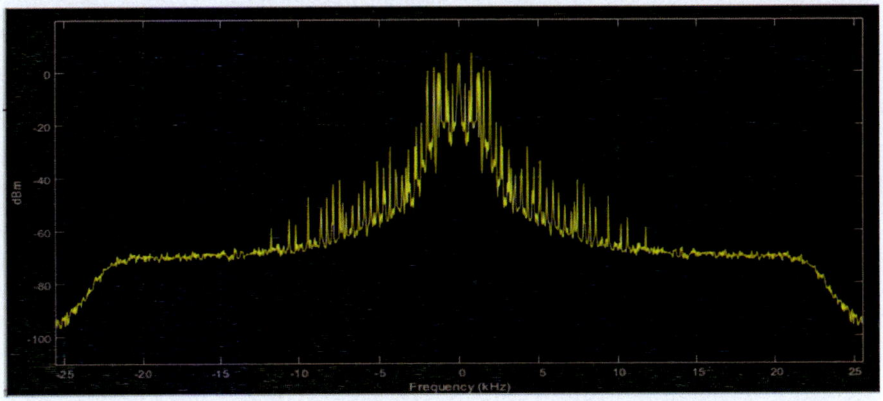

Fig. 11. Frequency spectrum of the ambulance siren signal.

From the preliminary analysis of the simulation results both algorithms lead to convergence. In particular, the FxLMS algorithm converges faster. In the future applications the algorithms will be experimentally tested.

4 Conclusions

In this paper a possible application of Active Noise Control (ANC) algorithms for the reduction of ambulance siren noise inside the vehicle is investigated. In particular, a general analysis of LMS and FxLMS is reported and a comparison of the two algorithms in reference to broadband noise and different fixed step-size is performed.

Furthermore, a preliminary analysis of the ambulance siren signal in reference to the two algorithms LMS and FxLMS is made and a possible preference for FxLMS is highlighted referring to the convergence speed.

On the basis of outcomes achieved in this paper, the ANC technique demonstrates to be a possible and promising application also in the reduction of noise due to the ambulance siren signal, potentially affecting the driver, the patient and the medical staff.

Next research activities will consist of the development of a prototype system for experimental tests to verify the algorithms and the optimal parameters for the siren case study. Moreover, in future steps the analysis of the applications with variable step-size will be tested to improve the FxLMS algorithm's convergence. However, authors consider that convergence could be a secondary problem when known signals have to be reduced as in the case of the ambulance siren noise.

References

1. Lueg, P.: "Process of silencing sound oscillations," U.S. Patent 2 043 416, June 9 (1936)
2. Goodwin, G.C., Sin, K.S.: Adaptive Filtering Prediction and Control. Prentice-Hall, Englewood Cliffs (1984)
3. Davisson, L.D., Longo, G. (eds.): Adaptive Signal Processing. ICMS, vol. 324. Springer, Vienna (1991). https://doi.org/10.1007/978-3-7091-2840-4

4. Cowan, C.F.N., Grant, P.M.: Adaptive Filters. Prentice-Hall, Englewood Cliffs (1985)
5. Honig, M.L., Messerschmitt, D.G.: Adaptive Filters: Structures, Algorithms, and Applications. Kluwer, Boston (1986)
6. Borchi, F., Carfagni, M., Martelli, L., Turchi, A., Argenti, F.: An active noise barrier system optimized for reducing outdoor stationary noise. In: Proceedings of the 22nd International Congress on Sound and Vibration, Florence, Italy, 12–16 July (2015)
7. Lapini, A., Biagini, M., Borchi, F., Carfagni, M., Argenti, F.: Design of active noise control systems for pulse noise, advances on mechanics, design engineering and manufacturing. In: Proceedings of the International Joint Conference on Mechanics, Design Engineering & Advanced Manufacturing, Catania, Italy, 14–16 Sept (2016)
8. Sakshi, G., Gupta, V.K.: A review on filtered-X LMS algorithm. Int. J. Signal Process. Syst. **4**(2), 172–176 (2015). https://doi.org/10.12720/ijsps.4.2.172-176
9. Ardekani, I.T., Abdulla W.: FxLMS-based active noise control: a quick review. In: Proceedings of, pp. 1–11 (2011)
10. Ardekani, I., Abdulla, W.: FxLMS-based active noise control: a quick review. In: Proceedings of 2011 Asia Pacific Signal and Information Processing Association Annual (APSIPA) Summit and Conference, Xi'an, China (2011)
11. Sachau, D., Jukkert, S.: Real-Time Implementation of the Frequency-Domain FxLMS Algorithm without Block Delay for an Adaptive Noise Blocker. Helmut-Schmidt-University, Germany (2014)
12. Meucci, F., Pierucci, L., Del Re, E., Lastrucci, L., Desii, P.: A real-time siren detector to improve safety of guides in traffic environments (2008)
13. Nataletti, P., Pieroni, A., Annesi, D., Di Giovanni, R., Morgia, F., Ruggeri, D.: Studio pilota dell'esposizione a rumore e vibrazioni degli addetti dei mezzi di soccorso del 118 e dei trasportati", Atti del convegno dBA2008, curato da e prodotto dal Dipartimento di Igiene del Lavoro, Laboratorio Agenti Fisici dell'Inail (ex Ispesl) (2008)
14. Weber, U., et al.: Emergency ambulance transport induces stress in patients with acute coronary syndrome. Emerg. Med. J. **26**(7), 524–528 (2009)
15. Sharma, M.K., Vig, R.: Ambulance siren noise reduction using LMS and FXLMS algorithms. Indian J. Sci. Technol. (2016). ISSN 0974-5645
16. Krishna, A., Ravinchandra, L., Fei, T.K., Yong, L.C.: Active noise reduction using LMS and FxLMS algorithms. J. Phys.: Conf. Ser. **1228**, 012064 (2019). https://doi.org/10.1088/1742-6596/1228/1/012064

A Graph-Based Multi-level Framework to Support the Designing of Collaborative Workplaces

Castrese Di Marino[1](✉), Andrea Rega[2,3], Fabio Fruggiero[4], Agnese Pasquariello[3], Ferdinando Vitolo[3], and Stanislao Patalano[3]

[1] Department of Management, Information and Production Engineering, University of Bergamo, Bergamo, Italy
castrese.dimarino@unibg.it

[2] Department of Neurosciences, Reproductive and Odontostomatological Sciences, University of Naples Federico II, Naples, Italy

[3] Fraunhofer JL IDEAS, Department of Industrial Engineering, University of Naples Federico II, Naples, Italy

[4] School of Engineering, University of Basilicata, 85100 Potenza, Italy

Abstract. In the paradigm of Industry 4.0, innovative workplaces characterized by Human-Robot Collaboration represent an important topic to improve productivity and adaptability of manufacturing plants. In this context, the design of a collaborative workplace is a challenging issue because of the high level of complexity due to multidisciplinary and non-homogeneity of its features, as well as the presence of human very close to the robot.

This work faces with the complexity of collaborative workplace and proposes a structured framework to support strategic decisions in designing. It suggests a clusterization of factors and effects, based on five domains involved in collaborative workplace, in order to better consider the human safety and working conditions. Consequently, the main elements of a collaborative workplace are highlighted in a matrix decomposed in relevant features and main incident factors, and a multi-level designing workflow is described to report collaborative performances. The proposed approach manages connections among the elements by means of the graph theory in the form of an adjacency matrix in order to show and manage the complexity of the problem. A user interface named Smart Graph Interface was developed to read and manipulate the contents of the adjacency matrix. Main results are reported on an assembly and sealing of a refrigerator, to spread out principal outcomes in terms of applicability and robustness.

Keywords: Human-robot collaboration (HRC) · Collaborative workplace design · Multi-level designing · User interface · Graph theory

1 Introduction

The technology driven progress of Industry 4.0 is forcing to adopt the interactive collaboration between humans and robots to improve flexibility and productivity [1]. In this context, the leftover approach of cooperation, designed over the principles of "how human

C. Rizzi et al. (Eds.): ADM 2021, LNME, pp. 641–649, 2022.
https://doi.org/10.1007/978-3-030-91234-5_64

is serving robot" or "is robot helping human", is taking over by the human and robot co-existence [2]. Thus, risk and safety issues [3], human physical and mental aspects [4], logistics, technological and operational requirements [5], product and process specification, have to be analyzed and designed in order to accomplish the environmental, social, and economic sustainability. Long over the state of art, there is currently the attempt of supporting the smart configurations of collaborative systems while proposing solution for human-centered allocation [6], flexible configuration [7] and automatic layout generation [8]. This requires a multi-level analysis arranged inside a decision support framework evolving over complex [9], dynamic rules [10]. In the collaborative workspace, different factors, and causal entities act. They evolve over domains (from their aggregate to detailed representation), regulated by parameters, configurations, and constraints, arranged to manifest benefits across their functionalities. Whenever collaboration between human and robots is arranged in systems, severe implications on human, robot/machine and environmental are required [11]. Those influence the design of configurational and operational systems to support the superior long-term performances in both employs and system outcomes.

The aim of this work is to propose and explain a consistent approach to support the designing of collaborative workplaces. Section 2 presents and discusses a five domains framework, decomposed in simpler elements organized in a multi-level workflow. The relations among the elements are underlined by means of the graph theory and a graphic interface is presented for this purpose. In Sect. 3, the methodology is tested on a state of art test case, and in Sect. 4 main outcomes are underlined.

2 Collaborative Workplace Framework

2.1 Decomposition: The Five Domains of a Collaborative Workplace

Collaborative workplaces include arrangement and assessment of a huge and non-homogenous number of entities/features/data/performances. Consequently, those require an architectural frame to support their causal interferences. Therefore, it is proposing a framework where the collaborative workplace is the combination of five domains (Table 1). Each domain meets a need, and it is arranged in groups and elements. Specifically, groups face to macro aspects, elements face specific functionalities.

The selected domains represent subject areas of the collaborative workplace:

- Logistic domain meets the management needs of the collaborative workplace. It is related to the management of spaces and flows.
- Technologic domain meets the technical needs. It is related to the active resources (act an action) and passive resources (suffer an action).
- Safety & ergonomics domain meets the safeguard and wellness of human in collaborative workplace. It involves ergonomic and environmental issues on a side, and the safety issue on the other side. It is strictly related to the human presence in the workplace and the collaboration with the robot.
- Process domain meets the working needs. It faces the aim of the workplace as working process, task sequencing and the interaction between human and robot.

- The economic domain meets the economic needs. It is related to the cost and the evaluation of the performances.

Table 1. The five domains decomposition Matrix for collaborative workplace.

Domains	Needs	Groups	Macro aspects	Elements	Functionalities
Logistic	Management	Layout	Management of spaces	Physical limits	Spatial constrain
				Workspaces	Space division
		Movement	Management of flow	Paths	Accessibility
				Feeding	Type of supplying
Technologic	Technic	Passive resources	Passive actor: suffer the action	Workpiece	Target to achieve
				Equipment	Supportive object
				Usable devices	Operator support
		Active resources	Active actor: act the action	Operator	Operator features
				Robot	Robot features
				AGV	AGV features
Safety & Ergonomics	Security & wellness	Human	Human wellness	Ergonomics	Operators constrain
				Environment	Working condition
		Safety	Human security	Minimum distances	Disposals constrain
				Perimeter	Workplace border
				Safeguarding devices	Safety device
Process	Working	Operation	The aim of the workplace	Type of work	Operation
				Task sequence	Relative operativity
		Interaction	Human-robot interaction	HRC	Type of interaction
				HMI	Communication
Economic	Economic	Value	Benefits vs. costs	Costs	Economic constraint
				Benefits	Key Performance Indicators (KPI)

The domains are decomposed in elements which report functionalities and cover principal aspects of the collaborative workplace:

1. Physical limits: available space within the workplace can be located;
2. Workspaces: layout and working areas;
3. Paths: definition of the access and exit points as well as the paths;
4. Feeding: type of supplying modality;

5. Workpiece: information about the piece to be worked and the components;
6. Equipment: support instruments and furniture;
7. Usable devices: devices under the control of the operator;
8. Operator: operator characteristics;
9. Robot: robot characteristics;
10. Autonomous guided vehicle: AGV characteristics;
11. Ergonomics: ergonomic requirements;
12. Environment: environmental working condition tolerable;
13. Minimum distances: minimum space among fixed and mobile resources;
14. Perimeter: physical or virtual workplace frontier;
15. Safeguarding devices: device not under control of the operator;
16. Type of work: working operation to be executed on the workpiece;
17. Task sequencing: set of operations to be performed by the active resources;
18. Human-robot collaboration: interaction modality between operator and robot;
19. Human machine interface: communication between operator and robot;
20. Costs: economic constraint;
21. Benefits: key performance indicators (KPI).

2.2 Designing Steps: Multi-level Designing Process of the Workplace

The designing multi-level process follows four steps as depicted in Fig. 1, in order to face all the aspects of a collaborative workplace. At the beginning, the "inputs", in the form of (i) constraints, (ii) customers' needs and (iii) strategic requirements has to be set and combined in order to arrange the aggregate contents scenarios. This is called *aggregate representation.* It presents the main workspace outcomes and collects all the relative "constraints" of the workplace. In the second step, it is planning the master "configurations" of the domains, basing on the selection of resources and the designing of tasks sequence for jobs. This reports the *main representation* where are designed the master sets. In the first two phases, the human presence is closely considered because the constraints - with the ergonomic requirements, the environment in terms of designing, the main safety and health - are very connected to the wellness and safety of the human. Indeed, the consideration of the ergonomics, or human factor, in the first stage, guides and limits the designer's choices in factors and elements. The third step provides the "manifestations" that come out of the previous statements. It includes the layout arrangement and its safety requirements. This is called *resulting representation* because it should be auto-generated across an expert knowledge-based system or through algorithms including meta-expert opinion implemented according to the needs of company and the designer. At the end, the "evaluation" of the workplace performances is set. This is called *evaluation representation.* The illustrated process recursively evolves updating the previous stages if some requirements are not met. Whenever inconsistency in results manifests or acceptance in arrangement is rejected, it is possible to return at the upper state that is logically and temporally preceding (e.g., the designer can modify type and amount of factors and element in order to optimize principal manifestation).

Consequently, the elements of the framework were grouped (Table 2) in constraints, configurations, manifestations, evaluations. This categorization, using the designing methodology exposed above, creates a flow from the constraints to the evaluation. The classification is the following:

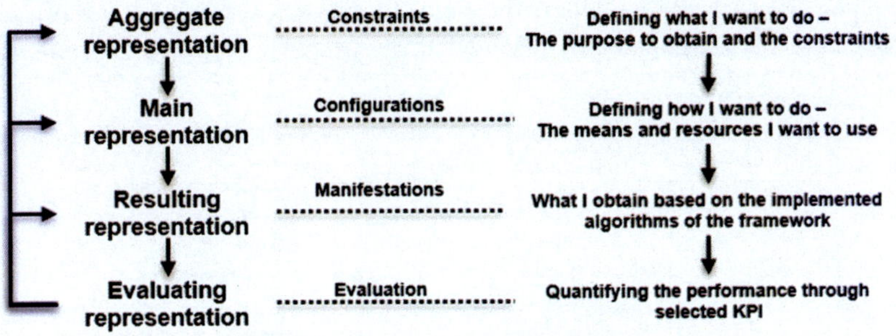

Fig. 1. Multi-level designing process

- Constraints: contain the requirements from the customer and needs of the company.
- Configurations: contain all the elements to be set.
- Manifestations: contain all the elements object of an optimization.
- Evaluation: consists of the evaluation of the performances.

Table 2. Aggregation of the elements according to the four levels of representation

Constraints	Configurations	Manifestations	Evaluation
Workpiece	Operator	Workspaces	Benefits
Type of working	Robot	Path	
Physical limits	AGV	Perimeter	
Ergonomics	Equipment	Safeguarding devices	
Environment	Usable devices		
Minimum distances	Task sequencing		
Costs	HRC		
	HMI		
	Feeding		

2.3 The Graph Based Representation of the Collaborative Workplace

The connection among elements is made explicit in order to face with the complexity of the problem and manage it at the same time. To this purpose, the graph theory was adopted and the relationship among the elements explained as per Di Marino et al. [9] by means of a multi-level graph. A graph is composed of a set of nodes connected by edges (G = {N, E}) [12] and presents a double representation: a matrix representation, by the adjacency matrix, where the relations among the nodes are explicated by the number 1 on the row (the row element has a direct connection with the element in the column), and a graphical representation where the elements are represented as nodes of a network and the relations are explicated by oriented arrows.

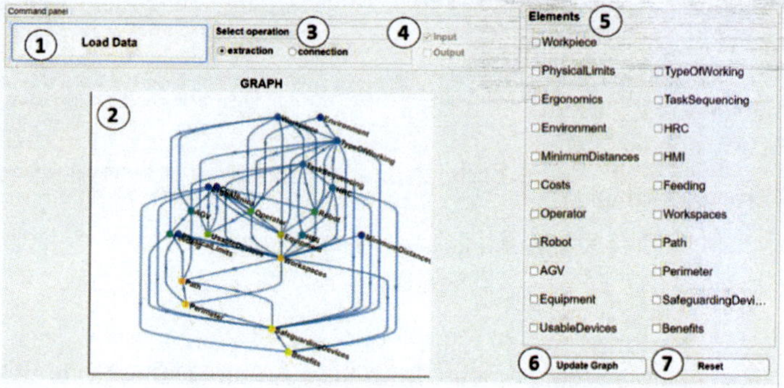

Fig. 2. Smart tool interface (SGI): 1) import the adjacency matrix; 2) graph displayed; 3) select the operation: extract a subgraph or highlight all the connections of a node; 4) set input and/or output; 5) set the element to display; 6) update the graph; 7) reset.

The information structured in the adjacency matrix or in the graph representation are not easily to be exploited, therefore a User Interface, named Smart Graph Interface (SGI), has been developed in Matlab (Fig. 2). Importing an adjacency matrix as a CSV file, the SGI reads the content of the matrix, recognizing the name of the elements, and depicts the corresponding digraph. At the same time, a list of checkboxes is generated on the side of the SGI. The user can exploit this information either by the "extraction" of a subgraph over the selection of a specific set of nodes or focusing on a single node to see all its direct "connections" (inputs and outputs). The first selection allows to investigate a reduced system, hiding uninterested elements; the second selection visualizes the elements directly connected to the selected node. The SGI provides support in designing and analysis of a collaborative workplace.

3 Application to a Multi Human-Robot Assembly

The proposed framework was applied to an existing use case in scientific literature. The use case consists in a collaborative workplace for the assembly and the sealing of a refrigerator [2]. The workplace employs two collaborative robots and two operators that work alongside on the parts to be assembled on a conveyor (Fig. 3). After the workpiece arrival, humans prepare the components whereas robots apply the sealant. Consequently, humans assemble the pipes and cables at the back of the workpiece. In the end, humans can manipulate the robots to apply any further necessary sealant.

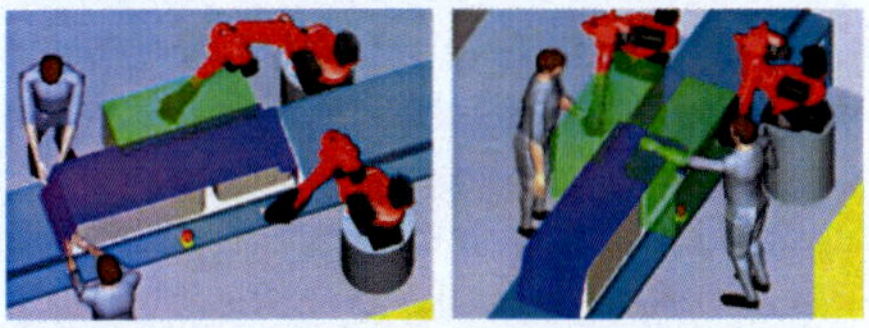

Fig. 3. Use case: assembly and the sealing of a refrigerator [2]

The designing process follows the steps explained in Fig. 1 and considers the connection established in the adjacency matrix and depicted in the SGI. The "aggregate representation" consists in the definition of the context and constraints. The technical specifications of the workpiece and the type of working (assembly) are firstly provided. Then, the constraints and the needs of both customers and company are given. Therefore, physical limits, cost and human working conditions are established. As a result, the context of the workplace is defined, limiting the possibilities choices of the next step, especially considering the human safety and working conditions.

The second stage, named "configuration representation", requires the selection of the active and passive resources (equipment and usable devices), the type of interaction between humans and robots, and the feeding mode. The task sequencing is established considering the constraints posed above. Two humans and two robots are set as active resources and the conveyor as feeding device. The passive resources are set as support for the active resources. The interaction modality is Power and Force Limiting (PFL) as indicate in ISO/TS 15066 [13] and the communication occurs by means of a starting button or direct contact. In this stage the collaboration is the core of the designing process. Indeed, the resources and the interaction modality limit the possible choices and guide the decision-making for the designer. Therefore, the collaboration is an opportunity over a strong constraint.

The third stage, named "resulting representation", is the direct effect of the statements carried out in the previous stage. For instance, the disposal of all the resources in the workplace is optimized with respect of the previous statements and constraints. Therefore, by the use of the SGI, all the inputs of the node "workspaces", that represents the working areas as defined in [14], are displayed (Fig. 4). Eleven related elements - used as inputs to optimize the layout - are recognized: type of working; task sequencing; HRC;

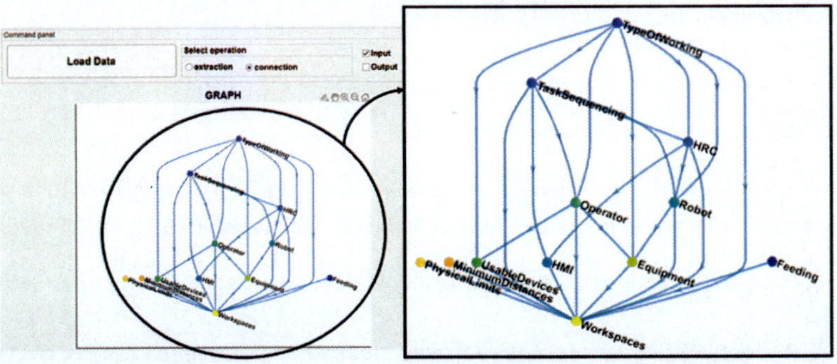

Fig. 4. Application of SGI: displaying all the inputs of the element "workplaces" as per the digraph on the right side

robot; operator; feeding; equipment; HMI; usable devices; minimum distances; physical limits. Consequently, the path, the safeguarding perimeter as well as the safeguarding devices are arranged according to implemented approach.

The last stage involves the analysis of the key performance indicators (KPIs) resumed in the node "Benefits". It provides a set of KPIs and evaluation functions about several aspects of the workplace.

4 Conclusion and Future Work

This work analyses the complexity of a collaborative workplace due to the high multidisciplinary and non-homogeneity of all the information and shows an approach to manage this complexity. Five domains involved in a collaborative workplace were identified and structured in a decomposition matrix, as well as all the elements that face specific aspects of the problem. Then, a multi-level designing workflow was presented. The elements were grouped, and the complexity of the relationship was managed by means of the graph theory. The relations among the elements of a collaborative workplace are highlighted in the human-robot collaboration perspective, and a tool, named Smart Graph Interface (SGI), was presented to display and manage them. In the end, outcomes for a case test of collaborative scenarios were reported using a combinations of multi-level workflow and SGI. The result is an example of designing process of a collaborative workplace starting from an aggregate point of view by adopting a collaborative perspective and considering the needs of the operator during the collaboration. The proposed structural approach confirms that it is possible to face with complexity in collaborative workplace while managing principal outcomes.

This approach can be used to select relevant groups while assigning main elements to relevant domain. According to the selected scenario, the user can modify the entity in order to come out alternative configurations in a *what if* approach. Afterwards, the framework is structuring to improve visual dependency among the nodes and optimize workflow on operations. Furthermore, an optimization engine is implementing in the SGI in order to auto generate and evaluate with local changes the designed workspace.

Acknowledgment. This study was developed with the economic support of MIUR (Italian Ministry of Universities and Research) under the remit of project ARS01_00861, 'Integrated collaborative systems for smart factory – ICOSAF".

References

1. Villani, V., Pini, F., Leali, F., Secchi, C.: Survey on human–robot collaboration in industrial settings: safety, intuitive interfaces and applications. Mechatronics **55**, 248–266 (2018)
2. Michalos, G., Makris, S., Tsarouchi, P., Guasch, T., Kontovrakis, D., Chryssolouris, G.: Design considerations for safe human-robot collaborative workplaces. Procedia CIRP. **37**, 248–253 (2015)
3. Awad, R., Fechter, M., van Heerden, J.: Integrated risk assessment and safety consideration during design of HRC workplaces. In: 2017 22nd IEEE International Conference on Emerging Technologies and Factory Automation (ETFA), pp. 1–10. IEEE, Limassol (2017)

4. Maurice, P., Padois, V., Measson, Y., Bidaud, P.: Human-oriented design of collaborative robots. Int. J. Ind. Ergon. **57**, 88–102 (2017)
5. Zanella, A., Cisi, A., Costantino, M., Di Pardo, M., Pasquettaz, G., Vivo, G.: Criteria definition for the identification of HRC use cases in automotive manufacturing. Procedia Manuf. **11**, 372–379 (2017)
6. Vitolo, F., Pasquariello, A., Patalano, S., Gerbino, S.: A multi-layer approach for the identification and evaluation of collaborative robotic workplaces within industrial production plants. In: Rizzi, C., Andrisano, A.O., Leali, F., Gherardini, F., Pini, F., Vergnano, A. (eds.) ADM 2019. LNME, pp. 719–730. Springer, Cham (2020). https://doi.org/10.1007/978-3-030-31154-4_61
7. Krüger, J., et al.: Innovative control of assembly systems and lines. CIRP Ann. **66**, 707–730 (2017)
8. Tsarouchi, P., et al.: A decision making framework for human robot collaborative workplace generation. Procedia CIRP. **44**, 228–232 (2016)
9. Di Marino, C., Rega, A., Vitolo, F., Patalano, S., Lanzotti, A.: A new approach to the anthropocentric design of human–robot collaborative environments. ACTA IMEKO. **9**, 80–87 (2020)
10. Fruggiero, F., Riemma, S., Ouazene, Y., Macchiaroli, R., Guglielmi, V.: Incorporating the human factor within manufacturing dynamics. In: IFAC- PapersOnLine, pp. 1691–1696. Troyes (2016)
11. Neumann, W.P., Dul, J.: Human factors: spanning the gap between OM & HRM. Neumann **30**, 923–950 (2010)
12. Deo, N.: Graph Theory with Applications to Engineering and Computer Science. Dover Publications, Made available through hoopla, United States (2017)
13. ISO/TS 15066:2016 Robots and robotic devices - Collaborative robots (2016)
14. Di Marino, C., Rega, A., Vitolo, F., Patalano, S., Lanzotti, A.: The anthropometric basis for the designing of collaborative workplaces. In: 2019 II Workshop on Metrology for Industry 4.0 and IoT (MetroInd4.0 IoT), pp. 98–102. Naples (2019)

Finite Element Model Proposed for Determine the Rotational Equilibrium Around of the Subtalar Joint Axis (SJA)

Rubén Lostado Lorza[1](✉), Fátima Somovilla Gomez[1], Saúl Íñiguez Macedo[2], Marina Corral Bobadilla[1], María Ángeles Martínez Calvo[1], Álvaro Pérez-Sala[3], and Ignacio M. Larráyoz[3]

[1] University of La Rioja, 26004 Logroño, La Rioja, Spain
ruben.lostado@unirioja.es
[2] Footwear Technology Center of La Rioja (CTCR), 26580 Arnedo, La Rioja, Spain
[3] Center for Biomedical Research of La Rioja (CIBIR), 26006 Logroño, La Rioja, Spain

Abstract. Finite Element Method (FEM) has been widely used combined with Computerized Axial Tomography (CAT) and a Magnetic Resonance Imaging (MRI) for the study of biomechanics of human body. Behavioral problems of the human foot and their associated pathologies have been studied with the FEM in recent years. Rotational Equilibrium around of the Subtalar Joint Axis (SJA) model is used to determine supinating and pronating moments and know if the footprint analyzed is neutral or not. This paper proposes a Finite Element analysis for determine the SJA and know the footprint for a 30-year-old male with 180 cm tall and 90 kg in weight. The Finite Element model was validated with the 3DSSPP software, and the footwear was obtained when the patient's foot was stepping on a semi-rigid surface. The contact forces obtained for the toe and heel of the patient were compared, and they showed that the results did not vary significantly. Subsequently, a load of 50 N was applied between the metatarsals and the cuneiforms of the middle and ring fingers, and the foot rotates in an anticlockwise and clockwise direction respectively. In conclusion, it can be said that the type of footprint of the patient studied is neutral, and therefore the patient will not suffer pronation and supination problems.

Keywords: Finite element method (FEM) · Subtalar joint axis (SJA) · Human biomechanics

1 Introduction

In recent years, the Finite Element Method (FEM) has been widely used for the study of biomechanics applied to different parts of the human body. Numerous researchers have modeled the human foot with this numerical method, and have considered each of its parts (mainly bone, cartilage and ligaments). Computerized Axial Tomography (CAT) and a Magnetic Resonance Imaging (MRI) are generally the techniques used to generate the geometry of these FE models after a segmentation process. Behavioral problems

C. Rizzi et al. (Eds.): ADM 2021, LNME, pp. 650–657, 2022.
https://doi.org/10.1007/978-3-030-91234-5_65

of the human foot and their associated pathologies have been studied with the FEM in recent years. The concept of Rotational Equilibrium around of the Subtalar Joint Axis (SJA) was described in 2001 [1]. In this work, the central axis of his proposed theoretical model is the spatial location of the SJA and its projection on the plantar surface of the foot, which allows dividing this region into two well differentiated areas. One internal to the shaft, where the ground reaction forces (generate supinating moments). An external one where these reaction forces generate pronating moments. For a healthy foot (without any important pathology, especially of the bone type or hallux valgus), the magnitude of the pronating moments must be equal to the magnitude of the supinating moments (See Fig. 1).

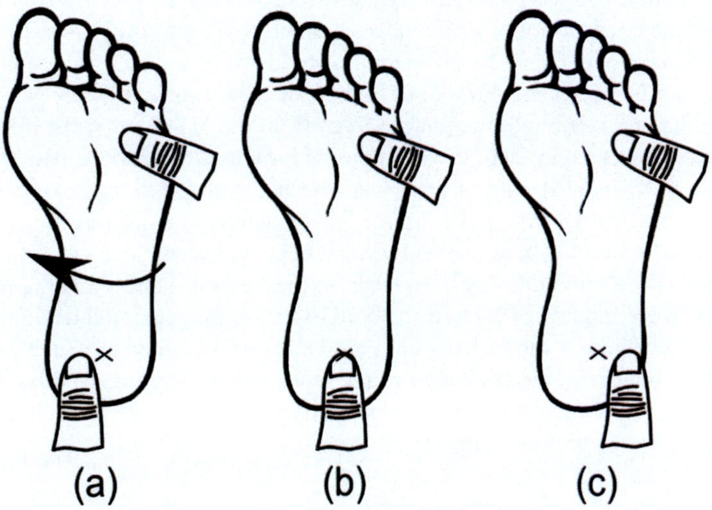

Fig. 1. (a) Subtalar joint supination motion. (b) Plotting the points of No rotation on the plantar foot represents the plantar location of the subtalar joint axis. (c) Subtalar joint pronation motion.

To date, no references have been found in which models of feet based on the FEM have been used to determine the rotational balance around the SJA. This paper proposes a FE analysis for determine the SJA and know the footprint for a 30-year-old male with 180 cm tall and 90 kg in weight. Firstly, a human foot FE model that included all hard and soft tissues, and considered the foot, ankle, and calf was developed. This model was generated from a medical CAT scan obtained from a patient's foot through a segmentation process. On a semi-rigid flat surface, the plantar contact pressures on the toe and on the heel of the foot were determined, as well as the reactions on both areas. Also, the 3DSSPP software [2] was used to determine these reactions on the toe and heel of the foot, and thus validate the proposed FE model. The results showed that the reactions obtained in the FE model on both toe and heel areas hardly differed from those reactions using the 3DSSPP software. Once the FE model was validated, the rotational balance around the (SJA) was determined. The results of the study showed that the type of footprint analyzed is neutral, and that the patient will not suffer from pronation and supination problems. The remainder of paper is structured as follow: in Sect. 2 the segmentation process is briefly introduced, in Sect. 3, a detailed explanation about the FE model is described. Section 4 describes and discuss the results obtained.

2 Segmentation Process and FE Models Proposed

The image segmentation was applied to the acquired Digital Imaging and Communication On Medicine (DICOM) dataset [3] from the CAT of a foot, ankle and calf of a selected patient using 3DSlicer software v4.10.1 [4]. In order to separate the cortical and the trabecular bone, the Hounsfield units (HU) threshold was applied. The HU is a quantitative scale used in CAT studies to describe the different levels of radiodensity in human tissues [5]. A threshold with a range of values from 700 Hu to 3070 Hu was applied in this case for the cortical bone, while values from 150 HU to 700 HU was considered for the trabecular bone. Once the different thresholds were applied, individual surfaces (each for the cortical and cancellous bone of the foot, ankle, and calf) were generated and exported in Standard Triangle Language (STL) from the 3D Slicer software. Once the geometry was generated in STL, it was imported into the simulation software based on the FEM, Mentat-Marc [6], said surfaces were meshed with 2D triangular elements with an edge size between 0.4 mm and 1.2 mm. This geometry was formed by the different surfaces that delimited the hard tissues (cortical bone and cancellous bone) for the foot, ankle and calf of the patient. Using these surfaces meshes with shell elements as the outer contour, the 3d FE models of each of the hard tissues were generated using tetrahedral linear elements. Figure 2 shows a detail of the FE model obtained in which the cortical and cancellous bone of some phalanges of the foot studied can be seen. Table 1 summarizes the Young and Poisson modulus, element types, number of elements considered and the area of the ligaments (mm^2).

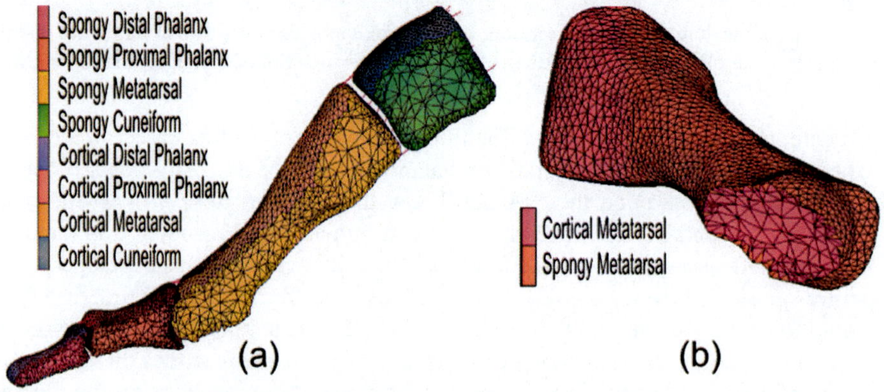

Fig. 2. Hard tissues considered: (a) Mesh density for the cortical bone and cancellous bone in the phalanx of the thumb. (b) Metatarsal mesh corresponding to the thumb.

Table 1. Mechanical properties, element types, number of elements and the area of the ligaments considered in the FE model.

Tissue type	Young modulus (MPa)	Poisson modulus	Element types	Number of Elements	Surface (mm^2)	References
Cortical bone	17000	0.25	Tetraedral	128412	–	[7]
Trabecular bone	150	0.3	Tetraedral	444324	–	[7]
Cartilage joints	1	0.4	Tetraedral	100857	–	[8]
Ligaments	260	0.4	Bar	151	18.4	[9]
Plantar fascia	350	0.4	Bar	5	290.7	[10]
Soft tissues	1.15	0.49	Tetraedral	548146	–	[11]

2.1 Modeling the Cartilaginous Joint

In this work, the modeling of cartilage was studied in 2 different ways: (1) considering the cartilaginous joint as 2 different bodies (as it really is) assuming that at the end of each cortical bone there is a cartilaginous body [12]. Each of these cartilaginous joints are in contact with each other, thus, for example the distal cartilage is in contact with the proximal cartilage (See Fig. 3a). (2) considering the cartilaginous joint as a single cartilage block and a single body, in such a way that mechanical contact is eliminated, notably reducing the computational cost of the FE model [13] (See Fig. 3b). Both proposals for the FE models were completed and simulated successfully considering in both cases a load of 90 kg and a contact mechanical contact on a semi-rigid flat surface. The plantar contact pressures on the toe and on the heel of the foot were determined, as well as the reactions on both areas. From both simulations it was observed that the computational cost of proposal (1) was much higher than the computational cost of proposal (2). This was due to the fact that proposal (1) had many more pairs of mechanical contacts than those of proposal (2), which generated a greater difficulty in the convergence of model (1) with respect to (2). It was for this reason that the FE model proposal (2) was selected to perform the SJA analysis.

2.2 Modeling of Ligaments

Usually, ligaments are modeled with the FE model in 3 different ways [13]: (1) in a 1D, considering truss elements, (2) in 2D, considering shell elements and (3) in 3D, considering space truss elements). In this case and in order to facilitate the convergence of the proposed FE model, all ligaments were molded with 1D truss elements. Figure 4 shows a detail of how the different ligaments were assembled in the proposed FE model. In this case, since the plantar fascia is one of the ligaments with the greatest capacity

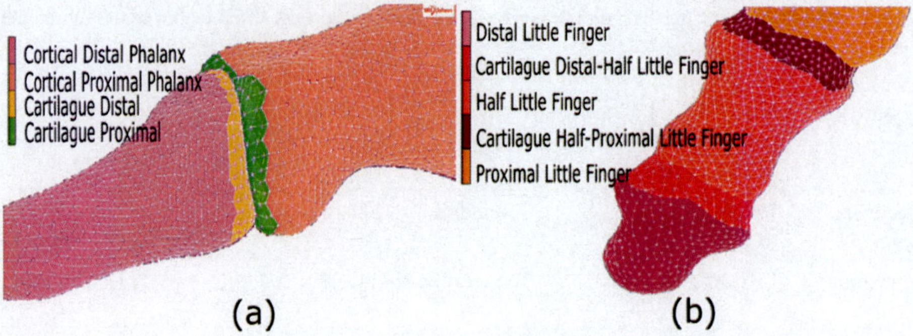

Fig. 3. Considerations proposed for the cartilage in the FE model: (a) Cartilaginous joint as 2 different contact bodies. (b) Cartilaginous joint as a single cartilage block.

and size in the human body, it was modeled with five truss elements, one element per phalanx (Fig. 4a). Also, the transverse metatarsal ligaments were modeled with truss elements, four elements per phalanx (Fig. 4b). The purpose of having truss elements in parallel in the proposed FE model was none other than to eliminate the concentrated stresses caused by a single element in the area of union with the cortical bone. One way to significantly reduce this localized stress is to arrange several equivalent truss elements in parallel, so that the stiffness of all these truss elements will be the same as the stiffness that a single element would have.

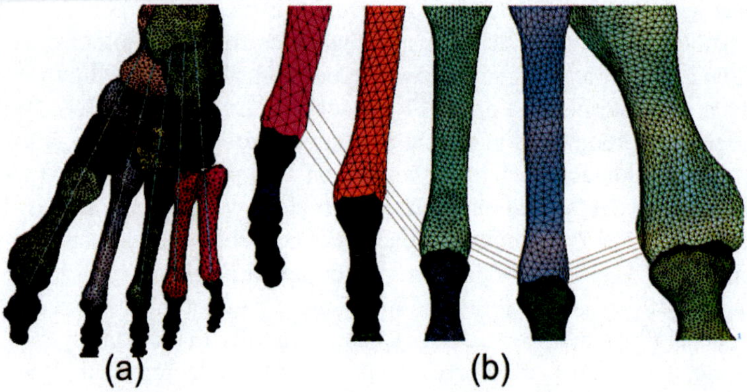

Fig. 4. Ligaments modeled in the proposed FE model: (a) Plantar fascia. (b) Transverse metatarsal ligaments.

2.3 Soft Tissue Modeling

Once the hard tissues were generated, the elements that formed the soft tissue were generated from the 2D shell elements that made up the outer skin of cortical bone and the inner skin of the foot, ankle and calf. In this case, the FEM modeling of these tissues was using tetrahedral elements with linear formulation. Figure 5 shows a detail of the process followed to generate said tetrahedral elements. In Fig. 5a are shown the 2D shell elements used as support to fill the 3D tetrahedral elements and generate the elements that made up the calf. Figure 5b shows the 2D shell elements used for generate the elements that made up the foot.

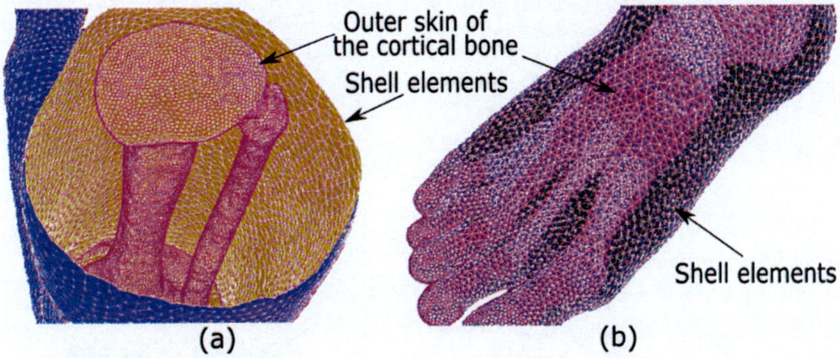

Fig. 5. 2D shell elements considered as support to fill the 3D tetrahedral elements and generate the elements: (a) Calf and (b) Foot.

3 Results

Once the FE models with both proposals were completed, they were simulated considering the characteristics of the individual (male of 180 cm and 90 kg of weight) assuming further that the entire body mass was applied to a single foot when it was contacting on a semi-rigid flat surface. As previously mentioned, proposal (1), despite successfully converging, was discarded because its computational cost was very excessive. The results obtained for the contact forces between the semi-rigid flat surface and the proposal (2) FE model showed values of 172.99 N for the toe and 712.42 N for the heel (See Fig. 6a). In order to validate the FE model, an analysis with the same characteristics of the individual studied was performed in a biomechanical model with the 3DSSPP software [2] (See Fig. 6b). This software provides an approximate simulation that includes posture data, force parameters and male/female anthropometry. The results obtained in this case for the toe and heel were, respectively, 172.2 N and 712.8 N respectively. From the results obtained, it was observed that the values obtained for both models (FE and biomechanical) hardly varied significantly, so it can be deduced that the model would be validated. Subsequently, the SJA was determined on the FE model. In this case, any load, large enough for the FE model undergo visible displacements (for example 50 N) was applied in the area of union of the metatarsals and the cuneiforms for each of the phalanges. The results showed that the foot rotated according to [1] (See Fig. 7). It is known that the axis of the subtalar joint (SJA) begins in the heel and ends in the toes, so that the patient's foot is not pronator or supinator, the axis must be kept in a more or less straight direction and not deviate towards the internal or external part of the foot since otherwise the patient would suffer pronation or supination problems.

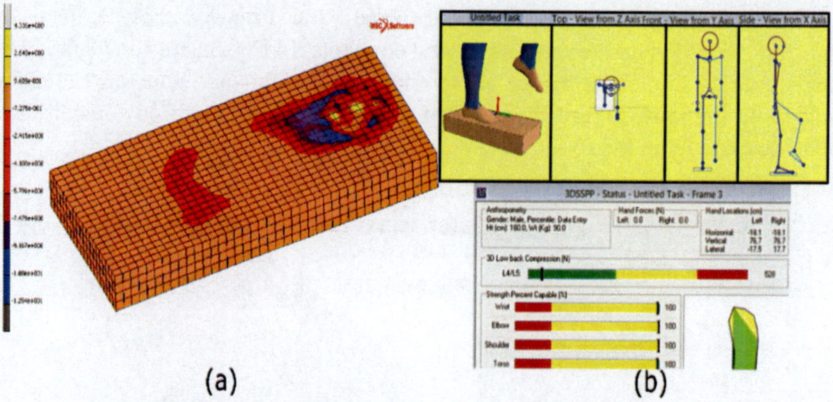

Fig. 6. (a) Contact pressures on the semi-rigid flat surface. (b) Results obtained from the analysis with the 3DSSPP software.

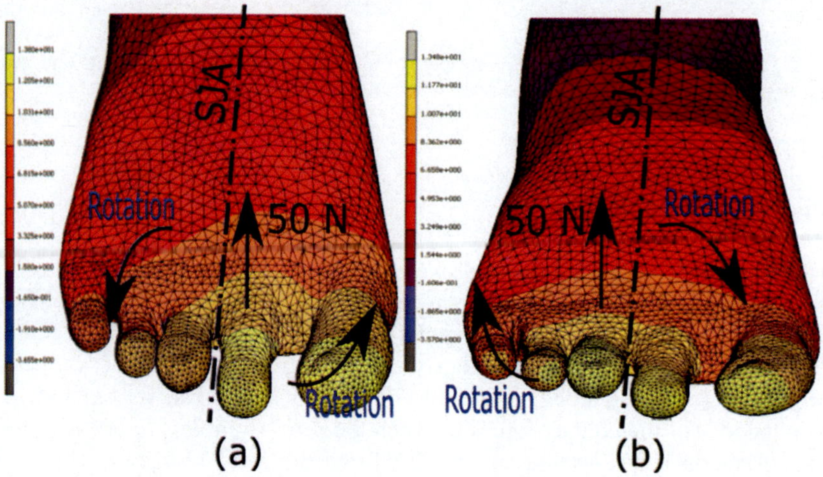

Fig. 7. Applications of forces on the different phalanges (in the area of union of the metatarsals and the cuneiforms) for the determination on the FE model of SJA: (a) On the middle finger. (b) On the ring finger.

From this figure it is observed that after applying a 50 N load on the middle finger, the foot rotates in an anticlockwise direction (Fig. 6a), while when applying that same load on the ring finger, the foot rotates clockwise (Fig. 6b). It can therefore be deduced that the type of footprint of the patient studied is neutral and therefore the patient will not suffer from pronation and supination problems.

4 Conclusions

Finite Element Method (FEM) has been widely used combined with Computerized Axial Tomography (CAT) and a Magnetic Resonance Imaging (MRI) for the study of biomechanics of human body. Numerous researchers have modeled with the FEM the human foot considering the bone, cartilage and ligaments. To date, the FEM has not been used to determine the rotational balance around the SJA. This paper proposes a FEM analysis for determine the SJA and know the footprint for a 30-year-old male with 180 cm tall and 90 kg in weight. The FE model was validated with the 3DSSPP software. Subsequently, a load of 50 N was applied between the metatarsals and the cuneiforms of the middle and ring fingers to determine the rotational balance around the SJA. In this case, foot rotates in an anticlockwise and clockwise direction respectively, so it can be stated that the type of footprint of the patient studied is neutral, and therefore the patient will not suffer pronation and supination problems.

References

1. Kirby, K.A.: Subtalar joint axis location and rotational equilibrium theory of foot function. J. Am. Podiatr. Med. Assoc. **91**(9), 465–487 (2001)
2. 3D Static Strength Prediction Program TM Version 7.0.0 User's Manual (2017)
3. Varma, D.R.: Managing DICOM images: tips and tricks for the radiologist. Indian J. Radiol. Imaging **22**(1), 4 (2012)
4. Fedorov, A., et al.: 3D Slicer as an image computing platform for the quantitative imaging network. Magn. Reson. Imaging **30**(9), 1323–1341 (2012)
5. Hurrell, M.A., Butler, A.P.H., Cook, N.J., Butler, P.H., Ronaldson, J.P., Zainon, R.: Spectral Hounsfield units: a new radiological concept. Eur. Radiol. **22**(5), 1008–1013 (2012)
6. Öchsner, A., Öchsner, M.: The Finite Element Analysis Program MSC Marc/Mentat. Springer, Singapore (2016). https://doi.org/10.1007/978-981-10-0821-4
7. Ozen, M., Sayman, O., Havitcioglu, H.: Modeling and stress analyses of a normal foot-ankle and a prosthetic foot-ankle complex. Acta Bioeng. Biomechan. **15**(3) (2013)
8. Athanasiou, K.A., Liu, G.T., Lavery, L.A., Lanctot, D.R., Schenck, R.C., Jr.: Biomechanical topography of human articular cartilage in the first metatarsophalangeal joint. Clin. Orthop. Relat. Res. **348**, 269–281 (1998)
9. Siegler, S., Block, J., Schneck, C.D.: The mechanical characteristics of the colateral ligaments of the human ankle joint. Foot Ankle **8**(5), 234–242 (1988)
10. Wright, D.G., Rennels, D.C.: A study of the elastic properties of plantar fascia. J. Bone Joint Surg. **46**, 482–492 (1964)
11. Antunes, P.J., Dias, G.R., Coelho, A.T., Rebelo, F., Pereira, T.: Non-linear finite element modelling of anatomically detailed 3D foot model. Report paper, 1 (2008)
12. Hassan, C.R., Qin, Y.X., Komatsu, D.E., Uddin, S.M.: Utilization of finite element analysis for articular cartilage tissue engineering. Materials **12**(20), 3331 (2019)
13. Somovilla Gómez, F., Lostado Lorza, R., Corral Bobadilla, M., Escribano García, R.: Improving the process of adjusting the parameters of finite element models of healthy human intervertebral discs by the multi-response surface method. Materials **10**(10), 1116 (2017)

Requirements Engineering in Complex Systems Design

Domenico Marzullo[1(✉)], Giuseppe Di Gironimo[2], Antonio Lanzotti[2], Rocco Mozzillo[3], and Andrea Tarallo[2]

[1] Department of Engineering and Architecture, University of Trieste, Trieste, Italy
dmarzullo@units.it
[2] Department of Industrial Engineering, University of Naples Federico II, Naples, Italy
[3] School of Engineering, University of Basilicata, Potenza, Italy

Abstract. The realization of nuclear fusion reaction as energy source is under investigation, among the scientific community, through the design and development of tokamak reactors. Among the several experiments worldwide, the ITER project is the major international experiment and it involves several research institutes from several countries. In such a project, a Systems Engineering (SE) approach is requested to organize and manage the design due to its highly integrated design, the safety requirements related to nuclear aspects and the complex procurement scheme. The SE discipline focuses the attention on the requirements which are crucial for every successful project, defining what the stakeholders want from a potential new system, namely what the system must do to satisfy stakeholders need. Correctly stating WHAT is needed for the system, it is possible to obtain its conceptual design (HOW) as much as possible complying the requirements. The incorrect definition of requirements often leads to the failing of a project. Stakeholders' needs are written in Natural Language that is generally ambiguous, imprecise, incomplete and redundant. Their transformation into SMART requirements is crucial to avoid design failure. However, it requires a great expertise, unless a specific procedure is assessed. To this end, this work presents a specific procedure based on "like-mind" processes to make systematic the SMART requirements definition and assessment from stakeholders needs. The procedure is based on a demand/response framework and it is developed to obtain ITER requirements. However, it can be easily extended to every project using its own specifications. A specific case study on ITER Remote Handling is presented in this paper as example of the conceived requirements transformation procedure.

Keywords: Systems engineering · Requirements engineering · SMART requirements · ITER tokamak

C. Rizzi et al. (Eds.): ADM 2021, LNME, pp. 658–667, 2022.
https://doi.org/10.1007/978-3-030-91234-5_66

1 Introduction

The development of innovative products requires a clear framework and tools to systematically deal with the concept design stage. In this design phase, the documentation and the traceability of design information play a main role for the development of a successful project and to avoid iterative re-design cycles, as well as to quickly explore many concepts determining those most likely to succeed. Systems Engineering (SE) discipline is widely recognized as the preferred mechanism for managing engineering activities in highly integrated environments. Both the international standard ISO/IEC 15288 [1] and the main Systems Engineering Handbooks (i.e. INCOSE Systems engineering handbook [2], NASA Systems engineering handbook [3]) define the requirements elicitation activities as drivers for many of the system life cycle processes. Several research have demonstrated that successful projects depend on the correct definition of the needs and requirements of the customers and stakeholders from the beginning of the early design stages. Without establishing detailed requirements, the risk of project failure would be unacceptably high.

Requirement elicitation is an iterative activity and benefits from continuous communication and validation with the customer. No design can be completed before establishment of the System Requirements Documents (SRD) reflecting all relevant design inputs. Stakeholders' needs are written in Natural Language that is generally ambiguous, imprecise, incomplete and redundant. Their transformation into the so-called *SMART* requirements is crucial to avoid design failure. However, it requires a great expertise, unless a specific procedure is assessed. To this end, this paper presents a specific procedure based on "like-mind" processes to make systematic the SMART requirements definition and assessment from stakeholders needs. Such approach identifies the main activities to be followed to obtain SMART requirements even without expertise. The procedure is based on a demand/response framework and developed to obtain ITER requirements. However, it can be easily extended to every project using its own specifications. The advantages are mostly related to simplify the work, save time and make more effective the transformation procedure as well as the final requirements. In the next section a general overview on the requirements engineering will be provided. Then in Sect. 3 the conceived procedure is presented and it is then exemplified through a case study discussed in Sect. 4.

2 Materials and Methods

2.1 Systems Engineering Discipline

SE is a methodical, disciplined approach for the design, realization and management (both operations and retirement) of a system [3]. A frequently used definition of a system is "a set of interrelated components working together toward some common objective", which implies a multiplicity of interacting parts that together perform a specific function [3]. SE is a holistic and integrative discipline, wherein the contributions of several engineering areas and many more disciplines are evaluated and balanced, to produce a coherent whole that represents homogeneously the different disciplines involved [5].

Among the various *Technical Processes* provided by SE, the *Stakeholder Requirements Definition Process* and the *Requirements Analysis Process* represent the crucial technical processes for the efficient development of a project, in particular during the conceptual design stage. Indeed, during the early design stage the requirements can be unclear, ambiguous and imprecise and they can be affected by lack of agreement among various stakeholders [5–7]. These situations are common at the early stage of the design process of complex systems, due to different experts involved in the collaborative design [8]. This work fits within these two technical processes aiming to provide a rigorous procedure for SMART requirements assessing from stakeholders' expectations.

2.2 Requirements Engineering and SMART Requirements

Requirements Engineering is a branch of Systems Engineering. It enables an agreed understanding between stakeholders, allowing the system's implementation and providing a basis of verifying designs and accepting solutions. Thus, requirements engineering is where the informal meets the formal [9].

The requirement should state "what" is needed, not "how". It means that in the earlier stage of *Stakeholders' Requirements Definition Processes* the important issue consists in defining what the stakeholders' needs, leaving out the possible solution which makes satisfy specified needs. Giving a solution in this phase could only constrain the design phase. The real issue of the requirements engineering can be found in the formal specification of the requirements. Stakeholders' expectations (needs) are generically written in natural language and they require to be translated in more detailed and not ambiguous forms. The essential characteristics for requirements can be seen in the word *SMART*. *SMART* is an acronym representing criteria to write good requirements. The letters S and M indicate *Specific* and *Measurable* respectively. Regarding the other three letters, there is not a common criterion to define them. In this work it is followed the interpretation provided by Fusion for Energy[1] in [10] with which the letters A is for *Achievable,* R is for *Relevant* and T is for *Traceable.* Following these five aspects users should formulate an effective requirement without misleading.

A requirement must be ***Specific***, because it must address **only one** aspect/function of the system design or performance being concise and simple. It must be ***Measurable***, because any characteristic required have to be expressed objectively and quantitatively. A requirement must be objectively verifiable (by test, analysis, inspection, or demonstration) to prove compliance. A desired capability that cannot be objectively verified should be written as a design goal, not a requirement. A requirement must be ***Achievable***, because it must be technically achievable at affordable costs within the relevant schedule conditions. A requirement must be ***Relevant***, because it must be needed to solve a problem or achieve an objective. A requirement must be ***Traceable***, because each requirement must have a unique ID and parents-children's relations must be clear and easily identifiable. Thus, requirements without a parent are referred to as orphans and their inclusion in the whole project context shall be assessed. Furthermore, each requirement must be linked to the stakeholders' need.

[1] Fusion for Energy (F4E) is the European Union's Joint Undertaking for ITER and the Development of Fusion Energy.

The SMART requirements must be assessed with two important processes: the requirements validation and the requirements verification.

Requirement's validation confirms the completeness, the compatibility and the correctness of the requirements. Requirement's verification is the process of design qualification and product's acceptance. Generally, the *verification plan* should be preliminary defined by the project team which should indicate the requirement verification method (test, demonstration, analysis or/and inspection).

3 Procedure for Writing SMART Requirements

The main flowchart of the conceived procedure for writing SMART requirements is shown in Fig. 1. The aim of the work has been the highlighting of several steps taken during transformation processes in view of the extraction of a Unified Modeling Language (UML) Activity Diagram, which is described in the current section.

The UML is a general – purpose visual modeling language that is used to specify, visualize, construct, and document the artifacts of a system. It captures decisions and understanding about a system that must be constructed [11].

According to the presented flowchart it is possible to make systematic the SMART requirements writing process, which is usually committed to system engineers with great expertise. The aim of this work is mostly related to simplify the work, saving time and making more effective the procedure as well as the final requirements. The benchmark of this work is the F4E released Requirements Writing Guidelines [10].

The procedure focuses the main actions to follow depicted through the main flow chart in Fig. 1. Some of such actions, defined "*invocation action*", have several sublevels needed to explain the source action. The invocation action can be recognized from the rake present on bottom right of the box action. To any of the actions presented in the main flowchart, a sub-flowchart is associated reporting the passages to perform such specific action. For sake of brevity, in this work only the main flowchart and the flowchart for one of the invocation actions are reported and discussed.

The first action is the "*Discussion with the stakeholders*". In this stage the stakeholders describe needs, wants, desires, expectations concerning the stakeholders' project. Generally, the output of such level are the requirements guidelines and the set of needs. The action "*Define a glossary*" suggests to list and to define the terms concerning the project. Creating the glossary means adding semantic definitions and synonyms of added terms; assuring that all those persons involved in the project understand a word in the same way, avoiding misunderstandings and fuzziness of communication. The next tasks are "*Select a stakeholders' need to be transformed*" and "*Record the stakeholders' need in a requirements depository*". The user shall select a first need to be analyzed, and then he shall record both the selected need and the need's source in a work file (e.g. file.xlsx). It is extremely important that every requirement is always traceable as well as linked to the corresponding need.

The next action "*Understand the stakeholders 'need*" is an invocation action and it maps to a specific flowchart. It is a brain work: the user shall reach a clear understanding of the selected need, identifying what the system should perform according to the stakeholders 'need, identifying what the system should perform according to the stakeholders 'need, performing a syntactic analysis, resolving all unknown terms (e.g. technical terms). It is important to check the list of words because if there are some ambiguous terms, stakeholders' clarifications must be integrated. Thus, the selected requirement must be marked as "to be updated", the analysis must be stopped and the next requirement can be considered. Concluding this task, from the main flow chart, it can be seen that if the need does not require stakeholders' clarifications, it is possible to move to the next action: "*Passage from the stakeholders' need to n SMART Requirement candidates*". This is an invocation action as well, characterized by three sub-levels as depicted in Fig. 2. Each stakeholders' need can contain several information. However, each SMART requirement shall hold a single information or feature that the final system shall have or perform. As a consequence, the user shall identify several *atomic requirements* to consider one at time. The user shall identify if the need expressed by atomic requirement is indeed a requirement rather than a goal, or a possibility, or an intention as well. The goal means an aim or objective to achieve; it is less strong than a requirement, but it is good that the system will achieve the fixed goal. If the atomic requirement is defined as requirement, the user shall transform it into a SMART requirement candidate following the invocation action "*Transform the stakeholders' needs into the SMART requirement candidate*". That candidate will be indeed a SMART requirement after several checks. Now, from the main flow chart, the user shall follow the action "*Record the SMART requirement candidate/possibility/goal/intention into requirements' depository*" and continue to formulate the next stakeholders' need. Some needs will require some stakeholders' clarification, and just after that the user will be able to continue the analysis. When all stakeholders' needs become SMART requirement candidate, the user can follow the next two actions of main chart "*Select a SMART requirement candidate*" and "*Check SMART requirement candidate*". A procedure for checking the whole set of requirements is also provided and it will be discussed in separate works by authors.

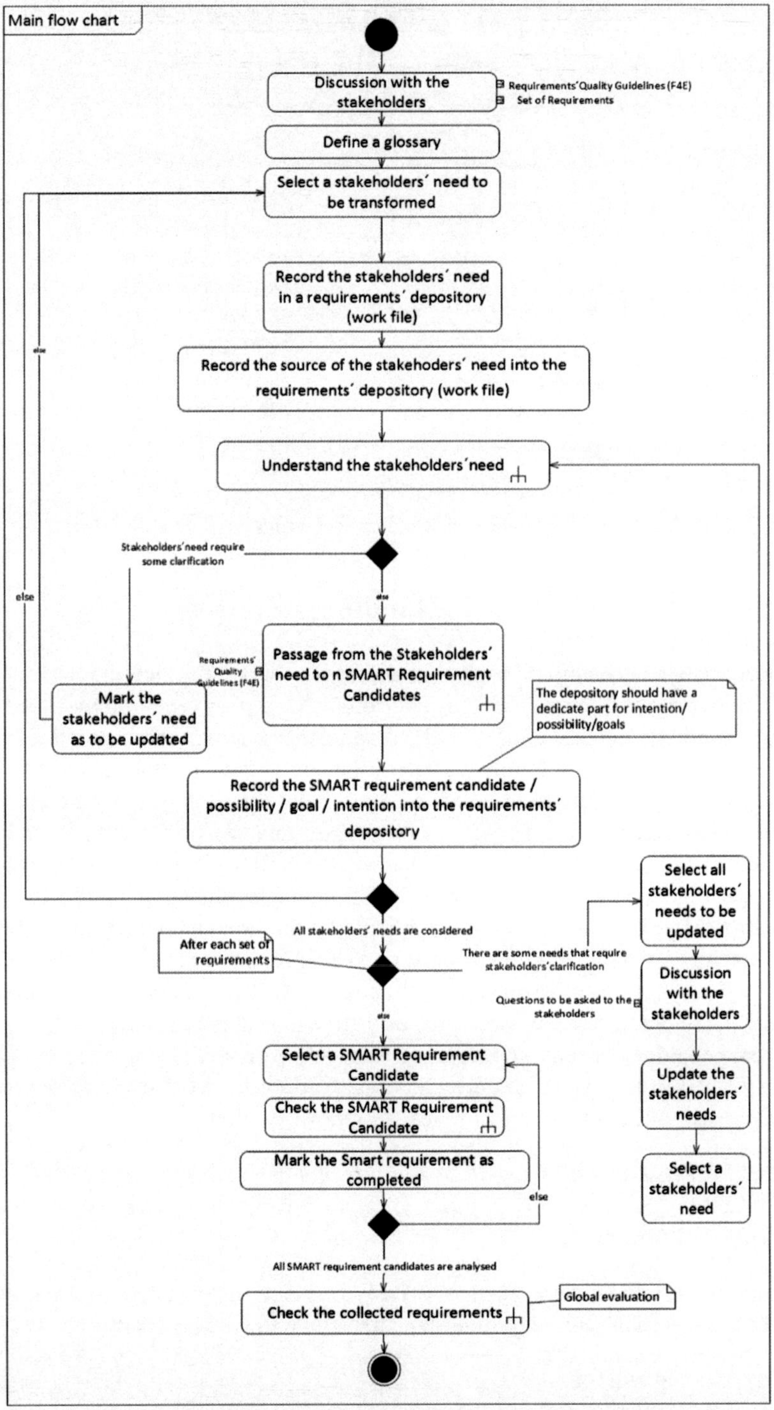

Fig. 1. Main flowchart for SMART requirements definition

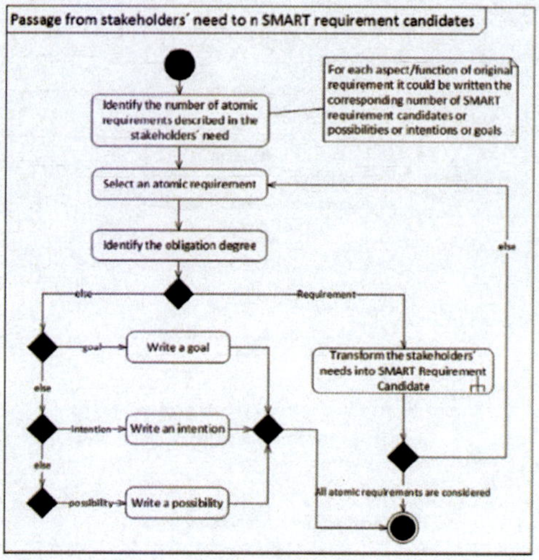

Fig. 2. Passage from the stakeholders' need to SMART requirements

4 Case Study: ITER Remote Handling Connector

The procedure has been applied to the requirements concerning the remote handling of a ITER component: the connector [12]. The connector is placed behind the divertor [13] cassettes, ensuring electrical energy to the diagnostic system mounted on the cassettes. The divertor area of ITER vacuum vessel consists of 54 modular cassettes, which must be replaced by means of Remote Handling equipment three times during the estimated 20 years of operation of the ITER. A clear representation of the connector location is depicted in Fig. 3. The connector must be uncoupled whenever the instrumented cassettes are removed from the ITER vessel. This would be done using a remote handling manipulator (MAM) and the tooling required for the connector operation should be minimal to simplify the deployment logistics. RH operations must be developed taking account the available space. To make easy the coupled (RH operation) between connector and cassette, the connector has three pins of different length and shape.

Ten stakeholders' needs, concerning the RH connector, have been analysed and.transformed into SMART requirements, according to the procedure described in Sect. 3. For sake of brevity, two examples are reported below.

> STK.01 "*Electrical bonding shall be reliable and corrosion-free over the lifetime ofthe ITER installation, with no paint or any other insulating surface coating on bonded metal surfaces.*"

Reading this stakeholders' need, it is easy to notice that it refers to three different aspects of the electrical bonding: reliable, corrosion and surface coating.

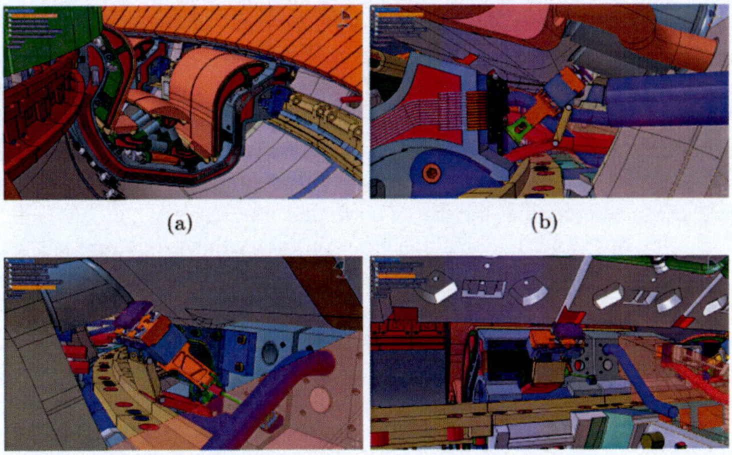

Fig. 3. (a) Divertor Cassette, (b) - (c) - (d) Connector behind the divertor

Considering that every requirement has to be specific, namely it must address only to one aspect/function of the system design/performance, for each aspect must be written a requirement. After that, it is important to make these three aspects measurable. So, the reliability has been expressed using failure for time and the corrosion free using the corrosion velocity, obtaining the three following **SMART requirements**:

REQ1. The electrical bonding must have a failure rate not greater than λ^2.

REQ2. The electrical bonding must have the corrosion velocity less than Y over the lifetime of the ITER installation (excluding disposal stage).

REQ3. The electrical bonding must have metal surface without paint or any other insulating surface.

> STK.05 "*Pickling is rarely specified for vacuum components, normally only for those to be used in rough vacuum, since the process attacks the metal surface and the oxide layer, tending to leave residues which are difficult to remove.*"

SMART Requirement

REQ4. If the pressure is less than 3×10^3 Pa the vacuum components must not have pickling as surface treatment.

REQ5. The vacuum components used in rough vacuum (1×10^5 to 3×10^3) may have the pickling as surface treatment. (POSSIBILITY)

From several stakeholders' need it is possible to obtain the rationales and comments for SMART requirements, which are collected with the requirement itself.

5 Conclusions

In this work, a specific procedure for SMART requirements definition based on several "like-mind" processes has been presented. The aim is to make systematic the process of

SMART requirements assessment and definition, saving time and reducing the required knowledge. It indeed can support users with no great expertise in systems engineering and requirement engineering. In this context, a specific tool has been designed in all parts looking at the many human processes made during requirement assessment. This procedure has been developed within ITER project and thus allows to obtain the requirements as required by ITER. However, changing the templates used and few particular requests (e.g. cross referencing, the use of particular modal verbs), it can be extended to every project. One of the main advantages introduced by this procedure is related to the unifying of the language used in the requirements' document, giving to every person that works at the same project the same understanding degree. On the other side, it should be noted that the process of requirements generation can be affected by variabilities and uncertainties depending on the type of project, so the proposed procedure should be verified in several different contest to be improved in its flexibility and robustness.

The implementation of a software supporting the user will allow to rapidly make SMART the stakeholders' needs. The software can be simply implemented following the flow chart and it is based on a demand/response framework.

Acknowledgments its Flexibility and Robustness. Authors would like to express their sincere appreciation to E. Manda and the team of VTT Technical research Centre of Finland – Tampere for their contribution to this work.

References

1. Arnold, S.: ISO 15288 systems engineering—system life cycle processes. International Standards Organisation (2002)
2. Haskins, C., Forsberg, K., Krueger, M., Walden, D., Hamelin, D.: Systems Engineering Handbook: INCOSE (2006)
3. Kapurch, S.J. (ed.): NASA systems engineering handbook. Diane Publishing (2010)
4. Kossiakoff, A., Sweet, W.N., Seymour, S.: Systems Engineering Principles and Practice. Wiley, New York (2011)
5. Marzullo, D., Di Gironimo, G., Dongiovanni, D.N., Lanzotti, A., Mozzillo, R., Tarallo, A.: Iterative and participative axiomatic design process to improve conceptual design of large-scale engineering systems. In: Rizzi, C., Andrisano, A.O., Leali, F., Gherardini, F., Pini, F., Vergnano, A. (eds.) ADM 2019. LNME, pp. 492–505. Springer, Cham (2020). https://doi.org/10.1007/978-3-030-31154-4_42
6. Di Gironimo, G., Lanzotti, A., Marzullo, D., Esposito, G., Carfora, D., Siuko, M.: Iterative and participative axiomatic design process in complex mechanical assemblies: case study on fusion engineering. Int. J. Interact. Des. Manuf. (IJIDeM) **9**(4), 325–338 (2015). https://doi.org/10.1007/s12008-015-0270-7
7. Legardeur, J., Boujut, J.F., Tiger, H.: Lessons learned from an empirical study of the early design phases of an unfulfilled innovation. Res. Eng. Design **21**(4), 249–262 (2010)
8. Marzullo, D., Bachmann, C., Coccorese, D., Di Gironimo, G., Mazzone, G., You, J.H.: Systems engineering approach for pre-conceptual design of DEMO divertor cassette. Fusion Eng. Des. **124**, 649–654 (2017)
9. Zowghi, D., Jin, Z. (eds.): Requirements Engineering. CCIS, vol. 432. Springer, Heidelberg (2014). https://doi.org/10.1007/978-3-662-43610-3
10. Serra, G.: Requirements Writing Guidelines (Private Communication) (2012)

11. Rumbaugh, J., Jacobson, I., Booch, G.: The Unified Modeling Language Reference Manual, 2nd edn. Addison-Wesley, Boston (2004)
12. Lyytinen, J., et al.: Development of the remote handling connector for ITER divertor diagnostic system. Fusion Eng. Des. **165**, 112243 (2021)
13. Marzullo, D., et al.: Progress in the pre-conceptual CAD engineering of European DEMO divertor cassette. Fusion Eng. Des. **146**, 942–945 (2019)

Virtual Prototyping and FEM Analysis of the Crystal Eye Detector

Fabrizio Renno[1], Stefano Papa[1(✉)], Domenico Marzullo[2], Felicia Carla Tiziana Barbato[3], and Giancarlo Barbarino[4]

[1] Università Degli Studi di Napoli Federico II, Napoli, Italy
stefpapa@unina.it
[2] Università Degli Studi di Trieste, Trieste, Italy
[3] Gran Sasso Science Institute (GSSI), L'Aquila, Italy
[4] Istituto Nazionale di Fisica Nucleare, Sezione di Napoli, Napoli, Italy

Abstract. The main aim of this article is to describe the design of a new sensor to study the electromagnetic field portions of gravitational waves. On August 17, 2017, the observation of the gravitational wave event started the era of multi-messenger astronomy. Therefore, new tools and optimal synchronization of the available telescopes are needed. The sensor that is designed is a cross-cutting technology, it is named Crystal Eye: a wide field of view in the energy field from 10 keV to 10 meV with a structure made of pixels. As the detector will be involved in the mission in 2023, the virtual prototype phase needed for optimization and production of the payload has been completed. Particular attention was paid to the results of the FEM analysis carried out to examine and predict the thermal and vibration behavior of the conceived mock-up during the launch phase and under strong temperature variations in the space environment.

Keywords: Virtual prototyping · Vibration and thermal analysis · Detector · FEM

1 Introduction

An established thesis in astrophysics is that impacts or mergers of two neutron stars produce considerably relativistic and collimated jets that supply energy to γ-ray bursts of short endurance [1–3]. This was only a theoretic concept until the X-ray part interested by the GW170817 gravitational-wave event was detected [4]. The so far considerable belief that gravitational-wave events from analogous fusions should be linked to γ-ray bursts, and that their prime part should be watched off axis, was so confirmed. Together, scientists were able to observe X-ray and, thereafter, radio frequencies which result to be coherent with a quick γ-ray blast viewed off-axis. The find of X-ray emissions that coincide with the transitory of a kilonova gives the observation link that was missing between the short γ-ray blasts and the gravitational waves from the neutron-stars fusions, and provide independent corroboration of the collimated nature of the γ-ray blast emission [5]. So, all the best and most versatile devices are needed to detect something even when and

C. Rizzi et al. (Eds.): ADM 2021, LNME, pp. 668–674, 2022.
https://doi.org/10.1007/978-3-030-91234-5_67

where no information is expected. The main currently active experiments in the space environment that can detect the γ-ray and X-ray range are Chandra and XMM Newton', unfortunately characterized by a small field of view. Although this feature allows to point a source and to obtain a good imagine of that, it reduces the observation of the Universe to a very small zone. The Fermi γ-ray blast monitor instrument (Fermi-GBM) is not very accurate because it is designed to be a wide sight experiment [6]. So, an enhanced and more efficient detector is needed.

Today, many tools (software and hardware) can be used to design, improve, and optimize products. So, for instance, CAD, CAE, CFD methodologies are "consolidated" instruments for the realization of Digital Mock-Ups (DMU) and for testing purposes. Furthermore, the use of the FEM methods can be very useful to analyze and foresee the thermal and vibration behavior of the mock-up during the launch or in the adverse conditions of the space environment due to the high temperature variations [7, 8]. In addition, the modern production technologies for the realization of the prototypes, such as Additive Manufacturing techniques, represent a fundamental tool to get enhanced physical prototypes. They are supported by design and optimization methods aimed at selecting adequate materials and shapes to improve performances and to reduce the weight of the final model, and so the costs. Traditionally, by means of conventional methods these kinds of products are made by many elements which need to be machined and jointed (by welding or screws). On the contrary, 3D printing, particularly with metals, is increasingly being used in the manufacture of components in aerospace industry (i.e. rockets parts, propulsion modules or engines elements) because it enables these components to be manufactured as a single piece in one-shot process [7]. Furthermore, the application of these new methodologies to the field of aerospace experimentation allows to test the weight reduction, the material efficiency and the behavior of the artifacts in this way produced, in unconventional operating conditions, providing further study data. The analysis of the responses of the prototypes can also contribute to the standardization of an innovative process for the design and development of products to support the experimental studies target of the Project.

2 The New Detector

2.1 Crystal Eye

As mentioned before, there are several currently active experiments in the space environment that can detect the gamma and X-rays, as for instance XMM-Newton' and Chandra [6]. They are a consequence of an old way of designing sky observatories and they have the main drawback related to the limited field of view and so, even if these detectors allow to get good results, they decrease the observation of the Universe to a very small zone. Therefore, the Crystal Eye proposal starts from a new approach for the Universe observation, and it will be the advanced version of the Fermi-GBM detector, designed for a mission on the International Space Station (ISS). It will be a new and crossover technology among all sky telescopes and monitors: an observatory with a really wide field of view (local observation equal to 2π), analyzing the range of energy from 10 keV to 10 meV with a structure full of pixels. Main target is to improve the field of view of the GRBs allowing a better localization due to the monitor with a low-cost mission spatial resolution. Furthermore, the compact dimensions (that allow an astronaut to move it easily) and the wide field of view with small pixels are the prerequisites chosen to make this device competitive with the other systems. Therefore, the main objectives of the program are to raise awareness of events with X-rays, low-power γ-rays and identify long-term variability X-ray founts, promote multi-wave length visualization of various entities, and search for scattering cosmic X-rays. The detector conceived is a hemispherical multi-level structure made by a double layer of LYSO pixels covered by a veto dome. The final model has been realized starting from a hemisphere with a 140 mm radius. Furthermore, it has been built with a full parametric approach to modify each feature of the virtual prototype in real time. In addition, it has been designed to be 3D printed. It will be characterized by 112 pixels on two scintillators layer designed to observe the down-going X and γ-rays and to reject the up-going ones (Fig. 1).

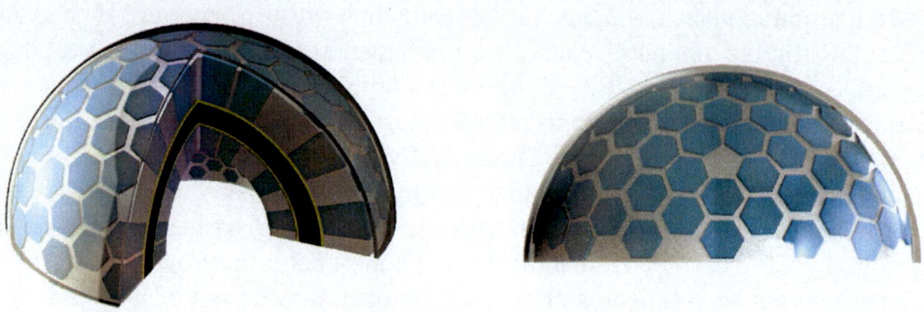

Fig. 1. CAD model of the Crystal Eye detector.

The smaller size and the larger number of contiguous pixels will improve the spatial resolution and, therefore, the localization of the source. Figure 2 shows the position of the pixels (characterized by hexagonal top and bottom surfaces). Each one consists of a hexagonal pyramid of LYSO read by adequate Silicon Photomultiplier (SiPM) matrices. A regular polygonal mesh with 112 hexagons and 4 pentagons defines their location assuring the symmetry with respect to the zenith. The external veto dome will be made by a scintillator material with a dig where an optical fiber will guide the collected light to the SiPMs. The wiring will be placed between the pixel layers and an adequate insulation.

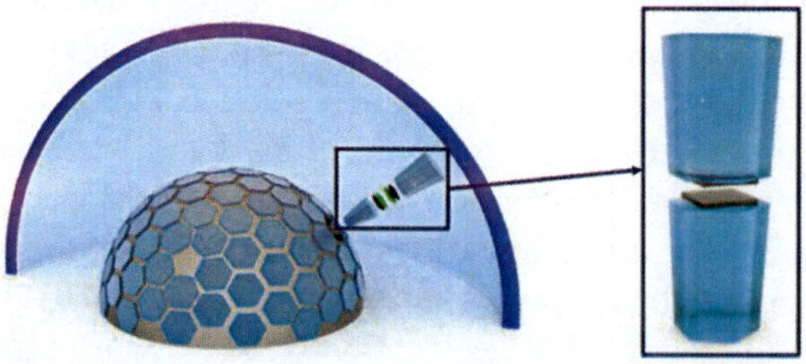

Fig. 2. Exploded view of the detection module of the Crystal Eye prototype.

2.2 Crystal Eye Pathfinder

A compact prototype of the conceived Crystal Eye model, based on only four pixels, will be launched on September 2023 aboard of the Space Rider, an uncrewed orbiting laboratory of European Space Agency (ESA) designed to give reusable access to the space. The Crystal Eye pathfinder was designed as a lightweight version of the complete detector that can provide basic results that act as a potential physical assessment. The system is characterized by an electronic board and four pixels (Fig. 3). Each one consists of a hexagonal pyramid of LYSO read by adequate Silicon Photomultiplier (SiPM) matrices as previously defined for the full detector. The two months Space Rider mission will follow the LEO orbit (as the International Space Station) and a Solid-State Drive will gather the data (also sent to the ground base).

Fig. 3. Crystal Eye pathfinder: lightest version of the full detector.

3 FEM Analysis

First aspect analyzed is related to the high shocks, stresses, and strains that the pathfinder will have to support especially during the launch and landing phases. Several tests have been carried out as the system shall withstand the mechanical loads which consists of static accelerations up to 6g. So, the boundary conditions and the constraints have been defined to analyze the behavior of the system in adverse conditions. Specifically, the screwed connection on the base of the pathfinder have been simulated by means of "body-to-ground" joints. Static analysis has been conducted out assuming 6g accelerations in all directions evaluating the total displacements and the Von-Mises stress distribution (Fig. 4) and no specific concerns have been identified at this stage (maximum equivalent stress < 80MPa).

Fig. 4. Equivalent stress and deformation analysis of the Crystal Eye pathfinder.

Afterwards, modal analysis was carried out to investigate natural frequencies of the system and to support the random vibration and the response spectrum analysis.

Specifically, three types of environmental loads have been applied at the interfaces: random vibroacoustic, sine and shock payload spectrums.

No critical loads close to the natural frequencies of the system have been identified. The natural frequencies of the system and the related participation factors are reported in Fig. 5: the system results quite rigid, presenting the first natural frequency at 225 Hz, taking the 57% of the total mass in the X direction. The modal analysis has been limited to the first 20 modes since higher frequencies are not considered interesting with respect to the applied loads (Fig. 5).

KINETIC ENERGIES, MODAL MASSES, TRANSLATIONAL EFFECTIVE MASSES SUMMARY

				EFFECTIVE MASS					
MODE	FREQUENCY	MODAL MASS	KENE	X-DIR	RATIO%	Y-DIR	RATIO%	Z-DIR	RATIO%
1	225.0	0.5244E-03	523.9	0.8701E-03	57.54	0.1839E-14	0.00	0.2353E-04	1.56
2	394.9	0.9731E-05	29.96	0.2132E-05	0.14	0.2622E-11	0.00	0.4217E-04	2.79
3	490.5	0.8166E-05	38.77	0.2719E-05	0.18	0.3611E-12	0.00	0.9278E-05	0.61
4	526.5	0.2182E-04	119.4	0.3384E-06	0.02	0.1034E-10	0.00	0.6714E-04	4.44
5	665.0	0.1109E-03	968.2	0.9470E-11	0.00	0.9202E-04	6.08	0.2564E-09	0.00
6	668.2	0.5322E-05	46.90	0.3151E-07	0.00	0.4836E-07	0.00	0.5450E-06	0.04
7	699.3	0.5225E-03	5044.	0.4430E-13	0.00	0.8754E-03	57.89	0.3783E-11	0.00
8	802.4	0.1059E-04	134.6	0.6834E-06	0.05	0.2116E-11	0.00	0.1365E-06	0.01
9	938.7	0.1510E-04	262.7	0.3381E-05	0.22	0.8847E-09	0.00	0.3412E-04	2.26
10	947.0	0.1302E-04	230.5	0.2294E-09	0.00	0.1923E-04	1.27	0.1805E-08	0.00
11	1123.	0.3342E-03	8318.	0.2594E-04	1.72	0.3746E-11	0.00	0.1150E-02	76.07
12	1184.	0.1561E-04	431.8	0.4628E-11	0.00	0.3020E-05	0.20	0.2537E-09	0.00
13	1307.	0.2205E-04	743.5	0.9113E-05	0.60	0.4568E-08	0.00	0.5450E-04	3.60
14	1315.	0.2401E-04	819.2	0.5246E-08	0.00	0.7271E-05	0.48	0.2921E-07	0.00
15	1371.	0.8008E-05	297.0	0.3566E-05	0.24	0.2348E-11	0.00	0.2996E-04	1.98
16	1449.	0.9405E-05	389.8	0.1758E-09	0.00	0.5787E-06	0.04	0.2527E-08	0.00
17	1459.	0.2491E-04	1047.	0.3033E-05	0.20	0.2538E-09	0.00	0.1297E-04	0.86
18	1493.	0.8510E-05	374.7	0.4306E-05	0.28	0.1218E-11	0.00	0.4511E-06	0.03
19	1578.	0.4938E-05	242.7	0.1962E-09	0.00	0.4637E-07	0.00	0.1853E-09	0.00
20	1659.	0.1226E-04	665.8	0.3679E-09	0.00	0.1496E-06	0.01	0.4990E-10	0.00
sum				**0.9254E-03**	**61.19**	**0.9978E-03**	**65.98**	**0.1425E-02**	**94.24**

Fig. 5. Results of the modal analysis of the Crystal Eye pathfinder.

Through the floor response spectrum analyses, problems related to critical hotspot at screwed connections have been identified. The physical issues have been solved revising the screwed connections layout, having more distributed reaction forces close to the center of gravity, while the artificial numerical hotspot have been checked locally decreasing the mesh dimension. The worst load condition has been identified in the shock environment payload.

The mechanical behavior of the pathfinder has been improved by means of some modification of the structure. So, the cut-outs of the top face of the base have been removed.

Afterwards, as the pathfinder will be exposed to high temperature variations for a two-months period, the thermal behavior of the system has been tested and evaluated. The optimal pathfinder thermal environment for operation and measurement shall vary between 15 °C and 35 °C. Considering the temperature limits of the electronic card (−20 °C and + 85 °C) and of the SiPM (−20 and + 35 °C) several analyses have been carried out contemplating unconventional operating conditions due to the Solar irradiance and the orbit that the Space Rider will describe. Thermal conduction with the interfacing systems has been deemed as well as radiation to the external environment. All the external surfaces of the pathfinder have been covered with reflective material to reduce the impact of the direct sunrays and so to decrease the thermal effects on the several electronic components in the system (Fig. 6).

Fig. 6. Results of the thermal analysis of the Crystal Eye pathfinder.

4 Conclusions and Future Works

The conceptual design and the virtual prototyping of the new Crystal Eye are described. Thanks to its architecture, the detector Crystal Eye will explore the gravitational waves events in terms of electromagnetic portion and will provide continuous monitoring and analysis of the Universe after these events with a better precision than other detectors. The Virtual Prototyping and FEM analysis phases have been performed with particular attention to the thermal and vibration behavior of the conceived mock-up during the launch phase and under high temperature variations in the space environment. The virtual tests have revealed good performances of the structure designed, in terms of responses to the adverse temperature conditions (from − 20° to + 85 °C) considering the worst cases related to the temperature limits of the several components, and the problems come out during these phases have been solved modifying the virtual prototype thanks to the parametric approach used from the very beginning of the project.

In the next months, the engineering phase will be terminated and the Physical Mock-Up of the compact model of the Crystal Eye detector will be quickly completed thanks to 3D printing. Therefore, the shock and vibration tests on the Physical Mock-Up will be performed respecting the time scheduled in the Space Rider project.

References

1. Eichler, D., et al.: Nucleosynthesis, neutrino bursts and -rays from coalescing neutron stars. Nature **340**(6229), 126–128 (1989). https://doi.org/10.1038/340126a0
2. Pendleton, G.N., et al.: The identification of two different spectral types of pulses in gamma-ray bursts. Astrophys. J. **489**(1), 175–198 (1997). https://doi.org/10.1086/304763
3. Rezzolla, L., Giacomazzo, B., Baiotti, L., Granot, J., Kouveliotou, C., Aloy, M.A.: The missing link: merging neutron stars naturally produce jet-like structures and can power short gamma-ray bursts. Astrophys. J. **732**(1), L6 (2011). https://doi.org/10.1088/2041-8205/732/1/L6
4. Troja, E., et al.: The x-ray counterpart to the gravitational wave event GW170817. Nature **551**(7678), 71–74 (2017). https://doi.org/10.1038/nature24290
5. Renno, F., et al.: AHP and multi-voting approaches for conceptual design of new detectors: the crystal eye case study. Eng. Lett. **28**(2), 470–476 (2020)
6. Abbott, B.P., et al.: Multi-messenger observations of a binary neutron star merger. Astrophys. J. **848**(2), L12 (2017).https://doi.org/10.3847/2041-8213/aa91c9
7. Sunil, C.J., et al.: 3D printing in aerospace and its long-term sustainability. Virtual Phys. Prototyping **10**(4), 175–185 (2015). https://doi.org/10.1080/17452759.2015.1111519
8. Papa, S., et al.: Towards the integration of thermal physics and geometrical constraints for a 3D-multiphysical sketcher. In: 2015 IEEE International Symposium on Systems Engineering (ISSE), Rome, Italy, pp. 248–252 (2015). https://doi.org/10.1109/SysEng.2015.7302765

Virtual Prototyping as a Supporting Tool for the Design of Complex Robotic Cells

Roberto Raffaeli[1](✉), Federico Neri[2], Margherita Peruzzini[2], Giovanni Berselli[3], and Marcello Pellicciari[1]

[1] DISMI-InterMech, Università di Modena e Reggio Emilia, 42122 Reggio Emilia, Italy
roberto.raffaeli@unimore.it
[2] DIEF-InterMech, Università di Modena e Reggio Emilia, 41125 Modena, Italy
[3] DIME, Università di Genova, 16145 Genova, Italy

Abstract. The fourth industrial revolution is characterized by flexible production systems that can respond to the demand for high variability and customization of the product. To maintain the efficiency of the production process, automated and flexible solutions are mandatory. This paper describes an approach to design Virtual Prototypes of robotic cells and support designer in the definition and simulation of the manufacturing system. The identified model is capable of replicating the performance of the cell under different aspects in a holistic manner: geometry, operating logic, performance, and physical behavior. The design approach is demonstrated on a robotic cell composed of two anthropomorphic robots for the flexible process of automatic assembly of mechanical parts. The resulting model proves to be straightforward, accurate and complete.

Keywords: Digital Twin · Physics-based simulation · Robotic assembly · Virtual Prototype

1 Introduction

Automation and simulation are two important pillars of Industry 4.0. The current business relies on the possibility of offering a vast variety of personalized products to customers, who could take part to the process itself of product definition [1]. However, always different products require adequate production systems, which are strongly based on automation and advanced collaboration between the human and the machine. Given small production lots, there is a need to automate the product design phases, quickly conceive new production cycles and machines' instructions, use advanced logistics. Therefore, along with the need of adequate product design approaches, process design becomes crucial, as well the necessity of flexible production systems.

In this context, robotic cells come to a role, as the inherent capability of highly reconfigurable execution cycles. However, for decades, robotic cells have been employed for repetitive tasks, worker substitution, high workload. Activity cycles have been defined in rigid manners, with limited adaptability. Changes to the product geometry

C. Rizzi et al. (Eds.): ADM 2021, LNME, pp. 675–682, 2022.
https://doi.org/10.1007/978-3-030-91234-5_68

or required operations traditionally require long setup times. Then, it is desirable to provide approaches, including traditional and new design tools, to allow the implementation of rapid changes in the configuration of manufacturing systems. Virtual Prototyping (VP) is a powerful tool to take advantage of, in order to carry out early design and set-up of real production systems in a virtual environment. In the robotic cells field, it is also advantageous to support tasks programming, that aims to reduce the stops of the production system and to implement Virtual Commissioning (VC).

The aim of this paper is to propose a holistic design approach for complex robotic cells that can enhance the design phase as well as the Offline Programming (OLP). In this context, the physically based simulation is presented as an additional strategic tool capable to make the simulation of the real system more intuitive and reality-compliant. To demonstrate the approach, the case of an advanced assembly robotic cell is presented.

2 State of the Art

2.1 Industry 4.0 and Advanced Simulation

The concept of Industry 4.0 has taken shape around some new needs dictated by the market, such as the request of higher levels of products customization and shorter development times. [2]. The technological support is a crucial aspect to be met to achieve the new business goal. Many new technological tools are indeed spreading on the industrial scene: Big Data, Internet of Things, Digital Twin, powerful simulation software are only some of applications at the base of Industry 4.0 paradigm [3].

Nowadays, advanced simulation and Cyber-Physical Systems (CPS) are appearing in production plants and machines are becoming "smart" [4]. Furthermore, it is possible to model the physical environment with advanced software applications, in order to obtain a virtual model of the real machines or products. The above-mentioned characteristics permit to realize CPS composed of three essential elements: the physical system, its virtual representation and the data flow between them. Such new trends demand for an extended definition of VP, which need to conveniently integrate all the aspects of the manufacturing systems [5].

2.2 Approaches for Programming Robotic Cells

A field where the VP concept finds an interesting application is the robotic cells design. New approaches for cells definition are mandatory given that the widespread of robotic manipulators in the industrial context. Currently, there are two main approaches for robotic cells programming, which are online programming and OLP [6]. The online programming is the traditional method, still very diffused, in particular among the small-medium enterprises. In fact, online programming is quite a simple methodology to apply, requiring low operators' skills [7] and investments. These peculiarities make the online programming the first choice for companies with limited budgets, even though it lack of flexibility and inability to implement complex paths [8]. Despite some higher programming skills are required, OLP could be the right choice to overcome those limitations. In fact, robot programs developed with OLP techniques could be easily modified and reused. Moreover, OLP offers the possibility to perform robot programming without stopping the production [9]. The programming environment is based on the virtual model of the robotic system, most of the times realized from CAD data. Overcoming the initial obstacles of investments and skills building, the OLP proves to be effective in robotic cells programming. Not surprisingly indeed, many computer applications for OLP have been launched in past few years [9].

2.3 Advanced Simulation and Physics Engines

The implementation of OLP strategies on robotic cells is a significant step in this direction of validating the system before its commissioning. In general, the idea of bringing the production system testing phase forward takes shape in the concept of VC, that is defined as a form of VP for the evaluation of production system functionality, performance and safety before the physical assembly. The VC technology is based on the realization of a virtual model of the production system. Both the real and the virtual systems can be divided in two macro-elements: the Control System (CS) and the Physical System (PS). Depending on the combination of those elements, VC can assume different denominations [10, 11]: i) Hardware-in-the-loop Commissioning: real CS and virtual PS are involved; ii) Software-in-the-loop Commissioning: CS and PS are both virtual; iii) Reality-in-the-loop Commissioning: virtual CS and real PS are involved. Although VP is proved to reduce times and costs of the design/commissioning phase of a production system [11], the introduction of VP in existing engineering processes is difficult because of various reason, such as high initial investments and implementation complexity [10]. Building up the virtual models requires time and efforts and several literature works focused on finding strategies to reduce the efforts [11]. Nonetheless, the studies targeted the geometry and kinematic models, meanwhile the dynamic aspect of the virtual model is quite a recent and little-investigated researching field, yet.

In this context, the tools to build virtual replicas face the necessity to introduce dynamics laws to capture realistic behaviors. Physics Engines are parts of simulation software that are able to model the motion of rigid bodies in a physical world. To simulate objects behavior, attributes such as mass, velocity and friction are used [12]. Most of the Physics Engines evolved from software libraries targeted to computer graphics applications such as Bullet, Open Dynamics and PhysX. Nowadays, some of the most widespread applications are Gazebo, USARSim and V-REP, whereas one of the oldest is STAR (Simulation Tool for Automation and Robotics), from which ABB RobotStudio (RS) has born afterwards. Recently, RS has been enriched with a physics engine developed by Algoryx Simulation AB, that is, AGX Dynamics. Another software including a physics engine is Visual Components.

3 Robotic Cell Design Approach

An approach for an optimal design of complex robotic cells supported by an advanced VP is here presented. Different software tools contribute to the design process of a robotic cell. The central element is represented by an OLP system. An example of a standard robot programming procedure has been proposed in [13]. However, such work does not include some important elements for an effective VP, for instance the Physics simulation. In fact, as it will be extensively explained in the following, it is fundamental to improve the cell virtual behavior in several applications and the match with the real counterpart. Figure 1 depitcs the proposed VP definition process. To begin with an OLP programming procedure, the geometries of manipulators, mechanical tools and all other devices that compose the real system need to be accurately modelled in a virtual environment. Geometrical definition is imported from CADs and completed by robot targets, Tool Center Point (TCP) and additional reference frames, which are useful for the following steps. The robot targets define the positions and orientations that the TCP shall reach. They can be both taught, by jogging the robot in the desired position, or created by inserting cartesian values and orientations. In addition, targets can also be automatically generated using auto-path functions. Once all the targets relative to a specific task have been created, the path that the robot should follow to reach all of them can be defined. Then, the target reachability and the absence of singularities in robot trajectories are verified. If any problem arise, rearrangements of the cell layout can be iteratively performed.

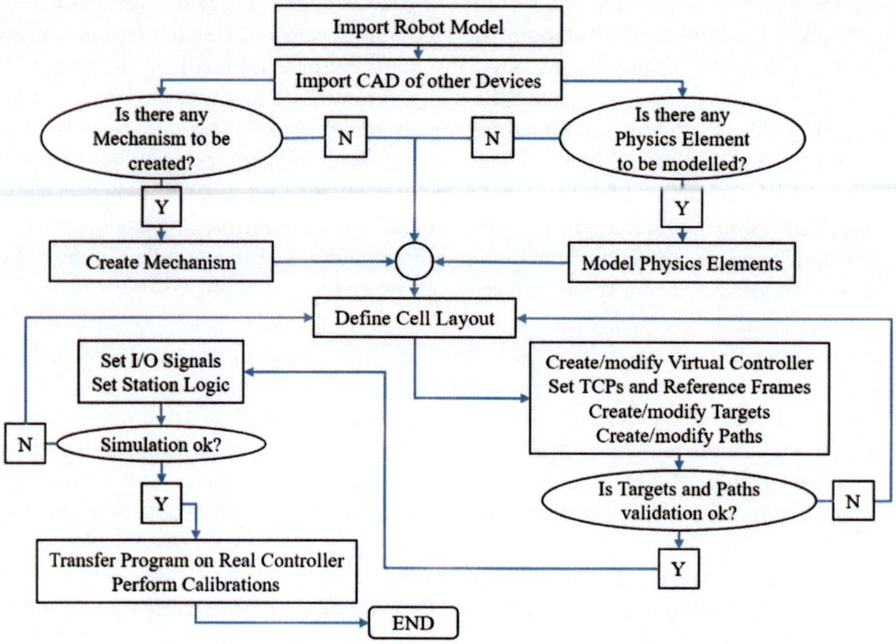

Fig. 1. Definition process of the Digital Twin of a robotic cell

To define the VP behavior, it is necessary to enrich the geometries of the devices with additional kinematic behaviors. For instance, a gripper should open and close its jaws to grab objects. Those functionalities are realized by the mechanism concept.

Kinematic mechanisms are usually enough for a realistic simulation, but many applications require more advanced physics-based behaviors. In fact, pursuing the best correspondence between the VP and the real system is useful to make the simulation more similar to reality and intuitive. Ultimately, physics-based behavior allows results to be obtained which can be easier transferred to the real system. After that, a control logic of the whole system is generated, including the communication layer among the control systems and the production management systems. According to this aim, program code is generated through OLP software function and saved on the robot virtual controllers. Once validated, it will be transferred on the real controller. Dedicated Programmable Logic Controllers (PLCs) need to be introduced to coordinate more complex scenarios composed of multiple robots. I/O signals are created on the virtual controller and connected with the logical representations of the cell's devices. In fact, actual control system of the single devices is simulated on the OLP software by a logical representation, that is employed to guarantee the right I/O connections. A relevant aspect of the proposed approach is to guarantee that the generated program for the virtual controller exactly matches the one sent to the real controller. The interaction with the devices in the cell must be carried out in the virtual environment with the same logic of the real one, even if the single device is represented by a logical "black-box" system. After that, the simulation of the process can be executed. Iteratively, the whole process is refined and validated. Finally, the elaborated program is ready to be transferred on the real controllers and PLCs.

4 The Assembly Robotic Cell Test Case

The procedure presented in Sect. 3 has been applied for the definition of the VP of an advanced automated assembly robotic cell. The details regarding the model building are neglected, since quite straightforward, while the discussion is here focused on the implementation of aspects related to the Physical simulation.

The test case is a demonstrative robotic cell situated in the SIR plant in Modena (Italy). The purpose of the assembly cell is to mount fasteners and bearings on a basement. The fasteners include studs, pins, helicoils and two different types of bearings. Once the assembly phase is ended, the finished product is scanned with a vision system to check the result. Among the devices included in the cell, two ABB robotic manipulators are found: an ABB IRB 1600–10/1.2, dedicated to pick-and-place operations; and an ABB IRB 2600–20/1.65, used for screwing. They both come with their IRC5 controller. The IRB 1600 (Robot 1) is equipped with a tool changer, a Schunk Quick-Change-Systems SWS-011. The device ensures automatic tool substitution, high repeatability performances and a fail-safe system. Five grippers can be mounted, each having a different type of fingers, according to its task. The grippers' models are parallel gripper Schunk PNG-plus 80–1 and 3-finger centric gripper Schunk PNZ-plus 80–1.

The IRB 2600 (Robot 2) is equipped with a torque-controlled screwing system which is composed by two main parts: the Bosch Rexroth electrical screwer and a pneumatic mechanism for fasteners grabbing and tool changing. Two different tools are provided to pick up studs and helicoils. The other fasteners, such as pins and bearings, are pressed into their seats on the basement using an electric press. The press is a Coretec Servo Press

CS20-350B, a model having the possibility to be programmed. Since there are different elements to press, the press is equipped with different punches, attached/detached to/from the press via a tool changer installed on the press itself. A Cartesian positioning table with 2 degrees of freedom is used to locate the basement in the right position under the press. The moving table is actuated by Bosch permanent magnet synchronous motors controlled by Bosch drivers. A vision system closes the process taking high quality pictures of the assembled product. It has a degree of freedom along the vertical axis thanks to a Bosch servomotor system: it is used to bring the camera at the right focal distance from the product. The control of the whole system is entrusted to a Beckhoff Embedded PC CX5140 equipped with Intel Atom quad-core processor. It is a DIN-rail-mountable PC running Windows 7 with TwinCAT real-time extension. The PLC has been programmed with an object-oriented approach.

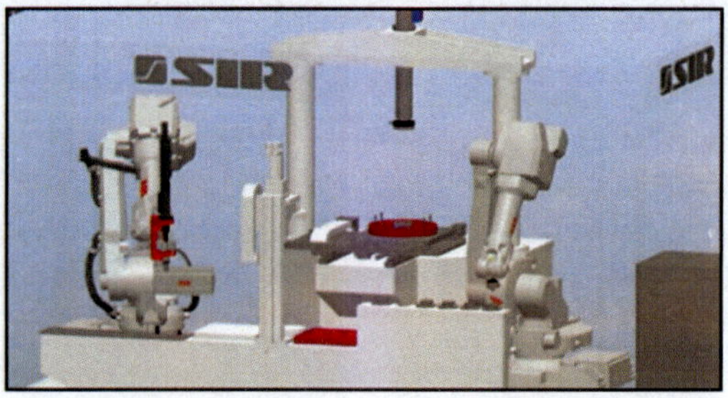

Fig. 2. The model of the assembly cell in RS environment

The cell (Fig. 2) has been modelled via SolidWorks CAD system and then imported on RS and kinematic mechanism have been created, such as the mobile table, the press and the grippers. However, some functionalities of the cell cannot be simulated using kinematic behaviors only. In particular, bearing pressing and the screwing of the studs has been modelled recurring to physical simulation. Thanks to the reliable CAD modelling, the targets of the robots can easily be set, as well as the paths. A harder task was to model the working logic of the devices, since RS does not offer the possibility of integrating any control system that is not an ABB robot controller. That required to realize specific smart components, that are RS objects that allow to recreate the working logic of a device. In the Station Logic environment, RS allow to connect inputs and outputs of the smart components with the I/O signals set on the virtual controllers. Thus, the simulation of the cell can be carried out.

4.1 Physics-Based Components

Physically based simulations have been added in the VP for modeling elements that are not moved by predefined kinematic rules, but that act according to the Physics laws. In this section, two applications are presented where also Dynamics is involved. The first example concerns the press and, specifically, the pressure of pins and bearings. The issue is to make sure that the fasteners are moved by the punch when their geometry collides with the geometry of the fasteners. Since the simulation with basic functionalities introduces complexity and non-realistic behavior, a different solution, involving the Physics simulation, has been applied. In particular, the simulation leverages on the possibility in RS software to change the physical behavior of the objects from kinematic to dynamic, making the fasteners react dynamically when the punch collides with them. This is also the case of the screwing of studs and helicoils. To better understand this case, an analysis of the screwer functioning and of the stud screwing sequence is given.

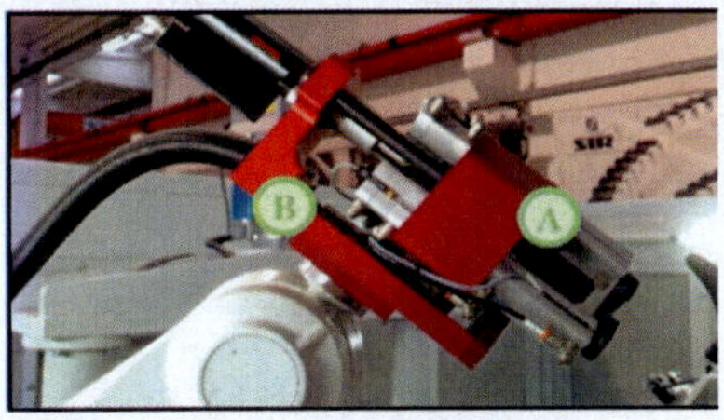

Fig. 3. The screwer: (A) Mobile element; (B) Fixed element

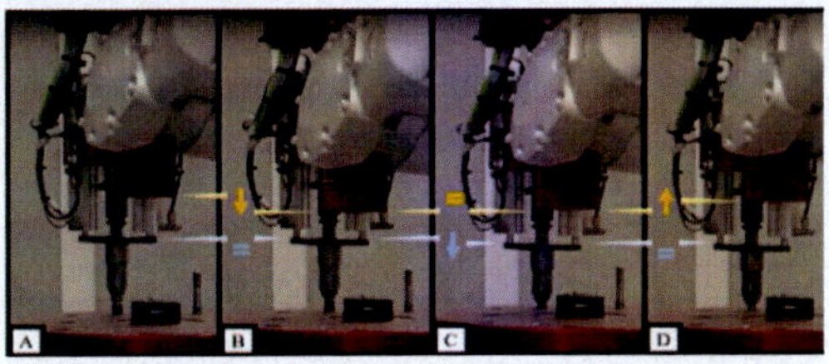

Fig. 4. Stud screwing sequence: (A) The stud hits the basement; (B) The robot reaches the target; (C) The screwing ends; (D) The robot moves away

In Fig. 3, the screwer is shown: the structure of the pneumatic mechanism is divided in two main elements, that are, the mobile part (A) and the fixed part (B). The mobile element slides on a vertical rail locked on the fixed element. This mechanism is design for the screwing phase. In fact, during the screwing of the studs/helicoils, the robot must maintain its position while the fasteners are screwed: the mobile element has to support the movement. Given that, the screwing phase in Fig. 4 is explained. In step A, the robot has not reached the target yet, but the stud has just touched the basement in correspondence of its seat. In step B, the robot reached the target and the screwer is starting to screw the stud. As it is highlighted by the signs in the pictures, the robot moves down, while the moving part of the screwer stood still. Proceeding with the next step, the robot is stationary in its position, but the mobile element slides down because of the screwing (step C). When the screwer ends its work, the robot starts to move towards the next target, but the mobile element does not move until it reaches its end-of-stroke (step D). What emerges is that there is a relative movement between the robot and the mobile element of the screwer and that the movement is due to physical contacts and gravity. Actually, a kinematic approach would be possible, but it will be fictitious and inefficient. Therefore, the Physics simulation is a good manner to properly model that movement. That way, the robot targets could have been properly set, gaining a very realist aspect of the simulation, too. Moreover, in order to solve a problem related to the coupling between the fasteners and their seats on the basement, a dedicated software component was developed by custom coding. Such component allows to create Physics joints during the simulation, permitting to set friction between the fasteners and their seats and efficiently perform the simulation itself. This result is relevant because it gives the possibility to ease the modelling process of the virtual models of many applications that request a dynamic interaction between objects.

5 Conclusions

The ambition to enhance product customization brings to the need of strategies allowing fast production systems design and re-configuration. In this paper, an approach for improving robotic cells design has been proposed leveraging the realization of advanced VPs which include several aspects of the robotic system: geometry, Kinematics, Dynamics, task executions, controlling logic, sensing. The relevance of physically based simulation in pursuing this goal has been highlighted, bringing as examples the real applications of a press and of a screwer operating in the assembling robotic cell. As a result, the correct functioning and coding of the devices has been simulated.

References

1. Raffaeli, R., Mengoni, M., Germani, M, Mandorli, F.: An approach to support the implementation of product configuration tools. In: Proceedings of the ASME IDECT-CIE 2009, vol. 5 (PART A), pp. 559–570 (2010)
2. Rojko, A.: Industry 4.0 concept: background and overview. Int. J. Interact. Mob. Technol. **11**(5), 77–90 (2017)
3. Lattanzi, L., Raffaeli, R., Peruzzini, M., Pellicciari, M.: Digital twin for smart manufacturing: a review of concepts towards a practical industrial implementation. Int. J. Comput. Integr. Manuf. **34**, 567–597 (2021)
4. Hozdić, E.: Smart factory for industry 4.0: a review. J. Mod. Manuf. Syst. Technol. **7**(1), 28–35 (2015)
5. Negri, E., Fumagalli, L., Macchi, M.: A review of the roles of digital twin in CPS-based production systems. Procedia Manuf. **11**, 939–948 (2017)
6. Holubek, R., et al.: Offline programming of an ABB robot using imported CAD models in the robotstudio software environment. Appl. Mech. Mat. **693**, 62–67 (2014)
7. Pan, Z., Polden, J., Larkin, N., Van Duin, S., Norrish, J.: Recent progress on programming methods for industrial robots. Robot. Comput. Integr. Manuf. **28**(2), 87–94 (2012)
8. Dietz, T., et al.: Programming system for efficient use of industrial robots for deburring in SME environments. In: ROBOTIK2012, 7th German Conference on Robotics (2012)
9. Gan, Y., Dai, X., Li, D.: Off-line programming techniques for multirobot cooperation system. Int. J. Adv. Rob. Syst. **10**(7), 282 (2013)
10. Lechler, T., Fischer, E., Metzner, M., Mayr, A., Franke, J.: Virtual commissioning-scientific review and exploratory use cases in advanced production systems. Procedia CIRP **81**, 1125–1130 (2019)
11. Lee, C.G., Park, S.C.: Survey on the virtual commissioning of manufacturing systems. J. Comput. Des. Eng. **1**(3), 213–222 (2014)
12. Gadaleta, M., Pellicciari, M., Berselli, G.: Optimization of the energy consumption of industrial robots for automatic code generation. Robot. Comput. Integr. Manuf. **57**, 452–464 (2019)
13. Qi, L., Yin, X., Wang, H., Tao, L.: Virtual engineering: challenges and solutions for intuitive offline programming for industrial robot. In 2008 IEEE Conference on Robotics, Automation and Mechatronics, pp. 12–17 (2008)

An Innovative Mechanical Solution to Better Understand Human-Robot Interaction Forces

Irene Pippo[1], Jacopo Zenzeri[1], Giovanni Berselli[2(✉)], and Diego Torazza[1]

[1] Robotics Brain and Cognitive Sciences, Istituto Italiano di Tecnologia, via Enrico Melen 83, 16152 Genova, Italy
irene.pippo@iit.it

[2] Department of Mechanics, Energetics, Management and Transportation, University of Genova, Via all'Opera Pia 15/A, 16145 Genova, Italy
giovanni.berselli@unige.it

Abstract. This paper describes the design and development of a new mechanical layout for the integration of additional sensors on robots that directly interact with humans. This specific work starts from the need to control human-robot interaction forces in order to ensure safety for users. The activity consisted on the conceptualization and design of a new mechanism, with the aim of including two sensors: a force/torque sensor to measure interaction forces and a linear encoder that let to know the real configuration of the device. The solution has been tested on a specific device named WristBot, whose limits are the inability to directly measure forces applied to it by humans and the impossibility to know precisely the end-effector position, with the consequences of relevant errors in calculating the interaction forces. The results obtained validate the designed mechanism comparing it with the old one mounted on the robot and prove the benefits of integrating additional sensors into the system.

Keywords: Industrial robotics · Human-robot interaction · CAD design · Mechanical optimization · Interaction forces

1 Introduction

The technological progress that characterized the last years has significantly improved our lives. The primary requirements of the first robots was speed, force and precision, but the unavoidable consequence was a lack of safety in human-robot interaction [1]. For this reason, robot and human did not cooperate directly. However, nowadays the robotics workspace is characterized by the operator increasingly working directly with robots: they support the human in many activities, from medical field to manufacturing [2]. Consequently, comes the need to accurately estimate their interaction forces.

This paper shows the design process of a new mechanical solution that includes two sensors in order to measure human-robot interaction forces. Without a force/torque sensor and a linear position sensor is not possible to evaluate the real human and robot effort during interaction activities. For this purpose, we tested the new mechanism on a human wrist rehabilitation robot, named WristBot, currently controlled with an impedance control, to highlight the benefits of this new mechanical solution.

C. Rizzi et al. (Eds.): ADM 2021, LNME, pp. 683–690, 2022.
https://doi.org/10.1007/978-3-030-91234-5_69

2 Materials and Methods

As previously said, emerging applications require robots to act safely in dynamic and unstructured environments. In order to avoid damaging robots and their surroundings (physical objects and/or human), special sensors are normally applied to control human-machine interaction, providing the desired safety [3]. Clearly, in end-effector robots it is important to have the appropriate control system.

In this work, a new mechanical solution has been applied to the human wrist rehabilitation robot WristBot [4, 5], an end-effector device developed in the Motor Learning, Assistive and Rehabilitation Robotics laboratory of the Istituto Italiano di Tecnologia (IIT). The robot is a three-Degrees of Freedom (DoF) manipulandum that allows wrist movements with three DoF: flexion/extension (FE), radial/ulnar deviation (RUD) and pronation/supination (PS). It is compatible with the full range of motion (ROM) of human subjects. Four brushless motors let to deliver torques to manipulate the wrist joint, guarantee low inertia and compensate weight. High-resolution incremental encoders measure angular displacements on the three axes, while an integrated virtual reality environment provides visual feedback to the user. The following Fig. 1 shows the total setup (robot and user), then the CAD assembly of the robot and a zoom of the old mechanism that has been implemented in this work.

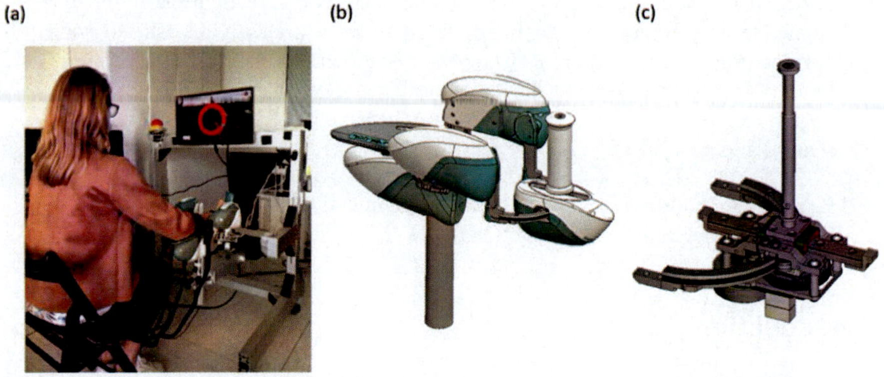

Fig. 1. (a) Robot WristBot and user. (b) WristBot CAD assembly. (c) Old mechanical solution CAD.

The limits that characterize this case study robot concern the fact that currently the only way to know the real force values generated by the robot is from motor currents. However, as the human wrist is not a perfectly spherical joint, the end-effector has a non-actuated degree of freedom. Therefore, we cannot know the real configuration of the device during its operation. In addition, it is impossible to accurately monitor the intrinsic dynamics of the robot, and this leads to a decrease of its transparency. For all these reasons, the proposed new mechanical solution has been implemented redesigning some robot mechanisms, with the aim of ensuring safety to the user monitoring interaction forces. This allows to quantify and separate human effort from robot effort, and to make the device more back drivable.

For these purposes, the mechanical solution adopted involves the addition of two sensors:

- a 6-axis Force/Torque (FT) sensor, to measure human-machine interaction forces on the end-effector;
- a linear position sensor, that allows to quantify precisely each robot configuration: given the end-effector position and the motor torque values, we can determine the force values applied by the device to the user.

Regarding the FT sensor, we have selected on the market the Mini 40 model by ATI Industrial Automation (Apex, USA). It measures all the six components of force and torque (Fx, Fy, Fz, Tx, Ty, Tz) using strain gauges with a high signal-to-noise ratio. This transducer is the best for a compromise between size and weight range. In addition, it is extremely tolerant to overloading. The following Table 1 gives a summary of its main mechanical parameters.

Table 1. ATI Mini 40 main parameters

Single-axis range of measure	Value
Fxy	± 810 N
Fz	± 2400 N
Txy	± 19 Nm
Tz	± 20 Nm

As linear position sensor, we have chosen a custom-made magnetic encoder developed inside IIT. The following Table 2 gives a summary of the main encoder parameters.

Table 2. Linear encoder main parameters

Parameters	Value
Resolution	0.5 μm
Maximum speed	40 cm/s
Pole length	1 mm
Magnetic input field amplitude	10–40 mT

The PCB layout has been designed using a CAD software and then it has been produced specifically for this case study: geometry, dimensions and layout of electronic components have been chosen to allow the best fit on the device.

In accordance with French's model of design process [6], the first step was to analyze and understand the problem, meeting the imposed constraints. For this specific case:

- maintain or minimize the mechanical range of motion (ROM) of the semicircular guide in the current configuration (62°);
- maintain the movement of the end-effector on the linear guideway (80.4 mm);
- in the new configuration the end-effector should be in the same position of the old one.

Criteria that will let to evaluate the successful result of the new solution are the minimum increase of weight due to the redesign of some components (the total weight of the replaced components in the old configuration was 171.5 g) and the minimum increase of size in all directions, which would cause a reduction of the current range of motion.

Once investigated the problem, we moved to the conceptual design phase and the idealization of several solutions. The primary choice was to determine where locate the two sensors in the assembly. The force/torque sensor must be located on the distal element of the robot kinematic chain, in this case on the end-effector. The magnetic encoder must be located also on the end-effector, in this way it is possible to measure its position during the movement on the non-actuated guideway.

In the original configuration, the end effector moved on a linear guideway by Hiwin (code MGNR12R0120HM). In the new solution, the two sensors take place on a trolley that moves with two recirculating ball bushings and two commercial mechanical rods. It was then necessary to design a new frame to assemble the commercial parts in the old configuration. For instance, Fig. 2 shows some drawings that helped to conceptualize the new mechanical solution.

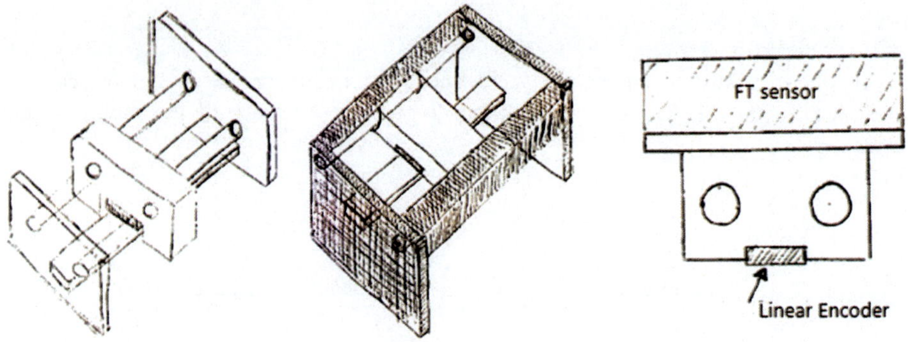

Fig. 2. Sample of drawings of different solutions that help to conceptualize the final design.

The new mechanical components have been designed using the CAD software Creo Parametric (PTC, USA). This software allows to take advantages of solid parametric modeling, designing each part individually and detailing it in assembly view. Figure 3 below shows the CAD assembly of the old configuration and the new mechanical solution. It is possible to see that the new solution meets both the requirements of dimension and weight: the total weight the new configuration is 120.7 g, the range of motion is still about 62° and the movement on the linear guideway is 84.7 mm.

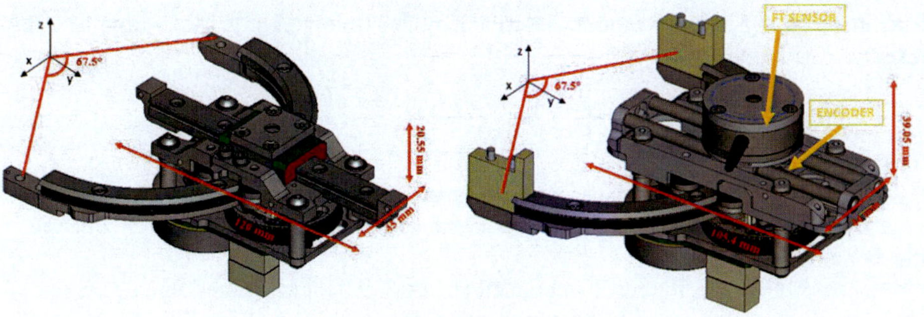

Fig. 3. CAD assembly of the old configuration mechanism (left) and new configuration mechanism (right), where dimensions and sensor position are marked.

The following Fig. 4 shows the comparison between CAD and real assembly of the new solution. When components do not require high stiffness and strength, additive manufacturing could be a good solution. In this case study FDM, SLS and PolyJet technique have been adopted.

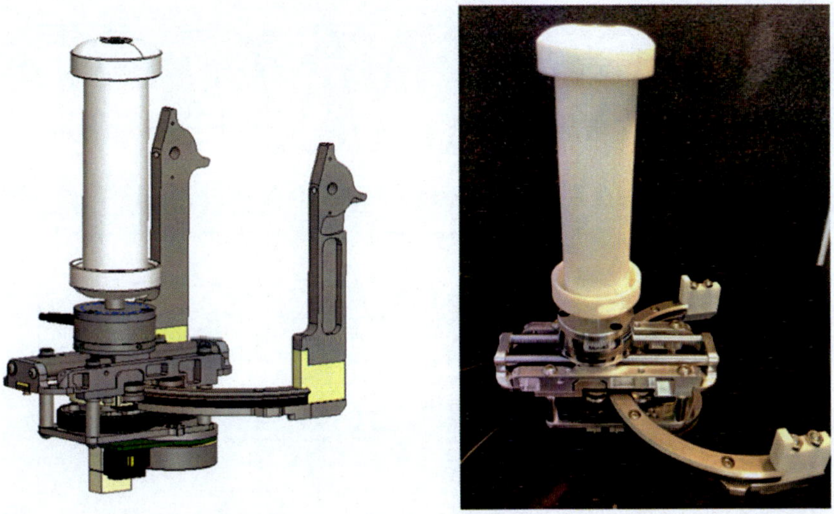

Fig. 4. CAD assembly (left) and real assembly (right) of the new mechanical solution.

3 Mechanical Testing

In order to check the impact of the sensors application in a real scenario, the new solution has been tested. The testing protocol requires the repetition of the same movement of the human wrist in flexion and extension while the other degrees of freedom are locked. During extension movements there is a visco-elastic force field that opposes to the motion. A spring-damper model, described by the following equation, simulates this resistive force field:

$$\tau = K \cdot (x_T - x) - B \cdot \dot{x} \quad (1)$$

where τ is the motor torque value, K = 900 mNm/rad is the spring stiffness, x_T = 0.84 rad (48°) is the final position of the spring, x and $\dot{x}$ are the current position and velocity of the human wrist on the flexion/extension plane and B = 120 mNms/rad is the damping coefficient.

Figure 5 illustrates the results obtained testing 3 patient (male and female 25 years). The first chart shows the end-effector position: the neutral position of the human wrist is 0°, the maximum flexion and extension point actually reached is around ± 55°. Then, the following charts represent the force values in the transversal (X) and longitudinal (Y) direction during the execution of the testing procedure. Force X has peak values when the user has to overcome the force field in extension movements. Due to the inclusion of the FT sensor in the device, now it is possible to detect the maximum force value applied by the user: 29.8 N. Y direction refers to the free linear movement of the handle: because of the human wrist is not a perfect spherical joint, during flexion and extension movements it moves also back and forward on the linear guideway.

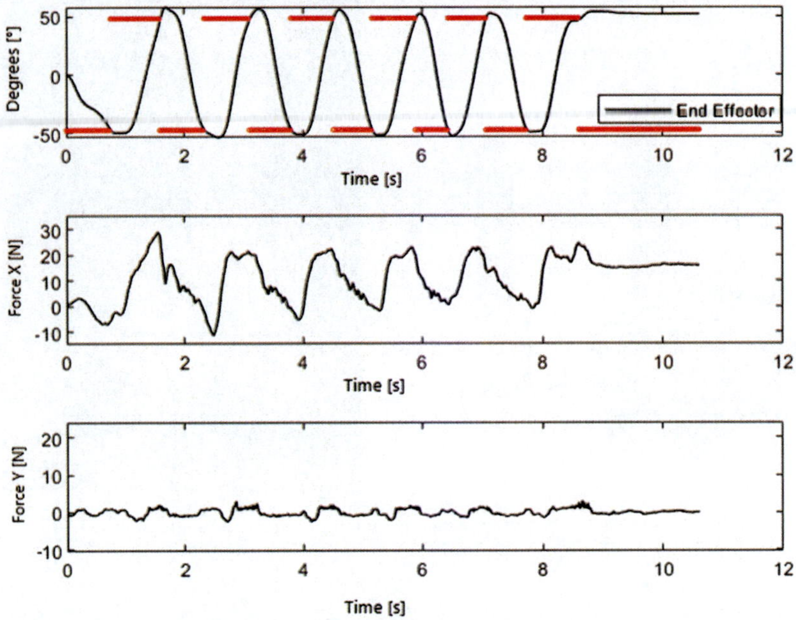

Fig. 5. Test results chart: end-effector position (upper), force along X direction values (middle), force along Y direction (lower) values.

In the old configuration, without sensors, it was not possible to detect these information, but for this experiment we have designed an adapter to mount the F/T sensor also on the old mechanism. The following Table 3 shows the comparison between the old mechanism and the new one analyzing force Y values. Despite of Fy on new mechanism is higher, the value remain acceptable thanks to recirculating ball bushings because does not negatively affect the perception of the user.

Table 3. Comparison between the two mechanism analyzing force Y values

	Maximum force Y [N]	Average [N]	Standard deviation [N]
Old mechanism	1.49	−0.22	0.57
New mechanism	3.04	−0.07	0.94

To identify robot effort, we need to know the real device configuration during the protocol execution. Known motor torque value, force value applied by the robot is given by the following equation:

$$F = {}^{\tau}/_{b} \quad (2)$$

where τ is the motor torque value and b is the distance between the rotation axis and the end-effector position.

Before encoder integration, the distance b was set considering the handle in the middle of its stroke. Figure 6 shows the comparison between hypothetical (estimated without encoder) and real force X values.

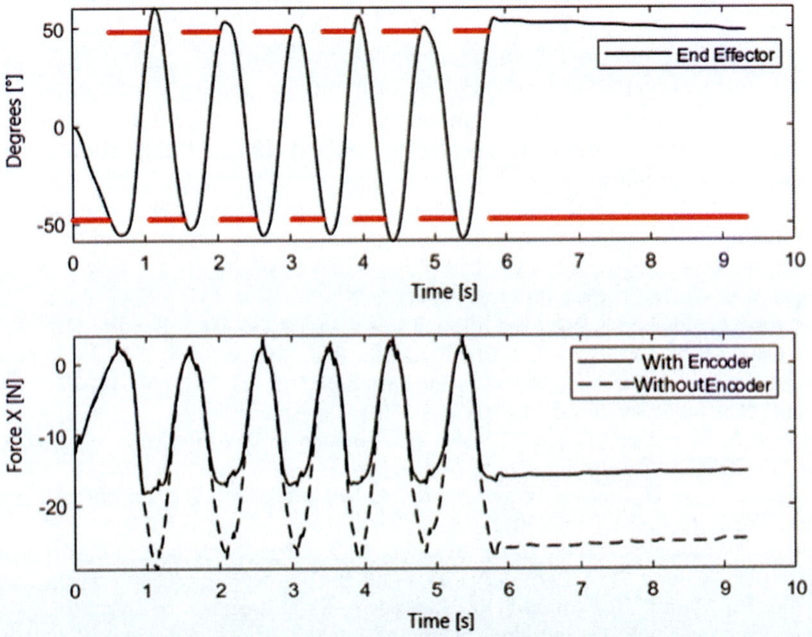

Fig. 6. Test results chart: end-effector position (upper) and comparison between force along X direction values in case of encoder and without encoder (lower).

Table 4 summarizes the maximum value estimated in the first case, the maximum real value calculated with the position sensor and the computation error leads by the difference between the two cases.

Table 4. Maximum force values estimated without and with encoder

Maximum estimated force [N]	Maximum real force [N]	Error [N]
−29.05	−18.22	10.83

4 Discussions

The aim of this paper was to point out the importance of controlling collaborative robots. Robots need to interact safely with human, which is why we have to exactly know the interaction forces between machines and users.

After testing the new mechanical solution on a specific device, the WristBot, it is possible to understand the improvement that this work has provided. The FT sensor applied on the end-effector allow us to measure the human effort during collaborative activities with robots. Moreover, analyzing Force Y values it is also possible to validate the new mechanism: the comparison between the old configuration and the new one suggests that the system satisfies all the requirements. However in the future it could be improved lowering forces along Y direction actually caused by friction.

Data collected using the magnetic encoder suggest how important is to precisely measure the real robot configuration, in order to correctly estimate the real force values applied by the robot to the user. This case study shows that the error between the maximum assumed force and the real one was about 37%.

In conclusion, the implementation of the case study shows the benefits of integrating these two sensors in robots that directly interact with human. A future development is to use the device with an admittance control scheme, to obtain a full transparency of the robot while generating forces.

References

1. Bragança, S., Costa, E., Castellucci, I., Arezes, P.M.: A brief overview of the use of collaborative robots in industry 4.0: human role and safety. In: Arezes, P.M., Baptista, J.S., Barroso, M.P., Carneiro, P., Cordeiro, P., Costa, N., Melo, R.B., Miguel, A.S., Perestrelo, G. (eds.) Occupational and Environmental Safety and Health. SSDC, vol. 202, pp. 641–650. Springer, Cham (2019). https://doi.org/10.1007/978-3-030-14730-3_68
2. Vysocky, A., Novak, P.: Human – Robot collaboration in industry. MM Sci. J. **2016**(02), 903–906 (2016)
3. Fumagalli, M., et al.: Force feedback exploiting tactile and proximal force/torque sensing. Auton. Robot. **33**(4), 381–398 (2012)
4. Masia, L., Casadio, M., Sandini, G., Morasso, P.: Eye-hand coordination during dynamic visuomotor rotations. PLoS ONE **4**(9), e7004 (2009)
5. Iandolo, R., Marini, F., Semprini, M., Laffranchi, M., Mugnosso, M., Cherif, A., et al.: Perspectives and challenges in robotic neurorehabilitation. Appl. Sci. **9**(15), 3183 (2019)
6. Cross, N.: Engineering Design Methods: Strategies for Product Design, 4th edn. Wiley, Hoboken (2008)

Finite Element Analysis of the Interaction Between an Endo-Surgical Tool and Colorectal Tissue for Setting up Force Feedback Evaluation in Virtual Reality-Based Applications

Robinson Guachi[1,2], Michele Bici[1], Fabiano Bini[1], Francesca Campana[1(✉)], and Franco Marinozzi[1]

[1] Department of Mechanical and Aerospace Engineering, DIMA – Sapienza University of Rome, Via Eudossiana 18, 00184 Rome, Italy
francesca.campana@uniroma1.it

[2] Department of Mechatronics, Universidad Internacional del Ecuador – UIDE, Av. Simón Bolívar, 170411 Quito, Ecuador

Abstract. Numerical simulations and Finite Element Analysis (FEA) have currently increased their applications in medical field for making preoperative plans to simulate the response of tissues and organs. Soft tissue simulations, such as colorectal simulations, can be adopted to understand the interaction between colon tissues and surrounding tissues, as well as the effects of instruments used in this kind of surgical procedures. This paper analyses through FEA the interaction between a surgical device and a colon tissue when it is fully clamped. Sensitivity analysis in the respect of the material mechanical behaviour, geometric approximation and the effect of thickness variation are investigated with the aim of setting up a virtual prototype of the surgical operation to aid mentoring and preliminary evaluation via haptic solutions. Through this investigation, the force feedback estimation that is necessary in many virtual-reality applications, may be estimated without discharging nonlinear effects that occur during clamping and that usually cannot be simulated efficiently to guarantee real-time solutions. Results are aligned with experimental data, confirming the reliability and right the set-up of FEA. Through them, the preliminary set-up of a haptic force feedback has been described and simulated through Simulink 3D animation, confirming the feasibility of the concept.

Keywords: Surgical simulation · FEA · Metamodeling · Virtual prototyping · Force feedback · Haptic device

1 Introduction

In recent years, virtual simulations have gained prominence in the medical field as a method for conducting preoperative planning and simulating tissues and organ responses. Many applications use Finite Element Analysis (FEA), especially to simulate real-world

C. Rizzi et al. (Eds.): ADM 2021, LNME, pp. 691–702, 2022.
https://doi.org/10.1007/978-3-030-91234-5_70

phenomena involving organs, surrounding tissues and surgical instruments (such as colorectal surgery simulations), as well as to design prosthesis, orthosis, analyse implants and simulations of surgery [1]. As a major drawback, FEA is not so efficient to achieve real-time simulation, as virtual reality-based applications ask for. Due to this, simplifications must be done to achieve interactivity, often reducing simulation's accuracy in terms of stress-strain distribution and force feedback. Furthermore, geometrical modelling and FEA may be principally used to set-up virtual investigations for the preoperative plan, on the base of the characteristics of the patient [2], but also, with the recent developments of additive technologies, to obtain printed 3D models useful for planning [3–5]. This kind of possibilities also gained importance in training and didactics applications, where learning by doing can be also associated to different levels of difficulties and accuracy. By this point of view, the linear elastic mechanical behaviour represents probably the most common assumption for simulating tissue deformation due to its low computational cost. Nevertheless, in soft tissues, well-known nonlinear effects are produced by hyper-elastic material behaviour and large displacements.

The aim of this paper is to investigate how nonlinear effects, evaluated through FEA, may be considered in a preliminary set-up of a virtual prototype of the colorectal surgery, where the response of the interaction between surgical tool and soft tissue is relevant. In Sect. 2, a brief state of the art about simulations of soft tissues both in virtual reality and FEA environment is presented, highlighting open issues, then, in Sect. 3, Materials and Methods, the investigated problem is presented. While, in Sect. 4, Results and Discussion about their implementation for force feedback are presented and finally, in Sect. 5, conclusions are outlined.

2 State of the Art

Surgical simulators based on virtual reality proves to be a good option to train new surgeons, to practice new processes, and to assist preoperative planning. For understanding tissues injury or critical physical behaviours during operations, accuracy of the mechanical mechanisms behind deformations is required. Many applications of virtual reality can be focused on the real phenomena that are present when there is an interaction between organs, surrounding tissues and surgical tools. In these pre-operative planning cases, stress accuracy is relevant similarly to the real-time feedback effect of the simulation, so that, equivalent mechanical models have been investigated [6].

The linear elastic mechanical behaviour is probably the most popular assumption to simulate the tissues deformation into virtual surgery simulators. It is based on Hooke‘s law and has the advantage of needing only two constants to determine its behaviour (under the hypothesis of isotropic and homogenous materials), and low computational cost. These characteristics enables real-time haptic rendering. Unfortunately, real soft tissues behave hyperelastically. Many material model formulations are present in literature, as reported in [7], and they are often implemented in the most adopted FEA codes [8]. FEA commercial software cannot simulate in real time although they may provide result with better accuracy. In [9], a condensation method to extend a FE model for real-time surgical simulations is presented. This method suggests reducing the FEA model only at the set of nodes which undergo stressed so that evaluated displacement may be rendered. A comparison between the proposed method and the conventional linear FE analysis presents similar results in terms of nodal displacement. Another research [10] presents a real time simulation, thanks to a pre-processing of elementary deformations derived from a finite element method wherein the bulk of the computations were performed during the pre-processing stage of the FE calculation. This method was tested in a real-time hepatic (liver) surgery simulation.

From the soft tissues point of view, better accuracy requests a nonlinear analysis that involves large displacement and a hyperelastic mechanical behaviour [11]. With the aim of a realistic behaviour in simulation for surgical training, in [12], researchers implemented the hyperelastic Mooney-Rivlin model. This work includes a proposal of a scheme for mesh adaptation based on an extension of the progressive mesh concept, so that real-time computation of deformation may be enabled.

Nowadays, it is possible to find software focused on the interactive computational medical simulations, as SOFA (Simulation Open Framework Architecture), which facilitates collaboration between specialists from various domains with the aim of approaching virtual reality [13]. To reach this goal, SOFA is organized in independent components in a scene graph data structure. Then, in each component it is possible to encapsulate a particular aspect regarding the simulation such as degrees of freedom (DOFs), forces and constraints, differential equations, main loop algorithms, linear solvers, collision detection algorithms, or interaction devices. In [14], SOFA is adopted to estimate in real-time, the position of the liver internal structures by evaluating nonlinear deformation of the liver, taken from images and vascular network simulated as vessels with simplified behaviour. Also in this case, the necessity of considering nonlinearity must be balanced with computational efforts.

From the virtual simulators point of view, it is possible to find several commercial solutions focused on surgical simulations. With the aim of a realistic simulation and a complete immersive experience, some surgical simulators provide a force feedback to the user by means of haptic systems. However, in many of these cases, the details of the adopted model for the soft tissues mechanical behaviour are not easily available. Linearization is often declared, tweaked on qualitative evaluation performed by few surgeons rather than actual material testing, highlighting the emphasis on producing models that are visually appealing. Recently, in [15], the problem of robot grasping during sutures is experimentally approached through haptic sensors suitable to assess the feeling at the finger. It investigates how visual and tactile information affects the evaluation of safety margin during tissue manipulations. In particular, two levels of indentation force are found assuming the presence or not of a visual feedback (respectively 75 mN and 87.5 mN).

3 Material and Methods

Surgical clamping is a manipulation task necessary during a laparoscopic process. It does not involve tissue rupture but deformation able to hold in contact walls of the organ. In a laparoscopic process made in Mini-Invasive Surgery (MIS), instruments must be inserted through a trocar to access the abdominal cavity without capability of direct manipulation of tissues, so that, a loss of sensitivity occurs. This is caused by the absence of a direct force-feedback resulting from the interaction between surgical instruments and soft tissues in conjunction with an indirect field of view, as described in [16]. The force feedback received during the operation is a composition of the reaction force of the tissue and of the mechanical characteristics of the surgical tool. Figure 1 shows a clamping tool in its opening position and the anatomical district of the large intestine and its FEA set-up when clamping is starting.

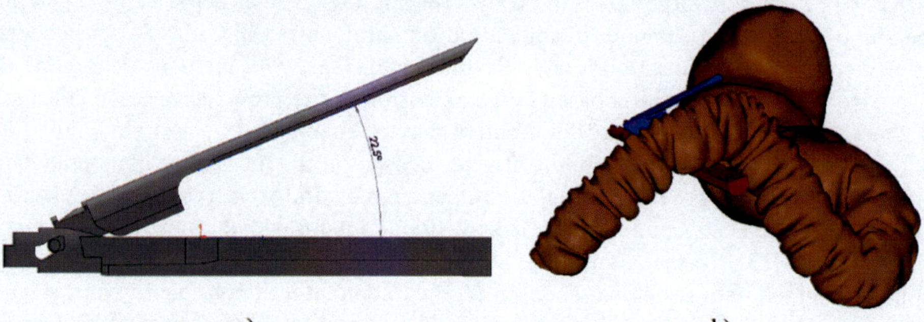

Fig. 1. Test case: clamping device (a) and interaction (b) with the anatomic district (geometric model named as "3D-SS").

In Fig. 2a, an example of the kinematic chain is sketched. Green dot lines show the final position when the grasp is close. Cam geometry, sketched in the detail, and spring stiffness (named as k) characterize the quality of the tool. Figure 2b describes the dynamic equilibrium, so that the force at the latch constrained in A, F_i, may be found as a function of the tissue reaction (F_g).

The clamping device may be assumed as rigid considering that, assuming a maximum allowable pressure of $8 * 10^{-2}$ N/mm^2 at the interface with the tissue [17], FEA results on the final part of the clamping device show a maximum deflection of about 0.4 mm. According to this, the relation among the momentum M_a induced by $F_i L_3$ is related to the F_g at the interface with the tissue by:

$$F_g = F_t' m/n \tag{1}$$

where F_t' stands for:

$$F_t' = \frac{F_i * L_3}{L_1} * cos\,\varphi * cos\,\beta - k(\Delta x) \tag{2}$$

where Δx represents the relative displacement of $C(\theta_i)$ from C_0 ($\theta_i = 22°$) and φ is function of θ_i that spans the range [0 to $\theta_i/2$]

$$sin\beta = \frac{\left(d\,cos\,\theta + sin\,\theta\sqrt{L_1^2 - d^2}\right) - d}{L_2} \tag{3}$$

$$\varphi = \frac{\theta_i}{2} \tag{4}$$

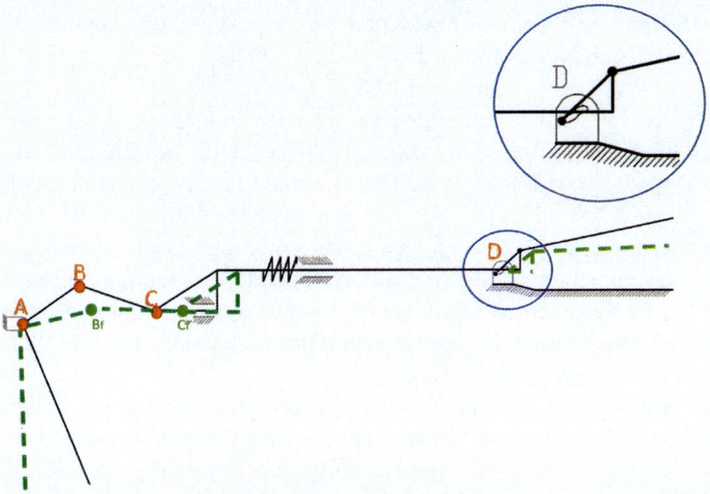

a

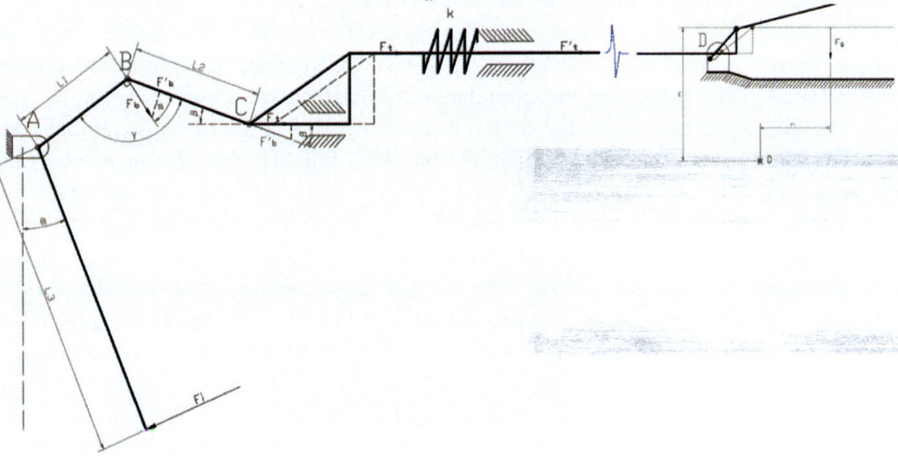

b

Fig. 2. a) Kinematic chain, b) Reaction force scheme.

The interaction with the soft tissue was investigated through FEA, applied on three different geometrical models, as discussed in [8]. The most accurate is the one shown in Fig. 1b, named 3D-SS, while an approximation of the real geometrical model is named 3D-S and finally the simplest is the cylindrical approximation of the region of interest, named CS, with diameter equivalent to the average one of the final segments.

FEA has been carried out in Altair HyperWorks suite using RADIOSS solver. RADIOSS have been chosen due to its advantages in handling contact algorithms and large displacements, since it is an explicit solver oriented to minimize computation time in FEA with high non-linearities [18]. Other parameters and input conditions have been set as described in the following:

- Mesh: Shell elements with three integration points along the thickness (named shell3N and shell4N) have been used to construct 3D-SS, 3D-S, and CS geometrical models.
- Load & Boundary conditions: it has been assumed the usage of a blue staple cartridge with a 1.5 mm closed height [19]. This gap has been adopted as final closure of the tissue under the clamping device, assuming no sliding at the interface. Concerning displacement constraints at the free edges of the models, rotations around all axes, as well as the displacement along the colon axial direction at the nodes along the free edges have been released, in order to reproduce constraints, close to those present in tissue when conducting colorectal surgery.
- Contact: The "Type 7 interface" into HyperWorks is the contact condition defined between the colon tissues and the surgical clamp, which is a multi-usage impact interface. To ensure that the simulation developed correctly and to establish self-contact and contact between colon tissues and the surgical clamp in the current case study, more than one "Type 7 interface" was used.
- Soft tissue's thickness: it is related on the healthiness of tissue. The colon wall thickness value is set in the range of 0.8 to 2 mm to consider all assumptions above mentioned, these values are based on data available in literature [20, 21].

  To highlight the role of the material model in the stress-strain behaviour, thus in the force feedback, both Mooney-Rivlin [22] (HE-MR) and linear elastic (LE) material models have been considered in this paper. In [2], the Constant values are set to $C_{10} = 0.085$ MPa and $C_{01} = 0.0565$ MPa. These values are sufficient to define the hyperelastic Mooney-Rivlin model when the used materials are incompressible. In the same way, several applications require a reasonable approximation of soft tissue mechanical properties. For this reason, some laboratory tests [23] have found an experimental approximation to the linear mechanical behaviour corresponding to $E = 5.18$ MPa.

4 Results and Discussion

Figure 3 highlights the comparison in terms of stress by changing geometrical model and tissue thickness, in the respect of LE and HE-MR material models.

The first consideration concerns the range of stress passing from LE to HE-MR. LE overestimates the maximum stress under the clamping device, despite the adopted geometrical model. The second one concerns with the irregularities of the 3D-S model in the respect of the others, considering that in many cases CS tends towards 3D-SS. The differences between the stress ranges are related to a major capability of stress distribution along the areas of the HE-MR model in the respect of LE. It means that the stress gradients are more gradual in the HE-MR case, as already discussed in [11]. Thickness increase always induces a stress increase.

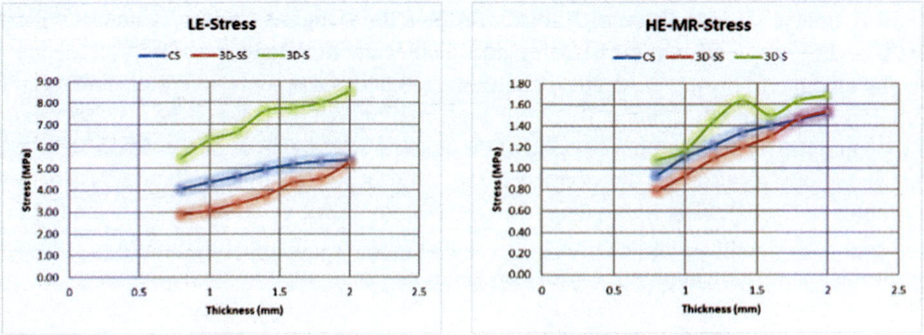

a) b)

Fig. 3. Trends of stress varying thickness and geometrical model for linear elastic material model (on the left), and for hyperelastic MR one (on the right).

Figure 4 compares the most stable solutions, CS and 3D-SS, in terms of contact pressure with the grasp, changing material model, thickness and geometrical model. Contact pressure increases with the changes of the initial thickness. The average values in the case of 3D-SS and CS models, assuming HE-MR is about 0.5 MPa, with a trend that can be reasonably seen as linear in the respect of the initial thickness changes. The LE material model, on the contrary, induces a major stiffness, thus higher contact pressure values.

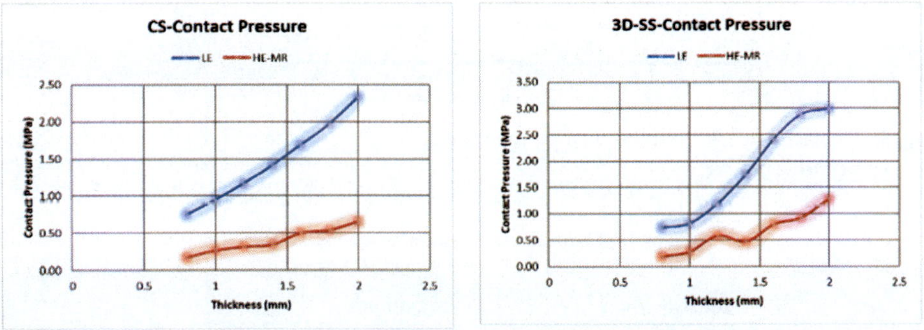

a) b)

Fig. 4. Trends of contact pressure varying thickness and material model: a) CS geometrical model and b) 3D-SS geometrical model.

It has to be noted that the contact pressure values are referred to a limited area of elements, taking the maximum values between those instantly in contact with the clamp. This fact is understandable looking also to the pressure force trend (Fig. 5).

FEA analysis made through Radioss follows the dynamics of the contact explicitly allowing the computation of force and displacements along time, so that, Force-Displacement curves may be derived changing geometrical models, materials and thickness. Figure 5 shows the contact force during clamping as a function of the displacement of the clamping tool, up to put in contact the upper and lower part of the colon section. This induces saturation of the contact force, after a plateau of less than 1 N (in the blue squared detail). This graph describes the reaction force on the grasp that the tissue applies under assigned displacements. It is consistent with the results found in [16] and the contact pressure found in the previous section.

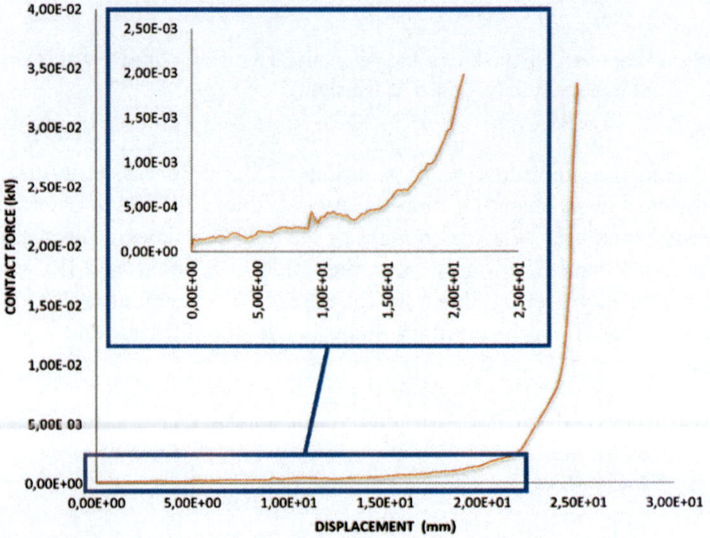

Fig. 5. Contact force vs vertical displacement, and, in the blue squared detail, a zoom of the plateau part of the curve.

Table 1. The kinematic chain's starting values.

Identification	Magnitude
L_1	21 mm
L_2	25.1 mm
L_3	54.1 mm
$\theta_{initial}$	22°
$\beta_{initial}$	19°
φ	$\theta_i/2$

Through the dynamic balance described in Sect. 3, the reaction force at the hand interface can be computed in accordance with the linkage lengths as reported in Fig. 2 and defined in Table 1.

Figure 6 shows the percentage values of F_g in the respect of F_i. This value has been adopted for a preliminary implementation of a virtual environment able to provide the force feedback during clamping. It has been designed in Matlab through Simulink 3D Animation and it will be interfaced with an Arduino's controller able to provide the feedback to a simplified clamping device achieved by additive manufacturing. Figure 7 shows the concept diagram of the system, where the force feedback function is based on the input provided from a post processing of the FEA collected in this work, starting from the CS model.

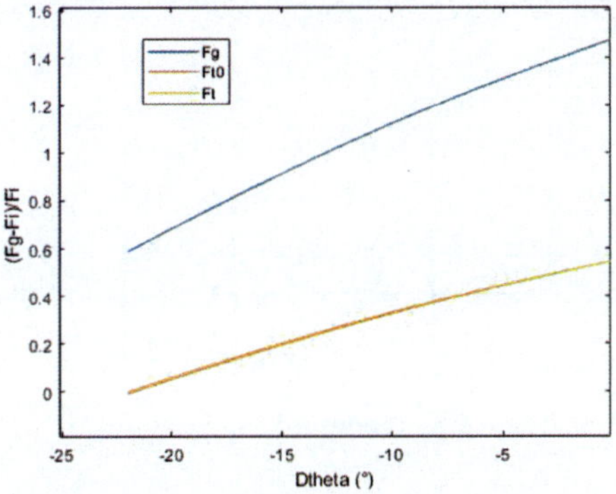

Fig. 6. Angular displacement (θ) vs F_g.

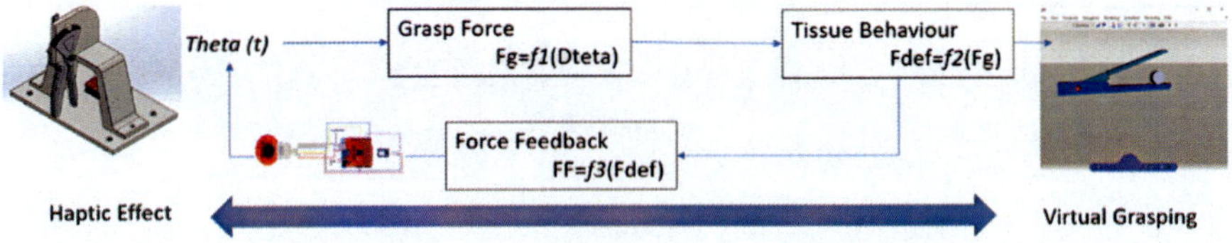

Fig. 7. Concept blocks of the proposed haptic device.

More in details, the VR sink implements a collision detection and the CS deformation derived from data collected in Fig. 5. The Simscape model of the grasping tool (Fig. 8) has been simplified having in mind the chance of implementing a haptic sensor through pulley, able to implement the force feedback on the base of an open loop impedance controller. Figure 9 captures the beginning of the contact and two simulation states along the contact force curve, assuming a half model of the CS, so that the visual correlation among theta and grasp displacement may be seen.

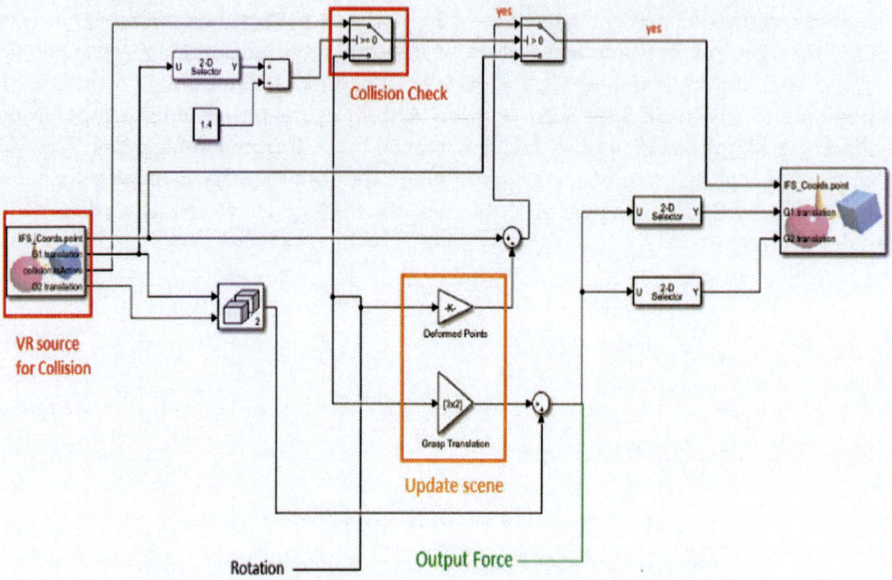

Fig. 8. Virtual Grasping modelling via Simulink 3D Animation.

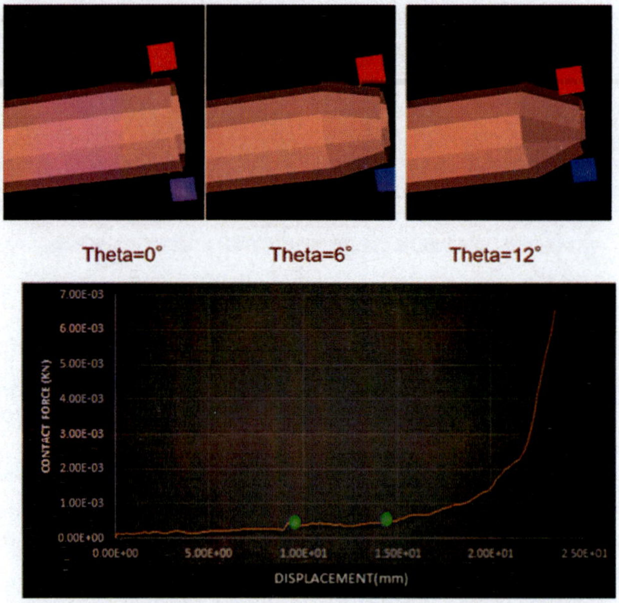

Fig. 9. VR results of the correlation among Theta-Displacements.

5 Conclusions and Future Works

This paper presents the stress-strain analyses of the surgical grasping of soft tissues in correlation to the force feedback that should be provided in haptic solution for virtual surgical experience. More in details, starting from previous analyses made about the role of hyperelasticity, accuracy of the geometric models and tissue initial thickness, FEA model based on Mooney-Rivlin hyperelasticity have been proved to achieve results aligned with values found in literatures. They confirmed a rather linear behaviour in the adopted range of simulations. Through these results, a haptic device with an open loop impedance feedback has been designed and preliminary presented. In particular, starting from a reasonable configuration of the grasping device, the force at the handling has been evaluated as a function of the imposed rotation and the related reaction force under the tissue has been estimated by the FEA. Then, concept blocks and test simulations through Simscape-Simulink animation 3D have been built. They confirmed the feasibility of the solution, starting from the CS geometrical model and the linearization made from the FEA results. The final goal is providing a VR environment able to support surgical testing through material behaviour information related to stress and strain and force feedback. Next developments will include the construction of the hardware and the testing of the controller that will be controlled through Arduino.

References

1. Shi, H.: Finite element modeling of soft tissue deformation (2007)
2. Mayeur, O., Witz, J.-F., Lecomte, P., Brieu, M., Cosson, M., Miller, K.: Influence of geometry and mechanical properties on the accuracy of patient-specific simulation of women pelvic floor. Ann. Biomed. Eng. **44**(1), 202–212 (2015). https://doi.org/10.1007/s10439-015-1401-9
3. Bici, M., et al.: Digital Design of Medical Replicas via Desktop Systems: Shape Evaluation of Colon Parts. J. Healthc. Eng. **2018**(3272596), 1–10 (2018)
4. Calì, M., Pascoletti, G., Gaeta, M., Milazzo, G., Ambu, R.: A new generation of bio-composite thermoplastic filaments for a more sustainable design of parts manufactured by FDM. Appl. Sci. **10**(17), 5852 (2020)
5. Ambu, R., Motta, A., Calì, M.: Design of a customized neck orthosis for FDM manufacturing with a new sustainable bio-composite. In: Rizzi, C., Andrisano, A.O., Leali, F., Gherardini, F., Pini, F., Vergnano, A. (eds.) ADM 2019. LNME, pp. 707–718. Springer, Cham (2020). https://doi.org/10.1007/978-3-030-31154-4_60
6. Misra, S.: Realistic tool–tissue interaction models for surgical simulation and planning. Doctor, p. 275 (2009)
7. Bustamante-Orellana, C., et al.: Biomechanics of soft tissues: the role of the mathematical model on material behavior. In: Botto-Tobar, M., León-Acurio, J., Díaz Cadena, A., Montiel Díaz, P. (eds.) ICAETT 2019. AISC, vol. 1066, pp. 301–311. Springer, Cham (2020). https://doi.org/10.1007/978-3-030-32022-5_29
8. Guachi, R., Bini, F., Bici, M., Campana, F., Marinozzi, F.: Finite element model set-up of colorectal tissue for analyzing surgical scenarios. In: Tavares, J.M.R.S., Natal Jorge, R.M. (eds.) ECCOMAS 2017. LNCVB, vol. 27, pp. 599–609. Springer, Cham (2018). https://doi.org/10.1007/978-3-319-68195-5_65
9. Bro-Nielsen, M.: Finite element modeling in medical VR. J. IEEE **86**(4), 490–503 (1998)
10. Cotin, S., Delingette, H., Ayache, N.: Real-time elastic deformations of soft tissues for surgery simulation. IEEE Trans. Vis. Comput. Graph. **5**(1), 62–73 (1999)

11. Guachi, R., Bini, F., Bici, M., Campana, F., Marinozzi, F., Guachi, L.: Finite element analysis in colorectal surgery: non-linear effects induced by material model and geometry. Comput. Methods Biomech. Biomed. Eng. Imaging Vis. **8**(2), 219–230 (2020)
12. Wu, X., Downes, M.S., Goktekin, T., Tendick, F.: Adaptive nonlinear finite elements for deformable body simulation using dynamic progressive meshes. Comput. Graph. Forum **20**(3), 349–358 (2001)
13. Allard, J., et al.: SOFA-an open-source framework for medical simulation. Stud. Health Technol. Inf. **125**, 13–18 (2007)
14. Haouchine, N., Dequidt, J., Peterlik, I., Kerrien, E., Berger, M.O., Cotin, S.: Image-guided simulation of heterogeneous tissue deformation for augmented reality during hepatic surgery. In: 2013 IEEE International Symposium on Mixed and Augmented Reality (ISMAR), pp. 199–208 (2013)
15. Cabibihan, J.-J., Alhaddad, A.Y., Gulrez, T., Yoon, W.J.: Influence of visual and haptic feedback on the detection of threshold forces in a surgical grasping task. IEEE Robot. Autom. Lett. **6**(3), 5525–5532 (2021)
16. Tholey, G., Desai, J.P., Castellanos, A.E.: Force feedback plays a significant role in minimally invasive surgery: results and analysis. Ann. Surg. **241**(1), 102–109 (2005)
17. Baker, R.S., Foote, J., Kemmeter, P., Brady, R., Vroegop, T., Serveld, M.: The science of stapling and leaks. Obes. Surg. **14**(10), 1290–1298 (2004)
18. Zienkiewicz, O.C., Taylor, R.L.: The finite element method. McGraw-hill, London (1977)
19. Chekan, E., Whelan, R.L.: Surgical stapling device - tissue interactions: what surgeons need to know to improve patient outcomes. Med. Devices Evid. Res. **7**, 305–318 (2014)
20. Egorov, V.I., Schastlivtsev, I.V., Prut, E.V., Baranov, A.O., Turusov, R.A.: Mechanical properties of the human gastrointestinal tract. J. Biomech. **35**(10), 1417–1425 (2002)
21. Liao, D., Zhao, J., Gregersen, H.: 3D Mechanical properties of the partially obstructed guinea pig small intestine. J. Biomech. **43**(11), 2079–2086 (2010)
22. Shahzad, M., Kamran, A., Siddiqui, M.Z., Farhan, M.: Mechanical characterization and FE modelling of a hyperelastic material. Mater. Res. **18**(5), 918–924 (2015)
23. Christensen, M.B., Oberg, K., Wolchok, J.C.: Tensile properties of the rectal and sigmoid colon: a comparative analysis of human and porcine tissue. Springerplus **4**(1), 1 (2015). https://doi.org/10.1186/s40064-015-0922-x

Experimental Methods in Product Development

Recent Findings About the Novelty Metric of Shah

Lorenzo Fiorineschi(✉), Francesco Saverio Frillici, Federico Rotini, and Monica Carfagni

Department of Industrial Engineering, University of Florence, Via Santa Marta 3, 50139 Firenze, Italy
lorenzo.fiorineschi@unifi.it

Abstract. The work presented in this paper aims at presenting the last findings about one of the most acknowledged and used novelty metrics in the field of engineering design. The a-posteriori novelty metric of Shah, Vargas-Hernandez and Smith is often considered in design research, but after two decades, there are still some critical issues to be solved. This paper highlights the issues identified in the last research works, by referring to the few published works that directly face the argument. It emerged that notwithstanding the presence of metric variants proposed by scholars, none of them is capable to overcome the problems. In particular, the assessment of set of ideas implementing heterogeneous numbers of attributes can lead to misleading results. In absence of a comprehensive solution, this paper provides a list of practical hints in order to allow the application of the metric (or the related variants) and ensuring the repeatability of the experiments.

Keywords: Novelty assessment · Creativity · Originality · Unusualness · Design

1 Introduction

Among the different definitions of creativity that can be found in the context of design (e.g. [1–3]), in the field of engineering it is often referred as the need to meet functional requirements in a novel way [4].

Focusing the attention on the concept of novelty, the related possible meanings depend on the considered assessment parameters and the reference used to establish what is novel or not. Shah and colleagues [5] identified two assessment approaches, namely the "a-priori" and the "a-posteriori" ones. The a-priori approach needs to identify a reference solution (or a reference set of solutions) to assess the novelty. Differently, in the a-posteriori approach, the reference is constituted by the set of ideas to be assessed. In the latter case, the assessment is fundamentally based on counting the occurrences of similar ideas that have been generated in the same session. Accordingly, novelty is considered as a measure of the "uncommonness" or unusualness or unexpectedness of a specific idea in relation to a group of ideas (normally from the same experimental activity or design task). Indeed, that of uncommonness is a concept relative to a specific set of ideas, which can be extremely useful for research purposes. Differently from a-priori approaches, the concept of uncommonness allows to assess the effect of treatments

C. Rizzi et al. (Eds.): ADM 2021, LNME, pp. 705–711, 2022.
https://doi.org/10.1007/978-3-030-91234-5_71

(e.g. stimuli, incubation, design methods, etc.), without the need to identify a subjective external reference. Maybe, for this reason, a-posteriori novelty metrics are often used in design research (e.g. [6–9]).

Among those available in literature, the a-posteriori novelty metric of Shah, Vargas-Hernandez and Smith [5] (hereinafter called SNM), is one of the most largely used and acknowledged among design researchers. However, besides the implicit limitation due to the relativeness of the specific novelty concept, additional flaws have been recently highlighted for SNM.

The aim of this paper is to highlight the issues recently highlighted for SNM, together with the possible solutions to overcome them. The outcomes of this work provide a first set of indications to support researchers for a correct application of the metric.

In particular, Sect. 2 shows a literature review of the currently acknowledged issues of SNM. Section 3 introduces pros and cons of the available SNM variants, while Sect. 4 reports if and to what extent the available variants can solve the SNM issues. Section 5 proposes an application guideline in order to avoid misleading assessments. Finally, conclusions are reported in Sect. 6.

2 SNM and the Acknowledged Issues

SNM requires the identification of key functions (or attributes) among the assessed ideas, and the identification of the partial solutions adopted to implement each attribute. Shah et al. [5] did not used a reference definition for attributes or functions but simply stated that their identification depends on the specific design and/or research objectives.

After the attribute identification, normalized weights are assigned to attributes by the evaluators. Consequently, novelty is scored by examining the ideas generated for each key attribute. In particular, the obtained scores are multiplied by the related weight and then summed togheter. The novelty score for each attribute (S_{ij}) is calculated by (1):

$$S_{ij} = \frac{T_{ij} - C_{ij}}{T_{ij}} \times 10 \tag{1}$$

where T_{ij} represents the total number of ideas (or solutions) generated for the key attribute i and design stage j. C_{ij} represents the number of times that the current solution appears for the attribute i and design stage j. The overall novelty of each idea M is obtained by Eq. (2):

$$M_{SNM} = \sum\nolimits_{i=1}^{m} f_i \sum\nolimits_{j=1}^{n} S_{ij} p_j \tag{2}$$

where f_i is the weight of the attribute i, m is the number of attributes, n is the number of design stages, and p_j is the weight assigned to the design stage j. Therefore, the greater is the amount of similar ideas, the lower is the novelty (see Eq. (1)).

Although the original paper of Shah et al. [5] does not mention particular issues, some limitations have been recently highlighted in the literature. For example, Brown [10] listed a set of issues concerning SNM, among which the subjectivity affecting the identification of key attributes, the subjectivity in identifying weights for each attribute, and the need to define a clear separation (in terms of design representations) between conceptual and embodiment design stages.

Indeed, nowadays it is clear that there are many different definitions of function [11, 12], which can lead to non-negligible subjectivity issues when applying this metric. Furthermore, the subjectivity issue of SNM has been recently investigated in depth [13], confirming that the identification of attributes and related solutions, as well as the weight assignment, are crucial processes in terms of subjectivity.

It has been also observed that SNM is unable to provide correct assessments when heterogeneous numbers of attributes characterize the set of ideas (e.g., when some ideas implement "n" functions, while others implement "m" functions) [14, 15]. This problem has been recently investigated through the adoption of a computational generator of sets of virtual ideas, in order to obtain extensive samples [16]. It emerged that when in presence of ideas implementing heterogeneous numbers of attributes, the adoption of different strategies to manage the problem can lead to sensibly different novelty scores.

Therefore, the main SNM issues can be resumed in the following points:

- Subjectivity in the Identification of the Attributes or Functions
- Subjectivity in the identification of the attributes weights
- Impossibility to perform comprehensive assessments when in presence of sets of ideas implementing heterogeneous numbers of attributes.

3 Available SNM Variants

The acknowledged variants of the metric proposed by Shah et al. share a similar approach, which consists in calculating the mean infrequency of the solution variants considered for the idea's attributes. However, non-negligible differences can be observed. For instance, ther are some SNM variants based on the genealogy tree (GT) [17–20]. The GT has been used by Shah et al. [5] to assess Variety, and consists in a hierarchical representation based on four items: physical principles (PPs), working principles (WPs), embodiment (EMB) and details (DETs) characterizing each implemented function [21].

In this work, the SNM variants using the GT are referred as "GT-based", while those based on general key attributes are referred as "attribute-based".

3.1 Attribute-Based Variants

Sluis-Thiescheffer et al. [14] observed that when used for a large set of ideas, SNM can lead to high novelty, even if similar solutions appear quite often. Accordingly, they proposed a binary metric (novel or not novel), where an "expectedness" threshold is considered as a reference to identify less frequent (novel) ideas. For example, Sluis-Thiescheffer et al. considered the 75th percentile for the frequencies of ideas generated for each attribute (i.e. the 25% of ideas having lower concurrencies are novel).

The contribution of Sluis-Thiescheffer et al. [14] can be considered a first attempt to face the impossibility to perform correct assessments of ideas when some attributes are not implemented. However, the selection of the percentile value is too subjective, and the metric can lead to misleading results in certain cases [22].

The proposal of Fiorineschi et al. [15] is claimed to allow the calculation of novelty also in presence of ideas with heterogeneous numbers of implemented attributes. However, it has been recently demonstrated that also this metric can lead to misleading results in certain situations [22].

3.2 GT-Based Variants

Vargas-Hernandez, Okudan and Schmidt [19] observed that the capability of SNM to accurately reflect changes on the set of ideas could be improved. Accordingly, they proposed two specidic metrics focused on the working principle level of the GT.

Differently, Peeters et al. [17] focused on the importance of abstraction levels, highlighting that an idea can reach higher novelty scores when the difference can be observed at higher abstraction levels. Accordingly, they use GT items in place of the SNM attributes, with the set of weights proposed by Nelson et al. (2009), but continue to apply very similar versions of Eqs. (1) and (2).

Johnson et al. (2016) proposed a new metric variant based on GT, with the additional "Strategy" level, intended to furtherly group solutions at an intermediate abstraction level between function and physical principle levels.

4 SNM Variants vs SNM Issues

4.1 Subjectivity in Attributes (or Functions) Identification

Unfortunately, none of the available variants proposes concrete solutions in order to overcome the problems of subjectivity in the identification of the attributes or functions.

However, a recent study found that the GT-based metric of Peeters [17] can lead to more stable results in terms of mean novelty scores [13].

4.2 Subjectivity in the Identification of the Attributes Weights

This issue still remains completely unsolved for attribute-based metrics, because the related literature contributions do not explicitly face this problem.

Differently, GT-based metrics implicitly can help because of the underpinning systematic construct used to build the GT. Indeed, each GT level is linked to a predefined weight. Actually, the identification of the weights for GT levels can still be affected by subjectivity. However, the presence of literature contributions that explain how to assign them and/or provide a reasonable set of weights can be used as a valid reference [5, 23].

4.3 The Problem of Ideas Implementing Heterogeneous Numbers of Attributes

As demonstrated in [22], a comprehensive solution is missing, which is capable to perform comprehensive assessments when in presence of sets of ideas implementing heterogeneous numbers of attributes. However, drastic modification of the examined sample of ideas should be avoided. Indeed, to overcome this issue, it would be hypothetically possible to select only those key attributes shared by the ideas to be assessed, i.e. considering the intersection of the key attributes characterizing each single idea. However, by following this approach, lot of important information can be lost about the actual novelty score, i.e. those concerning the discarded attributes.

Another trivial solution could be that of discarding ideas that implement unexpected numbers of attributes. However, the presence of unrequired attributes, as well as the absence of the required ones, should affect the assessment of "quality" and not of novelty, which are two distinct parameters in the set of metrics proposed by Shah et al.

5 Suggestions for Comprehensive Assessments

The absence of concrete solutions for the three SNM issues highlighted in this work should not discourage researchers from performing novelty assessments according to SNM and/or the related variants. Indeed, by following a comprehensive procedure, the negative effects of the mentioned issues can be kept under control.

In order to provide a set of suggestions for researchers involved in the application of SNM, the following list of suggestions is proposed:

- Carefully define the experimental objectives, to identify the most appropriated typology of attributes (e.g. related to form, functions and/or behaviors)
- Ensure the presence of at least two raters for the assessment procedure
- Perform a preliminary analysis of the sample to be examined (or an excerpt of it), in order to allow the involved raters to align their views about the possible attributes to be defined.
- Let the raters to perform the identification of the attributes independently
- Compare the set of identified attributes and extract a shared list to be considered for the assessment. In particular, when in presence of ideas implementing heterogeneous numbers of attributes, it is important to motivate the presence of them and to clearly highlight which is the strategy used to manage the problem.
- Assign a shared set of weights to the identified attributes
- Keep track of the rationale followed for the attributes and weights assignment. It is important to report it when publishing the experimental results, in order to allow peers to repeat the experiment and/or to verify the followed procedure.
- Allow fully independent assessments for the involved raters
- Evaluate the inter-rater agreement between the scores obtained by the involved raters. To that purpose, well-acknowledged literature approaches are available (e.g. [24, 25].

By following this simple set of suggestions, the three SNM issues, even if not solved, can be managed and, most importantly, acknowledged by the readers when the experimental results are published.

6 Conclusions

The last findings about the most acknowledged novelty metric in engineering design have been presented. The metric is the a-posteriori one in the set proposed by Shah et al.

[5], and it has been highlighted that notwithstanding the presence of metrics variants, the three main SNM issues are still unsolved.

Among the mentioned issues, recent studies highlight that when in presence of ideas implementing heterogeneous numbers of attributes, the application of the metric should be performed with particular care.

Indeed, while the subjectivity related issues can be somehow managed with planned alignment sessions and inter-rater agreement procedures, the presence of ideas with different attributes can be critical. Due to the absence of a solution capable to comprehensively overcome this flaw, a set of practical suggestions has been provided in this paper. The main objective of the proposed suggestions is to allow both comprehensiveness and repeatability of the experiments.

References

1. Amabile, T.M.: The Social Psychology of Creativity. Springer-Verlag, New York (1983). https://doi.org/10.1080/02640828708723925.
2. Moreno, D.P., Blessing, L.T., Yang, M.C., Hernández, A.A., Wood, K.L.: Overcoming design fixation: design by analogy studies and nonintuitive findings. Artif. Intell Eng. Des. Anal. Manuf. AIEDAM. **30**, 185–199 (2016). https://doi.org/10.1017/S0890060416000068
3. Sarkar, P., Chakrabarti, A.: Studying engineering design creativity - developing a Common Definition and Associated Measures. In: Gero, J.S. (ed.) NSF International Workshop on Studying Design Creativity 2008, Aix-en-Provence (2008)
4. Daly, S.R., Mosyjowski, E.A., Seifert, C.M.: Teaching creativity in engineering courses. J. Eng. Educ. **103**, 417–449 (2014). https://doi.org/10.1002/jee.20048
5. Shah, J.J., Vargas-Hernandez, N., Smith, S.M.: Metrics for measuring ideation effectiveness. Des. Stud. **24**, 111–134 (2003). https://doi.org/10.1016/S0142-694X(02)00034-0
6. Jansson, D.G., Smith, S.M.: Design fixation. Des. Stud. **12**, 3–11 (1991). https://doi.org/10.1016/0142-694X(91)90003-F
7. Linsey, J.S., Clauss, E.F., Kurtoglu, T., Murphy, J.T., Wood, K.L., Markman, A.B.: An experimental study of group idea generation techniques: understanding the roles of idea representation and viewing methods. J. Mech. Des. **133**(3), 1–15 (2011). https://doi.org/10.1115/1.4003498
8. Vargas-Hernandez, N., Schmidt, L.C., Okudan, G.E.: Systematic ideation effectiveness study of TRIZ. J. Mech. Des. **135**, 1 (2013). https://doi.org/10.1115/1.4024976
9. Fiorineschi, L., Frillici, F.S., Rotini, F., Tomassini, M.: Exploiting TRIZ tools for enhancing systematic conceptual design activities. J. Eng. Des. **29**, 259–290 (2018). https://doi.org/10.1080/09544828.2018.1473558
10. Brown, D.C.: Problems with the calculation of novelty metrics. In: Proceedings of the 6th International Conference on Design Computing and Cognition (2014)
11. Eckert, C., Alink, T., Ruckpaul, A.: Different notions of function : results from an experiment on the analysis of an existing product. J. Eng. Des. **22**, 811–837 (2011). https://doi.org/10.1080/09544828.2011.603297
12. Vermaas, P.E., Eckert, C.: My functional description is better! Artif. Intell. Eng. Des. Anal. Manuf. **27**, 187–190 (2013). https://doi.org/10.1017/S089006041300019X
13. Fiorineschi, L., Frillici, F.S., Rotini, F.: Subjectivity of novelty metrics based on idea decomposition. Int. J. Des. Creativity Innov. **8**, 1–17 (2020). https://doi.org/10.1080/21650349.2020.1811775

14. Sluis-Thiescheffer, W., Bekker, T., Eggen, B., Vermeeren, A., De Ridder, H.: Measuring and comparing novelty for design solutions generated by young children through different design methods. Des. Stud. **43**, 48–73 (2016). https://doi.org/10.1016/j.destud.2016.01.001
15. Fiorineschi, L., Frillici, F.S., Rotini, F.: Issues related to missing attributes in a-posteriori novelty assessments. In: International Design Conference - Design 2018, Dubrovnik, Croatia, pp. 1067–1078 (2018)
16. Fiorineschi, L., Frillici, F.S., Rotini, F.: Impact of missing attributes on a posteriori novelty assessments. Res. Eng. Design **31**(2), 221–234 (2020). https://doi.org/10.1007/s00163-020-00332-x
17. Peeters, J., Verhaegen, P.A., Vandevenne, D., Duflou, J.R.: Refined metrics for measuring novelty in ideation. In: Proceedings of IDMME - Virtual Concept 2010, Bordeaux, France, pp. 1–4 (2010)
18. Johnson, T.A., Caldwell, B.W., Cheeley, A., Green, M.G.: Comparison and extension of novelty metrics for problem-solving tasks. In: Proceedings of the ASME 2016 International Design Engineering Technical Conferences and Computers and Information in Engineering Conference IDETC/CIE 2016, Charlotte (2016). https://doi.org/10.1115/DETC2016-60319
19. Vargas-Hernandez, N., Okudan, G.E., Schmidt, L.C.: Effectiveness metrics for ideation: merging genealogy trees and improving novelty metric. In: Proceedings of the ASME 2012 International Design Engineering Technical Conferences & Computers and Information in Engineering Conference IDETC/CIE 2012, Chicago (2012). https://doi.org/10.1115/DETC2012-70295
20. Vargas-Hernandez, N., Schmidt, L.C., Okudan, G.E.: Systematic ideation effectiveness study of TRIZ. In: Proceedings of the ASME 2012 International Design Engineering Technical Conferences & Computers and Information in Engineering Conference IDETC/CIE 2012, pp. 1–10 (2012)
21. Pahl, G., Beitz, W., Feldhusen, J., Grote, K.H.: Engineering Design, 3rd ed. Springer-Verlag, London (2007). https://doi.org/10.1007/978-1-84628-319-2
22. Fiorineschi, L., Rotini, F.: A-posteriori novelty metrics based on idea decomposition. Int. J. Des. Sci. Technol. **23**, 187–209 (2019)
23. Nelson, B.A., Wilson, J.O., Rosen, D., Yen, J.: Refined metrics for measuring ideation effectiveness. Des. Stud. **30**, 737–743 (2009). https://doi.org/10.1016/j.destud.2009.07.002
24. Cohen, J.: Weighted kappa: Nominal scale agreement with pro-vision for scaled disagreement or partial credit. Psychol. Bull. **70**, 213–220 (1968)
25. Hayes, A.F., Krippendorff, K.: Answering the call for a standard reliability measure for coding data. Commun. Methods Meas. **1**, 77–89 (2007)

Mechanical Characterization of Power Electronics Solder Materials

Michele Calabretta[1(✉)], Alessandro Sitta[1], Angelo Alberto Messina[2], Salvatore Massimo Oliveri[3], and Gaetano Sequenzia[3]

[1] Automotive and Discrete Group, R&D Department, STMicroelectronics, 95121 Catania, Italy
michele.calabretta@st.com
[2] Italy Public Affairs, STMicroelectronics, 95121 Catania, Italy
[3] Dipartimento di Ingegneria Elettrica, Elettronica ed Informatica (DIEEI), Università degli Studi di Catania, 95125 Catania, Italy

Abstract. One of the main bottleneck for power semiconductor durability is the solder joint reliability. A proper design of the interconnections between silicon chip and printed control board is needed to fulfill the strict industrial and automotive requirements. Considering that solders are alloys with melting temperature lower than 450 °C, high-temperature package processes and costumer profile condition enhances the visco-plastic solder degradation, affecting the joint dimensional tolerances and reliability. The mechanical characterization of solder compounds and processes results fundamental to achieve reliability and geometric dimensioning and tolerancing targets. The presented work proposes an analytical-experimental methodology to characterize the mechanical constitutive equation of a specific solder compound widely used in semiconductor industries that is SnAgCu. Visco-plastic solder behavior with respect to environment temperature is experimental detected employing different uniaxial tensile tests considering some scenarios in terms of strain rate and temperature conditions. These outcomes are numerically post-processed to find out the Anand parameters of the analyzed solder according.

Keywords: Material characterization · Visco-plasticity · Anand model · Solder compound

1 Introduction

Nowadays there is a challenging request for high reliability and performance power devices both for consumer electronics and, especially, for automotive purposes. Semiconductor industries are therefore pushed to develop and provide more and more reliable package solutions. A main topic is the project of solder interconnections [1]. The scope of these components are to connect electrically and mechanically silicon chip (die). Solder comprises several compounds in which

C. Rizzi et al. (Eds.): ADM 2021, LNME, pp. 712–720, 2022.
https://doi.org/10.1007/978-3-030-91234-5_72

melting is performed below 450 °C. Industrial soldering process consists of melting the solder compound that have been placed between the parts to joint. During the cooling down intermetallic compounds are formed at the interfaces, resulting in a permanent connection. These joined components are constrained to deform together due to any temperature input. The differences in terms of coefficients of thermal expansion lead to out-of-plane deformation and thermal-stress inside package parts. Direct consequences are deviation from planarity with the related geometric dimensioning and tolerancing (GD&T) issues and the generation of thermal fatigue on solder joint due to the repetitive temperature input during qualification tests and mission profile conditions [2–4].

The knowledge of mechanical characteristic of solder joint results indeed crucial to predict and prevent the potential problems above mentioned [5,6]. Some aspects should be considered in this framework. The first one is that grain structure and mechanical properties of solder are determined by the considered temperature profile during soldering process [7]. The second aspect is that the viscosity behavior affects solder in their working conditions due to their low melting point [8,9]. The last point is that solder exhibits plasticity [10]. Several approaches have been proposed for solder characterization, based on creep [11], elasto-plasticity [12] and unified material model to account visco-plasticity [13]. Anand model is widely diffused for this purpose [10,14–16].

The present work proposes a methodology to study the visco-plasticity of solder alloy Sn96Ag3Cu0.5 (SAC 305). Experimental uniaxial tensile tests have been carried out to correlate stress, strain, time and temperature by means of solder Anand model. The theoretical basis has been developed in Sect. 2.1, the experimental plan in Sect. 2.2 and the outcomes have been presented in Sect. 3.

2 Materials and Method

2.1 Theoretical Framework

Anand's constitutive equations are defined by nine material parameters, shown in Table 1. Moreover, this formulation includes a scalar quantity s which is linked to the plastic flow material resistance. Basically, Anand theory formulates an internal scalar variable, s, that mathematically connects the plastic flow and the averages resistance to isotropic deformation. A strength of this model are that yield condition and kinematic plastic law are not needed. Furthermore, Anand formulation allows to consider the relation among mechanical properties, temperature and strain rate. Anand model can be formulated explicating $\dot{\varepsilon}_p$ that represents the inelastic strain rate:

$$\dot{\varepsilon}_p = A \exp\left(\frac{Q}{RT}\right) \left[\sinh\left(\xi \frac{\sigma}{s}\right)\right]^{1/m} \tag{1}$$

where A, Q, R, T, ξ, σ and m are, respectively, the pre-exponential factor, the activation energy, the constant of gas, the temperature (expressed in Kelvin), the multiplier of stress, the Von Mises stress and the sensitivity to strain rate of stress. The derivative of state variable $\dot{s}$ can be linked with saturation value of deformation resistance s^* and the strain hardening function h by:

$$\dot{s} = h(\sigma, s, T)\,\dot{\varepsilon}_p = \left[h_0 \cdot \left|1 - \frac{s}{s^*}\right|^a \cdot \operatorname{sgn}\left(1 - \frac{s}{s^*}\right)\right] \dot{\varepsilon}_p \tag{2}$$

$$s^* = \hat{s}\left[\frac{\dot{\varepsilon}_p}{A} \exp\left(\frac{Q}{RT}\right)\right]^n \tag{3}$$

where a, h_0, $\hat{s}$ and n are, respectively, the sensitivity to strain rate of hardening, the hardening constant,the saturation coefficient and the sensitivity to strain rate of saturation.

Table 1. Material parameters for Anand model.

Symbol	Description	Measurement unit
A	Pre-exponential factor	-1
Q	Activation Energy	K
$\hat{s}$	Saturation Coefficient	MPa
ξ	Multiplier of Stress	Dimensionless
m	Sensitivity to Strain Rate of Stress	Dimensionless
h_0	Hardening Constant	MPa
s_0	Deformation Resistance Initial Value	MPa
n	Sensitivity to Strain Rate of Saturation	Dimensionless
a	Sensitivity to Strain Rate of Hardening	Dimensionless

Anand model can be derived considering 1D case [16]. It is possible to write σ as function of s:

$$\sigma = c \cdot s = \frac{1}{\xi} \sinh^{-1}\left[\left(\frac{\dot{\varepsilon}_p}{A} \exp\left(\frac{Q}{RT}\right)\right)^m\right] \cdot s \tag{4}$$

σ can be correlated with ε and $\dot{\varepsilon}$:

$$\sigma = \frac{1}{\xi}\sinh\ ^{1}\left[\frac{\dot{\varepsilon}_p}{A}\exp\left(\frac{Q}{RT}\right)\right]^m \cdot \left(\hat{s}\left[\frac{\dot{\varepsilon}_p}{A}\exp\left(\frac{Q}{RT}\right)\right]^n - \left\{\left[\hat{s}\left[\frac{\dot{\varepsilon}_p}{A}\exp\left(\frac{Q}{RT}\right)\right]^n - s_0\right]^{(1\quad a)} + (a-1)\left[h_0\left(\hat{s}\left[\frac{\dot{\varepsilon}_p}{A}\exp\left(\frac{Q}{RT}\right)\right]^n\right)^{\quad a}\right]\varepsilon_p\right\}^{\frac{1}{1-a}}\right) \tag{5}$$

Considering that T and $\dot{\varepsilon}_p$ have been fixed during uniaxial tensile tests and recalling that plastic strain ε_p ideally tends to ∞ when saturation is reached, the ultimate tensile strength or saturation stress (σ^*):

$$\sigma^* = \frac{\hat{s}}{\xi}\left[\frac{\varepsilon_p}{A} \exp\left(\frac{Q}{RT}\right)\right]^n \cdot \sinh^{-1}\left[\frac{\dot{\varepsilon}_p}{A} \exp\left(\frac{Q}{RT}\right)\right]^m \tag{6}$$

Replacing σ^* as written in Eq. (5) with the expression reported above, stress-strain response after yielding results:

$$\sigma = \sigma^* - \left\{ (\sigma^* - c\, s_0)^{(1-a)} + + (a-1) \left[(c\, h_0) (\sigma^*)^{-a} \right] \varepsilon_p \right\}^{1/(1-a)} \tag{7}$$

Anand parameters shall be found out by uniaxial tensile tests performed at different conditions in terms of T and $\dot{\varepsilon}_p$. The general flow for extracting model parameters has been aligned with other researchers [16]. Parameters $\hat{s}$, ξ, A, Q/R, n and m can be initially obtained fitting Eq. (6) by nonlinear least-squares regression, recording σ^* at the different test conditions. According to the experimental stress-strain curves coming from uniaxial test, it has been observed that σ is quite constant at $\varepsilon = 0.25$, therefore σ^* has been extracted from stress-strain curve at this deformation value. The fitting of the first 6 parameters have been done considering Levenberg-Marquardt algorithm and blocking A to 100,000 according to previous works [10, 16]. The remaining Anand parameters (s_0, h_0 and a) are determined best-fitting σ-ε experimental curves by optimizing parameters in Eq. (7). Levenberg-Marquardt algorithm has been used also for this estimation.

2.2 Experimental Uniaxial Tensile Test

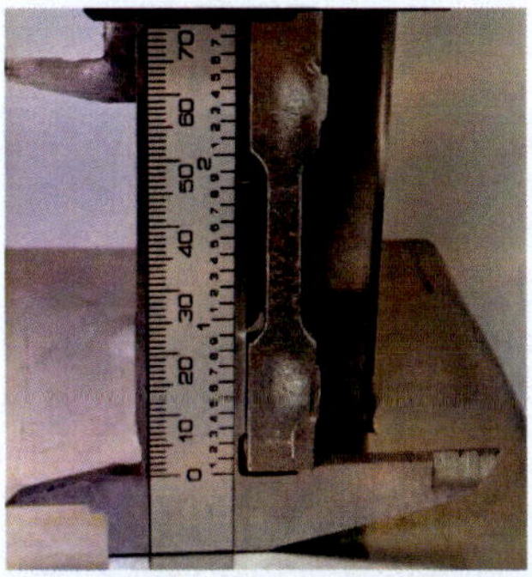

Fig. 1. Samples of SAC 305 compound for uniaxial tensile test.

Four different T (–40, 25, 75 and 125 °C) and three different $\dot{\varepsilon}_p$ (1×10^{-3}, 1×10^{-4} and $1 \times 10^{-5}\mathrm{s}^{-1}$) have been considered for uniaxial tensile tests. Three samples have been stressed for every test condition (36 tests). These samples have been formed in a glass mold by casting of solder alloys ingots of Sn96Ag3Cu0.5 provided by material supplier. The reflow profile has been selected in compliance with industrial soldering process in terms of maximum temperature and transient rate. Temperature profile inside the reflow oven has been continuously checked and regulated by dedicated thermo-couples. The dimension of the sample have been 5 mm for width and thickness 17 mm for the gauge length. The dogbone sample, shown in Fig. 1, has been selected according to the employed uniaxial tensile test experiment that usually considers such samples [10].

3 Results

As first step, a relation between elastic modulus and temperature has been found. The following linear relation has been found: $E_{SAC305} = 85267 - 128.5T$ MPa with R^2 equal to 0.996, while no significant correlation has been observed between elastic modulus and strain rate. Figure 2 shows experimental data and best-fit regression line.

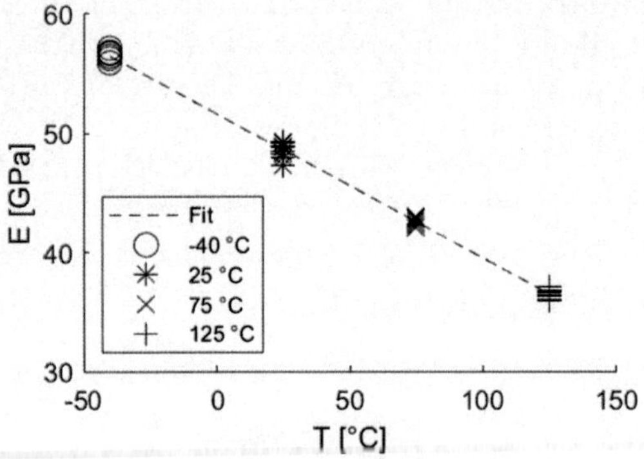

Fig. 2. Young modulus vs temperature for SAC 305.

According to the methodology explained in Sect. 2.1, Anand parameters have been extracted for the analyzed SAC solder. σ-ε curves after yielding at different T are depicted in Fig. 3. Experimental data and theoretical curves have been plotted together in order to graphically benchmark the predictability of Anand model for solder. Analyzing the difference between these curves, it could be observed that maximum error has been about 3.5% and it has been observed in case of $T = 125\,°\mathrm{C}$ and $\dot{\varepsilon_p} = 0.001\,\mathrm{s}^{-1}$. The obtained parameters for Anand model has been reported in Table 2.

The order of magnitude of such values is aligned with similar Anand characterization, e.g., [10,16]. Observing the plot reported in Fig. 3, it could be observed elasto-plastic stress decreases at T and lower $\dot{\varepsilon}_p$. For the sake of example and considering test at $0.001\,\mathrm{s}^{-1}$, σ^* ranges from 70 to 25 MPa, changing temperature from -40 to $125\,°\mathrm{C}$ respectively. The observed trend with $\dot{\varepsilon}_p$ and T is aligned with other researches on solder compounds [17]. Nevertheless, it is highlighted that similar stress-strain curves can be found by different parameters' set due to the customized non linear fitting procedure and algorithms and the degrees of freedom on some parameters' boundaries.

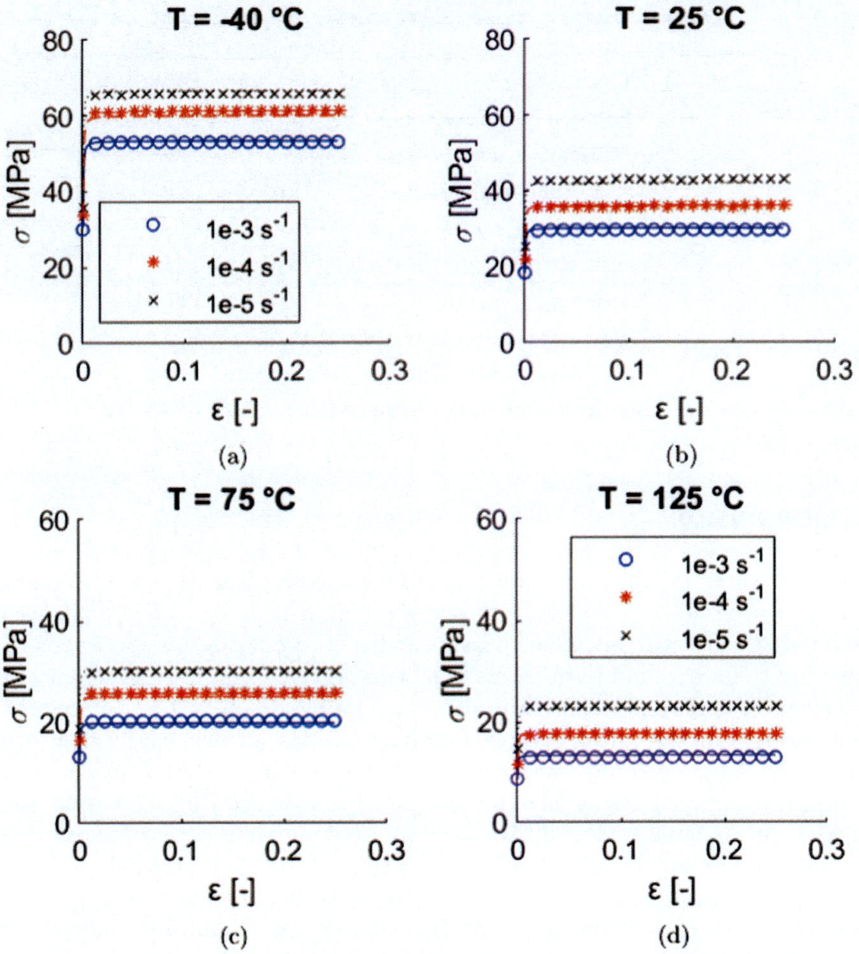

Fig. 3. SAC σ-ε curves at different strain rate for the four analyzed temperature conditions: $-40\,°C$ (a), $25\,°C$ (b), $75\,°C$ (c) and $125\,°C$ (d). They are reported both experimental (markers) and numerical (lines) data.

Table 2. Extracted Anand parameters for SAC solder.

Symbol	Description	Value
A	Pre-exponential Factor	100,000
Q	Activation Energy	9320
$\hat{s}$	Saturation Coefficient	73.29
ξ	Multiplier of Stress	5.09
m	Sensitivity to Strain Rate of Stress	0.16
h_0	Hardening Constant	326,680
s_0	Deformation Resistance Initial Value	46.41
n	Sensitivity to Strain Rate of Saturation	0.006
a	Sensitivity to Strain Rate of Hardening	1.74

4 Conclusion

In the present work, the structural constitutive law for lead-free SAC 305 solder alloy have been derived from uniaxial experimental tests in order to extract the visco-plastic constitutive model parameters of the considered specimens. The procedure to find Anand parameters has been described. Tensile tests have been performed at different strain rate and considering different test temperatures representative of the full operative conditions of the device in the field application [18]. Anand parameters have been found out by finding a mathematical relationship between stress, strain, time and temperature of the tested solder material. As depicted by Fig. 3, Anand analytical model fits quite accurately the experimental output curve of tensile tests within the 3.5% of error. This is aligned with other literature researches and it confirms the likelihood of Anand model to describe solder visco-plastic behavior. The aim of this activity is leaded by the current need of having a predictive model for one of the key bottleneck for the power device lifetime durability in application conditions. The obtained model is representing the starting point for having a predictive solder joint reliability model by using them as coefficients for a finite element model.

References

1. Goehre, J., Schneider-Ramelow, M., Geißler, U., Lang, K.D.: Interface degradation of al heavy wire bonds on power semiconductors during active power cycling measured by the shear test. In: 2010 6th International Conference on Integrated Power Electronics Systems, pp. 1–6. IEEE (2010)
2. Sitta, A., Renna, M., Messina, A.A., Sequenzia, G., D'Arrigo, G., Calabretta, M.: Thermal measurement and numerical analysis for automotive power modules. In: 2020 21st International Conference on Thermal, Mechanical and Multi-Physics Simulation and Experiments in Microelectronics and Microsystems (EuroSimE), pp. 1–3. IEEE (2020). https://doi.org/10.1109/EuroSimE48426.2020.9152204

3. Thoben, M., Sauerland, F., Mainka, K., Edenharter, S., Beaurenaut, L.: Lifetime modeling and simulation of power modules for hybrid electrical/electrical vehicles. Microelectron. Reliab. **54**(9–10), 1806–1812 (2014). https://doi.org/10.1016/j.microrel.2014.07.009
4. Calabretta, M., Sitta, A., Oliveri, S.M., Sequenzia, G.: Experimental characterization and modeling of power module warpage during assembly process. Int. J. Interact. Des. Manuf. (IJIDeM) **15**(1), 113–115 (2020). https://doi.org/10.1007/s12008-020-00735-z
5. Calabretta, M., Renna, M., Vinciguerra, V., Messina, A.A.: Power packages interconnections for high reliability automotive applications. In: ESSDERC 2019–49th European Solid-State Device Research Conference (ESSDERC), pp. 35–39. IEEE (2019). https://doi.org/10.1109/ESSDERC.2019.8901742
6. Knoerr, M., Kraft, S., Schletz, A.: Reliability assessment of sintered Nano-silver die attachment for power semiconductors. In: 2010 12th Electronics Packaging Technology Conference, pp. 56–61. IEEE (2010)
7. Basit, M.M., Motalab, M., Suhling, J.C., Lall, P.: The effects of aging on the Anand Viscoplastic constitutive model for sac305 solder. In: Fourteenth Intersociety Conference on Thermal and Thermomechanical Phenomena in Electronic Systems (ITherm), pp. 112–126. IEEE (2014)
8. Xie, X., Bi, X., Li, G.: Thermal-mechanical fatigue reliability of PbSnAg solder layer of die attachment for power electronic devices. In: 2009 International Conference on Electronic Packaging Technology & High Density Packaging, pp. 1181–1185. IEEE (2009)
9. Sitta, A., Calabretta, M., Renna, M., Cavallaro, D.: Solder joint reliability: thermomechanical analysis on power flat packages. In: Advances on Mechanics, Design Engineering and Manufacturing. LNME, pp. 709–716. Springer, Cham (2017). https://doi.org/10.1007/978-3-319-45781-9_71
10. Wilde, J., et al.: Rate dependent constitutive relations based on Anand model for 92.5 pb5sn2. 5ag solder. IEEE Trans. Adv. Packaging **23**(3), 408–414 (2000)
11. Sitta, A., et al.: An integrated approach to optimize solder joint reliability. In: 2020 21st International Conference on Thermal, Mechanical and Multi-Physics Simulation and Experiments in Microelectronics and Microsystems (EuroSimE), pp. 1–5. IEEE (2020). https://doi.org/10.1109/EuroSimE48426.2020.9152688
12. Calabretta, M., Sitta, A., Oliveri, S.M., Sequenzia, G.: Design and process optimization of a sintered joint for power electronics automotive applications. In: Rizzi, C., Andrisano, A.O., Leali, F., Gherardini, F., Pini, F., Vergnano, A. (eds.) ADM 2019. LNME, pp. 470–480. Springer, Cham (2020). https://doi.org/10.1007/978-3-030-31154-4_40
13. Durand, C., Klingler, M., Coutellier, D., Naceur, H.: Power cycling reliability of power module: a survey. IEEE Trans. Device Mater. Reliab. **16**(1), 80–97 (2016). https://doi.org/10.1109/TDMR.2016.2516044
14. Calabretta, M., Sitta, A., Oliveri, S.M., Sequenzia, G.: Power semiconductor devices and packages: solder mechanical characterization and lifetime prediction. IEEE Access **9**, 22859–22867 (2021). https://doi.org/10.1109/ACCESS.2021.3056281
15. Anand, L.: Constitutive equations for hot-working of metals. Int. J. Plast **1**(3), 213–231 (1985). https://doi.org/10.1016/0749-6419(85)90004-X
16. Motalab, M., Cai, Z., Suhling, J.C., Lall, P.: Determination of Anand constants for sac solders using stress-strain or creep data. In: 13th InterSociety Conference on Thermal and Thermomechanical Phenomena in Electronic Systems, pp. 910–922. IEEE (2012)

17. Zhang, L., Han, J.G., Guo, Y., He, C.W.: Anand model and fem analysis of SnAgCuZn lead-free solder joints in wafer level chip scale packaging devices. Microelectron. Reliab. **54**(1), 281–286 (2014). https://doi.org/10.1016/j.microrel.2013.07.100
18. ACEPACK™ 1 sixpack topology, 1200 V, 25 A, trench gate field-stop M series IGBT with soft diode and NTC. Datasheet, STMicroelectronics, Geneve, CH, November 2018. https://www.st.com/resource/en/datasheet/a1p25s12m3.pdf

3D-Printed Pure Copper: Density and Thermal Treatments Effects

Ilaria Rago[1,2], Marco Iannone[1], Francesco Marra[3], Maria Paola Bracciale[3], Laura Paglia[3], Donato Orlandi[4], Daniele Cortis[4], and Valerio Pettinacci[1](✉)

[1] National Institute for Nuclear Physics (INFN), Section of Rome, 00185 Rome, Italy
valerio.pettinacci@roma1.infn.it

[2] Department of Physics, Sapienza University of Rome, 00185 Rome, Italy

[3] Department of Chemical Engineering, Materials and the Environment, Sapienza University of Rome, 00184 Rome, Italy

[4] Gran Sasso National Laboratory, National Institute for Nuclear Physics (INFN), 67100 L'Aquila, Italy

Abstract. In the field of Additive Manufacturing (AM), interest in the production of pure copper parts is growing. The excellent thermal and electrical properties of this material make it irreplaceable in many industrial and scientific applications. Combining these aspects with the advantages provided by AM in terms of design and geometry optimization, the application potential is believed to be very broad. Notwithstanding the growing request of this kind of components, their processability via Selective Laser Melting (SLM) technology is still challenging due to the high reflectivity of copper at the emission wavelength of the conventional SLM laser sources (1064 nm) along with the thermal issues resulted from copper's high conductivity. Considering the relevance of physical and mechanical properties in the design, this work aims to verify the quality, in terms of microstructure/density, chemical and mechanical characteristics, of the pure copper parts produced via a standard SLM machine and the effect of post-processing thermal treatments on these features. The results of this study confirm the possibility to achieve a sufficiently high density and to improve mechanical properties with appropriate heat treatments. This aspect is crucial in the perspective of understanding the current readiness level of the AM technology and the process parameters to be set in order to properly obtain copper's properties for the desired thermomechanical performances.

Keywords: Additive Manufacturing · Pure Copper · Density · Thermal Treatments

1 Introduction

Additively Manufactured (AM) of pure copper combines the benefits of AM technologies with copper's unusual electrical and thermal properties, required for many applications [1, 2]. As a typical example, it is possible to mention micro heat exchangers, that with the appropriate application of design methodologies for AM, can be realized

C. Rizzi et al. (Eds.): ADM 2021, LNME, pp. 721–728, 2022.
https://doi.org/10.1007/978-3-030-91234-5_73

both with complex surface geometries (cooling fins and similar) able to guarantee a maximized convective exchange, and with a system of internal channels for the circulation of coolants. Similar applications are of great interest in the aeronautical and aerospace fields where space and masses tend to be minimized, but also in electronics where temperature is one of the main issues. However, despite the continuous demand of copper components realized via AM processes, their industrial production is still limited because of various processing challenges, especially in the case of laser-based methods. Two main issues limit the processability of pure copper via Selective Laser Melting (SLM): the very high reflectivity of the material when conventional SLM laser sources (IR 1064 nm) are employed [3, 4] and the remarkable thermal conductivity. Both factors reduce the energy available for the melting, producing highly porous surface that makes fail the electrical and thermal properties of pure copper.

To increase copper powder bed densification, different innovative approaches have been proposed [3–8]. Although the remarkable findings achieved by upgrading commercially available SLM machines, through the increment of laser power or the reduction of laser emission wavelength, the possibility to realize 3D-manufactured parts via SLM processes based on conventional 200 W lasers could be very interesting. Within this framework, all the efforts carried out to reduce the porosity of the printed pure copper components have led to density not exceeding 90% [9–14]. However, a striking improvement in densification of the copper parts via standard 200 W lasers was reached by employing fine-grained powders (10–15 μm) which promote the manufacturing of copper components showing a notably increased density (98.1%) [2, 15, 16].

In this scenario, the aim of the paper is to consolidate the knowledge and experience in the production of AM-pure copper parts, by employing a commercial SLM machine presenting a typical 200 W laser source, with the final purpose of producing highly dense copper parts characterized by desired geometry and properties. The chemical composition and microstructure of the SLM-manufactured copper materials are investigated by X-Ray Diffraction (XRD) and Scanning Electron Microscopy (SEM), respectively. Additionally, a mechanical characterization of the SLM samples is performed and the impact of different post-processing thermal treatments on the microstructural, chemical, and mechanical properties of the here-produced copper parts is evaluated.

2 Materials and Methods

In the printing experimental tests, a 99 wt% pure copper powder (particle size $5 \div 25\ \mu m$) was used. A SLM machine, produced by SISMA S.p.a. (MYSINT100 RM model), was employed to fabricate different copper-based samples (size of $25 \times 25 \times 2$ mm) under a nitrogen atmosphere, in order to avoid possible oxygen contamination during the printing process (see Fig. 1, on the left). The SLM machine is equipped with a 200 W fiber laser with a wavelength of 1 μm and a laser spot diameter of 30 μm. Usually, three samples per each printing job are produced. The samples geometry (vertical plate) is the best compromise between the needs of an efficient heat dissipation, the stability of the printing process, and an appropriate horizontal cross-section to observe the microstructure.

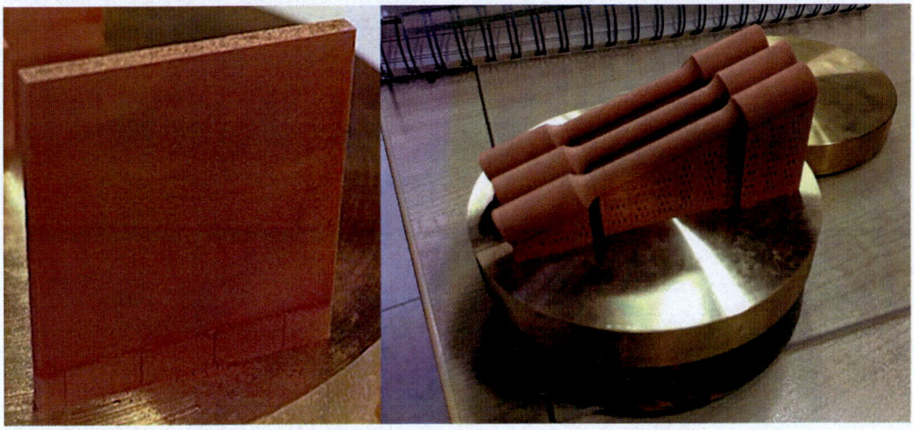

Fig. 1. Copper-based samples for SEM analysis (on the left) and for tensile test pursuant ASTM E8 (on the right).

The density of the here-fabricated samples was quantified by adopting the Archimedes method and by considering a theoretical density of copper as 8.93 g/cm^3. For this case, the optimum SLM processing parameters were identified with a laser power of 175 W, a scanning speed of 2000 mm/s, a hatch space of 40 μm and a layer thickness of 20 μm. No different parameters were used in the contour and core printing. The as-manufactured samples, after being detached from the substrate, were heated in an electric resistance furnace under argon atmosphere at 800 °C for 40 h to develop a specifically tailored post-treatment aiming to improve the density and, therefore, the properties of the copper-built parts. On the other hand, one specimen per each batch was excluded from these treatments to compare their as-built characteristics with those of their heat-treated counterpart, thus working as reference samples.

SEM imaging was conducted on the copper SLM-processed parts by collecting back-scattered electrons on a TESCAN MIRA3 operating at an accelerating voltage of 15 keV in order to explore the morphology, the porosity along with the presence of possible defects in the as-produced samples.

The crystalline phase analysis was performed by Xray diffraction (XRD, Philips Analytical PW1830) equipped with a Ni-filtered Cu Kα (1.54056 Å) radiation over 5–90° of diffraction angle (2θ) with a step size of 0.02°. The data were collected with an acceleration voltage and applied current of 40 kV and 30 mA, respectively. The crystalline phases in the resulting diffractograms were identified through the COD database (Crystallography Open Database – an open-access collection of crystal structures) [17].

The mechanical properties were evaluated through a mechanical testing machine (Instron 5584) equipped with a load cell of 150 kN and at a crosshead rate of 1 mm/min according to the ASTM E8 standard [18]. These tests were carried out on cylindrical samples with a diameter of about 6 ± 0.1 mm and a gauge length of about 30 ± 0.1 mm (see Fig. 1, on the right). These specimens were printed with dedicated jobs using identical process parameters of the square samples. The deformation was measured during the test with a contact type extensometer (Zwick-Roell MacroXtens) and each sample was tested until rupture. Porosity evaluation was carried out on digital images of cross sections (ASTM E-2109-01) at 200× magnification [19] using a Nikon Eclipse L150 optical microscope with the LUCIA Measurement image analysis software. For this purpose, the images were first digitally binarized through a grey-scale threshold, then the porosity was evaluated by the software by means the ratio between the dark and clear areas.

3 Results and Discussion

Common pure copper samples with dimensions of 25 × 25 × 2 mm were fabricated, as described in the previous section and, afterwards, morphological and chemical-structural characterizations were conducted. SEM imaging was performed on the as-grown specimens by considering both the central part and the border of sections under investigation. It has been found that the external region presents higher porosity compared to the core probably ascribed to the temperature gradient between the two zones with higher temperatures reached in the most internal regions. The resulting density of the as-built materials is approximately 86.93 ± 2.41%. Afterwards, these samples were subjected to a 40 h annealing treatment at 800 °C, leading to an evident improvement in terms of material density. Specifically, such post-processing heat treatment led to a value as high as 91.89 ± 2.07% (see Fig. 2).

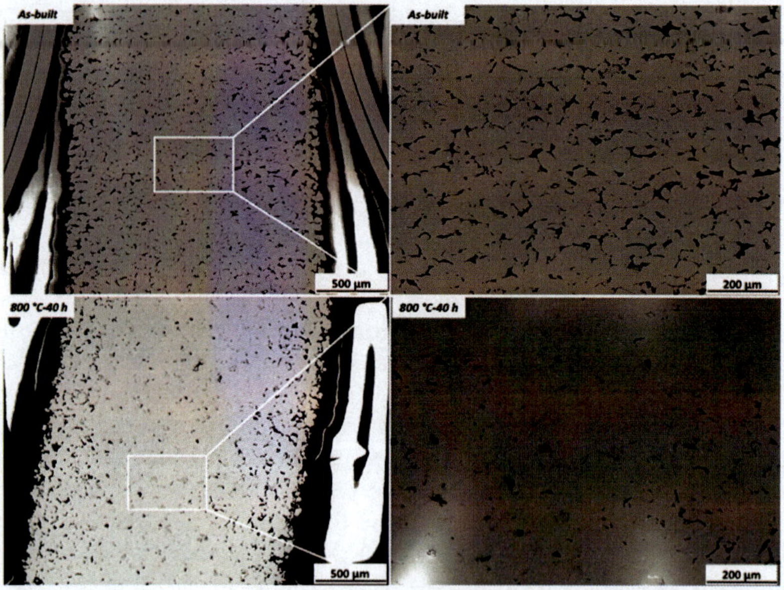

Fig. 2. SEM micrographs of 3D-built Cu-based samples before (As-built) and after the thermal treatment (800 °C for 40 h).

To investigate the structural parameters of the here-manufactured copper-based samples, before and after annealing, X-Ray Diffraction (see Fig. 3) analysis was carried out. The XRD diffraction patterns of untreated and treated samples at 800 °C for 40 h reveal the 2θ values of 43.4°, 50.5°, and 74.2° corresponding to the (111), (200) and (220) planes of FCC crystal structures of bulk copper and no impurities or changes due to thermal treatment can be detected [13].

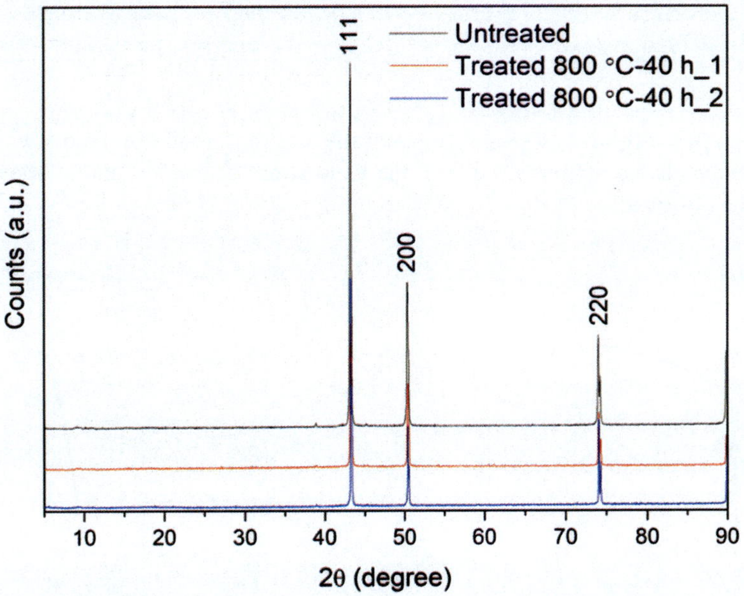

Fig. 3. XRD diffraction patterns of 3D-built Cu-based samples before (As-built) and after the thermal treatment (800 °C for 40 h).

The mechanical properties of the copper specimens, before and after the heat treatments for 40 h at 800 °C, were investigated in order to explore the effect of the post-processing. Table 1 summarizes the averaged tensile properties evaluated for the SLM-realized samples as well as for the heat-treated counterpart. Data are presented as the average plus or minus the Standard Deviation (SD) derived from at least three independent experiments. Observing the data in Table 1, it is clear that the annealing treatment, performed at 800 °C for 40 h, induced a fourfold increase of the elongation at break along with a rise in the tensile strength. The Young's modulus – calculated in the elastic part of the curve – seems to rise as well, but more likely it is the preload effect of the traction machine (given the extreme shortness of that segment). These findings are in agreement with data from Lykov and co-workers, who have reported an ultimate tensile strength, for copper parts realized via SLM, of 149 MPa for samples showing a relative density of 88.1% [12]. Further studies have disclosed improved tensile properties for SLM-realized copper parts characterized by densities surpassing 90% [9].

Table 1. Results of the mechanical tests performed on 3D-built Cu-based samples before (As-built) and after the thermal treatment (800 °C for 40 h).

Specimen ID	Young's Modulus (GPa)	Ultimate Tensile Strength (MPa)	Uniform Elongation (%)
Cu 175W 2000 mm/s T (As-built)	74.23 ± 1.97	166.95 ± 9.21	6.42 ± 2.58
Cu 175W 2000 mm/s (800 °C for 40 h)	78.36 ± 0.98	179.70 ± 0.17	30.87 ± 0.92

Furthermore, Fig. 4 shows an example of the different trend observed after the heat treatment: the elastic region is reduced, the yield stress decreases, and a larger plastic area can be observed.

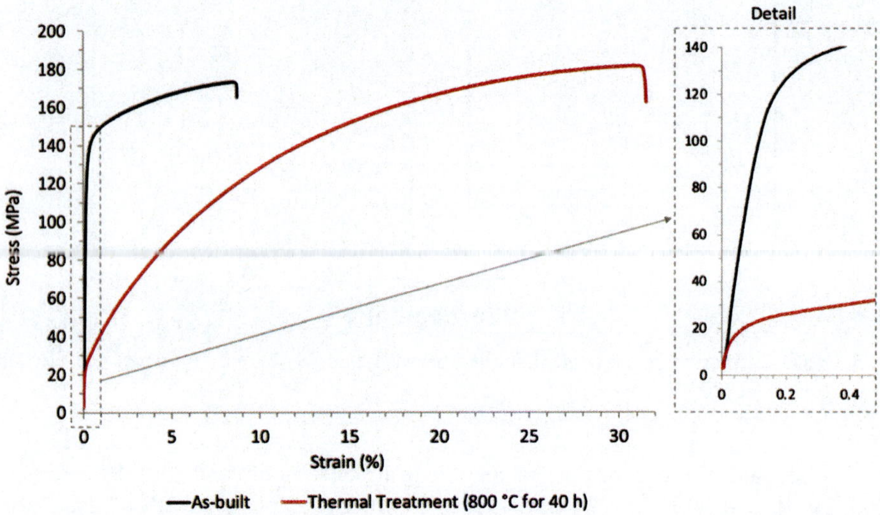

Fig. 4. Stress/Strain curves from tensile test on cylindrical samples before (As-built) and after the thermal treatment (800 °C for 40 h).

From a metallurgical point of view, it is believed that recrystallization generates new grains at the edges of existing grains as early as about 200–250 °C, and then with increasing temperature up to 800 °C – a level at which they are maintained for 40 h – these grains enlarge at the expense of the nearby original grains until they form a completely renewed crystalline structure devoid of residual stresses. Between the aforementioned edges, porosity zones are believed to be privileged areas of accretion; therefore, the overall effect is that of an increase in density. The parameters have been chosen from literature, but further studies will be necessary to optimize them in order to define the best microstructure possible.

4 Conclusion

Herein, the processability of pure copper powders via a commercial SLM machine, presenting a typical 200 W fiber laser with a wavelength of 1 μm and a laser spot diameter of 30 μm, was evaluated. The obtained results have revealed that, despite the high reflectivity of copper, highly dense pure copper parts can be produced by properly tuning both SLM process parameters and post-processing thermal treatments.

Moreover, the mechanical properties of the SLM-realized pure copper parts were investigated. Such analysis has disclosed that post-synthesis thermal process performed at 800 °C for 40 h leads to a strong increase in the elongation at break along with in the ultimate tensile stress and the general ductility of the material (plastic phase extended on the stress/strain curves). Compared to current literature, the work adds the experimental evaluation of that thermal treatments in order to affects the density and the related mechanical properties of pure materials, like copper, processed by SLM technology.

This study can be considered a starting point to additional improvements, both in terms of mechanical properties and density. For examples, future studies will be conducted to further increase the pure copper density and process repeatability, in particular working on the SLM machine printing parameters, such as the laser scanning speed which directly affects the energy density transmitted to the powder bed.

Acknowledgements. Special thanks to Angelo Corsi of the Mechanical Workshop Service of the Gran Sasso National Laboratories of INFN for the technical support provided during the phases of specimen printing, specimen removal and preparation of the printing platforms of the SLM machine.

References

1. Frazier, W.E.: Metal additive manufacturing: a review. J. Mater. Eng. Perform. **23**, 1917–1928 (2014)
2. Tran, T.Q., et al.: 3D printing of highly pure copper. Metals **9**, 756–780 (2019)
3. Jadhav, S.D., Dadbakhsh, S., Goossens, L., Kruth, J.P., Van Humbeeck, J., Vanmeensel, K.: Influence of selective laser melting process parameters on texture evolution in pure copper. J. Mater. Process. Technol. **270**, 47–58 (2019)
4. Colopi, M., Caprio, L., Demir, A.G., Previtali, B.: Selective laser melting of pure Cu with a 1 kW single mode fiber laser. Procedia CIRP **74**, 59–63 (2018)
5. Heussen, D., Meiners, W.: Green light for new 3d printing process Fraunhofer institute for laser technology (ILT) press release (2017)
6. Masuno, S., Tsukamoto, M., Tojo, K., Keita, A., Funada, Y., Yu, S.: Metal powder bed fusion additive manufacturing with 100 W blue diode laser. In: 36th International Congress on Applications Lasers Electro-Optics ICALEO, pp. 4–5 (2017)
7. Ikeshoji, T.T., Nakamura, K., Yonehara, M., Imai, K., Kyogoku, H.: Selective laser melting of pure copper. JOM **70**(3), 396–400 (2017). https://doi.org/10.1007/s11837-017-2695-x
8. Jadhav, S.D., Vleugels, H., Puyvelde, Y., Kruth, H., Vanmeensel, K.: Influence of carbon nanoparticle addition (and impurities) on selective laser melting of pure copper materials (Basel). Materials **12**, 2469–2486 (2019)

9. Sinico, M., Cogo, G., Benettoni, M., Calliari, I., Pepato, A.: Influence of powder particle size distribution on the printability of pure copper for selective laser melting. In: Proceedings of the 30th Annual International Solid Freeform Fabrication Symposium – An Additive Manufacturing Conference, pp. 657–667 (2019)
10. Kaden, L., et al.: A selective laser melting of copper using ultrashort laser pulses. Appl. Phys. **123**, 596–602 (2017)
11. Lodes, M.A., Guschlbauer, R., Koernerc, C.: Process development for the manufacturing of 99.94% pure copper via selective electron beam melting. Mater. Lett. **143**, 298–301 (2015)
12. Lykov, P.A., Safonov, E.V., Akhmedianov, A.M.: Selective laser melting of copper. Mater. Sci. Forum Trans. Tech. Publ. Ltd. **843**, 284–288 (2016)
13. Huiyan, X., Zhenhua, L., Baoren, T., Bo, R., Xin, L.: Preparation and properties of copper matrix composite reinforced with SLM Fe alloy lattice. Mater. Res. Express **6**, 056530–056536 (2019)
14. Trevisan, F., Calignano, F., Lorusso, M., Lombardi, M., Manfredi, D.: Selective laser melting of chemical pure copper powders. In: Euro PM2017, pp. 1–6 (2017)
15. Bajaj, P., Wright, J., Todd, I., Jägle, E.A.: Predictive process parameter selection for selective laser melting manufacturing: applications to high thermal conductivity alloys. Addit. Manuf. **27**, 246–258 (2019)
16. Jadhav, S.D., Vleugels, J., Kruth, J.P., Humbeeck Kim, J.V.: Vanmeensel Mechanical and electrical properties of selective laser-melted parts produced from surface-oxidized copper powder. MDPC **2**(2), e94–e102 (2020)

Check for updates

Design and Validation of a Block-on-Ring Test Bench

Diego Paderno(✉), Ileana Bodini, Gabriele Baronio, Stefano Uberti, Lorenzo Montesano, Simone Tonoli, Annalisa Pola, and Valerio Villa

Department of Mechanical and Industrial Engineering, University of Brescia, via Branze 38, 25123 Brescia, Italy
diego.paderno@unibs.it

Abstract. According to the activities carried out within the Proof of Concept laboratory, a machine for block-on-ring wear tests was designed and realized.

Consistent with the approach of the laboratory, the machine was developed with particular attention to the requirements of reliability, simplicity, as well as easy and fast production starting from existing devices integrated with ad-hoc made parts by the department equipment.

The experimental campaign was performed according to ASTM G99 and ASTM G77. The results obtained on commercial pin-on-disc device and on the realized block-on-ring machine, showed a good agreement, validating the proposed solution.

This appears particularly effective also based on an economic and technical comparison with commercial equipment complying with the same standard.

Keywords: Design and test · Design methods · Machine design

1 Introduction

In scientific research the availability of equipment for the qualitative and quantitative evaluation of the phenomena under investigation plays a key role. In many cases, the standard instruments are supported by newly designed devices, often born as an integration of the technologies already owned, in order to adapt them to the peculiarities of the analysis [1, 2]. Sometimes, proper instruments are designed, aimed at exploring specific issues or at understanding single aspects of a more complex research [3, 4].

The research activity on Proof of Concept devices carried out at our PoC laboratory since 2010 comes from the experience of the research group in the design of equipment for scientific research and from the skills developed by the previous laboratory Models for Design in the creation of prototypes and study models [5–7]. Therefore, a project, aimed at the development of devices able to measure or reproduce the studied phenomena, is accompanied by the approach to the creation of formal or functional models for understanding specific problems of the design process typical of the Design field. In many cases, this approach is characterized by the need of rapid results and low costs that allow to quickly clarify any problematic aspects and to continue with the project free from

C. Rizzi et al. (Eds.): ADM 2021, LNME, pp. 729–738, 2022.
https://doi.org/10.1007/978-3-030-91234-5_74

reserves. This means that, the expected adherence of the developed device compared to the ideal scientific instrument can be renegotiated during the design process. The partiality of the result, in fact, can reveal aspects not completely known at the beginning of the development process. In the field of Design, multiple study models, often simple, contribute to the definition of all the parameters that will lead to the development of the project and of a prototype as close as possible to the final product [8].

The activities of the PoC laboratory are of further importance when considered within a research institution, where the possibility of carrying out construction activities internally leads to many advantages. One of these, for example, is the reduction of the bureaucratic activity, in comparison with that needed to work in outsourcing, or the reduction of the costs.

The proposed approach is also focused on the realization of low-cost and quick-build equipment. This represents a significant and competitive advantage in the development of a research. However, the greatest benefit consists in the possibility of designing devices not in a perspective of "entrusting the construction activity", but rather in a constant dialogue between disciplines. This allows a prolific collaboration between specialized research sectors, devoid of design-construction skills, and research groups able to extrapolate from a scientific problem the concrete needs and design solutions. Additionally, this approach promote a constant comparison with the client group, in a perspective of mutual stimulus to identify the most satisfactory solution, which is often enriched with new potential during its evolution and construction [1, 2, 4, 9–11].

Finally, the fact that all the authors belong to the same academic field sometimes allows to develop new devices starting from the practical knowledge and the availability of equipment, working or disused, already in possession of the instrumental laboratories.

In this paper we illustrate the result of an activity in collaboration with the Metallurgy research group of our department, aimed at the construction of an experimental device for carrying out block-on-ring wear tests. Consistently with the developed method, the new instrument was realized by reusing a disused device intended for other purposes, and integrating it with commercial components, parts made ad hoc at the department laboratories, and acquisition equipment common to other instruments. In this way, it was realized in a short time and minimizing costs.

The experimental results obtained with the new device were compared with those obtained from other commercial instruments, representing the state of the art for this type of tests, showing the potential of the new machine.

2 Methods

2.1 Development of the Block-on-Ring Test Machine

The project was developed in a Department collaboration with the Metallurgy research group, which was evaluating the possibility of realizing a machine for block-on-ring tests by adapting an old self-built machine for dry-sand-rubber-wheel tests.

The machine had the standard configuration for this type of test: a vertical metal plate to support the whole system, a motor with a rotating rubber disc directly connected to the shaft and an L-shaped arm with a rotation fulcrum in the corner between the two sides. The two sides of the arm had a 2:1 ratio, with the aim of doubling the load on the short side compared to that applied on the long side.

On the short side, there was a support for fixing the sample, in order to place it in contact with the rubber disc on the horizontal axis of the rotation fulcrum, keeping the long side of the L horizontal.

The aim was to realize an instrument for the measurement of the force generated by the friction between the disc and the sample and of its evolution during the test, like commercial machines, in addition to the weight loss of the sample, and in accordance with ASTM G77-05(2010) standard [12].

An initial evaluation ruled out the possibility of making the new equipment by reusing the arm, because it would have been necessary to instrument the sample holder, which would also need to be mounted on a guide, thus leaving uncertainties related to the quality of the sliding of the components.

It was therefore decided to design and build a new arm (Fig. 1), which would allow to collect data as much as possible free from other factors.

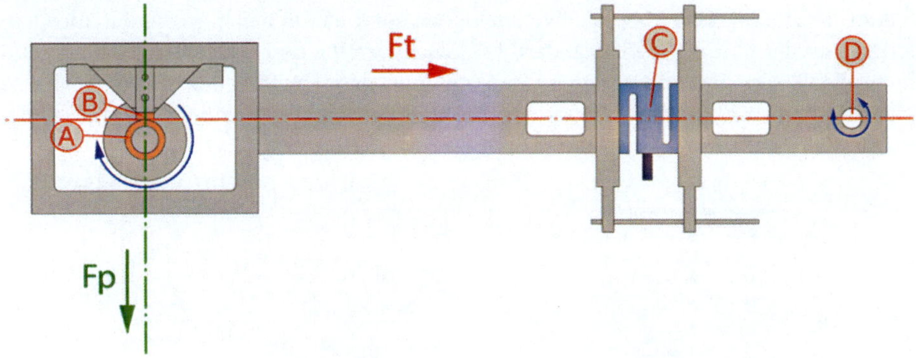

Fig. 1. Scheme of the machine with the new arm. (A) Ring; (B) Specimen; (C) Load Cell; (D) Arm Fulcrum.

Figure 1 shows the scheme of the machine: the design choice was to place on the same horizontal axis the point of contact between sample and disc, the fulcrum of rotation of the arm and the axis of the load cell provided for the measurement of forces. In this way the load cell directly measures the resulting friction force (F_t) as the product between the friction coefficient (μ) and the force perpendicular to the contact surface, which is the whole gravitational force in the designed configuration.

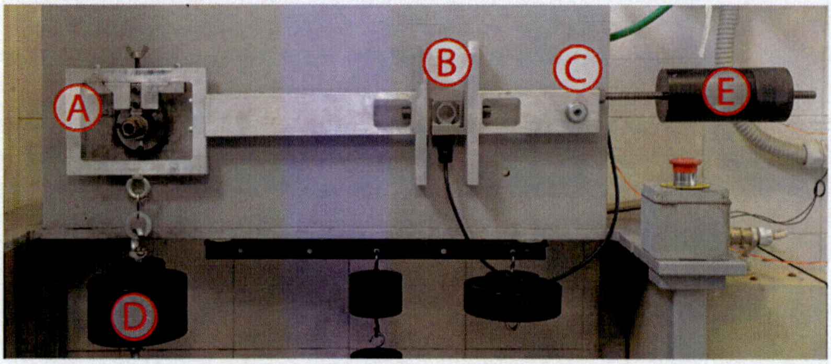

Fig. 2. Implementation of the block-on-ring machine. Below the detailed parts: (A) Testing zone; (B) Load cell; (C) Arm fulcrum; (D) Load; (E) Counterweight.

The structure of the arm was made of commercial size 6061 aluminum alloy profiles. The morphology was designed to be made as easily as possible with the equipment available within the department's laboratories.

Figure 2 shows the resulting block-on-ring machine. The arm is divided into two parts, between which the strain gauge load cell was mounted. This transducer was chosen to allow a full scale of 1 kN [13], like analogous commercial machines, and it allows to perform tests according to the standard [12]. Between the two segments of the arm and the load cell were placed two aluminum plates, connected by threaded rods to be able to make any pre-loading adjustments, needed for more accurate measurements by the load cell.

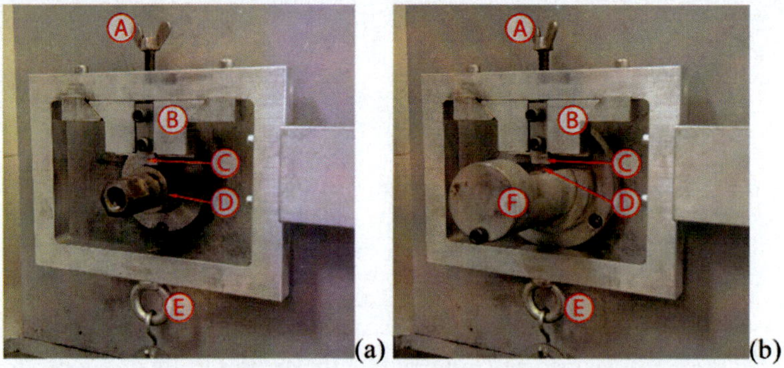

Fig. 3. Detail of the block-on-ring machine, without (a) and with (b) lubricant tank. They can be distinguished: (A) Sample-holder tightening screw; (B) Sample holder; (C) Sample; (D) Ring; (E) Weight support; (F) Lubricant tank.

At the opposite side of the fulcrum was fixed a rectangular element obtained from a solid profile, shown in details in Fig. 3. Once mounted, the vertical axis of this element coincides with the vertical axis of the drive shaft and the disc. In the upper part is mounted the sample holder, in the lower part is positioned the support to hang the weights necessary for the test.

Particular attention was paid to the assembly and disassembly procedure of the sample. In fact, the test involves monitoring the weight loss of the sample at regular intervals. To this aim, it is important to ensure the extraction and repositioning of the sample without the risk of damaging it with scratches or cuts due to the clamping system, resulting in weight loss able to compromise the accuracy of the collected data. This becomes even more relevant when, depending on the chosen test materials (sample and counterpart) and the test phase, the weight loss is small.

Once reassembled, the sample must be in the same position, to perfectly fit the worn track already present on the disc. These operations should be as fast as possible, given the frequency with which they must be carried out. To this aim, an intermediate element was added, on which the sample (of standard dimensions) is fixed. This element is inserted into a special housing with appropriate limit stops and an angular invitation for alignment with the vertical axis of the disc. Tightening takes place by means of a wing nut screw, so that the operator is not required to use tools.

To allow possible tests with cooling lubricants/liquids, a mountable tank was realized on the axis of the motor. This allows the disc to remain immersed in the lower part for the duration of the test.

A set of weights of increasing mass was also made to be able to compose various forces for performing tests in different conditions.

At the opposite side of the arm, beyond the fulcrum of rotation, an adjustable counterweight system was made to ensure the perfect balance of the arm, so that its weight would not add to that of the applied load.

Finally, it was arranged to connect the sample to a thermocouple, which allows to monitor the temperature trend of the sample during the test.

Thermocouples, load cell, and motor encoder data were all acquired through a commercial data acquisition system [14].

2.2 Experimental Tests

To validate the block-on-ring machine, friction coefficient was calculated through the results of both pin-on-disc and block-on-ring tests. Both types of test are based on sliding wear, and both induce adhesive, tribo-oxidative and in some cases abrasive wear.

Pin-on-Disc Test. The reference regulation is the ASTM G99-05 standard [15].

For the pin-on-disc wear test, two specimens are required: a pin with a known tip radius, usually a rigidly fixed sphere, and a flat disc. In the test configuration the axis of the pin test is perpendicular to the top face of the disc. The pin is pressed against the disc with a known force, generated by an applied load, usually by means of an arm or lever to which a weight is attached.

The amount of wear depends on several factors and the main parameters influencing the results of the test are: (i) the applied load; (ii) the relative linear speed between the contact surfaces; (iii) the sliding distance of the pin on the disc; (iv) the temperature of the pin and the disc.

The tests were carried out on ZnAl15Cu1 naturally aged alloy specimens, a recently developed alloy used for both casting and wrought products [16].

Nine cylindrical samples with a height of 8 mm were obtained from the extruded bar. The faces of the specimens were grinded with abrasive paper of progressively finer grit: P320, P600 and P800; obtaining a roughness Ra < 0.8 μm, as required by the standard. Tests were carried out using a 100Cr6 steel ball of 6 mm diameter as counterpart.

The test procedure included: (1) the assembly of the specimen and the pin to the test apparatus and their cleaning with ethyl alcohol, which had the purpose of eliminating any undesirable lubricating substances present on the surface. (2) The balancing of the arm, thanks to special counterweights. It is necessary that no loads due to the weight of the equipment act on the pin, but only the established test load that is added after balancing. (3) The definition of the test parameters between tests: the applied load and the sliding speed. All the tests were performed in air at room temperature. (4) The interruption every 5000 laps: the samples were observed under a digital stereographic microscope, the width of the worn track was measured to allow the calculation of the wear volume, and then the sample was reassembled. (5) The continuation of the test, until the next 5000 laps were reached. (6) The conclusion when 25000 laps were completed.

Each sample was tested with different test parameters, Table 1, which identifies each sample with the corresponding test parameters used. Three loads levels and three different sliding speeds were used, in order to evaluate if a variation of these parameters results in a variation of the friction coefficient of the alloy.

Table 1. Pin-on-disc samples identification.

Name	Load [N]	Sliding speed [cm/s]	Name	Load [N]	Sliding [cm/s]	Name	Load [N]	Sliding [cm/s]
P1.2	1	2	P3.2	3	2	P5.2	5	2
P1.4	1	4	P3.4	3	4	P5.4	5	4
P1.8	1	8	P3.8	3	8	P5.8	5	8

Block-on-Ring. The reference is the ASTM G77-05(2010) standard.

The test setup consists of a block, in the shape of a parallelepiped, which is pressed with a known force against the outer surface of a rotating ring. This test is often used to study wear under lubricated conditions. It is very flexible, any material that can be manufactured as blocks and rings can be tested, therefore, the potentially usable materials and combinations are infinite. The sizes of the ring and the specimen are fixed by the standard.

As with the pin-on-disc test, there are several parameters to consider during the test: (i) the load applied to the block; (ii) the rotation speed of the ring; (iii) the temperature of the block and the ring; (iv) the number of laps made by the ring (i.e. the sliding distance); (v) the environment within which the test takes place; (vi) the presence of lubrication. Generally, the tests are performed under the same conditions on different couplings to allow comparison; several tests are carried out varying the different parameters one at a time, for each coupling. The aim is to classify the different couplings according to their wear behavior.

For the block-on-ring test, the results obtained from the test are the weight loss, measured for the block and the ring separately, and the frictional force generated between the two surfaces, which is measured over the entire test period using a load cell.

In agreement with the pin-on-disc analysis, the tests were carried out on ZnAl15Cu1 naturally aged alloy specimens.

Nine parallelepiped-shaped specimens were prepared according to the standard. The tests were performed using a chromium steel ring with hardness of 62 HRC as counterpart. The test procedure included (1) the weighing of the specimen and of its holder separately; (2) the mounting of the specimen on the holder attached to the arm, taking care that the specimen and the ring are aligned; (3) the disassembly and weighing of the specimen; (4) the mounting of the specimen holder on the test arm; (5) the cleaning of the ring with abrasive paper of grit P500 and P800, to remove any residues of material from the previous tests; (6) the cleaning of the ring and of the sample contact surface with ethyl alcohol, to eliminate any undesired lubricating substances present on the surface; (7) definition of the applied load and the speed of rotation (temperature and atmosphere were those of the environment); (8) the interruption every 850 laps of the test: the sample holder, with the mounted sample, was unhooked from the arm, weighed and re-hooked; (9) the continuation until the next 850 laps; (10) the conclusion after 4250 laps. The number of revolutions is considerably lower than that of the previous test, because the circumference of the ring is much larger than that of the worn track in the pin-on-disc tests and the criterion used to make the pin-on-disc and the block-on-ring tests comparable is to maintain the same sliding distance.

As in the case of pin on disk tests, each sample was tested with three different loads and three sliding speeds to assess if the instrument can detect variation in the friction coefficient by changing the test parameters.

Table 2. Block-on-ring samples identification.

Name	Load [N]	Linear sliding [cm/s]	Name	Load [N]	Linear sliding [cm/s]	Name	Load [N]	Linear sliding [cm/s]
P2.2	19.6	6.3	P6.2	58.8	6.3	P10.2	98	6.3
P2.4	19.6	14.6	P6.4	58.8	14.6	P10.4	98	14.6
P2.8	19.6	31	P6.8	58.8	31	P10.8	98	31

Table 2 identifies each specimen through its test parameters: it shows the loads in N and the speeds in cm/s. For the sake simplicity, the specimens were named using two digits: the first represents the load expressed in kgf, i.e. 2, 6, 10, and the second represents the speed expressed according to the scale of the machine's potentiometer, i.e. v2, v4, v8.

For both pin-on-disk and block on ring tests, the friction coefficient was continuously recorded during the tests. The mean value was calculated at the end of the tests as the average of the recorded values in the steady state condition. Moreover, the standard deviation was obtained.

To test the repeatability of the measure, three samples were tested for each parameter set. In the corresponding results the mean value and the standard deviation are reported.

Comparison with a Commercial Bock-on-Ring Testing Machine.
The realized block-on-ring testing machine was compared with a commercial one in terms of technical data and costs.

3 Results and Discussion

3.1 Experimental Tests

The friction coefficient obtained through the two tests described in Sect. 3.1 were compared.

Data in Fig. 4 are presented to easily evaluate the effect of the sliding speed for constant loads. Each group of columns is characterized by the same applied load while each color represents a different slinging speed.

Figure 4(a) shows the steady-state friction coefficient μ from the pin-on-disc tests summarized in Table 1. Figure 4(b) shows the steady-state friction coefficient μ from the block-on-ring tests described in Sect. 3.1 and summarized in Table 2, with their standard deviation bars.

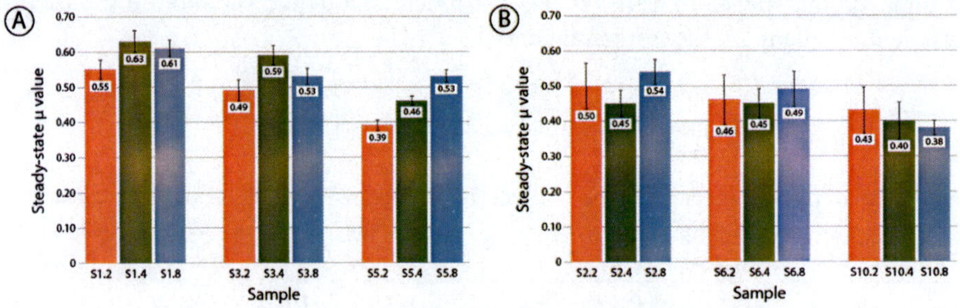

Fig. 4. Friction coefficient μ. (a) Pin-on-disc tests. (b) Block-on-ring tests

Considering the results in Fig. 4, it can be observed that each variation in the test parameters produces a variation in the measured friction coefficient in both the tests, notwithstanding the two different test machines produce different wear mechanism ad thus different material response and friction coefficient. It can be assessed that the block-on-ring machine can reveal the variation in the friction coefficient when the test parameters change as well as the commercial pin-on-disc apparatus.

3.2 Comparison with a Commercial Bock-on-Ring Testing Machine

Table 3 shows the comparison between technical data of a commercial block-on-ring testing machine [17] and of the proposed one. The same maximum load can be applied. The rolling speed depends on the available motor; both the machines have a lubricant container, which allows to perform tests in different conditions, and both the machines are provided with thermocouples. Finally, even if the built one doesn't allow Wear Depth Measurements, the comparison in terms of overall costs shows that the built block-on-ring testing machine is convenient, allowing satisfying test conditions and robust results. It is worth noting that all the cost related to the existing components (DAQ system and dismissed machine) as well as the working hours are not real costs in this work but only an indicative quotation for the complete machine realization.

Table 3. Comparison between a commercial block-on-ring testing machine and the presented built one.

	Commercial device	Built device
Max load	100 kN	100 kN
Max speed	5000 rpm	150 rpm
Lubricant container	Yes	Yes
Thermocouple	No	Yes
Wear depth measurement	Yes	No
Costs	~100 k€	~300 € new bought parts ~3000 € dismissed machines parts (considered as new) ~6000 € existing DAQ System (considered as new) ~3000 € of working hours

4 Conclusions

Consistently with the methodology used in our PoC laboratory, a device was developed for block-on-ring wear tests according to the ASTM G77-98 Standard.

To verify the correct functioning of the instrument, two campaigns of tests were carried out on the same materials, using the block-on-ring machine built and a commercial pin-on-disc test machine. The comparison between the results obtained showed the new instrument allows to reveal variations in the friction coefficient when the test parameters change as well as the commercial equipment.

From an economic point of view, the new device showed interesting competitiveness compared to commercial models, guaranteeing similar test conditions, lower cost, and consistent results. As a further development, results of the developed machine will be validated with those coming from commercial block-on-ring wear machine in other research facilities, considering the same test conditions (material, sliding speed, applied load etc.).

The Proof of Concept approach confirmed good results in terms of identifying the simplest and most effective solution, and the best compromise between available technologies, economic resources and results obtained.

Moreover, the development of devices for research in a multidisciplinary environment and between designers, manufacturers, and users allowed to identify the best requirements to be met and ensures the possibility of a constant evolution and improvement of the realized tools.

References

1. Uberti, S., Baronio, G., Cambiaghi, D.: Study & design of a special test bench for hydrostatic spindle housings. In: 11th International Design Conference, DESIGN 2010, pp. 1729–1740 (2010)
2. Bodini, I., et al.: A novel optical apparatus for the study of rolling contact wear/fatigue based on a high-speed camera and multiple-source laser illumination. Rev. Sci. Instrum. **87**(8), 083701 (2016)
3. Otten, J., et al.: Proof-of-concept of the shape memory alloy resettable dual chamber lift device for pedestrian protection with tailorable performance. J. Mech. Des. **135**, 061008–1 (2013)
4. Bontempi, E., et al.: A new method for municipal solid waste incinerator (MSWI) fly ash inertization, based on colloidal silica. J. Environ. Monit. **12**(11), 2093–2099 (2010)
5. Takai, S., Esterman, M., Midha, A.: An approach to study associations between design concepts and design outcomes. J. Mech. Des. **137**, 041101–1 (2015)
6. D. Paderno, G. Baronio: Aspectos de integracion entre modelacion virtual y maqueta en la enseñanza del diseño. In: Atti del Congresso nazionale XX INGEGRAF, Valencia (Spagna) 4–6 June (2008)
7. Mendel, J.: A taxonomy of models used in the design process. Interactions **19**(1), 81–85 (2012)
8. Fiorineschi, L., Rotini, F.: Unveiling the multiple and complex faces of fidelity. In: Proceedings of the Design Society: International Conference on Engineering Design, vol. 1, no. 1, pp. 1723–1732 (2019)
9. Zacco, A., et al.: Use of colloidal silica to obtain a new inert from municipal solid waste incinerator (MSWI) fly ash: first results about reuse. Clean Technol. Environ. Policy **14**(2), 291–297 (2012)
10. Pola, A., et al.: Effect of globular microstructure on cavitation erosion resistance of aluminium alloys. Solid State Phenom. **256**, 51–57 (2016)
11. Bonomi, G., et al.: Muon tomography as a tool to detect radioactive source shielding in scrap metal containers. In: International Journal of Modern Physics: Conference Series, vol. 27, p. 1460157 (2014)
12. ASTM International: ASTM G77-05: Standard Test Method for Ranking Resistance of Materials to Sliding Wear Using Block-on-Ring Wear Test (2010)
13. UB6 tension load cell datasheet. www.flintec.com. Accessed 25 June 2021
14. www.hbm.com. Accessed 5 June 2021
15. ASTM International: ASTM G99-05: Standard test method for wear testing with a pin-on-disk apparatus (2016).
16. Pola, A., et al.: On the ageing of a hyper-eutectic Zn-Al alloy. Metallurgia Italiana **107**(4), 37–41 (2015)
17. CETR tribometer UMT3 – triple configuration. www.cetr.com

Human-Related and User-Centered Design

The New Ergonomic Paradigm for Academic Learning Environments During Covid-19 Lockdown

Iolanda Fiorillo(✉), Rosaria Califano, and Alessandro Naddeo

University of Salerno, Via Giovanni Paolo II, 132, 84084 Fisciano, SA, Italy
ifiorillo@unisa.it

Abstract. The COVID-19 virus proliferated from its first emergence to a global phenomenon becoming a pandemic as declared by the World Health Organization on 11th March 2020. This sanitary emergency obliged people to suddenly face several changes in their way of life due to imposed countries' lockdown and social distancing. One of these rapid changes involved schools and universities and, consequently, students and professors: with the sudden lockdown of universities and colleges worldwide, they had to turn the traditional "in presence" classes into online courses. A published survey among Italian academic staff and students allowed to identify and highlight the factors that affect ergonomics of workstation (learning and teaching workplace), postural and environmental comfort and teaching/learning effectiveness.

This paper would like to give guidelines to set the right home environment and the timing of learning/teaching activities to improve global comfort, both for students and teachers, and improve learning effectiveness. Thirty-Five Master Degree students were involved in the experiment during the second-semester lessons (during Italian lockdown) and were asked to fill questionnaires and check their learning outcomes by answering multiple-choice tests. Data were statistically processed, and results were used to create guidelines to set the learning home-environment at best, where and when possible.

Keywords: Workplace design method · Comfort · eLearning effectiveness

1 Introduction

The COVID-19 (also named 2019-nCoV, or 2019 novel coronavirus) overgrew from its first emergence to a global phenomenon becoming a pandemic as declared by the World Health Organization (WHO) on 11th March 2020 [1]. This sanitary emergency obliged people to suddenly face several changes in their ways of life, such as the sudden lockdown of universities and colleges worldwide forced professors and students to turn the traditional "in presence" classes into online courses [2, 3]. The new eLearning approach implied several changes that positively or negatively impacted the teaching/learning effectiveness and wellbeing of all the people involved.

Certain problems arose, such as the complexity to arrange remote laboratories [4], to convert complex/traditional lessons in eLearning format [2, 5, 6], unavailability of

C. Rizzi et al. (Eds.): ADM 2021, LNME, pp. 741–752, 2022.
https://doi.org/10.1007/978-3-030-91234-5_75

resources, problems (accessibility and bandwidth) in Wi-Fi connections and limited knowledge, among students and teachers [6], of eLearning platforms and their limitations. However, on the one side, working or studying at home could have positive aspects. For instance, people could handle more over their work-life balance [7], saving more personal time since they don't need to reach the classroom or the workplace. On the other side, being forced to remain at home can lead to isolation feelings [7, 8]: being able to talk with other colleagues or classmates is not an aspect to neglect. The use of technology during eLearning can be a cause of distraction [9–11] since technological devices could be an easy alternative to cope with boredom during online lessons [10], especially for the "multitasking generation" [12]. Chatting with a friend, for example, can be a desirable activity that induces a positive effect that offsets the boredom of homework [13]. Furthermore, eLearning implied spending most of the time sitting, and this can affect learning [14–16]: uncomfortable body postures can drop students' interest and, consequently, can impact negatively learning, even during the most stimulating lessons [17, 18]. The eLearning approach presupposes using several devices, such as laptops, PCs, telephones, and smart TV, that increase risks for postural discomfort, especially for the neck [19]. Indeed, generally, it is suggested to modify the workstation layout or body postures to reduce postural diseases [20, 21]. Furthermore, several environmental factors such as lighting, noises, and temperature in an environment [22–24] could influence students' performance during online courses [25], learning [26], and perceived levels of (dis)comfort [27]. Previous work [28] investigated factors that affected learning/teaching effectiveness and general human comfort and wellbeing, after the sudden transition, due to COVID-19 in Italy, from the traditional "in presence" lessons into online lessons through eLearning platforms. The analysis of 700 filled questionnaires spread out among Italian universities has allowed evaluating which factors impacted, positively or negatively, the learning/teaching effectiveness, the all concerned people's wellbeing [28] and how. The difficulties to communicate with colleagues, the necessity to use several devices, the need to organize in our apartments, as much as possible, a workstation suitable for smart-working (teaching) or attending eLearning courses are an example of the several aspects that emerged from the survey [28]. As a principal result,

the most affecting factors on learning/teaching effectiveness and comfort/discomfort that need to be deeply investigated have been reported in a table [28]. The authors want to investigate the most important factors from the previous work through this work [28], such as cognitive factors, devices', postures', distraction's, and visual comfort. The investigation was carried out during an online lesson with a class of 34 students asking for comfort/discomfort and learning effectiveness through questionnaires. The paper is divided into three sections: the first one deals with the methods for experiments' design and data collection; the second one shows specific results from surveys that have been reported and summarized; the last one is about conclusion and suggestions for an eLearning workstation advice.

Thus, the research question was: which factors influence perceived comfort and learning effectiveness during eLearning? Starting from these considerations, is it possible to define guidelines to recreate, at home, a workspace suitable for online learning/teaching activities?

2 Materials and Method

2.1 Questionnaire

Questionnaires were developed with Google Forms platform and spread during the online lessons. The first questions were meant to cluster students according to gender, age, weight, height, left/right-handed, clothing, furniture (the type of desk and table) and layout of the workstation (position/number/screen dimension devices and notebook position). Then, Body Part Discomfort questions on a 5-point scale [29], global discomfort, global comfort, visual comfort (on 10-point scales) were asked to monitor students' wellbeing. These questions were submitted three times during the lesson. Moreover, since there was a break, students were asked to rate the global perceived comfort on a 10-point scale and select the actions taken during the break to monitor the break's influence on wellbeing. Finally, the last question (multiple choice) regarded the disturbing factors that emerged during the online lesson.

2.2 Participants

Thirty-Four Master Degree students (2 Females and 32 males) took part in the experiment. 31 of them were right-handed and only 3 left-handed. All participants signed the Informed Consent about the test's nature, following the University of Salerno's ethical standards. Demographic data of participants are gathered in Table 1.

Table 1. Demographic data

Criteria	Mean	Std. Deviation	Range
Age (years)	25.38	1.09	23–29
Weight (kg)	76.06	10.80	50–96
Height (m)	1.76	0.07	1.64–1.9

2.3 Experiment Protocol

Experiments were taken during an online lesson through the Microsoft Teams platform, from 3.30 pm to 6.30 pm in the second half of May with the mean temperature of 21 °C. The experiment protocol was the following:

1. Before the beginning of the lesson, students compiled the first part of the questionnaire (5 min). Information about personal data (Gender, Age, Weight, Height, Left/Right-Handed, Clothing), and the layout of the workstation (the type of chair, type of desk, audio devices, number of electronic devices, number of the screen, headphones usage) were acquired.
2. Protocol test was explained (10 min).

3. Forty minutes of online lesson.
4. Immediately after 40 min of the lesson, students were asked to evaluate the level of perceived discomfort (globally and for each body part) and perceived comfort (globally, acoustic, visual) on a numerical scale from 1 to 10.
5. Same as points 3–4.
6. Break of 10 min.
7. Immediately after the break, students had to report their actions during the break (coffee break, talk to someone, make a call, etc.) and their overall perceived comfort.
8. Same as points 3–4.
9. Furthermore, students had to report information about disturbing factors during the 3-h lesson (5 min).
10. At the end of the lesson, the professor carried out a test (20 min) to evaluate the learning effectiveness of topics covered during the lesson.

2.4 Data Analysis

The comfort/discomfort perception change over time was investigated by asking participants to fill the questionnaire three times along the experiment. The professor's final test on the lesson arguments was useful for evaluating the students' learning level. Collected data were analyzed using the software for statistical analysis SPSS16® to identify correlations between the factors detected by the questionnaire. The Spearman coefficient was used to assess correlations, as the variables had a non-normal distribution.

3 Results and Discussions

3.1 Discomfort and Comfort

Figure 1 shows results from the Body Part Discomfort questionnaires rated on a scale from 1 = "No discomfort" to 5 = "Extreme discomfort". The body parts with higher discomfort were neck, lumbar area and buttock. Moreover, the break was beneficial only for the neck, as seen with decreased perceived discomfort in the 40-min lesson after the

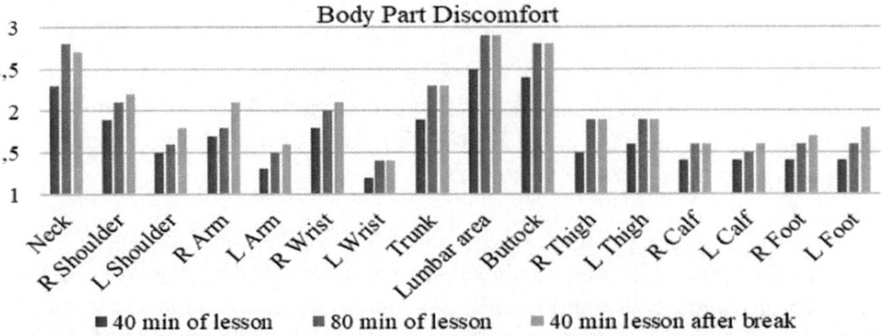

Fig. 1. Results from the Body Part Discomfort questionnaire rated 3 times during the experiment. The scale was from 1 = No discomfort to 5 = Extreme discomfort. R stands for Right, while L for Left.

break. Students that sat on office chair rated lower discomfort and higher global comfort than others.

Figure 2 shows global perceived Comfort and Discomfort results rated on a scale 1 to 10 (1 = No Comfort/No Discomfort; 10 = Extreme Comfort/Extreme Discomfort). The perceived global Comfort decreased after 80 lesson minutes, but the break helped students feel more refreshed, and the comfort increased. Then, the comfort decreased again after 40 min lesson. The global perceived discomfort had an opposite trend, and there were no data after the break. However, the break generally helped to stop the increase of perceived discomfort.

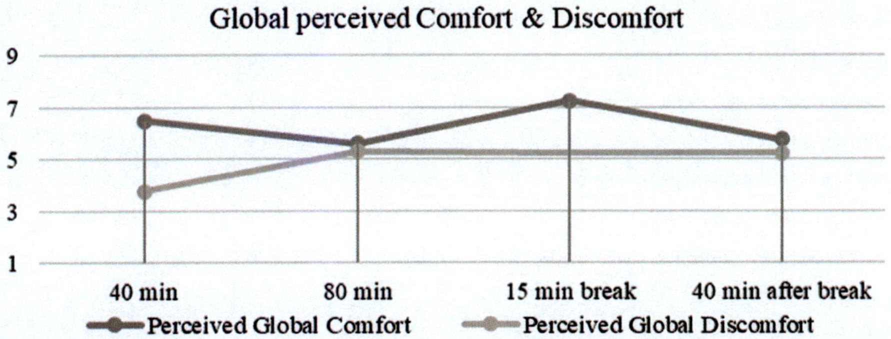

Fig. 2. Results from questionnaires about the global perceived comfort and discomfort. The scale ranked from 1 to 10.

The actions taken during the 15-min break are listed in Table 2: most students preferred to move from the workstation. Furthermore, the correlation analysis showed a lack of statistical correlations between the test scores (provided in the last part of the lesson) and the perceived comfort/discomfort.

Table 2. Actions taken during the 15-min break

Which of these actions have you made during the break?	Answers	Frequencies
Just a walk to stretch my legs	24	26,37%
Eat/Drink	23	25,27%
Satisfy physiological needs	13	14,29%
Social (WhatsApp, Instagram)	13	14,29%
Talk with someone	11	12,09%
Sit in the workplace	5	5,49%
Smoke a cigarette	1	1,10%

Sharing the work environment with other people and the presence of noises from inside and outside can affect the perceived wellbeing [28]. During the experiment, data about environmental disturbing factors were acquired to understand if they influenced the learning effectiveness.

Table 3 shows the disturbing factors reported during the entire lesson. Almost half of students reported no disturbing factors, while the others reported external noises and exogenous factors. The mean test score was 16.4 for students with disturbing factors and 16.7 for the others. The difference about the learning effectiveness was very minimal therefore it was not possible to draw valid conclusions.

Table 3. Disturbing factors came out during the entire lesson

Which of these disturbing factors came out during the lesson?	Answers	Frequencies
No disturbing factor	14	41,18%
Other people in the house	11	32,35%
External noises	11	32,35%
Lot of calls and messages received	9	26,47%
Connection problems	7	20,59%
Variation of environmental parameters	6	17,65%

3.2 Workstation

For the whole lesson duration, 73% of students sat on an office chair, 21% on a wooden chair, 3% on an armchair and 3% on a cantilever chair. Figure 3 shows the results of global perceived discomfort and comfort for office chair and wooden chair (since the others type were used only by one student): the office chair was the most comfortable.

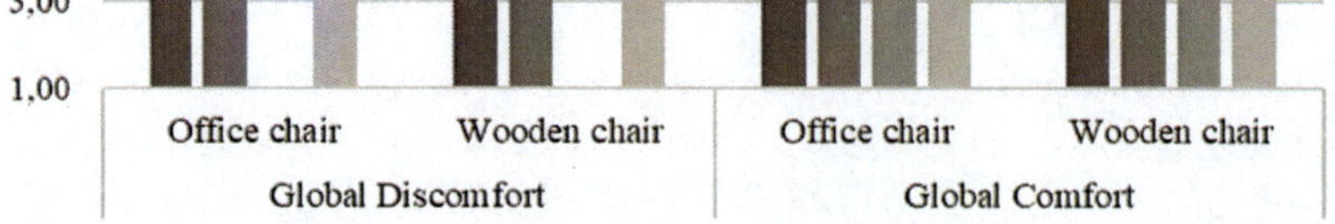

Fig. 3. Influence of perceived discomfort and comfort according to chair type. The scale was from 1 = No Comfort/No Discomfort; 10 = Extreme Comfort/Extreme Discomfort.

Moreover, 71% used a desk, 23% a wooden table, 6% a table with glass shelf. Figure 4 shows that the desk is more comfortable than the wooden table.

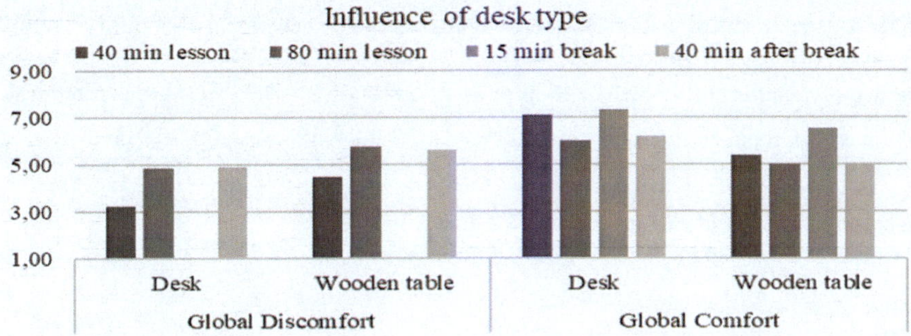

Fig. 4. Influence of perceived discomfort and comfort according to desk type. The scale was from 1 = No Comfort/No Discomfort; 10 = Extreme Comfort/Extreme Discomfort.

Regarding the workstation layout (Fig. 5), half of students positioned the laptop on the side, the notebook (or draft book) frontally and the smartphone beside; while the other half positioned the laptop frontally with the notebook in front of the laptop and the smartphone beside. Usually, those who used the second screen, place it on the left. 50% of students used laptop of 15″ screen size (18% used the 16″ screen size, 15% the 17″ screen size). Moreover, 10 students declared to use a second screen (tablet or TV) whose screen size was in the range of 15″–49″.

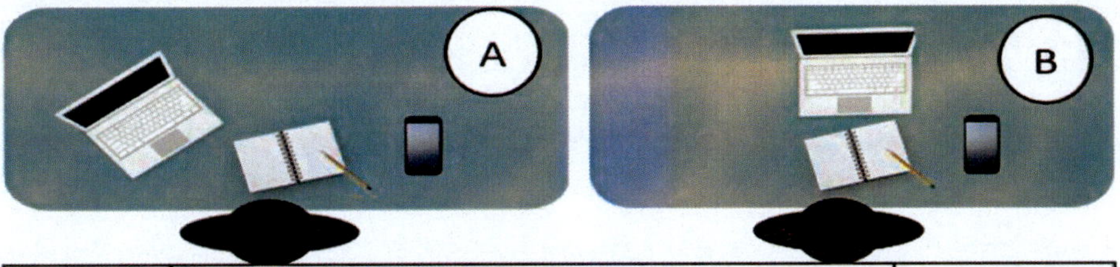

Layout	Global comfort			Test scores
	40 min	80 min	40 min after break	Average
A	7	6,16	6,64	16,4
B	6,14	5,21	5,29	16,6

Fig. 5. Workspace layouts with global comfort and test scores - A: laptop on the side, notebook in front and smartphone on the right; B: laptop and notebook in front, smartphone on the side.

3.3 Clothing

Most of the students wore sportswear (61,76%) and casual (26,47%) clothes that resulted in more comfortable (global perceived comfort: 6,3; global perceived discomfort 4,3) than other types of clothing. Table 4 shows Spearman correlation calculations results: there are negative correlations between discomfort and clothing since wearing sportswear the discomfort decreases. Instead, the positive correlations with perceived comfort mean wearing sportswear the comfort increase. There are no correlations after the break.

Table 4. Spearman correlation between clothing and perceived global discomfort/comfort

Spearman correlations between:		40-min	80-min	40-min after break
Clothing &	Global discomfort	−0,350*	−0,0405*	−0,175
	Global comfort	0,296	0,475**	0,319

* The correlation is significant at level 0.05 (2-queues)
** The correlation is significant at level 0.01 (2-queues)

3.4 Visual and Acoustic Comfort

Table 5 shows results from questionnaire regarding visual and acoustic comfort rated on a 10-point scale. While the visual comfort was constant, the acoustic comfort decreased.

Table 5. Results from questionnaire regarding visual and acoustic comfort

	40 min of lesson		**80 min of lesson**		**Break**	**40 min lesson after break**	
Comfort	*Average*	*SD*	*Average*	*SD*		*Average*	*SD*
Visual	7,4	1,89	7,4	2,03		7,2	2,16
Acoustic	7,7	1,50	7,1	1,80		6,8	1,87

In detail, 50% of students used headphones for the whole lesson duration and rated lower acoustic comfort levels than those attended without headphones. This statement was confirmed by negative correlations between headphones usage and acoustic comfort, as shown in Table 6.

Table 6. Spearman correlation between the usage of headphones and the acoustic comfort rated 3 times

		Acoustic comfort - 40 min	Acoustic comfort - 80 min	Acoustic comfort - 40 min after break
Headphones	Pearson correlation	−0,363*	−0,412*	−0,429*
	Sig. (2-tailed)	0,035	0,015	0,011
	N	34	34	34

* The correlation is significant at level 0.05 (2-queues)

Considering results showed in Fig. 6 and Fig. 7, the higher the number of devices used and size screen, the higher perceived comfort. However, while there is a correlation between the number of devices and test scores, there are no relevant influences of screen size on learning effectiveness.

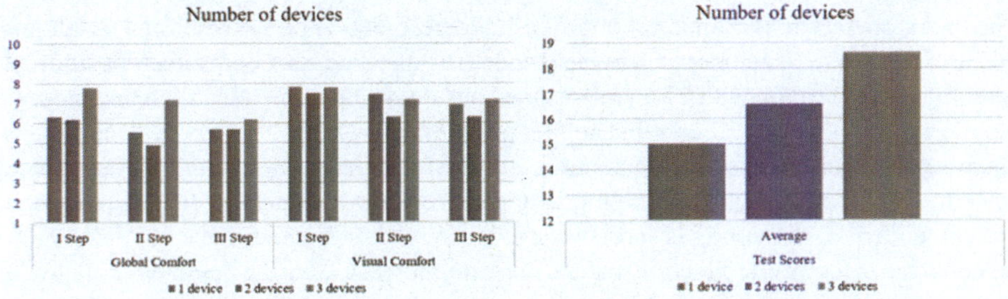

Fig. 6. Global comfort, visual comfort and average scores according to number of devices. I, II, III steps stand for 40-min lesson, 80-min lesson, and 40-min lesson after break.

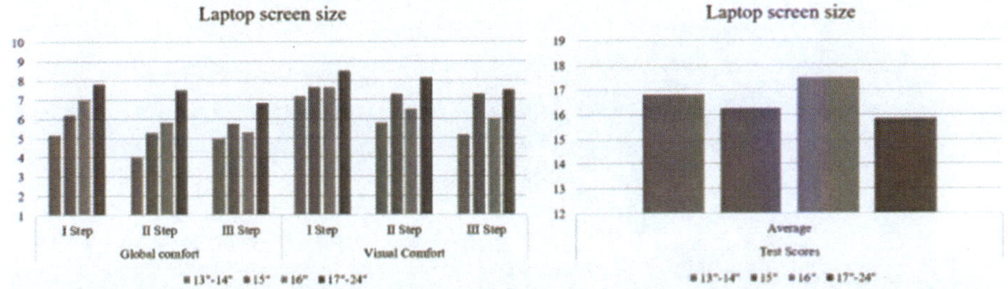

Fig. 7. Global comfort, visual comfort and average scores according to screen size expressed in inches. I, II, III steps stand for 40-min lesson, 80-min lesson, and 40-min lesson after break.

3.5 Trends

By proposing the same questionnaire at three different times, it is possible to analyze the comfort and discomfort trends over time. Data showed three different trends (Fig. 8) and the percentages of students who exhibited each or them. Different trends could mean there is no global trend equal for everyone since (dis)comfort is a subjective perception and is influenced by workstation's devices/furniture.

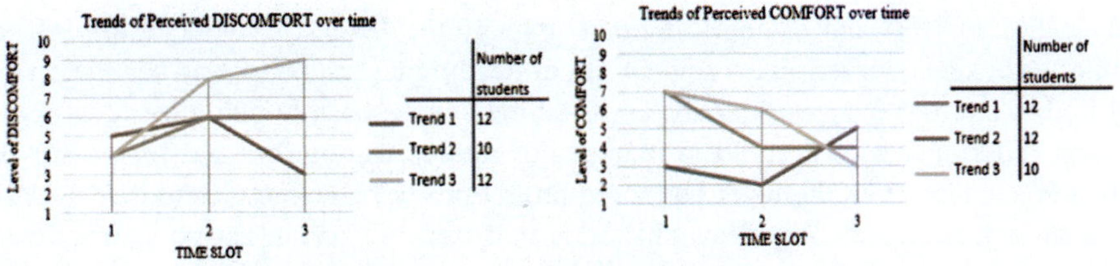

Fig. 8. Trends of perceived comfort and discomfort

3.6 Guidelines

Some guidelines (Fig. 9) could be identified to suggest the best workplace's configuration capable of reducing the perceived discomfort and increasing the work performance.

The best workstation is composed of office chair, designed to be suitable for a sedentary job, and the desk. Using more than one device (for instance a second screen positioned frontally) and a Laptop with 17 in. (positioned beside) is recommended for a better combination of high perceived comfort and learning effectiveness. The use of headphones is strongly discouraged since the acoustic comfort decreases over time, in accordance with the previous work [28]. Studying at home gives more choice on clothing, the one recommended for eLearning is sportswear. It is interesting to underline that who did not perceive the discomfort reduction and the comfort increase after the break did not have a conversation with someone. Thus, the absence of verbal communication with colleagues due to lockdown can lead to an emotional state of isolation and alienation that changes the subjective perception of comfort and discomfort.

Fig. 9. Ideal configuration for eLearning lessons

4 Conclusions

In this study, the existing condition that could influence perceived comfort/discomfort and learning effectiveness during eLearning lesson have been analyzed on a class of 34 Italian Master Degree students. During a 3-h lesson information about body part discomfort, global discomfort, global comfort, visual comfort, were asked to monitor students' wellbeing. The same questionnaire was proposed at three different times to analyze the comfort and discomfort trends over time. About the workstation info on number of devices, screen size, usage or not of headphones and desk and seat type were acquired. Acquired data were analyzed to define some possible guidelines to suggest the best workplace's configuration in order to reduce discomfort and increase work performance. However, the work has some limitations to be acknowledged.

As far as learning effectiveness, it has been analyzed only on one lesson, so no relevant results were obtained. Thus, further researches are recommended in order to understand the correlations between learning effectiveness and disturbing factors, especially on psychological stress.

The sample of 34 students belonged to the same class and was very homogeneous and relatively small. The workstation data (desk height, seat pan height, etc.) were acquired, but the data were so diversified that it was difficult to identify a statistical sample. Data about air quality, room temperature and lighting environment were acquired. However, these data were not analyzed. Students used different tools (mobile app), causing a lack of homogeneity among the data. Thus, further experiments in a controlled environment and objective measures (posture analysis) are suggested to analyze how these factors at home could influence comfort/discomfort and students learning.

References

1. World Health Organization: Coronavirus disease (COVID-19) pandemic
2. Mulla, Z., Osland-Paton, V., Rodriguez, M., Vazquez, E., Plavsic, S.: Novel coronavirus, novel faculty development programs: rapid transition to eLearning during the pandemic. J. Perinat. Med. **48**, 446–449 (2020). https://doi.org/10.1515/jpm-2020-0197
3. Girik Allo, M.: Is the online learning good in the midst of Covid-19 Pandemic? The case of EFL learners. J. Sinestesia **10**, 1–10 (2020)
4. Bright, C., Lindsay, E., Lowe, D., Murray, S., Liu, D.: Factors that impact learning outcomes in Remote Laboratories (2008)
5. IULM: Online teaching: the testimonials of students and teachers. https://www.iulm.it/it/news-ed-eventi/news/didattica-online-testimonianze-docenti-studenti
6. Reyes-Chua, E., Sibbaluca, B.G., Miranda, R.D., Palmario, G.B., Moreno, R.P., Solon, J.P.T.: The status of the implementation of the e-learning classroom in selected higher education institutions in region IV - a amidst the covid-19 crisis. J. Crit. Rev. **7**, 253–258 (2020). https://doi.org/10.31838/jcr.07.11.41
7. Robertson, M., Vink, P.: Examining new ways of office work between the Netherlands and the USA. Work **41**, 5086–5090 (2012). https://doi.org/10.3233/WOR-2012-1042-5086
8. Kurland, N., Cooper, C.: Manager control and employee isolation in telecommuting environments. J. High Technol. Manage. Res. **13**, 107–126 (2002). https://doi.org/10.1016/S1047-8310(01)00051-7
9. Xu, J.: Investigating factors that influence conventional distraction and tech-related distraction in math homework. Comput. Educ. **81**, 304–314 (2015). https://doi.org/10.1016/j.compedu.2014.10.024
10. Calderwood, C., Ackerman, P.L., Conklin, E.M.: What else do college students "do" while studying? An investigation of multitasking. Comput. Educ. **75**, 19–29 (2014). https://doi.org/10.1016/j.compedu.2014.02.004
11. Dietz, S., Henrich, C.: Texting as a distraction to learning in college students. Comput. Human Behav. **36**, 163–167 (2014). https://doi.org/10.1016/j.chb.2014.03.045
12. Wallis, C.: The multitasking generation. Time **167**, 48–55 (2006)
13. David, P., Kim, J.-H., Brickman, J.S., Ran, W., Curtis, C.M.: Mobile phone distraction while studying. New Media Soc. **17**, 1661–1679 (2015). https://doi.org/10.1177/1461444814531692
14. Fiorillo, I., Anzisi, F.J., Carbone, A., Califano, R., Naddeo, A.: A comfort evaluation tool for sitting postures: the case of Library chairs. In: WORK (2019)
15. Piro, S., Fiorillo, I., Anjani, S., Smulders, M., Naddeo, A., Vink, P.: Towards comfortable communication in future vehicles. Appl. Ergon. **78**, 210–216 (2018)
16. Naddeo, A., Fasulo, L., Vallone, M.: Ergonomic analysis and comfort driven redesign of student classroom combo desks at university of Salerno (Italy). J. Ind. Des. Eng. Graph. **10**, 33–44 (2015)
17. Castellucci, H.I., Arezes, P.M., Viviani, C.A.: Mismatch between classroom furniture and anthropometric measures in Chilean schools. Appl. Ergon. **41**, 563–568 (2010). https://doi.org/10.1016/j.apergo.2009.12.001

18. Dianat, I., Karimi, M.A., Asl Hashemi, A., Bahrampour, S.: Classroom furniture and anthropometric characteristics of Iranian high school students: proposed dimensions based on anthropometric data. Appl. Ergon. **44**, 101–108 (2013). https://doi.org/10.1016/j.apergo.2012.05.004
19. Honan, M.: Mobile work: ergonomics in a rapidly changing work environment. Work **52**, 289–301 (2015). https://doi.org/10.3233/WOR-152164
20. Hamberg-van Reenen, H.H., van der Beek, A.J., Blatter, B.M., van der Grinten, M.P., van Mechelen, W., Bongers, P.M.: Does musculoskeletal discomfort at work predict future musculoskeletal pain? Ergonomics **51**, 637–648 (2008). https://doi.org/10.1080/00140130701743433
21. Alessandro, N., Sandro, M.: Postural comfort inside a car: Development of an innovative model to evaluate the discomfort level. SAE Int. J. Passeng. Cars – Mech. Syst. **2**, 1065–1070 (2009). https://doi.org/10.4271/2009-01-1163
22. Ng, P.K., Jee, K., Yee Lim, S.: Development of ergonomics guidelines for improved sitting postures in the classroom among Malaysian University students. Am. J. Appl. Sci. **13**, 907–912 (2016). https://doi.org/10.3844/ajassp.2016.907.912
23. Michael, A., Heracleous, C.: Assessment of natural lighting performance and visual comfort of educational architecture in Southern Europe: the case of typical educational school premises in Cyprus. Energy Build. **140**, 443–457 (2017). https://doi.org/10.1016/j.enbuild.2016.12.087
24. Cajochen, C., et al.: Effect of daylight LED on visual comfort, melatonin, mood, waking performance and sleep. Light. Res. Technol. **51**, 1044–1062 (2019). https://doi.org/10.1177/1477153519828419
25. Marchand, G.C., Nardi, N.M., Reynolds, D., Pamoukov, S.: The impact of the classroom built environment on student perceptions and learning. J. Environ. Psychol. (2014). https://doi.org/10.1016/j.jenvp.2014.06.009
26. Higgins, S., Hall, E., Wall, K., Woolner, P., McCaughey, C.: The impact of school environments: a literature review. The Centre for Learning and Teaching, School of Education, Communication and Language Science, University of Newcastle (2005)
27. Galinsky, T., Swanson, N., Sauter, S., Hurrell, J.J., Schleifer, L.M.: A field study of supplementary rest breaks for data-entry operators. Ergonomics **43**, 622–638 (2000)
28. Naddeo, A., Califano, R., Fiorillo, I.: Identifying factors that influenced wellbeing and learning effectiveness during the sudden transition into eLearning due to the COVID-19 lockdown. Work, 1–23 (2021, Preprint). https://doi.org/10.3233/WOR-203358
29. Grinten, M.P.: Development of a practical method for measuring body part discomfort. Adv. Ind. Ergon. Saf. **4**, 311–318 (1992)

A System to Support the Design and Management of Customer Experience Based on a Customer-Centered Approach

Silvia Ceccacci(✉), Andrea Generosi, and Maura Mengoni

Department of Industrial Engineering and Mathematical Sciences, Università Politecnica delle Marche, Ancona, Italy
{s.ceccacci,a.generosi,m.mengoni}@univpm.it

Abstract. This article introduces at a conceptual level a system based on AI technologies, able to determine the customer profile, in order to support customer experience design and management accordingly to a customer-centered approach, by extracting information from video stream provided by the security cameras installed in a store. The system collects customer demographic and behavioral information (e.g., age, gender, time spent in determined areas of the store, time spent interacting with the salesperson, etc.) through Deep Learning algorithms, in a completely anonymous way, without saving bio-metric data. To predict the customer profile based on the collected data it exploits a Bayesian Belief Network (BBN). The paper describes the overall system architecture, details the method used to model the BBN and reports, through the description of a use case scenario, some examples of insights useful to guide the choice of possible actions to be taken to improve the customer experience strategy.

Keywords: Customer experience · Customer profiling · Machine learning · Predictive models · Video analysis

1 Introduction

Customer eXperience (CX) plays a key role in determining company success in a retail context, as it may influence customer purchase probability, customer satisfaction and customer loyalty [1]. The more a company can correctly identify what stimuli needs to be provided to ensure the best CX, depending on the nature of the touchpoint where the interaction takes place, the more it will be able to influence the customer to choose or repurchase products, to earn more money. Understanding in real time how individual elements contribute to determining CX is a topic of utmost interest in the retail sector, because it is the prerequisite for providing customized services. The adoption of new technologies to detect and analyze customer behavior opens the possibility to develop a new generation of adaptive management systems able to determine the customers' profile and providing suggestions in real time to improve the sale services. This can lead to a revolution in the retail sector.

C. Rizzi et al. (Eds.): ADM 2021, LNME, pp. 753–760, 2022.
https://doi.org/10.1007/978-3-030-91234-5_76

For example, as it is well known, shop assistants are fundamental in determining a good customer experience and persuading purchase [2]. In particular, the buyer-seller relationships are determined by several factors, including: the difference in sex, age and race between customer and salesperson, their character affinity [3], the customer's personality [4], the physical attractiveness and accent of the salesperson [5], and many, many others.

Defining the best behavior that a shopping assistant must deal to satisfy the customer in the best possible way, represents a big challenge: a salesperson must be able to assess very quickly, "who" is in front of him/her, and to decide what actions need to be taken. In this task, in the best case, he/she can rely on his/her own expertise about sales techniques, gained over years of sales or acquired during training courses. Otherwise, he/she can simply rely on his/her own sensitivity and intuition. Nowadays, the growing availability of sensors and intelligent devices connected to the Internet and powered by the pervasiveness of the Internet of Things, create an exponential growth of available data (e.g., the number of customers in a specific area, the time they spend in the store, their movement patterns, and their passing time in different areas, etc.). Valuable insights can be obtained by correlating this information with customer demographics to drive decisions for product placement, price, assortment optimization, promotion design, cross-selling, layout optimization, and staffing [6]. However, despite several applications based on big data are today available to enable the study of buying behavior of groups of customers, there are no systems able to automatically detect and analyze in real time customer behavior and provide useful insights to manage retailing services, in an adaptive way [7].

In this context the present paper proposes a new system to guide the definition and management of a proper CXS. It exploits Bayesian Belief Network (BBN) to define a customer model able to determine the customer personality according to the information extracted from video recorded images provided by the cameras installed in the store.

2 Research Background

Today, cameras became an integral part of our daily lives. We are all so accustomed to their presence, that we no longer care about them, without being fully aware of the large amount of information that can be extracted from the analysis of videos. In fact, through their video it is possible to collect a huge amount of information useful for an number of different applications. For example, through video processing it is possible to detect information about person profile (e.g., gender and age) and behavior (e.g., by analyzing eye-gaze and emotions), in an automatic way. Several methods based on image processing and machine learning techniques to determine user gender and age [8]. Methods based on regression and Convolutional Neural Network enable person eye-gaze tracking by using normal webcams or camera phones [9]. Many studies proposed methods to recognize patterns from facial expressions and to connect them to emotions. Most of them use Convolutional Neural Networks (CNN) and allows to recognize the Ekman & Keltner's primary emotions [8, 9]. By using multiple cameras, it is also possible to track people position and actions within an environment [10]. All this information allows the construction of a user model, which can be used to manage adaptive systems.

To manage user models, which are characterized by uncertain data, numerically approximate reasoning techniques can be used. Today, the Bayesian Belief Networks (BBNs) represent the most adopted reasoning technique to manage user models. They are one of the most complete and consistent knowledge tools for the acquisition, representation and use of data in conditions of uncertainty and they are considered one of the best available techniques for diagnosis and classification problems, have emerged as a powerful technique for handling uncertainty in context awareness domain, for the management of adaptive system, because they can learn from monitoring data [11]. BBNs combine principles from graph theory, probability theory, computer science, and statistics, and provide a graphical model in which each node represents a variable, and each link represents a causal influence relationship. Conditional dependencies are estimated using statistical and computational methods.

3 The Proposed System

The proposed system aims to support the definition of CXS according to a customer-centered design process defined in [7].

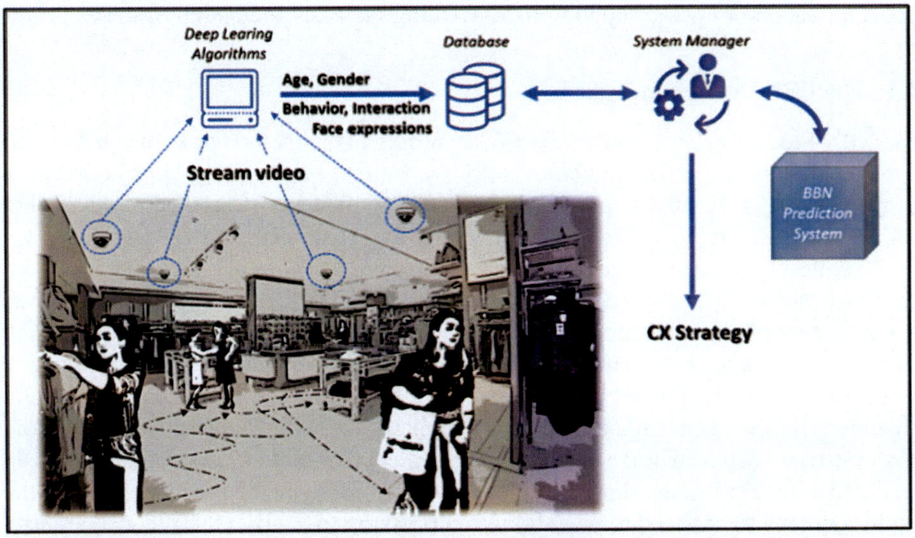

Fig. 1. The proposed system.

It is based on the following four main iterative activities: (1) Analysis of customers in the store, which implies the observation of behaviors of customer and the understanding of their emotional state during the interactions with products and staff in every touchpoint; (2) planning of strategies to improve the shopping experience, which aims at defining of all the actions to be implemented to maximize the shopping experience (e.g., varying the lighting conditions, the volume and type of music inside the store, discount offers, manage and shopping assistant behavior); (3) Implementation of the prototypal

CXS along the customer journey to test the achieved CX performance; (4) Testing and evaluation of resulted CX, by measuring the customers' satisfaction to assess the new CXS and define possible improvements.

Figure 1 describe the overall system architecture, which is characterized by three main modules, which are described in detail below: the data acquisition system, the BBN Prediction system, and the System Manager.

3.1 The Data Acquisition System

The data acquisition system takes in input the videos provided by the security cameras in the store and performs person tracking, facial recognition and facial expressions recognition through proper SW tools based on Deep Learning algorithms. It can be implemented by using the most widely used video capture technologies (webcams, IP cameras, CCTV, etc.). To detect customer age and gender, and real time emotions, it implements the algorithms respectively describes in [8] and [9], while the system proposed in [10] is used to track the customers' position and their movements inside the shop. In this way, it is possible to mark the persons with IDs and register their movement inside the shop, and easily determine the time they spend in the various areas of the shop, without saving biometric data. Collected data, are stored in a Database to enable statistics elaborations, and support proper training the BBN Prediction System.

3.2 The BBN Prediction System

The BBN Prediction system is developed, according to the method described below and implemented by using the simulation software Netica, produced by Norsys Software Corporation [12]. It aims to predict the personality of the customer based on his/her behavior in the store, to provide indication regarding customer expectation.

To collect the data necessary to model the customer behavior in a shop and to observe the different strategies adopted by expert salespeople, we carried out a video analysis in the context of a mono-brand fashion store characterized by mid-range products and target customer aged between 18 and 30. In particular, two salespeople (a woman and a man) with more than 5 years of expertise have been involved in the study: they were previously informed on methods and objectives and gave her informed consent. To avoid that customers behavior was affected by experimental condition, customers have been recruited as they exited the shop: they were informed about methods and purposes of the study, and they were asked to give us their permission to analyze their related security videos by signing the informed consent and was asked to answer the Personality Plus test adapted from [13]. A total of 60 customers (30 men and 30 women) were involved in the study. Observations were made in eight days. Through the video analysis the following data has be collected:

- Gender and Age: Two age groups have been considered: ≤24 and >24. We found a greater percentage of women that respond to Relater personality than men.
- Decision behavior: we considered the first 90 s of customer's stay in the store. By analyzing observation results, two main customers behavior have been identified: *Straight to the goal* and *Wandering*. The customer behave "straight to the goal" if

he/she move within a small number of areas in the store and he/she spent in each area more than 30 s (i.e., the customer goes straight to a particular shelf and looks at a particular product) The customer behave "wandering" if he/she moves in more than 2 areas and stays less than 30 s (i.e., the customer goes around the shop without focusing on a particular product).

- Interaction with salesperson: we considered three interactions level. An "High" level of interaction if the customer stays more than 5 min close to the salesman, "Medium" if the detected time is between 1 and 5 min and "Low" in the total time is less than 60 s. Interactions between customer and salesperson at the checkout are not considered.
- Personality: this is the dependent variable of this study. While the other data are collected through facial recognition SW, customer personality is analyzed by means of a questionnaire. Four customer personalities have been considered according to the four stereotypes defined in [13, 14], reported in Table 1.

Table 1. Considered customer personalities

Category	Personality	Expectation
Director	Demanding people who have a very clear goal, take their decision as quickly as possible and do not care about anyone's interest other than their own	They do not need/want any help from the salesperson, unless they ask for it
Thinker	Analytic people, who conduct research and analyze all the possibilities before deciding	They are unaffected by small talk provided by shop assistants within a store visit
Relator	People having a flat-type personality, who has a strong need to feel part of a group	Slow decision-makers and usually desires to include others in the decision-making process
Socializer	People who love to talk and make new friends	They want to build friendship with salesperson and are self-centered: they love to receive and give compliments

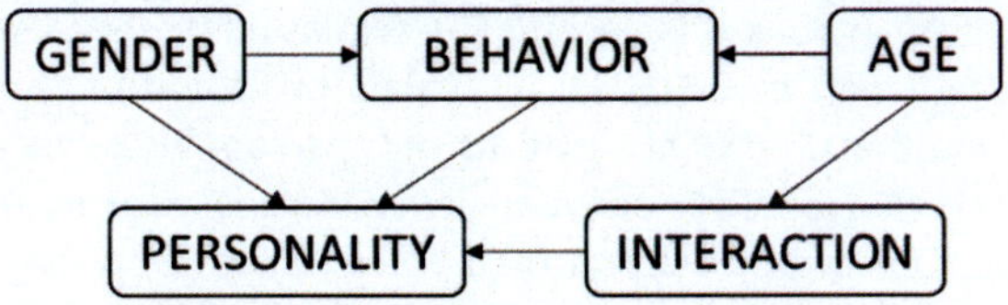

Fig. 2. The resulted BBN casual map

The BBN casual map (Fig. 2) has been defined based on experts' knowledge and prior data analysis (i.e., Correlation analysis and multicollinearity test).

3.3 The System Manager

The results generated by the BBN Prediction system are used as a knowledge based by a System Manager similar to that described in [7] to provide to the CX manager some suggestions on possible actions to improve the CXS, according to pre-defined objectives and constraints. The System Manager also provides real-time and historic information related to the customers' emotions collected within the shop. Such information can be useful to measure the effectiveness of the defined CXS and allow correcting it in a retroactive way. In particular, it provides the computation of a weighted average value (that goes from −100 to 100) based on the positivity or negativity of the registered emotions. Happiness and surprise are considered as positive emotions and all the others (i.e., fear, disgust, anger, and sadness) as negative emotions. In this way, two main thresholds are defined, which divide the continuous values in three ranges: positive, negative and neutral. When the emotional average value remains in a range from −15 to 15 the situation remains stable and the customer is mostly in a neutral mood. If the customer is in a positive mood (value between 15 and 100), or in a negative mood (value between −15 and −100) the CXS can be considered respectively effective or detrimental and then adjusted accordingly.

4 Use Case Scenario

Based on the data collected through the data acquisition system (i.e., customers' age, gender and behavior), the knowledge provided by the BBN can be used to provide indication to manage the customer service inside the shop. For example, at a conceptual level, we can focus on the salespersons' behavior observed during considering two main time-phases: approach and selling. Regarding the approach stage, during the observation study previous described, we observed two main shop assistant behaviors: Waiting (i.e., the shop assistant greets the customer and then waits until the customer asks directly for support) and Supporting (i.e., the shop assistant greets the customer and approach him/her to offer support). In the same way, in the selling phase, we identified two main behaviors: Answering (i.e., the shop assistant merely meets the customer's explicit requests. For example, it supplies the required size, looks for a particular product, etc.) and Proposing (i.e., the shop assistant tries to answer also to customers' inexplicit requests and attempts to sell other, or more, than what it is specifically requested by the customer).

Based on the prediction of the BBN, it is then possible to define a proper CXS. For example, the conditional probability of director personality is very low for younger women (Fig. 3). In this case, could be convenient that the salespersons adopt a supporting-proposing behavior. Otherwise, when a man enter in the shop, is very important to observe his behavior: if goes "straight to the goal" (Fig. 3) the condition probability that he corresponds to a director profile is high, so that the salesperson should wait until he asks for help. On the contrary, if he goes wandering (Fig. 3), the condition probability that he is a "Thinker" is high, so that in this case the salespersons should immediately approach and support him.

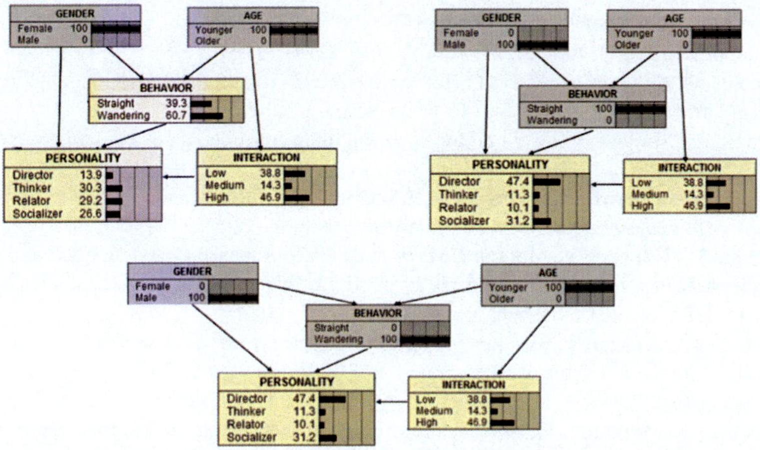

Fig. 3. BBN prediction in case of female customer aged ≤ 24 (on the top right). BBN prediction in case of male customer aged ≤ 24, who goes straight (on the top left). BBN prediction in case of male customer aged ≤ 24, who goes wandering (on the bottom).

5 Conclusion

The proposed system opens new frontiers in management of customer services in shops and opens new possibilities for scientifically defining the type of service to be provided according to the characteristics and behavior of customers. However, its commercial application today may be severely restricted by the privacy regulations in force today in most Western Countries. To overcome this issue, the proposed system collected customer data in a completely anonymous way, without saving biometric data, while video collected through the security system, which remain subjected to the respective regulations, are periodically discarded. However, the system performs customer behavior tracking and profiles them in order to improve their customer experience. Despite the customers are aware that a video surveillance system is active in the shop, they may not be fully aware that their actions are being tracked to gather information to improve the service. This may open to several ethical issues, so that the benefits to provide an optimized service must be weighed against the impact on user privacy and the fact that there is no obvious consent given for such a tracking. Future studies should be carried out to implement this system in a real use case to assess its accuracy and reliability.

References

1. Ceccacci, S., Giraldi, L., Mengoni, M. From customer experience to product design: reasons to introduce a holistic design approach. In: DS 87-4 Proceedings of the 21st International Conference on Engineering Design (ICED 17) Vol 4: Design Methods and Tools, Vancouver, Canada, 21–25 August 2017, pp. 463–472 (2017)
2. Lee, S., Dubinsky, A.: Influence of salesperson characteristics and customer emotion on retail dyadic relationships. Int. Rev. Retail Distrib. Consum. Res. **13**(1), 21–36 (2003)
3. Ainscough, T.L., Motley, C.M.: Will you help me please? The effects of race, gender and manner of dress on retail service. Mark. Lett. **11**(2), 129–136 (2000)

4. Hurley, R.F.: Customer service behavior in retail settings: a study of the effect of service provider personality. J. Acad. Mark. Sci. **26**(2), 115–127 (1998)
5. DeShields, O.W., Kara, A., Kaynak, E.: Source effects in purchase decisions: the impact of physical attractiveness and accent of salesperson. Int. J. Res. Mark. **13**(1), 89–101 (1996)
6. Gandomi, A., Murtaza, H.: Beyond the hype: big data concepts, methods, and analytics. Int. J. Inf. Manage. **35**(2), 137–144 (2015)
7. Ceccacci, S., Generosi, A., Giraldi, L., Mengoni, M.: Tool to make shopping experience responsive to customer emotions. Int. J. Autom. Technol. **12**(3), 319–326 (2018)
8. Generosi, A., Ceccacci, S., Mengoni, M.: A deep learning-based system to track and analyze customer behavior in retail store. In: 2018 IEEE 8th International Conference on Consumer Electronics-Berlin (ICCE-Berlin), pp. 1–6, September 2019
9. Generosi, A., Ceccacci, S., Faggiano, S., Giraldi, L., Mengoni, M.: A toolkit for the automatic analysis of human behavior in HCI applications in the wild. Adv. Sci. Technol. Eng. Syst. J. **5**(6), 185–192 (2020)
10. Hamdoun, O., Moutarde, F., Stanciulescu, B., Steux, B. Person re-identification in multi-camera system by signature based on interest point descriptors collected on short video sequences. In: 2008 Second ACM/IEEE International Conference on Distributed Smart Cameras, pp. 1–6. IEEE, September 2008
11. Gullà, F., Cavalieri, L., Ceccacci, S., Germani, M.: A BBN-based method to manage adaptive behavior of a smart user interface. Procedia CIRP **50**, 535–540 (2016)
12. Norsys Software Corp.: Netica Application. https://www.norsys.com/netica.html. Accessed 11 July 2019
13. Littauer, F.: Personality plus. Revell (1992)
14. Allesandra, T., O'Connor, M.: The Platinum Rule: Discover the Four Basic Business Personalities and How They Can Lead You to Success. Warner Business Books, Victoria (1998)

Evaluation of a Morphable Anthropomorphic Articulated Total Body Model

Giulia Pascoletti[1,4(✉)], Toon Huysmans[2,3], Paolo Conti[4], and Elisabetta M. Zanetti[4]

[1] Department of Mechanical and Aerospace Engineering (DIMEAS), Politecnico di Torino, 10129 Turin, Italy
giulia.pascoletti@polito.it

[2] Section on Applied Ergonomics and Design, Department of Human Centered Design, Faculty of Industrial Design Engineering, Delft University of Technology, Delft, The Netherlands

[3] imec-Vision Lab, Department of Physics, University of Antwerp, Antwerp, Belgium

[4] Department of Engineering, University of Perugia, 06125 Perugia, Italy

Abstract. In this work a new approach for the creation of Articulated Total Body (ATB) models for person-specific multi-body simulations is presented, with the main aim of overcoming limitations related to classical multi-ellipsoids ATB models, based on regression equations having only the weight and the height of the subject as input. The new methodology is based on a Statistical Shape Model (SSM), morphable according to up to 24 input parameters: the SSM was obtained from Principal Component Analysis (PCA), applied on a wide database of 3D human scans (CAESAR). The so obtained geometry can be segmented automatically to generate body segments with the respective inertial properties (mass, principal moments of inertia, and centres of mass location). The routine has been tested on a random set of 20 male subjects and the classical multi-ellipsoids models were compared to these in terms of inertial properties and 3D external geometry: the highest differences were registered at the abdomen and the thighs for what concerns the mass (60%), principal moments (75%) and centres of mass (50 mm) properties; the trunk, the shoulder and the calves are the most critical areas for the external geometry (average distance between the anthropomorphic and ellipsoids models equal to 50 mm). A contribution has been made to build person-specific multibody models. This is a valuable method since approximations made by multi-ellipsoidal models have resulted to be relevant at specific body areas, and personalised models can be a support to design and to forensic analyses.

Keywords: Principal component analysis (PCA) · 3D parametric human model · Articulated total body · Forensic biomechanics · Multibody analysis

1 Introduction

Articulated Total Bodies (ATB) are widely used to simulate human body kinematics in order to study accidents [1, 2], man-vehicle interactions [3, 4], sport movements [5, 6], and so on.

C. Rizzi et al. (Eds.): ADM 2021, LNME, pp. 761–772, 2022.
https://doi.org/10.1007/978-3-030-91234-5_77

These articulated models are usually made of solid bodies, representing human body's segments. Since ATBs are implemented in multibody codes, kinematic (linear and angular displacements, velocities and accelerations) and dynamic (forces and moments) information can be obtained with a limited computational effort [2].

This article is focused on ATB models to be employed for subject-specific analyses (see, for example, forensic biomechanics [7, 8] or subject's specific design [9, 10]). This family of ATBs are usually made of articulated ellipsoids [11, 12]: the geometry and the inertial properties of these ellipsoids are calculated through regression equations, based on surveys information, whose only inputs are the subject's gender, height and weight. As an alternative, anthropomorphic models can be built, where the geometry is obtained through laser scanning [13], CT scans [14], double X-ray exposure [15], photogrammetry [16]; as a next step, the anthropomorphic 3D geometry is segmented in order to identify the above-mentioned body elements and obtain an ATB model. In the latter models, inertial properties (that are the mass, the moments of inertia, the centres of mass) are obtained from segment's geometry, assuming a given density value [12, 17, 18]. In the work here presented, an approach that combines 3D parametric human geometry creation and segmentation is proposed. An anthropomorphic geometry has been created from a limited number of subject measures, morphing a 'standard' geometry; this approach could be implemented thanks to the existence of a wide 3D scans database (CAESAR database [19]), which has been analysed through a Statistical Shape Model (SSM) based on Principal Component Analysis (PCA) [20, 21], in order to establish the key measures to be given as an input to reproduce subject-specific geometry with good details [22, 23].

The performances of the two human modelling approaches are here compared, in order to assess if differences between these two models are significant, and if creating an anthropomorphic model does worth the effort. The anthropomorphic model was here used as reference because it has been proved to be able to replicate the actual human shape with an excellent accuracy (average error equal to 6.59 ± 1.28 mm [22]), while ellipsoidal models are affected by limitations coming from their rough approximation of the actual human shape.

2 Materials and Methods

A set of 20 male adult subjects has been considered, sorting the respective 3D scan from a subset of CAESAR database, referring to Dutch population (518 male human subjects) [19]; these subjects have been chosen randomly trying to cover a wide range of heights (1.62 to 2.05 m), weights (53.6 to 150 kg) and body mass indices BMI, calculated as the weight divided by the square of the height (18 to 40 kg/m^2).

In the following the Statistical Shape Model (SSM) is presented and the new anthropomorphic model is described in detail as well as the respective multi-ellipsoids model.

2.1 Sample Subjects and Statistical Shape Model

The Statistical Shape Model (SSM) implemented for the generation of anthropomorphic 3D geometries is described by the following equation:

$$\boldsymbol{S_r} = \overline{\boldsymbol{S}} + \boldsymbol{\Phi y} \tag{1}$$

where $\boldsymbol{S_r}$ is a new random shape vector (the desired anthropomorphic geometry in this case), $\overline{\boldsymbol{S}}$ is the mean shape based on the overall training data set ($\boldsymbol{S_k}$) and $\boldsymbol{\Phi y}$ is the chosen variability model; shape vectors $\boldsymbol{S_r}$, $\overline{\boldsymbol{S}}$ and $\boldsymbol{S_k}$ are vectors containing 3D human geometry points (nodes). The mean shape is computed based on human 3D scan data: the Generalised Procrustes Analysis was applied in order to align and scale all N shapes in the training set and compute the mean shape as:

$$\overline{\boldsymbol{S}} = \frac{1}{N}\sum\nolimits_{k=1}^{N} \boldsymbol{S}_k \tag{2}$$

The variability model of the SSM was based on the Principal Component Analysis (PCA), which provides the main modes of variation of the analyzed shapes as eigenvectors: these eigenvectors represent the so-called Principal Components (PCs) and are the column vector of the matrix $\boldsymbol{\Phi}$ in Eq. 1. Linear combination of the first PCs (that account for most of the shape variance) are added to the mean shape through the $\boldsymbol{y}$ vector in order to deform the mean shape $\overline{\boldsymbol{S}}$ and generate a new shape vector $\boldsymbol{S_r}$.

Usually, parameters in the $\boldsymbol{y}$ vector do not have a physical meaning; for this reason, these parameters are linked to physical features representing body characteristics (as height, weight, waist circumference, shoulder breadth, etc.). The relation between these features and the shape model parameters was sought by multiple multivariate linear regression [22]: having defined $\mathbf{f} = [\mathrm{f}_1, \mathrm{f}_2, \ldots, \mathrm{f}_F, 1]^{\mathrm{T}}$ as the feature vector of a specific body shape (that is the vector containing the F selected features), the PC weights $\boldsymbol{y}$ of a body shape with feature $\boldsymbol{f}$ can be generated as:

$$\boldsymbol{y} = \boldsymbol{Mf} \tag{3}$$

where $\boldsymbol{M}$ is the mapping matrix defining the relationship between $\boldsymbol{y}$ and $\boldsymbol{f}$.

For the subjects' set here considered, 10 measurements have been chosen as input features (Fig. 1); these parameters have been identified, among 24 available, as the measurements playing the highest influence on the actual human shape, according to statistical regressions [22]. The following figure shows the comparison between the average shape (blue shape) and two different body shapes belonging to the subjects' set considered.

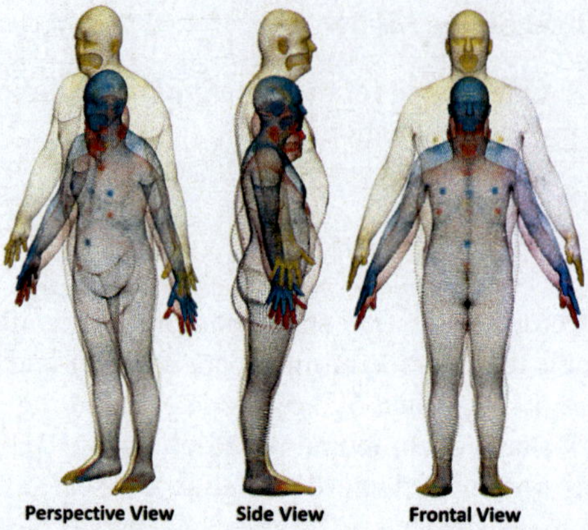

Fig. 1. Average shape (blue shape) and two different body shapes (yellow and red shapes) belonging to the set of 20 subjects

2.2 Anthropomorphic Model

The external geometry of the anthropomorphic model was obtained with a dedicated software [21] created at the Delft University of Technology and based on the Statistical Shape model described in Sect. 2.1; this code is able to generate the outer subject geometry for a maximum set of 24 anthropometric measurements. The standard model (average shape, Fig. 2b) and the subject-specific model (predicted shape, Fig. 2a and c) share iso topological meshes and include anatomical landmarks (blue and yellow nodes in Fig. 2) also derived from the CAESAR data.

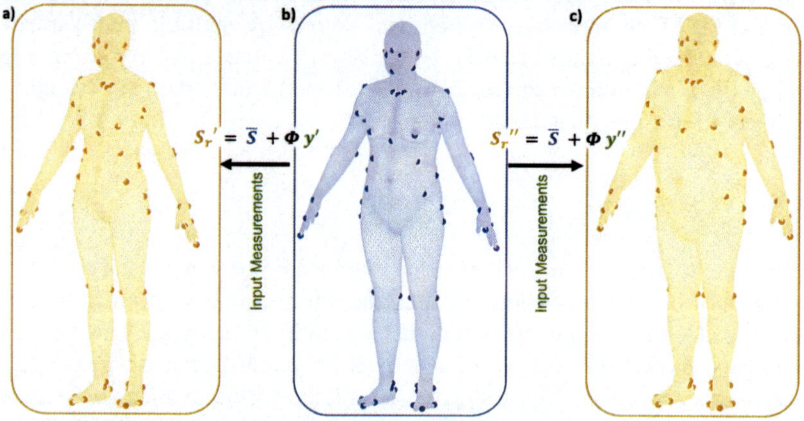

Fig. 2. a) Predicted shape for a female subject; b) Average shape; c) Predicted shape for a male subject. Blue and yellow nodes represent anatomical landmarks

The anthropomorphic 3D models were segmented into 15 components (Fig. 3b, c): this operation was made on the standard geometry, using anatomical segmenting planes (grey planes in Fig. 3a) built with reference to anatomical landmarks (Fig. 2b).

Mesh nodes belonging to each body segment were defined by a point cloud that represents the set of nodes which lies between two or more planes of segmentation.

These nodes were identified using the sign property of the dot product between two vectors ***a*** and ***b***: ***a*** is the normal vector of the segmenting plane and ***b*** is defined as the vector of the distance between a point lying on this plane and nodes of the average shape geometry (Fig. 3a).

All sets of nodes were so obtained on the average shape (Fig. 3b), along with the respective identification numbers; in this way the segmentation can be easily transferred to any predicted shape generated by the code (Fig. 3c) thanks to mesh iso-topology.

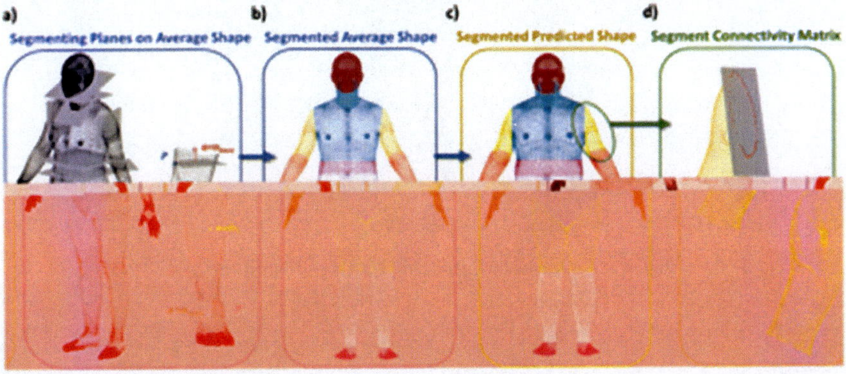

Fig. 3. a) Segmenting planes on the average shape and segments' nodes analytical definition; b) Segmented average shape; c) Segmented predicted shape; d) Segment connectivity matrix

A closed mesh representing a manifold geometry is required in order to create the multibody ATB model. For this reason, additional nodes and triangles have been added to the boundary regions of the segment, with the final aim of closing the open segment's portions (Fig. 3d).

In the next step, the inertial properties of each segment are calculated from the respective geometry, assuming an average constant density ρ, equal to:

$$\rho = \frac{M_{TOT}}{V_{TOT}} \tag{4}$$

where M_{TOT} is the subject weight in kg and V_{TOT} is the respective volume in m^3 calculated on the anthropomorphic geometry.

2.3 Multi-ellipsoid Model

The multi-ellipsoids model is made of 15 ellipsoids, articulated at anatomical joints position (Fig. 4). Given the input weight and height, the subject gender (all males), and its age range (all adults), 32 anthropometric measurements can be derived, according to regression equations obtained by USAF (United States Air Force) on 2420 male subjects [24]. The geometry (semiaxes length) and the inertial properties (centres of mass, weight, principal moments of inertia) of the 15 ellipsoids are then derived from these anthropometric measurements through regression equations.

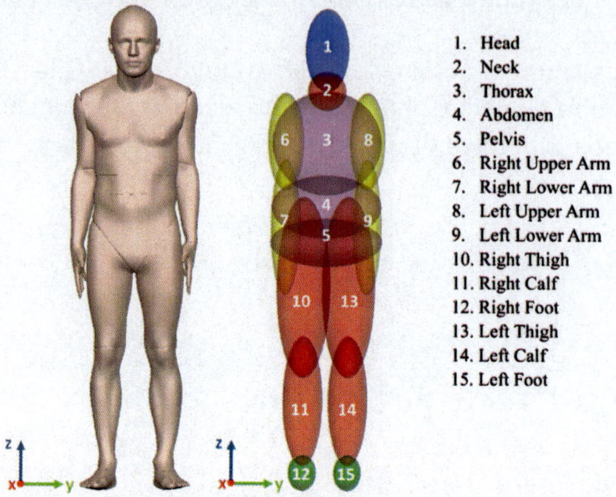

Fig. 4. Anthropomorphic and multi-ellipsoid model

2.4 Comparison Between the Ellipsoids Model and the Anthropomorphic Model

The comparison between the ellipsoids model and the anthropomorphic one, has included the analysis of the following properties:

- Inertial properties: mass, principal moments of inertia, centre of mass location
- 3D geometry

As stated in the Introduction section, the anthropomorphic model was used as reference for these analyses, due to its high and known accuracy in replicating the actual human shape (average error equal to 6.59 ± 1.28 mm [22]).

The reference system reported in Fig. 4 has been used: x is the posterior-anterior axis; y is the medio-lateral axis, directed from the plane of symmetry to the left of the subject; z is the vertical axis, directed upwards.

Inertial Properties. The following entities have been considered:

- Principal moments of inertia ($I_{xx_{i,j,k}}, I_{yy_{i,j,k}}, I_{zz_{i,j,k}}$)
- Masses ($M_{i,j,k}$)
- Centers of mass position ($CM_{x_{i,j,k}}, CM_{y_{i,j,k}}, CM_{z_{i,j,k}}$)

where i is the segment ($i = 1$ to 15), j is the subject being considered ($j = 1$ to 20), k is the model being assessed (*'E'* for ellipsoids model, *'A'* for anthropomorphic model).

Differences between values calculated for the ellipsoids model or for the anthropomorphic one have been analysed both in absolute (e_{abs}) and relative (e_{rel}) terms for all inertial properties; for example, considering I_{xx}:

$$e_{I_{xx},i,j,abs} = I_{xx,i,j,E} - I_{xx,i,j,A} \tag{5}$$

$$e_{I_{xx},i,j,rel} = \frac{\mu I_{xx}, i}{\mu A, I_{xx}, i} \cdot 100 \tag{6}$$

For both e_{abs} and e_{rel}, mean and standard deviation parameters have been considered.

The position of the centers of mass has been analyzed using the abdomen center of mass as a reference; the reason is that the whole body center of mass can be roughly approximated with the abdomen center of mass [25, 26]. Given the subject j, the 3D geometry of the ellipsoids' model was translated in order to make abdominal centers of mass coincide with the abdominal centers of mass of the anthropomorphic model. In the next step, the vertical distance (along z axis) between $CM_{z_{i,j,A}}$ and $CM_{z_{i,j,E}}$ was assessed for all segments. This analysis results in incremental errors moving from the abdomen center of mass to the farthest centers of mass; a second analysis was therefore performed where the reference center of mass changed from segment to segment according to Table 1. The vertical distance has been considered in place of the Euclidean distance because the second one has resulted to be heavily influenced by small difference in posture between the two models.

Paired-samples Wilcoxon tests have been used to assess, segment by segment, if differences between the anthropomorphic and ellipsoidal models were significant.

Table 1. Reference segments for relative centres of mass distance

Segment	Reference segment	Segment	Reference segment
1. Head	2. Neck	8. Left Upper Arm	3. Thorax
2. Neck	3. Thorax	9. Left Forearm	8. Left Upper Arm
3. Thorax	4. Abdomen	10. Right Thigh	5. Pelvis
4. Abdomen	–	11. Right Calf	10. Right Thigh
5. Pelvis	4. Abdomen	12. Right Foot	11. Right Calf
6. Right Upper Arm	3. Thorax	13. Left Thigh	5. Pelvis
7. Right Forearm	6. Right Upper Arm	14. Left Calf	13. Left Thigh

3D Geometry. 3D geometries have been compared considering, node by node, the Euclidean 'nearest neighbor distance'. The average distance d_{ave} has been calculated among all nodes as well as the 90% percentile distance d_{90}.

An F-test was performed in order to test if the variance of distances d_{ave} was significantly greater than the variance of distances measured between the anthropomorphic model and the true body shape, as reported in [22].

3 Results and Discussion

3.1 Inertial Properties

Figure 5 shows results of the absolute error analysis for the inertial parameters.

Ideally the absolute error for all the segments would be close to zero (black line in Fig. 5) if the ellipsoids model was able to provide a perfect estimate.

For most of the segments the approximation related to the ellipsoids model is very good, indeed the absolute error is low and approximately constant for all the subjects.

This approximation is not equally good for all segments: for example, data referring to the abdomen (blue circles) and the pelvis (red circles) are systematically below zero, pointing out that for these segments the relative inertial properties are systematically underestimated; on the contrary, data referring to the thorax (light blue circles) and the thighs (yellow and pink circles) are systematically overestimated. According to this evidence, ellipsoids model provides a very good approximation of inertial properties calculated from the actual geometry.

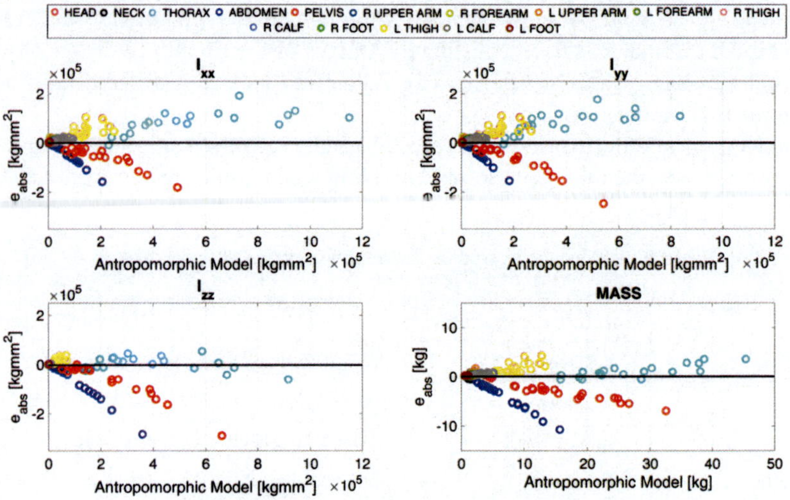

Fig. 5. Absolute error analysis results for inertial parameters: absolute errors are plotted versus the respective anthropomorphic model's inertial data

These findings are confirmed also by the analysis of relative errors (Eq. 6): in particular the abdomen remains critical also with reference to this parameter, whose average value is larger than 65% for the moments of inertia and about 60% for the mass estimation.

As a general rule, the ellipsoids model provides a better approximation of the mass compared to inertial moments.

Distances of centres of mass between the anthropomorphic model and the ellipsoids' one has evidenced major errors (>25 mm) for the upper arms ($\mu \approx 35$ mm), thighs ($\mu \approx 53$ mm), calves ($\mu \approx 54$ mm) and feet ($\mu \approx 100$ mm). However, when the nearest segments are taken as a reference, the head ($\mu \approx 35$ mm), the neck ($\mu \approx 34$ mm) and the forearms (μ ≈ 40 mm) and thighs (μ ≈ 34 mm) have resulted to be more critical: these are the key segments which determine errors observed in Fig. 6b.

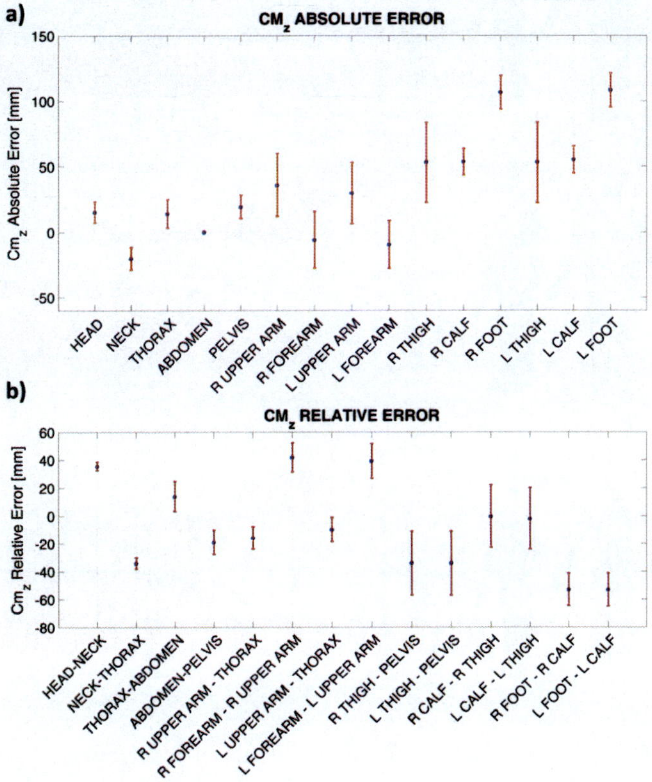

Fig. 6. Analysis of the centres of mass error distribution along the z coordinate with respect the abdomen (a) and with respect the nearest segment (b)

According to Wilcoxon paired tests, inertial parameters are significantly different between the anthropomorphic and the ellipsoids model ($p < 0.01$), exception made for the head mass, thorax moment of inertia I_{ZZ} and the right forearm CM_Z.

3.2 3D Geometry

The distribution of the Euclidean distance between respective nodes has resulted to be consistent among all 20 subjects, with the most critical areas located at shoulders, arms and calves (Fig. 7). A 90th percentile value equal to about 50 mm has been assessed and an average distance of 23 ± 7.39 mm has been computed; errors can get greater for obese subjects, especially in the trunk where they reach values larger than 130 mm.

Figure 7 a, b and c show the geometrical error distribution for a normal-weight, an over-weight and an under-weight subject respectively.

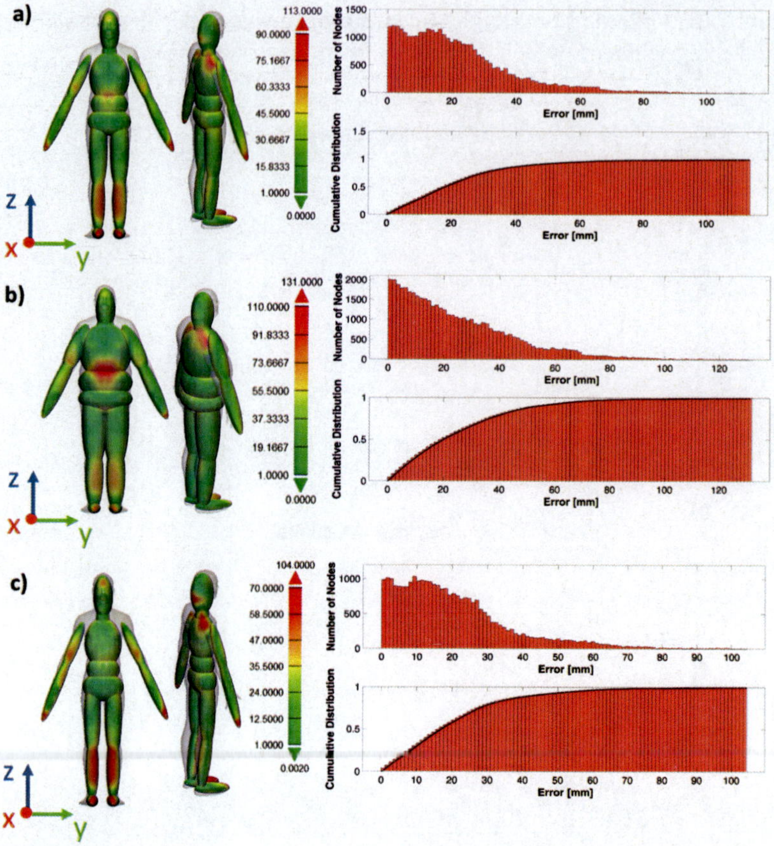

Fig. 7. Geometric distribution error analysis for: a) normal-weight subject; b) over-weight subject; c) under-weight subject

The F-test has shown that the variance of d_{ave} (that is the ellipsoidal geometry deviation from the anthropomorphic one) is significantly greater than the variance of distances measured between the anthropomorphic model and the true body shape. In other words, the ellipsoids model is affected by a significant added source of error, coming from the geometrical approximation through ellipsoids.

4 Conclusions

This work was finalised to assess if creating an ATB model from an anthropomorphic model did worth the effort providing significantly better accuracy in geometrical reconstruction and significantly different values of inertial total bodies. The analysis has given positive results: the statistical relevance of differences between the two analysed models has been proved. With reference to inertial properties, only a limited number of segments (head, thorax and right forearm) has exhibited weak significance of differences, coming from the associated absolute error being very limited. More in detail, with reference to the ellipsoids model, the abdomen and the thighs are the most critical segments for what concerns inertial properties, while when the external geometry is considered, the most critical areas are located at the shoulders, the calves, and, for obese subjects, at the trunk. The ellipsoids model errors concerning mass estimation can reach 60%, those concerning inertial moments 75%; the centres of mass can result to be displaced up to 50 mm; distances between the actual geometry and ellipsoids geometry can reach 130 mm for most obese patients, while in most cases they remain below 50 mm.

In this work the anthropomorphic ATB model has been used as reference for the performed analyses, due to its ability to more accurately replicate the actual human geometry. In light of this, it was assumed that also inertial properties computed from segments' geometries and density are most reliable than those provided by regression equations. The differences between the two models could have a significant impact on the outcome of multibody analyses, especially when the human subject geometry interacts with the external environment.

On the whole, the methodology here presented wants to give a contribution to the field of human body modelling, with particular reference to the subject-specific models where the novel approach was proved to provide a significantly more accurate method for segments' properties definition.

References

1. Cheng, Z., Rizer, A.L., Pellettiere, J.A.: Modeling and simulation of OOP occupant-airbag interaction. SAE Trans. **112**, 565–575 (2003). https://doi.org/10.4271/2003-01-0510
2. Haug, E., Choi, H.Y., Robin, S., Beaugonin, M.: Computational Models for the Human Body. Elsevier, Ansterdam (2004)
3. Shang, S., et al.: The predictive capacity of the MADYMO ellipsoid pedestrian model for pedestrian ground contact kinematics and injury evaluation. Accid. Anal. Prev. **149**, 105803 (2021). https://doi.org/10.1016/j.aap.2020.105803
4. Crocetta, G., Piantini, S., Pierini, M., Simms, C.: The influence of vehicle front-end design on pedestrian ground impact. Accid. Anal. Prev. **79**, 56–69 (2015). https://doi.org/10.1016/j.aap.2015.03.009
5. Shewchenko, N., Withnall, C., Keown, M., Gittens, R., Dvorak, J.: Heading in football. Part 2: biomechanics of ball heading and head response. Br. J. Sports Med. **39**, i26–i32 (2005). https://doi.org/10.1136/bjsm.2005.019042
6. LifeMOD Tutorials Bicycle Rider. http://www.lifemodeler.com/LM_Manual_2010/T_bike.shtml. Accessed 27 Jan 2021
7. Muggenthaler, H., Drobnik, S., Hubig, M., Schönpflug, M., Mall, G.: Fall from a Balcony-accidental or homicidal? Reconstruction by numerical simulation. J. Forensic Sci. **58**, 1061–1064 (2013). https://doi.org/10.1111/1556-4029.12148
8. Pascoletti, G., Catelani, D., Conti, P., Cianetti, F., Zanetti, E.M.: Multibody models for the analysis of a fall from height: accident, suicide, or murder? Front. Bioeng. Biotechnol. **7**, 419 (2019). https://doi.org/10.3389/fbioe.2019.00419
9. Cessna Aircraft Company: Methodology for seat design and certification by analysis (revision A) (2001)
10. Gavelin, A., Lindquist, M., Oldenburg, M.: Modelling and simulation of seat-integrated safety belts including studies of pelvis and torso responses in frontal crashes. Int. J. Crashworthiness **12**, 367–379 (2007). https://doi.org/10.1080/13588260701482906

11. TNO: MADYMO Human Body Models Manual (2012)
12. Gross, M.E.: THE GEBODIII Program User's Guide and Description (1991)
13. Corazza, S., Mündermann, L., Chaudhari, A.M., Demattio, T., Cobelli, C., Andriacchi, T.P.: A markerless motion capture system to study musculoskeletal biomechanics: visual hull and simulated annealing approach. Ann. Biomed. Eng. **34**, 1019–1029 (2006). https://doi.org/10.1007/s10439-006-9122-8
14. Andersson, B., Valfridsson, M.: Digital 3D facial reconstruction based on computed tomography (2005). www.ep.liu.se/exjobb/itn/2005/mt/010/
15. Hosseinian, S., Arefi, H.: 3D reconstruction from multi-view medical X-ray images - review and evaluation of existing methods. Int. Arch. Photogramm. Remote Sens. Spat. Inf. Sci. ISPRS Arch. **40**, 319–326 (2015). https://doi.org/10.5194/isprsarchives-XL-1-W5-319-2015
16. Remondino, F.: Three-dimensional reconstruction of static human body with a digital camera. In: Proceedings of SPIE - The International Society for Optical Engineering, pp. 38–45. SPIE Digital Library (2003). https://doi.org/10.1117/12.473090
17. Penmatsa, R., et al.: Estimation of mass and inertia properties of human body segments for physics-based human modeling and simulation applications. SAE Int. J. Passeng. Cars - Mech. Syst. **2**, 1626–1631 (2009). https://doi.org/10.4271/2009-01-2301
18. Ma, Y., Lee, K., Li, L., Kwon, J.: Nonlinear regression equations for segmental mass-inertial characteristics of Korean adults estimated using three-dimensional range scan data. Appl. Ergon. **42**, 297–308 (2011). https://doi.org/10.1016/j.apergo.2010.07.005
19. Robinette, K.M., et al.: Civilian American and European Surface Anthropometry Resource (Caesar). Final report, volume I: summary (2002). https://doi.org/10.1093/milmed/137.11.407
20. Pascoletti, G., Calì, M., Bignardi, C., Conti, P., Zanetti, E.M.: Mandible morphing through principal components analysis. In: Rizzi, C., Andrisano, A.O., Leali, F., Gherardini, F., Pini, F., Vergnano, A. (eds.) ADM 2019. LNME, pp. 15–23. Springer, Cham (2020). https://doi.org/10.1007/978-3-030-31154-4_2
21. Huysmans, T., Molenbroek, J.F.M.: DINED Mannequin. Tijdschrift voor Human Factors (45), 4–7 (2020)
22. Danckaers, F., Huysmans, T., Sijbers, J.: Adaptable digital human models from 3D body scans. In: DHM Posturography, pp. 459–470 (2019). https://doi.org/10.1016/B978-0-12-816713-7.00033-7
23. Danckaers, F., Huysmans, T., Lacko, D., Sijber, J.: Evaluation of 3D body shape predictions based on features, pp. 258–265 (2015). https://doi.org/10.15221/15.258
24. Grunhofer, H.J., Kroh, G.: A review of anthropometric data of German Air Force and United States Air Force flying personnel 1967–1968. Appl. Ergon. **7**, 110–111 (1976). https://doi.org/10.1016/0003-6870(76)90172-1
25. Mcconville, J.T., Churchill, T.: Anthropometric Relationships of Body and Body Segment Moments of Inertia. Report, vol. 105, pp. 7405–7409 (1980). https://doi.org/10.1073/pnas.0710346105
26. Young, J.W., Chandler, R.F., Snow, C.C., Robinette, K.M., Zehner, G.F., Lofberg, M.S.: Anthropometric and Mass Distribution Characteristics of the Adult Female (1983)

Preliminary Study of a High-Fidelity Simulator for the Management of Paediatric Tracheal Pathologies

Chiara Santarelli(✉), Luca Puggelli, Monica Carfagni, and Lapo Governi

Department of Industrial Engineering, University of Florence, Florence, Italy
{chiara.santarelli,luca.puggelli,monica.carfagni,
lapo.governi}@unifi.it

Abstract. In the biomedical field, high-fidelity simulation plays a fundamental role for medical and surgical staff as it allows them to simulate real scenarios from everyday clinical practice. The availability of patient-specific pathological models allows doctors to simulate surgical procedures before entering the operating room, thus reducing the risks associated with surgery. Such models are also employed in medical training to enhance the experience of trainees by confronting them with possible real-life emergency scenarios. To obtain a strict correspondence between simulator and reality - both in terms of geometry and haptic feedback - important decisions must be taken from the beginning of the design phase. High fidelity is a fundamental requirement of the simulators, since the design phase of the anatomical model, for what concerns geometries and materials. In this paper, a preliminary study for the development of a paediatric tracheal simulator is presented. This study is focused on the definition of some general geometric parameters and of the material composing the simulator. Mechanical characteristics of the trachea are investigated to identify a set of materials able to reproduce a realistic haptic feedback of the simulator. Materials are chosen in relation with the technological process that will be used for the simulator manufacturing: additive manufacturing or mould casting. To test the various materials, specimens are created and submitted to the judgement of a specialized medical team. The results showed Shore 40A silicone with a 1.75 mm wall-thickness value to be the best compromise for reproducing the haptic feedback of the trachea.

Keywords: Airway simulation · Soft tissue modelling · 3D printing · Haptic feedback · Paediatric trachea

1 Introduction

In the last decade, the use of life-size simulation manikins has grown exponentially in clinical practice. The use of high-fidelity simulators leads to relevant advantages in training and pre-surgery planning applications, compared with traditional approaches, which respectively consist of the direct practice of medical residents on real patients under expert supervision (in the case of training) and of an analysis limited to medical images (in the case of planning).

C. Rizzi et al. (Eds.): ADM 2021, LNME, pp. 773–784, 2022.
https://doi.org/10.1007/978-3-030-91234-5_78

In the first cases, medical residents have the possibility of gaining experience in routine procedures by replicating everyday clinical practices on realistic anatomical models. As a result, before operating on a patient, trainees can gradually practice more complex medical procedures to gain competency. This training approach, ensures greater patient safety as the risk of complications is reduced, compared to the traditional one [1, 2]. As previously mentioned, the advantages of using realistic simulators are not limited to training purposes. In fact, the availability of patient-specific simulators, i.e. 3D models designed based on the patient's anatomy, allows experienced surgeons to plan surgery before entering the operating room. In this way, different surgical strategies can be simulated and investigated with the aim of identifying the best one, thus reducing operating times and surgical risks. For both patient-specific and general training simulators, a close correspondence between 3D replica and patient's anatomy is crucial, both under geometrical and haptic feedback points of view [3]. Regarding the first aspect, diagnostic images of the patient, such as computed tomography (CT) and magnetic resonance imaging (MRI), are commonly used to retrieve the virtual model of the anatomical replica. The accuracy of CT-scan based 3D reconstruction has been validated in [4], in which an average accuracy error of 0.1 mm has been measured in a set of 3D reconstructions obtained from CT scans. From the digital model, it is possible to produce the physical replica by means of computer-based manufacturing technologies (e.g. Additive Manufacturing (AM)) that can easily handle the creation of complex physical models from 3D virtual geometries. The issue is more complex when referring to haptic feedback. This is a crucial aspect to make the simulated experience actually realistic, although it is considered of secondary importance as evident in many commercial simulators and specific studies in the scientific literature [5–7]. Anatomical structures are indeed composed of multiple organic materials, some of which are composed by fibres, thus their behaviour is anisotropic. For this reason, in the scientific literature the mechanical characterization is generally referred to general structures and "equivalent" materials, whose mechanical characteristics mimics the overall behaviour of the general anatomical structure or for the main substructure [8–10]. In addition, the manufacturing process, especially in the case of AM, limits the choice of the proper material with optimal characteristics.

In this paper, the attention is focused on these last aspects, related to the haptic feedback of an anatomical replica. In detail, a preliminary study for the development of a paediatric tracheomalacia (TM) management simulator is carried out. As widely recognized, TM refers to a weakness of the trachea, resulting in the collapse of the cartilage structure during respiration [11]. When the bronchi are also involved in the pathology, the term tracheabronchomalacia (TBM) is adopted. An effective clinical procedure in the management of TM involves the insertion of a stent into the area affected by cartilage restriction [12]. The simulator proposed in the present work, developed in the collaboration with AOU Meyer Children's Hospital located in Florence (Italy), is principally meant to simulate this treatment.

The paper is structured as follows: in Sect. 2, the tracheal mechanical characteristics will be discussed and a list of suitable materials to replicate them will be defined. In Sect. 3, the geometric characteristics of the trachea will be investigated, to define the simplified geometry of the specimens, that have been used to help the material choice, as explained in Sect. 4.

2 Materials

As mentioned above, the main aim of this paper is to devise a paediatric tracheomalacia (TM) management simulator for training and/or planning purposes. Therefore, the haptic feedback of the proposed simulator is of outstanding importance. For this reason the present paper aims to overcome the limitations of previous works [5–7] where tracheal simulators are developed but the haptic feedback is not generally taken into account. To cope with the issue of creating a simulator with optimal haptic feedback, it is necessary to select the materials, which allow to render the behaviour of actual trachea. In fact, a realistic haptic response is important in many simulation scenarios, starting from the merely endoscopic navigation to stent insertion procedure. Moreover, it becomes crucial to test stent stability and relative tracheal behaviour.

The complex mechanical behaviour of anatomical tissues can be simulated only through a proper choice of the materials to use, starting from an investigation on the mechanical characteristics of paediatric trachea (in particular, elastic modulus).

Focusing to scientific literature, a limited number of studies have dealt with estimating the elastic modulus of the trachea using different mathematical models [9, 13, 14]. After isolating several samples of tracheal cartilage, mechanical tests were carried out to establish the stress-strain curves and thus obtain an approximate value for the elastic modulus (Fig. 1). In [9] the stress-strain curves are retrieved for each main tracheal structure, such as cartilage, connective tissue, smooth muscle and so on.

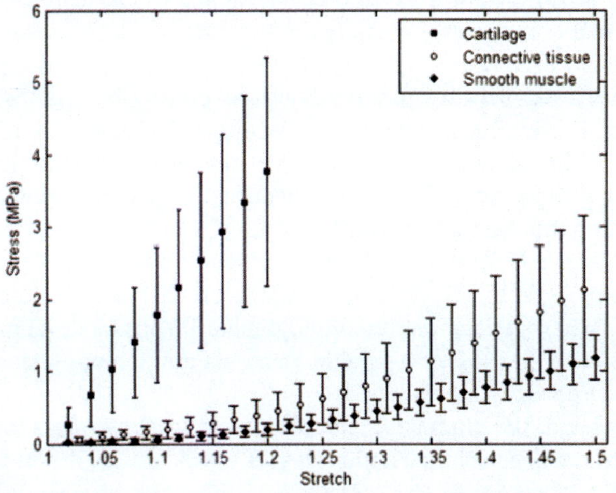

Fig. 1. The average stress-stretch curves for cartilage, connective tissue and smooth muscle samples [10].

In this case, authors define an average value of the elastic modulus for the three main tracheal components.

In general, all the literature papers agree with the fact that the stiffness of the trachea increases with the age of the patient. For this reason and for the different procedures adopted, not a precise value for tracheal stiffness is defined. Moreover, data are mostly referred to adult case studies. Therefore, a specific investigation on paediatric tracheal stiffness has been carried out, taking into consideration the lowest values for adult tracheal cartilage retrieved in other works (1–3.3 MPa) [13–16]. The choice of considering cartilage stiffness is motivated by the fact that cartilage rings are the structure responsible to maintain the trachea open during respiration, preventing collapse [11]. In order to simulate the resistance of trachea to inner navigation or to stent insertion, this aspect is therefore the principal to be considered.

As mentioned in the introductory section, the material choice is limited by the manufacturing technology that will be adopted to realize the complex simulator geometry. Since the production is limited to 1–5 items only for each specific geometry, the most suitable choice is AM, which can be employed in two different ways:

1. Direct 3D printing.
2. Silicone elastomer (SE) casting in 3D printed moulds.

In the first case, the final shape of the simulator is obtained directly from the 3D printer. The direct technique is increasingly used in the biomedical field [17–19] for the realization of prosthesis, orthosis or other medical devices. Unfortunately, the characteristics of printable materials generally do not match with the desired ones. In fact, an elastic modulus in the range [1 - 3.3] MPa is equivalent to a rubber like material, while printed ones are generally more rigid (at least plastic-like). In the last years, AM technologies and 3D printable materials were developed in this direction, with the aim of allowing the realization of physical models with characteristics that mimic human body's soft tissue. Among these, two of them are identified as potential suitable solutions to directly 3D printing the tracheal structure.

The first one is the *J750™ Digital Anatomy Printer* (DAP), a 3D printer based on *Polyjet™* technology specifically developed by *Stratasys 3D* for biomedical applications. The second one is a new generation stereolithographic 3D printer developed by *Formlab: Form 3B*. In this case, the softest available materials are photopolymers which - once solidified - have a rubber-like behaviour.

On the other hand, the alternative manufacturing strategy offers the possibility of using SE materials, whose characteristics vary in a wide range, depending on reagents composition and concentration. Consequently, in this case the material choice is way less limited than it is for direct 3D printing, while 3D modelling is more complex, since mould design must consider all the problems related to the part extraction.

With the exception of J750, in which materials are predefined for each anatomical tissue typology (e.g. specific materials for the reproduction of bone or cardiovascular tissues), both *Formlab Form3B* and soft SE materials are generally characterized by shore A hardness. An analytical/experimental relation between elastic modulus and

Shore hardness was retrieved in [13] and is provided by the following equation:

$$E = \frac{0.0981(56 + 7.62336S)}{0.137505(254 - 2.54S)} \quad (1)$$

where S represents the shore hardness value and E is the Young's Modulus (elasticity).

By using this equation and considering the range of elasticity [1–3.3] MPa Shore hardness results equal to a value between 30 and 58. Starting from this result it has been possible to select a list of potential proper materials for trachea simulator manufacturing. In the case of Formlab Form3B printer, the material "Elastic 50A" seems to be suitable for the application (50 Shore A). In the case of SE materials, it has been decided to consider four different Shore A hardness values: 30, 40, 50, 60. Since the elasticity value is merely indicative, as it is derived from adult age case studies, AOU Meyer Children's Hospital anaesthetists and otolaryngologists, by means of experimental tests on specifically designed specimens, validated the material choice. In fact, their experience and sensitivity can be considered the primary feedback about the simulator's realism.

3 Geometrical Characterization

In this chapter, the tracheal geometry is analysed to define a proper shape for the set of specimens that will be used to validate the material choice. From a geometric point of view, the tracheal cross-section can be described as a circle, partially collapsed in the posterior region (Fig. 2). For this reason, two dimensions generally define the cross-section of a healthy trachea: transverse and anteroposterior diameters, respectively referred to as T and AP in Fig. 1.

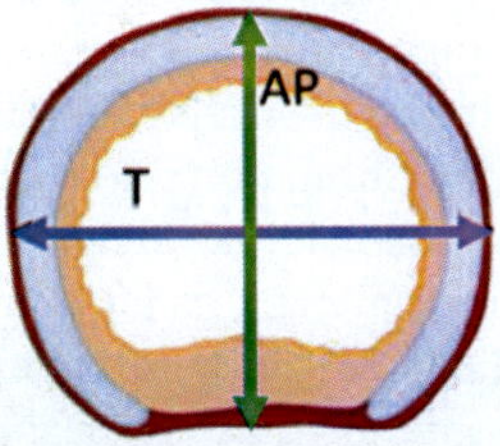

Fig. 2. Cross-sectional view of a tracheal geometrical model. AP and T are respectively anteroposterior and transverse diameters.

As already mentioned, most of the papers in the scientific literature are referred to adults' anatomy; the same is also true for what concerns tracheal anatomy. Therefore, information about tracheal dimensions are usually referred to adult case studies. Only a few works report statistical data referred to paediatric age [20–22]. One of the most complete statistics [23] is reported in Table 1.

Table 1. Geometric tracheal dimensions of paediatric patients [23].

Age (years)	0–2	2–4	4–6	6–8	8–10	10–12
Length (mm)	54	64	72	82	88	10
Anteroposterior diameters (mm)	5.3	7.4	8	9.2	11	12
Transverse diameters (mm)	6.4	8.1	9	9.3	11	12

As visible in the Table 1, anteroposterior and transverse diameters are very close in size; hence, trachea can be easily approximated to a hollow cylinder (Fig. 3).

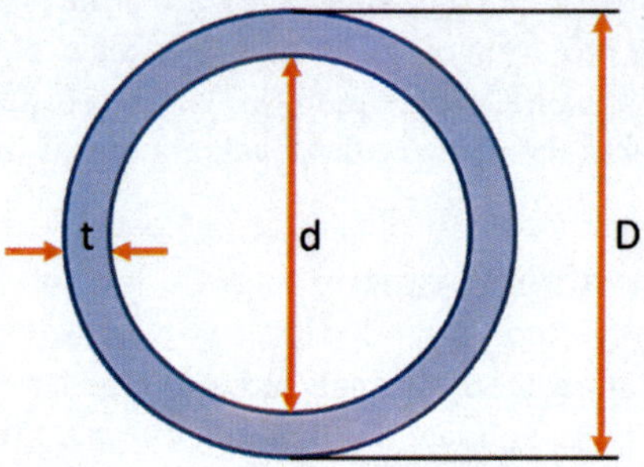

Fig. 3. Cross-sectional view of the simplified geometry. D and d represent outer and inner diameter respectively and t is the thickness.

In particular, the internal diameter value can be defined as the average between anteroposterior and transverse diameters. Focusing on 6–8 years old children, anteroposterior diameter is around 9.2 mm while transverse is 9.3 mm in size, therefore the internal diameter of the specimen was approximated to 9 mm, while length has been set at 82 mm (as defined in table). Unfortunately, no information is given about wall-thickness or external dimensions. This is probably because external dimensions or thickness are considered less important dimensions than internal ones, as far as trachea is generally inspected by means of laparoscopy. On the other side, the same cannot be said about the mechanical behaviour. As widely known, considering the trachea approximated with a circular tubular geometry, its reaction under external forces (bending moment or crashing) depends also on wall-thickness (or external diameter).

Therefore, to retrieve information about trachea wall-thickness in paediatric age (in the range 6–8 years old) AOU Meyer's otolaryngologists and anaesthesiologists were involved. Based on their background and experience, it has been possible to define an approximate thickness value. In particular, wall-thickness resulted roughly equal to 1.5 mm. This information was successively confirmed by analysing diagnostic images obtained from CT scans. In particular, 4 different patients' tracheal data have been measured at 3 different axial sections (A, B, C) indicatively at 20% of the length (from top to bottom), 50% and 80% as shown in Fig. 4. For each one, the average wall-thickness value is reported in Table 2. As noticeable, even if the sample is limited in size, collected data partially confirm the previously defined 1.5 mm wall thickness value.

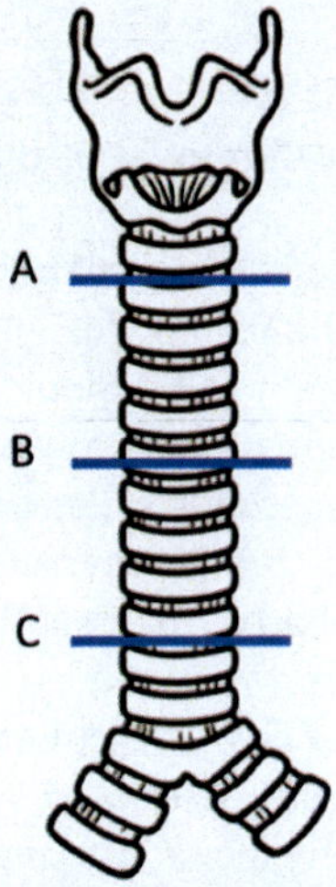

Fig. 4. Trachea sections for data collection.

Table 2. Tracheal wall thickness values of different patients measured at different sections.

Patient	#1	#2	#3	#4
Section A	1.24	1.33	1.44	1.52
Section B	1.56	1.54	1.29	1.56
Section C	1.43	1.62	1.63	1.26

However, it can be observed that a small difference in thickness causes a relevant difference in mechanical properties. For this reason, three wall thickness values are considered for the realization of the specimens: 1.25–1.50–1.75 mm, as shown in Table 3.

Table 3. Geometric dimension of the validation specimens.

Specimen	#1	#2	#3
Internal diameter	9	9	9
Wall thickness	1.25	1.50	1.75
Length	82	82	82

4 Validation

To define which is the best combination of wall-thickness and material, a set of 18 specimens was realized: 3 wall thicknesses for 6 different materials (2 for direct 3D printing, 4 silicone materials – respectively with Shore A hardness of 30 40 50 and 60).

As previously mentioned, two of them are designed to be directly 3D printed while the remaining four of them are thought to be mould casted. Therefore, the polygonal model (STL) of the specimen is required in the first cases, while the mould shells must be designed in the second ones. In particular, the design was carried out considering both the geometry to be reproduced and the characteristics of the material to be cast. The mould is composed by 4 parts: base, 2 half shells and a positioning cap, as shown in Fig. 5. The core is part of the base, while the two shells are maintained in the right position during moulding by base and cap. This solution allows to maintain the distance between shells and core as constant as possible, to obtain constant thickness along the moulded specimens. In addition, being the mould subdivided in two halves, it allows an easy extraction of the hardened specimens. All the parts of the moulds were realised by FDM 3D printing, in PLA material (Fig. 6).

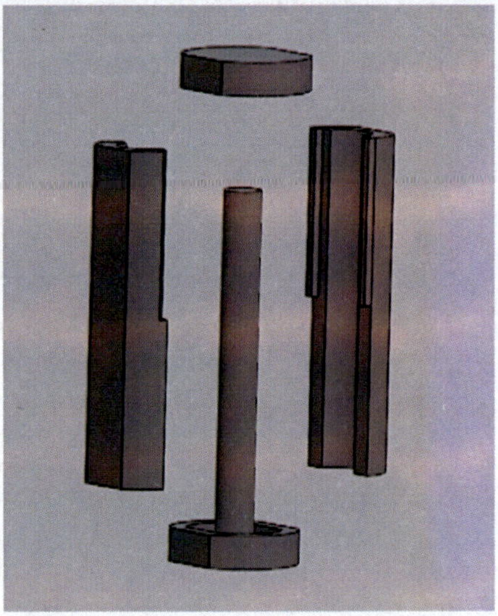

Fig. 5. Silicone casting mould prototype.

Once the moulds are printed, the silicone material can be prepared by mixing the silicone with the catalyst. In order to make the samples even more realistic, it is possible to modify the chromatic aspect of the simulator using special colorants that give the final product a high degree of realism. Before pouring the mixture into the mould, a degassing system was used to eliminate air bubbles in the mixture and ensure better results [24, 25]. To prevent the silicone from sticking to the mould surfaces and to facilitate the removal of the physical replica, an aerosol spray is employed as a release agent. After

Fig. 6. 3D printed silicone casting mould using PLA material.

the mixture was poured into the mould, a curing time of about 12 h had to be waited before removing the physical replica.

The 18 specimens were tested by team of physicians specializing in the treatment of upper and lower airways of the Meyer Children's hospital. Figure 7 shows the specimen realized with Formlab Form 3B direct printing (Fig. 7A) and with SE mould casting (Fig. 7B). The first phase of the test involved a tactile evaluation in which the specimens were manipulated and crushed to assess their elastic and tactile feedback. In a second step, a bronchoscope, an instrument used to investigate and treat tracheal pathologies, was used. Once lubricated, the instrument was inserted into the test specimens simulating a bronchoscopy and making the experience as close as possible to the real one. The whole procedure evaluated both the geometry and the haptic feedback of the simulator. Despite the evaluation in both steps is merely qualitative, the large background experience and the advanced specific sensitivity of the involved medical staff can be considered as valid reference. The various tests performed show that the materials used to reproduce the specimens with direct 3D printing are not suitable to simulate the flexibility of the paediatric trachea. Moreover, direct 3D printed materials have shown a tendency to change their physical properties over time, resulting gradually more rigid, more brittle and changing colour. In contrast, silicone models resulted more stable and appropriate at simulating both texture and flexibility. Specifically, the best real-world resemblance result was obtained with the 1.75 mm thick, 40 Shore A hardness specimen.

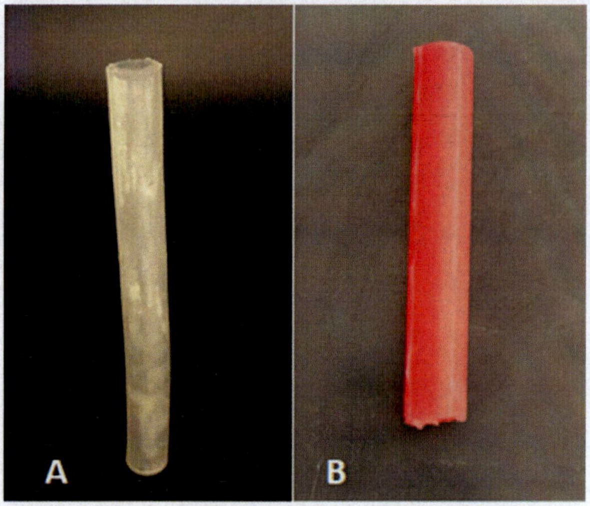

Fig. 7. Specimens realized by direct 3D printing and casting in SE mould for material validation. (A) *Formlab Form 3B* printed specimen with "Elastic 50A" material; (B) 40 Shore A SE specimen.

5 Conclusion

In this paper, a decision method to define the better design solution of a paediatric tracheal simulator is proposed. For what concerns geometry, as far as no information was retrieved in scientific literature about tracheal wall-thickness for the paediatric age, a possible range of this value has been retrieved. In particular, the range is defined thanks to the experience of AOU Meyer's otolaryngologists and anaesthesiologists together with information obtained from direct measurements on 3 different paediatric patients in the specific age range of interest (6–8 years old). As a result, three possible wall-thickness values were defined: 1.25 – 1.50 – 1.75 mm. Referring to the material choice, a range of values for the trachea elasticity (E = 1–3.3 MPa) was found. Since the simulator is meant to be realized using 3D printing (i.e. the model is directly 3D printed or obtained by casting in 3D printed moulds), the choice of the material is a-priori limited to those suitable for these manufacturing technologies. Hence, it was possible to draw a list of materials that fit the requirements of flexibility and elasticity for reproducing the mechanical resistance of tracheal tissues. In order to test both wall thickness and material behaviour, a set of 18 specimens was realized (6 different materials and 3 different wall thickness values). The specimens were tested by AOU Meyer's otolaryngologists and anaesthesiologists team, which simulated a tracheoscopy. Among the various solutions, silicone was the best material for simulating tracheal elasticity. More specifically, the specimen with 1.75 mm of wall thickness manufactured using silicone of Shore 40 shore A hardness resulted as the one that best simulates both the geometric and the haptic response of an actual paediatric trachea. In the light of the results obtained by this preliminary activity, it was possible to define thickness, material and - therefore - the technological process - for the modelling and the realization of the paediatric tracheal simulator. Future works will be addressed to develop a paediatric simulator for airway management, including the trachea and bronchi, which will be used by AOU Meyer's physicians for training and pre-surgery purposes.

References

1. Marker, S., Mohr, M., Østergaard, D.: Simulation-based training of junior doctors in handling critically ill patients facilitates the transition to clinical practice: an interview study 11 medical and health sciences 1117 public health and health services 17 psychology and cognitive sciences. BMC Med. Educ. **19**, 1–8 (2019)
2. Aggarwal, R., et al.: Training and simulation for patient safety. Qual. Saf. Health Care. **19**(2) (2010)
3. Rodr, E., Kypson, A.P., Moten, S.C., Nifong, L.W., Jr, W.R.C.: Robotic mitral surgery at East Carolina University : Int. J. 211–215 (2006)
4. Puggelli, L., Uccheddu, F., Volpe, Y., Furferi, R., Di Feo, D.: Accuracy assessment of CT-based 3D bone surface reconstruction. In: Cavas-Martínez, F., Eynard, B., Fernández Cañavate, F., Fernández-Pacheco, D., Morer, P., Nigrelli, V. (eds.) Advances on Mechanics, Design Engineering and Manufacturing II. Lecture Notes in Mechanical Engineering. Springer, Cham (2019). https://doi.org/10.1007/978-3-030-12346-8_47
5. Miyazaki, T., Yamasaki, N., Tsuchiya, T., Matsumoto, K., Takagi, K., Nagayasu, T.: Airway stent insertion simulated with a three-dimensional printed airway model. Ann. Thorac. Surg. **99**, e21–e23 (2015)
6. Cheng, G.Z., Estepar, R.S.J., Folch, E., Onieva, J., Gangadharan, S., Majid, A.: Three-dimensional printing and 3D slicer powerful tools in understanding and treating structural lung disease. Chest **149**, 1136–1142 (2016)
7. Wilson, C.A., et al.: Printed three-dimensional airway model assists planning of single-lung ventilation in a small child. Br. J. Anaesth. **115**, 616–620 (2015)
8. Xu, C., Brennick, M.J., Dougherty, L., Wootton, D.M.: Modeling upper airway collapse by a finite element model with regional tissue properties. Med. Eng. Phys. **31**, 1343–1348 (2009)
9. Safshekan, F., Tafazzoli-Shadpour, M., Abdouss, M., Shadmehr, M.B.: Mechanical characterization and constitutive modeling of human trachea: Age and gender dependency. Materials (Basel). **9**, 456 (2016)
10. Ozen, M., Sayman, O., Havitcioglu, H.: Modeling and stress analyses of a normal foot-ankle and a prosthetic foot-ankle complex. Acta Bioeng. Biomech. **15**, 19–27 (2013)
11. McNamara, V.M., Crabbe, D.C.G.: Tracheomalacia. Paediatr. Respir. Rev. **5**, 147–154 (2004)
12. Godoy, M.C.B., et al.: Multidetector CT evaluation of airway stents: what the radiologist should know. Radiographics **34**, 1793–1806 (2014)
13. Trabelsi, O., del Palomar, A.P., López-villalobos, J.L., Ginel, A., Doblaré, M.: Experimental characterization and constitutive modeling of the mechanical behavior of the human trachea. Med. Eng. Phys. **32**, 76–82 (2010)
14. Rains, J.K., Bert, J.L., Roberts, C.R., Pare, P.D.: Mechanical properties of human tracheal cartilage. J. Appl. Physiol. **72**, 219–225 (1992)
15. Lambert, R.K., Baile, E.M., Moreno, R., J Bert, P.D.P.: A method for estimating the Young's modulus of complete tracheal cartilage rings. (1991)
16. Roberts, C.R., Rains, J.K., Paré, P.D., Walker, D.C., Wiggs, B., Bert, J.L.: Ultrastructure and tensile properties of human tracheal cartilage. J. Biomech. **31**, 81–86 (1997)
17. Martelli, N., et al.: Advantages and disadvantages of 3-dimensional printing in surgery: a systematic review. Surg. (US) **159**, 1485–1500 (2016)
18. Tserovski, S., Georgieva, S., Simeonov, R., Bigdeli, A., Röttinger, H., Kinov, P.: Advantages and disadvantages of 3D printing for pre-operative planning of revision hip surgery. J. Surg. Case Reports. **2019**, 1–4 (2019)

19. Berman, B.: 3-D printing: the new industrial revolution. Bus. Horiz. **55**, 155–162 (2012)
20. Dave, M.H., Kemper, M., Schmidt, A.R., Both, C.P., Weiss, M.: Pediatric airway dimensions—A summary and presentation of existing data. Paediatr. Anaesth. **29**, 782–789 (2019)
21. Naina, P., Syed, K.A., Irodi, A., John, M., Varghese, A.M.: Pediatric tracheal dimensions on computed tomography and its correlation with tracheostomy tube sizes. Laryngoscope. **130**, 1316–1321 (2020)
22. Szelloe, P., Weiss, M., Schraner, T., Dave, M.H.: Lower airway dimensions in pediatric patients—A computed tomography study. Paediatr. Anaesth. **27**, 1043–1049 (2017)
23. Griscom, N.T., Wohl, M.E.B.: Dimensions of the growing trachea related to age and gender. Am. J. Roentgenol. **146**, 233–237 (1986)
24. Mussi, E., et al.: Current practice in preoperative virtual and physical simulation in neurosurgery. Bioengineering **7**, 1–14 (2020)
25. Kuo, C.C., Lai, M.Y.: Development of an automatic vacuum degassing system and parameters optimization for degassing process. Indian J. Eng. Mater. Sci. **18**, 405–410 (2011)

Trends in Human Factors Integration for the Design of Industry 4.0

Fabio Grandi[1], Margherita Peruzzini[1](✉), Roberto Raffaeli[2], and Marcello Pellicciari[2]

[1] Department of Engineering Enzo Ferrari, University of Modena and Reggio Emilia, via Vivarelli 10, Modena, Italy
margherita.peruzzini@unimore.it

[2] Department of Sciences and Methods for Engineering, viale Amendola 2, Reggio Emilia, Italy

Abstract. The fourth industrial revolution is evolving the machines as well as the abilities of people working in the factories. Human roles and tasks are changing, moving from highly physical tasks to decision-making and high-precision activities, asking for different competencies and creating new types of interactions with machines. This paper reviews the design and engineering methods for the inclusion of human factors in modern companies. Human factor integration (HFI) can play a key role in the design of factories with a great impact on social aspects and global process sustainability. The paper proposes a systematic view of the main tools to design human-centered industrial processes, with a specific focus on manufacturing, and discusses trends to achieve and effective HFI.

Keywords: Engineering methods · Human factors · Ergonomics · Human-centered design

1 Introduction

From 2011, when the term Industry 4.0 (I4.0) was firstly mentioned, the world started thinking factories in a novel way, according to the official guidelines [1]. Its purpose is to create a "highly flexible production model" based on real-time interactions between people, products, and devices during the production process [2]. In this context, machines and people should help each other in the execution of their respective tasks and to help decision-making and optimize the management of production processes [3]. So far, a lot of research on I4.0 machines and data management, while a very poor research focuses on people, even if they still play an important role in the development and implementation of the I4.0 concept. Indeed, the modern factory should also help to anticipate issues in human-machine interaction [4] and combine production needs with human needs, in a sustainable perspective [5]. For this purpose, human factors should be included into design and engineering practices to efficiently transfer ergonomic principles to human-machine system design [6]. Moreover, the introduction of new technologies in the intelligent factory is radically changing the role of human work, increasing the cognitive load (problem-solving, abstraction, managing) and changing the proportion between

C. Rizzi et al. (Eds.): ADM 2021, LNME, pp. 785–792, 2022.
https://doi.org/10.1007/978-3-030-91234-5_79

physical and cognitive activities [7]. Up to now, the main challenge for companies is to effectively combine human factors in the design of I4.0 systems with the final aim to mutually enhance the overall performance and the workers' well-being [8]. The aim of the study is defining the most promising trends in HFI in the design of I4.0. HFI is the process of applying HF and ergonomics within engineering process design, taking into consideration end user needs and limitations [9]. From the analysis of the recent scientific literature, the research grouped the tools into five levels with a different digital maturity for HFI and discussed the most interesting applications in the manufacturing area.

2 Review Methodology

A systematic literature review was conducted in order to give a complete overview of the main methods and tools to design human-centered industrial processes, following the PRISMA method [10]. The first step of the research consisted in the selection of two classes of keywords to be searched jointly, according to the paper's purpose. The first class includes at least one of the following keywords: *Human Factors, Ergonomics, Human Factors Integration.* The second class includes at least one of the following keywords: *Industry 4.0, Human Simulation, Digital/Virtual Manufacturing, Human-centered/-centric Manufacturing.*

Research has been carried out on the main scientific databases (i.e., Scopus, Web of Science, ScienceDirect) combining the two classes of keywords and considering different variants of word spelling and abbreviations, from 2011 to 2021. From these studies, a selection of 157 papers was extracted and further analyzed screening title and abstract, deleting papers published before 2011 and excluding those that do not deal with manufacturing topics. After this phase, a set of 100 papers was selected including articles on international journals (73), papers on international conferences (21) and book chapters (6). A quantitative analysis was carried out considering (i) year of publication, ii) quantity of papers, and iii) adopted tools for HFI, as reported in Fig. 1. The adopted tools were organized into 5 levels considering a growing digital maturity. Such tools are all about HFI in I4.0.

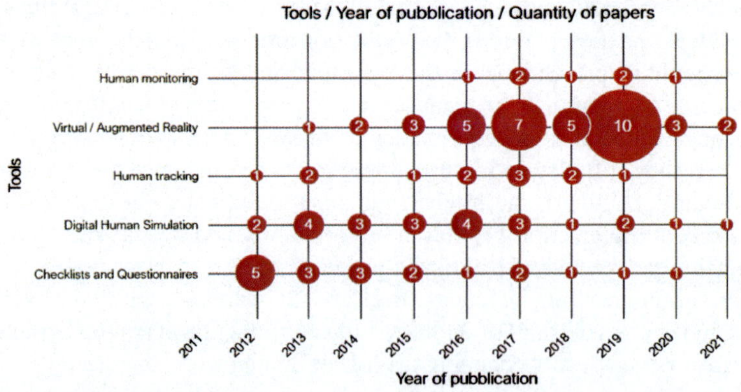

Fig. 1. Quantitative analysis on tools for HFI from the literature review

3 Tools for Human Factors Integration in I4.0

This section describes the tools emerged from the review in more detail. The tools were grouped with respect to digital innovation to express the digital maturity in pushing HFI in industry. In this context, digital maturity expresses how a certain class of tools is aligned with the concept of the digital factory, from level 1 (lower level of digitization) up to level 5 (highest level of digitization). The proposed levels of digital maturity for HFI in industry are defined as follows:

Level 1 - Checklists and Questionnaires (C&Q): this level is based on paper-based (no digital) or better excel-based or word-based tools (low digital). Checklists are usually adopted for the analysis of the workers' physical workload, used by experts to identify risks at work and evaluate the impact of ergonomic changes. There are a large number of checklists using different approaches (such as OWAS, RULA, EAWS, …) depending on the analysis goal. On the other hand, questionnaires are used to assess the workers' cognitive effort based on self-evaluation, such as NASA Task Load Index (NASA-TLX). Data from checklists and questionnaires can be combined to effectively highlight problems and suggest improvements to increase the operator comfort level, from a physical and cognitive point of view;

Level 2 - Digital Human Simulation (DHS): this level is characterized by fully digital simulation, using virtual workspaces and virtual mannequins to evaluate the system layout, task sequence, level of comfort, visibility and reachability. A semi-automatic ergonomic assessment can be carried out using the mannequins' postures representing the work cycle to support the identification of ergonomic issues;

Level 3 - Human Tracking (HT): this level uses different types of tools to track real uses into real spaces, observing real users and virtualizing them for posture assessment. Tracking and recording of the user movements allows postural analysis on a continuous flow of actions, closer to real activities. Two main approaches are widely recognized: real time motion capture, using specific systems (e.g. optical, inertial) and video analysis using images (e.g. RGB, depth images). These tools are able to detect pretty well awkward postures. Some specific tools refer to face reading to reveal users' emotions and stressful conditions using face recognition;

Level 4 - Immersive X-Reality (IXR): this level exploits high-end digital tools such as Virtual Reality (VR) and Augmented Reality (AR) to create multi-sensorial, immersive virtual scenarios in which operators can reproduce their task and take over possible issues. VR allows a complete immersion in a virtual scenario, while AR overlaps digital contents to the real world;

Level 5 - Sensorized Virtual eXperience (SVX): this level focuses on tracking the users while interacting within VR or AR scenarios, collecting biometric data to detect comfort problems or stress situations during task execution in an objective way. Human physiological parameters (human monitoring) can be carried out by wearable or environmental sensors, detecting both physical and mental workload.

Such tools can be used in the design of I4.0 considering different phases, as described in Fig. 2.

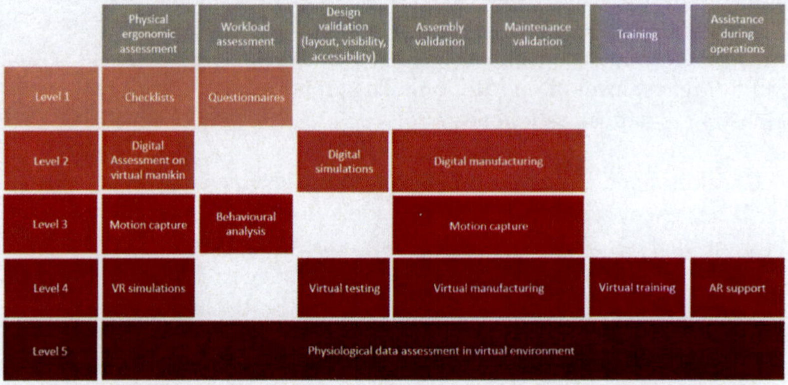

Fig. 2. HFI tools and application areas to support the design of I4.0

4 Industrial Applications, Limits and Future Trends

For each level, hereafter the most recent applications in the manufacturing field are presented and discussed. Regarding Level 1, C&Q have been used for a long time and are still frequently used for manufacturing ergonomics nowadays. In order to define the most suitable methods for factories, a multipath methodology to effectively encourage analysts in the selection of proper ergonomics assessment and measurement tools have been presented [11]. Alongside the well-known tools, specific checklists have been defined for specific purposes such as the Postural Ergonomic Risk Assessment (PERA) [12], that evaluates the postural ergonomic risk in short cyclic assembly operations. Indeed, to effectively implement questionnaires in modern factories in an easy way, web applications have been developed to assist experimenters in using NASA-TLX for the evaluation of subjective workload [13]. The main limitations of C&Q are related to the need of an ergonomics expert to collect and interpret information, analyze data, and suggest improvements, who strongly affects also the quality of final results [14], and the late feedback, since analysis is carried out observing users on the job, when the product or process already exist, and engineering changes are expensive and time-consuming [15]. The future of C&Q is assessment digitization of the collected data and the automation of analysis, in order to speed up the assessment process and to enlarge the number of people able to use these tools. As far as Level 2, DHS effectively helps the detection of static postures and exceptional strains in a more securely and fast way than in real-life assessments [16]. A set of applications have been recently developed for ergonomic virtual analysis: Gášová et al. [17] developed an app (i.e., CERAA, Ceit Ergonomics Analysis Application) for the evaluation of space limits and positions of workers in risky workplaces; similarly, Fantini [18] proposed a methodology to support the design of human work integrated within cyber-physical-systems and the assessments of different human-systems configurations in order to orient design choices in manufacturing context; also Baskaran et al. [19] used DHS not only to evaluate the ergonomic condition of workers in automotive assembly context, but also to enhance the collaboration between humans and robots using digital twins. In general, DHS is found to be applied for redesign purposes, using a corrective approach, but recent applications showed the effectiveness of these tools also in the design process, encouraging a preventive approach. In order to save time, DHS tools are now integrated in various commercial CAD platforms, to reduce file exchange problems and shorten the import-export time. The main limitations refer to the "static analysis" provided by DHS, based on static "snapshot" during task execution, without considering the dynamic aspects of human-system interaction [20].

About Level 3, HT is mainly used for acquiring postural data of workers, especially in assembly lines by motion capture and data post-processing [21], or supporting requirements definition in product-process design, considering manufacturing or maintenance operations [22]. In order to speed up the ergonomic analysis, specific motion analysis systems have been developed to perform posture assessment in an automatic way [23]. Recently, the OpenPose approach has been set for detecting bi-dimensional poses of many people from the analysis of a video image [24], using human body key points, such as hands or face. Therefore, it is difficult to apply HT in the preliminary stages of the design process, where most of the parts and equipment of the process have not yet been physically realised [25]. Relating to Level 4, referring to IXR technologies, different frameworks have been proposed to support HFI in recent years. Caputo et al. [26] used the digital twin of the workstation to optimize the development of new assembly lines, reducing time and assuring human safety. Similarly, Otto et al. [27] proposed the so-called "VR assembly assessment" (VR2A) to reach reliable validation results in production validation of the automotive and manufacturing industry. Other studies focused on training, as a fundamental aspect to leverage the workers' skills and to promote the HFI in industry. Specific VR-based training systems were defined to train operators to execute complex assembly tasks, enhancing the quality of user evaluations and results, improving the learning curve, and contemporary reducing the mental demands [28]. Focusing on AR, a recent study presented a detailed review identifying benefits, challenges and limitations to help factories to achieve process flexibility by improving inspection, logistics and support maintenance process monitoring [29]. Some methodologies have been recently proposed to support the creation of technical documentation [30] and quality inspection tasks [31]. As a result, review suggested that IXR tools can be used in the various design stages, especially for assembly and maintenance tasks. The main limitations are referred to the absence of specific industry standards in terms of data formats and CAD data exchange. The most common platforms to develop VR and AR applications are Unreal Engine 4 or Unity 3D, allowing 3D models import and creation of interaction features, but their use requires skilled programmers [32].

Concerning Level 5 about SVX, in the last few years different approaches have been developed in order to improve the HFI during the design process. The most interesting feature of SVX is the integration of human monitoring systems with immersive VR scenarios: this could allow an assessment of the process already in the early stages of the design process, preventing design errors and raising the HFI level of the system.

Lee et al. [33] examined the usability and reliability of wearable sensors for monitoring workers in everyday jobs, measuring the workers' individual level of workload and their physiological responses. Peruzzini et al. [34] presented a reference model for the analysis of UX merging digital modelling and physiological data collection, according to the Operator 4.0 paradigm. In addition, Grandi et al. [35] proposed a mixed reality (MR) application where humans are digitized and monitored to evaluate the quality of work. Ito and Kamat [36] developed a prototype monitoring system to prevent ergonomics issues in manufacturing industries. Such tools can be validly integrated with the design cycle, to give feedback on the project quality along the design. The main issues of such tools are related to the complexity and the cost of implementation, even if the recent technological developments and the diffusion on a large scale are gradually reducing the technology cost, year by year, improving the functionality.

5 Conclusions

This paper provided a comprehensive review of the most adopted methods and tools for HFI to support the design of I4.0, considering the scientific literature in the last ten years, and defined the main trends related to human-centered design for industrial purposes. The research defined five levels of digital maturity; for each level, the related tools are discussed. Moreover, the paper mapped the tools with the main application areas to support the design of I4.0 and highlighted the most significant trends of research in the manufacturing field. As a result, it provided a clear overview about the available technologies for HFI in industry and the main strengths and weaknesses. Human tracking and sensorized simulations seem to be necessary to extend the physical ergonomic analysis, including also the users' workload analysis considering also cognitive aspects. This aspect is crucial considering the new evolutions in tasks and job organization in the modern factories. For instance, the combined use of VR and AR simulations with the analysis of human physiological data can enrich task assessment and limit the effect of subjective impressions in the evaluation. Moreover, this approach could be effectively used to involve end users in the production process of products and services, improving the Industry-consumer relationship. However, there is a lack of standards for industry in this field, especially referring to data exchange formats, communication protocols and privacy data management.

References

1. Kagermann, H., Wahlster, W., Helbig, J.: Securing the future of German manufacturing industry: Recommendations for implementing the strategic initiative INDUSTRIE 4.0, Final Rep. Ind. 4.0 Work. Gr., no. April, pp. 1–84 (2013)
2. Zhou, K., Liu, T., Zhou, L.: Industry 4.0: towards future industrial opportunities and challenges. In: 12th International Conference on Fuzzy Systems and Knowledge Discovery, pp. 2147–2152, FSKD 2015. IEEE (2016)
3. Lucke, D., Constantinescu, C., Westkämper, E.: Smart factory - a step towards the next generation of manufacturing, manufacturing systems and technologies for the new frontier, no. Sfb 627, pp. 115–118. Springer (2008). https://doi.org/10.1007/978-1-84800-267-8_23

4. Ŝtefánik, I.A., Gregor, I.M., Furmann, I.R., Ŝkorík, I.P.: Virtual manufacturing in research & industry. IFAC Proc. Vol. **41**(3), 81–85 (2008)
5. Jovane, F., et al.: The incoming global technological and industrial revolution towards competitive sustainable manufacturing. CIRP Ann. **57**(2), 641–659 (2008)
6. Riley, M.W., Dhuyvetter, R.L.: Design cost savings and ergonomics. Proc. Hum. Factors Ergon. Soc. Ann. Meet. **44**(12), 2-511-2–514 (2000). https://doi.org/10.1177/154193120004401219
7. Kong, F.: Development of metric method and framework model of integrated complexity evaluations of production process for ergonomics workstations. Int. J. Prod. Res. **57**(8), 2429–2445 (2019)
8. Pacaux-Lemoine, M.P., Trentesaux, D., Rey, G.Z., Millot, P.: Designing intelligent manufacturing systems through human-machine cooperation principles: a human-centered approach. Comput. Ind. Eng. **111**, 581–595 (2017)
9. Cullen, L.: Human factors integration–Bridging the gap between system designers and end-users: a case study. Saf. Sci. **45**(5), 621–629 (2007)
10. Moher, D., Liberati, A., Tetzlaff, J., Altman, D.G.: Preferred reporting items for systematic reviews and meta-analyses: the PRISMA statement. PLoS Med. **6**(7), e1000097 (2009). https://doi.org/10.1371/journal.pmed.1000097
11. Ceccacci, S., Matteucci, M., Peruzzini, M., Mengoni, M.: A multipath methodology to promote ergonomics, safety and efficiency in agile factories. Int. J. Agile Syst. Manag. **12**(4), 407–436 (2019)
12. Chander, D.S., Cavatorta, M.P.: An observational method for postural ergonomic risk assessment (PERA). Int. J. Ind. Ergon. **57**, 32–41 (2017)
13. Pandian, V.P.S., Suleri, S.: NASA-TLX web app: an online tool to analyse subjective workload. arXiv preprint arXiv:2001.09963 (2020)
14. Jou, Y.T., Lin, C.J., Yenn, T.C., Yang, C.W., Yang, L.C., Tsai, R.C.: The implementation of a human factors engineering checklist for human–system interfaces upgrade in nuclear power plants. Saf. Sci. **47**(7), 1016–1025 (2009)
15. Peruzzini, M., Carassai, S., Pellicciari, M.: The benefits of human-centred design in industrial practices: re-design of workstations in pipe industry. Proc. Manuf. **11**, 1247–1254 (2017)
16. Schaub, K.G., et al.: Ergonomic assessment of automotive assembly tasks with digital human modelling and the 'ergonomics assessment worksheet'(EAWS). Int. J. Hum. Factors Model. Simul. **3**(3–4), 398–426 (2012)
17. Gašová, M., Gašo, M., Štefánik, A.: Advanced industrial tools of ergonomics based on Industry 4.0 concept. Proc. Eng. **192**, 219–224 (2017)
18. Fantini, P., Pinzone, M., Taisch, M.: Placing the operator at the centre of Industry 4.0 design: modelling and assessing human activities within cyber-physical systems. Comput. Ind. Eng. **139**, 105058 (2020)
19. Baskaran, S., et al.: Digital human and robot simulation in automotive assembly using siemens process simulate: a feasibility study. Proc. Manuf. **34**, 986–994 (2019)
20. Jayaram, U., Jayaram, S., Shaikh, I., Kim, Y., Palmer, C.: Introducing quantitative analysis methods into virtual environments for real-time and continuous ergonomic evaluations. Comput. Ind. **57**(3), 283–296 (2006)
21. Caputo, F., Greco, A., D'Amato, E., Notaro, I., Spada, S.: On the use of virtual reality for a human-centered workplace design. Proc. Struct. Integrity **8**, 297–308 (2018)
22. Peruzzini, M., Grandi, F., Pellicciari, M.: A Reference Model to Analyse User Experience in Integrated Product-Process Design. A Paradigm Shift, IOS Press, Transdisciplinary Engineering (2017)
23. Bortolini, M., Gamberi, M., Pilati, F., Regattieri, A.: Automatic assessment of the ergonomic risk for manual manufacturing and assembly activities through optical motion capture technology. Procedia CIRP **72**, 81–86 (2018)

24. Xie, L., Guo, X.: Object detection and analysis of human body postures based on tensorflow. In: 2019 IEEE International Conference on Smart Internet of Things (SmartIoT), pp. 397–401. IEEE (2019)
25. Manns, M., Otto, M., Mauer, M.: Measuring motion capture data quality for data driven human motion synthesis. Procedia CIRP **41**, 945–950 (2016)
26. Caputo, F., Greco, A., Fera, M., Macchiaroli, R.: Digital twins to enhance the integration of ergonomics in the workplace design. Int. J. Ind. Ergon. **71**, 20–31 (2019)
27. Otto, M., Lampen, E., Agethen, P., Langohr, M., Zachmann, G., Rukzio, E.: A virtual reality assembly assessment benchmark for measuring VR performance & limitations. Procedia CIRP **81**, 785–790 (2019)
28. Dobrescu, R., Merezeanu, D., Mocanu, S.: Process simulation platform for virtual manufacturing systems evaluation. Comput. Ind. **104**, 131–140 (2019)
29. de Souza Cardoso, L.F., Mariano, F.C.M.Q., Zorzal, E.R.: A survey of industrial augmented reality. Comput. Ind. Eng. **139**, 106159 (2020)
30. Gattullo, M., Scurati, G.W., Fiorentino, M., Uva, A.E., Ferrise, F., Bordegoni, M.: Towards augmented reality manuals for Industry 4.0: a methodology. Rob. Comput. Integr. Manuf. **56**, 276–286 (2019)
31. Grandi, F., Khamaisi, R.K., Peruzzini, M., Raffaeli, R., Pellicciari, M.: A reference framework to combine model-based design and AR to improve social sustainability. Sustainability **13**(4), 2031 (2021)
32. Liagkou, V., Salmas, D., Stylios, C.: Realizing virtual reality learning environment for Industry 4.0. Procedia CIRP **79**, 712–717 (2019)
33. Lee, W., Lin, K.Y., Seto, E., Migliaccio, G.C.: Wearable sensors for monitoring on-duty and off-duty worker physiological status and activities in construction. Autom. Constr. **83**, 341–353 (2017)
34. Peruzzini, M., Grandi, F., Pellicciari, M.: Exploring the potential of operator 4.0 interface and monitoring. Comput. Ind. Eng. **139**, 105600 (2020)
35. Grandi, F., Peruzzini, M., Zanni, L., Campanella, C.E., Pellicciari, M.: Digital manufacturing and virtual reality for tractors' human-centred design. Adv. Transdisciplinary Eng. **7**, 702–711 (2018)
36. Ito, T., Kamat, S.R.: A proposal of integrated worker-monitoring system towards ergonomic manufacturing environment. Int. J. Agile Syst. Manag. **12**(4), 437–452 (2019)

Evaluation of Mandibular Advancement Devices in Patients Suffering from Obstructive Sleep Apnea and Periodontitis

Marco Mandolini[1](✉), Manila Caragiuli[1], Giovanni Bruno[2], Alberto De Stefani[2], Antonio Gracco[2], Alida Mazzoli[3], and Agnese Brunzini[1]

[1] Department of Industrial Engineering and Mathematical Sciences, Università Politecnica delle Marche, 60131 Ancona, Italy
m.mandolini@univpm.it

[2] Department of Neuroscience, Section of Dentistry, University of Padua, 35100 Padua, Italy

[3] Department of Materials,Environmental Sciences and Urban Planning, Università Politecnica delle Marche, 60131 Ancona, Italy

Abstract. The Obstructive Sleep Apnea Syndrome (OSA) concerns episodes of complete or partial obstruction of the upper airway. Mandibular Advancement Device (MAD) is one of the most used systems for treating this syndrome. Clinicians frequently observe a combination of OSA and periodontitis. There is no research aiming to evaluate how periodontitis staging affects the overall mandibular and maxillary dental arches in the literature. Furthermore, no one has studied the combination between OSA and periodontitis and the effects of MADs on the patients in this condition. This paper aims to develop a numerical simulation approach based on FEM and evaluate the consequences (displacement and stress fields) of the periodontitis staging on PDL and teeth of patients suffering from OSA and treated with MADs. Simulations have been performed for evaluating Stage I of periodontitis. Results highlight a correlation between bone resorption and teeth displacement and periodontal ligaments stress (the higher the bone resorption, the higher the stress and displacement).

Keywords: Obstructive Sleep Apnea · Periodontitis · Mandibular Advancement Device · Finite Element Analysis · Numerical simulation

1 Introduction and Literature Review

People suffering from Obstructive Sleep Apnea (OSA) syndrome are characterised by episodes of complete or partial obstruction of the upper airways during sleep. Nowadays, OSA is treated by several devices, such as C-PAP (Continuous Positive Airway Pressure) and MADs (Mandibular Advancement Devices). MADs are non-invasive tools widely used worldwide for the treatment of OSA. Their effectiveness has remained undisputed over time.

Recently the need to investigate the side effects of this pathology has grown. Many articles nowadays look for OSA treatment's negative consequences to make practitioners more sensitive to the problem and encourage solutions and improvements. An issue

C. Rizzi et al. (Eds.): ADM 2021, LNME, pp. 793–800, 2022.
https://doi.org/10.1007/978-3-030-91234-5_80

related to the MAD is the stress accumulated on the temporomandibular joint (TMJ), caused by the mandible forward's forced movement. Scholars [1, 2] analysed the stress distribution over the TMJ through the Finite Element Analysis (FEA). Although deformations and loads induced by MADs increase with the mandibular advancement, these are lower than the physiological limits of TMJ. Numerical simulation models can help practitioners select suitable MAD concerning the patient's characteristics [3].

OSA is often coupled with other pathologies, and periodontitis is one of the most widespread. Periodontitis is an inflammatory disease that affects the oral cavity in several stages. The initial stage begins with a slight inflammatory activity at the gingival level (gingivitis). Then, periodontitis proceeds wearing down the periodontium. Some teeth' loss may occur at the most severe stages (staging III and staging IV) [4]. As the severity increases, more significant percentages of alveolar bone and teeth are lost.

Almost all the scientific papers presenting numerical simulations for evaluating the effects of periodontitis staging on periodontal ligaments and teeth are focused on just one tooth at a time. Bica et al. [5] evaluated the stress and displacement during the application of orthodontic forces in adult patients with different bone loss values. Geramy [6] assessed the displacement of the tooth with constant application of force (1 N amplitude) and with the reproduction of different stages of bone loss (from 1 to 8 mm). In a recent study [7], the initial stresses in teeth, PDLs, and bone were investigated in the presence of different levels of bone heights during the application of a force for intrusion and tipping on labial and lingual sides of the tooth. In this study, six 3D finite element models of a maxillary central incisor with PDL were designed. At each model, researchers applied a force on the tooth.

Zargham et al. [8] illustrated the effects of force magnitudes on the long-term tipping movements of teeth under different bone loss levels. Another factor that affects the stress over teeth and PDLs in periodontitis is reported in [9]. Zargham's work presents the effects of a maxillary central incisor labiolingual inclination. Here, the aim was to investigate whether labial tooth inclination and alveolar bone loss affect the moment per unit of force (M/F) in controlled tipping and consequent stresses on PDLs.

This research paper wants to answer the following research questions. What happens if OSA patients are also affected by another oral cavity pathology, like periodontitis? What are the consequences of MAD usage in this case? By analysing the literature review, it is possible to conclude that no research aims to evaluate how periodontitis staging affects the overall mandibular and maxillary dental arches. Furthermore, no one has studied the combination between OSA and periodontitis and the effects of MADs on the patients in this condition.

In conclusion, the paper presents a numerical simulation approach, based on FEM, for evaluating the consequences (displacement and stress fields) of the periodontitis staging on PDL and teeth of patients suffering from OSA and treated with MADs. The geometrical model considers the mandible (split into cortical and trabecular bones), periodontal ligaments, mandibular and maxillary arches, and a specific MAD (Avant®, by SomnoMed®). Simulations have been performed for evaluating Stage I of periodontitis.

2 The Numerical Simulation Model

The development of the finite element model can be summarised in the following steps: (i) *geometry acquisition,* (ii) *3D Model reconstruction,* (iii) *model pre-processing,* and (iv) *FEA of the numerical model.*

2.1 Geometry

The 3D virtual model includes mandibular bones, PDLs, teeth, and MAD. The anatomical bodies were reconstructed through a Reverse Engineering approach starting from the Cone Beam Computed Tomography (CBCT) scans of a female patient affected by OSA. Modelling tools allowed to rebuild the PDLs, that are too thin to be acquired from CBCT, and the mandibular bone by discriminating between cortical and cancellous tissue. The cancellous bone was obtained by the Boolean difference between the reconstructed mandible from CBCT scan and the cortical bone. The cortical bone thickness was set at 1.7 mm (average value recovered by analysing the CBCT slices). A thin layer of 0.2 mm was assumed around the tooth socket [10]. By offsetting the tooth-root surface of 0.3 mm, it was possible to obtain a thin homogeneous layer representing the PDL [11].

The SomnoMed Avant device is a MAD with two resin pins (on both the lateral sides of the inferior splint) that act as an insertion point for a rod that fits in central support located over the superior splint. The rod's length defines the advancement; the smaller its size, the greater the entity of mandibular advancement. Such a device was designed by digitising a physical prototype through a 3D laser scan and morphing the splints to fit the patient's teeth. Computer-aided design tools allowed to develop the flexible rod connecting the two MAD's splints, which is responsible for the protrusion of the mandible.

To mimic a bone reduction of 15% according to the first stage of periodontitis, as described by Tonetti et al. [4], a cutting surface was built by sweeping a profile curve through the curve connecting the teeth longitudinal axes' points corresponding to the 15% of the tooth root (Fig. 1a).

The reconstructed geometries were imported in Ansys software v.20 R2 (Canonsburg, PA, USA) to carry out the FEA. To reduce the simulation's complexity in terms of the number of elements and nodes, the authors decided to neglect the maxillary bones and take advantage of the symmetry condition to simulate only half of the model.

2.2 Materials

The material properties of the components representing MAD, teeth, cortical and cancellous bone were assumed to be homogeneous, isotropic, and linearly elastic, as listed in Table 1. The 3D model of the MAD consists of a flexible rod (nylon 6/6) and stiffer splints (acrylic). For what concerns the PDLs, a 2nd Ogden hyperelastic model was chosen for the best fitting of uniaxial experimental data of Natali et al. [12].

Table 1. Material properties of the modelled bodies

Material	Model	E MPa]	υ [-]	Coefficient values
MAD	Linear elastic	8300	0.28	
Rod	Linear elastic	1480	0.41	
Tooth	Linear elastic	18300	0.31	
Cortical Bone	Linear elastic	13700	0.3	
Cancellous Bone	Linear elastic	1370	0.31	
PDL	Hyperelastic Ogden 2nd order	–	–	$\alpha 1 = -3.476$; $\alpha 2 = 18.679$ $\mu 1 = -0.034004$ MPa $\mu 2 = 0.00088691$ MPa $d1 = 0$; $d2 = 0$

2.3 Boundary Conditions

Only half model has been considered assuming a sagittal symmetry plane that divides the model into two symmetric halves to reduce the computational effort during the simulation (Fig. 1b). This way, the model is constrained from moving in a direction normal to the symmetry plane (x-direction, i.e., left/right). Moreover, the maxillary PDLs have been constrained in all directions to prevent rigid body motion. Perfect bonding was assumed between the bones, the cortical bone and the PDLs and between the PDLs and the teeth since PDLs strongly bind the tooth root to the supporting alveolar bone. Thus, no sliding or separation is allowed between the bodies. Mutual sliding without separation was assumed between the two splints of the MAD. A continuous mesh through shared topology was created between MAD and teeth with shared nodes at the boundary where bodies meet. Frictionless, non-linear contact was applied between the rod's centre and the maxillary splint's central support and at the interface between the pin of the mandibular splint and the shoulder's rod. A cylindrical joint was assumed between the rod's end and the mandibular splint pin's shoulder to couple the relative movement between the rod and the splint. It enables lateral translation and axial rotation about the pin's longitudinal axis. A load of 5.59 N (measured by Cohen-Levy et al. [13]) was scoped to the condyle of the mandible to represent the restorative force of the temporomandibular joint at the maximum advancement of the MAD that is 9.5 mm (measured by Bruno et al. [14]).

2.4 Mesh

The simulation models share the same mesh settings to carry out a reasonable comparison supposing linear tetrahedrons with three degrees of freedom per node, suitable for complex geometries. The discretisation approach assumed smaller elements in the PDLs, teeth and thin-wall regions such as the bones' alveoli and greater elements size in the bones' peripheral areas. An element size of 0.2 mm was considered an acceptable trade-off between the mesh's quality in the PDLs and the workstation's computational effort. A dense and tight mesh was obtained in the bodies' thickness, ensuring at least two thickness elements. The curvature and proximity-based function allowed to generate a finer surface mesh in the alveolar zone of the cortical and cancellous bone to improve the accuracy of the mesh quality (Fig. 1c).

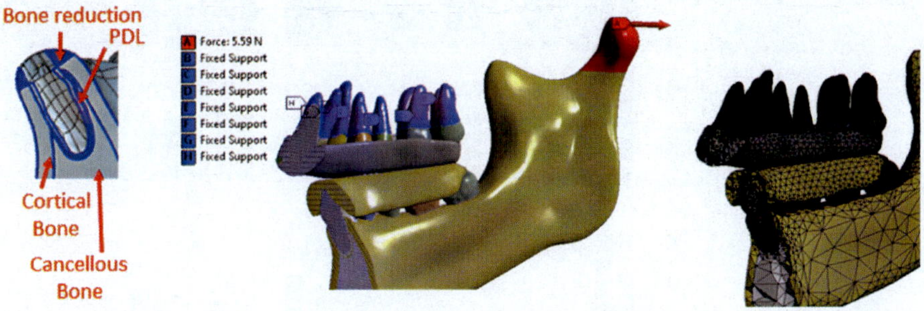

(a) (b) (c)

Fig. 1. Bone reduction caused by periodontitis (a), boundary conditions (b) and discretisation of the 3D healthy model (c).

The healthy 3D model consists of 1,300,732 nodes, 5,676,993 elements, with an orthogonal and skewness criteria of respectively 0.71 ± 0.15 and 0.29 ± 0.15.

3 Results and Discussion

Results in terms of von Mises stress and displacement have been presented only related to the mandibular arch due to the assumed constraint on the maxillary PDLs.

According to Fig. 2a, it is possible to note an averaged increment of the stress of about 9% in teeth affected by periodontitis. This general trend is supported by the findings of Kumar et al. [7], who investigated the initial stress in periodontally compromised models. The most stressed tooth is the II molar, followed by the canine. Simultaneously, the higher increment of stress is recorded in the canine, followed by the I molar, which are more exposed due to the noticeable reduction in bone tissue. Stress mainly affects the coronal side of teeth at the interface with the MAD.

The reduction of bone tissue due to the first stage of periodontitis entails lower support for teeth that are more prone to move due to the applied load, in agreement with Bica et al. [5]. Figure 2b shows a uniform increment of teeth mobility around 10%. The canine tooth is subjected to the lowest reduction in bone tissue. Thus it is less exposed than the other teeth and moves less (9%). Since molars have multiple roots, they provide a greater anchorage than single-rooted teeth. Thus, they are less prone to move than single-rooted teeth. It is worth noting that the applied load on the condyle causes a backward rotation of the mandible which brings the anterior cortical bone upwards and the posterior cortical bone downwards. Thus, the mobility of teeth interests mostly the apex of their root due to the MAD's presence that firmly keeps the teeth crown. For this reason, it is not possible to make a direct correlation with other studies which apply a load in a preferential direction to analyse the tipping or the translation/rotation of a tooth, such as in the study of Geramy et al. [6]. They found out 45% of increment in the displacement of the maxillary central incisor in alveolar bone loss.

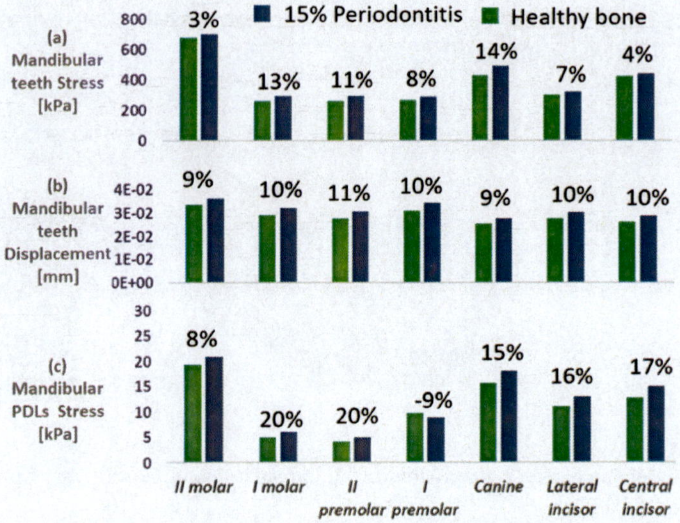

Fig. 2. Evaluation of the mandibular teeth stress (a), teeth displacement (b), and PDLs stress (c) for healthy (green) and diseased (blue) models.

Regarding the PDLs, by observing Fig. 2c, the trend reflects that one of the teeth. The most stressed PDLs are those associated with II molar tooth and canine tooth. The stress decreases by reaching lower values in the second premolar and first molar teeth. It is recorded an increase of stress in all PDLs of the periodontitis model except that one corresponding to the I premolar (–9%) that is the least loaded. Although the increase of stress is relatively uniform in all PDLs, the second molar is characterised by the lowest value (+8%) due to the limited tissue loss. The stress mainly involves the distal root of the PDL and the distal cervical edge due to the mandible downward and backward movement, which tends to move the entire MAD-PDL-bone block under the force's action.

4 Conclusions

The paper presented a numerical simulation approach, based on FEM, for evaluating the consequences (displacements and stresses) of the periodontitis staging on PDL and teeth of patients suffering from OSA and treated with MADs. The mandibular teeth stress increases from Stage 0 (healthy bone) to stage I (15% of bone resorption) of periodontitis up to 14% (9% average value). Mandibular teeth displacement has the same behaviour. From Stage 0 to stage I, the maximum displacement increment is 10%, and the trend is the same for all the teeth. The stress on mandibular periodontal ligaments tends to increase from Stage 0 to Stage I. The maximum increment value is about 20%. However, the behaviour is different among teeth, and the average stress increment is about 13%. This work results align with the articles describing bone resorption in single teeth since more significant deformation and stress values are documented in the most worn areas.

Future research aims to improve the 3D model of mandible (following the approach proposed in [15]) and maxilla to increase their resemblance with actual patients suffering from periodontitis. CT images of patients affected by different periodontitis stages will be analysed for better modelling mandible and maxilla vs. bone loss.

Acknowledgement. The authors are grateful to Dr Maddalena Barocci for her invaluable help in this study's practical development.

References

1. Brunzini, A., Gracco, A., Mazzoli, A., Mandolini, M., Manieri, S., Germani, M.: Preliminary simulation model toward the study of the effects caused by different mandibular advancement devices in OSAS treatment. Comput. Methods Biomech. Biomed. Engin. **21**, 693–702 (2018). https://doi.org/10.1080/10255842.2018.1511776
2. Crivellin, G., et al.: Strength distribution on TMJ using mandibular advancement device for OSAS treatment: a finite element study. Dent. Cadmos. **86**, 757 (2018) https://doi.org/10.19256/d.cadmos.09.2018.04
3. Bruno, G., et al.: Comparison of the effects caused by three different mandibular advancement devices on the periodontal ligaments and teeth for the treatment of osa: a finite element model study. Appl. Sci. **10**, 6932 (2020). https://doi.org/10.3390/app10196932
4. Tonetti, M.S., Greenwell, H., Kornman, K.S.: Staging and grading of periodontitis: framework and proposal of a new classification and case definition. J. Periodontol. **89**, S159–S172 (2018). https://doi.org/10.1002/JPER.18-0006
5. Bica, C., Brezeanu, L., Bica, D., Suciu, M.: Biomechanical reactions due to orthodontic forces. Finite Element Study. Proc. Technol. **19**, 895–900 (2015). https://doi.org/10.1016/j.protcy.2015.02.128
6. Geramy, A.: Alveolar bone resorption and the center of resistance modification (3-D analysis by means of the finite element method). Am. J. Orthod. Dentofac. Orthop. **117**, 399–405 (2000). https://doi.org/10.1016/S0889-5406(00)70159-4
7. Kumar, J., Reddy, G., Sridhar, M., Reddy, Tj., Reddy, P., Rao, S.: A finite element analysis of initial stresses and displacements in the tooth and the periodontium in periodontally compromised simulations: labial versus lingual force application. J. Dr. NTR Univ. Heal. Sci. **5**, 34 (2016). https://doi.org/10.4103/2277-8632.178976
8. Zargham, A., Geramy, A., Rouhi, G.: Evaluation of long-term orthodontic tooth movement considering bone remodeling process and in the presence of alveolar bone loss using finite element method. Orthod. Waves. **75**, 85–96 (2016). https://doi.org/10.1016/j.odw.2016.09.001
9. Choi, S.-H., Kim, Y.-H., Lee, K.-J., Hwang, C.-J.: Effect of labiolingual inclination of a maxillary central incisor and surrounding alveolar bone loss on periodontal stress: a finite element analysis. Korean J. Orthod. **46**, 155 (2016). https://doi.org/10.4041/kjod.2016.46.3.155
10. Borák, L., et al.: Bilinear elastic property of the periodontal ligament for simulation using a finite element mandible model. Dent. Mater. J. **30**, 448–454 (2011). https://doi.org/10.4012/dmj.2010-170
11. Hohmann, A., et al.: Periodontal ligament hydrostatic pressure with areas of root resorption after application of a continuous torque moment. Angle Orthod. **77**, 653–659 (2007). https://doi.org/10.2319/060806-234

12. Natali, A.N., Carniel, E.L., Pavan, P.G., Bourauel, C., Ziegler, A., Keilig, L.: Experimental–numerical analysis of minipig's multi-rooted teeth. J. Biomech. **40**, 1701–1708 (2007). https://doi.org/10.1016/j.jbiomech.2006.08.011
13. Cohen-Levy, J., Pételle, B., Pinguet, J., Limerat, E., Fleury, B.: Forces created by mandibular advancement devices in OSAS patients. Sleep Breath. **17**, 781–789 (2013). https://doi.org/10.1007/s11325-012-0765-4
14. Bruno, G., et al.: A procedure for analyzing mandible roto-translation induced by mandibular advancement devices. Materials (Basel). **13**, 1826 (2020). https://doi.org/10.3390/ma13081826
15. Pascoletti, G., Calì, M., Bignardi, C., Conti, P., Zanetti, E.M.: Mandible Morphing Through Principal Components Analysis. In: Rizzi, C., Andrisano, A.O., Leali, F., Gherardini, F., Pini, F., Vergnano, A. (eds.) ADM 2019. LNME, pp. 15–23. Springer, Cham (2020). https://doi.org/10.1007/978-3-030-31154-4_2

Limitations of Using Generative Design for Maxillo-Facial Reconstructive Surgery Applications

Stefano Filippi[1(✉)], Massimo Robiony[2], Alessandro Tel[2], and Francisco Daniel Montoya Buteler[1]

[1] DPIA Department, University of Udine, Udine, Italy
stefano.filippi@uniud.it, montoyabuteler.franciscodaniel@spes.uniud.it

[2] Maxillofacial Surgery Department, University of Udine, Udine, Italy
massimo.robiony@uniud.it, alessandro.tel@icloud.com

Abstract. Generative Design (GD) is an emerging approach to the development of industrial products. Its key-point is the automatism in defining the shapes of artifacts by obeying to parameters describing the environment the artifacts will interact with, their mechanical properties, their materials and the manufacturing processes that will be used to build them. This paper describes some GD experiences in medicine, aimed at developing artifacts for maxillo-facial surgery. Once highlighted a real surgical case, designers, engineers and surgeons worked together to set the parameters to enable GD generating the artifact to replace a part of a patient's jaw hit by a genetic disease. Three adoptions of the GD algorithms occurred, along with parameter refinement and integration time after time. Designers, engineers and surgeons reasoned together on the solutions; this led to considerations about the suitability of GD in maxillo-facial surgery and about the current limits of this approach, as well as to some hints for carrying on the research activities in the near future.

Keywords: Generative design · Maxillo-facial surgery · CAD in medicine

1 Introduction

Generative Design (GD) is an emerging approach to the development of industrial products. The novelty of this approach is that designers and engineers do not define shapes anymore; they just need to describe the environment in terms of interfaces with the surrounding, loads and constraints, safety factors, manufacturing processes, etc., and the GD algorithms generate the shapes accordingly [1]. GD focuses mainly on industrial products because of its origin; indeed, it can be seen as the natural, big evolution of topological optimization. For this reason, parameter setting, algorithmic functioning, etc., all of them work well when the problem to solve belongs to the industrial/mechanical domain. Nevertheless, also due to the youth of this design approach, this could be the right moment to explore other application contexts in order to highlight possible lacks

C. Rizzi et al. (Eds.): ADM 2021, LNME, pp. 801–808, 2022.
https://doi.org/10.1007/978-3-030-91234-5_81

and suggest more ways to describe the environment to the GD, aiming at generating more and more suitable and compliant shapes in heterogeneous domains.

This research aimed at verifying GD suitability in the maxillo-facial surgery context. Designers, engineers and surgeons together highlighted a surgical case where an artifact was required. Then, all the actors contributed to the setting of the process parameters (parameters, hereafter) to enable the first adoption of the GD algorithms. The results suggested some refinements of the parameters and the process reran. Three full cycles were considered as enough to start drawing some considerations about the expected results of the research. The following discussion took care of this.

This paper develops as follows. Section 2, materials and methods, summarizes the GD as well as the real surgical case this research focused on. Section 3 describes the research activities in detail. Section 4 presents the results of the research and discusses them. The Conclusions and References sections close the paper.

2 Background

2.1 Generative Design

Generative Design (GD) is an emerging approach to the shape definition of industrial products. It can be considered as the evolution of the topology optimization since topology optimization focuses just on material minimization in existing models to satisfy mechanical constraints while GD defines shapes from scratch, being able to consider requirements about materials, safety factors, manufacturing processes, etc. [1]. In other words, GD allows designers and engineers reasoning about the problems to solve rather than on how to solve them. Once described the problem by setting parameter values, the GD algorithms give a set of solutions, all of them obeying to those parameters. Then, designers and engineers apply their skill and knowledge in selecting those solutions embodying the optimum from their points of view. There are some software packages on the market implementing the GD, like Fusion 360 by Autodesk, PTC's Creo, Siemens' SolidEdge and NX, etc. This research exploited the GD section of Fusion360, a multi-purpose, cloud-based CAD tool [2].

2.2 Maxillo-Facial Surgery

Maxillo-facial surgery is a surgical specialty whose focus is the surgery of the splanchnocranium and the neck, with particular attention to the facial region. Compared with other specialties of the head and neck area, for instance otolaryngology, a large amount of maxillofacial surgery procedures is represented by re-constructive surgery. Reconstruction aims to restore anatomy and function when tissues are damaged by traumatic events, or iatrogenically sacrificed to excise pathological processes, as in the case of oncologic malignancies. Soft tissue reconstruction is performed using locoregional flaps and suturing techniques, whereas in case of wide defects for which substance loss coverage is not possible using nearby tissues, the use of a microvascular free flap is advocated. For bone reconstruction, consistent defects can be restored using microvascular free flaps as well, especially the fibula and the iliac crest free flap, although in the last years customized alloplastic implants provided a much more straightforward method to reconstruct bony defects, without the complications related to tissue harvesting. Currently, customized prostheses are designed starting from an anatomical model derived from the three-dimensional reconstruction of a CT scan [3]. Using the real anatomy as a template, customized implants are modelled to provide optimal fit to anatomical surfaces and therefore they are referred as "patient-specific" implants (PSIs) [4]. However, PSIs are generally designed using free-form techniques using a combination of design techniques based on polygons, clays or NURBS. However, such devices after design do not undergo any engineering validation, and therefore biomechanical features are strictly dependent of the designer's experience. There is documented research comparing custom-made artifacts to commercial ones, as described in [3].

3 Materials and Methods

Among the several specificities of the maxillo-facial surgery discipline, this research focused on a real case dealing with the surgery on the jaw of a patient hit by osteopetrosis, a genetic disease which altered the quality of bone, making it extremely fragile and subject to pathological fractures. As the disease is diffuse to all the bones of the patient, which in his past years underwent multiple surgical operations for limb fractures, the use of a fibula free flap to reconstruct the mandible was unsuitable. Therefore, the choice was to design and position a customized prosthesis to reconstruct the missing bone structure and cover it with a radial free flap to provide the soft tissue lining within the oral cavity. Figure 1 shows an overview of the case considered in this research: reconstruction of preoperative model showing areas of bone destruction in the mandibular body (left); definition of resection planes (right). Each GD adoption occurred in this research started from the STL file containing the triangle mesh describing the patient's anatomy as shown in Fig. 1. This data representation was made available by the surgeons who generated it thanks to the elaboration of CT scans.

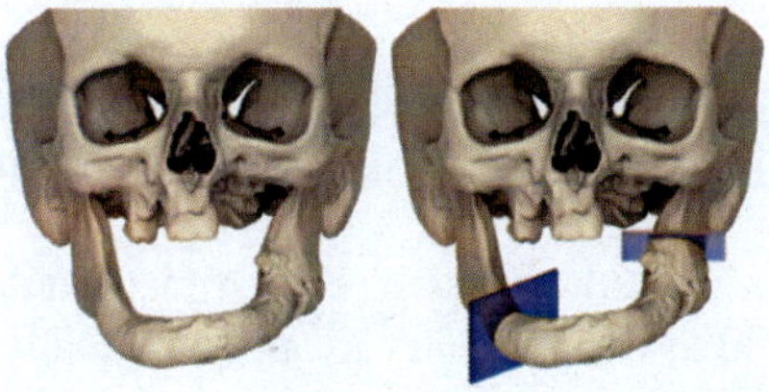

Fig. 1. Surgical case considered in this research.

Once selected the surgical case to consider, the setting of the general parameters took place as well as that of the parameters specific for the first GD adoption. Some research has been of real help during this setting, like that of Chen et al. [5]. The results of the first adoption suggested refining the parameters and rerunning the GD algorithms. In the end, three GD adoptions occurred before to get meaningful data to reason about and to draw some early considerations about the suitability of GD in the maxillo-facial surgery context. Every GD adoption occurred in the same way: the parameter setting allowed running the GD algorithm; then, the results were discussed by all the actors involved during the debriefing and the decision to consider the outcomes as satisfactory or not arose, as well as hints and suggestions for the new parameter setting. The following sections depict all of this.

3.1 Setting the General Parameters

In all the GD adoptions, values and points of application of loads and constraints were chosen to be consistent with what happens in situations where artifacts similar to the expected ones are used. Some references have been of great help for these settings. Commisso et al. [6] showed an interesting biomechanical model of the human jaw, complete with bones and muscles mathematical models. This approach represents a powerful methodology to analyze the functioning of the human jaw. Although a similar path would have been ideal to follow in the study taken in consideration in this article, this approach could not be implemented due to the lack of flexibility in the way that loads and constraints are implemented in the current GD software. Bekcioglu et al. [7] and Mazzocco et al. [8] performed FEM analysis of TMJ prostheses; they found that maximum stress is in the order of magnitude of 200–250 Mpa in the condyle area.

Based on this, designers, engineers and surgeons decided to adopt reasonable loads and constraints in order to have results that, in a first attempt, could be consistent with what happens in situations where artifacts similar to the expected ones are used. Specifically, all loads were considered as pressures with values between 200 and 250 Mpa. Along with the settings of loads and constraints, according to the surgeons, designers and engineers selected PLA (polylactic acid) and titanium 6Al–4V as the materials for all adoptions, the manufacturing processes were 3D printing and 5-axis machining to allow the GD algorithms generating both proof of concept (3D printing) and proof of product (5-axis machining) prototypes. Finally, the safety factor was set to two.

3.2 First GD Adoption

The first adoption occurred mainly to verify the feasibility and functioning of GD in the specific application context. Therefore, result quality was not the main concern.

Parameter Setting. In this case, the supports the artifact would have been fixed to the jawbone with were selected as the geometry to apply the loads and constraints to.

Results. As expected, result quality was very low. Nevertheless, the GD algorithms converged and gave more than one result. Figure 2 shows an overview of these results and the highlighting on one of them.

Debriefing. Clearly, the results were unacceptable from the surgical point of view. First, there was a significant concentration of mass near the supports; second, the shape invaded completely the oral cavity. Nevertheless, these results gave precious hints for tuning the

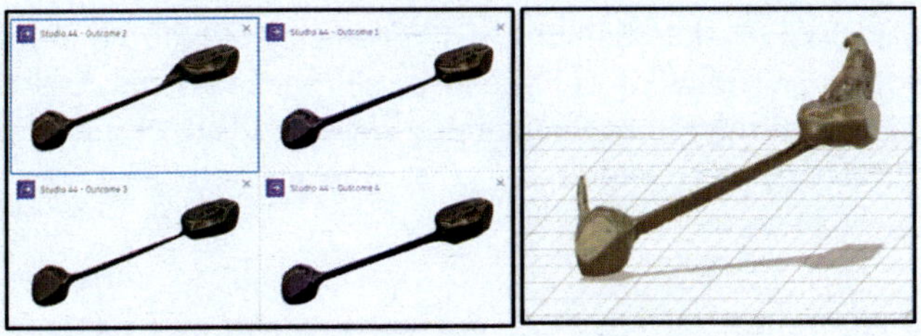

Fig. 2. Results of the first GD adoption.

parameters for the next GD adoption. Moreover, the research description considers this first attempt also to replicate as closely as possible the GD adoption in maxillo-facial surgery step by step, by almost every designer and/or engineer.

3.3 Second GD Adoption

Parameter Setting. The results of the previous GD adoption suggested two main parameter changes. The first change consisted of a review of the points of application of loads and constraints. As second change, geometries - two tube-like rails - were introduced to guide the generation of the shape; moreover, the supports to fix the artifact to the jawbone were set as obstacle geometry. Figure 3 (left) shows the parameter setting. Green geometry corresponds to the shape to keep and red geometry is the obstacle (space not to invade). Arrows represent the points of application of the loads.

Results. Figure 3 (right) shows one of the results of the second GD adoption. This result started replicating quite well the morphology of the artifacts currently used in this kind of surgery. The invasion of the oral cavity disappeared and the profile was compatible with the missing part of the jawbone.

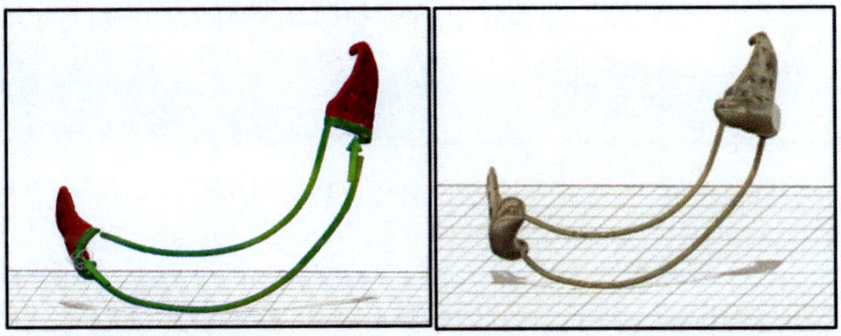

Fig. 3. Parameter setting (left) and one of the results (right) of the second GD adoption.

Debriefing. Nevertheless, weird concentrations of mass were already present and all the actors involved - designers, engineers and surgeons - agreed that such a shape would not offer enough support for the positioning of tissues, muscles, skin, etc., during surgery.

Quite clearly, the reason is that the preserving geometry introduced to force the shape to replicate the jawbone curvature well satisfies itself the loads and constraints applied; therefore, the GD algorithm thinks that nothing more needs to be added. Then, further changes to the parameters were required.

3.4 Third GD Adoption

Parameter Setting. The parameter setting for the third adoption consisted of a new refinement of loads and constraints. Along with this, the part of the jawbone affected by the disease (the one between the two planes in Fig. 1, to the right) was considered as starting shape for a refinement based on the mirroring of the remaining, right-side part of the jawbone, and on similarities with other patients' anatomies. The resulting shape, once made negative, was used as obstacle for the GD algorithms.

Results. This time, the GD algorithms did not converge. They were not able to give definitive solutions, maybe due to the complexity of the parameter values. The computation failed after few iterations (usually, the GD generates final shapes after 30 or more iterations). Then, designers and engineers considered the last partial solution before the failure. Clearly, in this case, the GD gave very scarce contribution, because this partial solution corresponds more or less to the parameter setting. Nevertheless, as shown in Fig. 4, the weird masses disappeared and the shape was thicker and more suitable to act as support for tissues, muscles, skin, etc. Both the digital model (to the left in Fig. 4) and the 3D print (to the right) of the result were available for reasoning and discussing during the debriefing.

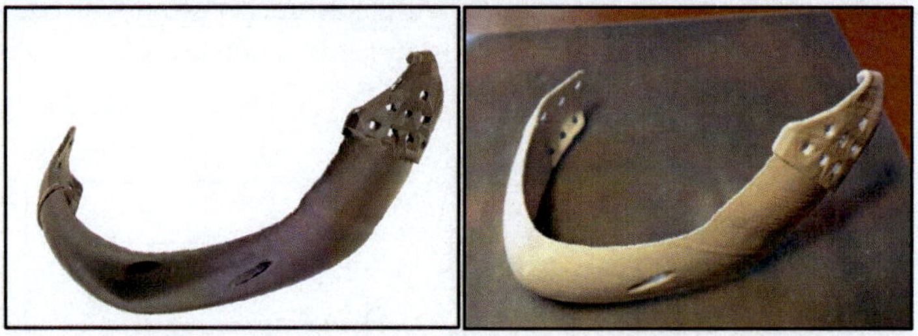

Fig. 4. Digital (left) and physical (right) representation of the result of the third GD adoption.

Debriefing. This third adoption generated better results than before and surgeons were slightly optimistic about their suitability in real surgery. These results were quite comparable to those obtained adopting the traditional approach to the generation of the artifacts. Shapes were close to each other, showing maximum deviation values lower than two millimeters. Nevertheless, from the designers and engineers' point of view, the setting of the parameters started becoming too challenging; moreover, and most important, the GD algorithms did not converge. Finally, the tight constraint setting determined indeed the resulting shape; in other words, almost all the degrees of freedom the GD exploits to give its contribution at best were locked. In turn, all of this made the benefit of using the GD fading drastically.

4 Results and Discussion

The results of the three adoptions of the GD algorithms show a clear trend towards better solutions. The results of the first adoption make appear it just as the way for the actors to get confident with the possible synergy between GD and maxillo-facial surgery. Clearly, these results do not have anything in common with something suitable for real surgery. Nevertheless, this adoption was of great help in focusing on the most important parameters, on the sensitivity of their values and on the pieces of information required to get better results. The results of the second adoption seem to be a good tradeoff between the effort in setting the parameters and the improvement of the shape against that obtained previously; nevertheless, this result was still unacceptable on the surgical side. The results of the third adoption reached a reasonable quality from the surgical point of view but the parameter setting was hard and time-consuming and the GD algorithms did not converge (in other words, the resulting model corresponded to the parameter setting; no shape generation/optimization came from the GD adoption). All of this pushed designers and engineers to consider classic design strategies as still preferable. As general considerations, at the end of this first experience in adopting the GD in the maxillo-facial surgery context, all the actors involved - designers, engineers and surgeons - agreed about judging this experience as quite unsatisfying. More specifically, from the surgeons' point of view, in maxillo-facial surgery more than in other surgical disciplines, esthetics and functionalities have the same importance. Current GD algorithms are great in taking care of the latter but they cannot deal with the former. From the designers and engineers' point of view, the number and definition of the spatial constraints needed to get acceptable shapes made the GD adoption hard and time-consuming. A classic modeling approach would have generated better results, earlier and more easily. Moreover, many problems arose in dealing with STL models, the only data format available from the surgical side, because current GD algorithms work with B-Rep models only. Clearly, transforming STL triangle meshes into B-Rep models is possible but current software packages hardly convert meshes made of hundreds of thousands of triangles or cannot manage B-Rep models with such numbers of triangles.

Consequently, all of this allows drawing some suggestions and some research directions. Clearly, GD is currently quite far from domains different from those mechanical or industrial in general. Parameters determining shape generations allow describing these contexts at best but they cannot deal with requirements present in different domains. For example, it would be great if GD could consider requirements about appearance, user experience, life quality, etc. In the meantime, waiting for this big improvement, software packages implementing GD should at least allow using triangle meshes as they are, without forcing the users to translate them into B-Rep models. Moreover, a tradeoff could be to use the GD approach for developing the maxillo-facial artifact functionalities while the management of appearance matters would occur separately. Once finished, a smart way to merge the partial results would be needed, to get something that, at the same time, shows the best accomplishing appearance and is the top for what concerns mechanical properties, manufacturability, etc.

5 Conclusions

This paper described the use of Generative Design (GD), an emerging approach to the development of industrial products, in developing artifacts for maxillo-facial surgery. The analysis of the solutions coming from three adoptions of the GD algorithms enabled by different parameter settings highlighted lights and shadows about this possible synergy. Among the others, if, on one hand, GD guarantees mechanical properties, manufacturability, safety factors, etc. of the solutions proposed, on the other hand appearance matters deserve and require much more attention. One of the suggested paths to follow would consist of starting reasoning about the artifacts from the appearance and functional points of view separately, and merging the outcomes afterwards.

Thus, the research perspectives will develop in this direction. There will be the attempt to find a smart way to join the contributions to the definition of the artifact shapes coming from the GD for what concerns the technological aspects and from the surgeons' experience and the bibliography regarding the appearance. Nevertheless, more adoptions of the GD will occur, focusing on different surgical cases, in order to refine even more the conditions needed by the GD algorithms to make better artifacts available.

References

1. Tyflopoulos, E., Tollnes Flem, D., Steinert, M., Olsen, A.: State of the art of generative design and topology optimization and potential research needs. In: DS 91: Proceedings of NordDesign 2018. Linköping, Sweden (2018)
2. Autodesk Homepage. https://www.autodesk.com/solutions/generative-design/manufacturing?us_oa=dotcom-us&us_si=e98789f9-9b2b-451b-9a4f-92f48b1b5314&us_st=Generative%20Design. Accessed 13 Mar 2021
3. Ackland, D. C., Robinson, D., Redhead, M., Vee Sin Lee, P., Moskaljuk, A., Dimitroulis, G.: A personalized 3D-printed prosthetic joint replacement for the human temporomandibular joint: from implant design to implantation. J. Mech. Behav. Biomed. Mater. **69**, 404–411 (2017)
4. Luo, D., Rong, Q., Chen, Q.: Finite-element design and optimization of a three-dimensional tetrahedral porous titanium scaffold for the reconstruction of mandibular defects. Med. Eng. Phys. **47**, 1350–4533 (2017)
5. Chen, X., et al.: Biomechanical evaluation of Chinese customized three-dimensionally printed total temporomandibular joint prostheses: a finite element analysis. J. Craniomaxillofac. Surg. **46**(9), 1561–1568 (2018)
6. Commisso, M.S., Martınez-Reina, J., Ojeda, J., Mayo, J.: Finite element analysis of the human mastication cycle. J. Mech. Behav. Biomed. Mater. **41**, 23–35 (2015)
7. Bekcioglu, B., Bulut, E., Bas, B.: The effects of unilateral alloplastic temporomandibular joint replacement on the opposite-side natural joint: a finite-element analysis. J. Oral Maxillofac. Surg. **75**(11), 2316–2322 (2017)
8. Mazzocco, A.G., Jardini, A.L., Nunes, E.L., Maciel Filho, R.: Custom-made temporomandibular joint prosthesis: computer aided modeling and finite elements analysis. Chem. Eng. Trans. **74**, 1489–1494 (2019)

Heart Sound Processing Model for a Mat-Shaped Device

Luca Salvati[1], Nicola Cappetti[1], Matteo d'Amore[2], Anita Fiorentino[3], Arcangelo Pellegrino[1], Pasquale Sena[2], and Francesco Villecco[1(✉)]

[1] Department of Industrial Engineering, University of Salerno, Via Giovanni Paolo II 132, 84084 Fisciano, Italy
fvillecco@unisa.it

[2] Department of Pharmacy, University of Salerno, Via Giovanni Paolo II 132, 84084 Fisciano, Italy

[3] Advanced Technology and Pre-Development Programs, Stellantis, Via Ex Aeroporto, 80038 Pomigliano d'Arco, NA, Italy

Abstract. The development and spread of human-centered products that are increasingly simple and affordable has seen their application areas increase over the years, as well as their effectiveness and reliability. The study of a model based on the interpretation of biosignals is able to provide real-time monitoring of physiological conditions and to offer useful support in both clinical and driving safety fields. Heart rate variability (HRV) analysis is useful in assessing the dynamics of the autonomic nervous system (ANS) and detecting the effects of numerous systemic diseases as well as changes related to the normal daily biological rhythm or to accidental situations of stress or fatigue, as well as comfort or vigilance. The conventional system for the acquisition of the electrocardiographic (ECG) signal uses a methodology that provides for the contact of electrodes with the human skin. This research proposes a detection platform without direct contact with the skin capable of acquiring cardiac signals in real time: through a digital signal processing algorithm able of filtering noise and identifying peaks, heart rate (HR), the interval between beats (RRI) and the characteristic indices of HRV in the time and frequency domain are determined. The parameters were compared with those of a conventional ECG using the Pearson correlation coefficient which produced an index ranging from 0.30 to 0.78 for the tachogram, managing to provide, in the cases less affected by noise, a correspondence in the results of the spectral analysis useful for the evaluation of sympatho-vagal balance.

Keywords: Human-centered-design · HRV · RRI · Non-invasive

1 Introduction

Human-centered design is based on satisfying the user's needs mainly related to performance, interaction, comfort and usability. This concept finds application in product development in the industrial sector: the diffusion of increasingly accessible and easy-to-use instruments, their application in the most diverse fields, is the context from which

C. Rizzi et al. (Eds.): ADM 2021, LNME, pp. 809–817, 2022.
https://doi.org/10.1007/978-3-030-91234-5_82

this study of a model able to provide support for research in both clinical and purely commercial fields was conceived.

In several recent works, especially from the design of aircraft seats and car seats but not limited to these fields [1, 2], there has been an increasingly application of bio-sensors, touch pressure measure devices, optical methods to collect data useful for improving both the posture and general comfort [3]. In the analysis of the cross relations between posture, seats design and comfort level there has been an increasingly development and application of some electro-cardio sensors which however are directly interacting with the user thus altering the comfort. To avoid this problem, have been recently studied some contactless sensors which enable to efficiently measure the physiological conditions of the subject thus leading to a more objective analysis of the comfort. In previous studies, a subject's condition monitoring systems were developed: they acquired biological signals using a polyvinylidene fluoride sensor [4], a microwave radar [5] or measurement sensors installed under a mat [6] which can also monitor arrhythmia and sleep apnea syndrome. A technology based on the acquisition and processing of biosignals, in this case the cardiac ones, allows the search for physiological parameters capable of providing an objective reading of perceptual states related to the sensations of fatigue, malaise and postural discomfort, concentration and sleepiness, as well as define the product ergonomic performance. The variation of the intervals between two consecutive beats, corresponding to the R waves in the electrocardiogram, is caused by the activity of the ANS and it is known as heart rate variability (HRV): it is the main tool available for non-invasive evaluation of ANS dynamics, it reflects the homeostatic interaction between perturbations in central cardiovascular functions and the dynamic responses of cardiovascular regulatory systems [7]. Physiological parameters HR and HRV are significant in the diagnosis of various clinical and functional conditions or for daily health monitoring [8, 9]. Therefore, reliable techniques are needed in clinical practice for the evaluation of the sympatho-vagal balance of ANS due to impaired autonomic balance in some cardiovascular diseases and pathophysiological conditions [10]: a widely used technique to evaluate ANS dynamics is to calculate the power spectral density of HRV [7]. It is known that the high frequency (HF) components of HRV are exclusively influenced by the parasympathetic system; conversely, the low frequency (LF) components of HRV are influenced by the sympathetic and parasympathetic nervous systems.

In this study we compare the HRV parameters obtained from the analysis of the data acquired through an electrocardiograph with those obtained from the use of a non-invasive system consisting of a mat-shaped device on which the subject's back is laid: it incorporates a microsensor capable of detecting the sound vibrations produced by the heart despite the subject wearing garments.

2 Methods of HRV Analysis

The study of HRV focuses mainly on the analysis in the time and frequency domain [7, 11–13]. However, cardiac activity is an integrated signal that is influenced not only by the two branches of the ANS, but also by a number of other subordinate physiological mechanisms and various external factors.

Time domain measures are basically divided into two classes: (a) measures of variability derived from RRI data; (b) measures of variability derived from the differences between adjacent RRIs. To the first subclass belong the mean RRI, the HR and the standard deviation of the RRI (SDNN), immediate but also the least informative. In the second subclass, the most explanatory parameter is the RMSSD: it is used to estimate the high frequency beat-to-beat variations that represent vagal regulation activity. A low RMSSD value indicates poor parasympathetic activity and difficulty in recovering from physical exertion or a situation of high emotional stress. Other parameters used to assess beat-to-beat variations include NN50, the number of adjacent RRIs that differ by more than 50 ms, and pNN50 (NN50/total number of RRIs). Since these parameters are highly correlated with RMSSD, they are also good estimators of vagal activity [14, 15].

The Poincaré plot is the scatter plot of the intervals between beats as a function of the previous intervals between beats. It is a geometric technique taken from non-linear dynamics. In the Spectral Analysis, the FFT technique allows to assign power in different bands to different underlying physiological functions. It is widely accepted that the power in the HF band (0.15–0.4 Hz) represents vagal activity, while the LF band (0.04–0.15 Hz) and that one of the very low frequencies VLFs ($\leq$0.04 Hz) are associated with sympathetic activity or both [16].

2.1 Detection System

This section describes the procedure for detecting the heartbeat acquired through a sensor integrated into a mat placed on the back of a seat at a height close to that of the heart (about 40 cm from the seat). To detect the HR and the characteristic HRV indices, an algorithm was developed that can identify the R peaks of the ECG wave by processing the heart sound signal coming from the seated subject. This signal is recorded through a capacitive microphone MP34DT06J sensor (64 dB SNR and –26 dBFS $\pm$ 1 dB sensitivity; 3 mm $\times$ 4 mm $\times$ 1 mm) embedded in a specific support made of 3D polyester material and inserted into a mat. This material acts as an oscillator with a center frequency of approximately 20 Hz and, through the phenomenon of stochastic resonance, allows the filtering of the acoustic signal around 20 Hz for the removal of artifacts [17, 18].

The acquisition of cardiac signals, both from ECG and from sensor, takes place at a sampling rate of 1000 Hz.

2.2 Signal Processing

In order to eliminate the noise components due to breathing or movement artifacts, three types of bandpass filter with different cut-off frequencies were tested to recognize which produced the cardiac signal most similar to that obtained with the electrocardiograph. Therefore for the first bandpass filter (b.p.f) the frequency bands 10–30 Hz, 15–80 Hz and 4–50 Hz were chosen and subsequently the signal was filtered between 0.8 and 1.8 Hz so that it assumed the analogous wave form to an electrocardiographic signal, having a frequency close to 1 Hz. The evaluation of the time interval between the peaks of this waveform allows an estimate of the parameters in the time and frequency domain typical of HRV analysis (Fig. 1).

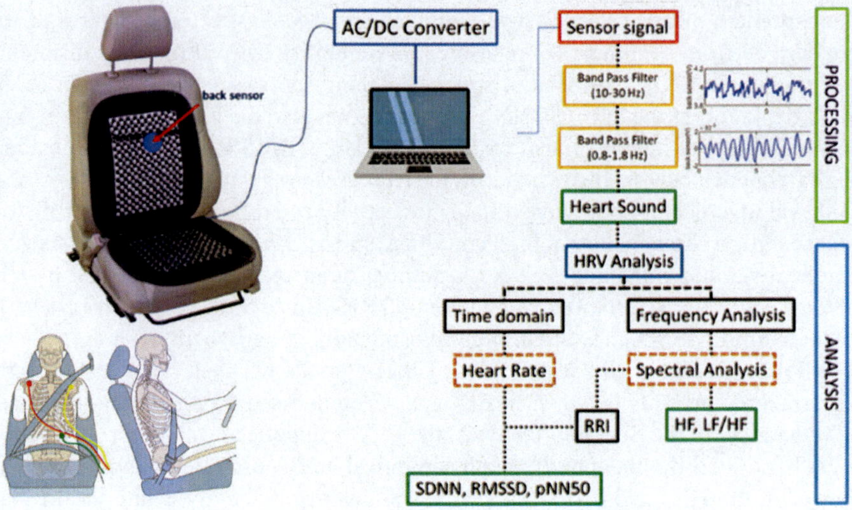

Fig. 1. Signal processing and postural scheme

2.3 Experiment Protocol

Ten male healthy subjects between the ages of 25 and 54 (mean age ± S.D.: 35.6 ± 9.4 years), informed in advance about the type of experiment and consenting, were made to sit for 5 min on a car seat fixed on the floor of the laboratory and equipped with the mat that incorporates the sensor. Participants were asked to sleep at least 7 h the night before the experiment, not to smoke and not to take caffeine, alcohol or energy drinks after waking up until the time of the experiment. The subjects were also asked to arrange the three ECG electrodes on themselves. The electrodes were placed in RA (right arm), LA (left arm) and LL (left leg) positions.

Due to the distance between body trunk and the mat, the sensor signal is much weaker (about ±5 mV) than the signal received by the electrodes applied to the subject's body (about ±0.5 V).

3 Results

A preliminary visual examination of the Poincaré Plot, relating to both the ECG signal and the HS (heart sound) signal processed by the algorithm, revealed that no signal loss was recorded during the detection (Fig. 2).

To establish which filter offered the best performance, the tachograms of the ECG and HS signals relating to each acquisition were compared. A Pearson correlation coefficient "r" was calculated for each comparison. Then the average relative to the analyzes carried out with the same b.p.f. was calculated. and the most performing filter turned out to be the one with a frequency range of 10–30 Hz, as shown in Table 1.

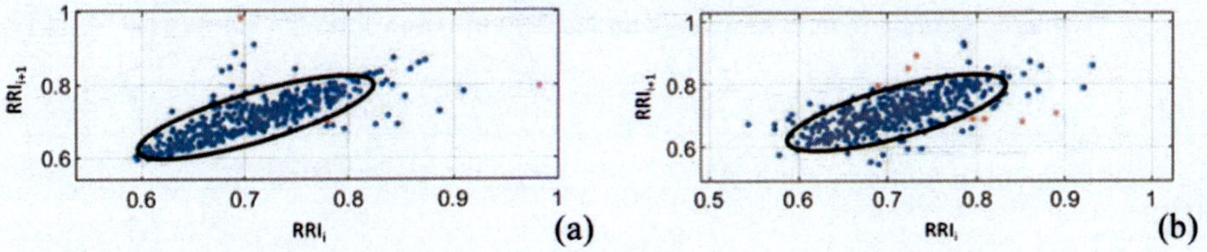

Fig. 2. (a) RRI Poincaré Plot; (b) HS peaks Poincaré Plot.

Subsequently, the measurements obtained with the sensor that showed the greatest and least correspondence with the tachogram obtained from the ECG are reported, respectively those of Subject 1 and Subject 4. With regard to Subject 1, in Fig. 3 the trend of the tacogram obtained from the HS signal (red line) reproduces with good approximation the trend of the RRI obtained from the ECG, so much that even the parameters in the time domain HR, RRI, SDNN, RMSSD, SD1, SD2 are equal or deviate by a few units (Table 2).

To further appreciate the similarity of the results obtained, an analysis in the frequency domain was also performed. Figure 4 shows the time course of the HF and

Table 1. Comparison of b.p.f. filters.

	Pearson correlation coefficient r		
	10-30 Hz	15-80 Hz	4-50 Hz
Subjects	r	r	r
1	0.78	0.89	0.28
2	0.68	0.18	0.17
3	0.62	0.56	0.47
4	0.30	0.31	0.22
5	0.38	0.37	0.15
6	0.34	0.58	0.47
7	0.43	0.25	0.15
8	0.52	0.14	0.44
9	0.68	0.31	0.37
10	0.63	-0.19	0.20
Mean	0.54	0.34	0.29

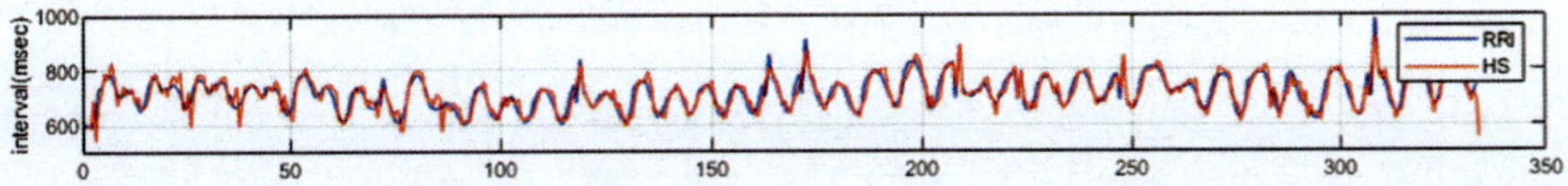

Fig. 3. Subject 1: overlap of the tachogram obtained with ECG (HR) and sensor (HS).

Table 2. Comparison of cardiac parameters in the time domain (Subject 1).

	ECG	HS
Mean RR|HR	709|85	709|85
SDNN	58.7	61.4
RMSSD	44.6	45.2
pNN50	11.7	21.6
SD1|SD2	31.5|76.5	31.9|80.5
SD1/SD2 ratio	0.41	0.4

LF/HF parameters for both ECG acquisition and sensor one: from a visual estimate it can be seen that in this case the algorithm can only approximate the actual LF/HF trend and it works better on the HF signal. The spectral diagram in Fig. 5 certainly offers a quantifiable estimate of the algorithm's efficiency.

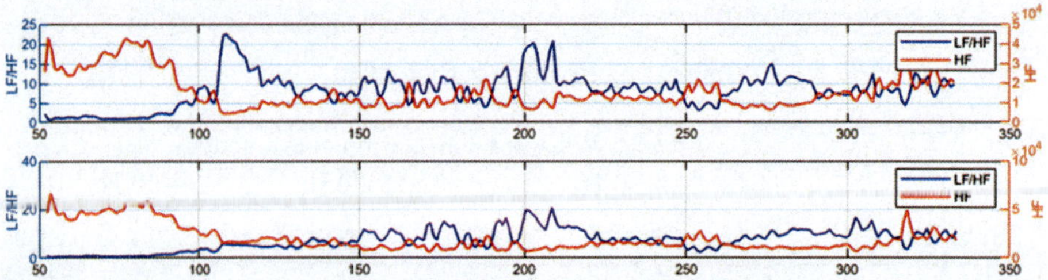

Fig. 4. Subject 1's HRV Frequency Analysis: comparison between ECG (top) and sensor (bottom) results.

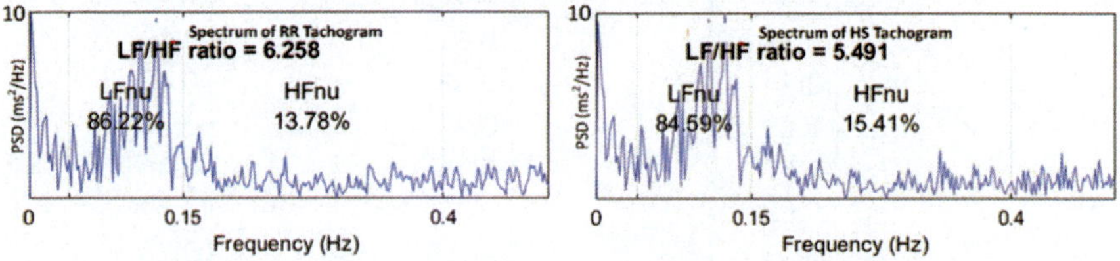

Fig. 5. Subject 1: spectral analysis of the RR tachogram from ECG and HS signals.

By observing the results obtained on Subject 4, it can be seen that the algorithm still manages to offer a good interpretation of the heart signal, however failing to provide close reproduction in all time segments (Fig. 6).

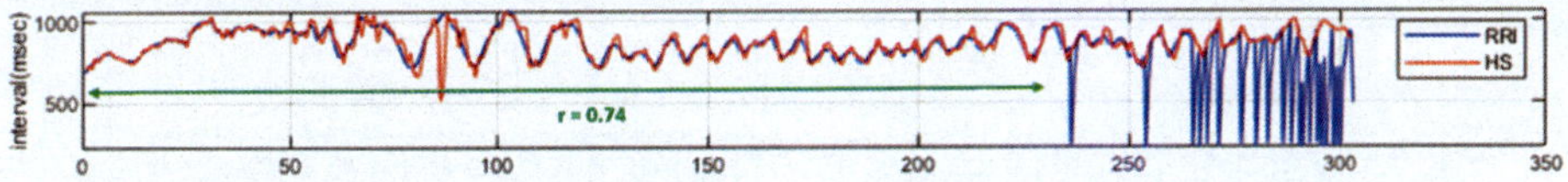

Fig. 6. Subject 4: overlap of the tachogram obtained with ECG (HR) and sensor (HS).

This loss of correct information is inevitably transferred to the calculation of the typical parameters of the HRV, as it is evident from the extent of the deviations between the values shown in Table 3.

Table 3. Comparison of cardiac parameters in the time domain (Subject 4).

	ECG	HS
Mean RR|HR	860|70	857|70
SDNN	79.0	84.8
RMSSD	35.9	56.2
pNN50	13.4	34.1
SD1|SD2	25.4|108.5	39.7|112.5
SD1/SD2 ratio	0.23	0.35

At such low correlation values, even the frequency analysis becomes unreliable (Figs. 7–8).

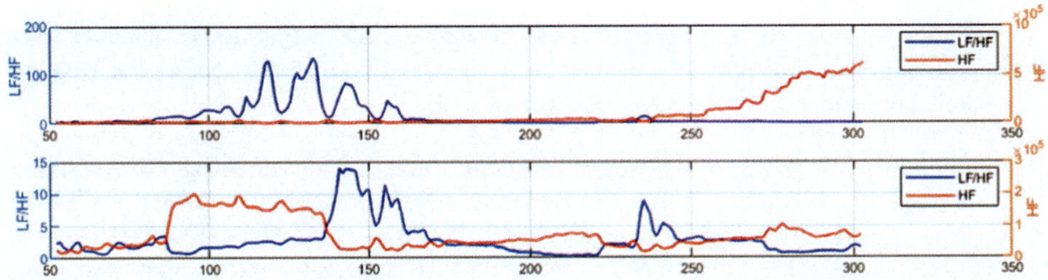

Fig. 7. Subject 4's HRV Frequency Analysis: comparison between ECG (top) and sensor (bottom) results.

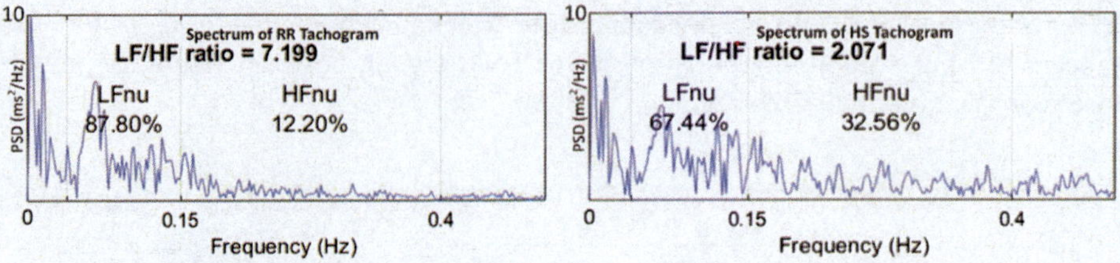

Fig. 8. Subject 4: spectral analysis of the RR tachogram from ECG and HS signals.

4 Conclusions

In this study, a non skin-contact ECG detection system is developed. The performance is validated through comparison with a clinic device. Even if this tool does not present the practicality of wearable models, it certainly offers the possibility of accessing more accurate information than what an average user or sportsman needs. The potential offered by such a system in both the clinical and commercial fields is considerable: the possibility of continuous monitoring of health state or a remote homecare service are just some of the prospects. Some more advantages are the possibility of monitoring in real-time states of stress, fatigue and drowsiness of a driver, passenger or worker. The further development will be the possibility to validate comfort indices in ergonomic and postural studies [19–21] by HRV analysis, using contactless sensors too. The non-invasiveness of this device is a breakthrough. However, the susceptibility of the results of such a device still represents a limit and above all it would be necessary to increase the number of tests in order to understand the source of these interpretative errors which appear suddenly and therefore could be attributed to movement artifacts or anomalies such as ectopic beats.

References

1. Fasulo, L., Naddeo, A., Cappetti, N.: A study of classroom seat (dis)comfort: relationships between body movements, center of pressure on the seat, and lower limbs' sensations. Appl. Ergon. **74**, 233–240 (2019)
2. Smulders, M., Naddeo, A., Cappetti, N., Van Grondelle, E.D., Schultheis, U., Vink, P.: Neck posture and muscle activity in a reclined business class aircraft seat watching IFE with and without head support. Appl. Ergon. **79**, 25–37 (2019)
3. Cappetti, N., Naddeo, A., Soldovieri, V.M., Vitillo, I., Fiorillo, I.: Perceived comfort and muscular activity: a virtual assessment of possible correlations. In: Rizzi, C., Andrisano, A.O., Leali, F., Gherardini, F., Pini, F., Vergnano, A. (eds.) ADM 2019. LNME, pp. 59–70. Springer, Cham (2020). https://doi.org/10.1007/978-3-030-31154-4_6
4. Wang, F., Tanaka, M., Chonan, S.S.: Development of a wearable mental stress evaluation system using PVDF film. J. Adv. Sci. **18**(1), 170–173 (2006)
5. Matsui, T., Ishizuka, T., Gotoh, S., Takase, B., Kikuchi, M., Ishihara, M.: A new instrument microwave radar to prevent exposure of rescue persons to toxic vapors and infectious bacteria. Jpn. J. Disaster Med. **9**(3), 279–285 (2005)
6. Watanabe, K., Watanabe, T., Matsui, Y.: Health monitoring system at home using an air mattress. J. Jpn. Soc. Appl. Sci. **14**(2), 4–8 (2000)

7. Task force of the European Society of Cardiology and the North American Society of Pacing and Electrophysiology: Heart rate variability. Standards of measurement, physiological interpretation, and clinical use. Eur. Heart J. **17**(3), 354_381 (1996)
8. Montano, N., et al.: Heart rate variability explored in the frequency domain: a tool to investigate the link between heart and behavior. Neurosci. Biobehav. Rev. **33**, 71–80 (2009)
9. Pumprla, J., Howorka, K., Groves, D., Chester, M., Nolan, J.: Functional assessment of heart rate variability: physiological basis and practical applications. Int. J. Cardiol. **84**, 1–14 (2002)
10. Grassi, G., Esler, M.: How to assess sympathetic activity in humans. J. Hypertens. **17**(6), 719–734 (1999)
11. Cerutti, S., Bianchi, A.M., Mainardi, L.T.: Spectral analysis of the heart rate varibility signal. In: Malik, M., Camm, A.J. (eds.) Heart rate variability, pp. 63–74. Futura Publ. Comp. Inc, Armonk, N.Y. (1995)
12. Kleiger, R.E., Stein, P.K., Bosner, M.S., Rottman, J.N.: Time domain measurements of heart rate variability. In: Malik, M., Camm, A.J. (eds.) Heart rate variability, pp. 33–45. Futura Publ. Comp. Inc., Armonk, N.Y. (1995)
13. Litvack, D.A., Oberlander, T.F., Carney, L.H., Saul, J.P.: Time and frequency domain methods for heart rate variability analysis: a methodological comparison. Psychophysiology **32**, 492–504 (1995)
14. von Borell, E., Langbein, J., Després, G., Hansen, S., Leterrier, C.: Heart rate variability as a measure of autonomic regulation of cardiac activity for assessing stress and welfare in farm animals – a review. Physiol. Behav. **92**(3), 293–316 (2007)
15. Salvati, L., d'Amore, M., Fiorentino, A., Pellegrino, A., Sena, P., Villecco, F.: Development and testing of a methodology for the assessment of acceptability of LKA systems. Machines **8**, 47 (2020)
16. Malliani, A., Lombardi, F., Pagani, M., Cerutti, S.: Power spectral analysis of cardiovascular variability in patients at risk for sudden cardiac death. J. Cardiovasc. Electrophysiol. **5**, 274–286 (1994)
17. Hagiyama, N., Hirano, H., Mito, A.: Unconstrained vital sign monitoring system using an aortic pulse wave sensor. Sci. Rep. **9**, 17475 (2019)
18. Salvati, L., d'Amore, M., Fiorentino, A., Pellegrino, A., Sena, P., Villecco, F.: On-road detection of driver fatigue and drowsiness during medium-distance journeys. Entropy (Basel) **23**(2), 135 (2021)
19. Naddeo, A., Cappetti, N.: Comfort driven design of innovative products: a personalized mattress case study. Work (Reading, Mass.) **68**(s1), S139–S150 (2021)
20. Fasulo, L., Naddeo, A., Cappetti, N.: A study of classroom seat (dis)comfort: relationships between body movements, center of pressure on the seat, and lower limbs' sensations. Appl Ergon. **74**, 233–240 (2019)
21. Naddeo, A., Cappetti, N., Vallone, M.: Range of rest posture of human lower limbs. In: Bagnara, S., Tartaglia, R., Albolino, S., Alexander, T., Fujita, Y. (eds.) IEA 2018. AISC, vol. 824, pp. 33–46. Springer, Cham (2019). https://doi.org/10.1007/978-3-319-96071-5_4

Design of Wearables for Biosignal Acquisition: A User Centered Approach for Concept Generation and Selection

Teodorico Caporaso(✉), Stanislao Grazioso, Giuseppe Di Gironimo, and Antonio Lanzotti

Fraunhofer Joint Lab IDEAS, Dipartimento di Ingegneria Industriale, Università degli Studi di Napoli Federico II, P.le Tecchio 80, 80125 Napoli, Italy
teodorico.caporaso@unina.it

Abstract. The presented work shows how a user centered approach might be used to generate and select the optimal design of smart garments for biosignal acquisition. Design is driven by human biosignal analysis, allowing the translation of subjective user's feelings into technical specification and the definition of customized criteria for concepts evaluation. So, different concepts are generated and, involving users again, the optimal one is chosen using multi criteria decision making based on Fuzzy AHP theory. A case study on a wearable system (i.e., electromyographic shorts) for football performance and risk injury analysis is shown.

Keywords: Wearable technologies · User centered design · Biological knowledge in engineering science

1 Introduction

Although conventional rigid sensors for biosignal acquisition are efficient, monitoring biosignals using soft wearable sensors is a novel interesting direction [1]. These sensors offer novel methods for recording and analysing body signals and can integrate garment and technology. These garments are wearable devices for biosignal acquisition that allow: (i) monitoring clinical and rehabilitation processes to provide technical support to the clinicians in a practical scenario [2]; (ii) monitoring industrial processes to prevent occupational injuries [3]; (iii) monitoring sport's gestures to optimize performance and decrease injury risk [4].

Recently the main criteria for designing wearables for biosignal acquisition have been underlined [5]. Based on [5], the main Customer Needs (CNs) could be synthetized as: (i) user friendly: simple for users to use in terms of hardware design and software interface; (ii) compact: wireless (following standardized IEEE protocols), preferably self-powered or with long battery life (low power and memory consumption); (iii) comfortable: non-invasive, easily wearable, using hygienic material (iv) safe: maintaining data and personal information security and confidentiality; (v) storing data: ability to

C. Rizzi et al. (Eds.): ADM 2021, LNME, pp. 818–826, 2022.
https://doi.org/10.1007/978-3-030-91234-5_83

store data, synchronized with cloud server, health records and consulting physician; (vi) reliable: standardized signal acquisition, it must be reliable and avoiding artifacts and operating conditions; (vii) multisignal analysis: signal analysis algorithm able to handle the non-stationary, non-linear and variable different type of signals; (viii) easy interpretation: easy to use and to understand by physicians (with small computational time); (ix) cheap: cost effective.

Although these guidelines are defined for wearables for biosignal acquisition, they are generally addressed for medical applications. In this context the novelty of the work is to present a "user-centered approach" for the design of a smart short for a specific sport gesture analysis in order to customize the available criteria for designing wearables for biosignal acquisition in this application. In particular a case study on a smart electromyographic short for football is developed, where the main biosignal of interest is the muscle activation signal. In traditional electromyography with rigid sensors, muscle activation analysis is carried out using surface electrodes. They allow: (i) precise positioning; (ii) the evaluation of single muscle excitation; (iii) higher accuracy for high excitations; (iv) higher sample rates [6]. At the same time, they do not allow dynamic activities assessment (such as the ones performed in sports as football) due to the different acquisition modules (with their weight and volume) and the difficulties to use them in conditions with high sweating and crash risk. To overcome these issues in the recent years, shorts with textile electrodes are designed. In this context, Sect. 2 presents the methodology of the "User centered design" approach with the aim: (i) to optimize the selection of CNs; (ii) to define specific Design Factors (DFs); (iii) to generate different concepts; (iv) to evaluate them based on criteria defined according to user's feeling. In Sect. 3, the proposed approach is applied for the design of electromyographic shorts with textile electrodes for football.

2 Methodology

A possible approach for the design of wearable for human biosignal acquisition based on "User centered design" is shown in Fig. 1. The starting point is the "human biosignal analysis", consisting in the selection of operational measurements that the system has to collect for the selected case study. Indeed, understanding the biomechanics and neuroscience of human movements is used for the design of wearable systems customized for humans [7]. Starting from that, a preliminary list of CNs is defined. Following, the selection of the main CNs is carried by emotional information from the user-end about their relationship with the product. In particular, this selection is carried out using specific questionnaires based on the KANO approach [8]. After the definition of the main CNs their translation into Technical Requirements (TRs) is defined through the Quality Function Deployment [9]. To translate CNs into TRs the House of Quality is built. This step is carried out according with the characteristics of the human biosignal under investigation.

The main TRs (with the highest technical importance rating) are summarized in DFs. Basing on the combinations of these DFs, Design Of Experiments (DOE) techniques can be applied [10]. Finally, the users are again involved in the concept selection process, through ELIGERE [11], an open-source software platform for multiple criteria decision-making method (based on Fuzzy AHP approach) that allows to support decision making in many engineering design scenarios. For the selection of the criteria, during the concept selection process the Kansei approach is adopted [12]. The semantic span is used in this approach to assess the criteria for the ranking of the alternatives. So, the optimal concept (with the ranking of criteria) is defined from the generated options by DOE.

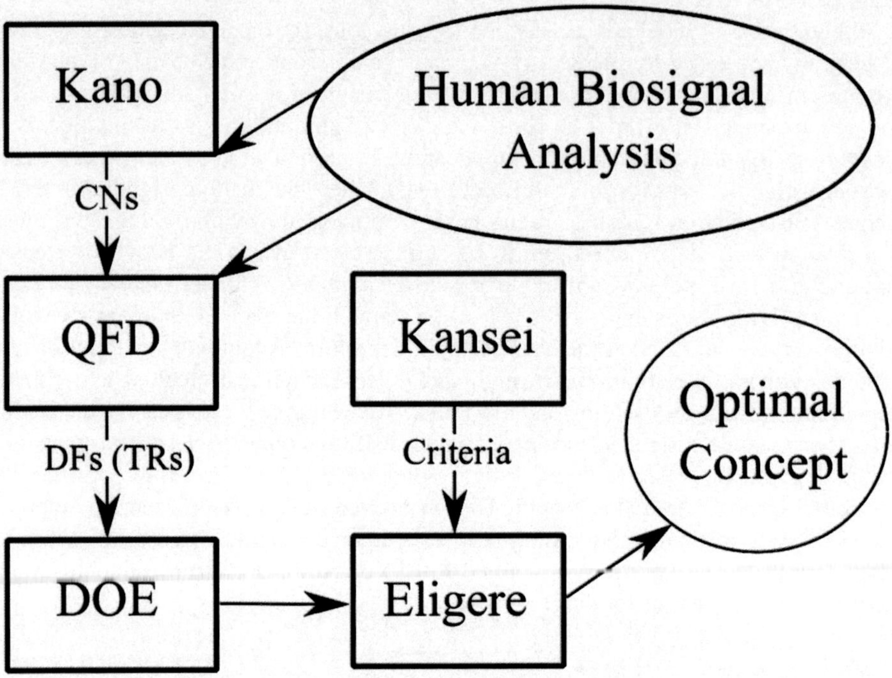

Fig. 1. Flow chart of the methodological approach based on "user centered design".

3 Case Study

A specific questionnaire was elaborated to register emotional user needs. The questionnaire was administered to fifty-three users from the football world. Interviewees consisted in thirty-nine athletic trainers, seven coaches, five footballers and four members of the medical staff (e.g., physician or physiotherapist). A percentage of 66% of them were pro. The structure of the questionnaire was composed by three parts. The first one was composed by sixteen questions based on the five Ws (i.e., Who, Whats, Why, Where and When) in order to collect user's demographic information and how the users interacts with the product. The second part of the questionnaire is an open question that allows to collect five words associated with the product by the users (i.e., Kansei words). Finally, the last part was composed by fourteen pairs of questions (i.e., functional and dysfunctional questions) to select the most important CNs (through the KANO approach).

3.1 Customer Needs Selection

The preliminary list of the CNs was composed by fourteen CNs. This list was composed by ten CNs suggested by [5]. We added four CNs considered useful for a sport device: (i) washable: that can be washed without shrinking or fading; (ii) waterproof: that does not allow water to penetrate; (iii) multisport: the concurrent or interleaved execution of two or more sport by a single CPU; (iv) bright: giving off or reflecting light.

Table 1. Kano CNs evaluation table and mean score (last column). For Kano CNs evaluation (i.e., A = Attractive; M = Must be; O = One dimensional; R = Reverse; Q = Questionable; I = Indifferent), the higher value for each row is underlined.

CNs	A	M	O	R	Q	I	Score
Washable	8	<u>23</u>	13	1	1	7	8.17
Multisignal analysis	8	<u>25</u>	13	0	0	7	8.25
Safe	9	<u>20</u>	19	0	2	3	8.57
Storing data	14	<u>21</u>	11	1	2	4	8.40
Cheap	14	10	<u>25</u>	0	0	4	8.17
Easy interpretation	14	11	<u>20</u>	0	0	8	8.55
User friendly software	<u>18</u>	3	12	7	2	11	6.98
Bright	11	3	2	5	0	<u>32</u>	4.87
User friendly hardware	<u>22</u>	5	4	0	1	21	5.87
Multisport	<u>25</u>	11	8	0	2	7	7.75
Compact	<u>20</u>	8	10	1	0	14	7.08
Reliable	<u>20</u>	7	18	0	1	7	7.79
Comfortable	6	<u>20</u>	<u>20</u>	0	3	4	8.96
Waterproof	12	<u>18</u>	17	0	0	6	8.79

Functional and dysfunctional questions related to all selected CNs were administered to all users. In addition, a 10-point ordinal subjective rating scale with extremes 1 (no relevance) and 10 (maximum relevance) was collected for each CN. In Table 1 the classification of the CNs based on the KANO approach and the mean score achieved by all CNs are reported. Considering CNs reporting a classification of Must Be (M) or One dimensional (O), eight CNs are selected (i.e., washable, multisignal analysis, safe, storing data, cheap, easy interpretation, comfortable and waterproof).

3.2 Design Factors Definition

In Fig. 2, a part of the House of Quality with the relationship between the eight main CNs and thirteen selected TRs is reported. The selected CNs reported a Score between 8.17 and 8.96 (see Table 1). So, for the assessment of the relative importance of each CNs (from 1 to 5) in the House of Quality we divided the range from 8.17 to 8.96 into five equal intervals (0.16 of wide range for each interval). For the definition of the list of TRs, we started for the main technical parameters presented in [5] for EMG wearable: (i) EMG sample rate; (ii) inertial data (i.e., dynamic range and sample rate); (iii) cardiac data (i.e., heart rate); (iv) number of sensors; (v) battery life; (vi) connectivity; (vii) memory capacity; (viii) dedicated graphical interface; (ix) app compatibility. In this phase, we also added four additional TRs connected with the additional CNs of the case study: (i) wash temperature; (ii) water resistance (i.e., International Protection (IP)); (iii) material stiffness; (iv) mass. From the House of Quality diagram, we selected eight main TRs (Technical Importance Rating ≥10). Then, three synthetic DFs were defined. The first one is "material", it includes the TRs stiffness of the material and mass. The second one is "sensors" it includes the TRs: battery life, water resistant, number of sensors, memory capacity. The third one is the "user interface" it includes the TRs: dedicated graphical interface and connectivity.

Relationships:
- • Strong= 3
- ⊙ Fair= 2
- ○ Weak= 1

Customer Needs (CNs)	Customer Importance / Technical Requirements (TRs)	Wash temperature	EMG sample rate	Inertial data (dynamic range, sample rate)	Cardiac data (heart rate)	Battery life	Memory capacity	Connectivity	App compatibility	Number of sensors	Mass	Dedicated graphical interface	Material Stiffness (Young's modulus)	Waterproof (IP)
Washable	1	•												⊙
Multisignal analysis	1		•	⊙	⊙	○	⊙							
Safe	3					○	⊙	○						
Storing data	2						•	•	○					
Cheap	1	⊙	⊙	○	○	○	⊙	○	⊙	•	⊙	⊙	⊙	⊙
Easy Interpretation	3								○			•		
Comfortable	5					○	○			⊙	⊙		•	
Waterproof	4													•
Technical Importance Rating Σ(CNs X TRs)	-	5	5	3	3	10	21	10	7	13	12	11	17	16

Fig. 2. Part of the House of Quality diagram. The higher values of the Technical Importance Rating are underlined.

In Table 2 for each DF two possible levels with different specification of TRs are reported. The limited number of levels chosen is based on the state of the art of main patents on electromyographic shorts [13, 14].

Table 2. Design factors and their levels, selected for the development of concepts.

Design factor	Level	
	0	1
A – Material	71% Polyamide micro - 29% Elastane	71% Polyamide micro - 29% Elastane & polymer reinforcement
B – Sensors	14 Textile electrodes & 2 inertial measurement units	8 textile electrodes & 1 inertial measurement unit
C – User interface	Simplified visualization of data	Detailed visualization of data

3.3 Criteria Selection

Through the preliminary questionnaire more than fifty words associated with the product were collected. The most frequent words were grouped by their semantic meaning. In particular, according to span the sematic space, a reduction of words based on the affinity diagram was performed. Six groups of words were obtained and six Kansei words were respectively assigned (Table 3).

Table 3. Kansei words based on the affinity diagram.

Ergonomic	Trendy	Affordable	Strong	Performing	User friendly
Cushy	Beautiful	Cheap	Breathable	Accurate	Handy
Soft	Short	Useful	Washable	Functional	Intuitive
Snug	Long	Adaptive	Waterproof	Effective	Smart
Elastic	Unique	Long-lasting	Safe	Reliable	Simple
Comfortable		Durable	Robust	Multifunctional	Easy to use
		Innovative			

3.4 Concept Generation and Choice of the Optimal

Table 4 shows the adopted 2^{3-1} fractional factorial design. We chose this type of planning in order to reduce the time required for the questionnaire. The generated concepts (in Table 4) present both the "Measurement Unit" (i.e., electromyographic short) and the "Control Unit" (i.e., mobile App). Seventeen users, the expert ones, were selected for a second questionnaire. The composition of the new sample was representative of the categories of the starting sample but with more years of experience in the field (i.e., over 90% more than five years).

The selected users filled a web-based questionnaire (developed for expert users) [15] divided in two parts: (i) preference section, consisting in a comparison about six criteria (defined in Sect. 3.3); (ii) suitability section, consisting in a comparison about

Table 4. Fractional factorial design of three factors defining four concepts.

Treatment	A	B	C	Concept
T_1	0	0	1	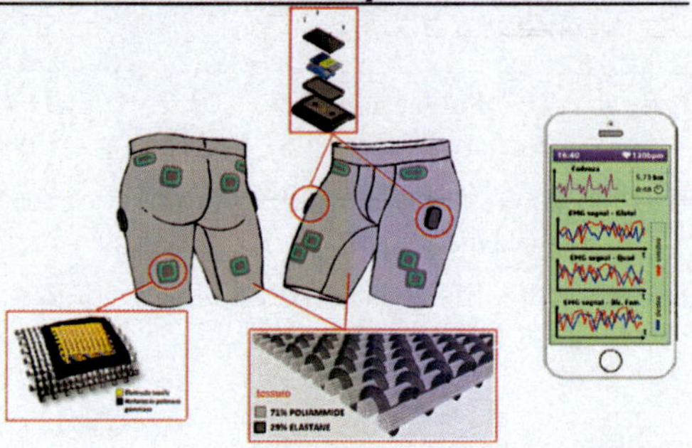
T_2	1	0	0	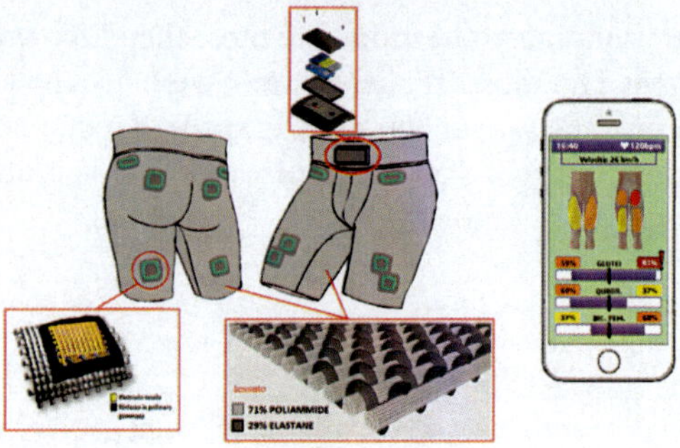
T_3	0	1	0	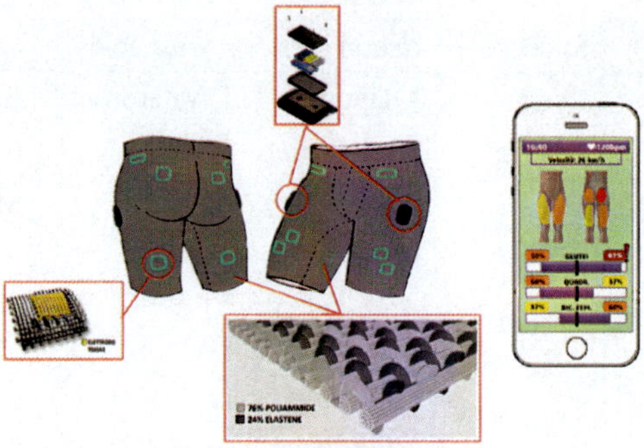
T_4	1	1	1	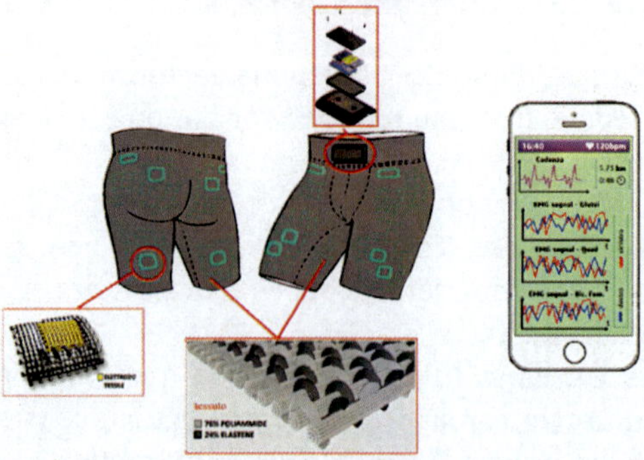

four treatments (shown in Table 4) with respect to each criterion separately. Figure 3 shows the output data from ELIGERE for this case study. Treatment T_1 is considered the best system solution among the four concepts (with a score of 34.3). Four criteria (i.e., ergonomic, affordable, strong and performing) are considered to have the same importance (with a weight of 18.4%).

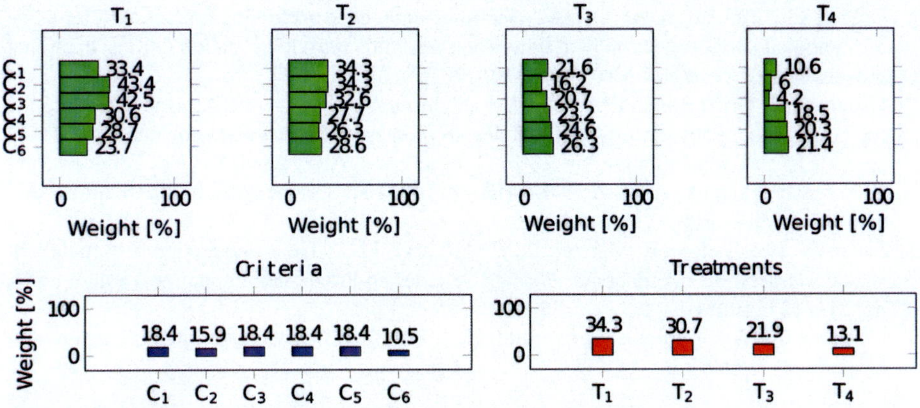

Fig. 3. ELIGERE results: on the top the score reported by the six criteria (i.e., C_1 = Ergonomic; C_2 = Trendy; C_3 = Affordable; C_4 = Strong; C_5 = Performing; C_6 = User Friendly) for each treatment. On the bottom: (left) the ranking of criteria; (right) the ranking of treatments.

4 Conclusions

In this work we presented a design methodology that could be used for improving design of wearable systems for biosignal acquisition. The proposed methodology was applied on a specific case study in sport field (i.e., electromyographic shorts for football). Results underlined the possibility to optimize and improve the suitable of the actual guidelines on design criteria for wearable systems for biosignal acquisition. A further improvement of the strategy that could be included in the design process is a 3D body scanning step, useful for design of tailored assistive devices [16, 17]. Indeed, for the presented case study, it could be adopted for the individuation of the anatomical landmarks [18]. This could guarantee: the optimization of the positioning of the textile electrodes on the body and the possibility of evaluating single muscle activation while guaranteeing good accuracy even when varying anthropometric measurements.

References

1. Shih, B., et al.: Design considerations for 3D printed, soft, multimaterial resistive sensors for soft robotics. Front. Robot. AI **6**, 30 (2019)
2. Tramontano, M., et al.: Effectiveness of a sensor-based technology in upper limb motor recovery in post-acute stroke neurorehabilitation: a randomized controlled trial. J. Biol. Regul. Homeost. Agents **34**(5 Suppl. 3), 165–174 (2020)

3. Conforti, I., et al.: Validation of a novel wearable solution for measuring L5/S1 load during manual material handling tasks. In: 2020 IEEE International Workshop on Metrology for Industry 4.0 & IoT, pp. 501–506. IEEE (2020)
4. Caporaso, T., Grazioso, S.: IART: Inertial assistant referee and trainer for race walking. Sensors **20**(3), 783 (2020)
5. Athavale, Y., Krishnan, K.: Biosignal monitoring using wearables: observations and opportunities. Biomed. Signal Process. Control **38**(2017), 22–33 (2017)
6. Colyer, S.L., McGuigan, P.M.: Textile electrodes embedded in clothing: a practical alternative to traditional surface electromyography when assessing muscle excitation during functional movements. J. Sports Sci. Med. **17**(1), 101 (2018)
7. Caporaso, T., Grazioso, S., Panariello, D., Di Gironimo, G., Lanzotti, A.: Understanding the human motor control for user-centered design of custom wearable systems: case studies in sports, industry, rehabilitation. In: Rizzi, C., Andrisano, A.O., Leali, F., Gherardini, F., Pini, F., Vergnano, A. (eds.) ADM 2019. LNME, pp. 753–764. Springer, Cham (2020). https://doi.org/10.1007/978-3-030-31154-4_64
8. Sauerwein, E., Bailom, F., Matzler, K., Hinterhuber, H.H.: The Kano model: how to delight your customers. In: International Working Seminar on Production Economics, vol. 1, no. 4, pp. 313–327 (1996)
9. Melemez, K., Di Gironimo, G., Esposito, G., Lanzotti, A.: Concept design in virtual reality of a forestry trailer using a QFD-TRIZ based approach. Turk. J. Agric. For. **37**(6), 789–801 (2013)
10. Park, G.J.: Design of experiments. Analytic Methods for Design Practice, pp. 309–391 (2007)
11. Grazioso, S., Selvaggio, M., Marzullo, D., Di Gironimo, G., Gospodarczyk, M.: Eligere: a fuzzy AHP distributed software platform for group decision making in engineering design. In: 2017 IEEE International Conference on Fuzzy Systems (FUZZ-IEEE), pp. 1–6. IEEE (July 2017)
12. Nagamachi, M.: Kansei engineering: a new ergonomic consumer-oriented technology for product development. Int. J. Ind. Ergon. **15**(1), 3–11 (1995)
13. Koivumaa, V., Martikka, M., Kylliainen, J., Remes, A., Tolvanen, P.: U.S. Patent Application No. 11/021,100 (2005)
14. Berg, J., Butt, H., Jayalath, D., Wiebe, C.: U.S. Patent No. 10,292,652. Washington, DC: U.S. Patent and Trademark Office (2019)
15. ELIERE Homepage. http://www.eligere.org. Accessed 21 Mar 2021
16. Grazioso, S., Selvaggio, M., Di Gironimo, G.: Design and development of a novel body scanning system for healthcare applications. Int. J. Interact. Des. Manuf. (IJIDeM) **12**(2), 611–620 (2017). https://doi.org/10.1007/s12008-017-0425-9
17. Grazioso, S., Selvaggio, M., Caporaso, T., Di Gironimo, G.: A digital photogrammetric method to enhance the fabrication of custom-made spinal orthoses. JPO: J. Prosthet. Orthot. **31**(2), 133–139 (2019)
18. Di Angelo, L., Di Stefano, P.: Bilateral symmetry estimation of human face. Int. J. Interact. Des. Manuf. (IJIDeM) **7**(4), 217–225 (2012). https://doi.org/10.1007/s12008-012-0174-8

A Workflow for the Numerical Evaluation of Hemodynamics in a Patient-Specific AAA After Stent-Graft Implantation

Michele Bertolini(✉), Marco Rossoni, and Giorgio Colombo

Department of Mechanical Engineering, Politecnico di Milano, Via la Masa 1, 20156 Milano, Italy
michele.bertolini@polimi.it

Abstract. An Abdominal Aortic Aneurysm (AAA) is a focal dilatation of the abdominal aorta, which if not treated can rupture with catastrophic internal bleeding. Besides open surgery, minimally invasive EndoVascular Aneurysm Repair (EVAR) has gradually taken place in clinical practice. In this paper, a workflow to evaluate the hemodynamic changes in a patient-specific AAA after stent-graft implantation is proposed. After the extraction from Computed Tomography (CT) images of the AAA 3D model, a patient-specific idealized stent-graft was obtained by means of CAD software. Models were then imported into the simulation environment, where a proper mesh was generated. Once calculations were completed, parameters of interest were extracted and a comparison between values before and after stent-graft implantation was conducted. Results show that, as expected, the introduction of the stent-graft makes the flow pattern smoother and more regular, with no vortices. This is because the stent provides a more physiological geometry for the blood flow, without the aneurysm enlargement. Anyway, simulations were conducted in a simplified way, being the main aim of the paper the implementation of the workflow by which parameters of interest can be obtained. As a development, the introduction of more refined boundary conditions, maybe considering data of the real patient, would increase the reliability of the simulations. A further step would be the passage from Computational Fluid Dynamics (CFD) to Fluid-Structure Interaction (FSI) simulations, so considering wall vessel compliance.

Keywords: AAA · Stent-graft · CFD

1 Introduction

An aneurysm can be defined as an abnormal irreversible dilatation of an artery. Depending on its shape, an aneurysm is usually classified as saccular, if it looks like sacs out of a side of a blood vessel, or as fusiform, if the whole circumference of the artery is affected. An Abdominal Aortic Aneurysm (AAA) is a focal dilatation of the abdominal aorta. Just like other aneurysms, if untreated AAA can rupture, causing catastrophic internal bleeding. The principal methods to repair an AAA are open surgery or EndoVascular Aneurysm Repair (EVAR) [1]. The first one is the traditional approach: it involves

C. Rizzi et al. (Eds.): ADM 2021, LNME, pp. 827–835, 2022.
https://doi.org/10.1007/978-3-030-91234-5_84

direct access to the aorta through a surgical incision into the abdomen. The second one, instead, requires a small incision over the femoral vessels, to access the lumen of the abdominal aorta. Then, a stent-graft is placed within the lumen of the AAA and expanded by ballooning, aiming at restoring the regular blood flow. This approach needs a close follow-up after the operation since stent-graft failures might occur due to blood leakage into the aneurismal sac or due to stent migration [2].

In the last years, Computational Fluid Dynamics (CFD) has been established as a powerful tool to perform simulations of complex systems in industry and academic research [3]. It has been widely used for cardiovascular research to visualize and analyze blood flow in arteries, opening the way for objective and quantitative pre-operative assessments (see e.g. [4]).

The present study investigates a fusiform abdominal aortic aneurysm, dealing with the hemodynamic changes after stent-graft implantation, exploiting the possibilities of modern CFD. The main goal of the paper, however, is not so much to provide clinically relevant results, but rather to implement and describe the workflow through which these results can be obtained. For this reason, idealized CFD simulations have been conducted, with simplified boundary conditions. Considering the state of the art, an easily accessible and versatile workflow for the fast evaluation of the fluid dynamics effects of the implantation of the designed patient-specific stent-graft is proposed.

The AAA 3D model has been extracted from Computed Tomography (CT) images while the 3D model of a patient-specific stent-graft has been modeled to adequately restore the blood flow. The two models, i.e. before and after the stent graft-implantation, have been then imported into the simulation environment, where a proper mesh was generated, and the boundary conditions applied. Once calculations were completed, the parameters of interest were extracted and a comparison between values before and after stent-graft implantation was conducted.

2 Materials and Methods

2.1 Anatomical Model Generation

The segmentation process has been conducted exploiting Mimics software (Materialise, Leuven, Belgium), version 21, starting from a stack of anonymised CT scans. The selected stack consisted of 523 slices in the axial plane, with a resolution of 511 $\times$ 403 pixels. The size of each pixel was about 0.75 mm, while the slice thickness was 1.25 mm. Dynamic Region Growing was first applied to the dataset, in order to extract a preliminary mask. After the specification of the region of interest (ROI) through the crop mask operation, manual editing was performed, to guarantee a satisfactory correspondence between the generated mask and anatomical borders in the slices. After other minor corrections, the resulting 3D model was calculated from the mask.

This was then exported to be post-processed in 3-Matic (Materialise, Leuven, Belgium), version 13, another software from the same suite. Here, the model was edited, in order to reach a watertight mesh. Subsequent operations were performed: a global smoothing step, in which a 1st order Laplacian algorithm with shrinking compensation was used, followed by a manual local smoothing of areas still presenting relevant irregularities. This is a crucial step, if we want a continuous, uniform and smooth model that could be "simulated". "Planar" vessel extremities were also obtained, on which easily apply boundary conditions. The product of this work is reported in Fig. 1. The resulting file was then exported in ANSYS Workbench (Canonsburg, USA), version 20.1.

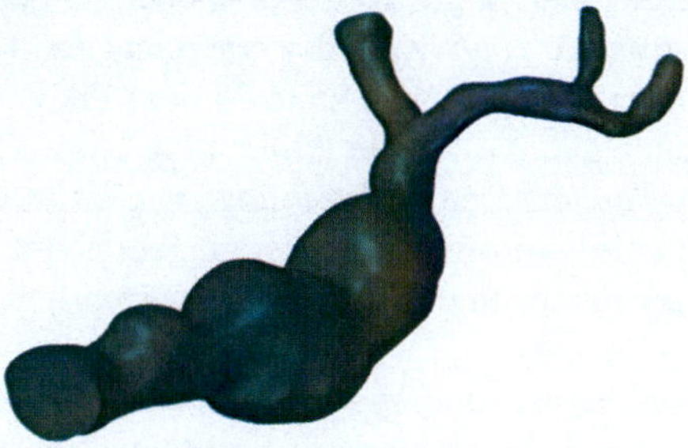

Fig. 1. Segmented model of a fusiform AAA.

2.2 Stent-Graft Generation

A stent-graft is usually composed of a set of metal rings and a fabric. The latter determines the geometry of the restored vessel. The idea, here, is to neglect metal rings, to focus on the shape of the fabric, which affects the path of the blood flow. Starting from the just segmented aneurysm, an idealised patient-specific stent-graft model was so created.

Exploiting Mimics software, the centreline of the aneurysm model was first extracted, together with the inlet and outlet sections profiles of interest, as can be seen from Fig. 2. The locations where to extract these profiles have been qualitatively chosen, based on the operator experience, and taking a commercial stent graft as reference (Fig. 2 on the right).

Fig. 2. Centerline extraction and inlet and outlet sections definitions for the stent-graft modelling (on the left); Sent graft geometry superimposed to the reconstructed geometry of the vessel (on the right).

The information about the centerline and the diameters were exported to SolidWorks software (Dassault Systèmes, Paris, France) as a reference for the 3D model of the stent. Here, a set of additional circular profiles was sketched, regularly along the centreline and perpendicular to it. The arbitrarily chosen diameters for these auxiliary profiles were set lower than the corresponding ones in the aneurysm model, so that the shape of an ideally healthy aorta was reconstructed. To give an idea, the diameter of the main branch of the restored vessel (i.e. the fabric) ranges from 25 to 28 mm. The sections were lofted: one loft for the principal branch and two for the bifurcating ones. The smoothness and the continuity of the geometry are ensured by the following constraints: (i) the loft has been forced to be normal to the inlet profile and (ii) a tangent constraint has been imposed where the two branches connect with the main one. The stent-graft geometry was finally obtained, as shown in Fig. 3 (left).

The upper part of the original aneurysm model was then attached at the inlet of the generated stent-graft model, to simulate the healthy part of the aorta just before the stent-graft itself, too. The resulting model was exported as a STEP file, to be read in ANSYS Workbench environment. The final 3D model is shown in Fig. 3 (right).

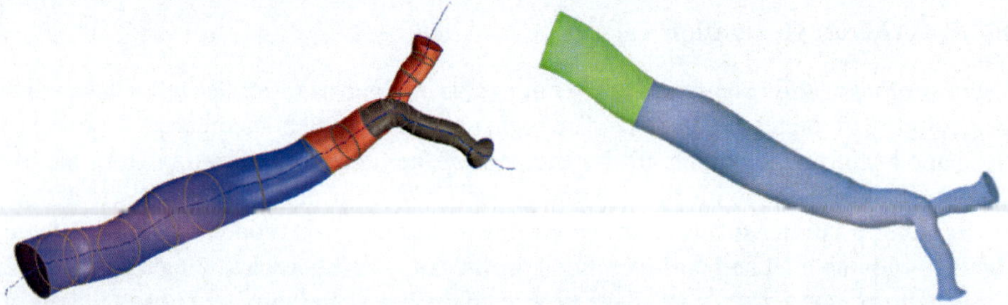

Fig. 3. Designed stent-graft model, by means of three loft features, highlighted in different colours (on the left). Stent-graft model with the inlet extension from the AAA model highlighted in green (on the right).

2.3 Simulations Set-Up

The obtained models were then imported in ANSYS Workbench (ANSYS, USA), 2020 R1. Here, the first important step to be performed was the mesh generation. With this regard, tetrahedral meshes were chosen for both models, to guarantee a fast and easy generation. A mesh independency study was performed: different tetrahedral elements sizes were tested, also imposing an inflation layer at the wall boundary made of seven prism layers, with a total thickness of 0.5 mm. Downstream from this study, a mesh with an average element size of 1 mm was chosen.

The result of the meshing phase, with previously defined parameters, for the fusiform model is shown in Fig. 4, where the grid at one outlet section is also highlighted. Indicatively, the total number of elements of the model is about 3.6 million elements.

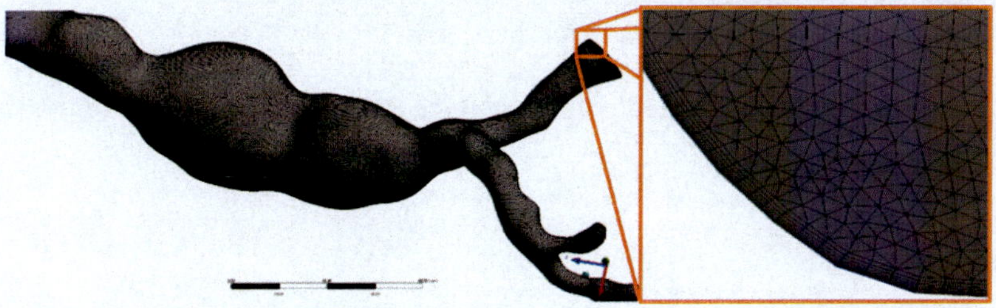

Fig. 4. Generated tetrahedral mesh for the AAA model, with a zoom on the boundary layer.

The simulation set-up was conducted in the ANSYS Fluent environment. The chosen solver type was Pressure-Based, while the steady in time simulation was set. Blood was modelled as an incompressible Newtonian fluid. Although blood is a biphasic fluid with non-Newtonian behavior, in large vessels this aspect can be neglected [5]. Thus, the viscosity was considered as constant and equal to 0.0035 kg/(m * s), while the density was imposed equal to 1060 kg/m^3 [5]. A laminar viscosity model was also considered and, to check this hypothesis, the Reynolds number was computed. Results are shown in the following Table 1.

Table 1. Main parameters for CFD simulation setting.

	Fusiform aneurysm
Inlet diameter [mm]	35
Inlet velocity [mm/s]	0.0347
Reynolds number	367

As expected, the obtained value is much lower than 2000, which corresponds to the laminar flow threshold.

As regards boundary conditions, arterial wall was considered as rigid, with a no-slip condition applied. On the inlet surface, a constant, normal-to-boundary velocity was set, equal to 0.0347 m/s, while at the outlet surfaces, a Gauge pressure equal to 0 Pa was fixed. Even though these are not a physiologically realistic hypothesis, these values are widely used in the literature to perform this kind of simulations [5].

Regarding the mathematical settings, a second-order scheme was set for the pressure discretisation, while a second order upwind for the momentum one. To check the convergence, residuals were fixed at 10^{-4} for the continuity and at 10^{-5} for the momentum in the three directions. Once calculations were completed, for each model mean velocity, maximum velocity and wall shear stress (WSS) fields were evaluated.

3 Results

The steady simulation was carried out for both the fusiform AAA case and the stented AAA case. In the following Fig. 5, the resulting flow patterns are reported.

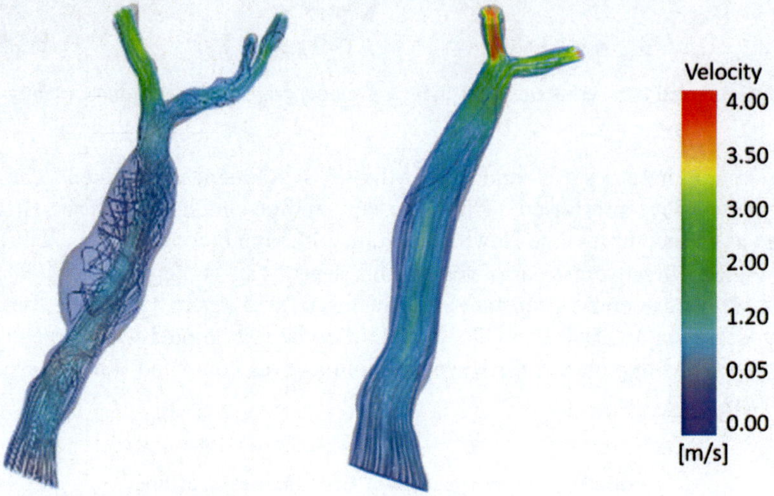

Fig. 5. Velocity streamlines for AAA (on the right) and stent-graft model (on the left), with the same boundary conditions.

As predictable, in the aneurysm region the velocity is very low. It can be noticed that the flow is chaotic and presents many turns both in circumferential and in longitudinal directions. Moreover, inside the aneurysm the main flux is not symmetrically distributed, due to inertial effects. On the contrary, the flow pattern is smooth and more regular after the stent-graft implantation. This is because the stent-graft provides a more "physiological" geometry for the blood flow since there are not sharp enlargements anymore.

The highest velocities are detected in downstream branches, after the iliac bifurcation, in both cases, where sections decrease. Since the flux is mainly oriented, the velocity inside the Right Common Iliac (RCI) artery is higher.

To carry out a more quantitative evaluation, five planes were considered along with the models, namely the inlets (plane 1), upstream healthy tract (plane 2), central zone (plane 3), iliac bifurcation (plane 4) and downstream branches (plane 5). The data confirms a significant increase in the average velocity of blood when the stent-graft is placed inside the AAA, as can be seen from the following Fig. 6.

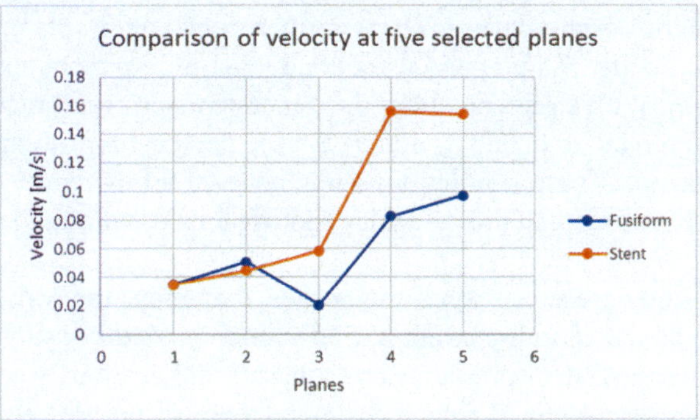

Fig. 6. Average velocities at each reference plane for AAA and stent-graft models.

Special attention must be paid to plane 3, which corresponds to the aneurysm sac: here, after stent-graft implantation, the average velocity is higher, due to the smaller diameter. On plane 1 there is the same value of velocity since it is imposed from the inlet boundary condition.

In Fig. 7, the wall shear stress (WSS) for both models is reported for comparison.

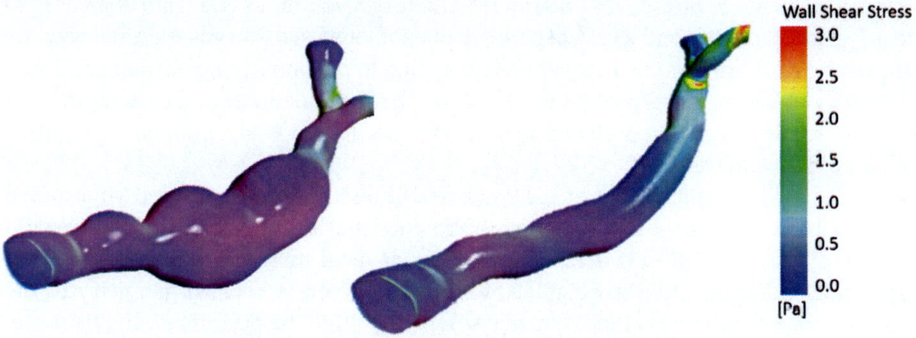

Fig. 7. Wall shear stress distribution for AAA (on the left) and stent-graft (on the right) models.

In both fusiform and stented aneurysm cases, the areas of highest wall shear stress are located after the bifurcation region. This is due to the impact of the blood flow against the wall of the vessel. In all the other zones, a lower WSS is detected, also because of the small inlet velocity imposed. This is especially true in correspondence with the aneurysmatic sac, where particularly low WSS values are detected. This has a specific influence on pathological conditions, as reported in the literature [5].

4 Conclusions and Developments

The aim of this project was to investigate the fluid dynamics of a fusiform abdominal aortic aneurysm, with a focus on the implementation of a workflow to quantitatively and qualitatively predicting hemodynamic changes after stent-graft implantation, exploiting a CFD commercial code. Simulations were conducted on a patient-specific model, as obtained after medical CTs segmentation; the generated stent-graft model was tailored on the basis of this model, to obtain a patient-specific device, too. As expected, obtained results for the stent-graft case simulations show better fluid dynamics, with respect to the AAA model. Flow is much more regular and organized, with higher velocities and lower vorticity.

Anyway, this study presents several limitations, that opens the way to future works. The main one is related to the adopted simplified boundary conditions. The input velocity profile is flat, while the output pressures are constant and arbitrarily equal to 0 Pa. By introducing specific user-defined functions, more realistic profiles could be adopted. Moreover, more refined models, e.g. lumped parameters models, could be useful to increase the accuracy, even if the complexity significantly increases. The introduction of proper viscosity and turbulence models would complete the picture.

The stent-graft model, generated from the AAA model, is idealized, too: it is not based on commercial models available on market and does not take into account the metal wireframe that enables these devices to be deployed and fixed to the arterial wall [6]. Anyway, from a merely fluid dynamic point of view, this is not so important.

In order to conduct a deeper analysis on the different working conditions of the vessel, transient simulations have to be introduced, together with some other indexes: Time Averaged Wall Shear Stress (TAWSS) and Oscillatory Shear Index (OSI) are often used to identify a threshold beyond which pathologic phenomenon starts to generate and become dangerous [7]. Furthermore, a deeper understanding of the underlying phenomena could be done by means of Fluid-Structure Interaction (FSI) simulations. In the literature, FSI is applied to cardiovascular problems in order to evaluate how wall compliance can affect the blood flow pattern [8].

Lastly, an important aspect of the work would be to provide easy and widespread access to these computer-based techniques. Provided that an automated workflow would be desirable to treat DICOM files, coming from medical images, CFD analyses could find a practical application in hospitals, too. Starting from a cleaned geometry of the anatomic zone of interest, a fluid dynamic simulation might be performed. In this sense, a relevant goal would be to find a good trade-off between mesh quality and the reliability of CFD results; if such a trade-off could be found, this would allow reducing as less as possible the time needed to run the simulation and so provide a quick response and support to the medical decision-making process.

In conclusion, from the implemented workflow and this preliminary CFD analysis we were able to extract some quantitative conclusions about the fluid dynamic advantages of implanting a stent-graft for AAA treatment. It has to be considered as a preliminary analysis, more methodological than able to provide new clinical results, but that anyway opens the way to more complex and accurate ones.

References

1. Kandail, H., Hamady, M., Xu, X.Y.: Patient-specific analysis of displacement forces acting on fenestrated stent grafts for endovascular aneurysm repair. J. Biomech. **47**, 3546–3554 (2014)

2. Li, Z., Kleinstreuer, C.: Analysis of biomechanical factors affecting stent-graft migration in an abdominal aortic aneurysm model. J. Biomech. **39**, 2264–2273 (2006)
3. Bluestein, D.: Utilizing computational fluid dynamics in cardiovascular engineering and medicine - what you need to know. Its translation to the clinic/bedside. Artif. Org. **41**, 117–121 (2017)
4. Algabri, Y.A., Chatpun, S., Taib, I.: An investigation of pulsatile blood flow in an angulated neck of abdominal aortic aneurysm using computational fluid dynamics. J. Adv. Res. Fluid Mech. Therm. Sci. **57**, 265–274 (2019)
5. Boyd, A.J., Kuhn, D.C., Lozowy, R.J., Kulbisky, G.P.: Low wall shear stress predominates at sites of abdominal aortic aneurysm rupture. J. Vasc. Surg. **63**, 1613–1619 (2016)
6. Colombo, G., Bartesaghi, S., Pallares, E.: A method to design custom-fit stent graft for abdominal aortic aneurysm. In: ASME International Mechanical Engineering Congress and Exposition, ASME, Houston, Texas, USA (2015)
7. Rana, M.S., Rubby, M.F., Hasan, A.B.M.T.: Study of physiological flow through an abdominal aortic aneurysm (AAA). Procedia Eng. **105**, 885–892 (2015)
8. Lin, S., Han, X., Bi, Y., Ju, S., Gu, L.: Fluid-structure interaction in abdominal aortic aneurysm: effect of modeling techniques. Biomed. Res. Int. **2017**, 1–10 (2017)

Engineering Education

A Systematic Approach to Innovative MCAD Education Based on Negative Knowledge Development and Formative Feedback

Ferruccio Mandorli(✉) and Harald E. Otto

Università Politecnica delle Marche, Via Brecce Bianche, Ancona, Italy
f.mandorli@staff.univpm.it, h.e.otto@univpm.it

Abstract. Within the field of industrial engineering, the use of systems for computer-aided design in mechanical engineering (MCAD) has been consolidated over the years. However, the potential benefits and advantages that these MCAD systems are capable of bringing to the product development process are still limited by, among other factors, the prevalent application of low-quality modeling strategies. This situation is reflected in the numerous MCAD models being created which can hardly be altered without causing unexpected and undesired side effects and errors, even though model alterability and reusability are essential requirements within the traditional design process. To improve this situation, MCAD curricula and education at institutions of higher education need to be restructured to further the development of skills and competency. This includes the development of MCAD modeling behavior in students, which is the outcome of understanding how to translate and correctly apply domain core concepts within strategic MCAD modeling. The aim of this paper is two-fold. Firstly, it attempts to provide an overview of efforts to restructure the MCAD course at the institution represented by the authors. Secondly, it tries to outline the individual stages of the development and implementation of the proposed innovative systematic approach to MCAD education, and to report on and summarize results achieved and experiences noted. A study of work previously published by the authors will facilitate a better understanding of this educational project, and such work is referenced below, but cannot be repeated, due to space limitations.

Keywords: MCAD skills and competency development · Learning from errors · Reflection on performance and outcome · Learning experience design

1 Introduction

Nowadays, systems for computer-aided design in mechanical engineering (MCAD) are widely used in the manufacturing industry. They play various roles in the product development process, ranging from aiding early phases such as analysis and simulation, and assisting planning and manufacture, to supporting the creation of technical documentation (cf. [1]). It is well known that product development, and design in particular, is an iterative process that requires continual adjustment (see also [2]). The MCAD models,

C. Rizzi et al. (Eds.): ADM 2021, LNME, pp. 839–850, 2022.
https://doi.org/10.1007/978-3-030-91234-5_85

which are used to perform tests and to provide feedback within this iterative process also need to be continually modified, in order to correctly represent the updated design solutions. Since their appearance on the market, dating back to the mid-1990s, MCAD systems have provided parametric functionality that allows the user to perform local modification and model re-generation. However, in order to be correctly re-generated, the CAD model must be created in compliance with a set of criteria which depends on how the MCAD systems have been implemented and operate [3]. Recent developments in MCAD systems have been increasingly focused on the long-standing goal of the integration of surface models with solid models. Those efforts have led to a kind of hybrid modeling that requires the user to be versed in and familiar with traditional techniques in both feature-based modeling and surface-based modeling.

A lack of understanding of the mechanisms that are the basis of how an MCAD system operates, can easily lead to what is commonly defined as 'poorly designed' MCAD models: that is, models that are almost impossible to alter without causing unexpected and undesired side effects and errors. In order to overcome such impediments, it is common practice to re-model from scratch instead of trying to alter a deficient model. This represents a somewhat paradoxical situation, because we are now encountering advanced MCAD technology which, instead of facilitating easy and efficient model alteration, seems rather to support the creation of models that are unalterable and difficult to reuse. Currently, the way CAD system developers and vendors are dealing with this problem can perhaps be best described as a kind of unsophisticated brute-force approach, where new modeling commands are added to existing MCAD systems, enabling a direct modification of the CAD model shape by means of stretching, turning, and twisting the geometric entities of the boundary model representation. This hybrid modeling approach, in the context of commercially available CAD systems, also referred to as *direct modeling* and *synchronous modeling* (cf. [4, 5]), might be capable of solving some specific problems, especially those related to the alteration of models that were exchanged between different CAD systems, and, as a result, have lost their feature history structure. However, from a methodological point of view, this is just a postponement of the problem of model reuse due to interpretation and alteration issues.

From an educational point of view, teaching students how to take advantage of all that such complex technology has to offer represents quite a challenging task. This is especially true in cases such as the introductory CAD course for students in their first year of undergraduate studies, where large numbers of students are attending the classes. In [16–22] various issues are reported, and these have been addressed during the restructuring project for the CAD course at the institution where the authors operate. The project, which has now been running for about a decade, has had to cope with various methodological and practical hurdles. The former include the integration of newly-designed course elements related to negative knowledge and expertise development, and learning from errors. The latter include the design and development of novel and innovative software tools for students and teachers to support CAD model assessment. To monitor progress and gain insight into various factors of those course-restructuring efforts, empirical studies analyzing the feedback from students, based on surveys and online questionnaires, are conducted throughout each academic year. In the remainder of this paper, the approach to and central principles of this restructuring project for the

CAD course are summarized. An overview of and insight into the structure and workings of the newly developed software tools is also provided.

2 Background, Objectives and Approach

2.1 Scope and Problems

The traditional approach to MCAD education is based on the teaching of system commands, the interaction with user interfaces, domain subject tutorials, guidelines, and best practices (cf. [6]), with the overall aim of developing enough domain knowledge, know-how, and skills to operate a modern MCAD system. However, learning outcomes achieved with this educational approach seem to fail in regard to the development of the strategic know-how and the basic domain expertise that are required to fully benefit from the functionality of such systems.

One of the major drawbacks of the traditional teaching approach is that, when students have to face new modeling situations not explicitly encountered during training, usually, due to their being novices, they do not recognize that certain strategies may lead to design and modeling situations best avoided. This is due to the fact that tutorials and best practices usually teach 'what to do' (positive knowledge). However, in many situations, being aware of critical circumstances that might lead to mistakes and errors, and thus also knowing 'what not to do' (negative knowledge), becomes equally important to achieving a desired outcome [7–9].

From an educational point of view, the achievement of learning outcomes related to the development of elements of negative knowledge can be supported with a learning-from-errors approach (see also [10, 11]). This educational approach requires from the student, among other tasks, the performance of a large number of modeling exercises, for which the teacher needs to provide timely and personalized formative feedback, and this feedback needs to be derived from analysis and assessment of the deficiencies and errors that were introduced into the models created by the student.

2.2 Goals and Approach

One central objective within the proposed educational approach has been the development and implementation of a teaching, performing, and assessment cycle, which is based on a continuous interaction between the teacher and the students. Here CAD modeling exercises are proposed to the students and formative feedback (cf. [12–14]) is generated by analyzing the results and outcomes produced by students during CAD laboratory and course assignments. The more exercises the students perform, the more feedback can be provided by the teacher, thus increasing the quality of results that can be achieved in terms of learning outcomes. However, to actually translate the advantages of this approach into educational practice and eventually benefit from it, first the fundamental hurdle of enabling a teacher to thoroughly analyze and assess a considerable number of MCAD models within a relatively short time period needs to be overcome (the duration of the CAD course is 48 h). This also harkens back to general efforts aimed at discovering a more efficient and effective means of supporting the creation and provision of a timely quality feedback loop, as well as offering a better use of resources.

Nowadays, within advanced MCAD systems, some very basic CAD model analysis is partially supported by functions embedded in the systems. However, those functions have not been designed to support analysis and assessment from an educational perspective, nor do they provide any statistical feedback in regard to model quality criteria or allow for comparative analysis among different models. In this regard, commercial tools are also available to perform CAD model analysis, but those tools just check the geometric quality of the model and are not able to provide any valuable feedback related to the modeling strategy adopted to create the model [15].

To overcome these shortcomings and hurdles as outlined earlier, the framework of the proposed educational approach includes a newly developed software tool, referred to as the *CAD model assessment* (CMA) tool. This CMA tool is specifically designed to assist and support the teacher during the individual stages of model analysis, formative assessment, and formative feedback generation (see Fig. 1).

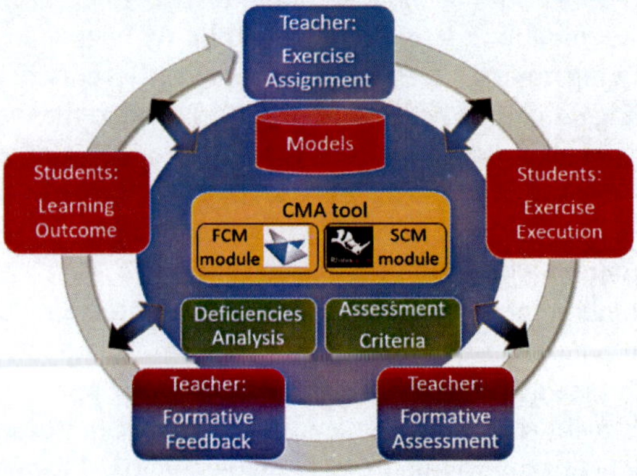

Fig. 1. Overview of structural components within the integrated MCAD course.

In particular, the CMA tool has been developed and tailored so that it can support the analysis and assessment of CAD models created by students in respect to, among other learning outcomes of the restructured MCAD course, the development of skills and competency to define and create both geometrically sound non-uniform rational B-splines (NURBS) based surface models and alterable as well as reusable parametric feature-based part models.

3 Development and Implementation of the Innovative MCAD Educational Framework

3.1 Point of Departure and Framework Development

The traditional approach to MCAD education is based mainly on positive knowledge, that is, teaching 'what to do'. However, another important aspect of strategic knowledge is knowing how to avoid serious mistakes. This is a central quality of professional

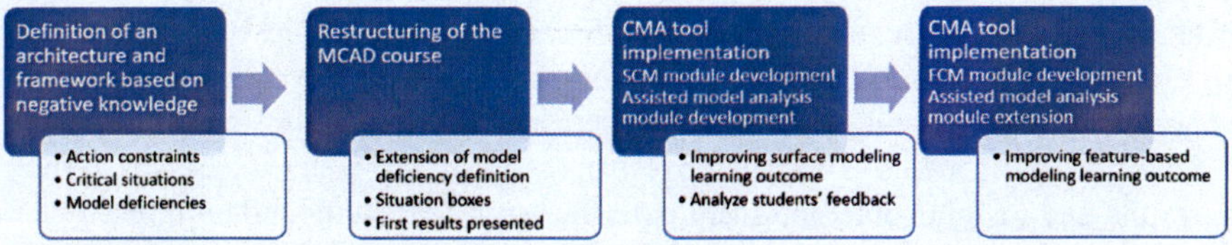

Fig. 2. Stages of and milestones in the implementation of the educational approach.

expertise, which is commonly referred to in the literature as *negative knowledge* (see again [7–9]).

Accordingly, the objective of the early stage of the educational project (see Fig. 2) has been the development of a framework for negative knowledge, which allows for direct translation of the proposed approach into practice, with the goal of improving learning behavior, skill acquisition and competency building for CAD education in mechanical engineering. This objective was approached by formulating negative knowledge as an element of strategic knowledge constraining actions within critical situations that would otherwise lead to errors and mistakes. In order to support the definition of what is to be avoided in respect to particular situations and contexts, and to have a qualitative measure to help express certain characteristics of situations during modeling, the concept of *CAD model deficiency* [16, 17] has been introduced. To systematically collect, synthesize, and document critical situations, which are an explicit representation of the negative knowledge relating to the modeling, use, and assessment activities, the concept of a *situation box* [18] has been introduced (see again Fig. 2).

3.2 Restructuring of the MCAD Course

Promising results from the newly developed educational approach were initially achieved with a limited number of volunteer students, as reported in [19]. However, to enable and support a full implementation of the new approach, it was necessary to speed up the process of CAD model analysis and assessment, and to assist the teacher during the individual steps of scaffolding and creating formative feedback. Then the new approach could reach all students of the MCAD course, which is currently a part of the curriculum for the Laurea degree in mechanical engineering at the institution represented by the authors, and which is attended by well over 100 students every year. This led to the earliest efforts dedicated to the design and development of the first prototype of the CMA tool. These efforts were focused on the identification of shape deficiencies within NURBS-based MCAD models. The tool was used experimentally to assist the teacher during the analysis and assessment of hybrid surface/solid models, which had been created by students during a set of shape engineering exercises, as reported in [20].

With the newly developed educational framework now fully implemented and successfully running in its third year, the authors were able to extend their work to better include the viewpoint of students: that is, obtaining further valuable input and translating it into course improvements and adjustments. This was approached by systematically investigating the opinions and views of students in regard to the learning experience and the learning outcomes that had been achieved within the restructured MCAD course.

This led to the design and execution of an empirical study that was based on data obtained from student feedback through online questionnaires, which, structured as surveys and tests, were related to learning outcomes, test performance, self-assessment, and opinions on teaching/learning methods used in the restructured MCAD course (see also [21]). Positive results and insights obtained from both the empirical study and an interim course evaluation encouraged a continuation of the CMA tool development, among other efforts. This was then extended to support the identification and analysis of feature deficiencies within parametric feature-based MCAD models, as reported in [22].

3.3 Requirements for and Development of the CMA Tool

In general, assessment of the quality of NURBS surfaces requires examination of both the layout of the curves and the patches of the CAD model that was created by using those curves as input. For example, this includes, among other requirements, an analysis of the degree and number of control points of curves and surfaces, checking for the occurrence of curve self-intersections, and the examination of critical points within patch boundaries with a tendency to lead to degenerated surface patches, etc. (for more details see [20]). Assessment of the quality of a parametric feature-based CAD model requires a thorough examination of the CAD model structure in terms of modeling features used, the modeling sequence generated, feature dependencies and constraints used during profile definition. This often involves, among other tasks, examining a considerable number of elements and determining the level of correctness based on a set of multi-dimensional criteria (cf. [22]). In pursuit of creating a software tool aimed at supporting a systematic and efficient CAD model analysis for teachers, CAD domain specific requirements have been compiled and translated into system specific requirements, leading to the development and implementation of a software tool framework and system architecture as follows.

The software tool design is based on a modular open system structure (MOSS), as shown in Fig. 3. Currently, the system contains two modules, the surface-based CAD model assessment (SCM) module and the feature-based CAD model assessment (FCM) module. The MOSS operates through the NURBS and feature-based MCAD models, which are based on the geometric entity (CMGE) repository and the feature entity (CMFE) repository. Those, in turn, facilitate the import from and export to not only different modeling environments, but also sets of linked geometric entities (GEs), feature entities (FEs), deficient geometric entities (DGEs), and reference feature entities (RFEs). Then the DGEs and RFEs are used to identify and record deficiencies in the respective MCAD models. Within the MCAD models, the geometric entity (CMGE) inventory and the feature entity (CMFE) inventory are then compiled, together resulting in the model entity analysis reports and various charts that are provided for assisting the teacher during the analytic and visual analysis process.

The newly developed software tool features a technical architecture that leverages API-based functionality provided by commercially available MCAD systems to support the modular and highly cohesive system architecture shown in Fig. 3. Within the current implementation, the modeling environment deploys a commercially available parametric feature-based solid modeling system, namely *SolidEdge* from Siemens AG, and a NURBS-based surface modeling system, namely *Rhinoceros* from Robert McNeel and

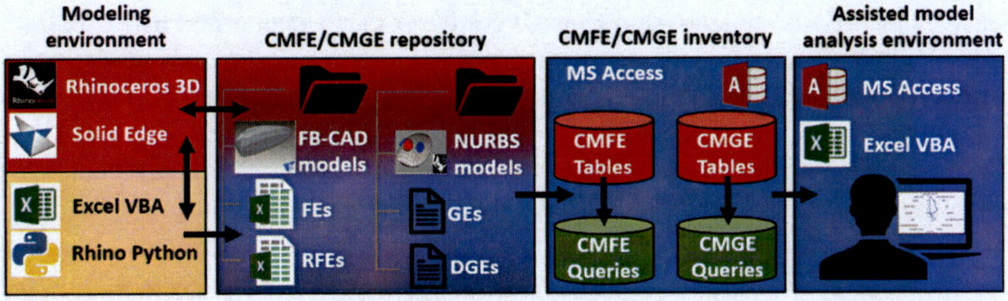

Fig. 3. Architecture of the FCM/SCM modules integrated into the CMA software tool.

Associates. At present, the CMFE repository is compiled by extracting MCAD data from the *SolidEdge* part models using VBA (Visual Basic for Applications) functions, while the CMGE repository is compiled by extracting geometric data from *Rhinoceros* using the Rhino Python framework. This extracted data is then further processed and stored in structured text and Microsoft *Excel* files. Next, those structured files are imported into the Microsoft *Access* RDBMS (relational database management system) by means of macros, to facilitate the creation and built-up of the CMFE and CMGE inventories. Assessment criteria used for the MCAD model deficiency analysis are specified and implemented using a set of domain-specific filter functions and Microsoft *Access* queries.

4 Results and Experiences from Within Educational Practice

Efforts to restructure the MCAD course at the institution represented by the authors have resulted in numerous promising outcomes as well as eye-opening experiences and repercussions as a matter of course. In what follows, a summary is provided based on a selection of representative examples of those considered to be important and valuable from the viewpoint of educational research and practice. Further results, details, in-depth analyses, and comprehensive discussions are provided by the authors in [16–22].

The overall theoretical evaluation and a specific more practice-oriented examination of empirical data related to learning outcomes, performance, and self-assessment, which have been obtained from course work, CAD laboratory exercises, and a series of online-based tests, questionnaires and surveys, showed promising results and outcomes as follows. Students developed a better understanding of central concepts related to the geometric usability of CAD models. This development was accompanied by an increased capability of recognizing critical modeling situations that would have led to errors, thus helping to avoid mistakes typically made by novices. Also, confidence in domain subject knowledge and strategy formation increased substantially. This observation, among other issues, was reflected in data from self-assessment regarding CAD modeling skills and abilities, and laboratory exercises, which correlated with the suggestion that students had advanced in both positive and negative knowledge. The opinion poll administered within a recently conducted survey, regarding the two teaching methods integrated in the newly designed course structure, showed that the majority of students found teaching based on positive knowledge and teaching based on negative knowledge equally important.

However, most of the students who showed good test performance and had a high self-rated competency score found learning about errors to be more useful, while most of the students with poor test performance and a low self-rated competency score found learning from tutorials more useful.

Promising results, and outcomes that suggested that the MCAD course had improved as a result of restructuring, were achieved not only in regard to learning and skill development, as outlined above, but also in regard to aspects of teaching. For example, application of and support from the CMA tool helped teachers to significantly reduce the time that is usually required for manually analyzing and assessing a large number of CAD models created by students. This, in turn, facilitated the creation of timely and high-quality formative feedback provided to students, and this feedback could be structured with meaningful and personalized content in regard to the specific mistakes and errors committed. Hence, it supported students in reflecting on and learning from their previous mistakes and in overcoming skills deficiencies. The CAD model assessment quality was also improved in terms of accuracy, consistency, reliability, and efficiency. For example, as each CAD model is based on hundreds of parameters such as sketch and profile relations, and dimensions, even small mistakes can be detected by software tools such as the CMA tool. Such minor errors may, in many cases, be overlooked in an exclusively manual analysis. Moreover, due to the complexity and considerable volume of data that needs to be examined during the analysis and assessment of large numbers of CAD models, undertaking this without software tool assistance usually becomes a time-consuming task that can be affected by fatigue and influenced by subjectivity.

Use of the CMA tool to assist with CAD model assessment also allowed for the expansion of knowledge regarding CAD model deficiencies that resulted from errors committed by students, which, however, were neither noticed by the teacher nor encountered previously in exercises. Note that the expansion of knowledge on MCAD model deficiencies is made possible in this manner, due to the integrated semi-automatic CMA tool for MCAD model analysis and assessment, where man and machine cooperate to produce results based on effective and efficient use of their distinct and outstanding capabilities. In what follows, some examples of cases actually encountered within educational practice are presented for the domain subjects of surface-based 3D modeling and parametric feature-based MCAD modeling.

The main learning goal of the exercise presented and discussed as a representative example for the domain of surface-based 3D modeling was to create a closed poly-surface model of the subject of design, which was a plastic mineral water bottle as shown in Fig. 4(a). This surface-based 3D model needs to be good enough to enable it to be used within exercises that are scheduled later in the course, such as the modeling of simplified parts/components related to molding cavities within blow molding processes. The errors most commonly committed by students during the modeling of the top patch are related to using a wrong input curve and subsequently selecting a wrong system command to create the patch surface. Using a silhouette curve as the input curve and then employing the CAD system command that creates a surface from a boundary curve is a common example of this mistake. Another error that is frequently committed by students during this exercise is forgetting to actually model the top patch.

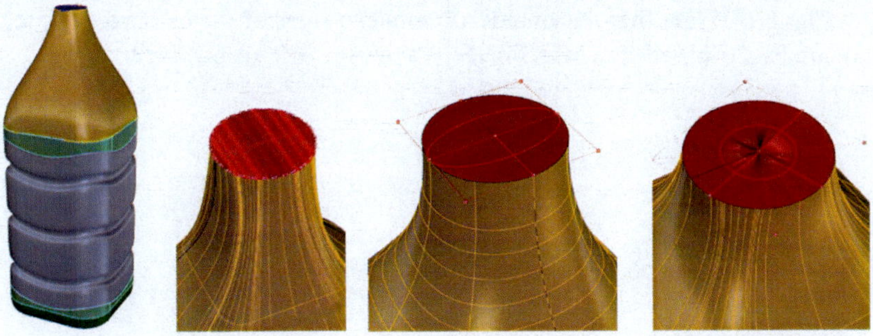

Fig. 4. Reference model and examples of deficient top patch surfaces.

In the restructured MCAD course, software-tool-assisted CAD model assessment of surface-based 3D models is now supported by the SCM module. As a result, it has become possible to detect and identify CAD model deficiencies that were neither noticed by the teacher nor encountered previously in exercises. Such model deficiencies are the result of solutions noticeably different from what is generally found or expected, which some students have used for the modeling of the top patch surface. A selection of examples of these newly encountered CAD model deficiencies is as follows. The flawed top patch surface shown in Fig. 4(b) has been modeled as a non-planar boundary patch using the faulty open boundary of the bottle rim. The defective top patch surface shown in Fig. 4(c) has been modeled by applying an unusual and obscure strategy, resulting in a patch with 1 joint edge and 2 boundary edges employing two sets of overlapping control points consisting of 2 and 4 control points, respectively. Another faulty top patch surface, shown in Fig. 4(d), has been modeled by applying the revolution of a straight segment, resulting in a patch with 3 boundary edges, 1 joint edge and 8 overlapping control points at its center.

The main learning goals of the exercise presented and discussed as a representative example for the domain of parametric feature-based MCAD modeling were as follows. Firstly, according to the approximate geometric appearance the part is expected to have, to plan and organize an appropriate modeling strategy. Secondly, based on a partial specification, to create a feature-based model of the subject of design, which was a bolted yoke clamp fastening mechanism with stiffening ribs and a rectangular base mount. Thirdly, based on the characteristics of the expected final assembly, to alter the previously created model to fit as a part within the assembly as shown in Fig. 5(a). Within this CAD domain context, the teacher was assisted in the analysis and assessment of MCAD models by the FCM module, which also resulted in the detection and identification of various CAD model deficiencies that had not been encountered previously in exercises. For example, a part of the yoke clamp head of the fastening mechanism was created with two revolved protrusion features based on a profile as shown in Fig. 6(a) and, even more surprisingly, with two rib features as shown in Fig. 6(b). Another example of feature-related deficiencies which were the result of solutions noticeably different from what is generally found or expected was the creation of the two lateral stiffening ribs of the fastening mechanism by using an extruded protrusion feature based on a profile as shown in Fig. 6(c). Note that the final CAD model shapes of the examples depicted in Fig. 6(b) and Fig. 6(c) are shown in Fig. 5(b) and Fig. 5(c), respectively.

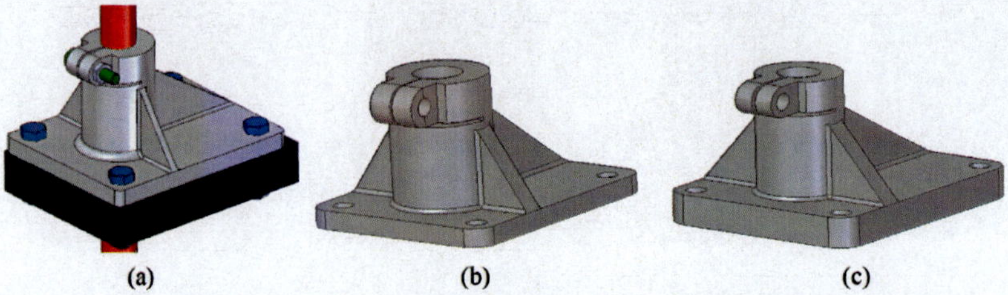

Fig. 5. Reference assembly and examples of deficient feature-based MCAD models.

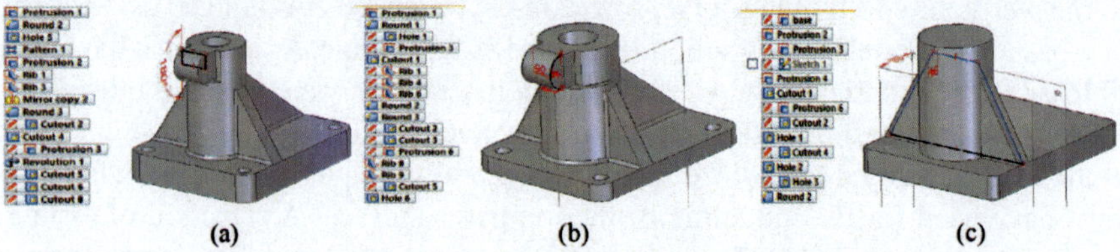

Fig. 6. Examples of deficient feature-based MCAD models and their feature history.

At this point, perhaps it should be made explicit that simply obtaining the numbers and types of features used to create a model by analyzing, for example, the feature history manually or with the assistance of a tool, is not sufficient to correctly reason about the actual presence of those features in the CAD model. When the examples presented above are re-examined, it becomes evident from Fig. 5(b) that only 3 stiffening ribs were actually present in the model, even though 5 rib features are listed in the feature history shown in Fig. 6(b). On the other hand, in the feature history shown in Fig. 6(c), no rib features are listed, but the final CAD model shape includes 3 stiffening ribs, as shown in Fig. 5(c).

5 Conclusions and Future Work

An overview with examples has been presented of efforts to restructure an actual MCAD course using an innovative approach based on the integration of positive and negative knowledge. Development of the restructured course has taken into consideration the need for frequent and methodical formative assessment and feedback. These efforts have been aimed at improving teaching quality and learning experiences throughout the entire course. However, a recent course evaluation showed that there is still room for improvement regarding both student performance during exercises and the development of learners' understanding of core concepts.

This has led to a second round of course improvements, which are currently in progress. However, this time the objective of improving the learning experience has been focused on increasing student engagement, particularly during exercise work related to CAD laboratory and course assignments. This has involved transferring back to the students some of the responsibility for learning and for achieving the required learning outcomes. As this requires assessment and feedback to be even more timely and individual, students need to be systematically supported in their efforts to improve self-assessment and self-adjustment. Efforts currently in progress aim to achieve this by helping students to bring core concepts to life through computer-based simulation, by supporting CAD model analysis through a dedicated student software tool, and by introducing new innovative key metrics for success that can be used by both students and teachers.

References

1. Bi, Z., Wang, X.: Computer Aided Design and Manufacturing. John Wiley & Sons, Hoboken, NJ, USA (2020)
2. Hirz, M., Dietrich, W., Gfrerrer, A., Lang, J.: Modeling techniques in CAD. In: Hirz, M., Dietrich, W., Gfrerrer, A., Lang, J. (eds.) Integrated Computer-Aided Design in Automotive Development, pp. 241–308. Springer, Heidelberg, Germany (2013). https://doi.org/10.1007/978-3-642-11940-8
3. Mandorli, F., Borgo, S., Wiejak, P.: From form features to semantic features in existing MCAD: an ontological approach. Adv. Eng. Inf. **44** (2020). Paper-No., 101088
4. Ault, H., Phillips, A.: Direct modeling: easy changes in CAD? In: Proceedings of the ASEE Engineering Design Graphics Division Conference, January 24–26, Daytona Beach, FL, USA, pp. 99–106 (2016)
5. Wang, L., Zhang, J.Y., Yang, W., Fan, Q.: Advantages analysis of synchronous modeling technology based on solid edge. WSEAS Trans. Comput. **14**, 248–252 (2015)
6. Camba, J.D., Contero, M., Company, P.: An analysis of strategies for design reusability. Comput. Aided Des. **74**, 18–31 (2016)
7. Minsky, M.: Negative expertise. Int. J. Expert. Syst. **7**(1), 13–19 (1994)
8. Gartmeier, M., Bauer, J., Gruber, H., Reid, H.: Negative knowledge: understanding professional learning. Vocat. Learn. **1**(2), 87–103 (2008)
9. Parviainen, J., Eriksson, M.: Negative knowledge, expertise and organisations. Int. J. Manag. Concepts Philos. **2**(2), 140–153 (2006)
10. Harteis, C., Bauer, J., Gruber, H.: The culture of leaning from mistakes: how employees handle mistakes in everyday work. Int. J. Educ. Res. **47**(4), 223–231 (2008)
11. Harteis, C., Bauer, J.: Learning from errors at work. In: Billett, S., Harteis, C., Gruber, H. (eds.) International Handbook of Research in Professional and Practice-based Learning. SIHE, pp. 699–732. Springer, Dordrecht (2014). https://doi.org/10.1007/978-94-017-8902-8_26
12. Irons, A.: Enhancing Learning through Formative Assessment and Feedback. Routledge, Abingdon, UK (2008)
13. Hattie, J., Timperley, H.: The power of feedback. Rev. Educ. Res. **77**(1), 81–112 (2007)
14. Marsh, P.: What is known about student learning outcomes and how does it relate to the scholarship of teaching and learning? Int. J. Scholar. Teach. Learn. **1** (29) (2007). Paper-No.: 22
15. Gonzáles-Lluch, C., Company, P., Contero, M., Camba, J.D., Plumed, R.: A Survey on 3D CAD model quality assurance and testing tools. Comput. Aided Des. **83**, 64–79 (2017)
16. Mandorli, F., Otto, H.E.: Negative knowledge and a novel approach to support MCAD education. Comput. Aided Des. Appl. **10**(6), 1007–1020 (2013)

17. Otto, H.E., Mandorli, F.: Integration of negative knowledge into MCAD education to support competency development for product design. Comput. Aided Des. Appl. **14**(3), 269–283 (2017)
18. Otto, H.E., Mandorli, F.: A framework for negative knowledge to support hybrid geometric modeling education for product engineering. J. Comput. Des. Eng. **5**(1), 80–93 (2018)
19. Mandorli, F., Otto, H.E.: Innovation in MCAD education toward competency development using negative knowledge: from theoretical framework to practical implementation. Comput. Aided Des. Appl. **15**(3), 337–352 (2018)
20. Otto, H.E., Mandorli, F.: Surface model deficiency identification to support learning outcomes assessment in CAD education. Comput. Aided Des. Appl. **16**(3), 429–451 (2019)
21. Mandorli, F., Otto, H.E.: Systematic support of learning from errors and negative knowledge development in mcad education: empirical analysis of student feedback. Comput. Aided Des. Appl. **17**(2), 384–406 (2020)
22. Otto, H.E., Mandorli, F.: Parametric feature-based solid model deficiency identification to support learning outcomes assessment in CAD education. Comput. Aided Des. Appl. **18**(2), 411–442 (2021)

The Problem-Solution Network for Systematic Conceptual Design: Lesson Learned

Lorenzo Fiorineschi(✉), Federico Rotini, and Monica Carfagni

Department of Industrial Engineering, University of Florence, Via Santa Marta 3, 50139 Firenze, Italy
lorenzo.fiorineschi@unifi.it

Abstract. The need for innovative products led scholars to develop design methods to improve or at least to support creativity of designers. Among the different alternatives, the Problem Solution Network (PSN) was developed to solve some of the issues that characterize the German systematic approach for conceptual design. The objective of this paper is to report and discuss pros and cons of the use of the PSN for didactic purposes in an engineering design course. Both, feedback from students and teachers' impressions, have been gathered in about ten years. The key information is then reported and discussed in this paper, highlighting that as for other academic design methods, the key problem still resides in motivating students to better understand the learned approach.

Keywords: Conceptual design · Engineering education · Creativity

1 Introduction

The modern marketplace leads industry to heavily push on innovation, thus enforcing the need for novel products. To prepare tomorrow's designers to this challenge, teaching them methodologies and tools capable to creatively fulfill the requirements of any design task is becoming even more important. Taught approaches should promote comprehensive design space explorations, thus overcoming counterproductive effects from design fixation [1, 2]. Indeed, students usually tend to start from a specific desired solution, working backward through iterative modifications to the first conceived ideas [3]. In this way, unfortunately it is quite improbable to perform comprehensive design space explorations.

However, abstraction can be successfully used to overcome design fixation [4, 5], since it allows to solve design problems without the need to directly refer to subjective preconceptions about the final product. Among the approaches that exploit the benefits of abstraction, the Functional Decomposition and Morphology (FDM) is acknowledged as one of the most taught and widespread [6]. However, some researchers point out that conceiving function structures is impossible without fixating on a reference structural solution [7, 8]. This FDM issue somehow implicitly leads students (and designers), to work backward from their first ideas, hence hindering the exploration of the design space. In order to overcome the flaws ascribed to FDM, the Problem Solution Network (PSN)

C. Rizzi et al. (Eds.): ADM 2021, LNME, pp. 851–860, 2022.
https://doi.org/10.1007/978-3-030-91234-5_86

approach has been proposed [9], and recently upgraded with an underpinning construct to better exploit abstraction [10]. The PSN does not require to work on function structure, and is based on the co-evolution of problems and solutions [11]. The PSN allows the decomposition of the overall problem at different abstraction levels, by keeping track of the different possible solutions.

The objective of this paper is to share the lesson learned from the considered teaching experience, by showing both pros and cons observed in using the PSN approach for teaching creative conceptual design in an engineering design course. More precisely, students' feedback and teachers 'observations are reported and discussed, finding that the main PSN tool can be efficiently used for understanding the actual process followed by students in facing the design task. Nevertheless, the industrial use of the PSN still needs to be investigated in depth.

2 Background

One of the most debated issue is that related to the spread of academic design methods in industry [6, 12, 13], which is acknowledged to be also rooted in an insufficient training of students [14, 15]. More recently, a comprehensive list of requirements has been published, concerning what the methods should fulfil for being accepted in industry [16]. Additionally, it has been observed that some key differences exist between experts and students in performing design activities [17, 18]. In particular, non-negligible differences have been observed in terms of problem scoping and information gathering performed by designers with different expertise [19]. In particular, while experts rely in problem decompositions, students can find several difficulties in doing such a fundamental activity. As a further example, it has been observed that students need to learn a procedural knowledge to understand problem solving approaches [20]. Furthermore, students are used to believe that mathematics is the unique language of engineering, often neglecting other types of human cognition that are involved in the design process [21]. Consequently, a first step to motivate students in learning methods, is to make them aware about the drawback of their design approach. In other words, taught methods should be considered a valid procedural support for novices, which can allow them to rapidly reach higher expertise levels.

The classical and maybe the most widespread teaching approach is characterized by frontal lessons supported by tutorials and practical exercises [22]. In such a case, text-based material constitutes an important part of the different media exploited for sharing knowledge with students. This is justified, because it has been observed that learning theory of methods from books improves the problem solving skills [23].

However, other approaches can be potentially used for an effective learning experience, as for example, inductive approaches like problem-based learning [24] or project based learning [25]. Unfortunately, specific situations may hinder a successful application of these approaches, for example when the theoretical knowledge is required to apply a specific method, but the available didactic resources are not sufficient for performing both theoretical lessons and practical exercises.

3 The Problem-Solution Network

The PSN approach relies on three main phases that substantially provide a further nesting of the design process. In this case, the conceptual design process is subdivided in three distinct phases, i.e., the Concept Generation (CG), the Concept Composition (CC) and the Concept Selection (CS). In CG the design space is explored in terms of problems and solutions sequences, which can be represented by a hierarchical network organized in terms of abstraction levels (Fig. 1).

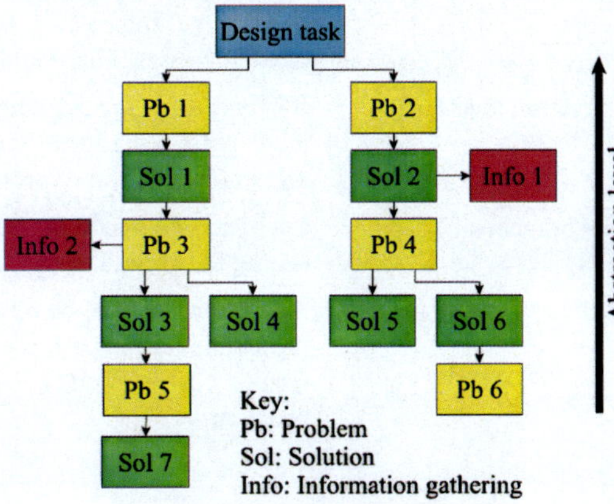

Fig. 1. Generic PSN

In the CC, different solutions combinations available from the CG network are evaluated to obtain different concepts. Eventually, the third phase is the Concept Selection (CS), where concept variants are selected.

The following paragraphs introduce CG and CC, while CS is fundamentally based on acknowledged approaches available in literature (e.g. Concept selection matrices [26], Selection charts [27] or QFD-like matrices [28]).

3.1 Concept Generation

Problem's formulation and solution's identification are performed in CG by the sequential application of the Analysis-Synthesis-Evaluation (ASE) process. In particular, the problem is analyzed (analysis), a solution idea is proposed (synthesis) and then evaluated by the designer (evaluation) in order to decide whether to consider it or not [9]. Each solution can potentially lead to new problems at lower abstraction levels, leading to different branches composed by problems and solutions (see Fig. 1). Such a network allows to visualize the design rationale followed during the conceptual design activity. Moreover, pink boxes (see Fig. 1) are used to keep track of the info-gathering activities.

Problems are formulated in the general form: "How to verb – noun?", thus allowing to consider any form of design problems among functional and non-functional ones. Moreover, the six rules summarized in Table 1 constitute the fundamentals of the approach.

Table 1. PSN rules for CG

PSN rule	Description
Main task formulation	The main task is formulated as the first problem box of the network
First-level problem formulation	The first level of solution-independent problems are related to the main functions expected for the system
Solution-independent problem decomposition	"problem-problem" decomposition can be done only if the passage is not influenced by any fixation on a particular solution
Correct sequence of abstraction levels	It is necessary to reach the highest abstraction level, both in the identification of solutions and in formulation of problems
Independency of the PSN branches	Each branch is independent on the others during his growth
Completeness of the PSN	A minimum of one complete problem-solution path (i.e. ending with a solution) is required for each "first-level" problem

3.2 Concept Composition

A variety of possible solutions can be found for each problem, each of them capable to generate additional problems and related solutions, and so forth. The PSN is also a morphological tool for selecting the most promising branches in order to create different sets of problem-solution sequences. This peculiarity is a valid support for composing the overall concept variants. More precisely, potential combinations can be selected at different abstraction levels (from abstract to concrete solutions), to support the identification of a limited set of ideas. However, it is necessary to follow three CC fundamental rules when selecting branches:

- All the "first level problems" must be solved to avoid incomplete concepts.
- Any solution is an input of an "OR" logical port (i.e. one of them or a combination of them is sufficient to solve the parent problem);
- Problems constitute the inputs of an "AND" logical port (i.e., each of them must be solved in order to accept the parent solution).

Eventually, CC can lead to some problems between the solutions that come from different PSN branches, or from the combination of two solutions coming from the same parent problem.

3.3 Teaching Approach

The reference course is inserted in the first year of the Mechanical Engineering master degree of the University of Florence, and is focused on the fuzzy front end phases of the design process. More specifically, after a first part focused on the product planning phase, the central part of the course (about 30 h) is focused on systematic conceptual design. Some theoretical notions about creativity and conceptual design are presented, and then students are taught about FDM and related flaws. After that, the PSN is introduced by means of both theoretical lessons (about five hours) and tutorials (about two hours), and a practical exercise lasting about four hours. Then, some well-known procedures for concept selection are also presented [26, 29].

The practical exercise consists in developing a product concept, according to the information generated in the first part of the course. To this aim, design teams of 2/4 students are formed, and the teacher directly follows them during the exercise. After completing the task, design teams present their work and discuss about the performed experience in an interactive session of about two hours. In particular, students can interact each other, and the teacher moderates the discussion.

Then, for the final exam, students are asked to apply the taught tools in a brand-new project, directly proposed by students themselves. The desired outcome is the identification of a preferred concept.

Unfortunately, this specific approach has not been followed in the last academic year, due to the severe restrictions caused by the COVID-19 pandemic. In particular, the practical exercise has been removed, due to the impossibility for the teacher to follow students in a comprehensive way. Accordingly, also the final exam is no longer based on a project, but it is delivered through an oral proof.

4 Gathered Impressions

4.1 Feedback from Students

The impression from students have been gathered in about ten years both during the course and at the final exam. Different approaches have been used to gather students' feedback. The main approach was that of anonymous questionnaires [10], used to evaluate both the teacher performance as well as to receive specific suggestions about how to improve the course (e.g. the amount of theoretical lessons vs practical in class exercises). An excerpt of the questionnaire administered in 2018 and 2019 can be found here: https://rb.gy/xukmls. However, also direct impressions have been received during the final exam (mainly from students obtaining negative evaluations for their work).

It is always difficult to screen information from students due to their natural tendency of asking for more and more examples, which however is in contrast with the need of teaching them the required theory in a predefined and limited amount of time.

Concerning the PSN, both positive and negative feedback have been received from students. Table 2 summarizes this information by clustering it into four main factors, i.e.: the capability perceived by students in supporting the exploration of the design space, the usability of the learned approach, the perceived learning load, and aspects related to the application of the approach.

Table 2. Main feedback received from students

Argument	Positive feedback	Negative feedback
Design space exploration	The PSN helped in exploring additional solutions	-
Approach	The presence of a predefined set of rules helped in organizing the work	The need to follow a predefined set of rules appears quite boring when the designer already has his/her own idea
Learning load	-	The required theoretical lessons are cumbersome if compared with an intuitive design approach
Project workload	The PSN helped to work in groups, since allows to present ideas in a very abstract form	Expressing ideas in an abstract form can be quite difficult when approaching the PSN for the first time
	The independency rule allows to easily subdivide the workload among the team's subjects	

4.2 Observations from Teachers

Teachers observed students during both courses and exams, and they kept track of the most recurring issues during the exams. The main aspects teachers are reported in Table 3. More in particular, they took in consideration the task of defining the system's function, as it is acknowledged as a key step in conceptual design, the adherence of the applied approach with the approach taught in the course, as the compliance with the rules in developing and combining concepts is an indication of the correctness of the application.

For the last year, the COVID-19 pandemic made impossible to repeat the course with the same didactical approach. Differently, the course moved towards a more emphasis on theory, thus strongly reducing practical exercises. Indeed, it was impossible to manage the latter remote lessons. Therefore, the observations reported in Table 3 concern normal in-class lessons with the didactical approach shortly introduced in Sect. 3.

Table 3. Main observations performed by teachers

Argument	Most recurring issues
Function definition	Students often fail in formulating correct functions according to the EMS model
	Students often fail to understand which are the actual actions that the designed system should implement
PSN rules	The most transgressed rule is that of the independency of branches
Concept composition	Students often fail in modelling the problems related to the concept composition phase

5 Discussions and Conclusions

5.1 Lesson Learned

The feedback and observations shortly listed in Sect. 4 suggest that the PSN approach can be efficiently used to teach creative conceptual design to engineering students. Indeed, the main issues gathered from students are somewhat common when teaching academic design approaches (see Sect. 2). In particular, the teachers involved in the considered course found the PSN a very useful tool when judging students projects. Indeed, problem-solution co-evolutionary nature of the PSN, together with its specific sets of rules, allows to easily find out possible flaws or lacks in the idea generation processes adopted by students.

More precisely, besides possible formal errors (e.g. function definitions, problem formulation, etc.), the PSN allows to easily highlight incongruences in the formulation of the problem-solution series. For example, it is possible to easily understand if the students have learned how the proposed solution works, simply by observing the problem-solution sequences used to describe it. Additionally, it is possible to understand how and to what extent students performed the design space exploration. Indeed, the PSN branches show the different solution alternatives considered for each problem (horizontal exploration), as well as the problems formulated for each identified solution (vertical exploration).

However, most of the problems observed by the teachers concern the actual understanding that students have about the theoretical underpinnings of the PSN approach. It seems that due to the particular verification procedure (practical project), most of the students tend to learn the minimal instructions required to compile a "sufficient" PSN, hacking the procedure by performing a reverse modeling of PSNs starting from intuitively conceived ideas. By operating in this way, the creativity support of the PSN is nullified, and the procedure appears as an useless burden for users.

The highlighted problems cannot be neglected, and as for other academic design approaches, they probably are a concurrent cause hindering the diffusion in industry. Indeed, to the authors opinion, the benefit/cost ratio characterizing the adoption of academic conceptual design methods still needs to be comprehensively evaluated and discussed. Without such a kind of information, it is quite improbable to convince firms to spend time and efforts in learning academic methods. For this reason, students are the most important resource for methods diffusion. Indeed, being the designers of the future, they will have the opportunity to import the learned approaches in industry.

5.2 Limits and Research Hints

As stated in Sect. 3, the restrictions due to the COVID-19 pandemic led teachers to move towards a complete oral exam, thus fundamentally based on theory. Notwithstanding the importance of practical exercises in teaching design methods, this can be an important opportunity to understand how the exam procedure can impact on students' motivations. Indeed, one of the possible problems that hinder the diffusion of the methods, is due to insufficient understanding of the approach [27]. The main issues highlighted in Sect. 4 confirm that students have some problems in understanding the fundamental rules and formalisms, but it is unclear if the problem is due to the particular teaching procedure, or if it is a consequence of the particular examination procedure.

The possibility to completely change the examination approach will allow to observe if the same lacks about the theoretical underpinning are still present. If not, or if sensibly reduced, it is possible to assume that the project-based approach leads some students to neglect theoretical notions. This could be an important information in order to develop more efficient teaching procedures.

However, the authors are conscious that what reported in this work is probably affected by their subjectivity and therefore cannot be generalized. In order to obtain a comprehensive set of hints for educational purposes, it is necessary to develop a structured course assessment approach, involving specific investigation activities (e.g. active monitoring of students during the exercises, recorded interviews, in addition to the questionnaires already administered in this case). This work should be considered for what it is, i.e. a lesson learned to share the impressions from the teaching experience and to provide a first base for discussions.

References

1. Jansson, D.G., Smith, S.M.: Design fixation. Des. Stud. **12**, 3–11 (1991). https://doi.org/10.1016/0142-694X(91)90003-F
2. Purcell, A.T., Gero, J.S.: Design and other types of fixation. Des. Stud. **17**, 363–383 (1996). https://doi.org/10.1016/S0142-694X(96)00023-3
3. McNeill, N.J., Douglas, E.P., Koro-Ljungberg, M., Therriault, D.J., Krause, I.: Undergraduate students' beliefs about engineering problem solving. J. Eng. Educ. **105**, 560–584 (2016). https://doi.org/10.1002/jee.20150
4. Goldschmidt, G.: Avoiding design fixation: transformation and abstraction in mapping from source to target. J. Creative Behav. **45**, 92–100 (2011). https://doi.org/10.1002/j.2162-6057.2011.tb01088.x
5. Vasconcelos, L.A., Crilly, N.: Inspiration and fixation: Questions, methods, findings, and challenges. Des. Stud. **42**, 1–32 (2016). https://doi.org/10.1016/j.destud.2015.11.001
6. Tomiyama, T., Gu, P., Jin, Y., Lutters, D., Kind, C., Kimura, F.: Design methodologies: industrial and educational applications. CIRP Ann. **58**, 543–565 (2009). https://doi.org/10.1016/j.cirp.2009.09.003
7. Chakrabarti, A., Bligh, T.P.: A scheme for functional reasoning in conceptual design. Des. Stud. **22**, 493–517 (2001). https://doi.org/10.1016/S0142-694X(01)00008-4

8. Kroll, E.: Design theory and conceptual design : contrasting functional decomposition and morphology with parameter analysis. Res. Eng. Des. **24**(2), 165–183 (2013). https://doi.org/10.1007/s00163-012-0149-6
9. Fiorineschi, L., Rotini, F., Rissone, P.: A new conceptual design approach for overcoming the flaws of functional decomposition and morphology. J. Eng. Des. **27**, 438–468 (2016). https://doi.org/10.1080/09544828.2016.1160275
10. Fiorineschi, L.: Abstraction framework to support students in learning creative conceptual design. J. Eng. Des. Technol. **16**, 616–636 (2018). https://doi.org/10.1108/JEDT-02-2018-0017
11. Cross, N., Dorst, K.: Creativity in the design process: co-evolution of problem–solution. Des. Stud. **22**, 425–437 (2001). https://doi.org/10.1016/S0142-694X(01)00009-6
12. Eder, W.E.: Design modeling-a design science approach (and why does industry not use it?). J. Eng. Des. **9**, 355–371 (1998). https://doi.org/10.1080/095448298261499
13. Frost, R.B.: Why does industry ignore design science? J. Eng. Des. **10**, 301–304 (1999). https://doi.org/10.1080/095448299261218
14. Jagtap, S., Warell, A., Hiort, V., Motte, D., Larsson, A.: Design methods and factors influencing their uptake in product development companies: a review. In: International Design Conference - DESIGN 2014, pp. 231–240 (2014)
15. Geis, C., Bierhals, R., Schuster, I., Badke-Schaub, P., Birkhofer, H.: Methods in practice – a study on requirements for development and transfer of design methods. In: DESIGN 2008 - 10th International Design Conference, pp. 369–376 (2008)
16. Guertler, M.R.: How to design methods for application - empirical insights from industry. In: International Design Conference - Design 2018, pp. 1161–1172 (2018)
17. Björklund, T.A.: Initial mental representations of design problems: differences between experts and novices. Des. Stud. **34**, 135–160 (2013). https://doi.org/10.1016/j.destud.2012.08.005
18. Ozkan, O., Dogan, F.: Cognitive strategies of analogical reasoning in design: differences between expert and novice designers. Des. Stud. **34**, 161–192 (2013). https://doi.org/10.1016/j.destud.2012.11.006
19. Cardella, M.: Engineering design processes : a comparison of engineering design processes : a comparison of students. J. Eng. Educ. **96**, 359–379 (2007). https://doi.org/10.1002/j.2168-9830.2007.tb00945.x
20. Barak, M.: Teaching engineering and technology: cognitive, knowledge and problem-solving taxonomies. J. Eng. Des. Technol. **11**, 316–333 (2013). https://doi.org/10.1108/JEDT-04-2012-0020
21. Dym, C.L., Agogino, A.M., Eris, O., Frey, D.D., Leifer, L.J.: Engineering design thinking, teaching, and learning. J. Eng. Educ. **94**, 103–120 (2005). https://doi.org/10.1109/EMR.2006.1679078
22. Cascini, G., Graziosi, S., Montagna, F., Rotini, F.: On the factors affecting design education within a multi-disciplinary class. J. Integr. Des. Process. Sci. **21**, 21–44 (2017). https://doi.org/10.3233/jid-2017-0001
23. Atman, C.J., Bursic, K.M.: Teaching engineering design: Can reading a textbook make a difference? Res. Eng. Des. – Theor. Appl. Concurrent Eng. **8**, 240–250 (1996). https://doi.org/10.1007/BF01597230
24. Northwood, M.D., Northwood, D.O., Northwood, M.G.: Problem-based learning (Pbl): from the health sciences to engineering to value-added in the workplace. Global J. Eng. Educ. **7**, 157–163 (2003)
25. Bell, S.: Project-based learning for the 21st century: skills for the future. Clearing House J. Educ. Strategies Issues Ideas. **83**, 39–43 (2010). https://doi.org/10.1080/00098650903505415
26. Pugh, S.: Total Design. Integrated Methods for Succesfull Product Engineering. Addison Wesley Publishing Company, Reading, Massachusetts (1991)

27. Pahl, G., Beitz, W., Feldhusen, J., Grote, K.H.: Engineering Design 3rd ed. Springer-Verlag, London (2007). https://doi.org/10.1007/978-1-84628-319-2
28. Akao, Y.: Quality Function Deployment: Integrating Customer Requirements into Product Design. Productivity Press, Cambridge (1990)
29. Ulrich, K.T., Eppinger, S.D.: Product Design and Development. Mc Graw Hill Irwin, New York (2012). https://doi.org/10.1016/B978-0-7506-8985-4.00002-4

A New Table to Manage 1D Tolerances Stack-Up

Sergio Rizzuti(✉) and Luigi De Napoli

Università Della Calabria, DIMEG - Ponte Pietro Bucci 46/C, 87030 Rende (CS), Italy
{sergio.rizzuti,luigi.denapoli}@unical.it

Abstract. Analyzing an assembly and recognizing how the components can mate with each other in order to satisfy the functionality for which they have been designed is not a trivial task. Teaching such a problem to engineering students requires they are familiar with a set of components and how much larger or thin can be the errors intrinsically related to the technologies employed in their production. A set of steps are necessary to reasoning about the right identification of a chain of dimensions that influences a functionality. Then can be useful to have a table where collect the data, especially when the number of parts is relevant. The paper presents a new format for the table that can be used in all kinds of problems that may occur in design: analysis, analysis with constraints, and synthesis. The way how to employ such a table and the steps to solve each problem is discussed with known examples.

Keywords: Tolerance stack-up · Modeling · Tolerance analysis and synthesis

1 Introduction

The search for a stack-up inside an assembly is a fundamental step to push students to understand the main functionalities of a device. The correct identification of a stack-up, that involves several dimensions, requires the selection of those basic dimensions that are influent on the functionality. A stack-up may assume really heterogeneous forms, 1D, 2D, or 3D depending on the relative orientations of the dimensions that it takes into account. 2D and 3D stack-ups require more sophisticated tools to be managed, 1D may appear more simple, even computation requires a certain time to investigate the results and reasoning about them. Several tools of computer-aided tolerancing analysis [1–4] are integrated into CAD systems and give results in terms of Worst Case, Statistic Tolerances, general, or RSS. This approach is mandatory to guide professionals during product development, when it is not convenient or impossible to apply for a physical prototype. Different is the task when a stack-up is introduced to students for the first time when they have not a well-structured knowledge of all factors that influence the form and the dimension of all components inside an assembly. The identification of the best stack-up of dimensions is not trivial and many are the abilities that must be strengthened, first of all how to recognize the most influential dimension inside every single part involved in a stack-up. The paper presents a table that collects data selected by the designer to create a stack-up. The main difference from others such a kind of table, built collecting the extreme value of each dimension, like in the ASME tradition, is that

C. Rizzi et al. (Eds.): ADM 2021, LNME, pp. 861–871, 2022.
https://doi.org/10.1007/978-3-030-91234-5_87

this one employs both the nominal dimension and its variations. The table is intended for 1D studies and can be used even for problems of analysis, analysis with constraints, and synthesis. After its description, it will be used to revisit well-known cases taken from literature [5, 6] (see Fig. 1 and Fig. 7).

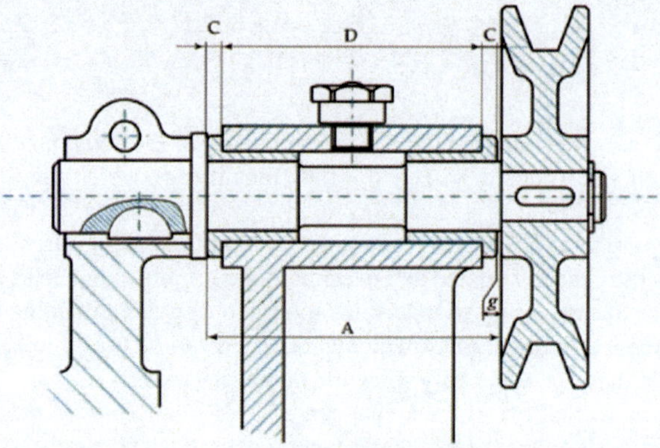

Fig. 1. Support for a pulley that guides a crank.

2 The Importance of Modeling

One of the most intriguing topics in the course of Machine Drawings at the first level of Mechanical Engineering studies is the identification and solution of a stack-up of dimensions affected by errors, since students still have not complete knowledge of all the influences that concur on the final form of the components, from materials, technologies, mechanical behavior. Most of the consideration hidden in the correct and complete dimensioning of each component will appear more clear only when all the questions about the functionalities are well understood. At the same time, the knowledge of dimensioning cannot be left to a purely cosmetic application of dimensions on drawings, also taking into account the new need for geometric dimensioning and tolerancing. In any case, pure dimensions or geometric dimensions must be put on drawings only after the study of the device's functionality. Therefore, when designers start to reasoning about functionalities they must identify the so-called "functional conditions", or dimensional characteristics that appear on the assembly when all components are mounted to each other. Before performing any kind of study the first step consists of identifying a "functional condition", then find all the components involved in it, and identify on each component what is(are) the dimension(s) that may influence the functional condition. In a certain sense, a model must be built for studying a certain functionality. For each problem, the functional condition, that will be named "g" in the following, must be put in relation to a set of other entities, or dimensions. Each dimension must not be identified only by a number, but in conjunction with its variation, which depends on the technological process by which can be obtained. This latter is influenced also by the costs for their manufacturing. So the term functional has sense, in that g can be seen as a function that depends on other ones, and designers must manage not only geometrical problems. The best choice will lie in finding the best solution that gives the best tradeoff between functionality and cheapness.

When studying any phenomenon, the importance to create a model has been put in evidence by Vito Volterra, a distinguished Italian mathematician, in 1901 said: "... Il tradurre nel linguaggio dell'aritmetica o della geometria i fatti della natura è piuttosto schiudere il varco alle matematiche... Plasmare dunque concetti in modo da potere introdurre la misura; poter misurare quindi; dedurre poi le leggi; risalire da esse ad ipotesi; dedurre da queste, mercé l'analisi, una scienza di enti ideali sì, ma rigorosamente logica; confrontare poscia colla realtà; rigettare o trasformare, man mano che nascono contraddizioni fra i risultati del calcolo e il mondo reale, le ipotesi fondamentali che han già servito; e giungere così a divinare fatti e analogie nuove...." [7].

3 Strategy to Identify 1D Tolerance Stack-Up

A functional condition g can be associated with a set of dimensions affected by errors. The first step is to identify which are the dimensions and the tolerances by which g may depend, as in Eq. 1, where n is the number of dimensions involved in the stack-up

$$g = f(Q_1, Q_2, Q_3, \ldots, Q_n; t_1, t_2, t_3, \ldots, t_n) \tag{1}$$

This is the more difficult step, absolutely, in which students frequently fail to identify what is useful and what it is not. From the pedagogical point of view, the problem is only suggested and the functional condition is shown, underlying the problem in the assembly or describing a more global problem of functionality. Students have to research what are the parts involved in the chain and which dimensions in each part. The rule suggested [8], starts from the identification of g between two parts, and assigning to this dimension a verse (arbitrary and positive), like a vector. From the origin of such a vector and moving backward along all the involved parts, the associated dimensions are identified. The search ends when the last dimension reaches the vector arrow of g. In this way, it has been formed a "chain of dimensions" or stack-up. The model proposed to students is typically a 1D chain of dimensions that develops along only one direction. The nature of the functional f is typically an algebraic sum in 1D stack-up and vector sum in 2D and 3D cases. The most proper model in which the dimensions' variability t occurs on each components is decided by designer/technologist/metrologists on the basis of the manufacturing by which the components will be produced and of the operating condition of the device. Many models of stack-up exist: the so-called "worst case" (Eq. 2a) or "statistical" (general or RSS) (Eq. 2b) tolerance analysis. In the worst case (Eq. 2a) t_g and t_i measure the whole range, in the statistical cases (Eq. 2b) these terms measure half of the range.

$$t_g = \sum t_i \tag{2a}$$

$$t_g = \sqrt{\sum t_i^2} \tag{2b}$$

Now, the worst-case model, which is introduced to students, will be treated.

3.1 New Table to Manage Functional Problems as Worst Case

In Fig. 2, the table shows the ordered organization of data when reasoning about the fulfilment of a functional condition. In the upper part of the table, the dimensions involved in the stack-up are classified. Having algebraic sum of dimensions, these must be classified, in order, as positive (concordant with the sign of g) or negative (discordant with the sign of g). Accordingly, the extreme of their variations dimensions must be inserted in the relative columns on the right. In the last column are put the errors related to each dimension. In the lower part of the table shown in Fig. 2, a set of elements linked to the *functional condition* are reported, that is in strict relation with the data of the upper part: the nominal value of g, that may assume the meaning of "clearance" or "interference", its extremes of the variations dimension. On the last column on the right, a set of cells are present in which are collected: the sum of dimensions' variability t_i of all dimensions involved in the stack-up; the variations of g (t_g) obtained as the difference of its variation extremes; finally, a question to check if Eq. 2a is satisfied. The last three rows at the bottom collect if there is a mandatory requirement to g ("required g") if it has been satisfied ("actual g") and how much different these values are, that can be used as starting point to review some elements contained in the table. Let see now how to use the Table in two identified cases [5, 6].

PARTS	DIMENSIONS	VARIATIONS		TOLLERANCE
CONCORDANT	$Q_i^{(+)}$	i_i: LOWER	S_i: UPPER	t_i
DISCORDANT	$Q_i^{(-)}$	S_i: UPPER	i_i: LOWER	t_i
				$\sum t_i =$
FUNZIONAL CONDITION (g)	g: NOMINAL DIMENSION	i_g: LOWER VARIATION	S_g: UPPER VARIATION	$t_g = s_g - i_g$
				$\sum t_i = t_g$?
		g minimum	g maximum	
REQUIRED g				
ACTUAL g				
DIFFERENCE (Δ)				

Fig. 2. Content of the new table schema.

4 Case Studies: Reexamination of a Typical Problem

The mechanism represented in Fig. 1 [5] is a support for a pulley (on the right) that guides a crank (on the left). The motion is transmitted by a shaft that rotates freely inside two bushings. The functional problem that emerges is that the pulley should not touch the external surface of the bushing. The *functional condition* g is associated with a space between bushing and pulley that must be guaranteed. Following the dimensioning in Fig. 1, g is in relation to a set of dimensions: A on the shaft, D on the support, and C on both bushings, each with their tolerances.

So Eq. 1 can be written, as

$$g = f(A, C, D, t_A, t_B, t_C) \tag{3}$$

The dimensions in Fig. 1 are the following, according to UNI EN 22768/1-f:

$$A = 84^{\pm 0.15};\ C = 4^{\pm 0.05};\ D = 76^{\pm 0.15}$$

4.1 The Problem of Analysis

A problem of analysis occurs when assembling a set of components already manufactured of whom dimensions and variability are known. The procedure starts with the identification of the sequence of dimensioning, properly writing Eq. 3.

$$g = A - 2C - D$$

a)

PARTS	DIMENSIONS	VARIATIONS		TOLLERANCE
CONCORDANT	$Q_i^{(+)}$	i_i: LOWER	S_i: UPPER	t_i
A	84	- 0.15	0.15	0.3
DISCORDANT	$Q_i^{(-)}$	S_i: UPPER	i_i: LOWER	t_i
C	2 x 4	2 x 0.05	2 x (- 0.05)	2 x 0.1
D	76	0.15	- 0.15	0.3
				$\sum t_i = 0.8$
FUNZIONAL CONDITION (g)	g: NOMINAL DIMENSION	i_g: LOWER VARIATION	S_g: UPPER VARIATION	$t_g = S_g - i_g$
	0	- 0.4	0.4	0.8
				$\sum t_i = t_g$!
		g minimum	g maximum	
REQUIRED g				
ACTUAL g				
DIFFERENCE (Δ)				

b)

PARTS	DIMENSIONS	VARIATIONS		TOLLERANCE
CONCORDANT	$Q_i^{(+)}$	i_i: LOWER	S_i: UPPER	t_i
A	84	0.35	0.65	0.3
DISCORDANT	$Q_i^{(-)}$	S_i: UPPER	i_i: LOWER	t_i
C	2 x 4	2 x 0.05	2 x (- 0.05)	2 x 0.1
D	76	0.15	- 0.15	0.3
				$\sum t_i = 0.8$
FUNZIONAL CONDITION (g)	g: NOMINAL DIMENSION	i_g: LOWER VARIATION	S_g: UPPER VARIATION	$t_g = S_g - i_g$
	0	0.1	0.9	0.8
				$\sum t_i = t_g$!
		g minimum	g maximum	
REQUIRED g		0.1		
ACTUAL g		0.1		
DIFFERENCE (Δ)		0		

Fig. 3. Table filled in case of tolerance analysis (a), and with constraint, after modifications (b).

Then the table shown in Fig. 2 is filled, inserting dimensions and orderly their variations in the right columns, distinguishing if the dimension is concordant or discordant with g (see Fig. 3a). Therefore, the g is computed summing algebraically the number written in the dimension column, and accordingly its lower and upper variation ig and sg. Then tg is evaluated and this result is verified in terms of Eq. 2a.

A first consideration can be done on the sizes of g and t_g, in terms of functionality, thinking if these values are reasonably acceptable.

The Problem of Analysis with Constraints. This kind of problem is frequently present when a particular constraint is assigned to the *functional condition.* In the previous problem, the designer recognizes that in the device may occur that pulley is in contact with the bushing. This is because the lower variation of g, or as it can be said its mini-mum value $g_{min} = g + i_g$, is "negative", and that this corresponds not to a "clearance" but to an "interference". For this problem in the bottom part of the table must be added such condition, as "required g", in the column that characterizes it. In this particular problem suppose that designer has required a $g_{min} = 0.1$ to guarantee a clearance. Pre-vious computations resulted a different value of $g_{min} = -0.4$. Copying this value in "actual g" emerges a difference between what is prescribed and what is obtained. This difference, named Δ in the table, must be corrected operating a translation of the dimensions' variation of one or more sizes in the assembly parts (Fig. 4).

	g minimum	g maximum	
REQUIRED g	0.1		
ACTUAL g	- 0.4		
DIFFERENCE (Δ)	0.5		

Fig. 4. Detail of the table in case of analysis with constraint.

At this point, a decision must be taken. The designer must decide on which components operate some changes. The difference Δ can be assigned to only one component or subdivided among several ones. Supposing that the bushings have been bought and the support has been made outsourcing, the difference can be absorbed by shaft A, shifting its lower variation: $i'_A = i_A + \Delta = -0.15 + 0.5 = 0.35$. Accordingly, the upper variation of A must be translated $s'_A = s_A + t_A = 0.35 + 0.3 = 0.65$.

This update must be made without altering the tolerances assigned to all components because otherwise technologies would have been changed and so the related device costs. The problem is now summarized as in Fig. 3b.

4.2 The Problem of Synthesis

A problem of synthesis occurs when a set of components of whom the best manufacture must be chosen, must be assembled considering the need to satisfy a specific functional condition. The problem is reversed from the analysis. Now t_g is known and all t_i should be identified considering the best subdivision of that global variation. Furthermore, a set of constraints could be posed that may alter the initial hypothesis. Reconsidering the same problem of Fig. 1, where the designer has established a variation to t_g equal to 0.8 mm, suppose now to buy the bushings with a size C = 4h11. The problem now is how a)

PARTS	DIMENSIONS	VARIATIONS		TOLLERANCE
CONCORDANT	$Q_i^{(+)}$	i_i: LOWER	s_i: UPPER	t_i
A	84			0.35
DISCORDANT	$Q_i^{(-)}$	s_i: UPPER	i_i: LOWER	t_i
C	2 x 4	2 x 0.075	0	0.15
D	76			0.30
				$\sum t_i = 0.8$
FUNZIONAL CONDITION (g)	g: NOMINAL DIMENSION	i_g: LOWER VARIATION	s_g: UPPER VARIATION	$t_g = s_g - i_g$
	0			0.8
				$\sum t_i = t_g$!
		g minimum	g maximum	
REQUIRED g				
ACTUAL g				
DIFFERENCE (Δ)				

b)

PARTS	DIMENSIONS	VARIATIONS		TOLLERANCE
CONCORDANT	$Q_i^{(+)}$	i_i: LOWER	s_i: UPPER	t_i
A	84	-0.175	0.175	0.35
DISCORDANT	$Q_i^{(-)}$	s_i: UPPER	i_i: LOWER	t_i
C	2 x 4	2 x 0.075	0	0.15
D	76	0.15	-0.15	0.30
				$\sum t_i = 0.8$
FUNZIONAL CONDITION (g)	g: NOMINAL DIMENSION	i_g: LOWER VARIATION	s_g: UPPER VARIATION	$t_g = s_g - i_g$
	0	- 0.475	0.325	0.8
				$\sum t_i = t_g$!
		g Minimum	g Maximum	
REQUIRED g				
ACTUAL g				
DIFFERENCE (Δ)				

c)

PARTS	DIMENSIONS	VARIATIONS		TOLLERANCE
CONCORDANT	$Q_i^{(+)}$	i_i: LOWER	s_i: UPPER	t_i
A	84	0.25	0.6	0.35
DISCORDANT	$Q_i^{(-)}$	s_i: UPPER	i_i: LOWER	t_i
C	2 x 4	2 x 0.075	0	0.15
D	76	0	-0.3	0.30
				$\sum t_i = 0.8$
FUNZIONAL CONDITION (g)	g: NOMINAL DIMENSION	i_g: LOWER VARIATION	s_g: UPPER VARIATION	$t_g = s_g - i_g$
	0	0.1	0.9	0.8
				$\sum t_i = t_g$!
		g minimum	g maximum	
REQUIRED g		0.1		
ACTUAL g		0.1		
DIFFERENCE (Δ)		0		

Fig. 5. The table partially filled in case of tolerance synthesis (a), when the functional condition is checked (b), and with solved constraint (c).

much larger can be the variations admissible for A and D. The table collects the data as reported in Fig. 5a.

The designer move from the requirements to the upper part of the table. From Eq. 2a and adopting UNI EN 22768/1-f

$$t_g = t_A + 2t_C + t_D$$

$$0.8 = t_A + 2 \cdot 0.075 + t_D$$

$$t_A + t_D = 0 \cdot 65$$

The manufacturing processes and their costs may suggest that D may be obtained with a finer variation with respect to A. So $t_A = 0.35$ and $t_D = 0.30$ may be assigned.

If there are no other conditions imposed to "required clearance", the dimensions' variation of size A and D may be equally distributed by the classical $\pm\, t_i/2$ and the table would appear as in Fig. 5b. If further restriction should be imposed to the "required clearance", reasoning similar to those made in Sect. 4.1 must be done. If the same condition of Sect. 4.1 is assigned, now the difference appears as in Fig. 6.

	g Minimum	g Maximum
REQUIRED g	0.1	
ACTUAL g	-0.475	
DIFFERENCE (Δ)	0.575	

Fig. 6. Detail of the table in case of synthesis with constraint.

The decisions on how to distribute such difference may be several. Surely bushing must not be touched, because their dimensions' variations are guaranteed. The difference Δ could be subdivided between A and D, as

$$\Delta = \Delta_A + \Delta_D = 0 \cdot 425 + 0 \cdot 15$$

Then the lower variation of A must be raised and upper variation of D lowered:

$$i'_A = i_A + \Delta_A = -0 \cdot 175 + 0 \cdot 425 = 0 \cdot 25$$

$$s'_D = s_D - \Delta_D = 0 \cdot 15 - 0 \cdot 15 = 0$$

Accordingly, s_A and i_D must be shifted:

$$s'_A = i'_A + t_A = 0 \cdot 25 + 0 \cdot 35 = 0 \cdot 6$$

$$i'_D = s'_D - t_D = 0 - 0 \cdot 3 = -0 \cdot 3$$

The problem is now summarized as in Fig. 5c.

4.3 The Problem of Analysis with Least Material Condition

The problem is proposed in [6] and represents the assembly of two components that are relatively positioned by a junction between a slot and a tab. In Fig. 7 the assembly and dimensioning of the parts are reproduced, in altered scale. The stack-up analysis asks to define the maximum and minimum dimension of the external surfaces of both parts when mounted. The *conditional function g* is the distance between these two surfaces. The dimensions involved on both parts are the axes of slot and tap, and the dimension of these. In order to find the extreme values of g, two different assemblies must be considered, considering that the contact between slot and tab may happen on the sur-faces on the left or on the right. Two different stack-up must be built.

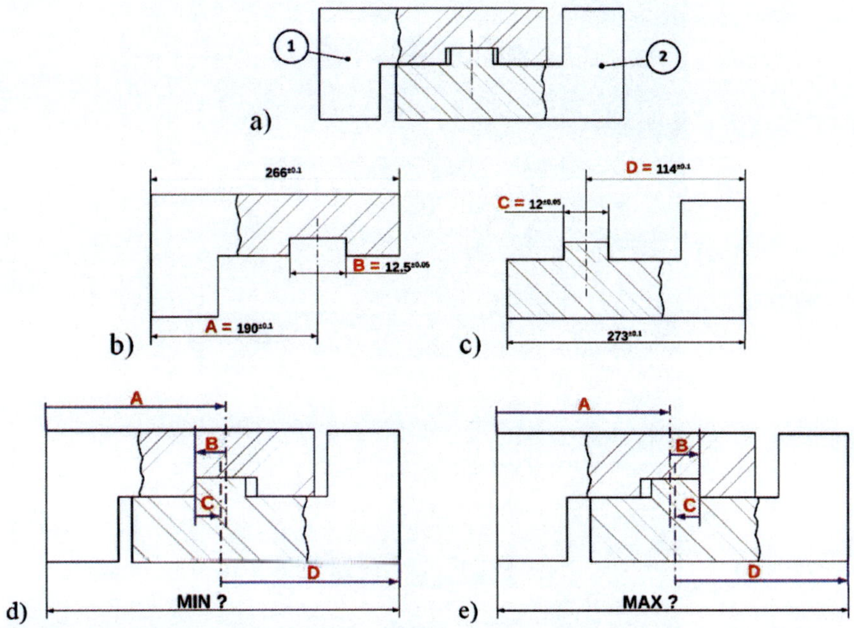

Fig. 7. a) Assembly; b) Part 1; c) Part 2; contact on the left (d) and right (e) surfaces.

Anyway, the wider variability occurs when slot and tab are in least material condition.

Minimum Distance. The case is relative to Fig. 7d when the contact is on the left sur-faces of the slot-tap. Table in Fig. 8a collects the data associated to the functional con-dition g = A − (B/2) + (C/2) + D. The minimum value of *g* must be evaluated as *gmin* = *g* + *ig*. The other value *gmax* has not a real sense, because its value does not represent anything other than a further possibility, but not an extreme value.

Maximum Distance. The case is relative to Fig. 7e when the contact is on the right surfaces of the slot-tab. Table in Fig. 8b collects the data associated to the functional condition g = A + (B/2) − (C/2) + D. The maximum value of *g* must be evaluated as *gmin* = *g* + *sg*. The other value *gmin* has not a real sense, because its value does not represent anything other than a further possibility, but not an extreme value. The ability in this kind of problem, and every time a joint by a dowel pin is involved, consists in identify correctly both stack-ups.

a)

PARTS	DIMENSIONS	VARIATIONS		TOLLERANCE
CONCORDANT	$Q_i^{(+)}$	i_i: LOWER	S_i: UPPER	t_i
A	190	-0.1	0.1	0.2
C/2	6	-0.025	0.025	0.05
D	114	-0.1	0.1	0.2
DISCORDANT	$Q_i^{(-)}$	S_i: UPPER	i_i: LOWER	t_i
B/2	6.25	0.025	-0.025	0.05
				$\sum t_i =$
FUNZIONAL CONDITION (g)	g: NOMINAL DIMENSION	i_g: LOWER VARIATION	S_g: UPPER VARIATION	$t_g = s_g - i_g$
	303.75	-0.25		
				$\sum t_i = t_g$?
		g minimum	g maximum	
REQUIRED g		303.5		
ACTUAL g				
DIFFERENCE (Δ)				

b)

PARTS	DIMENSIONS	VARIATIONS		TOLLERANCE
CONCORDANT	$Q_i^{(+)}$	i_i: LOWER	S_i: UPPER	t_i
A	190	-0.1	0.1	0.2
B/2	6.25	-0.025	0.025	0.05
D	114	-0.1	0.1	0.2
DISCORDANT	$Q_i^{(-)}$	S_i: UPPER	i_i: LOWER	t_i
C/2	6	0.025	-0.025	0.05
				$\sum t_i =$
FUNZIONAL CONDITION (g)	g: NOMINAL DIMENSION	i_g: LOWER VARIATION	S_g: UPPER VARIATION	$t_g = s_g - i_g$
	304.25		0.25	
				$\sum t_i = t_g$?
		g minimum	g maximum	
REQUIRED g			304.5	
ACTUAL g				
DIFFERENCE (Δ)				

Fig. 8. Minimum distance (a), and Maximum distance (b).

5 Conclusion

One of the most important topics that must be taught in a course of Machine Drawing in mechanical engineering studies is the right identification of a *functional condition* in an assembly. The training in using a table like that presented in the paper may guide the reasoning for the solution of the problems typology shown. The first step in studying a *functional condition* is the search for the dimensions of the components involved. The second step is related to the choices that must be done in order to satisfy the conditionality imposed on the *functional condition*. The usage of the table is valid not only when the study is performed on drawings, but also, and overall when a CAD model has been built. The software tools [1–4] that are available at the moment may give a result to a stack-up of dimension tolerances, which may involve sometimes also geometric tolerances. The problem of identifying the right sequence of dimensions and geometric conditions continues to be the main problem. On CAD models the difficulties are greater because the nature of geometric modeling superimposes the possibility of identifying some measurements when for example during modeling other dimensions and relations have been employed on each component.

Furthermore, CAD systems give a lot of results and they must be checked properly, because often these results do not have sense, being only the management of dimensions and variation that are far from the characterization of a *functional condition* to be solved.

References

1. Sigmetrix Homepage. https://www.sigmetrix.com/. Accessed 17 Mar 2021
2. Siemens-Tecnomatix Homepage. https://www.plm.automation.siemens.com/global/en/products/manufacturing-planning/dimensional-quality.html. Accessed 17 Mar 2021
3. 3DCS Homepage. https://www.3dcs.com/. Accessed 20 Mar 2021
4. Eventive Homepage. https://www.enventive.com/. Accessed 20 Mar 2021
5. Chirone, E., Tornincasa, S.: Disegno Tecnico Industriale, vol. 2. Ed. 2020, Il Capitello, Torino (2020)
6. Meadows, J.D.: Tolerance stack-up analysis. In: Geometric Dimensioning and Tolerancing Handbook: Applications, Analysis & Measurement Chap. 23. Ed. ASME Press, Meadows (2009)
7. Volterra, V.: Sui tentativi di applicazione delle matematiche alle scienze biologiche e sociali. Discorso inaugurale. Annuario della R. Università degli studi di Roma, pp. 5–28 (1902)
8. Manfé, G., Pozza, R., Scarato, G.: Disegno Meccanico, vol. 2. Principato Ed, Milano (1987)

A Service-Learning Experience Applied in Corporate Identity Design of Newly Created Companies

Natalia Muñoz López[1] (✉), Anna Biedermann[1], José Luis Santolaya Sáenz[1], and Camille Bertrand[2]

[1] Universidad de Zaragoza, María de Luna 3, 50018 Zaragoza, Spain
{nataliam,anna,jlsanto}@unizar.es
[2] OTRI Universidad de Zaragoza, Pedro Cerbuna 12, 50009 Zaragoza, Spain
camille@unizar.es

Abstract. The presented project is based on a collaboration between two public entrepreneurship incubators, Ceminem and Zaragoza Activa, and the subject of Graphic Design Applied to Product. In this project, students develop various design proposals for the graphic image of newly created companies that participate in the business incubation program. In this way, students actively learn not only the contents of the subject but also the treatment with the client, being their first experience, and in parallel they acquire knowledge about entrepreneurship. Entrepreneurs from different centers participate in a networking work focused on branding. The service-learning methodology will facilitate the development of contact networks for students and their first clients, as well as reinforce a user-centered work attitude. Participants will meet in the different centers to carry out the activities together and get to know their offer first hand. Innovation is based on: introducing students to the entrepreneurship ecosystem, turning a university exercise into a real job that includes the requirements of working with a client-user, evaluating the satisfaction of all parties involved. The results show that all participating companies decided to use one of the designs performed by students and both entrepreneurs and students are very satisfied with the experience.

Keywords: Service-learning · Entrepreneurship · Knowledge transfer · Branding

1 Introduction

Today's society requires that the knowledge developed and transmitted by educational institutions should not be for accumulation but rather serve to stimulate the society's transformation [1]. That is why various authors indicate the need to include the new pillar of education: "Learn to be entrepreneur" [2–4] to the four basic pillars of teaching presented in the Report "Education in the XXI century" [5]. Entrepreneurship competence plays an important role in strengthening capacities that enable creativity, innovation, independent action and a critical view of the environment, as well as the resilient ability

C. Rizzi et al. (Eds.): ADM 2021, LNME, pp. 872–878, 2022.
https://doi.org/10.1007/978-3-030-91234-5_88

to take risks and achieve objectives. Creativity and innovation are especially needed in the current situation [6, 7].

Batista Hernandez et al. [1] explain that teaching entrepreneurship is to build in the student an attitude towards the social, ethical, intellectual and political values that constitute the basis for independence in thinking and acting. At the same time, it favors the student's search for the new solutions, the creation and development of their potential, preparing them to assume a disposition towards the future that provides optimism and progress of personal and collective potential. The competence of learning to be an entrepreneur composes a novel reinforcement to the system of competence training; develops skills that enable the effective performance of the individual in society. It opens the educational process by turning social environments into pedagogical settings. Developing entrepreneurship in summary is to increase students' ability to solve problems with initiative and creativity, leadership in the development of work and willingness to be agents of change.

The development of this competence is linked to the capacity that society has to create new companies, an action that has a direct impact on the economic and social progress of the communities, as stated by various international institutions such as the OECD or the EU [8, 9]. A special focus is placed on university entrepreneurship as the one that is capable of generating quality and long-term jobs [10]. Nonetheless Ripollés [2] reviewing the university programs for entrepreneurship development concludes that, in most cases, those programs are based on master classes, specialized readings, practical cases or talks. They are conventional teaching methods that fail to give the student a necessary active role [11].

The environment in which companies incubate is very important because it has an immediate influence on the formation of the startup itself. An environment conducive to creating a wide network of contacts helps create symbiosis between enterprises and also in the development of the commercial network. The enterprise itself can only be developed in society, in which it is possible to have a network of contacts. Learning to build this network - Networking- is an opportunity for training social skills. Effective networking requires dialogue, information exchange to find ways to help each other [12].

The entrepreneurial ecosystem, on the other hand, consists of both the environment prone to the creation and development of startups and also of the community, group of individuals and institutions that create this environment. In Zaragoza, there are various institutions specialized in supporting the entrepreneur in the different stages of development of their project, whether public or private. This support usually includes training and consulting. Analyzing the offers of these centers we have detected the opportunity to collaborate with these institutions, offering their participants a service they need and networking opportunities between various institutions. At the same time, we provide our students with teaching that reserves an active role for them and introduces them to the entrepreneurial ecosystem of the city and promotes the role that the University can play in this field.

The project presented in this paper has the nature of an intervention that gets students involved in socio-community activities and is also a productive project since the results are being applied. During the second semester of the second course of the four-year degree in Engineering in Product Design and Development at the University of Zaragoza, students attend the compulsory subject of Graphic Design Applied to Product. They learn to make multiple elements of visual communication using the vector design tools, layout and image editing software. The course aims to facilitate the use of these tools, given that on average more than 70% of students never used them beforehand and the development of compositional criteria, knowledge of color and type fonts, as well as the characteristics of different environments of graphic design application. One of these environments and at the same time content to develop in the subject is Branding - the process of creating a corporate identity. This work requires multiple analysis before proceeding to the development of brand concepts. With the aim of bringing this process to professional experience since 2012, the role of the clients is taken by the entrepreneurs incubated in a municipal program of support for entrepreneurship- Semillero de Ideas. In order to broaden the scope of the project, it has been decided to also count on the collaboration of entrepreneurs incubated at CEMINEM within the entrepreneurship support programs of the University of Zaragoza itself.

2 Objectives

The objectives of this collaboration are as follows:

- To provide a design experience similar to the professional one in which it is the client who establishes the design specifications, there are deadlines imposed to do the work and there is professional competence.
- To provide quality designs to support startups selected in public calls, offering them a design service at no cost at the critical moment of development of their business activity. The scope of this service is clearly delimited in the bases of the collaboration.
- To create the situation in which students relate with entrepreneurship cases directly.
- To facilitate a space for exchange between the entrepreneurs incubated in the different institutions, supporting the connections within the ecosystem of entrepreneurship in Zaragoza.
- To increase students' perception of themselves as professionals capable of solving real design needs.

To bring students closer to public institutions with validated entrepreneurship programs.

3 Methodology

The methodology applied in this case is Service Learning methodology, based on the needs, problems or real challenges that can be found in the near and far context. Service learning is a methodology that is not only limited to identifying the needs and acquiring knowledge but to involve the students in providing a service to the community. With the term service, we mean doing an activity linked to the future career of these students. This activity provides a benefit that is freely delivered to the community [13]. In the presented case, the university community performs a service to the communities related to entrepreneur incubators and in this way, supports the development of entrepreneurship.

The structure of this project follows the learning process defined in the teaching guide of the subject, as well as the design methodology applied in the Grade. This project represents 20% of the 6 ECTS course of Graphic Design Applied to Product, which means 12 h of work in class and 15 h of individual work. The designs are realized in couples and students can decide freely if they want to offer their design to the entrepreneurs´ use. If they do so, and their design is chosen by the company to be used, they get one score more over the academic assessment of their work. If they decide they don't want their design to be used by entrepreneurs, they note it at the project documentation delivery and still they can get the maximum academic score if the result of their work is accurate.

The collaboration is structured in the following phases:

- Drafting the bases of collaboration that define the tasks, deadlines and obligations of both parties (entrepreneurs and students). Registration of business projects interested in participating in the experience.
- Random assignment of entrepreneur projects to groups of students.
- Online joint meeting instead of at Zaragoza Activa headquarters as originally planned due to COVID-19 between entrepreneurs from the two centers and designers.
- Sessions of practical online classes -development of the graphic image of business projects.
- During this process, entrepreneurs and students communicate freely to exchange opinions on the designs that are being developed.
- Online presentation of the different proposals instead of at CEMINEM head office as scheduled, which is the second meeting of the entrepreneurs of both institutions and students. Two proposals per couple are presented.
- Final decision of the entrepreneurs on the design to be used.
- Development of brand manuals and their applications.
- Evaluation of the works by the teachers independently and subsequent sum of a point in the case of the chosen designs. In the academic evaluation, the following ponderation of the score is applied: research (5%), concepts development (10%), final brand design (30%), corporate identity manual (20%), visit card (20%), applications (10%), presentation (5%).

The activity is developed within the framework of the PIIDUZ_19_500 Program for Teaching Innovation at the University of Zaragoza.

4 Results

In the 2019 course, nine companies participated in the experience, four of them incubated in Semillero de Ideas (Zaragoza Activa) and five of them incubated in CEMINEM. 79 students participated in the experience. They were grouped in 35 pairs and 3 groups of 3 students (38 teams in total) to work on the branding of randomly selected company. The companies defined the specifications for their brand design: the values and the message to be transmitted, the definition of the target public. The deadlines were defined on the bases of collaboration, giving to the students a month and a half for concepts preparation and another month for completion of the documentation in the brand manual and brand applications.

All of the students decided to offer the companies their designs, none of them wanted to use the exercise just for academic reasons. Altogether entrepreneurs have been able to choose from 79 design proposals. All of the nine participating entrepreneurs have decided to use the corporate identity proposed by the students for their companies. The chosen brand designs can be seen in Fig. 1, and not chosen alternatives can be seen in Fig. 2.

Entrepreneurs have expressed their satisfaction with the experience (100% have confirmed that they are satisfied with the development of the collaboration), a vision also shared by students (Fig. 3). 78.6% of them consider that the projects made for a real client are necessary for preparing for the future career, although the latter have also expressed that working with real clients made the task more difficult in comparison to other academic practices (opinion of 10%). The average of the academic grade obtained by the students is 7.68 out of 10.

100% of the companies also confirmed that they are satisfied with the designs presented by the students. 85.7% of companies confirmed that they would participate in other experiences like this where students develop a service for their company, and 14.3% would participate in the similar experience with some modifications. 28.6% of the members of the companies didn't know of the existence of the Grade in Industrial Design and Product Development Engineering, before the collaboration. 100% of the companies would recommend other entrepreneurs to participate in the next edition of this collaboration.

At the beginning of the course only 6.9% of the students had some skills using vector design software, at the end of the experience 85.4% of the students confirm having fluency in using vector design software. After the experience 26.8% of the students are convinced that entrepreneurship will be their professional choice, 39.0% of the students see entrepreneurship as the future possible option, and 26.8% of the students doesn't discard it. Only 7.3% of the students are not interested in entrepreneurship. So, it has been achieved that students know at first hand the real meaning of entrepreneurship and they feel motivated to undertake. 82.9% of the students consider that the projects realized for the real clients are motivating for them and let them have the perception of themselves as professionals capable of solving real design needs, and 14.6% partially agree with this affirmation.

Restrictions caused by the pandemic have made impossible visits to any of the entrepreneurship centers as expected which would have facilitated a space for exchange between students and entrepreneurs incubated in different institutions. However, all the nine companies agreed that the pandemic has not affected the collaboration, just one of them indicates that the collaboration would be more interactive with face-to-face contact. From the point of view of the students, the perception of the influence of the pandemic restrictions is more critical. 36,58% of the students believe that the suspension of classes and face-to-face meetings has negatively affected the experience, and 24.4% partially agree with this affirmation. 26.8% of the students believes that the restrictions have affected their academic results, and 36.6% of the students partially agree with this affirmation.

Fig. 1. Chosen brand designs.

Fig. 2. Alternative designs.

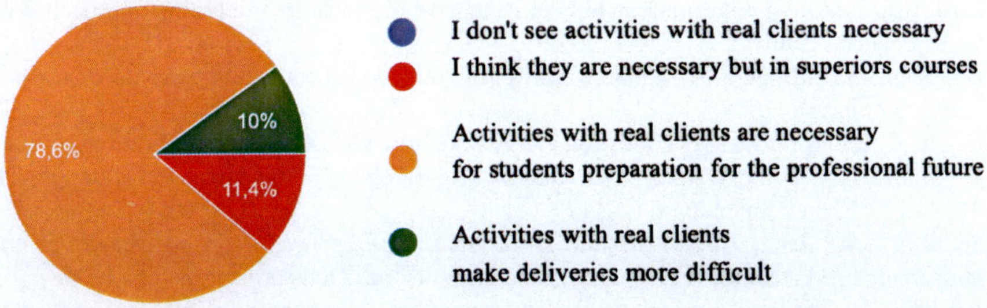

Fig. 3. Students opinion about activities with real clients within academic works.

5 Conclusions

Isenberg [14] defines six domains that describe the state of the entrepreneurial ecosystem: politics, finance, culture, support services, human capital, and markets. Our project affects the cultural domain by developing tolerance to risk and failure, encouraging innovation and creativity, and helping to define the creation of companies as a desirable form of career. At the same time, it generates services that support the visibility of companies and forms part of students' training. The project generates new synergies between institutions dedicated to entrepreneurship, the entrepreneurs themselves and students of the Degree in Engineering in Industrial Design and Product Development. The fact that all of the nine participating companies have decided to use the design proposed by the students, reflects that the proposed practices responded to the real need of branding design the start-up companies have and proved as well that the second-year students have the abilities to develop graphic designs that can satisfy the client's needs. The facilitation of the exchange between the entrepreneurs incubated in the different institutions, supporting the connections within the ecosystem of entrepreneurship in Zaragoza has been limited due to the pandemic restriction.

References

1. Hernández, N.B., Estupiñán, J.: Competencia de emprendimiento como sustento de la formación integral e inserción social del estudiante. Revista Órbita Pedagógica (2017)
2. Ripollés, M.: Aprender a emprender en las universidades. Arbor **187**(Extra_3), 83–88 (2011)
3. Ferreyra, H.A.: Enseñar a "aprender a Emprender": orientaciones metodológicas. Novedades Educativas **148**, 14–16 (2003)
4. Ferreyra, H.A.: Nuevo tópico formativo: "aprender a emprender." Revista Novedades Educativas **259**, 56–58 (2012)
5. Delors, J.: Informe a la UNESCO de la Comisión Internacional Sobre la Educación Para el Siglo XXI, Presidida por JACQUES DELORS La Educación Encierra un Tesoro. UNESCO, Paris (1996)
6. The World of Higher Education after Covid-19. https://www.yerun.eu/wp-content/uploads/2020/07/YERUN-Covid-VFinal-OnlineSpread.pdf. Accessed 26 June 2021
7. The Future of Jobs Report 2020. http://www3.weforum.org/docs/WEF_Future_of_Jobs_2020.pdf. Accessed 26 June 2021
8. OCDE (Organisation for Economic Co-operation and Development). http://www.oecd.org. Accessed 26 June 2021
9. EU (European Commission). http://ec.europa.eu. Accessed 26 June 2021
10. Bird, B.J.: Learning entrepreneurship competences: the self-direct learning approach. Int. J. Entrepreneurs. Educ. **2**(1), 41–59 (2002)
11. Bogotá, C.D.C., Bogotá, A.M.: Networking: El poder de los contactos personales para hacer crecer la empresa (2008)
12. Bager, T.: Entrepreneurship education and new venture creation: a comprehensive approach. In: Hindle, K., Klyver, K. (eds.). Handbook of Research on New Venture Creation, pp. 299–315. Edgar Elgar, Cheltenham (2011)
13. Serrano, L.R., Roig, A.E.: Aprendizaje-Servicio (ApS): Claves Para su Desarrollo en la Universidad. Educación Universitaria, 1st edn. Ediciones Octaedro, Barcelona (2018)
14. Isenberg, D.: The big idea: how to start an entrepreneurial revolution. Harv. Bus. Rev. **88**(6), 41–50 (2010)

Students Creativity Assessment Through Gray's "Forward Flow" Tool: A Preliminary Study

Ileana Bodini(✉), Mariasole Bannò, Diego Paderno, Gabriele Baronio, Stefano Uberti, and Valerio Villa

Department of Mechanical and Industrial Engineering, University of Brescia, via Branze 38, 25123 Brescia, Italy
ileana.bodini@unibs.it

Abstract. The promotion of creativity in education is intended to address many of the political challenges and goals for the development of a country, but among all of them the role of creativity in technology and economics is seen as crucial in helping nations to achieve higher employment, better economic performance and to cope with global competition.

This study, currently in its preliminary phase, aims to compare the creativity levels of first-year students at a medium-sized Italian university. The degree courses in Engineering Management, Mechanical Engineering, Law and Motor Sciences were analyzed.

In this analysis, students' creativity is measured with the "Forward Flow" test by Gray et al. This test implements a new metric that uses latent semantic analysis to measure the evolution of thoughts over time. Operationally, students are asked to produce a sequence of semantically related words from a given initial word. Latent semantic analysis calculates the semantic distance between words by examining the frequency with which they appear together in a very large collection of documents. Studies conducted on the test, found in the literature, reveal that Forward Flow can predict the creativity of college students. According to these studies, even that membership in real-world creative groups (e.g. professional actors or entrepreneurs) is statistically predicted by scores on the Forward Flow test, even when controlling for divergent thinking.

Keywords: Engineering education · Creativity

1 Introduction

Although Plato had already expressed interest in creativity as a component of human capital and creativity is found in the traditions of almost every culture in the world [1–3], it is only since the mid-1990s that we have witnessed a marked interest in education [4, 5] so much so that the most advanced countries in the world have declared their political interest in the education of creativity [6, 7].

C. Rizzi et al. (Eds.): ADM 2021, LNME, pp. 879–888, 2022.
https://doi.org/10.1007/978-3-030-91234-5_89

This attention, over the years, has shifted from being on the margins to being an indispensable aspect of youth education [8–10]. Indeed, the promotion of creativity in education is meant to address many of the policy challenges and goals for a country's development: the role of creativity in the economy is seen as crucial [11] especially in helping states achieve higher employment, better economic performance [12] and to cope with global competition.

For these reasons, creativity cannot be ignored, nor can its development be left to chance [13], but on the contrary it must be among the important long-term goals of the university education plan [14]. Creativity should be considered a fundamental skill [8] that must be developed to prepare future generations [15, 16]. Developing the creativity of young people during their education, including university education [17, 18], is in fact part of building the human capital on which, according to Adam Smith and subsequent economists, the wealth of nations depends [19].

The idea of the present contribution is to verify the level of this non-technical competence at the entrance to different degree courses of a medium-sized Italian university, and then to test its levels in different degree courses. An attempt will be made to identify possible factors able to intervene both positively and negatively in the creativity training pathway. In this paper we propose the first step of this study and some very preliminary results.

2 The Concept of Creativity

The first and broadest definition of creativity includes not only characteristics of intellect, but also motivational aspects and personality qualities [20]. Among others, Morawski [21] defines creativity as the result of the interaction of intellectual skills, creative talents, personality traits and the sociocultural and economic environment. Creativity according to Pfeiffer [22] refers to the generation of new and useful ideas and the ability of employees to create new practical ideas. Other authors, instead, decry creativity as closely related to economic, managerial, and entrepreneurial aspects [23, 24].

Thus, in line with the above definitions, we define creativity as the ability to think creatively, as the ability to generate ideas constantly seeking new solutions, the ability to choose the optimal solution, perseverance in overcoming difficulties, concepts or new associations and connections with existing ideas or concepts [1, 9].

This paper refers to the concept of the creativity process, which we can trace back to two styles of thinking (i.e., convergent, and divergent) [25]. Guilford [26] laid the groundwork for understanding the roles that convergent and divergent thinking play in the creativity process. Although divergent thinking is often associated exclusively with creativity, it is important to recognize that convergent thinking is also critical, particularly in the context of problem solving, whatever its nature. For example, engineers recognize convergent thinking as a feature of the design process. Indeed, the core of engineering design includes two fundamental phases: a creative synthesis phase (i.e., divergent thinking), followed by a logical analysis phase (i.e., convergent thinking). Divergent cognition [27] involves not only generating many possible ideas or solutions, but also involves seeing connections between disparate information (e.g., recognizing patterns, relating different concepts, combining unrelated ideas). A particularly interesting aspect of divergent thinking is the process of creating associations beyond the traditional, conventional, orthodox approach. Mednick [28] states that when a known stimulus is presented in an atypical situation, people usually select a common and familiar response; they do this out of habit and lack creativity. In the present paper, an attempt will be made to measure and thus understand creativity specifically declined as divergent thinking.

3 Methods

In this analysis, creativity is measured through the Forward Flow test by Gray et al. [29] that can measure the degree of imagination and divergent thinking. The test is free and accessible at www.forwardflow.org. This test can predict creativity in the laboratory and in the real world using latent semantic analysis to capture the semantic evolution of thoughts over time, i.e., how much present thoughts diverge from past thoughts. This method is quite different from those already employed by authors in the engineering field [30].

The experiment is carried out in the following way: a "respondent", whose purpose of the test is kept hidden, is asked to produce a sequence of semantically related words, starting from a given initial word, the same for all respondents. Latent semantic analysis calculates the semantic distance between two words by examining the frequency with which they appear together in texts. For example, "snow" and "white" have minimal semantic distance (they are close in meaning), while "snow" and "carburetor" have greater semantic distance. Forward Flow then measures the average semantic distance between any given word and all preceding words. It can be interpreted as the ability of a stream of consciousness to flow forward, leaving previous thoughts behind and thus as the ability to diverge. Having a low Forward Flow indicates that thoughts are returning to previous thoughts (e.g., happy, smile, dentist, teeth, smile, happy); having a high Forward Flow indicates that thoughts are continuing to flow by diverging from the past (e.g., happy, smile, dentist, doctor, hospital, helicopter).

Studies conducted on the test have revealed that Forward Flow predicts creativity in college students and in representative samples of Americans, even when controlling for intelligence [29, 31, 32]. Studies also reveal that the Forward Flow score correctly predicts the creativity of people recognized as creative (e.g., being professional actors or entrepreneurs).

The experiment was conducted by testing students enrolled in different degree courses: Engineering (i.e., Management, Mechanics and Automation), Law and Motor Sciences. Participants receive an online questionnaire, in a Google Form, and, for two successive times, they are given an initial word (i.e. "Candela" the first word, "Neve" the second) followed by 9 blank lines with instructions for completion. The words are in Italian because they are addressed to Italian students, thinking that it would be easier and more natural to find semantic proximity in their mother tongue. The Forward Flow tool, however, works with sequences in English: it was therefore necessary to proceed with the translation, which respected the meaning of the words chosen. We consider this step valid because it is not a psychological test. We have also asked directly to the author of the test, Gray, who has confirmed that the translation is an acceptable step, not able to influence the result of the test. Important aspects for the success of the test are that the participants do not know the purpose of the test (i.e. to measure creativity) and that they are guaranteed that the results will be anonymous. However, the questionnaire asks for personal information (i.e. age, gender), information on high school studies (i.e., school, high school grade), information on university studies (i.e., degree course, year of course), and finally it asks if they have had work experience and what is their preferred subject of study.

A database with Forward Flow results and personal information was compiled, and possible statistical correlations were searched. Figure 1 shows and example of the data collected in the resulting database.

#	Flow "CANDELA"	Flow "NEVE"	Age	Gender	Degree Program	Course Year	Current Academic Degree	Hobby	Sport	High School	Graduation grade	Preferred Subject	Work Experiences (Yes/No)
ML42	0.77	0.58	19	M	Motor Science	1	Bachelor	Gym	Soccer	Technical Institute	82	Humanistic Subjects	None
GG29	0.88	0.79	21	F	Motor Science	1	Bachelor	Drawing	Golf	Technical Institute	62	Stem	Yes
OS47	0.81	0.84	20	M	Motor Science	1	Bachelor	Cooking	Soccer	Technical Institute	63	PE	None
LE77	0.83	0.77	19	M	Motor Science	1	Bachelor	Sport	Rugby	Scientific High School	96	Medicine	Yes
AA57	0.89	0.89	20	M	Motor Science	1	Bachelor	Bodybuilding	Soccer	Gymnasium and Human Sciences	75	Humanistic Subject	Yes
AS79	0.81	0.87	18	M	Motor Science	1	Bachelor	Videogame	Basket	Scientific High School	80	Humanistic Subject	None
ML24	0.76	0.64	19	F	Motor Science	1	Bachelor	Cinema	Kick boxing	Scientific High School	80	Humanistic Subject	Yes
DM50	0.69	0.81	19	F	Motor Science	1	Bachelor	Music	Athletics	Linguistic High School	80	Humanistic Subject	Yes
FL00	0.76	0.71	19	M	Motor Science	1	Bachelor	Fitness	Pallavolo	Linguistic High School	78	PE	None
CS39	0.87	0.79	21	M	Law	1	Bachelor	Fishing	Martial Art	Scientific High School	78	Humanistic Subject	Yes
GM10	0.81	0.76	20	F	Law	1	Bachelor	Photography	Basket	Gymnasium and Human Sciences	80	Humanistic Subject	None
DR39	0.84	0.81	18	F	Law	1	Bachelor	calligraphy	kickboxing	Gymnasium and Human Sciences	89	Humanistic Subject	Yes
AM63	0.69	0.80	19	F	Motor Science	1	Bachelor	Trekking	Vollyball	Linguistic High School	100	Stem	Yes
AA32	0.86	0.69	19	F	Engineering	1	Bachelor	Listening to music	Gym	Technical Institute	97	Humanistic Subject	Yes
FS46	0.67	0.85	20	F	Engineering	3	Bachelor	Reading	Gym	Scientific High School	80	Stem	Yes
GD08	0.70	0.69	20	M	Engineering	3	Bachelor	MotoGp, Formula 1	Calcio	Technical Institute	88	PE	None
RS95	0.88	0.68	21	M	Engineering	3	Bachelor	Photography	Gym	Technical Institute	73	Stem	Yes
DE47	0.75	0.74	22	M	Engineering	5	Master Degree	Trekking	Swimming	Scientific High School	72	Stem	Yes
sI87	0.83	0.70	22	M	Engineering	5	Master Degree	MTB	Gym	Scientific High School	73	Stem	Yes

Fig. 1. Example of some data in the compiled database.

4 Results

221 male and female students responded to the questionnaire. Up to December 2020, there were 205 responses useful for analysis purposes. Table 1 shows the statistics describing the entire sample. Figure 2, on the other hand, shows the distributions of the results of the two Forward Flow tests.

The sample consists of 131 boys and 74 girls who, in half of the cases, come from a High School, either Scientific (about 40%) or Gymnasium (more than 13%). About a third came from a Technical School and only a very small percentage from an Art School (about 2%) or a Vocational School (2.43%). The average age of the interviewees is around 20, although there are also cases of those over 30. Most of those interviewed have worked (about 2 out of 3).

As far as their university career is concerned, the sample is made up of over 38% of students enrolled in Engineering and for the most part attending a bachelor's degree (85.53% of cases).

A first analysis of the sample shows a distribution of values in response to Forward Flow with a concentration between 0.67 and 0.88, with few responses that are very creative (9 responses) or not very creative (15 responses) (Fig. 2).

Table 1. Statistics describing the entire sample.

Variable	Mean value/%	Standard deviation	Min	Max
Forward Flow 1	0.76	0.07	0.50	0.93
Forward Flow 2	0.73	0.8	0.51	0.93
Age (years)	20	4	18	55
Women (dummy)	36.09%	0.48	0	1
Technical Institute	33.17%	0.47	0	1
Vocational School	2.43%	0.15	0	1
Art School	1.95%	0.14	0	1
Gymnasium and Human Sciences	13.17%	0.34	0	1
Scientific High School	40.48%	0.49	0	1
Other High School	8.80%	0.29	0	1
Maturity Grade (in hundredths)	80.85	11.53	60	100
Engineering (dummy)	38.53%	0.48	0	1
Bachelor Degree Course (dummy)	85.53%	0.34	0	1
Work Experience (dummy)	64.39%	0.48	0	1

All the following analyses are intended to test the robustness of the test and were conducted on the two tests performed for the words "Candela" and "Neve" respectively. The results are stable for both answers, although candle was interpreted both as candle (in 89.6% of the cases) and as car spark plug (in the remaining 10.4% of the cases).

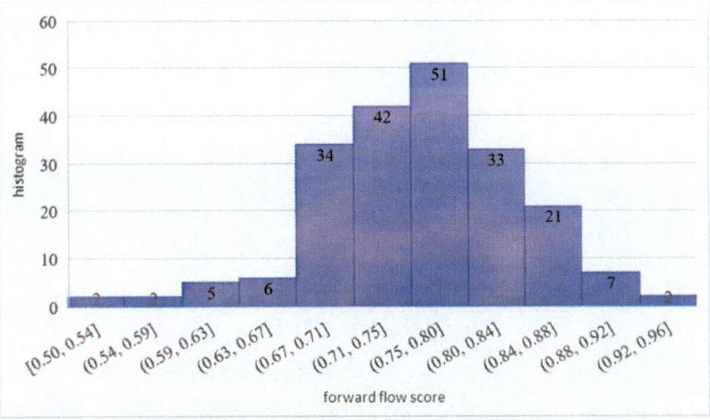

Fig. 2. Forward Flow Scores Distribution by classes of values.

Checking the distribution of Forward Flow scores by high school grade, gender, and age, no significant differences emerge (t-tests of differences between averages are not significant). Nor do having had work experience or having a preference in a specific discipline seem to affect creativity (Fig. 3).

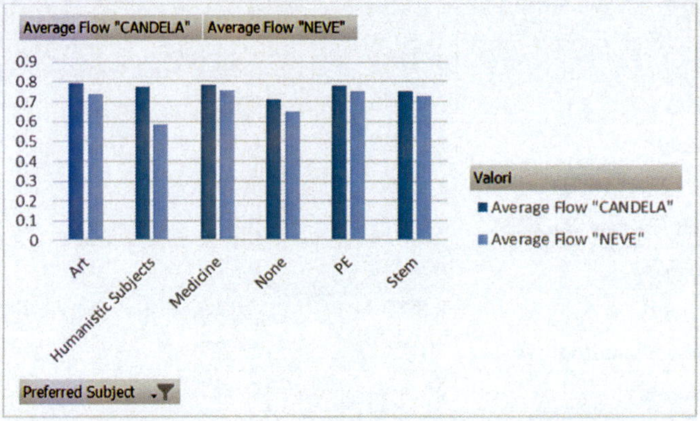

Fig. 3. Distribution of response scores to Forward Flow by preferred subject of study.

On the other hand, the analyses in relation to the course of study just completed (Fig. 4) and the degree course undertaken (Fig. 5) are interesting. The results show a statistically and clearly higher creativity for students who have completed their high school diploma at the Art School (t-test significant for $p < 0.01$), while there are no statistically significant differences for the other paths (Fig. 4). As far as the degree pathway is concerned, it clearly emerges that those who are pursuing engineering studies are clearly characterized by statistically lower values (t-test significant for $p < 0.01$). On the other hand, no stable and significant differences emerge between the pathway in Law and in Motor Sciences (Fig. 5).

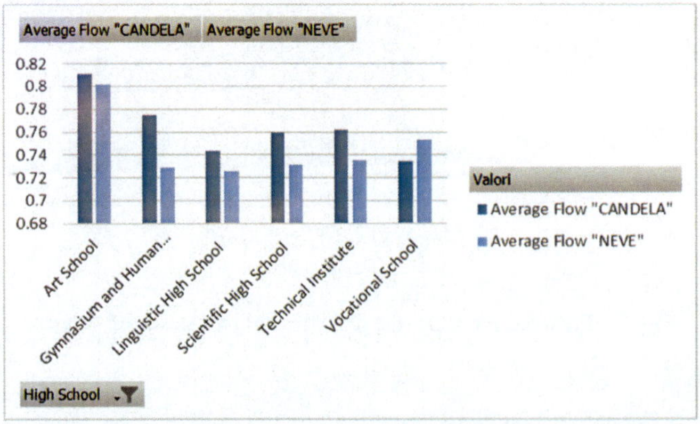

Fig. 4. Distribution of Forward Flow response scores by school of origin.

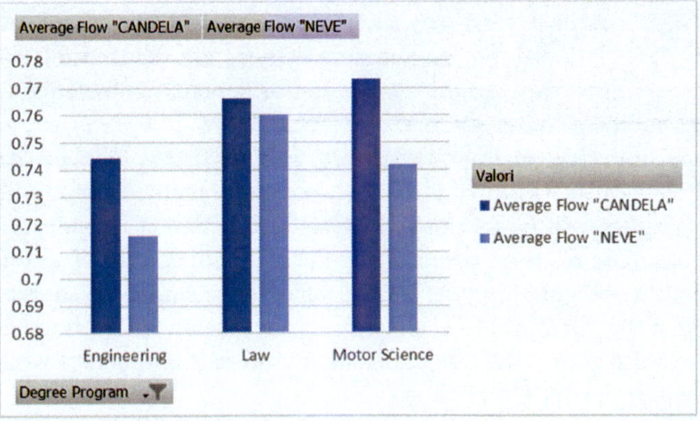

Fig. 5. Distribution of Forward Flow response scores by degree program.

The comparison of the results in Fig. 3, Fig. 4 and Fig. 5, suggests that a significant element capable of influencing creativity is the course of study, in fact the high school course and even more the degree course have a significant influence. Personal preference, on the other hand, identified through the preferred subject request, does not bring out any distinctive trait. In the future it would be interesting to verify whether creativity, as many authors already claim, can be taught and is not an innate characteristic.

5 Conclusions

Knowledge, because of university education, has not been sufficient for some time [33–35]. This is because it is difficult if not impossible to know what knowledge will be needed in the future [10]. On the other hand, formal education, which is both a right and a need [36], has been criticized in recent years for not being able to form freely creative and original thinkers [37]. One of the reasons why educational systems have been considered obstacles to the development and release of creative potential is that teaching focuses primarily on the acquisition of knowledge [38].

In response to these now evident needs, there has been a change in educational policy around the world and efforts have been made to combine creativity and knowledge for years now [9], and although Italian universities generally lag behind in the development of didactic activities expressly aimed at teaching and stimulating creativity, in some areas, including ING-IND/15, work has been underway for some time to study creativity, stimulate it and promote it as an added and indispensable value to the work of the engineer [17, 18, 30].

From an initial analysis, the proposed study seems to highlight how some courses of study in which creativity should perhaps be cultivated as a necessary skill seem instead to negatively influence student development in this area. While a high school course focused on the arts seems to stimulate creativity, the continuation in engineering studies seems to penalize it. Such evidence, while preliminary and exploratory, is consistent with what Kazerounian and Foley [39] already stated more than a decade ago: "If creativity is so central to engineering, why is it not a part of the engineering curriculum at every university?" There may be several reasons why our educational tracks ignore creativity. Among others, it is believed that engineering degrees are focused on strong specializations and teaching focuses on the acquisition of concrete knowledge [40]. Finally, educators lack specific training on creativity [25, 41].

One unexpected result of this preliminary work is that it seems that engineering students are less creative (following the Forward Flow method) than other examined students: one purpose will be a critical analysis of the method proposed in this paper which is only based on measure creativity on the ability to diverge. The study will continue by increasing and enlarging the sample surveyed, refining the source words from which to generate the Forward Flow, to verify its consistency as an effective measure of creativity. One of the aims will be to understand if this definition suits well mechanical and management engineers. Other definitions and model will be considered, and creativity will be measured also following other methods. Authors will also try to understand if less creative students enroll in mechanical and management engineering or if Mechanical and Management Engineering negatively affect students' creativity. The tests, considering different definitions of creativity, will be administered to young people leaving high school and to the same after embarking on a degree course, following them over time, to see if and how their creativity evolves. If necessary, other tests will be proposed and solutions will be sought to "teach" creativity, focusing on mechanical and management engineers, with whom authors work.

References

1. Cropley, A.J.: Creativity in the Classroom: The Dark Side of Creativity. Cambridge University Press, Cambridge (2010)
2. Cropley, D.H.: Promoting creativity and innovation in engineering education. Psychol. Aesthet. Creat. Arts **9**(2), 161–171 (2015)
3. Craft, A.: An analysis of research and literature on creativity in education. Qualifications Curriculum Authority **51**(2), 1–37 (2001)
4. Acar, O.A., Tarakci, M., van Knippenberg, D.: Creativity and innovation under constraints: a cross-disciplinary integrative review. J. Manag. **45**(1), 96–121 (2019)
5. Wilson, N.C., Stokes, D.: Managing creativity and innovation. Journal of Small Business and Enterprise Development (2005)
6. Turner-Bisset, R.: Performativity by stealth: a critique of recent initiatives on creativity. Education 3–13 **35**(2), 193–203 (2007)
7. Feldman, D.H., Benjamin, A.C.: Creativity and education: an American retrospective. Camb. J. Educ. **36**(3), 319–336 (2006)
8. Craft, A.: Fostering creativity with wisdom. Camb. J. Educ. **36**(3), 337–350 (2006)
9. Dickhut, J.E.: A brief review of creativity. Internet Available. www.personalityresearch.org/papers/dickhut.html (2003)
10. Sharif, R.: The relations between acculturation and creativity and innovation in higher education: a systematic literature review. Educ. Res. Rev. **28**, 100287 (2019)
11. Burnard, P., et al.: Documenting 'possibility thinking': a journey of collaborative enquiry. Int. J. Early Years Educ. **14**(3), 243–262 (2006)
12. Davies, D., Jindal-Snape, D., Collier, C., Digby, R., Hay, P., Howe, A.: Creative learning environments in education-a systematic literature review. Thinking Skills Creativity **8**, 80–91 (2012)

13. Nesta, L., Mangematin, V.: Industry life cycle, knowledge generation and technological networks. In: Networks, alliances, and partnerships in the innovation process. Springer, Boston (2002). https://doi.org/10.1007/978-1-4615-1151-9_11
14. Duderstadt, J.J.: A choice of transformations for the twenty-first century university. Chronicle of Higher Education, vol. 46, no. 22 (2000)
15. Downey, G.L., et al.: The globally competent engineer: working effectively with people who define problems differently. J. Eng. Educ. **95**(2), 107–122 (2006)
16. Shaheen, R.: Creativity and education. Creative Educ. **1**(03), 166 (2010)
17. Borgianni, Y., Maccioni, L., Fiorineschi, L., Rotini, F.: Forms of stimuli and their effects on idea generation in terms of creativity metrics and non-obviousness. Int. J. Des. Creativity Innov. **8**(3), 147–164 (2020)
18. Fiorineschi, L., Rotini, F.: Novelty metrics in engineering design. Journal of Engineering Design (2021). https://doi.org/10.1080/09544828.2021.1928024
19. Walberg, H.J.: Creativity and talent as learning. The nature of creativity. Contemporary psychological perspectives, vol. 340 (1988)
20. Groborz, M., Necka, E.: Creativity and cognitive control: explorations of generation and evaluation skills. Creat. Res. J. **15**(2–3), 183–197 (2003)
21. Morawski, M.: Business model used in companies representing creative industries. In: Global Opportunities for Entrepreneurial Growth: Coopetition and Knowledge Dynamics within and Across Firms. Emerald Publishing Limited (2017)
22. Pfeiffer, R.S.: The scientific concept of creativity. Educ. Theor. **29**(2), 129–137 (1979)
23. Hamid, H.: The strategic position of human resource management for creating sustainable competitive advantage in the VUCA. World. J. Hum. Resour. Manag. Labor. Stud. **7**, 1–4 (2019)
24. Shalley, C.E., Zhou, J., Oldham, G.R.: The effects of personal and contextual characteristics on creativity: Where should we go from here? J. Manag. **30**(6), 933–958 (2004)
25. Cropley, D.H.: Creativity in engineering. In: Corazza, G.E., Agnoli, S. (eds.) Multidisciplinary Contributions to the Science of Creative Thinking. CTFC, pp. 155–173. Springer, Singapore (2016). https://doi.org/10.1007/978-981-287-618-8_10
26. Wreen, M.: Creativity. Philosophia **43**(3), 891–913 (2015). https://doi.org/10.1007/s11406-015-9607-5
27. Boden, M.A.: Agents and creativity. Commun. ACM **37**(7), 117–121 (1994)
28. Mednick, M.T., Mednick, S.A., Mednick, E.V.: Incubation of creative performance and specific associative priming. J. Abnorm. Soc. Psychol. **69**, 84–88 (1964)
29. Gray, K., et al.: "Forward flow": A new measure to quantify free thought and predict creativity. Am. Psychol. **74**(5), 539 (2019)
30. Borgianni, Y., Cascini, G., Rotini, F.: Assessing creativity of design projects: criteria for the service engineering field. Int. J. Des. Creativity Innov. **1**(3), 131–159 (2013)
31. Rossiter, J.R.: Forward flow—An alternative interpretation: Comment on Gray et al. (2019) (2020)
32. Kenett, Y.N., et al.: Clarifying what forward flow is (and isn't): Reply to Rossiter (2020) (2020)
33. Guilford, J.P.: Traits of creativity. In: Anderson, H.H. (Ed.): Creativity and its Cultivation. Harper, New York (1959)
34. Weick, K.E.: Drop your tools: on reconfiguring management education. J. Manag. Educ. **31**(1), 5–16 (2007)
35. World Economic Forum: The future of Jobs Report 2020 (2020)
36. Carnoy, M.: Paper commissioned for the EFA Global Moni-toring Report 2005: The quality imperative, UNESCO (2005)
37. Muneta, M.L.M., de Avila, M.L., Romero, G., Mindán, J.F.: Searching for the most creative engineer. Int. J. Eng. Educ. **31**(1), 354–360 (2015)

38. Davies, T.: Incorporating creativity into teachers practice and self-concept of professional identity. J. Educ. Change **14**(1), 51–71 (2013)
39. Kazerounian, K., Foley, S.: Barriers to creativity in engineering education: A study of instructors and students perceptions (2007)
40. Sternberg, R.: Creativity as a habit. In: Creativity: A Handbook for Teachers (2007)
41. Craft, A.: Creativity in schools: Tensions and dilemmas. Psychology Press (2005)

Reasoning About Technical Drawing Online Teaching During COVID-19

Barbara Motyl[1(✉)], Stefano Filippi[1], Gabriele Baronio[2], and Valerio Villa[2]

[1] DPIA – Università degli Studi di Udine, Udine, Italy
barbara.motyl@uniud.it
[2] DIMI – Univeristà degli Studi di Brescia, Brescia, Italy

Abstract. This work is part of the historic collaboration between the design and methods research groups of the universities of Udine and Brescia. In particular, it presents the results of a survey on the perceptions of engineering students on the online teaching methods activated for the technical drawing courses during the COVID-19 pandemic. 111 students at the University of Udine participated in the survey and the results of the analysis of the responses showed that, in general, online teaching methods are not comparable to face-to-face ones; however, they have been appreciated as they allowed regular teaching during this critical period. Furthermore, opinions and scores from the new teaching methods were more than positive regarding both the availability of the recordings of the lessons and the introduction of generalized corrections of exercises. The authors planned to extend the survey also to the University of Brescia to collect further pieces of information in order to constantly improve this teaching paradigm.

Keywords: Technical drawing · Online teaching · Engineering education · COVID-19 · Students survey

1 Introduction and Background

Since the World Health Organization has declared the COVID-19 outbreak as a global pandemic, all nations applied lockdowns and restrictions to control the spread of this disease. These decisions affected all aspects of life and, inevitably, all levels of education as well. In higher education, such measures resulted in a sudden shift from face-to-face teaching towards online teaching. All over the world as well as here in Italy, many institutions of all sizes and types opted to cancel face-to-face classes, including labs and other learning experiences, and moved their courses online to help preventing the spread of the disease. Due to the contingent need to move online, as suggested in [1–3], rather than talking about online learning, it seems more correct to refer to these activities as "emergency remote teaching", mainly to avoid undermining the reputation that online learning has earned in recent years comparing to face-to-face learning [1]. Choices of this type have been made also in our universities of Udine and Brescia to ensure the continuity of the education proposals.

C. Rizzi et al. (Eds.): ADM 2021, LNME, pp. 889–897, 2022.
https://doi.org/10.1007/978-3-030-91234-5_90

Traditionally, the technical drawing courses offered at the University of Udine are provided to both management and mechanical engineering students of the second year of the bachelor's degree in engineering. They are 48-h courses and mainly consist of theory lessons and laboratory lessons dedicated to hand drawing exercises. All of them are carried out in the classroom. Six to eight hours are dedicated to the use of the Auto CAD 2D drawing software. Usually, two to four hours of optional classroom exercises are added, in which exam preparation exercises are proposed. The exercises, assigned weekly, are collected, and corrected to be commented in the classroom in subsequent lessons. Switching to the online mode, the 48-h courses were organized in modules of 4 h per week. During these 4 h, each week, students have been offered an argument and assigned exercises. While the online lessons were delivered through the Microsoft TEAMS platform, for reasons of habit and management practicality the exchange of materials (slides of the lessons, files or other) and the collection of drawings in the form of images or scans (jpeg or pdf) were made using the Moodle system. Instead of the extra exercises in the classroom, three online exercises have been proposed in the form of guided exercises with the teacher present online and available to provide clarifications or answer questions. As for the lessons in the classroom, the exercises collected were regularly corrected and commented during the lessons and then made available as downloadable materials. All the online activities (lessons, exercises correction, etc.) were recorded and the records were immediately available. Thus, given the particularity of the technical drawing subjects offered in our courses and the high number of students who attend them, it seemed appropriate to investigate how this situation has been experienced by the students to get understand strengths and weaknesses and to guarantee a margin for improvement. Authors were interested in understanding what students thought about their online learning experience. Thus, the goal of this work is twofold. First, to understand if they have experienced difficulties following the lessons due to their technological equipment and/or low skills. Second, to collect their perceptions on online lesson experiences (in comparison with previous modalities). Furthermore, we were also interested in understanding whether the change in some well-established practices relating to the teaching of technical drawing, such as carrying out practical exercises in the classroom, has introduced further difficulties. All this to understand if there are opportunities for improvement and to review the contents of the courses or reschedule the activities in the future.

In the following sections, first, the survey method used will be presented, based on an on-line questionnaire, and subsequently the results obtained at the Udine university will be discussed. Finally, conclusions and ideas for future developments will be highlighted.

2 Methodology

To carry out this survey, we expanded and reformulated the questionnaire already in use for our technical drawing courses. The questionnaire, implemented through a Google form, is composed of 23 questions. The 17 questions refer to technical drawing topics already used and discussed in [4–6]. Together with these, six new questions were introduced to collect students' perception of the remote teaching situation due to the

COVID-19 pandemic. These questions were formulated thanks to a literature search on current surveys about COVID-19 effects on online teaching and learning [7–9].

Questionnaire Design

For the purposes of this paper, only the questions relating to the COVID-19 pandemic will be considered. There are three closed questions on the technological equipment available to the students and on the Internet connection methods used (Q17, Q18, Q19). Two more 5-point Likert-type questions allow students expressing their opinions and experiences during online lessons or about connection difficulties (Q20, Q21). The first part of the questionnaire was used to collect the personal data of the sample of students interviewed. For reasons of practicality and ease of reading, the results obtained, the questions statements and the answers will be reported in the third section.

Participants

All 296 students enrolled in the virtual class of the Technical Design course in Microsoft TEAMS have been invited to participate to the survey during the last week of the course (December 2020). The response rate to the questionnaire was 37.5% (111 students filled the questionnaire). These are mainly students enrolled in the second year of Management Engineering or Mechanical Engineering courses (80.2%), with the addition of some cases of students of other years (11.7% third year, 8.1% other). Table 1 shows the personal data relating to the interviewed students.

Table 1. Data relating to the sample of students interviewed.

Data	Value in number	Value in %
Students of Management Engineering	62	55.9
Students of Mechanical Engineering	49	44.1
Male	74	66.7
Female	37	33.3

3 Results and Discussion

This section reports the results of the analysis of the answers obtained for each COVID-19-related question.

Students' Technological Equipment and Internet Connection Availability (Q17-Q18-Q19)

The students were asked what kind of technological equipment and Internet connection they have, using some closed questions (Q17, Q18 and Q19). Table 2 reports the results obtained for these three questions.

All students have a PC or laptop (100%) paired with a smartphone (95%), while about 48% also have a tablet. The most available types of Internet connections are

Table 2. Questions about the technological equipment and Internet connection.

#	Statement	Response %
Q17	Which of the following digital devices do you own?	
	Desktop or laptop computer	100
	Smartphone	95.5
	Tablet	47.7
	Other	3.6
Q18	What kind of internet connection do you have at home?	
	Fiber	36.9
	ADSL	44.1
	Mobile hotspot	24.3
	Other	11.7
Q19	Which device do you use the most to take a lesson?	
	Desktop or laptop computer	96.4
	Tablet	2.7
	Smartphone	0.9
	Other	0

ADSL (44%) and Fiber (37%), followed by mobile hot spots (24%) and other 12%. If used in combined mode, 24.3% of the students declare to combine Fiber and ADSL with Hotspot. Furthermore, 96.4% of the students stated that they mainly use the PC/laptop as the device to follow/participate to the lessons.

Table 3. Students' perceptions on their technology equipment and skill for online teaching.

Q20 Statements		Values					Mean	SD
		1	2	3	4	5		
S20-1	I own a sufficient number of hardware and software devices (PC/tablet/smartphone/internet/programs) to follow an online lesson	0	0	3	26	82	4.71	0.51
S20-2	I have a sufficient level of knowledge and skills to manage online learning activities	1	1	5	36	68	4.52	0.71
S20-3	Tools and software used for the online lessons are enough easy to use	0	0	8	44	59	4.46	0.63
S20-4	The information/instructions to follow the online lessons are quite clear	0	2	9	49	51	4.34	0.71

Students' Opinions on Hardware and Software Skills (Q20)
Considering the students' opinions on their technological equipment and hardware and software skills needed to follow online lessons (Q20), Table 3 reports their perceptions classified on a Likert scale from 1 to 5, where 1 = strongly disagree and 5 = strongly agree. Likert scores were reported as both frequency and mean and relative standard deviation (SD). The internal reliability for this question composed by four sub questions was verified considering Cronbach's α. The obtained value for α is 0.77 and this can be considered a good level of reliability [10]. From the analysis of the answers, the students' opinion on their level of technological equipment (S20-1) and their knowledge (S20-2) is more than positive and stands at rather high average values (over 4.00), also witnessed by low SD values. Statement S20-1 scored and average of 4.71 while statement S20-2 scored an average of 4.52. Most of the students declare that the software used for the online lessons are enough easy to use (S20-3, average 4.46) and that the information and instructions to follow the online lessons are quite clear (S20-4, average 4.34).

Students' Perception of Online Teaching (Q21)
With the question Q21, with its sixteen statements, students were asked to express their opinions or to rank their experience of online teaching using a Likert scale from 1 (strongly disagree) to 5 (strongly agree). Also, for this question, composed by 16 sub questions, the internal reliability was verified considering Cronbach's α [10]. The obtained value for α is 0.84 and this can be considered a very good level of reliability. Table 4 shows the frequency of the values next to the mean value and the standard deviation. The analysis of the answers to the questions on students' perception of online teaching highlighted, as expected, that the students' opinion is not entirely positive.

In fact, the average score for statement S21-1, regarding the comparison of online and face-to-face lessons is 2.92, suggesting that most of the students prefer face-to-face teaching. With reference to the statement S21-2 or participation and involvement during online lessons compared to participation in traditional lessons, the students scored their experience with an average of 3.32. Statement S21-3 scored an average of 2.86 due to the presence of demotivation towards participation in online lessons. The evaluation of the lack of direct contact with other students or friends (S21-4) received an average score of 2.17. However, the score attributed to the presence of difficulties in contacting or communicating with the teacher (S21-5) obtained an average of 3.65. Overall, the teacher's ability to engage students in online classes compared to face-to-face lessons averaged 3.58 in statement S21-6. On the other hand, the acquisition of greater flexibility in the management of lessons (S21-7) received an average score of 3.93 and the experimentation of new teaching methods (S21-8) received an average score of 3.65. These perceptions are considered positive and personal growth aspects.

Furthermore, students essentially maintain a neutral attitude with an average score of 3.05 towards the home environment as suitable for learning (S21-09). Considering the student-teacher interaction they score an average of 3.49 (S21-10) and towards the possibility of continuing to use online teaching methods in the future an average score of 3.32 for (S21-11) both moderately above the neutral value. Considering the students' opinion related to the technical drawing course offered online, most of the students (89.2%) are satisfied with the lessons and materials of the course made available by the teacher, being their responses scored as 4.29 in average (S21-12).

Table 4. Students' perceptions of their online teaching experience.

Q21 Statements		Values					Mean	SD
		1	2	3	4	5		
S21-1	Online lessons are as effective as traditional classes/face-to-face lessons	11	32	36	19	13	2.92	1.15
S21-2	The level of my participation in online lessons is comparable to that for face-to-face lessons	9	21	25	37	19	3.32	1.19
S21-3	I did not experience demotivation and lack of interest for learning due to health emergency	19	28	22	34	8	2.86	1.23
S21-4	The lack of direct contact with other students did not reduce my interest in lectures	37	38	19	14	3	2.17	1.23
S21-5	Communicating with the teacher during the online lessons was easy	4	12	31	36	28	3.65	1.08
S21-6	Teacher's ability to interest/engage students online is comparable to face-to-face lessons	2	13	31	47	17	3.58	0.94
S21-7	By online lessons I gained flexibility in. time management	1	5	18	64	23	3.93	0.79
S21-8	Online lesson mode made me experiment new learning experiences	1	12	27	56	15	3.65	0.88
S21-9	Home environment is suitable for participating in online lessons	17	19	27	37	11	3.05	1.23
S21 10	I am satisfied with the student-teacher interaction experienced during online teaching	6	12	31	46	16	3.49	1.04
S21-11	I would like to continue to take online classes even after the health emergency	15	13	29	30	24	3.32	1.30
S21-12	I am satisfied of the online lessons and of the materials available for the technical drawing course	1	0	10	55	45	4.29	0.70
S21-13	I had the opportunity to ask questions or clarify concerns during the online lessons	0	2	31	48	30	3.95	0.79
S21-14	Presence of the recordings of the lessons is useful	0	0	1	20	90	4.84	0.39
S21-15	Teacher's generalized comments of the exercises are useful	0	0	4	15	92	4.79	0.49
S21-16	Online guided exercises in extra lesson hours are useful	0	0	18	37	56	4.34	0.74

Moreover, in (S21-13) the average of 3.95 testifies that the students appreciated the possibility of obtaining clarifications from the teacher during the lessons, while all the students interviewed greatly appreciated the possibility of using the recordings of the lessons (S21-14, with an average score of 4.84). Most students also welcomed, in statement (S21-15), the introduction of generalized corrections of the exercises proposed weekly with the teacher's comments (average score 4.79) and the possibility, in S21-16, of carrying out guided exercises in extra lesson hours with the presence of the teacher (score average 4.34).

What emerges from this analysis, looking at the mean scores obtained in the various questions' statements, is that students suffered most from the lack of interaction and contact with each other (see the mean values for S21-1, S21-3, S21-4). In fact, one of the things that students missed the most was the opportunity of contact and exchange views with each other. In this regard, it could be hoped that in the future the platforms used for conducting online classes (for example: TEAMS, Zoom, Google Classroom,...) will try to reduce these distances, for example by making the modes already available such as TEAMS together, or Zoom Immersive view more customizable. It would be desirable for these tools to allow individual participants to be able to virtually sit next to their friends, rather than randomly, to make these online classroom experiences even more similar to the actual experience of a real classroom. On the other hand, considering the higher average scores obtained in the two questions Q20 and Q21, some considerations can be drawn, below. For question Q20, it can be assumed that students did not report difficulties from the point of view of their technological equipment or personal knowledge. In fact, on average, the number of devices in their possession and their level of knowledge and use of digital technologies were quite high, this also because they are students of courses with a high scientific and technological content. As for the averages obtained for the responses to question Q21, what was most appreciated was the quality of the materials made available by the teacher and the possibility of using video recordings of the lectures, so it is planned to continue to provide for the availability of these materials in future courses.

4 Conclusions and Future Developments

This survey aimed to investigate the opinions of engineering students on online teaching methods and on the changes of the technical drawing course to move online. As might be expected, the students rated their online learning experiences as not comparable to face-to-face ones in terms of actual engagement and modality. However, this survey found that engineering students are essentially well equipped to take online classes and have good quality internet connections. Some positive aspects also emerged related to the fact that the students declared that they had acquired greater autonomy and experience in the use of videoconferencing systems and greater flexibility in managing time and online lessons. Moreover, the level of participation and the possibility of interacting with the teacher were also appreciated. It should also be noted that, regarding the specific technical drawing course, the strategies implemented by the teachers to maintain a good level of interaction and feedback with the students have been highly appreciated. In particular, the possibility of having the recordings of the lessons and the corrections of the various exercises was much appreciated. Some limitations may be represented by the fact that this is the first survey since the beginning of the COVID-19 pandemic. In addition, the number of questions used in the questionnaire should be expanded and better organized. Moreover, it would be significant to interview teachers about implementing online courses to identify the various strategies adopted. For this reason, we planned to extend this survey also to the courses of the University of Brescia to assess whether there are substantial differences, leaving then the possibility to all interested parties to be involved in the research. It would also be interesting to check the progress of the final exams between the pre and post pandemic situation. However, it should be noted that the examination methods used both within the same location and between different locations have been changed, so this analysis should be carried out with due caution.

Acknowledgments. The authors would like to thank all the engineering students of the University of Udine, who participated in the survey during the Academic Year 2020–21.

References

1. Hodges, C., Moore, S., Lockee, B., Trust, T., Bond, A.: The difference between emergengy remote teaching and online learning. Educause Rev. **27**, 1–12 (2020)
2. Gomez Recio, S., Colella, C.: The world of higher education after COVID-19. How COVID-19 has affected young universities. https://www.yerun.eu/wp-content/uploads/2020/07/YERUN-Covid-VFinal-OnlineSpread.pdf. Accessed 20 July 2021
3. Govindarajan, V., Srivastava, A.: What the shift to virtual learning could mean for the future of higher ed. Harvard Business Review 31, (2020). https://hbr.org/2020/03/what-the-shift-to-virtual-learning-could-mean-for-the-future-of-higher-ed. Accessed 20 July 2021
4. Motyl, B., Filippi, S., Baronio, G., Speranza, D.: Investigating students' perception on the outcomes of technical drawing university courses. In: Roucoules, L., Paredes, M., Eynard, B., Morer Camo, P., Rizzi, C. (eds.) JCM 2020. LNME, pp. 379–385. Springer, Cham (2021). https://doi.org/10.1007/978-3-030-70566-4_60
5. Speranza, D., Baronio, G., Motyl, B., Filippi, S., Villa, V.: Best practices in teaching technical drawing: experiences of collaboration in three Italian Universities. In: Eynard, B., Nigrelli, V., Oliveri, S., Peris-Fajarnes, G., Rizzuti, S. (eds) Advances on Mechanics, Design Engineering and Manufacturing. Lecture Notes in Mechanical Engineering, pp. 903–913. Springer, Cham (2017). https://doi.org/10.1007/978-3-319-45781-9_90
6. Metraglia, R., Villa, V., Baronio, G., Adamini, R.: High school graphics experience influencing the self-efficacy of first-year engineering students in an introductory engineering graphics course. Eng. Des. Graph. J. **79**(3), 16–30 (2015)
7. Dost, S., Hossain, A., Shehab, M., Abdelwahed, A., Al-Nusair, L.: Perceptions of medical students towards online teaching during the COVID-19 pandemic: a national cross-sectional survey of 2721 UK medical students. BMJ Open **10**(11), e042378 (2020). https://doi.org/10.1136/bmjopen-2020-042378
8. Mishra, L., Gupta, T., Shree, A.: Online teaching-learning in higher education during lock-down period of COVID-19 pandemic. Int. J. Educ. Res. Open **1**, 100012 (2020). https://doi.org/10.1016/j.ijedro.2020.100012

9. Almekhlafy, S.S.A.: Online learning of English language courses via blackboard at Saudi universities in the era of COVID-19: perception and use. PSU Res. Rev. **5**(1), 16–32 (2020). https://doi.org/10.1108/PRR-08-2020-0026
10. Lovelace, M., Brickman, P.: Best practices for measuring students' attitudes toward learning science. CBE—Life Sci. Educ. **12**(4), 606–617 (2013). https://doi.org/10.1187/cbe.12-11-0197

Engineering Methods and Applications in Cultural Heritage

Automatic Segmentation of Sweep Features of Constant Radius from Archaeological Artifacts

Luca Di Angelo[1], Paolo Di Stefano[1], and Anna Morabito[2(✉)]

[1] Department of Industrial Engineering, University of L'Aquila, L'Aquila, Italy
[2] Department of Engineering for Innovation, University of Salento, Lecce, Italy
annaeva.morabito@unisalento.it

Abstract. The sweep features of constant radius are particularly significant for the purposes of the historical-archaeological investigation and classification of ancient artifacts. The paper focuses on the automatic recognition of this specific class of features from triangulated 3D models experimentally acquired from cultural heritage objects. This is not a trivial problem. The ancient artifacts, although repeatable, are unique handmade pieces with a geometry commonly characterized by complex and non-analytical shapes. Their surfaces are also usually damaged and worn, so that the related geometric properties are altered or lost.

The methodology proposed here is inspired by the one previously developed by the authors for the automatic segmentation of fillets, rounds and grooves from high-density triangulated models of mechanical components. The paper, in particular, focuses on the aspects of this methodology that must be tuned to allow the recognition of the sweep features of constant radius from archaeological finds. The methodology has been implemented and finally applied to an archaeological find acquired by a laser scanner.

Keywords: Computer methods in archaeology · 3D archaeology · Feature recognition · Laser scanning · 3D digital models

1 Introduction

The semantic decomposition of the archaeological artifact into high-level features is a fundamental operation for the analysis, interpretation, and classification of the find. These features are elements of interest for the archaeological context and may be useful for further cultural processes involving the object, such as comparisons and similarity analysis, object reconstruction, documentation, and detail annotation of the artifact. Archaeological feature recognition is traditionally performed by domain specialists and aims at identifying the surface portions of the object characterized by specific properties related to geometry, texture, color, pattern, decorative motifs, etc. Most of these features concern the shape of the object and often require, in order to be properly characterized, a quantitative evaluation of certain dimensional parameters. For example, the evaluation of the maximum diameter of a potsherd needs the preliminary extraction of the external wall

C. Rizzi et al. (Eds.): ADM 2021, LNME, pp. 901–908, 2022.
https://doi.org/10.1007/978-3-030-91234-5_91

(that is a shape feature) and the identification of its axis of revolution. Unfortunately, the human ability to evaluate these quantitative elements is scarcely repeatable and largely depends on the experience and skills of the expert. In recent years, several computerized methods have been proposed in order to support the semantic decomposition of the archaeological find and the dimensional characterization of its features. These methods make use of the 3D digital data captured from the surface of the object by a 3D scanner. The high-density acquisition of this data, in the form of point clouds or polygonal models, allows the capture of small details from the object, thus making it possible to analyze and document archaeological finds more precisely and efficiently than using drawings or traditional photographs. Now the archaeologist can work directly on the 3D digital model of the find by segmenting the features of interest manually or automatically (with a feature recognizer) and adding the 3D annotations. The automatic segmentation process of archaeological features makes use of a feature recognizer that segments the find into surface patches based on certain rules. The most challenging aspect, in implementing this recognizer, lies in the need to define a set of rules that are in accordance with the way in which archaeologists perform the conceptual categorization of ancient pottery. The knowledge's coding of the domain specialist into a series of mathematical algorithms is certainly not a trivial problem. Firstly, complex and non-analytical shapes usually characterize the geometry of an archaeological artifact. The finds, also, even if repeatable, are unique handmade pieces with surfaces usually damaged and worn, so that their geometric properties are often altered or lost. Finally, the ancient artefact is often found in the form of fragments. This paper is focused on the automatic recognition of a specific class of ceramic features: the Constant Radius Sweep (CRS) features. They are detail features often present on the surfaces of the potsherds such as engravings, working signs and decorations. In some cases, these features are axially symmetric too. From the geometric point of view, the CRS features have an approximately constant cross section, characterized by one or more circular arcs, which is swept along a curve called sweep curve. The CRS features can be created by fingers action or by tools either intentionally, as in the case of inscriptions and decorative motifs, or involuntarily, such as the working signs left during the creation of the artifact. A CRS feature can also derive from the application of sweeping geometries of constant radius section in clay. The sweeping, applied to a section composed of approximately circular arcs, results in concave or convex traces on the ceramic surface. Examples of concave features are engravings, graffiti, impressions/stampings, etc. Convex features are plastic and molded reliefs, etc. Figure 1 depicts some examples of ancient pottery including these detail features. These features are particularly important both for the "identification" of the archaeological find, such as the assigning of a potsherd to a specific ceramic type, and for understanding if a modal or attribute-level variability of the cultural heritage (CH) object is culturally meaningful or not.

Recently some methods have been proposed which, through the calculation of the shape descriptors of the find surface, aim to emphasize the visualization of these detail features through the algorithmic generation of colored and enhanced images [1–3] and without developing any methodology aimed at their automatic and selective recognition. For some time, the authors of this paper have been working on the recognition of geometric features for the automatic semantic decomposition of several categories of objects, from mechanical components [4] to archaeological finds [5]. A feature recognizer, aimed at the automatic segmentation of constant radius secondary features from mechanical parts (such as fillets, rounds and grooves), has been developed and described in [6]. Recently, an original 3D informational database has been also proposed for the automatic archiving of features coming from the semantic decomposition of archaeological pottery finds [7]. In this work, the recognizer developed for the automatic segmentation of secondary features from mechanical components is modified to adapt it to the specific problems of the CRS feature recognition in the archaeological context. The paper, therefore, discusses on the aspects of the methodology that must be tuned to allow the recognition of these features from archaeological artifacts.

Fig. 1. Some examples of ancient pottery including CRS features.

2 The Methodology for the Automatic CRS Feature Recognition

Figure 2a) depicts the discrete model of the archaeological artefact used here as case study. This is the tessellated model of an ancient jar (*olla* in Latin), acquired by a 3D scanner FARO (Edge ScanArm) with a laser line probe (accuracy $2\sigma = 0.0688$ mm, sample rate $S_r = 0.25$ mm) positioned at the end of an anthropomorphic arm. This technology allows minimizing the number of acquisitions required for the 3D reconstruction of the object, reducing the errors inevitably associated with the registration process. The iterative closest point algorithm (with a maximum value for the mean distance between point clouds equal to 0.05 mm) has been applied to perform this process. The point cloud has been then subjected to decimation and filtering.

As highlighted by the Fig. 2, the olla shows various embossed decorations, mostly placed on the body and between the neck and shoulder of the pot. These are positive decorations obtained, in most of the cases, by adding clay to the already formed surface of the vase. Among these decorative prominences, only one, placed on the neck of the vase, is characterized by axial symmetry. From the geometric point of view, the decorations show a roughly triangular cross-section, the vertices of which are rounded off with circular arcs of approximately constant radius along the decorative element. As can be seen from the detail representation shown in Fig. 2b), the convex vertex of the triangular profile is rounded with an appreciably smaller radius than the fillets of the other vertices (characterized by radii approximately equal to each other). The sweep curves, along which the triangular profile moves, are quite circular or of complex shape.

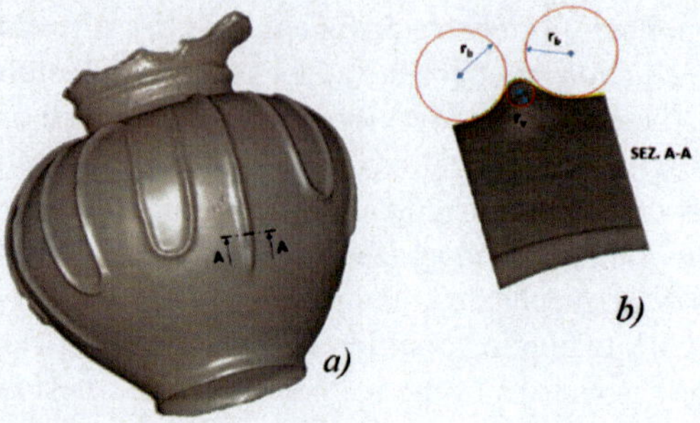

Fig. 2. a) The discrete model of the olla used as case study. b) The triangular cross-section of the olla's embossed decorations.

Various CRS features of a given characteristic radius R_i, therefore, can be identified and associated with each decorative prominence.

The methodology for the automatic segmentation of the CRS features consists of the following principal phases:

- Estimation of geometrical differential properties at each node **p** of the tessellated model;
- Evaluation of the characteristic radius R_i^* (related to the inverse of the principal curvature maximum in absolute value (k_1)) for each category of CRS features;
- Automatic selection of the nodes potentially attributable to each category of CRS features;
- Identification and segmentation of the CRS features.

The curvature estimation from discrete surface models is not trivial because they contain only low-level geometric information of the object, such as coordinates of the vertices and normals to the triangular facets. Added to this is the further source of uncertainty due to the small extension of the CRS features with respect to mesh sampling. In the paper, the *5-coefficients paraboloid* method has been implemented for the evaluation of the principal curvatures k_1 and k_2. The preliminary estimation of the characteristic radii R_i^* is carried out by an iterative analysis of the recurrences of the r_1-values (with $r_1 = \frac{1}{k_1}$) estimated at the regular points of the model. This is necessary because the histogram peaks tend to widen and to cover other potentially useful peaks in the case of experimentally acquired tessellated models [8]. The blurring effect of the peak is particularly evident in the case of archaeological artefacts, mainly due to the surface irregularities (i.e. small fluctuations of curvature produced by the action of time and weather on the objet surface). Figure 3 a) depicts the histogram of the r_1-values. Once the radius R_i^* has been estimated as the mode of the histogram evaluated at the iteration *i*, the methodology first identifies all the CRS features with characteristic radius equal to R_i^* and then removes them in order to evaluate the histogram of the next iteration. The preliminary recognition of the CRS features is performed by the fuzzy approach described in the following section. More details about the iterative methodology for R_i^* evaluation can be found in [9]. Three values of the characteristic radius R_i^* have been identified from the histograms respectively obtained at the iteration 1, 2 and 3 (Fig. 3a), b), c));

these values are: $R_1^* = 11.5$ mm, $R_2^* = 3.5$ mm, and $R_3^* = 17.5$ mm. No significant peak could be detected from the histogram of iteration 4 (in Fig. 3d)); the residual relative frequencies of this histogram are due, in fact, to surface roughness and measurement noise.

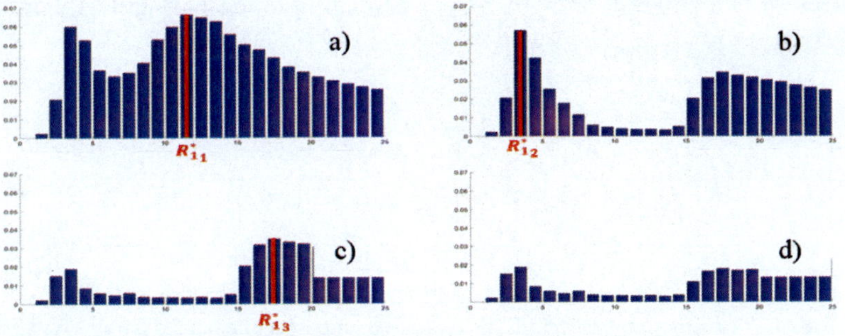

Fig. 3. The histogram, with possibly associated peak radius, obtained at: a) Iteration 1, b) Iteration 2, c) Iteration 3 and d) Iteration 4.

2.1 The Fuzzy Approach

To manage the r_1-values variability, the methodology for CRS feature recognition implements a fuzzy approach. This approach quantifies the membership of each vertex of the tessellated model to the following categories of points:

- *Non-regular points*, usually located on the fracture surfaces of an archaeological sherd.
- *Points potentially belonging to CRS features of radius R_i^*.*
- *Points belonging the remaining regular surface of the vessel.*

The implementation of this approach requires the definition of a set of membership functions, measuring the possibility that a generic node belongs to one of the above categories. The three membership functions are respectively denoted with μ_s, μ_{R_i}, $\mu_{\neg R_i}$. Since the generic node of the discrete model can belong unequivocally only to those three categories, these functions must satisfy the following *normality condition* at each node of the model.

$$\mu_s + \mu_{R_i} + \mu_{\neg R_i} = 1 \tag{1}$$

The possibility μ_s that a point is regular or not is quantified by a local indicator, called sharpness indicator (*SHI*) [10]. The a_s and b_s values of μ_s (Fig. 4a)) are identified by considering that, in an archeological artefact the transition zones from one surface feature to another are very gradual so that non-regular points are located on the fracture surfaces of a fragment and/or originate from measurement noise and surface roughness. For this reason, it is recommended here to select a higher value for a_s and to reduce the slope of the ramp function μ_s, compared to the case of the mechanical parts. The possibility of attributing a point **p** to a CRS feature, with characteristic radius R_i and sweep radius R_s (usually variable along the sweep curve and related to the radius $r_2 = \frac{1}{k_2}$), is measured respectively by the membership functions μ_{R_i} and μ_{R_s}. This possibility depends on the local quality of the tessellation, which is measured by the factor of *curvature approximation* γ, defined as the maximum value of the tangent of the dihedral angle between two adjacent triangular faces belonging to the 1- ring neighborhood of the generic point **p**.

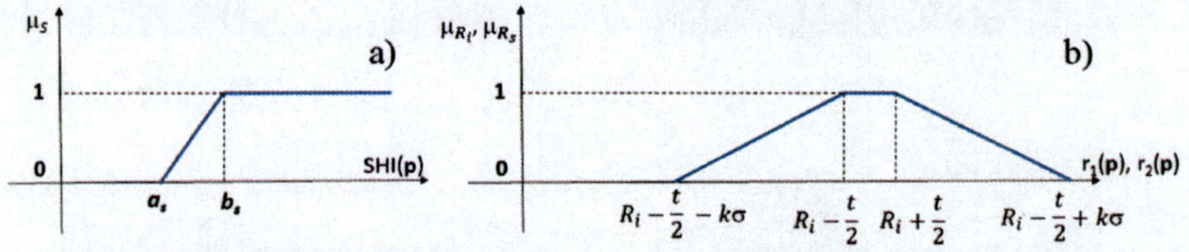

Fig. 4. The function types with the relative parameters chosen for μ_s and μ_{R_i}.

Both these functions are trapezoidal-shaped, as shown in Fig. 4b). The t-values for μ_{R_i} and μ_{R_s} have been obtained by setting the coarsest tolerance class (from tables II and I of the UNI ISO 2768 standard), given the high variability of the curvature radii of the archaeological object. The σ-values have been empirically determined by specific tests consisting of synthetic case studies that included CRS features tessellated with γ values in the range 0.01–1. Several tests have been performed, moreover, to establish the optimal value for the slope of the trapezoid's inclined sides in Fig. 4b).The CRS features are identified by a region growing process. During this phase a node **P**, neighboring to the growing region $\mathbf{F_j}$, is added to it if recognized similar. This is performed by an algorithm based on the fuzzy concept of dissimilarity [11] between the analyzed point $\mathbf{P_k}$ and the growing region $\mathbf{F_j}(R_i)$ in terms of μ_s, μ_{R_i}, $\mu_{\neg R_i}$ and μ_{R_s}. The region growing starts at the seed node $\mathbf{P_0}$, where a value of the membership function μ_{R_i} higher than the assigned threshold value μ_0 is reached. By using the dissimilarity function D, the node $\mathbf{P_k}$, on the boundary of the growing region $\mathbf{F_j}$, is compared with each point belonging to its 1-ring neighborhood. When D is lower than a threshold value given for D_0, the node $\mathbf{P_k}$ is added to the region $\mathbf{F_j}$. The growing algorithm stops when dissimilar nodes are met or all the nodes are analyzed. The threshold values coefficient μ_0 and D_0 are parameters of the methodology that require being empirically set.

3 The Experimental Results

Figure 5a) and b) show the nodes of the model allocated by the methodology to the various categories of points (described in the Sect. 2.1) when D_0 is respectively 0.5 and 0.3 and μ_0 is equal to 0.9. The dissimilarity coefficient D_0 is an important parameter of the recognition process as it can influence the extent of the grown region. Increasing D_0 means allowing "less similar" points to be allocated to the growing region. The effect is, therefore, that more points are attributed on average to a growing region, although the total number of points belonging to the various categories (reported in the two relative tables) does not increase significantly. Figure 5c) shows the results by setting $\mu_0 = 1$, while D_0 is equal to 0.3. The threshold value μ_0 is a parameter useful for the identification of seed points during the region growing. This parameter can affect the total number of points allocated to the various node categories during the growing step. Figures 5b) and c) show, as expected, that lowering the μ_0-value increases the number of the points allocated to the various node categories. The two segmentations, shown in the Fig. 5b) and c), although different from each other, can both be considered valid.

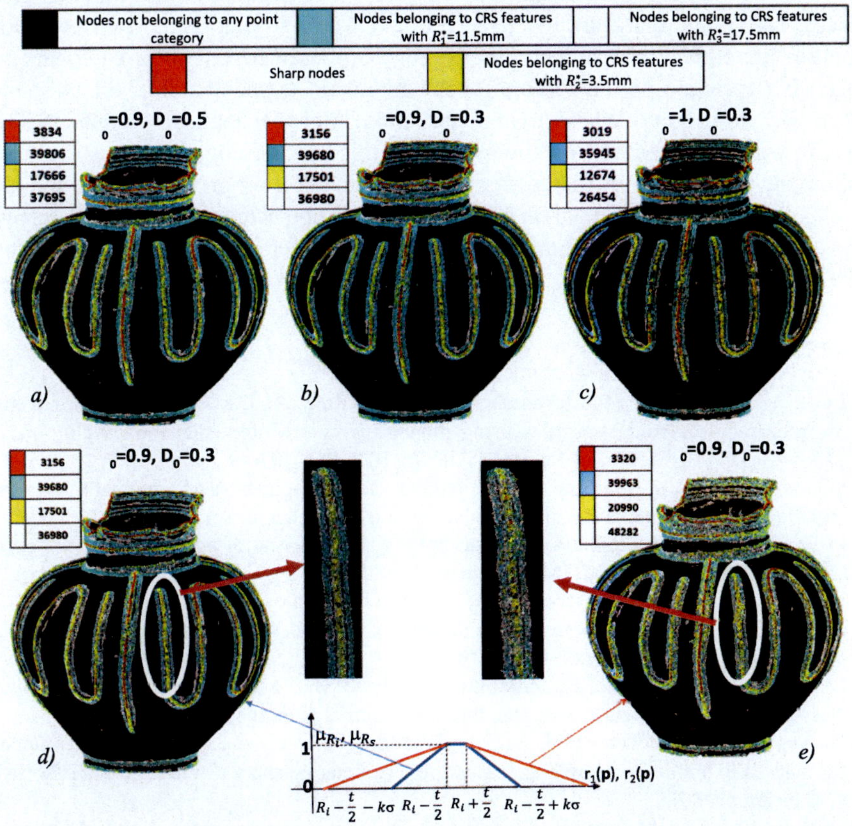

Fig. 5. The results of CRS feature recognition process (in terms of points allocated to the various point categories) by varying μ_0 and D_0.

Finally, Fig. 5d) and e) show what happens, in terms of recognition results, when the slope of the inclined sides of the membership functions μ_{R_i} and μ_{R_s} is halved. This modification of the membership functions has, as foreseen, the effect of making the identification of the various categories of nodes more blurred. This is well highlighted from the two enlargements (Fig. 5d) and e)) including the same decorative detail.

4 Conclusions

The aim of the paper is to describe and implement a new recognizer suitable for the automatic segmentation of sweep features of constant radius, which are quite common in archaeological artifacts. It is a specific class of ceramic features, often present on the surfaces of the archaeological objects in the form of engravings, working signs and decorations. Recognizing these constant radius features in archaeological artifacts is not a trivial problem. The ancient finds, even if repeatable, are unique handmade pieces with a geometry often characterized by complex and non-analytical shapes. Their surfaces, additionally, are usually damaged and worn, so that the related geometric properties are altered or lost. The methodology proposed here is inspired by the one previously developed by the authors for the automatic segmentation of fillets, rounds and grooves from high-density triangulated models of mechanical components. The paper, in particular, focused on the aspects of this methodology that must be tuned to allow the recognition of the constant radius features from archaeological artifacts. The experimentation carried out on an archaeological find acquired with a laser scanner showed the effect that some important parameters of the methodology have on the results of the recognition process.

References

1. Lucena, M., Fuertes, J.M., Martínez-Carrillo, A.L., Ruiz, A., Carrascosa, F.: Classification of archaeological pottery profiles using modal analysis. Multimedia Tools Appl. **76**(20), 21565–21577 (2016). https://doi.org/10.1007/s11042-016-4076-9
2. Gilboa, A., Tal, A., Shimshoni, I., Kolomenkin, M.: Computer-based, automatic recording and illustration of complex archaeological artifacts. J. Archaeol. Sci. **40**, 1329–1339 (2013)
3. Wilczek, J., et al.: Computer-assisted orientation and drawing of archaeological pottery. J. Comput. Cult. Herit. (JOCCH) **11**, 22 (2018)
4. Di Angelo, L., Di Stefano, P., Morabito, A.E.: Recognition of intrinsic quality properties for automatic geometric inspection. Int. J. Interact. Des. Manuf. (IJIDeM) **7**(4), 203–215 (2012). https://doi.org/10.1007/s12008-012-0173-9
5. Di Angelo, L., Di Stefano, P., Morabito, A.E., Pane, C: 3D Metrology for ancient pottery classification and reconstruction, Handbook of Cultural Heritage Analysis (2021)
6. Di Angelo, L., Di Stefano, P., Guardiani, E., Morabito, A.E.: A 3d informational database for automatic archiving of archaeological pottery finds. Sensors (Switzerland) **21**(3), 1–18 (2021). art. no. 978
7. Di Angelo, L., Di Stefano, P., Morabito, A.E: Fillets, rounds, grooves and sharp edges segmentation from 3D scanned surfaces. CAD Comput. Aided Des. **110**, 78–91 (2019)
8. Di Angelo, L., Di Stefano, P., Morabito, A.E.: Segmentation of secondary features from high-density acquired surfaces. In: Eynard, B., Nigrelli, V., Oliveri, S., Peris-Fajarnes, G., Rizzuti, S. (eds.) Advances on Mechanics, Design Engineering and Manufacturing. Lecture Notes in Mechanical Engineering, pp. 1043–1051. Springer, Cham (2017). https://doi.org/10.1007/978-3-319-45781-9_104
9. Di Angelo, L., Di Stefano, P., Morabito, A.E., Pane, C.: Constant radius geometric features segmentation in archeological pottery. In: IMEKO International Conference on Metrology for Archaeology and Cultural Heritage, MetroArchaeo 2017, pp. 69–74 (2019)
10. Di Angelo, L., Di Stefano, P.: C1 continuities detection in triangular meshes. Comput. Aided Des. **42**(9), 828–839 (2010)
11. Cox, E.: The Fuzzy Systems Handbook. AP Professional, Cambridge

A Review on Computer-Based Methods for Archeological Pottery Classification and Reconstruction

Luca Di Angelo(✉), Paolo Di Stefano, Emanuele Guardiani, and Caterina Pane

University of L'Aquila, Piazzale Pontieri, 1, 67040 Monteluco di Roio, AQ, Italy
{luca.diangelo,paolo.distefano}@univaq.it, {emanuele.guardiani, caterina.pane}@graduate.univaq.it

Abstract. The knowledge of the history, economy, and art of an archeological site is based on information that can be taken from ceramics analysis, classification, and reconstruction. The traditional methods used by the archeologists for their investigation are time-consuming, not reproducible, and repeatable, and the results depend on the subjectivity, specialization, personal skills, and professional experience of the operator. An as consequence, only a few indicative samples that have characteristic components are analyzed with wide uncertainties.

In order to overcome these limitations, in the last years, some automatic methods for studying archeological pottery's findings are proposed in the literature. To help all the researchers involved in this field, this paper aims to provide a complete and critical analysis of the state-of-the-art until the end of 2020 of the published methods on pottery classification and reconstruction from a 3D discrete manifold model. For this purpose, papers in English by the Scopus database are collected by using the following keywords: "computer methods in archaeology", "3D archaeology", "3D reconstruction", "automatic feature recognition and reconstruction", "3D puzzling". The list is completed by additional references found through the reading of selected papers. The 35 selected papers are divided into three groups: Geometric model fragment processing, 3D information systems for archaeological pottery visualization and documentation, 3D puzzling of archaeological fragments.

The results of the present review are focused on the presentation of the pros and cons of the techniques used on these different issues.

Keywords: 3D archaeology · Computer methods in archaeology · Surface segmentation · Automatic feature recognition · Computer-based methods for sherd classification and reconstruction

1 Geometric Model Fragment Processing

The geometric model processing methods published in the literature concern three different issues:

C. Rizzi et al. (Eds.): ADM 2021, LNME, pp. 909–919, 2022.
https://doi.org/10.1007/978-3-030-91234-5_92

- *Axis identification*; that, being archaeological potteries characterized by an axially symmetric geometry, is fundamental for the following dimensional evaluation steps;
- *Segmentation and recognition* of the most important *geometrical* and *morphological* fragments' features.
- *Evaluation of the characteristic dimensions* of the geometric and morphological features that can be found on the pottery's fragments.

1.1 Axis Evaluation

The axis evaluation is a fundamental topic in the archaeological field since it is used for the analysis of the fragment in order to retrieve the original pottery shape, and it is at the base of the computer-based systems developed for ancient pottery classification. Because of that, the axis estimation needs to be very accurate: it affects the classification and the data archiving of the archaeological finds. To find the axis starting from discrete geometric models is difficult because:

- the geometry of an archaeological vessel is only approximately axially-symmetric since they are handmade objects with handles, decorations, inscriptions, feet, and, for the specific case of a sherd, fracture surfaces.
- the axially symmetric surface of an archaeological pot is generally characterized by several defects, due for example to extensive wear for weathering, encrustations, chipping and other damage

These factors inevitably influence the quality of the axis estimation and consequently the quality of the measurements and evaluations that derive from it. The working principles, the strengths, and weaknesses of the most important published methods [1–6] are summarized in Table 1.

1.2 Profile Evaluation

The pair axis/profile is the base of the geometry of an axially symmetric sherd [7]. An axially symmetric surface, in fact, can be always generated by rotating a planar curve, the profile, around the axis of rotation (or axis of symmetry). The profile definition is the first step in the documentation, classification, or reconstruction of archaeological pottery. The profile, in fact, is the feature that contains information needed to perform these tasks. Archaeological methods for extracting the profile consist of hand-drawn sketches, which are an interpretation of the fragment's geometry. These sketches could result in an inaccurate representation, living aside the information hold into the fragment.

Several computer-aided and automatic methods were thus proposed to assist archaeologists in profile extraction. Typically, the state-of-the-art explore two main solutions: a first attempt profile identification and the final representative profile. The exception to this trend is the approach proposed by Kampel et al. in [8] that identifies the representative profile of a sherd as that in correspondence of fragment maximum height. In Table 2, the most important characteristics of the analyzed methods are summarized.

Table 1. Comparison of axis detection methods.

Methods	Strength	Weakness
Normal-based methods [1]: based on the property of axially symmetric surfaces for which the normal vectors at any point of the surface always intersect the axis of symmetry	Quick estimation without requiring any preliminary evaluation	Sensitive to the measurement noise, outliers, surface roughness and all those factors that make the surface not perfectly axially symmetric
Osculating sphere [2]: exploits that the center of the maximal sphere tangent to the surface (osculating sphere), evaluated at each node of the tessellated surface, is located on the axis of the axially symmetrical surface	It provides a better solution compared with the *Normal-based* method	Sensitive to the measurement noise, outliers, surface roughness and all those factors that make the surface not perfectly axially symmetric A preliminary axis estimation is required It fails when the fragments shape is close to spherical
Symmetry line-based [3] is based on the property for which the symmetry line of any curve resulting from the planar section of a complete axially symmetrical surface always intersects the surface axis	It provides a robust and accurate axis estimation also in presence of singularities and noise	Only for complete axially symmetric surfaces
Circle and line fitting [4, 5], exploit the property whereby each section of an axisymmetric surface perpendicular to the axis is a circle, whose center is on the axis	Recommended for metrological applications	An accurate results for objects with incomplete circular span, especially when affected by micro and macro irregularities (such as roughness and local deformations) that make the surface not perfectly axially symmetric, (such as in the case of fragments of archaeological vessels) It fails when the fragments shape is close to spherical

(*continued*)

Table 1. (*continued*)

Methods	Strength	Weakness
Thickness versor intersection-based [6] is based on that the minimum wall thickness line always intersects the axis of revolution	The method seems to be robust in the axis identification also on an object affected by extensive wear, encrustations and all the typical damage that can be found in a common archaeological sherd	Only for thin-walled axially symmetric surfaces

1.3 Features Detection

An effective computer-based method should automatically recognize the fragment features similarly the archaeologists perform their conceptual classification of ceramic. The shifting of the archaeological way of studying pottery into an automatic method is not a trivial process. This is because the characteristic elements of potteries' sherds not associable to analytical surfaces; so, specific operative rules, applicable to the above-mentioned feature recognition, for as wide as possible set of types of pottery fragments are required.

Moreover, this process is made difficult because the 3D discrete models are obtained from finds generally discovered damaged and worn so that their geometric properties have been lost. For all these reasons, in the last years, the automatic methods published in the related literature have proposed several new shape descriptors that, even if easily calculable didn't present an obvious correspondence with those used by archaeologists to study finds [8, 11–14].

Table 2. Comparison of profile detection methods.

Method	First attempt profile identification	The final representative profile	Strength	Weakness
Hlavackova-Schindler et al. [9]	A section of the surface with a half-plane passing through $\boldsymbol{\alpha}$	The profile is divided in several parts and each of them are approximated by a cubic B-spline curve	Universal approximation property	Relatively simple mathematical definition The number and complexity of the curves approximating the profile is not known a priori

(*continued*)

Table 2. (*continued*)

Method	First attempt profile identification	The final representative profile	Strength	Weakness
Karasik and Smilansky [12]	Projection on the plane z-ρ ($z \equiv \boldsymbol{\alpha}$ and ρ is the distance from $\boldsymbol{\alpha}$) of all the points of the surface (excepted those recognised belonging to fractured surface)	The centre of mass of the points within rectangles with width inversely proportional to the mean curvature at the point and oriented perpendicularly to the line locally approximating the profile	Procedure valid of each type of profile	The final result, depending on the width of the rectangles (calculated using the differential geometrical properties), is affected by the noise of the profile
Di Angelo et. al. [10]	The radii of the circles centred on $\boldsymbol{\alpha}$ approximating the slices perpendicular to $\boldsymbol{\alpha}$	Points smoothed by a parabola approximating the neighbourhood of each point profile with a defined width	Ease of implementation and speed of calculation Procedure valid of each type of profile	The optimal dimensions of neighbourhood are not known in advance
Kampel et al. [8]		Profile in correspondence of fragment maximum height	Ease of implementation and speed of calculation	Evaluation affected by local and global surface imperfections

Only recently, the authors published in [15] an automatic method following the previously mentioned requirements (Fig. 1). The method, applied to hundreds of real archaeological fragments, correctly recognized, in most cases, all the major features here exposed, failing in the reconstruction of chips between two different pottery features, fractured surface with some encrustations present, and the boundary between adjacent features when it is blurred by damages and wears.

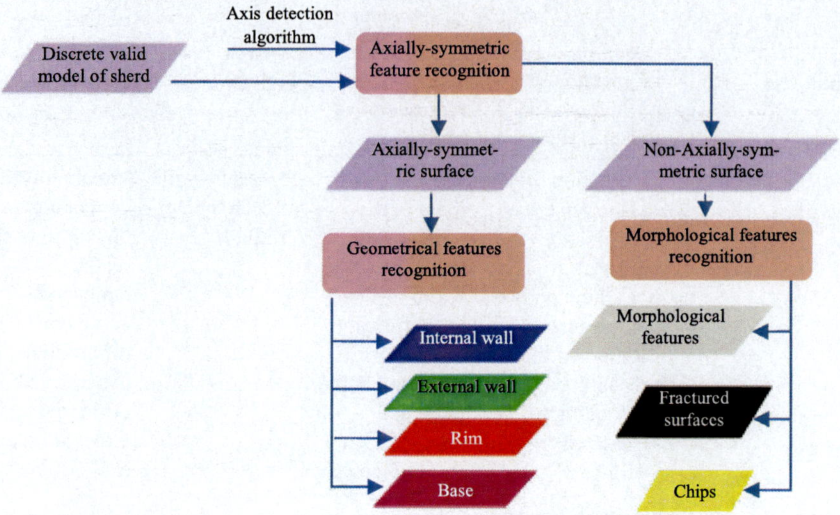

Fig. 1. Flow-chart of the features recognition method proposed in [15].

1.4 Secondary Features Detection

Ancient pottery often carries some detailed features, not necessarily axially symmetric, full of historical and archaeological meaning useful for classification aims. These features result in traces on potsherd surface, they can be left by finger or by tools intentionally (inscriptions and decorative motifs), or unintentionally, (working marks). From a geometric point of view, these detailed features are left by a sweeping action consisting of negative or positive traces on the ceramic surface. These features are characterized by a cross-section approximated to one or more circular arcs and may belong to the axially symmetric portion of a fragment. These detail features, moreover, are of limited extension as regards the mesh sampling, so the evaluation of geometrical differential properties is more difficult to carry on and affected by greater uncertainties.

In [16], Solbrig et al. proposed a method that locates the ridges and valleys of triangle meshes of decorated fragments by finding the local minimum and maximum of principal curvatures. These ridges are detected in correspondence of convex decoration i.e., cordons, pointed annexes, and embossed decorations. Valley lines are identified from concave ornamentations, such as grooves, channels, engraved lines, and impressions. Ridge and valley lines detection, however, does not allow obtaining information on the general shape and dimensions of these features. This last, however, is a fundamental aspect for classification or for trying to identify the tools used to shape them.

In [17] a computer-based method focused on the automatic detection and measurement of decorations with a constant radius in archaeological pottery is proposed. The method provides the hypothesis which considers the values of radii a priori unknown, and it follows two major steps:

- constant radius features segmentation: it is performed by means a fuzzification of principal curvatures at the nodes of discrete geometric models;

– radii measurement.

The results reported show the method can be used to analyze decorations of real 3D scanned objects both negative, such as engraving, graffiti, excised decoration, impression/stamping, burnishing, roulette, and in reliefs, such as barbotine, applied, and molded relief.

1.5 Dimensional Analysis of Fragments

Figure 2 shows the typical dimensional features used by the archaeologists to drive the sherd classification. To the best of the authors' knowledge, Di Angelo et al. [10] propose the only automatic method, published in the literature, suited to evaluate these dimensional features of fragments. The method implements specific rules applied on the profile detected (whose method is summarised in Table 2) starting from the results of features segmentation and recognition of the sherd (whose method is detailed in [15]). The proposed experimentation shows the method's robustness.

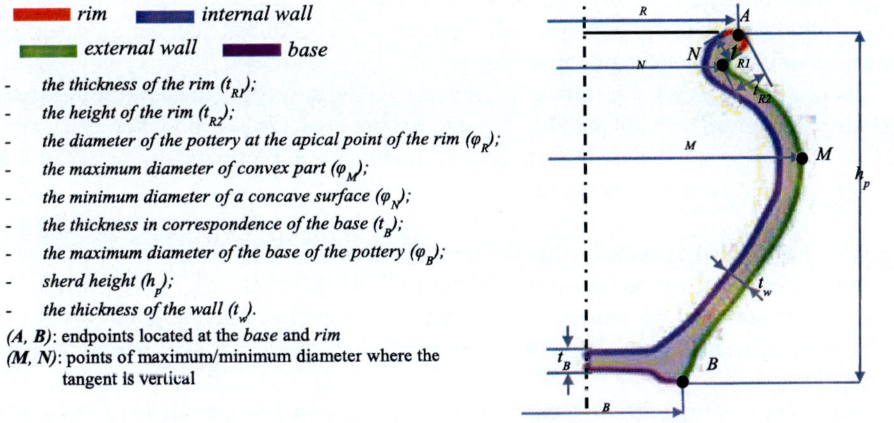

Fig. 2. Dimensional features and characteristic points.

2 3D Information Systems for Archaeological Pottery Visualization and Documentation

The available databases relative to archeological ceramic have different approaches and aren't built on a common structure and format, so that data could be difficult to share. By analyzing the published databases/structures, relative to the shape and morphology of sherds, it results in report digitization of out-of-print fascicules ([19], http://www.beazley.ox.ac.uk/pottery/default.htm, https://www.cvaonline.org/cva/) or information performed by skilled operators [20]. In Table 3 the most important available databases/structures, which start from a 3D scanned model of the sherd, are listed and compared in the voices outlined. Furthermore, annotations about quantity, origin, class, shape, typology, decoration, production, chronology, and bibliography can be added to all structures.

3 3D Puzzling of Archaeological Fragments

Several computerized methods have been developed to help archaeologists in sorting and separating all the shards part of the excavated strata, which are mixed and could be difficult to reassembly also because of the large quantity of ceramic discovered. Han et Hahn in [5] proposed the distribution of principal curvatures for grouping sherds, preliminarily clustered by an operator according to their position along with the pottery, with matching color and texture. A more general method is proposed by Biasotti et al. in [26], in which the similarity is qualified by the following criteria translated into computational tools: overall fragment size, thickness, material texture, shape continuity, color, and decorations. Even though this can be applied to real cases, showing the possibility to help archaeologists in estimating their findings from a quantitative point of view, they are still far from finally resolve the problem of assessing similarity. This is because of several problems, such as keeping results consistent across different resolutions and the supervision of damaged fragments.

The reassembly of an object from fragments is called mosaicing in literature and it can be generally viewed as a 3D puzzle: recently, Eslami et al. [28] proposed an exhaustive review of the state-of-the-art. The published methods for automatically reassembling pottery fragments can be divided in:

- *global methods*: they assume that the overall shape of the broken object is known a priori [30]; orientation and position of fragments along the α axis is defined by matching the geometry of the axially-symmetry pottery template; their angular position is determined by matching the contour curves and fractured surfaces of overlapping fragments
- *local methods*: they rely on accurately aligning only two fragments at the time and the reassembly of the object is obtained gradually, by fixing a sherd and finding the roto-translation (rotation around α and translation along α) transformation of the other one which minimizes an objective function. Typically, matching sherds include the contour curve and fractured surfaces matching ([31] and [32]). Willis and Cooper in [33] proposed to minimize a distance function of points coordinates and normal of contour curves.

Table 4 summarizes the strengths and weaknesses of two general strategies. Both approaches can provide a global relaxation [34] in order to minimize simultaneously all local alignment errors.

Table 3. Comparison of 3D information systems for archaeological pottery.

Structure/databases	Voices outlined
The Pottery Informatics Query Database (*PIQD*) [21]	The morphology of the ceramic vessel (coming from both 2D scans of illustrated ceramics profile and 3D acquisitions) is converted in measurements of radii, tangent and curvature of its representative profile as a function of arc-length
Computer-Assisted Drawing of Archaeological Pottery (*DACORD*) [23]	- general layout (layout are represented in the plane y-z, where z-axis coincides with the axis of symmetry); - overall form (the selected profile section contains information for further fragment classification, e.g., rim, foot, handles, distance measurements, volume estimation, maximum sector angle); - fragment rendering; - vessel regularity (evaluated as distance between model points and axis or model points and the 3d object obtained by rotating the profile around the axis); - 3D reconstruction (the 3d object obtained by rotating the profile around the axis) - Illustration exportation (archeological illustrations, 3D reconstruction)
A 3D Informational Database for Automatic Archiving of Archaeological Pottery Finds [22]	- General information: archeologist, warehouse and site references, - Fragment rendering; - Fragment features segmentation: Texture based features, Axially-symmetric features (rim, wall, base), non - Axially-symmetric features (handle, fractured surfaces, chips), Constant radius features - Fragments dimensional features

Table 4. Comparison of methods for sherd reassembly.

Reassembly methods	Strength	Weakness
Global methods	Simplification of the process of aligning each sherd	- The shape of pottery has to be known
Local methods	It is more general	- The alignment process can be laborious (hundreds of possible configurations); - Accumulation of errors; - Strongly affected by the physical degradation of fractured surfaces

References

1. Pottmann, H., Peternell, M., Ravani, B.: An introduction to line geometry with applications. Comput.-Aided Des. **31**, 3–16 (1999)
2. Cao, Y., Mumford, D.: Geometric structure estimation of axially symmetric pots from small fragments. In: Signal Processing, Pattern Recognition, and Applications, IASTED International Conference (2002)
3. Di Angelo, L., Di Stefano, P., Morabito, A.E.: A robust method for axis identification. Precis. Eng. **39**, 194–203 (2015)
4. Halir, R.: An automatic estimation of the axis of rotation of fragments of archaeological pottery: a multi-step model-based approach. In: Proceedings of the 7th International conference in Central Europe on computer graphics, Visualization and Interactive Digital Media (WSCG 1999) (1999)
5. Han, D., Hahn, H.S.: Axis estimation and grouping of rotationally symmetric object segments. Pattern Recogn. **47**, 296–312 (2014)
6. Di Angelo, L., Di Stefano, P.: Axis estimation of thin-walled axially symmetric solids. Pattern Recogn. Lett. **106**, 47–52 (2018)
7. Willis, A., Orriols, X., Cooper, D.B.: Accurately estimating sherd 3D surface geometry with application to pot reconstruction. In: IEEE Computer Society Conference on Computer Vision and Pattern Recognition Workshops, vol. 1, Article no. 4624519 (2003)
8. Kampel, M., Mara, H., Sablatnig, R.: Robust 3D reconstruction of archaeological pottery base on concentric regular rills. In: Proceedings of the 6th International Workshop on Image Analysis for Multimedia Interactive Services, Montreau, Switzerland, 13–15 April 2005, pp. 14–20 (2005)
9. Hlavackova-Schindler, K., Kampel, M., Sablatnig, R.: Fitting of a closed planar curve representing a profile of an archaeological fragment. In: Proceedings VAST 2001 Virtual Reality, Archeology, and Cultural Heritage, pp. 263–269 (2001)
10. Di Angelo, L., Di Stefano, P., Pane, C.: An automatic method for pottery fragments analysis. Measurement **128**, 138–148 (2018)
11. Mara, H., Sablatnig, R.: Orientation of fragments of rotationally symmetrical 3d-shapes for archaeological documentation. In: Third International Symposium on 3D Data Processing, Visualization, and Transmission, pp. 1064–1071. IEEE (2006)
12. Karasik, A., Smilansky, U.: 3D scanning technology as a standard tool for pottery analysis: practice and theory. J. Archaeol. Sci. **35**, 1148–1168 (2008)
13. Cohen, F., Liu, Z., Ezgi, T.: Virtual reconstruction of archaeological vessels using expert priors and intrinsic differential geometry information. Comput. Graph. **37**, 41–53 (2013)

14. Gilboa, A., Tal, A., Shimshini, L., Kolomenkin, M.: Computer-based, automatic recording and illustration of complex archaeological artifacts. J. Archaeol. Sci. **40**(2), 1329–1339 (2013)
15. Di Angelo, L., Di Stefano, P., Guardiani, E., Pane, C.: Automatic shape feature recognition for ceramic finds. J. Comput. Cult. Heritage (JOCCH), **13**(3), 1–21 (2020)
16. Solbrig, T., Brunner, D., Brunnett, G.: Finding ornamentations on bronze age vessels to support the automated classification. In: EVA 2006 Berlin Conference Proceedings, pp. 72–78 (2006)
17. Di Angelo, L., Di Stefano, P., Morabito, A.E., Pane, C.: Measurement of constant radius geometric features in archaeological pottery. Measurement **124**, 138–146 (2018)
18. Anichini, F., Fabiani, F., Gattiglia, G., Gualandi, M.L.: A database for archaeological data recording and analysis, MapPapers. **1en**-II, 21–38 (2012)
19. Classical Art Research Centre: Beazley Archive Pottery Database (BAPD). http://www.beazley.ox.ac.uk/pottery/default.htm
20. Verga, F., Fabbri, F.: Un approccio globale, archeologico, archeometrico, informatico, alla schedatura della ceramica a vernice nera della Valle del Tevere. Archeologia e Calcolatori **22**, 51–65 (2011)
21. Smith, N.G., Karasik, A., Narayanan, T., Olson, E.S., Smilansky, U., Levy, T.E.: The pottery informatics query database: a new method for mathematic and quantitative analyses of large regional ceramic datasets. J. Archaeol. Method Theory **21**(1), 212–250 (2014)
22. Angelo, L., Di Stefano, P., Guardiani, E., Morabito, A.E.: A 3D informational database for automatic archiving of archaeological pottery finds. Sensors **21**(3), 978 (2021)
23. Wilczek, J., et al.: Computer-assisted orientation and drawing of archaeological pottery. J. Comput. Cult. Heritage (JOCCH) **11**(4), 22 (2018)
24. Willis, A.R., Cooper, D.B.: Computational reconstruction of ancient artefacts: from ruins to relics. IEEE Sig. Process. Mag. **25**(4), 65–83 (2008)
25. Grosman, L., Smikt, O., Smilansky, U.: On the application of 3-D scanning technology for the documentation and typology of lithic artifacts. J. Archaeol. Sci. **35**(12), 3101–3110 (2008)
26. Biasotti, S., Moscoso Thompson, E., Spagnuolo, M.: Experimental similarity assessment for a collection of fragmented artifacts. In: Eurographics Workshop on 3D Object Retrieval, EG 3DOR, pp. 103–110 (2018)
27. Marie, I., Qasrawi, H.: Virtual assembly of pottery using Moire surface profile measurements. J. Archaeol. Sci. **32**(10), 1527–1633 (2005)
28. Eslami, D., Di Angelo, L., Di Stefano, P., Pane, C.: Review of computer-based methods for archaeological ceramic sherds reconstruction. Virtual Archaeol. Rev. **11**(23), 34–49 (2020)
29. Lu, Y., Gardner, H., Jin, H., Liu, N., Hawkins, R., Farrington, I.: Interactive reconstruction of archaeological fragments in a collaborative environment. In: Proceedings of 9th Bien nial Conference on Australian Pattern Recognition Society on Digital Image (DICTA 2007), pp. 23–29. IEEE (2007)
30. Liu, J., Zhou, M., Geng, G., Xiamixiding, R.: Broken pottery relic reassembling based on mixed feature vector. Boletin Tecnico/Tech. Bull. **55**(5), 1–9 (2017)
31. Kampel, M., Sablatnig, R.: 3D puzzling of archaeological fragments. In: Proceedings of the 9th Computer vision Winter Workshop, Piran, Slovenia, 4–6 February 2004, pp. 31–40 (2004)
32. Zheng, S.Y., Huang, R.Y., Wang, Z., Li, J.: Reassembling 3D thin fragments of unknown geometry, cultural heritage. ISPRS Ann. Photogram. Remote Sens. Spatial Inf. Sci. **2**(5), 1–7 (2014)
33. Willis, A.R., Cooper, D.B.: Bayesian assembly of 3D axially symmetric shapes from fragments. In: Proceedings of the 2004 IEEE Computer Society Conference on Computer Vision and Pattern Recognition, pp. 82–89 (2004)
34. Garcia Castaeda, A., Brown, B., Rusinkiewicz, S., Funkhouser, T., Weyrich, T.: Global consistency in the automatic assembly of fragmented artefacts. In: Proceedings of the 12th International Symposium on Virtual Reality, Archaeology and Cultural heritage VAST, Prato, Italy (2011)

A Photogrammetry Reconstruction of Emissario Romano in Capistrello

Luca Di Angelo, Paolo Di Stefano, and Emanuele Guardiani(✉)

Department of Industrial and Information Engineering and Economics, University of L'Aquila, Piazzale E. Pontieri, Monteluco di Roio, 67100 L'Aquila, AQ, Italy
emangua@univaq.it

Abstract. UAV Photogrammetry is a quite widely use technology for 3D reconstruction of territory, thanks to its satisfying results in many fields of applications and its low cost. Most of researches that addressed this topic use Ground Control Points (GCPs) for scaling and georeferencing the acquired point-cloud. In this paper, due to the particular morphology of the surveyed territory that makes very difficult to define GCPs in the region of interest, an innovative technique based on geometrical features was used for orienting, scaling and georeferencing the 3D point-cloud. The case is the ancient Roman outlet of the Fucino emissary, sited in a very steep and not accessible area, near Capistrello (AQ) city, Italy. The obtained results show that the proposed methodology, although less accurate than standard one using GCPs, provides quite good results, so that it can be used where the use of GCPs is difficult or impossible.

Keywords: UAV photogrammetry · Archaeology · Feature-based alignment

1 Introduction

Fucino's emissary was an ancient structure of Roman's era, built for controlling the water drainage of Fucino Lake, which was sited about 100 km east of Rome and it was one of the largest lake in Italy (Fig. 1). Its construction is considered as one of the most engineering challenges of antiquity: the tunnel was built underground, 15 m under the water level of the lake and the gallery was the longest tunnel ever made until the 19th century. Today the lake does not exist anymore since it was drained definitely in the 19th century.

The necessity of creating such a complex structure was justified by the endorheic nature of Fucino Lake, which led to a significance water level changing during years, causing several discomforts for the local population. The tunnel, whose length is about 5650 m, was destroyed in the 19th century, when it was replaced by a more larger and deeper modern tunnel excavated on the same path as the ancient one. [1]. In order to build the Fucino emissary, more than thirty wells were built along the gallery path. The functions of these wells were

C. Rizzi et al. (Eds.): ADM 2021, LNME, pp. 920–929, 2022.
https://doi.org/10.1007/978-3-030-91234-5_93

Fig. 1. Geographic map of Roman Emissary, about 100 km east of Rome.

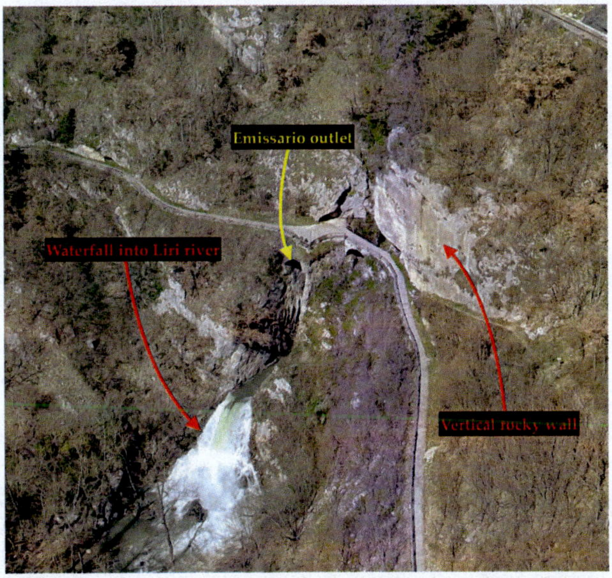

Fig. 2. Aerial view of emissary outlet and some of other natural and artificial elements in the area.

to permit the extraction of removed material and to define the layout and depth of the tunnel. The gallery of the emissary ends near Capistrello town, in the narrow canyon where flows the Liri river, in which the evacuated waters ends up. In Fig. 2 a photo of the tunnel outlet, the waterfall of the emissary, the Liri river and the steep surrounding area is shown. Watching the same photo, it can be observed to the right of the Emissary exit a large rocky wall, worked to create a vertical flat surface. The function of this rock face and the reason why the Romans carved it has never been clarified. It can not be excluded that the flat rock face in Fig. 2 was used by the ancient roman engineers as reference element to survey the territory and to define the tunnel path. A similar hypothesis could be evaluated by territory survey, geometric 3D reconstruction of the ancient emissary outlet and the measurements of its significant elements. At this purpose, a reconstruction of the territory, using photogrammetry techniques, is presented. This approach is performed by means of photos acquired by an Unmanned Aerial Vehicle (UAV). This reconstruction is compared and referenced with data of the emissary obtained by traditional survey technology (Fig. 3).

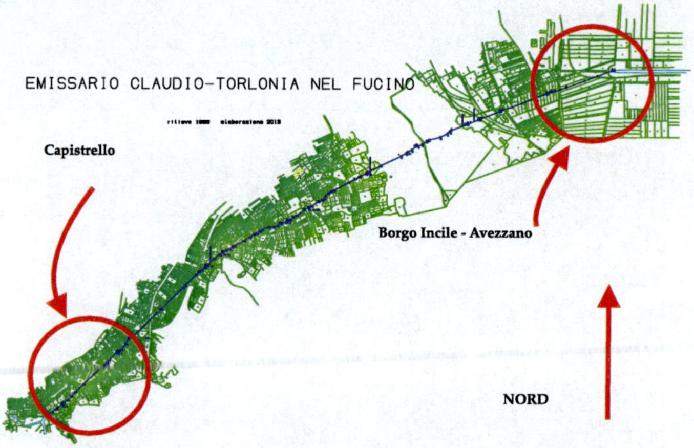

Fig. 3. Topographic map of Emissario Claudio-Torlonia.

The map surveyed and the 3D point-cloud reconstruction of the outlet of the emissary are both merged together to obtain a 3D model of the opera. At this purpose, it was necessary to identify some geometric features of the opera that were used as reference elements of the Emissary outlet. At the end the results obtained are compered with measured data and then critically discussed and evaluated.

2 State of Art

Photogrammetry is the art, science, and technology of obtaining reliable information about physical objects and the environment through processes of recording, measuring, and interpreting photographic images and patterns of recorded radiant electromagnetic energy and other phenomena [2]. Thanks to the large diffusion of Unmanned Aerial Vehicle (UAV), capable to acquire geo-referenced and high-resolution photos with low effort and cost, UAV photogrammetry is largely used in geomatic applications. Archaeology and architecture is one the field where these technologies are applied. One of the most interesting possibilities offered by using UAV photogrammetry in archaeological applications is the capability of recognizing, after the reconstruction process, some specific features which could be not visible by ground-based observers.

Accuracy of the final 3D model is one of the most challenging aspects of UAV photogrammetry. A 3D point-cloud is reconstructed by associating the same geometric point in different shots acquired from different positions of the camera. This reconstruction is performed by recognising, in each shot, the matching image patterns. Since the technical features of the image sensor and of the optical lens are known, the single point position in the space can be evaluated through a simple geometric triangulation: Structure from Motion (SfM) and Multi-View Stereo (MVS) are, nowadays, the most diffuse photogrammetry techniques [3] adopted for generating a 3D point-cloud, starting from digital images. These last are discrete by nature; since the atomic element of image grid is the pixel, an unavoidable limit to 3D accuracy of the model is related to the distance of two neighboring pixels on the camera sensor. Furthermore, bigger is the distance between camera and reconstructed object, whose is generally known as Ground Simple Distance (GSD), less will be the accuracy of the obtained 3D model. GSD is only one of the limits afflicting the photogrammetry reconstruction process. Other reasons which may compromise the model accuracy could be related to noise in photos, that makes difficult for algorithm to recognise the matching patterns in the textures of the images. Moreover, camera lens can generate a deformation of acquired scene, making a mismatch between real position and recognized position of points. In addition to this, some photogrammetric techniques base their evaluation on camera position using UAV position data. In most cases this introduces a misalignment in geo-referenced coordinate system, due to low performances of UAV integrated Global Navigation Satellite System sensors [4]. The literature concerning the analysis of accuracy in UAV photogrammetric reconstruction is voluminous and, consequently, cannot be fully mentioned here. Furthermore, each UAV reconstruction is characterized by proper contextual conditions, so that each case requires a specific approach. Ferrer-González et al. [5] tried to identify the optimal number of Ground Control Points (GCPs) in a large map reconstructed by Structure of Motion (SfM) technique. For this methodology, the use of GCPs is needful for geo-referencing the point cloud. They compared the obtained point-clouds through Root Mean Square Error (RMSE) and the Multiscale Model to Model Cloud Comparison (M3C2). They found that a number between 9 and 11 GCPs per kilometer gives rise the best accuracy results. Uysal et al. [6] performed an accuracy analysis on their reconstruction of a partly rocky hill near Afyon Kocatepe University campus (Turkey), approximately 40000 m^2 wide and having an average elevation of 1005 m above sea level, georeferenced using 27 GCPs. The results evidenced a RSME of some centimeters for each GCP, which can be considered a good result, especially for archaeological applications. The previous studies evidence the capability and accuracy of UAV photogrammetry when the use of GCPs is allowed, but very low knowledge is available about how to adopt UAV photogrammetry when no GCPs can be assigned. This is our case of study, where the territory is extremely steep and inhospitable. Thence, in the next section, an innovative methodology proposed by the Authors for scaling, aligning and georeferencing the point-clouds is introduced.

3 Methodology

GCPs are the most adopted solution for georeferencing the point-cloud, but their use it's not always possible, due to morphological features of the territory, so that excessive cost and operative difficulties in surveying the GCPs are encountered. In these situations, in order to scale the acquired point-cloud, could be useful to use, as dimensional references, geometric features identified in the region: for example, vertical rocks, walls, or artificial arches. These kinds of geometric features can be adopted to find the vertical orientation of the point-cloud. Some of them are also Control Points (CPs) on the ground surface, whose relative distances can be easily measured by using classical meteorological tools like yard stick or laser, so that they can be used as reference elements for scaling the point cloud to the real size. The aim of the proposed research is to define and test a procedure to make a 3D reconstruction of the Emissary outlet, so that it is scaled to the real dimensions, and to georeference it, avoiding the use of GCPs. The resulting model was obtained in three different acquisition stages, performed at different times. The first one was conducted in 2019; during this stage, a large area, including the outlet of the Emissary and the Well n.2, sited in the center of Capistrello city, was acquired. In order to increase the density of point-cloud and to acquire more details of the Emissary outlet, two other missions were scheduled between february and march 2021. In Fig. 4, the surveyed area during the three missions is evidenced. A DJI Mavic Pro, which belongs to Micro UAV quadricopter category, was used for the survey. In Table 1, some significance data of the optical sensor mounted on it are indicated.

Table 1. Optical sensor data of the UAV DJI Mavic Pro, used for the survey.

Focal length	35 mm
FoV	78°
Pixel dimension	10 μm
Resolution	12 Mpixel
Sensor dimensions	6.17 mm × 4.55 mm

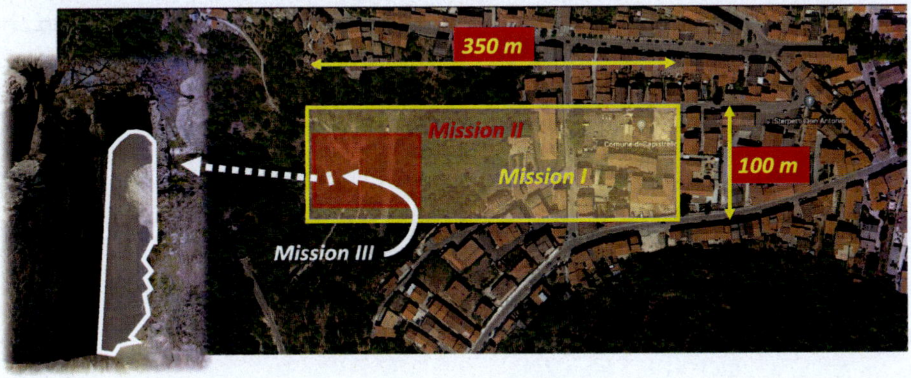

Fig. 4. Detail of acquired area during the three missions (source Google Earth).

The 3D reconstruction was performed adopting Agisoft Metashape Professional (v 1.6), which is one of the most powerful and diffuse software for photogrammetric processing. 247 photos were taken during the first mission. Due to the large area to survey, the photos were acquired by UAV camera manually, to maintain the control of battery usage. Near the Emissario outlet and Well n.2, a major number of shots were performed. The flight altitude respect to the ground not exceeded 50 m, but due to the irregularity of terrain and different camera angulation used during the survey, no consideration can be done a priori about the minimum point-cloud resolution. Metashape was able to align 241 of the 247 provided shots for creating a first set of tie points. For the second and third mission, although still manually operated, the drone was configured for taking a shot on every horizontal and vertical translation, respectively of 1 m for the second and 2 m for the third mission, using Pix4Dcapture app for iOS. 301 photos were aligned by Metashape for the second reconstruction; 381 for the third one. In Table 2, some significance data of the 3D reconstruction process, performed into Agisoft Metashape Professional, are reported. The ultra high preset was adopted, both for tie and dense point-clouds.

Table 2. Some output data of 3D reconstruction performed into Agisoft Metashape Professional.

	Mission I	Mission II	Mission III
RMS reprojection error	1,29 px	1,58 px	1,41 px
Mean key point size	3,22 px	2,52 px	2,18 px
Tie points	222339	281393	5871063
Total elaboration time	3 h 17 m	13 h 0 m	7 h 14 m

In order to direct the point-clouds according to the vertical direction associated with the z axis, some geometric features oriented vertically (2 planes) and one oriented horizontally (the axis of a cylinder) were identified in the point-cloud, using Geomagic Control 2014. In Fig. 5, the geometrical features recognised in the built or worked parts of the emissary outlet are evidenced. The plane n.1 is associated to the vertical rock face, worked with a chisel; The plane n.2 is associated to the vertical worked rock wall that delimits the west part of the exit channel; The cylinder n.1 is associated to the vault of the building, made of stone masonry. The z-axis was identified according the following alignment procedure:

1. y-axis of the point-cloud aligned with axis of Cylinder.
2. yz-plane aligned with Plane 1.
3. xz-plane aligned with Plane 2.

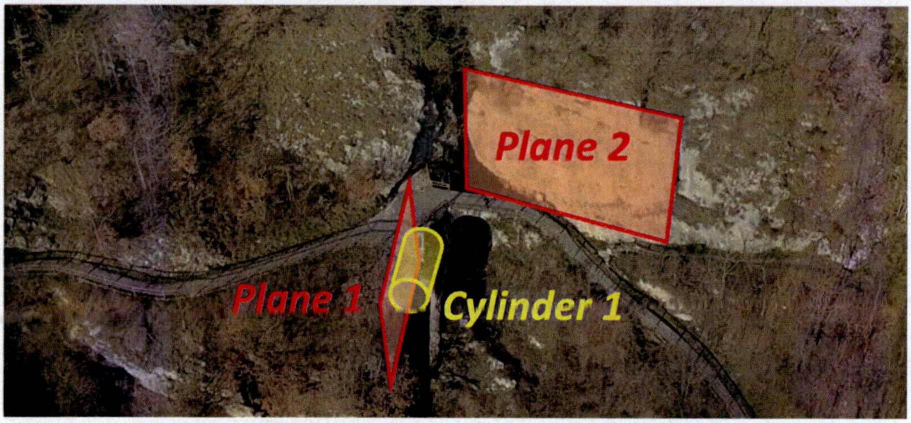

Fig. 5. Identified geometrical features on Emissario exit.

A set of measurements, indicated in Fig. 6 using a progressive numeration, was defined using traditional measuring instruments. These measures were used to determine the scale factors of the point-clouds along the three main principal directions. Their value is reported in Table 3.

Table 3. Measurements of segments in Fig. 6.

Segment	Length [m]	Segment	Length [m]
A	7.27	E	9.30
B	13.20	F	8.90
C	15.60	G	14.36
D	7.03	H	3.59

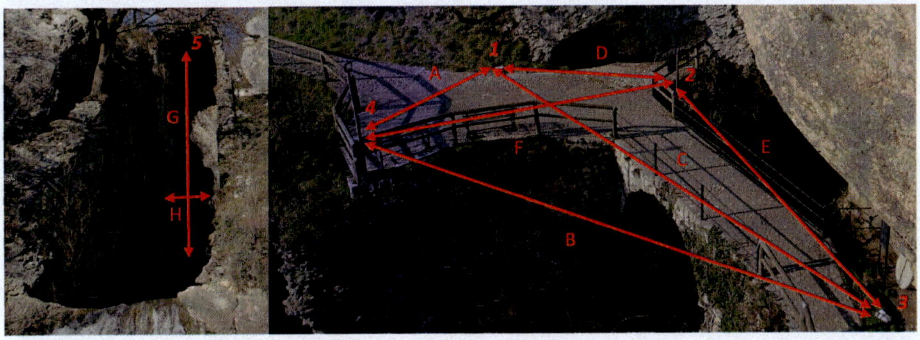

Fig. 6. Measurements of CPs.

The elements chosen to evaluate the corrective scale factors were selected strategically, in order to be as orthogonal as possible. Furthermore, only some CPs near of the Emissary outlet were considered, since this is the area of main interest, for which an accurate reconstruction was required. This makes difficult to evaluate distortions of the point-cloud far from the investigated area. The evaluation of scale factor along each coordinated direction was realized through a statistic analysis of corresponding measurements performed on the point-cloud by Geomagic software. These repeated measurements were made necessary due to the finite size of CPs reference marks that, being defined by many points, did not have a unique identification. The results evidenced very similar values of scale ratio along the three principal coordinated directions, with standard deviations of few millimeters. The point-clouds were scaled, taking fixed the points centroid, by using an unique value of scale factor f, equal to $1,075$. The last phase of the proposed method consists in aligning the resulting point-cloud and a pre-existing 2D topographic survey, already mentioned in the Introduction (Fig. 3). This phase is necessary for georeferencing of the point-cloud and verifying the dimensional accuracy of distant elements in the 3D reconstruction. This process was conducted by the use of a specific software for geomatic applications, Thopos 2020, thanks to which the horizontal projection of the resulting point-cloud (upon the xy-plane) was aligned with the topographic survey. The alignment process was performed using some remarkable points of the topographic polygon: the Well n.2 and the end of the tunnel. The results of the process are shown in Fig. 7. It can be observed a satisfying overlay between the 3D point-cloud and the 2D topographic survey, which is considered as the reference element for the alignment process. In order to verify the validity of the proposed methodology, a measurement between the Emissary outlet and Well n.2 was performed (red dot-line in Fig. 7), both into point-cloud and 2D map. The value obtained from this last is 208,85 m, while the same elements, measured into the horizontal projection of the 3D point-cloud, have a distance of 208,70 m. This can be considered as a good result, considering the extension of the area, whose longest side is about 350 m. Moreover the alignment along the vertical direction was performed by using natural features, shown in Fig. 5, with no-guarantee of their effective vertical alignment. Least but not last, a little uncertainty of few centimeters is related to the effective position of well n.2 on the 3d point-cloud.

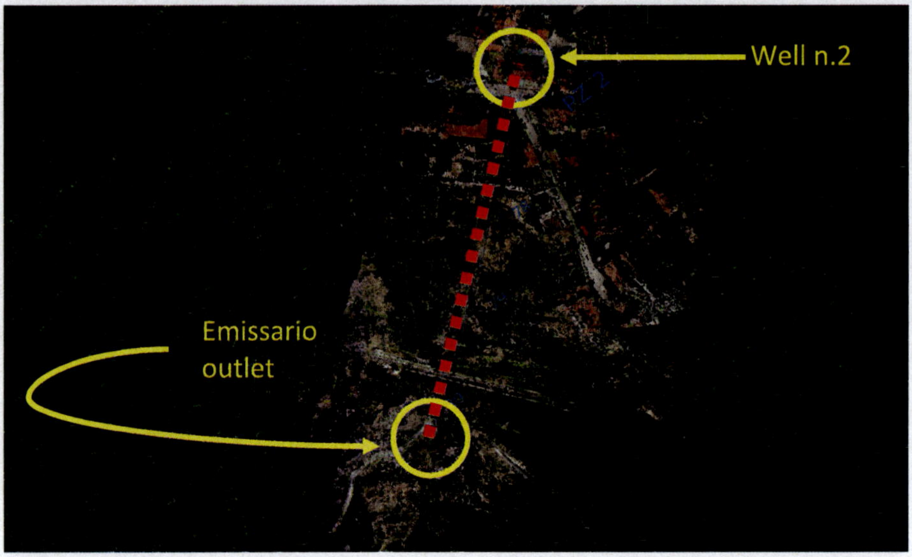

Fig. 7. Results of alignment process between 3D point-cloud and 2D topographic survey.

4 Conclusions

The survey of the structures of the outlet of the ancient emissary of the Fucino lake was made. This part of emissary has the particularity of not being accessible to direct measurements, since it is currently used and, consequently, a flowing watercourse that plummet into the Liri river (the waterfall is about 10 m high) is present. Furthermore, the Emissary outlet is in a very steep place inside a deep canyon with natural vegetation that obstructs the full view of the ancient works. In this context it was quite impossible to use referenced GCPs points. In order to perform a georeferenced survey, it was useful to use photogrammetry technology by using photo shot by a drone. The point-clouds were scaled and oriented by a non conventional approach based on the use of properly identified geometric features. Although the accuracy, in terms of dimensions and consistency with the few topographical surveys available, can not be compared with traditional UAV surveys adopting GCPs, the results evidenced the validity of the proposed methodology. Therefore, the illustrated procedure can be a valid alternative for UAV territory surveys when the use of GCPs is prevented. In particular, this method can be applied in most of anthropized contexts were, generally, a significance number of redundant features can be recognized. Buildings, roads, railways or bridges are only some of the most diffused human constructions which could be exploited for a feature-based alignment.

References

1. Di Angelo, L., Di Stefano, P., Guardiani, E., Morabito, A.E., Pane, C.: 3D virtual reconstruction of the ancient Roman Incile of the Fucino Lake. Sensors (Switzerland) (2019). https://doi.org/10.3390/s19163505
2. Wolf, P.R., Dewitt, B.A.: Elements of Photogrammetry with Applications in GIS. 3rd edn. (2000)
3. Westoby, M.J., Brasington, J., Glasser, N.F., Hambrey, M.J., Reynolds, J.M.: "Structure-from-Motion" photogrammetry: a low-cost, effective tool for geoscience applications. Geomorphology **179** (2012). https://doi.org/10.1016/j.geomorph.2012.08.021
4. Santise, M.: UAS PHOTOGRAMMETRIC BLOCKS: ACCURACY, GEOREFERENCING AND CONTROL, Ph.D. thesis (2016). https://doi.org/10.13140/RG.2.2.21597.41449
5. Ferrer-González, E., Agüera-Vega, F., Carvajal-Ramírez, F., Martínez-Carricondo, P.: UAV photogrammetry accuracy assessment for corridor mapping based on the number and distribution of ground control points. Remote Sens. **12**(15) (2020). https://doi.org/10.3390/RS12152447
6. Uysal, M., Toprak, A.S., Polat, N.: DEM generation with UAV Photogrammetry and accuracy analysis in Sahitler hill. Measure. J. Int. Measure. Confederation **73**, 539–543 (2015). https://doi.org/10.1016/j.measurement.2015.06.010

A Methodological Proposal for the Comparison of 3D Photogrammetric Models

Andrea di Filippo[1], Francesco Villecco[2](✉), Nicola Cappetti[2], and Salvatore Barba[1]

[1] Department of Civil Engineering, University of Salerno, Fisciano, Italy
[2] Department of Industrial Engineering, University of Salerno, Fisciano, Italy
fvillecco@unisa.it

Abstract. The field of simplification of geometric surfaces still lacks a formal and universally recognized definition of the error, which should involve both the approximation of the shape and the conservation of the other attributes of the mesh (starting from the colour). In order to solve this problem, we propose a hypothesis of methodological comparison that allow the evaluation of differences between two homologous surfaces, quantified employing the Hausdorff distance. The main advantage of this method is the independence from sampling techniques used to produce the mesh, without losing its characteristics of objectivity and generality. The Hausdorff distance geometrically represents the distance between two sets A and B in a suitable metric space, and it is defined as the maximum between the excess of A over B and the excess of B over A. This value is then compared with the average length of the diagonals of the "bounding boxes" of the homologous models, i.e. the parallelepipeds corresponding to the minimum volume that completely envelops each set; this results in an effective representation of error in relative terms.

Keywords: Hausdorff distance · Mesh · Accuracy assessment · Cultural Heritage

1 Introduction

Many applications of the digital survey - especially for Cultural Heritage -, such as multi-image photogrammetry and laser scanning, involve the manipulation of geometric objects [1], usually represented by triangular meshes because of their perfect compatibility with rendering [2].

The need for a better realism with higher geometric fidelity has produced over time a continuous increase in the complexity and size of the mesh, putting a strain on the storage, transmission and hardware processing capacity [3]. The use of a higher number of faces guarantees, in fact, a more detailed representation of the model, while causing a loss of performance due to the growing computational demand. Therefore, it is often advisable to define a compromise between the visual quality of the model and the processing times, appropriately evaluating the rank of the 3D graphic content, especially when the final use is designed for mobile devices [4].

C. Rizzi et al. (Eds.): ADM 2021, LNME, pp. 930–937, 2022.
https://doi.org/10.1007/978-3-030-91234-5_94

The idea behind this study is to make a comparison between homologous models obtained downstream of methodologically consolidated and recognized processes, such as photogrammetry.

In spite of intensive use of meshes - triangular or more generally polygonal - in geometric modelling and computer graphics, there is no agreement on the most appropriate method to estimate the differences between models of the same object, in the absence of a common criterion for formalizing the error [5]. This is because the approaches to mesh generation can be very different and affect aspects of the three-dimensional structure of the model, such as curvature, distance, or non-geometric attributes (colour, texture), which are not very compatible with each other.

For these reasons, we propose a hypothesis of a comparison method that does not require prior knowledge of algorithms for generating and simplifying the models produced, but still guarantees the possibility of quantifying the approximation error.

The measurement of the difference between two surfaces is intuitively related to the definition of a "distance" between them; it is therefore a question of identifying a metric able of summarizing, through a single indicator, the differences between the two objects.

2 The Choice of a Suitable Metric

When we talk about distance, we generally refer to the minimum; for example, if a point x is placed at a distance d from a polygon P, it is assumed that d is the distance between x and the closest of the points of P. Try to apply the same logic to two polygons A and B; if they are separated precisely by a distance d, it is generally understood that d is the minimum distance between each point of A and each point of B. Operationally, we are defining a function of minimum, being the distance d between A and B expressed by the following relation [6]:

$$d(A, B) = \min_{a \in A}\left\{\min_{b \in B}[d(a, b)]\right\} \tag{1}$$

It should be noted that the replacement of the infimum with the minimum in the definition of distance is possible because the polygons are closed sets. However, this definition of distance might be quite unsatisfactory for some applications: let's consider, for example, the two triangles ABC and DEF (Fig. 1a).

It is possible to state that they are close by considering their minimum distance, the one that separates the red vertices. However, one would expect that a 'small' distance between the polygons would indicate that no point of the first is 'particularly' distant from the second and vice versa. In this sense, the two triangles considered are not so close since their furthest points, highlighted in blue, can be very distant from the other polygon.

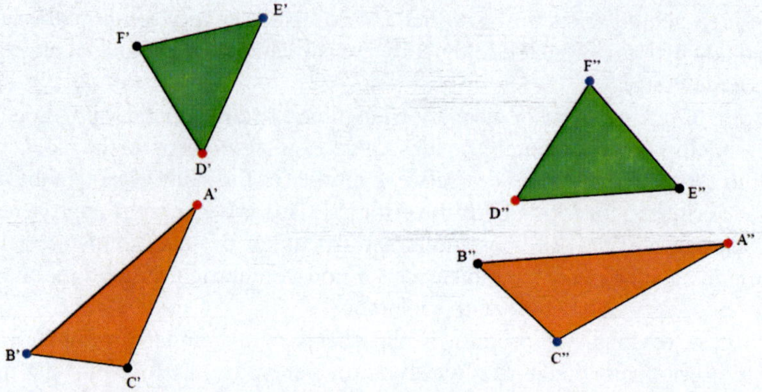

(a) (b)

Fig. 1. Two spatial configurations for a pair of triangles with the same minimum distance.

In the second example (Fig. 1b) we have the same two triangles, separated by what should be the same minimum distance as in the case reported in Fig. 1a, but in a different configuration. It is evident that the concept of minimum distance has a very low information content since the distance value has not changed compared to the previous case, although the two objects have a very different spatial configuration.

Despite its apparent complexity, Hausdorff distance can capture these details, which are ignored by the minimum distance [7]. First, we define the excess of a set A over a set B as the following maximum function:

$$e(A, B) = \max_{a \in A} \left\{ \min_{b \in B} [d(a, b)] \right\} \tag{2}$$

a and b are points of sets A and B respectively and d(a, b) is any metric between these points, generally coinciding with the Euclidean one (the same argument applies to the previous relation regarding the replacement of the supremum with the maximum). Since e(A, B) $\neq$ and e(B, A) (the asymmetry is a property of the maximum functions, while the minimum ones are symmetric), the Hausdorff distance can be obtained using the maximum of the two values.

Going back to the triangles of the example, some of the points are to be considered close, but not all. The Hausdorff distance provides an interesting measure of their proximity, indicating the maximum distance of each point of the first triangle (or polygon) from the other one (Fig. 2): therefore, it can be more effective than the minimum distance, which refers to a single point of each triangle (or polygon) and completely neglects their spatial configuration, to which, instead, the Hausdorff distance is sensitive. The graphic representation of the excess of ABC over DEF and vice versa - materialized in the circumferences of Fig. 2 - is an example of how much the metric chosen for comparison is capable of expressing the proximity of the figures.

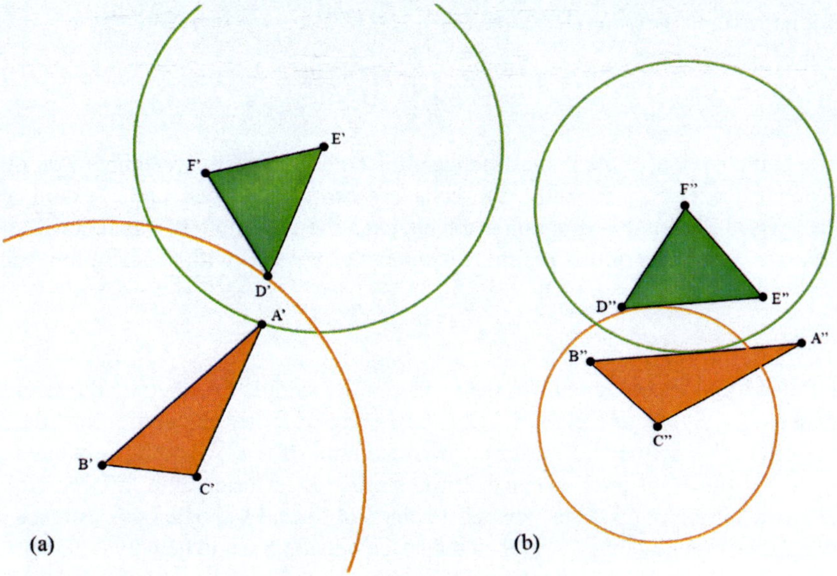

Fig. 2. Representation of the excesses of each triangle over the other for the two configurations.

3 A New Approach to Model Comparison

3.1 Surface Comparison

Special pairs of homologous models have been prepared for the experiments using the photogrammetric technique. In each pair, one model was processed starting from frames acquired with an entry-level reflex camera and the second with images acquired via smartphone. Both datasets were then processed in the Agisoft Metashape environment [8]. In the orientation, we employed the high option to search for key points on original resolution images. Calibrated scale bars appropriately placed in the scene prior to acquisition were used to solve the external orientation. These are characterized by targets positioned at a known distance, useful to define a local reference system common to the two homologous models. Tie points were then filtered to keep their reprojection error less than one pixel and orientation optimization was then performed. We then moved on to dense cloud and mesh processing. Prior to the comparison, the correct overlap of the models was improved with an algorithm derived from the ICP [9]. We employed a simplified approach based on sampling the two models and calculating the distance of a point from a surface. The bounding box of the reference mesh was partitioned into cubic cells following a regular pattern. Subsequently, a list of the faces of the reference mesh that intercept each cell was drawn up. For each point p, belonging to the sampled mesh and the analysed cell, the distance between p and all the faces of the reference mesh contained in its own cell was calculated. The adjacent cells were then processed, progressively increasing the distance from p, until it is verified that all the unprocessed cells are further away than the current closest face. Given a point p and a surface S, the following metric was essentially defined:

$$d(p, S) = \min_{p' \in S} d\left(p, p'\right) \tag{3}$$

In the specific case d(p, p′) is the Euclidean distance between two points in a three-dimensional space. Subsequently, the cells adjacent to the one under examination were processed. The approximation error between the two models was defined as the maximum of the values of the minimum distances associated with the sampling cells:

$$e(S_1, S_2) = \max_{p \in S_1} d(p, S_2) \tag{4}$$

There follows that it corresponds to the excess of the sampled mesh over the reference one. Since $e(S_1, S_2) \neq$ and $e(S_2, S_1)$ it is necessary to repeat the entire procedure by reversing the roles of the two meshes. The maximum of the two values obtained will correspond to the Hausdorff distance. If the surface S_1 is orientable, it is possible to extend the definition of distance between a point p of S_1 and S_2 so that this distance d′ is positive if the closest point $p' \in S_2$ is in the space outside S_1, and negative in otherwise (Fig. 3). In other words, if N_p is the vector normal to S_1 in the sampled point p and $p' \in S_2$ is the closest point, the sign of the distance sought is the same as $Np \bullet (p' - p)$. This definition of signed distance was introduced because the tested algorithm employs it to distinguish between the positive and negative distance between two surfaces. The distinction is aimed at a correct graphical representation of the values obtained, while from a computational point of view the process uses only positive distances.

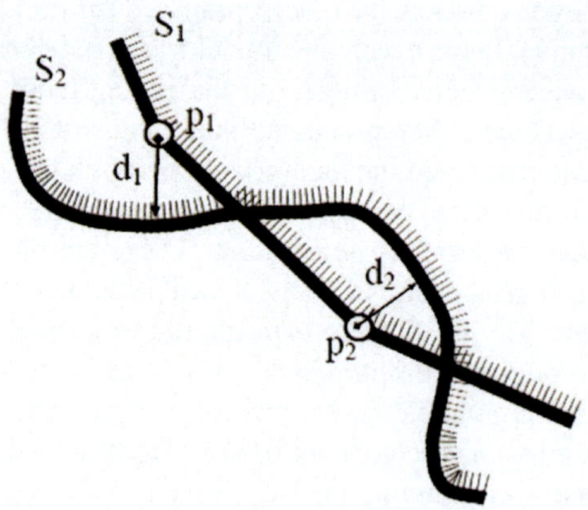

Fig. 3. Oriented distance computed with the employed algorithm [5].

At this point, it is necessary to make some clarifications. The evaluation of the distance of a point from a surface, based on the proposed comparison method, is partly dependent on the algorithm used to generate the mesh [10]. To overcome this drawback and keep the fluctuations of the result produced by the changing topology, only the vertices detected with the photogrammetric technique and used for the construction of the mesh were selected. This expedient guarantees the possibility of generalizing the proposed approach for comparison to any pair of models that approximate a given surface, regardless of the generation and simplification techniques; in fact, it is very likely that the points furthest from the target mesh fall right at the vertices.

3.2 Analysis of Results

To verify the applicability of the proposed methodology we compared the homologous photogrammetric models of a vase found in ancient Corinth, 24.81 cm high: the first obtained with a 24.2-million-pixel APS-C sensor and the second with a 1/3.2″ sensor with 8 million of pixels, implemented in a smartphone. Then we calculated the reciprocal excesses, obtaining a value of 2.8 mm by comparing the first with the second and 3.1 mm in the reverse case. This last value represents the Hausdorff distance, which describes the differences between homologous surfaces. A further step forward was to relate the approximation error between the two surfaces, with a characteristic dimension of the modelled object. In this regard, we selected the average length of the diagonals of the "bounding box" of the homologous models, equal to 31.73 cm (Fig. 4b). The "bounding box" constitutes, as anticipated, the minimum volume of parallelepiped shape that entirely encloses the modelled surface.

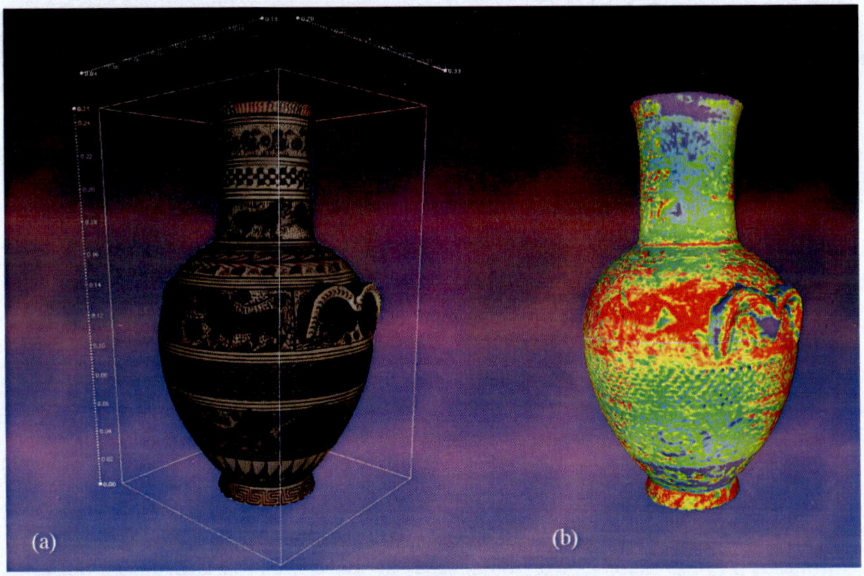

Fig. 4. "Bounding box" of the reference model (a) and colour representation of deviations between homologous models (b).

The Hausdorff distance in the case study was less than 1% of the average length of the diagonal. Up to now, we have limited ourselves to an estimate of the global error. To know the effective distribution of this error on the surface, we selected, among the homologous models, the mesh that gives the maximum value between the two possible excesses and we applied to the surface a colour filter based on the same parameter associated with its vertices (remember that only the vertices detected by photogrammetry are employed). After setting the range of values to be represented, a map of red-green-blue surfaces was generated, where red indicates the minimum value (set equal to − 3.1 mm) and blue the maximum (set equal to 3.1 mm). What is represented by the map, therefore, is none other than the distance, from the reference mesh, of the sampled points; since the mesh is a non-continuous surface, the major deviations between the two surfaces are observable and concentrated in the areas characterized by a sudden variation of the discrete curvature (Fig. 4b).

To further substantiate the choice of Hausdorff distance we can consider that the distributions of the surpluses for the two models, computed as positive distances, are not normal, which happens very often in comparisons between point clouds or meshes. Using the average distance, as often happens in the literature, would not allow an effective description of the error but rather an underestimation of it.

4 Conclusions

The main goal of this study was to develop a novel approach to the comparison of homologous models generated downstream of heterogeneous processes. The need therefore emerged to verify that these models met certain rigorous criteria.

To quantify the approximation error, we employed the Hausdorff distance, a particular metric capable of providing an interesting measure of the proximity of two surfaces as it depends, unlike the minimum distance, on the spatial configuration of the same.

For a simpler interpretation of the results, this value was compared with a characteristic size of the modelled mesh, made to correspond to the average length of the diagonals of the "bounding box" relating to the models.

The approximation error was found to be less than 1% of the characteristic size of the surface, allowing us to conclude that the models developed with the use of mobile devices can 'approximate' in sufficient detail homologous photogrammetric models produced downstream of consolidated and scientifically recognized processes. However, these conclusions required further investigation. In fact, even if the approximation error between two polygonal surfaces is contained, they could still present topological differences. The topology of a mesh, in fact, refers to how its constituent elements relate and influences its ability to preserve the distinctive characteristics of the object it represents, such as depressions, protrusions, or through holes. It is trivial to observe how the topological variations, as much as the ability of the mesh to approximate a continuous surface, will directly depend on the processing and simplification algorithms.

A limitation of the methodology presented has therefore emerged, as it is not always possible to trace the algorithms of generation. To partially overcome this drawback, it was decided to select, for sampling aimed at the comparison, only the vertices of the mesh (i.e., those points used for the construction of the mesh itself). In fact, likely, the points of the sampled surface most distant from the target mesh fall precisely in correspondence with the latter, so as not to underestimate the approximation error.

A second observation, concerned the topology of the models taken as a sample. This approach is particularly suitable for the comparison of closed meshes while, in the case of open models, on the other hand, the irregularities along the edges of the perimeter could excessively influence the estimate of the Hausdorff distance if not properly treated.

Despite these problems, the proposed metric offers an effective representation of the approximation error, certainly more representative of the average distance, often used in the literature without an effective finding of normality in the distribution of errors. Future studies and insights will aim to clarify the mechanisms through which a meshing algorithm can influence the approximation error between two surfaces (also from laser scanning), to be able to isolate the component attributable solely to the specifications. characteristics of the algorithm.

References

1. Guidi, G., Russo, M., Angheleddu, D.: 3D survey and virtual reconstruction of archeological sites. Digit. Appl. Archaeol. Cult. Heritage **1**(2), 55–69 (2014)
2. Grilli, E., Remondino, F.: Classification of 3d digital heritage. Remote Sens. **11**(7), 847 (2019)
3. Denning, J.D., Kerr, W.B., Pellacini, F.: Meshflow: interactive visualization of mesh construction sequences. In: Hoppe, H. (eds.) ACM SIGGRAPH 2011 papers, pp. 1–8. Association for Computing Machinery, New York (2011)
4. Schütz, M.: Potree: rendering large point clouds in web browsers. Wien (2015)
5. Morena, S., Barba, S., Tordesillas A.A.: Shining 3d einscan-pro, application and validation in the field of cultural heritage, from the Chillida-Leku museum to the archaeological museum of Sarno. Int. Arch. Photogram. Remote Sens. Spat. Inf. Sci. XLII-2/W18, pp. 135–142 (2019)
6. Gregoire, N., Bouillot, M.: Hausdorff distance between convex polygons. Computational Geometry Web project (1998)
7. Mezhenin, A., Zhigalova, A.: Similarity analysis using Hausdorff metrics. In: CEUR Workshop Proceedings, vol. 1, pp. 1–8 (2019)
8. Luhmann, T., Robson, S., Kyle, S., Boehm, J.: Close-range Photogrammetry and 3D Imaging. Walter de Gruyter, Berlin (2013)
9. Chetverikov, D.: The trimmed iterative closest point algorithm. In: Object recognition supported by user interaction for service robots. IEEE (2002)
10. Edelsbrunner, H.: Geometry and Topology for Mesh Generation. Cambridge University Press, Cambridge (2001)

A CAD-Based Framework for Interactive Analysis in the Restoration of Bronze Statues

Paolo Cicconi[1], Michele Bici[2], Olimpia Colacicchi Alessandri[3], Giorgio D'Ercoli[4], and Francesca Campana[2](✉)

[1] Dipartimento di Ingegneria, Università degli Studi Roma Tre, Rome, Italy
[2] Dipartimento di Ingegneria Meccanica e Aerospaziale, Sapienza Università di Roma, Rome, Italy
francesca.campana@uniroma1.it
[3] Museo Nazionale Romano, Rome, Italy
[4] Istituto Superiore per la Conservazione e il Restauro, Rome, Italy

Abstract. Nowadays, restoration is a multidisciplinary work that gathers knowledge and skills from different areas (technical, artistic, historical, architectural, ...). In the field of ancient bronze statues, technical knowledge may also concern with materials behaviour and its preservation, surface quality, non-destructive diagnostics for integrity, a better understanding of the manufacturing technology, and of details, sometimes hidden, in not directly accessible sections of the artefact. This knowledge, got from different domains, can support restorers in their decision-making process. In many cases, they summarise it on pictorial views of the artefacts, or on images derived from the 3D model that is experimentally acquired through reverse engineering, to reference information on the interested areas. The aim of this paper is to explore the advantages related to a CAD-based framework able to gather the technical domains involved in the restoration of historical artifacts. Doing so, CAD functionalities and related benefits may be extended to cultural heritage applications as tools oriented for restoration, according to a life cycle perspective of the restorer's activities and the artefact preservation and fruition. The proposed CAD-based framework has been implemented to manage the investigation for restoration and conservation of bronze statues. The approach has been applied to the Principe Ellenistico, part of the collection of Palazzo Massimo, one of the sites of Museo Nazionale Romano (in Rome). The obtained results show that the CAD-based framework may speed-up the investigation processes without losing accuracy and restorers' good practices.

Keywords: CAD-CAE · Virtual prototyping · Cultural heritage · Design for restoration · Principe Ellenistico

1 Introduction

In the field of Cultural Heritage, 3D model acquisition is now a common practice that helps archival, dissemination and investigations carried out during restoration, as highlighted by Scopigno et al. in [1]. Many recent works demonstrate that different acquiring

C. Rizzi et al. (Eds.): ADM 2021, LNME, pp. 938–950, 2022.
https://doi.org/10.1007/978-3-030-91234-5_95

methodologies can be used to achieve a virtual model (photogrammetry, laser scanner, phase shift methods, etc.). Accuracy of the digitized shape is the most relevant aspect together with the easiness of acquisition and the surface texturing reproduction, especially if virtual reality applications are investigated for divulgation or analysis purposes [2, 3]. Photogrammetry and laser scanner are widely applied in many works. 3D acquisition of buildings commonly supports architectural investigations at different levels, including restoration [4, 5]. Close-range photogrammetry is an emerging technique also for artefacts [6], and, in [7], a recent application is discussed with specific attention to inaccuracy due to reflection and occlusion-points. In [8], a multi-sensor network is investigated to widespread low-cost acquisition for sharing data, so that, more detailed models may be achieved thanks to collaborative work through data cloud sharing.

Reverse Engineering techniques may also help fragment re-compositions and mechanical design of supports as shown in [9, 10]. When the focus of the restoration is the static behaviour of the statue or its integrity, the 3D model can be converted in a model for structural analysis by Finite Element Analysis (FEA), as conveniently demonstrated in applications on marble statues (e.g. David of Michelangelo [11], the marble of Alba Fucens [12]) and bronze statues (such as Marco Aurelio and Minerva di Arezzo [13, 14]). FEA is used also to evaluate stress concentration and to map critical areas for material weakness or static stability. In [15] and [16], activities related to the Neptune Fountain in Bologna are presented. In these studies, FEA and experimental measurement are gathered on the 3D model to support restoration decisions. Measurements of different kinds (radiography, ultrasound, acoustic emission, etc.) are often necessary for monitoring, for maintenance purposes and for a better understanding of the artefact [17, 18]. They may be punctual data or mapped, related to the multidisciplinary aspects involved in the diagnosis of the problems. From the 70s of the last century, innovative non-destructive diagnostic methods allowed restorers to find new strategies for preventive conservation and monitoring with non-destructive procedures [19, 20]. In parallel, technological evolution of materials and mechanical design made possible new reversible assembly techniques of fragments and lacuna reconstruction. Due to this, new support designs were developed, and engineering solutions started to increase their help to restores. For example, in [21], the 3D reconstruction of an ancient statue of Zeus Enthroned is integrated through a process of Digital Design by Additive Manufacturing to achieve the legs of the throne.

In [22, 23], the integration of different information in a unique tool to support restorers is discussed. In [22], a demo tool is addressed to demonstrate how a "reality-based" model of the Pietà Rondanini may help data archiving and statue exploration for preservation purposes. To achieve a "reality-based" tool, the main requirements of its development are connected to the accuracy of the acquisition, surface texture acquisition and rendering. Data organization has been divided in metadata (local extraction of model data) and "paradata" (results of analysis and investigation by the "reality-based" model). Technical investigations interlaced in the model are physical analysis on single referenced points and areas, like measurements.

The present paper further investigates this topic, presenting a CAD-based framework able to integrate data from sensors and other activities usually planned during the investigation of bronze statues. The originality of the work is related to the analysis of the methodological approach usually applied by restorers. So, the CAD-based framework is proposed as a design for restoration tool in the respect of the organization of the knowledge stratification derived from the different fields of expertise applied in Cultural Heritage (historical, artistic, technical, economics, etc.). Doing so, the operational activities may be anchored, to the other relevant aspects of the lifecycle of an artefact (e.g. exposition monitoring and fruition management), thanks to a Product Lifecycle Management approach that is connate in industrial engineering CAD and takes into account all the actors and stakeholders involved in Cultural Heritage sectors [24].

Starting from the requirements related to the workflow adopted for bronze statue restoration, we implemented the tool, analysing its usability on the preservation activities of the Principe Ellenistico of the Museo Nazionale Romano (placed in Rome, at Palazzo Massimo). In Sect. 2, the concept design of the framework is discussed to define the basic requirements and the different domains to be collected; in Sect. 3, the application is presented and then, in Sect. 4, it is discussed. Finally, in Sect. 5, some conclusions and perspectives will be defined.

2 Framework Definition

2.1 Fields of Expertise and Data Collection

Operations connected to restoration and conservation of ancient bronze statues involve many activities both when they are related to a primary intervention after an archaeological finding, and when they concern a planned intervention to preserve the integrity or to change the exhibit status (e.g. change of room, style of the exposition, and so on). Recent examples of these scenarios are the restoration of the Satiro Danzante of Mazara del Vallo, found in 1997 underwater in the Mediterranean Sea [25] and the Vittoria Alata di Brescia, recently restored to be exposed in a new location in Capitolium of Brescia.

Evaluation and decisions are made through an experimental assessment that involves: (a) technical expertise and diagnostic measurements, (b) historical, artistic and archival studies; (c) visual inspection. More in detail:

(a) Technical expertise and diagnostic measurements. Technical expertise can be distinguished in:

- Analyses on the statue integrity which involve fragment analysis (if any) and their assembly condition, on the structural integrity of the cast and related joins, on the structural integrity of the material, on the static support;
- Surface assessment (material composition, corrosion, finishing);
- Architectural aspects for exhibition (position, anti-seismic devices, lighting).

(b) Historical, artistic, and archival studies. They concern with repertory photos, reports (even ancient chronicles or literature), drawings of related studies or previous restorers. They can be organized both by keywords and chronologically, so that research and extraction may be possible in accordance with the criteria selected by the user.

(c) Visual inspection pertains to the professional expertise of the restorer. It merges knowledge and evaluation criteria based on the morphological and surface recognition of the details.

(d) The framework here proposed aims to collect data and knowledge related to (a), (b), and (c) into virtual annotations anchored to the digital mock-up of the statue.

(e) From the visualization point of view, each of these fields is associated to different kinds of data set that involve points or areas of the statue, as reported in Table 1.

Table 1. Fields of expertise and related data types

		3D model	Surface map /patch	Section/photo	Point/path	Graph/table	Docs/images	3D annotation
Technical expertise	**Structural Integrity**							
	Fragment analysis	X		X			X	
	Radiography			X				
	Endoscopy		X		X			
	Acoustic Emission				X	X		
	Inner part 3D acquisition	X	X	X				
	Thickness measurements				X			
	Weight, center of mass and inertia characteristics				X	X		
	FEA results		X		X	X		
	Support Design	X						
	Surface assessment							
	Polarization Resistance (corrosion rate)				X	X		
	Eddy current				X	X		
	Architectural aspects for the exhibition							
	Assembly model	X						
	FEA results		X		X	X		
	Rendering	X	X					
Historical, artistic, and archival studies				X			X	X
Visual inspection		X	X	X				X

2.2 Bronze Statue Restoration Workflow

Figure 1 presents a classic design workflow to support the restoration of bronze statues as from in-field experience. Central activities are **ACTION PLANNING** and **STATUE RESTORATION** that are the results of investigations related to the light-blue blocks in the figure. The initial input of the process derives from historical, artistic, and archival studies collected in reports, photos, analysis, etc., as described in 2.1. This phase is here called **EARLY ANALYSIS**.

In **DEFECT RECOGNITION**, restorers correlate technical data from additional 3D shape acquisitions (e.g., inner surfaces), visual inspection, and experimental data related to discontinuities. **STRUCTURAL EVALUATION** represents the following step that is also interlaced to data from the support design that develops and investigates the interaction effects in the assembly statue-inner supports, to understand loads and static. At this step, shells characteristics, as thickness distribution and joining, are evaluated together with experimental data concerning the position of centre of mass. In parallel, **SURFACES EVALUATION** investigates the external defects related to the skin of the statue, chemical analysis, and corrosion effects. Visual inspection also occurs at this stage. Another input related to surface evaluation includes the architectural aspects derived from the exhibition design. Aspects such as lights and environmental parameters may affect the final aspect of the surfaces during exhibition.

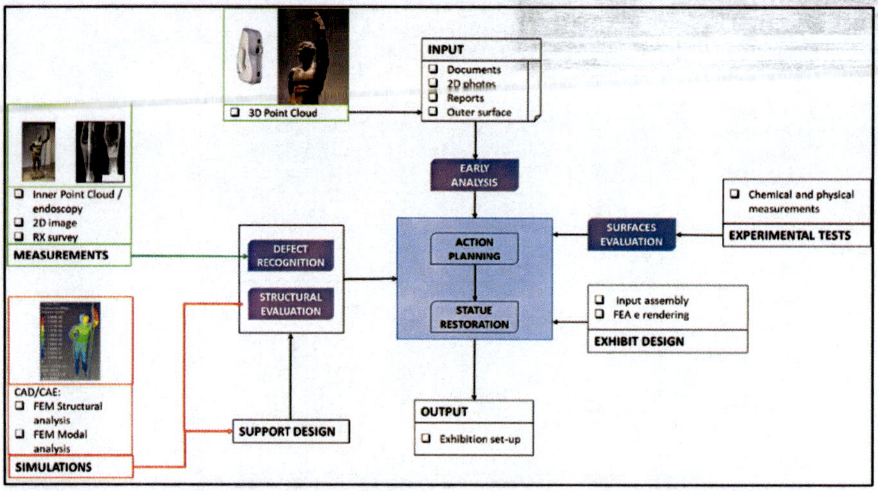

Fig. 1. Traditional restoration design framework

The lateral disposition of the blue blocks has been selected to point out that from the time scheduling point of view, also restoration has the necessity of speed up the process adopting parallel investigations. Unfortunately, this may increase the efforts made in the action of planning and restoration, since it may produce high volume of information to be organized and compared.

2.3 CAD-Based Framework

To aid the data analysis, we propose a CAD-based framework as a data hub to support action planning (Fig. 2). The **3D MODELING** phase, via CAD tool, processes a 3D representation of the statue converting a point cloud into an STL file for the CAD elaboration. The **3D MODELING** allows the digital representation of defects and other relevant information that can be tagged into the CAD tool, through the **3D ANNOTATIONS** step. Through them, the Early Analysis of Fig. 1 is collapsed within the input step in which 3D models and documentation are processed. Documentation is uploaded and classified through keywords and tagged through 3D-Annotation where necessary. 3D Annotations are also applied during defects and surfaces evaluation to gather visual inspections. **DATA and RESULTS** phase regards point paths, surface map, and the reference sections used for structural and surface evaluation. They include RX, computed tomography, thickness measure, surface chemical, and physical investigations, etc. A knowledge-based engine is also proposed in the workflow to support the definition of DATA and RESULTS through a collection of rules and best practices to be formalized in an application tool. This knowledge base, represented in Fig. 2 by the KBE tool, can propose past experiences about material selections, procedures, logistics, etc.

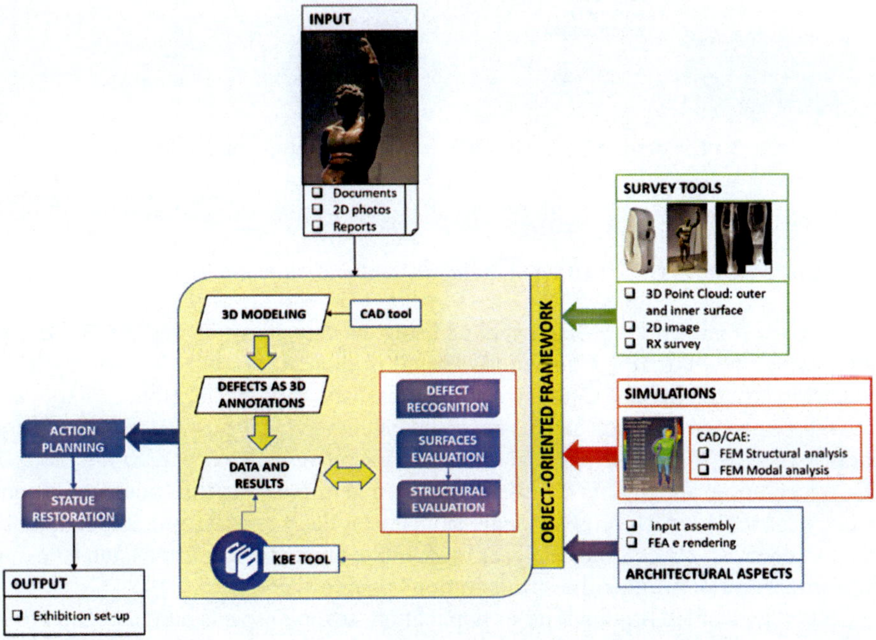

Fig. 2. Restoration workflow with CAD-based framework

According to this, the CAD-based framework has been developed as an Object-Oriented Framework (OOF). Different digital information has been integrated and managed with a class-diagram organization. Figure 3 shows the OOF, highlighting the main objects of the analysed structure (STATUE, REPORT, DEFECT, SOLUTION) with the main fields and functions (SET SOLUTIONS). The SET SOLUTIONS function may allow the restorer to analyse the possible solutions for producing the planning of the restoration actions through the KBE tool.

As a CAD platform, Rhinoceros® has been selected for the 3D representation of the bronze statue with the related information. The automatic elaboration of the geometrical and non-geometrical information has been performed using scripts in Python.

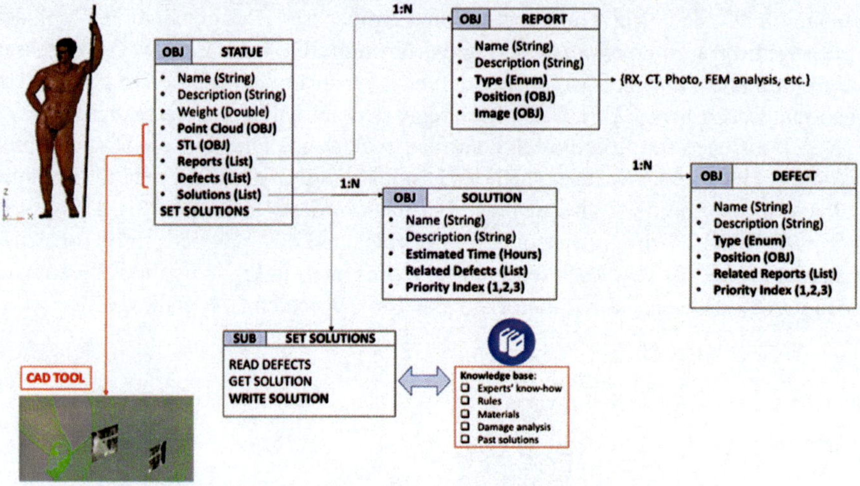

Fig. 3. The description of the OOF related to the proposed framework.

3 Application: The Principe Ellenistico

The described framework has been applied using, as case study, an ancient Hellenistic bronze statue, the "Principe Ellenistico", found in the end of 19th Century in Rome. It is datable around the 2nd Century B.C. and probably brought in Rome as spoils of war. This prince, two meters tall, is represented in a typical pose as a sovereign or warlord. Since its finding, it has undergone to several restoration operations firstly in order to recompose and join its fragments and then to improve its exhibition conditions. Currently conserved in Museo Nazionale Romano, at Palazzo Massimo, it is frequently subject to controls, checks and analyses useful to evaluate its conditions and to assess possible interventions, under the "preservation" point of view.

As already detailed, in this kind of application, several aspects and data need to be overlaid, to accomplish a complete and exhaustive evaluation that could drive to the optimal intervention to be planned and made.

Concerning archival data, reports of previous restorations represent the core of the input documentation preliminary investigated, together with works on the finding and the timeline of different exhibition both in the museums of Rome and temporary expositions. They usually involve manoeuvring, thus inspections and possible measurements. Repository includes documents collected as pdf files, photos and archive references.

Fig. 4. a) The Principe Ellenistico of Museo Nazionale Romano (Rome, Italy); b) 3D tessellated model through laser scanner; c) simplified model for FEA.

A 3D model has been acquired through a portable laser scanner by MCM s.r.l. and assembled as a single object of about 200×10^3 points (Fig. 4b), another acquisition has been provided more recently in 2019 through close-range photogrammetry [7]. Figure 4c shows the reduced model used for structural evaluation made by FEA [20]. Reduction consists of a point filtering that provided about 20×10^3 points tessellated through triangular elements.

From the structural integrity point of view, a wide range of data have been collected. First, the information about fragments and previous actions to guarantee solidity of legs and left arm. Figure 5 summarizes part of these data related to legs. Figure 5a, on the left, shows details of the left leg as from an old report and from the tessellated 3D model obtained through laser scanner. The 3D model can highlight the conjunction plates as depicted in the old document made to map criticalities. On the right, the 3D annotations associated to that detail are shown, as derived from the application of the CAD-based framework. Figure 5b shows the sketch related to the inner support and the composition of the RX images related to the same view and to a 3D rotation, to highlight the advantages of RX in the case of inner details reconstruction. Figure 5c shows an example of data visualization related to measurements useful for structural and surface evaluation. Figure 5c, on the left, represents thickness distribution as derived from ultrasonic measurements, they are reported as a path of points associated to spikes. On the right, a detail of stress distribution as derived from FEA. Similar representation may be derived for other measurements like corrosion rate.

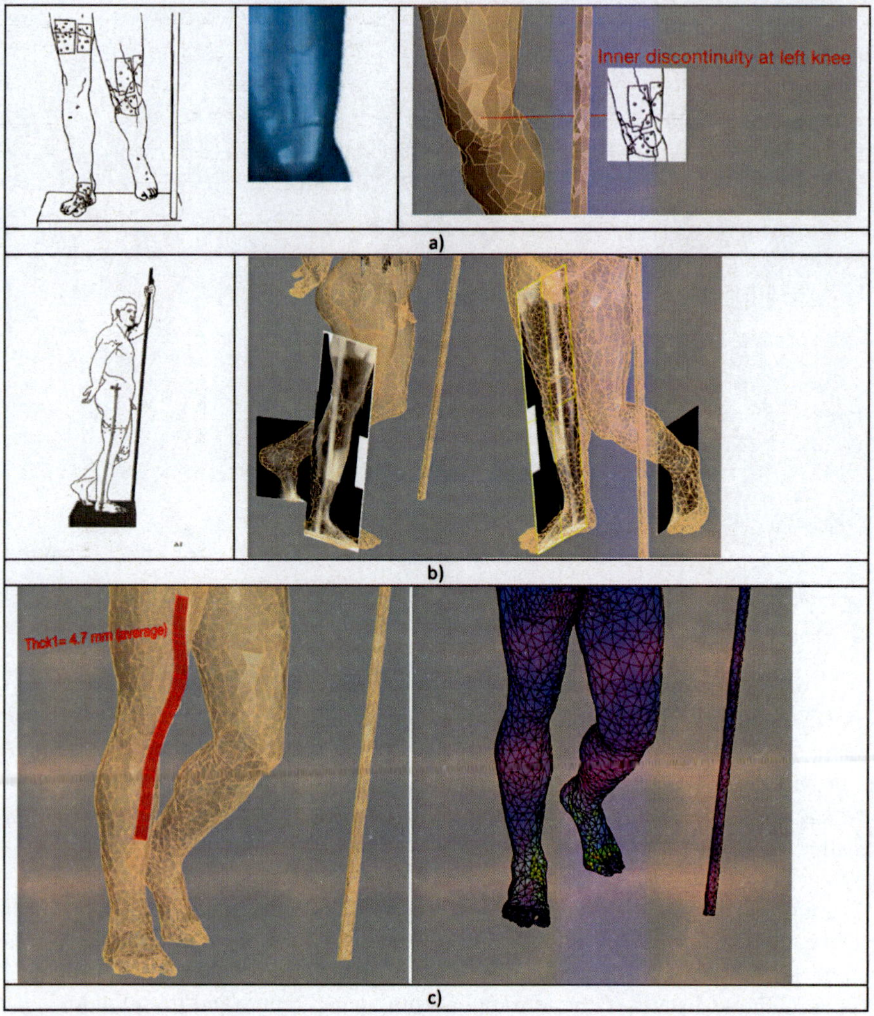

Fig. 5. Legs: a) discontinuities on the 3D model (on the left), 3D annotations (on the right); b) RX image as sections of the 3D model; c) visualization of the measured thickness path (on the left) and stress map evaluated by FEA (on the right).

Figure 6a investigates the upper part of the statue, showing fragment recognition of the left arm together with a reference image and associated RX. Figure 6b shows the superimposition of data in the CAD-based framework, including original fragment subdivision.

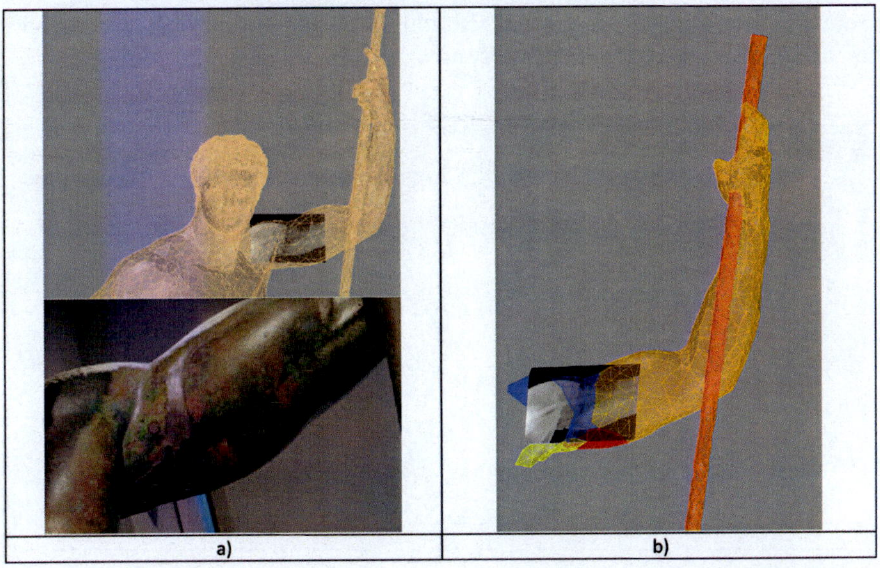

Fig. 6. Arm's fragment assessment: a) RX section (at the top), visual inspection of fragments (at the bottom); b) data overlapping on the 3D model through the CAD-based framework.

4 Discussion

The application made on the Principe Ellenistico may explain how the CAD-based framework assists the data visualization derived from different fields of knowledge. The 3D model overcomes the adoption of pictorial maps as shown in Figs. 5a and 5b, adopting an interactive digital mock-up that allows a virtual navigation according to the different layers of expertise. According to [26], polygonal modelling has been selected as visualization strategy, since, as requested from restorers, main requirements for what concerns model visualization are precision, interactivity, and visual consistency.

The relevance of the precision requirement is not only related to the capability of seeing discontinuities, as shown in Fig. 5a, but also the chance of benefit with image overlapping between RX and 3D model. It may assist reverse modelling of supports, as shown in Fig. 7a, where FEA of the inner support is derived through sketches design from RX images (Fig. 5b). It assumes importance in the respect of structural integrity assessment and support design. In fact, reverse engineering of the inner parts of statues is not always possible and RX coupled with 3D modelling from sketching are a good approach to rebuild geometries. Doing so, FEA can become the tool for structural assessment of the statue, allowing the correlation among material weakness, stress distribution and fragment statics. An example of this analysis is provided by means of the right part of Figs. 5c and 7b. In the first one, the stress distribution on the right foot shows relevant changes that can be correlated to the lines of the edge among fragments, confirming the critical conditions of that part. This imposed to increase stiffness and displacement control on the support arrangement at the basement, but also a corrosion evaluation driven by corrosion rate measurements. In this context, the CAD-based framework assumes an innovative aspect, since it may extend also to bronze statues the concept of design of restoration, already present in the preservation of buildings [27].

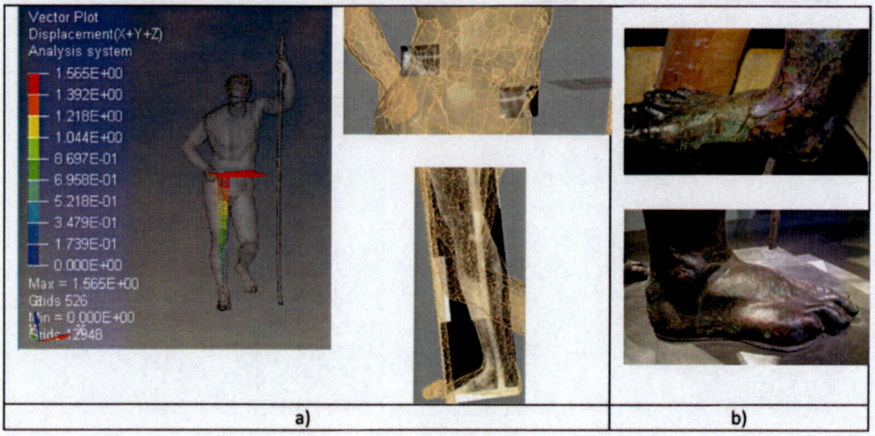

Fig. 7. Leg and support analysis: a) FEA (on the left), RX at their reference positions (on the right); b) Details of the left foot.

5 Conclusions

During bronze statues restoration, researchers plan activities and solutions through an intensive analysis anchored to surfaces and specific parts of the statue. So that, usually, the overall view of the data is stratified on pictorial views of the statue, usually called criticality maps. 3D models in CAD-CAE environments derived from an engineering design approach may increase the capability of the restorers of investigating details from multiple points of view, thanks to multi-layer data visualization, integration of FEA analysis, component design, and rendering of the exhibition set-up and full data management.

This paper presents a CAD-based framework able to fulfill these requirements, as discussed on the application made on the ancient bronze of the Principe Ellenistico, exhibited in Museo Nazionale Romano in Rome (Italy). In particular, the data structure, gathered from practical experience, technical measurements, simulations, and past reports, has been organized and implemented in an Object-Oriented Framework. The proposed design framework has been implemented using Rhinoceros® and scripts in Python. The results show a good level of usability towards data collection, visualization, and interaction towards simulations. Moreover, the proposed approach improves the analysis of the inner characteristics of the bronze statue by mixing the CAD geometry with the 3D annotations. The use of the digital mock-up is an innovative method in the field of restoration. Future applications will concern other important bronze statues such as the Pugilatore and the Vittoria Alata, so that a KBE tool may be developed from the collected experience.

References

1. Scopigno, R., et al.: 3D Models for Cultural Heritage: beyond plain visualization. IEEE Comput. **44**(7), 48–55 (2011)
2. Gherardini, F., Santachiara, M., Leali, F.: Enhancing heritage fruition through 3d virtual models and augmented reality: an application to roman artefacts. Virtual Archaeol. Rev. **10**(21), 67–79 (2019). https://doi.org/10.4995/VAR.2019.11918
3. Santachiara, M., Gherardini, F., Leali, F.: An augmented reality application for the visualization and the pattern analysis of a roman mosaic. In: IOP Conference Series: Materials Science and Engineering, vol. 364, no. 1, pp. 012094 (2018). https://doi.org/10.1088/1757-899X/364/1/012094
4. Inzerillo, L., Santagati, C., Di Paola, F.: Image-based modeling techniques for architectural heritage 3d digitalization: limits and potentialities. Int. Arch. Photogrammetry Remote Sens. Spat. Inf. Sci. **5**(W2), 550–560 (2013)
5. Anzani, A., Caramel, C.: Design and restoration: a holistic approach. In: I. Management Association (ed.) Architecture and Design: Breakthroughs in Research and Practice, pp. 1–17. IGI Global (2019). https://doi.org/10.4018/978-1-5225-7314-2.CH001
6. Samaan, M., Héno, R., Pierrot-Deseilligny, M.: Close-range photogrammetric tools for small 3d archeological objects. Int. Arch. Photogrammetry Remote Sens. Spat. Inf. Sci. ISPRS Arch. **40**(5W2), 549–553 (2013)
7. Bici, M., Gherardini, F., Campana, F., Leali, F.: A preliminary approach on point cloud reconstruction of bronze statues through oriented photogrammetry: the "principe ellenistico" case. In: IOP Conference Series: Materials Science and Engineering, vol. 949, no. 1, (2020). https://doi.org/10.1088/1757-899X/949/1/012117
8. Alsadik, B.: Practicing the geometric designation of sensor networks using the crowdsource 3d models of cultural heritage objects. J. Cult. Herit. **31**, 202–207 (2018). https://doi.org/10.1016/J.CULHER.2017.11.001
9. Bici, M., Guachi, R., Colacicchi, O., D'Ercoli, G., Campana, F.: Posture evaluation for fragment re-alignment of ancient bronze statues: the case study of the principe ellenistico. In: Cavas-Martínez, F., Eynard, B., Fernández Cañavate, F.J., Fernández-Pacheco, D.G., Morer, P., Nigrelli, V. (eds.) Advances on Mechanics, Design Engineering and Manufacturing II. LNME, pp. 323–335. Springer, Cham (2019). https://doi.org/10.1007/978-3-030-12346-8_32
10. Gherardini, F., Santachiara, M., Leali, F.: 3D virtual reconstruction and augmented reality visualization of damaged stone sculptures. In: IOP Conference Series: Materials Science and Engineering, vol. 364, no. 1 (2018). https://doi.org/10.1088/1757-899X/364/1/012018
11. Borri, A., Grazini, A.: Diagnostic analysis of the lesions and stability of michelangelo's David. J. Cult. Herit. **7**(4), 273–285 (2006)
12. Bagnéris, M., et al.: A complete methodology for the mechanical diagnosis of statue provided by innovative uses of 3d model. application to the imperial marble statue of Alba-La-Romaine (France). J. Cult. Herit. **28**, 109–116 (2017). https://doi.org/10.1016/J.CULHER.2017.05.002
13. Accardo, G., Amodio, D., Bennici, A., Cappa, P., Santucci, G., Torre, M.: Strain fields on the statue of marcus aurelius. Exp. Mech. **30**(4), 372–376 (1990). https://doi.org/10.1007/BF02321507
14. Fontana, R., et al.: Three-dimensional modelling of statues: the Minerva of Arezzo. J. Cult. Herit. **3**(4), 325–31 (2002). ISSN 1296–2074
15. Apollonio, F.I., Ballabeni, M., Bertacchi, S., Fallavollita, F., Foschi, R., Gaiani, M.: Digital documentation and restoration tools reusing existing imagery: a multipurpose model of the neptune's fountain in Bologna. Appl. Geomatics **10**(4), 295–316 (2018). https://doi.org/10.1007/s12518-018-0210-x

16. Apollonio, F. I., Gaiani, M., Basilissi, W., Rivaroli, L.: Photogrammetry driven tools to support the restoration of open-air bronze surfaces of sculptures: an integrated solution starting from the experience of the neptune fountain in Bologna. Int. ISPRS Arch. **42**(2W3), 47–54 (2017). https://doi.org/10.5194/ISPRS-ARCHIVES-XLII-2-W3-47-2017
17. Menningen, J., Siegesmund, S., Krompholz, R., Rieffel, Y.: The marble sculptures of general bülow and scharnhorst in Berlin: comparative, non-destructive analysis of the weathering state after 12 years of exposure. Environ. Earth Sci. **79**(11), 1–23 (2020). https://doi.org/10.1007/s12665-020-08984-7
18. Pascale, G., Lolli, A.: Crack assessment in marble sculptures using ultrasonic measurements: laboratory tests and application on the statue of david by Michelangelo. J. Cult. Herit. **16**(6), 813–821 (2015). https://doi.org/10.1016/J.CULHER.2015.02.005
19. Bici, M., Campana, F., Colacicchi, O., D'Ercoli, G.: Metodi diagnostici e programmazione di interventi sui grandi bronzi: il caso del principe ellenistico del museo nazionale romano. atti del convegno il restauro dei grandi bronzi archeologici – laboratorio aperto per la vittoria alata di brescia presso opificio delle pietre dure, Firenze (2019)
20. Bici, M., Campana, F., Colacicchi, O., D'Ercoli, G.: CAD-CAE methods to support restoration and museum exhibition of bronze statues: the "principe ellenistico". In: IOP Conference Series: Materials Science and Engineering, vol. 364, no. 1, art. no. 012014 (2018). https://doi.org/10.1088/1757-899X/364/1/012014
21. Di Paola, F., Milazzo, G., Spatafora, F.: Computer aided restoration tools to assist the conservation of an ancient sculpture, the colossal statue of zeus enthroned. international archives of the photogrammetry. Remote Sens. Spat. Inf. Sci. ISPRS Arch. **42**(2W5), 177–184 (2017). https://doi.org/10.5194/ISPRS-ARCHIVES-XLII-2-W5-177-2017
22. Mandelli, A., Achille, C., Tommasi, C., Fassi, F.: Integration of 3d models and diagnostic analyses through a conservation-oriented information system. Int. Arch. Photogrammetry Remote Sens. Spat. Inf. Sci. ISPRS Arch. **42**(2W5), 497–504 (2017).. https://doi.org/10.5194/ISPRS-ARCHIVES-XLII-2-W5-497-2017
23. Turco, M.L., Rinaudo, F., Piumatti, P., Calvano, M., Spreafico, A., Patrucco, G.: The digitisation of museum collections for research, management and enhancement of tangible and intangible heritage. In: Proceedings of the 2018 3rd Digital Heritage International Congress, Digital Heritage 2018 held jointly with the 2018 24th International Conference on Virtual Systems and Multimedia, VSMM 2018, Art. No. 8810128 (2018). https://doi.org/10.1109/DIGITALHERITAGE.2018.8810128
24. Settembre Blundo, D., Ferrari, A.M., Fernández Del Hoyo, A., Riccardi, M.P., García Muiña, F.E.: Improving sustainable cultural heritage restoration work through life cycle assessment based model. J. Cult. Herit. **32**, 221–231 (2018)
25. Donati, P.: Satiro da Mazara del Vallo: intervento di restauro. A cura di Petriaggi R., Il Satiro Danzante, Catalogo della Mostra presso la Camera dei Deputati (aprile - giugno 2003) e della Mostra presso i Musei Capitolini (giugno - luglio 2003), Roma, pp. 90–96 (2003)
26. Foni, A.E., Papagiannakis, G., Magnenat-Thalmann, N.: A taxonomy of visualization strategies for cultural heritage applications. J. Comput. Cult. Herit. **3**(1), 1–21 (2020). https://doi.org/10.1145/1805961.1805962, Art. No. 1
27. Cruz Ramírez, L.C.: Parametric design in the restoration project. https://editorialrestauro.com.mx/parametric-design-in-the-restoration-project/. Accessed 21 Mar 2021

Design of the New Inner Frame for the *Vittoria Alata* di Brescia: How Engineering Design May Support Ancient Bronze Restoration

Michele Bici[1](✉), Annalena Brini[2], Francesca Campana[1], Sergio Capoferri[3], Riccardo Guarnieri[3], Francesca Morandini[4], and Anna Patera[2]

[1] Dipartimento di Ingegneria Meccanica e Aerospaziale, Sapienza Università di Roma, Roma, Italy
michele.bici@uniroma1.it
[2] Ministero della Cultura, Opificio delle Pietre Dure, Firenze, Italy
[3] Capoferri S.p.A., Adrara San Martino, Bergamo, Italy
[4] Comune di Brescia - Fondazione Brescia Musei, Brescia, Italy

Abstract. Mechanical design and engineering can support projects with high-added value in the Cultural Heritage field, such as restoration of artefacts like statues and architectonical decorations. Many examples have been carried out, along the recent past, in the field of ancient bronze restorations (as for Marco Aurelio, Satiro Danzante di Mazara del Vallo and Principe Ellenistico). Engineering design techniques help the assessment of structural problems through physical measurements and FEA simulations; the digital acquisition of surfaces represents a fundamental base for CAD modelling, and the inner frame design helps for guaranteeing stability and manoeuvrability requirements for transport and exhibition. Workflow peculiarities and requirements to accomplish the restorers' activities and investigations may highlight best practices and rules. The design of the new inner frame of the *Vittoria Alata* of Brescia, an ancient roman bronze statue, represents a recent example of this kind. Its design workflow was provided in the loop of the restoration program, and it was assessed considering structural integrity, surfaces protection, inner inaccessibility, and dimensions. The solutions adopted are the result of a collaborative process with restorers to evaluate each proposed concept, in compliance with the studies and the constraints highlighted during the investigations. CAD-CAE tools applied starting from the 3D acquisition, helped the development and its verification, reducing the efforts during the manufacturing phase and final set-up. This paper aims to discuss the obtained result demonstrating how structural analysis and mechanical design anchored to 3D acquisitions may help restoration of bronze statues.

Keywords: CAD-CAE · Structural analysis · Cultural heritage · Design for restoration · *Vittoria Alata*

1 Introduction

Mechanical design and engineering are increasingly supporting restoration of artefacts like statues, architectonical decorations and paintings. Engineering design techniques

C. Rizzi et al. (Eds.): ADM 2021, LNME, pp. 951–962, 2022.
https://doi.org/10.1007/978-3-030-91234-5_96

may help the assessment of structural problems through physical measurements and Finite Element Analysis (FEA) simulations; the digital acquisition of surfaces is currently fundamental for CAD modelling, and, often, inner frame design is carried out guaranteeing stability and manoeuvrability requirements for transport and exhibition.

Measurements involve surface corrosion, material composition and integrity, technological information related to fabrications and previous restorations. Major examples of this are provided by relevant projects such as the one made on the Michelangelo's David [1] or applications for deeper investigations such as the Roman bronze statue of the Pugilatore at Museo Nazionale Romano [2]. Peculiar studies and design activities involve seismic preservation and monitoring, as demonstrated in [3]. Simulations for structural integrity are also increasing their role: besides the architectural field, original applications were carried out to evaluate stability of many statues such as the roman equestrian bronze of Marco Aurelio [4] or the imperial marble statue of Alba-la-Romaine (France) [5]. This work also demonstrates that many different CAD investigations may be provided starting from digital acquisition, such as measurement report and collection [6] and posture restoration. Further works connected to this topic are related to fragment recognition and assembly of potteries [7] or statues [8]. Finally, the applications related to digital design and additive manufacturing to aid transportability and manoeuvring [9], exhibition and lacuna reconstruction [10] deserve to be mentioned.

Fragment alignment is naturally linked to the design of inner frames to guarantee stability. Many ancient statues were found in pieces and the original solutions for their postures and stability is often unknown, in some cases also the final aspect may be under discussion, when lacunas are relevant. Examples of this are provided by the statues of the Satiro Danzante di Mazara del Vallo [11] and la Minerva di Arezzo [12, 13]. In the first case, the final design accomplished the requirements of fragment positioning with regulations through precision screws and spherical joints, and material compatibility through contact surfaces made by compatible resin. In the second case, the favourable accessibility of the inner parts allowed the design of a wireframe able to align the original sherds with the reconstruction of the lacunas that can be removed easily. Other important bronze statues, like Principe Ellenistico [8] and the *Vittoria Alata* of Brescia present structural peculiarities related to their integrity status and to the restoration made after their finding in the nineteenth century. They are quite continuous along the body, thus there are not easy to be accessed inside. The support devices inside both the statues consisted of frames with square section beams linked to the bronzes by compact concrete fillings and, only in the case of the Principe Ellenistico, bronze plates with rivets. The fillings acted as rigid stiffening between bronzes and basements where the frames were inserted. The filling naturally provided solidity among sherds, original or not, and reduction of local bending loads, but it resulted to be not able to guarantee about surface corrosion. In the last century, this old kind of arrangement was amended by new restoration criteria that prefer showing original fragments by themselves without invasive new reconstructions and diffuse fillings.

In this paper, we are going to report the efforts to design and set-up the new inner frame for the *Vittoria Alata* di Brescia (Fig. 1) for its new exhibition in the Capitolium of Brescia. It has been done after a restoration programme carried by the Opificio delle Pietre Dure (OPD) of Florence that included the removal of the nineteenth century filling.

The preliminary assessment defined not only the requirements related to the new architectural exhibition, designed by Juan Navarro Baldeweg, but also the structural effects induced in the bronze shell by the filling removal. It was obtained through experimental analysis and FEA simulations as discussed in Sect. 3 after the presentation of the statue and its restoration project (Sect. 2). Then, in Sect. 4, relevant concepts and final development are discussed, so that, in Sect. 5, main conclusions are outlined.

2 The *Vittoria Alata* and its New Exhibition Set-Up

The *Vittoria Alata* of Brescia (Fig. 1), datable around the middle decades of the 1st Century A.D., represents a precious and rare archaeological artefact in bronze, which has fortunately escaped to the common practice of re-use (for further information about the statue's history, the restoration intervention and the new exhibition set-up [14, 15]).

Fig. 1. A view of the new exhibition of the *Vittoria Alata* in Brescia's Capitolium (Courtesy of Archivio Fotografico Musei Brescia, Fotostudio Rapuzzi).

The statue's details brought out by the restoration, and the analyses about the used manufacturing technologies, suggest that it was made by a high-level atelier, which could rely on a mature artist for the choice of the model and the harmonious construction of the figure, and on qualified and well-organized workers.

The statue was discovered in 1826, during a campaign of archaeological investigations carried out in the Capitolium of the ancient city of Brescia, where it had been intentionally hidden, in antiquity, with its wings and arms detached and neatly placed near the body. At that time, specialists recognised the female figure as a winged victory (*Vittoria Alata*) in the pose of writing the name of the victor in battle on a shield, not found in the excavation, which the divinity held on her raised left leg. This interpretation guided the first nineteenth-century restoration aimed at displaying the large bronze statue. To give stability to the entire structure and allow the wings and arms to be attached, an ingenious metallic structure, with joining mechanisms for the originally detached parts, was inserted inside the statue. This supporting structure was held in place by a massive filling, which remained in place until 2018, when, after careful studies and inspections, it was decided to remove this inner system as it was no longer able to safely accomplish the function which it was designed for.

The urgent restoration work has become the hub of a wider process of research, conservation and valorisation [16, 17]. The project has been promoted by the Municipality of Brescia and the Fondazione Brescia Musei, with the direct involvement of the Opificio delle Pietre Dure in Florence, the Soprintendenza for Archeologia Belle Arti e Paesaggio for the districts of Bergamo and Brescia, the Lombardy Region, and thanks also to the support of private individuals. It has led to huge progresses in the knowledge of this extraordinary workpiece and its peculiar history. In addition to archaeologists, restorers and scientific experts from the involved institutions, external specialists and technicians from various universities and research institutes contributed to the implementation of this ambitious project in accordance with a modern interdisciplinary research approach.

At the end of the restoration, the *Vittoria Alata* has been exhibited in the Capitolium of Brescia, in a space specially designed by the Spanish artist and architect Juan Navarro Baldeweg, who was able to enhance the statue with a particularly monumental setting in which intentional references to antiquity can be seen. The statue has been placed in an elevated position on a cylindrical base, inside which is concealed the connection with an anti-seismic platform placed at the floor level. The Botticino marble, used for this supporting cylindrical base, evokes the shafts of the columns, while bricks and mortar cladding of the walls are inspired by Roman architectural materials (Fig. 1).

The *Vittoria Alata* posed an interesting challenge both for the restoration purposes and for the related studies and research. A thorough examination of the state of preservation and preliminary investigations carried out with the acoustic emission technique, conducted to the decision of emptying the statue of filling and old supporting structure, designing a suitable new support that would allow the various parts of the statue to be safely assembled and positioned high up on the new basement connected to the anti-seismic platform.

The critical areas, initially observed through an examination of the statue, were the conspicuous dimpling of the lower neck, the evidence of the welding point, the consequent stretching of the metal on the back of the neck and the series of cracks running through almost the entire circumference. Moreover, several cracks and missing parts of the upper part of the chiton, at the insertion of the left arm, and the detachment of a portion of the roll of folds of the cloak, were very visible, in addition to some other small breaks distributed. Thermography also revealed a problem of discontinuity in the lower back part of the robe, and radiographic examinations revealed numerous cracks and breaks that were hidden under the layers of bronze alteration products and of previous restorations. The usage of X-rays brought into a sharper focus a large area of possible weakness in the left side of the statue due to both the thinning of the alloy in that area, and to its progressive mineralisation and suffering due to the imperceptible change in posture caused by the natural process of degradation of the organic filling materials that secured the 19th century support to the statue and by the different balance caused by the weight of the wings. This weakness was also made evident by acoustic emission surveys repeated in successive years [18].

At the end of the emptying process, conducted stratigraphically with the criteria of the archaeological micro-excavation [19], a photogrammetric relief of the internal surfaces was made possible, also having the chance of linking it to the 3D relief of the external surfaces, becoming useful as a support and reference system for specific mappings [6]. This system of sequential images made possible the revealing of further points of discontinuity, such as the precariousness of the attachment of the bust in the upper part of the joints to the lower part of the figure, which was therefore reinforced by the assembly and welding of the horizontal folds of the cloak.

All these structural weaknesses were therefore considered, studied and used for the mapping of the critical points, which had to be taken into account for the construction of the new support that would allow the new exposure in full safety.

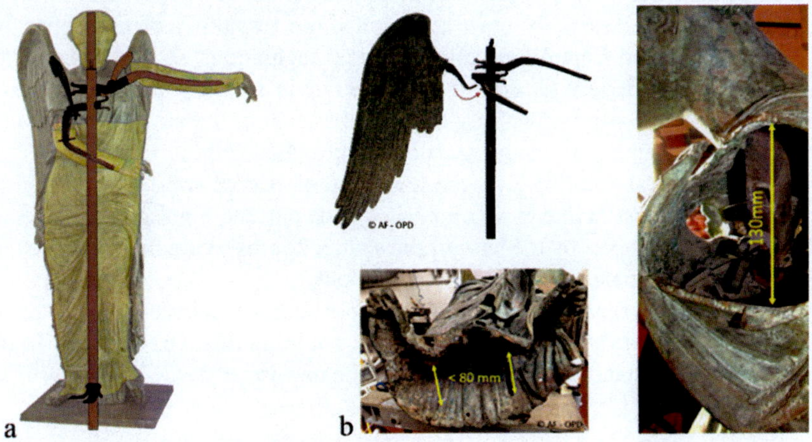

Fig. 2. (a) The old frame and the filling (light yellow); (b) The old mechanisms for wings and arms connections and the useful openings of the statue (Courtesy of Archivio fotografico dell'Opificio delle Pietre Dure di Firenze).

The 19th-century inner frame, originally designed to support the statue and drowned in the mentioned filling, is presented in Fig. 2. It consisted with a central square-section beam with transverse elements to connect wings and arms in the respect of the central body. Transverse elements were fastened by means of screws and regulated through spacers. Wings were hooked, one for each transverse element. Arms were engaged through pins with rectangular sections. The stability between the inner frame and the bronze shell was guaranteed, as said, through a conglomerate filling, that presented also organic composition, that guaranteed adhesion and solidity between the internal structure and the bronze shell of the statue. It also furnished a structural advantage on the central body, increasing its stiffness. The central beam was inserted in the base of the neck through a prismatic guide embedded in the filler of the head. At the base, a positioning cross ensured the correct place of the central beam, before filling the statue. The central beam supported the load entirely, as confirmed by the slight and discontinuous imprint of the bottom perimeter on the basement put in place in the 1950s. No specific regulation for the relative positioning of arms and wings were defined except the use of shims.

3 Design Requirements, Analyses and Concept Design

From the previous sections, it can be deduced how the new frame design could be challenging and the high number of different requirements to be accomplished. Main functionality of the new inner frame is withstanding the weight of the central body, wings and arms, respectively of about 217 kg, 30 kg (left wing), 38 kg (right wing), 12 kg (left arm), 8 kg (right arm), allowing the connection to the new anti-seismic base provided by THC and a height from the floor of 1.6 m as requested by the museal set-up. Moreover, material integrity of the ancient bronze must be prevented without overload at critical points defined by diagnostics, as reported in Sect. 2. The set-up must be able to be disassembled and it must be made of materials compatible at the bronze interface, to guarantee surfaces. Assembly must accomplish the feasibility of manoeuvring the statue both in OPD and Capitolium, allowing final regulations on site. Other relevant constraint is the limited room for assembly manoeuvre inside the statue. Access towards the inside of the statue is possible only from the base, from the arms openings and on the shoulders where areas for wings insertion are provided (Fig. 2b).

From the structural point of view the final map of critical areas, derived by visual inspections and diagnostics like radiography and eddy current, highlighted the necessity of: 1) not providing support by the base of the statue; 2) preserving the weak area of the left hip; 3) preventing new opening of the neck crack.

The filling removal represented an additional element of weakness since its presence increased the overall stiffness of the central body, avoiding relative rotations. To define a reference structural state for the evaluation of the new inner frame concept, FEA was adopted.

3.1 Preliminary Structural Evaluation by FEA

FEA model was based on the 3D acquisition made by a structured light device by Mattia Mercante [20]. For facilitating the computation, the initial high-density cloud has been filtered to achieve a suitable coarse triangular mesh. Loads have been distributed as gravity loads after an experimental evaluation of thickness distribution and of the central body centre of mass. The thickness measurements have been made by Eng. Giorgio D'Ercoli (ISCR, Roma), through calliper in all the accessible areas of the statue (Fig. 3a) and the measurement of the centre of mass position has been achieved by the Apice Srl using dynamometers and winches in one of the operations of movement during the restoration.

Crossing these results with the 3D model of the external surface of the central body of the statue and its weight, two distributions of thickness emerged, one for the torso, with a rather uniform trend, on average about 6 mm, and a more variegated one of about 8 mm in average, in correspondence of the legs. In Fig. 3b, it is reported a side view of the loads diagram to the resulting centres of mass for the arms (P_b), the wings (P_a) and the central body (P_{cc}). The reference structural state has been defined as the one

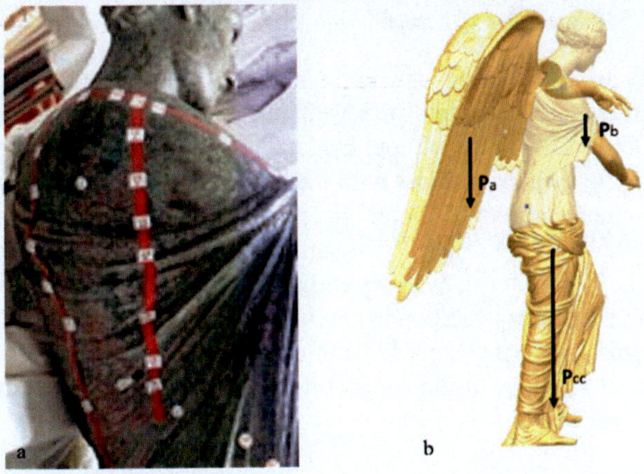

Fig. 3. (a) Experimental point distribution for thickness measurements on the upper part; (b) Simplified load distribution

achieved simulating the presence of filling, reproducing the condition of the statue before the restoration intervention. The action of filling has been simulated as a lock of relative rotations. The wings have been applied in contact with the statue assuming a rotation of the connection hooks. Their critical effect on the left hip area has been detected as a stress state related to a bending moment induced by the wings, that in the ancient restoration may touch the back of the statue (Fig. 4). The final maximum displacement on the head has been assumed as a sizing criteria for the new inner frame, so that, the stiffness effect of the filling may be reproduced in accordance to stress distribution at the neck and the left hip, not greater than the reference case.

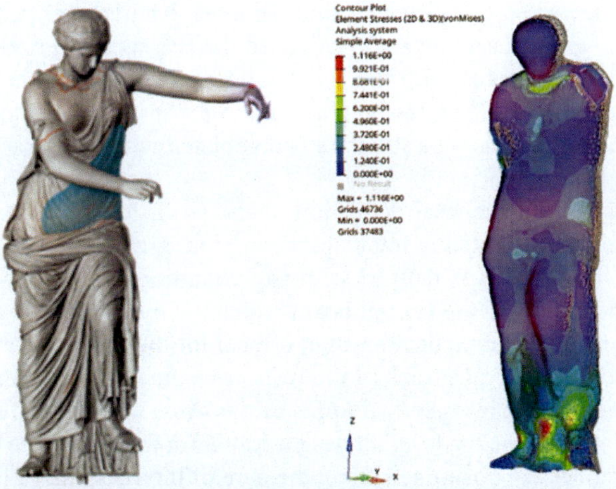

Fig. 4. Left: critical area map (in blue, results of the Eddy Current test about material weakness; in red, cracks); right: preliminary FEA displacement results (mm), in grey undeformed shape.

3.2 Digital Mock-Up of the Concept

The internal surface of the statue was not completely available, due to inaccessibility, and, above all, it was not a simple offset of the outer, being more like an irregular pipe with a very rough finishing without evidence of the external folds of the drapery. According to this, the contact surfaces with the inner frame have been derived by offset and smoothing operations from the outer one, guided by the knowledge of the thickness measures and the visual inspection of the surfaces. The functional components of the concept are load support interfaces and stabilization interfaces to limit the oscillations of the bronze with respect to the support, regulations of the interfaces in the respect of problems of alignment with the supposed contact surfaces and in the respect of contrast tension. The contact surfaces of the support interfaces were identified in the left thigh and in the right shoulder (Fig. 5).

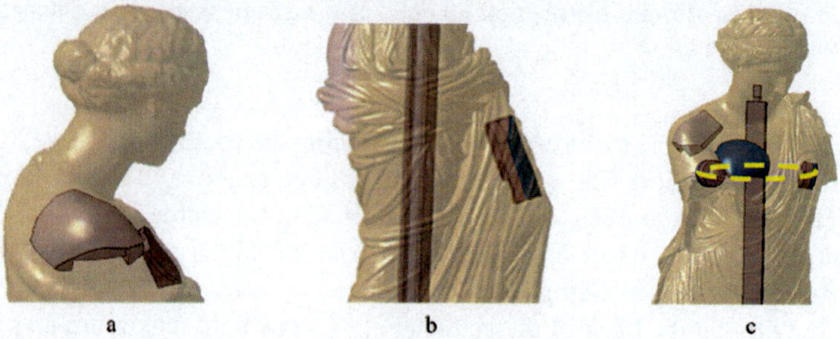

Fig. 5. Preliminary layout of the contact interfaces: (a) right shoulder; (b) left leg; (c) stabilization at the breast.

The stabilization interfaces have been set in two cross sections at the height of the breast and at the lower base. In each section there are 3 horizontal pads adjustable in translation by screws. The section at the height of the breasts also provides for angular adjustment in correspondence with the spherical graft of the struts in the three discs placed at approximately 120°, at the height of the right breast, of the lower left armpit hollow and on the right back. The structure is completed with a contrast support at the height of the thigh.

To limit, as possible, the final deflection of the head, for the central beam, an H section was opted. It is oriented to intercept the original graft left in the head, by using an oriented pin. The graft was defined with gap condition to avoid overstressing the neck. Correct orientation of the central beam is defined by the stabilization interfaces at the base, assigned to specific positions not critical for the mineralization of the area. Regulation of the interface positions in the respect of the contact of internal surfaces of the statue is guaranteed by spherical joints of the struts in the metallic anima of the interfaces. Regulation of the loads by screws positioned at the base since the struts of the thigh and the shoulder are positioned along the core of the H beam, or directly near by the interfaces for the stabilization ones. The wings are engaged from the upper openings by means of prismatic shanks that are housed in special crosspieces connected to the central beam. For the support of the arms a pantograph arrangement has been selected so that proper tension can be regulated to guarantee stabilization.

Regarding the materials, it was decided to make the internal structures in stainless steel with contact interfaces in compatible resin (PE 664 hardener IPE 764 (100:50)). The realization of the pads has been examined since the concept phase, due to the obstacles related to the positions to be reached; the need to guarantee reversibility and stability of position during forming; the necessity of moving the statue from laying down to stand up.

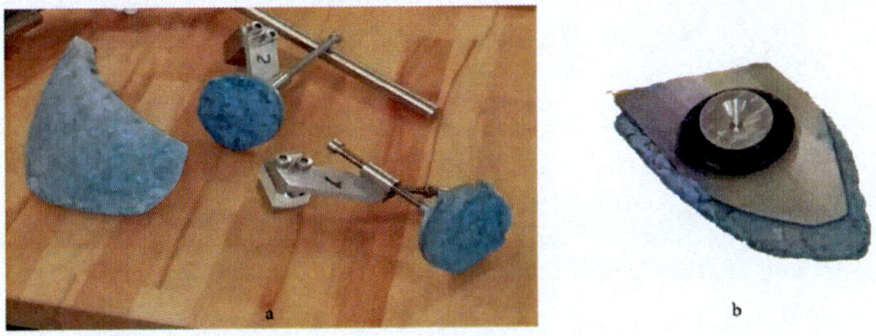

Fig. 6. (a) Resin bronze-frame interfaces; (b) particular of the back of the left leg interface.

To reduce position errors, a direct molding process has been defined. It works after covering the resin with a film suitable to avoid leaks. Markers positioned on the internal surface have been adopted as fastening elements of the pads during the forming phase, as well as the repetition of multiple tests, have ensured good precision of execution and repeatability of positioning. Figure 6 shows right shoulder, left leg and stabilization interfaces with part of the steel supports.

During the restoration, diagnostics and inspections confirmed that the central body consists with 2 parts joined by weldment along the bottom of the hip. Weldment integrity has some weakness, so that in the absence of supports at the base of the drapery, we opted for the insertion of steel cables (Fig. 7) capable of linking the lower part of the garment, which was connected to the rest of the body with internal welding, currently not entirely safe. The positions of the three cables are at different heights to avoid interference with internal surface of the bronze.

4 Preliminary and Final Set-Up

Preliminary set-up was made at OPD to test the solutions developed from the concept design. This was made maintaining the statue laying down, horizontally. Testing included: verification of the manoeuvring of the central beam and its insertion in the guide of the neck; marking the areas of the contact interfaces and verifying the room for assembly. Thanks to markers, breast, shoulder and thigh interfaces have been placed prior to assembly struts and central beam. Figure 7b shows the "as built" model provided by Capoferri S.p.A.. Figure 7c reports the alignment of the initial interface design with the "as built" model. The alignment of the two projects was done by bringing to coincidence (by minimizing the relative distances) the three points of comparison of the lower

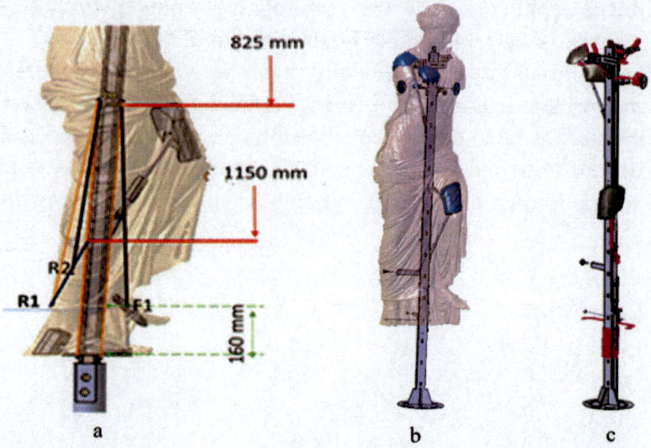

Fig. 7. (a) Cables set-up for safety of the shell at the bottom; (b) Mock-up as built; (c) Alignment between "as built" and final design (in purple).

position of the central pole on the statue and, subsequently, by rotating the planes of the statue-interface contact at the points related to armpit and breast, until they coincide. The differences between the two central poles are of a few millimetres and, despite some differences in the positions of the thigh strut, it can be concluded that the two models comply with a good approximation. Finally, in Fig. 8, the gap among the statue and the basement is shown to demonstrate that accomplish of the request about preservation of the bottom part.

Fig. 8. Detail of the final set-up to show the suspension of the statue.

5 Conclusions

The restoration of the new exhibition set-up for the *Vittoria Alata* of Brescia included the design and set-up of a new inner frame suitable to support its parts (central body, arms and wings) and preserve material integrity. The openings for manoeuvring and the wide lengths represented the major obstacles to define feasible solution. CAD-CAE analysis starting from 3D digitalization of the outer surfaces were the starting point of the work to assess the structural integrity and to evaluate assembly drawbacks. Concepts were defined considering these drawbacks and, after digital mock-ups, physical prototypes were built to assess possible limits and tolerances of interface manufacturing and mounting. Co-operation with restorers and integration among simulation and practical set-up optimized the scheduling and decision making, demonstrating how mechanical integrated product-process design may help restoration in the cultural heritage field.

References

1. Borri, A., Grazini, A.: Diagnostic analysis of the lesions and stability of Michelangelo's David. J. Cult. Herit. **7**(4), 273–285 (2006)
2. Mercuri, F., et al.: Interface thermal conductance characterization by infrared thermography: a tool for the study of insertions in bronze ancient Statuary. Infrared Phys. Technol. **90**, 31–39 (2018)
3. Siami, A., Karimi, H.R., Cigada, A., Zappa, E., Sabbioni, E.: Parameter optimization of an inerter-based isolator for passive vibration control of Michelangelo's Rondanini Pietà. Mech. Syst. Signal Process. **98**, 667–683 (2018)
4. Accardo, G., Amodio, D., Bennici, A., Cappa, P., Santucci, G., Torre, M.: Strain fields on the statue of Marcus Aurelius. Exp. Mech. **30**(4), 372–376 (1990)
5. Bagnéris, M., et al.: A complete methodology for the mechanical diagnosis of statue provided by innovative uses of 3D model application to the imperial marble statue of Alba-la-Romaine (France). J. Cult. Herit. **28**, 109–116 (2017)
6. Cicconi, P., Bici, M., Colacicchi Alessandri, O., D'Ercoli, G., Campana, F.: A CAD-based framework for interactive analysis in the restoration of bronze statues. In: Lecture Notes in Mechanical Engineering: Proceedings of ADM2021 International Conference, Accepted for Publishing (2021). https://doi.org/10.1007/978-3-030-91234-5_70
7. Eslami, D., Di Angelo, L., Di Stefano, P., Pane, C.: Review of computer-based methods for archaeological ceramic sherds reconstruction. Virt. Archaeol. Rev. **11**(23), 34–49 (2020)
8. Bici, M., Guachi, R., Colacicchi, O., D'Ercoli, G., Campana, F.: Posture evaluation for fragment re-alignment of ancient bronze statues: the case study of the Principe Ellenistico. In: Cavas-Martínez, F., Eynard, B., Fernández Cañavate, F.J., Fernández-Pacheco, D.G., Morer, P., Nigrelli, V. (eds.) Advances on Mechanics, Design Engineering and Manufacturing II. LNME, pp. 323–335. Springer, Cham (2019). https://doi.org/10.1007/978-3-030-12346-8_32
9. Fatuzzo, G., Sequenzia, G., Oliveri, S.M., Barbagallo, R., Calì, M.: An Integrated Approach to Customize the Packaging of Heritage Artefacts. Lecture Notes in Mechanical Engineering, pp. 167–175. Springer, Cham (2017).https://doi.org/10.1007/978-3-319-45781-9_18
10. Balletti, C., Ballarin, M., Guerra, F.: 3D printing: state of the art and future perspectives. J. Cult. Herit. **26**, 172–182 (2017)
11. Petriaggi, R.: Il Satiro Danzante di Mazara del Vallo. Diario di un Restauro (Rome – Istituto Centrale per il Restauro) (2004)
12. Pucci, E., Sarri, S.: L'Arringatore e la Minerva di Arezzo: il Restauro per il miglioramento delle condizioni statiche ed espositive. In: Morandini, F., Patera, A. (eds.). Il Restauro dei Grandi Bronzi Archeologici. Laboratorio Aperto per la *Vittoria Alata* di Brescia, Atti del Convegno, Firenze, 27–28 May 2019, pp. 175–182. Edifir, Firenze (2020)
13. Fontana, R., et al.: Three-dimensional modelling of statues: the Minerva of Arezzo. J. Cult. Herit. **3**(4), 325–331 (2002)
14. Morandini, F., Patera, A.: Il restauro dei grandi bronzi archeologici. Laboratorio aperto per la *Vittoria Alata* di Brescia, Atti del Convegno, Firenze, 27–28 May 2019. Edifir, Firenze (2020)

15. Morandini, F., Patera, A.: La *Vittoria Alata* di Brescia. Non ho mai visto nulla di più bello. Skira Editore, Milano (2021)
16. Patera, A., et al.: La *Vittoria Alata* di Brescia: Avvio del progetto di restauro e prospettive di ricerca su un grande bronzo di età romana. OPD Restauro **30**, 78–89 (2018)
17. Morandini, F., et al.: The Brescia winged victory: ongoing diagnostic work, conservation treatment and restoration. In: Baas, P. (ed.) Proceedings of the XXth International Congress on Ancient Bronzes. Resource, Reconstruction, Representation, Role, Tübingen, 17–21 April 2018, pp. 35–43. Oxford (2019)
18. Caneva, C., Pampallona, A.: Rilievo delle criticità strutturali di un bronzo complesso con la tecnologia della Emissione Acustica. Caso della *Vittoria Alata*. In: Morandini, F., Patera, A. (eds.). Il restauro dei grandi bronzi archeologici. Laboratorio aperto per la *Vittoria Alata* di Brescia, Atti del convegno, Firenze, 27–28 May 2019, pp. 141–144. Edifir, Firenze (2020)
19. Brini, A., Casu, S., Patera, A.: La rimozione del riempimento ottocentesco della *Vittoria Alata*. In: Morandini, F., Patera, A. (eds.). Il Restauro dei Grandi Bronzi Archeologici. Laboratorio Aperto per la *Vittoria Alata* di Brescia, Atti del convegno, Firenze, 27–28 May 2019, pp. 49–60. Edifir, Firenze (2020)
20. Mercante, M.: Applicazioni del Rilievo 3D durante il Restauro della *Vittoria Alata*. In: Morandini, F., Patera, A. (eds.). Il Restauro dei Grandi Bronzi Archeologici. Laboratorio Aperto per la *Vittoria Alata* di Brescia, Atti del Convegno, Firenze, 27–28 May 2019, pp. 155–160. Edifir, Firenze (2020)

Author Index

C. Rizzi et al. (Eds.): ADM 2021, LNME, pp. 963–966, 2022.
https://doi.org/10.1007/978-3-030-91234-5